Birds of
ECUADOR

HELM FIELD GUIDES

Birds of ECUADOR

Juan F. Freile
and
Robin Restall

H E L M
LONDON • OXFORD • NEW YORK • NEW DELHI • SYDNEY

To Elvia, Rubén, Gabi, Ele, Martín, Emilia, Bernardo and Marimba

HELM
Bloomsbury Publishing Plc
50 Bedford Square, London, WC1B 3DP, UK
29 Earlsfort Terrace, Dublin 2, Ireland

BLOOMSBURY, HELM and the Helm logo are trademarks of Bloomsbury Publishing Plc

First published in Great Britain 2018

A catalogue record for this book is available from the British Library

Library of Congress Cataloguing-in-Publication data has been applied for

ISBN: PB: 978-1-4081-0533-7; ePDF: 978-1-4729-2564-0; ePub: 978-1-4729-2565-7

4 6 8 10 9 7 5

Designed by Julie Dando, Fluke Art
Maps on p. 13 and p. 32 by Brian Southern

Printed and bound in India by Replika Press Pvt. Ltd.

CONTENTS

PREFACE

For me, living in one of the richest bird countries on the globe, is a privilege comparable to being a champion footballer in the World Cup. It is both a huge honour, and an enormous responsibility.

Over the years, all Ecuadorian birdwatchers and ornithologists – myself included – have hoped to possess, either in the field or at home, a book that illustrates all the birds that occur in our country. The first-ever Ecuadorian ornithologist Gustavo Orcés no-doubt perused the pages of the eternal bible – *Distribution of Bird Life in Ecuador* (1926) – by Frank M. Chapman. Similarly, the prolific Fernando Ortiz Crespo pursued the birds of Ecuador, travelling with a copy of the *Landmark Birds of South America* (1966) by Rudolphe Meyer de Schauensee.

By 1991, Ortiz and Juan Manuel Carrión had published a small but great book (*Introduction to the Birds of Ecuador*, in Spanish). Their contribution to Ecuadorian ornithology made several blossoming naturalists and field biologists (again, myself included) to fall in love with birds, as had Steve Hilty and William Brown five years earlier with their outstanding *Birds of Colombia*. In Orcés and Ortiz's times, the desired field guide seemed a distant daydream. Nevertheless, Robert Ridgely and Paul Greenfield made this dream come true: in 2001, their magnificent two-volume *Birds of Ecuador* was launched, following a lengthy but fruitful two-decade 'pregnancy'.

For present-day ornithologists, this milestone book pointed the way in bird observation, identification, research, and, indirectly, conservation. Since Ridgely & Greenfield's book, a new stage in Ecuadorian ornithology has commenced, with more Ecuadorian birders in the field, more collaborative research between residents and visitors, and an increasing body of information on the natural history, distribution and conservation of the avifauna of mainland Ecuador. A latest important output of this new 'era' is the recent publication of a compact field guide by McMullan & Navarrete (2013).

Furthermore, Ridgely & Greenfield's, masterpiece also highlighted three crucial facts about Ecuadorian birding and ornithology: (1) populations of dozens of species are in sharp decline; (2) knowledge of the ecology and distribution of many birds is still insufficient; and (3) plumages of several species have not been illustrated and are, therefore, not well known – not even after the exceptional effort made by Paul to portray nearly 1,600 species in almost 2,500 individual images.

The task to illustrate all 1,640 species currently known to occur in mainland Ecuador (i.e., not including the Galápagos Islands) in as many plumages as possible (sexes, ages, breeding stages, morphs, subspecies) proved as great a challenge as when, in the early 2000s, Robin Restall put his health at serious risk in his efforts to paint almost 6,500 images for the *Birds of Northern South America*. Those images for Ecuadorian birds were subsequently adapted, corrected, some improved and redrawn, several others added, and compiled into the field guide you hold now.

Thirty years ago, Ortiz Crespo and his contemporary colleagues would have been sceptical as to the possibility of this achievement. I have been fortunate to live in a time of harvest, reaping what scores of colleagues have sowed. The praiseworthy work of Chapman, Ortiz Crespo and Carrión, as well as Ridgely & Greenfield, the founders of Ecuadorian ornithology and 'ornithophily' have inspired this field guide and were, naturally, prime sources of information. This book is also the result of the collective effort of dozens of ornithologists, birders, park guards, conservationists, naturalists and the like, from William Jardine and the first contributions to Ecuadorian ornithology in 1849, to the current revisions of the Ecuadorian avifauna undertaken by the Committee for Ecuadorian Records in Ornithology (CERO), to those young birders announcing what they observed yesterday afternoon via the country's electronic discussion group: Aves_Ecuador@yahoogroups.com, and other social media.

I have tried to synthesise and update in this book what the ornithological community of mainland Ecuador has hitherto learned concerning our feathered neighbours. Four attributes represent the backbone of this book: (1) a taxonomic update and revision following the South American Checklist Committee (SACC) of the American Ornithologists' Union; (2) an update of knowledge concerning the distribution and status of birds (i.e., new country records, range extensions, etc.); (3) new information on the natural history and vocalisations published by leading ornithologists working on Ecuadorian birds (e.g., Niels Krabbe, Harold Greeney, the late Paul Coopmans, John Moore, Olaf Jahn, Rudy Gelis, Alejandro Solano, Tatiana Santander, Diego Cisneros, Ana Ágreda, Patricio Mena, Juan José Alava, Ben Haase, Boris Tinoco, Jordan Karubian, Lelis Navarrete, Esteban Guevara, among others); and (4) the illustration of as many plumages as possible for every species (i.e., almost all subspecies, sexual dimorphism, age variation, morphs, seasonal variation, flight) known to occur in mainland Ecuador up to July 2107.

Our greatest wish was to illustrate the entire range of plumage variation found in Ecuador. Even though we have included almost 4,200 individual images in this book, we know that several plumages await discovery, description and, consequently, portrayal. My hope is that this new field guide to the birds of Ecuador will become a new tool, complementary to the keystone works already mentioned, in the better understanding of the avifauna of my home country, and for the everlasting survival of the Neotropical birds that manage to co-exist with us in tiny Ecuador.

ACKNOWLEDGEMENTS

Thanks to all fellow birders, ornithologists and conservationists, who have contributed in many ways; there are just too many people to mention them all. Those who have largely contributed with information, bird records, sharing birding journeys, ID discussions, input, etc. over recent years are marked with an asterisk: Adam Baty, Adrián Azpiroz*, Adrián Soria, Adriana Lara, Alejandro Aguayo, Alejandro Solano* (for many hundreds of birds seen together and thousands of topics fervently discussed!), Álex Boas, Ana Ágreda, Ana Charpentier, Andrea Nieto*, Andrés Cuervo, Andrés León, Andrés Vásquez, Andrew Spencer*, Ángel Paz, Arne Jent Lesterhuis, Aster Real Wallace, Ben Haase* (for always replying to my queries about seabirds and waders), Bert Harris, Beto González, Boris Herrera, Boris Tinoco* (for raising the bar in Ecuadorian ornithology), Borja Milá*, Brandt Ryder, Byron Palacios*, Carlos Cajas, Carlos Camacho, Carlos Morochz, Carlos Rodríguez*, Carlos Vinueza, Carmen Bustamante, Carola Bohórquez, Carolina Toapanta, Caroline Dingle, Catherine Graham, Catherine Vitts*, César Garzón*, César & Elicio Tapia, Charles Hesse, Charlie Vogt, Christian Devenish, Chris Canaday*, Chris James, Chris Sharpe, Clare Sobetski, Dan Lane*, Dana Houkal, Daniel Cadena, Daniel Martínez*, Daniela Bahamonde, David Brewer* (for kindly sharing unpublished Vireonidae information), David Díaz*, David Parra, David Pearson, Derek Kverno, Diana Esther Arzuza, Diego Calderón Franco, Diego Castro, Diego Cisneros Heredia*, Domingo Cabrera, Dušan Brinkhuizen* (for shaking the wasps' nest), Ecuador Experience, Eduardo Carrión Letort*, Eliana Montenegro, Elisa Bonaccorso* (for her patience, support and friendship), Esteban Guevara* (p.e.d.q.), Evelyng Astudillo, Fabián Cupuerán, Fabián Granda, Fabián Rodas*, Félix Man Ging*, Fernanda Salazar, Fernando Angulo, Fernando Castillo, Fernando Donoso, Fernando Nieto, the late Fernando Ortiz Crespo*, Fidel Chiriboga, Forrest Rowland, Francis Commerçon, Francisco Cuesta*, Francisco Sornoza* (for the hillstar), Freddy Cáceres, Freddy Gallo, Frederik Brammer, Fredrik Ahlman*, Fundación EcoCiencia, Fundación Numashir, Gabriel Bucheli*, Gabriela Castañeda, Galo Buitrón*, Galo Real*, Gary Stiles*, Giovanni Soldato, Glenda Pozo, Gonzalo Nazati, Gustavo Jiménez, Guy Kirwan*, Harold Greeney* (for all of his bird breeding and natural history data, and his deadly-hot curry), Héctor Cadena*, Heimo Mikkola, Henry Hernandez, Hernán Vargas*, Hugolino Oñate, Huilo Vaca, Irina Muñoz*, Itziar Olmedo, Jaime Chaves* (for being an excellent friend during my early years in ornithology and birding), Jaime García-Domínguez, Jaime Salas*, Janeth Lessmann, Jarol Vaca*, Joep Hendriks, Johan Ingels, John Gray, John Moore* (for generously providing collections of audio recordings), Jon Fjeldså, Jonas Nilsson* (for many new country records), Jordan Karubian* (for his friendship and steadfast encouragement), Jorge Bedoya, Jorge Correa, Jorge Olivo, José Cabot (for reviewing accounts of 'Variable Hawk'), Jose DeCoux, José Hidalgo, José Illanes, José Luis Yanza, José María de Zabala, José María Loaiza* (for the Red-ruffed Fruitcrow!), José Navarrete, José Simbaña*, Juan Carlos Valarezo, Juan Guayasamin, Juan José Álava*, Juan Luis Parra, Juan Manuel Aguilar, Juan Manuel Carrión*, the late Juan Mazar Barnett, Julio Loor, Karen Terán, Keith Barnes, Kerem Boyla* (for 'databasing' my life), Lani Miller, Latin Roots Travel, Laurent Raty, Lelis Navarrete*, Leonardo Ordóñez*, Lloyd Kiff, Lucho Carrasco*, Luis Agila, Luis Daniel Montalvo*, Luis Miguel Renjifo, Luis Saigaje, Luis Suárez, Luke Browne, Manuel Plenge, Manuel Sánchez* (for the chip chops), Marcelo Luque, Marcelo Mosquera, Marco Jácome, Marco Vinicio Salazar, Marcus Braun, María Laguna, Mariano Muñoz, Mark Robbins*, Markus Tellkamp*, Mathieu Siol, Martín Obando, Martin Riesing, Martin Schaefer, Matthias Heinz, Mauricio Guerrero, Mauricio Guerrón, Mauricio Vargas, Melina Costantino, Mercedes Rivadeneira, Mery Juiña, Milton Orozco, Mitch Lysinger* (for the 'San Isidro' Owl), Mónica González, Mort Isler, Murray Cooper* (for his generosity with hundreds of photos, enjoyable field trips and friendship), Nacho Areta,

Nancy Hilgert*, Neblina Forest Birding Tours, Nick Athanas*, Nicole Büttner, Niels Krabbe* (for sharing his massive knowledge of the birds of Ecuador), Norby López, Normand David, Olaf Jahn*, Orfa Rodríguez, Orlando Carrión*, Óscar Laverde, Óscar Tapuy, Ottavio Janni, Pablo Menéndez, Paco Hernández*, Pancho Prieto*, Paola Gastezzi, Paolo Piedrahita*, Patricio Mena Vásconez, Paula Enríquez, the late Paul Coopmans* (for brilliantly responding to so many e-mails), Paul Greenfield* (for sharing his vast experience as birder and bird artist), Paúl Tito*, Paul Van Gasse, Paulo Catry, Paulo Pulgarín, Pedro Astudillo, Peter Ginsburg, Pierre Ives Henry*, Randy Vickers, Rasmus Boegh, Red Aves Ecuador, Redmar Woudstra, Renata Durães, Richard Mellanby, Richard Waldrop, Rob Clay, Rob Williams*, Robert Ridgely* (for sharing data and knowledge, and for endless e-mail discussions of species distribution and status), Roberto Zamora, Roger Ahlman* (for his massive collection of photographs and so many new and surprising bird records), Rolando Hipo, Ronald Navarrete*, Rudy Gelis*, Ruth Muñiz, Sandra Loor Vela, Santiago Lara, Santiago Torres, Santiago Varela*, Santos Patiño, Scott Olmstead*, Scott Walter, Sebastian Herzog, Sergio Basantes, Sergio Córdoba, Sergio Seipke, Servio Pachard, Stefan Kreft, Steve Bright, Steve Hilty (for his initial support and encouragement), Tarsicio Granizo, Tatiana Santander* (for IBA co-work), Thierry Garcia, Thomas Donegan, Tino Mischler, Tjitte de Vries* (for reviewing accounts of 'Variable Hawk'), Tom Schulenberg*, Tom Smith, Trotsky Riera, Tuomas Siemola, Van Remsen*, Vanessa Barham, Vincent Mouret*, Vinicio Pérez, Vítor Piacentini, Will Cresswell, Xavier Amigo* (for many conversations about music, great birding trips and the Pink-throated Brilliant!), Xavier Muñoz*, Willan Poveda and Zoltan Waliczky. If I forgot someone, my most sincere apologies.

For all I have learnt from them: the late Fernando Ortiz Crespo, Tjitte de Vries, the late Paul Coopmans, Niels Krabbe, Elisa Bonaccorso, Bette Loiselle, John Blake, Ben Haase, Robert Ridgely, Lelis Navarrete, Paul Greenfield, Borja Milá, Jordan Karubian and Alejandro Solano.

To Robin Restall, for his outstanding effort in illustrating *all* the birds of northern South America, and his enthusiasm for this project. To Nigel Redman for permitting me to produce this field guide; Julie Dando for designing it; Steve Hilty, David Ascanio and Gustavo Rodríguez for partial or concurrent work during this project; Jim Martin, Jane Lawes and Jenny Campbell for final editorial work; and Gustavo Rodríguez and Esteban Garcés for their superb work in image editing.

The maps would never have been as neat as they appear now without the invaluable support provided initially by Elisa Bonaccorso, Ana Charpentier and Itziar Olmedo. Subsequently, the kind collaboration of Fernando Espíndola, Alejandro Aguayo, Jorge Bedoya, Esteban Garcés and Johana Simbaña was so crucial that I will always be indebted to them. Fernando's and Johana's work was especially helpful. Curators and collection managers of the following museums and natural history collections are also acknowledged for data shared in respect of this or other projects: American Museum of Natural History (AMNH), New York City; Academy of Natural Sciences of Philadelphia (ANSP); Natural History Museum, Tring (NHMUK); California Academy of Sciences (CAS), San Francisco; Canada Museum of Nature (CMN), Ottawa; Cambridge University Museum of Zoology (CUMZ); Delaware Museum of Natural History (DMNH), Wilmington; Denver Museum of Nature and Science (DMNS); Escuela Politécnica Nacional (EPN), Quito; Florida Museum of Natural History (FLMNH), Gainesville; Field Museum of Natural History (FMNH), Chicago; Instituto de Biología de la Universidad Autónoma Nacional de México (IBUNAM), Mexico City; Instituto de Ciencias Natural, Universidad Nacional de Colombia (ICN), Bogotá; Illinois State Museum (ISM), Springfield; Instituto Nacional Mejía (INM), Quito; Kansas University Natural History Museum (KU), Lawrence; Los Angeles County Museum (LACM), Los Angeles; Louisiana State University Museum of Zoology (LSUMZ), Baton Rouge; Museo Nacional de Ciencias Naturales de Madrid (MCNM); Museo Nacional de Costa Rica (MNCR), San José; Museum of Science, Boston (MOS); Manchester Museum (MMU); Museum of Comparative Zoology, University of Harvard (MCZ), Cambridge, MA; Museo Ecuatoriano de Ciencias Naturales (MECN), Quito; Museo de Historia Natural Javier Prado, Universidad Mayor San Marcos, Lima (UNMSM); Museu de Zoologia Universidade de São Paulo (MZUSP); Muséum d'Histoire Naturelle, Neuchâtel (MHNNL); Muséum d'Histoire Naturelle, Geneva (MHNG); Muséum d'Histoire Naturelle, Paris (MNHN); Natural History Museum, Luxembourg (MNHNL), Luxembourg City; Museo di Storia Naturale, Università di Pisa (MSNTP), Calci; Michigan State University Museum (MSU), East Lansing; Museum of Vertebrate Zoology, Berkeley (MVZ); Moore Laboratory of Zoology (MLZ), Los Angeles; North Carolina Natural Science Collections (NCSM), Raleigh; Naturhistoriska Riksmuseet, Stockholm (NRM); New York State Museum (NYSM), Albany; Osaka Museum of Natural History (OMNH); Oxford University Zoology Museum (OUZM), Oxford; Museo de Zoología, Pontificia Universidad Católica del Ecuador (QCAZ), Quito;

Naturalis, Leiden (RMNH); Royal Ontario Museum (ROM), Toronto; Royal Saskatchewan Museum (RSM), Regina; Santa Barbara Museum of Natural History (SBMNH); Texas A & M University (TAMU), College Station; Texas Cooperative Wildlife Collection (TCWC), College Station; Universidad de Costa Rica (UCR), San José; University of Arizona (UAZ), Tucson; University of California, Los Angeles (UCLA); Museum of Zoology University of Michigan (UMMZ), Ann Arbor; University of Missouri, St. Louis (UMSL); University of Nebraska Science Museum (UNSM), Lincoln; University of Washington Burke Museum (UWBM), Seattle; Western Foundation of Vertebrate Zoology (WFVZ), Camarillo; Yale Peabody Museum (YMP), New Haven; Zoology Museum University of Copenhagen (ZMUC). Again, Elisa Bonaccorso and her team are thanked for compiling most of this information.

Last but not least, to my much-loved family and friends for their permanent support and inspiration. Thanks go to Rubén Freile, Elvia Ortiz, Gabi Freile, Elena Freile, Bolívar Ullauri, Santiago Oliva, Martín, Emilia & Bernardo Ullauri, María Chiliquinga, my Freile Ortiz aunts, uncles, and cousins; Martín, Alegría and Sergio Bustamante, Gabriela Granda, Hugo Mogollón, Ernesto Pfafflin, Cristina Wasner and family, Amaranta Valencia, Oso Barreto, Palli, Lola, Tito, Julián & Lucas Guarderas, Mateo Vinueza, Tina Gleich, Xavier Cisneros, Nicole Marchán and family, Esteban Garcés, Omar Torres, Selene Báez, Andrés Vallejo and family, the Ecuador Terra Incognita crew (Gabi, Mariela, Carlitos, Victoria, Mónica and Nadya), Nanis & Pancho Cordovez, Silvana Sáenz, Juan & Nina Guayasamin, Elisa Bonaccorso, Jaime Chaves, the late René Fonseca, the late Carlos Boada, Santiago Espinosa and family, Lorena Endara, Chichi Terán, Melissa Moreano, Fede Brown and Antonia, Blanca Ríos, Xavier Amigo, Geovanna Lasso and Julia, Fabricio & Amaru Guamán, Antonia Manresa, Inti, Manuela, and Iara Arcos, Emiliano Granda, Polo and Margarita, Alejo Solano, Agus Arcos, Vane Wurth, Nina Duarte, the Torres brothers, Mare Arcos, Lia & Copal, Ana Cobano Gael and Saru, Gabriela Castañeda, Cata Carrasco, Ailín Blasco, Leo Santillán, Miguel & Gabriel Torske, Pauli Lasso, Ana María Vintimilla and Damián, the Durán sisters and Aitano, Javier & Gael Carrera, Fer Meneses, Rogelio, Johana & Mishelle Simbaña, Martha Guamán, the Tola Chica community, Karla Muñoz, Gene and Camila, the Tyod crew, Mimi Foyle, Jaime West, the Mashpi community, the Ortiz-Acosta family, George Fletcher and family, Murray Cooper and family, the Puerto Cabuyal community, Anita Torres, Raúl, María Elena, Mafe & Negro Moscoso Rosero, Alegría Acosta, Martina Avilés, Narci and Karina Pastrana and family, Marimba, Tami and Piñu. Lastly, my endless gratitude and affection to Paola Moscoso for the good days.

Juan Freile

INTRODUCTION

More than 6,000 birdwatchers every year cannot be wrong: Ecuador is a fantastic destination for birders worldwide. Full of birds from north to south, every corner of this tiny country offers superb opportunities for watching birds. Irrespective of your location: the suburbs of a large city, or the canopy tower at an Amazonian lodge, if you keep your eyes and ears wide open, you will enjoy as many birds as only one of the richest Neotropical countries can supply.

Let the roads (generally good) take you throughout the country in a handful of hours, from lush Amazonian jungles to misty cloud forests on its Andean slopes; from the startling dry forests of the south-west to the super-wet Chocó rainforests of the north-west; or from the cold and diverse northern *páramos* to the southern Andes packed with endemics. All of these areas will certainly fill your mind with the colours and sounds of in excess of 1,600 species. What could be better?

This guide aims to be your field partner to the great diversity of birds found in mainland Ecuador, and as much plumage variation as possible. Literally, thousands of different plumages are shown here. We left virtually no subspecies out, either those living and breeding here, or those that visit Ecuador only seasonally, occasionally or even accidentally. This field guide, the ideal counterpart to the monumental *Birds of Ecuador* (Ridgely & Greenfield 2001, 2006) presents brief descriptions of distribution, habitat, plumage, behaviour, voices, similar species, and status of all species currently known to occur in Ecuador, excluding the superb Galápagos archipelago. In all, 1,636 species are illustrated – including five species no longer considered to occur in Ecuador due to prior misidentifications, taxonomic changes or distributional revision, four species that probably don't occur in Ecuador, and one taxon of uncertain taxonomic status. The updated mainland Ecuador bird list in the appendix includes 1,640 species (with 53 species requiring supporting evidence in the form of a specimen, published photograph or audio-recording; four species are not illustrated). At least 17 species suggested by Restall *et al.* in *Birds of Northern South America* to occur in Ecuador were not included due to lack of evidence, as well as another 30+ species assigned to the Ecuadorian avifauna by different authors without any supporting documentation.

In the following chapters, we summarise the geography, climate, vegetation, diversity, conservation and current state of ornithology in Ecuador, along with an explanation of this field guide's contents, how to use it, and how it was produced.

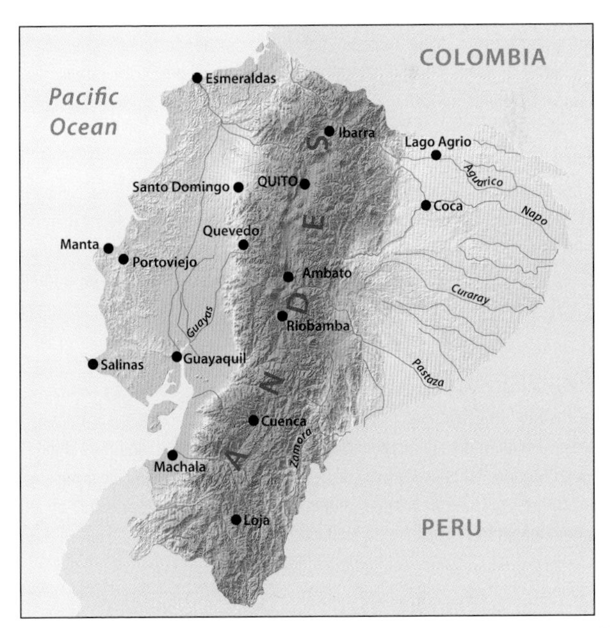

Topographic map of Ecuador, with major rivers and main cities. Green indicates areas below 100m elevation; ochre shades indicate areas above that elevation and brown shades indicate areas above 1000m.

HOW TO USE THIS BOOK

TAXONOMY AND NOMENCLATURE

In tandem with the current flow of taxonomic revisions, this field guide follows the South American Classification Committee (SACC) taxonomy nearly universally (up to December 2015). What makes SACC taxonomy robust is the fact that all taxonomic and nomenclatural changes to be adopted – hence, the resulting South American checklist – require strong supporting evidence, based on scientific studies published in peer-reviewed journals. Is this mere scientific arrogance? Not really. It is simply an evidence-led process, just like those followed by detectives, journalists or attorneys trying to solve their own cases. It is a matter-of-fact approach; 'seeing is believing'. In the Appendix that closes the book is an updated and revised checklist of the birds of mainland Ecuador, up to December 2015, as validated by the CERO Committee. In the Appendix, but not necessarily the plates, we have adopted the December 2015 SACC sequence of genera and species within families.

Additionally, a shorter annotated list precedes the checklist of the birds of Ecuador, presenting a selection of species that might eventually be found in Ecuador (Potential species list), particularly long-distance migrants. These 40 species are known to range close to Ecuador's current boundaries, some just a few km away, and are included as an alert for observers. We acknowledge that other species might well also show up, though.

Plate and species accounts sequences primarily follow recent suggestions for field guide arrangements at family-group level made by Howell *et al.* (2012), coupled with SACC taxonomic sequence. We have tried to adopt Howell *et al.*'s suggestions as much as possible, but as their article appeared when our species accounts were already planned, following wholeheartedly would have necessitated entirely reorganising the book (which was not plausible). Likewise, we failed to adopt the latest changes in families and orders presented by SACC, as they are so revolutionary that another major reorganisation would have been needed (see the Howell *et al.* 2012 discussion about taxonomy in field guides).

English names also follow SACC, with a single exception that will be explained below. Unfortunately, Spanish names are not included in the species accounts due to space constraints.

SPECIES ACCOUNTS

Species accounts and plates for 1,640 species, starting with tinamous and ending with introduced passerines, form the bulk of the book. Short headings precede each set of species accounts, in which general information about the species included on each plate is presented. Each species account contains: English and scientific names, measurements, distribution, elevation, habitats, plumage description, subspecies, behaviour, voice, similar species, and status. Emphasis is laid on field identification; however, due to space constraints, descriptions are brief and focused mainly on typical adult plumage (both sexes if needed), with additional notes on ages or morphs if relevant. In albatrosses, petrels, shearwaters and storm-petrels (the Procellariiformes), all barely vocal away from the breeding grounds, voice descriptions are replaced by flight descriptions, as flight is a relevant identification character. Flight is also described for species that are mostly seen in flight, like swifts and diurnal raptors.

Descriptions of plumage, behaviour and similar species are based on the senior author's field experience, published literature, some unpublished sources, and experiences shared by fellow birders and ornithologists.

Measurements

Taken mostly from published sources, measurements (in cm) are primarily taken from museum specimens, so do not necessarily reflect real sizes. However, they do provide a general idea for comparisons across species. It must be noted that it is not easy to accurately appreciate size differences in the field, and they should only rarely be used as a sole identification criterion. Additional measurements are provided for specific groups; wingspan for diurnal raptors and Procellariiformes, bill lengths for many hummingbirds (when this represents an identification feature) and tail length in some relevant instances.

Distribution and elevation

Descriptions of distribution are not detailed as they might easily fill a full page for a single species. They are simple descriptions of the whereabouts of species in Ecuador, in terms of geographic allocation (N, S, E, W, etc.) and altitudinal regions (tropics, foothills, subtropics, etc.). See Geography and Climate on page 21 for descriptions of altitudinal regions. Aside from typical terms for geographic and altitudinal allocation, some descriptors are used for more specific examples: adjacent lowlands for areas immediate to foothills (versus extreme lowlands); wet NW for the Chocó lowlands and foothills (versus arid SW); Andean slopes (when a species' distribution covers several altitudinal zones); Inter-Andean slopes for slopes above the central Andean valleys; highlands for all areas in the Andes above the subtropics. Also others like inland and offshore.

In some cases, distribution is combined with altitudinal range or habitat, mainly if number of records is few, hindering the definition of range. Where appropriate, localities are used for more specific distributions or scanty knowledge, including regions (e.g., Santa Elena Peninsula, Río Napo, Cordillera del Condor, Zumba), provinces or localities (e.g., Mar Bravo, Salinas). Likewise, in a few instances of species with one or just a few records, names of observers are included.

Do not expect precise wordings for distribution descriptions. Although terms for geographic allocation and altitudinal regions are standardised throughout the book (see above), we have used equivalent words for cases with few records (scattered, few, scanty, single record, recorded once) and discontinuous ranges (sparse, local, spottily). We have also introduced some terms more appropriate for status descriptions, but considered adequate for describing the distribution of migratory (austral, boreal, intra-tropical migrants) or casual species (vagrant, wanderer, transient, seasonal), as well as for describing historical (formerly, historically, rarely), expected (spreading) or even current status (widespread versus uncertain).

Altitudinal ranges are taken from published literature, with Ridgely & Greenfield as a base (owing to their exceptional revision of pre-2000 ornithological literature and thousands of museum specimens), but with updates, revisions and corrections based on post-2000 literature compiled by Solano & Freile (2012), a collection of unpublished field data by reliable observers, data reported to CERO, and further revision of museum specimens. Extralimital records often accompany ranges; i.e., those seemingly outside the 'normal' range. In some cases, several upper or lower elevations are mentioned. Needless to say, elevation ranges are not absolute; they can change regionally (lower in the southern Andes), locally or seasonally. They can also vary with time, following deforestation (species of open areas spread into newly open lands) and shifts are even expected due to global climate change. Therefore, we urge readers to contact the senior author if they have new records to report; even better, we encourage readers to publish their records or to report them to the Committee for Ecuadorian Records in Ornithology (Comité Ecuatoriano de Registros Ornitológicos, CERO at: www.ceroecuador.wordpress.com).

Non-Ecuadorian distribution is not included.

Habitat

Brief habitat descriptions are based on the general classification presented in the Habitat section. It is not a specialised classification, but one that aims to guide observers to the habitats where birds are more frequently found. Vertical stratification of forests is also included (understorey, subcanopy, canopy), as well as if a species occurs in borders or the forest interior. Different types of coastal habitats are described (e.g., rocky or sandy beaches, mudflats, estuaries, etc.).

Plumage description

As is conventional in field guides, males are described first, followed by females when a species is sexually dimorphic. Likewise, adults are described first, as well as breeding versus non-breeding plumages. However, for migrants not expected to occur in breeding plumage, non-breeding plumage is described in more detail. For females, juveniles and other 'variants', only those characters that differ from males, adults or breeders, respectively, are described. Subspecies are not fully described; differences are given primarily if remarkable, as in the case of races that might deserve species status. Subspecies are not mentioned in plumage descriptions, but a geographic reference is employed.

Generally, plumages are described from front to rear and from upper- to underparts, with soft parts being mentioned first and tail last. Birds often seen in flight have separate descriptions of plumage as seen from below; in a few cases (e.g., some seabirds, swifts) also from above. Key identification features are *italicised*, whereas in few cases, **bold** is used to highlight plumage phases or morphs.

Subspecies

Subspecies (or races) are generally only mentioned, not discussed, with references in parentheses to geographic range. In most species, subspecific designation follows Clements *et al.* (2016), with deviations to the comprehensive review made by Robert Ridgely in vol. 1 of the *Birds of Ecuador* when more appropriate (or by Moore *et al.* 2013 in a couple of instances). Moreover, updates and modifications are incorporated following an exhaustive revision of recent literature (including recently described taxa, e.g., *Amazilia amazilia azuay*). Roughly 1,300 subspecies currently recognised occur in Ecuador (not counting monotypic taxa that comprise *c.*300 species), plus *c.*70 that are probably present and at least a dozen seemingly undescribed forms. An additional note is added when subspecies might merit species status, or if there is taxonomic controversy, but no decisions are taken.

Behaviour

Regrettably, behaviour and habits are described in a very succinct form due to space constraints. Nonetheless, we have tried to summarise important aspects for field identification or for finding birds in the field. Information included herein varies across families and species groups, as different traits are more important for some but less for others. Accordingly, feeding behaviour, association with mixed-species flocks, foraging substrates and strata, flight and social habits are sometimes described.

Voice

Voice descriptions are based on the senior author's field experience and a meticulous auditory revision of the Moore *et al.* series, Krabbe & Nilsson (2003) DVD, and www.xeno-canto.org. Songs are usually described first, followed by the most commonly heard calls; in some cases, only calls are mentioned. Unfortunately, space constraints do not permit extensive descriptions of 'all' vocalisations uttered by a species. As conventional in field guides, voice descriptions are intended as phonetic representations of bird vocalisations, using qualitative modifiers and onomatopoeic similes, using Standard English phonetics. As the senior author is not a native English speaker, however, any slip from English pronunciation must be excused! Abbreviations in parentheses after some voice descriptions denote published sources (see Abbreviations pp. 19–20).

The following variants are used for characterising phonetics:

Accent: stressed syllables (*chép*).
Capital letters: even more stressed syllables (*weED-weED*).
Exclamation mark: highly stressed syllables often ending abruptly (*wéeoo!*).
Question mark: inflected syllables, primarily at end of a phrase (*kee-woop?*).
Vowels: denote pitch, in following order from low to high: u-o-a-e-i (*tseE-tsu-tsu*).
Dieresis: resonant syllables, notes or phrases (*cäo-cäo-cäo*); for more resonant sounds, a series of *ooo, uuu, hmm, mmm* are used.
No vowels: more mechanical sounds with 'no discernible vowels' (*grrrk-g'kk*).
sh, dz, z, zz, zh, jz: increasingly buzzy.
kk, kh, j, dj, tj: increasing harshness.
ch, tch: chipped sounds.
nn, nh, wh, jn: nasal sounds (also an *h* at end of syllable).
wr, kr, kh, kk: metallic sounds.
s, th, ts, ps, sw, w, ch, d, y, l, f, b, v, z, followed by a vowel: for softer sounds, whistles or chips.
k, j, g, c, q, t, tch, p, r, h, followed by a vowel: for rougher sounds.
trrr, prrr, krrr, frrr, brrr: for fast rattles, trills and similar sounds.
tzzzz, fzzzz, bzzz: for fast shrill or buzzy sounds.

Increasing length of pauses between notes or phrases is indicated as follows:

Consecutive syllables: fast voices with no discernible separation (*dududu*); often rattled, trilled or rolled sounds.

Apostrophe: very short pause (*ti'di'di*); rattled, trilled or shrill sounds. An apostrophe at the end of a phrase denotes slightly rising terminus.

Hyphen: short pause (*chip-chip-chip*).

Comma: longer pause (*shee-u, shee-u*).

Space: even longer pause (*swee swee*); sometimes replaced by two consecutive points (*keee..e..uu*).

Suspension points: phrase or syllable series continues in a similar way (*sweép-sweép-sweép...*); in one case (Scaly-breasted Wren), increasing number of points denotes increasing distance between notes.

Vocal variation is not fully described, but mentioned in a few cases when geographic variation is remarkable and might involve separate species. Please note that vocal transcriptions are only 'the next best thing' to nothing. Do not expect bird vocalisations to 'sound' exactly like their transcriptions, as local, racial, seasonal, age-related or individual variation in bird voices can make them sound quite different. Also, what sounds like a *tchip* to me, might sound like *tseep* for you, as everyone's ears operate differently. Although vocal descriptions are useful in the field, nothing compares to learning vocalisations. Luckily, one of the largest published audio collections in the Neotropics corresponds to the Ecuadorian avifauna, recently summarised by Moore *et al.* (2013). As important as learning how to identify birds visually, is learning to identify them by ear. In fact, many Neotropical birds are much more often heard than seen. It is a matter of enjoying sounds as much as colours and plumage patterns.

Similar species

Similar species that overlap in distribution are described here, by necessity in summary fashion. In some cases, potentially or locally overlapping species, or species that do not overlap, are also included, whilst in very few instances species not yet found in Ecuador (though not wholly unexpected) are compared – especially for tricky species. When two species are very similar, comparisons are made directly within plumage descriptions, whereas **SS** is used for other potentially confusing species. In complex groups like Procellariiformes, raptors or waders, readers are referred to specialised guides and monographs. No references are provided to plates where similar species are presented in cases where the relevant species do not appear on the same plate.

Status

Relative abundance, conservation status, migratory status and months present in Ecuador are summarised. Defining the abundance or population density of birds require specifically designed protocols within a given timespan. Such information is not available for Ecuadorian birds as a whole, but only for few localities where intensive research has been undertaken. And, even at well-known sites, abundance can change with seasonal availability of food resources, differential habitat preferences for breeding and non-breeding seasons, and other aspects of a species' natural history. Furthermore, birds tend to be more numerous in preferred habitats, easier to detect at certain seasons (e.g., more vocal, flocking behaviour, gathering at rich food sources), and more abundant at the centre of their ranges than at borders.

Here, a rough estimation of relative abundance is presented, as an average of species abundance in areas where they are regularly – or mostly – found. Do not expect these abundances to hold up for all species in every corner of Ecuador, but take relative abundances as a reference for your expectations.

All in all, the following abundance and status categories are employed:

Common: seen daily in moderate to large numbers.

Fairly common: seen daily or nearly so in moderate to low numbers.

Uncommon: seen every few days, numbers moderate to low.

Rare: small numbers; might go undetected in short to mid-length visits.

Local: patchy populations.

Casual: occasional visitor, not following any regular or seasonal pattern.

Vagrant: just one or few records, outside regular range or migratory route.

Accidental: just one or few records, well outside regular distribution or migratory route; very low likelihood of being recorded again.

Erratic: unpredictable visitor.

Transient: in-transit migrants that remain short periods in the country.

For conservation status, IUCN classification is followed, both at global and national scales (Granizo *et al.* 2002, BirdLife International 2016). In some cases, a suggestion as to potential threat status is given.

Migratory status and months when present in Ecuador follow major published sources and an extensive review of recent literature, well-supported records by reliable birders, and records submitted to CERO; no discussions are made of species' migration. Species are classified as winter residents, transients, vagrants and casuals.

DISTRIBUTION MAPS

Maps were developed using Maxent and Quantum GIS 1.8 software, through the outstanding abilities of Fernando Espíndola, in Quito. After initial range modelling, maps were edited 'manually' using Adobe Photoshop CS5.1 to better depict species distribution. In most cases, ranges depicted do not exclude already deforested areas. Ranges illustrated are not as 'precise' as niche models can be, in part due to scale. It should be noted that distributions can change over time or across regions. For instance, Andean species tend to range lower in the southern Andes – which are also lower than the northern Andes.

Yet, we believe that our maps represent a very good approach to actual and updated species distributions. Major sources of information include: (1) a database constructed by Elisa Bonaccorso, Alejandro Aguayo, Itziar Olmedo, Ana Charpentier, Jorge Bedoya and the senior author (with major contributions by Niels Krabbe and Pancho Cuesta), which includes locality data for *c.*80% of Ecuadorian museum specimens worldwide; (2) an extensive review of literature, especially a very thorough revision of post-2000 material; and (3) countless reliable unpublished records (up to December 2016) also compiled by the senior author (mainly with Alejandro Solano-Ugalde and CERO). All in all, we have compiled >70,000 individual records. Range maps published by previous authors (Krabbe *et al.* 1998, Ridgely & Greenfield 2001, and others) were used as comparative material. The maps represent graphic hypotheses of species distributions and might be improved or corrected after further knowledge is acquired. Updates and corrections should be sent to the senior author.

Key to the distribution maps

As is conventional in a field guide, green denotes breeding resident species. Boreal migrants are shown in blue; austral migrants in ochre; intra-tropical migrants or wanderers from other tropical areas in red. Areas where the ranges of resident and migrant populations of species overlap are depicted using a green background (for the resident population), with a hatched foreground colour corresponding to its migratory origin. Altitudinal migration is not indicated on the range maps, as knowledge of these movement patterns in Ecuador is too deficient. Some maps include only locality dots, mostly for species with scattered records. Localities marked as empty dots reflect historical records. In accordance with colours used for ranges, dots for little-known species – including historical records – are shown in the corresponding colour. A question mark is used if a distribution is not well understood. Subspecies ranges are not indicated on the maps.

breeding resident species

boreal migrants

boreal migrant overlap

austral migrants

austral migrant overlap

intra-tropical migrants or wanderers

? prescence uncertain

• locality dot (for species with scattered records only)

○ historical records

All maps include country limits, three shaded areas representing the 1,000m altitudinal contour line, and seven major rivers: Aguarico, Napo, Curaray, Pastaza, Zamora, Guayas and Esmeraldas. Mainland areas outside species' ranges appear in grey, whereas oceanic areas are in pale blue. Ecuadorian oceanic territory (20 nautical miles) is not fully included in the map, as limits are not easily measured in the field and because our knowledge is extremely scanty.

PLATES

Some 1,646 species are illustrated on 291 plates: 139 plates cover non-passerines and 152 the passerines; plates 140–226 show the suboscine passerines; 227–243 various oscine families; and 244–291 the nine-primaried oscines.

Most plates include the same taxonomic families or orders. A few exceptions exist though. For example, seedsnipes (Charadriiformes) appear on a plate with single-species families of Gruiformes (trumpeter, Sungrebe, Sunbittern and Limpkin).

Plate organisation follows a simple rule: not to pack too many (or too few) images on a single plate. Most plates depict 5–7 species, linked by: (1) taxonomic relationships; (2) similarity; (3) geographic ranges; and (4) page set-up convenience in very few cases. Again, a few exceptions exist. Plate 66 shows only two 'species'. This plate is also exceptional on taxonomic grounds, as this is the only case in which we almost abandoned SACC taxonomy, with the purpose of highlighting the remarkable case of 'Variable Hawk', which is considered a single, variable species by SACC but treated as two species (Red-backed and Gurney's Hawks) by José Cabot and Tjitte de Vries based on their exhaustive ongoing research. Whether the approach by SACC or Cabot & de Vries is the correct one will only be elucidated by further research. Likewise, Plate 242 includes only four vireos as the complex subspecific variation in Red-eyed Vireo requires much explanation; additionally, a handful of additional plates show four or eight species.

Overall, 4,200 individual images illustrate nearly all of the subspecies recorded in Ecuador, as well as morphs, much sex-related, age-related, moult-related and some individual variation. Even this huge number of images is insufficient to illustrate all of the variability one might encounter in the field; variability we are still far from fully understanding. Consequently, any comments and suggestions for improving the illustrations (as well as species accounts and maps) are welcomed.

Scale is the same within each plate but not across plates; however, inset images, particularly of birds in flight, are necessarily presented at a smaller scale, as should be obvious.

ABBREVIATIONS

The following abbreviations and symbols are employed; months are also abbreviated:

Alt:	alternative English name
ad.	adult
AMNH	American Museum of Natural History.
NHMUK	Natural History Museum, Tring.
C	central
c.	*circa*, approximately
cf.	compare with
CR	Critically Endangered status
DD	Data Deficient status.
E	east
EN	Endangered status
imm.	immature
juv.	juvenile
LE	Locally Extinct (in Ecuador)
N	north
NE	north-east
NT	Near Threatened status

NW	north-west
prov.	province/provinces
S	south
SE	south-east
sec.	second(s)
sp.	species (unidentified species)
SS	similar species
SW	south-west
VU	Vulnerable status
w	wingspan
W	west

Species lacking solid evidence of their presence in Ecuador (specimens, published photographs, archived sound-recordings or video-tape) are marked with an asterisk (*);

introduced species are marked i;

species no longer included on the Ecuadorian bird list – or species that should be removed – owing to updated knowledge of their distribution and systematics, and corrections to previous identifications (but retained here for comparative purposes) are marked †;

uncertain information is marked with a question mark (?).

Abbreviations of published sources for voice descriptions

C *et al.*	Curson *et al.* 1994. *New World Warblers.*
ChD	Chantler & Driessens. 1995. *Swifts: A Guide to the Swifts and Treeswifts of the World.*
FK	Fjeldså & Krabbe. 1990. *Birds of the High Andes.*
Harrison	Harrison. 1987. *Seabirds of the World.*
H *et al.*	Hayman *et al.* 1986. *Shorebirds. An Identification Guide.*
HB	Hilty & Brown. 1986. *A Guide to the Birds of Colombia.*
HBW	*Handbook of the Birds of the World* series.
JB	Jaramillo & Burke. 1999. *New World Blackbirds.*
KG	Kirwan & Green. 2011. *Cotingas and Manakins.*
KN	Krabbe & Nilsson. 2003. *Birds of Ecuador DVD.*
LJ	Jonsson. 1992. *Birds of Europe.*
MB	Madge & Burn. 1988. *Waterfowl. An Identification Guide to the Ducks, Geese and Swans of the World.*
NN	Cleere & Nurney. 1998. *Nightjars: A Guide to Nightjars and Related Nightbirds of the World.*
R *et al.*	Restall *et al.* 2006. *Birds of Northern South America: An Identification Guide.*
RG	Ridgely & Greenfield. 2001. *Birds of Ecuador.*
RT	Ridgely & Tudor. 2009. *Birds of South America. Passerines.*
S *et al.*	Schulenberg *et al.* 2007. *Birds of Peru.*
TvP	Taylor & van Perlo. 1998. *Rails: a Guide to the Rails, Crakes, Gallinules and Coots of the World.*

GEOGRAPHY AND CLIMATE: AN OVERVIEW

Ecuador has traditionally been classified in four geographic regions; three in its continental portion, and the Galápagos archipelago, which is not included herein. Being a tiny country, its geography is rather 'simple' compared to that of neighbouring Colombia and Peru. The Andes cordillera, running north to south, represents the main feature of Ecuadorian geography, and divides the country's western and eastern lowlands.

The eastern lowlands (or Amazon) descend smoothly from 500–600m to *c.*200m elevation, with hilly terrain dominating areas between large rivers, mainly south of the Napo. North of this river terrain is more level. Two large rivers transect the eastern lowlands, namely the Napo – the largest Amazon tributary in Ecuador – and Pastaza. Meanwhile, the following fairly large rivers form other important drainages: Aguarico – north of the Napo – and Curaray – north of the Pastaza. Both run east and empty into the Napo. Further south, the Santiago, Morona, and Zamora Rivers run south and empty into the Marañón River, in northern Peru.

Westwards, the Andes are preceded by isolated mountains (Sur Pax, Sumaco-Pan de Azúcar massif, and Reventador) and ridges (Galeras, Kutukú and Cóndor). These mountains reach up to 3,000m elevation and the ridges up to *c.*2,500m. Mountains in the north are linked to the Andes, whereas the Kutukú and Cóndor cordilleras in the south are topographically, structurally and historically rather different (having a separate geological history to the Andes). These cordilleras somewhat resemble the distant tepui massifs of the Guianan Shield, being covered in stunted shrubs and forests.

The Andean region of Ecuador (*Sierra*) comprises two separate and parallel (north–south) cordilleras, the inter-Andean valleys between them, and small east–west ridges connecting the eastern and western Andes (*nudos*).

The eastern slope of the east Andes rises rather steeply from 500–600m in the Amazonian foothills to 5,000+m in the snow-capped Cayambe, Antisana, Tungurahua, Altar and Sangay volcanos. Several deep valleys intersect the eastern Andes; these are, from north to south: the Pastaza, Paute and Zamora. Of these depressions, the Pastaza presents a drier climate and consequently drier vegetation than the neighbouring slopes. The Zamora Valley has been hypothesised as an important barrier to the distribution of several species or subspecies, with some examples also in the Pastaza. Furthermore, a distinct valley opens southwards in extreme south-east Zamora Chinchipe province (in the Mayo River drainage), draining towards the dry Marañón valley of northern Peru and supporting a distinctive semi-dry vegetation.

The inter-Andean slopes of both Andean chains are less steep and drier than the outward-facing slopes (and are largely deforested now). Similarly, natural habitats in the inter-Andean valleys are almost totally degraded. These valleys are settled at elevations above 1,500–2,000m, and drain either east or west; no valley drains into both tropical regions. From Chimborazo to the north, the valleys are wider and some *nudos* are peaked by gigantic volcanos (Chimborazo, Cotopaxi) or tall mountains (Sincholagua, Rumiñahui, Carihuairazo). The topography of the southern Andes is more complex, but lower, rarely surpassing 4,000m. Also, the southern valleys are narrower.

The western slope of the west Andes, facing the Pacific lowlands, descends abruptly from *c.*5,000m (at the summit of Illinizas and Cotacachi volcanos) to 500–600m, in the western foothills. West-opening river depressions also intersect the west Andean chain, most being narrower and much drier than those in the east. Therefore, vegetation differs more notably from humid slopes. These river valleys are the Mira, Guayllabamba, Chimbo, Chanchán, Jubones, Puyango and Catamayo. Of these, the Mira, Chimbo and Chanchán apparently represent important limits to the current distributions of some species.

The western lowlands (*Costa*) are quite level. The large Daule-Guayas River basin, running north–south, forms a large and flat floodplain in the centre of this region. This floodplain is regularly inundated and was formerly covered by an important set of natural wetlands (now mainly converted to agricultural fields). To the north, the Esmeraldas River forms an important drainage that emerges from the Guayllabamba Valley (around Quito). Further north, topography becomes rougher, the climate wetter, and forests are still quite extensive. The Santiago, Cayapas and Mira Rivers are the most important drainages in the north-west. South of Guayas province, the Andes lie very close (<20km) to the coast.

Running parallel to the coast, low-elevation ridges separate the western lowlands from the sea. These coastal ridges (collectively the *Cordillera de la Costa*) are discontinuous but in some areas represent a genuine barrier for humidity originating at sea. This results in a foggy climate and humid vegetation atop them. The highest portions – reaching 800–900m elevation on some ridges – are in southern Esmeraldas / northern Manabí provinces (Cordillera Mache Chindul), and in southern Manabí and northern Santa Elena provinces (Cordillera Chongón Colonche).

The coastline has few remarkable features, the most important being the Santa Elena Peninsula in the southwest (an important stopover locality for boreal and austral migrants) and the Gulf of Guayaquil, where large tracts of mangroves exist. There are few islands along the Ecuadorian coast, the most important being Isla de la Plata (where an endemic subspecies thrives), Islote Pelado, Isla Santa Clara, and Isla Puná, the largest.

WEATHER

Climate in Ecuador is mostly not seasonal and, normally, mild to warm. Lying right on the equator, the country does not experience autumn and spring, whereas summer and winter are never as marked as at temperate latitudes. Except on volcanos and high ridges above 4,000+m elevation, snow is an unknown climatic event. There follows a more detailed overview of climate in mainland Ecuador's geographic regions.

The *Costa* (western lowlands)

Two markedly different climates exist in the Pacific lowlands: very humid to the north, and increasingly dry to the south. However, as for vegetation (see Habitats below) there is a north–south gradient of humidity / dryness and seasonality. Climate is strongly correlated with the movements of two sea currents (the cold Humboldt's Current from south to north, and Tropical Warm Current from north to south).

Rainfall reaches a peak in the interior of northern Esmeraldas, where *c.*4,000mm of rain can fall annually. Here, seasonality is not marked, with rain most of the year. Southwards, the climate becomes more seasonal and rains less extreme.

Often, the ocean becomes warmer in December / January through April / May, resulting in a wet or humid season in the central to southern coastal region. Meanwhile, May to December are the drier months. Curiously, hot and sunny days are more frequent in the wet season, because during the dry season oceanic fog prevails almost daily, particularly from Manabí southwards.

Extreme rains occur during ENSO (El Niño Southern Oscillation) events, a climatic phenomenon of increasing oceanic temperatures that provokes heavy rains every 3–16 years. At the opposite extreme are normal years on the Santa Elena Peninsula, the westernmost point of Ecuador. Here, rain is a true rarity.

Temperature in the Pacific lowlands is *c.*20–25°C on average, with little variation throughout the year in the north and interior, but a 5–10°C decrease on the southwest coast (on and around the Santa Elena Peninsula).

The *Sierra* (Andes)

Seasonality is weak, especially on outer Andean slopes. Generally, March to May are the wettest months, while July to September are the driest. Rains on the outer slopes are prevalent for most of the year, with annual rainfall reaching up to 3,000 mm in parts. Rain shadows occur locally due to the complex Andean topography. These result in drier climates in some valleys, especially in the south, where orientation of mountain ranges expose one slope to more humidity, while the other receives little rain.

Temperature variation throughout the year is insignificant, but daily variation is. So, while monthly variation averages *c.*3°C, temperatures can range from *c.*0–5°C at midnight to 20–25°C at midday. Furthermore, temperate decreases with elevation, approximately 0.5°C every *c.*100m. Inter-Andean valleys are warmer and drier than the slopes and *nudos*, and extremely cold temperatures, albeit rarely below -5°C, occur above the snowline (5,000m elevation).

The Amazon lowlands

Mean annual temperatures, as well as rainfall, are basically uniform year-round, *c.*25–30°C on average, with up to 3,000–3,500mm of rain (even up to 4,000mm near the Andes). Seasonality is not marked, but June to August are somewhat wetter. Likewise, temperature varies little during each day. During the austral winter (June–September) southern cold fronts can arrive, lowering temperatures to *c.*15°C. These fronts are called *surazos*, meaning 'heavy southerns'.

HABITATS

Habitat diversity in Ecuador is overwhelming, rather unbelievably so for a country smaller than Britain, where, on the contrary, ecosystem diversity is quite low. A recent country-level classification of ecosystem types led by the Ministry of Environment described 109 major ecosystems grouped into 8–10 main types (i.e., forests, wetlands, steppes, etc.). Here, we summarise those ecosystem types and provide brief descriptions in terms relevant to birdlife. Ecosystems are described from west to east of the country.

Humid forest

Characterised by tall canopy, with emergents surpassing 40m. Understorey is fairly open if pristine conditions dominate. Found in rainy areas in the north-west, primarily on hilly terrain, although seasonally flooded forests occur along the lower parts of large rivers (mostly the Santiago). In the interior north-west, where rainfall is heavier, forest is even wetter. Overall, the wet season spans much of the year (± 8 months). Current deforestation levels are alarming, with most humid forests having vanished from southern Esmeraldas, Santo Domingo, Manabí, Pichincha and Los Ríos provinces.

Deciduous and semi-deciduous forests

In the centre-west to south-west, forests become increasingly drier. Semi-deciduous forests occur from northern Manabí southwards, and close to the Andes in Guayas and neighbouring provinces. This forest is half deciduous, half humid; many trees shed their leaves in the dry season. Canopy reaches 20–25m and understorey becomes denser, more tangled, in the dry months. Southwards, dry forests replace semi-deciduous vegetation. In dry forests, the canopy is lower, diversity too, while seasonality and rates of leaf loss are more marked. *Ceiba* and *pretino* trees dominate some areas. In increasingly drier areas to the south (El Oro and Loja provinces) and near the coast, dry forests are replaced by drier scrubland, and in some areas even barren thorn scrub. Habitat loss throughout the region is extreme, especially in El Oro, interior Guayas and Manabí, and especially Santa Elena.

Secondary dry forest dominated by *Ceiba* trees, Guayaquil (Martín Bustamante)

Foothill and subtropical forests

Very humid forests occur from the lower portions of the Andes to *c.*2,500–2,800m altitude, on the outward-facing slopes of both west and east Andean chains. Presence of dense fog is very frequent (these are often called cloud forests). Canopy reaches 20–25m in foothills, but is lower with increasing elevation. Vegetation diversity is notable, mixing Andean, tropical and subtropical species; diversity and the sheer number of epiphytes are noteworthy, with their increasing prevalence at higher altitude. Understorey is very dense at edges and gaps. Eastern foothills and subtropics are largely forested compared to their western counterparts, but the agricultural frontier is expanding fairly rapidly. A similar forest type occurs at lower elevations atop coastal mountain ridges that receive humidity from sea fog. However, vegetation composition has important differences compared to the Andes. Some Andean bird species inhabit these low-lying cloud forests in coastal ranges.

Mature cloud forest in steep terrain, Bellavista (Juan Freile)

Montane Andean forests

Above the subtropics, forest becomes smaller in stature (up to 10–15m tall), denser and more impenetrable; moss and epiphyte load is impressive. In re-growing areas, like landslides, gaps and edges, understorey is extremely tangled, with bamboo thickets dominating in places. Fog is frequent and provides a prevailing humidity – coupled with regular rains. Upslope, forest becomes even shorter, more tangled, branches are twisted, and mosses very abundant throughout. In pristine areas, this elfin forest gradually grades into bushy *páramo* (and other *páramo* types), but in degraded areas – where fires are frequently used to 'improve' croplands – the habitat shift is sharp. Above the treeline, woodlands dominated by *Polylepis* and *Gynoxys* trees provide additional habitats for a number of specialist birds.

Páramos

Located above treeline and dominated by grasses, but with remarkable local differences caused by weather or terrain. Drier *páramos* have sparse, mostly grassy vegetation (in drier areas interspersed with barren terrain). More humid areas are covered in bushy grassland, *Espeletia* (very locally) and *Chuquiraga* stands, bamboo thickets (even more locally) and damp grasslands. In the south, the treeline occurs at lower elevations and *páramos* tend to be shrubbier. Habitat alteration in Ecuadorian *páramos* is extensive, resulting in mostly open grasslands of low diversity and low stature. Upwards, near the snowline, *páramos* become nearly barren.

Tussock grass dominating semi-humid *páramo*, Papallacta (Martín Bustamante)

Inter-Andean valleys

Dry valleys opening into the western Andes were formerly covered in dry montane woodland and scrub, but are currently largely replaced by urban and agricultural areas. Other valleys are more humid, with montane scrub in currently remote creeks and slopes; the remaining areas are largely degraded. Slopes of both Andean cordilleras and *nudos* have (or formerly supported) montane forests, fairly similar but less humid to those on outer slopes.

Isolated mountain ridges

Two isolated mountain ranges in the south-east are characterised by low-stature, stunted, twisted forests on their upper slopes and ridges (increasingly dwarf with elevation). As ridgetops are often flat and soils rocky, sandy and poor, vegetation becomes scrubby, almost *páramo*-like. Cordillera Napo Galeras in the north (Napo province) has fairly similar vegetation on its top.

Terra firme Amazon forest from a canopy tower, Añangu (Juan Freile)

Terra firme forest

Very tall forest (canopy *c.*30m), with even taller emergents. Grows on well-drained, often hilly terrain in the Amazon. Understorey is regularly open, but frequent treefall gaps provide dense and tangled habitats for some birds. Forest loss is less extensive than in the superficially similar humid forest of the western lowlands.

Oxbox lake surrounded by *várzea* and *terra firme* Amazon forest, Yasuni National Park (Martín Bustamante)

Várzea and *igapó* forests and river islands

Two distinctive forest types grow in areas subject to seasonal or temporal flooding. Areas flooded by black waters originating in Amazonia are known as *igapó*, those inundated by white waters from the Andes as *várzea* (both forest types are more simply referred to as *várzea*). Diversity and forest height are lower than in *terra firme* forest (poorer in *igapó*), but the understorey is denser. *Igapó* is more widespread in the Aguarico and Lagartococha drainages. Other flooded areas are dominated by Moriche palms (known as *morichales* or *moretales*).

Vegetation along the shores of large rivers is dominated by dense lower growth, where *Gymnerium* cane and *Tessaria* shrubs govern. *Cecropia* and other fast-growing trees, typical of second-growth areas, dominate river shores. Similar vegetation, often lower in stature but with very dense understorey, grows on river islands (in the Napo, Pastaza and Aguarico). Lastly, oxbow lakes form in meanders left behind as rivers change their course. These lakes support special successional vegetation and an interesting selection of birds. Habitat loss is quite frequent in riparian areas, less so in more remote areas with *igapó* and *várzea* forests.

Coastal wetlands

The lower drainage of the Guayas River was formerly dominated by extensive wetlands (less extensively in other minor river basins). Currently, most marshes and swamps, very important for waterbirds, have been converted into ricefields and other crops. Few still support an interesting avifauna, and are covered in dense reedbeds and other riparian habitats. A handful of natural wetlands remain, but several artificial ones are very important for waterfowl and waders, including salt-evaporating lagoons. Seasonality in water level is an important feature of coastal wetlands.

Andean wetlands

Páramos are, by default, water sponges. Therefore, thousands of lakes, bogs and marshes exist throughout the Andes. Most are shallow and some have muddy shores and dense reeds. Some wetlands also occur in the inter-Andean valleys, nowadays mostly in Imbabura province and locally in Chimborazo province, where quite large stands of reeds persist.

Beaches and mangroves

Sandy beaches dominate the Ecuadorian coastline, but muddy shores are frequent in river estuaries, where mangroves are also prevalent (or formerly were so). Mangrove forest growing in salty waters has low diversity, fairly low canopy (though some extraordinary forests grow up to 40m!), and an impenetrable understorey. Only found in estuaries along the coast, the largest remnants are in the San Lorenzo area, Muisne estuary, Gulf of

Permanent freshwater coastal wetland, La Segua (Juan Freile)

Guayaquil, and on the Jambelí Islands. Rocky beaches are infrequent and sparse, more regular in southern Esmeraldas, southern Manabí and Santa Elena. Cliffs are rare and never prominent, the most relevant ones are on islands like Santa Clara, La Plata and Pelado. Heavy urban and touristic development, and shrimp farming, have modified many parts of the coastline. Sandy shores also characterise large Amazonian rivers, representing important nesting and feeding habitat for several waterbirds.

Offshore waters

Probably the least-known habitat in Ecuador, and the entire Neotropics. Some sandy banks, rocky and coral reefs, and shallower areas (known as *bajos*) occur around the continental shelf off Ecuador. Areas beyond the continental shelf are even less well known and in much need of further exploration.

Man-made habitats

A number of different habitats have been created by humans; from diverse croplands, gardens and parks to extensive monoculture stands of banana, oil palm, potato, rice, maize, sugarcane, pastures, shrimp ponds, and exotic trees. Habitat modification is more prevalent in the western lowlands than the eastern. Inter-Andean valleys are more altered than Andean slopes. Some man-created habitats support fairly diverse avifaunas, albeit mostly of common to fairly common species, tolerant of habitat modification.

Andean bogs in lake shore, Papallacta (Martín Bustamante)

BIRD DIVERSITY PATTERNS

SPECIES RICHNESS

I'm sitting in front of a lush cloud forest, it is midday, and the only audible sounds are the unrelenting ticking of a Speckled Hummingbird and a distant rushing river. It seems that this forest, in the Intag Valley of north-west Ecuador, is almost devoid of birds.

However, just three hours earlier, a frenzied mixed-species flock comprising nearly 30 species was packed into a single large, moss-covered, fruiting tree. We remained glued to our binoculars, ears wide open and necks aching. This type of birdwatching is frequent in nearly all of the country's ecosystems. Packed with 1,640 species (1,683 when the Galápagos birds are included), Ecuador is the fourth most diverse bird country in the world, harbouring only a few hundred species fewer than Colombia or Peru – both four times larger than Ecuador – and even the gigantic Brazil.

Generally, bird diversity is greater in forested ecosystems than in non-forested areas, in tropical ecosystems as opposed to Andean, in humid ecosystems rather than in dry ones, and east of the Andes versus the west. A more detailed assessment of bird diversity patterns reveals that the Amazon lowlands support *c.*650–700 species. Many range throughout nearly all of the western Amazon basin (Colombia, Ecuador and Peru), and most occur in the entire Ecuadorian Amazon, but some are limited to the northern or southern banks of the Napo River, the largest tributary of the Amazon to rise in the Ecuadorian Andes.

Contrastingly, on the other side of the Andes – in the Pacific lowlands – patterns of bird diversity are notably different. The northern part is wet to very wet, forming part of the southern portion of the Chocó biogeographic region of western Colombia. Here, *c.*450–500 species occur. Southwards, rainforests are progressively replaced by drier, more seasonal ecosystems, with very dry woodlands and xeric scrub in the south-west, which abuts the Peruvian coastal desert just a few km from the international border. This wet–dry gradient is reflected by avian diversity: *c.*250 species in the central-west, but fewer than 200 species in the extreme south-west. Although the eastern lowlands considerably surpass the Pacific lowlands in overall diversity, what is remarkable in the west is endemism. The wet Chocó harbours nearly 60 endemics, while the dry south-west supports 50–60 endemic species. Few of these are entirely confined within the political boundaries of Ecuador, but many are nearly endemic as they barely cross the Colombian or Peruvian borders. In addition, it is also noteworthy that *c.*150 aquatic and marine species – including *c.*100 migrants – occur on the sandy and rocky beaches, estuaries, wetlands and mangroves of coastal Ecuador.

Elucidating bird diversity patterns in the Andes is more complex. The avifauna of the two main Andean cordilleras in Ecuador is fairly similar, as is also the total number of species, although diversity is marginally greater in the eastern Andes. Nevertheless, a significant number of species are confined to just one Andean cordillera (*c.*50 species are limited in Ecuador to the western Andes, and *c.*110 to the east). Meanwhile, the southern Andes – south of Azuay province – with a lower but more broken topography, hold *c.*45 species that do not range further north. Consequently, if we sum all of the birds that occur across the country's long Andean backbone, a total of 700+ species results.

ENDEMISM

'Political' endemics to Ecuador – i.e. the total number of species wholly confined to the current boundaries of the country – are very few (excluding the 26 Galápagos endemics), because Ecuador itself is tiny and birds do not recognise human-designed country frontiers. Only seven species are endemic to mainland Ecuador: El Oro Parakeet, Esmeraldas Woodstar, Violet-throated Metaltail, Black-breasted Puffleg, El Oro Tapaculo, Pale-headed Brush-finch and the enigmatic Turquoise-throated Puffleg (of which a couple of specimens of uncertain origin are labelled as being from Bogotá, Colombia). However, in terms of biogeographic boundaries, which are very dissimilar to geopolitical ones, the number of endemic species, also known as restricted-range species, is overwhelming: more than 300 species, confined to eight regions of biological endemism – or biogeographic regions – occur in Ecuador. These regions are: Chocó, Tumbesian, Eastern Andes between southern Colombia and northern Peru, Amazonian lowlands in the Napo region, North Central Andes, South Central Andes, Amazon-Andean Ridgetops, and Marañón Valley. Of these biogeographic endemics, at least 130 species are almost confined to Ecuador; i.e., 80–90% of their distribution lies within Ecuador.

El Oro Parakeets, (*Pyrrhura orcesi*), El Oro foothills, Ecuador (Melvin Grey/Nature Picture Library)

Male Club-winged Manakin, (*Machaeropterus deliciosus*), Milpe Cloudforest Reserve, Ecuador (Tim Laman/Nature Picture Library)

MIGRATION

Approximately 220 migratory species visit Ecuador, of which *c.*125 are regular visitors and *c.*95 are casual, accidental or vagrants. The chances of seeing some of these latter again are low.

The bulk of migratory species (*c.*130) recorded in Ecuador are boreal migrants that breed in North America; 57 winter in Ecuador, whereas *c.* 30 occur only as transients and *c.* 50 are accidentals or vagrants. The first migrants arrive from mid to late September, rarely earlier (on southbound autumn migration), while the last depart around mid to late April (spring migration). Overwintering individuals are regular in some common species (e.g., Whimbrel, Spotted Sandpiper).

Although Ecuador does not possess highly significant stopover or transit sites (unlike northern Venezuela or Central America), it does receive a regular suite of migrants annually and astonishing surprises can be expected locally. By far the most important area for migrant shorebirds is the Santa Elena Peninsula, where many thousands of birds congregate (especially Wilson's Phalaropes). This has resulted in the designation of the saline ponds of Santa Elena as a Ramsar Site, Important Bird Area, and part of the Western Hemisphere's network of important shorebird sites. Curiously, a site that is becoming increasingly important for migrant landbirds (primarily passerines) is the tiny Quito Botanical Garden, nestled in the heart of the capital's commercial district.

Fewer austral migrants visit or pass through Ecuador, but it should be borne in mind that austral migration is still poorly understood, especially at these latitudes. Some 25 austral migrants have been recorded in Ecuador, 60% of them in the eastern lowlands. Austral migration is far less well known than boreal, so arrival and departure times are not well understood. Roughly, the first austral migrants are found in late March to early April, with departing birds – or southbound migrants – recorded in mid September to early October. The vast majority of austral migrants are passerines.

A separate set of 'austral' migrants includes seabirds that arrive in a less predictable manner. These species (12) are strongly associated with the Humboldt Current, seemingly moving northwards according to variations in sea temperature. Chilean Flamingo is a remarkable example as it has not been found breeding in Ecuador, but is a year-round resident (in variable numbers) on the Santa Elena Peninsula, where even immatures are often seen.

Two additional sets of migrants occur in Ecuador, both of them very poorly known. Intra-tropical migrants (i.e., those species that breed in one tropical area and migrate to another) are found in the southern Andes, and the western and eastern lowlands (6–8 species). For example, Snowy-throated Kingbird breeds in the arid south-west in the first half of the year, and migrates to the more humid north-west.

Pelagic seabirds form the least known and most seldom-seen group of migrant or wide-ranging species. With 35 species recorded to date, much work is needed to better understand migration patterns of boreal, austral and tropical migrant seabirds, as well as the wide movements of a fair number of species. If there is an area where a number of new country records are waiting to be discovered, it is out there in our oceans.

Lastly, an undetermined number of forest birds (e.g., doves, hummingbirds, toucans, flycatchers, cotingas and even antbirds) are believed to engage in altitudinal movements. This type of 'migration' is mostly related to resource availability and weather conditions. Frugivores and nectarivores, for example, 'follow' fruiting and flowering plants, whereas *páramo* species move downslope during harsh weather. Again, much work is needed to understand patterns of seasonal / altitudinal movements by birds that are usually considered resident.

CONSERVATION

As everywhere in the world, in Ecuador conservation problems are currently increasing. Natural habitats in most of the Pacific lowlands or the Andean valleys no longer persist, whilst formerly remote rainforests of interior Esmeraldas or distant Amazon jungles are being exploited at an ever-faster rate. Timber extraction, large-scale agriculture, oil extraction, slash-and-burn agriculture, water pollution, unsustainable fishing, and mining concessions all threaten natural areas, at different scales and in different regions. Habitat loss and over-exploitation are seriously imperilling biodiversity and mankind alike.

Consequently, no fewer than 130 bird species that occur in Ecuador are currently on the brink of global extinction, whereas some 250 species might disappear from the country in the near future. With some exceptions, the entire Ecuadorian avifauna is experiencing a significant decline over most regions of our country. Many examples can be cited: Pale-headed Brush-finch, Esmeraldas Woodstar, Grey-headed Antbird, Violet-throated

Metaltail and others endemic or near-endemic to Ecuador have a total global populations barely exceeding 100 individuals. Black-faced Ibis, Peruvian Thick-knee, Andean Condor, Southern Pochard and others are just a handful of birds away from vanishing in Ecuador. Bird conservation in the country is far from 'rosy'.

But not all is 'gloom and doom'. Starting with a large national network of protected areas administered by the Ecuadorian government that cover roughly 20% of the country's area, several conservation initiatives are presently underway. Privately run reserves have flourished over the last decade, totalling more than 70,000ha and still increasing. Many are strategically located where major populations of threatened, endemic or rare birds are concentrated. Similarly, many communities protect vast areas in the high Andes, coastal mangroves and lowland rainforest, where management mainly derives from ancestral land-use practices. Several sustainability initiatives are also flourishing in Ecuador; from alternative energy sources and ecological food provisioning, to waste management and sustainable use of natural resources, including shade coffee, cacao, permaculture crops, tropical fruits, fair-trade practices, and nature tourism.

Birdwatching, of course, may become a major source of sustainability, and your support can be crucial. At present, birding facilities are already widespread throughout Ecuador, both for independent birders and tour operators. However, birding tourism in Ecuador is still in its infancy. Many areas, aside from the already famous destinations, have an outstanding potential as new birding paradises. Nevertheless, site custodians and their visitors need to establish guiding principles regarding use of playback or feeders, to avoid major disturbance to birds. How much birdwatching can accomplish for conservation in Ecuador will depend on what my fellow countrymen accomplish, and what you, international birders, can contribute.

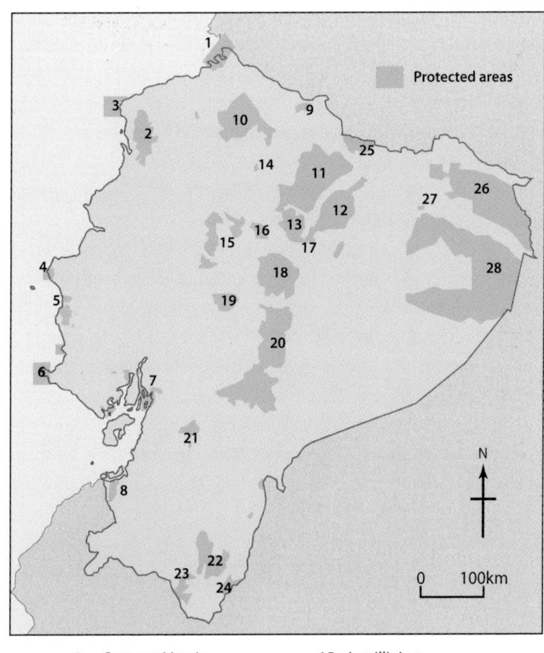

1	Cayapas-Mataje	15	Los Illinizas
2	Mache-Chindul	16	Cotopaxi
3	Galera-S. Francisco	17	Colonso-Chalupas
4	Pacoche	18	Llanganates
5	Machalilla	19	Chimborazo
6	Puntilla S. Elena	20	Sangay
7	Manglares-Churute	21	Cajas
8	Arenillas	22	Podocarpus
9	El Angel	23	Yacuri
10	Cotacacahi-Cayapas	24	Cerro Plateado
11	Cayambe-Coca	25	Cofán-Bermejo
12	Sumaco-Galeras	26	Cuyabeno
13	Antisana	27	Limoncocha
14	Pululahua	28	Yasuni

Main national protected areas in mainland Ecuador; some new and small areas are not mentioned due to space limits.

BIRD TOPOGRAPHY

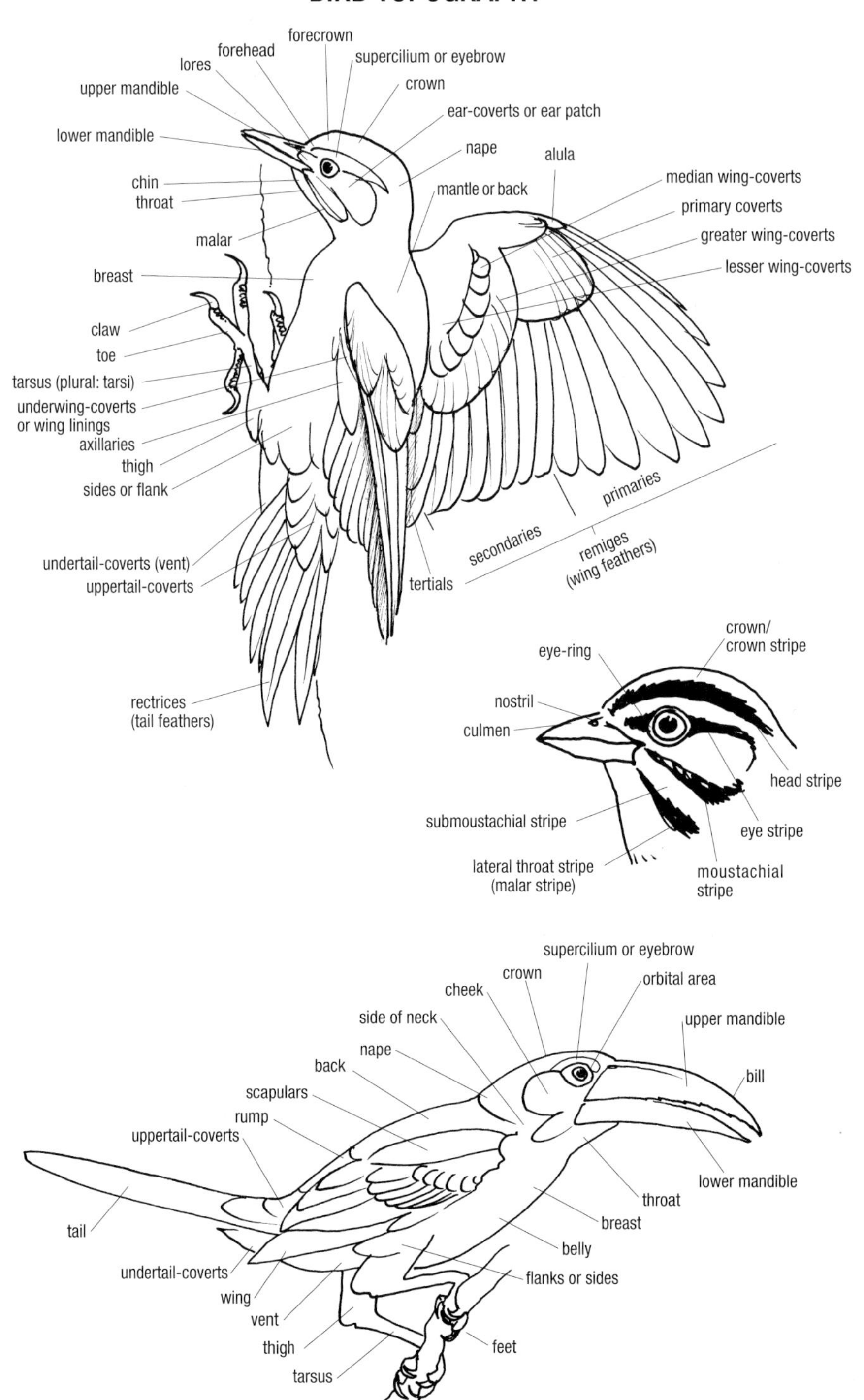

Two fairly large Andean tinamous confined to forests and only rarely seen (*Nothocercus*); and the heavier *Tinamus*, with two brownish, forest-based, regularly-heard species in lowlands, and two dark, hardly ever seen, montane species. All tinamous prefer to escape on foot.

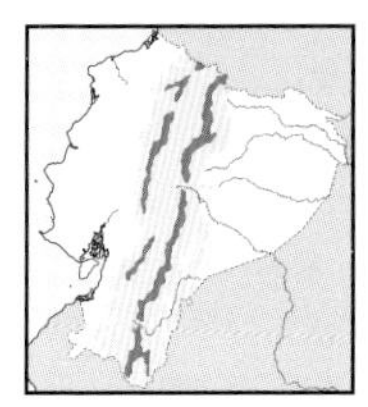

Tawny-breasted Tinamou *Nothocercus julius* 35–40cm

E & NW Andean slopes (2,300–3,400m). In temperate to montane forest. Greyish bill and legs. Mainly rufous-brown, *richer on head, cinnamon-rufous belly, white throat*. Fine dusky vermiculations on upperparts and vent, wings with faint buffy spots. Monotypic. Sometimes at borders and adjacent clearings at dawn or in overcast conditions. **Voice** Loud, two-noted *trrraur, trrraur, trraur*... or *trreea-trreea-trreea* ... (up to 40 times), fading after short *trru-trru-trru*. **SS** Confusion unlikely if well seen; pattern resembles Great Tinamou but no overlap; cf. Highland Tinamou's voice (less raspy, simpler, nasal). **Status** Uncommon.

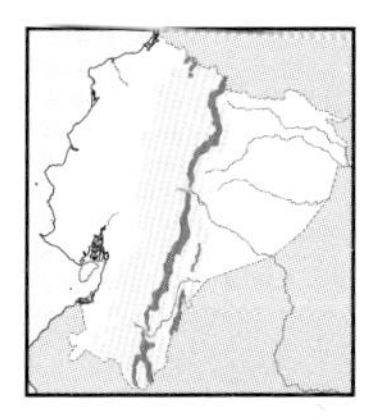

Highland Tinamou *Nothocercus bonapartei* 35–40cm

E Andean slope (1,600–2,200m). Montane forest. Grey legs. *Crown and cheeks blackish-brown*, otherwise mostly rufous-brown, *more tawny on throat*, richer on breast. Fine dusky vermiculations on back and vent, faint buffy barring on wings. Race *plumbeiceps*. Solitary, primarily in forest, very retiring and difficult to see even when close as is unobtrusive in dense undergrowth; *blue* eggs. **Voice** Far-carrying, deep, nasal *kaw'-kaw'-kaw'*...or *kä-kä-kä*... (up to 20–30 times), simpler and less trilled than previous species. **SS** From Tawny-breasted Tinamou by throat colour and voice. **Status** Rare, probably overlooked.

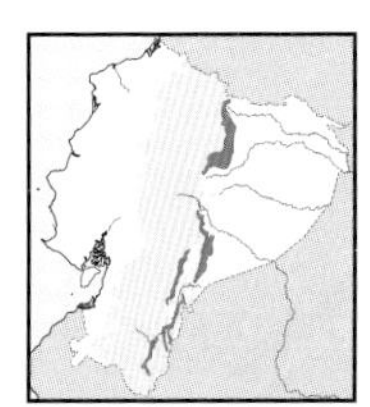

Grey Tinamou *Tinamus tao* 42–46cm

E Andean slope (400–1,600m). Inside foothill to lower subtropical forest, locally adjacent *terra firme* in lowlands. Blackish bill, pale mandible, bluish-grey legs. Large, *mainly grey*, with *blackish head and neck, white stripe on neck-sides, whitish throat*. Upperparts and wings coarsely banded and vermiculated greyish-olive, tawny vent. Race *kleei*. Confined to forest interior, typically shy and elusive, even seems to vocalise little; *greenish-blue* eggs. **Voice** Single, low-pitched *duuu*, at long intervals. **SS** Paler than Black Tinamou, with more striking head pattern; Black gives three descending, melancholic notes. **Status** VU globally, rare, local, overlooked?

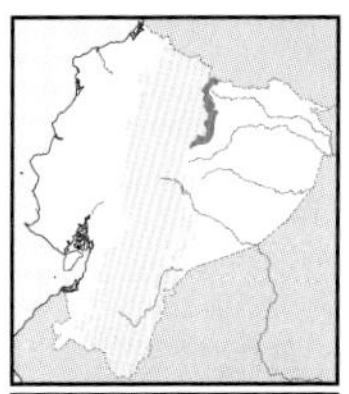

Black Tinamou *Tinamus osgoodi* 40–46cm

NE slope, few records (1,000–1,400m). Inside montane and remote subtropical forest. *Mostly blackish*, quite contrasting *rufous-chestnut vent*. Legs bluish-grey. Race *hershkovitzi*? Poorly known, rarely encountered as occurs in remote areas, but might emerge when overcast, or at dawn and dusk. **Voice** Melancholic, descending, quavering whistles, *wooo-woo-duU*, last note fading, superficially resembling Great Tinamou but sadder, simpler, weaker. **SS** Darker than Grey Tinamou, without white on neck. **Status** Rare, local, VU globally.

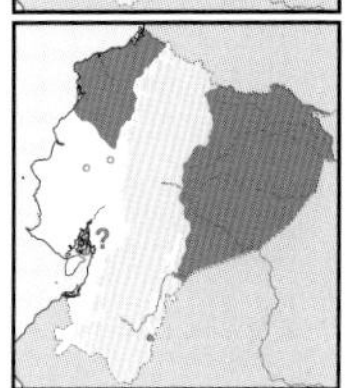

Great Tinamou *Tinamus major* 38–46cm

E & NW lowlands to foothills (to 1,200m). Inside forest, mostly *terra firme*. Greyish legs. Olive-brown above, *richer rufous head* (E), *dark brown crown* (W). Greyish-brown below, paler belly, *whitish throat* and vent. Faint blackish barring on upperparts and wings. Races *peruvianus* (E), *latifrons* (W). Retiring, difficult to see but very vocal; prefers open understorey; *turquoise* eggs. **Voice** Lovely, tremulous, whistled *whooo, wHO-oo-oo'oo'o, wHO-oo-oo'oo'o*... second note higher, then fading, sometimes notes more separated. **SS** Unmistakable in W, in E see smaller White-throated Tinamou (brownish crown, buff wing spots; voices differ); cf. higher-pitched, less resonant, shorter voice of Little Tinamou. **Status** Fairly common, but NT globally.

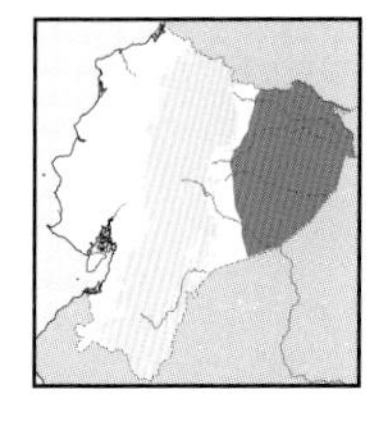

White-throated Tinamou *Tinamus guttatus* 32–36cm

E lowlands (to 400m, locally higher). Inside *terra firme* forest. Grey bill, *greenish legs. Blackish-brown crown, white throat*, freckled tawny face and neck. Upperparts rufous-brown barred dusky, *wings with faint but noticeable buffy spots*. Breast greyish-brown, paler belly. Monotypic. Typical shy tinamous, often found in close proximity to Great Tinamou, sometimes near water; turquoise eggs. **Voice** Low-pitched, mellow, paused, two-noted whistles, *whooo...whüü'a*, 1–2 sec between notes. **SS** Similar to Great Tinamou but smaller, buff wing spots, darker head; simpler voice. **Status** Uncommon; NT globally.

juvenile/immature
Tawny-breasted Tinamou
variant
Highland Tinamou
Grey Tinamou
Black Tinamou
latifrons
peruvianus
paler variant
Great Tinamou
variants
White-throated Tinamou

PLATE 2: 'UNIFORM' TINAMOUS

Six species of medium-sized to small tinamous, all forest-based, but at least Little and Undulated tolerate some degree of disturbance. Little pattern to plumage but notable vocal differences; one species confined to lowlands of NW, three to E, one widespread, one confined to extreme SE.

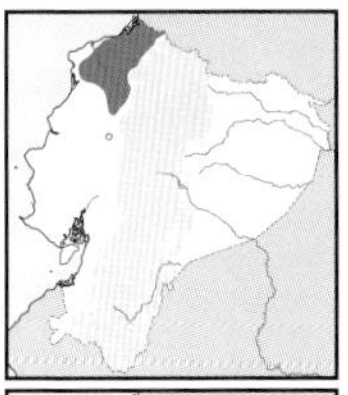

Berlepsch's Tinamou *Crypturellus berlepschi* 28–32cm

NW lowlands (to 400m, locally higher). Inside wet primary forest, old second growth. ***Orange eyes, reddish legs***, pinkish mandible. Mostly ***sooty-black***, paler belly. Monotypic. Probably favours tangled streamside vegetation, but poorly known and seldom seen; ***pale turquoise***, blotched eggs (B. Palacios). **Voice** Short, very high-pitched whistle, *tdeeee*, not tinamou-like. **SS** Only dark tinamou in range, darker than darkest Little Tinamou, with red legs; simple voice piercing and distinctive. **Status** Rare to locally uncommon, EN in Ecuador.

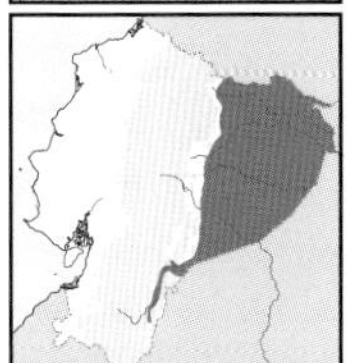

Cinereous Tinamou *Crypturellus cinereus* 29–31cm

E lowlands (to 600m, locally 900m). Inside *várzea* and riparian forest and woodland, damp areas by streams in *terra firme*. Bill grey above, pale horn below, greenish legs, ***orange-brown eyes***. ***Uniform*** olive-brown to greyish-brown, paler below, throat with ***faint white streaks***. Monotypic. Primarily confined to tangled vegetation and very difficult to spot; favours wet forest areas; ***violet*** to ***salmon*** eggs. **Voice** Single, short, clear, tremulous, evenly-pitched whistle, *puüü*, every 3–5 sec. **SS** Only plain-brown Amazonian tinamou, but see higher-elevation Brown Tinamou (more richly coloured). **Status** Fairly common.

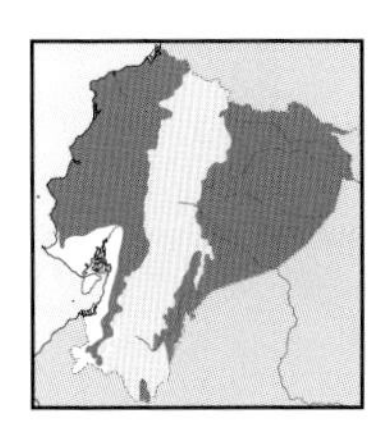

Little Tinamou *Crypturellus soui* 21–24cm

E & W lowlands and foothills (to 1,200m). Forest interior, second growth, borders, dense shrubby clearings. Greyish bill, greenish legs. ***Small***, plain but variable; rufous to olivaceous-brown, more ***ochraceous or buffy below. Blackish head, pale throat***, often greyer or buffier; female ***more richly coloured***. Races *nigriceps* (E), *harterti* (W). Numerous by voice, but difficult to see; might emerge at borders, in adjacent clearings and crops, but rarely flushes; ***brown*** eggs. **Voice** Clear, slightly high-pitched, quavering *wuëë-ee-üyrr...üyrr-üyrr...wuëëë-uüür...* increasing in volume; quite variable; resembles Great Tinamou but less resonant, higher-pitched as progresses, ending abruptly. **SS** By size and habitat; cf. locally Brown and Cinereous Tinamous (simpler, weaker voices). **Status** Common.

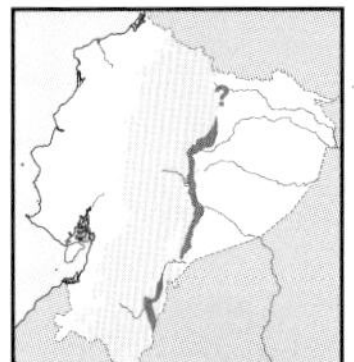

Brown Tinamou *Crypturellus obsoletus* 25–28cm

E Andean slopes (500–1,100m). Inside humid montane and foothill forest. Drab tinamou, mostly rufous-brown, ***head and neck more greyish***, vent faintly barred black. Race *chirimotanus*? Poorly known, mainly away from edges; also very difficult to spot or flush; eggs ***brown***. **Voice** Tremulous *du-du-du-duut...du-du-du-du-ut...du-du-du-dudududu...* slightly rising, ending abruptly. **SS** From Little Tinamou by uniform greyish, not white throat; voices differ markedly. **Status** Rare, local, NT in Ecuador.

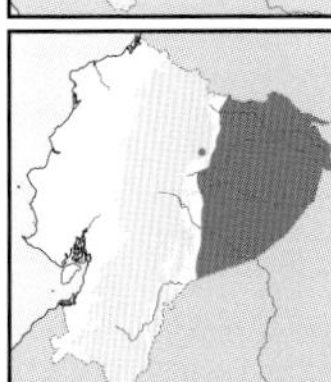

Undulated Tinamou *Crypturellus undulatus* 25–30cm

E lowlands (to 600m). Inside forest, mostly *várzea* and riparian, also second growth, borders, shrubby clearings, river islands. Blackish bill, pale below, greenish-grey legs. ***Dusky-brown head and neck*** fading into greyish-brown breast and belly, tawnier vent finely barred dusky, ***white throat. Slightly vermiculated lower belly***. Race *yapura*. The easiest to see forest tinamou, often venturing into open areas; sometimes attracted to whistled imitations; ***pink*** eggs. **Voice** Distinctive 3–4 mellow whistles, *wuu, wuu-wuu-uah?*, transcribed as *ti-na-moou?* **SS** White throat fairly conspicuous, also readily identified by voice. **Status** Fairly common.

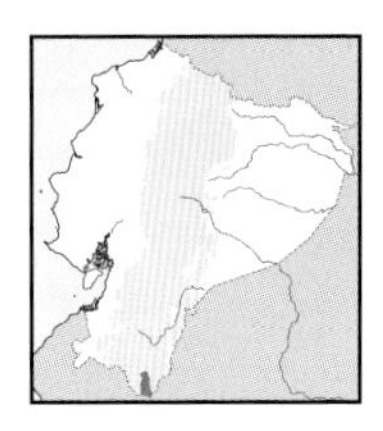

Tataupa Tinamou *Crypturellus tataupa* 24–27cm

Extreme SE in Zumba region (650–950m). Semi-humid forest, woodland, scrub. Grey and ***reddish*** bill, ***purplish-red legs. Head, neck and breast grey, paler throat***, whitish belly, vent ***barred blackish and white***. Upperparts uniform brown. Race *inops*. Less shy than many tinamous, often at borders and in adjacent clearings, but still more often heard than seen. **Voice** Descending series of short 'police' whistles, longer at first, then faster and shorter: *drreeep, drreeep, drreeep, drreeep, dree, dree-dree-dree-dree, dru-dru-dru...dreep, dru-dru-dru-dru*; sometimes shorter. **SS** Only occurs with Little Tinamou (dark legs and bill, grey foreparts); voices differ. **Status** Uncommon, local.

Berlepsch's Tinamou
Cinereous Tinamou
adult
juvenile
♂
Little Tinamou
dark variant
harterti
♀
nigriceps
♂
♀
juvenile
SE dark variant
adult
Brown Tinamou
Undulated Tinamou
Tataupa Tinamou

PLATE 3: 'PATTERNED' TINAMOUS

Three small to medium-sized *Crypturellus*, two of them confined to Amazonian forests, one capable of persisting in marginal woodland and scrub, all more boldly patterned than congeners on previous plate. Beautifully patterned Andean *Nothoprocta* of open habitats are seldom seen.

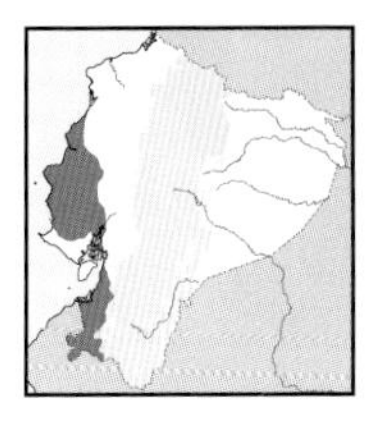

Pale-browed Tinamou *Crypturellus transfasciatus* 25–27cm

CW to SW lowlands, to subtropics in Loja (mostly to 800m, 1,600m in S). Deciduous to semi-deciduous forest, woodland, dense scrub. ***Pink to pale orange legs***. Dusky crown, ***contrasting whitish supercilium***. Whitish throat, *grey breast, pale buff below*, barred flanks and vent. Brown above ***barred blackish*** including wings (♀) or ***barred only on lower back and rump*** (♂). Monotypic. Very vocal throughout day, difficult to see; persists in agricultural areas with dense patches of vegetation, often in open. **Voice** Liquid, abrupt, piercing *ooo-iin!* **SS** Unique in most range, cf. Andean Tinamou on SW slopes (spotted breast, curved bill, streaked back). **Status** Fairly common but NT globally, VU in Ecuador.

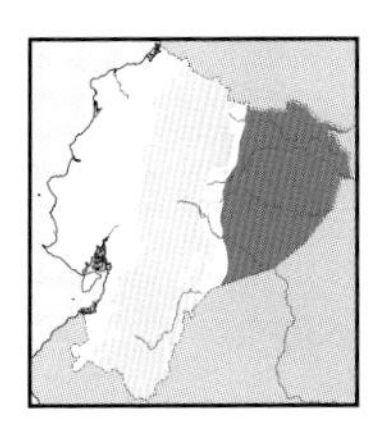

Variegated Tinamou *Crypturellus variegatus* 25–30cm

E lowlands (to 400m). Inside *terra firme* forest. Black bill, greenish-yellow legs. ***Blackish hood, pale throat, rich cinnamon-rufous neck, upper mantle and breast***, paler belly, barred flanks to vent. ***Blackish back finely barred cinnamon***. Monotypic. Retiring inhabitant of forest floor, when spotted freezes and warily walks away; ***purplish*** eggs. **Voice** Melancholic, long *wuuuuuu* whistle, then a fast, ascending, bouncing *wuu, wuu, wuu, wuu....* **SS** Rufescent breast and neck, and black hood separate from rarer, duller Bartlett's Tinamou, voices differ; other Amazonian *Crypturellus* plain above. **Status** Uncommon.

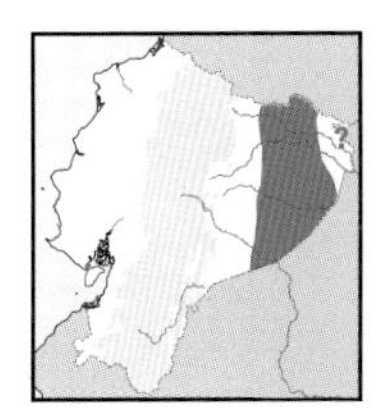

Bartlett's Tinamou *Crypturellus bartletti* 23–28cm

E lowlands (to 400m). Inside *terra firme* forest. Black, fairly short bill, greenish-yellow legs. Variable. ***Blackish-brown to rufescent above***, darker head, ***back to tail densely barred buff***. Whitish throat, ***dusky-brown to rich tawny breast***, paler belly, barred flanks. Monotypic. Poorly known, seldom seen; recorded at few sites, mainly in drainages of Napo and Aguarico Rivers; ***brown*** eggs. **Voice** Penetrating, abrupt whistles, *puuüü'... puuüü'-puuüü'* at long intervals, followed by faster, less spaced, rising series of short, clear *puu-puu-puu* whistles. **SS** Variegated Tinamou (more richly coloured, clear-cut contrasting hood); whistles resemble Cinereous Tinamou but louder, intervals longer. **Status** Rare, overlooked?

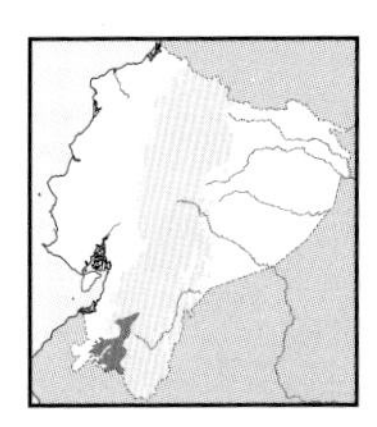

Andean Tinamou *Nothoprocta pentlandii* 26–30cm

SW highlands (1,000–2,300m). Montane scrub, adjacent clearings and agricultural fields. ***Decurved bill, pale yellow legs***. Dusky freckled head, ***bold whitish supercilium***, whitish throat, ***greyish-buff breast spotted buff***, pale buff belly to vent. Upperparts strikingly ***streaked white and barred buffy-brown and blackish***. Race *ambigua*. Not easy to see as prefers open areas with tall grass; often in open areas but shy. **Voice** Not tinamou-like *pee-dui!* or *we-wui* at long intervals. **SS** No overlap with similar Curve-billed; Pale-browed has straighter bill, unstreaked back, unspotted breast; voice and habitat differ. **Status** Rare, local.

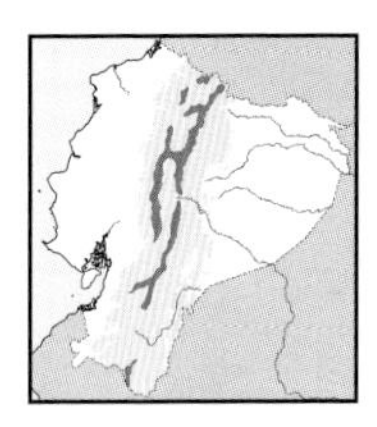

Curve-billed Tinamou *Nothoprocta curvirostris* 28–30cm

Andean highlands S to Azuay, one record from Zamora Chinchipe (3,000–3,900m). *Páramo* grassland and woodland, nearby agriculture. ***Decurved bill, pale yellow legs***. Dusky-brown freckled crown and face, ***bold whitish supercilium***. ***Whitish throat, tawny-ochraceous underparts***, dense dark spotting on neck, whitish spots on breast. Upperparts strikingly ***streaked white, mottled and banded chestnut and black***. Nominate race. Shy and calm, walks in *páramo*, often well concealed but sometimes in adjacent open crops or seldom-used roads; ***pink*** eggs. **Voice** Soft series of *pee-pee-pee* whistles (FK). **SS** Unique in range; cf. seedsnipes and snipes when flushed (chunkier, 'tail-less', broader and shorter wings. **Status** Rare.

Pale-browed Tinamou
♂
♀
Variegated Tinamou
adult
juvenile
Bartlett's Tinamou
dark variant
Andean Tinamou
adult
juvenile
Curve-billed Tinamou

Medium-sized to large guans, clad in brown and most streaked over head and neck, all typically reminiscent of domestic fowl; guans have low tolerance to forest disturbance and are threatened by intensive hunting. Arboreal and frugivorous.

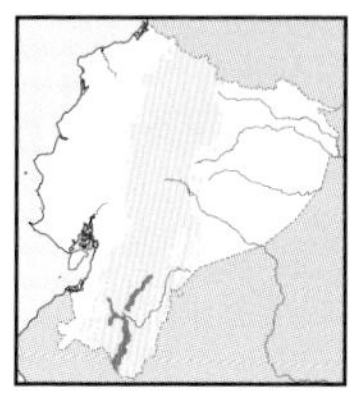

Bearded Guan *Penelope barbata* 55–61cm

S subtropics to highlands (1,900–3,100m). Forest, woodland, borders. Blackish bill and orbital skin, mid-sized dewlap. Head brown densely ***speckled silvery-white***, including bushy crest and cheeks. Dusky-brown neck to breast *streaked white*. Otherwise olive-brown, paler below, ***brownish tail tipped rufous***. Monotypic. Singles to small groups, arboreal, mostly confined to forests; vocal at dawn, including wing-whirring. **Voice** Various cackles, whistles and honks, *kléj-kléj-clá-clá...ca'a'a'a'w... cau-cau-cau... cléh-cléh-cléh....* **SS** Andean Guan (little overlap) has darker head and less streaked neck. **Status** Rare, declining; VU globally, EN in Ecuador.

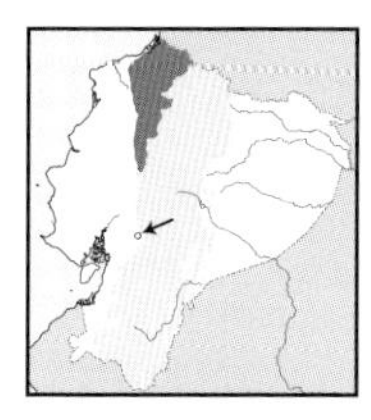

Baudó Guan *Penelope ortoni* 56–63cm

NW lowlands and foothills (below 1,300m). Wet forest, borders. Greyish facial skin, large dewlap, bushy crest. Mostly dusky-brown, darker head and neck. ***Foreneck to upper belly markedly streaked white***; plain brown wings and tail. Monotypic. Poorly known, very shy; singles to small groups, may follow army ants. **Voice** Not very vocal, a frog-like *whääk..whääk..whääk...* at dawn; soft rising whistles, weak cackles, *jón-jón-jónk-jo'o'o'on....* **SS** Larger, sympatric Crested Guan has more reddish belly and rump, more crested, very noisy when approached. **Status** Rare, likely declining; EN globally and in Ecuador.

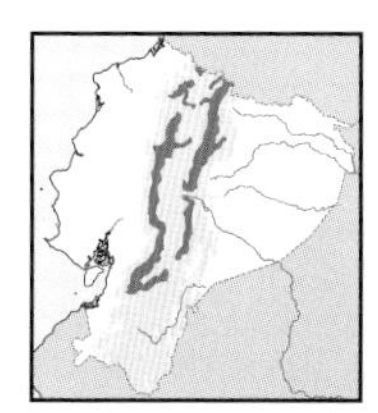

Andean Guan *Penelope montagnii* 53–59cm

E, W and inter-Andean highlands (2,500–3,600m). Humid forest, woodland, borders. Pinkish bill, slaty facial skin, small dewlap. Dark brown head and neck, ***densely chevroned white to breast***. Otherwise olive-brown, paler, more ***rufescent below and on rump; bronzy tail***. Darker race *atrogularis* (W); *brooki* (E) has more silvery head and neck. Singles to small groups at borders, noisy and numerous where undisturbed. **Voice** Various cackles, honks, whistles, *dik-dik-dik-dik-di'i'i'ik..cláh-cláh-cláh..jhen-jhen-jhen..cláh-cláh....* **SS** In Azuay cf. Bearded Guan (more silvery-white head speckling, tail tipped rufous). **Status** Rare, locally uncommon.

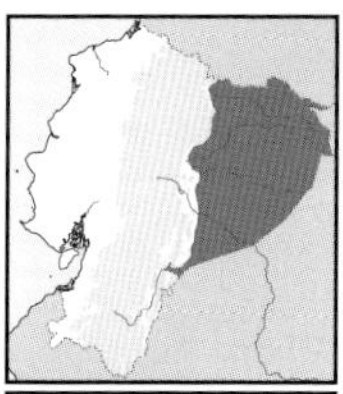

Spix's Guan *Penelope jacquacu* 76–84cm

E lowlands and lower foothills (to 1,000m). *Terra firme* and *várzea* forest, borders. Greyish bill and facial skin, ***large dewlap***. Mostly olive-brown, darker head, paler belly, ***rufous vent***. ***Neck, breast and mantle finely streaked white***, bronzy tail. Nominate race. Singles or pairs in forest, often near ground; fairly vocal before dawn, also wing-whirring. **Voice** Loud, deep honks, *kewëër-kewëër-kewëër..ánk-ánk-ánk...kerrow-kerrow-kerrow...*, loud yelps *oohN-ooHN-oOHN...whan-whan-whan'whan'whan...* and others. **SS** Only *Penelope* in range. **Status** Rare to locally uncommon.

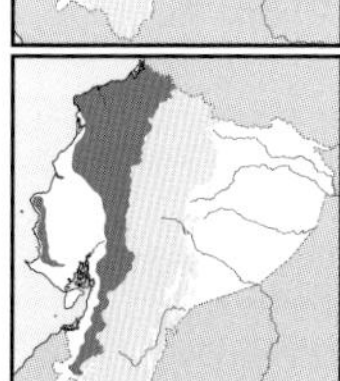

Crested Guan *Penelope purpurascens* 84–91cm

W lowlands to lower subtropics, south to Loja / El Oro Andean slopes (below 1,500m). Humid forest, borders, locally in drier forests. Large, *crested, large dewlap*. Mostly olive-brown, darker head, paler belly. ***Rufescent lower belly to vent and rump; neck and breast finely streaked white***, bronzy tail. Race *aequatorialis*. Singles to small arboreal groups, more often in canopy; abruptly flee when threatened. **Voice** Various loud honks, yelps, include *kAw-kAw-kAw..kw'e'e'e'e-Ónh-Ónh-Ónh..., kweee, kweee* whistles, *kOnh-kOnh-kOnh...wOnk-wrrrrr....* **SS** Smaller Baudó Guan lacks rufescent below, less crested. **Status** Rare, likely declining; EN in Ecuador.

Bearded Guan
Baudó Guan
Andean Guan
brooki
atrogularis
Spix's Guan
Crested Guan

PLATE 5: CHACHALACAS AND GUANS

Three monotypic guans, but *Pipile* and *Aburria* are likely closely related and may represent a single genus. Chachalacas are the smallest cracids, more tolerant to forest disturbance than guans.

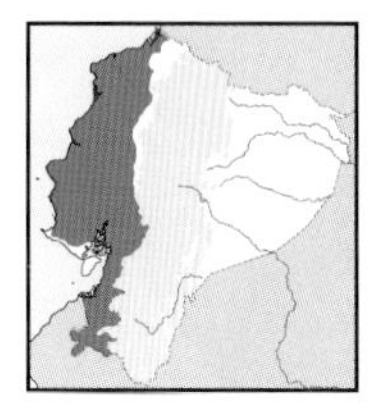

Rufous-headed Chachalaca *Ortalis erythroptera* — 56–66cm

W lowlands to SW subtropics (to 1,800m in SW). Semi-deciduous and deciduous forest, woodland, borders. Greyish bill and orbital skin, red bare throat. ***Rufous head and neck*** grading into brown nape, wings and upperparts, ***rufous flight feathers***. Bronzy tail, ***outer feathers edged chestnut***. Monotypic. Small groups in all forest strata, might persist in small wooded patches in degraded areas; noisy at dawn. **Voice** *cha-cha-cáw, kra-kra-kraú* and others, transcribed as *go-to-work, go-to-work… I-don't-go, I-don't-go…*, also whistled *tleeeú, kle-kle-kle…*. **SS** Only chachalaca in range. **Status** Uncommon and declining; VU.

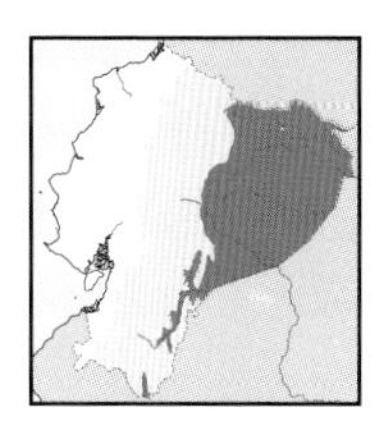

Speckled Chachalaca *Ortalis guttata* — 49–60cm

E lowlands and foothills (to 1,100m). *Terra firme* and *várzea* forest borders, secondary and riparian woodland. Greyish bill and orbital skin, bare red throat. Mostly olive-brown, *neck and* ***breast obviously speckled whitish. Greyish-buff belly***, buffier on vent. Bronzy tail, ***outer feathers edged chestnut***. Nominate race. Arboreal, small groups might persist in borders and degraded areas; rarely in canopy or on ground; occasionally at feeding stations. **Voice** Vocal; several sing simultaneously: *cha'cha-la'cák, cha'cha-la'cák…, chá-chá-lakák*, and other whistles, squeals and cackles. **SS** Confusion unlikely. **Status** Uncommon.

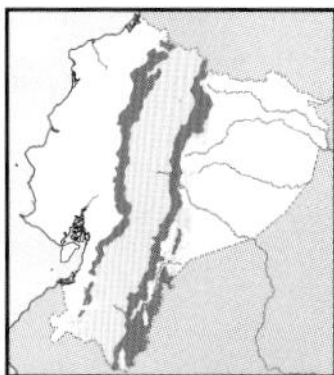

Sickle-winged Guan *Chamaepetes goudotii* — 51–65cm

Subtropics on both Andean slopes (900–2,600m). Montane forest, borders. Black bill, ***large blue facial skin***, pink legs. Blackish to dusky-brown upperparts, ***rufous-chestnut underparts***, tail bronzy-green to black. Larger *tschudii* race (E) has ***chestnut extending to chest***, longer tail; race *fagani* (W) has ***shorter, bronzier tail***. Singles to small groups, mostly arboreal, often at borders; wing-whirring displays; can be attracted to feeding stations. **Voice** Not very vocal; soft, *whee-át, whee-át…* first note whistled; thin *keeeee'u* calls when foraging. **SS** Blue facial skin diagnostic. **Status** Uncommon.

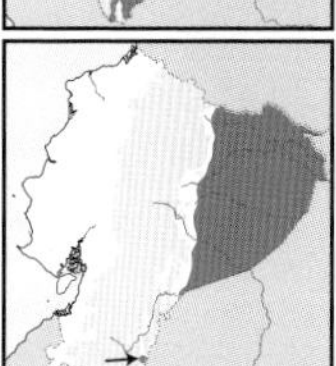

Blue-throated Piping-Guan *Pipile cumanensis* — 68–74cm

E lowlands (to 400m). *Terra firme* and *várzea* forest, borders. Handsome guan with ***bluish-white bill tipped black, white facial skin, blue dewlap***. *Clean white crest*, often held low but sometimes raised. ***Mostly glistening black, large white patch in wing-coverts***. Nominate race. Pairs to small groups, swiftly hopping and running between canopy and borders, often high; regular at river edges. **Voice** Piercing, rising series of 6–8 brief piping whistles: *fuit-fuúít-fuuult-fuuUlt-fuuÚÍT…*, weak *teeui* calls. **SS** Unmistakable. **Status** Uncommon, confined to forest, rarer in settled areas.

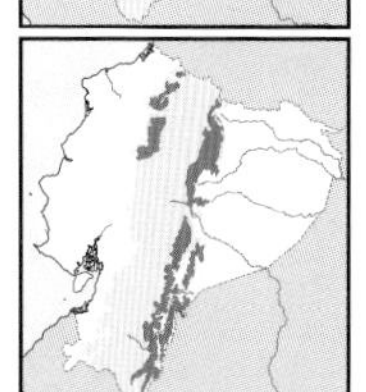

Wattled Guan *Aburria aburri* — 72–81cm

Subtropics on E & NW Andean slopes (1,200–2,100m). Montane forest, borders. Unique, ***mainly glossy black*** with ***bluish bill tipped black, long orange dewlap, yellow towards tip; bright yellow legs***. Monotypic. Pairs to small groups, difficult to observe during day; more often heard at dusk and dawn; presumably moves seasonally following fruiting trees. **Voice** Loud, trilled *ar-aaaar'rrrreeeééE*, mostly at dusk, also on moonlight nights, thin *buít..buít..buít* and louder *f'tiiiw* calls. **SS** Unmistakable. **Status** Rare, rather local; NT globally, VU in Ecuador.

Rufous-headed Chachalaca
Speckled Chachalaca
fagani
Sickle-winged Guan
tschudii
Blue-throated Piping-Guan
Wattled Guan

Three large curassows, all rare inhabitants of deep forest interior; both *Crax* are extremely scarce. Severe hunting depletes curassow populations around towns and settlements. The smaller, similarly retiring Nocturnal Curassow, tolerates some human pressure.

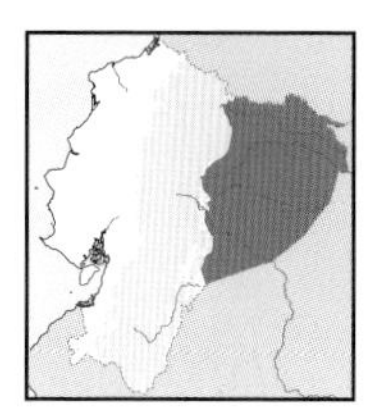

Nocturnal Curassow *Nothocrax urumutum* 50–57cm

E lowlands (below 1,000m). *Terra firme* and *várzea* forest interior, often near water. Unique, small curassow with ***orange-pink bill, many-coloured orbital skin*** (mainly yellow, bluish and lavender). ***Mostly chestnut***, black bushy and curved crest, coarse dusky barring above, pale chestnut belly. Outer tail feathers dusky, tipped buff. Monotypic. Extremely difficult to see, resting birds can be seen low inside forest, even in tree holes; primarily nocturnal. **Voice** Far-carrying, very deep, slow *mm-mmm.. mm, mm-mm-mmmm, óH!*, last note often not heard. **SS** Confusion unlikely. **Status** Uncommon, easily overlooked if not vocalising.

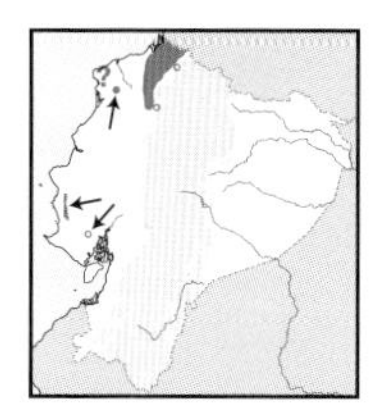

Great Curassow *Crax rubra* 87–92cm

NW lowlands and lower foothills, locally to SW (to 700m). Humid forest interior (formerly in more deciduous forests). Male mostly ***glossy black, erect and curly crest, bright yellow knobs at bill base. Belly to vent contrastingly white***. ♀ has ***black-and-white erect crest, black-and-white speckled head and neck***. Otherwise mainly ***chestnut***, tail prominently ***banded chestnut, buff and black***. Nominate race. Pairs to small groups on forest floor; very retiring, slowly walks away. **Voice** Very deep booming *óom..mmm-mmm-mmM*; loud, piping *pfíp-pfíp-feeeéE* whistles. **SS** Unmistakable. **Status** Very rare, nearly extinct in Ecuador, where CR; VU globally.

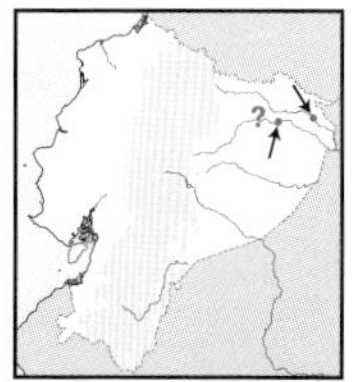

Wattled Curassow *Crax globulosa* 82–89cm

E lowlands (below 300m), two old records of uncertain location; possibly *várzea* and riparian forest. Mostly glossy black, ♂ with ***bright red knob and wattle at bill base***, ♀ has ***bright red skin alone***. Short curly black crest. ***Contrasting white belly*** (♂) or ***chestnut*** (♀); ***long uniform tail***, slaty legs. Monotypic. Unknown in Ecuador and poorly known overall; regularly near water, but shy. **Voice** Long, thin, falling whistles, *déééeeuuuu, TEéeeeeeeeuuu*. **SS** Salvin's lacks bill wattles and has white-tipped tail. **Status** Uncertain, possibly extirpated; EN globally, CR in Ecuador.

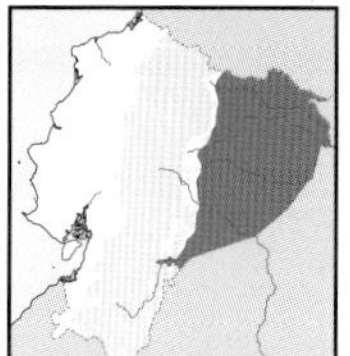

Salvin's Curassow *Mitu salvini* 75–89cm

E lowlands (below 400m). *Terra firme* and *várzea* forest interior. ***Bright red and strikingly arched bill***, legs ***coral-pink***. Mostly glossy blue-black, short ruffled crest often held flat. ***White belly to vent, long tail tipped white***. Sexes alike. Monotypic. Mostly terrestrial in pairs that rest low in trees; vocalises mainly at dusk; silently walks away. **Voice** Deep, slow booming: *mm-mmmm...mmH..mmmh...mmmm-mmm-mmmü*. **SS** Extremely rare, perhaps extirpated Wattled Curassow (red bill wattles, female has chestnut vent); simpler voice than Nocturnal Curassow. **Status** Rare; VU in Ecuador.

Nocturnal Curassow
♀
♂
Great Curassow
♀
♂
Wattled Curassow
Salvin's Curassow

PLATE 7: WOOD-QUAILS

Chunky quails of forest floor, always in family coveys, seldom observed but regularly heard. All are clad in cryptic plumages, but at least two show vivid bare facial skin. Reminiscent of tinamous, but chunkier, shorter-necked and strong-billed. Two strictly Amazonian, two in W lowlands, two on Andean slopes (one each side of Andes).

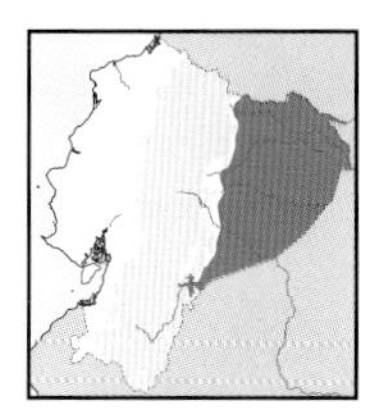

Marbled Wood-Quail *Odontophorus gujanensis* 25–28cm

E lowlands (to 900m). Inside *terra firme* forest and old second growth. Broad *red orbital skin. Mostly dark brown*, darker crown, greyish throat, neck and nape. Slightly paler and black-banded breast and belly, buffier rump. Wing-coverts have blackish spotting, very short tail. Sexes alike. Race *buckleyi*. Family groups on forest floor, often in dense tangles; two or more coveys often head at once. **Voice** Far-carrying, rollicking duets: *cororó-kuó, cocoró-kuó, cocoró-kuó... coö-cow, coöw-cow...*, slower *coo-coo-coo-coo...* calls. **SS** Locally sympatric Starred Wood-Quail (paler overall, greyer neck, yellow facial skin, longer crest). **Status** Fairly common, but NT globally.

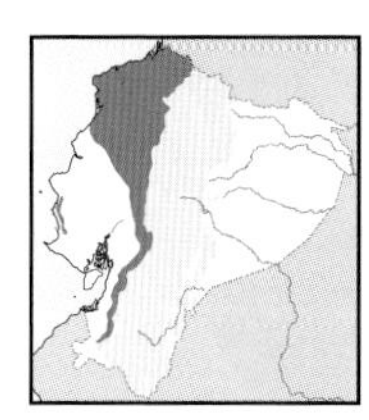

Rufous-fronted Wood-Quail *Odontophorus erythrops* 23–26cm

W lowlands to lower subtropics S to El Oro and Loja (mainly to 1,200m). Primary and old secondary forest. Purplish orbital skin. Dusky-brown above, *rufous front, face and underparts, throat black with white crescent*, vent and flanks barred dusky, black dots on wing-coverts. ♀ similar but duller. Race *parambae* (S to Chimborazo), nominate (S to El Oro). Persist in small forest fragments, but not into open or degraded areas; two or more coveys often head simultaneously. **Voice** Far-carrying *kloö-kloö-kloö-kluö-kluö-kluö...* running into a fading *clé-klö'clé-klö'clé-klö...*, *klökle-klöke-klöke...*; sweet *chüee* calls. **SS** Dark-backed Wood-Quail (little overlap) lacks black-white throat; faster voice. **Status** Uncommon; VU in Ecuador.

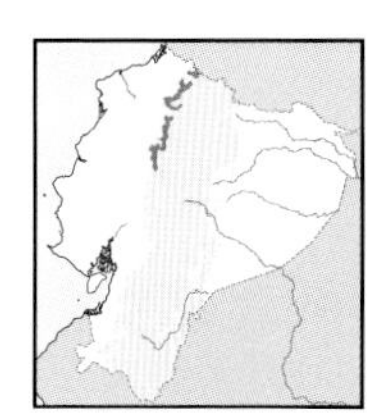

Dark-backed Wood-Quail *Odontophorus melanonotus* 23–26cm

NW subtropics (1,200–2,700m). Montane primary and old secondary forest. Purplish orbital skin. *Dusky-brown above faintly vermiculated blackish. Rufous throat to lower breast*, belly has dusky barring. Sexes alike. Monotypic. Family groups forage on steep slopes, ravines, borders where not disturbed; vocal mostly at dawn and dusk, climbs in row along oblique tree limbs to roost nocturnally. **Voice** Fast, sweet duets: *cloró'cloré, cloró'cloré, cloró'cloré...* or *cö-ro-co-ré, cö-ro-co-ré, cö-ro-co-ré ...*, also weak *veet-ve'e'e't* whistles. **SS** Rufous-fronted Wood-Quail (contrasting black-and-white throat, with less throaty voice). **Status** Uncommon, small range; VU globally and in Ecuador.

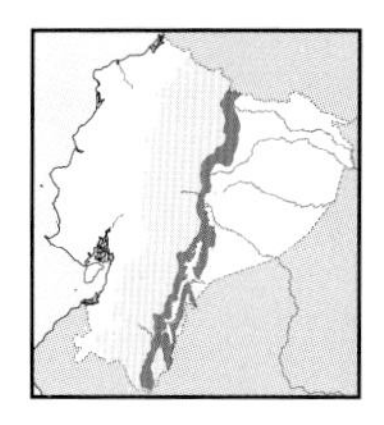

Rufous-breasted Wood-Quail *Odontophorus speciosus* 24–27cm

E foothills to subtropics (800–1,800m). Montane primary and old secondary forest. Purplish orbital skin. Dark brown upperparts, crown and cheeks, *whitish-flecked supercilium and malar stripes, black throat*, paler rump, faint white streaks on mantle. *Rufous underparts*, brown lower belly finely barred dusky; black dots and small whitish spots in wing-coverts. ♀ with *greyish-brown underparts, rufous confined to chest*. Race *soederstroemii*. Retiring, forest floor dweller, difficult to see but readily heard; regularly on steep slopes and ravines. **Voice** Similar to Dark-backed Wood-Quail: *keeree-keeree-keeree... keeroró-keeroró-keeroró-keeroró...*, sweet *viit* whistles. **SS** Little overlap with Marbled Wood-Quail (red orbital skin, no rufous below). **Status** Rare, NT globally and in Ecuador.

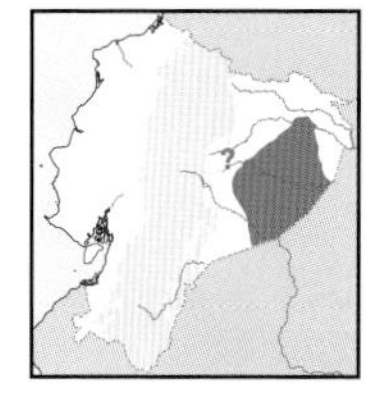

Starred Wood-Quail *Odontophorus stellatus* 24–27cm

Local in SE lowlands (to 400m). Primary forest. *Yellow orbital skin. Rufous crest, grey nape, face, throat and mantle*. Tawny lower back and rump, grading into dusky tail. Brown wings, whitish spots in coverts, dusky bars on flight feathers, black dots on scapulars. *Underparts rufous, streaks of white stars* in breast, brown belly with dusky barring. ♀ has brown crest, less grey in foreparts. Monotypic. Little known in Ecuador; elsewhere, similar to other wood-quail, but perhaps scarcer. **Voice** Slower than Marbled Wood-Quail; a repeated, rather trilled *koo-ko'ö'ö'ö, ko-koo'ö'ö'ö, ko-ko'ö'ö'ö...*, *kör-kooorr, kör-kooorr...* **SS** See Marbled (red orbital skin). **Status** Very rare, NT in Ecuador.

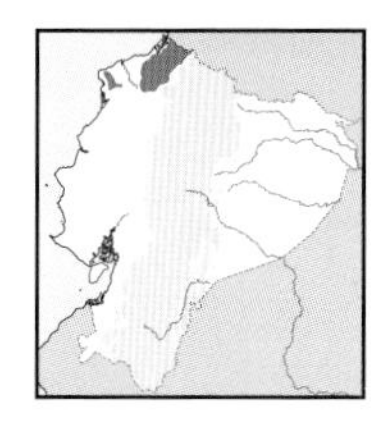

Tawny-faced Quail *Rhynchortyx cinctus* 18–19cm

Extreme NW lowlands (to 600m). Very humid primary forest. Smaller bill, grey eye-ring. *Rufous-tawny head*, thin black ocular line; *white throat, grey upper mantle to lower breast*, tawny-cinnamon belly. Brown above, obviously spotted / barred black on back and wing-coverts. ♀ has brown crown, *long whitish supercilium*, white speckling on throat; chocolate breast, belly *barred black and white*. Race *australis*. Mostly in pairs, very shy, run and freeze when threatened; rarely at borders. **Voice** Tinamou-like whistled *kiiio, kuuo, kuuooo, kuuo-kwe-kwe'kwe'kwe...kuu-kwo-kwi..kuu-kwo-kwi-wiwi....* **SS** Confusion unlikely. **Status** Uncommon, EN in Ecuador.

Marbled Wood-Quail
erythrops
parambae
Dark-backed Wood-Quail
Rufous-fronted Wood-Quail
♀
♂
Rufous-breasted Wood-Quail
♂
♀
Starred Wood-Quail
♂
♀
Tawny-faced Quail

Thickset birds of freshwater wetlands, including the odd-looking screamer, the only goose in Ecuador, and the only native duck domesticated by pre-Columbian Americans. All currently rare and local.

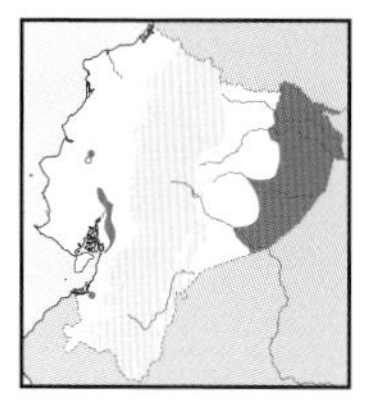

Horned Screamer *Anhima cornuta* 84–92cm

E & SW lowlands (to 300m). Marshes, lakes, river islands. Unique, massive, ***mostly black***, heavy bill, long greyish legs and odd ***antenna-like horn***. Sharp spurs at bend of wings. ***Crown and foreneck speckled white***, belly white. ***White patch in wing-coverts*** conspicuous in flight. Monotypic. Pairs to small groups feed on vegetable matter and nest on vegetation platforms, often seen atop trees and bushes; mostly in remote areas. **Voice** Very deep, loud *güü-guul-güüp...; güiy-giyü..güiy-giyü..güiy-giyü*, with odd, nasal *yaa-yaan-yaa-yaah...* honking. **SS** Unmistakable. **Status** Rare, sharp decline in W; EN in Ecuador.

Orinoco Goose *Oressochen jubatus* 56–66cm

Local in E lowlands (to 300m, once in extreme SW). Sandbars on major rivers. Short bill, pink legs. ***Greyish to buffy-white head to breast, tawny-rufous flanks, belly***, mantle and scapulars, rest of upperparts dark. In flight, ***metallic green secondaries with broad white inner patch***, blackish underwing. Monotypic. Pairs to small groups on sandbars, grazing on riparian grasses; males fight fiercely pre-breeding; rather reluctant to fly. **Voice** Nasal *iuunk!, aaanh* honks, high *zreee* whistles. **SS** Confusion unlikely. **Status** Rare, local, VU in Ecuador, mostly confined to Río Pastaza; NT globally.

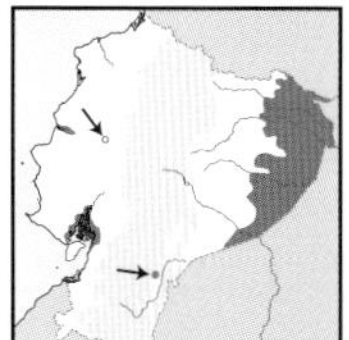

Muscovy Duck *Cairina moschata* 70–84cm

E lowlands, locally in SW (below 300m). Remote lakes, slow-moving rivers and marshes. Pale eyes, ***red facial wattles*** (♂), bicoloured bill. ***Mostly black glossed green***, slightly combed crest. ***Wing-coverts broadly white***. In flight, ***white under- and upperwing***. Juvenile browner, little white in wing-coverts. Monotypic. Wary, small groups roost high above water; up-ending, dabbling and grazing on shores. **Voice** Soft hisses and quacks: *oohg-oohg...oonh-oonh...fii'd'd'd*. **SS** See escaped, paler domestic Muscovy (little too much white throughout, red soft parts). **Status** Rare, local, declining; EN in Ecuador.

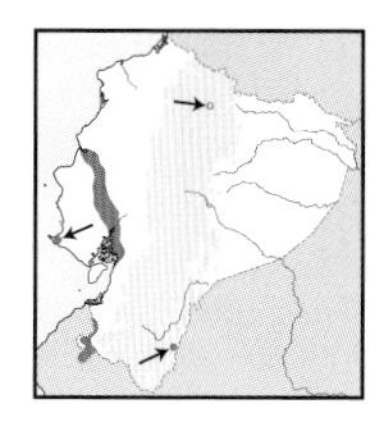

Comb Duck *Sarkidiornis melanotos* 60–75cm

Locally in SW lowlands (to 500m, once at 900m in SE; historical highland records). Rivers, marshes, ricefields. ***White below***, glossy black-green above, ***black bill, odd bill knob*** (♂); black flanks. In flight, metallic green coverts, ***dark underwing***. Juvenile dusky-brown, paler below, ***dusky ocular stripe***. Race *sylvicola*. Active at night, may move seasonally with water levels; small groups graze river shores, often with whistling-ducks in ricefields; buoyant swimmer. **Voice** Some grunts, hisses or honks; i.e., *gggak-gggak-gggak...*. **SS** Confusion unlikely. **Status** Rare, local, VU in Ecuador.

Horned Screamer
Orinoco Goose
♂
♀
Muscovy Duck
♀
♂
Comb Duck
♂
♀
juvenile

PLATE 9: DUCKS I

Four miscellaneous ducks: whistling-ducks tend to wander around following floods; congregations can be impressive. Torrent Duck is the sole species of duck in cold rushing Andean rivers and streams. Brazilian Teal is an accidental from open Amazonian lands.

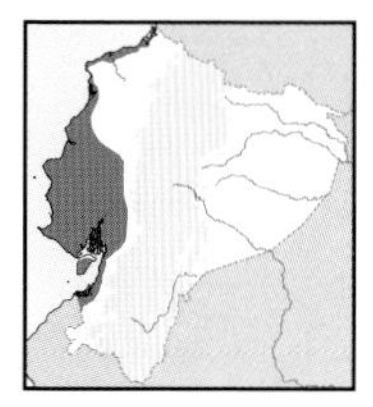

Fulvous Whistling-Duck *Dendrocygna bicolor* 45–54cm

W lowlands (to 100m). Freshwater marshes, lake edges, ricefields. Tall-standing duck with ***grey bill and legs***. Mainly ***fulvous***, browner crown and upperparts, ***long fluffy whitish flank feathers, white vent***. Back feathers edged fulvous, ***dark wings in flight, white uppertail-coverts***. Monotypic. Gregarious (sometimes several thousands), often with Black-bellied; feeds by dabbling, up-ending, sometimes diving or walking on shore. **Voice** Shrill whistles, *kee'deeo-k'daoo-kee'deeo...*, *weED-weED....* **SS** Black-bellied more patterned, has white in wing; juv. Comb Duck has all-dark wings, dark rump, much paler below, stockier. **Status** Locally common.

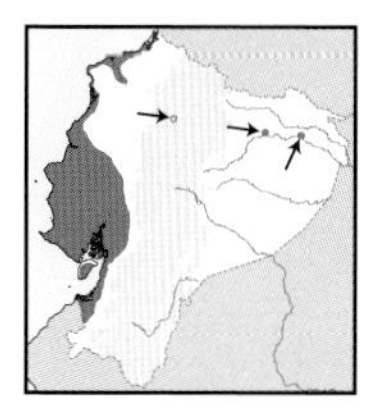

Black-bellied Whistling-Duck *Dendrocygna autumnalis* 46–53cm

W lowlands, local in E; once in highlands (to 200m). Freshwater marshes, ricefields. ***Pink bill and legs***. Chestnut-brown above, ***greyish face and throat***, cinnamon-brown chest, ***black belly***. In flight, ***conspicuous white band in mid wing, dark underwing***. Juv. duller, dark soft parts. Race *discolor*. Active at night, more regular grazing on ground and perching in trees than congener, but also swims and congregates with other species. **Voice** High-pitched, weak, whistled *k'wee-k'wee-k'wee...*, song ending in shrill *pi-TEW-ti'di'di'diti*, emphatic *tEw-tEw-tEw-tEw* in flight. **SS** Fulvous Whistling-Duck (plain wings, uniform underparts, grey bill); Southern Pochard (dark bill, shorter neck, whitish remiges from below). **Status** Common.

Brazilian Teal *Amazonetta brasiliensis* 36–40cm

One record from fishponds in NE lowlands (R. Ahlman). Small. Dark crown, ***pale cheeks***. Warm brownish, speckled below; green speculum and ***white trailing edge*** to upperwing. ♂ has pinkish bill and legs; ♀ grey bill, bold ***white face marks***. Nominate race? Dabbles, alone or pairs in shallow water; tends to wander. **Voice** Piercing *tuwee-tuwee* whistles (MB) and soft *uuk-uuk...* (HB). **SS** Confusion unlikely. **Status** Uncertain, possibly wanderer.

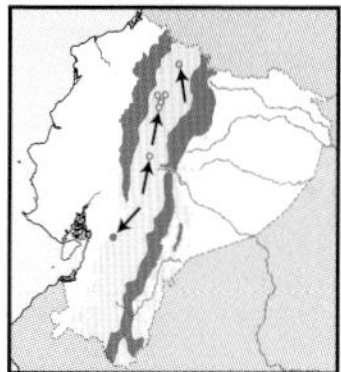

Torrent Duck *Merganetta armata* 38–42cm

Andean foothills to highlands (700–3,200m, locally higher). Fast-flowing rocky rivers and large streams. Handsome black-and-white, streaky ♂ with ***bright red bill and legs***, long stiff tail. ♀ ***slaty above, cinnamon below***. Juv. ♂ drabber, no black head-stripes, greyer head. Race *colombiana*. Pairs or family groups swim and dive in turbulent rivers, resting atop boulders; more prone to swim than fly. **Voice** High-pitched *week-week*, but often silent. **SS** Unique. **Status** Uncommon in clean-water rivers.

juvenile
adult
adult
Fulvous Whistling-Duck
juvenile
adult
Black-bellied Whistling-Duck
♂
♀
Brazilian Teal
♂
♀
♂
♀
juvenile
Torrent Duck

Four species of dabbling ducks, including two residents in the highlands and our only common boreal migrant. Cinnamon Teal is now possibly extinct as resident, but recent records of visitors from N and S. Sexual dimorphism and eclipse plumage of males more marked in migratory species. Wing pattern important for species recognition.

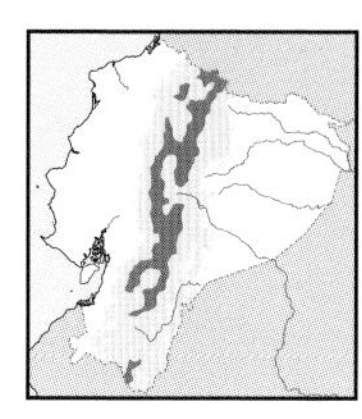

Andean Teal *Anas andium* 38–43cm

Andean *páramos* (3000–4000m, locally higher or lower). Lakes, marshes, ponds. ***Bill grey.*** Mostly brown, ***densely speckled head. Breast to flanks pale tawny*** with dusky spots. In flight pale underwing, ***dark green speculum bordered by tawny stripes.*** Nominate. Pairs to small groups in surprisingly small ponds to open lakes, rests on lakeshores; feeds by dabbling or up-ending; flight often low. **Voice** Soft, fast, harsh *krree-rriik* in flight; *güaak, gük-gük-gük...* quacking; fast *quik'qui-quikquikquik* on take-off. **SS** Paler Yellow-billed Pintail has yellow bill; female Blue-winged Teal has pale supercilium, blue wing patch. **Status** Locally common.

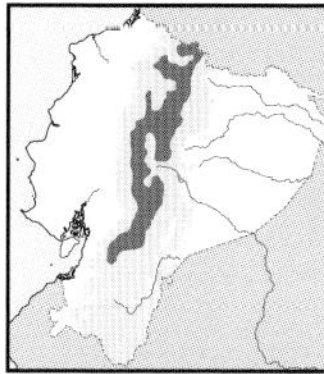

Yellow-billed Pintail *Anas georgica* 48–57cm

Highlands (2,200–4,000m). Lakes and *páramo* ponds, locally on inter-Andean lakes. ***Bill mostly yellow. Pale warm brown, heavily speckled dusky below.*** Upperparts darker, scaled buff, ***long dusky tail.*** In flight dark underwing, ***black speculum bordered by two whitish-buff stripes.*** ♀ duller. Race *spinicauda.* Pairs to small groups, often with Andean Teal, feed by dabbling, up-ending, grazing or even diving. **Voice** Low *prrrrp* whistle, *prrrp-prrp-pirp-pirp,* but not very vocal. **SS** Andean Teal (dark bill, darker overall); cf. ♀ Blue-winged and Cinnamon Teals, and Northern Pintail. **Status** Uncommon.

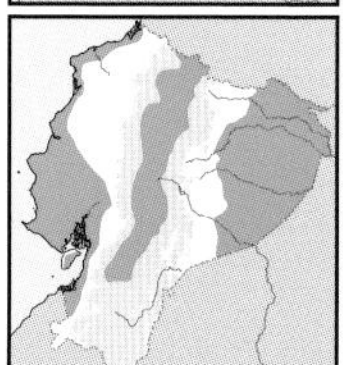

Blue-winged Teal *Anas discors* 35–41cm

Boreal migrant to W lowlands, locally in highlands, rarely to E (to 3,200m, locally higher). Freshwater wetlands. Dark bill. ***Bluish-grey head, white facial crescent.*** Dark brown above edged buff, tawny below spotted dusky. ♀ and eclipse ♂ dusky-brown, buffier below; whitish supercilium, ***dusky eyestripe, whitish loral spot.*** In flight white underwing, ***blue upperwing-coverts,*** green speculum outlined white. Monotypic. Up to dozens dabble in open water or rest on muddy shores. **Voice** Thin *tsee-tsee..pfa-pfa-pfa* whistles. **SS** ♀ Cinnamon Teal (warmer brown, weaker facial pattern); juv. trickier (Cinnamon warmer brown). **Status** Uncommon (Oct–Apr).

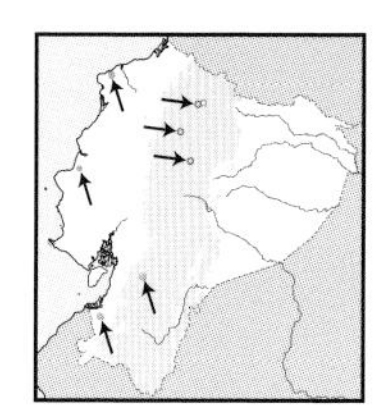

Cinnamon Teal *Anas cyanoptera* 35–45cm

N Andes; lowlands in W (sea level, 2,500–2,800m). Freshwater wetlands. ***Rich cinnamon-chestnut,*** darker above. ♀-plumaged birds ***dull brown,*** faint dusky ocular line, ***darker lores.*** In flight white underwing, ***blue upperwing-coverts,*** green speculum. Resident race *borreroi* (Andes, extinct; dark spotting below), boreal migrant *septentrionalium;* larger *orinomus* likely in SW. Might consort with Blue-winged Teal, feeds by dabbling, up-ending, sometimes dipping head. **Voice** Similar to Blue-winged Teal. **SS** ♀-plumaged Blue-winged Teal (darker overall, bolder face pattern; wing pattern same); Andean Teal smaller, darker, shorter-billed, no blue in wing. **Status** Extinct in highlands, accidental migrant expected Dec–Apr.

Andean Teal
♂
♂
♀
♀
Yellow-billed Pintail
♀
juvenile
♂
adult
♂
breeding
♀
Blue-winged Teal
♀
♂
♂
eclipse
Cinnamon Teal
♀
♂

White-cheeked Pintail is the only dabbling duck currently breeding in the lowlands, plus four accidental visitors from the N Hemisphere, the shoveler being more regular than the other species on this plate.

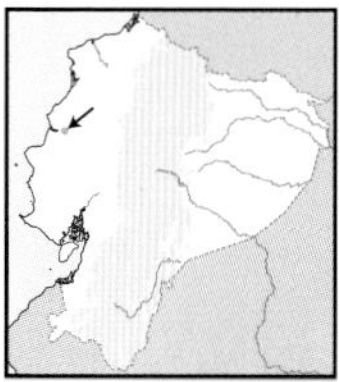

Green-winged Teal *Anas crecca* 35–37cm

One record from La Segua (R. Ahlman). ***Small. Chestnut-and-green*** head of breeding male unmistakable. Featureless eclipse ♂ and ♀ have blackish crown, grey-brown face with ***faint eyestripe, yellowish bill base***. In flight white underwing, ***green speculum***. Race *carolinensis*. Feeds by dabbling and up-ending in shallow water, often near cover. **Voice** Soft, liquid *preep-preep* (MB). **SS** Blue-winged Teal (larger, pale face, blue upperwing-coverts); cf. Cinnamon Teal. **Status** Accidental boreal visitor (Dec).

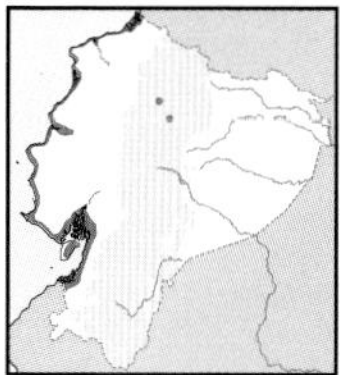

White-cheeked Pintail *Anas bahamensis* 44–50cm

Coasts and neighbouring lowlands (below 50m; once at 3,800m). Open wetlands. ***Bill bright red at base***. Warm brown, darker head, ***contrasting white cheeks and throat***. Underparts spotted dusky, long buffy tail. ♀ duller. In flight ***bright green speculum broadly bordered buff on trailing edge***, some white in underwing-coverts. Race *rubrirostris*. Pairs to small groups (larger numbers locally) often swimming, dabbling, up-ending or head-dipping in open water, sometimes on muddy shores. **Voice** Soft, nasal *gyaak-gyaak-gyaak*... in flight; thin *fiip-fiip* whistles. **SS** Confusion unlikely. **Status** Locally common.

American Wigeon *Anas americana* 45–56cm

One record at a freshwater wetland in NW lowlands. *Short bluish bill* tipped black. ***Heavily speckled greyish-buff head*** contrasts with darker (♂) or buffier (♀ and eclipse ♂) underparts. ♀ upperparts are blackish edged buff. ***White patch in wing-coverts*** prominent in ♂, also white belly and black vent. Monotypic. In regular range, pairs or loose to fairly large groups graze next to shallow water; also dabble. **Voice** Unlikely to be heard in Ecuador. **SS** Contrasting head, prominent white wing patch and white belly distinctive. **Status** Accidental (Dec).

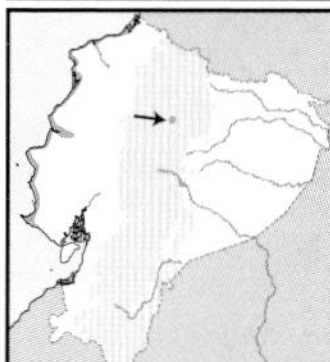

Northern Shoveler *Anas clypeata* 43–52cm

Local in W near coast (to 10m; twice in *páramo* at 3,800m; D. Brinkhuizen, M. Sánchez). Shallow brackish to freshwater wetlands. ***Long-necked***, with large ***spatula-like bill***. ♂ has ***glossy green head, contrasting white chest and rufous underparts***. ♀-plumaged birds brown speckled dusky, bicoloured bill. In flight ***white underwing***, bluish upperwing-coverts, ***green speculum outlined white*** (♂), narrow white stripe in coverts, duller speculum (♀). Monotypic. Singles or small groups seen in Ecuador; elsewhere mostly in pairs to small groups, dabbling in shallow water. **Voice** Not heard in Ecuador. **SS** Unique. **Status** Accidental boreal winter visitor (Dec–Mar).

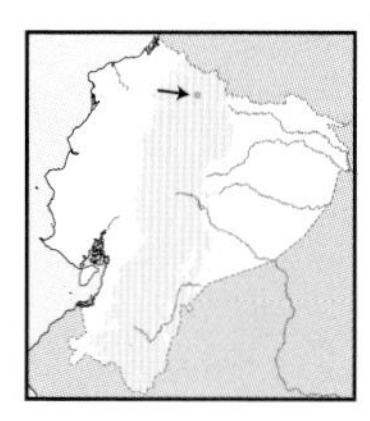

Northern Pintail *Anas acuta* 50–56cm

One record from Lago San Pablo, 2,400m (K. Terán *et al.*); one in NW coast (R. Ahlman). ♂ unmistakable in breeding plumage. Eclipse ♂, ♀ and juv. better identified by ***slim, longish neck, paler and warmer*** than upperparts and sides to flanks (speckled dark). Bill ***dark*** in all plumages (cf. Yellow-billed Pintail); ***green*** wing speculum (cf. Blue-winged and Cinnamon Teals); ***no*** blackish head, ***dark*** underwing-coverts (cf. Andean Teal). Nominate race. Singles observed with other ducks; elsewhere forms large flocks, dabbling and up-ending. **Voice** Mellow *proop-proop* (MB). **Status** Accidental boreal winter visitor (Nov–Feb).

♂
breeding
♀
♀
Green-winged Teal
♂
moulting eclipse
♂
♂
♂
White-cheeked Pintail
♀
♂
breeding
♀
♀
♂
American Wigeon
♂
♀ / ♂ non-breeding
♂
moulting eclipse
♀
Northern Shoveler
♂
breeding
♀
♂
♂
juvenile
Northern Pintail
♂
moulting eclipse
♀

Miscellaneous ducks: one extremely rare and declining in freshwater lakes of W and Andes; two secretive diving resident species not easily spotted; two rare boreal winter visitors; Lesser Scaup becoming more regular.

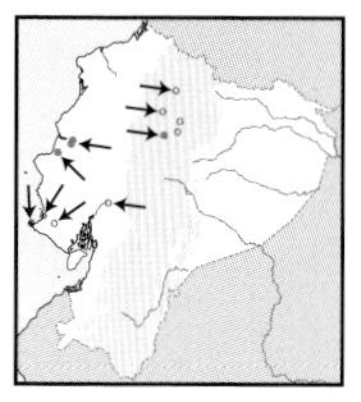

Southern Pochard *Netta erythrophthalma* 46–48cm

Few records from SW lowlands and N highlands (W to 100m; formerly to 3,200m). Freshwater marshes, lakes. *Greyish bill, red eyes* (brown in ♀). *Head to back blackish-maroon*; otherwise mostly *dark brown*. ♀ has dusky head and upperparts, *white facial crescent* to throat, paler underparts. In flight *extensive white in flight feathers*. Nominate. Small diving flocks, apparently move seasonally following water level changes; patters surface in taking-off. **Voice** *perrr-perrr-perrr* in flight (MB). **SS** Black-bellied Whistling-Duck (white in upperwing but longer-necked, black underwing); see ♀ Lesser Scaup. **Status** Very rare, declining, CR in Ecuador.

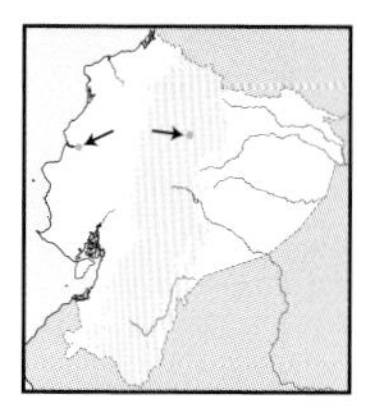

Ring-necked Duck *Aythya collaris* 40–45cm

Two recent records; La Segua marshes (T. Seimola), Micacocha (D. Brinkhuizen). ♀-plumaged ducks brown, *paler face, whitish eye-ring and foreface*; faint *pale ring on bill*. Otherwise, recalls Lesser Scaup, but *drabber* brown. ♂ less vermiculated above. Note head shape with *bump on rear crown* (more crested, *higher-crowned* in scaup). In flight *no white* in flight feathers. Monotypic. Mostly dives, also dabbles and up-ends (MB). **Voice** Not expected to vocalise in Ecuador. **SS** ♀ Southern Pochard (warmer brown, bolder face marks; juv. has whiter foreface, more rounded head). **Status** Accidental boreal vagrant (Jan–Feb).

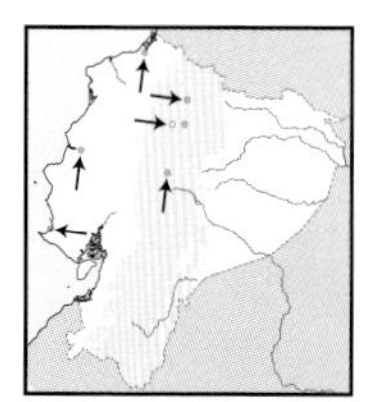

Lesser Scaup *Aythya affinis* 38–43cm

Few records from N highlands and SW lowlands (sea level, 2,600–2,800m). Freshwater and brackish wetlands. Greyish bill, *yellow eyes. Head glossy black-purple*, slightly crested, *breast black*. Grey above, densely vermiculated, *white below*. ♀ *brownish*, contrasting *white foreface*, white belly. In flight extensive *white in underwing*, white bar across *mid flight feathers above*. Monotypic. Dives in open water; numbers in Ecuador increasing (up to 95 birds recently counted). **Voice** Mostly silent. **SS** See Ring-necked Duck and ♀ Southern Pochard (paler facial marks, more rounded head). **Status** Boreal vagrant to rare visitor (Mar, May, Nov, Dec–Feb).

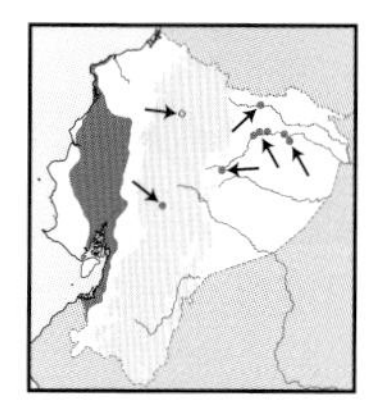

Masked Duck *Nomonyx dominicus* 33–36cm

W lowlands, local in Andes and E (mostly to 300m). Freshwater marshes and ponds with abundant aquatic vegetation. Small, chunky. *Bill bright bluish tipped black. Black head, chestnut overall, black spotting on upperparts and sides*; black stiff tail. ♀ and eclipse ♂ dull brown, *two blackish lines on face*. In flight *white axillaries and square white patch in secondaries* (above). Monotypic. Mainly small parties, often remains hidden, avoids open water; dives. **Voice** Rarely gives *cruee-cruee...* and other hisses. **SS** Ruddy Duck in highlands (black *hood*, white underwing, no head-stripes in ♀). **Status** Uncommon, local.

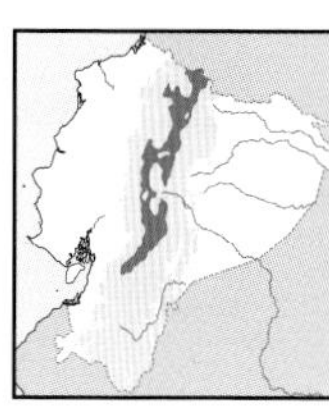

Ruddy Duck *Oxyura jamaicensis* 41–43cm

Andes (2,100–4,000m). Open freshwater lakes. Fairly long, *bright bluish bill. Black hood*, otherwise mostly *rich chestnut*, black stiff tail. *Belly whitish mottled dusky*. ♀ *dark brown*, brown bill, *two vague* horizontal lines on face. In flight extensive *white underwing and black upperwing*. Race *ferruginea*; *andina* with white cheeks (locally in N). Social, but groups not compact; regularly swims buoyantly on open water, tail cocked; dives. **Voice** Mostly silent; a mechanical, nasal *üäk* or *üüü* in display. **SS** Masked Duck (black mask, white patch in secondaries); little if any overlap. **Status** Uncommon.

Southern Pochard
♀
♂
♂ juvenile
♀
♂
♀ juvenile
♀
♂
♂ 1st-winter
♂ breeding
♂ eclipse
♀ breeding
juvenile
Ring-necked Duck
♀
♂
♀ non-breeding
♂ 1st-winter
♀ breeding
Lesser Scaup
♂ breeding
juvenile
♀
Masked Duck
♂ non-breeding
♂ breeding
ferruginea
♀
♂
♂
Ruddy Duck
andina

PLATE 13: GREBES AND PENGUIN

Three resident breeding grebes, plus one only recently recorded. Silvery is confined to fairly remote Andean wetlands (currently extinct at many accessible ones). The distribution of Least and Pied-billed is chiefly dependent on seasonal/geographic variation in water level. Just one penguin has to date been reported in mainland Ecuador, but Galápagos Penguin *Spheniscus mendiculus* could occur.

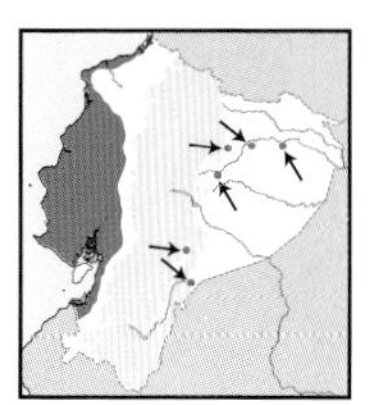

Least Grebe *Tachybaptus dominicus* 22–25cm

W lowlands, spottily in E (to 700m). Ponds, lakes, estuaries, ephemeral wetlands, dams. ***Small; thin, pointed, slightly up-curved bill, yellow eyes.*** Crown and throat black, ***slate-grey cheeks to chest***, grey-brown underparts. *Blackish above*, white fluffy flank feathers. Non-breeders *paler*, no black throat. Young have dusky stripes over white head, eyes already ***yellow***. Race *eisenmanni* (W), *brachyrhynchus* (E). Small loose groups, sometimes pairs or singles; tends to prefer wetlands with floating and riparian vegetation, persist temporarily in surprisingly small ponds. **Voice** Rarely vocal but breeders utter a trilled *tirririri* song and louder *hónk* calls. **SS** Co-occurring larger, browner Pied-billed Grebe has heavier bill. **Status** Fairly common.

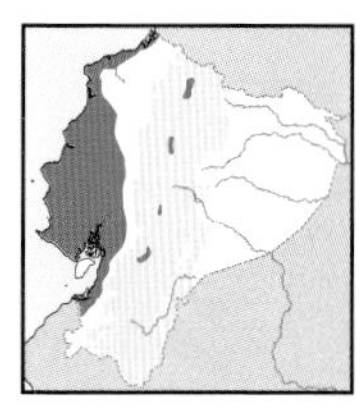

Pied-billed Grebe *Podilymbus podiceps* 28–33cm

W lowlands, locally in Andes (below 200m in W, 2,000–3,200m in Andes). Ponds, lakes, dams. Larger than previous species; ***heavy, whitish bill*** with ***blackish ring*** near tip in breeder; ***dark eyes. Greyish-brown***, richer brown below, ***black crown and throat***. Non-breeder paler, ***white throat, no ring on bill***. Young has dusky stripes on throat and neck-sides, dark eyes. Race *antarcticus*. Often associates with Least Grebe but favours open, deeper water; mostly singles but several may gather at certain sites. **Voice** Deep, hollow *kuh-kuh-kok'kok'kok..ku-kow, kow, kow....* **SS** Bill shape unique among Ecuadorian grebes; from gallinules or coots by behaviour, structure, etc. **Status** Common.

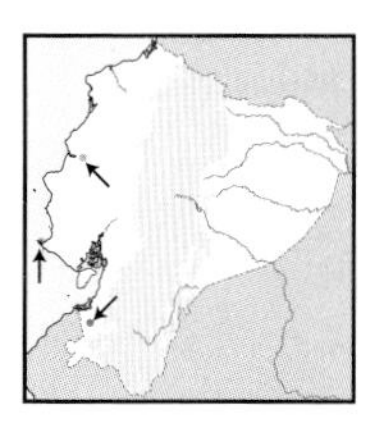

Great Grebe *Podiceps major* 70–78cm

Few records at SW wetlands (Santa Elena Peninsula, Tahuín and La Segua). Large, ***long-necked, thin*** pointed bill, small crest. Blackish head and upperparts, ***grey face, reddish neck and breast***, white belly, cinnamon undertail-coverts. Non-breeder (more regular in Ecuador) ***paler***, with whitish underparts and pale bill. Presumed race *major*. Elegant plumage, shape and swimming behaviour, with neck often held straight; recorded in Mar at Ecuasal saline ponds, recently found breeding in SW Ecuador. **SS** Readily identified from other grebes. Imm. Neotropic Cormorant or Anhinga have very different bill shapes. **Status** Mainly vagrant (Jan–Jul, Sep, Dec).

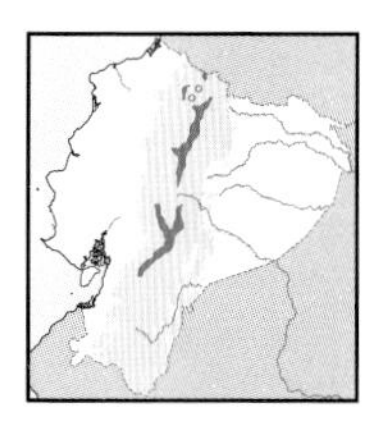

Silvery Grebe *Podiceps occipitalis* 27–31cm

Andes (2,200–4,100m). Open lakes in *páramo*, locally (formerly?) also on inter-Andean lakes. Slender, thin, pointed bill, ***red eyes. Silvery-grey head***, with long ***silvery-white plumes on cheeks***. Grey above, ***snowy-white below***. Non-breeders ***lack cheek plumes***. Race *juninensis*; likely valid species. Somewhat less frequent diver than other grebes, picks aquatic invertebrates from surface; formerly in large groups, but numbers so small now that mostly seen in small parties. **Voice** Not very vocal but thin *wit-wit-wit* or *dzit-dzit-dzit* calls. **SS** Only grebe in range; local overlap with Pied-billed Grebe (heavier bill, browner and duller plumage). **Status** Rare, local, declining, total population just hundreds; VU in Ecuador, NT globally.

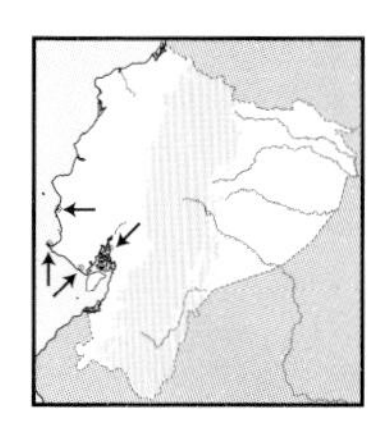

Humboldt Penguin *Spheniscus humboldtii* 65–70cm

Few records on SW beaches at Santa Elena Peninsula. Iris brown, heavy black bill with extensive ***flesh-pink** area at base*. Handsome ***black-and-white pattern***, with white line from supercilium to pectoral band and sides. ***Broad black pectoral band arches down to lower underparts***. Juv. ***brown*** above, with greyer cheeks, ***brown chest, whitish underparts***; pink bill skin paler. Monotypic. Accidental; weak penguins sporadically found on shore, presumably carried by sea currents or even by fishermen (?). **SS** Although not yet recorded, Galápagos Penguin could occur in Ecuadorian waters as it has wandered to Colombia and Panama; smaller, lacks extensive pink in face. **Status** Vagrant (Oct, Dec).

adult
non-breeding
adult
breeding
juvenile/
immature
Least Grebe
adult
non-breeding
adult
breeding
juvenile/
immature
Pied-billed Grebe
Great Grebe
juvenile
adult
non-breeding
adult
breeding
adult
breeding
juvenile/
immature
adult
non-breeding
Silvery Grebe
Humboldt Penguin
adult
juvenile

PLATE 14: FRIGATEBIRDS AND CORMORANTS

Large aquatic or marine birds that show little resemblance among families apart from heavy bills. Magnificent Frigatebird is common along the entire Ecuadorian coast, whilst Neotropic Cormorant is also common and widespread. Identification of the two frigatebirds is difficult. Guanay Cormorant probably breeds in extreme S.

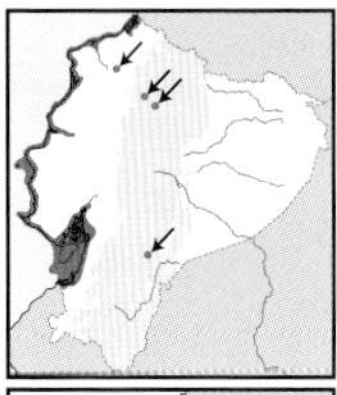

Magnificent Frigatebird *Fregata magnificens* 97–107cm

Entire coast, sometimes at inland wetlands (twice in highlands). ♂ large, mainly black, red gular pouch (breeders), ***blackish to greyish legs***, very long wings and tail; ***little to no contrast in upperwing***. ♀ has ***broad white pectoral band, black throat***, some white in axillaries. Juv. has ***entire head to belly white, dark spurs*** on breast-sides. Intermediate plumages frequent. ***Blackish to bluish orbital ring***. Monotypic. Chases seabirds and boats to rob prey; mostly seen in flight, patrolling high overhead. **Voice** Screams, chatters, cackles, etc. at breeding colonies. **SS** Great Frigatebird. **Status** Common.

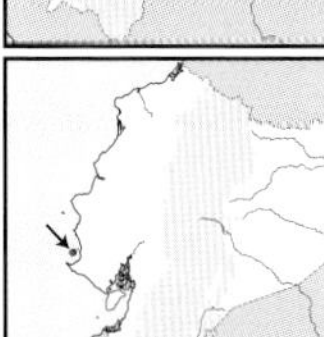

* Great Frigatebird *Fregata minor* 89–97cm

One May record from SW coast (Olón; B. Haase). ♂ very similar to Magnificent, not readily distinguished in field, but ***orbital ring red, pale band on upperwing***. ♀ ***white from chin to breast-sides***, juv. with ***dull tawny head***, white belly. Also intermediates. Leg colour probably not a safe character as Magnificent-plumaged birds with red legs have been seen. **SS** Wings ***broader and tail shorter*** than Magnificent, but hard to discern in field. Race *ridgwayi*. **Status** Accidental visitor to coasts, but habits and behaviour similar to other frigatebirds.

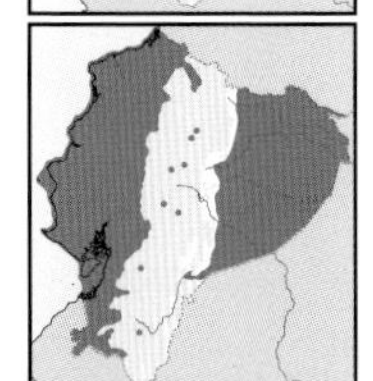

Neotropic Cormorant *Phalacrocorax brasilianus* 64–69cm

E & W lowlands, locally in inter-Andean valleys (mainly to 800m, to 2,400m in Andes, locally higher). Rivers, freshwater lakes, ponds, estuaries, coastline. Mostly ***glossy black***, bill greenish-yellow, ***whitish at base***, bluish eyes. Juv. ***Dusky-brown above, whitish below***. Nominate. Highly gregarious, often seen perched atop snags drying wings; strong diver; flight steady, fast and direct. **Voice** Some raucous grunts but mostly silent. **SS** Confusion unlikely; cf. longer-billed Anhinga. **Status** Common.

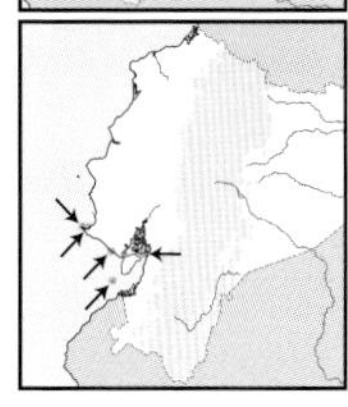

Guanay Cormorant *Phalacrocorax bougainvillii* 69–74cm

Visitor to SW coast; possibly breeding at Isla Santa Clara (F. Sornoza). ***Glossy black above, white throat and below***. Bill greenish-olive, ***red facial skin***, pinkish legs. Non-breeders duskier and browner, juv. duller, ***dirty white below***. Monotypic. Erratic and irruptive visitor to SW coast (N to Santa Elena Peninsula), movements probably linked to warm-water incursions during El Niño events; feeds mostly on anchovies; highly gregarious at colonies, regular offshore. **Voice** Nearly silent. **SS** Commoner Neotropic Cormorant (smaller, blacker ad., juv. much duller, no pinkish legs, yellower bill). **Status** Irregular and local (Mar, Aug, Nov); NT globally.

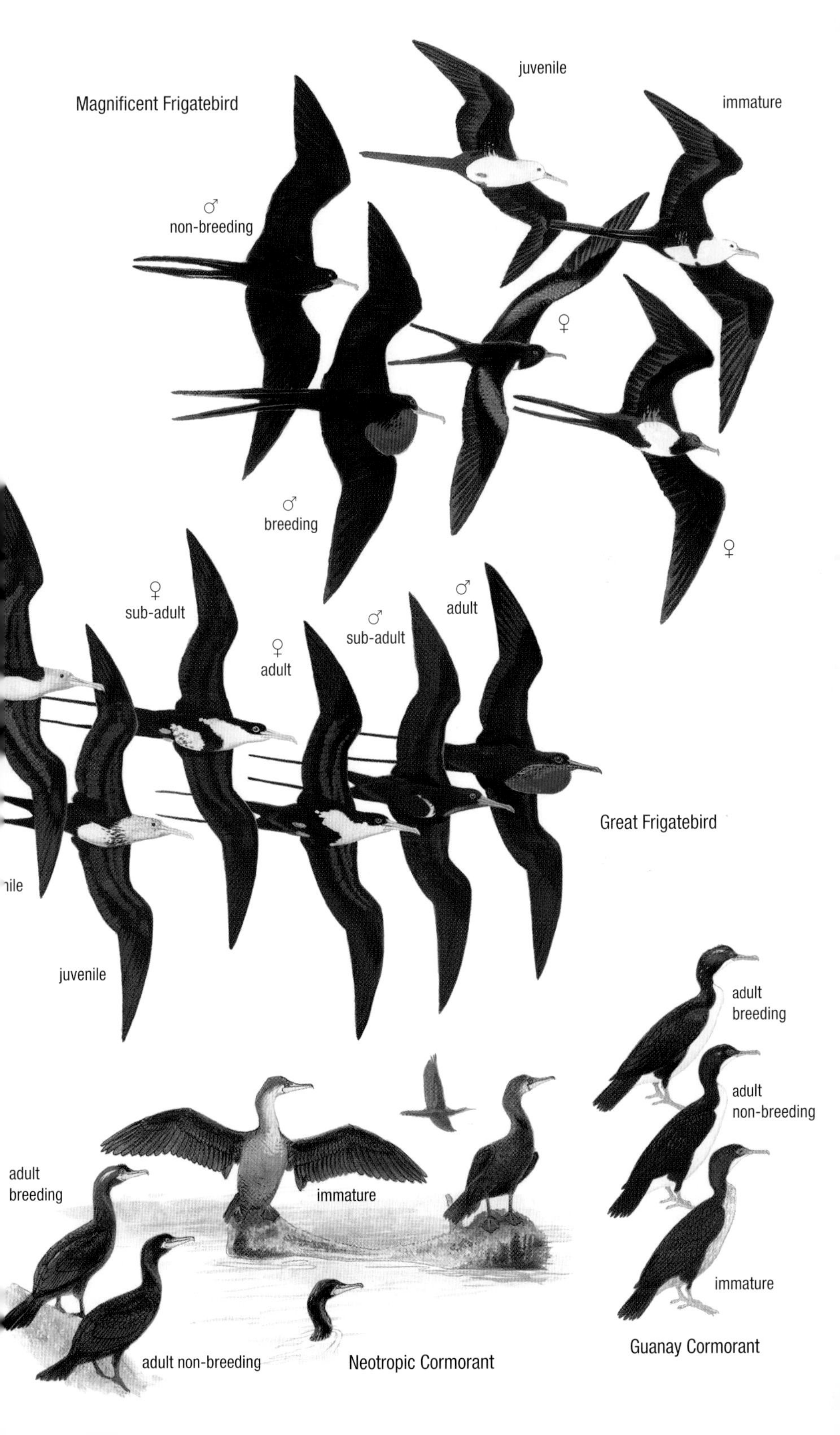

Magnificent Frigatebird
juvenile
immature
♂
non-breeding
♀
♂
breeding
♀
♀
sub-adult
♀
adult
♂
sub-adult
♂
adult
Great Frigatebird
ile
juvenile
adult
breeding
adult
non-breeding
adult
breeding
immature
immature
adult non-breeding
Neotropic Cormorant
Guanay Cormorant

PLATE 15: ANHINGA, PELICANS, TROPICBIRD AND GANNET

Identification of the two pelicans demands care. Anhinga is the only species confined to freshwater bodies. Tropicbird is pelagic and, consequently, rarely seen from shore. Plus a gannet tentatively identified as Cape Gannet.

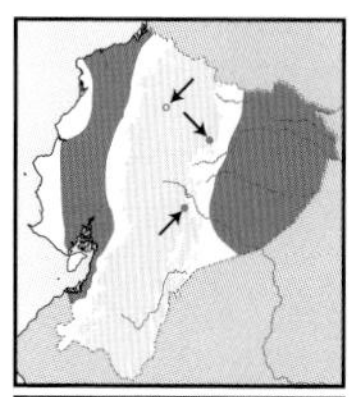

Anhinga *Anhinga anhinga* 83–89cm

E & W lowlands (to 400m). Rivers, freshwater lakes and ponds. ***Long spear-like yellowish bill, long slender neck.*** ♂ ***glossy black*** with conspicuous ***silvery streaks in upperwing-coverts.*** Tail long, tipped brown. ♀ has ***buffy-white head to upper breast***; juv. similar but duller. Nominate. Mostly singles or pairs perched with open wings to dry; often seen soaring with fanned tail and extended neck; can swim partially submerged. **Voice** A mechanical *gk-gk-gk-gk*…, but often silent. **SS** Confusion unlikely, Neotropic Cormorant shorter-billed, shorter-necked, darker, etc. **Status** Uncommon.

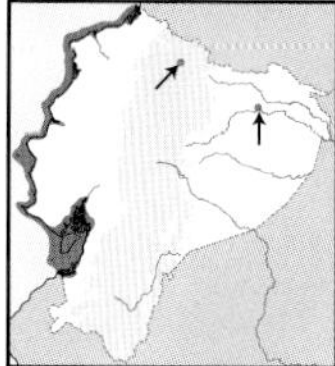

Brown Pelican *Pelecanus occidentalis* 117–132cm

Widespread along coast (once at 200m and 2,200m). Bill ***mostly grey, orangey tip. Head and neck-sides dull white, tinged buff in breeders.*** Chestnut hindneck, ***grey upperparts***, blackish flight feathers, dusky-brown underparts. Juv. duller and duskier, whitish belly. Race *murphyi*. Highly aerial, always seems to be 'going somewhere else'; gregarious at breeding and roosting sites, less so while feeding; plunge-dives. **Voice** Silent. **SS** Very similar but larger Peruvian Pelican (more coloured bill, large whitish wing patches). **Status** Common.

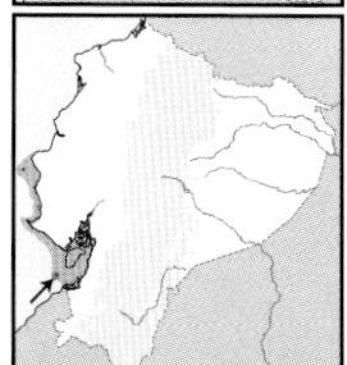

Peruvian Pelican *Pelecanus thagus* 137–152cm

Locally on SW coast, mainly to S Manabí. ***Larger*** than commoner Brown Pelican, with long colourful bill (horn, pinkish, orange). Head and foreneck white tinged yellowish-buff in breeders, chestnut hindneck. Blackish flight feathers, ***whitish patch in upperwing-coverts.*** Dusky-brown below. Juv. recalls Brown but clear ***whitish wing patches.*** Monotypic. Often consorts with Brown, tends to dive less; more regular offshore. **SS** When together, size separates from Brown; also white wing patches, more colourful head. **Status** Austral migrant, year-round; small breeding colony at Isla Santa Clara, NT globally.

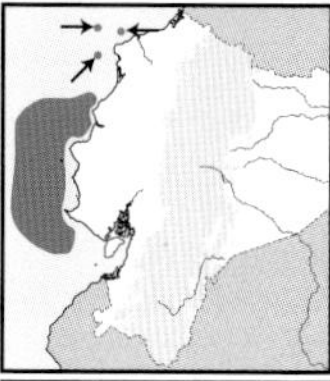

Red-billed Tropicbird *Phaethon aethereus* 43–48cm

Offshore, more regular around Isla de la Plata. ***Mainly white***, barred upperparts, ***black outer primaries and ocular line; bright red bill, very long tail-streamers***; bill ***yellower*** in juv., ***streamers lacking.*** Race *mesonauta*. Nests in rocky crevices; often seen flying near nest sites in spectacular displays; more wary and solitary at sea, where flies high above water; plunge-dives. **Voice** Guttural *tk-tk-tk-tk-tcheeer*, other cries in display (RG). **SS** Large red-billed terns (Royal, Elegant, Caspian); other tropicbirds could occur. **Status** Uncommon at breeding sites, even rarer elsewhere.

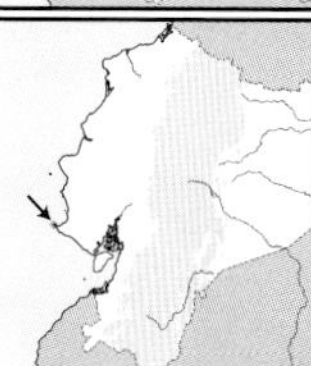

* Cape Gannet *Morus capensis* 85–90cm

One record from Santa Elena Peninsula (B. Haase). Larger than Ecuadorian boobies, mostly white except ***yellowish nape.*** Recalls Nazca and Masked Boobies but bulkier, ***no*** black mask, ***bluish bill, narrower*** black secondaries, ***white tertials.*** Juv. recalls ad. Peruvian Booby but darker, much ***larger***, with ***darker bill.*** Monotypic. Prone to wandering during non-breeding periods, but further records not really expected. Flight powerful and purposeful, with heavier flaps than boobies. **SS** Red-footed Booby juv. has dark head; see larger, narrower-winged albatrosses and paler shearwaters and petrels. **Status** Accidental visitor (May); accepted as *Morus* sp. by CERO.

Anhinga
juvenile
♀
♂
Brown Pelican
immature
adult non-breeding
adult breeding
Peruvian Pelican
adult non-breeding
adult breeding
Red-billed Tropicbird
adult
juvenile
Cape Gannet
immature
adult

PLATE 16: BOOBIES

Boobies are familiar in the Galápagos Islands but in mainland Ecuador only one (Blue-footed) is commonly seen from shore; the other species are rarer, more pelagic, or mostly confined to SW. See also gannet on previous plate.

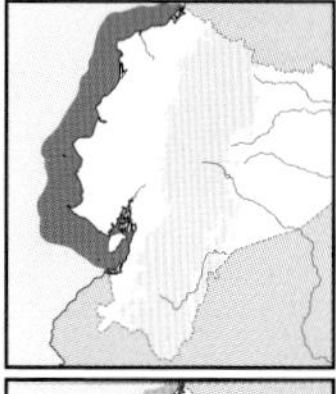

Blue-footed Booby *Sula nebouxii* — 76–84cm

On coast, nesting locally in rocky islets and islands. ***Bright blue legs***, yellow iris. ***Head densely streaked dusky-brown, white upper mantle and rump***; underparts white, underwing blackish and pale. ♀ larger, with ***larger pupils***. Juv. has ***solid brown head to upper breast, white nuchal patch***, grey legs. Nominate. Coastal, spectacular plunge-dives; direct and purposeful flight in long lines. **Voice** Various grunts, bottle whistles and hisses. **SS** Peruvian Booby whiter overall, no blue legs, no white rump; Brown Booby (yellow bill, no white rump, no blue legs). **Status** Common.

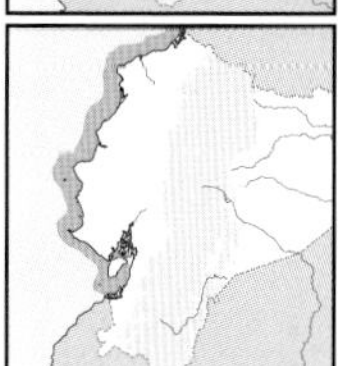

Peruvian Booby *Sula variegata* — 71–76cm

Mainly SW coast. Greyish bill and face, ***greyish legs. Mostly white, back to rump dusky-brown heavily scaled white***. Dark underwing with ***white central stripe***. Upperwing-coverts ***finely spotted white***. Juv. has ***head and underparts mottled yellowish-brown***. Monotypic. Irregular visitor, sometimes numerous but occasionally not seen for years; often with Blue-footed Booby; plunge-dives. **Voice** See previous species. **SS** From Blue-footed by contrasting white head, underwing pattern, upperwing spots, no white rump; see accidental Cape Gannet. **Status** Irregular year-round visitor.

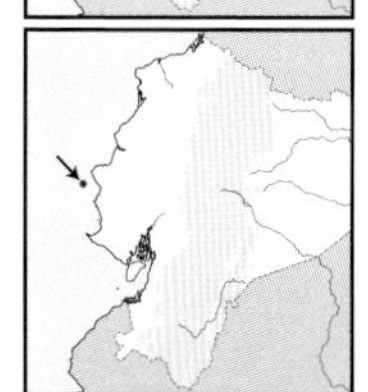

Masked Booby *Sula dactylatra* — 80–85cm

Three records on Isla de la Plata, apparently breeding (Oct–Feb) in very small numbers. Much like Nazca Booby, but ***bill yellow to pale yellow, especially at base***, possibly ***larger*** black mask. Some Nazca have white central remiges (S *et al.*), none in Masked. Some juv. Masked have broader white nuchal collar, browner upperparts. Seemingly Nazca has ***shorter, shallower bill***. Race *personata*. Feeds much further offshore than Nazca, apparently engages in long-distance movements. **Voice** As other boobies. **SS** Juv. recalls rare Brown Booby but more extensive white below, no straight division on breast; bill darker. **Status** Accidental.

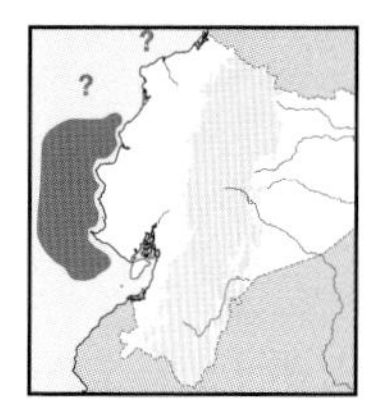

Nazca Booby *Sula granti* — 79–87cm

On S coast (C Manabí to Santa Elena Peninsula), rarer northwards. ***Orange bill***, richer at base. ***Snowy-white***, black mask, tail, ***greater wing-coverts and flight feathers***. Juv. has ***dark*** bill, ***solid blackish-brown hood and upperparts, white nuchal collar***. Monotypic. More pelagic than most Ecuadorian boobies; colonies on open ground and cliffs, likely only at Isla de la Plata. **Voice** Bottle whistles, grunts, hisses in display. **SS** Juv. recalls rare Brown Booby, but brown less extensive on underparts (with nuchal collar); see much rarer Masked Booby and Cape Gannet. **Status** Common at breeding site, rarer elsewhere.

Red-footed Booby *Sula sula* — 66–74cm

Offshore around Isla de la Plata, S to Santa Elena Peninsula. Bill bluish-grey, ***pink basal patch, red legs***. Dark morph ***dusky-brown, paler below***. White morph (rare in Ecuador) ***all white***, black flight feathers, ***black carpal patch***. Duller juv. has ***dark bill, greenish-grey legs***. Race *websteri*. Nests colonially in trees and bushes; regular in open seas. **Voice** As other boobies. **SS** Compare white morph with larger, orange-billed Nazca Booby (no red legs); juv. Brown Booby (larger, not uniformly dark below). **Status** Uncommon at breeding site, rarer elsewhere.

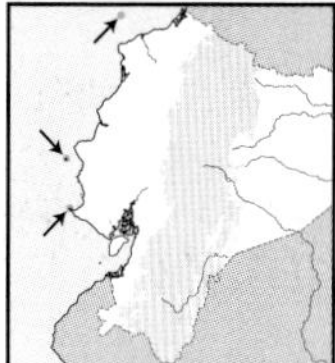

Brown Booby *Sula leucogaster* — 66–74cm

Few confirmed records along coast (Jun, Aug–Sep, Nov–Dec). Yellowish bill and legs. ***Solid dark brown above, to upper breast. White below***, including axillaries and broad central stripe in underwing. Juv. ***similarly patterned but underparts whitish-brown***, bill and legs duller. Race *etesiaca*. Pelagic to near coast, also plunge-dives; numerous elsewhere but accidental or irregular visitor to Ecuadorian waters. **Voice** As other boobies. **SS** Resembles juv. Nazca Booby but lacks nuchal collar; underwing pattern differs; juv. Red-footed more uniform below, cf. Peruvian and Blue-footed. **Status** Uncertain.

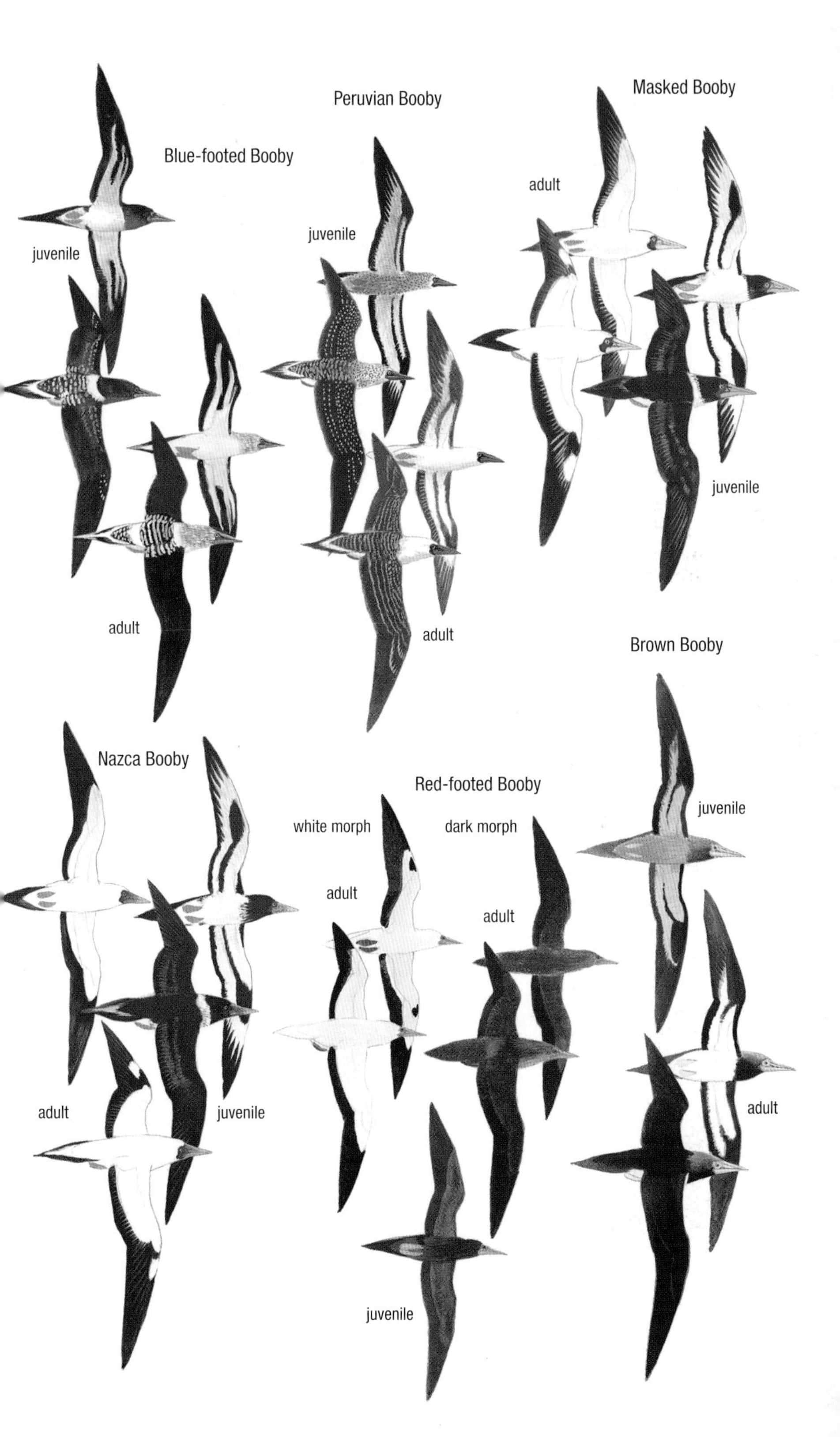
Peruvian Booby
Masked Booby
Blue-footed Booby
adult
juvenile
juvenile
juvenile
adult
adult
Brown Booby
Nazca Booby
Red-footed Booby
juvenile
white morph
dark morph
adult
adult
adult
juvenile
adult
juvenile

PLATE 17: ALBATROSSES AND GIANT PETREL

Only one of the four albatrosses is regularly recorded in Ecuadorian waters. Highly pelagic and great wanderers, spending consecutive months at sea. Identification can be tricky, but note bill, head and underwing patterns (see Onley & Scofield 2007 for details). Also, the largest procellariids, highly pelagic and nomadic, rare in Ecuadorian waters.

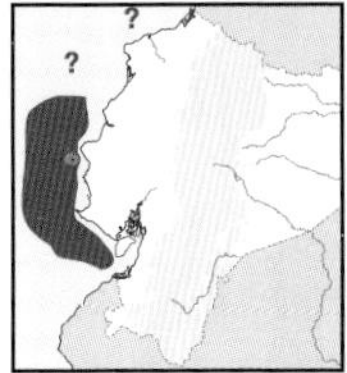

Waved Albatross *Phoebastria irrorata* 88–92cm / w 230–240cm

Small breeding population on Isla de la Plata (mostly Apr–Nov), regular offshore (Santa Elena / Manabí prov.). *Long yellow bill. White hood, crown and neck tinged yellowish.* Most upper- and underparts *finely waved dusky*, pale rump. Monotypic. Pelagic, singles or up to 30 congregate near fishing boats, but do not follow ships. Flies low over water on stiff, outstretched wings; deep wingbeats in light winds. **SS** Only regular albatross in Ecuador; identified by size, bill length and colour, hood, underwing and wavy underparts. **Status** Rare (year-round), CR globally, EN in Ecuador.

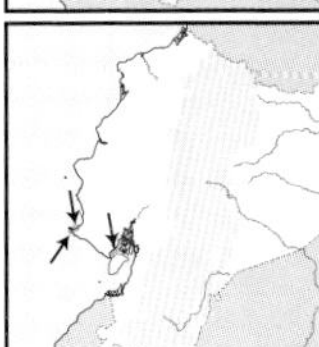

Black-browed Albatross *Thalassarche melanophris* 82–86cm / w 210–250cm

Three SW records; Santa Elena Peninsula and Gulf of Guayaquil. Possible passage off SW coast. *Yellow-orange bill, short black eyebrow. White head*, blackish above, *white rump to uppertail.* White *underwing, broad black wing edges* from below. Juv. has *greyish-yellowish bill tipped dark*, blacker underwing, grey collar. Nominate? Highly pelagic, habitual ship follower. Flight as Waved Albatross. **SS** Juv. with Salvin's Albatross (whiter underwing); Grey-headed Albatross *T. chrysostoma* not yet recorded, but might wander to Ecuadorian waters (darker bill, darker head, broader collar in juv.); see Kelp Gull. **Status** Accidental (May, Oct, Nov).

Buller's Albatross *Thalassarche bulleri* 75–80cm / w 210–215cm

One record of dead specimen, Ayangue, Santa Elena prov. (R. Carvajal & P. Amador). *Yellow-and-black bill. Pale grey hood*, whitish front, black orbital area. White underparts, dark grey upperparts, white underwing with *narrow black margins*. Duller juv. Monotypic. Highly pelagic, follows ships, rarely in tropical waters. Flight effortless, right above surface. **SS** Other mollymawks, especially Salvin's (bill greyish-horn, narrower black leading edge) and Grey-headed Albatross (darker hood, broader leading edge, blacker bill; not yet in Ecuador). **Status** Accidental (Apr).

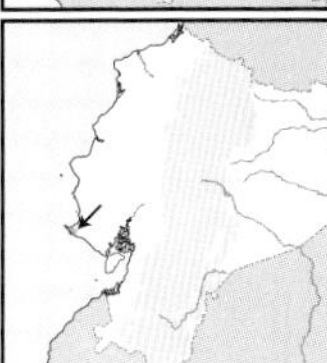

* Salvin's Albatross *Thalassarche salvini* 90–95cm / w 240–250cm

Single record, Mar Bravo, Santa Elena Peninsula (R. Carvajal). Identification not 100% certain; juv. not safely separated from White-capped Albatross *T. cauta*. *Greyish-horn bill tipped dark. Pearl-grey hood, whitish forehead.* White underparts and rump. Underwing has *very narrow black margins, black thumb mark at base of leading edge*. Juv. has *darker bill tipped dusky*, darker face. Ad. White-capped *lacks dark bill tip, head whiter*, narrower tips to primaries; juv. *whiter-headed*. Monotypic. Follows ships, rarely into tropical waters. Flight: soars frequently, wings held stiffly outstretched. **SS** Buller's and Grey-headed Albatrosses (not recorded in Ecuador). **Status** Accidental (Apr).

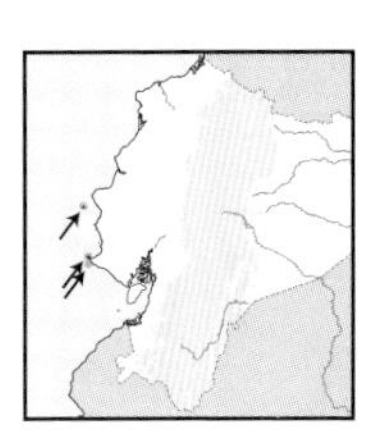

Southern Giant Petrel *Macronectes giganteus* 85–100cm / w 150–210cm

Few records in SW: off Salinas and Isla de la Plata. *Sturdy, humpbacked. Heavy yellow-horn bill tipped olive, conspicuous tubular nostrils. Dirty white head.* Otherwise mostly dull brown (or dirty white in white phase), heavily mottled / barred. Broad wings, *whitish leading edge, dusky underwing-coverts*. Juv. *blackish*, becoming paler with age. Monotypic. Scavenger, tends to wander over Humboldt Current, especially juv. Flight fast, on stiff wings, light soaring, agile chasing. **SS** Bill, silhouette and size separate Waved Albatross; see Northern Giant Petrel *M. halli* (not yet recorded in Ecuador). **Status** Casual transient or vagrant (Apr, Jul, Aug).

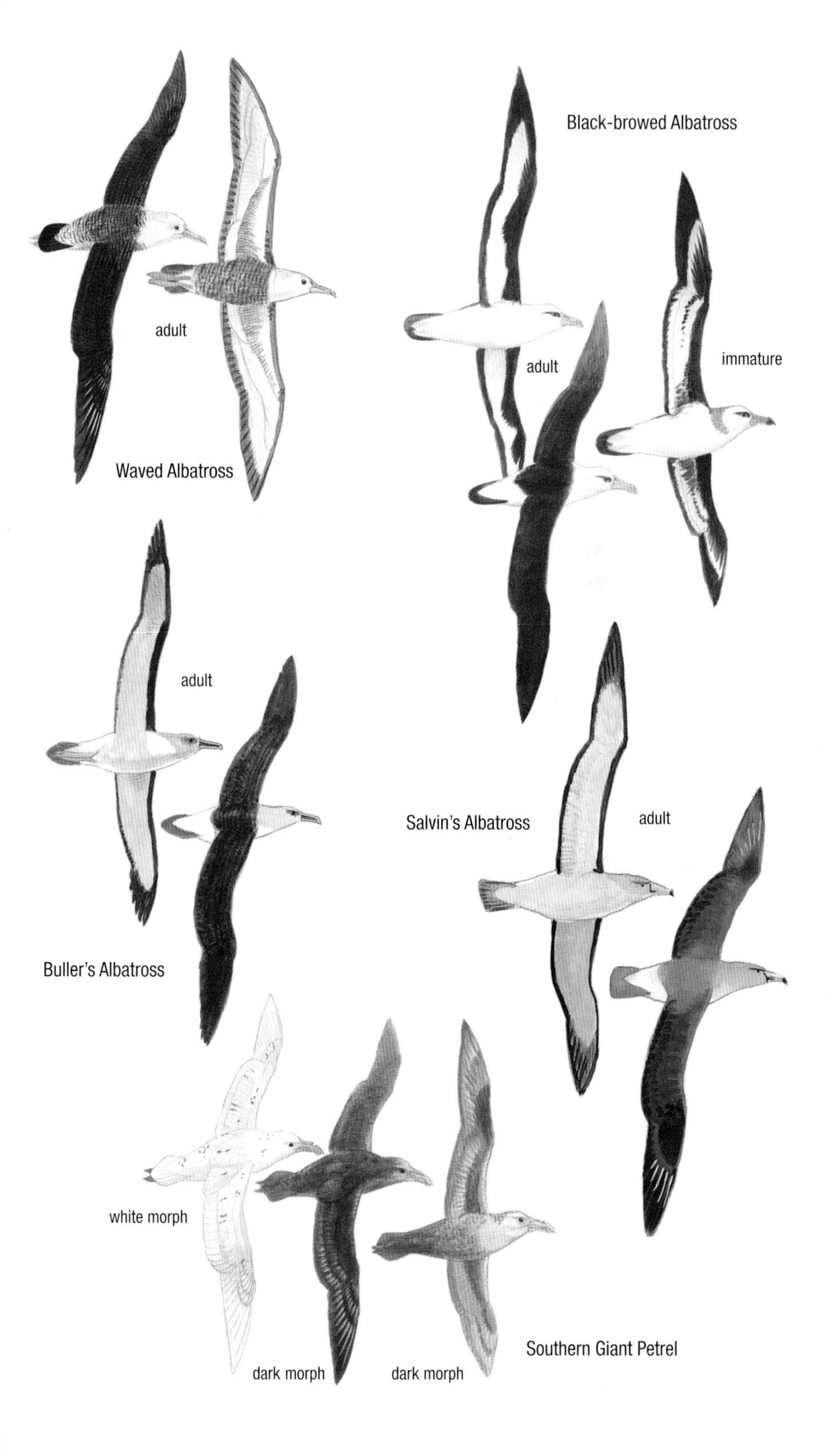
Black-browed Albatross
adult
immature
adult
Waved Albatross
adult
adult
Salvin's Albatross
adult
Buller's Albatross
white morph
dark morph
dark morph
Southern Giant Petrel

PLATE 18: FULMAR AND PETRELS

Poorly known and seldom-seen pelagic petrels, most appearing seasonally or wandering far offshore. Most are not easily identified, but note underwing patterns and flight behaviour. Only two species regular in Ecuador.

* Southern Fulmar *Fulmarus glacialoides* 45–50cm / w 115–120cm

One record from Santa Elena Peninsula. *Short pinkish bill tipped dusky.* White head and *underparts*, pearl-grey above, whiter tail. *White patch near wingtip, white below, very narrow* dark trailing edge. Monotypic. Flocks at sea, do not follow ships but often attracted to chum. Flight: short fluttering flaps on stiff wings, accomplished dynamic gliding, wings sometimes slicing waves. **SS** Whiter than other procellariids, with pink bill and distinctive wing pattern; recalls large gull but see bill, robustness, flight pattern. **Status** Accidental (Nov).

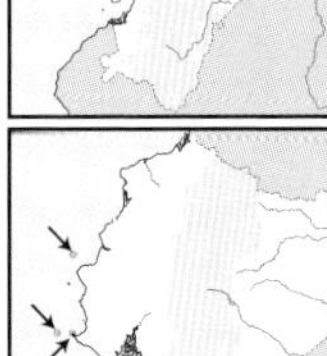

Cape Petrel *Daption capense* 36–40cm / w 85–86cm

Few records off SW coast, once N to Manta. Complex *checkered black-and-white pattern above*, underwing *largely white with broad black margins* and black smudges near axillaries. Possibly nominate. Several congregate, though not in Ecuadorian waters, following offshore ships, not near coast. Flight: bouncing glides. **SS** Confusion unlikely; no other checkered petrel in Ecuador; Sabine's Gull has distinctive V-shaped wing pattern. **Status** Casual transient (Jul–Aug, Nov, Feb).

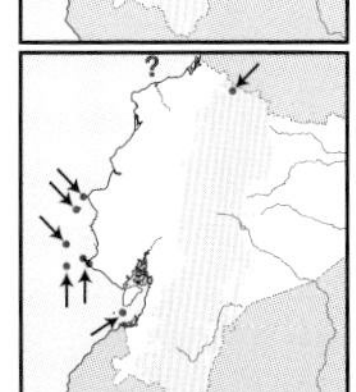

Galápagos Petrel *Pterodroma phaeopygia* 41–43cm / w 90–102cm

Off SW coast (one exceptional record at 1,400m in W Carchi). Long wings. *White forehead*, black cap, *dark rump*. White below. Underwing white, *narrow blackish trailing edge, blackish diagonal bar, dusky smudging on sides*. Monotypic. Mostly singles picking food from surface, rarely following ships. Flight fast, bounding, in long glides, little flapping, swooping over waves, then steeply downwards. **SS** Only pied Ecuadorian gadfly petrel; underwing distinctive. Juan Fernández Petrel likely in Ecuador (no diagonal bar on underwing, whiter sides). **Status** Rare visitor (Feb–Oct), Galápagos endemic, CR.

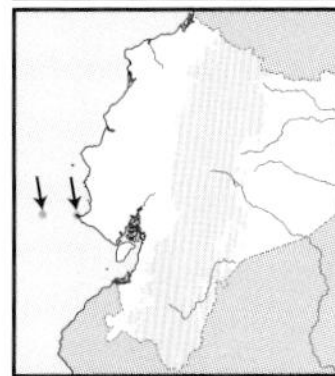

* White-chinned Petrel *Procellaria aequinoctialis* 53–58cm / w 134–147cm

Two records on SW coast, one off Santa Elena Peninsula, one dead (K. Barnes). *Greenish-ivory bill, pale-tipped*. Blackish-brown, *small white chin patch*, dark underwing. Monotypic. Elsewhere, frequent ship follower often in small flocks. Flight strong, purposeful, slow wingbeats and long glides. **SS** Very similar to Parkinson's Petrel, but larger, heavier, lacks dark bill tip (difficult to see in field); Sooty Shearwater (paler, slenderer bill, pale underwing patches). **Status** Accidental (Aug 1983 and 2007).

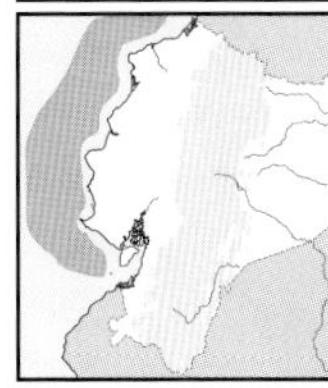

Parkinson's Petrel *Procellaria parkinsoni* 46–48cm / w 115–118cm

Regular off coast, possibly more numerous in SW, apparently overlooked northwards. Ivory bill *tipped dark*, blackish feet. Mostly *dark chocolate*, paler outer primaries below. Monotypic. Small groups at fishing boats, readily attracted to chum but rarely follows ships; often in mixed flocks. Slow wingbeats, long, lazy, low glides. **SS** Smaller, slimmer than White-chinned Petrel (pale bill tip), but field identification unsafe; Sooty Shearwater (greyer, slenderer bill, whitish underwing). **Status** Uncommon visitor (year-round), VU globally.

Cape Petrel
Southern Fulmar
Galápagos Petrel
White-chinned Petrel
Parkinson’s Petrel

All but one of shearwaters reported in Ecuadorian seas are rare vagrants or overlooked transients. Identification intricacies might obscure understanding of distribution and seasonal patterns. Sooty is only species regularly seen from shore.

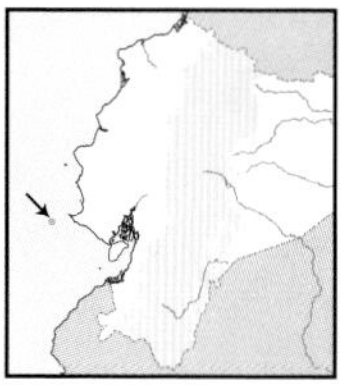

Wedge-tailed Shearwater *Ardenna pacifica* 45–47cm / w 97–99cm

Single record off Santa Elena Peninsula (B. Haase). Recalls Pink-footed Shearwater but ***longer, wedge-shaped tail*** and shorter, narrower wings provide different silhouette, and more buoyant flight. Bill *narrower*, throat and neck *less smudgy*. Dark morph has ***much less*** whitish in underwing than Sooty. Bill *narrower* than bulkier Parkinson's Petrel. Monotypic. Rarely attracted to boats, often follows dolphins. Flight very leisurely, flap-flap-glide. **SS** Buller's Shearwater (whiter underwing, bolder upperparts). **Status** Accidental (Oct).

Buller's Shearwater *Ardenna bulleri* 44–46cm / w 95–97cm

Single record from SW coast (Palmar). ***Blackish cap***, grey back and rump, whitish underparts. ***Bold dark M and pale V pattern above. Mainly white underwing, very narrow*** black margins. Monotypic. Unknown in Ecuador; elsewhere gregarious diver, often follows ships. Flight leisurely, long, effortless glides followed by graceful wingbeats (flap-flap-glide). **SS** Upperwing pattern distinctive; from below cf. Galápagos (much smaller, duskier underwing, different flight), Wedge-tailed (broader dark trailing edge) and Pink-footed Shearwaters (duskier underwing, pale soft parts). **Status** Accidental, possibly rare transient far offshore, VU globally (Jan).

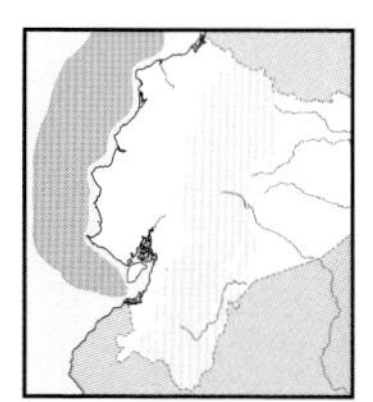

Sooty Shearwater *Ardenna grisea* 40–46cm / w 95–105cm

Regular off coast, more numerous off Santa Elena Peninsula. ***Blackish feet. Sooty-brown***, paler throat. Blackish axillaries, variable ***whitish on underwing. Narrow wings***. Monotypic. Small groups at fishing boats, but do not follow ships, regularly seen from shore and in breakers. Flight: fast, flap-flap-flap-glide, often zigzagging. **SS** Only mainly sooty-brown Ecuadorian shearwater, with whitish flash in underwing; slenderer than *Procellaria* petrels, bill black and slenderer; see dark-morph Wedge-tailed (darker underwing, longer tail, slenderer bill) and Short-tailed Shearwaters (not yet in Ecuador). **Status** Uncommon year-round visitor.

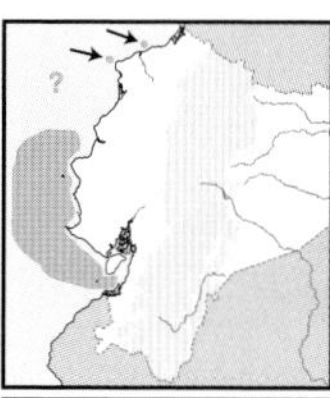

Pink-footed Shearwater *Ardenna creatopus* 45–48cm / w 109–117cm

Regular off coast, mostly Santa Elena Peninsula. ***Slender pink bill tipped dark, pinkish feet***. Sooty-brown above, darker cap, ***whitish below smudged dusky on sides***. Underwing whitish, ***broad dusky edges; short, rather rounded tail***. Monotypic. May congregate around fishing boats, some follow ships; dives. Flight: laboured flapping and long leisurely glides. **SS** Only bicoloured shearwater regular in Ecuador; underwing darker than Buller's; Wedge-tailed has longer, wedge-shaped tail, whiter underwing, smaller head, slenderer body. **Status** Uncommon visitor, commoner from Jul; VU globally.

* Galápagos Shearwater *Puffinus subalaris* 28–31cm / w 63–70cm

Few records off SW coast (also northwards?). ***Sooty-brown above including cap***, white below, ***dark undertail-coverts***. White underwing, broad black leading edge, dusky trailing edge, dusky smudges on coverts. Monotypic. In Galápagos congregates around boats and over fish shoals, mostly seen swimming. Flight fast, fluttery, few shallow flaps and short glides low over water. **SS** Much smaller than Ecuadorian shearwaters, but note other *Puffinus* in the Pacific (Manx recently recorded in Galápagos); white underparts and tail shape separate from storm-petrels. **Status** Rare visitor (Feb–Mar, May, Dec), possibly overlooked; endemic Galápagos breeder.

Wedge-tailed Shearwater
Buller's Shearwater
dark morph
pale morph
Sooty Shearwater
Pink-footed Shearwater
Galápagos Shearwater

Small pelagic seabirds rare in Ecuador, although this might reflect poor pelagic coverage rather than true scarcity, at least for some. Except a few species (e.g., White-faced), storm-petrels are difficult to identify and to observe well; note rump patterns, tail shape, flight type.

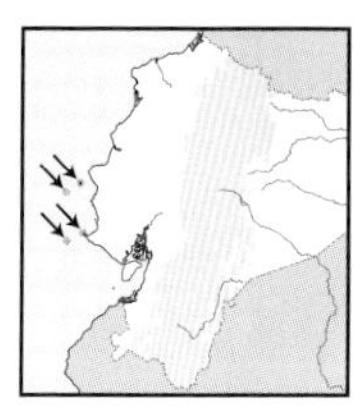

Wilson's Storm-Petrel *Oceanites oceanicus* 17–19cm / w 38–42cm

Scattered records off SW coast, N to Isla de la Plata. Blackish-brown, pale upperwing band, ***U-shaped white rump band almost reaching vent***. Black ***square-ended tail, legs protruding***. Yellow-webbed feet. Race *oceanicus*, perhaps *exasperatus*. Might congregate at and follow ships; regularly patters surface with wings in V, switches to new location and patters again. Flight direct, purposeful, fluttery wingbeats, little gliding. **SS** Smaller Elliot's (white belly patch visible only when banks sideways, pale band in underwing-coverts); all *Oceanodroma* have longer, not squared tails, legs never protruding beyond tail, larger rump band, more arched wings in flight (see Band-rumped). **Status** Rare austral winter visitor (Jun–Oct), possibly overlooked.

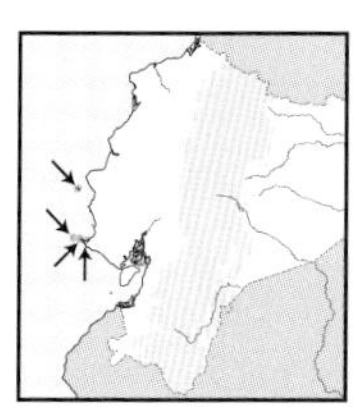

Elliot's Storm-Petrel *Oceanites gracilis* 14–16cm / w 36cm

Few records off SW coast, N to Isla de la Plata, more *inshore* than previous species. Closely resembles Wilson's including rump band, but ***belly to vent white, paler band in underwing-coverts***. Nominate, possibly *galapagoensis* (smudgy belly patch). Might congregate at sea at ships or whales, often near coast; patters surface. Flight direct and purposeful, shallow fluttery wingbeats and little gliding, lighter than Wilson's. **SS** Square tail and protruding legs separate from *Oceanodroma* storm-petrels; Black-bellied *Fregetta grallaria* and White-bellied *F. tropica* (not yet in Ecuador, but likely) much whiter below. **Status** Rare visitor (year-round?), possibly overlooked, DD globally.

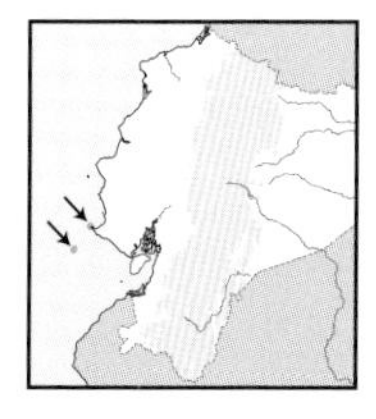

White-faced Storm-Petrel *Pelagodroma marina* 18–21cm / w 42–43cm

Two records off SW coast. Greyish-brown above, ***extensive white in face and long supercilium, dusky face stripe. White underparts, pale grey rump, white underwing***. Pale U-shaped band in upperwing-coverts. Forked black tail, ***legs protruding***. Race *maoriana*. Elsewhere may congregate at sea but does not follow ships; kicks forwards with feet, sometimes 'walking' and 'jumping' over surface. Flight erratic, permanently banking and weaving; stronger wingbeats and short glides, legs dangling. **SS** Ringed (bolder head pattern, grey breast-band, broader and paler upperwing band, dark underwing, longer forked tail, white rump, legs not protruding). **Status** Casual transient or vagrant (Jun–Jul).

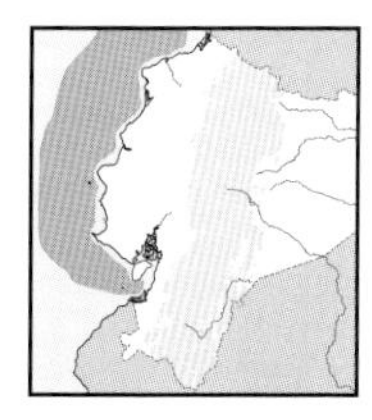

Wedge-rumped Storm-Petrel *Oceanodroma tethys* 17–19cm / w 37–40cm

Regular off coast, more numerous in SW but possibly overlooked N. Blackish-brown with ***large white rump patch extending as wedge to mid tail and vent-sides***. Pale band in upperwing-coverts. Fairly long, ***slightly forked black tail***. Race *kelsalli*, possibly *tethys*. Several may congregate around fishing boats, often near coasts; patters water with wings rather horizontal. Flight fast, erratic, repeatedly banks and weaves, quick, deep wingbeats; when feeding more bounding, legs dangling. **SS** Large white rump patch distinctive; Band-rumped (narrower rump band, more laboured flight) and *Oceanites*. **Status** Fairly common visitor (Apr–Feb).

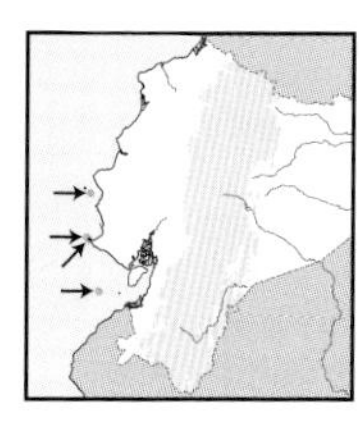

* Band-rumped Storm-Petrel *Oceanodroma castro* 19–22cm / w 44–49cm

Two records off Santa Elena Peninsula. Blackish-brown, *faint* pale upperwing band. ***Narrow white band on lower rump and uppertail-coverts*** extending to vent-sides. ***Fairly long, barely notched black tail***. Monotypic. Very pelagic, not congregating like other storm-petrels, does not follow ships, often patters with horizontal wings. Flight strong, buoyant and steady, much banking and switching, deep wingbeats, often gliding on flat wings. **SS** Narrow rump band separates from Wedge-tailed. Wilson's and Elliot's have shorter squared tails, legs protruding; Leach's *O. leucorhoa* (not yet in Ecuador, but likely), has narrow dark line dividing rump (white less noticeable), more prominent upperwing band, more forked tail. **Status** Very rare transient (May–Feb).

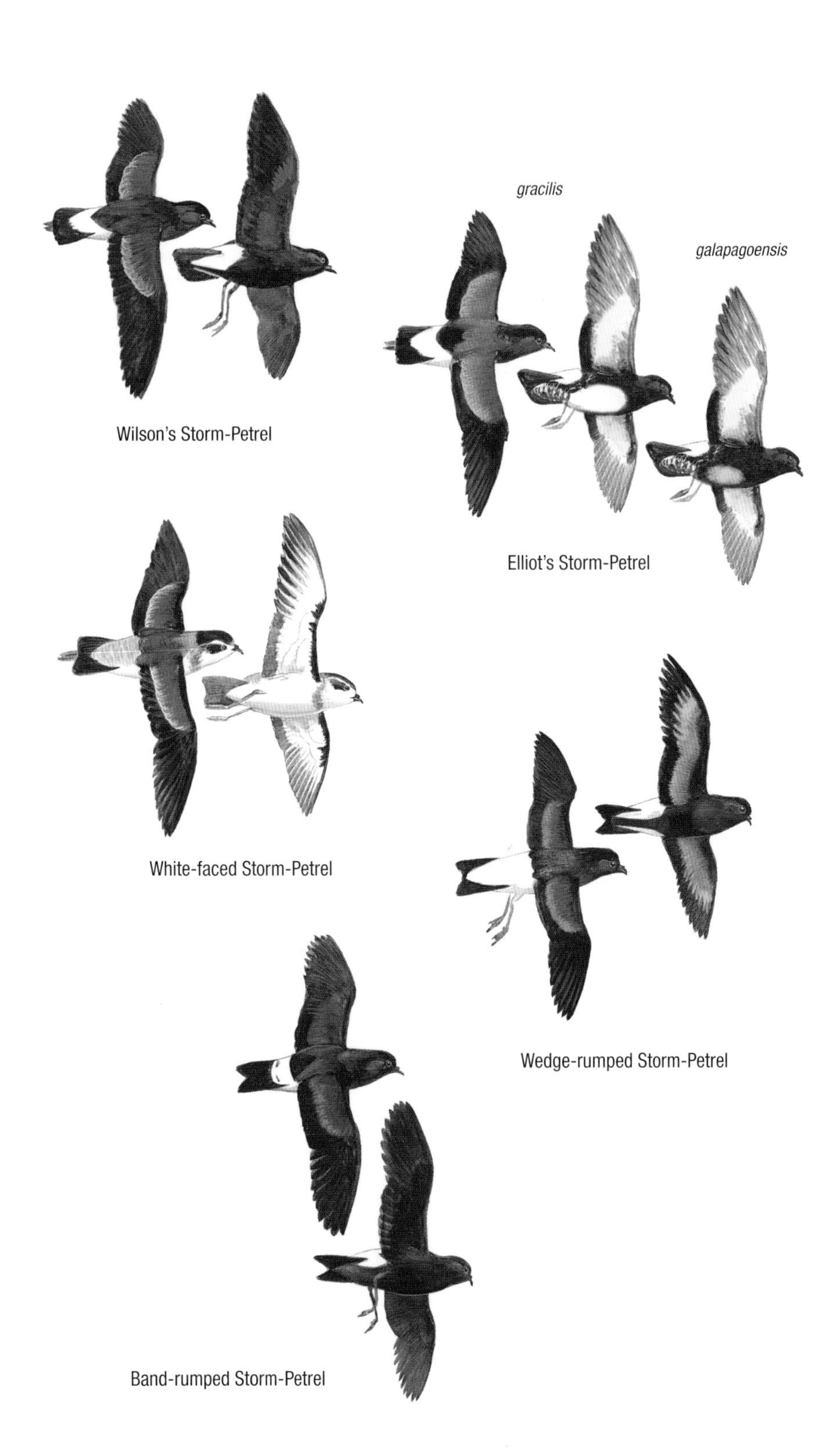

Wilson's Storm-Petrel

Elliot's Storm-Petrel

White-faced Storm-Petrel

Wedge-rumped Storm-Petrel

Band-rumped Storm-Petrel

PLATE 21: STORM-PETRELS II

One distinctive plus four dark, confusing storm-petrels. Least, Markham's, Ashy and Black are separated by size, upperwing pattern and flight. Observers should consult specialised guides (Harrison 1987, Onley & Scofield 2007) for seabird identification.

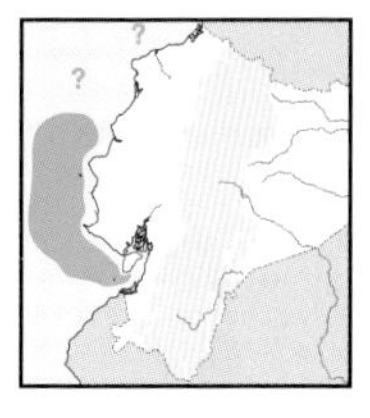

Least Storm-Petrel *Oceanodroma microsoma* — 14–15cm / w 32–34cm

Apparently regular off C-SW coast, possibly also N. *Small.* Mostly blackish-brown, *vague* pale band on upperwing-coverts, dark underwing. *Rather short wedge-shaped tail, short bowed wings.* Monotypic. Flies low over water, not usually following ships. Flight fast, bat-like, constant, direct, deep wingbeats; when feeding more erratic, wings held over back, little pattering. **SS** Smallest dark storm-petrel, wedged-shaped tail distinctive, shorter wings, wing band fainter than Markham's, Ashy and Black; Leach's *O. leucorhoa* (not yet in Ecuador) has long forked tail, bolder upperwing band. **Status** Rare boreal winter visitor (Dec–Jun).

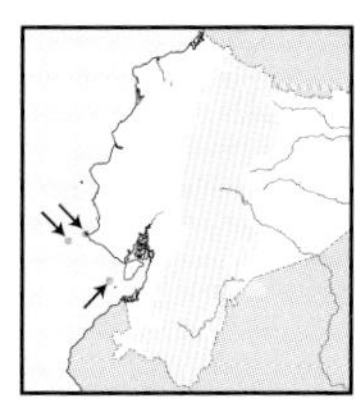

Markham's Storm-Petrel *Oceanodroma markhami* — 21–23cm / w 49–54cm

Scattered records off Santa Elena Peninsula. Blackish-brown, greyer cast above, *pale band in upperwing-coverts extending to carpals. Long forked tail.* Monotypic. Poorly known, apparently does not follow ships, but may congregate with other seabirds at fishing boats. Flight fast, purposeful, *light shallow* wingbeats, *frequent gliding.* **SS** Black Storm-Petrel (blacker overall, *wing band does not reach carpal*, stouter bill, heavier legs); flight differs: Markham's is faster, more controlled and fluttery; cf. much rarer Ashy (paler underwing); smaller Least (wedge-shaped tail); also dark-rumped form of smaller Leach's (faster, bounding, more erratic flight; not yet in Ecuador). **Status** Very rare (Jan–Feb, May, Jul, Dec), DD.

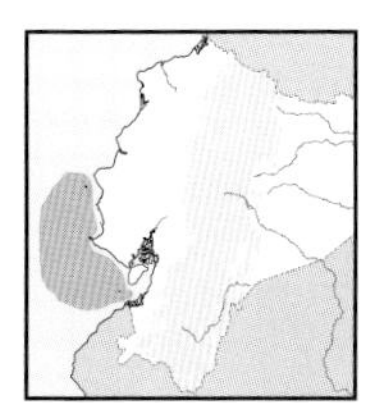

Ringed Storm-Petrel *Oceanodroma hornbyi* — 22–24cm / w 50cm

Irregular off C-SW coast, off Santa Elena Peninsula to Isla de la Plata. *Blackish cap, white face, throat and narrow nuchal collar, grey breast-band, white belly.* Grey above, *broad whitish band in upperwing-coverts, dusky underwing. Long forked tail.* Monotypic. Rarely follows boats, presence erratic. Flight erratic, rather deep wingbeats interspersed by short glides, little hovering and bouncing low over water. **SS** Confusion unlikely; White-faced Storm-Petrel (no grey breast-band or whitish upperwing bar, short square tail, white rump, long protruding legs). **Status** Rare visitor (Jun–Jan), possibly year-round.

†Ashy Storm-Petrel *Oceanodroma homochroa* — 18–20cm / w 42cm

Two unconfirmed records off Santa Elena Peninsula and Isla de la Plata. Entirely *ashy*-brown, *faint band in upperwing-coverts reaching carpals*, dark flight feathers, *pale suffusion to underwing.* Long forked tail. Monotypic. May congregate at sea, often around fishing boats, flying low over water, no pattering. Flight fluttery but direct, shallow wingbeats, little gliding, wings not reaching horizontal; rapidly changing direction. **SS** *Pale underwing* distinctive in good views, but otherwise difficult to separate from slightly larger Black and Markham's (and from dark Leach's; not yet in Ecuador); butterfly-like flight diagnostic. **Status** Uncertain, possibly overlooked irregular visitor (Jan, Sep); possibly misidentifications (B. Haase).

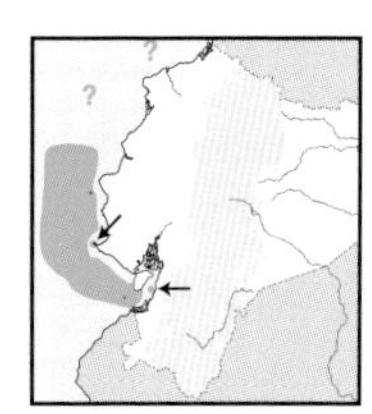

Black Storm-Petrel *Oceanodroma melania* — 21–23cm / w 43cm

Regular off coast. Entirely *sooty-blackish*, pale band in upperwing-coverts *not reaching carpals.* Long, deeply forked tail. Monotypic. Tends to follow ships and congregate at fishing boats. Flight buoyant, graceful and direct, *deep* wingbeats, occasional glides on flat wings often *lifting head.* **SS** Not safely separated from Markham's (*square forehead, more slender bill, paler legs* dangling less conspicuously); Markham's flight lighter, more graceful; see also rarer, smaller, shorter-winged Ashy (fainter upperwing band, paler underwing, shallower wingbeats); Least (smaller with wedge-shaped tail); also dark-rumped form of Leach's Storm-Petrel (not yet in Ecuador). **Status** Apparently uncommon visitor; status complicated by identification problems; year-round?

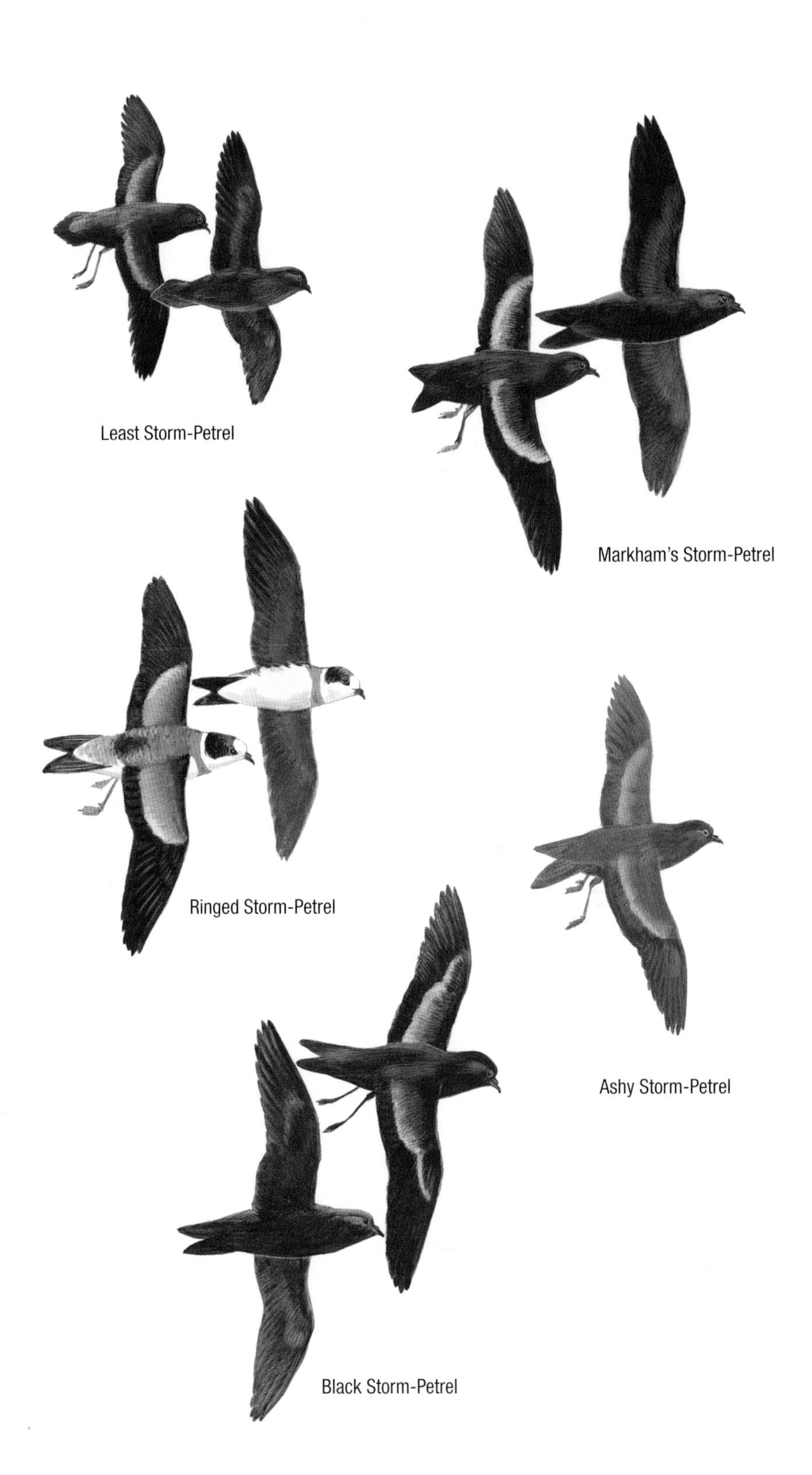
Least Storm-Petrel
Markham's Storm-Petrel
Ringed Storm-Petrel
Ashy Storm-Petrel
Black Storm-Petrel

Powerful seabirds that occur in Ecuador as transients and non-breeding visitors. Flight strong and fast, often chasing other seabirds. Larger species have broader wings. Complex identification due to: (1) very complex plumages with few differences among species; (2) polymorphism (Parasitic and Pomarine); (3) non-breeding plumages or immatures more expected in Ecuador; (4) observation conditions not always ideal.

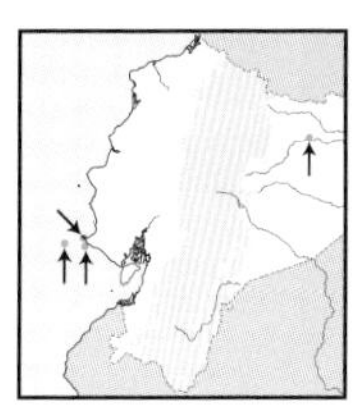

Long-tailed Jaeger *Stercorarius longicaudus* 54cm

Few records off Santa Elena Peninsula, one specimen Mar Bravo (B. Haase), once Río Napo (C. Hesse). Shorter bill than congeners. *Cold* brownish-grey above, blackish cap, whitish nuchal collar, *paler grey chest-band* (absent in breeders). Imm. similar, with distinct *white flecking on rump*. Juv. has *pale fringes* to upperwing-coverts and mantle, *prominent barring in tail-coverts, white shafts* to outer primaries. In flight: *narrow white flashes* in primaries, dark underwing, *pale upperwing-coverts, dark flight feathers*, long tail-streamers. Race *pallescens*? Possibly overlooked in Ecuador; flight buoyant, agile, more gliding, chases small seabirds. **SS** *Smaller, more delicate, shorter-headed* than congeners, almost resembles large tern; see heavier bodied Parasitic (larger, less rounded head, broader wings, warmer plumage, *more prominent white flash in primaries*, less graceful flight). **Status** Rare boreal winter transient (Jan, Jul, Oct–Nov). [Alt: Long-tailed Skua]

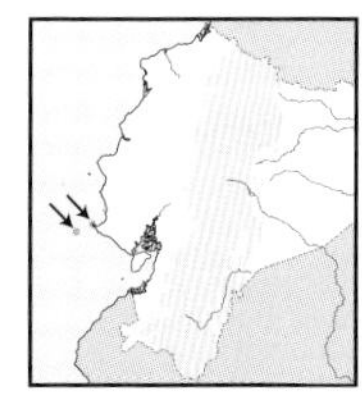

Parasitic Jaeger *Stercorarius parasiticus* 45cm

Offshore, often near coast at Santa Elena Peninsula; once in Esmeraldas. Bill slighter than Pomarine, *duller*. Generally greyish-brown above, *darker cap*, creamy nuchal collar, whitish below, *vermiculated grey breast-band* to belly-sides. Imm. warmer than Pomarine with *smaller cap, pale fringes* to upperwing-coverts, *less barred* tail-coverts. In flight: *darker underwing*, prominent white *crescent* at base of primaries. Monotypic. Piratic, mainly offshore but more frequently near coast than congeners, rapid flight, swiftly chases small seabirds. **SS** *Slimmer* body than Pomarine, more *slender neck, smaller head*, weaker bill; *flight more flicking*. Long-tailed smaller, *daintier*, with *colder plumage, paler upperwing-coverts, fainter 'windows' in wing*. **Status** Uncommon boreal winter visitor (year-round). [Alt: Arctic Skua]

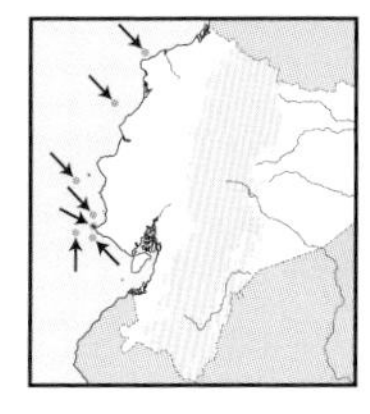

Pomarine Jaeger *Stercorarius pomarinus* 56cm

Offshore, often near coast at Santa Elena Peninsula. *Bicoloured bill*, spoon-shaped tail projections (breeding). *Stocky*. Blackish-brown above, *large dark cap*, creamy nuchal collar, whitish below, *barred chest-band* to belly-sides, *dark underwing*. Imm. has *upper- and undertail-coverts barred dark*. Juv. has *plain head, no* pale nuchal collar, *densely barred underwing*, often with *double pale patch, plain upperwing, barred tail-coverts*. In flight: *small, pale, star-shaped patch near wingtips*. Monotypic. Solitary, piratic, rarely near coast, flying high and aggressively attacking fishing seabirds. **SS** *Stockier, larger-headed, heavier-chested, wings broader especially at base* than other jaegers, *larger cap*; bill stronger, flight more *steady, active* and purposeful. **Status** Uncommon boreal winter visitor (mostly Sep–May); possibly year-round. [Alt: Pomarine Skua]

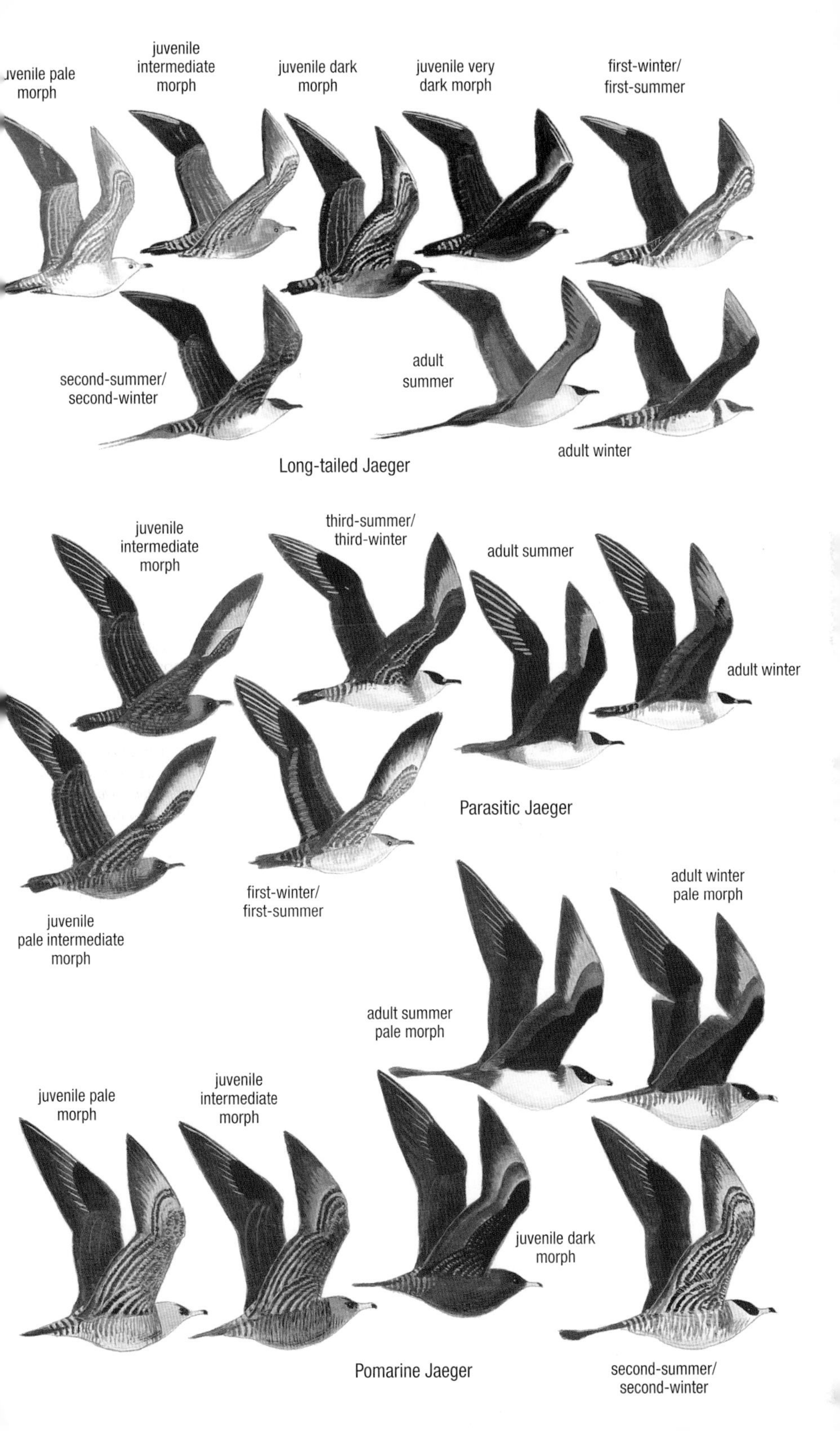

ıvenile pale morph
juvenile intermediate morph
juvenile dark morph
juvenile very dark morph
first-winter/ first-summer
second-summer/ second-winter
adult summer
adult winter
Long-tailed Jaeger
juvenile intermediate morph
third-summer/ third-winter
adult summer
adult winter
juvenile pale intermediate morph
first-winter/ first-summer
Parasitic Jaeger
adult winter pale morph
adult summer pale morph
juvenile pale morph
juvenile intermediate morph
juvenile dark morph
Pomarine Jaeger
second-summer/ second-winter

The two powerful skuas are very rarely seen off the SW coast. Flight strong and fast. Complex identification due to: (1) very complex plumages with few differences among species; (2) polymorphism (Parasitic and Pomarine); (3) non-breeding plumages or immatures more expected in Ecuador; (4) observation conditions not always ideal.

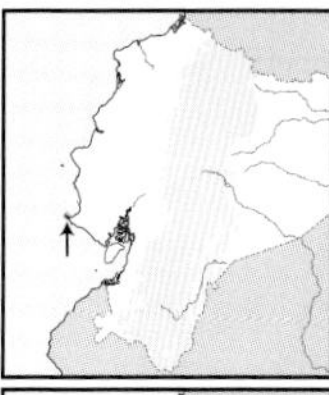

* Chilean Skua *Stercorarius chilensis* 58cm

Two records from Santa Elena Peninsula. ***Bicoloured bill.*** Brown above, ***dark cap, cinnamon to cinnamon-rufous*** below, ***dark breast-band.*** In flight: ***cinnamon underwing-coverts,*** white 'window' at base of primaries. Monotypic. Follows Humboldt Current in winter, sometimes approaching coasts and entering harbours. **Voice** Unlikely to be heard in Ecuador. **SS** Larger, more reddish than congeners, dark cap and breast-band distinctive; older individuals duller below; cf. Brown Skua *S. antarcticus* (not yet in Ecuador). **Status** Rare or accidental austral winter visitor (Apr–Aug).

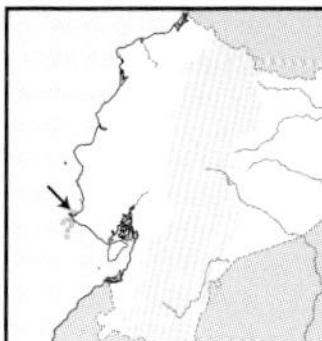

* South Polar Skua *Stercorarius maccormicki* 53cm

Few records off Santa Elena Peninsula. ***Bicoloured bill.*** Polymorphic: dark brown above, ***pale greyish to blackish-brown head and underparts, whitish necklace in nape, pale front.*** In flight: conspicuous white window in base of primaries, can extend to coverts, ***short, blunt tail.*** Monotypic. Solitary and more pelagic than Chilean Skua; flight strong, purposeful. **SS** Slightly lighter than Chilean, with pale nuchal patch and frontal blaze; ***no cinnamon below, no dark cap.*** Tail appears shorter, more blunt-tipped than jaegers, but see juv. Pomarine (smaller 'window' in primaries). **Status** Accidental austral vagrant (Jun, Sep, Nov?).

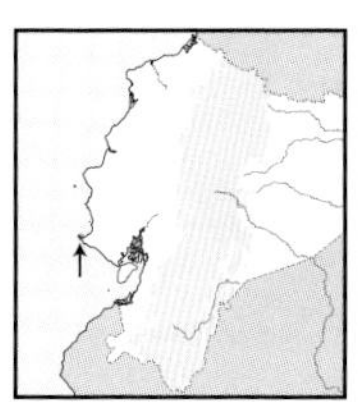

Lesser Black-backed Gull *Larus fuscus* 56–61cm

One record Santa Elena Peninsula. ***Heavy yellow bill,*** whitish eyes, yellow legs. ***Slate-grey mantle.*** Non-breeder densely streaked brownish on head to breast-sides. Juv. ***dark brown*** with whitish speckling / scaling, ***black bill,*** long wing projection. In flight: ***black primaries*** contrast with slate coverts; juv. has ***dark trailing edge, broad dark terminal tail-band, pale uppertail-coverts.*** Possibly race *graellsii.* Monotypic. Sandy beaches. **Voice** Not heard in Ecuador. **SS** Kelp Gull (blacker back; juv. has less contrasting rump); see rare Herring (bulkier, shorter wing projection, stouter bill, less contrasting rump) and California juv. (darker uppertail, paler bill). **Status** Vagrant (Jan).

adult
immature
juvenile
Chilean Skua
juvenile intermediate morph
adult pale morph
adult intermediate morph
adult pale intermediate morph
adult dark intermediate morph
adult dark morph
South Polar Skua
first-winter
second-winter
adult winter
first-winter
second-winter
adult winter
Lesser Black-backed Gull
adult winter

PLATE 24: LARGER GULLS

Large, primarily marine and coastal gulls. All species are only transients or accidentals. White heads and bright soft parts characterise breeders; non-breeders and imm. more likely in Ecuador. Identification complicated by complex non-breeding, imm. and juv. plumages, gulls often require 3+ years to attain full ad. plumage. We urge readers to consult specific monographs (Grant 1986, Harrison 1987, Malling Olsen & Larsson 2003, Howell & Dunn 2007).

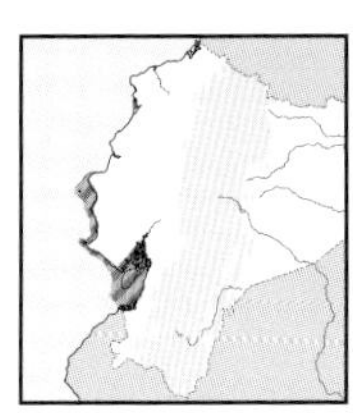

Kelp Gull *Larus dominicanus* 58–65cm

SW coast N to S Manabí, spottily further N. ***Stout yellow bill, red mandible spot, pale eyes, greenish legs.*** Blackish back. Juv. ***dark brown densely speckled / scaled whitish***; some have whiter head. In flight ad.'s ***tail white, broad*** white trailing edge; juv. ***dark underwing and tail***, dark upperwing-coverts, ***pale uppertail-coverts***. Monotypic. Singles or small groups, dominant over other gulls; breeds locally. **Voice** Loud nasal *kyA-ká-kyA-kyA-kyA...kháá-khaa-khaa...kayeéw....* **SS** Lesser Black-backed Gull (slate-grey back, longer wing projection); immature Belcher's (longer and paler bill, more solid dark head); ad. with broad black subterminal tail-band; paler California and Ring-billed are pink-billed; see much rarer Herring Gull. **Status** Uncommon local breeder, rare austral visitor (year-round).

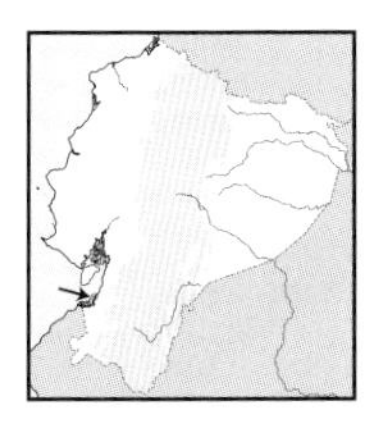

* Belcher's Gull *Larus belcheri* 51–53cm

One confirmed record SW coast. ***Heavy yellow bill, black mandible ring, broad red tip. Black back.*** Non-breeder: ***duller*** soft parts, ***dark eyes, incomplete dark hood***, blackish upperparts. Juv. has ***dark brown hood***, brownish above mottled buff, bill pale yellow tipped dark. In flight broad white trailing edge, ***broad black subterminal tail-band.*** Monotypic. Presence in Ecuador possibly associated with climate; often tame in harbours. **Voice** Not heard in Ecuador? **SS** Bulkier Kelp Gull (stouter, all-tail white in flight, or much less white in third-year bird; juv. lacks dark hood). **Status** Casual austral visitor (Jan).

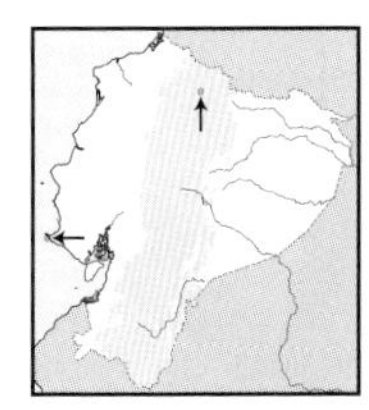

California Gull *Larus californicus* 52–54cm

Two records (Santa Elena Peninsula and Yaguarcocha Lake). ***Rather slender yellow bill***, dark ring near tip, relatively large ***yellowish legs, dark eyes. Medium grey back.*** Juv. ***pale brown, dense whitish mottling***, bill often ***pink tipped dark***, pinkish legs, ***long wing projection.*** In flight: juv. ***more uniform pale upperparts*** than Herring Gull, tail more uniformly dark. Unknown race. Singles on sandy beaches. **Voice** Not heard in Ecuador. **SS** Larger juv. Herring Gull (stouter pink bill, shorter wing projection, narrower wings, paler panel on inner primaries); Ring-billed (slimmer bill, whiter overall appearance, more contrasting rump and greater coverts); see Lesser Black-backed Gull. **Status** Vagrant (Feb–Mar).

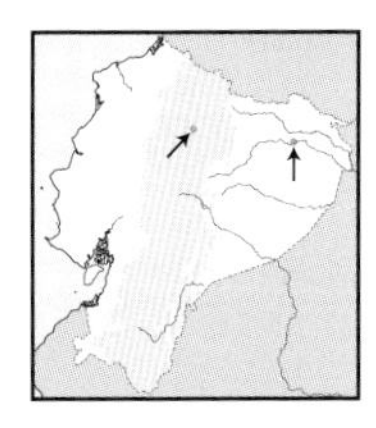

Herring Gull *Larus argentatus* 56–61cm

Three records (N coast, E lowlands and highlands). ***Heavy yellow bill***, red mandible spot, ***pale eyes, pinkish legs. Pale grey mantle.*** Non-breeder: ***hood streaked dusky.*** Juv. ***brown speckled / mottled whitish***, some with whiter head, ***blackish bill tip.*** Ad. in flight: ***two outer white mirrors in black primaries***; juv. ***flight feathers blackish, broad blackish terminal tail-band, scaled rump.*** Race *smithsonianus*. Sandy beaches and harbours. **Voice** Not heard in Ecuador. **SS** Lesser Black-backed Gull darker grey above; juv. has more contrasting pale rump, blacker inner flight feathers; more slender California (longer wing projection, pink bill) and Ring-billed Gulls (paler pink bill); much hybridisation on breeding grounds. **Status** Vagrant (Feb; Dec–Apr).

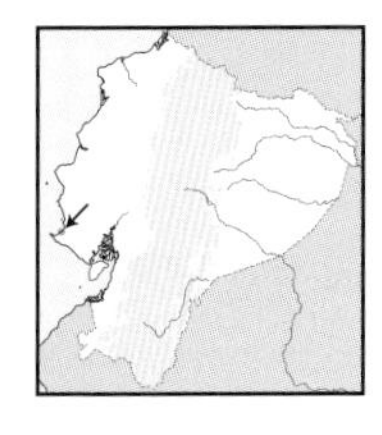

Glaucous-winged Gull *Larus glaucescens* (not illustrated) 70–71cm

One record San Pablo, Santa Elena Peninsula (B. Haase, P. Hernández). ***Large, pale*** grey mantle, yellow bill, pink legs, ***grey wingtips, primaries white below.*** Juv. black bill, ***pale*** brownish-grey with little speckling / mottling, ***buffy-grey primaries, brownish tail and rump.*** In flight: ***grey primaries***, creamy underwing in juv. Monotypic. Accidental in Ecuador; elsewhere, congregates on coasts and rubbish dumps. **Voice** Not expected to vocalise in Ecuador. **SS** Resident or regular gulls smaller, even Kelp Gull, which is blacker above; see Herring and Lesser Black-backed Gulls (more contrasting back, more speckled / mottled juv. but less 'dirty'); much hybridisation. **Status** Vagrant (Jan), unlikely to occur again.

adult
Kelp Gull
second-winter
third-winter
second-winter
third-winter
adult
non-breeding
juvenile
breeding
Belcher's Gull
second-winter
third-winter
California Gull
non-breeding
second-winter
adult summer
first-winter
second-winter
third-winter
adult winter
Herring Gull

A blend of gulls some with complex plumage variation related to age, requiring 3–4 years to attain ad. plumage. Better to consult specific monographs (Grant 1986, Harrison 1987, Malling Olsen & Larsson 2003, Howell & Dunn 2007). Swallow-tailed and Sabine's Gulls are rarely seen from shore.

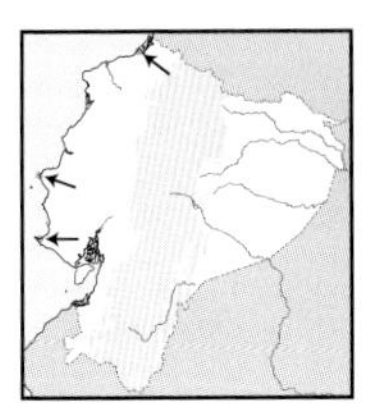

Ring-billed Gull *Larus delawarensis* 45–50cm

Few records, Santa Elena Peninsula. ***Yellow bill, black subterminal ring***. Mainly white, ***pale grey mantle***; non-breeder has ***head to breast-sides finely streaked dusky***. Juv. spotted and checkered brownish, ***bill pinkish tipped dark***. In flight ad. has black outer primaries, ***large white 'windows'***, white tail; juv. has ***pale greater wing-coverts*** and ***uppertail-coverts***, dark subterminal tail-band. Monotypic. More likely in beaches and harbours. **Voice** Not expected to be heard. **SS** Imm. of larger California and *much* larger Herring Gulls; Herring's bill darker, heavier, duskier belly; darker California has heavier and larger bill, blacker tail, darker wing-coverts and inner flight feathers, longer wing projection. **Status** Vagrant (Apr–Oct).

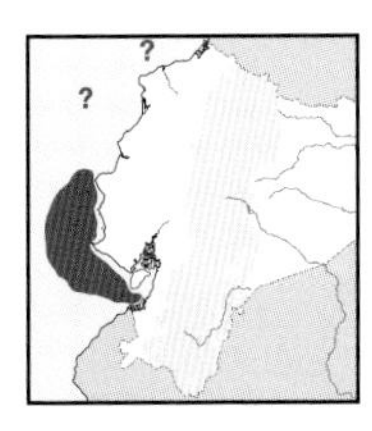

Swallow-tailed Gull *Creagrus furcatus* 54–57cm

Few records off SW coast. ***Long black bill tipped yellowish, red eye-ring, large dark eyes. Slate-black hood, pearl-grey neck to upperparts***, white below. Non-breeder replaces black hood with ***black orbital area***. Juv. boldly ***mottled*** grey, brownish and white above, ***red*** orbital area. In flight: ***deeply forked white tail, tricoloured*** upperwing. Monotypic. Crepuscular and nocturnal, breeds in Galápagos and Malpelo, from where few wander to continental waters, mostly singles or small groups. **Voice** Not expected to vocalise in mainland Ecuador. **SS** Confusion unlikely but see much smaller Sabine's. **Status** Rare offshore visitor (possibly year-round).

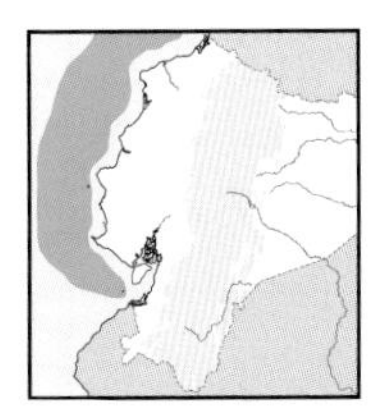

Sabine's Gull *Xema sabini* 33–35cm

Offshore, often near coast of Santa Elena Peninsula. ***Small. Black bill tipped yellow***. Slate-grey hood, grey above. Non-breeder: greyish confined to mid rear crown. Juv. rather similar but greyer mid-crown and hindneck, ***scalier upperparts***. In flight: ***forked tail***, bold ***tricoloured upperwing*** pattern. Nominate. Small parties in tern-like low buoyant flight, often following fishing boats. **Voice** Not expected to vocalise in Ecuador. **SS** Confusion unlikely given size and upperwing pattern; Swallow-tailed Gull much larger, deeper forked tail, darker wing-coverts, large eyes, longer bill. **Status** Rare boreal winter visitor (Jun–Mar).

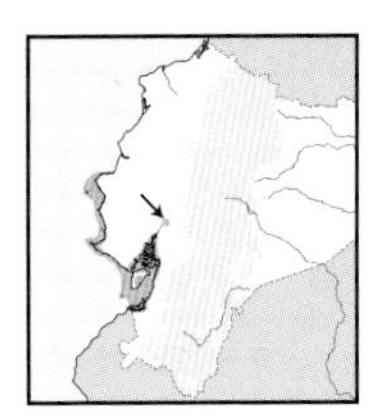

Grey Gull *Leucophaeus modestus* 44–46cm

SW coast, N to C Manabí. ***Black bill and legs***. Mostly ***grey, silvery hood***. Non-breeder ***nearly lacks hood***. Juv. ***Brownish-grey, some pale upperwing-coverts***. In flight: ***narrow white trailing edge to secondaries***, black primaries, ***grey tail, narrow white terminal band***. Monotypic. Small groups on sandy beaches running back and forth with waves. **Voice** Yelping *kyaow*, laughing *kye'kye'kye'kyeeeeaaaow*, growling *krrrryaa* (S *et al.*). **SS** Confusion unlikely; juv. has distinctive black soft parts, contrasting black primaries; Lava Gull from Galápagos (not recorded mainland Ecuador) slimmer and darker. **Status** Uncommon austral visitor (mostly May–Sep).

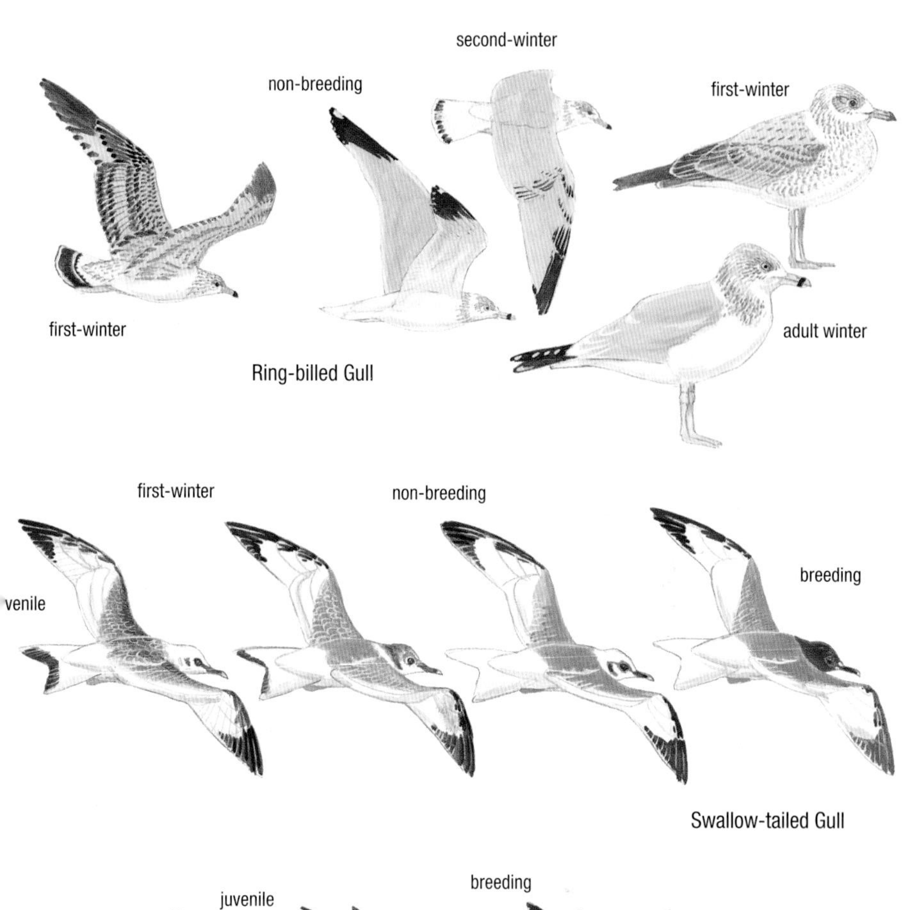
second-winter
non-breeding
first-winter
first-winter
adult winter
Ring-billed Gull
first-winter
non-breeding
venile
breeding
Swallow-tailed Gull

breeding
juvenile
juvenile
second-winter
first-winter
non-breeding
second-winter
Sabine's Gull
non-breeding
Grey Gull
breeding

Generally 'hooded' gulls, all but one primarily coastal and marine; recently separated from ubiquitous genus *Larus*. Grey-hooded is a common resident, Laughing and Franklin's are common boreal visitors. Identification might be challenging in non-breeding plumages (expected in Ecuador). The Andean is an attractive resident breeder of paramos.

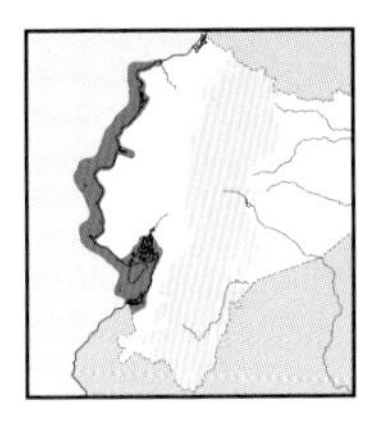

Grey-hooded Gull *Chroicocephalus cirrocephalus* 42–44cm

SW coast N to C Manabí, scattered records further N. ***Crimson bill, legs and narrow eye-ring, whitish eyes. Pearl-grey hood***, white underparts and nuchal collar, ***mid-grey upperparts***. Non-breeder: traces of grey hood, ***dusky ear-coverts spot***. Juv. has ***dusky bill, greyish-brown blotched crown***, dusky subterminal tail-band. In flight: ***bold white wedge-shaped panel in upperwing***, black primaries, small ***white subterminal 'window' in outer primaries, dark underwing***. Nominate. Breeds Santa Elena Peninsula, regular elsewhere on SW coast, often inland. **Voice** Loud, rough *crhäw-crhäw-crhëw*.... **SS** Non-breeding Laughing and Franklin's Gulls lack extensive white in upperwing; note subterminal tail-band in juv.; Andean Gull (larger 'window', smaller bill, paler juv.). **Status** Uncommon.

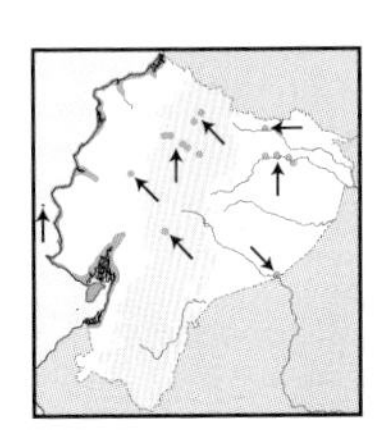

Laughing Gull *Leucophaeus atricilla* 38–40cm

Coasts, scattered inland records (to 3,000m). Beaches, brackish and freshwater wetlands. ***Longish black bill, black legs*** (deep red in breeder), ***white eye-crescents***. Black hood, slate-grey above. Non-breeder lacks hood; smudgy ***greyish patch in rear-crown to ear-coverts***. Juv. dark greyish-brown, mottled above, faint eye-crescents. In flight: ***blackish primaries merge into grey coverts and inner flight feathers; black subterminal tail-band*** in juv. Race *megalopterus*. Singles to flocks near shore or inland, locally numerous. **Voice** Laughing *craa-craa-craa-craa-cruaa-cruua'a*.... **SS** Smaller non-breeding Franklin's (bill *shorter, less drooped at tip*; bolder eye-crescents, more extensive grey in rear crown, white 'windows' in outer primaries, paler underwing); see non-breeding Grey-hooded and rarer Bonaparte's Gulls. **Status** Common boreal winter visitor (year-round).

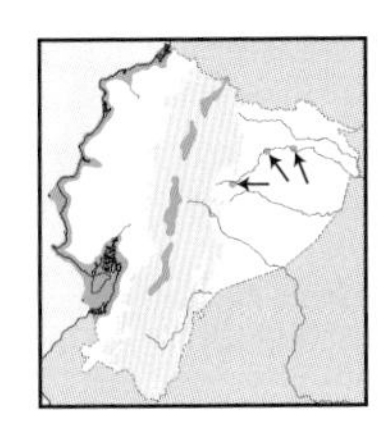

Franklin's Gull *Leucophaeus pipixcan* 35–38cm

Coastline, scattered inland records (to 3,000m). Mainly beaches, also brackish and freshwater wetlands. Resembles commoner Laughing Gull, but bill ***shorter***, less droop-tipped, wings proportionately broader and more ***blunt-tipped***. Eye-crescents ***bolder***. Non-breeder has ***darker*** and ***more extensive*** patch on rear crown to ear-coverts. Juv. fairly similar to non-breeder but ***duskier***. In flight: ***white apical dots in primaries, white band separating black primaries from grey primary bases***; juv. has dark subterminal tail-band ***not reaching outer feathers, darker*** underwing. Monotypic. More coastal than Laughing Gull, more frequent at sea. **Voice** Loud *wa-wak...wak-wak*.... **SS** Compare non-breeding Grey-hooded (black primaries tips); Andean (different wing pattern, no subterminal tail-band); smaller Sabine's (bold upperwing pattern); and smaller, accidental Bonaparte's Gulls (different wing pattern, smaller bill). **Status** Uncommon boreal winter visitor (mainly Oct–Feb; May–Jul).

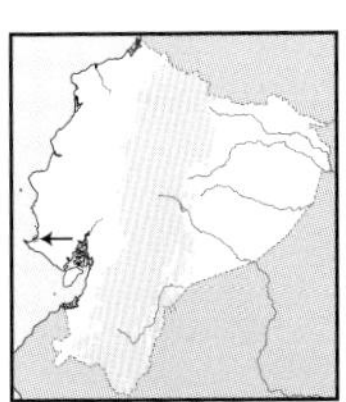

Bonaparte's Gull *Chroicocephalus philadelphia* 31–32cm

One record SW coast at Pacoa (B. Haase). ***Small with thin, black bill***. Head mostly ***white*** in non-breeder, bold ***dark ear-coverts patch***. White below, pearl-grey above. Pinkish legs. In flight: ***translucent white outer primaries, broad black trailing edge***. Monotypic. Quick wingbeats with light, almost tern-like flight; surface-feeder, legs dangling. **Voice** Nasal *cheerp*, shorter *chirp* (R *et al.*). **SS** Size, wing pattern and behaviour separate from commoner Laughing, Franklin's and Grey-hooded Gull. **Status** Accidental boreal winter visitor (Nov).

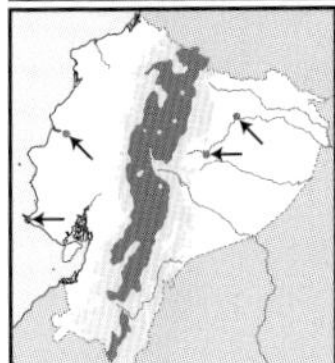

Andean Gull *Chroicocephalus serranus* 46–48cm

Highlands, scattered E & SW lowlands records (2,500–4,200m). Lakes, marshes, ploughed fields. ***Reddish bill and legs, bold white eye-crescents. Glossy black hood***, pale grey upperparts. Non-breeder and juv.: ***small dark ear-coverts patch, white eye-crescent***. Juv. has dark bill, upperparts mottled brownish-grey. In flight: ***large white panel on leading primaries, bold white 'window'*** in black primaries. Monotypic. Pairs to small groups, often pass over inhospitable habitat. **Voice** Strident, tern-like *keEy-keEyr-keEyr*.... **SS** Unmistakable in range, but see Laughing and Franklin's Gulls (less white in upperwing). Andean might wander to SW coast, where cf. Grey-hooded Gull (slimmer, darker above, smaller 'window'). **Status** Fairly common.

Grey-hooded Gull
breeding
juvenile
first-winter
first-summer
second-winter
first-summer
first-winter
breeding
Laughing Gull
non-breeding
juvenile
juvenile
first-winter
second-winter
non-breeding
Franklin's Gull
first-summer
non-breeding
first-winter
Bonaparte's Gull
non-breeding
first-winter
non-breeding
breeding
juvenile
first-winter
first-summer
Andean Gull

Terns mostly characterised by dark or yellow bills. All but the Amazonian Yellow-billed are marine and coastal. Two very local breeders (Bridled and Gull-billed); the other three are only visitors or transients. Sandwich is not uncommon, whereas Peruvian shows up every several years. Convoluted taxonomy; all formerly in genus *Sterna*, now split into several genera.

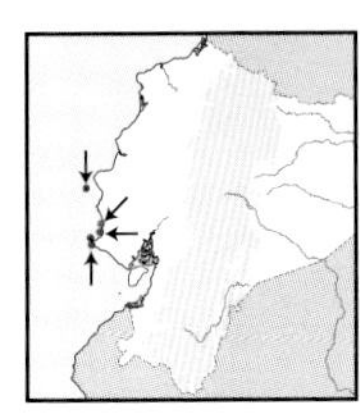

Bridled Tern *Onychoprion anaethetus* 36–38cm

Scattered records on SW coast in Santa Elena and Manabí prov. ***Black bill and legs.*** Black cap, ***white V in forecrown, greyish-brown upperparts***, whitish underparts. Forecrown ***darker*** in imm. In flight: ***deeply forked tail***, whitish underwing contrasting with blackish flight feathers. Possibly race *nelsoni*. Oceanic, rarely near shore, infrequently plunges but seizes prey from surface. **Voice** Nasal *kyaar!* (RG). **SS** Confusion unlikely owing to deeply forked tail, head pattern, black bill. Similar Sooty Tern apparently erroneously assigned to mainland Ecuadorian avifauna. **Status** Uncommon, local, tiny breeding colony on Isla Pelado, Santa Elena prov. (Apr–Sep).

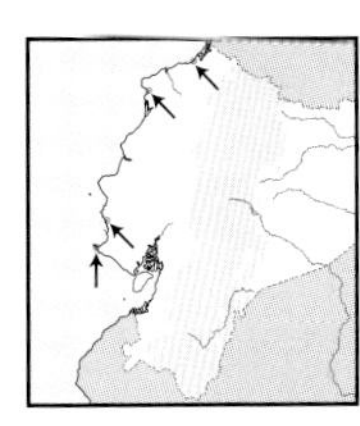

* Least Tern *Sternula antillarum* 23cm

Few records along coasts. ***Black bill yellow at base***, olive legs. ***Small. White front***, dark mask, pale grey above, ***pearly below, white tail***. Juv. browner above, dark carpal bar, ***bill browner***. Non-breeder has ***all-dark bill***. In flight: blackish outermost primaries forming wedge. Unknown race. Plunge-dives in shallow waters near coast, flight fast and dashing on shallow, hurried wingbeats. **Voice** A high-pitched *kee-jEÉT!* **SS** Closely resembles Amazonian Yellow-billed Tern (no known overlap, but Least could wander to E lowlands); see bill colour and blacker primaries; also Peruvian Tern (darker bill, greyer underparts, grey tail); moulting Least Tern might have dark-and-yellow bill. **Status** Casual visitor (Aug–Nov).

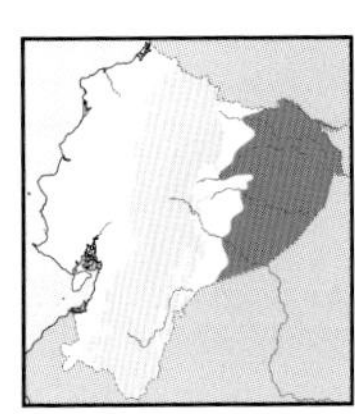

Yellow-billed Tern *Sternula superciliaris* 23–25cm

E lowlands on major rivers and adjacent lagoons (200–400m, once in highlands). ***Yellow bill and legs.*** Black cap, ***white front, grey upperparts***, white underparts. Non-breeder has ***mottled crown***. Juv. has duller soft parts, upperparts scaled brown. In flight: recalls Least Tern, but all primaries blackish. Monotypic. Singles to small parties in fast flight and dashing plunges, often resting on sandbars, logs, stumps; nests colonially with Large-billed Tern and Black Skimmer. **Voice** Sharp, high-pitched *keék-keék-keék-keék, keék-keérriik, kiirriík....* **SS** Very similar to Least Tern but no range overlap in Ecuador. Large-billed has massive bill, distinctive upperwing pattern. **Status** Fairly common breeder.

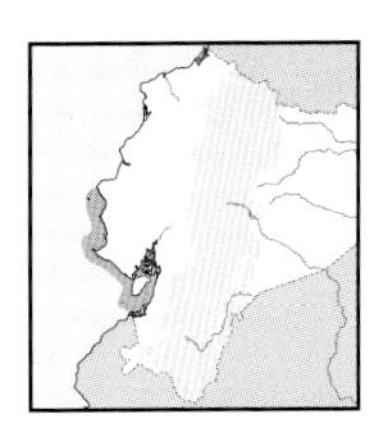

Peruvian Tern *Sternula lorata* 23–24cm

SW coast N to C Manabí. ***Black bill yellow basally***, yellowish legs. Black cap, ***white front***, grey upperparts, ***dirty grey underparts***, white lower face. Non-breeder whiter below, ***breast clouded grey***, crown mottled. In flight: black wedge in primaries. Monotypic. Pairs to small groups near coast, fast hovering and plunge-diving in fairly shallow waters; sometimes goes unrecorded for several years. **Voice** Mellow, high-pitched *churi* (RG) or *chirrii*. **SS** Underparts darker grey than rarer Least Tern, including underwing; yellow limited to bill base; moulting Least Tern might have little to no yellow. **Status** Rare austral visitor, rather erratic (May–Oct), large flocks on Jambeli (R. Ahlman); EN globally.

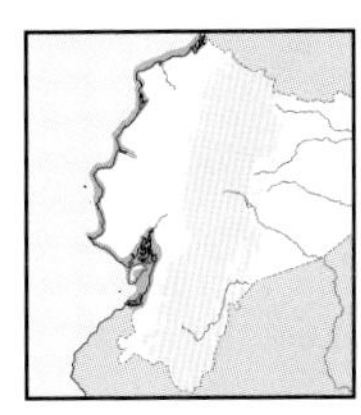

Gull-billed Tern *Gelochelidon nilotica* 35–39cm

Coast and adjacent wetlands, scattered records N Manabí and Esmeraldas. ***Stout black bill***, black legs. ***Black cap*** extends to nape, ***pale grey above***, white below. Non-breeder only shows ***dusky postocular to ear-covert patches***. Juv. crown ***washed brown***, brownish wing-coverts. In flight: ***rather short*** grey tail. Possibly races *aranea* and *vanrossemi*. Mainly small groups, seize prey from surface, often feeds inland; breeds locally, numbers possibly increasing. **Voice** Commonly, a raspy, high-pitched *ka'wee-krrra'wek...* also *ka-we'ék, ka-we'ék...wek-wek-wek....* **SS** Sandwich Tern longer-billed (yellow tip difficult to see), legs shorter, less sturdy; slenderer body and longer tail give less thickset jizz; non-breeder has black nape (feeding behaviour differs). **Status** Fairly common.

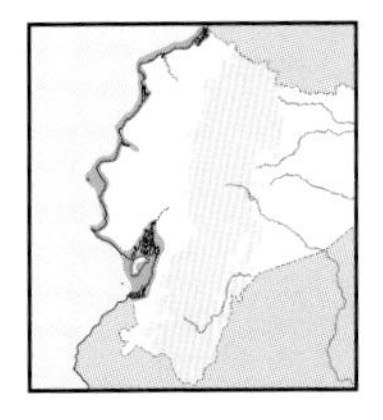

Sandwich Tern *Thalasseus sandvicensis* 41–43cm

Coasts, adjacent wetlands. ***Long slender black bill tipped yellow***, black legs. ***Black cap, shaggy crest***, pale overall. Non-breeder lacks bushy crest, ***black confined to rear crown and nape***. Juv. has brown-fringed upperparts. In flight: rather long, deeply forked ***white*** tail. Race *acuflavidus*. Small to fairly large flocks near coast and offshore, plunge-diving; often consorts with other terns. **Voice** High-pitched, shrilled *chiirrk, chrík!* **SS** From Gull-billed by bill length, leg size, more extensive black in nape (non-breeder), slimmer body, feeding behaviour; Elegant and Common Terns show at least some reddish in bill (Common also in legs). Peruvian and Least much smaller. **Status** Uncommon boreal winter visitor (year-round).

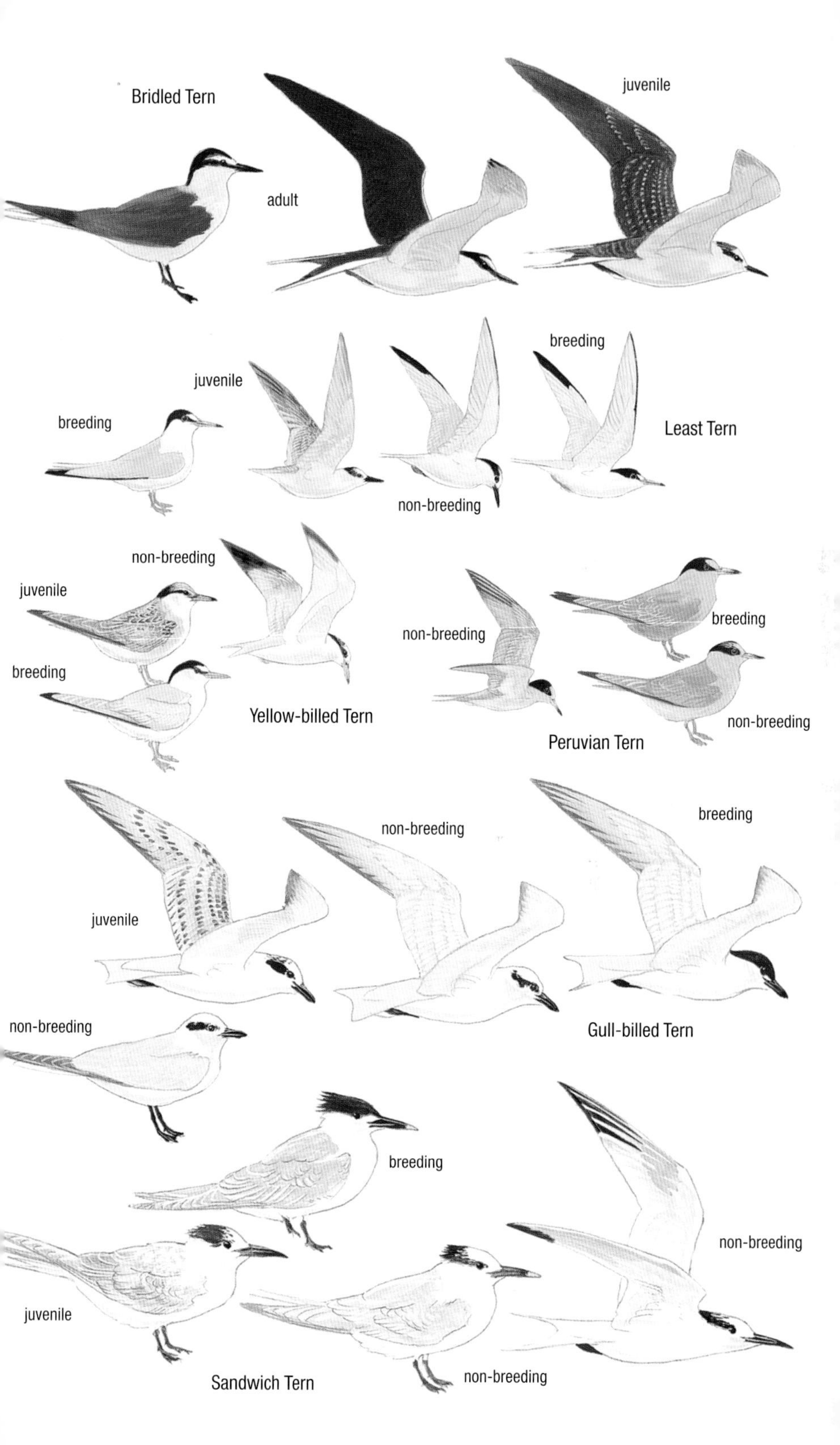
Bridled Tern
adult
juvenile
juvenile
breeding
non-breeding
breeding
Least Tern
non-breeding
juvenile
breeding
Yellow-billed Tern
non-breeding
breeding
non-breeding
Peruvian Tern
breeding
non-breeding
juvenile
Gull-billed Tern
non-breeding
breeding
juvenile
non-breeding
non-breeding
Sandwich Tern

Six migratory terns. Although red bills characterise breeders, imm. and non-breeders with duller/darker bills are more expected in Ecuador. Identification can be difficult, particularly of Common, Arctic and South American, whose differences are subtle. Clear observations precluded by their fast flight and light conditions. Note bill shape, size and colour, wing pattern and, to some extent, flight. With practice, Royal and Elegant readily separated by bill size.

Caspian Tern *Hydroprogne caspia* 53–56cm

Few records (C Manabí and Jambelí archipelago, El Oro). *Big. Heavy, blood-red bill tipped dark. Black shaggy crest streaked white* in non-breeder, pale grey above. Juv. similar but *upperparts mottled brown*. In flight: *large blackish wedge in primaries* (from below). Monotypic. Ecuador records involve singles on coast and inland wetlands; more likely along coast. **Voice** Elsewhere a harsh, almost strident *krrr-áárk!* (Harrison). **SS** Smaller Royal Tern (slimmer, more orange bill, paler underwing, no blackish outer wedge, black limited to hindcrown and nape); Elegant Tern (much smaller, slimmer, orangey bill, crest more as Royal). **Status** Accidental (Jan, Jul).

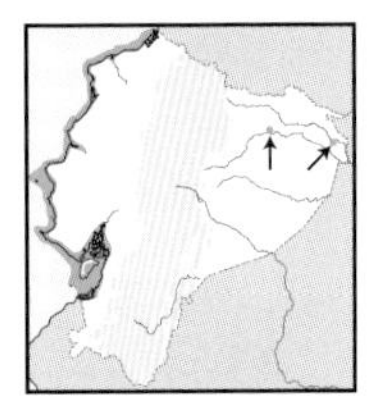

Common Tern *Sterna hirundo* 33–38cm

Coasts, scattered records inland. *Red bill tipped dark. Long black crest*, mid-grey above, pearl-white below. Non-breeder: *dark bill, basally reddish, black from postocular to nape*. Juv. recalls non-breeder but *dirtier*. In flight: *blackish primary tips*, dusky outer web of outermost rectrices; non-breeder and juv. show *dark carpal bar, grey secondaries*. Nominate. Graceful flight, deep wingbeats, more often coastal. **Voice** Short *kiirrt*. **SS** Resembles Arctic and South American Terns; in flight, Arctic shows *whiter primaries* with narrow black edge, white tail; juv. Common has more patterned upperwing with dark carpal bar and grey secondaries; legs shorter. See South American Tern. **Status** Fairly common boreal winter visitor (year-round).

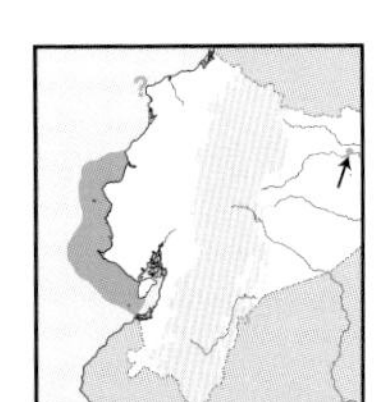

Arctic Tern *Sterna paradisaea* 33–36cm

Offshore, scattered records Santa Elena Peninsula. *Red bill*. Long black crest, *pale grey underparts*. Non-breeder: dark bill, no black in forecrown and lores. In flight: *white primaries with very narrow blackish trailing edge, 'translucent' remiges long white outermost tail feathers*. Juv. *lacks carpal bar, white secondaries*. Monotypic. Mostly small groups passing offshore, flight and behaviour recalls Common Tern. **Voice** Piercing *kiii'rrr* (B. Haase). **SS** Resembles 'dirtier' Common (shorter legs); wings appear longer, protruding beyond tail when perched; South American Tern (longer, heavier bill, broader black trailing edge). **Status** Uncommon northbound (Apr–May) and southbound transient (Aug–Nov).

South American Tern *Sterna hirundinacea* 42cm

Santa Elena Peninsula. *Coral-red bill*. Long black crest, pale grey above, *white below*. Non-breeder and juv. lack black in forecrown, soft parts duller, marbled mantle, *dark carpal bar*. In flight: *white underwing appears translucent* in good light, *blackish wedge in outer primaries*. Monotypic. Rather erratic, coastal, active and noisy; flight and feeding behaviour as Common and Arctic Terns. **Voice** Harsh, drawn-out *kyarrr-kyarrr-kyarrr...* (RG). **SS** From Common and Arctic by whiter underparts, wing pattern and bill colour; migrating Common and Arctic have dark bills in Ecuador; bill heavier, wings shorter. **Status** Rare or casual austral visitor (Mar–Oct); local breeder.

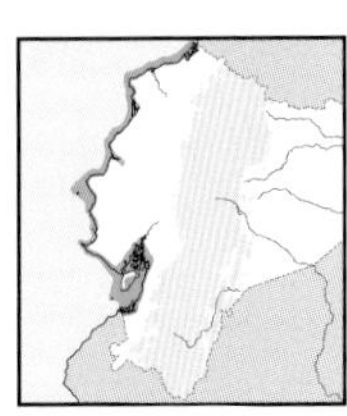

Elegant Tern *Thalasseus elegans* 41–43cm

Coasts, adjacent inland wetlands. *Rather long, slender, reddish-orange to orange-yellow bill. Long shaggy black crest (from postocular area to hindcrown and nape in non-breeder)*. Mid-grey above. Juv. recalls non-breeder but *upperparts spotted / mottled brown*. In flight: *slim body*, underwing as Royal. Monotypic. Transient, more regular on coast than offshore; large flocks at roosts, often with other terns; plunge-dives. **Voice** Similar to Royal but higher-pitched, also laugh-like *kree-kje-je'je'je'je'je*. **SS** Resembles larger Royal Tern but bill shape and length distinctive (tip slightly down-curved); focus on crest pattern: in Elegant appears to encircle eyes while Royal's orbital area is whiter. **Status** Uncommon boreal winter visitor (year-round); NT globally.

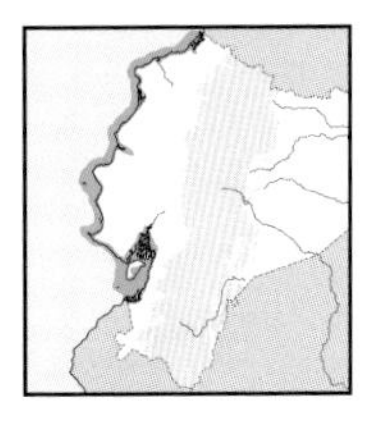

Royal Tern *Thalasseus maximus* 48–51cm

Coasts, adjacent inland wetlands. *Fairly heavy orange to yellow-orange bill. Black shaggy crest*, grey above. Non-breeder: *black confined to shaggy hindcrown and nape patch*. In flight: *thin blackish edges to outermost primaries*; juv. shows *dark carpal bar, pale band in median upperwing coverts*. Nominate. Numerous at sea and near coasts; hurried flight, dashing plunge-dives, often with other terns at roosts. **Voice** High-pitched, shrill *chiirr'rk!-chiir'rk, chirr-kikikiki....* **SS** Larger, heavier-billed than Elegant Tern, but size evident if together; hindcrown patch reaches eyes in Elegant; bill slimmer, slightly curved down; normally Royal looks stockier. See rarer Caspian Tern. **Status** Uncommon boreal winter visitor (year-round).

juvenile/first-winter
Caspian Tern
breeding
non-breeding
non-breeding
non-breeding
Common Tern
first-winter
non-breeding
first-winter
non-breeding
Arctic Tern
non-breeding
first-winter
breeding
non-breeding
reeding
immature
Elegant Tern
South American Tern
juvenile
breeding
non-breeding
breeding
Royal Tern

PLATE 29: TERNS III AND SKIMMER

Miscellaneous terns and the unique, spectacular skimmer. Large-billed Tern shares its Amazonian range with the smaller, shorter-billed Yellow-billed Tern (Plate 27). Inca Tern visits Ecuador infrequently from southern seas, while noddies are accidental visitors. Black Tern is a cosmopolitan tern infrequently seen offshore, even more exceptionally inland.

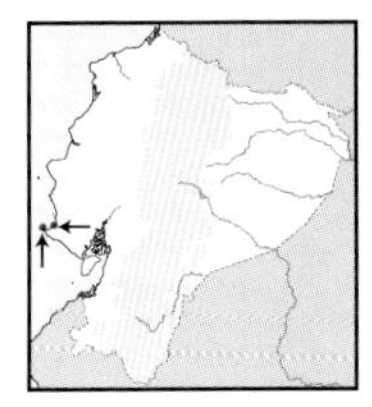

Brown Noddy *Anous stolidus* 42–43cm

Two certain records SW coast. Rather heavy bill, deep basally, *white partial eye-ring*. Mostly *dark brown, silvery cap* well demarcated by black loral and orbital areas, *two-toned underwing*. Possibly races *galapagensis* and *ridgwayi*. Highly pelagic, flocks above fish shoals, feed by surface-snatching, not diving. **Voice** Not expected to vocalise. **SS** B. Haase believes a Sep record pertains to Black Noddy, which also occurs in tropical seas and has reached Ecuadorian waters. Subtle differences; Black's bill longer, less deep basally, less robust; also paler underwing in Brown. **Status** Uncertain, probably vagrant (Feb, Sep).

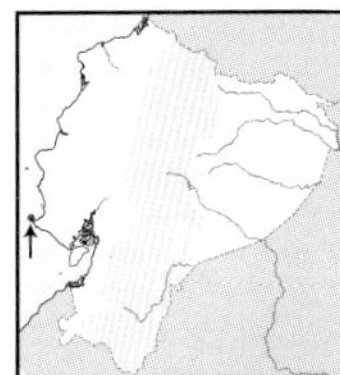

* Black Noddy *Anous minutus* 35–39cm

Single record SW coast at La Chocolatera (B. Haase). Very similar to Brown Noddy but smaller, *darker*, with white cap *sharply separated* from head-sides and neck; bill longer and thinner. Race unknown. Fluttery flight, highly pelagic. **Status** Accidental (Jul).

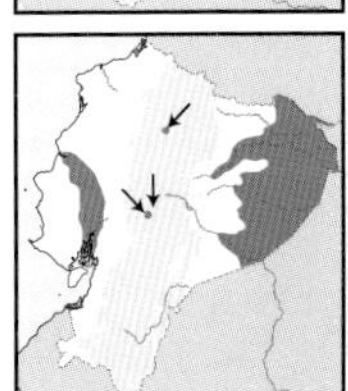

Large-billed Tern *Phaetusa simplex* 38–40cm

E lowlands along major rivers, adjacent lagoons; old records in highlands and SW lowlands (mostly below 250m, once 3,000m). *Large yellow bill. Black cap, dark grey above, rump and tail*, white nuchal collar and underparts. In flight: bold *grey, white and black upperwing pattern*, forked tail. Nominate. Singles to small groups perch on logs, stumps and sandbars; plunge-dives; breeds colonially. **Voice** Nasal *nyah-nyah-kha-kha-kha...* in flight; noisy moans at colonies. **SS** Much larger than sympatric Yellow-billed Tern, bill much heavier, upperwing pattern unmistakable. **Status** Uncommon in E, much rarer elsewhere.

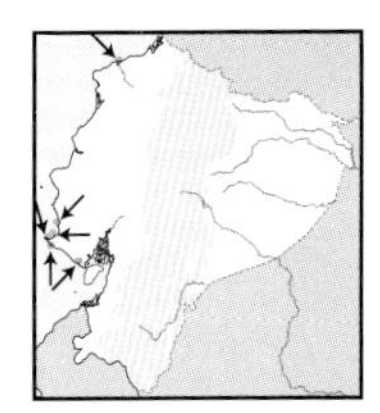

Inca Tern *Larosterna inca* 41–43cm

Few records on SW coast (Guayas and Santa Elena), once Esmeraldas; mainly rocky areas and ports. *Heavy scarlet bill, red legs, yellow wattle in gape*. Mostly *slate-grey, long curled white whiskers*. In flight *conspicuous white trailing edge*, slightly forked tail. Juv. duller. Monotypic. Graceful flight on shallow wingbeats with darting manoeuvres; irruptive in Ecuadorian waters, not present all years, presence perhaps associated with El Niño events. **Voice** Unlikely to be heard in Ecuador. **SS** Unmistakable by dark plumage and red soft parts. **Status** Rare to irregular austral visitor (mainly May–Oct, possibly Dec onwards); NT globally.

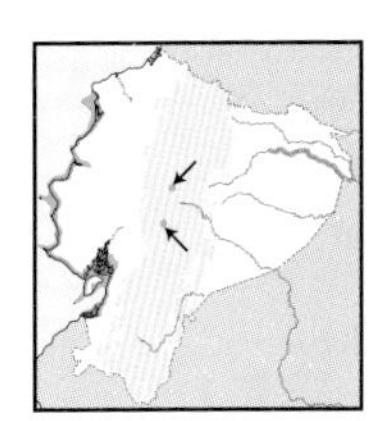

Black Tern *Chlidonias niger* 23–25cm

Coasts and adjacent wetlands N to C Manabí, once in E lowlands (Rio Napo), once in Andes (Yambo). Black bill; grey above, white below, *blackish cap extends to postocular and ear-covert areas*, grey extending as *small patches to breast-sides*. In flight: pale grey underwing, *short tail only slightly notched*. Breeder mostly black, *lower underparts white*. In moult *mottled black and white*. Race *surinamensis*. Buoyant graceful flight, rarely dives but seizes prey from surface; singles or small groups in Ecuador, but elsewhere larger flocks. **Voice** Short, high *kik-kik-kik...tiiik* (B. Haase). **SS** Size, blackish cap and short, nearly square tail distinctive. See Sabine's Gull (striking upperwing pattern). **Status** Rare boreal winter visitor (possibly year-round).

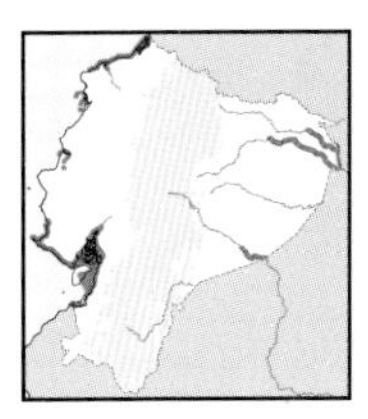

Black Skimmer *Rynchops niger* 42–46cm

SW coast to adjacent wetlands, locally N Esmeraldas (San Lorenzo, Atacames), E lowlands along major rivers (below 300m). *Long knife-like bill, mandible longer, red basally*, red legs. *Black above, white below*. In flight: broad white trailing edge, long forked tail, whitish underwing, *very long wings*. Juv. duller, mottled blackish and white, soft parts duller. Race *cinerascens*. Singles to small groups in graceful flight, skimming over water in low flight; often perching on stumps; breeds colonially. **Voice** Loud, penetrating, nasal *augh!, augh!...* **SS** Unmistakable. **Status** Uncommon, rather erratic in W; VU in Ecuador.

juvenile
ridgwayi
adult
Black Noddy
Brown Noddy
adult
juvenile
galapagoensis
adult
immature
non-breeding
breeding
juvenile
juvenile
Large-billed Tern
breeding
immature
adult
Inca Tern
non-breeding
breeding
juvenile
adult
juvenile
Black Tern
immature
adult
Black Skimmer

Resident and migratory waders with short, stout bills, chunky bodies and strong legs. Lapwings are boldly patterned and have prominent wing spurs. Complicated identification of golden plovers (Pacific very rare in Ecuador). The dotterel was recently re-found in Ecuador after decades.

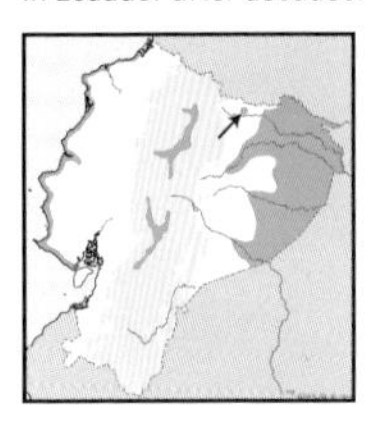

American Golden Plover *Pluvialis dominica* 24–28cm

Scattered records in lowlands and highlands (to 3,900m). Grassy fields, muddy beaches, banks. *Short, rather slim bill.* Non-breeder: *greyish-brown* above *spotted whitish and pale ochre*, duller breast mottled whitish, *whitish-tawny supercilium. Smoky-grey underwing, dark upperwing and rump*; wings *project beyond tail* at rest. Monotypic. Mainly singly, feeding as Black-bellied Plover, some defending winter territories. **Voice** Whistled *kweét, kweed'lee*. **SS** Commoner Black-bellied (black axillaries, white rump, white upperwing stripe); very similar to accidental Pacific Golden Plover (less mottled breast, more yellowish above, slightly slimmer, wings *do no project* beyond tail). **Status** Rare transient (mainly Feb–Apr, Aug–Nov).

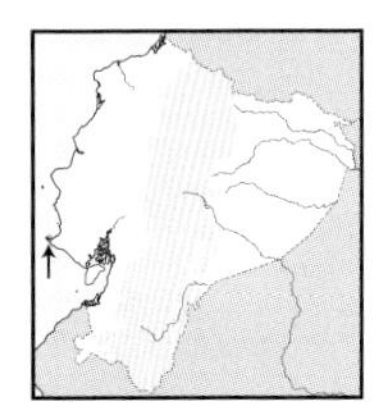

* Pacific Golden Plover *Pluvialis fulva* 23–26cm

Two sight records on Santa Elena Peninsula. Saltmarshes. Very like American Golden Plover in all plumages. Non-breeder mainly *yellowish-buff* above, but beware moulting birds. Breast mottling *yellower*. Somewhat smaller and *slimmer*. Monotypic. Behaves as congeners. **Voice** Fast, loud *tu-ee, chu-wit* (H *et al.*). **SS** Difficult to identify from American even by voice; differences are subtle (more yellowish-buff overall, sparser breast mottling, tawnier supercilium, *shorter wings* not projecting beyond tail); tail length, often considered a useful character, not always helpful. **Status** Accidental (Jan, Apr).

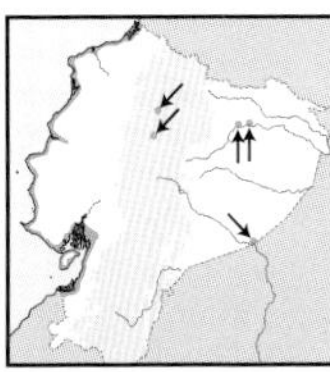

Black-bellied Plover *Pluvialis squatarola* 27–30cm

Mostly on coasts, sparse records elsewhere (to 300m). Sandy and muddy beaches, mudflats. Rather *stout black bill.* Non-breeder: *pale grey* to greyish-brown above *spangled whitish, white rump. White stripe in upperwing, black axillaries.* Monotypic. Mostly solitary, rather lethargic, performing rapid runs after prey. **Voice** Sad, upslurred *keéwl!, keéwl-klüe'e.* **SS** Larger and plumper than rarer American and Pacific Golden Plovers (tawny or yellowish above), white rump and black axillaries distinctive in flight. **Status** Fairly common (year-round). [Alt: Grey Plover]

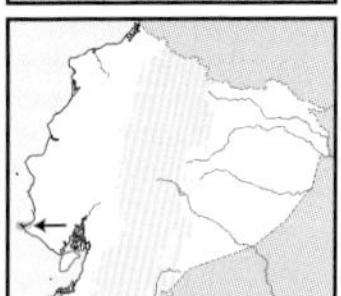

Tawny-throated Dotterel *Oreopholus ruficollis* 25–28cm

Single old record (Jan–Feb 1898) on Santa Elena Peninsula. Open semi-desert fields. Bold facial pattern. *Pale grey neck-sides to nape and breast, tawny-orange throat, black patch on white belly, white underwing.* Race *pallidus*. Unknown in Ecuador; elsewhere, gathers in small groups on barren terrain, running and flying fast and direct. **Voice** Tremulous *whee-tur-tur...* (H *et al.*). **SS** Confusion unlikely. **Status** Long lost from Ecuador; recently re-located (Jun 2015; B. Haase).

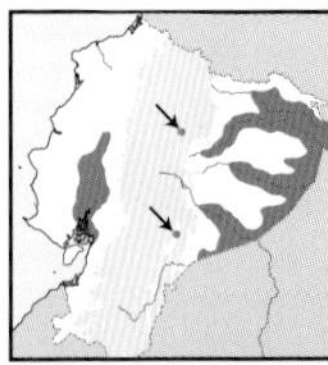

Pied Lapwing *Vanellus cayanus* 21–24cm

E lowlands, locally in SW lowlands (below 400m, occasionally higher). Large-river edges, sandbars. Beautiful *head pattern, black pectoral band*, white underparts, *black-and-white band in scapulars*, white rump. *Greyish-white and black wing pattern* in flight. *Red legs and eye-ring*. Monotypic. Loose pairs run after prey; apparently wander following water levels. **Voice** Querulous *kyeep?* or *kee-woop?-kee-woop?* in flight. **SS** Confusion unlikely; Southern Lapwing similarly patterned in flight but lacks scapular band, head pattern, blacker forewing, etc. **Status** Uncommon, VU in Ecuador.

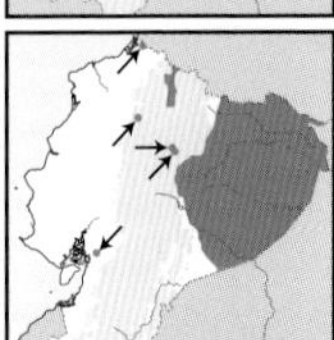

Southern Lapwing *Vanellus chilensis* 31–38cm

E lowlands, local in W lowlands, Andean slopes and valleys (to 2,600m). Grassy fields, river margins, sandbars. *Coral-pink bill tipped black, pink legs, long hairy crest.* Black-and-white foreface, *greyish head, black breast*, white below, *dark upperwing-coverts*. Race *cayennensis*. Noisy pairs to small flocks, buoyant flight; numbers apparently increasing. **Voice** Loud, unrelenting *khén-khén-khén... péu-péu-péu....* **SS** Greyish head and hairy crest distinctive; from Andean Lapwing by underparts and head. Pied Plover more boldly patterned. **Status** Uncommon, spreading.

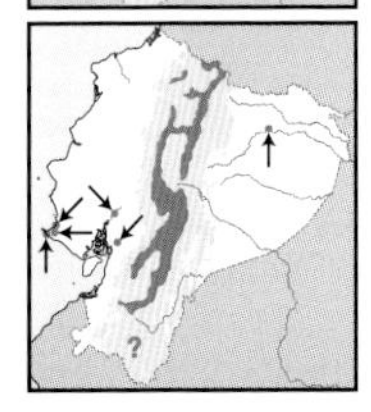

Andean Lapwing *Vanellus resplendens* 33–36cm

Andean highlands S to N Loja (mostly 3,500–4,400m, wanders to lowlands). Open *páramo*, short grassy fields, marshes. *Coral-pink legs and dark-tipped bill. Pale grey head to breast, whiter foreface. Bronzy-olive above.* Monotypic. Loose pairs or groups in floppy flight, loud and conspicuous. **Voice** Querulous *glee-glee-glee-e'e'e'e... wlee'e'e'e'week-glee-glee..., wík-wík-wík....* **SS** From Southern Lapwing by underparts and head; legs shorter, do not protrude beyond tail (no known overlap). **Status** Fairly common.

American Golden Plover
breeding
non-breeding
to breeding
non-breeding
juvenile
non-breeding
breeding
Pacific Golden Plover
juvenile
adult
Tawny-throated Dotterel
non-breeding
breeding
juvenile
non-breeding
Black-bellied Plover
juvenile
adult
Pied Lapwing
juvenile
adult
Southern Lapwing
adult
Andean Lapwing
juvenile

Small, chunky plovers – except Killdeer – with short stout bills, and short pink to yellowish legs. Most are coastal, run-and-stop feeders, but species identification can be tricky. Best identification features are bill length and colour, leg colour, extent of dark breast-band and neck collar. Three residents, one with migratory and resident populations, two boreal migrants.

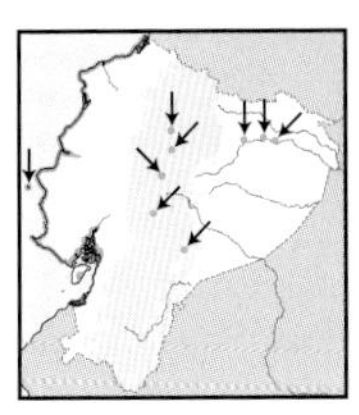

Semipalmated Plover *Charadrius semipalmatus* 17–19cm

Boreal winter visitor to coast, adjacent lowlands, few in E lowlands and highlands (to 3,700m). Muddy shores, lagoon edges. ***Short stubby bill, orange basally*** (non-breeder only ***pale base***); ***orange legs*** and narrow eye-ring. Non-breeder has dull brown head, upperparts and ***narrow pectoral band, white front, eyebrow, nuchal collar*** and underparts. In flight: white stripe on upperwing, ***white flanks***, white underwing. Breeder has black in face and pectoral band. Monotypic. Small, loose flocks with other waders. **Voice** Inflected *chee-vít?..chee-ví?...*. **SS** Larger Wilson's Plover (heavier black bill, no eye-ring, fleshy legs); Snowy Plover (dark legs and bill, broken breast-band). **Status** Common transient (year round).

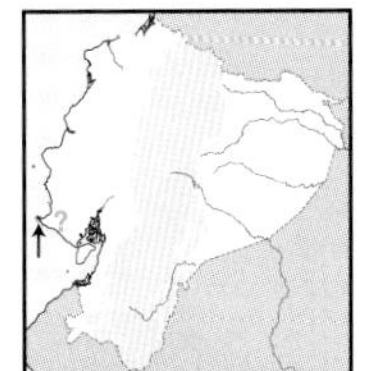

Piping Plover *Charadrius melodus* 17–19cm

One certain record from SW (Salinas). ***Short stubby black bill, orangey-flesh legs. Pale sandy-silver head, cheeks, upperparts and broken breast-band***, white foreface, faint eyebrow, ***nuchal collar*** and underparts. In flight: white stripe on upperwing, white underwing and tail, ***broad black patch above***. Breeder: ***black nape to pectoral band, orange bill base***, black frontal band. Monotypic. Deliberate, steady movements on sandy beaches. **Voice** Not heard in Ecuador, elsewhere sweet, descending *peep-lo*. **SS** Paler above than Semipalmated Plover, legs less brightly coloured, no eye-ring, incomplete breast-band. Snowy has longer, slenderer bill, greyish legs. **Status** Accidental (Oct).

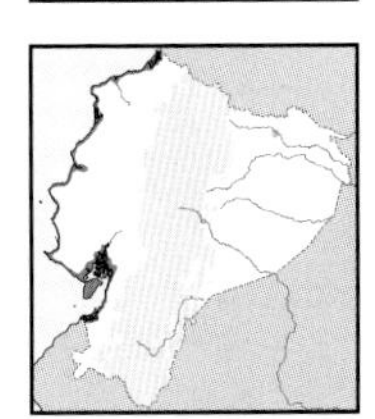

Wilson's Plover *Charadrius wilsonia* 18–20cm

Coasts. Muddy shores, lagoon edges, mudflats, shrimp ponds. ***Heavy black bill, dark flesh legs***. Dusky-brown crown, cheeks and ***complete breast-band, more rufescent upperparts***. White forecrown, short eyebrow, nuchal collar and underparts. Breeder has bolder, black frontal and breast-bands, ***chestnut wash on hindneck***. In flight: white wing stripe, ***white rump-sides***. Race *beldingi*. Pairs to small groups with other waders, compact group when threatened, running and stopping; rather tame. **Voice** Weak *fwít-fwít-fwít* whistles, sometimes more emphatic. **SS** Heavy bill distinctive; cf. Collared Plover (slender bill, no nuchal collar, yellowish legs); Semipalmated (shorter bill, bright legs); Snowy and Piping Plovers (paler above, short-billed). **Status** Uncommon.

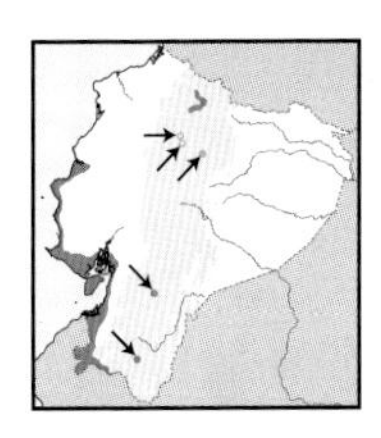

Killdeer *Charadrius vociferus* 23–26cm

Coasts to adjacent lowlands in SW, Chota Valley; boreal visitor to N highlands (to 1,200m, migrants higher). Open sandy fields. ***Large heavy black bill, rosy eye-ring, dark legs***. Brown crown to cheeks, more rufescent upperparts. ***Bold black double breast-band***, narrow black frontal band. White front, eyebrow, throat to ***nuchal collar*** and underparts. ***Cinnamon-rufous lower back and rump, black subterminal tail-band. Long graduated tail***. Race *peruvianus*, possibly migratory *vociferus*. Pairs to small wary groups, rather noisy and bold. **Voice** Very loud *keel-deé...chá-chá-chá...keél-keél-deé-dee-dee-dee...*. **SS** Confusion unlikely; smaller but heavy-billed Wilson's Plover has single breast-band. **Status** Uncommon resident, vagrant boreal visitor.

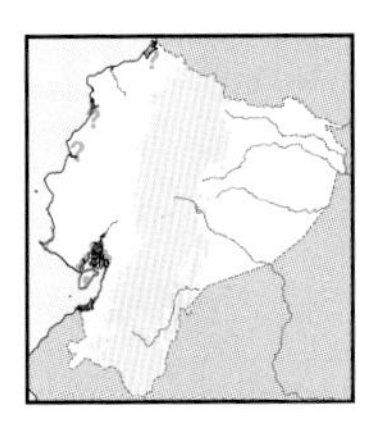

Snowy Plover *Charadrius nivosus* 15–16cm

Coasts in SW (sea level). Xeric vegetation, saltmarshes. ***Short, slender black bill, grey legs. Sandy-brown crown, cheeks and upperparts***, narrow black frontal band. White forecrown, face, eyebrow, ***nuchal collar*** and underparts, ***blackish patches on breast-sides***. In flight: white wingbar, ***white rump and tail-sides***. Race *occidentalis*. Mainly pairs, rather unwary, counting on cryptic plumage; feeds faster than most congeners. **Voice** Dry *churr-pirr'uirr, pirr'uirr...chüirr-pirr'uirr*, rather rhythmic; soft *kuwe'éet* calls. **SS** Semipalmated and Collared Plovers have complete breast-bands, Collared lacks nuchal collar, both species' legs more colourful. Rarer Piping Plover is paler silvery-grey above, pinkish legs, stubby bill. **Status** Uncommon; NT globally.

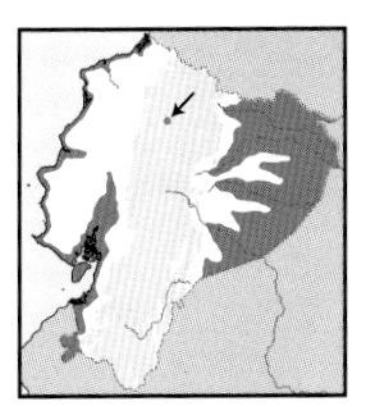

Collared Plover *Charadrius collaris* 14–15cm

Coasts to adjacent W lowlands, E lowlands (to 500m). Muddy shores of estuaries and lagoons, shrimp ponds, large-river margins, sandbars. ***Slender black bill, yellowish legs***. Brown above, ***cinnamon tinge to nape, hindcrown and neck-sides***. Black forecrown and ***single*** breast-band, ***black lores, white front***, eyebrow and rest of underparts. ***White flanks and tail-sides*** in flight. Monotypic. Tame pairs when breeding, small groups when not, often away from wader flocks and from water. **Voice** Sharp *kíip!* or *chíp*, sometimes running into fast jumble. **SS** Only plover with single breast-band and no white nuchal collar. See Semipalmated, Snowy and Piping Plovers. **Status** Fairly common.

Semipalmated Plover
juvenile
non-breeding
breeding
juvenile/non-breeding
Piping Plover
breeding
Wilson's Plover
juvenile
♀
♂
adult
Killdeer
Snowy Plover
non-breeding
♂
breeding
juvenile
immature
♂
Collared Plover

PLATE 32: MISCELLANEOUS WADERS

Distinctive waders in separate, rather species-poor families. Oystercatchers are strictly coastal; the stilt occupies brackish to freshwater wetlands; jacana in freshwater wetlands with abundant floating vegetation; and the nocturnal thick-knee is not tied to water at all. The avocet is only accidental to Ecuador. Note variation in Black-necked Stilt, especially in extreme SW.

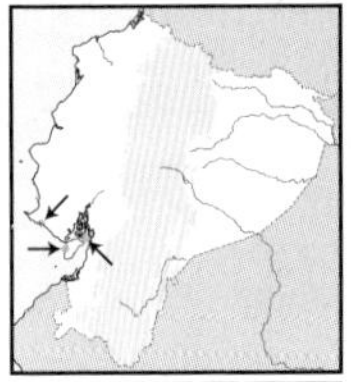

Blackish Oystercatcher *Haematopus ater* 43–45cm

Few records from SW coast at Chanduy and Puná. Sandy and rocky beaches. Mainly ***blackish, coral-red eye-ring*** and ***yellowish-tipped bill***, dull legs, yellow eyes. Monotypic. Mostly sedentary resident on rocky beaches in breeding grounds. **Voice** Like other oystercatchers, a loud, explosive *wheep!* and piping *pip-pip-pip-pip-pip*. **SS** Confusion unlikely, only resident oystercatcher is black and white; American Black Oystercatcher *H. bachmani* (not in South America to date) very similar, bill more slender. **Status** Vagrant.

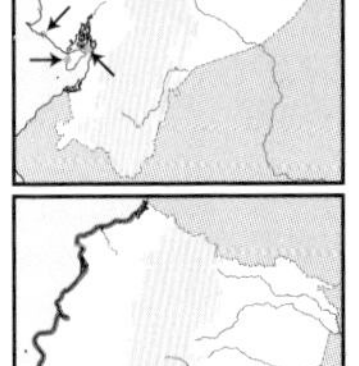

American Oystercatcher *Haematopus palliatus* 40–44cm

SW coast, locally northwards (sea level). Rocky or sandy beaches. ***Long coral-red bill***, red eye-ring, yellow eyes, ***short pink legs***. Bold. ***Black head to mantle and breast, white below***, dark brown above. ***White rump, white stripe in flight feathers, white underwing*** and ***black tail*** conspicuous in flight. Race *pitanay*. Scattered pairs or small parties, often wary and noisy when flushed; opens shells and turns stones, shells to feed. **Voice** Piercing *kleee!*, *khleep!*, often running into *klee-klee'kee'kee'kee*.... **SS** Unmistakable. **Status** Rare.

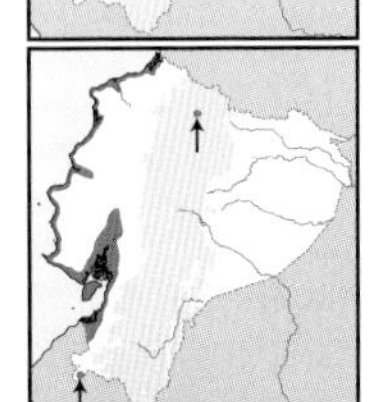

Black-necked Stilt *Himantopus mexicanus* 35–40cm

Coast to adjacent W lowlands (below 100m; once at 2,200m). Marshes, shrimp ponds, salt ponds, ricefields, pond margins. ***Needle-like black bill, long thin rosy-red legs***. Boldly ***pied, white lower back to rump***. *Melanurus* race has white head and ***nuchal collar***, black nape. Race *mexicanus* (most of range), *melanurus* (extreme SW) suggested species status. Small to large flocks wade in fairly deep water, wary when flushed, flocks compact when threatened, circling around, landing and flying again. **Voice** Strident, endless *kék-kék-kék, kee-kék!* yelps to high-pitched, faster *kikikikikik*... and others. **SS** Unmistakable. **Status** Common.

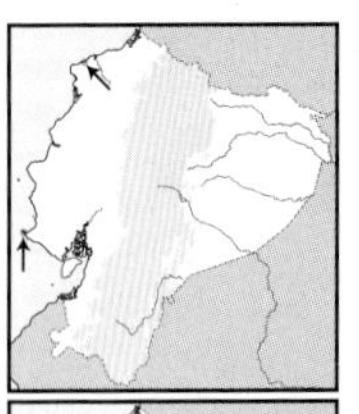

American Avocet *Recurvirostra americana* 40–45cm

Two records from coast. Salt ponds. ***Narrow upturned bill, very long thin grey legs. Pearl-grey head to upper breast and mantle, mainly white, black back, sides and flight feathers***, white underwing. Monotypic. Elsewhere in wintering range congregates in large flocks, wading in deep water sweeping bill sideways, also swims. **Voice** Not heard in Ecuador, a penetrating *kleet* reported (H *et al.*). **SS** Unmistakable, superficially similar Black-necked Stilt has bright rosy legs, straight bill, etc. **Status** Accidental (Jan, Jul–Sep).

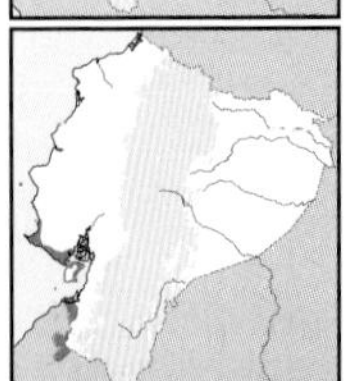

Peruvian Thick-knee *Burhinus superciliaris* 38–42cm

SW lowlands (below 200m). Open arid fields, xeric open scrub and agricultural fields. ***Stout yellowish bill tipped dark, very large yellow eyes, long yellowish-green legs. Sandy-greyish upperparts and neck to breast, bold white supercilium outlined black above.*** White throat and belly to vent. Black flight feathers with small ***white flash*** in inner primaries, white underwing. Monotypic. Nocturnal to crepuscular, by day resting motionless in grass; does not flush, rather walks slowly and warily. **Voice** Loud chattering *we-re-ké-ké...wer'ëëë...wer-keu-keu-ke-ke*..., querulous *we're-ë-ë-ë*.... **SS** Unmistakable, especially by bill, eyes, legs. **Status** Rare, declining, VU in Ecuador.

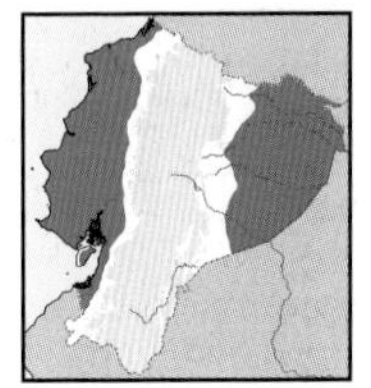

Wattled Jacana *Jacana jacana* 20–23cm

E & W lowlands (to 300m). Shallow marshes, ponds, lagoon margins, ricefields. ***Yellow bill, bright red wattle at base. Black*** head to mantle and underparts, ***chestnut*** (E) ***to rufous*** (W) ***upperparts, greenish-yellow flight feathers***. Long greenish legs with ***very long toes***. Juv. lacks wattle, bill pink basally, dull brown above, ***bold white supercilium, whitish underparts***. Race *intermedia* (E), *scapularis* (W). Polyandrous, congregates for breeding, but singles to loose groups when feeding; walks over floating vegetation, flies heavily when flushed, legs dangling, wings held outstretched after landing. **Voice** Loud *cla-ka'kee, cla-ka'kee*..., *kla-kla-kla-kla*...*kya-kya*... cackles, clicks and chatters. **SS** Unmistakable. **Status** Common.

adult
immature
adult
American Oystercatcher
juvenile
adult
immature
Blackish Oystercatcher
mexicanus
juvenile
adult
melanurus
adult
breeding
Black-necked Stilt
American Avocet
breeding
non-breeding
juvenile
Peruvian Thick-knee
Wattled Jacana
scapularis
adult
intermedia

Most snipes are resident in Ecuador. They dwell in densely vegetated damp places, most in highlands but two migratory or seemingly migratory species range elsewhere. All species very difficult to observe due to shy behaviour and, when seen, most are difficult to identify. Note bill proportions, belly and mantle patterns, but especially voices.

Imperial Snipe *Gallinago imperialis* 29–31cm

Scattered records on both temperate slopes of Andes (2,700–3,800m). Wet forest, adjacent woodland. *Long bill* (9cm), *slightly drooped at tip. Dark chestnut, black* streaks on neck-sides to breast, chevrons and bars in upperparts. *White belly barred black.* Monotypic. Crepuscular, seldom seen by day; flight display includes high circling and shallow dive, with whirring sound produced by tail. **Voice** Loud, raucous *wak-wak-khak, wak-wak-khan...whak-ti-túk-tuk-tuk-tuk... wok-ka-ti-ków-ti-ków-kuk-kuk-kuk...* in flight display; *whak* calls on perch. **SS** Confusion unlikely, deeper chestnut than Andean Snipe, with belly boldly barred; more forest-based than other snipes. **Status** Rare, local, NT globally.

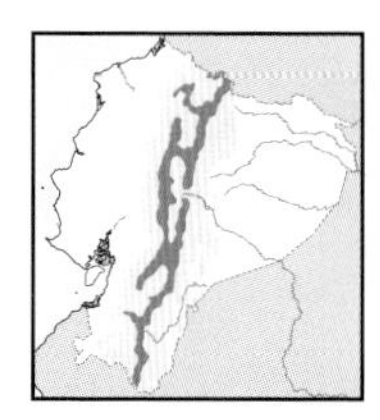

Andean Snipe *Gallinago jamesoni* 29–30cm

Andean highlands (3,100–4,400m). Wet *páramo*, bogs, *Polylepis* woodland. *Long bill slightly drooped at tip* (7.8cm). *Dark brown and chestnut above, no obvious* pale longitudinal V on back; *barred underparts.* In flight: *no white* trailing edge. Monotypic. Singly at dusk, abruptly flushes but flight short; display flight high circling rise-and-dive, with bellowing sound produced by tail. **Voice** Loud *weék'o-weék'o-weék'o...* or *weék-weék-weék'kee'kee'kee...* in flight display; rattled *tak'k'k'k'k'k'k* from ground or when flushed. **SS** From Noble Snipe by barred underparts, also less 'snipe-patterned' above. Imperial darker chestnut, belly barred white and black. **Status** Uncommon resident.

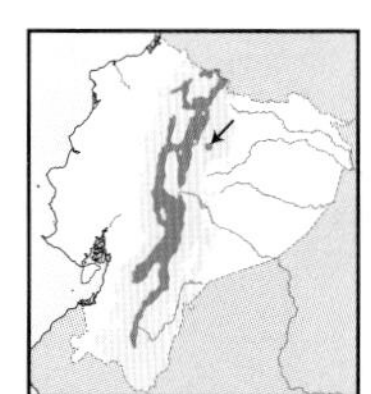

Noble Snipe *Gallinago nobilis* 30–32cm

Andean highlands S to N Loja (2,900–4,100m; locally breeding at 2,000m). Wet grassy *páramo*, marshes, bogs. *Very long bill* (9.4cm), short dark flesh legs. *Dark brown and chestnut above, narrow buff V on back,* head to breast coarsely streaked, *white unmarked belly.* In flight: *no* white trailing edge to secondaries. Monotypic. Singles or pairs skulk in dense vegetation, abruptly flying when almost stood on; apparently performs display flights. **Voice** Quite melodious *yAk-yAk-yAk-yAk* from ground, *ghughughughughughu* winnow; fast *kyAk-yak'yak'ky-yak'ky-yak* when flushed. **SS** Larger than Puna (overlap?), Wilson's and South American Snipes, bill longer, heavier; Andean Snipe (barred underparts, barred tail, darker, less 'snipe-patterned' upperparts). **Status** Uncommon; NT globally.

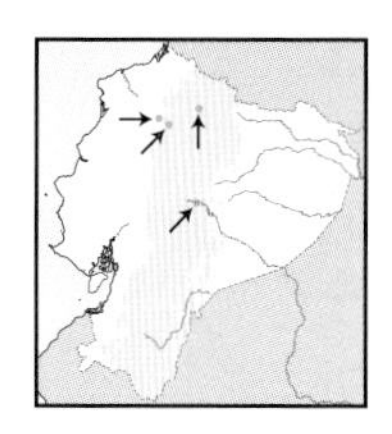

Wilson's Snipe *Gallinago delicata* 25–27cm

Scattered highland records (1,200–2,600m, possible elsewhere). Wet grassy fields. Long straight, *dark-tipped* bill (6.4cm), *greyish legs.* Variegated buff, chestnut and black above, *narrow whitish V on back, buff head to breast coarsely streaked dusky, white belly.* In flight: *narrow white trailing edge* to secondaries, *white tail-sides.* Monotypic. Singles or small parties probe in damp mud, rather dowitcher-like; zigzag flight when disturbed. **Voice** Long harsh *sscrraaai!* when flushed (R *et al.*). **SS** Nearly unidentifiable versus South American Snipe, except by voice, season and, to some extent, range. South American has *whiter underwing, longer bill and legs, shorter wings.* **Status** Rare boreal winter visitor (Oct, Dec–Feb).

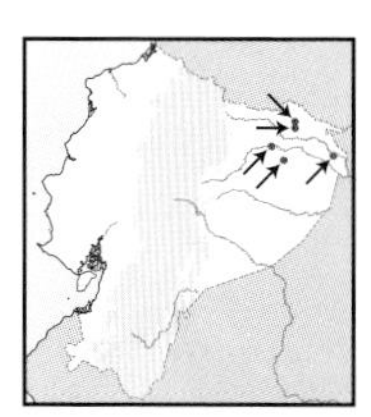

South American Snipe *Gallinago paraguaiae* 24–27cm

Few records in E lowlands (mainly below 250m). Wet grassy fields, bogs, lagoons. Near-identical in plumage to Wilson's Snipe but *paler underwing,* more white tail-sides. Bill slightly longer (7cm). Nominate. Apparently wanders to Ecuador; singles remain concealed in dense vegetation, abruptly flushing when disturbed, flight possibly less zigzagging than Wilson's. **Voice** Nasal, rather raspy *kriiaaa* in sudden flight, *kreea-kreea-kreea...* from ground. **SS** Not safely identified from Wilson's; focus on bill length as compared to nape–bill base length. **Status** Very rare, possibly wanderer (mostly Jan, Apr, Jul), perhaps largely overlooked.

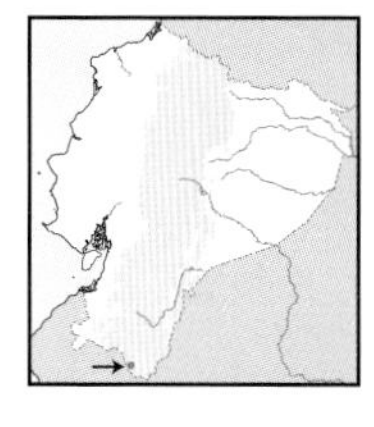

* Puna Snipe *Gallinago andina* 23–25cm

One record from extreme SE (3,300m). Wet grassy *páramo.* Longish bill *shorter* than similarly plumaged Wilson's and South American Snipes (5.4cm). *Short yellow legs.* Plumage resembles South American but *more buffy* back V, *more white in outer tail feathers, edge to outer primary white.* Monotypic. Poorly known in Ecuador; in Peru prefers bogs and damp stream sides, where displays and nests; circling and diving flight display. **Voice** Clear *dyák-dyák-dyák-dyák...* from ground, harsher when flushed; hoarse *shushushushush...* in flight display. **SS** Leg colour and size distinctive; Andean Snipe much larger, longer-billed, no yellow legs, no white belly, darker brown-chestnut above, barred tail. **Status** Uncertain, possibly very rare resident.

Imperial Snipe
Andean Snipe
Noble Snipe
Wilson's Snipe
South American Snipe
Puna Snipe

PLATE 34: LARGE WADERS

Mid-sized to large migratory waders. Dowitchers extremely difficult to identify; Long-billed is only accidental. Godwits, Whimbrel and Willet are tall, slender and elegant waders. Godwits are rarely found inland in Ecuador, whilst Whimbrel and Willet are among the commonest beach waders.

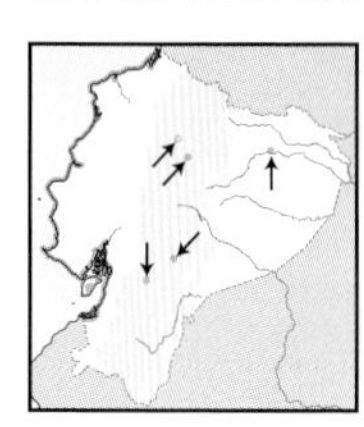

Whimbrel *Numenius phaeopus* 38–46cm

Coasts and scattered inland records (to 3,600m, mostly below 100m). Sandy and muddy beaches, pools, mudflats, brackish lagoons, shallow ponds. ***Long decurved bill***, rather short legs. ***Head striped whitish and dusky***. Race *hudsonicus*. Singles to small parties feed on mud or sand by picking, also probe, often with other waders. **Voice** Distinctive, steady, musical whistles *bee-bee-bee-bee... bëë-bëë-bee'bee'bee*. **SS** Only Ecuadorian wader with decurved bill. Note bill length versus other *Numenius* (e.g. Long-billed Curlew). **Status** Common boreal winter visitor (year-round). **Note** There is a recent, undocumented sight record of Long-billed Curlew *Numenius americanus* (***very long bill, cinnamon underparts, no head stripes***) – not illustrated.

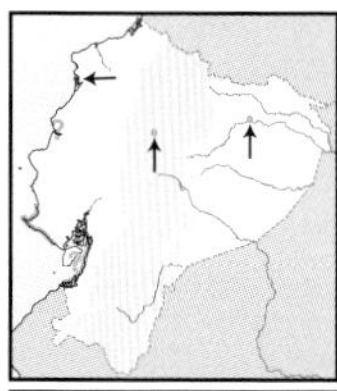

Hudsonian Godwit *Limosa haemastica* 37–42cm

Scattered SW, highland and E lowland records (to 3,800m). Lagoons, shallow pools, mudflats. ***Very long upturned bill, pink basally***, long legs. ***Pale grey***, white belly, short white supercilium. In flight: ***white wing stripe and rump, black tail, black underwing-coverts***. Monotypic. Might join other waders, feeding in deep water, probing deeply in mud. **Voice** Not vocal in winter; weak *teeu* calls reported. **SS** Willet has shorter straight bill, bolder wing stripe, grey tail, more patterned underwing. See rarer Marbled Godwit. **Status** Rare boreal winter visitor (Jan–May, Jul–Nov).

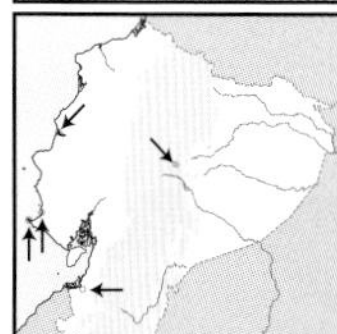

Marbled Godwit *Limosa fedoa* 42–48cm

Few SW records (Santa Rosa, Santa Elena, Bahía de Caráquez). Coastal lagoons, shallow ponds, mudflats. ***Very long upturned, mostly pink bill***, long legs. ***Largely cinnamon***, dusky speckling / barring above, short whitish eyebrow. In flight: ***black outer primaries, cinnamon-buff inner flight feathers, underwing-coverts, rump and tail***. Race *fedoa*. Behaviour similar to Hudsonian Godwit; winters mostly in Middle America. **Voice** Mostly silent on wintering grounds; nasal voice. **SS** Confusion unlikely due to long pink bill, cinnamon-buff plumage, long legs. Hudsonian Godwit (greyer, bolder wing pattern). **Status** Rare boreal winter visitant (Jan–Mar, Aug–Sep).

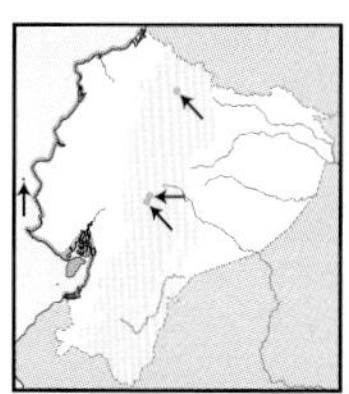

Willet *Tringa semipalmata* 35–41cm

Mostly coasts, few highland records (to 3,000m; once at Rio Napo). Sandy and muddy beaches, mudflats. ***Long straight bill, bluish basally, long bluish-grey legs***. Grey above short supraloral stripe, dark lores. In flight: ***bold white wing stripe, white rump, grey tail***. Races *inornata, semipalmata*. Small groups probe and pick, following tides; often with Whimbrels. **Voice** Loud, repeated *kEE-kip-kip-kip, wee-wee-wee...*. **SS** Bill and leg separate from similar-sized Greater Yellowlegs. Hudsonian Godwit has longer upturned bill, shorter wing stripe, black tail. **Status** Common boreal winter visitor (year-round).

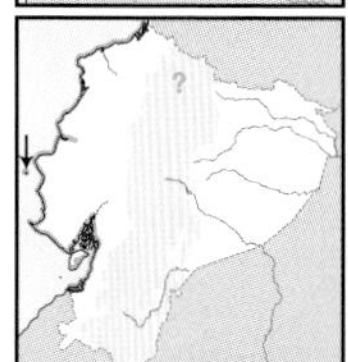

Short-billed Dowitcher *Limnodromus griseus* 25–29cm

Coasts. Brackish wetlands, mudflats, estuaries. ***Long snipe-like bill***, fairly short greenish legs. Mostly grey, paler below, white belly, short white supraloral. In flight ***extensive white in lower back to rump, tail barred black and white***. Possibly races *hendersoni, caurinus* and *griseus*. Large flocks feed in chest-deep waters, rapidly probing with sewing machine-like movements. **Voice** Rapid, mellow *tdu'tdu...tdu'tdu'tdup* on take-off. **SS** From other regular waders by very long straight bill, extensive white in rump, feeding behaviour. Near-identical Long-billed by plumage and bill length; note voices (see below). See Stilt Sandpiper (yellowish legs, shorter bill curved at tip, white uppertail-coverts). **Status** Common boreal winter visitor (year-round).

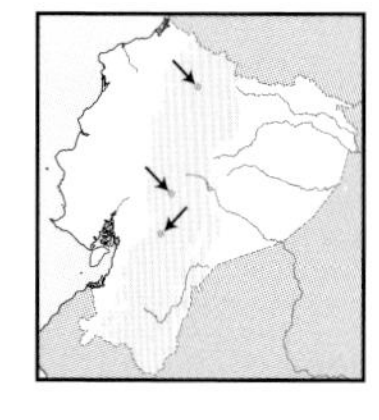

Long-billed Dowitcher *Limnodromus scolopaceus* 27–30cm

Three record from Andes (2,600–3,000m, once at Rio Napo). Freshwater marshes. Closely recalls Short-billed Dowitcher; field identification tricky even in good light. ***White dominates tail barring in Short-billed, black in Long-billed***. Long-billed somewhat darker above and on breast. Bill and legs average longer in Long-billed, but little aid in field. Juv. has tertials narrowly edged reddish-buff but fewer inner markings than Short-billed. Short-billed juv. ***brighter rufous-buff*** below. Monotypic. Poorly known in Ecuador, possibly vagrant but identification issues might obscure status. Feeds with head partly submerged, body more horizontal. **Voice** High-pitched, ringing *kík!* or *keék*, sometimes in accelerating series (H *et al.*). **SS** Voices only safe identification character, but note feeding behaviour. **Status** Accidental boreal winter visitor (Dec, Feb, Mar).

Whimbrel
juvenile
Hudsonian Godwit
non-breeding
non-breeding
Marbled Godwit
non-breeding
breeding/
early spring
Willet
juvenile
non-breeding
non-breeding
juvenile
non-breeding
non-breeding
Short-billed Dowitcher
Long-billed Dowitcher

Long-legged, elegant and slender; typical *Tringa* are rather common in coastal and highland wetlands. Commonest wader in Ecuador in variety of habitats from muddy estuarine shores and semi-forested streams, to *páramo* lakes, is Spotted Sandpiper. Also, three strictly coastal species, stocky, short-legged, mostly on rocky beaches.

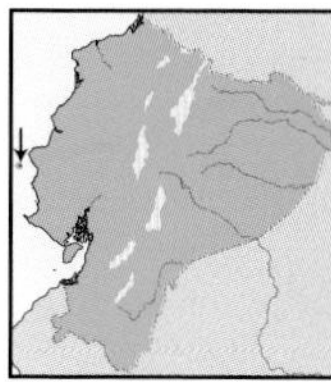

Spotted Sandpiper *Actitis macularius* 18–20cm

Widespread (to 3,800m). Rivers, streams, pools, lagoons, mudflats. ***Short greenish legs***. Greyish-brown above, white below, ***white supercilium, brownish patches on breast-sides***. Breeder: ***dark spotting below***. In flight: ***white wing stripe, dark rump and tail***. Monotypic. Singly, feeds leaning forwards, constantly bobbing rear parts. **Voice** Loud, slightly rising *peét'vít...peét-veét!...peét-veét-veét!* **SS** Note constant teetering and characteristic flight: shallow wingbeats, gliding on bowed wings. Solitary Sandpiper has legs and bill longer. **Status** Common boreal winter visitor (year-round).

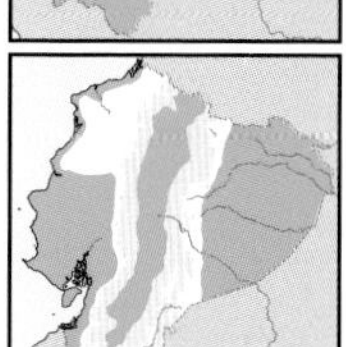

Greater Yellowlegs *Tringa melanoleuca* 29–33cm

E & W lowlands to foothills, inter-Andean valleys (to 4,500m). Brackish to freshwater ponds and lagoons, mudflats. ***Very long, slightly upturned bill, very long yellow legs***. In flight: white rump, barred tail. Monotypic. Singles or small parties with other waders; wades rather deeply, feeds by sweeping or picking. **Voice** Ringing whistles, *tew-tew-tee...tew-tew-tew-tew...* **SS** From Lesser Yellowlegs by bill: ***1.5 times head length in Greater***, slightly upturned; straight, equalling head length in Lesser. **Status** Fairly common boreal winter visitor (year-round).

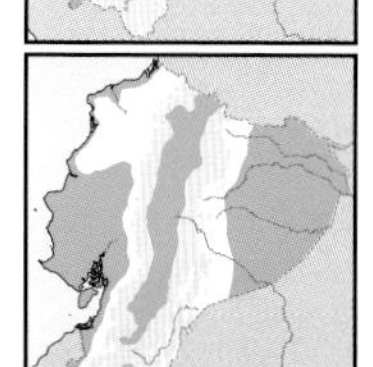

Lesser Yellowlegs *Tringa flavipes* 23–25cm

E & W lowlands to foothills, inter-Andean valleys (to 3,800m). Brackish to freshwater ponds and lagoons, mudflats. ***Long straight bill, very long yellow legs***. Plumage lie Greater Yellowlegs, separated by ***bill length and shape***. Monotypic. Singles or small parties pick from surface, elegant, high-stepping walk, bobs head when nervous. **Voice** Simple *tewp...tewp-teep*, less ringing than Greater. **SS** See also Solitary Sandpiper (smaller, darker above, short greenish legs), Stilt Sandpiper (shorter, greener legs, streakier underparts, decurved bill at tip, long supercilium), Upland Sandpiper (short bill). **Status** Uncommon boreal winter visitor (year-round).

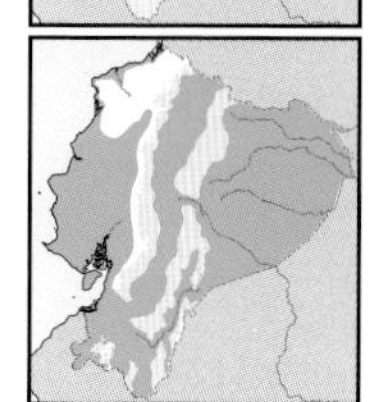

Solitary Sandpiper *Tringa solitaria* 18–21cm

Widespread (to 3,600m). Shallow freshwater ponds. ***Straight needle-like bill, greenish-yellow legs. Dark*** brownish-grey above to chest-sides, ***short white supraloral stripe, pale eye-ring***. In flight: ***all-dark upperparts, no wing stripe, dark grey underwing***. Races *solitaria* and *cinnamomea*. Mostly singles or small parties feeding slowly and deliberately, nodding head when nervous. **Voice** Sharp *peet'weet!* (H *et al.*). **SS** Smaller, shorter-legged than yellowlegs. Spotted Sandpiper paler above, bill and legs shorter, longer supercilium, white wingbar and underwing; constant bobbing. **Status** Uncommon boreal winter visitor (Jul–Apr).

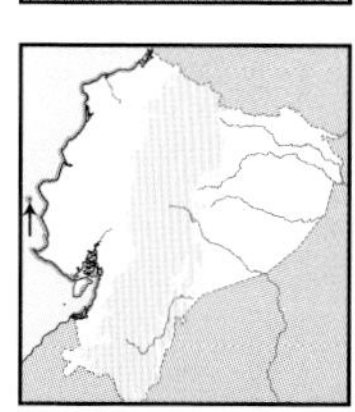

Wandering Tattler *Tringa incana* 26–29cm

Rocky coasts. ***Straight grey bill, short yellow legs***. Mostly ***pale grey, obvious white supercilium***. Breeder boldly barred / scaled below. In flight: ***plain upperwing and upperparts, dark grey underwing***. Monotypic. Singly on tideline, probing crevices and holes, constantly bobbing rear. **Voice** Rarely heard in noisy habitat, but gives plaintive short trill. **SS** Darker grey, stockier Surfbird has short bill, greener legs. Spotted Sandpiper shorter-billed, browner above, no grey chest; Solitary Sandpiper slimmer, greener legs, browner above. **Status** Uncommon boreal winter visitor (year-round).

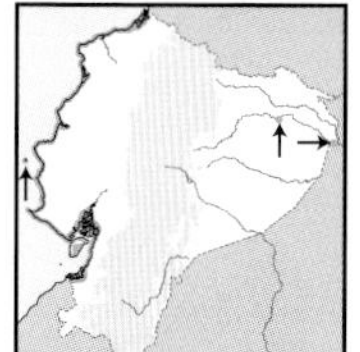

Ruddy Turnstone *Arenaria interpres* 21–25cm

Coasts, few records inland (below 200m). Sandy and rocky beaches, brackish wetlands. ***Short wedge-shaped bill, short orange legs***. Chunky, ***rufous to rufescent-brown upperparts, black to dusky chest, white belly***. Complex ***black-and-white head pattern***. In flight: *white back, scapulars and rump, broad black tail-band*, white wing stripe. Race *morinella*. Small groups purposefully turn over seaweed and rocks for prey. **Voice** Fast staccato *tuk-kl-tuk*; softer *puk'puk'puk...* calls. **SS** Confusion unlikely; see Surfbird (short yellow legs, no rufous above). Black Turnstone not recorded in mainland Ecuador. **Status** Fairly common boreal winter visitor (year-round).

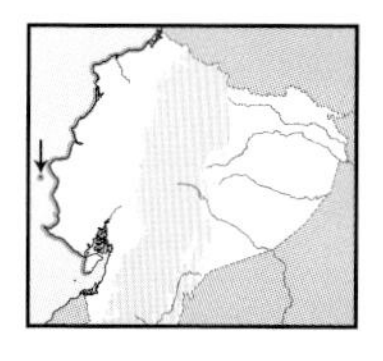

Surfbird *Aphriza virgata* 23–25cm

Rocky coastal beaches. ***Short conical bill, short greenish-yellow legs***. Stocky. ***Grey head, upperparts and chest***. Breeder ***boldly streaked*** where non-breeder grey, ***cinnamon scapulars***. In flight: ***white rump and uppertail-coverts, broad black tail-band***, white wingbar. Monotypic. Small groups often with Ruddy Turnstone, search for prey on seaweed and rocks at tideline. **Voice** Seldom vocal. **SS** Wandering Tattler greyer, yellower legs, longer bill. **Status** Rare boreal winter visitor (mostly Aug–Apr).

non-breeding
breeding
Spotted Sandpiper
non-breeding
Greater Yellowlegs
non-breeding
Lesser Yellowlegs
non-breeding
breeding
Solitary Sandpiper
juvenile
Wandering Tattler
non-breeding
juvenile
non-breeding
breeding
Ruddy Turnstone
juvenile
non-breeding
Surfbird

PLATE 36: MEDIUM-SIZED SANDPIPERS II

Mid-sized sandpipers with longish curved bills, long legs or both. Only Pectoral and Stilt Sandpiper are regular on Ecuadorian coast (rarer in highlands). The other four species are casual visitors. Identification can get complicated by gradual variation between non-breeding and breeding plumages, but breeders are less likely in Ecuador.

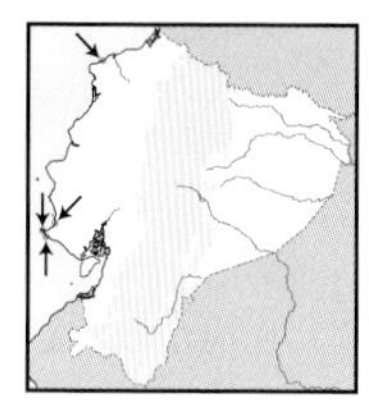

Red Knot *Calidris canutus* — 23–25cm

Few records (mostly Santa Elena Peninsula). Mudflats. ***Bulky***, fairly ***large, heavy, nearly straight bill, olive legs***. Dirty grey above, scaled appearance with ***dark-centred back feathers***. Whitish below, ***chest to flanks finely scaled dusky***. In flight: ***long-winged***, narrow white wing stripe, ***pale grey rump and tail***. Race *rufa*. Compact flocks probe and pick in shallow water; in Ecuador small parties or singles. **Voice** Mostly silent. **SS** Larger, more robust than most *Calidris*, recalls Curlew Sandpiper but bill shorter, straighter, no white rump. Pectoral has sharply delimited breast, yellow legs, longer neck. Rarer Dunlin darker above, longer bill drooped at tip, black legs. **Status** Rare boreal winter visitor (mostly Aug–Apr); NT globally.

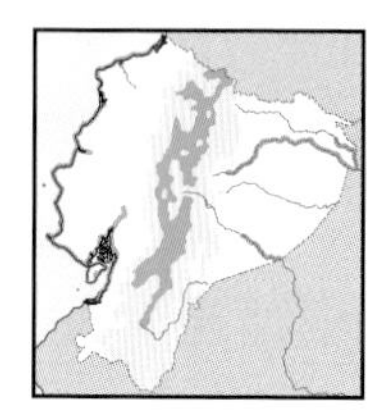

Pectoral Sandpiper *Calidris melanotos* — 20–23cm

Coasts to adjacent lowlands, Andes, E lowlands (to 3,800m). Grassy marshes, semi-vegetated muddy shores, sandbars, freshwater wetlands. ***Blackish bill faintly decurved, long neck, yellow legs. Brownish*** above finely streaked, buffier breast streaked dusky, ***sharply demarcated from white belly***. In flight: vague wing stripe, ***broad black rump centre***. Monotypic. Small, rather loose flocks probe and pick; more purposeful than smaller congeners. **Voice** Harsh, throaty *kiirr'k*. **SS** Sharply demarcated chest and yellow legs distinctive; Stilt Sandpiper has longer, more curved bill, darker legs, bolder supercilium. See also uncorroborated Sharp-tailed and short-billed, long-necked Upland Sandpipers. **Status** Uncommon boreal winter transient (mainly Jul–Apr).

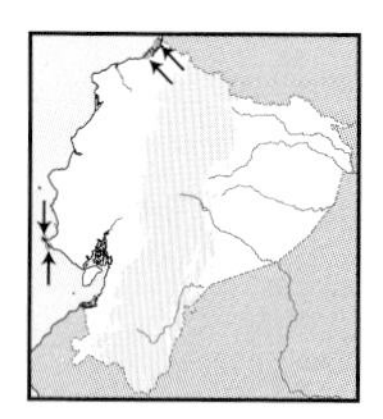

Dunlin *Calidris alpina* — 18–22cm

Few records on coastal mudflats. ***Long, black, decurved bill, longish black legs. Dirty grey*** above, paler breast, white belly, short white eyebrow. In flight: clear white wing stripe, ***white rump-sides***, dark tail. Possibly race *pacifica*. Small groups (large in regular wintering grounds) probe and pick in shallow water. **Voice** Harsh *treeep* trills (H *et al.*). **SS** Hunchbacked silhouette fairly distinctive; cf. rare Curlew Sandpiper and Red Knot; former has noticeably curved bill, clearer supercilium, longer legs; Red Knot (greyer, olive legs, straighter bill); smaller Western Sandpiper plainer grey above, bill finer, shorter. **Status** Rare boreal winter visitor (Dec–Feb, Apr).

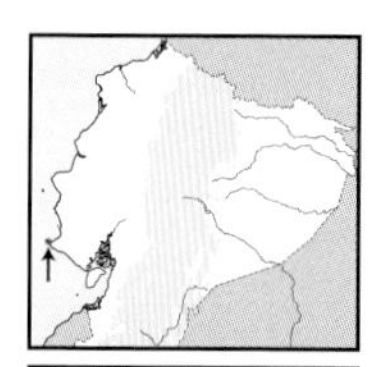

* Sharp-tailed Sandpiper *Calidris acuminata* (not illustrated) — 18–21cm

Single uncorroborated record (Santa Elena Peninsula: B. Haase). Brackish mudflats. Recalls Pectoral Sandpiper but shorter-billed, longer-legged. Breast ***not*** as sharply demarcated, ***darker crown, cleaner and bolder supercilium, more obvious eye-ring***. In flight show ***less white*** in rump-sides. Monotypic. Behaviour also recalls Pectoral. **Voice** Softer than Pectoral, *pheep* or *treet*, often faster sequence (H *et al.*). **SS** See Pectoral and other medium-sized *Calidris*. **Status** Accidental boreal vagrant (May).

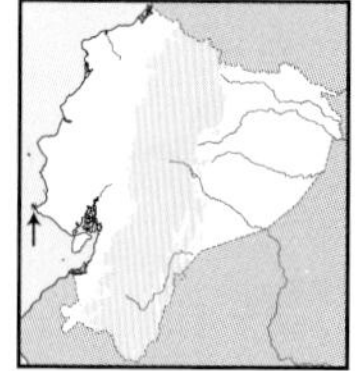

Curlew Sandpiper *Calidris ferruginea* — 19–23cm

Two records (Santa Elena Peninsula). Mudflats. ***Long, decurved bill***, shortish black legs. Pale grey above, more rufescent tone; ***long white supercilium***, blurry streaking on chest-sides. In flight: ***white wingbar, broad white rump patch***, pale grey tail. Monotypic. Probes wet mud, picks prey from surface, also wades in deeper water than most *Calidris*. **Voice** Clear *krirrup*. **SS** Similar-sized Stilt Sandpiper has greenish legs, straighter bill, streakier rear underparts; Red Knot's supercilium less bold, bill straighter, darker legs, more scaly underparts. Much smaller Western Sandpiper (finer, shorter, less decurved bill, no obvious eyebrow); White-rumped (narrower white rump patch, weaker wing stripe). **Status** Accidental boreal vagrant (Jul–Aug, Jan).

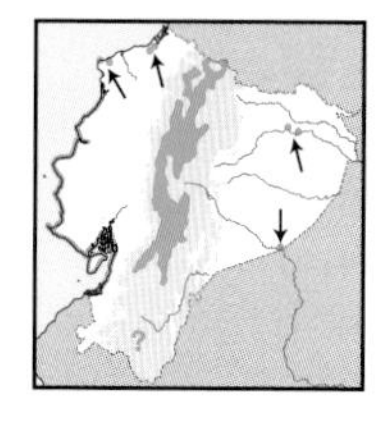

Stilt Sandpiper *Calidris himantopus* — 19–23cm

Scattered records from coast, highlands and lowlands (to 3,800m). Estuaries, mudflats, brackish lagoons, muddy shores. ***Long blackish bill, straight at base drooping at tip, long greenish legs***. Dirty grey above, blurry greyish streaking on flanks, ***long white supercilium***. In flight: virtually ***no wingbar, narrow white band in uppertail-coverts, legs partially project beyond tail***. Monotypic. Small parties wade in belly-deep water, dowitcher-like, with purposeful movements. **Voice** Soft, rattled *kir'irt... kir'kit-kit*. **SS** Pectoral Sandpiper distinctive by clearly demarcated breast, brighter yellow legs. Recalls Short-billed Dowitcher, but bill shorter, more curved, brighter legs, smaller white rump not extending onto back. **Status** Uncommon boreal winter visitor (mostly Jul–Apr).

Red Knot
juvenile
non-breeding
juvenile
non-breeding
Dunlin
Pectoral Sandpiper
non-breeding
juvenile
juvenile
non-breeding
non-breeding
Curlew Sandpiper
Stilt Sandpiper

Commonly seen stints are more regular at coastal mudflats and estuaries. Sanderling is quite distinctive but care needed with other species. Identification complicated by patterns of moult between breeding, non-breeding and juvenile plumages. Note size, leg colour, bill shape and proportions, rump and uppertail patterns, and in some, wing pattern. Further complexity is provided by their flocking behaviour.

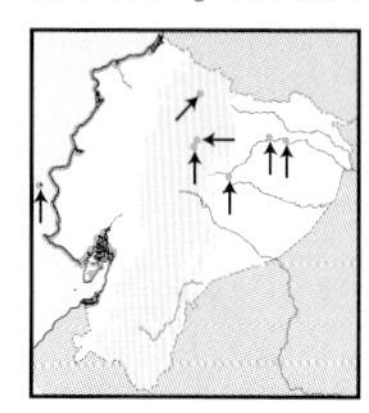

Sanderling *Calidris alba* 20–21cm

Coasts, few inland records (to 300m, once 2,200–2,350m). Mostly sandy beaches, sometimes mudflats. Black straight bill, black legs, ***lacks hind toe. Plain pearl-grey above, immaculate white below***. In flight: ***bold white wingbar*** contrasts with blackish wingtip, white rump-sides, ***pale grey tail***, white underwing. Marbled breeding plumage. Monotypic. Small compact flocks dash back and forth with wave line picking prey. **Voice** Liquid, soft *pwíp* or *twíp!* **SS** Extensive white below and wingbar distinctive, as behaviour. Pale individuals of smaller and slimmer Western and Semipalmated Sandpipers less white below, with fainter wingbars. **Status** Common boreal winter visitor (year-round).

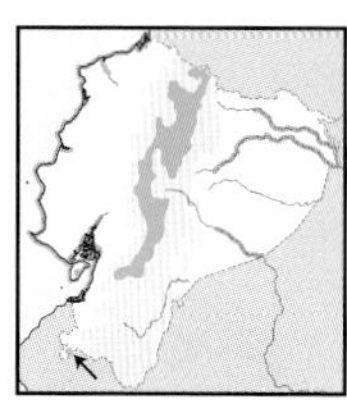

Least Sandpiper *Calidris minutilla* 13–15cm

Coast to adjacent lowlands, E lowlands, highlands (to 3,800m). Shallow ponds, mudflats, lagoons, muddy shores. ***Greenish-yellow to pale orange legs, fine black bill, slightly drooping tip. Brownish*** above with dark streaking, breast streaked dusky, white below. In flight: vague wing stripe, white rump-sides. Monotypic. Flocks with other waders, less fond of water's edge or bare mud, usually in semi-vegetated areas; rather tame, rising steeply if threatened. **Voice** High-pitched, shrill, rising *kreéyp* or *trre'eép*. **SS** Smallest 'stint', yellowish legs and fine bill distinctive; looks hunchbacked and short-necked, browner than Western and Semipalmated. **Status** Common boreal winter visitor (year-round).

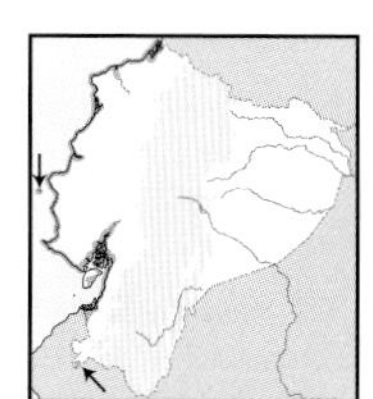

Semipalmated Sandpiper *Calidris pusilla* 13–15cm

Mainly mudflats, brackish wetlands on coast. ***Straight black bill, deep-based, blunt-tipped***, black legs. Brownish-grey above, white below, brownish-grey streaking on chest-sides. In flight: ***narrow wingbar***, white flanks, black tail. Monotypic. Flocks with other waders, including Western Sandpiper, probing muddy shores. **Voice** Harsh *chürp* or *kürp-kür'r'r'r*, also sharper *kyip!* **SS** Difficult from Western by plumage. Western's bill longer (though some overlap) and droop-tipped. Least has yellowish legs, finer bill; larger White-rumped and Baird's have dusky-streaked breasts, longer bills, longer legs. **Status** Fairly common boreal winter visitor (year-round); NT.

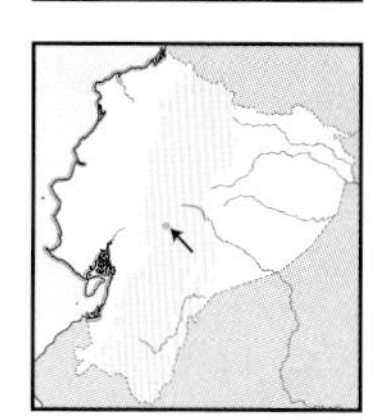

Western Sandpiper *Calidris mauri* 14–16cm

Mainly coastal, once at 3,000m. Mudflats, brackish wetlands. ***Rather long bill drooped at tip***, black legs. Brownish-grey above, streaky chest-sides, white below. In flight: as Semipalmated. Monotypic. With other waders on muddy shores, probing, often wading deeper than Semipalmated. **Voice** Shrill, sharp *jeet*. **SS** Very subtle plumage differences in non-breeding Western and Semipalmated; ***note bill shape rather than length***: deeper base in Semipalmated, drooped tip in Western. Least has yellowish legs, shorter, finer bill; from larger Dunlin (longer bill, darker grey upperparts); see White-rumped and Baird's (straighter bill, browner upperparts, streakier chests); Baird's has grey tail, vague wing stripe. **Status** Fairly common boreal winter visitor (year-round).

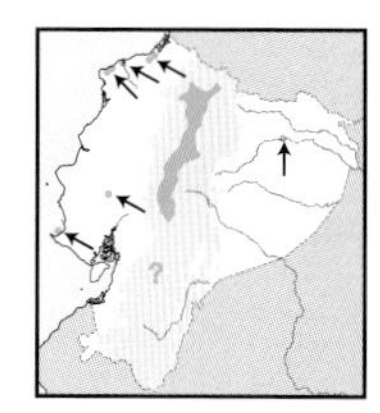

White-rumped Sandpiper *Calidris fuscicollis* 16–18cm

Scattered records (100–3,800m) at marshes and muddy shores. ***Black, nearly straight bill, yellowish base***, blackish legs, ***long wings extend beyond tail tip. Brownish***-grey above ***coarsely streaked dusky***, long pale supercilium, dusky breast streaking ***extends to flanks***. In flight: ***contrasting white uppertail-coverts***, *weak* wing stripe. Monotypic. Flocks favour muddy areas with other waders. **Voice** High, squeaky *ëeet-ëeet-ëet-et-et*, short *teep* calls. **SS** At rest recalls Baird's but bolder supercilium, less buffy and less streaked above; white rump conspicuous in flight; cf. much rarer Curlew Sandpiper with longer, more curved bill, whiter underparts and eyebrow, broader white rump; latter only reported in SW lowlands. **Status** Rare boreal winter transient (Jul–Oct, Mar).

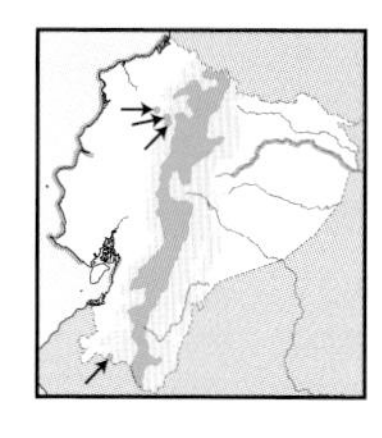

Baird's Sandpiper *Calidris bairdii* 15–18cm

Coast, Andes, scattered E lowland records (to 4,000m). Shallow ponds, marshy lagoons, mudflats. ***Straight black bill***, black legs, ***long wings. Buffy-brown above***, head and ***chest buffy-brown streaked dusky, back edged pale***. In flight: ***dark rump, vague*** wingbar. Monotypic. Small flocks with other waders, less fond of water's edge, more often in vegetated zones. **Voice** Low *prëeet...prëet-ëet'ëet*. **SS** Long-winged as White-rumped (plainer upperparts, bolder supercilium, streaky flanks, conspicuous white rump in flight); see smaller Western and Semipalmated Sandpiper, both greyer, Western's bill drooped at tip, Semipalmated deeper at base. **Status** Uncommon boreal winter transient (mostly Jul–Nov, Apr–Jun).

juvenile
Sanderling
non-breeding
juvenile
non-breeding
Least Sandpiper
juvenile
non-breeding
Semipalmated Sandpiper
juvenile
non-breeding
Western Sandpiper
juvenile
non-breeding
White-rumped Sandpiper
juvenile
non-breeding
Baird's Sandpiper

Three slender, short-billed sandpipers found mostly in grassy fields, not necessarily near water; Ruff is accidental, Buff-breasted rarer than Upland. Phalaropes are distinctive, two of them are mainly found at sea. Regularly swim.

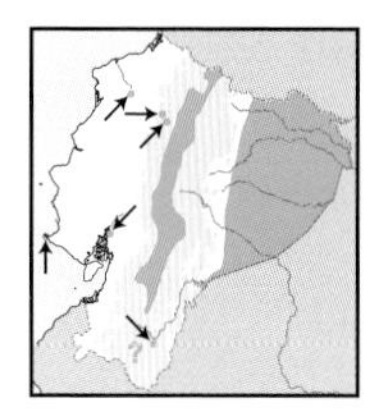

Upland Sandpiper *Bartramia longicauda* — 28–32cm

Andes and E lowlands (200–4,000m). Dry and marshy grassland. ***Short straight bill***, pale mandible, ***yellow-ochre legs, long neck, long tail protrudes beyond folded wingtips, small head. Buffy*** head, neck and upperparts streaked dusky, ***white belly***. In flight: dark rump and flight feathers, buff rump-sides, ***buffy underwing and axillaries***. Monotypic. Singles or small groups, graceful flight, often holds wings aloft on landing, walks nodding head. **Voice** Rather mellow *fuü-fii-fii-füüt, fui-it'it'it....* **SS** Smaller, rarer Buff-breasted has uniform buffy-ochre underparts, shorter neck, shorter tail. **Status** Rare boreal winter transient (Aug–Oct, Mar).

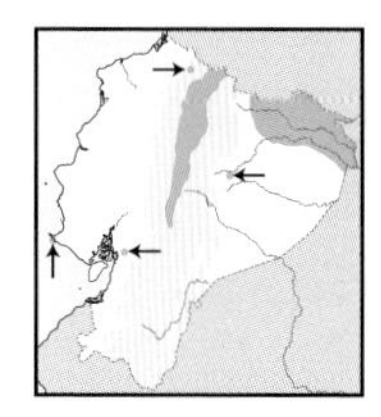

Buff-breasted Sandpiper *Tryngites subruficollis* — 18–20cm

N Andes, few lowland records (to 4,000m). Meadows, grassy marshes, sandbars. ***Short straight black bill, short yellow legs. Ochre-buff face and underparts, brown crown and upperparts coarsely streaked buff***, faint pale supercilium. In flight: ***bold white underwing***. Monotypic. Singles or small flocks, not necessarily near water, high-stepping walk, rather tame. **Voice** Not very vocal, a low growling *pr-r-r-reet* reported (H *et al.*). **SS** Nearly unmistakable; see larger, more elegant Upland with long neck, small head, long wings, white belly. **Status** Rare boreal winter transient (Jul–Oct, Mar–Apr), NT.

* Ruff *Philomachus pugnax* — 22–32cm

One uncorroborated record in W lowlands (R. Ahlman). ***Short bill with pale base***, rather ***short yellowish legs, small head. Humpbacked.*** Upperparts ***scaled***, neck streakier; underparts mostly whitish. In flight, ***V-shaped white rump-band***. Juv. similar but buffier. Monotypic. Elsewhere, regular in grassland, ricefields and similar habitat, often walking and pecking. **Voice** Not expected to vocalise. **SS** From Buff-breasted Sandpiper by whitish in face, scaled back, pale bill base, longer neck, longer bill, white rump; Pectoral Sandpiper has dull streaked breast sharply cut; Stilt Sandpiper is longer-legged, longer-billed, not scaled above. **Status** Accidental (Aug).

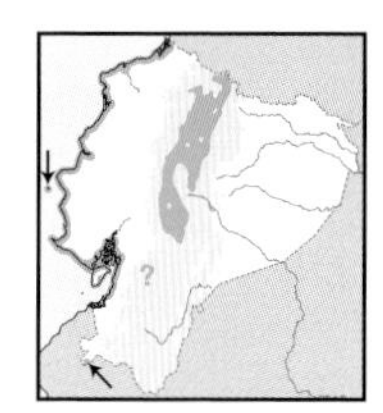

Wilson's Phalarope *Phalaropus tricolor* — 22–24cm

Coast and Andes (to 3,800m). Freshwater and brackish wetlands. ***Long slender black bill, yellow legs. Plain grey above, white below, pale grey postocular patch***. In flight: ***no obvious wing stripe, white rump***. Breeder shows combination of bluish-grey, black, chestnut, buff and brown. Monotypic. Large flocks spin in water, pick prey from surface, wade in mud; compact flocks in flight. **Voice** Not very vocal, a soft *aangh* reported (H *et al.*). **SS** Finer and longer-billed than Red-necked and Red Phalaropes (fainter ear-covert patch, somewhat paler upperparts, no wing stripe, white rump); more coastal than congeners, only phalarope inland. See taller Lesser Yellowlegs (brighter and longer legs, no postocular patch, less white below, pale eye-ring). **Status** Common boreal winter visitor (year-round).

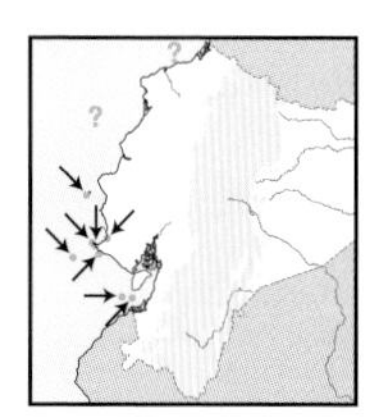

Red-necked Phalarope *Phalaropus lobatus* — 18–19cm

Offshore and coasts. ***Black needle-like bill***, bluish-grey legs. ***Sooty-grey above, some white feather edges give shaft-like appearance***, black postocular patch and crown. In flight: white wing stripe and ***rump-sides***. Monotypic. Rare on coast, never inland, migrates at sea, where forms rather large and compact flocks; flies low over water. **Voice** Not very vocal; single *twik* in flight reported (H *et al.*). **SS** Wilson's larger, with fainter face pattern, plainer upperparts, no wing stripe, yellow legs, longer bill; very similar to Red (bill shorter and thicker, plainer upperparts, pale edges to median/lesser coverts); note Red-necked has more erratic flight with fast wingbeats and rapid twists. **Status** Rare boreal winter visitor (year-round).

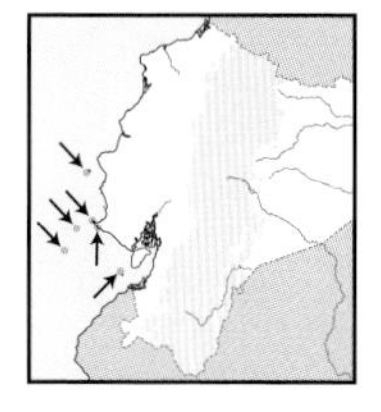

Red Phalarope *Phalaropus fulicarius* — 20–22cm

Offshore and coasts. ***Short thickish bill, pale yellow base***, bluish-grey legs. ***Plain grey above***, white below, blackish postocular patch and crown. In flight: resembles Red-necked in wing stripe but ***rump darker***. Monotypic. Small flocks gather at upwelling zones, seizing tiny prey from surface; less numerous than congeners. **Voice** Not very vocal. **SS** Bill shorter, deeper-based, less pointed than congeners; upperparts plainer than Red-necked, ***heavier ear-coverts patch***. Not safely identified at sea, but note plainer upperparts and slower wingbeats, less twisting flight. **Status** Rare boreal winter visitor (Jun–Feb, Apr; year-round?). [Alt: Grey Phalarope]

juvenile
non-breeding
Buff-breasted Sandpiper
Upland Sandpiper
♀
juvenile
♂
juvenile
♀
non-breeding
Ruff
non-breeding
juvenile
moulting
♀
breeding
Wilson's Phalarope
adult moulting
♂
breeding
♂
breeding
first-winter
♂
moulting
non-breeding
♀
breeding
Red-necked Phalarope
♀
moulting
moulting
♂
breeding
first-winter
Red Phalarope
non-breeding
♀
breeding

PLATE 39: MONOTYPIC GRUIFORMES AND SEEDSNIPES

Four odd Gruiformes, three living in or near water, one in Amazonian forests away from water. Limpkin and Sunbittern distantly recall herons. Sungrebe, as its name suggests, recalls a slender grebe even in behaviour, but plumage pattern differs notably. The trumpeter recalls a strange, tall, hunchbacked curassow. It is often kept as a pet locally owing to its fierceness in defending territories. The superb seedsnipes occupy open rather barren terrain.

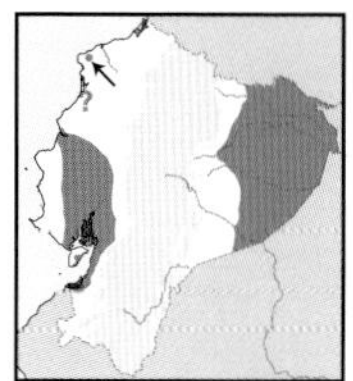

Limpkin *Aramus guarauna* 60–70cm

Lowlands of SW, locally in NE (to 400m). Freshwater marshes, edges and adjacent agricultural fields. ***Long, slightly curved, bicoloured bill***, long blackish legs, long neck. Mostly ***dusky-brown, head to mantle profusely streaked white***. Sexes alike. Nominate. Solitary marsh dweller, easy to see where not persecuted but rather wary; feeds largely on snails; in flight legs often held dangling. **Voice** Loud *ca-rra-ouu, ca-ca-rraoo..kha-kha-kha....* **SS** Superficially resembles juv. ibises (Scarlet, White and Glossy) or night-herons, but see bill shapes. **Status** Uncommon.

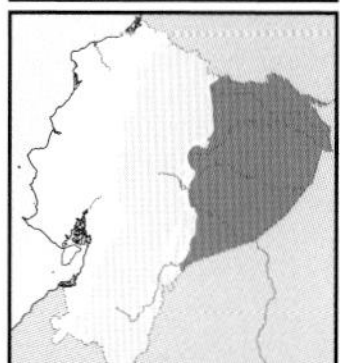

Grey-winged Trumpeter *Psophia crepitans* 45–62cm

E lowlands (to 700m, locally higher). Inside primary *terra firme* forest. Unmistakable ***hunchbacked appearance***, short, ***thick*** bill, ***short*** and strong greenish legs. ***Head to mantle and underparts glossy black***, glossy purple on foreneck. Head and neck feathers short, ***velvet-like***. ***Pale ochraceous lower back, pale grey rear parts***, including wings. Sexes alike. Race *napensis*. Small groups on forest floor, mostly taking fruit, also invertebrates and small vertebrates; territorial, cooperative breeders, shy around settlements. **Voice** Deep booming *hhummmm, hummm, hmmm-hmmm-hmmm-hmmm*; harsh *jjkk-jjkk* in alarm. **SS** Unmistakable. **Status** Rare, mostly in remote areas, VU in Ecuador, NT globally.

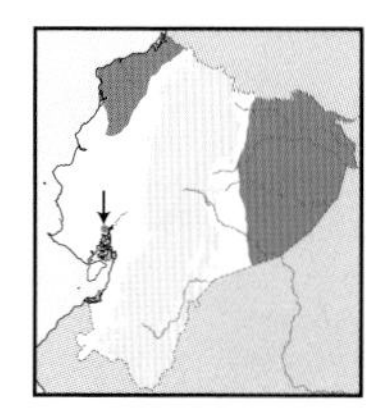

Sungrebe *Heliornis fulica* 28–30cm

E & W lowlands (to 500m). Slow-flowing rivers, streams, freshwater lakes, ponds with abundant vegetation. Unique, with ***red bill*** (♀) or bicoloured (♂), ***yellow-and-black legs***. Crown to neck-sides ***black, white supercilium and neck stripe, buff ear-coverts patch*** (♀), white throat and foreneck. Otherwise dusky-brown above, whitish below. Monotypic. Moves neck back and forth when swimming, often under overhanging vegetation cover; not good fliers but might perch on branches. **Voice** Honking *kro-kro-khu-guA...*, *ko-koo-koo...go-koo, khoo-khoo-khoo*; *gk-gk-gk* grunts in alarm. **SS** Unmistakable. **Status** Rare, rarer in W.

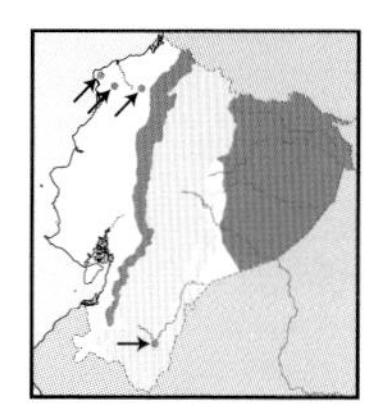

Sunbittern *Eurypyga helias* 42–48cm

E lowlands to foothills, W Andean foothills and subtropics (to 1,000m in E, 500–1,700m in W). Forest streams, oxbow lakes, lagoons. ***'Horizontal'*** slender body, ***long bicoloured bill. Head black, white face stripes***. Neck to underparts tawny, densely vermiculated dusky. ***Brown upperparts banded tawny***, long tail barred grey-white, with ***two black-and-chestnut bands***. Open wings with ***chestnut, black and buff 'sunspots'***. Race *helias* (E lowlands), *meridionalis* (E foothills), *major* (W). Solitary, walks along forested streams, flies to high perches when threatened, leaning back and forth. **Voice** High-pitched whistle *kuuúiiik* or *kuu'úuuk*. **SS** Unmistakable. **Status** Uncommon.

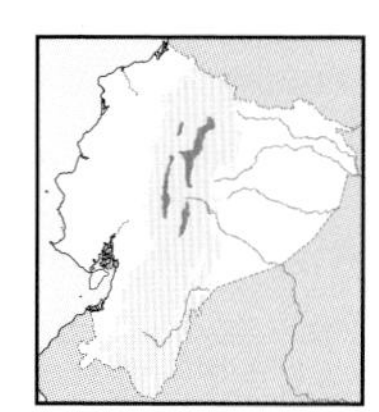

Rufous-bellied Seedsnipe *Attagis gayi* 27–31cm

Highlands from Pichincha S to Azuay (4,000–4,600m). Open *páramo* grassland and mostly barren terrain. ***Plump***, short yellowish legs, short conical bill. Complex, striking head, breast and upperparts: ***blackish, buff and chestnut vermiculations, scalloping, edging and freckling***. Rest of underparts ***rich rufous***, including ***underwing-coverts***. Endemic *latreillii* race possibly valid species. Pairs to small groups very inconspicuous, especially when they crouch and freeze, allowing close approach. **Voice** Calls *wulla-wulla-wull-wull-wull...* in alarm, *wull-wull-wü-wü* in flight. **SS** Confusion unlikely; Curve-billed Tinamou has very different silhouette. **Status** Rare but overlooked.

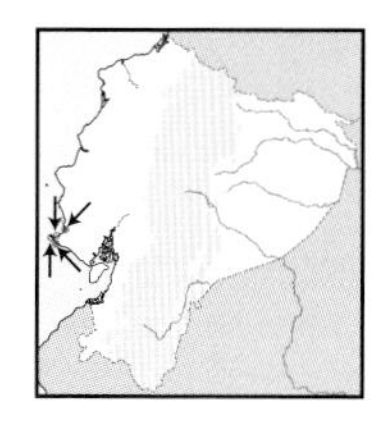

Least Seedsnipe *Thinocorus rumicivorus* 16–19cm

Local on Santa Elena Peninsula. Open barren ground. Small with ***short conical yellow bill***, very short yellow legs. ***Grey foreparts***, white throat outlined ***black, black median line on breast***. Pale brown above, with complex scalloped pattern. Belly to vent ***pure white, black underwing-coverts***. ♀ lacks grey and black line. Race *cuneicauda*. Hard to see; pairs to small groups walk, crouch and freeze; long unrecorded in Ecuador until rediscovery (L. Navarrete, Feb 2003); might only be an accidental visitor. **Voice** Calls *juk-juk* in flight (RG). **SS** Unmistakable. **Status** Very few records (Jan–Feb, Apr–May, Jul); very rare visitor possibly extinct as resident.

Limpkin
Grey-winged Trumpeter
♂
♂
Sungrebe
♀
helias
major
meridionalis
Sunbittern
♂
♀
Rufous-bellied Seedsnipe
Least Seedsnipe

PLATE 40: RAILS AND WOOD-RAILS

Wood-rails prefer cover, two live inside humid forests not necessarily near water. They are rarely seen, but their voices are loud and distinctive. Rufous-necked might emerge from mangroves with tides, as does Grey-necked in equivalent habitats of Amazonian lowlands. All wood-rails walk with head upright. *Rallus* rails move in more typical rail fashion in marshes and muddy mangrove understorey.

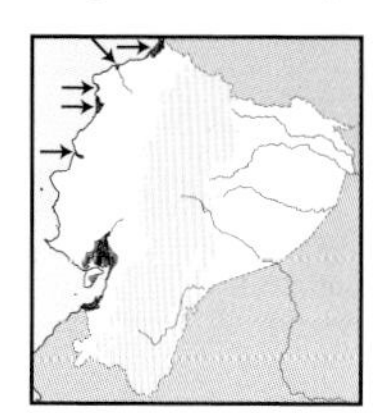

Clapper Rail *Rallus longirostris* — 31–38cm

Scattered records in mangroves. Dusky and pink, slightly curved bill, pinkish legs. Chestnut crown to hindneck, *greyish-buff face and neck grading into buffy-brown breast. Short whitish supercilium, white throat. Greyish*-brown flanks to vent, *banded white*. Olive-brown above, profusely streaked grey. Race *cypereti*. Poorly known and rare in Ecuador; solitary, forages on muddy floor, might emerge on exposed banks; diurnal and crepuscular. **Voice** Not well known in Ecuador, elsewhere a rapid, loud, accelerating *kák* or *chák* series. **SS** Sympatric wood-rails are boldly coloured, larger, unbarred below; Virginia Rail only in highlands. **Status** Rare, local, VU in Ecuador.

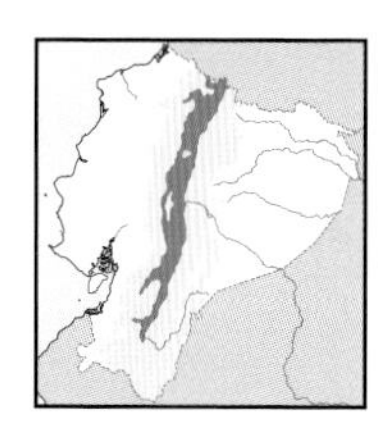

Virginia Rail *Rallus limicola* — 20–25cm

Andes (2,200–3,800m). Freshwater marshes, reedbeds, swampy areas. Smaller than previous species. Bill mostly *orange-pink. Grey face*, short whitish supraloral stripe, *white throat. Greyish-buff neck to belly*, vent banded black and white. Dusky-brown above, back densely streaked fuscous, wings unstreaked, with *rufous in coverts*. Juv. *sootier, browner* below. Race *aequatorialis* (probably separate species). Skulks in dense reeds and tall grass, rarely emerging in open, mainly early in morning or when overcast, difficult to flush. **Voice** Fast, descending, nasal *keh-keh…kr*, more metallic than Clapper Rail. **SS** Only rail in range, might co-occur with very different Sora or even Plumbeous Rail. **Status** Rare.

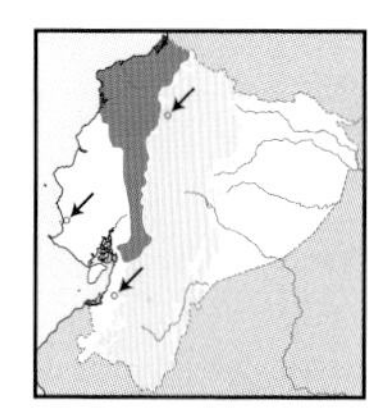

Brown Wood-Rail *Aramides wolfi* — 32–36cm

W lowlands to foothills (to 1,300m, but mostly lowlands). Swampy areas, forested streams, secondary woodland. Long yellowish bill, pink legs. *Greyish-brown head and neck*, whitish throat. Otherwise *brown*, flight feathers rich rufous, *blackish belly to vent*. Monotypic. Singles or pairs in muddy and swampy forested areas; territorial, apparently monogamous, favours secondary forest for nesting. **Voice** Nasal *kijui-kijui-kijui…*, repeated *kui-có, kui-có….* **SS** Sympatric Rufous-necked Wood-Rail richer rufous, sharper vocalisations, more metallic voice, less cackling. **Status** Rare, local, declining; VU globally, EN in Ecuador.

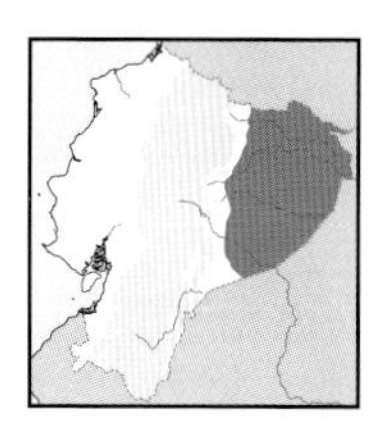

Grey-necked Wood-Rail *Aramides cajaneus* — 33–40cm

E lowlands (to 400m). Swampy areas, marshes, streams and lagoons. Long yellowish bill, pink legs. *Grey head to upper mantle and chest*, whitish chin. Contrasting *cinnamon upper belly*, black belly and vent. Olive-brown upperparts, *rufous-chestnut flight feathers*. Juv. duller, *sootier*. Nominate. Emerges to feed on muddy banks and adjacent shoreline, solitary or in pairs; walks rather than flies when threatened. **Voice** Loud cackles: *klee-ko-keé, kow-kow-kow, klö-kleyee-klö-kleyee-klö…, gö-chi-kokó-chi-kokó-chi-kokó…, kre-kó, kre'e-kó-kre'e-kó…* and others. **SS** Confusion unlikely; cf. rarer Red-necked Wood-Rail (extensive grey below, chestnut neck sides). **Status** Uncommon.

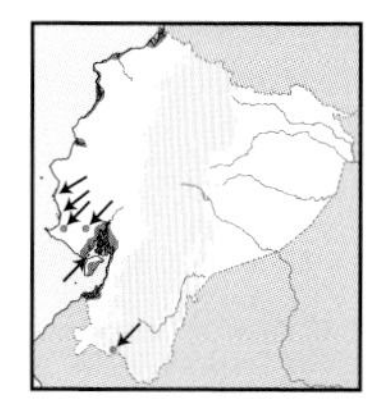

Rufous-necked Wood-Rail *Aramides axillaris* — 28–31cm

Local in coastal mangroves, locally breeds at 1,400m along dry forest streams. Long yellowish bill, pink legs. *Head neck, breast to upper belly and flanks rich rufous, white throat*, black vent, rump and tail. Grey upper mantle, rest of upperparts olive-brown. Duller juv. Monotypic. Singles or pairs often emerge from mangroves onto muddy banks and shores, more in open than Brown Wood-Rail, often walking over low mangrove branches. **Voice** Penetrating *kyór-kyór…kiy-kor, kiy-kor…* or *pik-pik-pik…* series. **SS** Darker Brown Wood-Rail or Clapper Rail (barred belly and flanks, smaller). **Status** Rare, probably declining due to extensive mangrove destruction; EN in Ecuador.

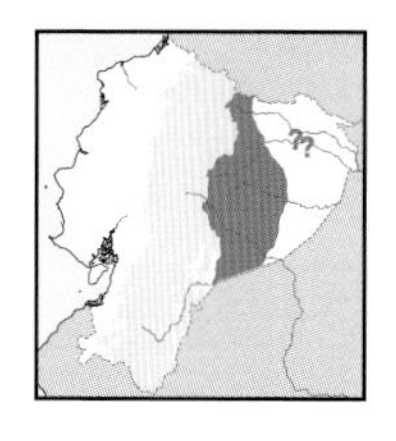

Red-winged Wood-Rail *Aramides calopterus* — 31–35cm

Locally in E lowlands and (mainly) foothills (250–1,200m). Streams in humid forest. Long greenish-yellow bill, pink legs. Upperparts mostly dark brown, *rich rufous-chestnut neck-sides, wing-coverts and flight feathers*. Whitish chin, *bluish-grey underparts*, blackish belly to tail. Juv. duller, *browner*. Monotypic. Seemingly very rare and local, not observed outside forest; nest reported from hilly terrain surrounding permanent swamp. **Voice** Apparently a sharp *keeeéEE-yut* or more tremulous *keeEE-yuuk*; also *klá, klá* calls. **SS** Commoner Grey-necked Wood-Rail (entire head and neck grey; voice not clacking). **Status** Rare and local.

immature
Clapper Rail
adult
Virginia Rail
Brown Wood-Rail
Grey-necked Wood-Rail
adult
juvenile
Rufous-necked Wood-Rail
Red-winged Wood-Rail

PLATE 41: CRAKES

Small rails in dense and tangled marshes, reedbeds, lake margins (one mainly along slow-flowing forest streams). All notably difficult to see. Very vocal, but explosive voices superficially similar. Laterally-compressed to facilitate skulking movements through dense vegetation, bills short and stout; body and head held forwards when walking. Some recent taxonomic changes based on molecular studies.

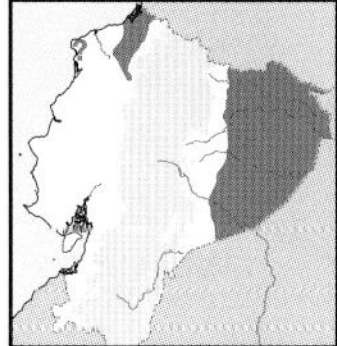

Uniform Crake *Amaurolimnas concolor* 20–23cm

E & NW lowlands (to 300m). Slow-flowing forest streams and swampy areas. ***Bill greenish-yellow***. Mostly ***uniform brown,*** paler lores, throat and underparts. Race *guatemalensis* (W) more olive above; *castaneus* (E). Very difficult to see as skulks in dense vegetation, often in pairs, difficult to flush. **Voice** Loud whistles: *dooee- dooee-DOOee, DOOEE-DOoee-dooee-dooee*…; sharp *tsjék* calls (O. Jahn). **SS** In E, Chestnut-headed Crake has bicoloured bill, chestnut-and-brown body; voices differ; wood-rails are longer-necked, longer-legged. **Status** Rare, local, likely overlooked.

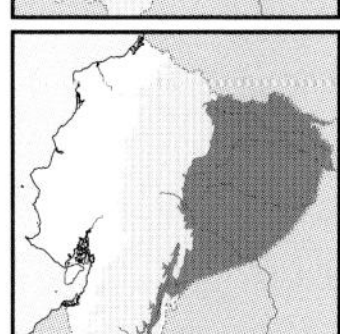

Chestnut-headed Crake *Anurolimnas castaneiceps* 19–22cm

E lowlands (to 1,000m). Swampy areas near slow-flowing forest streams. Bill ***black above, greenish-yellow below. Rufous-chestnut head to lower breast***, upperparts and ***belly brown***. Juv. duller. NE race *coccineipes* (red legs), SE race *castaneiceps* (brownish legs). Difficult to see as favours dense vegetation and nearly impossible to flush, walks erect. **Voice** Loud, far-carrying, rather melodious, long *tii-tuú, tii-tuú..., tii-tuu-doó, tii-tuu-doó...* duets. **SS** Uniform Crake concolorous throughout, voice more whistled, not in duets. **Status** Uncommon, probably overlooked.

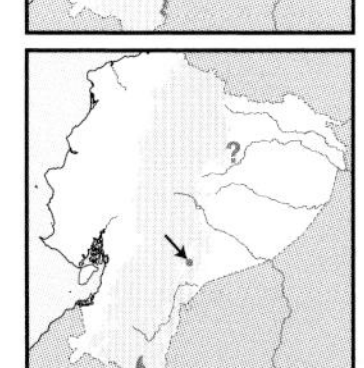

Russet-crowned Crake *Anurolimnas viridis* 16–18cm

Locally in SE Andean foothills (600–1,200m). Dense second growth, shrubby clearings. ***Blackish bill***, red eyes. ***Chestnut crown, grey mask***, whitish chin. Olive-brown above, cinnamon-rufous below, slightly paler, ***unmarked*** belly. Nominate. Very difficult to see as skulks in dense undergrowth, tends to walk over roots, twigs, logs, but difficult to flush. **Voice** Rattling *tkitkitki...* lasting *c.*4 sec. **SS** No bands on belly and vent as other crakes; see larger Chestnut-headed (bicoloured bill, brown belly); voice like other crakes in range. **Status** Rare, local, likely overlooked.

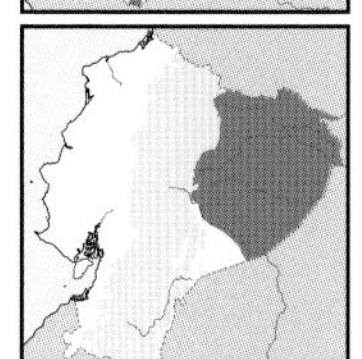

Black-banded Crake *Anurolimnas fasciatus* 17–20cm

E lowlands to foothills (to 1,000–1,100m). Damp areas near streams with dense, tangled vegetation, river islands. ***Black bill***, red eyes, reddish legs. ***Uniform rufous-chestnut head to lower breast***, olive-brown above, ***tawny to cinnamon belly and vent boldly banded black***. Monotypic. Very secretive and difficult to see or flush. **Voice** Long descending, explosive rattle (8–10 sec) *t'tr'tr'tr'trtr...*; sharp *tép* calls. **SS** Bill colour and belly pattern separate from Rufous-sided and Russet-crowned Crakes; voice more musical than sympatric crakes, longer, less explosive, slower. **Status** Rare, local, likely overlooked.

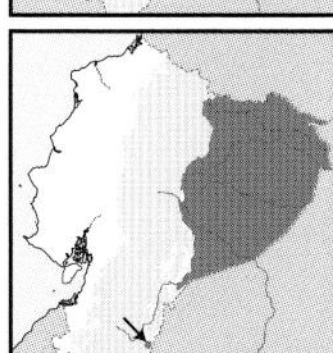

Rufous-sided Crake *Laterallus melanophaius* 15–18cm

E lowlands to foothills (to 1,350m). Swamps around oxbow lakes, damp grassland with tall vegetation. ***Bill greenish-yellow tipped black, greenish legs***. Olive-brown above, ***rufous-chestnut face to breast-sides***, white throat to mid-belly, ***flanks white boldly banded black***. Juv. duller below. Race *oenops*. Remains in dense thickets but often emerges in open, especially in early mornings or when overcast. **Voice** Explosive, descending, long rattling *ch'chrrrrrr*; also high-pitched *tréein-tréein....* **SS** Underparts differ from other crakes in E, rattling is longer. **Status** Uncommon, likely overlooked.

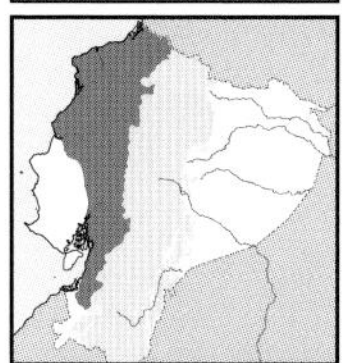

White-throated Crake *Laterallus albigularis* 14–16cm

W lowlands to lower subtropics (to 1,400m, locally higher). Damp grasslands, vegetation-fringed marshes. Bill greenish-yellow tipped black, ***greenish legs***. Olive-brown above, ***rufous-chestnut face, nuchal collar and breast, white throat***. Belly to vent ***boldly banded black and white***. Juv. duller. Nominate. Secretive, very difficult to flush, but sometimes in open, especially in early mornings or to cross roads. **Voice** Long, explosive, descending rattling *chrrrrr...*, also high-pitched *treeeng* alarm calls. **SS** Grey-breasted Crake has grey foreparts, voice shorter and harsher. **Status** Common.

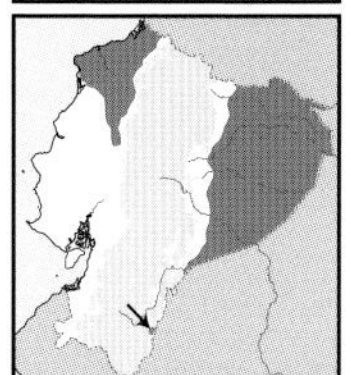

Grey-breasted Crake *Laterallus exilis* 13–15cm

E & NW lowlands (to 850m). Swampy tall vegetation around lakes, damp grassland, river islands. Bill greenish-yellow tipped black, greenish legs. ***Grey head to breast***, darker crown, ***whitish throat. Rusty mantle***, olive-brown upperparts. White mid-belly, ***flanks to vent banded black and white***. Juv. duller, ***greyer above***. Monotypic. Incredibly difficult to see and flush, skulks and runs away; rarely in open. **Voice** Short, descending, harsh rattling *t't-trrrrrrr*; also sharp *tít-tít-tít...* calls. **SS** Only grey *Laterallus*; voice harsher and shorter than other rattling crakes; cf. locally Ash-throated Crake (larger, paler-billed, redder legs). **Status** Uncommon.

coccineipes
castaneiceps
Uniform Crake
Chestnut-headed Crake
Russet-crowned Crake
Black-banded Crake
Rufous-sided Crake
White-throated Crake
♂
juvenile
♀
Grey-breasted Crake

PLATE 42: CRAKES AND RAILS

Three different genera that share predominantly grey plumage and brightly coloured bills, including the migratory and distinctive Sora, and its tiny, very rare and local congener Yellow-breasted Crake. *Mustelirallus* crakes are small, plump and reminiscent of *Laterallus* and *Anurolimnas*, but have different flanks pattern and bright red patches at bill base. *Pardirallus* have very long bills, two species with red patches at base.

* Yellow-breasted Crake *Porzana flaviventer* — 12–14cm

Two uncorroborated records from La Segua Marsh. Short *black* bill. ***Black crown and ocular stripe, white supercilium***. Brown above, back streaked white. ***Neck-sides to breast pale tawny-buff***, whitish throat, flanks boldly banded ***black and white***. Nominate? Secretive and difficult to observe, but can briefly emerge in open, walking on floating vegetation; flies short distances when flushed, legs dangling, head drooping. **Voice** Sharp *peep*, falling *zee-zee-zee…zerrr*, loud ringing *clureeo*, *krerrr…* and other calls (R *et al.*). **SS** Confusion unlikely. **Status** Very rare, local.

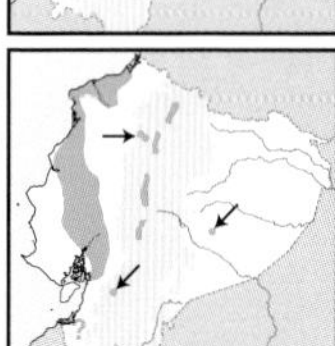

Sora *Porzana carolina* — 20–25cm

Migrant to W lowlands to foothills, inter-Andean valleys (to 2,800m). Freshwater marshes, ponds, lakes, swampy grasslands. ***Stout yellow bill tipped black***. Brown above, streaked black on back. ***Grey face, neck-sides and breast, black mask and throat***, whitish belly, flanks banded white and dusky. ♀ has less black on face and throat. Juv. duller, ***buff*** replaces grey, no black mask. Monotypic. Regularly in cover, might emerge in early morning and near dusk; quite furtive. **Voice** Sharp, abrupt *keeK* or *kiú* alarm calls. **SS** Confusion unlikely. **Status** Uncommon boreal winter visitor (Dec–Mar).

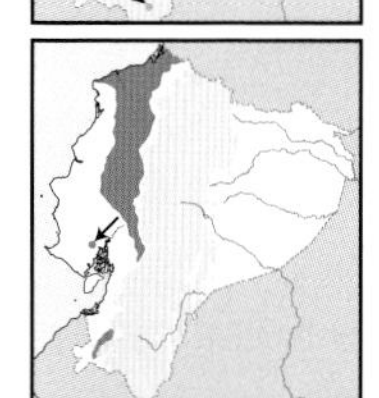

Colombian Crake *Mustelirallus colombianus* — 18–20cm

NW lowlands, locally S along Andean foothills (to 500m, locally 1,100m). Damp grassland, pond margins. ***Bill greenish-yellow, red basally. Head to upper belly grey, whitish throat, tawny lower belly to vent***. Nape and upperparts olive-brown. Nominate. Skulking and furtive, but might emerge at open muddy areas early morning; apparently tends to wander. **Voice** Slow, descending *tuee-tuee-tuee…* lasting *c.*10 sec (RG). **SS** Though no overlap reported cf. Paint-billed Crake (barred flanks and vent, paler bill); voice resembles Uniform Crake, but shorter notes. **Status** Rare, overlooked, DD globally.

Paint-billed Crake *Mustelirallus erythrops* — 18–20cm

Locally in W lowlands (to 300m). Damp grassland and marshy areas. ***Bill greenish-yellow, red basally. Head and underparts grey***, whitish throat, ***flanks to vent banded black and white***. Nape and entire upperparts olive-brown. Race *olivascens* or *erythrops*. Very furtive but might briefly emerge in muddy opens areas near dawn or dusk. **Voice** Long, series of vibrating, guttural notes, sometimes ending in brief trill; sharp *twak* in alarm (R *et al.*). **SS** Colombian Crake has plain flanks and vent, faster voice. **Status** Rare and local, likely overlooked.

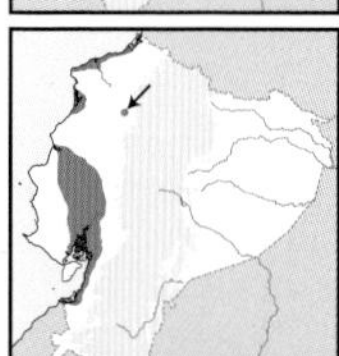

Spotted Rail *Pardirallus maculatus* — 25–30cm

Locally in W lowlands (mostly sea level). Dense reedbeds, overgrown freshwater marshes. ***Long yellowish bill, red basally***. Handsome, brown with ***dense and bold white streaking, freckling and spotting***, rustier wings, ***belly to vent barred black and white***. Juv. duller, fainter streaking / spotting / barring. Nominate. Can emerge to feed in muddy areas adjacent to dense cover; otherwise secretive and difficult to flush. **Voice** Various grunts and screeches: *skeey*, *skeuun*, *krkreich*, *gek*. **SS** Ad. unmistakable, cf. duller juv. with other rails. **Status** Rare, local, likely overlooked.

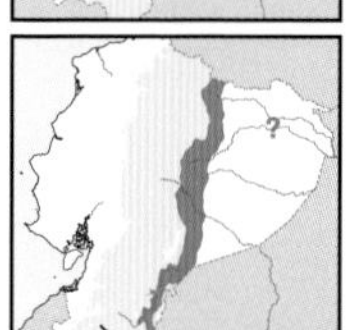

Blackish Rail *Pardirallus nigricans* — 27–30cm

E Andean foothills to subtropics, locally in adjacent lowlands (400–1,650m). Tall damp grassland and swampy ponds. ***Long greenish-yellow bill***. Grey head to underparts, ***throat white, black vent***. Dark olive-brown above. Juv. ***browner*** overall. Nominate. Skulking, but more prone to emerge in open muddy areas; flies short distances when flushed. **Voice** Loud, abrupt *drruu-ee-ee-eeT*, 'paired' with soft guttural *bu-bu-bu-bu…*; also sharp metallic *tiid-diit* and mournful *keeewaa* calls (R *et al.*). **SS** Not sympatric Plumbeous Rail has red-and-blue bill base, dark throat, paler vent. **Status** Uncommon.

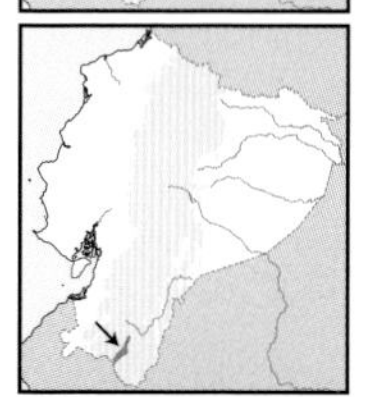

Plumbeous Rail *Pardirallus sanguinolentus* — 28–32cm

Locally in S inter-Andean valleys (1,500–1,900m). Wet grassland, damp areas, overgrown wet pastures. Bill long, ***greenish yellow, red and blue basally. Dark grey face and underparts, blacker lower belly to vent***. Dark brown above. Juv. duller, ***browner, bill and legs dusky***. Race *simonsi*. Singles or pairs emerge into open near cover more often than congeners; seen in gardens and crops. **Voice** Duet comprising repeated *rrueé-reeT* or *pu-rueet* squeals followed by soft *boo* hoots (given by ♀). **SS** No other grey rail in range. **Status** Rare and local, likely overlooked.

Yellow-breasted Crake
♂
subadult
juvenile
Sora
Colombian Crake
adult
juvenile
Paint-billed Crake
adult
juvenile
Spotted Rail
adult
juvenile
Plumbeous Rail
juvenile
adult
Blackish Rail

The more aquatic group of rallids, all are good swimmers and more conspicuous than others, except the recently discovered Ash-throated Crake. Common Gallinule is common and widespread in W lowlands and inter-Andean valleys, while at least one *Porphyrio* is also numerous in W. American Coot almost certainly extirpated from Ecuador. All swim with tails cocked. Coots are good divers.

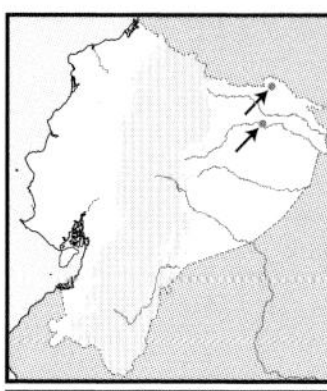

Ash-throated Crake *Porzana albicollis* 21–24cm

Two records in NE. Thick greenish-yellow bill, red eyes. Brownish above ***streaked dark; grey cheeks and most underparts***, whiter chin; barred flanks and vent. Duller juv. Race *typhoeca*? Flies short distance if flushed, legs dangling; often climbs vegetation but otherwise unobtrusive. **Voice** Duet includes loud cackling and prolonged trills; sharp *tuk* and longer *bewrewt* calls (TvP). **SS** Grey-breasted Crake (only grey crake in range) much smaller, darker-billed, yellow-legged, unstreaked above; more trilled voice. **Status** Uncertain, very local.

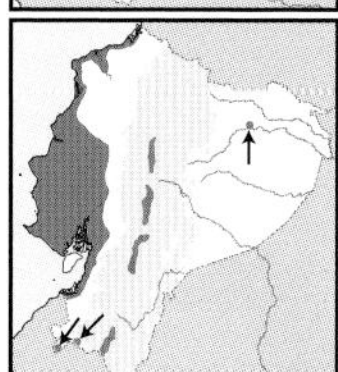

Common Gallinule *Gallinula galeata* 30–36cm

W lowlands, inter-Andean valleys, locally E lowlands (to 300m in lowlands, 2,200–3,200m in Andes). Freshwater ponds and marshy lakes. ***Bright red frontal shield and bill, yellow bill tip***, yellow legs. Mostly plumbeous-grey, browner upperparts, ***white stripe on flanks, white undertail-coverts*** with black centre. Imm. browner and duller. Sexes alike. Race *pauxilla* (W, Andes), possibly *galeata* (E). Regularly swims or walks on floating vegetation, cocking tail; singles to sizeable flocks even at small ponds. **Voice** Long *kruunk…kruunk, kreeen, kreh, kreh…kitik, kitik* and other cackles. **SS** Heavier Slate-coloured Coot (red less extensive, no white flanks); flanks and vent diagnostic. **Status** Common.

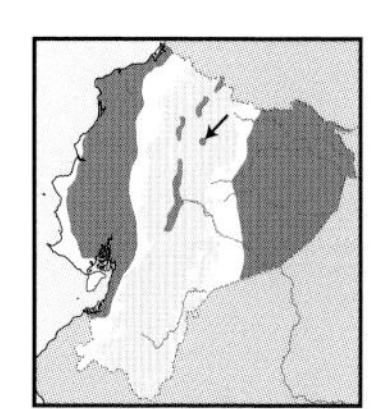

Purple Gallinule *Porphyrio martinicus* 27–32cm

W & NE lowlands, locally (wanderers?) to highlands (mostly below 500m, locally to 3,300m, once at 4,100m). Freshwater marshes, ponds, vegetation-fringed lakes, slow-flowing rivers. Unmistakable, ***deep lavender-blue***, green back, ***white vent. Bill bright red, tipped yellow***, bluish frontal shield. Juv. drab, bill greenish-yellow, ***ochraceous-buff head and underparts, white*** throat, belly and ***vent***. Dusky-green above, bright blue patch in scapulars. Monotypic. Walks on floating vegetation and shores, tail cocked, also swims; singles to sizeable flocks, tends to wander widely. **Voice** Various cackles and grunts: *kyup-kuyp-khukhukhukhu, cuhr-kú-kúk*, fast and descending *kukukuku…*; ghost-like *goó..goó-goó…*. **SS** Smaller juv. Azure Gallinule (paler below, more mottled above, bill lighter). **Status** Common.

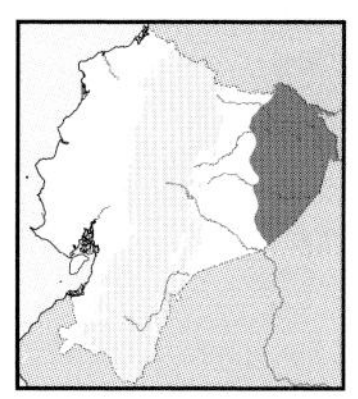

Azure Gallinule *Porphyrio flavirostris* 23–26cm

E lowlands, mostly along lower Napo and Aguarico Rivers (to 300m). Edges of oxbow lakes. Yellow bill, frontal shield and legs. ***Pale sky-blue face to breast-sides***, white throat to vent. Dusky-brown crown, ***mottled upperparts, bright bluish wing-coverts and flight feather fringes***. Juv. duller, buffier, white throat to vent, ***bluish wing-coverts and flight feather fringes***. Monotypic. Elusive, less gregarious than Purple Gallinule, often walks on floating vegetation, rarely swims. **Voice** Rather silent, but short, nasal *grrá...grrá…grrá…* reported. **SS** Larger Purple Gallinule more often in open, possibly overlap locally (no mottled back, less blue in wings); voice more nasal and cackled. **Status** Uncommon, local.

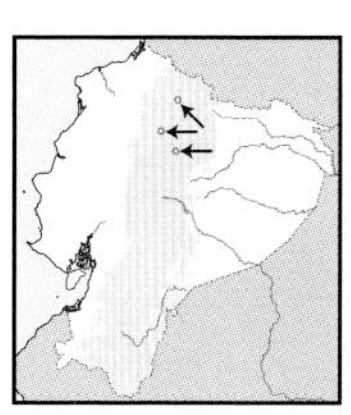

American Coot *Fulica americana* 35–40cm

Local and historically in N highlands (around 2,200m). Collected 80 years ago in Yaguarcocha Lake, Imbabura. Closely recalls common Slate-coloured Coot, but whitish bill shorter, less stout, with ***dark red and small frontal shield, dark broken ring near bill tip***. Race *columbiana*. Probably behaves like Slate-coloured, but not currently found in Ecuador where formerly resident. **Voice** Elsewhere (R *et al.*) various calls: *pulk* or *poonk, pow-uur* or *cooanh, punk-kuk-kuk* or *ka-pow, ka-pow*. **SS** More numerous and polymorphic Slate-coloured Coot; red frontal shield, always show more extensive red and yellow in bill, no ring. **Status** Feared extinct in Ecuador.

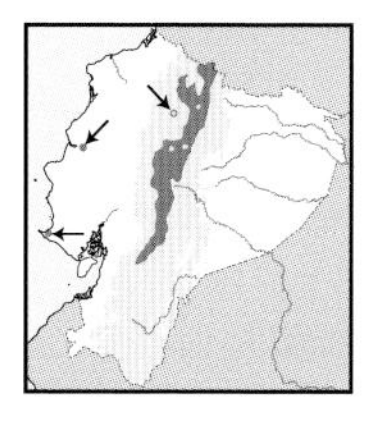

Slate-coloured Coot *Fulica ardesiaca* 39–44cm

Andes, locally (wanderers) to W lowlands (2,200–3,900m, locally higher). Freshwater lakes with abundant reeds. Bill colour variable, but more numerous form has ***whitish bill and frontal shield; yellow shield and white bill or red shield and yellow bill*** forms are less numerous. Slate-grey, some with little white in vent-sides. Juv. duller, ***browner, duskier-billed***. Race *atrura*. Swims on open water, suddenly diving for brief periods; flight strong, must paddle before take-off. Tends to congregate in sizeable groups, also walks or rests on shore and in dense reeds. **Voice** Repeated *kituuk-kituuk* and descending *chrrrr-chrrr…*. **SS** Probably extinct and smaller American Coot has smaller frontal shield. **Status** Uncommon.

Ash-throated Crake

juvenile

Common Gallinule

juvenile

adult

adult

Purple Gallinule

adult

Azure Gallinule

adult

juvenile

juvenile

ult

juvenile

American Coot

white-fronted
morph

yellow-fronted
morph

adult

red-fronted
morph

juvenile

Slate-coloured Coot

Cryptic-plumaged species, most occurring in dense reedbeds and vegetation-fringed ponds in lowlands. Tiger-herons, Zigzag Heron and bitterns are seldom seen and might present identification pitfalls, especially in juv. plumages. Fasciated Tiger-Heron is the only species occurring along rushing rocky Andean streams.

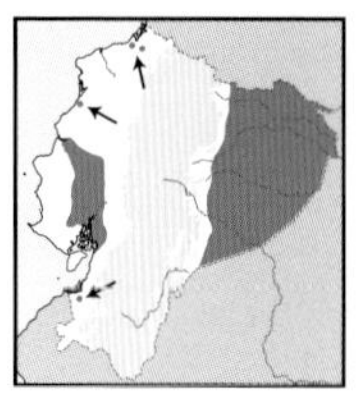

Rufescent Tiger-Heron *Tigrisoma lineatum* 66–76cm

E & W lowlands (below 500m). Oxbow lakes, forest streams, flooded forest. ***Rufous-chestnut head to chest-sides, brown upperparts*** faintly vermiculated dusky. White throat and narrow median stripe on belly. Juv. ***rich buff barred black***, whitish underparts, ***mottled / barred*** on sides and flanks, ***four tail-bands***. Race *lineatum*. Solitary and difficult to observe though sometimes surprisingly tame, remaining motionless for long periods. **Voice** Low *hon'hon'hon'hon'hon-HNnnn*, groaning, frightening *huuuoOO* calls. **SS** Juv. nearly identical to Fasciated Tiger-Heron, but more barring on flanks, heavier bill; Pinnated Bittern not barred but vermiculated or blotched. **Status** Rare.

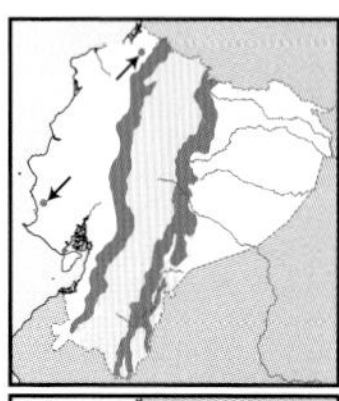

Fasciated Tiger-Heron *Tigrisoma fasciatum* 61–70cm

E & W Andean foothills to subtropics, SW Cordillera de la Costa (600–2,200m, locally lower). Shady swift-flowing streams and rivers. Mostly ***sooty-grey*** faintly vermiculated and barred buff. Whitish throat and median stripe to buff belly. Juv. closely resembles Rufescent Tiger-Heron but bill ***shorter, lighter***, only ***three*** tail-bands (difficult to see in the field), ***less barring*** on flanks. Race *salmoni*. Solitary, shy and wary, often seen standing on boulders or rocky shores. **Voice** In alarm *kwóK*, but mostly silent. **SS** Might overlap locally with previous species, but habitat often differs. **Status** Rare.

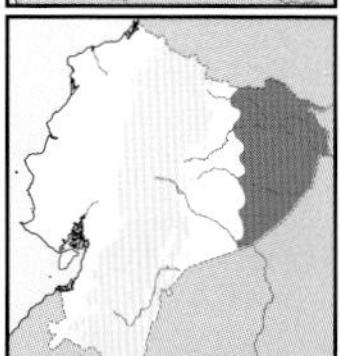

Zigzag Heron *Zebrilus undulatus* 30–33cm

E lowlands (below 300m). Shady lagoons and oxbow lakes, forest streams. Short bill, pale eyes, ***short greenish legs. Chunky***. Blackish crown, ***sooty-brown above densely barred buff, buff loral spot***. Buff throat to breast thickly vermiculated dusky-brown, paler belly. Juv. similar, but more ***cinnamon on face***, lighter, less marked underparts. Monotypic. Solitary, very difficult to see in dense undergrowth, remaining motionless on low perches. **Voice** Far-carrying, hollow *kUOown* repeated every 4–6 sec. **SS** Unmistakable. **Status** Rare but overlooked; NT globally.

Pinnated Bittern *Botaurus pinnatus* 64–76cm

Locally in W lowlands (below 50m). Reedbeds, marshes, vegetation-fringed ponds. ***Long, thick, mostly yellow bill***, greenish legs. ***Mostly buff with complex dusky*** vermiculations, streaks, mottles and bars. White throat, ***whitish foreneck to belly, dense brown streaking below***. Race *pinnatus*. Solitary, skulking, very rarely in open; when alarmed freezes stretching neck towards sky, when flushed flies low and hides in dense vegetation. **Voice** Booming, very low-pitched *ooonk*. **SS** Juv. night-herons plumper, darker, streaked. Juv. Rufescent Tiger-Heron heavily banded above. **Status** Rare, local, VU in Ecuador.

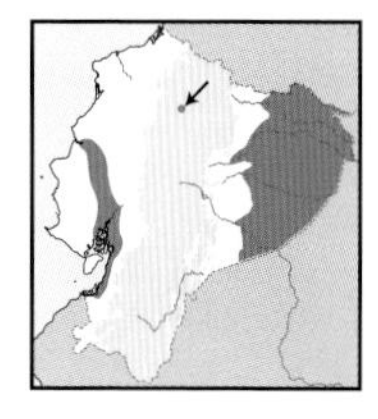

Least Bittern *Ixobrychus exilis* 28–32cm

Locally in E & SW lowlands, vagrants in highlands (below 300m, casual above 2,500m). Freshwater marshes, oxbow lakes. Ochraceous-buff face and neck-sides, ***black crown and upperparts*** (***brown*** back in ♀), ***prominent buff patch in wing-coverts***. Whitish-buff below, streaked ochraceous-buff in ♀. Juv. duller. Race *limoncochae* (E), possibly *peruvianus* (SW), *erythromelas* or *bogotensis* (wanders to highlands). Solitary, wary, skulking; when spotted might freeze like Pinnated; flies with legs dangling. **Voice** Low-pitched, hollow, trembling *göähh..göähh...*. **SS** See Stripe-backed Bittern. **Status** Rare and local; probably seasonal in W (Sep–Apr).

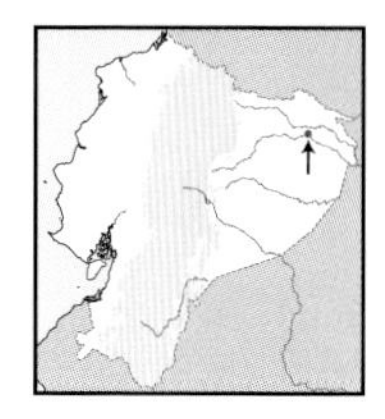

* Stripe-backed Bittern *Ixobrychus involucris* 30–34cm

One record (Mar) in NE lowlands (below 300m). Oxbow lakes. Resembles Least Bittern but ***upperparts black, markedly striped buff***, black crown, buffy face and neck-sides, ***duller*** wing patch. Monotypic. Poorly known in Ecuador; elsewhere also skulking and solitary, difficult to flush. **Voice** Far-carrying low *huuu*, lower-pitched than Least (R *et al.*). **SS** From Least Bittern by wing and back pattern. **Status** Uncertain, possibly wanderer.

Fasciated Tiger-Heron
adult
immature
adult
Rufescent Tiger-Heron
immature
adult
juvenile
Zigzag Heron
Pinnated Bittern
eruvianus
♂
erythromelas,
limoncochae
♂
juvenile
♀
bogotensis
♂
Least Bittern
Stripe-backed Bittern

Three mainly nocturnal herons with large eyes and heavy bills – notably so in Boat-billed. *Butorides* herons are small and plump, one a rare boreal migrant, the other widespread and numerous.

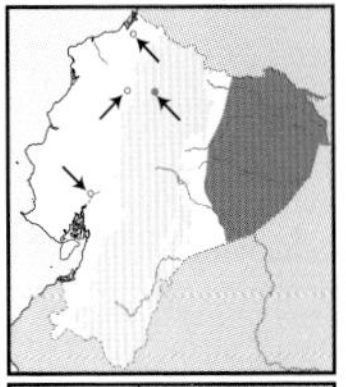

Boat-billed Heron *Cochlearius cochlearius* 45–53cm

E lowlands, locally (historically?) in W (to 400m, once at 2,900m). Freshwater lakes, sluggish rivers, swampy forest. ***Broad, massive, blackish bill***, large eyes. White forecrown, ***black elongated crest; pearl-grey above***, black mantle. White throat to breast, ***rufous-chestnut central belly***. Juv. brown, paler below. Nominate. Mainly nocturnal and solitary, feeds by scooping in shallow water, roosts concealed. **Voice** Low-pitched *cuuah* when flushed, *coon-coanh-cooan....* **SS** Black-crowned Night-Heron has very different bill shape, all-white underparts, yellow legs. **Status** Rare.

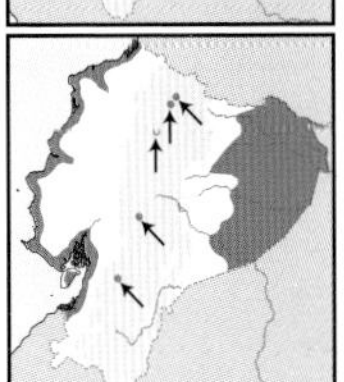

Black-crowned Night-Heron *Nycticorax nycticorax* 56–65cm

Coasts, adjacent lowlands, NE lowlands, locally in highlands (below 400m, to 3,300m in Andes). Freshwater and saline wetlands. Rather long ***stout black bill***, yellow legs (***red*** in breeder). White forecrown, ***black crown and upperparts, white neck, nuchal collar and underparts***, pearl-grey wings. Juv. brown with ***prominent whitish streaking above, whitish spots in wing-coverts***, whitish underparts, dusky streaking. Race *hoactli*. Diurnal and nocturnal, feeding on muddy banks and in shallow water. **Voice** Abrupt *WOk!* or *uAK!* when flushed. **SS** Juv. Yellow-crowned has longer legs, mainly dark mandible, is greyer, less spotted. **Status** Uncommon.

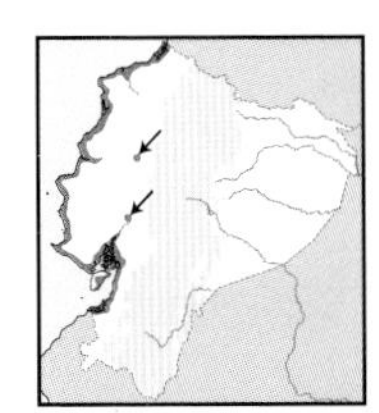

Yellow-crowned Night-Heron *Nyctanassa violacea* 56–68cm

Coasts, adjacent lowlands, locally inland (below 200m). Mangroves, tidal banks, abandoned shrimp ponds. ***Long stout blackish bill, large orange-brown eyes***. Yellow forecrown, ***white crown; black face and throat, broad white moustachial***. Mostly ***slaty blue-grey***, upperparts and wings black edged and streaked grey. Juv. resembles Black-crowned, but ***greyer*** above, ***more finely streaked, longer legs, darker bill***, wing spotting less prominent. Race *caliginis*. Mainly nocturnal and solitary, sometimes gather with other waders to feed. **Voice** Nasal *kwAK!*, higher-pitched than previous species. **SS** Compare juv with Pinnated Bittern in W. **Status** Fairly common.

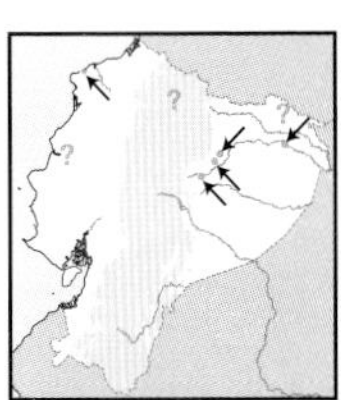

Green Heron *Butorides virescens* 38–43cm

NE lowlands, locally in NW (to 400m). Oxbow lakes, forest streams (E), mangroves (W). Very similar to commoner Striated Heron, not always safely separated. Neck to breast-sides ***chestnut, richer*** than Striated. Juv. nearly identical to Striated but somewhat ***browner in face and neck***. Nominate; complex status. Casual winter visitor, mostly solitary, hunting like Striated. **Voice** Sharp *kwóp* when flushed (R *et al.*). **SS** Striated shows variable amount and tone of rufous-chestnut on neck, but mostly ***grey***, juv. not safely separated. **Status** Rare boreal migrant (Dec–Mar).

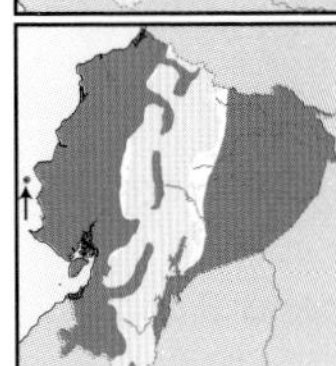

Striated Heron *Butorides striata* 35–43cm

E & W lowlands, locally in inter-Andean valleys (to 500m in lowlands, locally 2,200–2,800m). Freshwater wetlands, marshes, shady streams. Black crown, ***grey to rufescent-grey neck-sides to flanks***. White throat and narrow central stripe outlined rufous; greyish-green upperparts. Juv. has ***grey to brownish-grey neck to sides, underparts variably and densely streaked brown***, browner above. Race *striata*. Solitary, active crouching hunter, numerous in several habitats, often flushes abruptly. **Voice** Ringing *kyOg!-kyOg!-kyOg!* **SS** See previous species; shier Least and Stripe-backed Bitterns (smaller, bold wing or back patterns), Zigzag Heron (chunkier, shorter-legged, darker, browner). **Status** Common.

Boat-billed Heron
adult
juvenile
non-breeding
juvenile
adult
juvenile
breeding
Yellow-crowned Night-Heron
Black-crowned Night-Heron
adult
juvenile
adult
juvenile
intermediate
Striated x Green
Green Heron
Striated Heron

Slender egrets and herons that often congregate in wetlands, mostly freshwater. The three familiar white species are numerous and rather widespread; Cattle is native to Africa but has rapidly spread across the Neotropics. See bill and leg coloration for species recognition. Tricoloured and Little Blue are often together at muddy banks.

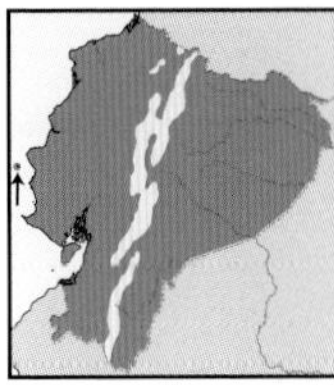

Cattle Egret *Bubulcus ibis* 46–52cm

W & E lowlands to foothills, inter-Andean valleys (to 2,800m, locally higher). Grassland. ***Yellow, rather stout bill***, yellow facial skin, ***yellowish-green legs***. Crown, chest and back tinged ***tawny-orange*** when breeding, soft parts redder. Nominate. Gregarious, highly mobile, regularly around but not always cattle, less often in wetlands. **Voice** Moans and grunts. **SS** Snowy Egret (more slender, longer, thinner black bill, black legs); see also imm. Little Blue Heron. **Status** Common, widespread, recent coloniser, possibly boreal migrant to E.

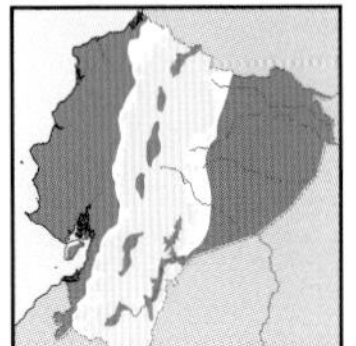

Great Egret *Ardea alba* 90–105cm

E & W lowlands, locally in inter-Andean valleys (to 2,800m, locally higher). Freshwater and saline wetlands. ***Long yellow bill, dark legs***, elongated white feathers in breast and lower back. Juv. with ***duller*** soft parts. Race *egretta*. Highly gregarious, also with other egrets, herons and waders; elegant flight with deep wingbeats. **Voice** Hoarse *gahrrr!*, more nasal *gaah!* in flight. **SS** Smaller Snowy Egret, Little Blue Heron (juv.) with thin black to blackish bill, and Reddish Egret. **Status** Common.

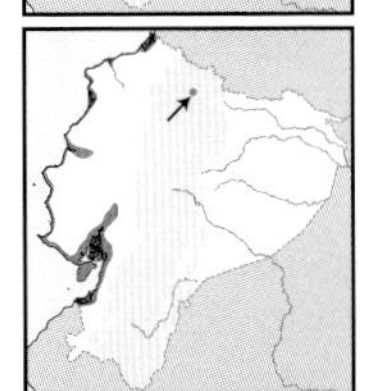

Tricoloured Heron *Egretta tricolor* 58–72cm

Coast and adjacent W lowlands (below 100m, once at 2,200m). Mangroves, tidal banks, estuaries. ***Long bicoloured bill, yellow facial skin. Blue-grey head to chest and upperparts, white rump and belly***. Juv. similarly patterned but ***duller, browner face and neck***. Race *ruficollis*. Active on tidal flats, often chases prey or fishes using wings as shade; often wades in belly-deep water. **Voice** Throaty *gaaanh* calls in flight. **SS** Little Blue Heron (solid blue-grey below); Reddish Egret (brownish-white, dull bill). **Status** Uncommon.

Reddish Egret *Egretta rufescens* 66–80cm

Recently recorded on a Río Napo island, 200m (D. Lane) and Pacoa saltpans (D. Brinkhuizen, J. Nilsson). Mostly ***brownish***, darker above. Juv. pale ***greyish to brownish-white***, somewhat mottled or dirty; ***dull greenish*** soft parts. Nominate. Mostly alone, walking and running along shores. **Voice** Soft grunts and groans (R *et al.*). **SS** Resident egrets much whiter, yellow soft parts; Capped and Whistling Herons are smaller, more colourful; Great Egret (taller, less sturdy). **Status** Rare vagrant.

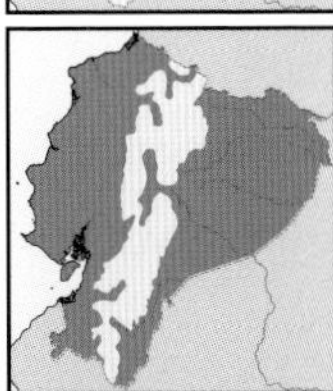

Snowy Egret *Egretta thula* 53–68cm

E & W lowlands, locally in inter-Andean valleys (below 500m, locally to 2,200–3,300m). Freshwater and saline wetlands. ***Long, slender black bill, yellow facial skin, black legs, yellow feet***. Elongated plumes on crown, breast and back of breeder. ***Paler*** soft parts in juv. Nominate. Active, gregarious, often chases prey or forces it to escape. **Voice** Raspy, fastidious *crruaag, craag*. **SS** Juv. Little Blue Heron (darker legs and bill, some greyish blotching); Great Egret (larger, yellow bill); Cattle Egret (stouter yellow bill); also Reddish Egret. **Status** Common.

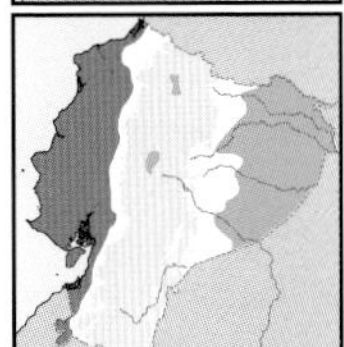

Little Blue Heron *Egretta caerulea* 55–76cm

E & W lowlands, locally in inter-Andean valleys (to 2,600m). Freshwater and saline wetlands, rivers. ***Grey bill tipped black, greyish facial skin and legs. Maroon-purple head to chest, bluish grey rear parts***. Juv. ***white*** blotched ***dusky, blackish tips*** to primaries; ***darker soft parts***. Monotypic. Singles to small groups, more patient hunter than other herons. **Voice** Mostly silent; raspy *rrraaag-rrraaag* in flight. **SS** Tricoloured Heron (white-bellied); juv. Snowy Egret (all white, yellow bill base and feet). **Status** Uncommon (W), likely boreal visitor (E and Andes).

non-breeding
breeding
Cattle Egret
non-breeding
breeding
Great Egret
non-breeding
juvenile
Tricoloured Heron
juvenile
non-breeding
white morph
non-breeding
Reddish Egret
non-breeding
breeding
Snowy Egret
juvenile
immature
Little Blue Heron
non-breeding

Slender herons, most in bolder shades than herons on previous plate. Great Blue is becoming a more regular boreal visitor; Whistling possibly expanding after deforestation.

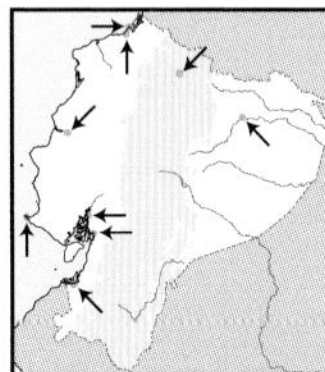

Great Blue Heron *Ardea herodias* 100–130cm

Casual in W lowlands and inter-Andean valleys (to 2,200m). Freshwater and saline wetlands. Long stout bill, greenish legs. ***White crown***, black ***crown-sides, greyish-buff neck*** grading into grey chest and upperparts, ***rufous thighs***. Juv. duller, pale grey above, ***buffy below***, white throat. Nominate, possibly *cognatus* from Galápagos. Casual visitor from northern latitudes, mostly singles but with other waders. **Voice** Hoarse *guk-huk*... **SS** Whiter Cocoi Heron has more black in crown (all ages), white thighs (rufous in Great Blue not always evident). **Status** Rare (Jan–Mar, Jul).

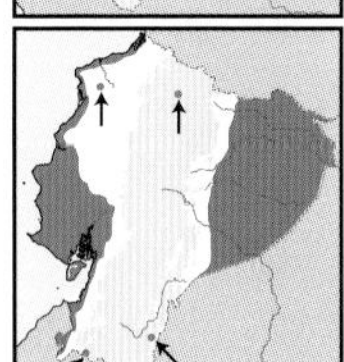

Cocoi Heron *Ardea cocoi* 100–130cm

SW & E lowlands, locally in NW (to 400m; locally to 900m and in highlands). Freshwater and saline wetlands. ***Long yellow, stout bill***, greenish legs. ***Black elongated crown, white throat to chest***, pale grey upperparts, ***white flanks and thighs***. Juv. mostly ***pale grey, black crown***, white throat and foreneck stripe, ***white thighs.*** Monotypic. Mainly solitary in shallow waters, sometimes gather at mud banks, abandoned shrimp ponds, etc. **Voice** Hoarse, grumpy *krooak*. **SS** Much rarer Great Blue Heron (little black in crown, rufous thighs). **Status** Fairly common.

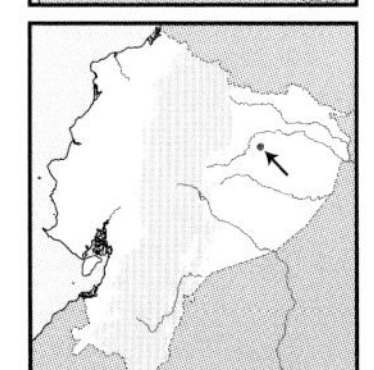

* Whistling Heron *Syrigma sibilatrix* 50–60cm

Two records in NE lowlands (300m). Freshwater ponds. Long bill, ***mostly orange-red, blue-grey facial skin***. Black crown, ***rich tawny neck to chest***, white below. ***Grey upperparts, whitish rear parts***, rufous-buff wing-coverts edged black. Juv. paler, including soft parts. Presumed race *fostersmithi*. Singles to small groups (though only seen twice in Ecuador) in pastures, not necessarily tied to water, possibly spreading. **Voice** High-pitched *wueee!* (R *et al.*). **SS** Confusion unlikely. **Status** Vagrant (Jan, Aug).

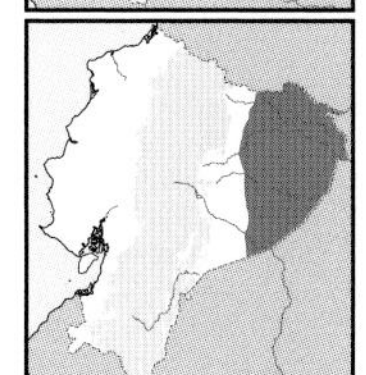

Capped Heron *Pilherodius pileatus* 53–60cm

E lowlands (to 400m). Rivers and lagoons. Long, pointed, ***pale blue bill, pale blue facial skin. Creamy-white forecrown, throat, face, neck and underparts***, whitish belly. ***Contrasting black crown, pale*** grey upperparts. Non-breeder ***mostly white below***, juv. mainly white with ***dusky mask and crown***. Monotypic. Mainly solitary in grassland, not necessarily tied to water; rather wary, spears and pursues prey, sometimes wading in shallow water. **Voice** Mostly silent, occasionally croaks when flushed (HB). **SS** Confusion unlikely. **Status** Uncommon.

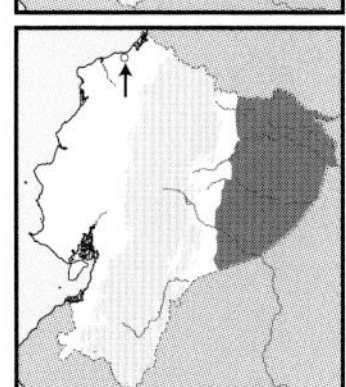

Agami Heron *Agamia agami* 64–74cm

E lowlands (below 300m). Flooded forest, forest streams and lagoons. ***Very long*** needle-like bill, ***long neck***. Black crown, ***elongated silvery crest, chestnut neck, mantle and underparts***, green above. White throat and median stripe ending in ***long silvery plumes***, median stripe edged black. Juv. ***dull brown*** above, whitish below variably ***streaked dull brown***. Monotypic. Solitary, wary, rarely emerges in open; spears fish from thickets or overhanging vegetation. **Voice** Mostly silent, weak *grreu* heard. **SS** Confusion unlikely; bill shape and length readily separates juv. from other herons. **Status** Rare, VU globally.

juvenile
adult
Great Blue Heron
adult
juvenile
juvenile
adult
Cocoi Heron
adult
juvenile
Whistling Heron
adult
immature
juvenile
Agami Heron
breeding
non-breeding
juvenile
Capped Heron

PLATE 48: IBISES

Generally rare, ibises mostly dwell in or near water. Three species apparently only wanderers, accidental or seasonal visitors. Resemble more familiar egrets and herons, but bill shape and size of legs distinctive. Two species, notably the *páramo*-inhabiting Black-faced, are experiencing a sharp decline.

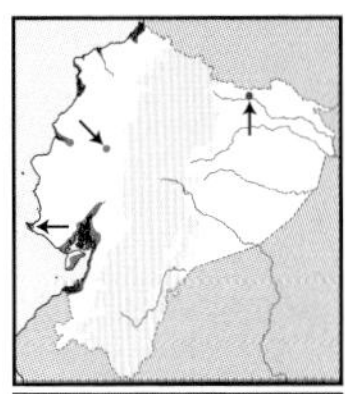

White Ibis *Eudocimus albus* — 56–70cm

Local on coasts, rarer inland (mostly sea level). Mangroves, tidal banks, marshes. ***Long decurved pink bill, red facial skin, red legs. Snowy***, outer primaries tipped black. Juv. has ***pale pink bill and legs, pink facial skin. Head to chest buff mottled dusky, white below***. Upperparts brown, ***white rump and uppertail-coverts***. Monotypic. Congregate on muddy banks, wade and probe mud; graceful flight with neck and legs extended, shallow wingbeats. **Voice** Nasal honks and whistles. **SS** Juv. Scarlet Ibis shows some red; if not, not safely identified. **Status** Locally uncommon.

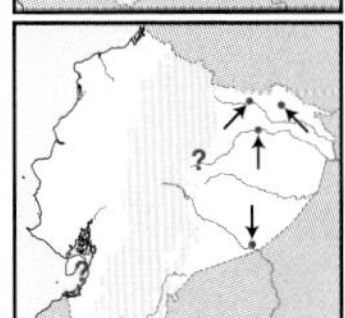

Scarlet Ibis *Eudocimus ruber* — 56–70cm

Local in NE lowlands, scattered uncorroborated records in SW (below 300m). Rivers and lakes. ***Long decurved pink-horn bill, rosy-red facial skin and legs***. Strikingly ***red***, black tips to outer primaries. Juv. nearly identical to White Ibis but some show ***some red***. Monotypic. Poorly known in Ecuador, where probably vagrant and solitary or few birds; otherwise, behaves much like White Ibis. **Voice** Mostly silent. **SS** Limpkin. **Status** Uncertain, perhaps increasing (Jan, Mar, May).

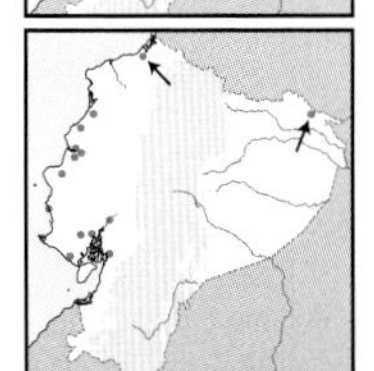

Glossy Ibis *Plegadis falcinellus* — 55–65cm

Locally in W lowlands (to 50m). Freshwater marshlands. ***Long decurved blackish bill, blackish legs. Mostly bronzy-green***, greener wings, ***white outline to bill***. Monotypic. Mostly in small flocks wading and probing mud; tends to wander widely; flies with neck and legs outstretched, legs protruding beyond tail. **Voice** Soft nasal *wehp-ehp*. **SS** Unmistakable in range (but an unidentified dark ibis, possibly Green, was observed in NW Esmeraldas by A. Solano & J. Freile). **Status** Uncertain, possibly casual; recorded Oct–Mar, Jul–Aug (year-round?).

* Puna Ibis *Plegadis ridgwayi* (not illustrated) — 60–65cm

One record from Lake Limpiopungo, 3,800m (Y. Potaufeu). Reddish bill. Mostly ***dark purplish***-brown, ***green gloss in wings***, some white streaks on head. Monotypic. Wades and probes mud with hunched posture, probably wanders widely. **Voice** Nasal *wut, cwurk* calls (FK). **SS** Recalls Glossy Ibis, but lacks white outline to bill. **Status** Accidental wanderer (Jan).

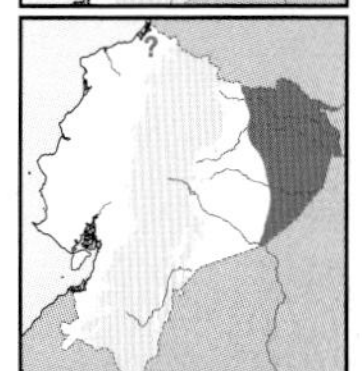

Green Ibis *Mesembrinibis cayennensis* — 52–58cm

E lowlands (to 300m). *Várzea* forest, forested and swampy streams, lake edges. ***Long, decurved, greenish-grey bill, short greenish legs***. Mostly ***dark metallic green***, bronzy upperparts, some ***long ruffled feathers on foreneck***. Juv. duller and darker. Monotypic. More forest-based and less gregarious than other ibises, flight erratic and bouncy, quick wingbeats; mainly solitary and wary, often perches on riparian trees. **Voice** Loud, rapid *corOcorOcorOcorO...*, rolling *coo'or* calls. **SS** Dark plumage and short legs distinctive, but see Bare-faced Ibis. **Status** Uncommon.

Bare-faced Ibis *Phimosus infuscatus* — 46–51cm

Local in NE lowlands (to 300m; locally higher). Riverbanks, grassy marshes. Long, slender decurved ***rosy*** bill, ***large red facial skin***, pinkish legs. ***Mostly dark bronzy***. Juv. duller overall, including bill. Race *berlepschi*. Infrequently seen, in small groups or solitary; elsewhere probes in mud, often near large rivers; flight fast and steady. **Voice** Mostly silent. **SS** Bare red face distinctive, see Sharp-tailed Ibis *Cercibis oxycerca* (not yet in Ecuador). **Status** Rare, accidental visitor (Dec, May), likely spreading.

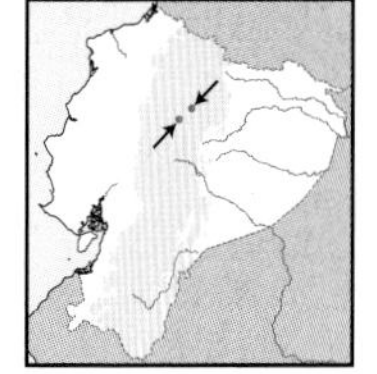

Black-faced Ibis *Theristicus melanopis* — 71–76cm

Locally in N Andes (3,800–4,300 m). Grassy *páramo*. ***Long, decurved, blackish bill***, reddish legs. ***Ochraceous crown, tawny neck to chest, black belly to tail***. Upperparts mostly ***brownish-grey***, paler wings, ***black flight feathers***. Juv. heavily ***streaked dusky*** in head to underparts. Race *branickii*, possibly full species. Not necessarily near water, often in small flocks or pairs, walking and probing in open *páramo*; nests in crevices. **Voice** Far-carrying, nasal *whíp, turt-turt..turt-turt* in flight. **SS** Unmistakable. **Status** Very rare, local, declining, CR in Ecuador, NT globally.

adult
White ibis
breeding
juvenile
adult
adult
Scarlet ibis
immature
non-breeding
breeding
juvenile
juvenile
Glossy ibis
adult
Green ibis
adult
juvenile
Bare-faced ibis
adult
juvenile
Black-faced ibis

The two stork species, among the largest Ecuadorian birds, resemble large herons but bills are much heavier and heads featherless. The only flamingo species in mainland Ecuador, not yet found to breed in the country, is likely to be resident.

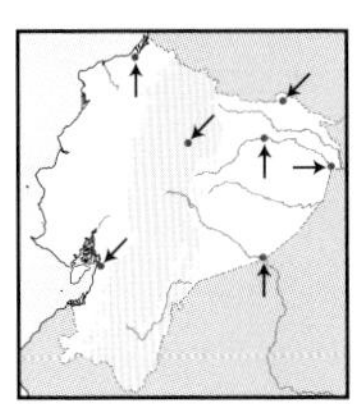

Jabiru *Jabiru mycteria* 120–150cm

Few records in E lowlands (below 300m, once at 2,000m, once in NW, Dec 2016, C. Vogt). Sandbanks, rivers, marshy grassland. The 'giant' of Ecuadorian birds, massive ***upturned black bill***, bare black head and neck, ***bare red band*** at neck base. Entirely white. Juv. duller, greyer. Monotypic. Only vagrant to Ecuador; elsewhere mostly in pairs; soars like a vulture but outstretched neck distinctive. **Voice** Silent, but bill claps at nest. **SS** Unmistakable. **Status** Rare visitor (Aug–Sep, Dec–Feb).

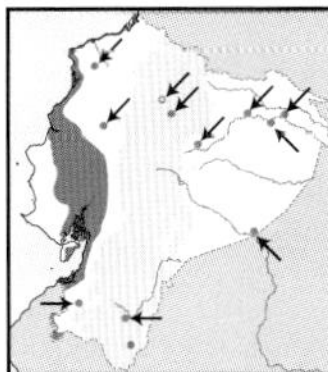

Wood Stork *Mycteria americana* 85–110cm

Mainly W lowlands but wanders to highlands and E lowlands (below 100m, wanders higher). Marshes, freshwater wetlands, abandoned shrimp ponds. Massive, ***slightly decurved black bill, bare dusky head and neck, black mask***. Mostly ***white, flight feathers black***. Juv. duller overall. Monotypic. Excellent flier, might soar high overhead with neck, legs and wings outstretched; congregates at feeding areas where preys on vertebrates and invertebrates; sometimes soars over forested areas but only on passage. **Voice** Some grunts and growls at colonies. **SS** Confusion unlikely. **Status** Uncommon in SW, vagrant elsewhere.

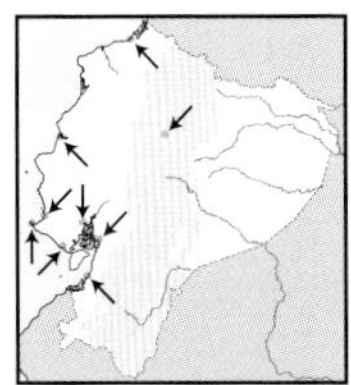

Chilean Flamingo *Phoenicopterus chilensis* 100–110cm

Locally in SW (sea level, once at 3,800m). Salty wetlands and mudflats on Santa Elena Peninsula; once N Esmeraldas (M. Sánchez). Unmistakably ***pink***, with ***greyish pink-and-black arched bill, greyish legs, reddish-pink 'knees' and feet***. Black flight feathers. Juvenile smaller, no pink. Monotypic. Nomadic at saline wetlands where gathers in groups (once 200); wary, when approached, runs before taking flight. **Voice** Low honks and grunts. **SS** Unmistakable, other flamingos unlikely in mainland Ecuador. **Status** Year-round but irregular, non-breeding visitant; NT globally.

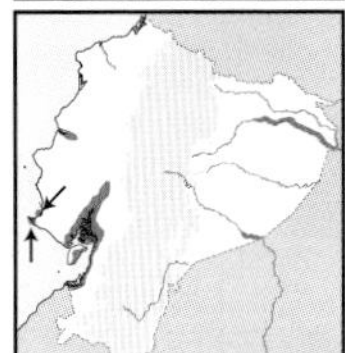

Roseate Spoonbill *Platalea ajaja* 71–80cm

Local in NE lowlands and SW coast (to 300m in E). Rivers (E), mangroves, tidal flats, marshes (W). Long, ***spatula-like*** bill, ***dark featherless head***, pink legs. White neck, ***pink upperparts and belly, scarlet wing patch and underwing-coverts***, pink to yellowish tail. Juv. duller, little pink overall, greyish legs, ***feathered head***. Monotypic. Gregarious at feeding and roosting sites, feeds by moving bill sideways; neck outstretched in elegant flight. **Voice** Mostly silent, some grunts. **SS** Unmistakable. **Status** Rare to uncommon.

immature
adult
Jabiru
adult
Wood Stork
juvenile
Chilean Flamingo
adult
immature
adult
immature
juvenile/1st-winter
Roseate Spoonbill

adult
juvenile
Plumbeous Kite
Plate 58
♂
♀
immature
Snail Kite
Plate 60
first-winter
adult
Mississippi Kite
Plate 58
Swallow-tailed Kite
Plate 58
immature
Slender-billed Kite
Plate 60
Osprey
Plate 60

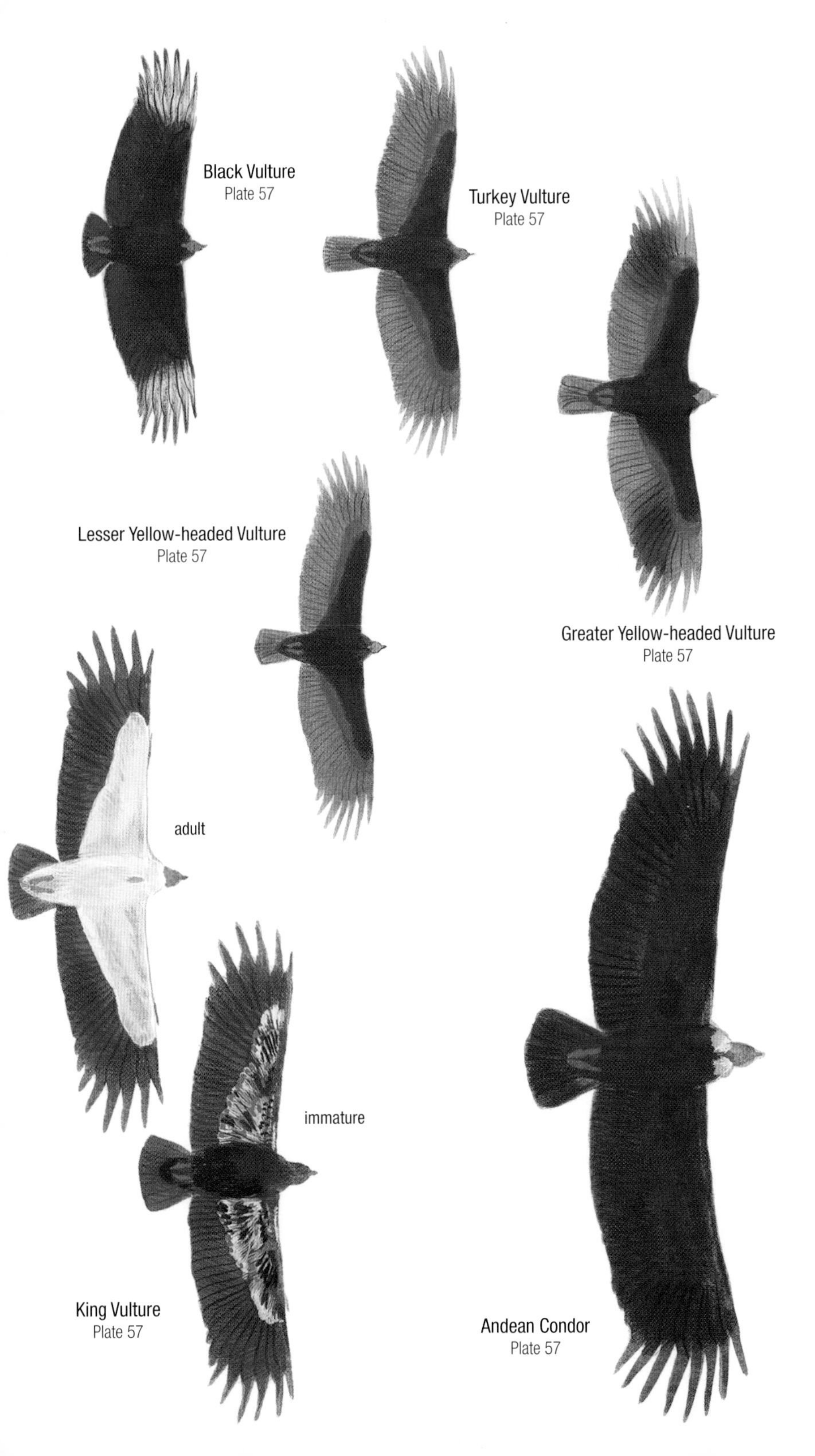
Black Vulture
Plate 57
Turkey Vulture
Plate 57
Lesser Yellow-headed Vulture
Plate 57
Greater Yellow-headed Vulture
Plate 57
adult
immature
King Vulture
Plate 57
Andean Condor
Plate 57

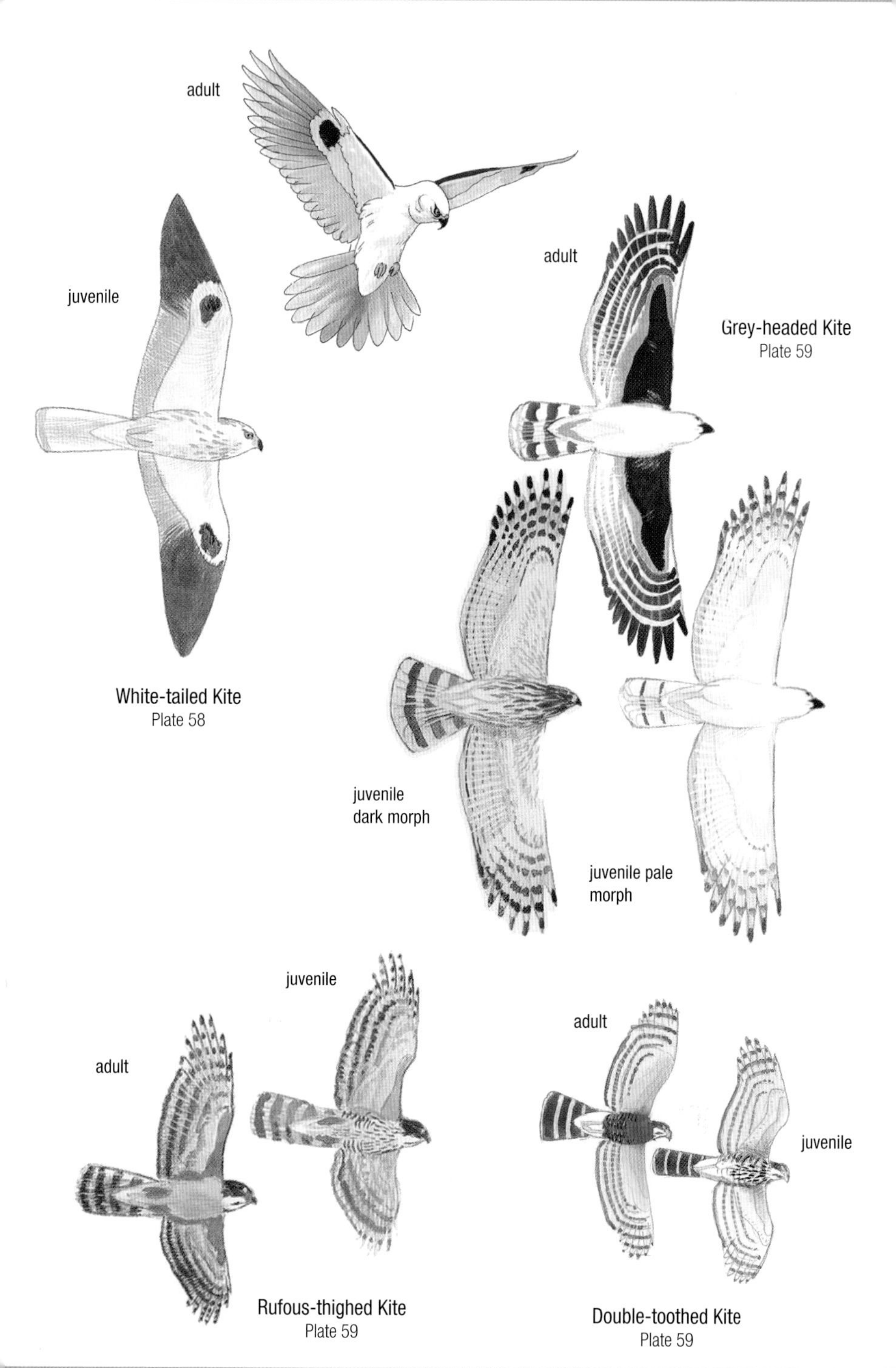
adult
juvenile
White-tailed Kite
Plate 58
adult
Grey-headed Kite
Plate 59
juvenile
dark morph
juvenile pale
morph
juvenile
adult
Rufous-thighed Kite
Plate 59
adult
juvenile
Double-toothed Kite
Plate 59

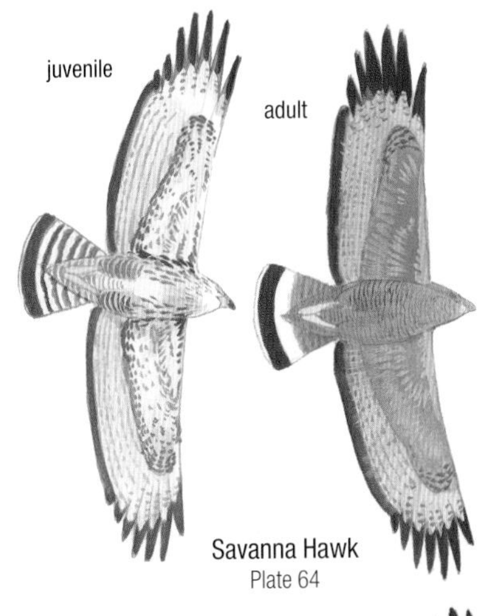

Savanna Hawk
Plate 64

Black-collared Hawk
Plate 64

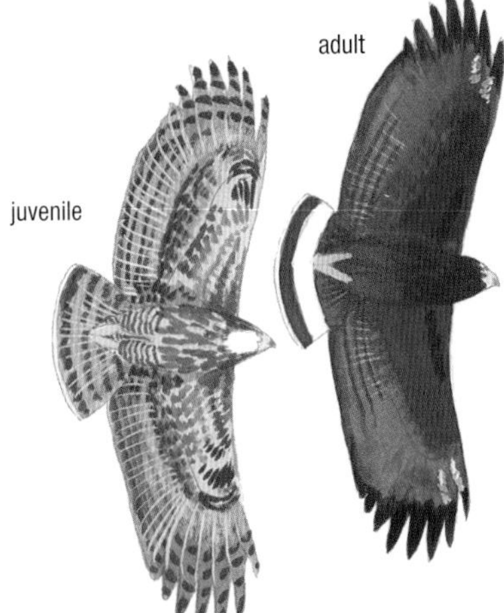

Common Black Hawk
Plate 63

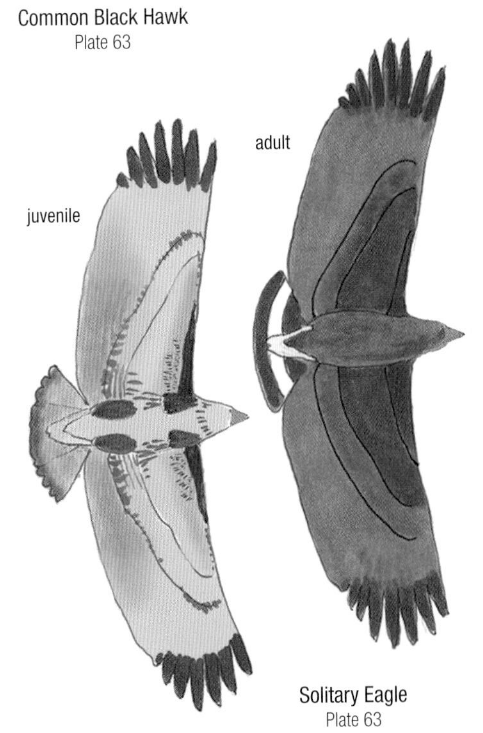

Solitary Eagle
Plate 63

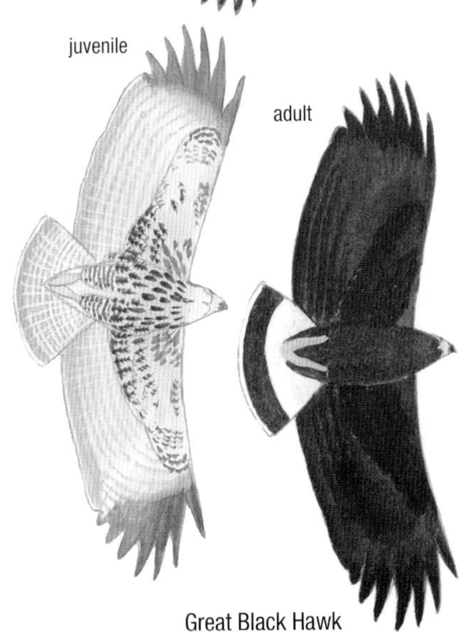

Great Black Hawk
Plate 63

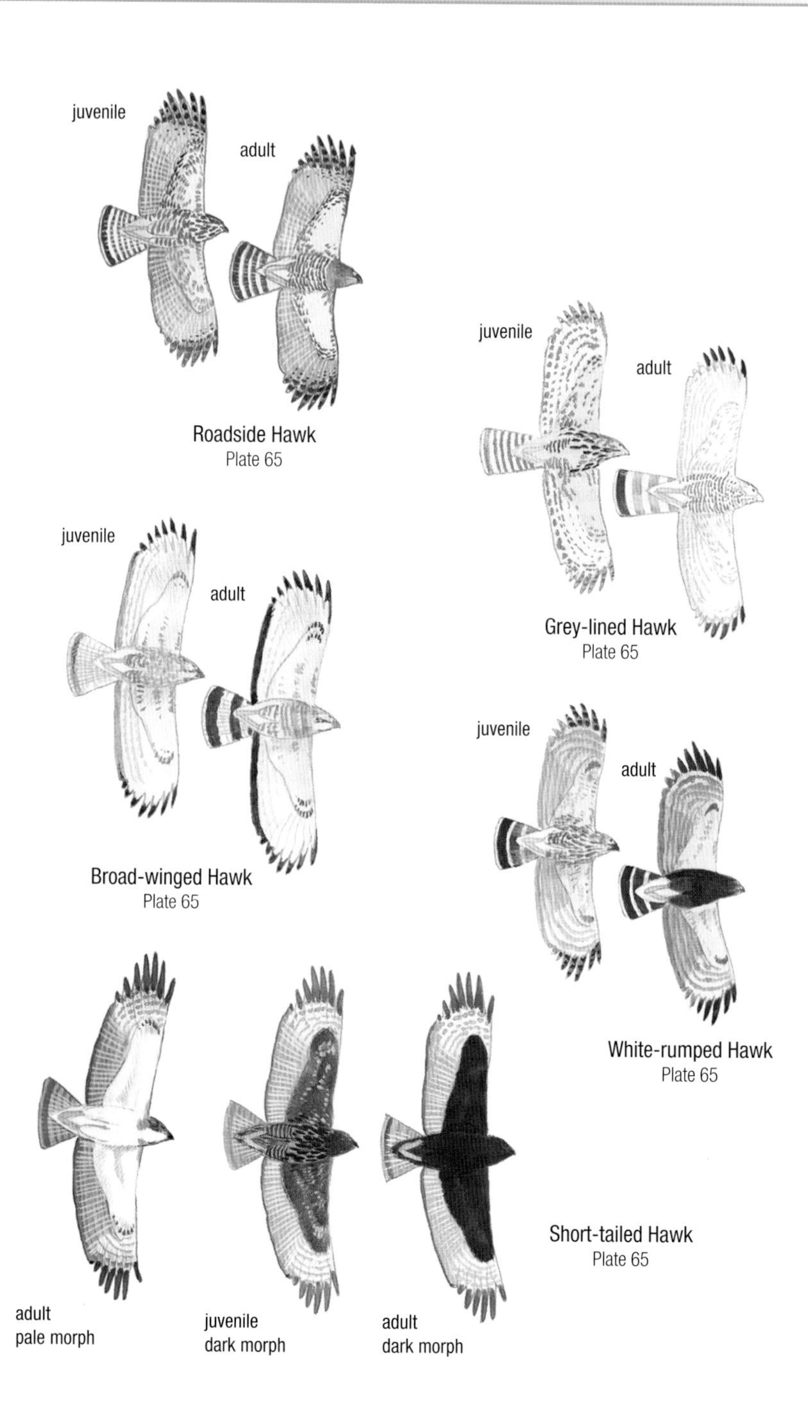
juvenile
adult
Roadside Hawk
Plate 65
juvenile
adult
Grey-lined Hawk
Plate 65
juvenile
adult
Broad-winged Hawk
Plate 65
juvenile
adult
White-rumped Hawk
Plate 65
Short-tailed Hawk
Plate 65
adult
pale morph
juvenile
dark morph
adult
dark morph

White Hawk
Plate 62

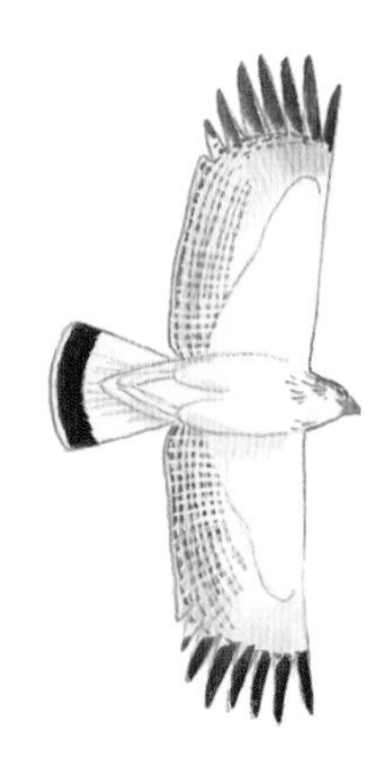

Grey-backed Hawk
Plate 62

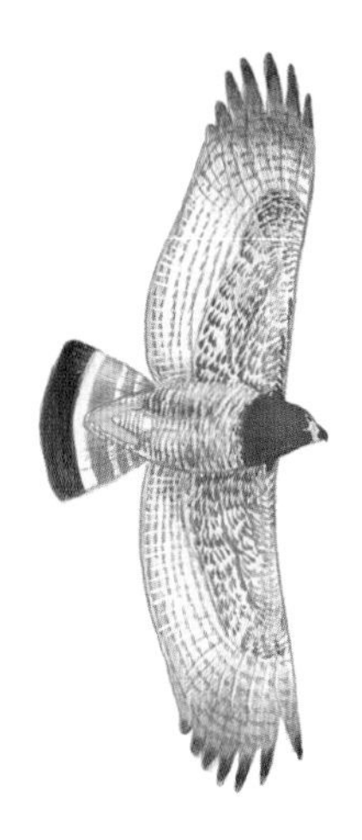

Barred Hawk
Plate 64

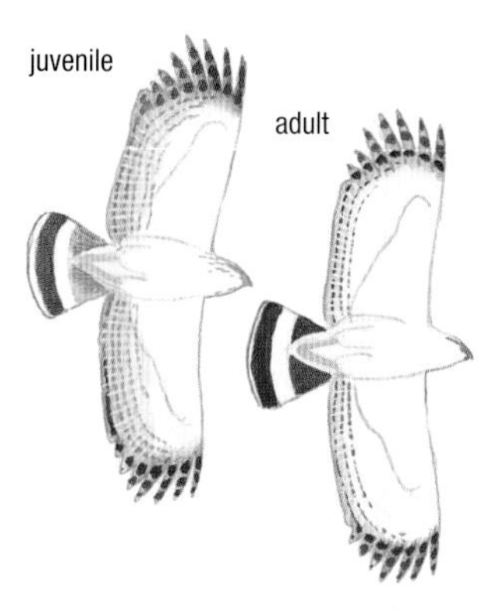

Black-faced Hawk
Plate 62

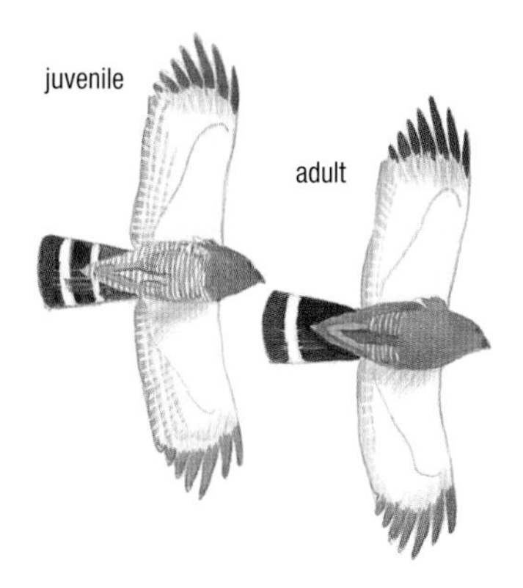

Plumbeous Hawk
Plate 62

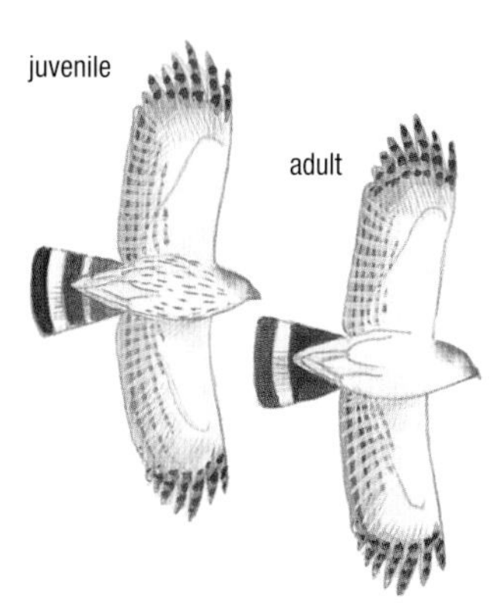

Semiplumbeous Hawk
Plate 62

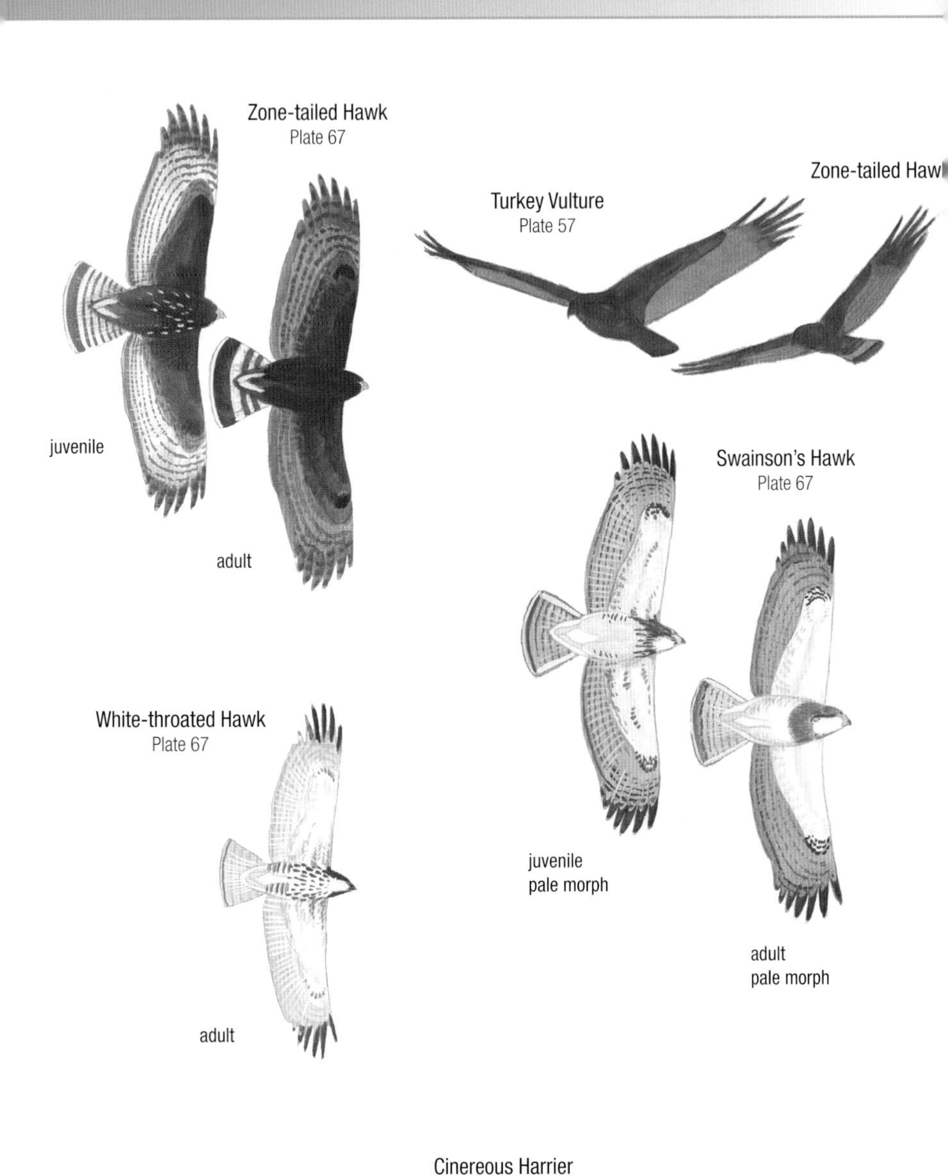

Cinereous Harrier
Plate 60

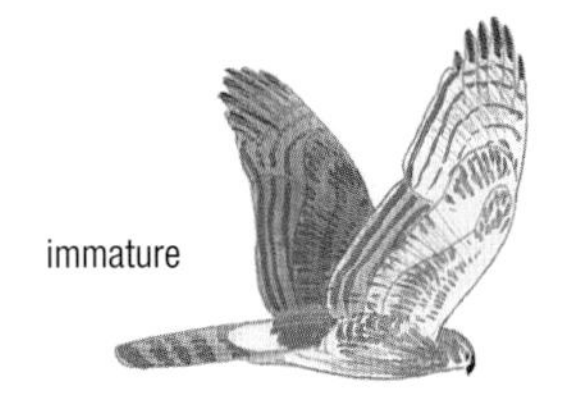

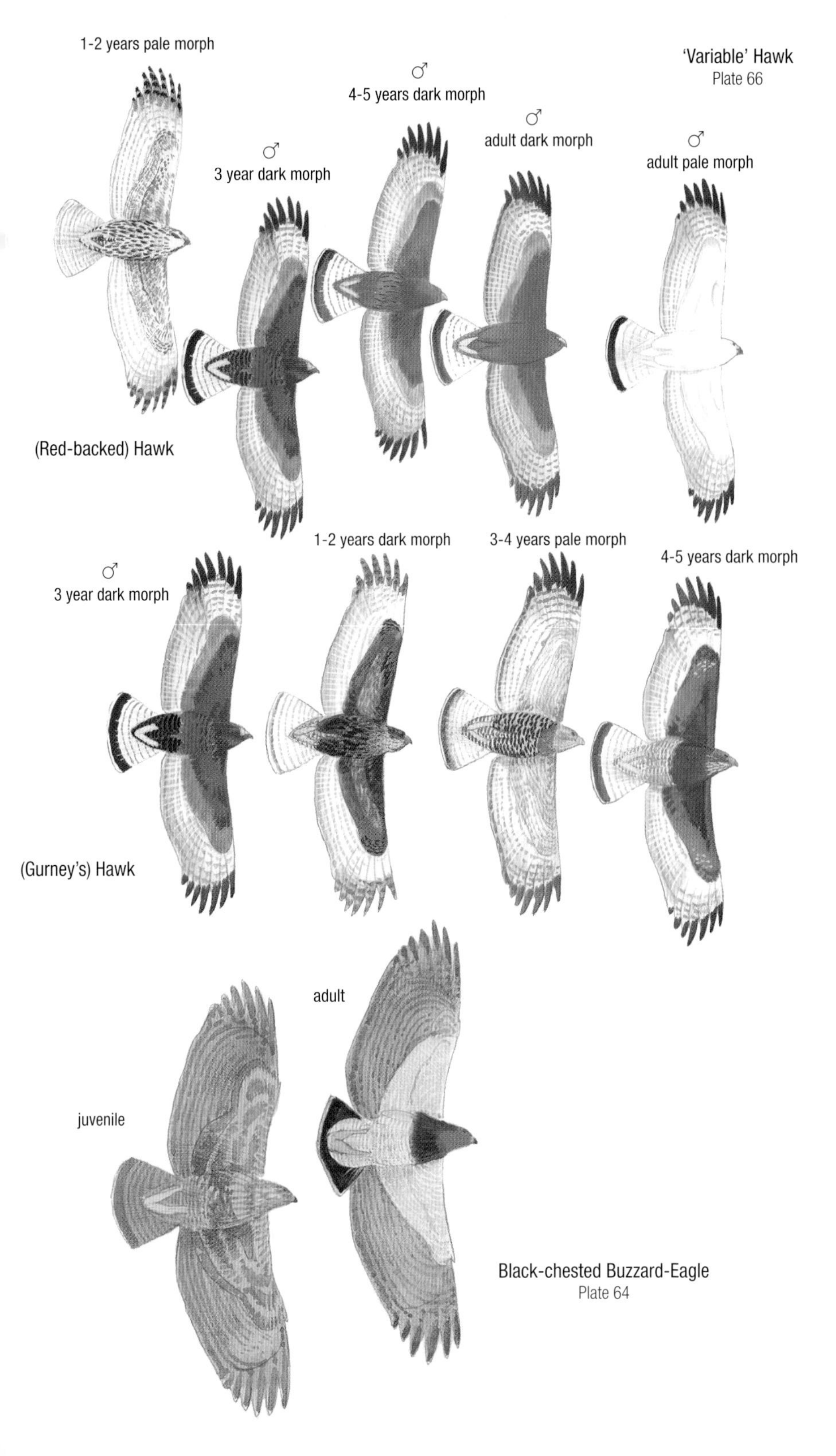

1-2 years pale morph
'Variable' Hawk
Plate 66
♂
4-5 years dark morph
♂
adult dark morph
♂
3 year dark morph
♂
adult pale morph
(Red-backed) Hawk
1-2 years dark morph
3-4 years pale morph
4-5 years dark morph
♂
3 year dark morph
(Gurney's) Hawk
adult
juvenile
Black-chested Buzzard-Eagle
Plate 64

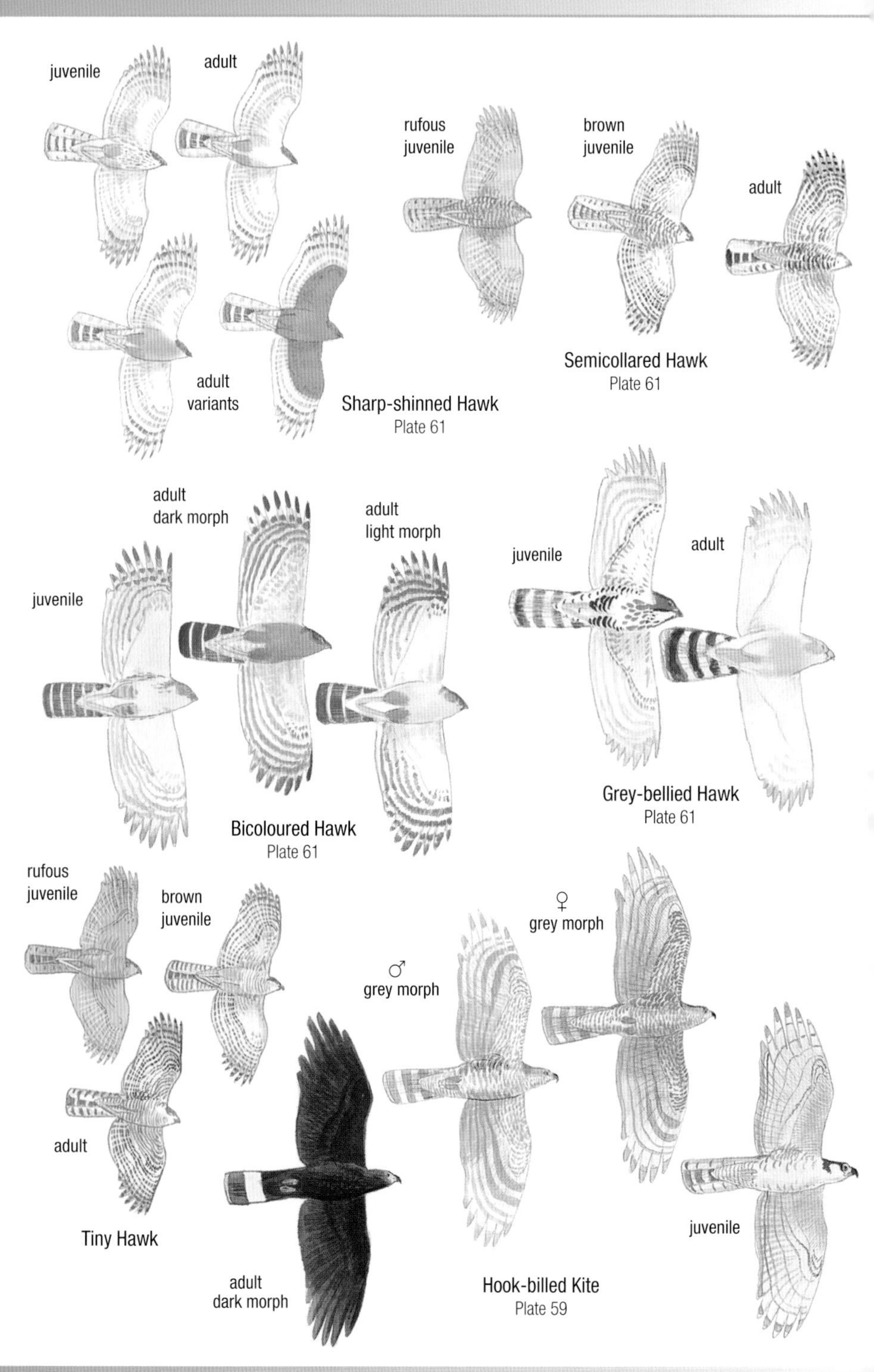
juvenile
adult
adult
variants
Sharp-shinned Hawk
Plate 61
rufous
juvenile
brown
juvenile
adult
Semicollared Hawk
Plate 61
adult
dark morph
adult
light morph
juvenile
Bicoloured Hawk
Plate 61
juvenile
adult
Grey-bellied Hawk
Plate 61
rufous
juvenile
brown
juvenile
adult
Tiny Hawk
♂
grey morph
♀
grey morph
adult
dark morph
juvenile
Hook-billed Kite
Plate 59

adult

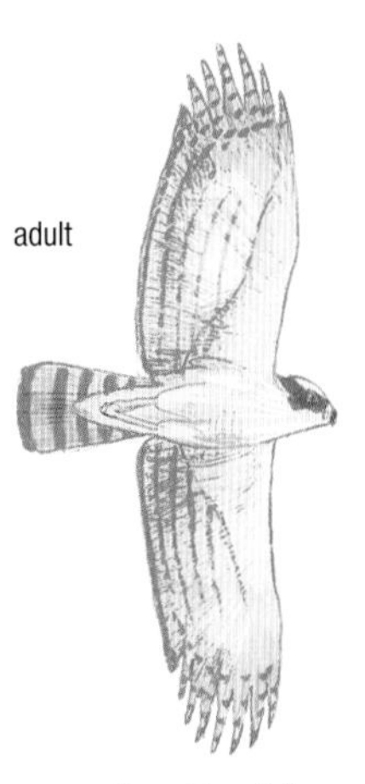

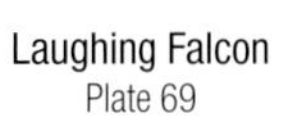

Laughing Falcon
Plate 69

juvenile adult

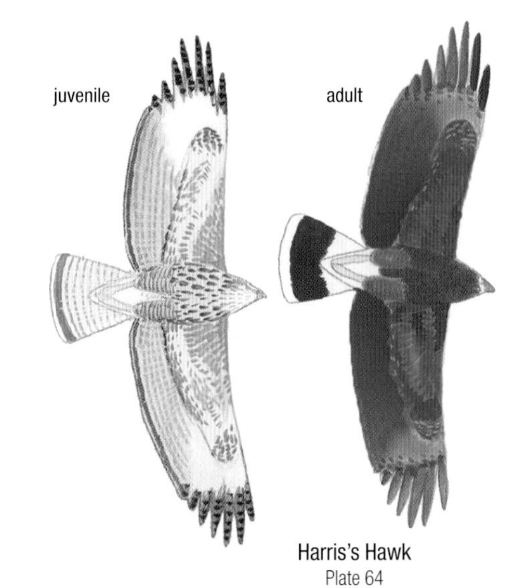

Harris's Hawk
Plate 64

juvenile adult

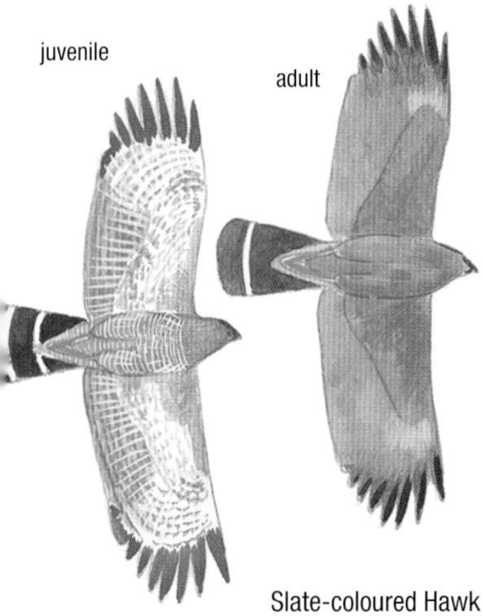

Slate-coloured Hawk
Plate 63

juvenile adult

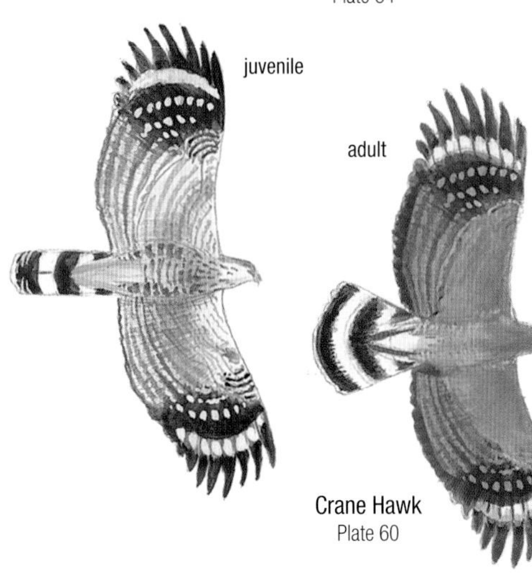

Crane Hawk
Plate 60

adult juvenile

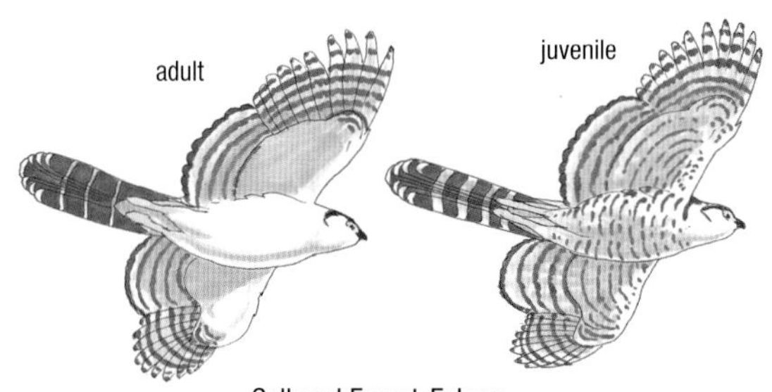

Collared Forest-Falcon
Plate 70

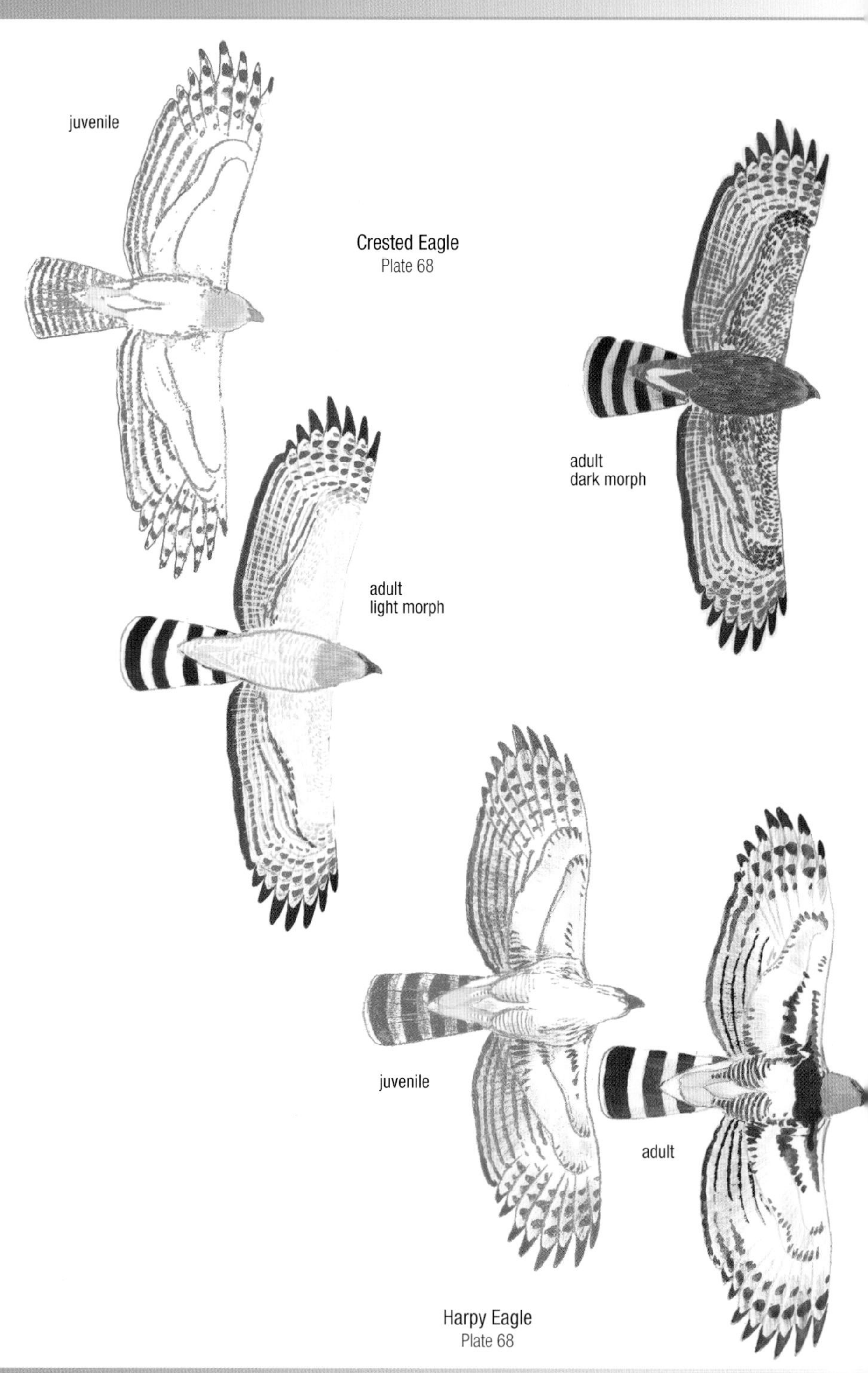
juvenile
Crested Eagle
Plate 68
adult
dark morph
adult
light morph
juvenile
adult
Harpy Eagle
Plate 68

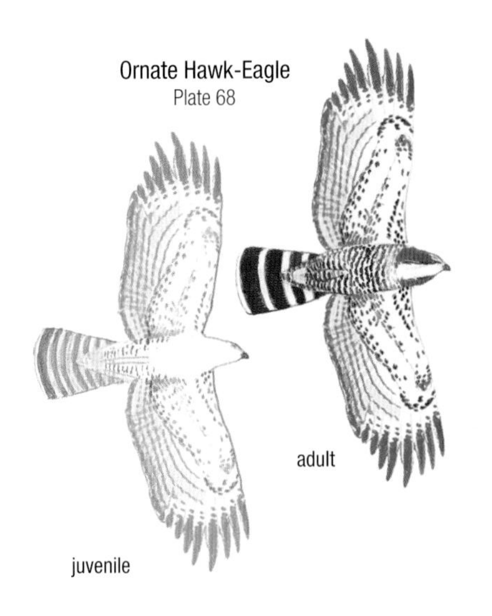

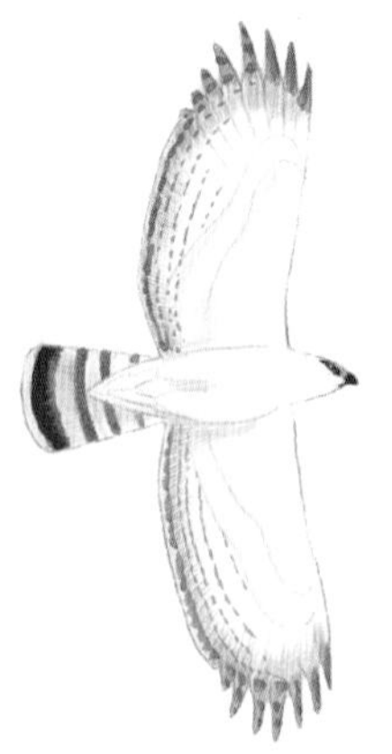

Black-and-white Hawk-Eagle
Plate 67

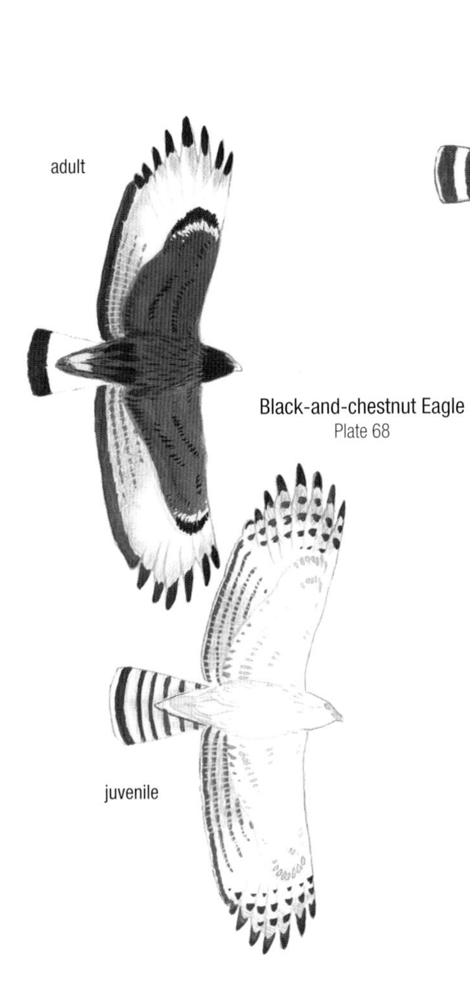

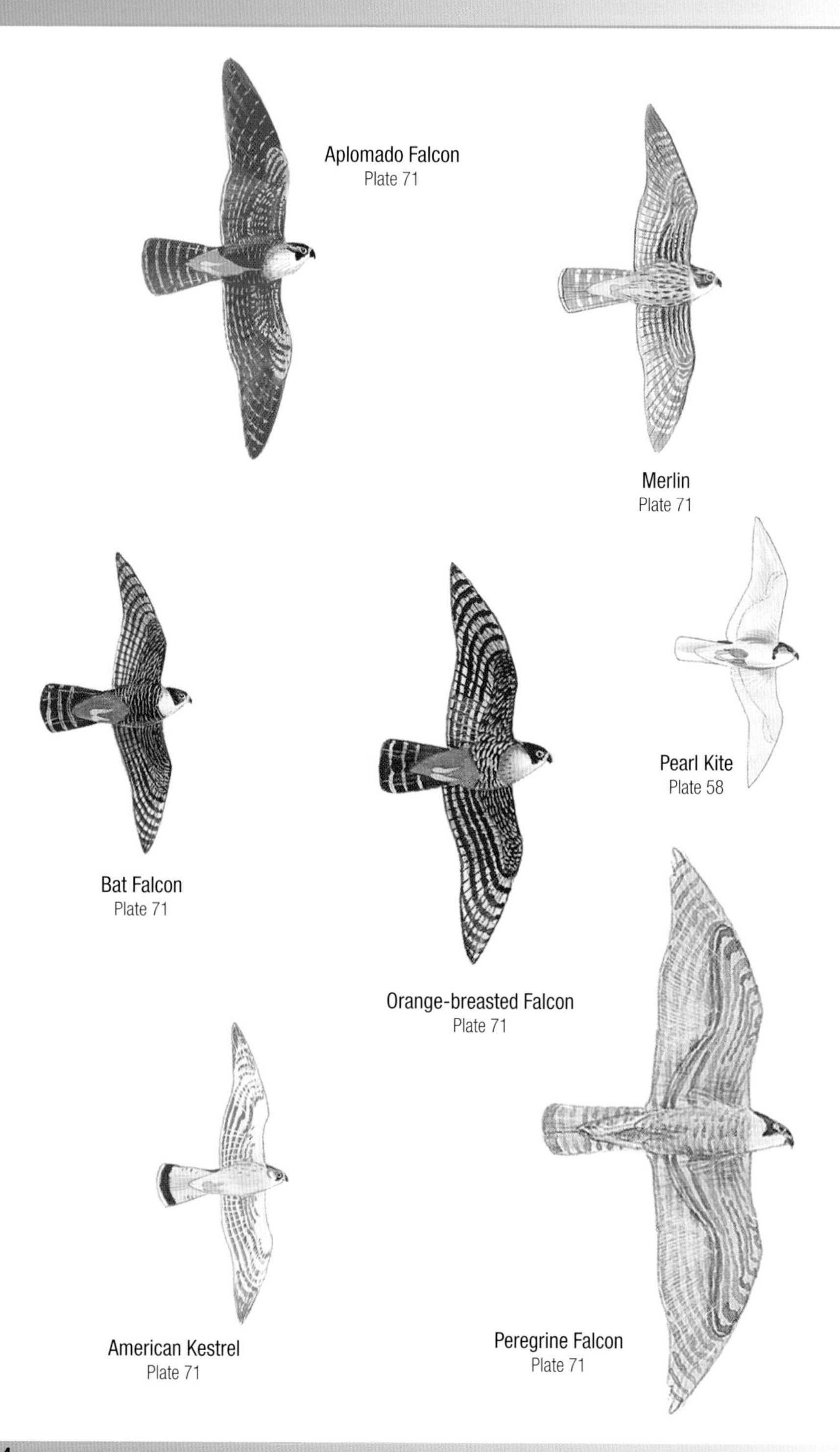
Aplomado Falcon
Plate 71
Merlin
Plate 71
Pearl Kite
Plate 58
Bat Falcon
Plate 71
Orange-breasted Falcon
Plate 71
American Kestrel
Plate 71
Peregrine Falcon
Plate 71

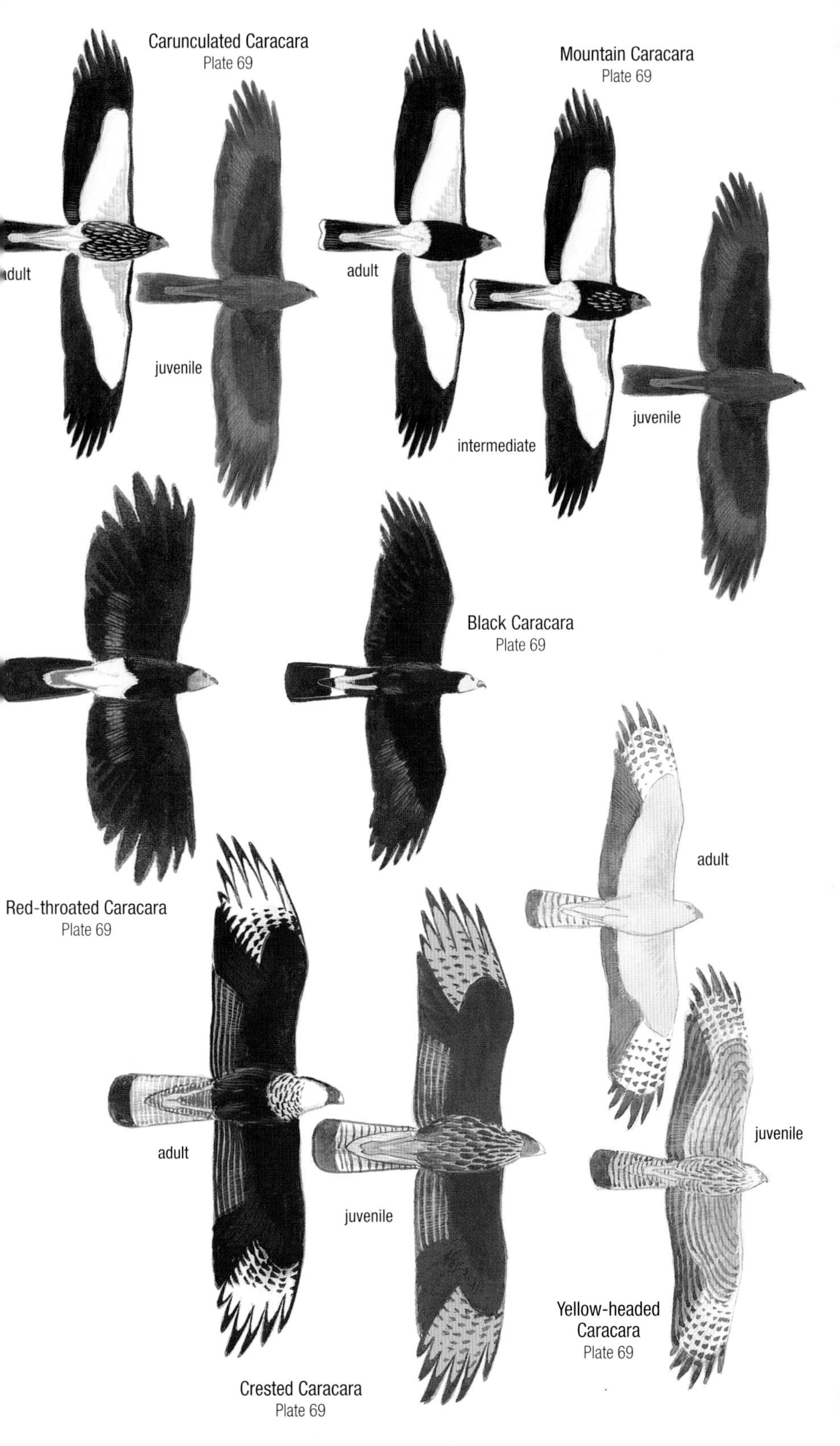
Carunculated Caracara
Plate 69
adult
juvenile
Mountain Caracara
Plate 69
adult
intermediate
juvenile
Black Caracara
Plate 69
Red-throated Caracara
Plate 69
adult
juvenile
Yellow-headed Caracara
Plate 69
adult
juvenile
Crested Caracara
Plate 69

PLATE 57: VULTURES

Black and Turkey Vultures are among the most ubiquitous birds in Ecuador; acquire familiarity with both to improve raptor and vulture identification skills. King Vulture and Andean Condor are becoming scarcer.

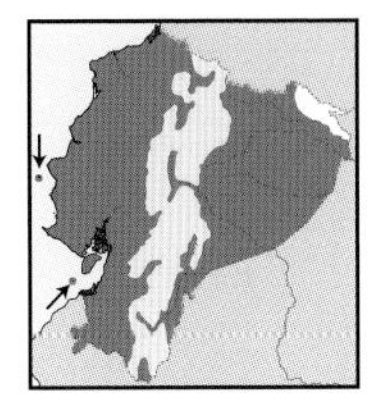

Turkey Vulture *Cathartes aura* 64–76cm / w 160–180cm

W lowlands to subtropics, inter-Andean valleys, locally in E lowlands to foothills (to 2,500m, locally higher). Open and semi-open areas. ***Bare red head***, yellow to brown on nape. Mostly sooty-black. In flight, ***wing-coverts black, flight feathers pale grey, long tail***, wings long, broad. Juv. has head black. Four races: *falklandicus* (offshore islands), *jota* (W), *ruficollis* (E), possibly *meridionalis* (boreal migrant). Tilting flight, marked dihedral, few flaps; frequent around habitation, taking advantage of roadkill. **Voice** Mostly silent. **SS** Greater Yellow-headed Vulture in E (see underwing); Black Vulture (short tail, silvery 'windows' in wingtips). Lesser Yellow-headed recently found in far E (smaller, yellow head, white primary shafts). **Status** Common. See also Plate 50.

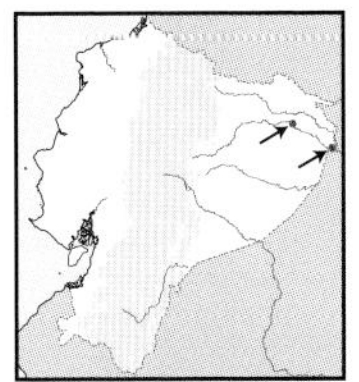

Lesser Yellow-headed Vulture *Cathartes burrovianus* 55–60 cm / w 150–165cm

Two records in E lowlands along Rio Napo (J. Nilsson, N. Athanas). Much like larger Greater Yellow-headed, but flight feathers ***entirely pale below,*** outer primaries with ***white shafts*** from above. Possibly race *urubitinga*. Not well known in Ecuador; elsewhere, tends to tilt while soaring not very high and seldom with other vultures, often seen gliding low over marshland. **Voice** Mostly silent. **SS** Combines underwing pattern of Turkey Vulture with overall appearance of Greater Yellow-headed. **Status** Uncertain, possibly vagrant or spreading? See also Plate 50.

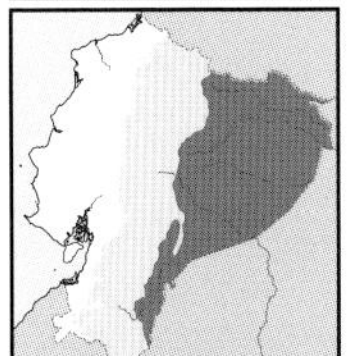

Greater Yellow-headed Vulture *Cathartes melambrotus* 74–80cm / w 165–180cm

E lowlands to foothills (to 1,300m). Above forest, lakes, rivers, adjacent clearings. ***Bare yellow head, bluish*** crown, ***orange*** nape, ***black*** mask. Mostly sooty-black. In flight, ***wing-coverts black, pale grey flight feathers except blackish inner primaries***, long tail, long, broad wings. Juv. has head blackish. Monotypic. More forest-based, flight less tilting, less dihedral than Turkey Vulture; finds food inside forest, not regularly with other vultures. **Voice** Silent. **SS** Bicoloured flight feathers from below separates from Turkey (if head is not seen well); Lesser Yellow-headed Vulture *C. burrovianus* recently found in Ecuador. **Status** Uncommon. See also Plate 50.

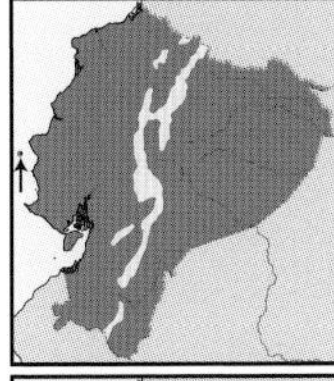

Black Vulture *Coragyps atratus* 56–66cm / w 130–160cm

Widespread up to subtropics and inter-Andean valleys (to 3,000m). Open and semi-open areas. ***Bare blackish head***, blackish legs, all sooty-black. In flight, underwing mainly black, ***silvery-white 'windows' in outer primaries, tail short and wedge-shaped***. Monotypic. Common over modified habitats benefitting from road kills and human garbage; flies with fast wing flaps, short soaring. **Voice** Grunts and sneezes. **SS** From other vultures by head colour and wing pattern; might even resemble some dark raptors (black hawks), but see head shape. **Status** Common. See also Plate 50.

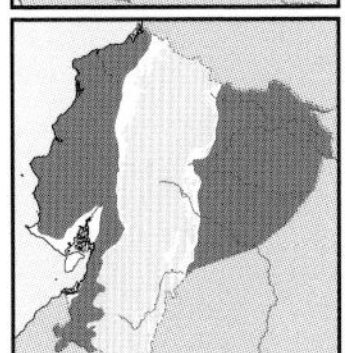

King Vulture *Sarcoramphus papa* 71–81cm / w 170–200cm

E & W lowlands (to 500m). Humid to semi-deciduous forests, adjacent clearings. ***Multicoloured bare head and neck, orange bill, white eyes***. *Immaculate* ***white***, dusky ruff, ***black rump, tail and flight feathers***. In flight, ***black-and-white underwing***, long and broad wings, ***short tail***. Juv. ***sooty*** overall, but some colours in head, ***whitish*** in underwing-coverts. Monotypic. Forest-based, singles to small groups depend on but dominate other (olfactory) vultures. **Voice** Grunts and hisses. **SS** Juv. with sooty-black Turkey and Greater Yellow-headed Vultures; flight more steady. **Status** Uncommon, rarer in W. See also Plate 50.

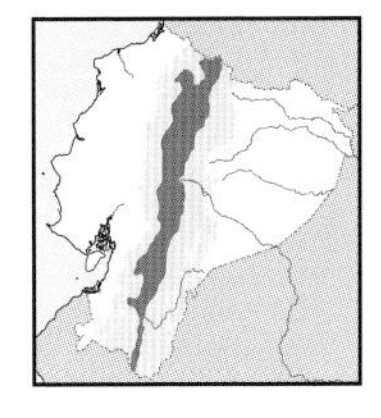

Andean Condor *Vultur gryphus* 100–130cm / w 275–320cm

Local in highlands (2,000–4,500m). Mainly remote *páramos*. ***Huge***! Pink bare head, protruding ***wrinkled comb*** in male. ***Pure white ruff. Black,*** with ***extensive white in wings***. ♀ ***lacks comb***, head duskier. Juv. ***brown***, lacks comb, dusky-brown head, ***pale brown ruff***. In flight, contrasting white flight feathers above, long and broad wings, short tail. Monotypic. Now rarely seen, solitary, pairs or (if lucky!) small parties soaring high overhead, with outstretched primaries. **Voice** Silent. **SS** Unmistakable. **Status** Very rare, declining, CR in Ecuador, NT globally. See also Plate 50.

uvenile
adult
jota, falklandicus
Lesser
Yellow-headed Vulture
ruficollis
meridionalis
boreal migrant
Turkey Vulture
Greater
Yellow-headed Vulture
3rd year
King Vulture
Black Vulture
adult
juvenile
2nd year
♂
♀
Andean Condor
♂
immature
♂

Two largely white and distinctive kites, plus two mainly grey species that are more complicated to identify, and one of the smallest diurnal raptors. Kites have long pointed wings, more reminiscent of a typical falcon than an accipitrid. All live mostly in open areas, but Swallow-tailed and Plumbeous regularly fly over extensive forest. Mississippi is a rare boreal winter visitor, perhaps only vagrant, but probably overlooked.

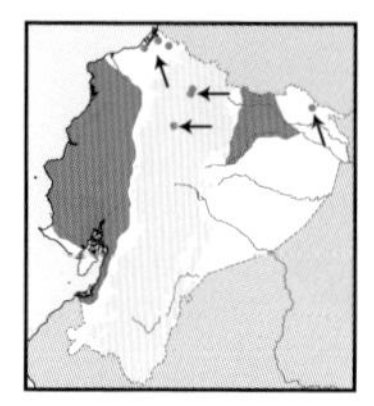

White-tailed Kite *Elanus leucurus* 40–42cm / w 90–100cm

Locally in N lowlands to foothills (to 1,200m). Open terrain with scattered trees. Falcon-like with long pointed wings, square tail. Mostly ***grey-white*** with ***blackish shoulders***, black orbital area, ***red eyes***. Juv. similar but ***brownish-grey*** above edged pale, ***streaky head and chest***. In flight: ***white underwing-coverts, black carpal patches***. Nominate. Deep graceful wingbeats; keeps wings raised in gliding, in V when soaring; regularly hovers. **Voice** Weak piping and chirping. **SS** *Ictinia* kites darker, no black shoulder; Plumbeous has bold rufous windows in primaries; tiny Pearl has shorter wings, no carpal patches, no black shoulder. **Status** Rare to uncommon, probably spreading. See also Plate 51.

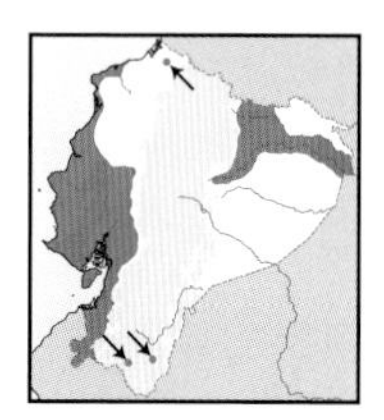

Pearl Kite *Gampsonyx swainsonii* 22–26cm / w 45–55cm

SW lowlands, NE lowlands to foothills (to 800m in E, locally to 1,650m). Arid land, spreading with deforestation. ***Tiny. Narrow pointed wings***, longish square tail, short legs. Blackish above, ***pure white below, narrow white nuchal collar***, cheeks and flanks variably ***tinged buff to rusty-orange***. Juv. duller brown above, edged pale, paler head, rusty-buff below. In flight: ***all white below***, hooded. Race *magnus* (SW), possibly *swainsonii* (E, NW). Flight recalls small falcons, even hovering briefly, but soars and glides more; often perches atop posts, logs, etc. **Voice** High-pitched *kipkipkipkipkip...* becoming *kiririririri* in flight. **SS** Confusion unlikely; in SW occurs with similar-sized American Kestrel, but pattern strikingly different. **Status** Uncommon, spreading?

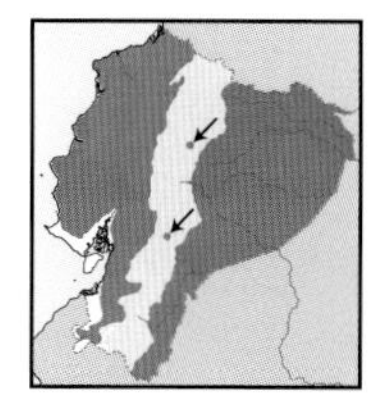

Swallow-tailed Kite *Elanoides forficatus* 56–61cm / w 120–135cm

E & W lowlands to subtropics (mostly to 1,600m, locally higher). Humid to montane forest, borders, adjacent fields. Unique ***black-and-white*** pattern with ***long deeply forked tail***. Nominate (boreal migrant) has blue-black gloss above, *yetapa* (resident) is green and bronze. Graceful, agile flight with much gliding and soaring, often low over canopy; dozens may gather to hunt aerial insects, regularly consorting with Plumbeous Kite. **Voice** Loud, piping, rather nasal *këy-këy-këy..., key-key-key-kiy-kip-kip....* **SS** Unmistakable. **Status** Common. See also Plate 50.

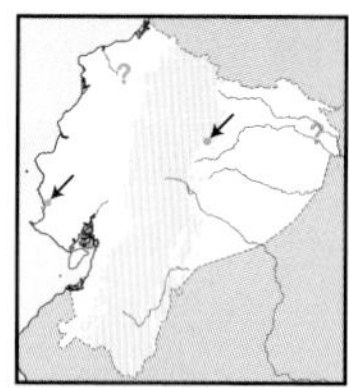

* Mississippi Kite *Ictinia mississippiensis* 35–38cm / w 75–85cm

Uncertain, few records from E & W foothills (below 1,000m). Open areas? Closely resembles Plumbeous Kite, but ***lacks*** tail-bands, no rufous in primaries (in flight), ***white secondaries***, paler grey overall with ***more contrasting head***. Juv. difficult to separate unless seen very well; focus on pale rufous patches in Plumbeous primaries, clearer tail-bands, sparser and darker streaking below (no hint of rusty streaks). Monotypic. Few migrants move through Ecuador, with regular route apparently further E; buoyant flight as Plumbeous but usually more direct migrating flight. **Voice** Not vocal in wintering grounds? **SS** Beware of paler-headed Plumbeous. **Status** Possibly rare transient (Apr, Oct, Dec, Jan). See also Plate 50.

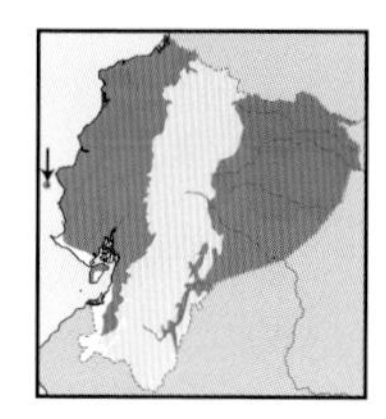

Plumbeous Kite *Ictinia plumbea* 34–38cm / w 70–85cm

E & W lowlands to foothills (to 1,000m). Humid forest, woodland, adjacent fields. ***Long, pointed, narrow wings***, square tail. ***Mostly grey***, paler head, more slate above. ***Short orange legs***. In flight: ***rufous windows in primaries***, two white tail-bands. Juv. dusky above edged pale, ***little rufous in primaries, creamy below streaked dusky***, brownish-grey secondaries tipped white. Monotypic. Buoyant flight, much gliding and high soaring; may congregate at swarms of flying insects. **Voice** Rather silent, occasionally utters a mournful, descending shrill *yeé-kweee*. **SS** Rarer Mississippi Kite (which see). **Status** Fairly common. See also Plate 50.

White-tailed Kite
adult
immature
swainsonii
magnus
Pearl Kite
adult
adult
adult
juvenile
forficatus
yetapa
Swallow-tailed Kite
juvenile
Mississippi Kite
adult variant
adult
immature
juvenile
Plumbeous Kite

Four kites, two forest-based (*Harpagus*) and two more tolerant of disturbed habitats. Grey-headed and, especially, Hook-billed remarkable for their plumage variation related to age and colour morphs. Hook-billed regularly soars, the other species soar little, especially both *Harpagus*, which often follow troops of monkeys.

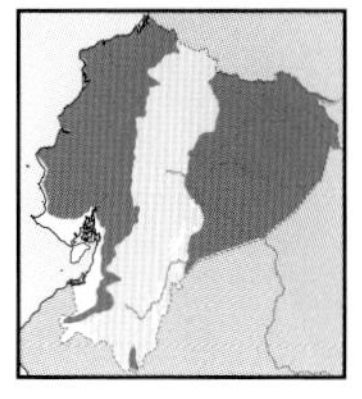

Hook-billed Kite *Chondrohierax uncinatus* 38–44cm / w 80–100cm

E & W lowlands, to foothills (E), subtropics (W), extreme SE in Zumba region (to 800m in E, to 1,800m in W; locally higher). Humid forest, woodland. ***Heavy bill, short yellow legs, white eyes. Greenish-yellow cere and lores, yellow loral spot.*** Grey morph: ***ashy-grey finely barred below***, broad grey mid tail-band, narrow white tip. Dark morph: blackish, two broad white tail-bands. ♀: browner above, ***more rufescent cheeks and collar, below barred pale rufous and whitish***, two grey tail-bands. Juv. has dark eyes, ***whitish collar and underparts***. Dark juv. Blackish-brown, yellow in face. In flight: boldly ***barred remiges***, bold tail-bands, ***very rounded wings***. Nominate. Glides on bowed wings, no long soars, often low above canopy. **Voice** Brief, fast, ascending-descending *te-khe'kha'kha'kha...*. **SS** Heavy bill, yellow lores, broad S-shaped wings distinctive; variable underwing / undertail pattern might cause confusion. **Status** Rare. See also Plate 54.

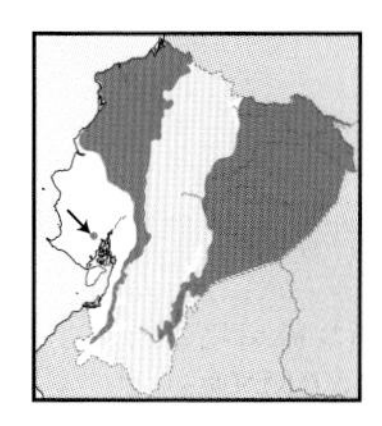

Grey-headed Kite *Leptodon cayanensis* 45–58cm / w 90–110cm

E, W lowlands to lower subtropics, rarer in SW (to 1,600m). Humid forest, borders, adjacent fields. ***Pale bluish-grey soft parts. Slate-grey above, pale grey cap, white underparts.*** Juv. has ***yellow bare parts; mostly white, black crown***, dusky upperparts. Dark juv. ***dull greyish above, streaked greyish-white below***. In flight: ***long-winged, black underwing, grey-and-white barred remiges, long tail***, three white bands below, white tip. Juv.: white underwing, ***remiges*** as ad. Nominate. Rarely soars above canopy, flap-and-glide flight; unobtrusive. **Voice** Nasal, loud, rising-descending or rising *küa-küa-küa-küa...*; nasal *nnyëëuuw!* **SS** White *Leucopternis* hawks (broader-winged, shorter-tailed, shorter-legged); pale juv. resembles Black-and-white Hawk-Eagle (orange cere, black mask, broader wings, feathered legs); see juv. Hook-billed Kite (heavy bill, short yellow legs, dark carpal patch). **Status** Uncommon. See also Plate 51.

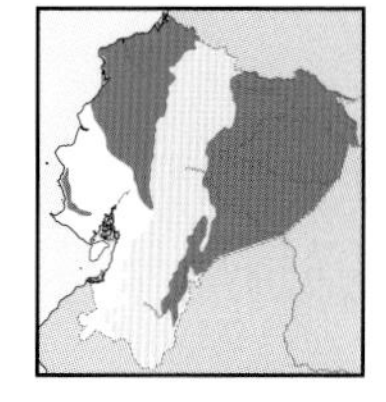

Double-toothed Kite *Harpagus bidentatus* 31–36cm / w 60–70cm

E & NW lowlands to subtropics (to 1,800m). Humid forest canopy, borders. Grey above, ***bluish-grey head***, underparts rufous to barred white and rufous, ***white throat, black mesial throat stripe, fluffy white vent***, whitish tail-bands. Juv. streaky below, ***mesial throat stripe***. In flight: ***short, rounded wings***, long tail, ***curved inner trailing edge, boldly barred primaries*** (less prominent in juv.). Nominate (E), *fasciatus* (W). Sluggish below canopy, often follows monkeys; briefly soars over forest, sometimes high. **Voice** Flycatcher-like, whistled *whee-yee..típ!, wheee-yee...típ, whee-yee....* **SS** *Accipiter* hawks, especially Bicoloured (longer-legged, fiercer expression, straighter trailing edge, no mid-throat stripe, no fluffy white vent); rare Rufous-thighed Kite (noticeable rufous thighs, but juv. only faintly rufous). **Status** Uncommon. See also Plate 51.

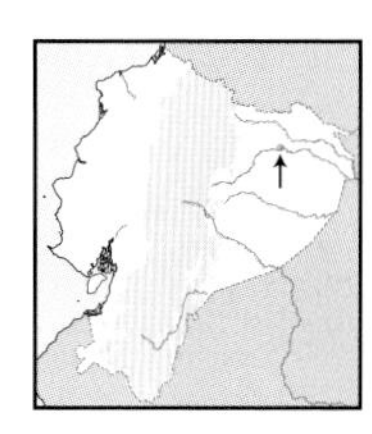

* Rufous-thighed Kite *Harpagus diodon* 31–36cm / w 60–70cm

Uncertain, single record in E lowlands (300m). *Terra firme* forest? Recalls Double-toothed Kite but ***head blacker, greyer*** tail-bands, ***grey below***, fainter throat stripe, ***rufous thighs evident***, pale rufous underwing-coverts. Juv. more similar, not always safely separated, but underwing often ***washed rufous-buff***, greyer tail-bands, ***pale rufous thighs, barred secondaries in flight***. Monotypic. Elsewhere apparently follows mixed-species flocks and army ants. **Voice** Weak, piping *whee-yeép, whee-whee-yép...yép-yép*. **SS** Larger Bicoloured Hawk (no mesial throat stripe, longer tail and feet, fiercer expression; juv. duller than Rufous-thighed, little rufous in thighs, less streaked below, pale-edged wing coverts). **Status** Rare, possibly accidental (Jul). See also Plate 51.

♂
grey morph
Hook-billed Kite
dark morph
immature
grey morph
♀
grey morph
immature
light morph
adult
immature
dark morph
Grey-headed Kite
bidentatus
juvenile
fasciatus
bidentatus
juvenile
♂
♀
♂
♀
Double-toothed Kite
adult
juvenile
adult
juvenile
adult
juvenile
Rufous-thighed Kite

A potpourri that includes the aquatic Osprey – sometimes seen soaring elsewhere – two smallish dark kites associated with water, and two distinctive raptors: Crane Hawk in forest and at edges and the harrier in open *páramo*. Cinereous Harrier has a unique buoyant flight that separates it from other raptors; it is apparently declining due to habitat loss.

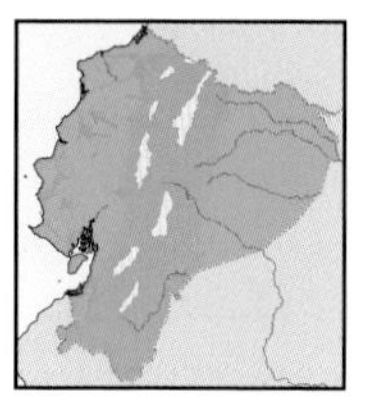

Osprey *Pandion haliaetus* 54–59cm / w 125–170cm

Widespread but commoner in lowlands (to 3,000m). Freshwater bodies. Long, pointed, angled wings, ***M-shaped***, prominent head. Dull brown above and on mask, ***white head and below***, pale-barred tail. Long greyish legs. In flight: ***white underwing, black carpal patches***, prominent ***M-shaped wings***. Race *carolinensis*. Always near water, perches upright atop stumps, logs, posts; slow, rather deep wingbeats, powerful hovering. **Voice** Loud, yelping *cleeyé!-cleeyé!..* rising and falling, inflected. **SS** Confusion unlikely as wing shape, underwing pattern, flight, behaviour, and habitat distinctive. **Status** Uncommon boreal migrant (year-round). See also Plate 50.

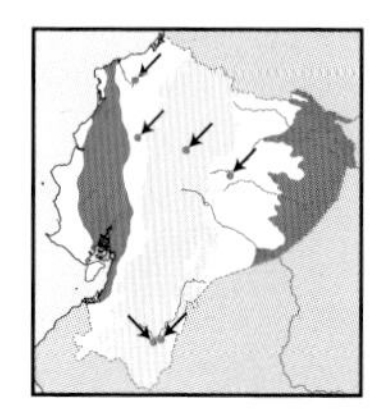

Snail Kite *Rostrhamus sociabilis* 40–45cm / w 100–115cm

SW lowlands, locally in NW & E lowlands (below 400m, not above 50m in W). Freshwater lakes, marshes, rice fields. ***Deeply hooked bill. Slate-grey, broad white tail base*** and narrower tip. ***Orange bill base and slender legs***. ♀ dull brown, ***blotched whitish below***, buffy eyebrow and face, tail as ♂, duller soft parts. Juv. similar but ***streaky head and underparts, yellowish*** soft parts, paler tail. Nominate. Floppy flight, glides and soars on bowed wings; social, perches low above water on posts or stumps. **Voice** Throaty *kra-khakhakhakha, kru- kr'u'hu'hu'hu*. **SS** Smaller Slender-billed Kite (E) entirely dark, shorter tail; juv. darker, tail shorter, barred black and white, wings shorter. Bill shape and tail pattern separate from black hawks, Slate-coloured and Crane Hawks. **Status** Uncommon, local, VU in Ecuador. See also Plate 50.

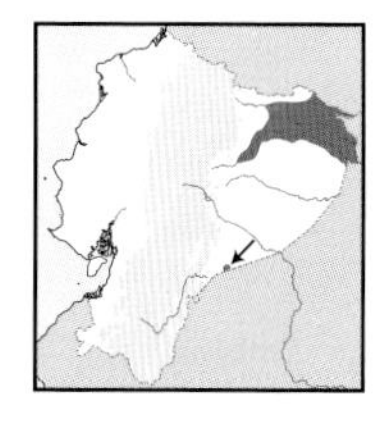

Slender-billed Kite *Helicolestes hamatus* 37–41cm / w 80–90cm

NE lowlands (below 300m). *Várzea* and *igapó* forest, swamps. Wings and tail shorter than similar Snail Kite, ***deeply hooked bill***. Entirely ***slate-grey, bright orange bare parts***. Juv. similar but upperparts edged pale, flight feathers obscurely barred below, 2–3 white tail-bands. Monotypic. Easily overlooked as usually perches in cover, foraging by still-hunting mostly from low perches. **Voice** Very nasal, upslurred *whyëëeeeaa!* **SS** Larger Snail Kite (longer wings, more floppy flight, extensive white in tail; juv. also lacks tail bands); cf. also Slate-coloured Hawk (bulkier, smaller head, single white tail-band; juv. has strongly barred underwing). **Status** Rare, rather local. See also Plate 50.

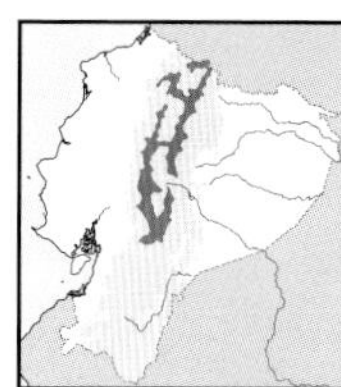

Cinereous Harrier *Circus cinereus* 43–50cm / w 90–115cm

N Andes S to Cañar (3,000–4,000m, locally higher?). Open grassy and dry *páramo*, adjacent fields. Longish wings, tail and legs. ***Grey*** (♂) or brown-mottled (♀) ***upperparts*** to chest, ***belly barred rufous and white***. Juv. as ♀ above, coarsely ***streaked*** brown below, except belly. In flight: ♂ has ***striking white underwing, black primaries and subterminal trailing edge, bold white rump***; ♀ and juv. mottled / streaked underwing-coverts. Monotypic. Slow graceful flaps low above ground, dihedral soaring, little gliding; perches on ground, boulders, etc. **Voice** Harsh, chattering *ghegheghe....* **SS** Few sympatric hawks and Black-chested Buzzard-Eagle are broader-winged, shorter-tailed, stockier, different flight and behaviour; see juv. Carunculated Caracara. **Status** Rare, rather local, NT in Ecuador.

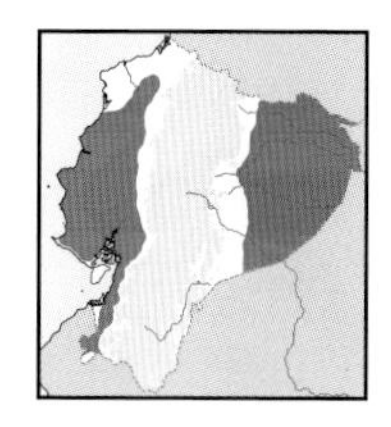

Crane Hawk *Geranospiza caerulescens* 46–51cm / w 75–110cm

E & SW lowlands (to 400m). Humid to deciduous forest, borders. Slim, small-headed, long-tailed, long-legged. ***Grey cere, orange legs***. Mostly ***grey, two broad white tail-bands***. Juv. ***paler grey***, upperparts mottled dark, underparts somewhat ***streaked / barred grey-whitish to creamy***, whiter face, barred thighs. In flight: small projecting head and long tail provide curious silhouette; ***white arc in primaries***. Nominate (E), *balzarensis* (W). Often singly, agile in canopy and borders, inspecting crevices and nests; slow floppy flight, with short glides and brief flat-winged soaring. **Voice** Loud, whistled, short *whéeeeoo!* **SS** Confusion unlikely, but cf. black hawks, Slate-coloured and Bicoloured Hawks. **Status** Rare. See also Plate 54.

Osprey
♂
♀
immature
juvenile
♂
♀
Snail Kite
immature
adult
Slender-billed Kite
♀
immature
♂
Cinereous Harrier
♂
♀
immature
adult
nmature
adult
balzarensis
juvenile
Crane Hawk
caerulescens

PLATE 61: *ACCIPITER* HAWKS

Small to mid-sized, forest-based hawks that predate birds in swift, agile flight. One species thrives near treeline, sometimes even venturing into open *páramo* and adjacent agricultural fields. Identification complicated in some species, and even between this and other genera of similar-sized forest raptors. *Accipiter* hawks have pale eyes and fierce expression.

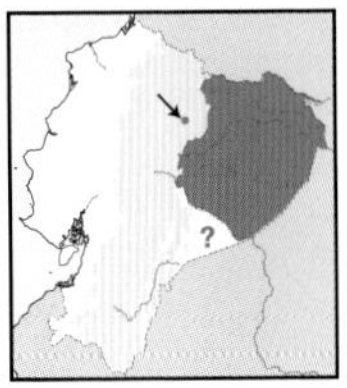

Grey-bellied Hawk *Accipiter poliogaster* — 38–51cm / w 70–85cm

E lowlands locally to subtropics (mostly below 400m, once at 1,950m). *Terra firme, várzea* forest, borders. Robust, heavy bill, short sturdy legs. ***Black and pale grey***, yellow in face. Juv. 'mimics' Ornate Hawk-Eagle: ***rufous cheeks, nuchal collar and breast-sides***, black crown, ***malar stripe***, chest mottling and ***belly barring***. In flight: whitish underwing, capped appearance. Monotypic. Shallow wingbeats, glides on flat wings, no soaring; chases birds agilely through forest interior. **Voice** Weak, short series of clear *wéyk* notes. **SS** Bicoloured Hawk (darker below, rufous thighs); Slaty-backed Forest-Falcon (graduated tail, no cap, shorter legs, less yellow in face); juv. at least 50% smaller than Ornate Hawk-Eagle (feathered tarsi, long crest). **Status** Rare, NT globally. See also Plate 54.

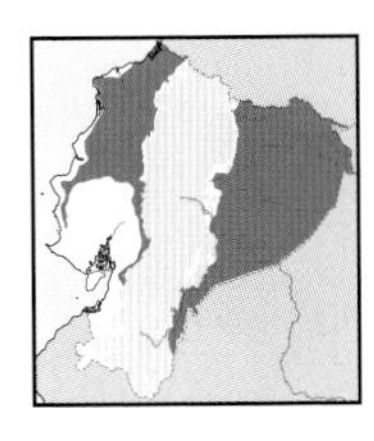

Tiny Hawk *Accipiter superciliosus* — 21–28cm / w 40–50cm

E & W lowlands to foothills, not in arid SW (to 900m). Humid forest, borders. ***Tiny***. Greyish-brown above, ***white barred dusky below, plain whitish throat***, pale cheeks, dark crown. ♀ and juv. browner above, ***below thickly barred brown to rufescent-brown and whitish, plain throat***. In flight: barred underwing, more mottled in juv. Nominate (E), *fontainieri* (W). Rapid shallow wingbeats, little gliding, no soaring; perches in subcanopy and dashes after flying birds. **Voice** Sweet, shrill *wék-wék-wék-wék... wé-wé-wék*. **SS** Slightly larger Semicollared Hawk (pale nuchal collar, sparser barring below, mottled cheeks; little range overlap); Double-toothed Kite (black throat stripe) and longer-tailed forest-falcons. **Status** Rare.

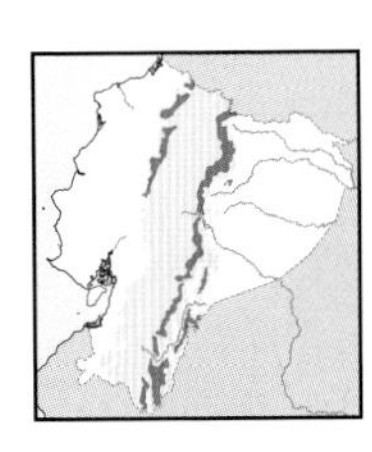

Semicollared Hawk *Accipiter collaris* — 24–27cm / w 45–55cm

Local on W & E Andean slopes (1,500–2,200m). Montane forest, borders. Small. Sooty-brown above, ***darker crown, mottled / streaked cheeks, pale nuchal collar***, whitish below ***evenly barred dusky, except white throat***. Juv. ***rufescent-brown above, pale rusty nuchal collar***, darker crown, rufescent barring below. In flight: ***whitish throat and nuchal collar***, juv. has up to six pale tail-bands. Monotypic. Rapid shallow wingbeats, *soars* briefly, dashes through canopy after prey. **Voice** Piercing, short *tikekeké*. **SS** Separated from Sharp-shinned Hawk by barred underparts, white throat, shorter tail, mottled cheeks; more similar to lower-elevation Tiny Hawk (no nuchal collar, plainer cheeks, heavier barring below). **Status** Rare, NT globally. See also Plate 54.

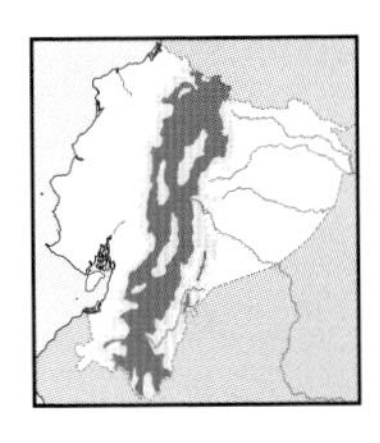

Sharp-shinned Hawk *Accipiter striatus* — 28–33cm / w 40–65cm

Andes and inter-Andean valleys (1,700–3,500m). Montane forest, borders, adjacent clearings. Slim. Grey to sooty-brown above, below ***variably*** barred brownish to rufescent or ***plain creamy, white or dark grey***. Neck-sides and cheeks mottled pale except in dark morph. ***Rufous thighs*** in pale morph. Juv. plain chestnut, finely streaked brown or heavily blotched dull rufous below. In flight: ***long tail, slightly notched***, short rounded wings, barred underwing. Race *ventralis* (species status?). Buoyant flight, little gliding but regular soaring; dashes after prey from perch, acrobatic chases. **Voice** Thin, high-pitched, slow yelping *keé..keé..keé..keé*. **SS** Occurs above most *Accipiter*; recalls larger Bicoloured (larger head, fiercer expression, dark cap, pale nuchal collar); Semicollared (shorter tail, whitish nuchal collar and throat, barred underparts). **Status** Uncommon. See also Plate 54.

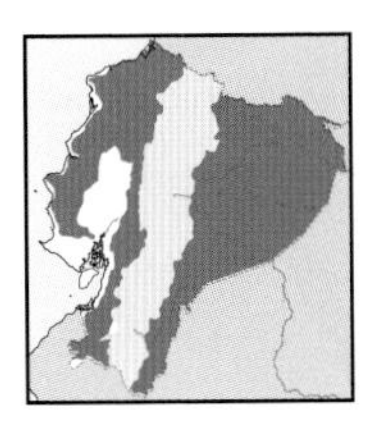

Bicoloured Hawk *Accipiter bicolor* — 33–46cm / w 60–80cm

E & W lowlands to subtropics (to 2,500m). Humid to deciduous forest. ***Slate-grey above***, darker cap, paler below, ***rufous thighs, white vent***. Juv. brown above, darker crown, ***creamy to tawny below extending to narrow nuchal collar, rufous thighs***. In flight: ***rufous thighs, white vent***, mottled underwing, longish tail. Nominate. Shallow wingbeats interspersed by brief glides, often soars in slight V; perches motionless and dashes after prey. **Voice** Loud, cackling *káh-káh-káh...*. **SS** Recalls rarer, smaller Rufous-thighed Kite (black mesial throat stripe, dark hood; behaviour differs); Grey-bellied Hawk (paler below, darker above, no cap, no rufous thighs); cf. juv. with forest-falcons (longer graduated tail, different head pattern). **Status** Rare, overlooked? See also Plate 54.

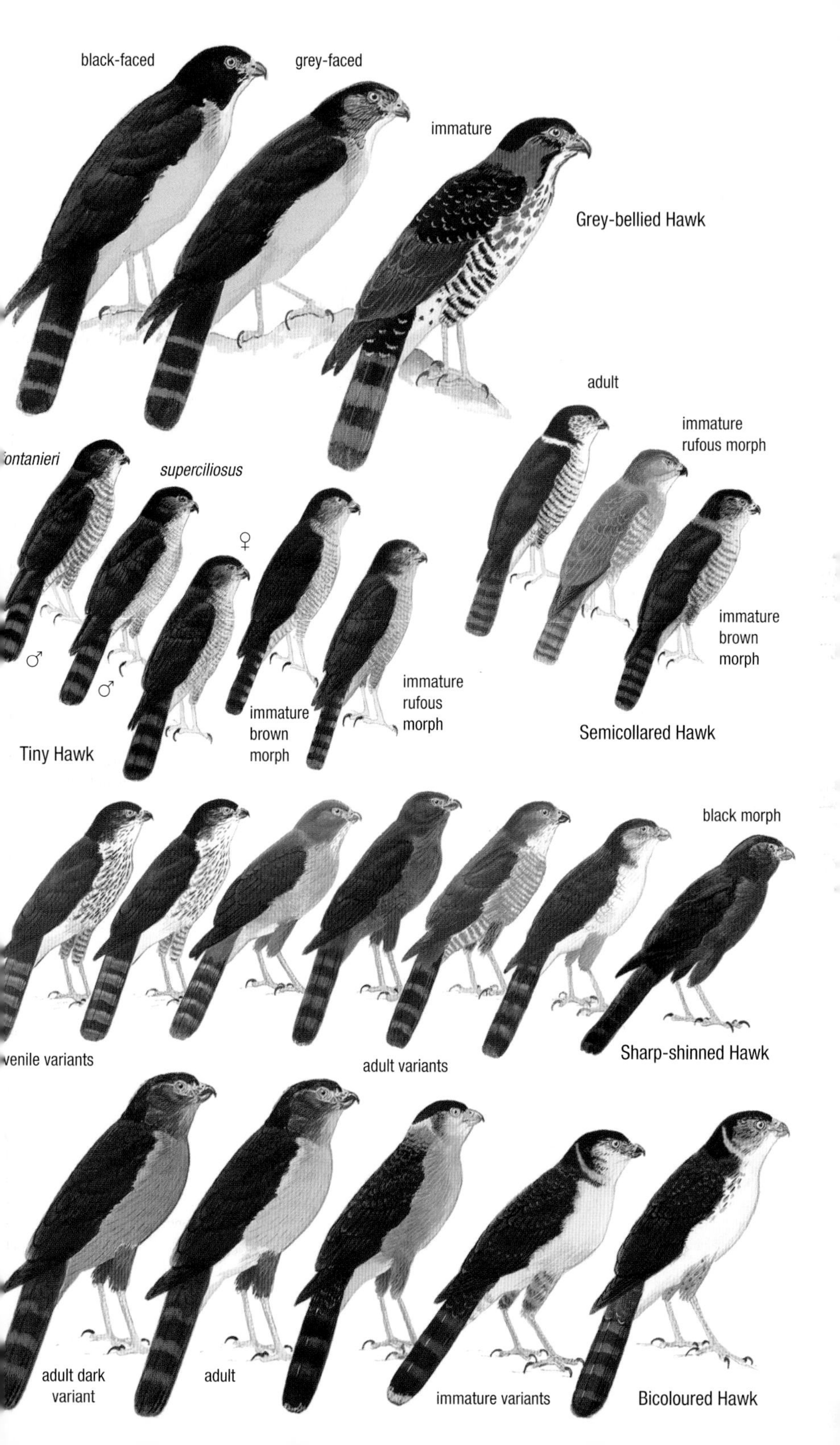
black-faced
grey-faced
immature
Grey-bellied Hawk
adult
immature
rufous morph
fontanieri
superciliosus
♀
♂
♂
immature
brown
morph
immature
rufous
morph
immature
brown
morph
Semicollared Hawk
Tiny Hawk
black morph
juvenile variants
adult variants
Sharp-shinned Hawk
adult dark
variant
adult
immature variants
Bicoloured Hawk

PLATE 62: HAWKS

Fairly large forest hawks, all are broad-winged, short- and broad-tailed. Recent taxonomic revision divided former genus *Leucopternis* into several genera and moved others to *Buteogallus*; few species remain in *Leucopternis*. All primarily in lowlands. White and Grey-backed regularly soar over semi-open areas; the remaining species rarely, if ever, abandon forest cover. Three western species apparently experiencing severe population declines due to habitat loss.

Plumbeous Hawk *Cryptoleucopteryx plumbea* — 36–38cm / w 70–80cm

NW lowlands to subtropics (to 1,700m). Humid forest, borders. ***Orange bare parts. Dark grey***, fine barring on lower belly and thighs. Shortish tail, ***single white median tail-band***. Juv. more obviously barred below. In flight: rather bulky, short rounded wings and tail, ***white underwing***, black primary tips. Monotypic. Mostly seen perched atop exposed branches, never soars. **Voice** Loud, whistled *weeeéáaa...weeeéáaa..wëëa-wëëa....* **SS** Semiplumbeous Hawk entirely white below; cf. perched Plumbeous Kite (long wings exceed tail tip, dark cere). **Status** Rare, overlooked? VU. See also Plate 52.

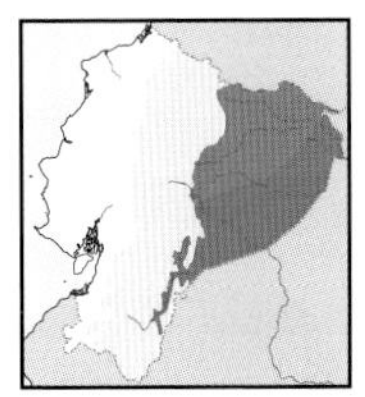

White Hawk *Pseudastur albicollis* — 43–49cm / w 100–120cm

E lowlands to foothills (to 1,100m). *Terra firme* canopy. Large, broad wings, ***grey cere***. ***White head to underparts***, black mask, ***upperparts mostly blackish***. Tail black tipped and based white. Juv. tinged buff, little crown streaking. In flight: stocky, ***white below, single black subterminal tail-band, black tips to flight feathers***. Nominate. Long soars on flat wings, perches on high bare branches. **Voice** Loud *shreeey...shree-shree-shree*. **SS** Black-faced Hawk (orange bare parts, streaky head, white median tail band); Black-and-white Hawk-Eagle (heavier, feathered legs, orange cere, broader mask, barred tail). **Status** Rare. See also Plate 52.

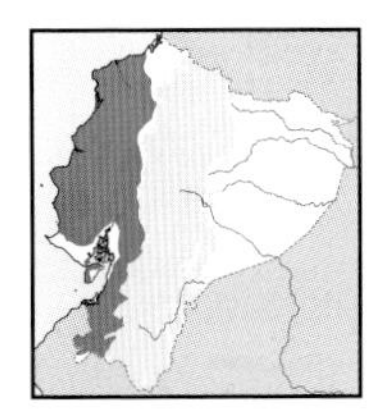

Grey-backed Hawk *Pseudastur occidentalis* — 46–52cm / w 105–115cm

SW lowlands to foothills, locally towards wet NW (to 1,300m). Humid to deciduous and montane forest. ***Blackish upperparts, heavy head streaking*** to chest-sides. ***White below***, mantle streaked white, tail white basally. In flight: broad rounded wings, ***white below, broad black subterminal tail-band and 'fingers'***. Monotypic. Long soars on flat wings, often perches for long periods on exposed branches. **Voice** Loud, husky scream *skeeeeey'*. **SS** Unmistakable in range; cf. locally Black-and-white Hawk-Eagle, Semiplumbeous Hawk, Grey-headed Kite. **Status** Uncommon, local, declining? EN. See also Plate 52.

Semiplumbeous Hawk *Leucopternis semiplumbeus* — 33–36cm / w 50–65cm

NW lowlands to foothills (to 1,500m). Humid forest, borders. Small ***dark grey and white***, short rounder wings, ***orange bare parts***, single white median tail-band. Juv. has narrowly streaked head, two tail-bands. In flight: compact, ***hooded***, dark-barred remiges. Monotypic. Perches on high, exposed perches, never soars; often low inside forest. **Voice** Drawn-out, high-pitched *wuu'eeeeee..wuu'eeeee*; also long *eeeeeaarr*. **SS** Grey-backed Hawk (heavier, broad wings, different tail pattern); Grey-headed Kite (larger, grey-headed, longer tail, striking underwing pattern); see Plumbeous and Slaty-backed Forest-Falcons. **Status** Rare, EN in Ecuador. See also Plate 52.

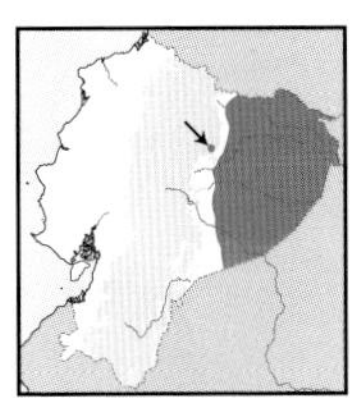

Black-faced Hawk *Leucopternis melanops* — 37–42cm / w 65–80cm

E lowlands (to 900m). Mostly *terra firme* canopy. Blackish above mottled white, ***head streaked black and white, underparts white. Black mask, orange cere***, single broad white median tail-band. Juv. creamier below. In flight: short broad wings, mostly white below, remiges distally barred dusky. Monotypic. Fast shallow wingbeats, little soaring, mostly inside forest perching below canopy. **Voice** High-pitched, slurred, short *kee'u?* every 5–6 sec; shorter *keyp* in flight. **SS** Larger White Hawk lacks head streaks, no orange cere, different tail pattern. Black-and-white Hawk-Eagle much larger, no head streaks, plain upperparts, stronger-legged. **Status** Rare. See also Plate 52.

adult
juvenile
Plumbeous Hawk
adult
immature
White Hawk
adult
immature
Grey-backed Hawk
adult
juvenile
Semiplumbeous Hawk
adult
immature
Black-faced Hawk

Three large, predominantly dark raptors plus a former *Leucopternis*, which is smaller, more forest-based. Black hawks confined to lowlands, one exclusively in mangroves. The large and rare Solitary Eagle is confined to extensively forested mountains and slopes.

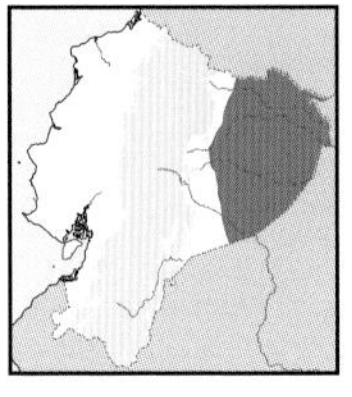

Slate-coloured Hawk *Buteogallus schistaceus* 41–46cm / w 85–95cm

E lowlands (to 400m). *Várzea* and *terra firme* borders. Broad wings, rounded short tail. ***Slate-grey, orange bare parts***. Black tail, ***single median white band***, narrow white tip. Juv. narrowly barred on belly and thighs. In flight: juv. has mottled underwing, finely-barred secondaries, two tail-bands. Monotypic. Rarely soars, mostly perches on large branches at various heights. **Voice** Loud, slurred *wuyeeeé...*; also fast *kikikikikí....* **SS** Slender-billed Kite (no tail-band, bill notably longer); dark-morph Hook-billed Kite (broader tail-band, heavier bill, tricoloured facial skin); black hawks (blacker, bulkier, yellow bare parts); Zone-tailed Hawk (longer wings, patterned underwing). **Status** Uncommon. See also Plate 51.

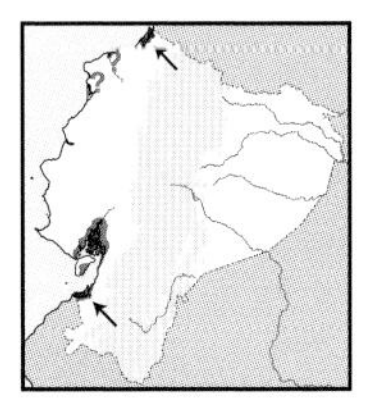

Common Black Hawk *Buteogallus anthracinus* 42–48cm / w 95–105cm

SW coast, locally in extreme NW (sea level). Mangroves. ***Broad wings***, short tail. ***Blackish, broad white band near tail base***. Juv. ***brownish above*** streaked / mottled buff, darker ocular area, ***paler eyebrow and malar***, streaky crown, streaked / blotched below, barred thighs. In flight: ***broad white tail-band, rufous-tinged secondaries, pale primary patch***. Juv. heavily streaked blotched below, ***buff primary patch***, heavily barred, longish tail. Feeds on crabs taken on muddy banks, often soars on nearly flat wings. **Voice** Fast, ringing, shrill *kleeklee-klee-klee-keekeke....* **SS** Unique in habitat, but Great Black Hawk occurs in forest adjacent to mangrove; juv. Great Black Hawk (sturdier, paler-faced, denser tail barring, more forest-based; ad. blacker, tail white basally); cf. Zone-tailed Hawk. **Status** Rare, VU in Ecuador. See also Plate 51.

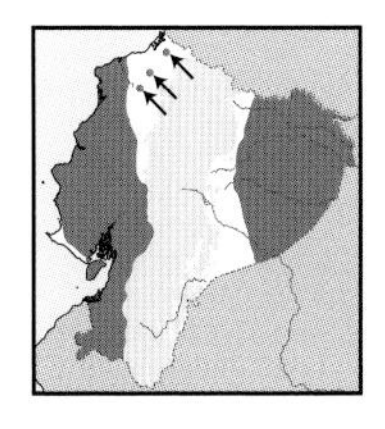

Great Black Hawk *Buteogallus urubitinga* 53–61cm / w 115–135cm

E & SW lowlands, to foothills in SW, locally in SE (to 400m in E, to 1,400m in W). Humid forest borders, often near water. Stocky. ***Blackish, white uppertail-coverts and tail base***. Juv. ***dull brown above blotched buff***, creamier below streaked dusky, ***pale face***, pale-tipped tail. In flight: ***white tail base*** distinctive; juv. streaked / blotched, blackish outer primaries, ***dark carpal arcs***, heavily-barred remiges and tail. Nominate. Strong wingbeats, long soars on flat wings, often seen perched low near water. **Voice** High-pitched, descending scream *kieeeeeeeeeeeu*; fast *kikikiki....* **SS** Local Common Black Hawk (no habitat overlap but close); Slate-coloured Hawk (slate-grey, orange bare parts); montane Solitary Eagle (heavier, greyer, broader-winged); Black Hawk-Eagle (heavier, large-headed, crested, longer- and broader-winged). **Status** Uncommon. See also Plate 51.

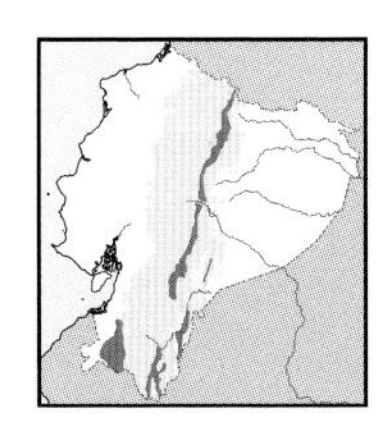

Solitary Eagle *Buteogallus solitarius* 68–74cm / w 160–180cm

Local in Andean foothills and subtropics (900–1,800m). Montane forest, borders. ***Large. Dark slate-grey, single white median tail-band***. Juv. dark brown above, ***head streaked buff-brown, buff supercilium, plain throat***, underparts variably streaked / blotched dusky, more ***solid flanks and thighs***. In flight: long, ***broad rounded wings***, bulging secondaries, ***very short tail***; juv. ***dark in sides to thighs and vent*** pale remiges, white panel on primary bases, ***nearly unbarred pale tail***. Nominate. Pairs soar high, apparently on flat wings; sometimes perch atop trees. **Voice** Melancholic, loud *kluuu-eet... kluuüüü...kyéét!...klee-klee-klee-klee....* **SS** Juv. Black-chested Buzzard-Eagle darker, less contrasting black primaries, darker tail and secondaries; juv. Great Black Hawk (smaller, pale-faced, longer-tailed; little range overlap). **Status** Rare, VU in Ecuador, NT globally. See also Plate 51.

adult
immature
Slate-coloured Hawk
adult
juvenile
Common Black Hawk
juvenile
adult
Great Black Hawk
juvenile
adult
Solitary Eagle

PLATE 64: HAWKS AND BUZZARD-EAGLE

Mixture of highland and lowland species, including the largest *páramo* raptor (buzzard-eagle) and Black-collared Hawk of lagoon edges in extreme E lowlands. Taxonomic affinities of Black-collared with kites recently unravelled. Savanna and Harris's share dry and open habitats in W.

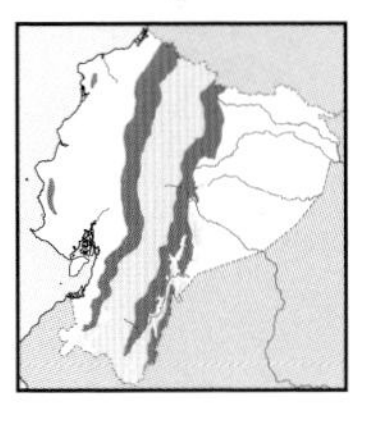

Barred Hawk *Morphnarchus princeps* 52–61cm / w 110–124cm

E & W Andean foothills to subtropics, locally atop coastal cordilleras (700–2,200m, to 500m in coastal ranges). Montane forest, borders. Yellow bare parts. ***Bluish-grey above to chest, evenly barred grey and white below***, single white median tail-band. Juv. similar but pale-edged above. In flight: robust, broad-winged, short-tailed, ***bicoloured underparts***. Monotypic. Often soars on flat wings, with up to four birds interacting, calling loudly. **Voice** Loud *kwee'kwee'kwee'kwee...*, *kweéaa!..kweé'aay...*. **SS** Much heavier Black-chested Buzzard-Eagle (paler, with pale shoulders); see Grey-lined Hawk. **Status** Uncommon, VU in Ecuador. See also Plate 52.

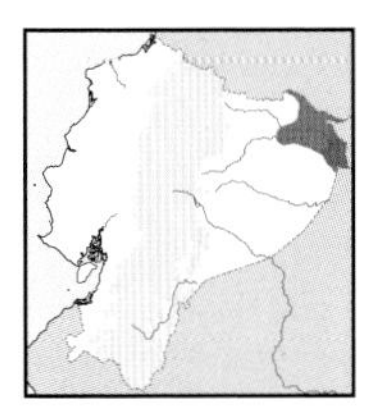

Black-collared Hawk *Busarellus nigricollis* 46–50cm / w 115–140cm

Local in lowlands of extreme NE (below 250m). *Várzea* and *igapó* forest edges. Bulky, short legs and tail. ***Whitish hood, cinnamon-rufous body***, incomplete ***black collar***. Duller juv. has ***streaky head and collar***, creamier below, barred belly. In flight: ***long and very broad rounded wings***, bulging secondaries, black-and-rufous underwing, black subterminal tail-band. Paler, barred tail in juv., dark carpal arcs. Nominate. Slow strong flaps, flies low over water, glides and soars on bowed wings; often perches low above water. **Voice** Low, raspy *khree-heee, kre-he, krhe-hë-hë*. **SS** Stockier than water-associated kites. **Status** Rare, local. See also Plate 51.

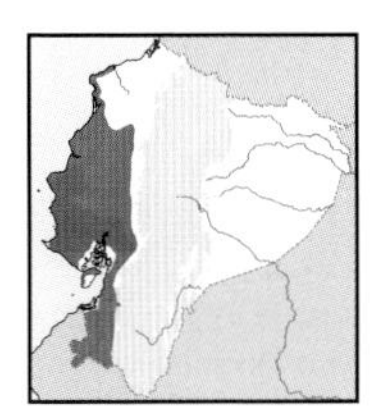

Savanna Hawk *Buteogallus meridionalis* 53–61cm / w 120–140cm

SW lowlands to foothills (to 1,000m, locally higher in SW). Open and semi-open terrain. ***Lanky. Mostly rufous, richer shoulders***. Black tail, ***single white median band***. Juv. dark brown above, ***whitish face and throat***, heavily streaked below, barred thighs and vent. In flight: ***long-winged***, longish-necked, conspicuous tail-band, black edges to remiges. Juv.: black primaries, dark trailing edge, barred remiges, ***barred tail, dark subterminal band***. Monotypic. Deep wingbeats, glides on cupped wings, long soars on flat wings; often perches low near water. **Voice** Brief, high-pitched *kéeeua*. **SS** Juv. Harris's Hawk darker, white rump, drabber underwing pattern; juv. Snail Kite longer-billed, paddle-shaped wings, different flight. **Status** Uncommon. See also Plate 51.

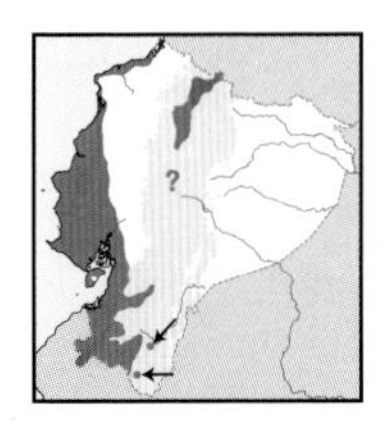

Harris's Hawk *Parabuteo unicinctus* 48–54cm / w 90–120cm

SW lowlands, inter-Andean valleys (to 1,700m in SW, 1,500–2,500m Andes). Deciduous scrub, adjacent fields. ***Lanky***, long legs, ***wings*** and tail. Dark brown, ***rufous shoulders and thighs, white rump, vent and tail base***. Juv. duller, ***pale rufous in shoulders and thighs***, heavy buff streaking below. In flight: ***bold rufous underwing-coverts and thighs, white wing 'window' and tail base*** (also distinctive in juv.). Race *harrisi*. Fast shallow wingbeats, glides on cupped wings, floppy flight, often in groups. **Voice** Throaty *krrrreaaa!, krrra-krrra-krrra...*. **SS** Shorter-tailed Snail Kite (no rufous wings, different bill shape); Savanna Hawk (more rufous overall, different tail pattern, no white tail base). **Status** Uncommon. See also Plate 54.

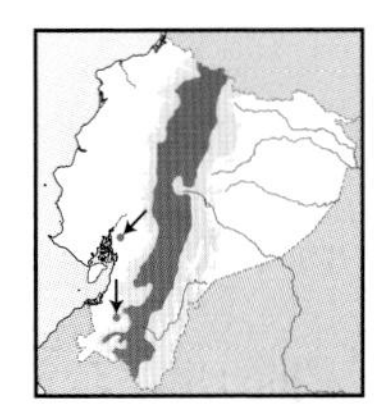

Black-chested Buzzard-Eagle *Geranoaetus melanoleucus* 60–69cm / w 150–185cm

Andes, locally to valleys and lower slopes (2,000–4,000m; locally lower or higher). Mostly open *páramo*. ***Slate-grey above and hood, paler shoulders***, barred belly. Juv. ***dark brown above, paler shoulders, almost solid belly streaking***. In flight: ***broad-based wings, stubby tail, clear-cut contrast*** below, ***grey remiges***; juv. ***longer tail, finely barred***, underwing blotched / streaked dusky-buff, pale 'window' in primary bases. Race *australis*. Slow strong wingbeats, long soars on flat wings; usually in pairs or trios. **Voice** Weak, doubled *wa-keé..wa-keé; ki'ii-kikikiki...*. **SS** Much larger than Barred Hawk; cf. juv. Solitary Eagle (fainter tail, contrasting black 'fingers', creamier underparts, more solidly streaked breast-sides); Black-and-chestnut Eagle (longer, evenly-broad wings, longer-tailed, feathered tarsi, plainer underparts); also smaller (Red-backed) and (Gurney's) Hawk. **Status** Uncommon. See also Plate 53.

adult
immature
Barred Hawk
adult
immature
Black-collared Hawk
adults
immature
Savanna Hawk
adult
immature
Harris's Hawk
adult
immature
juvenile
Black-chested Buzzard-Eagle

'Typical' hawks often seen soaring over non-forested habitats. All formerly classified in *Buteo*, but some species recently transferred to other genera. Identification intricate, especially juv. and imm. Head, underwing and undertail patterns diagnostic. Some species on this plate, especially Roadside and Short-tailed, are common and widespread. See also Plates 66–67.

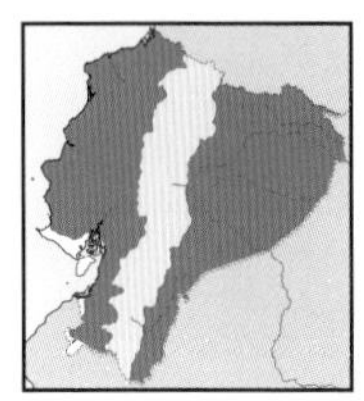

Roadside Hawk *Rupornis magnirostris* 33–38cm / w 65–90cm

E & W lowlands to subtropics (to 3,000m). Roadsides, clearings, agriculture. The most numerous hawk. ***Greyish above to chest, breast to thighs evenly barred dull rufous, rufous in primaries***. Juv. ***browner***, mottled buff, head streaked buff, underparts whitish ***streaked brown to lower breast***. In flight: ***rufous 'window' in primary bases***, clear chest–belly contrast; wing 'window' ***paler*** in juveniles. Nominate. Shallow stiff wingbeats, glides on bowed wings, frequent soaring; assumes exposed perches. **Voice** High-pitched, nasal *nýäääaa* scream; fast *nyä'nyä'nyä...* in flight. **SS** Broad-winged Hawk (longer wings, in flight blackish wing edge, paler underwing, no rufous 'window', dark malar); Grey-lined Hawk (paler belly barring, different underwing pattern). **Status** Common, widespread. See also Plate 52.

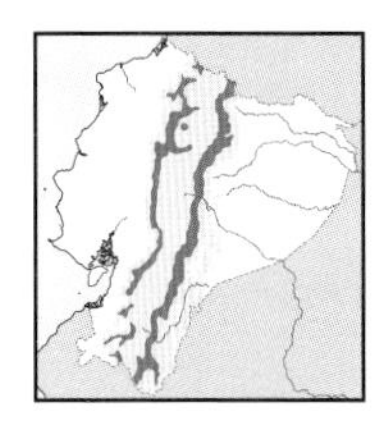

White-rumped Hawk *Parabuteo leucorrhous* 33–38cm / w 70–80cm

E & W Andes, subtropics to temperate zone (2,000–3,200m). Montane forest borders. ***Yellow-orange eyes***. Mostly ***black, rufous-barred thighs, white rump and vent***, black tail, two white tail-bands and narrow tip. Juv.: ***head to upper mantle and underparts boldly streaked*** dusky-buff, ***rump as ad.*** In flight: ***white underwing***, dark carpal arcs, white vent; juv. almost ***solidly streaked below, pale buff underwing***, dark carpal arcs, buffier vent. Monotypic. Fast shallow wingbeats, soars and glides low over canopy, perches atop emergent trees. **Voice** High-pitched *pyeee'u*, more nasal *pyh'h'h'h*. **SS** Dark-morph Broad-winged Hawk (dark underwing, silvery 'window' in primaries). **Status** Rare. See also Plate 52.

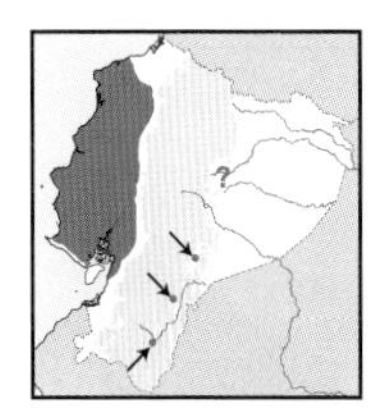

Grey-lined Hawk *Buteo nitidus* 40–45cm / w 75–95cm

W lowlands, not in wet NW, SE foothills (to 450m in W, 600–1,000m in SE). Humid to deciduous forest, woodland. Compact. ***Pale grey, prominently barred below***, white rump and vent. Juv. brown above, whitish streaky head, ***prominent supercilium and moustachial***, little streaking below. In flight: short rounded wings, contrasting black tail, ***pale underwing***, narrow black primary tips, ***dark subterminal band in pale tail***. Juv. tail longer, ***dusky malar***, streaked / mottled underwing, pale remiges. Nominate. Infrequently soars, stiff wingbeats, often seen perched at forest edge. **Voice** Nasal, high-pitched *kéeeuu!*, more drawn-out *keee..e..uu, wee'uu-wee'uuu-wee'üüü....* **SS** Juv. Broad-winged Hawk (heavier streaks below, plainer face, prominent dark trailing edge, unbarred thighs); Roadside (streakier head, rufous wing patches). **Status** Rare to uncommon. See also Plate 52.

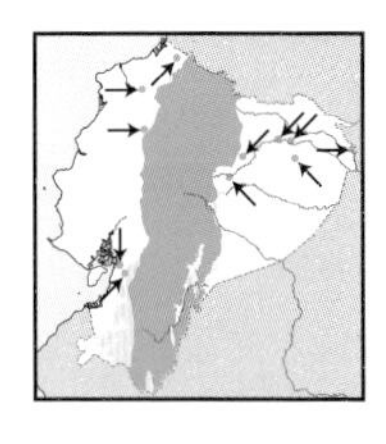

Broad-winged Hawk *Buteo platypterus* 38–43cm / w 75–95cm

Andean foothills and subtropics (mostly 800–2,800m). Montane forest, adjacent clearings. Stocky, ***longish wings***. Dark brown above ***edged pale***. Streaky head, ***broad malar stripe, underparts coarsely streaked / barred brown and whitish***; banded tail. Juv. streakier above, ***broad malar***, paler underparts, ***4–5 narrow pale tail-bands***. In flight: ***dark malar*** accentuates pale throat, ***pale underwing, whitish remiges, black trailing edge***, dark carpal arcs. Juv. fainter underwing, black ***'fingers'***, paler barred tail. Nominate. Stiff fast wingbeats, flat-winged glides, frequently soars; often perches at forest edges. **Voice** Piercing *pt'teeeey*. **SS** Juv. Grey-lined Hawk (less streaky, no dark wing edges), Short-tailed Hawk (hooded, two-toned underwing, paler undertail), White-throated Hawk (hooded, contrasting white throat, creamier below, streaked / blotched underwing, darker remiges), Roadside Hawk (rufous 'windows', evenly-barred belly). **Status** Uncommon boreal migrant (Oct–Apr). See also Plate 52.

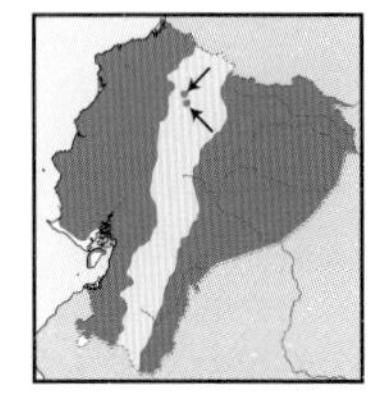

Short-tailed Hawk *Buteo brachyurus* 39–43cm / w 85–105cm

E & W lowlands to subtropics (to 1,600m). Forest borders, clearings. Stocky, long-winged. ***Blackish above, white below*** (also all-blackish morph), ***whitish forehead***. Juv. ***buffy below, streaked head to breast-sides***; dark juv. mottled whitish below, ***whitish forehead***. In flight: ***hooded***, plain white below (***black underwing*** in dark morph), ***black carpal arcs*** and 'fingers', ***silvery base to primaries***. Juv. ***buffier***, paler tail, ***hooded***. Nominate. Soars high in shallow V, often with vultures. **Voice** Thin, high-pitched, descending *peeeeuu*. **SS** Grey/white morph of Red-backed Hawk (no carpal arcs, bolder subterminal band); also Grey-lined, White-throated and Swainson's Hawks; juv. less streaked below than congeners, blackish primaries and carpal arcs distinctive; cf. dark morph with Zone-tailed Hawk. **Status** Uncommon. See also Plate 52.

magnirostris
juvenile
adult
adult
Roadside Hawk
immature
adult
dark morph
White-rumped Hawk
adult
juvenile
immature
Grey-lined Hawk
Broad-winged Hawk
adult
immature
Short-tailed Hawk
adult
dark morph
immature
dark morph
immature
light morph
adult
light morph

PLATE 66: 'VARIABLE' HAWK

A complicated taxon formerly in genus *Buteo*. Currently treated as a single, variable species, but originally described as two species: Red-backed and Gurney's Hawks. Identification thorny, as is recognising age-related 'morphs'. Most literature has failed to accurately depict plumage variation, which we intend to illustrate here following recent work by J. Cabot, T. de Vries and C. Pavez.

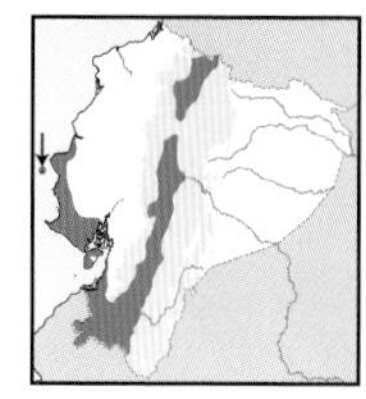

(Red-backed) Hawk *Geranoaetus polyosoma* 45–55cm / w 105–125cm

SW lowlands and inter-Andean valleys (below 1,000m in SW, 2,000–3,500m in Andes). Open arid scrub, agricultural fields. Two races: *peruviensis* (SW), *polyosoma* (Andean valleys). Apparently consistent size differences between Red-backed and Gurney's Hawks (wing length, cubito-radius, notches in pp10 and 8); Red-backed *smaller*. Red-backed has *shorter wingspan and narrower wings* than Gurney's, but some overlap exists between ♀ Red-backed and ♂ Gurney's. Seemingly, ad. plumage is attained after six years, sexual dimorphism as third-year. Predominant **pale morph**: **first-year** dark brown above, pale *ochraceous-buff supercilium, cheeks and underparts, dark malar*. **Second-year** *rustier* above, more uniformly barred rusty in lower underparts. **Third-year** ♂: *some reddish in mantle, whiter underparts* less rusty-barred on belly to thighs, faint tail-bands; ♀ more obvious *brick-red dorsal patch*. **Fourth-year** ♂ greyish above, *brick-red mantle, barred flanks*; ♀ darker above, more solid *brick-red dorsal patch*, underparts mostly white barred dusky. **Fifth-year** ♂ with *reddish intermixed feathers on mantle*; ♀ similar to fourth-year. **Ad.** ♂ grey above, mostly white below, *flanks and thighs faintly barred grey, unbarred* median and lesser wing-coverts; ♀ as fifth-year. **Rare dark morph**: **first-year** *wholly sooty-brown*, creamy spots / streaks *especially in flanks*. **Second-year** dark brown, few creamy blotches. **Third-year** ♂ dark brownish mantle *intermixed reddish*, underparts intermixed *reddish down to mid-belly, no barring below*, tail already shows *dark subterminal band*; ♀ *brick-red in mantle and most underparts*. **Fourth-year** ♂ *plumbeous-grey above*, little reddish, *reddish breast-band*; ♀ darker grey above, *brick-red mantle*; *underparts tinged sooty-brown and reddish*. **Fifth-year** ♂ almost entirely *slate-grey*, ♀: *grey on head*, contrasting *brick reddish median underparts* and *sooty blackish intermixed with reddish on lower underparts*. **Ad.** ♂ *more solidly* grey, *white undertail*; **ad.** ♀ as fifth-year. Some ad. show pale barring on belly, throat and belly, or belly to thighs. In flight: white tail, *bold black subterminal band* as third-year; pale remiges, primaries tipped blackish; juv. tail only *faintly barred*. Wing outline *more projected and pointed*, wings not uniformly broad as Gurney's (has longer secondaries). Red-backed flies on rather *flat wings*, not markedly dihedral as Gurney's. Red-backed soars above dry open areas, often singly or in pairs. **Voice** Short, high-pitched *kluëee..kluëee...*; sometimes faster *kleeee-kle'kle'kle'kle....* **SS** White-throated Hawk (well-marked white throat); Short-tailed Hawk (hooded, blacker above, whitish remiges); might overlap locally with Black-chested Buzzard-Eagle; cf. juv. (heavier, thickly streaked below, almost solid belly, broader wings); dark Swainson's Hawk (dark remiges, pale vent). **Status** Fairly common.

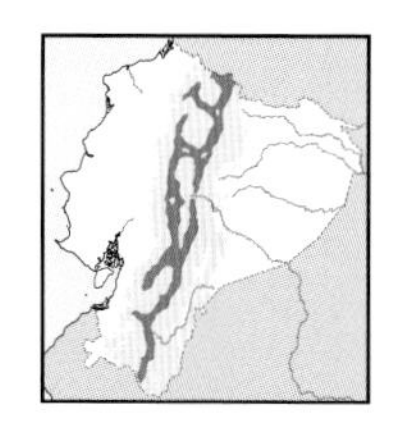

(Gurney's) Hawk *Geranoaetus polyosoma poecilochrous* 50–70cm / w 130–150cm

Andes (3,000–4,000m). Open grassy *páramo*. *Larger* than Red-backed; *larger wingspan* and *broader wings*, but some overlap exists. Apparently, ad. plumage attained after six years, and sexual dimorphism as third-year. **First- to third-year pale morph** similar to Red-backed but *darker mantle, underparts often marked brown, bolder* malar, *more extensively pale* on head-sides and supercilium; breast solidly tawny as second-year, *belly barred markedly rusty to rufous-chestnut*. **Fourth- and fifth-year**: greyish with some reddish (♂) or all brick-red (♀) back, dark grey head, *pale cheeks*; brick-red lower breast (*little rufous* in ♂, which disappears with age), *barred lower underparts*. **Ad.** ♂ *medium to slate-grey* above, white below, *fine dark vermiculations on mid-breast*, pale cheeks. **Ad.** ♀ *loses reddish chest-band*. **Dark-morph: juvenile** *tawny cheeks and broad malar streak, tawny* throat, upper chest streaked dusky, breast and flanks vermiculated dusky, barred lower belly, *vent extensively white*. **Second-year** *plainer* below, breast more solidly dark, barred belly to vent; *sootier* head and neck. **Third-year** dark greyish-brown, some reddish in breast and mantle (♂), ♀ *more extensively reddish* on breast, neck and mantle. **Fourth-year** ♂: *reddish dorsal patch* and *grey-and-white barring* on lower underparts; ♀: *reddish pectoral band*. As **fifth-year**, ♂ *slate-grey* above, *markedly barred* grey and white below; **ad.** ♀ retains *reddish mantle*, but loses breast-band; underparts *dark grey, markedly barred on belly to thighs*. In flight: white tail, *black subterminal band* as third-year; *heavy barring on inner remiges and coverts*, primaries tipped blackish; juv. tail *faintly barred*. Gurney's primaries form a *blunter, more rounded outline*; wings *evenly broad*. Gurney's flies in marked *dihedral* and tend to polyandry; often seen soaring over open areas, in groups. **Voice** Much as Red-backed. **SS** Pale morph similar in pattern to Red-backed but *ground colour creamier to tawny*. Fourth-year and older Red-backed has *dark head-sides* (pale in Gurney's); White-throated Hawk (well-marked white throat); juv. Black-chested Buzzard-Eagle (heavier, densely streaked below, almost solidly in belly, broader wings). **Status** Uncommon. [Alt: Puna Hawk]

(Red-backed) Hawk
dark morph
♂ adult
♀ 4–5 years
♂ 4–5 years
♂ 3 years
♀ 3 years
1–2 years
pale morph
♂ adult
♂ 4–5 years
♀ 4–5 years
♀ 3 years
2 years
1 year
(Gurney's) Hawk
dark morph
♂ adult
♀ 5 years
♀ 3 years
2 years
1 year
pale morph
♂ adult
♀ adult
♀ 4–5 years
♀ 3 years
1–2 years

Three additional *Buteo* hawks, two of them larger and more distinctive. Zone-tailed is actually more reminiscent of a vulture, owing to silhouette and flight behaviour. Swainson's is a boreal winter transient or vagrant, White-throated is presumably an austral migrant. The two forest-based hawk-eagles are seldom seen and poorly known.

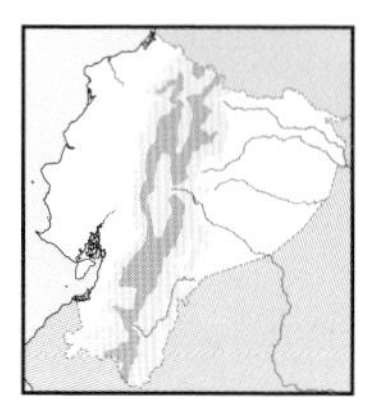

White-throated Hawk *Buteo albigula* 39–43cm / w 85–100cm

Andes, locally into inter-Andean valleys (2,200–3,200m). Montane forest, adjacent fields. Stocky, longish wings and tail. Blackish-brown above, ***mostly white below, dark cheeks provide hooded appearance***; streaked flanks, ***thighs barred rufous***. Juv. pale-fringed above, ***buffier below and more profusely streaked***. In flight: ***obvious hood, dark rufous breast-sides, pale underwing with few markings***, dark carpal arcs, narrow tail-bands, grey-barred remiges. Monotypic. Stiff wingbeats, soars on flat wings, apparently engages in long-distance movements, never numerous. **Voice** High-pitched, short *wyeéea*. **SS** Short-tailed Hawk (less streaky sides, bolder tail-bands, silvery wing 'window'); see larger, pale-morph (Red-backed) and (Gurney's) (no well-demarcated throat), and rarer Swainson's Hawk. **Status** Rare. See also Plate 53.

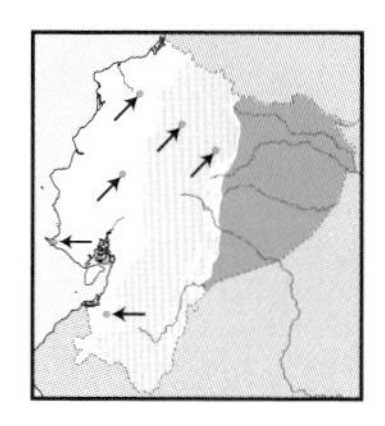

Swainson's Hawk *Buteo swainsoni* 48–56cm / w 115–135cm

Locally in SW & E lowlands, and Andes (to 2,500m). Mostly clearings. ***Long, narrow pointed wings***. Blackish-brown above, ***brown chest-band, white throat, dark malar***, white rump. Belly whitish, nearly unbarred. Dark morph mostly blackish-brown, tail greyer, ***whitish rump***. Rarer rufous morph: belly barred rufous, ***contrasting white throat***. Juv. paler, streaked head, ***pale supercilium, dark malar, white throat***, underparts variably streaked, pale tail. In flight: ***well-demarcated throat***, dark subterminal tail-band, ***dark grey remiges***, darker 'fingers', ***pale rump***. Monotypic. Soars in V, moderate wingbeats, frequent hovering. **Voice** Unlikely to be heard in Ecuador. **SS** Longer-winged than congeners, dark remiges, ***white rump U***, breast-band (when present), clear-cut throat; see Short-tailed, (Red-backed) (shorter wings) and Zone-tailed Hawks. **Status** Rare boreal transient (Nov–Apr). See also Plate 53.

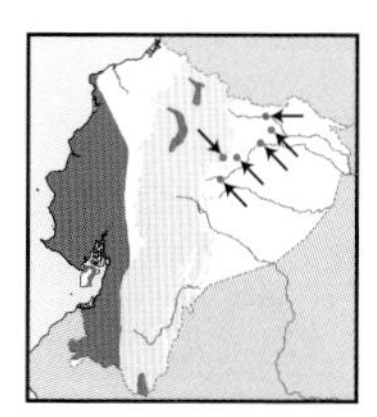

Zone-tailed Hawk *Buteo albonotatus* 47–56cm / w 115–140cm

W lowlands, to SW subtropics; locally in NE lowlands, foothills and Andes (to 1,500m, locally higher). Woodland, adjacent clearings. ***Lanky, longish-winged. Blackish, white forecrown***, 2–3 grey tail-bands. Juv. browner variably ***spotted white***, several narrow dark tail-bands. In flight: ***two-toned underwing pattern***, longish tail. Juv. has ***paler remiges*** and tail. Monotypic. Slow wingbeats, soars and glides in deep V, often tilting sideways; often seen with Turkey Vultures. **Voice** Short, raspy *yeep-yeep*.... **SS** Wing shape, bicoloured pattern and flight behaviour recall Turkey Vulture; cf. Broad-winged (whiter remiges, blackish trailing edge, broader, shorter wings) and Short-tailed Hawks (silvery primary 'window'); see black hawks. **Status** Rare. See also Plate 53.

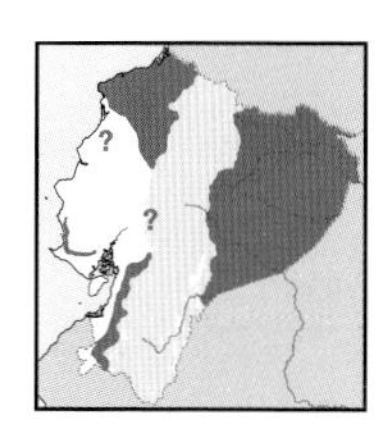

Black Hawk-Eagle *Spizaetus tyrannus* 58–66cm / w 115–145cm

E lowlands to foothills, NW lowlands, locally to SW and foothills (to 1,200m in E, 1,700m in W). Humid forest, borders. ***Strong feathered tarsi, short bushy crest. Black, lower underparts barred white, white crest base. Orange eyes***. Juv. ***creamy-white head***, neck- and chest-sides streaked dusky, ***barred lower underparts***; narrower tail-bands. In flight: broad wings ***pinched-in at base, blackish coverts, grey and white barred remiges***. Juv. has ***streaked / barred*** underwing, ***mostly whitish head***. Race *serus*. Often soars with flat wings slightly held forward, perches high in canopy and edges. Might soar high. **Voice** Loud *wër-wë-wo-wëe'er, wëer-wë-weer, wo-wor-wo'wo-weeé-e*.... **SS** Much rarer dark-morph Crested Eagle (bulkier, evenly barred below, longer crest, bare legs); juv. Ornate Hawk-Eagle whiter; also dark Hook-billed Kites (proportions differ). **Status** Rare. See also Plate 55.

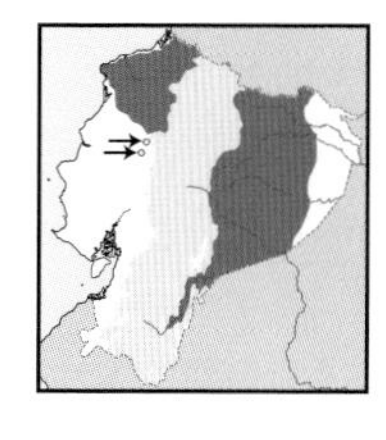

Black-and-white Hawk-Eagle *Spizaetus melanoleucus* 53–61cm / w 110–135cm

Locally in E & NW lowlands to foothills (to 1,400m). Humid forest, borders. ***Strong feathered tarsi***, short crest, ***orange cere, yellow eyes. White head to underparts, black above, on crest and mask***; 3–4 blackish tail-bands. Juv. ***browner***, fainter tail-bands. In flight: relatively long, ***evenly-broad wings, white coverts***, broader distal tail-band. Monotypic. Strong wingbeats, long glides on nearly flat wings, perches below canopy. **Voice** Piping *whee-whi, whee-whi, whee-whi-whi-whi'whi'whi*.... **SS** Juv. Ornate Hawk-Eagle (black-blotched underwing, axillaries and thighs, shorter wings, more rounded trailing edge, longer crest); Black-faced Hawk (smaller, rounded head, bare legs, bold tail pattern); also White Hawk. **Status** Rare, local. See also Plate 55.

White-throated Hawk
adult
rufous morph
adult
dark morph
adult
light morph
Swainson's Hawk
immature
light morph
immature
rufous/dark morph
immature
Zone-tailed Hawk
adult
adult
Black-and-white
Hawk-Eagle
adult
immature
immature
Black Hawk-Eagle

PLATE 68: EAGLES AND HAWK-EAGLES

True eagles are the heaviest and most outstandingly strong raptors in forested habitats. Harpy and Crested are the largest Neotropical raptors, both mainly hunters of arboreal mammals and large birds. The four *Spizaetus*, two on the preceding plate, are smaller but still powerful, have feathered tarsi and conspicuous crests. All are seldom seen.

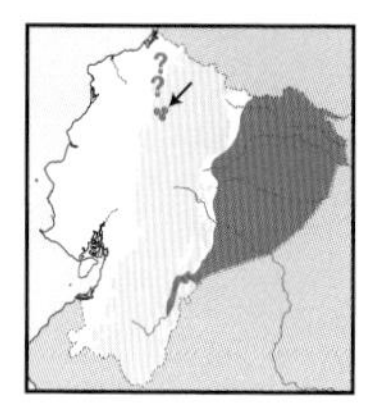

Crested Eagle *Morphnus guianensis* 70–85cm / w 140–155cm

Local in E lowlands, even more local in NW foothills to subtropics (mostly below 300m in E, 100–2,200m in NW). *Terra firme* forest. ***Large, single-pointed crest. Pale grey head to chest***, black crest, blackish upperparts, ***white below, narrowly barred rufescent***. Tail evenly banded. Rarer dark morph ***densely barred*** black and white below. Juv. ***mostly whitish***, mottled upperparts, fainter tail. In flight: bulky, large-headed, ***broad rounded wings, whitish underwing, barred remiges***. Juv. paler remiges, ***faintly banded tail***. Monotypic. Strong wingbeats, occasionally soars low, perches high in canopy, rarely in riparian edges. **Voice** Loud *wyeér-wyeér-wyeér-wyeér..weyeé* screams. **SS** Harpy Eagle (bulkier, forked crest, stronger legs, broader wings; juv. with broader tail-bands); juv. Black-and-chestnut Eagle has blacker 'fingers', shorter tail. **Status** Rare, local, VU in Ecuador, NT globally. See also Plate 55.

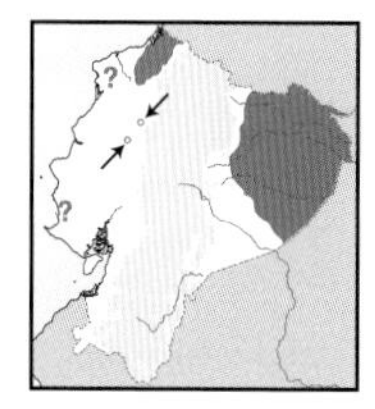

Harpy Eagle *Harpia harpyja* 89–99cm / w 175–200cm

Locally in E & NW lowlands (to 400m, formerly higher, once at 1,000m in NE). Humid forest. Huge, large-headed, ***strong-legged, forked crest. Grey head***, darker crest, ***broad black chest-band***; dark grey above, ***white below***. Juv. ***white head and underparts***, greyer-brown crest and upperparts, blackish remiges. In flight: ***very broad wings***, square tail, broad black and white tail-bands, ***underwing mottled black, remiges strongly barred***; juv. fainter underwing and tail pattern. Monotypic. Inconspicuous except when crossing rivers; rarely soars. Often perches in large branches below canopy, sometimes in riparian edges. **Voice** Rather weak *weeeea, weeee-eeeaa* whistles. **SS** Young has grey chest-band as Crested but size differs; juv. Crested has fainter tail-bands. **Status** Rare, declining? VU in Ecuador, NT globally. See also Plate 55.

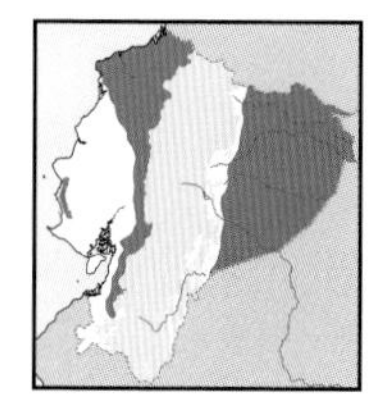

Ornate Hawk-Eagle *Spizaetus ornatus* 58–68cm / w 110–125cm

E lowlands, locally in W lowlands and foothills (to 500m in E, 1,000m in W). Humid forest, borders. ***Strong feathered legs, long erectile crest. Black crest and moustachial, rufous cheeks to chest-sides, white throat, densely barred below***. Blackish upperparts. Juv. ***whitish***, barred flanks and thighs, browner upperparts, some with faint moustachial. In flight: short rounded wings ***pinched-in at base***, clear-cut ***white throat***, bold barring below; juv. has sparse barring on underwing and thighs. Nominate (E), *vicarius* (W). Deep wingbeats; low, flat-winged soaring; perches below canopy, often hunts inside forest. **Voice** Piping *weer-wee'pee-pee..weer-pe-peer..weer-weer-wee-pee'pee'pee*. **SS** Juv. Grey-bellied Hawk (obviously smaller, crestless, unfeathered weaker legs); juv. Black Hawk-Eagle (thickly barred / streaked below, bolder remiges); juv. Black-and-white Hawk-Eagle (whiter, fainter remiges, straighter trailing edge). **Status** Rare NT globally. See also Plate 55.

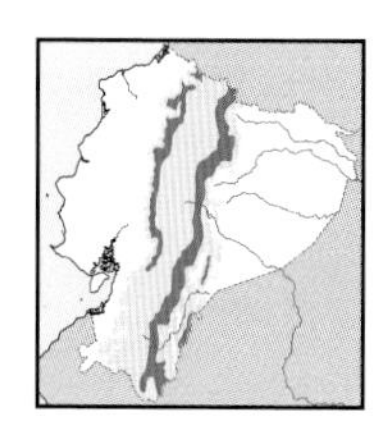

Black-and-chestnut Eagle *Spizaetus isidori* 66–74cm / w 150–165cm

E & NW Andean subtropics to temperate zone, locally in SW (1,500–3,000m). Montane forest, borders. ***Long pointed crest. Black above and on hood, chestnut below, pale grey tail, broad black subterminal band***. Juv. ***brownish-grey above, mottled, shaft-streaks whitish; greyish to buffy-white below***, some streaks / mottles on sides, three black tail-bands. In flight: broad ***rounded wings, pale grey primary bases***, black carpal arcs; juv. ***pale*** below variably streaked dusky, ***whitish primary bases***, more tail-bands. Monotypic. Deep flaps, flat-winged glides, soars high; inconspicuous. Often perches in exposed branches above canopy or in large, bare branches in forest edges. **Voice** Short nasal *kwyaaa* scream. **SS** Juv. of lower-elevation Ornate and Black Hawk-Eagles (wing shape and remiges pattern differ); Solitary Eagle (much shorter tail) and Crested Eagle (bare legs). **Status** Rare, VU in Ecuador, EN globally. See also Plate 55.

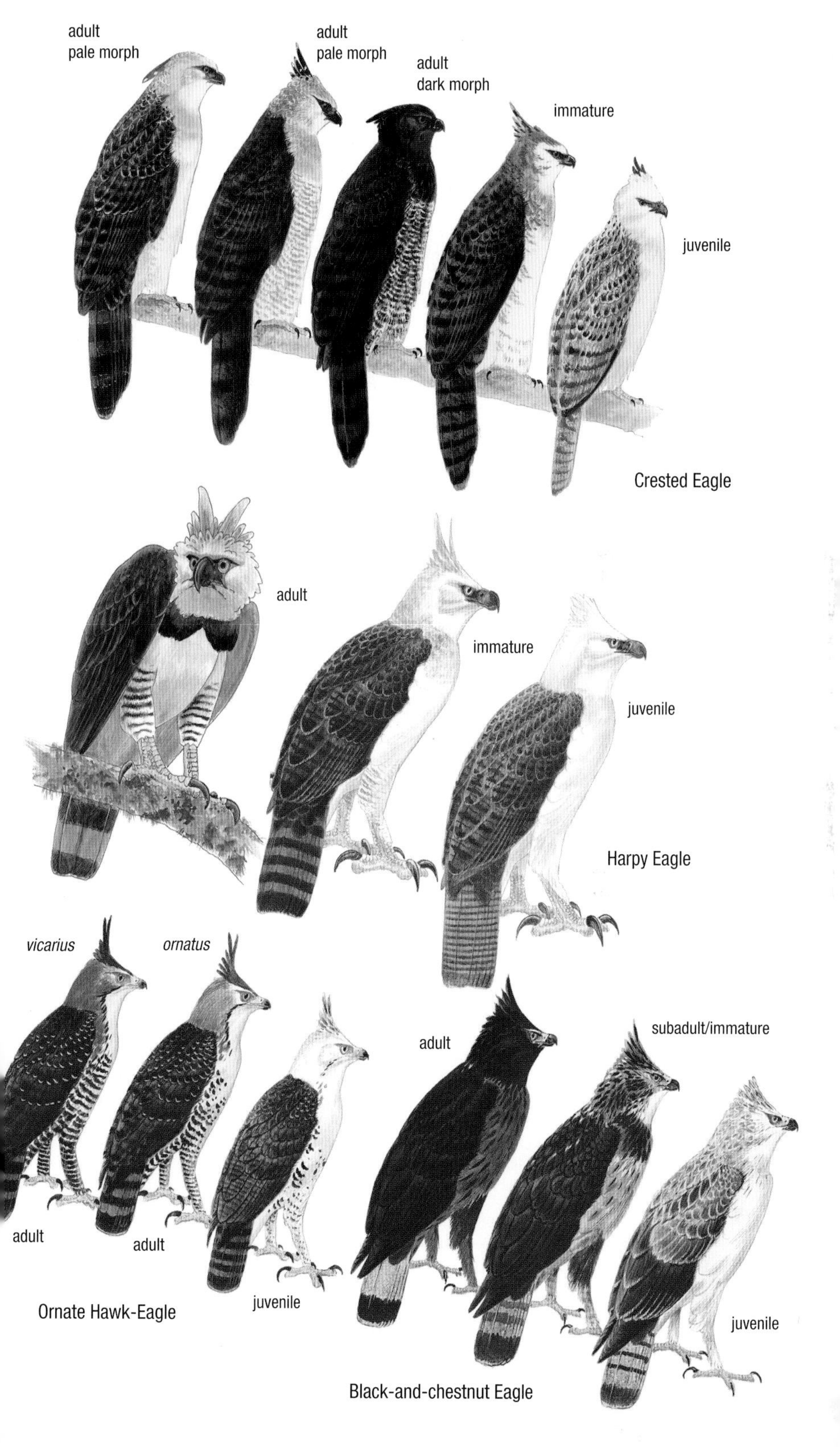
adult
pale morph
adult
pale morph
adult
dark morph
immature
juvenile
Crested Eagle
adult
immature
juvenile
Harpy Eagle
vicarius
ornatus
adult
adult
juvenile
Ornate Hawk-Eagle
adult
subadult/immature
juvenile
Black-and-chestnut Eagle

PLATE 69: LAUGHING FALCON AND CARACARAS

Caracaras are fairly large, found in open and semi-open country, and omnivorous, with less strong bills and talons than falcons and other raptors. *Phalcoboenus* caracaras occur in highlands, the remaining four species in lowlands. Laughing Falcon is a unique, fairly primitive falcon, well known for its remarkable, far-carrying, onomatopoeic voice.

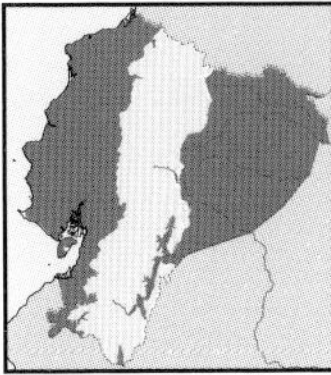

Laughing Falcon *Herpetotheres cachinnans* 46–52cm / w 75–90cm

E & W lowlands to lower foothills (mostly below 800m). Humid to deciduous forest, woodland, adjacent fields. Brown and *pale buff*, tail banded buff and brown, *contrasting black mask*. Juv. duller, paler-edged above. In flight: *large-headed*, short rounded wings, *obvious mask*, underwing mostly buff. Races *cachinnans* (E), *fulvescens* (W). Slow steady flight with shallow flaps and short glides; often long perches motionless. **Voice** Loud, onomatopoeic laughing *gua'cow-gua'cow...gua'co-wa...*, long *güa-güa-güa... güacow-güacow...*. **SS** Unmistakable, but cf. forest-falcon voices. **Status** Fairly common. See also Plate 54.

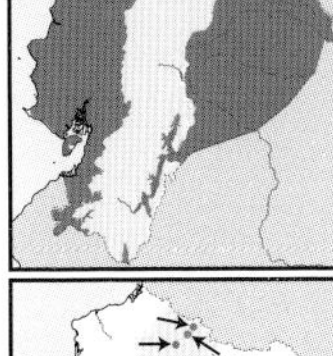

Crested Caracara *Caracara cheriway* 51–59cm / w 110–130cm

SW lowlands, inter-Andean valleys (mainly to 2,000m, locally higher). Dry open terrain, xeric scrub. Large-headed, heavy-billed; *long fingered wings, pale red facial skin. Blackish above and on crown; throat and neck dirty white*. Juv. *duller*, dirty *buff* replaces white, streaked below. In flight: projecting head, *white window in primary bases, broad* black terminal band. Juv. has dusky subterminal band. Monotypic. Steady flight on stiff wings, short glides on arched wings; often on ground or conspicuous perches. **Voice** Throaty *grrrk-g'kk-grrk...*. **SS** Confusion unlikely. **Status** Fairly common. See also Plate 56.

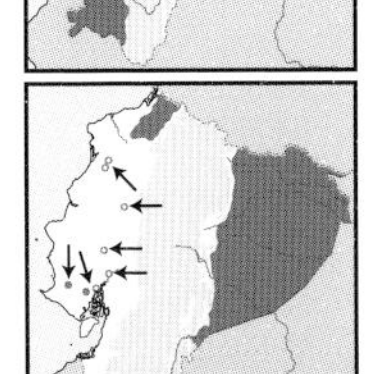

Red-throated Caracara *Ibycter americanus* 51–56cm / w 95–125cm

E & NW lowlands to lower foothills, locally to SW (to 800m). Forest canopy, borders. Slim, long-tailed, *bare red face and legs*. Mostly *black, lower belly and vent white*. Juv. has duller soft parts, black less glossy. In flight: small-headed, long-tailed, longish pointed wings, *contrasting white lower belly and red throat*. Monotypic. Laboured, rather deep wingbeats, never soars, short glides; groups at wasp nests. **Voice** Raspy, raucous, very loud *kloklá-KA-KA..kra-kra-KRAW...klö-klö-KRAW-graw-graw...*. **SS** Black Caracara almost all black, yellow to orange soft parts. **Status** Fairly common. See also Plate 56.

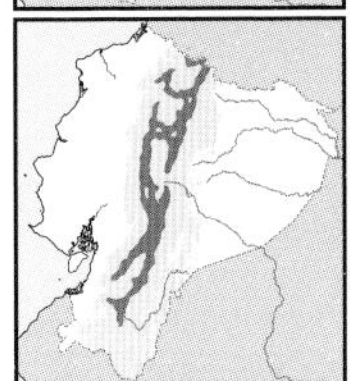

Carunculated Caracara *Phalcoboenus carunculatus* 51–56cm / w 110–120cm

Andes S to N Loja (3,000–4,700m). Grassy *páramo*, adjacent fields. *Bare red face*. Black above, *heavily streaked black and white below, white vent, rump and broad tail tip*. Juv. dull brown, *paler head and underparts*. In flight: *pied underwing*; juv. *pale window in primaries, pale buff rump*. Monotypic. Fast flight, strong wingbeats, erratic soaring; singles to groups on ground. **Voice** Harsh *khraa-raa-raa-raa...* barks. **SS** No overlap with Mountain Caracara; Crested differs in proportions, bill shape, underwing, undertail; juv. Cinereous Harrier (different shape, flight, behaviour). **Status** Uncommon. See also Plate 56.

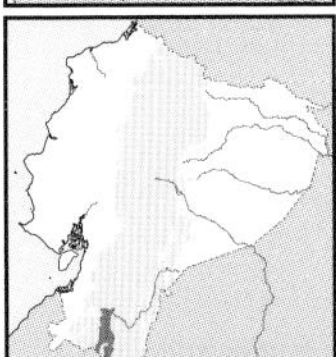

Mountain Caracara *Phalcoboenus megalopterus* 51–56cm / w 110–125cm

Locally in S Andes, in Loja / Zamora Chinchipe (2,900–3,500m). Grassy *páramo*. Replaces Carunculated Caracara in S. Differs in glossy black extending from upperparts to *chest*, *plain white belly*. Juv. largely similar but somewhat *rustier* in tone, *paler rump*, rustier 'window' in primaries. Monotypic. Strong wingbeats, erratic glides and soaring; small groups on ground. **Voice** Harsh, duck-like *kreaa-kra-kra-kra-kra* barks. **SS** Juv. *Buteo* hawks (different proportions, flight, behaviour, etc). **Status** Uncommon, rather local. See also Plate 56.

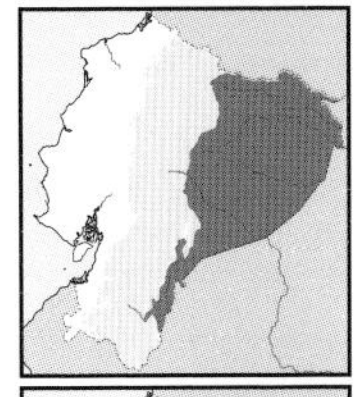

Black Caracara *Daptrius ater* 40–43cm / w 90–100cm

E lowlands to foothills (to 1,300m). Forest borders, open and semi-open areas. Slim, longish tail, *orange to yellow bare face and legs*. Entirely *glossy black, narrow white band near uppertail base*. Juv. duller, more tail-bands. In flight: note *partially hidden tail-band*, pale orange soft parts, longish wings. Monotypic. Laboured continuous flaps, short glides; often pairs or small parties at fruiting trees, carrion or canopy. **Voice** Harsh, irritating *krrryaaaaa* screams. **SS** Red-throated Caracara (white throat, red face) differs notably in voice and behaviour. **Status** Common. See also Plate 56.

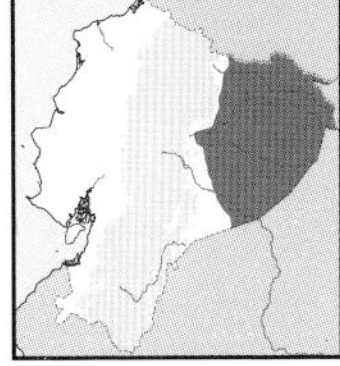

Yellow-headed Caracara *Milvago chimachima* 40–45cm / w 80–95cm

E lowlands (to 300m, locally higher, spreading?). Rivers, sandbars. Small. *Creamy-yellow*, dark brown back, tail and wings. Juv. *dull brown, coarse buff streaks on head and underparts*. In flight: yellowish underparts, *rump* and tail; *dusky secondaries, creamy underwing-coverts, white primaries* dark-tipped. Juv. underparts streaked / barred brown. Nominate. Buoyant floppy flight, short glides; often perches on ground, sandbars, fallen logs and other exposed perches. **Voice** Weak, nasal *wreeeaau!* **SS** Confusion unlikely but see Laughing Falcon (paler buff, rounded wings, mask, behaviour and habitat). **Status** Uncommon. See also Plate 56.

lvescens
cachinnans
adult
adult
juvenile
Laughing Falcon
juvenile
adult
Crested Caracara
juvenile
adult
Red-throated Caracara
juvenile
adult
Carunculated Caracara
juvenile
adult
Mountain Caracara
juvenile
adult
Black Caracara
juvenile
adult
Yellow-headed Caracara

Small forest raptors, extremely difficult to observe even at close range. Their loud barking or yelping voices are remarkable, and the best identification clue. Primarily in lowlands, but some species reach foothills and subtropics. All are shy. Longer-tailed, more sluggish than forest-based hawks and kites.

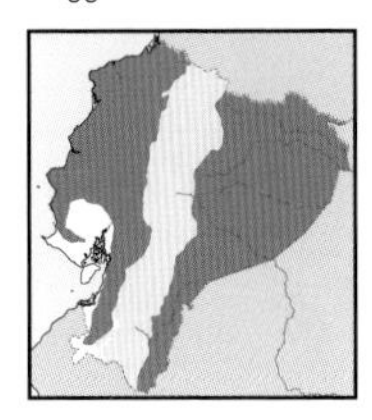

Barred Forest-Falcon *Micrastur ruficollis* 33–38cm / w 45–60cm

E & W lowlands to subtropics (mostly to 2,000m). Humid to montane forest. ***Yellow facial skin***. Blackish above, ***greyish-buff throat. Underparts barred dusky, 3–5 narrow pale tail-bands***. Juv. greyish-brown above, ***narrow pale nuchal collar, whitish to buff below variably barred, pale rump spots***. Race *interstes* (W), *zonothorax* (E slope), *concentricus*? (E lowlands). Crepuscular, often at borders. **Voice** Nasal, descending *kyo!* every 2–3 sec, often in 4–7 series, *kyo-kyó-kokokoko*. **SS** In E, Lined Forest-Falcon (orangey face, white eyes, sparser barring, fewer tail-bands, less ringing voice; juv. by voice); in W, Plumbeous (greyer above, sparser barring, one tail-band, orangey face, higher-pitched voice). **Status** Uncommon.

Plumbeous Forest-Falcon *Micrastur plumbeus* 31–36cm / w 50–55cm

NW lowlands to lower foothills (to 700m, locally higher). Wet forest interior. Recalls Barred Forest-Falcon but ***bare parts orange-red, paler eyes. Ashier-grey above, one white tail-band; sparser barring below***. Juv. has sparser barring below, ***no rump spots***. Monotypic. Often follows army ants; difficult to detect, chases prey from perch. **Voice** Nasal *kéw..kéw..kéw...kew-kew-kew'ke'ke'ke'ke*, lower-pitched, more serial than Barred. **SS** Some Barred can have orangey cere and orbital area; juv. Barred browner above, buffier below, yellower bare parts; voices differ. Collared larger, plain underparts, nuchal collar; see Semiplumbeous Hawk (different tail pattern, plain white below, etc.). **Status** Rare, local, VU globally, EN in Ecuador.

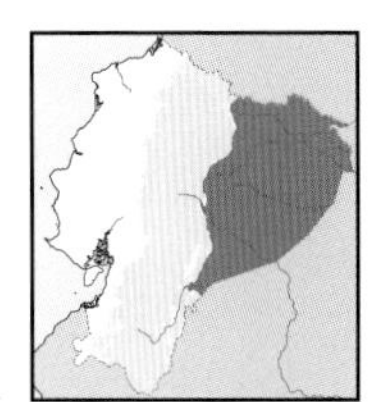

Lined Forest-Falcon *Micrastur gilvicollis* 32–38cm / w 50–60cm

E lowlands to foothills (to 1,500m). Mostly *terra firme* forest. ***Orange bare parts, white eyes. Ashy-grey above***, white below ***sparsely barred grey***, plain throat, ***1–2 narrow white tail-bands***. Juv. as Barred Forest-Falcon but lacks rump spots. Monotypic. May follow army ants; otherwise, retiring and weary. **Voice** Doubled nasal *caw-cáw*, also *caw-cáw-caw-caw..coo* or *caw-cou, caw-cou-cóu* similar to Barred but mellower, lower-pitched. **SS** Barred largely sympatric (E), has dark eyes, yellow bare parts, 3–4 tail-bands, more barred underparts, less nasal voice, single notes more spaced; see also juv. Slaty-backed (larger, dark-eyed, scaly below, cat-like voice). **Status** Uncommon.

Slaty-backed Forest-Falcon *Micrastur mirandollei* 41–45cm / w 65–70cm

E & NW lowlands (to 400m in E; below 100m in W). *Terra firme* forest. ***Yellow bare parts. Ashy-grey above***, blacker tail, ***3–4 narrow whitish tail-bands, uniform white below***. Juv. browner-grey above, ***dirtier below, chest to flanks coarsely mottled or scaled***. Monotypic. Usually at army ant swarms; poorly known. **Voice** Rising, nasal chant of cat-like notes, *caah..cööw..cööw..cööw...coow-coów-coow*, sometimes faster, shorter. **SS** Barred Forest-Falcon (barred below, slower voice); Lined (pale eyes, orangey bare parts, barred below, fewer tail-bands); Collared and Buckley's (larger, pale nuchal collar; more laughing, resonant voices); cf. Bicoloured and Grey-bellied Hawks. **Status** Rare.

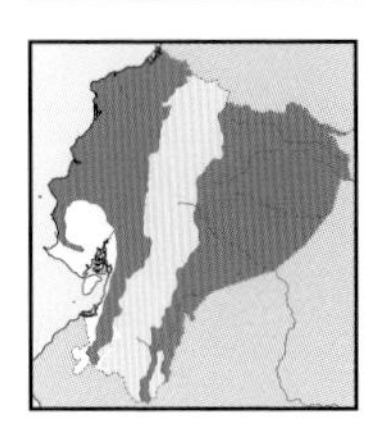

Collared Forest-Falcon *Micrastur semitorquatus* 52–58cm / w 70–85cm

E & W lowlands to subtropics (to 2,000m). Humid forest, borders. Slender, ***dull greenish cere***. Blackish above, black ***crescent on cheeks, white to buff below and nuchal collar, 4–5 white tail-bands***. Rare dark morph in E has sooty underparts. Juv. browner above, ***variably waved dusky below***. Nominate (E), *naso* (W). More often at borders than congeners; perches at various heights. **Voice** Nasal, hollow *cööw* or *cä-cAUP* every 3–4 sec; also rising *cäo-cäo-cäo-coow-coów-coów...cö-cö*. **SS** Buckley's Forest-Falcon similar but has white wing spots, juv. paler below; best told by voice (more hollow in Collared); other *Micrastur* paler above, lack nuchal collar; cf. Bicoloured Hawk. **Status** Uncommon. See also Plate 54.

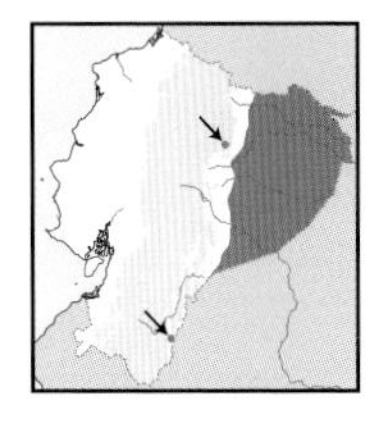

Buckley's Forest-Falcon *Micrastur buckleyi* 41–46cm / w 60–70cm

E lowlands, locally to foothills (to 1,200m). Mostly *várzea* forest, borders. Recalls Collared Forest-Falcon but smaller, ***shorter-legged***, some ***white spots in secondaries***, possibly ***fewer tail-bands***. Poorly known juv. has ***sparse wavy barring*** mostly on flanks. Best differentiated by voice. Monotypic. Poorly known, apparently behaves much like Collared though shyer. **Voice** Three-noted *ka-kwá.. kou, kawk-káw..kou* at long intervals; also rising *kawh-kawh-kawn...cö-cö* faster than Slaty-backed, or laughing *kawh-kawh-kawh*. **SS** See eastern *Micrastur*, especially juv.; Buckley's three-noted voice distinctive, less resonant than Collared, less repetitive than Slaty-backed, Barred and Lined. **Status** Rare, overlooked?

interstes
zonothorax
adult grey morph
adult grey morph
adult rufous morph
immature
immature
immature
immature
Barred Forest-Falcon
adult
juvenile
'lumbeous Forest-Falcon
adult
juvenile
Lined Forest-Falcon
juvenile
adult
Slaty-backed Forest-Falcon
Collared Forest-Falcon
dark immature
morph
pale morph
dark morph
pale immature
♀
♂
juvenile
Buckley's Forest-Falcon

PLATE 71: FALCONS

Typical falcons occur in range of habitats, but mostly in semi-open and open areas. Long pointed wings provide a boomerang shape, permitting very agile and fast flight, including impressive stoops and dives. Two migratory species and two rather common residents; Aplomado and Orange-breasted are rarer. Sexual size dimorphism very marked.

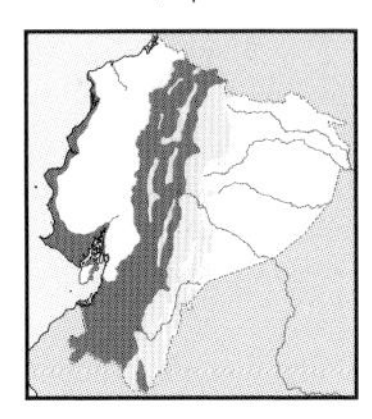

American Kestrel *Falco sparverius* — 26–29cm / w 50–60cm

Inter-Andean valleys, SW subtropics to lowlands (to 3,200m). Open areas, xeric scrub, towns. Small, *longish pointed wings. Blue-grey crown, blackish malar and crescent behind white cheeks. Rich rufous above*, blue-grey wings, orange-buff below. *Rufous tail, black subterminal band.* ♀ *barred above, streaked below, several dark tail-bands.* Races *aequatorialis* (N), *peruvianus* (S). Active fast-flapping flight, soars and glides on level wings, rapid hovering. **Voice** High-pitched shrill *klee-klee-klee-kleé*.... **SS** Slightly larger Merlin (no rufous, streaked below); Bat Falcon (much bolder); Pearl Kite (all white, rufous thighs); Aplomado Falcon (larger, different head pattern). **Status** Common, spreading. See also Plate 56.

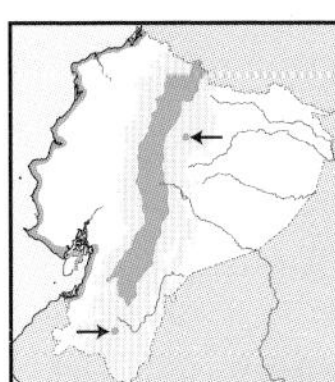

Merlin *Falco columbarius* — 27–32cm / w 55–75cm

Boreal winter visitor to W lowlands and Andes (to 3,600m). Open areas, *páramo*. *Short, broad-based wings. Dusky-grey above* vaguely streaked dusky, white throat, *broad dark malar, buffy eyebrow. Coarsely streaked dusky underparts*; blackish tail, *several pale bands*. ♀: browner above. In flight: *dark*, obvious *buffy streaking below*, dark subterminal tail-band. Nominate. Fast powerful wingbeats, no hovering; dashes after prey. **Voice** Not expected to vocalise. **SS** Peregrine much larger, bolder head pattern; Aplomado larger, bolder face pattern, rufous-orange below; American Kestrel (more rufescent, bolder head, paler underwing); Sharp-shinned Hawk (longer legs, rounded head, broader-based wings). **Status** Uncommon (Oct–Mar). See also Plate 56.

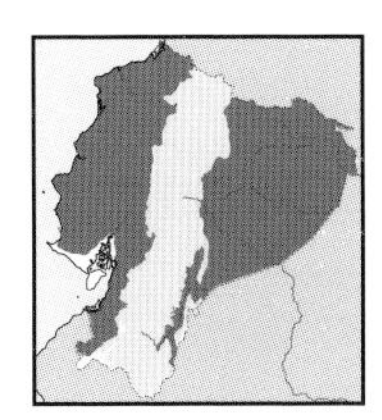

Bat Falcon *Falco rufigularis* — 24–29cm / w 50–65cm

E & W lowlands to foothills (to 1,000m, locally higher). Humid to deciduous forest, borders. *Long, narrow wings. Blackish above, whitish-buff throat and partial nuchal collar, orange-buff wash on neck-sides and chest. Heavily barred below, rufous belly.* Juv. duller above, more rufescent below. In flight: contrasting throat, underwing dark streaked / spotted. Nominate (E), *petoensis* (W). Swift-like dashing flight, strong wingbeats, brief glides and soars; mostly pairs on high perches; often crepuscular. **Voice** Loud *kyé-kyé-kyé-kyé*.... **SS** Orange-breasted Falcon larger, somewhat larger-headed; richer and *broader* orange-rufous chest-band, sparser wavy barring below, *barred vent*. **Status** Uncommon. See also Plate 56.

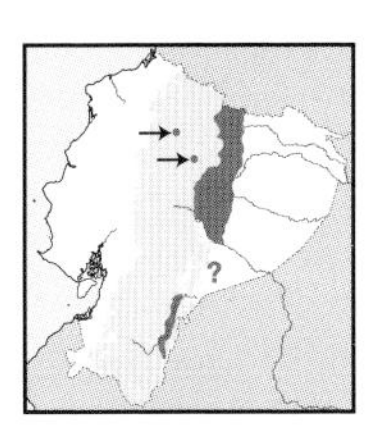

Orange-breasted Falcon *Falco deiroleucus* — 36–41cm / w 70–85cm

E foothills locally to adjacent lowlands (to 1,400m, once 2,900m). Humid forest, borders. Sturdy, large-footed, *broad-based wings. Larger* than Bat Falcon, orange-rufous *more extensive on chest to neck, clear-cut white throat*, belly and *vent more obviously barred*. Juv. browner-black above, duller underparts, streaky chest. In flight: white tail tip *broader* than Bat Falcon, broader-based wings. Monotypic. Fast powerful wingbeats, swift flight, flat-winged glides and soars (rarely); does not stoop on prey. **Voice** Loud barking *kyeea-kyeea-kyeea-kyeea*..., lower-pitched than Bat Falcon. **SS** See higher-elevation Aplomado and migratory Peregrine Falcons. **Status** Rare, local, VU in Ecuador, NT globally. See also Plate 56.

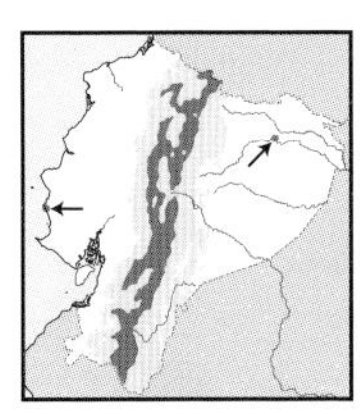

Aplomado Falcon *Falco femoralis* — 36–43cm / w 75–100cm

Andes, inter-Andean valleys, locally in lowlands (mostly 3,000–4,100m). Grassy *páramo*, adjacent fields. *Long narrow wings. Slate-grey above, black moustachial, dark postocular line*, long *buff supercilium, buffy face and throat. Underparts rufous-orange.* Juv. duller, chest streaked dusky. In flight: *buff V on nape*, bold head and underparts, *white trailing edge*. Race *pichinchae*. Shallow fast wingbeats, buoyant flight, flat-winged glides and soars; often perching near ground. **Voice** Occasionally loud, quavering *kyaáá*. **SS** Peregrine (larger, darker upperparts, simpler head pattern, broader-based wings); also lower-elevation Orange-breasted and Bat Falcons (simpler head patterns). **Status** Uncommon. See also Plate 56.

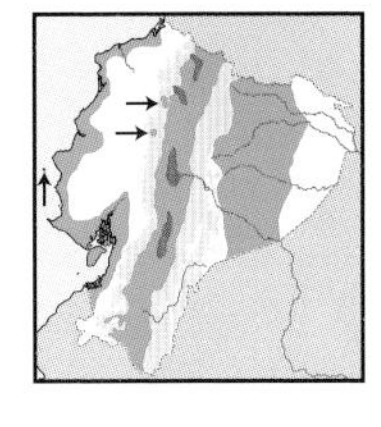

Peregrine Falcon *Falco peregrinus* — 38–48cm / w 80–115cm

Resident locally in inter-Andean valleys (mostly 2,000–3,200m), migrant to coast and lowlands (below 100m). Coasts, open areas. *Robust.* Boreal migrant *anatum*: *slate above, broad black triangular moustachial, whitish neck-sides*, underparts *buffy-white barred blackish*. Boreal migrant *tundrius*: *larger cheek patch, whiter, plainer chest.* Resident *cassini* looks *hooded*, darker belly / breast. Juv. duller, more streaked below. In flight: *broad-based wings*, prominent *moustachial or hood*. Powerful wingbeats, spectacular stoops; glides and soars on slightly depressed wings. **Voice** High-pitched, cackling *klyaa-klyaa-klyaa*.... **SS** Aplomado (smaller, bolder facial pattern, slimmer wings); Orange-breasted Falcon (more striking). **Status** Rare resident; uncommon year-round transient, VU in Ecuador. See also Plate 56.

aequatorialis
peruvianus
♀
♂
♂
American Kestrel
♂
♀
Merlin
petoensis
rufigularis
♂
immature
♀
Bat Falcon
♂
immature
♂
ımmature
Orange-breasted Falcon
♂
immature
Aplomado Falcon
♀
♂
juvenile
♂
tundrius
atum
♂
♀
♀
cassini
Peregrine Falcon

Some of the largest owls in Ecuador, including the distinctive, long-legged, pale, dark-eyed Barn Owl. Head shapes vary from the very long erectile tufts of Crested to the rounded heads of *Pulsatrix*. Aside from Barn, all are forest-based, but Crested and Spectacled are fairly common in forest borders and adjacent clearings.

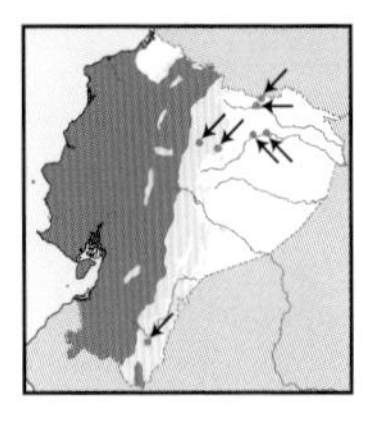

Barn Owl *Tyto alba* — 36–40cm

W lowlands to foothills, inter-Andean valleys, locally in E lowlands (to 3,200m, locally higher). Agricultural areas, town, cities. Whitish bill, ***brown eyes, long feather legs, whitish, heart-shaped facial disk*** outlined dark. Greyish to buff above with some dark and white speckling, ***white to rich tawny below sparsely spotted brown. Underwing-coverts white***, barred flight feathers, short, faintly barred tail. Race *contempta*. Nocturnal, active hunter; low, buoyant flight; regularly seen in city parks. **Voice** Harsh *sh-shhhhhhhhrp*, other hisses, screeches and loud bill snap. **SS** Confusion unlikely owing to facial disk, paleness, long legs, brown eyes, large rounded head. **Status** Common.

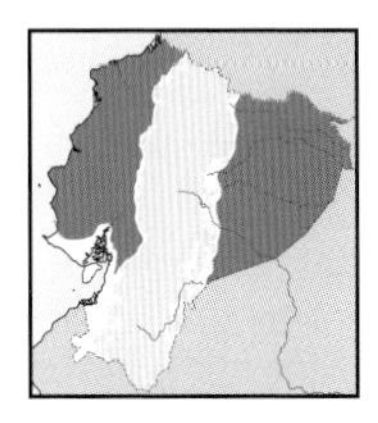

Crested Owl *Lophostrix cristata* — 36–42cm

E & W lowlands to foothills (to 800m, locally higher). Humid forest borders, woodland, adjacent clearings with tall trees. Very ***long white ear-tufts*** contrasting with dark brown head, ***rufous-chestnut cheeks***. Dark brown above, ***whitish spots in scapulars and wing-coverts***, dark tawny breast grading into ochraceous-buff belly, fine dusky vermiculations. Nominate (E), darker *wedeli* (W). Nocturnal, regularly in shady borders, often in pairs resting under palm fronds; vocalises intermittently at night, especially during full moonlight. **Voice** Harsh, croaking, rather short *gggoOORR* or *kkkworRRR*. **SS** Unmistakable; voice harsher than other owls. **Status** Uncommon.

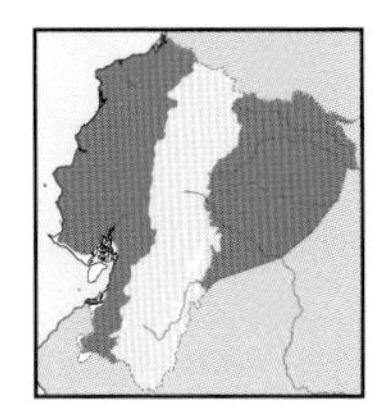

Spectacled Owl *Pulsatrix perspicillata* — 43–48cm

E & W lowlands to foothills (to 1,400m, locally higher). Humid to deciduous forest, borders, woodland, adjacent clearings with trees. Handsome! ***White spectacles*** and neck collar, ***yellow eyes. No ear-tufts***, breast and upperparts dark brown, ***tawny to buff below***, rather long banded tail. Juv. ***mostly whitish***, contrasting ***blackish mask***, brown wings and tail. Nominate (E), darker *chapmani* (W). Nocturnal, vocalises intermittently at night, especially in moonlight; regularly roosts near water and in bamboo stands. **Voice** Bouncing, fading hoots, *BUop-Buop-buop-buop-buop...*, higher-pitched in ♀. **SS** Confusion unlikely over most range; see darker Band-bellied Owl (little overlap in E foothills), with higher-pitched, faster, more accelerating voice. **Status** Uncommon.

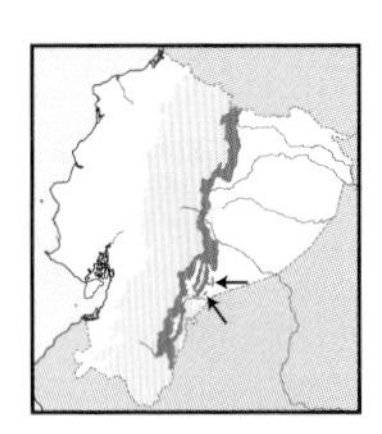

Band-bellied Owl *Pulsatrix melanota* — 36–38cm

Foothills to subtropics on E Andean slope (900–1,500m). Montane forest, borders, adjacent clearings with trees. Dark brown head, ***white spectacles and broad collar, brown eyes***. Dark brown upperparts, some white spotting in scapulars, coverts and flight feathers. ***Brown breast-band*** with whitish speckling, ***belly obviously barred white, rufous and black***, banded tail. Race *melanota*. Nocturnal, vocalises mostly on moonlit nights; forest-based but can emerge into clearings; little known. **Voice** Bouncing, accelerating and fading hoots, *BUap-Buap-buap-buap...*, higher-pitched than Spectacled Owl; also *buuaa!* lament. **SS** Unmistakable; lower-elevation Spectacled Owl has conspicuous face pattern, pale underparts; Mottled Owl (striped below, no white throat, hooting voice). **Status** Rare, locally uncommon.

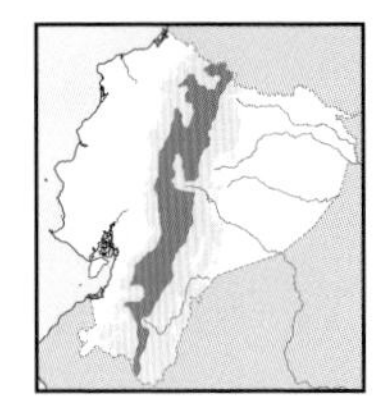

Great Horned Owl *Bubo virginianus* — 48–57cm

Andes (3,200–4,500m; locally to 2,200m). *Páramo*, woodland, *Polylepis* stands, remote agricultural fields. ***Very large, long bushy ear-tufts***, yellow eyes, strong feet. Sooty-brown above mottled dusky and white, ***broad white neck collar, whitish underparts clearly barred dusky-brown***, darker breast. Race *nigrescens*. Nocturnal, but also active at dawn and dusk; mostly solitary and retiring, vocalises on moonlit nights, often rests in crevices or high in *Polylepis* trees. **Voice** Deep, resonant hoots, *boo-boó-boo..boo-boó*; *ha-ha-nhá...* nasal calls. **SS** Confusion unlikely, other large highland owls (Stygian and Short-eared) differ in overall size, ear-tuft size, unbarred underparts; voice deeper than most highland owls, Stygian lower-pitched, notes more separated, simpler, less pronounced. **Status** Rare, local.

pale morph
Barn Owl
wedeli
cristata
rufous morph
grey morph
Crested Owl
adult
juvenile
Spectacled Owl
juvenile
adult
Band-bellied Owl
Great Horned Owl

PLATE 73: SCREECH-OWLS

Formerly included in the widespread genus *Otus*. All strictly nocturnal, forest-based, rarely seen during daylight. Species identification by plumage notably difficult; intraspecific variation (2–3 colour morphs) intricate. Voices are more reliable identification clues. Screech-owls utter two song types (one territorial, one for courtship) and various calls. Taxonomy still complex in some.

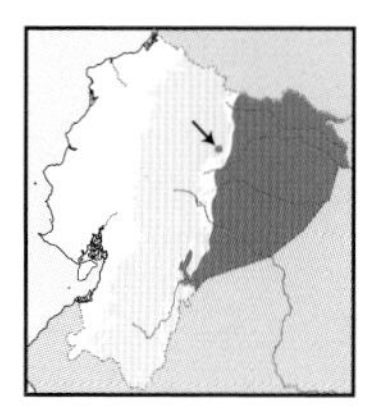

Tropical Screech-Owl *Megascops choliba* 23–25cm

E lowlands (mostly to 700m). Riparian woodland, borders, adjacent clearings. ***Yellow eyes.*** Cinnamon to greyish-brown above, white dots in scapulars and wing-coverts, ***whitish brows, face clearly outlined black***. Buffy to pale grey below ***streaked blackish in herringbone pattern***. Race *cruciger*. Regular in semi-open areas; vocalises just after dusk and before dawn; roosts in dense tangles. **Voice** Slow, rolled *poo-ru-ru-ru-ru-rúk, poo-ru-ru-ru-ru-rúk…*; short purring *purrr-kúk, purrr-ru-kúk*. **SS** Vermiculated Screech-Owl (foothills) darker face, conspicuous white in wings; voice longer, more trilled, ascending. Tawny-bellied (dark eyes, darker overall, pale hindneck, voice more tooting, longer). **Status** Uncommon.

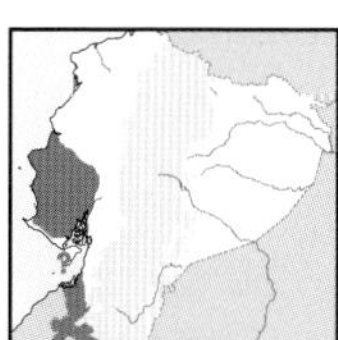

Peruvian Screech-Owl *Megascops roboratus* 19–22cm

SW lowlands to foothills, extreme SE in Zumba region (to 1,200m). Deciduous to semi-deciduous forest, borders. Brownish-grey to cinnamon-rufous above, ***darker crown. Whitish brows, face outlined black***. Paler underparts ***markedly herringboned blackish***. Nominate (SE), *pacificus* (SW) possibly valid species. **Voice** Ascending, low-pitched *purrrRR* trill, sometimes more stuttered; higher-pitched, more inflected and fading in W. **SS** Vermiculated Screech-Owl (shorter tufts, more white in wing, simpler pattern below, higher-pitched, less trilled voice; more humid habitats). **Status** Uncommon.

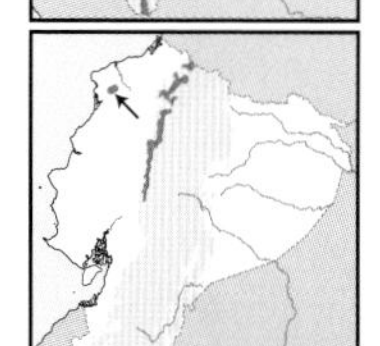

Colombian Screech-Owl *Megascops colombianus* 25–28cm

NW Andean slope, local in coastal ranges (1,300–2,300m, 500–600m in Mache-Chindul). Montane forest, borders. ***Dark eyes***. Brown to rufous-brown above, ***paler brows, buffy underparts faintly streaked and cross-hatched dusky***. Monotypic. Strictly nocturnal, vocalises soon after dusk, before dawn or on moonlit nights, readily responds to playback. **Voice** Long *bu* series, notes evenly pitched lasting up to 15 sec; also *bap-bap, bababa…*. **SS** Voice more hooting, longer than Vermiculated Screech-Owl; higher-elevation White-throated (yellow eyes, white throat, voices differ). **Status** Rare, NT.

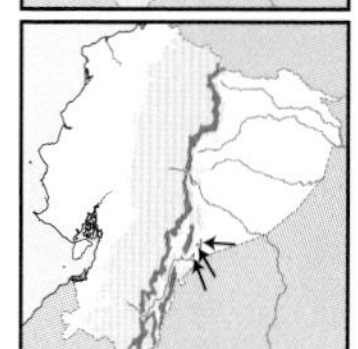

Rufescent Screech-Owl *Megascops ingens* 25–28cm

Locally on E Andean slopes (1,200–2,300m). Montane forest, borders. Closely recalls Colombian Screech-Owl but both morphs paler, ***more patterned below***, rather ***large whitish dots in scapulars***. Nominate. Poorly known, apparently on outlying ridges, strictly nocturnal, vocalises soon after dusk or before dawn. **Voice** Long rising series of evenly-pitched *bu* notes lasting 10–12 sec; or accelerating *bup-bup-bup-bup-bup-bup, bubububu…*. **SS** Smaller, plainer Cinnamon has shorter, lower-pitched, less accelerating, more evenly-pitched voice; Vermiculated voice more trilled. **Status** Rare, local.

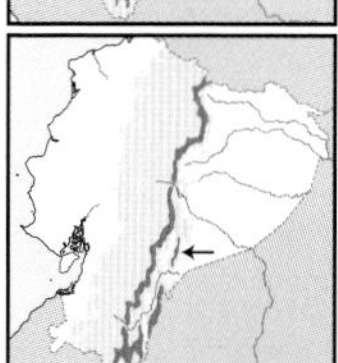

Cinnamon Screech-Owl *Megascops petersoni* 22–24cm

Local on E Andean slopes (1,700–2,200m). Montane forest. Dark eyes. ***Cinnamon-brown*** above lightly vermiculated dusky, narrow ***buffy nuchal collar, buff dots on scapulars. Whitish brows***, faint facial rim; paler underparts faintly herringboned and cross-hatched dusky. Monotypic. Broadly overlaps with Rufescent Screech-Owl, vocal interactions reported, Cinnamon at higher elevations. **Voice** Rather short series of low-pitched *too* notes, only slightly accelerating. **SS** Larger Rufescent (longer, slower, higher-pitched, more accelerating voice); Vermiculated voice more trilled. **Status** Rare, local.

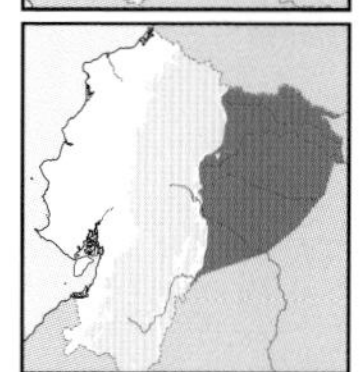

Tawny-bellied Screech-Owl *Megascops watsonii* 21–23cm

E lowlands (to 500m, locally higher). *Terra firme* and *várzea* forests. ***Orange-brown to amber eyes***. Greyish-brown above faintly mottled / vermiculated, slightly paler face outlined black, ***ochraceous-buff to greyish-buff underparts with blackish herringbone pattern***, plainer breast. Nominate. Rarely in borders and clearings; strictly nocturnal but may vocalise before dusk. **Voice** Long (20 sec) series of rapid, evenly-paced *too* notes increasing in volume, sometimes shorter, more staccato. **SS** Tropical Screech-Owl (paler below, yellow eyes, white in scapulars, more trilled voice). **Status** Uncommon.

Vermiculated Screech-Owl *Megascops guatemalae* 20–22cm

E & NW Andean foothills, also coastal ranges (500–1,400m, below 1,000m in W). Humid forest, borders. Yellow eyes. Greyish-brown to rufescent above, faintly vermiculated, ***white dots in scapulars and wing-coverts. Narrow*** black facial rim; ***paler underparts, narrow dusky barring, very faint streaking***. Races *napensis* (E), *centralis* (W); both possibly separate species. Strictly nocturnal, retiring; tends to stay in cover. **Voice** Fast, descending, tremulous trill lasting 8–10 sec (E), or short, inflected, purring trill ending abruptly (W). **SS** See Peruvian Screech-Owl. **Status** Rare.

Tropical Screech-Owl
ufous
brown
dark grey-brown
pacificus
rufous
grey
roboratus
rufous
grey
Peruvian Screech-Owl
brown
rufous
brown
Colombian Screech-Owl
grey-brown
brown
grey-brown
rufous
Rufescent Screech-Owl
Cinnamon Screech-Owl
grey-brown
brown
rufous
Tawny-bellied Screech-Owl
centralis
ous
brown
grey
juvenile
Vermiculated Screech-Owl
napensis
rufous
brown
grey

Ciccaba owls, plus the largest and most distinctive screech-owl, White-throated, which might belong in a separate genus. *Ciccaba* owls are fairly large, round-headed, and dark-eyed; one strictly montane species lives close to treeline, the remaining species range from lowlands to lower subtropics. Two handsome black-and-white species and two more cryptic species clad in brown, rufous and buff.

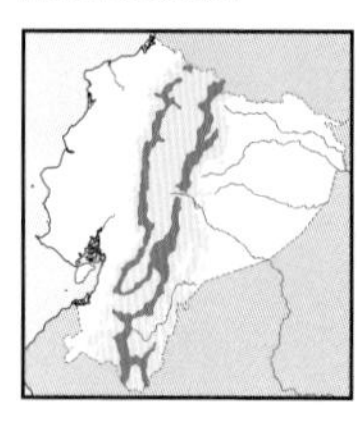

White-throated Screech-Owl *Megascops albogularis* — 25–28cm

Andes (2,500–3,500m). Temperate forest, borders, woodland. ***Yellow eyes***. Dark brown above, whitish mottling above, whitish brows. ***Contrasting white throat***, brown breast, ***buff to tawny belly*** coarsely herringboned in brown. Race *macabrus* (W), nominate (E). Strictly nocturnal, often vocalises in full dark; retiring, rarely in open. **Voice** Fast, chattering *woo-doo, woo-doo, woo-do-woo-woo-woo*; also long series of high-pitched *woo-woo-woo*... and more rhythmic *doo-woodoo, do-woodoo*.... **SS** No other yellow-eyed screech-owl in highlands; Rufous-banded Owl strikingly different, voice more hooting, shorter, simpler; Great Horned Owl voice shorter, hooted, resonant. **Status** Fairly common.

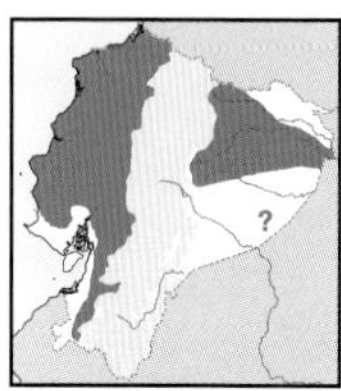

Mottled Owl *Ciccaba virgata* — 30–34cm

W lowlands to subtropics, locally in E lowlands (to 2,000m in W, lower in E). Forest canopy, borders, woodland. Brown head, breast and upperparts, somewhat vermiculated, whitish scapular dots. ***Whitish brows, face rim and throat, underparts buffy coarsely streaked brown***, barred tail. Nominate (W), *superciliaris*? (E); possibly separate species. Strictly nocturnal, vocalising in full dark, readily responds to playback. **Voice** Slow, evenly-pitched, emphatic 4–6 *kgoó* notes, fourth or fifth note more emphatic, higher-pitched in W; also cat-like *wyaaa!* and others. **SS** From screech-owls by size, rounded head, streaked underparts; higher-elevation Rufous-banded (more rufescent, unstreaked underparts); Black-and-white's voice is faster, more emphatic. **Status** Uncommon.

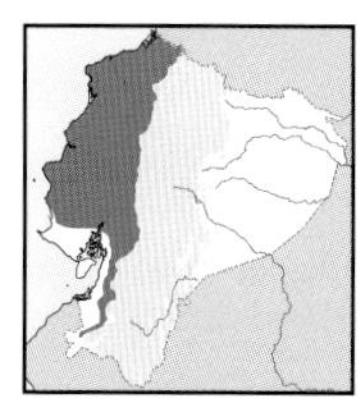

Black-and-white Owl *Ciccaba nigrolineata* — 36–40cm

W lowlands to foothills (to 1,400m, locally higher). Humid forest, borders, woodland. ***Dark eyes, yellow-orange bill and legs***. Black crown, ***heart-shaped mask*** and upperparts, ***white freckled broad face outline, white-and-black banded nuchal collar and underparts***, black tail with narrow white bands. Monotypic. Strictly nocturnal but can show up near lamps, day-roosts high in forest canopy or bamboo. **Voice** Rapid series of low *koop-koo-koo-koo-koo...kOÓ*, shorter, shorter and higher-pitched in ♀; loud *keywoo* and other calls. **SS** Confusion unlikely; voice faster, more emphatic, higher-pitched than Mottled Owl. **Status** Uncommon.

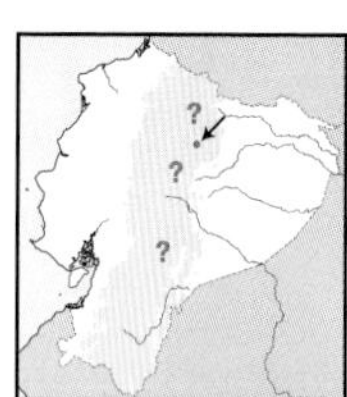

'San Isidro' Owl *Ciccaba* sp. — 36–40cm

Locally in E subtropics (2,000m). Forest borders and adjacent treed clearings in San Isidro region. Closely ***resembles Black-and-white and Black-banded Owls***; not safely separated but no overlap. Taxonomic status unclear, appears ***intermediate*** in plumage between Black-and-white (W lowlands to slopes) and Black-banded in E lowlands; possibly only a highland population of Black-banded. **Voice** Similar to Black-and-white and Black-banded (possibly lower-pitched?). **SS** Banding on underparts of Black-banded looks ***denser***, wings have ***bolder*** white barring, upperparts ***coarsely barred*** but very similar voice. **Status** Unknown, possibly overlooked separate taxon?

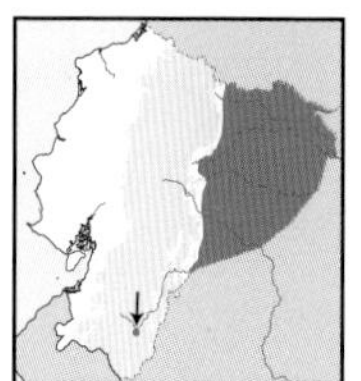

Black-banded Owl *Ciccaba huhula* — 36–40cm

E lowlands to foothills (to 1,000m, locally to 1,500m in S). Mostly *terra firme* forest canopy, borders. ***Dark eyes, yellow-orange bill and legs***. Black above ***faintly barred white, heart-shaped black mask, extensive white freckling on face rim; underparts black densely banded white***, black tail narrowly banded white. Nominate. Strictly nocturnal, forest-based, vocalises in full dark; day-roosts in tall canopy. Might respond to Black-and-white Owl's voice. **Voice** Song *boo-boop-boop-boop... boÓ*, similar to Black-and-white, but rather simpler, mellower; also cat-like *keyoow* and other calls. **SS** Unmistakable in lowlands; cf. very similar 'San Isidro' Owl on Andean slopes (no known overlap); Mottled Owl voice slower, evenly pitched. **Status** Uncommon.

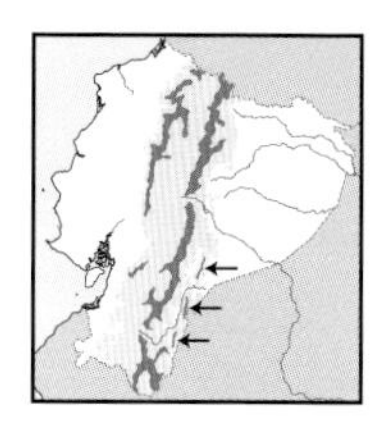

Rufous-banded Owl *Ciccaba albitarsis* — 34–38cm

Subtropics to temperate zone on E & NW Andean slopes (1,900–3,300m, locally higher). Montane forest, borders. ***Dark eyes*** and bill. Dark brown above coarsely ***banded rufous***. Rufous face, ***large whitish eyebrows***, narrow black face outline. ***Broad rufous breast-band mottled / barred blackish. Tawnier belly*** herringboned dusky, rufous and white. Monotypic. Strictly nocturnal, sometimes vocalises on moonlit nights; day-roosts in dense undergrowth to midstorey, reluctantly flushes. **Voice** Rather slow *bu-bu-bu-bu.. bUU*; fast *prrr* and paced *bu, bu, bu* calls. **SS** Lower-elevation Mottled Owl (no known overlap) streaked below, browner above, voices differ; White-throated Screech-Owl (prominent white throat, yellow eyes). **Status** Uncommon.

Mottled Owl
grey
brown
brown
rufous
brown
grey-brown
virgata
superciliaris
superciliaris
juvenile
White-throated Screech-Owl
adult variants
Black-and-white Owl
'San Isidro' Owl
adult variants
Black-banded Owl
Rufous-banded Owl

PLATE 75: PYGMY-OWLS

The smallest owls, active by day, or moonlit nights, dusk and dawn. Identification by plumage highly complex owing to intraspecific variation (colour morphs) and little interspecific differentiation. Again, voice is *the* identification clue; fortunately, voices are distinctive. All are forest-based, except Peruvian. Two recently described species replace commoner Andean at lower elevations; pay attention to range and habitat for identification.

Cloud-forest Pygmy-Owl *Glaucidium nubicola* 14–15cm

Local in foothills and subtropics of W Andes (1,400–2000m, 900–1,000m in SW). Montane forest canopy, borders. ***Greyish-brown head heavily spotted white***, long whitish eyebrows, narrow black and white 'false eyes' on nape. Brown above, ***faint white dots on scapulars, wing-coverts faintly edged buffy-white***. Whitish below obviously streaked dusky-brown, sides mottled richer brown. ***Short banded tail.*** Rufous morph more subdued overall. Monotypic. Diurnal and crepuscular, regularly vocalises on moonlit nights; apparently low population densities. **Voice** Paired *pu-pu, pu-pu…* lasting up to 20 sec. **SS** Very similar to upper-elevation Andean Pygmy-Owl (longer-tailed, upperparts spotted / barred); Andean voice longer, not paired. **Status** Rare, VU.

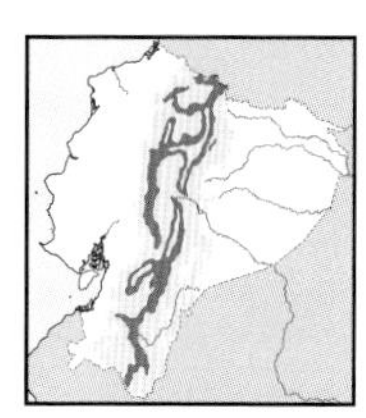

Andean Pygmy-Owl *Glaucidium jardinii* 14–15cm

Andes (2,000–3,500m). Montane forest, woodland, borders, adjacent clearings. ***Brown head heavily spotted white***, long whitish eyebrows, black and white 'false eyes' on nape. Brown above, ***faint white dots on scapulars, back and coverts faintly barred. Whitish throat***, streaked dusky-brown below. ***Fairly long*** banded tail. Wings ***relatively long***. Rufous morph mostly ***rich rufous***, plainer wings. Monotypic. Mainly diurnal and crepuscular, regularly vocalises on moonlit nights; often mobbed by passerines, moving tail sideways. **Voice** Stuttering *weer-weer…* followed by *du-du-du-du…*, introductory notes often missing. **SS** Lower-elevation Cloud-forest and Subtropical Pygmy-Owls might overlap locally; best told by voice. **Status** Uncommon.

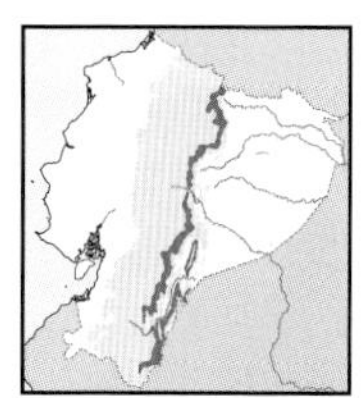

Subtropical Pygmy-Owl *Glaucidium parkeri* 14–15cm

Locally in E Andean foothills to subtropics (1,100–2,000m). Montane forest canopy, borders. ***Greyish-brown head densely spotted white***, long whitish eyebrows, white collar, narrow black and whitish 'false eyes' on nape. Brown upperparts, ***contrasting whitish dots in scapulars and wings***. Whitish below profusely streaked dusky-brown, breast-sides more splotched. ***Short*** banded tail. Monotypic. Diurnal and crepuscular; rarely leaves forest canopy. **Voice** Mellow, 3–4-noted *too-too, too?, too-too-too, too?* **SS** Best told from Andean Pygmy-Owl by 3–4-noted, inflected voice and less streaked underparts; Ferruginous Pygmy-Owl (larger, streaked head, longer-tailed). **Status** Rare, local.

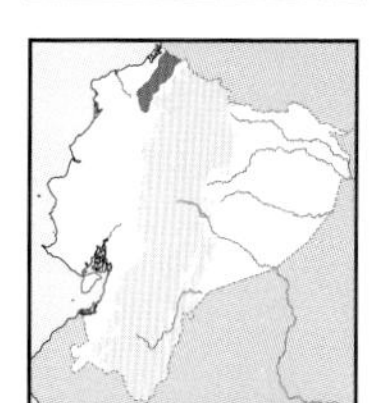

Central American Pygmy-Owl *Glaucidium griseiceps* 14–15cm

Local in NW lowlands (100–600m). Very humid forest canopy, borders. ***Greyish head with faint white dots***, white eyebrows, narrow black and whitish 'false eyes' on nape. ***Rufescent-brown above***, faint whitish dots on scapulars and wings, whitish below, breast-sides splotched rufescent-brown, ***belly streaked rufescent-brown***. Black tail banded white. Unknown race, presumably ***undescribed***. Mainly in forest canopy, rarely emerges in adjacent clearings; diurnal and crepuscular, seldom seen. **Voice** Spaced, fairly slow and ascending series of 3–5 ringing *doo* or *dew* notes, sometimes longer. **SS** No other pygmy-owl in range. **Status** Rare, local, VU in Ecuador.

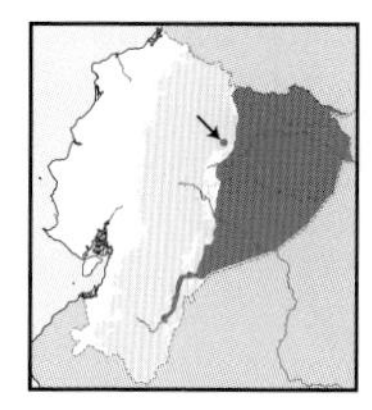

Ferruginous Pygmy-Owl *Glaucidium brasilianum* 16–17cm

E lowlands to foothills (to 1,200m). Mostly *várzea* forest, borders, woodland. Brown to rufescent head with faint ***whitish shaft-streaks***, white throat, grey-black 'false eyes' on nape. ***Greyish-brown to rufous above***, whitish dots on scapulars and wings. Whitish below, sides splotched and streaked brown to rufous. ***Fairly long, faintly barred tail.*** Race *ucayalae*. Mainly crepuscular, more in semi-open and canopy, where often perches exposed, easily attracted using playback. **Voice** Long, steady series of ringing *puu* notes lasting 30+ sec; other trills and whines. **SS** Only pygmy-owl in range; in foothills cf. Subtropical (dotted crown, scapulars and wings, black tail, shorter voice). **Status** Uncommon.

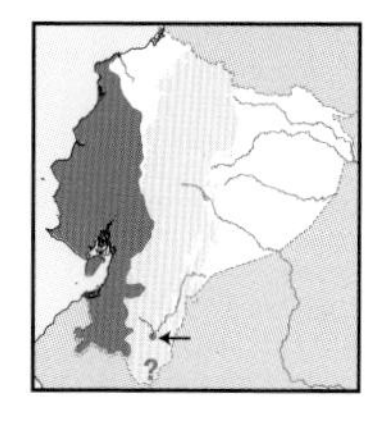

Peruvian Pygmy-Owl *Glaucidium peruanum* 16–17cm

SW lowlands to subtropics, extreme SE in Zumba region (to 2,400m in W, 900m in SE; locally higher). Deciduous forest, woodland, scrub, semi-open areas. Rufous morph has ***streaked crown***, faint wing dots. Brown morph with ***streaks and spots on crown, more evident wing spots***. Monotypic, but complex variation. Diurnal, crepuscular but also nocturnal; conspicuous in semi-open areas, often perching on stumps. **Voice** Long, fast series of staccato *tooi* notes lasting 7–30+ sec; also agitated *chw-chw-chw…kik-kik-kik!* **SS** Locally Andean Pygmy-Owl in S subtropics (dotted crown, black tail, slower and less staccato voice). **Status** Common.

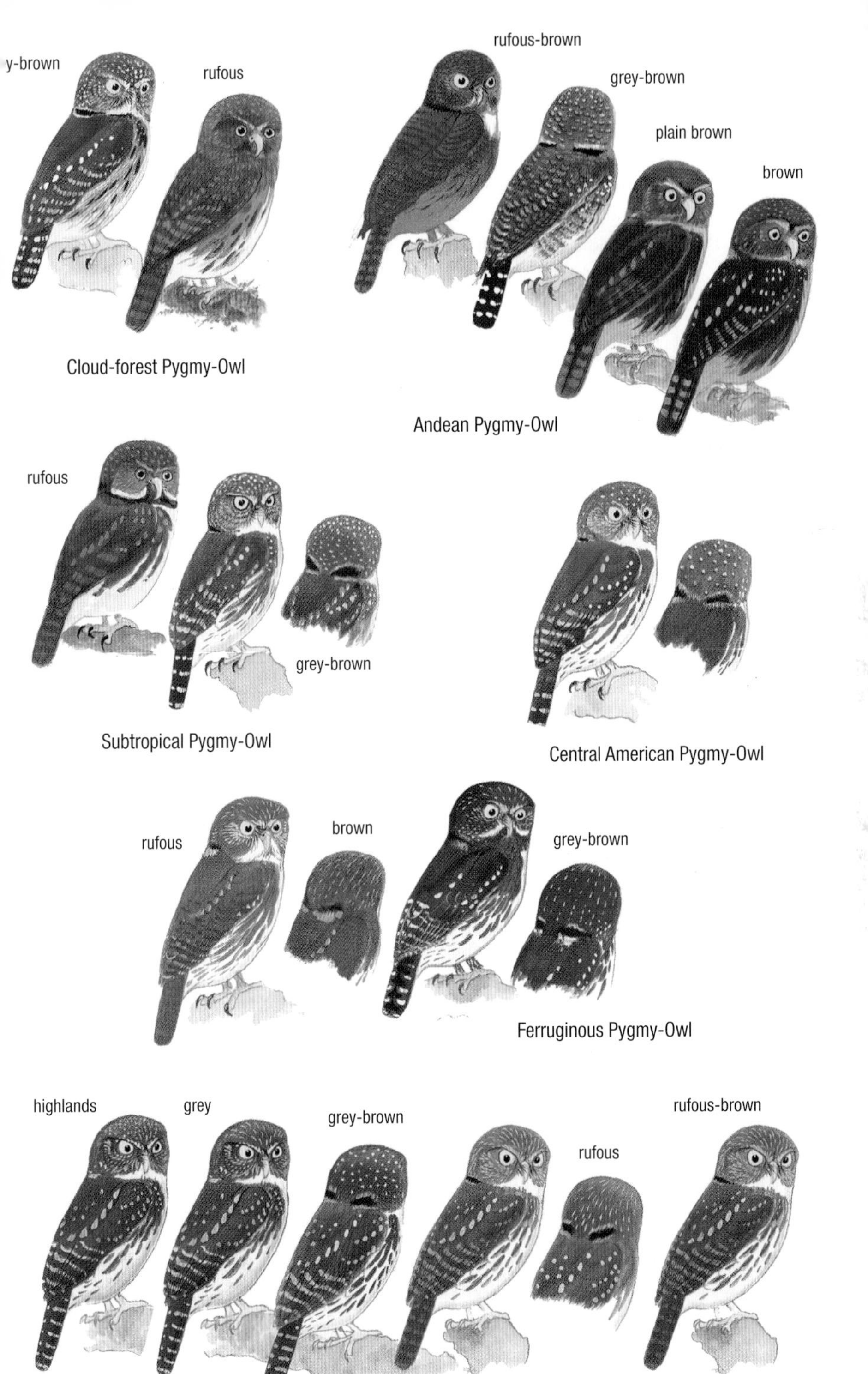
y-brown
rufous
Cloud-forest Pygmy-Owl
rufous-brown
grey-brown
plain brown
brown
Andean Pygmy-Owl
rufous
grey-brown
Subtropical Pygmy-Owl
Central American Pygmy-Owl
rufous
brown
grey-brown
Ferruginous Pygmy-Owl
highlands
grey
grey-brown
rufous
rufous-brown
Peruvian Pygmy-Owl

PLATE 76: MISCELLANEOUS OWLS

'Miscellaneous' owls, including the only terrestrial and sociable species (Burrowing). This small, long-legged owl ferociously defends nesting and roosting sites. *Asio* are large owls with complex underparts patterns, all three preferring non-forest habitats. *Aegolius* is perhaps the most distinctive and possibly one of the rarest owls in Ecuador; its habitat, distribution and natural history are very poorly known.

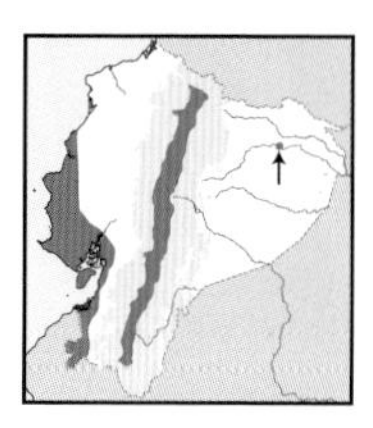

Burrowing Owl *Athene cunicularia* — 20–26cm

SW lowlands, inter-Andean valleys, Río Napo islands (below 50m SW, below 300m in NE, 1,500–3,000m in highlands). Arid open and semi-open terrain. ***Long feathered legs. Markedly streaked whitish on head, spotted on upperparts and wings; underparts barred and spotted dusky-brown*** and white; short tail. Race *pichinchae* (highlands); paler, less spotted *punensis* (SW); *carrikeri*? (NE). Mainly diurnal, terrestrial, stands next or close to burrows, often in pairs or groups, bobs up and down when excited. **Voice** Harsh *keaa* or *kreea* screams; shrieking *kee-kikikikikiki* and other screeching. **SS** Unmistakable. **Status** Uncommon to fairly common.

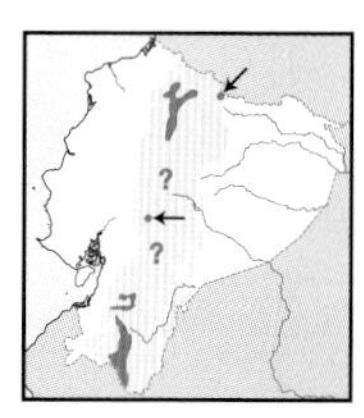

Buff-fronted Owl *Aegolius harrisii* — 19–20cm

Locally along Andes (1,900–3,100m). Montane forest, woodland, adjacent clearings, forested gorges. Small and beautiful. Brown above, ***white spots on scapulars, wing-coverts and flight feathers. Buff to tawny below, creamy face*** with ***two black stripes above eyes***, facial disk narrowly outlined black. Buff feathered legs, brown eyes. Nominate. Poorly known in Ecuador, where rarely seen, population densities probably low; crepuscular and nocturnal, not very vocal; forest-based but might emerge into adjacent areas. **Voice** Long, quavering *tur'r'r'r'r'r'r* trill; soft, high-pitched *tu* notes reported. **SS** Unmistakable (if seen), trilled voice differs from pygmy-owl whistles, and other owls' hooting. **Status** Very rare but probably overlooked; VU in Ecuador.

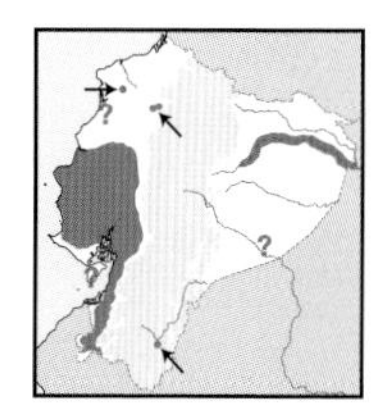

Striped Owl *Asio clamator* — 32–38cm

SW lowlands to foothills, locally in N lowlands (to 700m, below 300m in E but locally higher). Woodland, adjacent clearings, agricultural fields; river islands and riparian woodland in E. ***Dark eyes, long bushy ear-tufts. Buffy-white face outlined black. Cinnamon to tawny-brown*** above, streaked / mottled dusky. ***Buffier below markedly streaked brown***, more spotted breast. Banded tail. Nominate. Active from dusk to night but not very vocal; roosts in or near ground; possibly spreading. **Voice** Rather nasal hoots in slow series, *ghoo...ghoo....* **SS** Mottled Owl lacks ear-tufts; cf. higher-elevation darker, pale-eyed Short-eared and Stygian Owls (no overlap); voice simpler, at longer intervals than other hooting owls. **Status** Rare, more local in E.

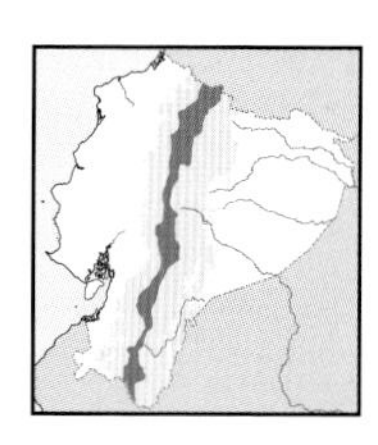

Stygian Owl *Asio stygius* — 41–46cm

Local in Andes (1,700–3,100m, locally lower). Semi-open terrain, agricultural fields, forest borders, woodland. Large, with ***long bushy ear-tufts, orange-yellow eyes; freckled white eyebrows***, dusky facial disk. Blackish-brown head and upperparts, some whitish spots on back. Underparts coarsely ***blotched / streaked brown and buff (herringbone pattern)***. Rather long banded tail. Race *robustus*. Active from dusk, often near houses; roosts high in trees near trunk. **Voice** Deep, very low-pitched *boo* at 5+ sec intervals, followed by nasal *nha* (♀). **SS** Short-eared Owl has nearly imperceptible tufts, no herringbone pattern below, voices strikingly different; voice higher-pitched in Great Horned Owl. **Status** Rare.

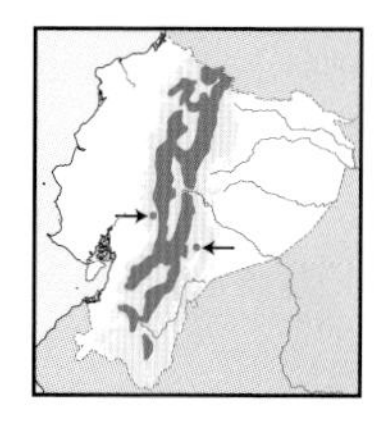

Short-eared Owl *Asio flammeus* — 36–40cm

Andes S to Azuay (3,000–4,000m, locally down to 2,200m). *Páramo*, dry valleys, wooded borders, adjacent agricultural fields, wooded gorges. ***Tiny bushy ear-tufts, yellow eyes.*** Brown face, ***freckled whitish brows and face***. Brown to dusky-brown above, streaked and spotted whitish-buff. ***Tawny to buff below streaked dusky.*** Race *bogotensis*. Regularly seen perched on stumps or fence posts, or gliding low over grassland and crops; mainly crepuscular but often diurnal. **Voice** Mostly silent but *shrie-shrie* growls often heard. **SS** Darker, longer-tufted Stygian tends to prefer less open situations, roosts in trees, not in tall grass. **Status** Uncommon.

grey
pale
normal
dark
punensis
pichinchae
carrikeri
Burrowing Owl
normal
morph
pale
morph
uff-fronted Owl
juvenile
Striped Owl
relaxed
alarmed
Stygian Owl
Short-eared Owl

Charismatic nightbirds more often heard than seen, but seen more frequently than nightjars or nighthawks. Oilbird is widespread throughout Ecuador, being especially numerous around caves, especially in E foothills. Potoos are like feathered branches or stumps and thus easily overlooked. However, as usually loyal to roost sites, once located, daylight observations are feasible. Potoos' eyes reflect light.

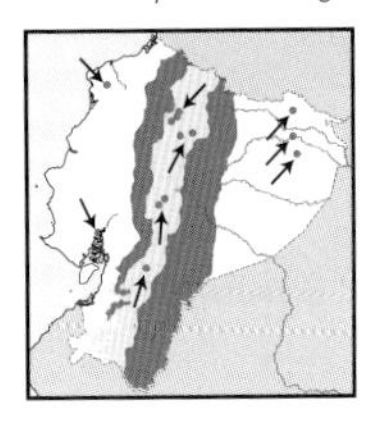

Oilbird *Steatornis caripensis* 43–48cm

Spottily throughout Andean slopes, locally in inter-Andean valleys and lowlands (500–2,400m, occasionally to 3,000m, accidental higher or lower). Canopy, borders, adjacent clearings, woodland. Heavy, ***hooked rosy-horn bill***, large dark eyes (reflect red). Unique, ***mostly rufous, sparse white dots outlined black on wings, whitish diamonds on underparts. Long graduated tail.*** Monotypic. Normally roosts in caves and deep gorges, sometimes numerous; noisy flying and screaming at unison; plucks fruits while hovering; lengthy foraging movements. **Voice** Phantasmagorical screams inside caves; in flight utters raucous screams, snarls, clicks and growls, *kkrra-krra-krra, krrkrkrr....* **SS** Unmistakable; Rufous Potoo similarly patterned but much smaller, behaviour totally different. **Status** Uncommon, local.

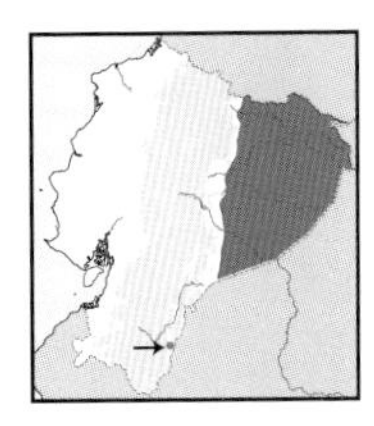

Great Potoo *Nyctibius grandis* 48–54cm

E lowlands (to 400m, locally to 900m). *Terra firme* and *várzea* canopy, borders. Large, ***big-headed***, large brown eyes (reflect bright reddish-orange). Mostly ***greyish-white*** to ***pale*** brownish-grey head and upperparts, ***thickly speckled / mottled dusky. Buffy-white*** below faintly vermiculated dusky, somewhat mottled black on breast; rather long barred tail, ***wings reach near tail tip.*** Nominate. Perches atop stumps or exposed canopy branches, and sallies for flying prey. **Voice** Far-carrying, scary, *WYeeeeoou*; harsh *waak* and other calls. **SS** Much larger than other potoos; Long-tailed darker, with prominent white malar, longer tail, mellower voice. **Status** Uncommon.

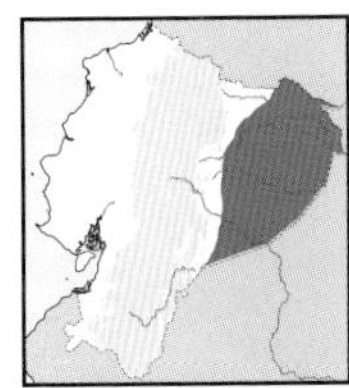

Long-tailed Potoo *Nyctibius aethereus* 42–50cm

E lowlands, mostly NE (to 700m). *Terra firme* forest subcanopy, borders. Large brown eyes reflect bright red. ***Brown upperparts*** streaked and mottled blackish, contrasting ***whitish malar***, buffy-white eyebrows. ***Pale buff*** wing-coverts, barred flight feathers. Underparts rufous-brown, paler on belly, intricately striped, mottled and vermiculated blackish. ***Long broad tail barred blackish-buff, wings do not reach mid tail.*** Race *longicaudatus*. Roosts in forest subcanopy to midstorey, rarely emerging from forest, vocalising just after dusk, before dawn and on moonlit nights. **Voice** Far-carrying, throaty *guaOOurr* every 5–7 sec. **SS** Larger, longer-tailed than Common Potoo and darker, smaller, longer-tailed than Great Potoo. **Status** Rare.

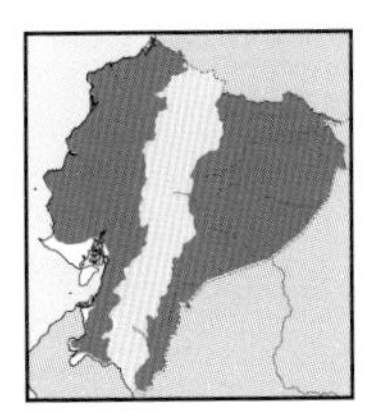

Common Potoo *Nyctibius griseus* 34–40cm

E & W lowlands to subtropics (to 2,000m). Humid forest borders, woodland, adjacent clearings. ***Yellow eyes*** reflect reddish-orange. Variable. Rufous to greyish-brown above complexly streaked, mottled and vermiculated blackish and buff. Black shoulders, brown mottled / marbled wings. ***Buffy below somewhat streaked dusky***, with ***narrow blackish malar and black breast spots. Short tail*** barred black and grey. Nominate (E), *panamensis* (W). Regular on stumps and fence posts in semi-open; not high above ground; active since dusk, often heard in full moonlight. **Voice** Descending, mournful *CUuua-cuua-cuu-cuu* or longer *WUA-WUa-woo-wú-wu-wu.* **SS** In E Andes, Andean Potoo (white wing patch, simpler voice); cf. higher-pitched, simpler voice of Rufous-bellied Nighthawk. **Status** Common.

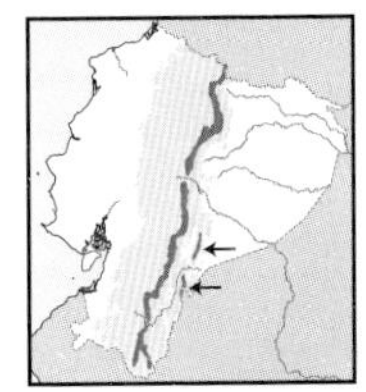

Andean Potoo *Nyctibius maculosus* 38–41cm

Local in E Andean subtropics (1,800–2,300m). Canopy to mid-strata in montane forest and borders. ***Yellow eyes*** reflect red. Similar to Common Potoo but ***darker brown*** overall, especially on head and breast. ***Contrasting whitish patch in wing-coverts.*** Monotypic. Poorly known, but observed foraging low over forest and adjacent clearings; regularly vocalises in full moonlight. **Voice** Far-carrying, querulous *Gwuuuaan* every 5–10 sec. **SS** Little overlap with variable Common Potoo, which sometimes has white in wings; voices differ. **Status** Rare but possibly overlooked.

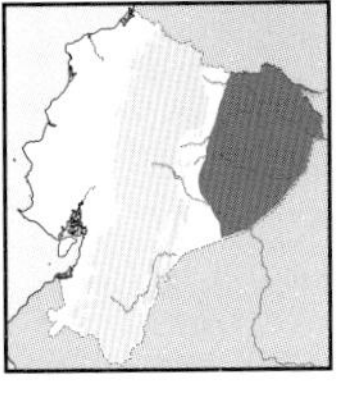

Rufous Potoo *Nyctibius bracteatus* 21–25cm

Local in E lowlands (to 500m). Understorey of *várzea*, seasonally flooded and adjacent *terra firme* forests. ***Small*** and plump, yellow eyes. Mainly ***rich rufous, white dots outlined black on scapulars, tertial tips and sparsely on breast, belly and vent.*** Monotypic. Mainly on low perches inside forest, regular on palm stumps or horizontal twigs; apparently preens and moves body in breeze during day imitating a moving leaf. **Voice** Short, trembling, slightly descending laugh, *wru-huhuhuhuhuhu.* **SS** Confusion unlikely if seen; voice vaguely resembles trill of a screech-owl. **Status** Rare, local, but perhaps overlooked.

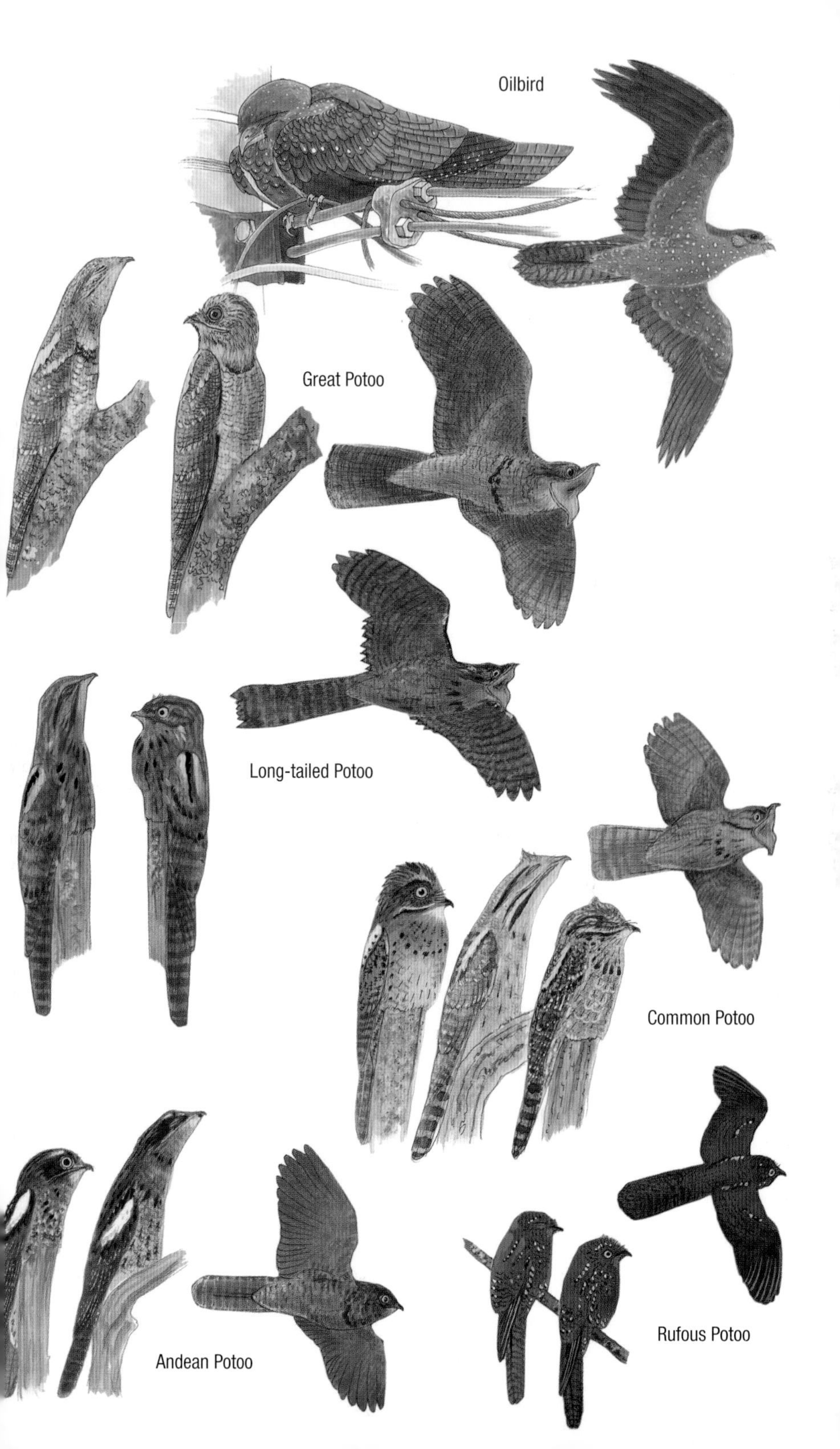
Oilbird
Great Potoo
Long-tailed Potoo
Common Potoo
Andean Potoo
Rufous Potoo

PLATE 78: NIGHTHAWKS

Nighthawks are crepuscular, feeding on airborne insects in graceful flights. All but one have long, narrow, pointed wings that facilitate acrobatic flight. Nacunda has broad wings and glides like a diurnal bird of prey; apparently highly nomadic. Wings of *Lurocalis* are broader-based than *Chordeiles*. Most roost concealed on ground, but *Lurocalis* prefer arboreal perches. One migratory species.

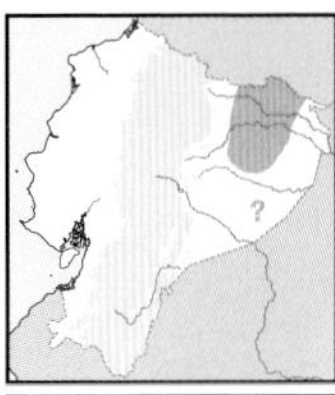

Nacunda Nighthawk *Chordeiles nacunda* 27–30cm

Locally in NE lowlands (below 300m). Canopy and borders. Dusky-brown above heavily vermiculated; brown breast thickly vermiculated / barred buff, ***underparts white, including underwing-coverts***. Long ***broad, rounded wings, broad white band on primaries***. Rather long tail with ***narrow black subterminal band, broad white tips***. Nominate. Active at dusk, high above open areas and waterbodies; graceful raptor-like flight, with glides; might perch on sandbars. **Voice** Mostly silent in Ecuador. **SS** Unmistakable. **Status** Uncertain, possibly rare austral migrant (Jun–Aug, Nov).

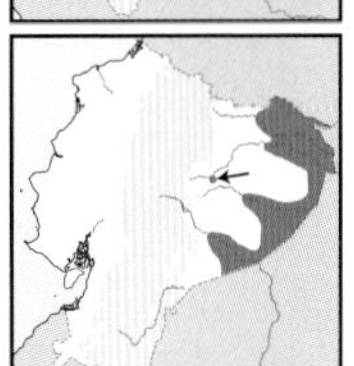

Sand-coloured Nighthawk *Chordeiles rupestris* 20–22cm

E lowlands (to 300m). Sandbars along large rivers, small tributaries, oxbow lakes. ***Very pale***. Greyish-brown upperparts speckled / vermiculated dusky. ***Brown mottled breast-band, pure white underparts, including underwing-coverts and flight feathers***, which are ***tipped black, outer primaries black***; long tail ***tipped black***. Nominate. Groups at dusk, sometimes numerous; erratic flight with deep flaps; often perches on bare branches and driftwood piles. **Voice** Gives *wo-wo…tr, t-wo-wo…tr, t-wo…* or *wo-wo-wo-g-g-wo-wo*; also sharp trill. **SS** Unmistakable in flight; cf. perched Ladder-tailed Nightjar and Lesser Nighthawk. **Status** Rare, locally uncommon.

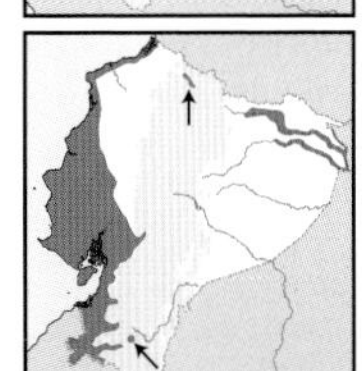

Lesser Nighthawk *Chordeiles acutipennis* 19–21cm

W arid lowlands, locally in inter-Andean valleys and E lowlands (to 800m lowlands, 1,200–2,000m in valleys). Open xeric scrub and grassland. Brownish above speckled / vermiculated blackish-grey; ***white throat, brownish underparts and underwing narrowly speckled / barred buff-grey***. Long pointed wings, ***broad white band near wingtip***; narrow white subterminal tail-band. ♀: ***buffy*** wing-band. Races *aequatorialis* (W), *acutipennis*? (E). Crepuscular, low buoyant flight; often perches on ground below trees. **Voice** Long, resonant *purrr,rrrrr,rrrrr…*; loud *wuaag, wuaag…* calls. **SS** Paler Common Nighthawk has more basal wing-band, longer outer primaries. **Status** Uncommon, local in E.

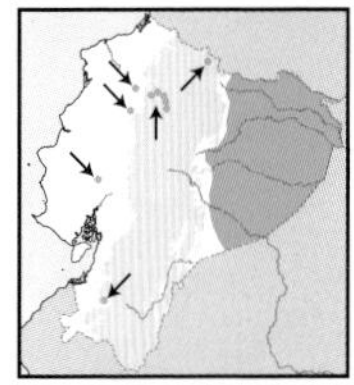

Common Nighthawk *Chordeiles minor* 23–25cm

E lowlands, locally (transients?) in highlands and W lowlands (to 3,500m). Forest borders, semi-open areas, canopy. Mostly blackish coarsely spotted / speckled whitish, white throat, barred underparts, whitish vent. ***Long, narrow, pointed wings***, barred underwing, ***broad white band in primaries more basal*** than previous species. Broad white subterminal tail band. Race uncertain; five races might pass through Ecuador. Mostly seen buoyantly flying high overhead, deep wingbeats; crepuscular. **Voice** Nasal, loud *beeent* and other calls. **SS** Lesser Nighthawk (less pointed wings, more distal underwing patch). **Status** Uncommon boreal transient and winter resident (Aug–May).

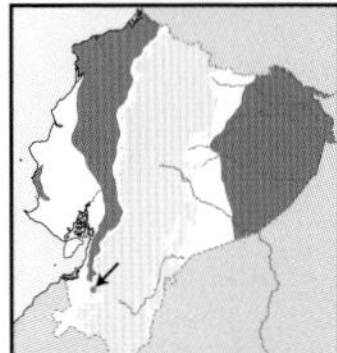

Short-tailed Nighthawk *Lurocalis semitorquatus* 19–21cm

E & W lowlands to foothills, more local in W (below 900m). Canopy, borders, *várzea* in E. Upperparts ***dark brown*** speckled / vermiculated rufous and tawny, darker underparts barred whitish and cinnamon. Long, narrow wings ***project beyond tail*** when perched; ***dusky underwing, short tail***. Races *nattereri* and *semitorquatus* (E), *noctivagus* (W). Active at dusk, in bat-like flight above canopy, gaps or waterbodies; often perches on heavy branches and forages in flocks. **Voice** Sharp *too-eert* or *quirrrt* in flight. **SS** *Chordeiles* have shorter wings, shorter tail, white band in wing. **Status** Uncommon.

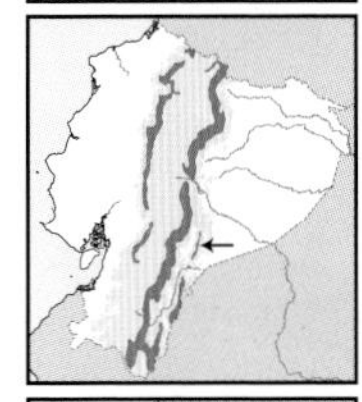

Rufous-bellied Nighthawk *Lurocalis rufiventris* 21–23cm

E & NW Andean slopes (1,500–3,300m, mostly below 3,000m). Montane forest canopy, borders. Blackish upperparts to chest speckled / spotted brown and buff, ***rufous-chestnut underparts, including underwing-coverts***. Long, rather broad, ***pointed wings, short tail***. Monotypic. Pairs to small groups, bat-like flight above canopy, sometimes lower at borders. **Voice** Ascending-descending, rapid *kwa-kwE-KWe-kwe*; also loud *kuoj-kuoj*. **SS** Rufous underparts unmistakable; voice vaguely reminiscent of Common Potoo but higher-pitched, ascending-descending. **Status** Uncommon.

Band-tailed Nighthawk *Nyctiprogne leucopyga* 18–20cm

Local in NE lowlands (below 200m). Vicinity of water. Very dark; ***small white streak on throat-sides***. Narrow, long, pointed wings, ***no white patch in primaries***. Rather long forked tail, ***black with broad white band near base***. Race *exigua*? Erratic flight over water, fluttering wingbeats, short glides with wings held in V; often perches on horizontal branches. **Voice** Steady *kwoit, ku-kwoit-kwoít…kwa-kwoít…*; guttural *kurk* calls. **SS** From *Chordeiles* by unmarked primaries, smaller size, single tail-band. **Status** Rare, local.

Nacunda Nighthawk
juvenile
♀
♂
adult
Sand-coloured
Nighthawk
Lesser Nighthawk
♂
♀
Common
Nighthawk
semitorquatus
nattereri
Short-tailed
Nighthawk
Rufous-bellied Nighthawk
Band-tailed Nighthawk

The commonest and most widespread nightjar (Pauraque), plus forest-based poorwills and three long-tailed species. Sexual dimorphism reaches its peak in montane *Uropsalis* nightjars. Tails are used in aerial displays that need further study. Pauraque, poorwills and Ladder-tailed Nightjar prefer to roost on ground, while *Uropsalis* rest both on ground and on low stumps or trunks.

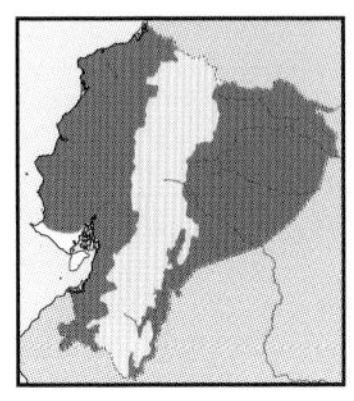

Common Pauraque *Nyctidromus albicollis* 24–28cm

E & W lowlands to subtropics (to 1,800m). Forest borders, clearings, woodland, gardens. Grey to rufescent-brown above, *black-buff spots on scapulars and wing-coverts, orange-chestnut cheeks.* White throat crescent, buff to cinnamon below barred dusky. *Broad white band near primary tips;* long rounded tail, *white inner web to outer feathers.* ♀: chestnut crown, *less contrasting cheeks, buff wing-band,* little white in tail. Nominate. Regularly perches at roadsides, zigzagging flight to ground when flushed; sometimes fly-catches from perch. **Voice** Loud, whistled *wuu-u-WIluu*; also *week-week-week...* calls. **SS** Become familiar with this species to identify rarer nightjars; tail longer than most, but see Ladder-tailed. **Status** Common, widespread.

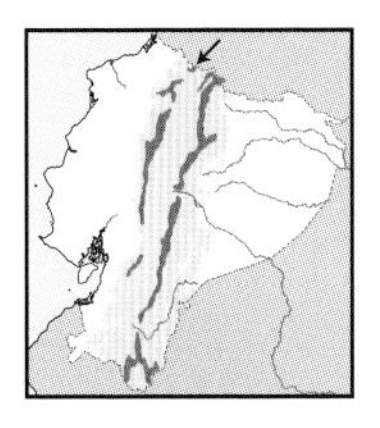

Swallow-tailed Nightjar *Uropsalis segmentata* 22–24cm, tail 45–50cm

E & NW Andean slopes, more local in W (2,200–3,500m). Montane forest borders, adjacent clearings, often near cliffs. Mostly blackish *thickly spotted rufous-tawny, narrow whitish throat crescent. Outer tail feathers incredibly long, rather arched at tip, white shafts.* ♀: lacks tail-streamers but plumage similar, *barred forked tail.* Nominate. Active from dusk when striking ♂ might be seen sallying to open from low perches; perches near ground. **Voice** Sharp *prr-wuiíírrrr* tremulous trill. **SS** Darker Lyre-tailed Nightjar (longer, looser, white-tipped tail, less scaly overall, lacks throat crescent, has chestnut nuchal collar); voices differ. **Status** Rare.

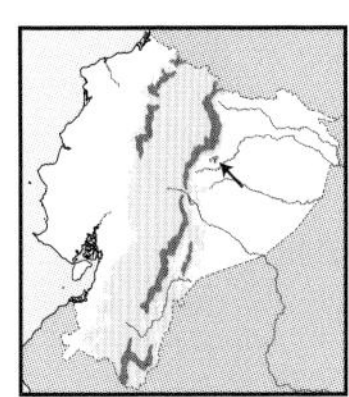

Lyre-tailed Nightjar *Uropsalis lyra* 23–26cm, tail 63–70cm

Local in E & NW Andean foothills to subtropics (800–2,000m, locally higher). Montane forest, borders, often near cliffs and gorges. Mostly dark brown, *less marked* than previous species but has *rufous-chestnut nuchal collar* and bold *tawny spots* on wing-coverts. *Outer tail feathers even longer and more incredibly arched, looser* than Swallow-tailed, *no white shafts, whitish tips.* ♀ lacks tail-streamers but plumage similar, *barred forked tail.* Nominate. Active from dusk, sallying from cliffs and rocky ledges, often near water. **Voice** Musical, rising *war-du-wyeu, war-du-wyeu...*; clear *chip-chip-chip* calls. **SS** Longer-tailed than previous species, with nuchal collar but no whitish throat crescent. **Status** Rare.

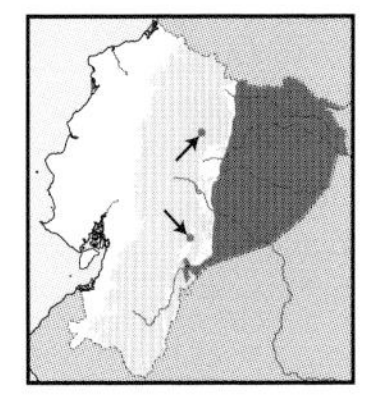

Ladder-tailed Nightjar *Hydropsalis climacocerca* 22–27cm

E lowlands (to 400m, locally higher; once breeding at 1,600m). River islands, rivers and banks. *Pale greyish above,* bold *buff wing spots; white throat and malar,* brownish underparts densely barred. Long wings, *broad* white band near primary tips. Long forked (*M-shaped*) tail, *white inner webs to most tail feathers.* ♀: *sandier brown, buff* wing-band, *forked tail* simply patterned. Nominate. Regular beside open water sallying for airborne insects in bowed flight, spreading tail; perches on ground or driftwood piles. **Voice** Squeaky *chweeít* or *cheeíp.* **SS** Pauraque bolder patterned above, has white crescent not entire throat, shorter tail. **Status** Uncommon.

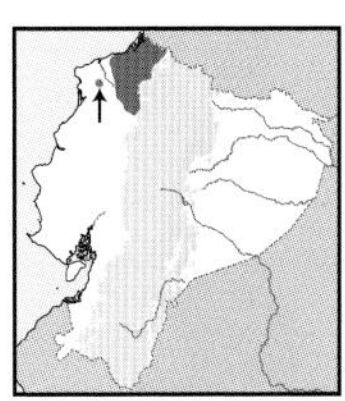

Chocó Poorwill *Nyctiphrynus rosenbergi* 20–22cm

NW lowlands (to 1,100m). Humid forest undergrowth, borders. *Sooty-brown upperparts,* tinged cinnamon-chestnut on wings, *2–3 white dots on upperwing-coverts; large white throat crescent,* blackish below barred chestnut on breast. Short rounded wings, *unmarked* primaries, rounded tail *narrowly tipped whitish.* Monotypic. Forest-based but more likely at forest edges than Ocellated Poorwill; often perches low or on ground. **Voice** Loud *duu-duú-kwíOU...* increasing in pitch; also abrupt *wí-woK...wí-wok, kwek-kwek....* **SS** Only occurs with larger Pauraque but habitat and voice markedly different. **Status** Rare, NT globally, VU in Ecuador.

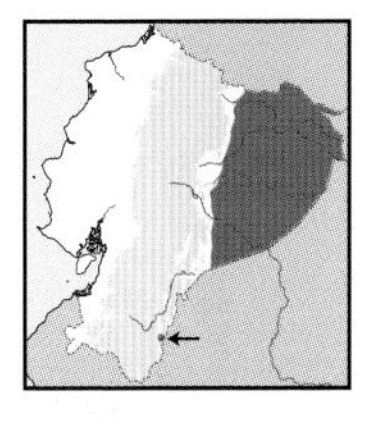

Ocellated Poorwill *Nyctiphrynus ocellatus* 20–22cm

E lowlands (to 500m, once 1,400m). *Terra firme* forest undergrowth. *Very dark. Black dots edged buff in scapulars, white throat crescent, white spots on belly-sides.* Short rounded wings, *unmarked* flight feathers. Rounded tail, *narrow white tips to outer feathers.* ♀: more *rufescent overall,* buffy-white tail tips. Nominate. Seldom leaves forest floor; perches on ground or low horizontal branches. **Voice** Melancholic, trilled *dweeyrr-dú,* once every *c.*5–7 sec; soft *wah-wah-wah* calls (NN). **SS** Blackish Nightjar (complex pattern throughout, no white throat crescent, less white in tail, white wing-band; voices differ). **Status** Rare.

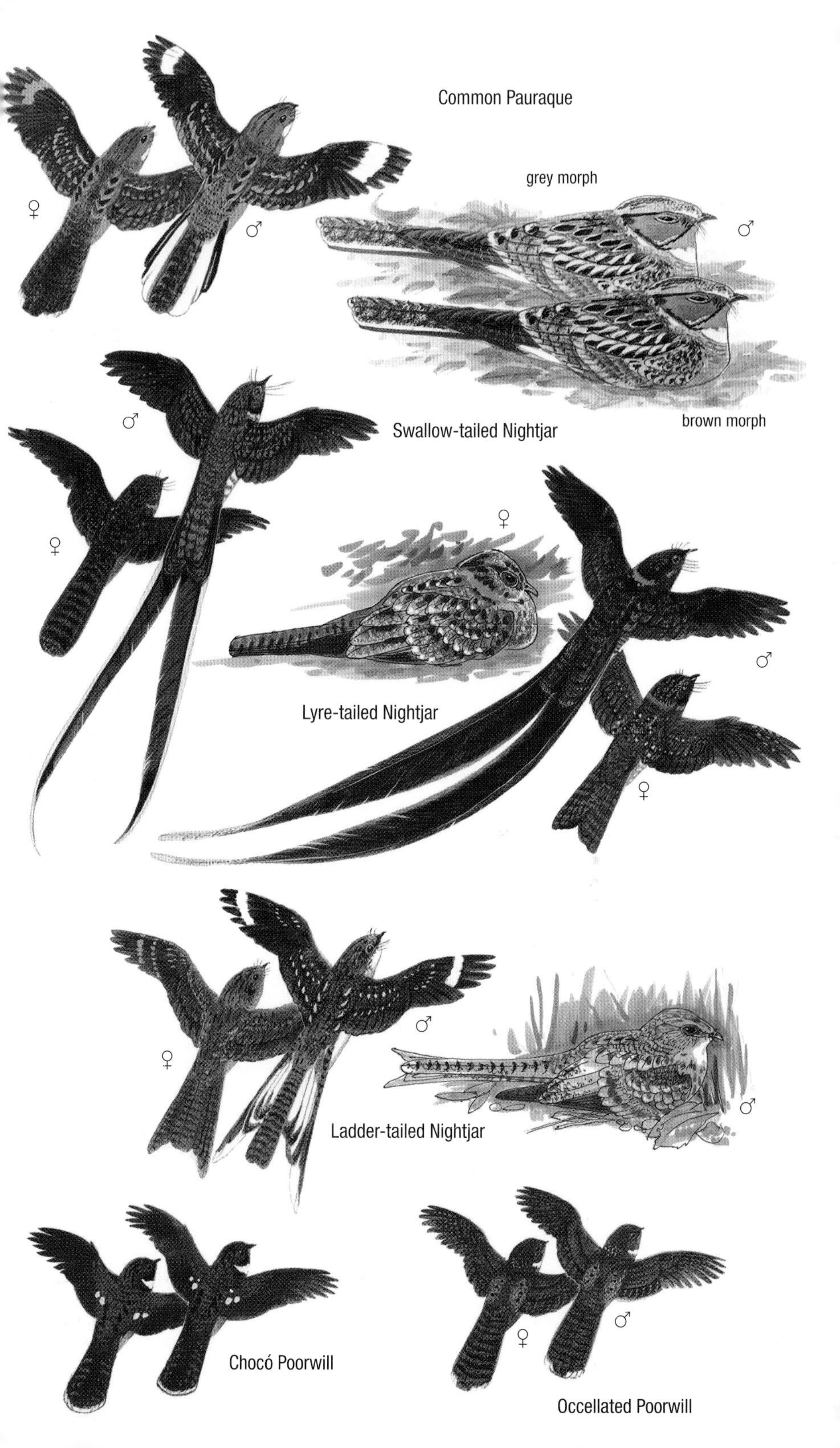
Common Pauraque
grey morph
♀
♂
♂
brown morph
♂
Swallow-tailed Nightjar
♀
♀
Lyre-tailed Nightjar
♂
♀
♂
♀
Ladder-tailed Nightjar
♂
Chocó Poorwill
♀
♂
Occellated Poorwill

More nocturnal than nighthawks, with more rounded wings and longer tails suited for agile zigzagging flight. Not confined to forest, but most rely on some cover. Nest and roost on ground. Two numerous in Andes and arid coast, respectively; the remaining are rare, local and poorly known. Recent taxonomic shake-up transferred all former *Caprimulgus* to different genera.

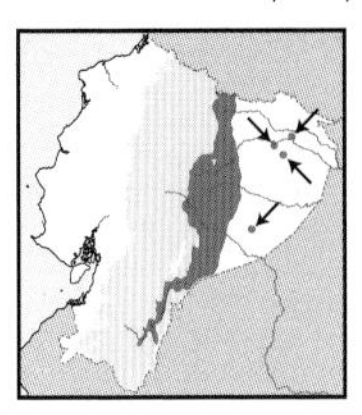

Blackish Nightjar *Nyctipolus nigrescens* 19–21cm

E lowlands and foothills (to 1,200m, but not below 300m?). Rocky outcrops in *terra firme* forest, borders. Blackish above heavily vermiculated / mottled tawny-grey or cinnamon. ***Brown below thickly barred buff, white patches in throat-sides. Short white band in inner primaries***; long tail with ***white tips to some inner feathers***. ♀ ***lacks*** white in wings and tail. Monotypic. Non-forest species, mostly sallying from ground; might permit close approach before abruptly flushing. **Voice** Short, high-pitched *ptrrrrrr* (police whistle); also louder *wroug*, and soft *vit-vit-vit* calls. **SS** Darker Ocellated Poorwill; Pauraque (larger, white tail); see Spot-tailed Nightjar. **Status** Rare, local.

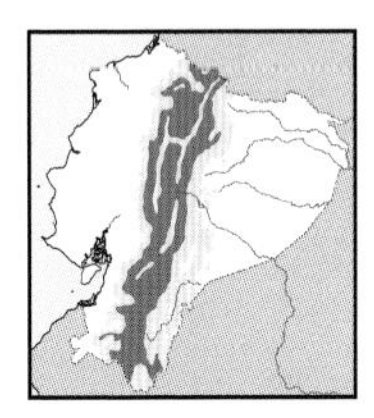

Band-winged Nightjar *Systellura longirostris* 20–23cm

Widespread in Andes (1,800–3,700m). Montane forest borders, clearings, semi-open areas, treeline, *páramo*, city parks. Mostly ***dark brown coarsely*** streaked / mottled buff, rufous and grey, bold ***buff spots on scapulars. White throat crescent, rufous nuchal collar***, brownish barred / spotted underparts. ***White band near tip of primaries***; long rounded tail, ***white tips to outer feathers***. ♀ has ***buff*** in throat, and wing-band, no white in tail. Race *ruficervix*. Regular at roadsides; often feeds around lights; jerky flight when flushed. **Voice** High-pitched *psseee'eeET*, thin *cheet* calls. **SS** ♀ Swallow- and Lyre-tailed Nightjars (darker, one lacks buff throat crescent, one lacks rufous nuchal collar); see also much paler White-tailed Nightjar. **Status** Fairly common.

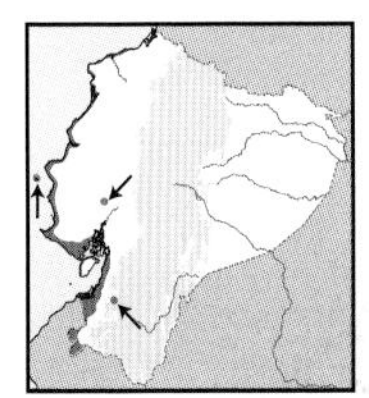

Scrub Nightjar *Nyctidromus anthonyi* 18–20cm

W arid lowlands (to 650m). Xeric scrub, semi-open areas, grasslands. Greyish-brown above streaked / speckled black and buff, ***narrow buff-white supercilium, blackish dots*** on scapulars. ***White throat, dusky breast densely barred buff. White band near wingtip***; long rounded tail, white inner webs to outer remiges. ♀: ***buff throat*** and ***band in primaries***. Monotypic. Regular near coast, perching below shrubs; vocal probably only in wet season. **Voice** Far-carrying *treeow-treeow-treeow…*; also *treu, tew* calls. **SS** Confusion unlikely; Pauraque larger, whiter tail, orangey cheeks; perched Lesser Nighthawk (longer wings, shorter tail, large white throat patch). **Status** Uncommon, seasonally commoner.

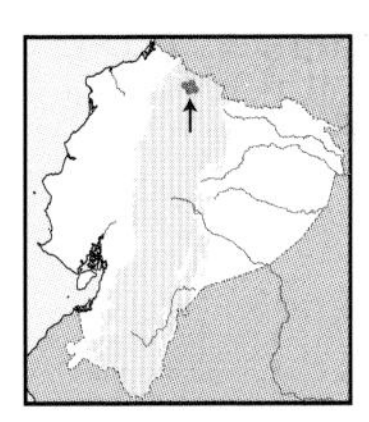

White-tailed Nightjar *Hydropsalis cayennensis* 19–22cm

Local in arid N inter-Andean valleys (1,400–2,000m). Arid scrub, adjacent clearings. ***Dusty greyish-brown*** above coarsely streaked / mottled dusky, ***tawny nuchal collar***, whitish supercilium. White throat and malar, ***buff breast*** densely spotted, ***white belly***. White band near wingtip. Long, slightly forked, ***mainly white*** tail. ♀: buff supercilium, malar, ***throat*** and belly, ***no white*** in tail and wings. Race *aperta*. Nocturnal, sallies from rocks or low perches; may roost along quiet roads. **Voice** High-pitched, slurred *tst-tseeeeeur?*, descending; high *see* calls. **SS** Longer-tailed Pauraque and darker Band-winged Nightjar; Band-winged voice longer, higher-pitched, less inflected. **Status** Common locally.

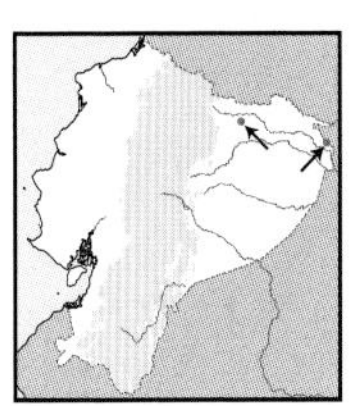

Spot-tailed Nightjar *Hydropsalis maculicaudus* 19–21cm

Very local in NE lowlands (to 350m). Mainly in grassy marshes. ***Rufous-buff supercilium, buffy throat, broad blackish malar, rufescent nuchal collar***. Sooty brown above finely barred / scaled dusky, ***broad buff dots on scapulars***. Blackish breast with ***large cinnamon-buff spots***, cinnamon-rufous below ***thinly barred dusky***. Spotted flight feathers ***lack band***; rounded tail, ***outer feathers broadly tipped white***. ♀: similar but ***no white in tail***. Monotypic. Poorly known in Ecuador, may increase following deforestation; perches on or near ground. **Voice** Sharp, high-pitched, steady *pit-sweeét*; rapid *t-seet* calls (NN). **SS** Blackish Nightjar (no white in tail, white moustachial). **Status** Uncertain.

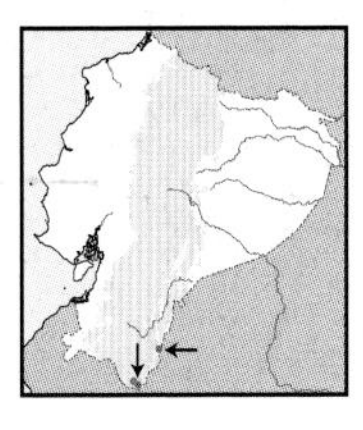

Rufous Nightjar *Antrostomus rufus* 25–28cm

Extreme SE in Zumba region (700–1,100m). Deciduous to semi-deciduous forest, borders, woodland. Mostly ***rufescent-brown*** streaked / vermiculated black, buff and grey, bold ***dark spots on scapulars. Whitish-buff throat crescent***, rufescent underparts barred / speckled buffy. Rounded wings, ***unmarked*** flight feathers; long tail, ***inner webs to outer feathers with broad white subterminal spots***. ♀ ***lacks*** white in tail. Race unknown. Forest-based, mainly on floor but sometimes low perches when foraging; regularly flies low. **Voice** Fast, long-repeated *chok, weel-weel-weeoo, chok, weel-weel-weeo…*. **SS** Only occurs with Pauraque (different tail pattern, white throat, orangey cheeks). **Status** Uncommon but very local.

♀
♂
Blackish Nightjar
♂
♀
♂
♀
Band-winged Nightjar
♂
♀
Scrub Nightjar
White-tailed Nightjar
♂
♀
Spot-tailed Nightjar
♂
♀
Rufous Nightjar

PLATE 81: GROUND DOVES

Small terrestrial doves of agricultural fields, parks and gardens, most rather conspicuous with the notable exception of Maroon-chested, a rare montane dove apparently tied to seasonal seeding of bamboo. At least two *Columbina* tolerant of habitat disturbance are probably spreading following deforestation. The two *Claravis* are more arboreal.

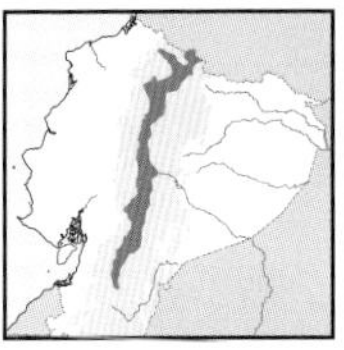

Common Ground Dove *Columbina passerina* 16–18cm

Andes S to Azuay (1,300–3,000m). Arid country, scrub, agricultural areas. Bicoloured bill and pink legs. Mostly pale vinaceous, duskier above. ***Dense scalloping on breast and neck-sides***. Some rufous spots on wing-coverts, ***rufous primaries*** conspicuous in flight. Outer tail tipped white. ♀: more greyish. Race *quitensis*. Pairs to small groups often in roadsides; short zigzagging flight when threatened. **Voice** Monotonous, raspy, fast *guoo'oop-guoo'oop-guoo'oop…*. **SS** Only *Columbina* with spotted breast, little if any overlap with congeners. **Status** Uncommon, familiar.

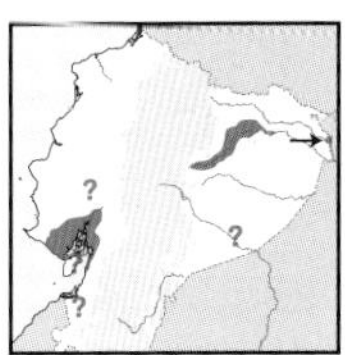

Plain-breasted Ground Dove *Columbina minuta* 14–16cm

Local in E & SW lowlands (to 500m). Open grassy country. Pinkish, dark-tipped bill, pink legs. Mostly ***pale vinaceous***, greyer crown and uppertail-coverts. Upperparts more olive, dark spots on wing-coverts. ***Rufous primaries and underwing*** conspicuous in flight. Outer tail feathers tipped white. ♀: mainly greyish-olive. Races *minuta* (E), *amazilia* (W). Habits like Common Ground Dove, also regular at roadsides. **Voice** Hollow, fast-repeated *whuup-whuup-whuup…*. **SS** Larger, more rufous Ruddy Ground Dove in E; in W see black-winged Ecuadorian. **Status** Rare, local, likely increasing.

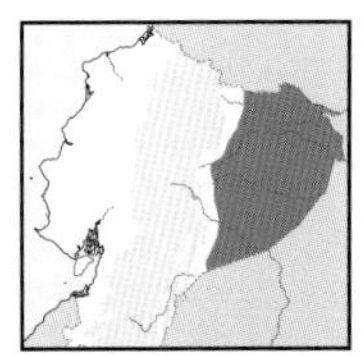

Ruddy Ground Dove *Columbina talpacoti* 16–18cm

E lowlands (to 500m). Open country, agricultural areas, towns, river islands. ***Black bill***, pink legs. ***Mostly ruddy, grey crown***, whitish throat, vinaceous cheeks. Some black spots on wing-coverts, rufous primaries, tail mainly black. ♀: ***paler*** overall. Nominate. Pairs to small groups forage in roadsides, crops, gardens; suddenly fly to cover when threatened. **Voice** Long-repeated, throaty *k'guu-k'guu-k'guu…*, faster than Blue Ground Dove. **SS** Only ruddy terrrestrial dove in range; cf. smaller, duller Plain-breasted (bicoloured bill). **Status** Fairly common, likely increasing.

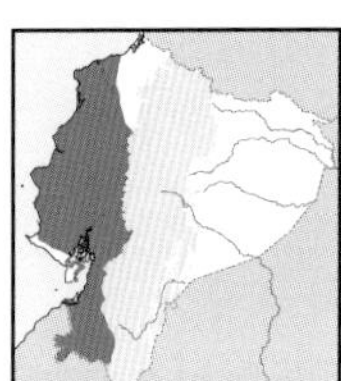

Ecuadorian Ground Dove *Columbina buckleyi* 16–18cm

W lowlands and foothills, to SW subtropics (to 2,000m). Open and semi-open areas. ***Dark bill***. Mostly vinaceous-buff, greyer upperparts, ***duller wings with conspicuous black spots. Primaries and underwing-coverts black***. Tail somewhat large, narrow white tips to outertail feathers. ♀: ***more olive-brown*** above, whiter throat and mid belly. Nominate. Pairs to small groups in roadsides, gardens, crops, parks; sometimes with Croaking Ground Dove. **Voice** Slow, short, rather moaning *guu'u, guu'u…*. **SS** Plain-breasted Ground Dove (smaller, rufous primaries), Croaking (yellow bill); Blue (bluer male, rufous-rumped female). **Status** Fairly common.

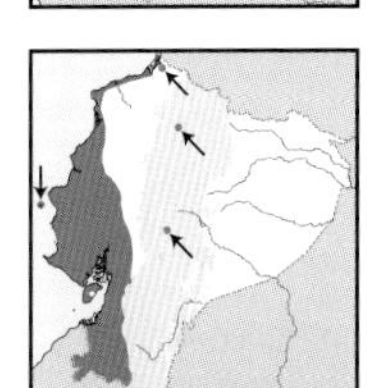

Croaking Ground Dove *Columbina cruziana* 16–18cm

W lowlands, to SW subtropics and dry Andean valleys (to 2,200m). Arid country, agricultural areas, towns. ***Obvious yellow-and-black bill. Grey head***, pale vinaceous below, brownish above. Black spots on wing-coverts, ***maroon line on scapulars; black primaries, grey underwing***. ♀: ***duller***. Monotypic. Mainly in groups, sometimes dozens, favouring arid land; regular in town parks and streets; locally flocks with Ecuadorian Ground Dove. **Voice** Distinctive, metallic *gruau*, *wreeao* or *wruui*. **SS** Bill colour readily separates from Ecuadorian; voice unique. **Status** Common, familiar.

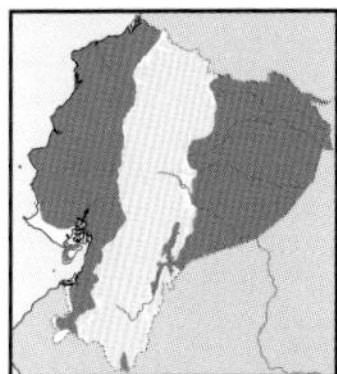

Blue Ground Dove *Claravis pretiosa* 20–22cm

E & W lowlands (to 1,000m). Forest borders, woodland, shrubby clearings. Handsome, ***bluish-grey, white foreface***. Wings darker with ***contrasting black bars and spots***. ♀: ***mostly brown, rufous rump and uppertail***, pale belly, ***rufous bars and spots on wings***. Monotypic. Retiring ground dweller, primarily in forest and woodland; less prone to fly when threatened, prefers to run and hide. **Voice** Far-carrying, slow, mellow *boop* or *goóp* series. **SS** Smaller *Columbina* (greyer, more vinaceous, less obvious black dots). **Status** Uncommon.

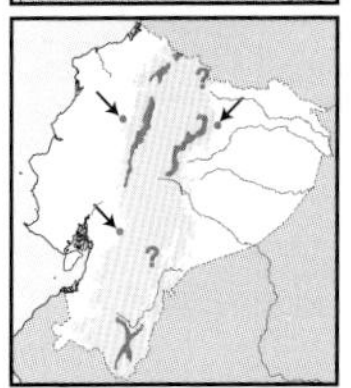

Maroon-chested Ground Dove *Claravis mondetoura* 21–23cm

Local in foothills to temperate zone on both Andean slopes (500–3,500m). Montane forest understorey, bamboo stands. Handsome. ***Grey upperparts, contrasting maroon chest***, white lower underparts. ***Purplish-black wingbars, outer tail feathers white***. ♀: mainly ***dark brown, paler fore-face and mid-belly. Maroon wingbars, outer tail tipped white***. Nominate. Retiring in forest, probably erratic following bamboo seeding, nests in bamboo stands; rarely seen. **Voice** Rhythmic, long-repeated *guoo-óop, guoo-óop…* or *ghoo-doó-doop…*. **SS** Confusion unlikely due to chest and tail, also habitat. **Status** Rare, local, likely nomadic; NT in Ecuador.

Plain-breasted Ground Dove
♂
♂
♂
amazilia
♂
minuta
♂
♀
♂
♀
♀
Common Ground Dove
juvenile
♀
♂
♂
♂
Ecuadorian Ground Dove
♀
♀
Ruddy Ground Dove
♀
♂
♂
Croaking Ground Dove
Blue Ground Dove
♀
♂
Maroon-chested Ground Dove

PLATE 82: DOVES

One ground dove and six predominantly terrestrial doves of open areas, including one confined to *páramo*. Most *Leptotila* are forest-based, Ochre-bellied being particularly tied to forest understorey in SW; and the most threatened dove in Ecuador. Eared is among the commonest Ecuadorian birds. *Zenaida* and *Metriopelia* are gregarious, *Leptotila* mainly solitary.

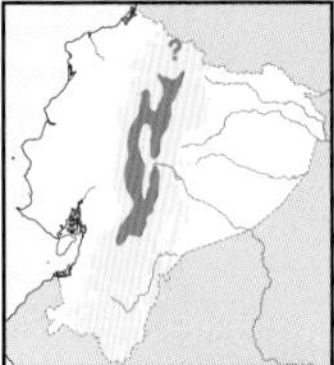

Black-winged Ground Dove *Metriopelia melanoptera* 22–24cm

Highlands from Pichincha to Azuay (3,300–4,300m). Open *páramo* and adjacent agriculture. ***Black bill, pale orange orbital skin, black legs***. Olive-brown above, buffier below, with prominent ***white wing bend. Flight feathers, underwing and tail contrasting black***. ♀: browner. Race *saturatior*. Small groups forage in open country, where might 'disappear' in vegetation; permit close approach, but suddenly flush with whirring wings. **Voice** Mostly silent, a weak *trru-iiioo* rolling; loud wing whirr. **SS** Unique in *páramo*; see barely similar Eared Dove. **Status** Uncommon.

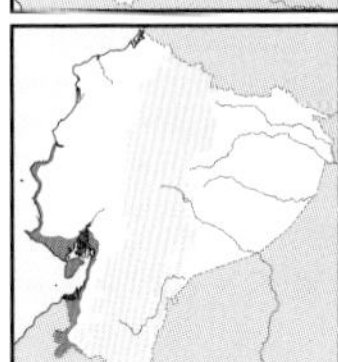

West Peruvian Dove *Zenaida meloda* 28–29cm

SW lowlands (to 700m in S). Arid second growth, shrubby clearings, riparian woodland. Black bill, ***blue orbital skin***. Olive-brown above, paler face, black ear-coverts spot, ochraceous breast to upper belly, greyish-olive below. ***Contrasting white band in wing-coverts***, conspicuous in flight; tail tipped white. Monotypic. Small groups sometimes congregate at waterholes; often perch in exposed shrubs and trees. **Voice** Rhythmic, sad *gooo..goo-oó-ooop. . .goo..goo-ooóp*, but phrasing variable. **SS** Only dove with prominent white in wings and blue orbital skin. **Status** Uncommon.

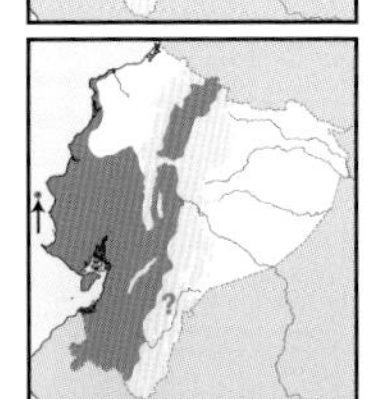

Eared Dove *Zenaida auriculata* 25–26cm

W lowlands, SW slopes, Andean valleys (to 3,200m). Open country, agricultural fields, gardens, parks, towns. Black bill, pink legs. Dull ochraceous to rufescent, more vivid breast, pink iridescence on neck-sides. ***Black ear crescent, black spots on wings, tail tipped and edged white***. Race *hypoleuca*. Familiar dove of human-modified habitats, congregate to forage; flight fast and direct, regularly perches on wires, houses, etc. **Voice** Hollow *coo-Úoo, coo-coo-guo....* **SS** Familiarity with this common Andean dove enables comparison with other terrestrial species. **Status** Common.

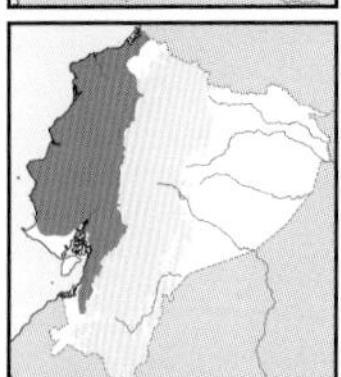

Pallid Dove *Leptotila pallida* 26–28cm

W lowlands (to 800m). Humid to semi-deciduous forest, borders, woodland. Black bill, ***red orbital skin***, pallid eyes. ***Forecrown silvery-grey***, dusky-grey mid crown to nape. Upperparts olive-brown, ***ruddier rump and uppertail-coverts***. Pale vinaceous breast, tail tipped white. Monotypic. Mostly terrestrial, singles under cover; flies low, fast and direct, may nervously perch in shrubs or trees. **Voice** Hollow, low-pitched *whooo...whooo...whooo*, ending abruptly; notes longer, less hollow than Ruddy Quail-Dove. **SS** White-tipped Dove (more white in tail tips, no silvery forecrown). **Status** Uncommon.

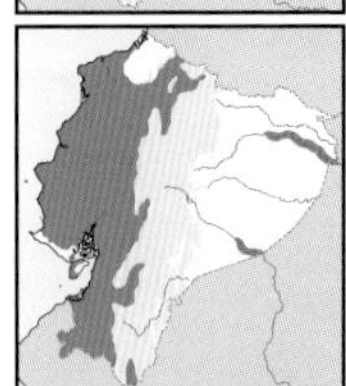

White-tipped Dove *Leptotila verreauxi* 26–28cm

W lowlands to subtropics, Andean valleys, extreme SE, locally along Río Napo (to 3,300m, below 300m in E). Semi-deciduous to humid forest borders, woodland, adjacent clearings, river islands in E. Olive-brown above, paler forehead and throat, ***more vinaceous below, whitish belly. Chestnut underwing, outer tail feathers broadly tipped white***. Race *decolor* (W, Andes), possibly *decipiens* (E lowlands). Mostly singles on ground; regular at roadsides and hedgerows. **Voice** Soft two-noted *cuu-guuuú*; phrasing varies, often rolled first *purr-purr-purrpu..cuu-guuú*. **SS** Pallid Dove (W), Grey-fronted Dove (E), both have ***silvery*** forecrown. **Status** Uncommon; spreading?

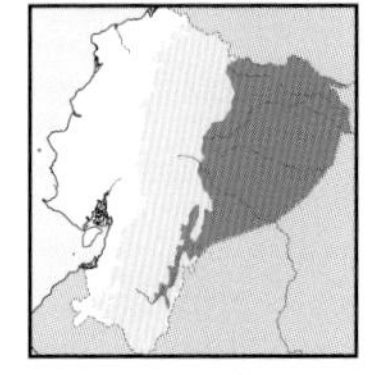

Grey-fronted Dove *Leptotila rufaxilla* 26–28cm

E lowlands and foothills (to 1100m). Mostly in *várzea* forest and woodland, river islands, local in *terra firme*. Black bill, red orbital skin, pallid eyes. ***Silvery forecrown, blue-grey mid crown***, olive-grey upperparts, ***brighter rump and uppertail-coverts***. Pale vinaceous below, whitish throat and mid belly, tail tipped white. Race *dubusi*. More often heard than seen as dwells on ground, seldom at borders. **Voice** Fading, sad *whoooo*, less hollow than Ruddy Quail-Dove. **SS** Locally overlapping White-tipped Dove (no silvery front, more in open areas). **Status** Uncommon.

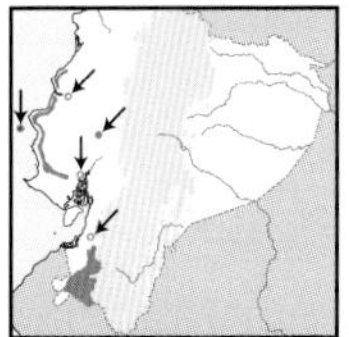

Ochre-bellied Dove *Leptotila ochraceiventris* 26–28cm

SW lowlands to foothills (to 1,700m in Loja). Humid to semi-deciduous forest and woodland. Attractive, rare dove with black bill, purplish orbital skin and pale eyes. Mostly brown, paler front ***fading into rusty crown and nape, mantle more lavender***. Whitish throat, ***vinaceous breast, buffy-ochraceous belly***, tail tipped white. Monotypic. Seldom seen, often congregates at waterholes; seemingly undertakes seasonal movements. **Voice** Throaty *grrooww* every 7–10 sec. **SS** Confusion unlikely due to rich plumage and voice. **Status** Rare, likely declining; VU globally, EN in Ecuador.

adult
juvenile
Black-winged Ground Dove
West Peruvian Dove
♂
♂
juvenile
♀
Eared Dove
♀
juvenile
juvenile
adult
Pallid Dove
juvenile
adult
decipiens
White-tipped Dove
decolor
juvenile
adult
Grey-fronted Dove
adult
juvenile
Ochre-bellied Dove

PLATE 83: PIGEONS

Large, arboreal, mostly forest-based but tolerate some habitat disturbance. Few plumage differences between some species, particularly Ruddy and Plumbeous. Voices better identification clues as most are more often heard than seen.

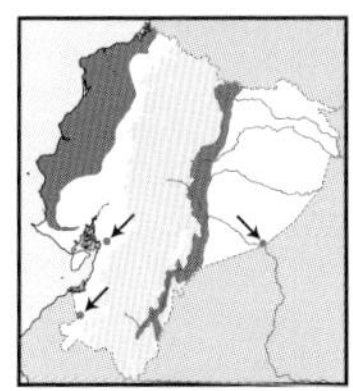

Scaled Pigeon *Patagioenas speciosa* 33–36cm

W lowlands and foothills, E foothills (to 1,200m). Mature forest canopy, borders, secondary woodland. *Attractive*, with *red bill, orbital skin and eyes*. Head chocolate-brown, *neck and chest finely spotted / scaled black and white, fading into buffy scalloping below and on mantle*. Chestnut upperparts, brighter rump. Juv. duller. Monotypic. Often singly or in pairs; small groups gather at fruiting trees. **Voice** Low-pitched, rhythmic *whooo..whoo-oo, whu-whu-whooo, whu-whu-whooo....* **SS** Red bill is, even under poor light conditions, distinctive. **Status** Uncommon.

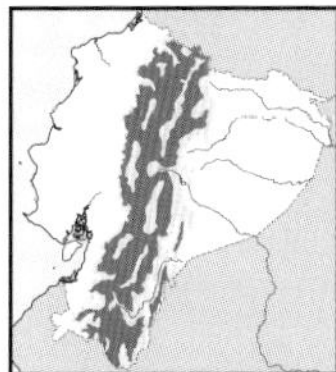

Band-tailed Pigeon *Patagioenas fasciata* 35–37cm

Andean subtropics to *páramo* edge (1,500–3,000m, locally lower or higher). Forest canopy, borders, woodland, adjacent clearings. Large, *yellow bill*, pale eyes. Mostly vinaceous-grey, *white hindneck band*. Tail ashy-grey, *terminal half paler, black median tail-band*. Juv. duller. Race *albilinea*. Mostly small groups, fly fast high overhead; congregate at fruiting trees, also move seasonally with fruiting periods; might appear above towns. **Voice** Weak, mellow, low-pitched *cuu'oo-cuu'oo-cuu'oo...* or *cu-hooo...* in short series; gravely *ggrra* in flight. **SS** Confusion unlikely. **Status** Fairly common.

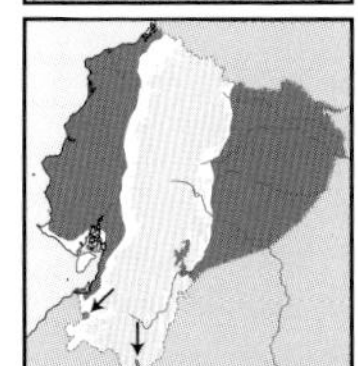

Pale-vented Pigeon *Patagioenas cayennensis* 28–31cm

E & W lowlands (to 500m, locally higher). Forest borders, woodland, adjacent clearings, near water in E. Black bill, narrow red eye-ring, brown eyes. *Mostly vinaceous*, glossy green hindcrown and ear-coverts, whitish throat. *Rump, uppertail-coverts and belly ashy-grey*. Pale grey tail, *paler terminal half*. Race *andersoni* (E), *occidentalis* (W). Non-forest, perches on exposed branches; singles to small groups at fruiting trees and roosts. **Voice** Rhythmic *wu'oooh..guk-wuu-woöo, guk-wuu-woöo; wuooö* calls. **SS** Pale belly and rump readily seen; less forest-based than most congeners. **Status** Uncommon.

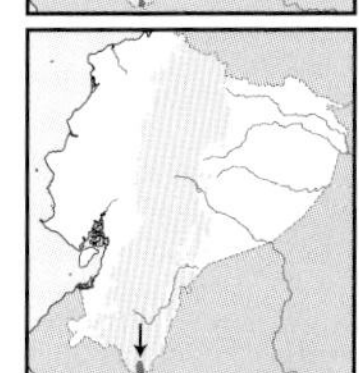

* Peruvian Pigeon *Patagioenas oenops* 31–33cm

Foothills in extreme SE (650–750 m). Semi-deciduous forest canopy, borders, woodland. *Reddish-and-grey bill*, purplish eye-ring, *pale eyes*. Mainly *rich vinaceous*, more chestnut upperparts, *outer wing-coverts, rump, uppertail-coverts and belly grey, uniform tail*. Monotypic. Singles to small groups perch on exposed branches, fly low over canopy; groups may congregate at fruiting trees. **Voice** Slow, melancholic *wuuüü, wa-wooo, wa-wooo*. **SS** From Pale-vented by reddish bill, *no pale* subterminal tail-band and darker vent. **Status** Rare, local, likely declining; CR in Ecuador, VU globally.

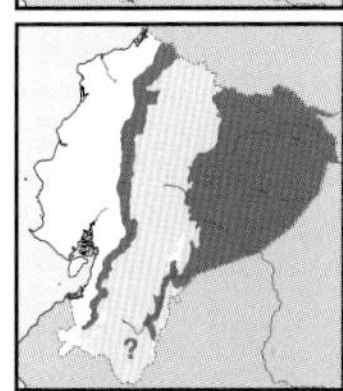

Plumbeous Pigeon *Patagioenas plumbea* 29–32cm

E lowlands to foothills, W foothills to subtropics (to 1,000m in E, 600–2,000m in W). Canopy and borders. Bill black, *pale iris*, pink legs. *Dark vinaceous-grey overall*, darker upperparts, *browner wings*, paler belly. Race *chapmani* (W), probably *bogotensis* (most E) and *pallescens* (SE). Flies low over forest canopy, may flock with Ruddy Pigeon; mostly solitary or small parties. **Voice** Three-noted, rhythmic *woóp-woo-whooó*, slower than Ruddy Pigeon; also *guo'rrrr* growls. **SS** Ruddy less grey, dark eyes; better told by voice; also Dusky Pigeon locally. **Status** Fairly common.

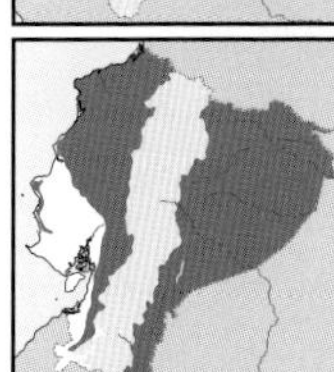

Ruddy Pigeon *Patagioenas subvinacea* 28–31cm

E & W lowlands to subtropics (to 1,900m). Canopy and borders. Black bill, *brown eyes*, pink legs. *Mostly vinaceous*, more chestnut back to uppertail-coverts, *browner wings*. Races confusing, need thorough revision; *ogilviegranti* (E), *berlepschi* (W). Singles or pairs in canopy, fly low over forest, may flock at fruiting trees. **Voice** Four-noted, mellow, rhythmic, syncopated *wüp, woo-whoó-wuu*, like *ru.. ddy-piii-geon*; also *grouw* growls. **SS** Greyer Plumbeous Pigeon (paler belly, darker wings); best told by voice; in W see Dusky Pigeon (grey head, three-noted voice). **Status** Fairly common, but VU globally.

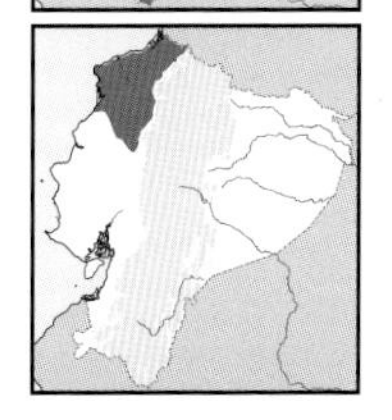

Dusky Pigeon *Patagioenas goodsoni* 27–28cm

NW lowlands and foothills (to 800m). Humid forest canopy, borders. *Black bill*, reddish eye-ring and eyes. *Grey head to chest*, whitish throat. *Ruddy mantle grading into brown back to uppertail-coverts*. Brown wings, vinaceous-grey breast and belly. Monotypic. Singles to small groups, fly low over forest, shy when perched; sometimes congregate at fruiting trees with Ruddy. **Voice** Three-noted rhythmic *wooa?..dup-dup*, shorter than Ruddy, last two notes brief; growls like Ruddy and Plumbeous. **SS** Ruddy, Plumbeous and less forest-based Pale-vented Pigeons (bolder heads, less rufescent back). **Status** Uncommon, VU in Ecuador.

juvenile
♀
♂
Scaled Pigeon
♀
♂
♂
♀
Band-tailed Pigeon
juvenile
adult
adult
andersoni
occidentalis
Pale-vented Pigeon
Peruvian Pigeon
bogotensis?
chapmani
pallescens
Plumbeous Pigeon
ogilviegranti
berlepschi
Ruddy Pigeon
Dusky Pigeon

PLATE 84: ROCK PIGEON AND QUAIL-DOVES

Terrestrial quails-doves plus the introduced, familiar Rock Pigeon. The thickset, short-tailed quail-doves are hardly ever seen as skulk on forest floor and only escape when hard-pressed. Heard much more frequently than seen. All but one in lowlands. Recent taxonomic shake-up divided the genus *Geotrygon*.

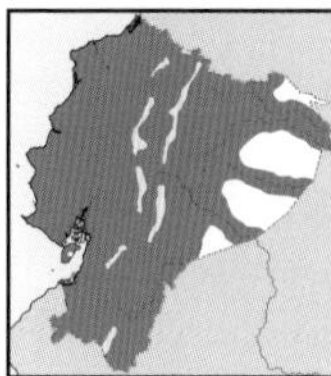

[i]Rock Pigeon *Columba livia* 33–36cm

Widespread in cities, towns, even small villages (to 4,000m), more numerous in Andean valleys. Remarkable plumage variation, with grey, brown, tawny, black and white morphs. *Two grey bands in wings and white rump* often present. Gregarious, particularly where food available (parks, plazas, haciendas, gardens), often semi-captive; flight powerful, sometimes almost falcon-like. **Voice** Moaning *coorh.* **SS** Only pigeon in human-modified habitat, unlikely to be confused with plainer and wilder native pigeons. **Status** Abundant, but not seen away from settled areas.

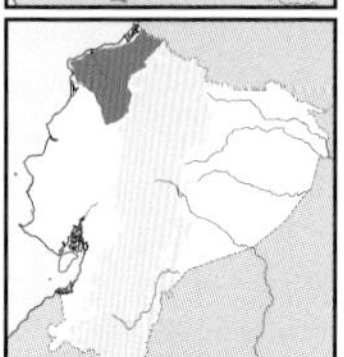

Purple Quail-Dove *Geotrygon purpurata* 23–24cm

W lowlands to foothills (200–700m). Humid mature forest, borders. Dark purplish bill. *White front, face, throat*, broad black malar, *sapphire-blue crown.* Tawny neck, bright green nape-band. Purplish-maroon upperparts, *rich purple rump. Ashy breast, white below.* Monotypic. Singles or pairs on hill slopes, seldom seen, tend to freeze or run when threatened; may become habituated to feeding stations. **Voice** Weak, 'high-pitched', sad *ghoou* every 2–3 sec (W). **SS** Duller and darker Olive-backed Quail-Dove. **Status** Rare, VU in Ecuador, EN globally.

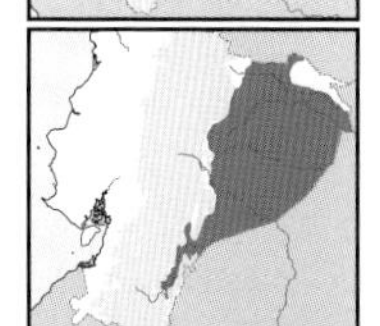

Sapphire Quail-Dove *Geotrygon saphirina* 23–24cm

E lowlands and foothills (to 1,100m). Humid mature forest. Dark purplish bill. *White front to throat*, broad black malar. *Greyish crown* grading to *green on nape*. Tawny neck, bright green nape-band. Purplish-maroon upperparts, *rich purple rump. Ashy breast, white below.* Nominate. Singles or pairs in swampy areas, seldom seen, tend to freeze or run when threatened. **Voice** Trembling, paused, 'high-pitched' *kgouuu....* **SS** Confusion unlikely; cf. White-throated Quail-Dove. **Status** Rare.

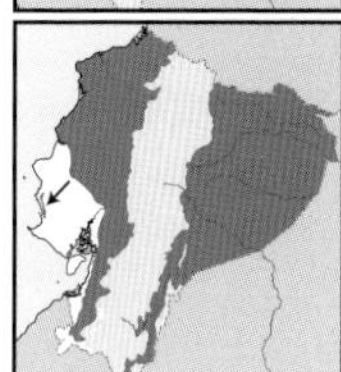

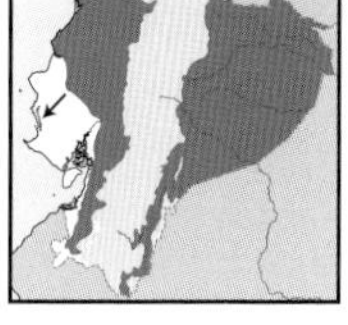

Ruddy Quail-Dove *Geotrygon montana* 22–24cm

E & W lowlands and foothills (to 1,300m). Humid forest ground, secondary woodland. *Reddish bill and orbital ring*, yellow eyes. *Mostly ruddy, buffy-white stripe on lower face, whitish throat.* Wings darker, *glossed purplish above*, breast washed vinaceous, paler below. ♀: similar but duller, face stripe browner. Juv. scaled above. Nominate. Seldom seen, retiring; like other quail-doves, walks away from observers out of sight; rarely emerges into clearings. **Voice** Soft, descending, moaning *ooooo*, lower-pitched than Pallid Dove. **SS** Face pattern, no black malar, ruddy upperparts diagnostic. **Status** Uncommon, likely overlooked.

†Violaceous Quail-Dove *Geotrygon violacea* 22–24cm

Single record from *terra firme* forest in Cuyabeno region (250m); later identification of tape-recordings confirmed vocalisations belonged to Grey-fronted Dove. This small quail-dove has extensive *pinkish-white in face, purplish bill and orbital ring, glossy lavender nape and mantle*, brown wings, whitish underparts. ♀: duller, less purplish on mantle. **Voice** Short, hollow *ho-ooo*, last part falling, mellower than other quail-doves (RG). **SS** Pale face separates from other quail-doves. **Status** Erroneously identified, this species was recently excluded from Ecuador's bird list.

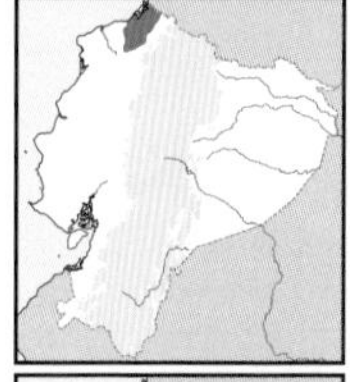

Olive-backed Quail-Dove *Leptotrygon veraguensis* 23–24cm

NW lowlands (to 300m). In humid forest. Black bill, dark red orbital skin, yellow eyes. *Front, lower face and throat white*, broad black malar stripe. *Mainly olive-brown, whitish mid belly, buff flanks and vent.* ♀ has *buffier* face, juv. duller. Monotypic. Singles or loose pairs on forest floor, regularly near streams; freeze or timidly walk away and hide when threatened. **Voice** Seldom vocalises, a frog-like, deep *ghoú'* or *ghun.* **SS** Purple Quail-Dove has similar facial pattern but more colourful crown, whiter underparts, richer upperparts. **Status** Rare, probably overlooked; EN in Ecuador.

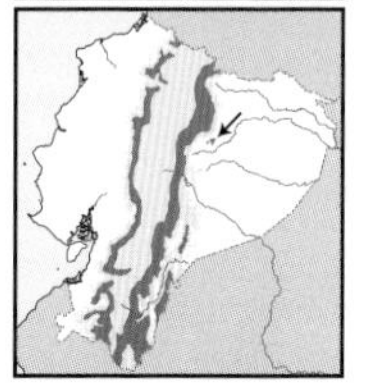

White-throated Quail-Dove *Zentrygon frenata* 29–32cm

Subtropics on both Andean slopes (1,300–2,600m). Montane forest interior. Black bill, pale brown eyes. *Buff face, bluish-grey crown and nape*, black malar stripe, *white throat, grey and black stripes* to neck and chest. Purplish-brown above, vinaceous below, ruddier rear parts. Juv. scaled above and on breast. Race *bourcieri* (most range), paler *subgrisea* (W Loja). Singles or loose terrestrial pairs, sometimes in adjacent clearings in dull conditions; walks away from observers; attracted to feeding stations. **Voice** Short, hollow, sad *hooop* every 3–5 sec. **SS** Only truly montane quail-dove; cf. more rufescent, smaller Ruddy. **Status** Uncommon.

Rock Pigeon
♂
♀
wild Rock Pigeon type
typical dark feral city bird
Purple Quail-Dove
juvenile
adult
Sapphire Quail-Dove
juvenile
♂
♀
Ruddy Quail-Dove
♀
♂
Violaceous Quail-Dove
immature
moulting
♀
venile
♀
♂
Olive-backed Quail-Dove
juvenile
♂
White-throated Quail-Dove

PLATE 85: MACAWS

Large macaws are uncommon birds of wild areas, progressively declining and disappearing from disturbed areas. Calls are loud and raucous, heard over long distances, while their silhouettes with long pointed tails are distinctive, even far away. Slow flight with steady, incessant wingbeats. Blue-and-yellow, Scarlet, Red-and-green and Great Green are often kept as pets even in towns and cities.

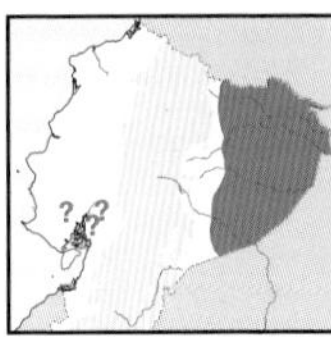

Blue-and-yellow Macaw *Ara ararauna* — 75–87cm

Lowlands of E and (formerly?) W (to 500m). Canopy and borders of *terra firme* forest, swampy areas and *várzea*. Very large ***bicoloured*** macaw, ***blue above, rich yellow below***. Monotypic. Pairs or small groups often seen flying fast high overhead; moves long distances from roosting to feeding grounds; perched birds difficult to locate. **Voice** Loud, raucous *krraAh!-krrrA...* in flight; more 'musical' calls when perched. **SS** Unmistakable in good light. **Status** Uncommon.

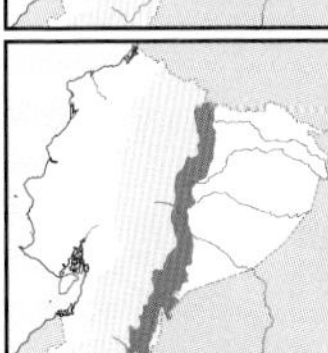

Military Macaw *Ara militaris* — 66–80cm

Foothills and subtropics on E slope (600–1,500m, rarely below, once at 3,000m). Primary forest canopy, borders. Facial skin white with black streaks, iris yellowish-white. Entirely ***grass-green, red forecrown***, bluish nape, wing-coverts, remiges, rump and vent. ***Long, mostly blue tail with red base***. Nominate. Long-distance flights; mostly in small groups flying high overhead; roosts and seemingly nests on cliffs and probably large trees. **Voice** Loud *wraaah!-krra...* as other macaws. **SS** Very similar Great Green Macaw only in W. **Status** Rare; EN in Ecuador, VU globally.

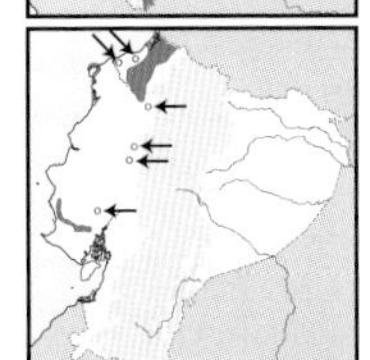

Great Green Macaw *Ara ambiguus* — 76–90cm

W lowlands and (formerly?) foothills (formerly to 800m, now to 200m). Canopy and borders of humid to deciduous forests. Black bill, pale iris, whitish-pink facial skin with black lines. ***Mostly green, red forecrown***, bluish wings, nape and rump. ***Tail red at base, greenish-blue terminally***. Near-endemic race *guayaquilensis*. Pairs or small groups, now rare; noisy, probably engages in long flights; tame for a macaw. **Voice** Powerful *k'krraaak-krraaA...* 'screams'. **SS** Unique in range; cf. much smaller Chestnut-fronted Macaw. **Status** Very rare and local; CR in Ecuador, EN globally.

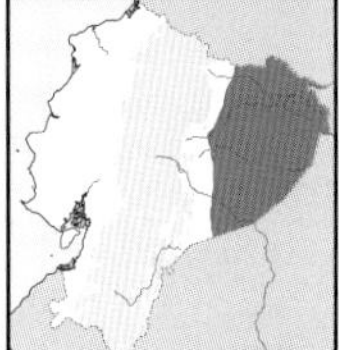

Scarlet Macaw *Ara macao* — 80–96cm

E lowlands (to 400m). Canopy and borders of (mostly) *terra firme* forest. Bill black below, horn above, pale iris, ***white facial skin. Scarlet-red with tricoloured wings***, blue rump, flight feathers and tail tip. Nominate. Pairs or small groups fly high overhead, often with Blue-and-yellow Macaw; tail undulates in flight; regular at clay licks, moves long distances in search of food. **Voice** As other large macaws, raucous *ggraAK-krrrAg!...*, various calls on perch. **SS** Darker Red-and-green Macaw (green in wings, streaked facial skin). **Status** Uncommon, declining; NT in Ecuador.

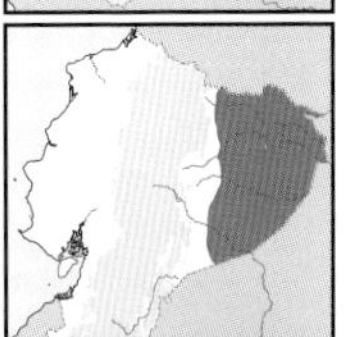

Red-and-green Macaw *Ara chloropterus* — 80–95cm

E lowlands (to 500m). Canopy and borders of *terra firme* forest in fairly remote areas. Black-and-horn bill, pale iris, ***red lines*** on white facial skin. ***Darker red*** than Scarlet Macaw, ***green replaces yellow in wing***. Blue in flight feathers, rump and tail tip. Monotypic. Mostly pairs, sometimes with other large macaws; probably long movements. **Voice** Louder, more powerful than other macaws. **SS** Scarlet has yellow in wings, plain facial skin. **Status** Rare and local, likely declining; VU in Ecuador.

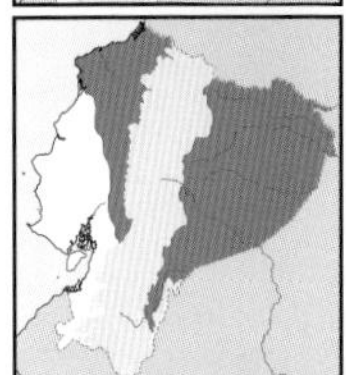

Chestnut-fronted Macaw *Ara severus* — 40–50cm

E & W lowlands and foothills (to 1400m, locally higher). Humid forest canopy, borders, *várzea*, adjacent clearings with tall trees. ***Small*** macaw, ***mostly green, dark chestnut front*** and throat. Blue flight feathers, red shoulders. ***Underwing and undertail reddish*** in flight. Monotypic. Pairs or small groups, more in open than larger macaws; active at midday, unusual for a macaw. **Voice** Less raucous than larger macaws, harsh *wráAH!-wráAH!...* in flight; strident chatters and other calls when perched. **SS** Size and reddish underwing separate from other macaws. **Status** Uncommon, declining in W?

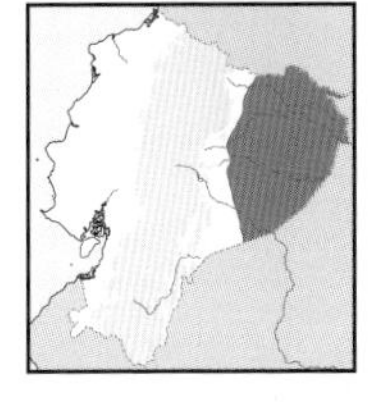

Red-bellied Macaw *Orthopsittaca manilatus* — 44–50cm

E lowlands (to 400m). Mostly *várzea* forest, swampy *Mauritia* palm stands, swampy areas in *terra firme* forest. ***Yellow facial skin, dark iris. Mostly green***, bluish in crown, throat, breast, flight feathers and tail. ***Reddish patch on mid belly; greenish-yellow underwing and undertail*** in flight. Monotypic. Usually in groups, flies rather slowly high overhead; roosts and perhaps nests in *Mauritia* and other palm stands. **Voice** High-pitched, far-carrying *kree-greee-géék...* and other calls when perched. **SS** From Chestnut-fronted Macaw by underwing, undertail and facial skin. **Status** Uncommon.

Blue-and-yellow Macaw
Military Macaw
Great Green Macaw
adult
juvenile
Chestnut-fronted Macaw
♀
juvenile
♂
Scarlet Macaw
Red-and-green Macaw
Red-bellied Macaw

PLATE 86: LARGE PARROTS AND PARAKEETS

All are like small macaws given their long tails and relatively heavy bills. Yellow-eared Parrot is a distinctive Andean species not seen in Ecuador for nearly two decades. Golden-plumed and Scarlet-fronted Parakeets are also Andean, and very rare. The others occur in the lowlands, two in E, one in W. Parakeets are gregarious and noisy, their flight is direct, fast, with no undulation and little rolling, often high above ground.

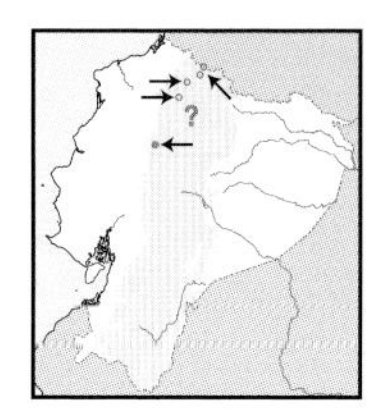

Yellow-eared Parrot *Ognorhynchus icterotis* 43–46cm

NW Andes (1,800–3,200m). Canopy, borders, adjacent clearings. Large, macaw-like parrot with *long* pointed tail. ***Bill black***, red iris. Grass-green above with ***striking yellow front, face and ear-coverts. Pale greenish-yellow below***, reddish undertail. Monotypic. Highly dependent on wax palms (*Ceroxylon*) for feeding, roosting and nesting; moves in pairs or small groups, probably engages in long-distance displacements. **Voice** Loud, raucous, nasal *raänh-raä-raänh...* in flight, *raäh-raänh-raä... rä'rä'rä-ränh* and others when perched. **SS** Smaller Golden-plumed Parakeet lacks extensive yellow in face. **Status** Very rare, perhaps extinct in Ecuador, last seen in 1998 in W Cotopaxi. CR.

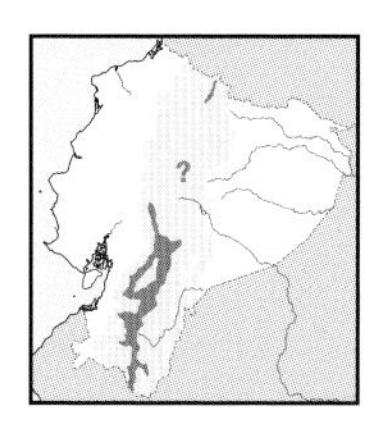

Golden-plumed Parakeet *Leptosittaca branickii* 34–38cm

Temperate forest on E slope and S Andes (2,400–3,400m). Humid montane forest canopy, borders, adjacent clearings. ***Long tail***, grey bill, bluish orbital skin, red eyes. ***Mostly green***, thin ***golden-yellow facial streak***. Reddish scalloping on ***pale*** green belly, yellowish underwing, ***reddish-orange undertail***. Monotypic. Small groups (seasonally larger) move noisily between forest patches, often landing on open perches in trees in clearings; somewhat associated with Andean palms, where nests. **Voice** High-pitched, throaty, shrill *kreeeh-kreeeh-kreeh!..kreaa-rhee....* **SS** Unique in range; cf. rarer Yellow-eared Parrot. **Status** Rare and local, fragmented range; EN in Ecuador, VU globally.

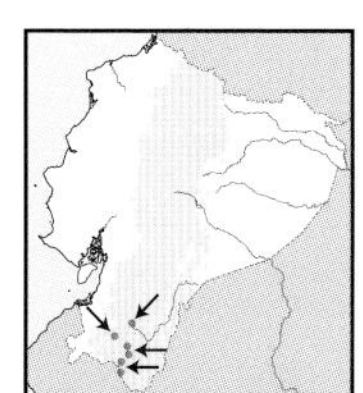

Scarlet-fronted Parakeet *Psittacara wagleri* 36–40cm

Foothills and subtropics in S Andes (1,000–2,500m). Canopy, borders, adjacent clearings with scattered trees. *Ivory*-white bill, white orbital ring, brown eyes. Mostly bright green, with ***red crown, wing bend and forewing*** (some show ***red in thighs*** too). Race *frontatus*. Poorly known in Ecuador where seems to be currently rarer than before; elsewhere, forms noisy flocks while foraging for fruit and nesting in crevices; roosts in cliffs. **Voice** Strident, high-pitched *chee-yk*. **SS** Lower-elevation Red-masked Parakeet (more extensive red in face covering entire ***mask***, red in ***underwing***); juv. more similar. **Status** Very rare and local, probably declining in Ecuador where CR.

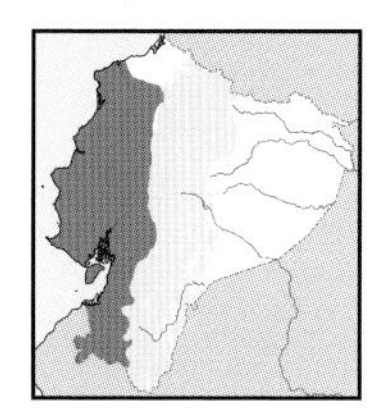

Red-masked Parakeet *Psittacara erythrogenys* 33–36cm

SW lowlands to lower subtropics (to 2,200m, locally higher). Canopy, borders, adjacent clearings, especially deciduous and semi-deciduous forest. Horn-white bill, broad white orbital ring. Green overall, ***contrasting red mask*** (varies with age). Red shoulders and ***underwing*** (in flight); greenish-yellow undertail. Monotypic. Mostly flocks, sometimes (seasonally?) large; noisy groups conspicuous in flight, often visiting maize crops and fruiting trees in agricultural land. **Voice** Very vocal; varied raspy calls and chatters: *kreE-kreE-kraa-keé...creEEh-creEEh..., sqüeey-sqüeey-keéy....* **SS** Imm. can recall Scarlet-fronted Parakeet, but see underwing; unique in most of range. **Status** Locally fairly common, but declining; VU in Ecuador, NT globally.

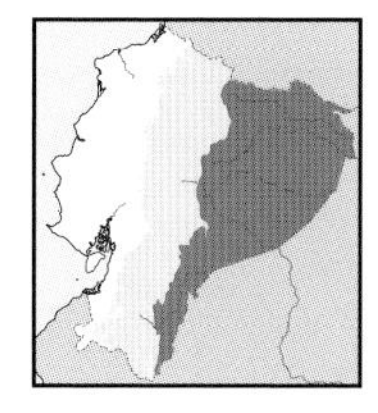

White-eyed Parakeet *Psittacara leucophthalmus* 32–35cm

E lowlands and foothills (to 1,100m, locally to 1,700m). Canopy and borders of *várzea* and riparian forest, adjacent clearings. Ivory-white bill, ***broad white orbital skin. Grass-green*** with few, variable red blotches in front, face and neck. Red wing bend, ***underwing mostly red; greenish-yellow undertail***. Race *callogenys*. Tends to avoid extensive *terra firme* forest; often seen flying fast overhead in small noisy groups; more prone to occupy open areas than other Amazonian parakeets. **Voice** Loud, raspy *kre'e'e'e'et...jeet-eet-eet-eet...* and other calls. **SS** Confusion unlikely in good light; see red underwing. **Status** Uncommon.

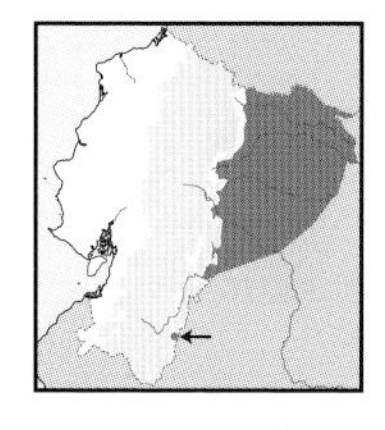

Dusky-headed Parakeet *Aratinga weddellii* 25–28cm

E lowlands, locally to foothills (to 900m). *Várzea* and flooded forest borders, adjacent clearings. ***Black bill***, bordered by pink line, ***broad greyish-white orbital skin, pale iris. Dusky-grey head***, more bluish on nape, neck and throat. Green chest and upperparts, paler belly, ***slaty-grey undertail and underwing***. Monotypic. Mostly small flocks (up to 15) at palm stands, where can be tame; do not fly high; regular at clay licks. **Voice** High-pitched, thin, nasal *jéék-jéék-jéék'éék-jéék'éék...*; thin perched calls recall White-eyed Parakeet. **SS** Confusion unlikely; see undertail and underwing in flight. **Status** Fairly common.

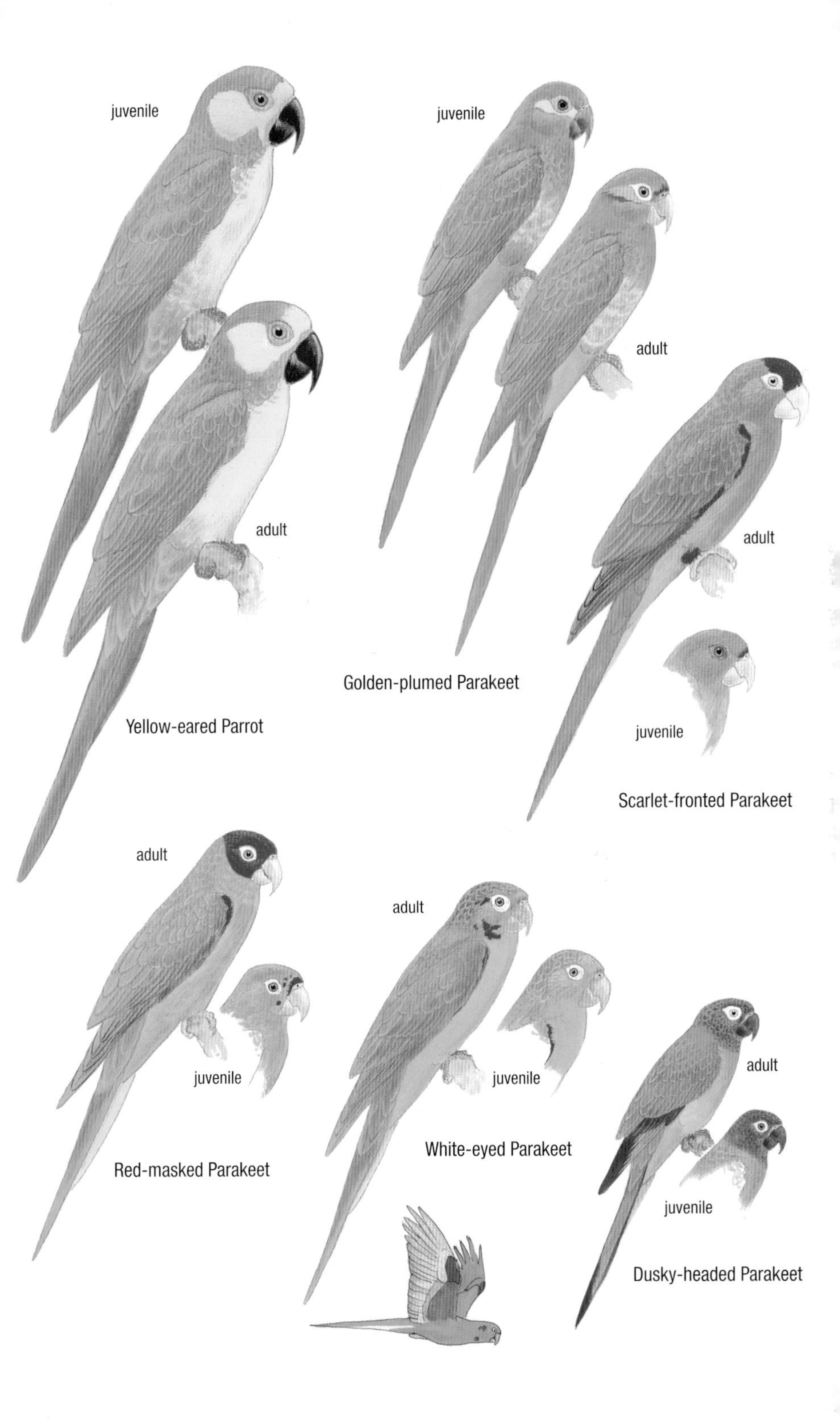
juvenile
adult
Yellow-eared Parrot
juvenile
adult
Golden-plumed Parakeet
adult
juvenile
Scarlet-fronted Parakeet
adult
juvenile
Red-masked Parakeet
adult
juvenile
White-eyed Parakeet
adult
juvenile
Dusky-headed Parakeet

Smaller parakeets of forested habitats, *Pyrrhura* resemble larger parakeets but are smaller; tails less pointed, maroon to reddish, bills smaller. Small compact flocks dart through canopy and subcanopy. Flocks oriented head-to-tail, and flight often undulating, with obvious rolling. Two local and range-restricted species, one endemic to Ecuador. Barred Parakeet is a small, chunky, highly nomadic Andean parakeet hardly ever seen perched. Flight fast, direct, with deep and strong wingbeats, shallow rolling, and no undulating.

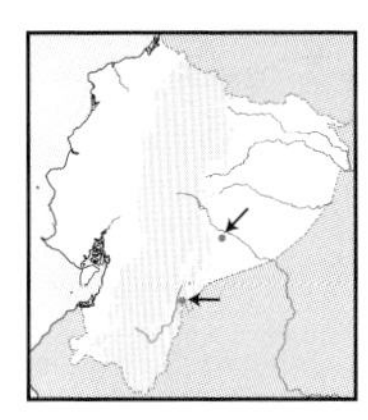

Rose-fronted Parakeet *Pyrrhura roseifrons* 22–25cm

Foothills in extreme SE (300–900m). Montane forest canopy and borders in Cordillera del Cóndor region. ***Black*** bill, dull whitish orbital skin. ***Dusky-brown head, white speckling at sides,*** blue wash in front. ***Maroon cheeks, creamy-white ear-coverts, dusky-brown throat densely chevroned buffy-white.*** Maroon mid belly, red rump; ***green underwing, blue primaries.*** Race *peruviana*, but taxonomy complex. Recently found in Ecuador at a clay lick; as other *Pyrrhura*, mostly in canopy, not high overhead; flocks of 8–15 seen. **Voice** Ringing *pirk-piirk-piiik*, several calls when perched. **SS** White-necked Parakeet and SE race of Maroon-tailed Parakeet (red in underwing and green head). **Status** Very rare and local.

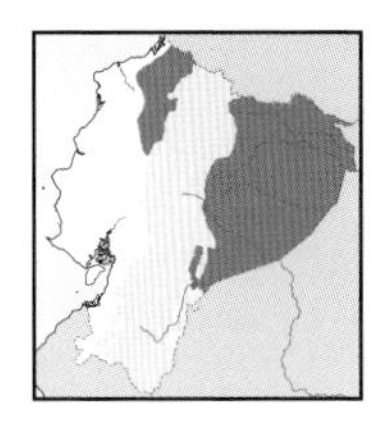

Maroon-tailed Parakeet *Pyrrhura melanura* 23–25cm

E lowlands and foothills, NW foothills to subtropics (to 1,200m E, locally higher, 500–2,200m W, locally lower). Humid forest canopy, borders, adjacent clearings. ***Grey*** bill, white orbital skin (***grey*** in W). Mostly green, ***throat and breast scalloped buffy to whitish, red wing-coverts*** (conspicuous in flight), ***maroon tail.*** Races *pacifica* (NW, narrow breast scaling, dark orbital skin), *melanura* (E lowlands, coarse whitish scaling), *souancei* (NE slopes, broader whiter scaling, dark orbital skin), *berlepschi* (SE, more solid white scaling); possibly several species. Small flocks fly through canopy, never high overhead, noisy in flight. **Voice** High-pitched, shrill or chattering, *kreE-kreE-kreéy...skreéy-skreéy...* in flight, screechy chatter on perch. **SS** White-necked (solid white breast), Rose-fronted Parakeet (dusky-brown head) in SE. **Status** Fairly common.

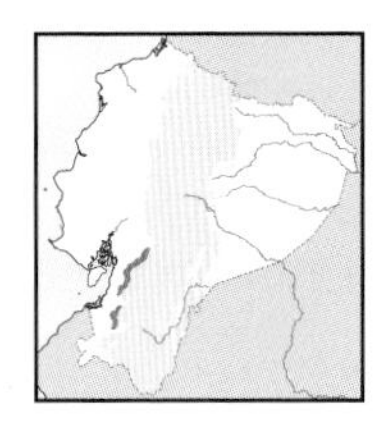

El Oro Parakeet *Pyrrhura orcesi* 22–24cm

SW foothills to lower subtropics (600–1,200m, locally 300–1,500m). Canopy, borders, adjacent clearings. Greyish bill, white orbital skin. Mainly bright green, ***red front, plain underparts; greater primary-coverts red*** (conspicuous in flight), some blue in primaries, maroon tail. Monotypic, recently (1988) described. Flocks of 3–15 in canopy, not high overhead; sometimes several flocks join at fruiting trees, but separate to roost in palms; nests communally, at natural crevices in trees; incubation duties shared by several individuals; some altitudinal movements following fruiting events. **Voice** High-pitched, shrill *krëEY-krëEY...* in flight, *kree'ee'ee'eey* and other calls on perch. **SS** Only *Pyrrhura* in range. **Status** Rare, very local, declining; EN, endemic.

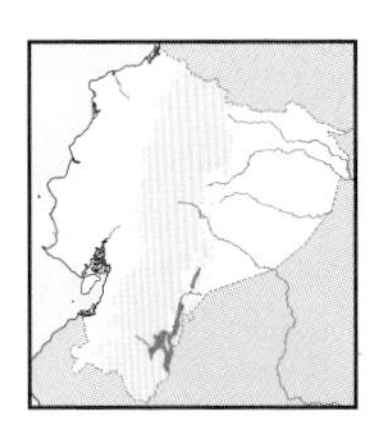

White-necked Parakeet *Pyrrhura albipectus* 23–25cm

SE foothills and subtropics (900–1,800m). Canopy, borders, adjacent clearings. Black bill, white orbital skin. Blackish-brown crown, ***green cheeks, orange ear patch. White throat, breast and nuchal collar,*** some yellow tinge on upper belly, green lower belly. Bright green above, dusky tail. ***Greater primary-coverts red*** (in flight). Monotypic. Small flocks fly through canopy, never high overhead, noisy in flight but almost invisible when perched. **Voice** As other *Pyrrhura*, a thin shrill chatter in flight, *kree'ee-kree'ee..skreey-kreeye...*; weaker calls when perched. **SS** Race *berlepschi* of Maroon-tailed Parakeet can look almost white-breasted; see head pattern; cf. Rose-fronted Parakeet (darker head, no red in wings). **Status** Uncommon, local, VU.

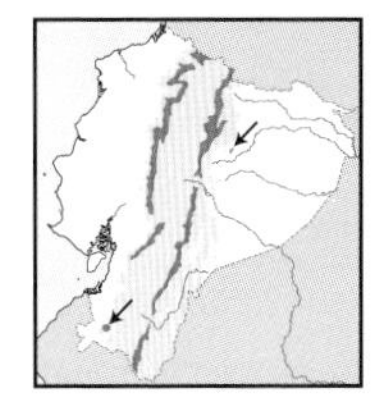

Barred Parakeet *Bolborhynchus lineola* 16–18cm

Subtropics to temperate zones on both Andean slopes (1,700–3,100m). Forest, borders, adjacent treed clearings. ***Small, chunky,*** pale bill, narrow whitish eye-ring. Bright green head, ***olive-green upperparts,*** paler underparts. ***Dense, coarse dusky banding, mostly above;*** barred wings. ***Dusky, short, pointed tail.*** Race *tigrinus*. Compact flocks more often heard than seen, as fly rapidly and directly, high overhead; seasonal movements following fruiting trees and seeding bamboo, where perches for long periods; noisy in flight, usually silent on perches. **Voice** A shrill *chiirr-shirr...shirii-chii-chii-chiir-irr* and other chattering calls. **SS** Unique in range and habitat by shape and size. **Status** Uncommon, local, overlooked?

e-fronted Parakeet
melanura
juvenile
adult
souancei
berlepschi
pacifica
Maroon-tailed Parakeet
♀
juvenile
♂
El Oro Parakeet
juvenile
adult
White-necked Parakeet
Barred Parakeet

PLATE 88: PARROTLETS AND SMALL PARAKEETS

Small to petite green parakeets, rather chunky, with wedge-shaped tails. *Forpus* occurs in open, semi-open and edges in E lowlands (two) and W; gregarious and noisy. Pacific is spreading north following deforestation. Flight often low, fast, deeply undulating and weaving. *Brotogeris* are more forest-based and very gregarious; flight faster, less weaving.

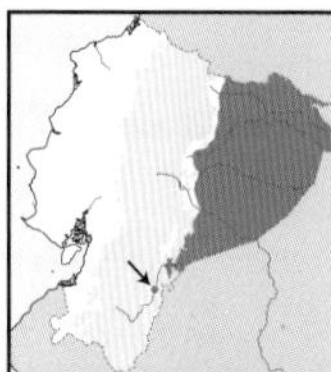

Blue-winged Parrotlet *Forpus xanthopterygius* — 12–13cm

E lowlands (to 500m, locally to 1,000m). *Várzea* and riparian forest borders, woodland, adjacent clearings. ***Sparrow-sized, whitish bill.*** Darker green above. ***Bright blue in greater wing-coverts, secondaries and rump.*** ♀: ***lacks blue parts***. Race *crassirostris*. Small groups tend to perch in open, noisy in flight, can 'disappear' when perched. **Voice** Endless, high-pitched chattering *tjeet-jeetjeetjeetzitzit*.... **SS** Darker Dusky-billed Parrotlet has bicoloured bill, deeper blue parts; differs in habitat. **Status** Uncommon and local.

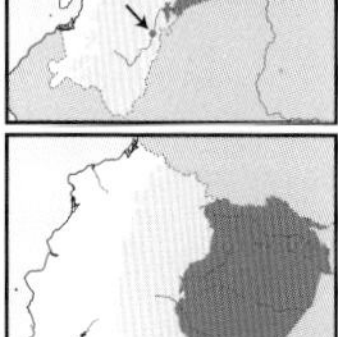

Dusky-billed Parrotlet *Forpus modestus* — 12–13cm

E lowlands (to 900m). Canopy and borders of *terra firme* forest, adjacent clearings. ***Sparrow-sized***, with ***bicoloured bill*** in both sexes. ***Dark green, brighter on head***, paler on belly. ***Deep blue*** greater wing-coverts, secondaries and rump. ♀: similar, but ***no blue*** in wings and rump. Race *sclateri*. Pairs or small flocks move through canopy, not high overhead; rarely descends, usually in borders and clearings. **Voice** Very high-pitched, whistled *ziiiit-ziit-ziriit-ziiit*.... **SS** Paler and brighter Blue-winged Parrotlet has pale bill. **Status** Uncommon but local.

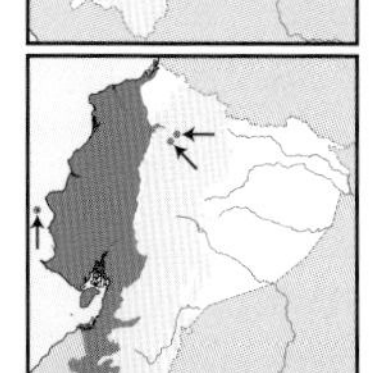

Pacific Parrotlet *Forpus coelestis* — 12–13cm

W lowlands and foothills (to 800m, locally higher, spreading). Borders, woodland, shrubby clearings, gardens, parks, crops. ***Sparrow-sized***, mostly ***bright lemon-green***, more olive above. ***Bluish nape and postocular stripe. Deep blue greater wing-coverts, secondaries and rump.*** ♀: similar but ***no blue***. Monotypic. Numerous, well adapted to modified habitats; groups up to 40–50, noisy, fast-flying. **Voice** High-pitched, continuous, fast chattering *tzit'tzit'tzizitzit...t'tzitzitzitzitzi*. **SS** Unique in range and habitat. **Status** Common, widespread.

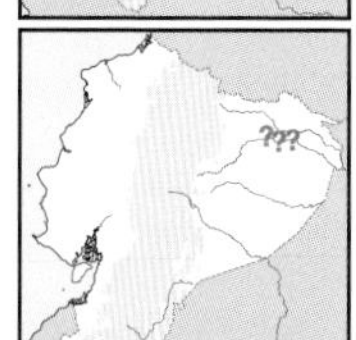

* Tui Parakeet *Brotogeris sanctithomae* — 17–19cm

Local in E lowlands (below 300m). *Várzea* and riparian forest and borders. ***Dark bill, pale eyes.*** Mostly green, ***yellow forehead; no blue in wings and tail***. Nominate. Poorly known in Ecuador, few records along Río Napo, maybe regular on river islands; elsewhere in noisy canopy flocks. **Voice** High-pitched, near-constant *skreek-kee-keekee*..., also other semi-musical calls. **SS** Cobalt-winged Parakeet (extensive blue in wings). **Status** Uncertain, undocumented visual records.

[i] Canary-winged Parakeet *Brotogeris versicolurus* — 22–25cm

Feral, local in SW lowlands and probably SE foothills (0–10m, 1,000m). Mostly green, ***yellow inner coverts, blue outer primaries, rest of flight feathers white***. Long, pointed tail. Monotypic. Presumed escapes observed near Guayaquil and Zamora, feral populations probably established and breeding in Guayaquil; elsewhere, fly high and tend to wander widely. **Voice** *screek-weecha-screek-weecha* in flight (HB). **SS** Bold wing pattern separates from other parakeets. **Status** Apparently introduced; previously believed part of Amazonian Ecuador avifauna, but no corroborated records.

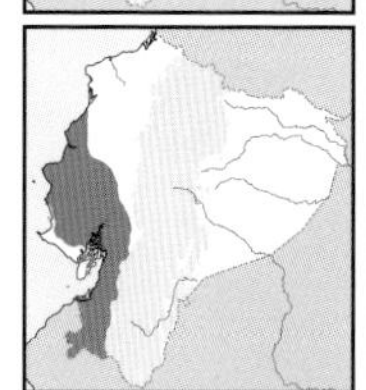

Grey-cheeked Parakeet *Brotogeris pyrrhoptera* — 19–21cm

SW lowlands to foothills (to 1,000m, rarely higher). Deciduous to semi-deciduous forest, borders, woodland, clearings with trees. Pale horn bill. ***Bluish crown and nape, pale grey face. Orange underwing-coverts***, deep blue in primaries. Monotypic. Small flocks often perch in open; fly high above ground, constantly vocalising; sometimes flocks with Red-masked Parakeet. **Voice** Shrill chatters: *kleet-leét-leét...klee-klee-leey-kee...sklee-kee-skleek*... and others. **SS** Only medium-sized parakeet in range. **Status** Uncommon but declining; popular in illegal trade; VU in Ecuador, EN globally.

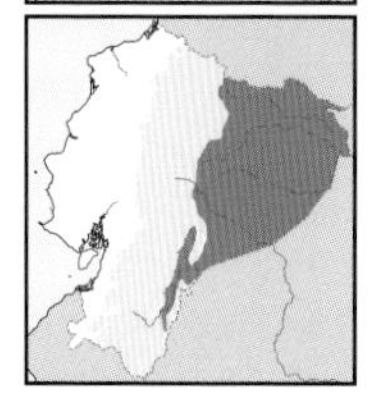

Cobalt-winged Parakeet *Brotogeris cyanoptera* — 19–21cm

E lowlands to foothills (to 1,000–1,100m). Canopy, borders, woodland, adjacent clearings. Pale bill. Mostly green, paler below. Small ***orange chin patch***, yellow front, bluish crown and nape. ***Deep blue flight feathers, short pointed tail with blue in central feathers***. Nominate. Noisy flocks, sometimes dozens, mostly in canopy but lower at borders and in shrubby clearings; numerous at clay licks. **Voice** Shrill *s's'sklínk-liiínk-sklíin-klíin-kílíkílíkílíkílí*... and other more musical notes. **SS** Rare Tui Parakeet has yellow forehead, no blue in wings. **Status** Common and widespread.

♂
♀
Blue-winged Parrotlet
♀
♂
Dusky-billed Parrotlet
♀
♂
Pacific Parrotlet
Tui Parakeet
juvenile
adult
Canary-winged Parakeet
juvenile
adult
Grey-cheeked Parakeet
Cobalt-winged Parakeet

Touit are small, chunky, with short square tails and small bills. Predominantly green, with brightly coloured underwing patches. Form small, rather compact flocks 'invisible' when perched, but quite distinctive and noisy in flight. Flight fast and direct, wingbeats deep and steady, with little to no undulation. Black-headed is a unique Amazonian parrot, very vocal, which flies fast and direct, with little rolling, wings whirring.

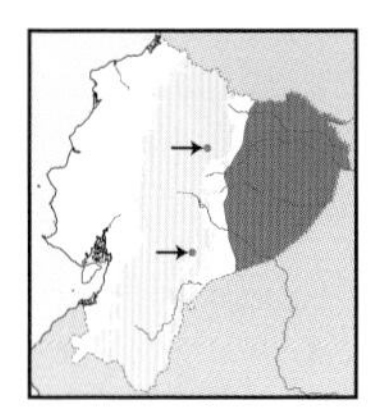

Scarlet-shouldered Parrotlet *Touit huetii* 15–16cm

Local in E lowlands to foothills (to 400m, locally to 1,400m). *Terra firme* forest canopy, borders. Whitish bill, ***bold white orbital ring***. Mostly green, paler below. ***Black front***, yellowish-brown crown, ***lavender-blue foreface. Scarlet shoulders and underwing-coverts***, blue wing-coverts. ***Short, square tail with maroon outer rectrices*** (***yellowish-green*** in ♀ and juv.). Monotypic. Compact flocks rather silent, moving through canopy or in high flight; seemingly erratic in Ecuador, where records are few; sometimes at clay licks. **Voice** In flight *tou-eet-eet-eet, tou-eeet..tou-eet-eet...*. **SS** From *Brotogeris* by tail; Sapphire-rumped Parakeet (plainer head, no white eye-ring, purple rump, no red in underwing). **Status** Rare, local, likely erratic, perhaps overlooked; VU globally.

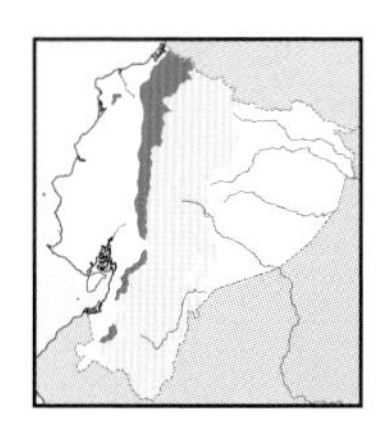

Blue-fronted Parrotlet *Touit dilectissimus* 14–17cm

W foothills, locally to lowlands (500–1,400m). Canopy and borders of humid forest. Yellowish-white bill, bold greyish orbital skin. Mostly green, paler below, brighter rump and wings. ***Bluish forecrown, red orbital line***, bright green cheeks, yellowish malar and throat. ***Scarlet wing-coverts, bright yellow wing bend and underwing-coverts. Short, square tail, orange-yellow outer rectrices***. ♀: ***less red in face and wing-coverts***. Monotypic. Small groups discreetly move through canopy, sometimes seen at borders or in compact flocks in zigzagging fast flight. **Voice** Thin, high-pitched *tu-deee* or *tuueeet, teéw..teéw...*. **SS** Unmistakable in range by tail shape, size, underwing and voice. **Status** Rare, quite local (overlooked or erratic?); NT in Ecuador.

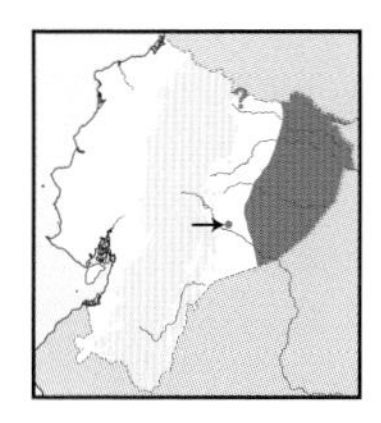

Sapphire-rumped Parrotlet *Touit purpuratus* 16–18cm

Local in E lowlands (to 300m). Canopy and borders of (mostly?) *terra firme* forest. Whitish bill, ***grey orbital ring, pale iris***. Mostly green, paler below, brighter crown, wings and mantle. ***Violet-blue rump*** (difficult to see); ***dark brown scapulars and tertials***. Tail mainly reddish, rectrices narrowly edged and tipped black. ♀ and juv. have ***narrow green*** tail tip. Race *viridiceps*. Pairs or small groups often missed when perched, or in zigzag flight just above canopy; might freeze when threatened, suddenly flying fast after few pre-flight calls. **Voice** Nasal, loud *nyaAh-nyaAh-nyaAh...* or *neyaá-neyaá...*. **SS** No red in underwing like Scarlet-shouldered Parrotlet; larger than *Forpus*, with square tail. **Status** Rare, likely overlooked.

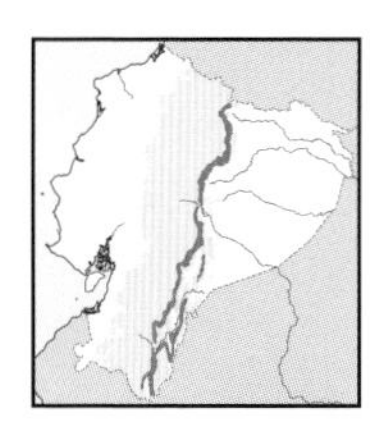

Spot-winged Parrotlet *Touit stictopterus* 16–18cm

E Andean subtropics (1,100–1,800m). Montane forest canopy, borders, adjacent clearings. Yellowish-white bill, grey eye-ring. Mostly bright green, contrasting ***dusky-brown wing-coverts, with buffy-white spotting***. Orange patch in greater coverts, ***short, green, square tail***. ♀: similar but ***wing-coverts green, yellower face***. Monotypic. Small flocks fly low above canopy, but can cross open areas between forest patches; twisting flight helpful for identification. **Voice** Harsh, shrill *zzrreet-ddreet-zzreet...*, *dree-dree-deet..dee-deet...* when perched. **SS** Unique in range, but difficult to see; Barred Parakeet has long pointed tail. **Status** Rare, range-restricted, possibly declining; VU.

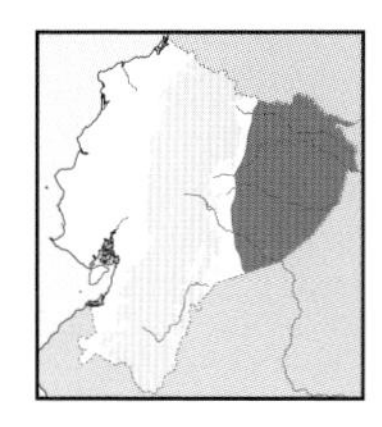

Black-headed Parrot *Pionites melanocephalus* 21–23cm

E lowlands (to 400m). Canopy and borders of *terra firme* and *várzea* forest, river islands. Handsome, mid-sized parrot, with ***black bill, dark grey orbital ring, bright red eyes. Black crown, rich orange-yellow cheeks, throat and upper breast, more buffy-orange on nape***; lower belly and thighs rich orange-yellow, white ***underparts***. Sexes alike. Race *pallidus*. Small flocks mostly fly through or low above canopy, sometimes perch on exposed limbs; quite tame when foraging or perching; feeding flocks include some vigilant birds while rest forage. **Voice** Notably vocal; wheezy *creeích-creeích-creeích...*; various ringing, semi-musical calls when perched. **SS** Confusion unlikely. **Status** Fairly common.

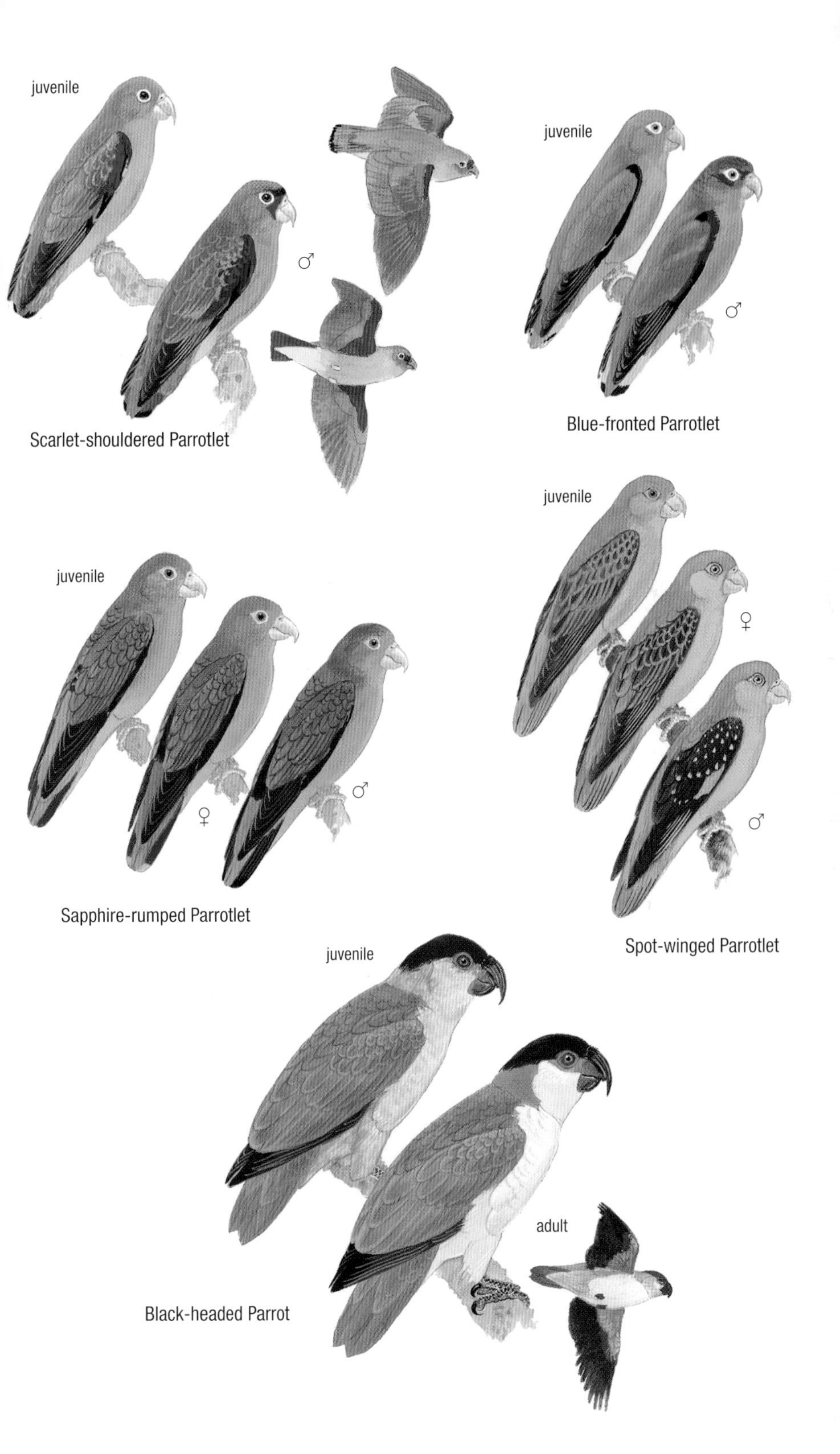

juvenile
♂
Scarlet-shouldered Parrotlet
juvenile
♂
Blue-fronted Parrotlet
juvenile
♀
♂
Sapphire-rumped Parrotlet
juvenile
♀
♂
Spot-winged Parrotlet
juvenile
adult
Black-headed Parrot

PLATE 90: PARROTS I

Two genera, predominantly green, with remarkable head and wing patterns. *Pyrilia* in lowlands, *Hapalopsittaca* exclusively montane, reaching close to *páramo*. Gregarious, but flocks small. Flight fast, low above canopy in *Pyrilia*, often high in *Hapalopsittaca*, wingbeats being deep, steady and powerful, with no undulation but marked rolling. Status in Ecuador of Saffron-headed and Rusty-faced uncertain, both requiring confirmation. The monotypic Short-tailed Parrot occurs on river islands and in riparian forests.

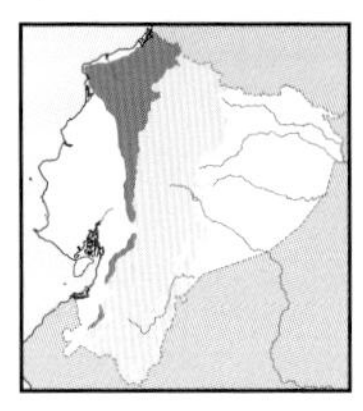

Rose-faced Parrot *Pyrilia pulchra* 21–23cm

NW lowlands and foothills, S on Andean slope to El Oro (up to 1,400m). Humid forest canopy, borders, adjacent treed clearings. Medium-sized, chunky, with ***pale bill, orbital ring and iris. Rosy face*** bordered by thin black line. ***Head coppery-brown, to nape and chest***. Dark green above, paler below, ***blue underwing-coverts***, red axillaries. Monotypic. Small flocks silent when perched, sometimes on exposed branches or below canopy, tends to fly high, twisting. **Voice** Harsh, far-carrying *shrëék-kEk-shrëék-shrëik-kEk...* or *skrEEk-skrEEk....* **SS** Larger *Pionus* show red vent in flight; rosy face unique; juv. Saffron-headed Parrot looks fairly similar. **Status** Uncommon, possibly declining; VU in Ecuador.

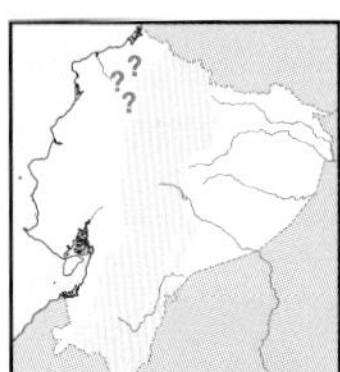

* Saffron-headed Parrot *Pyrilia pyrilia* 21–23cm

Lower NW foothills; two uncorroborated observations (to 450 m). Humid forest canopy, borders. Chunky, with whitish bill, white orbital ring, ***dark eyes. Entire hood bright sulphur-yellow***, more orange on ear-coverts. Green upperparts, brownish chest, pale green below. ***Orange wing bend***, some blue in upperwing-coverts, ***red underwing-coverts***. Juv. duller. Monotypic. Unknown in Ecuador; elsewhere habits and behaviour like other *Pyrilia*. **Voice** In flight *che-week*, also *cheeyk-cheeyk...* (HB). **SS** Saffron head unmistakable, but duller juv. similar to juv. Rose-faced Parrot (but see underwing). **Status** Uncertain, possibly very rare wanderer to Ecuador but perhaps misidentified; VU globally.

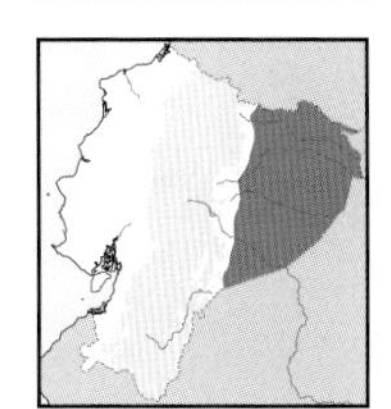

Orange-cheeked Parrot *Pyrilia barrabandi* 24–26cm

E lowlands (to 400m). Canopy and borders mostly of *terra firme* forest, also *várzea*. Unique, handsome, with slaty bill, ***creamy-white*** orbital ring, dark eyes. ***Black head, orange-yellow cheek patch***. Otherwise mostly green, olive-brown chest, ***orange*** shoulders and thighs. ***Bright red wing bend and underwing-coverts***. Nominate; race *aurantiigena* erroneously reported for Ecuador. Small flocks seen mostly in flight; perch quietly in canopy; flight fast, often twisting, regular at clay licks. **Voice** Soft, often disyllabic *chowít..chowít-ché*; soft *chuuit* on perch. **SS** Confusion unlikely; see smaller, more striking Black-headed Parrot. **Status** Uncommon in forested areas, NT globally.

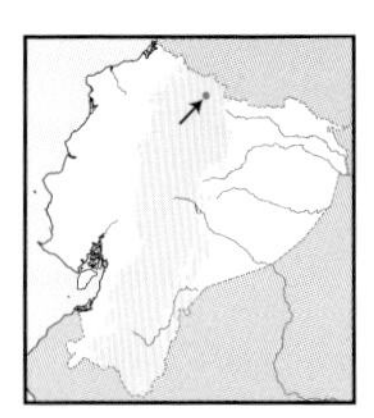

* Rusty-faced Parrot *Hapalopsittaca amazonina* 21–23cm

Temperate forest in extreme NE Andes; one record at 3,300m. Montane forest canopy, borders. Whitish bill, grey orbital ring, amber eyes. Mostly green, paler below, yellower breast. ***Rusty face, with olive-yellow ear-coverts and yellow plumes on ear-coverts. Bright red shoulders***, blue primaries, some red in underwing-coverts. Tail ***dull red, square***, with blue tips. Race *velezi*? Not well known; elsewhere small shy flocks perch in canopy, flight fast and direct, high overhead. **Voice** In flight *chek-chek-chek...* (HB). **SS** Red-faced Parrot (no overlap) has green tail and redder face; Indigo-winged Parrot *H. fuertesi* not in Ecuador but comes quite close Colombia. **Status** Uncertain, very rare or even accidental wanderer; CR in Ecuador, VU globally.

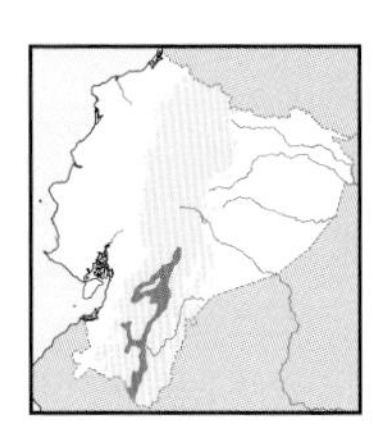

Red-faced Parrot *Hapalopsittaca pyrrhops* 21–23cm

Temperate forest in E and inter-Andean slopes of S Andes (2,700–3,500m). Montane forest canopy, borders, adjacent treed clearings. Whitish bill, grey orbital ring, pale iris. Mostly bright green, more olive breast. ***Red foreface and chin, olive cheeks with yellow plumes on ear-coverts***, crown and neck-sides washed blue. ***Shoulders bright rosy-red***, some red in underwing-coverts. ***Square tail, green tipped blue***. Monotypic. Small flocks fly high overhead, flight fast and direct; unobtrusive while perched. **Voice** Weak *kree-kekekee-kekee*; perched birds utter a high-pitched *Eek-Eek-Eek...*, more guttural *thruut-thru-thruut* in flight. **SS** Confusion unlikely; see very rare and local Rusty-faced Parrot. **Status** Rare and local, fairly nomadic; declining, EN in Ecuador, VU globally.

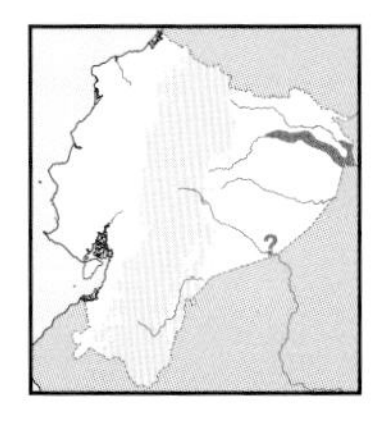

Short-tailed Parrot *Graydidascalus brachyurus* 21–24cm

NE lowlands (to 300m). Canopy and borders of *várzea* and riparian forest, woodland and river islands. ***Plump, mostly green with heavy blackish bill, large grey orbital skin, red eyes***. Paler underparts, ***short, square tail***. Red bases of outer tail feathers. Juv. dark eyes. Monotypic. Very noisy in flight, audible at considerable distance; small flocks fly high overhead; on landing, flocks twist and roll, flight fast and direct, deep wingbeats; often perch on exposed branches. **Voice** Loud, harsh *shhreeyk'k'k-shreeyk'k... jeEk-jeEk-jeEk!...*; perched, a harsh *shreey-sh'hre-sheedle'e'e- sheedle'e'e....* **SS** Plump body and very short tail separate from *Pionus*; *Amazona* larger, more coloured. **Status** Rare, local.

♂
♀
juvenile
Rose-faced Parrot
adult
juvenile
Saffron-headed Parrot
adult
juvenile
Orange-cheeked Parrot
juvenile
adult
Short-tailed Parrot
adult
juvenile
Rusty-faced Parrot
Red-faced Parrot

Compact parrots with short square tails, reminiscent of *Amazona* but smaller. Red-fan is rare and local in Amazonia. *Pionus* occur in loose pairs to fairly large flocks – for example at maize crops. One strictly montane, living near treeline; one on Andean slopes, one in W foothills and lowlands, one in E & W lowlands. Flight fast, with deep strong wingbeats. All have characteristic red vents.

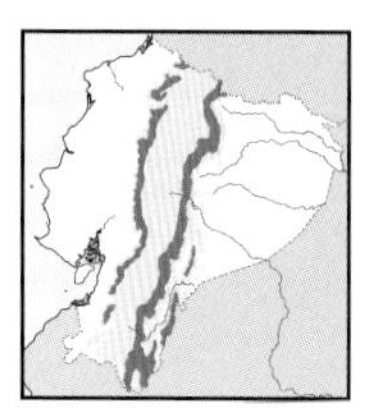

Red-billed Parrot *Pionus sordidus* 27–30cm

E & W Andean upper foothills to subtropics (1,200–2,400m). Montane forest canopy, borders, adjacent clearings with standing trees. ***Coral-red bill.*** Mostly green, bluish-tinged head, ***crown somewhat scaled.*** Variable ***blue throat and chest-band.*** Juv. duller. Race *corallinus.* Small noisy flocks similar in behaviour to Blue-headed Parrot; deep wingbeats. **Voice** High-pitched chatter *ee'waaan-kee'iin-keeyiiin-eeiiin...* in flight, louder and shriller when perched. **SS** Low-elevation Blue-headed (possibly overlaps locally) has entirely blue head, little red in bill; upper-elevation Speckle-faced Parrot (also overlaps locally) has paler head. **Status** Uncommon.

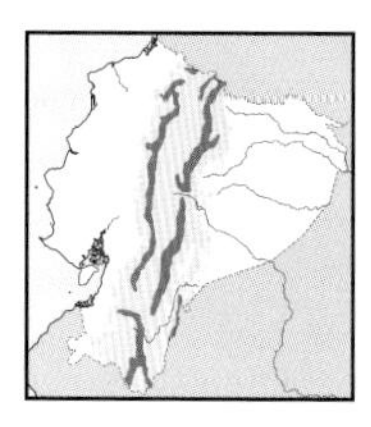

Speckle-faced Parrot *Pionus tumultuosus* 28–30cm

Subtropical and temperate forests on both Andean slopes (1,500–3,200m). Montane forest canopy, borders. ***Bill pale yellow. Head and neck violet-pink densely speckled white; white forecrown and throat, dusky below.*** Race *seniloides* possibly valid species. Small groups less conspicuous than other *Pionus*, silent in trees or flies through dense cloud cover; apparently makes seasonal or nomadic movements. **Voice** Thin, high-pitched, shrill *kree-yaá..yaan-kree-yaán...creé-eek-eek...*, other chatters when perched. **SS** Locally sympatric Red-faced Parrot lacks red vent; see also lower-elevation Red-billed Parrot; pale head and yellow bill distinctive. **Status** Rare to seasonally uncommon.

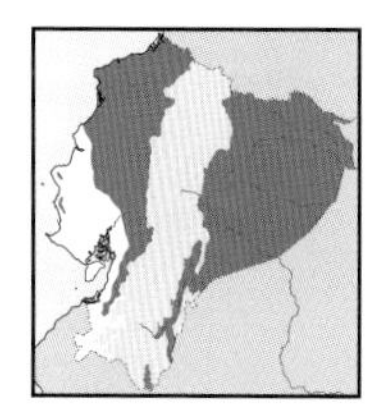

Blue-headed Parrot *Pionus menstruus* 25–29cm

E & W lowlands and foothills (to 1,100m). Humid forest canopy, borders, woodland, adjacent treed clearings. ***Dusky bill with pink spot on maxilla. Bright blue hood, black patch on ear-coverts, red throat patch.*** Race *rubrigularis* (W, more red in throat), *menstruus* (E, deeper blue head). Conspicuous and familiar in varied habitats including clearings and agricultural land; noisy flocks in direct flight, often perch on exposed branches, regular at clay licks. **Voice** High-pitched *keéwEnk-wEnk-wEnk... keewénk-wEn-wEn-wEn...shwEEnk-shwEEnk...*; several chattering calls when foraging and perching. **SS** Upper-elevation Red-billed Parrot show some blue on head, but bill all red. **Status** Common.

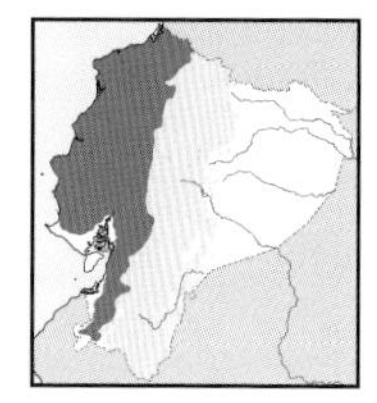

Bronze-winged Parrot *Pionus chalcopterus* 27–29cm

W lowlands to foothills (to 1,400–1,500m). Humid and semi-deciduous forest, borders, woodland, adjacent treed clearings. ***Pale yellow bill,*** pinkish orbital ring. Mostly ***purplish-blue, rosy-white pectoral patch,*** bronzy-green back and ***bronze wings. Flight feathers vivid blue above, turquoise below.*** Juv. duller. Monotypic. Apparently engages in seasonal movements, at least in semi-deciduous forests; flocks usually small, but dozens can gather at feeding sites. **Voice** In flight *këëink-këëink-kék'kék'kék...; chëëik-gëik-gëëk-gë...* and other calls when perched. **SS** Confusion unlikely due to dark blue plumage; Blue-headed Parrot occurs sympatrically, seasonally/locally, but differences noticeable. **Status** Uncommon, VU in Ecuador.

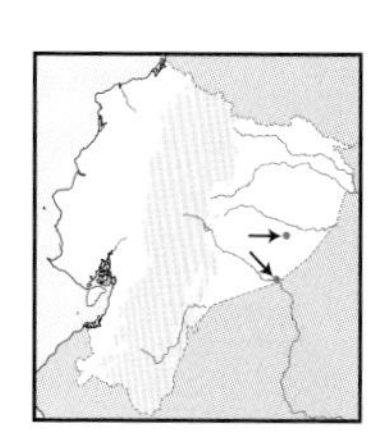

Red-fan Parrot *Deroptyus accipitrinus* 34–36cm

Local in lowlands of extreme SE (200m). *Terra firme* canopy, borders. Spectacular: ***long-tailed*** with blackish bill and orbital ring, yellow eyes. ***White forecrown, brownish head with white spotting and streaking, red nape fan with bluish edging.*** Nominate. Pairs to small groups often perch high and exposed; wing silhouette and flight differ from typical parrots, including short glides. **Voice** Loud, raucous *klék-klék-klék*...or *chEk-chEk-chEk...* in flight; *klëyk-klëyk-kek'ek'ek'eyk'eyk'eyk...* when perched. **SS** Unmistakable. **Status** Rare, local.

adult
variant
Red-billed Parrot
adult
juvenile
juvenile
adult
Speckle-faced Parrot
juvenile
adult
juvenile
adult
Blue-headed Parrot
juvenile
adult
Bronze-winged Parrot
adult
juvenile
Red-fan Parrot

Amazona are large, mainly Amazonian in distribution, characterised by square tails and brightly-coloured underwing patches – mostly seen in flight. Loud voices and strong but shallow wingbeats distinguish them from smaller *Pionus* in flight.

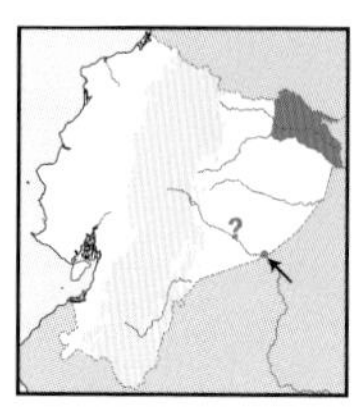

Festive Amazon *Amazona festiva* 33–36cm

Local in E lowlands (to 250 m). Canopy and borders of *várzea* and riparian forest, woodland. ***Dark grey*** bill and orbital ring. Mostly green, ***reddish front, lores and above eyes***, bluish postocular stripe. ***Red rump and lower back***, blue flight feathers, ***no speculum***. Juv. with little to no red on rump. Nominate. Associated with water, often perches on exposed branches; noisy in flight, less so when perched. **Voice** Repeated, nasal *rowg-rowg-rowg-rowg*... or *räng-häng-rräng-räng*.... **SS** Red rump and no red speculum unique; cf. larger Mealy (banded tail, green rump, etc.). **Status** Uncommon, local; NT globally.

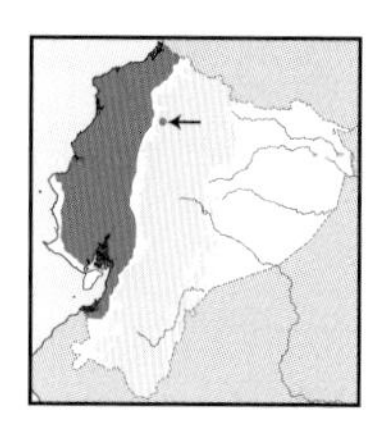

Red-lored Amazon *Amazona autumnalis* 31–35cm

W lowlands (to 700m). Deciduous to humid forest canopy, borders, mangroves. Dusky-and-horn bill, whitish orbital ring. Mostly green, ***red forecrown and supercilium, bluish crown***, bright pale green cheeks. ***Red speculum*** in secondaries conspicuous in flight. Broad yellowish-green tail-band. Race *salvini* (NW), endemic race *lilacina* (SW) possibly valid species. Pairs to small groups, shy when foraging, noisy in flight, seen passing from roosts to feeding grounds; concentrates in mangroves for roosting. **Voice** Loud, harsh *cheék-clor'Ak-clo'o'o'k*...*chEk-chEk-chEk*... or *cheek'yA-cheek'yA-cheek'yA-yák-klyaa*..., other calls when perched. **SS** Larger, plainer Mealy Amazon; louder voice. **Status** Rare, local, declining; EN in Ecuador and globally.

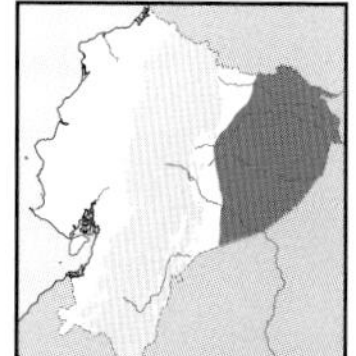

Yellow-crowned Amazon *Amazona ochrocephala* 34–38cm

E lowlands (to 400m). *Terra firme* canopy and borders, also *várzea*. ***Grey bill*** with horn patch on maxilla, white orbital ring. Mainly green, ***lemon-yellow crown, red shoulders and speculum in secondaries***, deep blue flight feathers. Tail yellowish-green distally. Race *nattereri*. Pairs to small groups mostly seen in flight from roosts to feeding areas; regular at clay licks. **Voice** Guttural *cu-gäow, cu-gäow*... in flight; various throaty calls *grr-graow* when perched. **SS** Mealy Amazon larger, plainer, voices differ; Orange-winged Amazon (yellow in face not in crown, orange wing speculum). **Status** Uncommon.

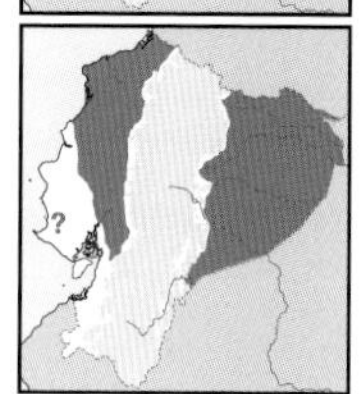

Mealy Amazon *Amazona farinosa* 38–41cm

E & W lowlands (mostly to 700m). Humid forest canopy, borders, mainly *terra firme* in E. Horn bill, ***broad*** white orbital ring. Mostly green, little yellow in mid crown (variable), some scaling on nape and mantle. Red wing bend, ***prominent red speculum in secondaries***. Distal tail half ***yellowish-green***. Nominate. Very social, mostly in small flocks, noisy in flight, quiet when perched; regular at clay licks. **Voice** Varied and loud repertoire, *chyök-chyök-chyök*...*kák-kák-kAkAkA*...*klëya'a'a'aa, tAyk-kay-yee, tAyk-ka-yee*..., *klá-klá-klyá-klyá*... in flight. **SS** Larger than other *Amazona*, voice distinctive (cf. Red-lored Parrot in W). **Status** Uncommon (E), now rare in W; NT globally.

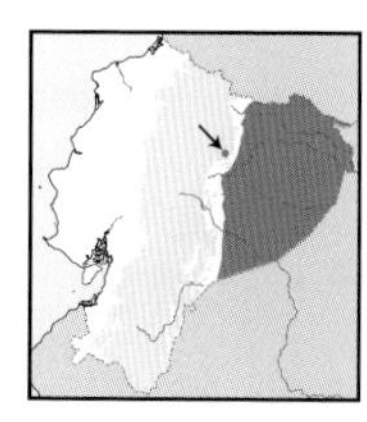

Orange-winged Amazon *Amazona amazonica* 31–33cm

E lowlands (to 500m). Canopy and borders of *várzea* and riparian forest, woodland, adjacent clearings. Grey-and-horn bill, greyish orbital ring. Mostly green, ***yellow cheeks and forecrown, bluish crown and lores. Orange speculum in secondaries*** conspicuous in flight. Distal half of tail yellowish-green. Monotypic. Tolerates edges and fairly open habitats, dozens often congregate at island roosts; noisy in flight, less regular at clay licks. **Voice** High-pitched *ké-wEEk, ké-wEEk, ké-wEEk*...; many loud chattering calls when perched: *kék-wEk*...*wekékEk*.... **SS** Orange speculum separates from other *Amazona*; see Yellow-crowned. **Status** Uncommon, widespread.

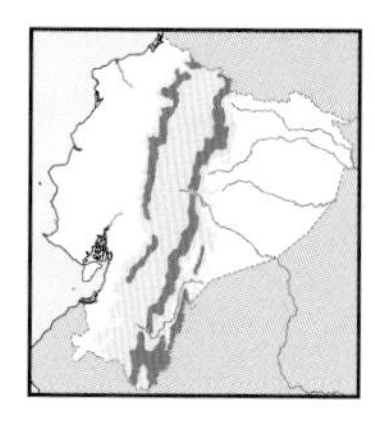

Scaly-naped Amazon *Amazona mercenarius* 31–34cm

Subtropical to temperate zones on both Andean slopes (mostly 1,200–2,600m). Montane forest canopy, borders and woodland. Grey-and-horn bill, white orbital ring. Mainly green, ***scaled crown, nape, neck-sides and mantle. Wings mostly uniform*** (little to no coloured speculum). ***Broad reddish subterminal*** tail-band. Race *canipalliata* (N), possibly *mercenarius* (S, with red speculum). Often seen flying high between roosts and feeding grounds; quiet when feeding. **Voice** In flight *ka-ka-ka-kaklee*.. *ka-klee*...*kla-kla-klee*; *kla-klee-kay-kay-kay-klëëërr*... calls and chatters when perched. **SS** Unique in habitat, larger than other montane parrots, with louder voice. **Status** Uncommon, but apparently declining in deforested areas.

Festive Amazon
lilacina
juvenile
adult
salvini
Red-lored Amazon
adult
adult
juvenile
Yellow-crowned Amazon
adult
adult
variant
juvenile
Mealy Amazon
juvenile
adult
Orange-winged Amazon
adult
juvenile
Scaly-naped Amazon

Three forest cuckoos, predominantly chestnut to rufous, with long graduated tails, coloured facial skin and bills, plus the bolder and very local Dwarf Cuckoo. Anis are social birds of semi-open and open areas, two are common and widespread in degraded areas of lowlands to foothills, while Greater is rare in W lowlands, commoner in E, but confined to water-side vegetation.

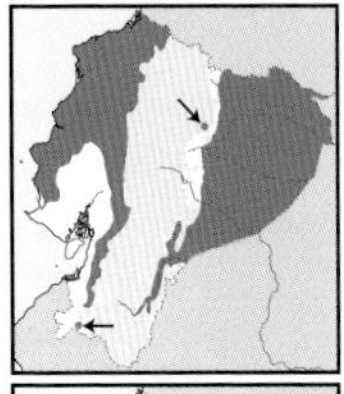

Little Cuckoo *Coccycua minuta* 25–28cm

E & W lowlands, locally to subtropics (to 1,900m). Borders, woodland, adjacent clearings. *Small* version of Squirrel Cuckoo. ***Yellow bill, red orbital skin. Chestnut upperparts, head and breast*** (more cinnamon in W), ***greyish-brown belly*** (paler in W). Nominate (E), *gracilis* (W). Not easy to see; shy, forages low in dense vegetation, briefly emerging at borders. **Voice** Mournful *tdeea, tdeea-ku?* song; harsh *tch'weaah, t'chuw* calls; also descending, bouncing *wii-a-de-de-de...* and other calls. **SS** Size separates from *Piaya* species. **Status** Uncommon.

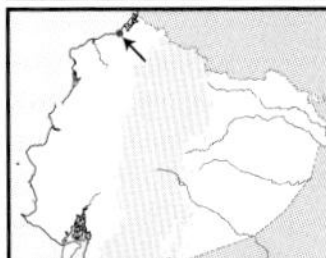

Dwarf Cuckoo *Coccycua pumila* 20–22cm

Few records from coastal scrub in NW (J. Nilsson ; R. Ahlman). Grey above, ***chestnut throat to chest, creamy-yellow below***. Monotypic. Sluggish and shy, seen mostly inside vegetation, rather nervous. **Voice** Calls *churr* or grating *trr, trr, trr...* (HB). **SS** Dark-billed Cuckoo has grey crown and tawny belly, but otherwise very different. **Status** Uncertain, possibly accidental visitor (Jun, Dec) or local breeder.

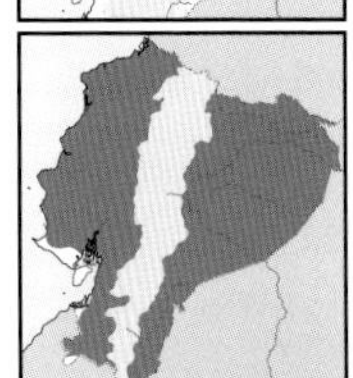

Squirrel Cuckoo *Piaya cayana* 40–46cm

E & W lowlands to subtropics (to 2,500m). Forest, woodland, borders, adjacent clearings. ***Yellow bill and yellow*** (W) ***or red*** (E) ***orbital skin. Uniform chestnut above, grey below***. Races *mesura* (E), *nigricrissa* (W). Lively, forages in upper strata, hopping on large branches with tail held loose; often conspicuous when crossing open areas, vocalising abruptly. **Voice** Loud, ringing *keéy-keéy-keéy-keeké...*; also loud, nasal *chiik-kwaa?*, bouncing *tchk-waaaa*; sweeter *duík..duík..duík* territorial song. **SS** Black-bellied Cuckoo (grey crown, red bill), Little Cuckoo (much smaller). **Status** Common.

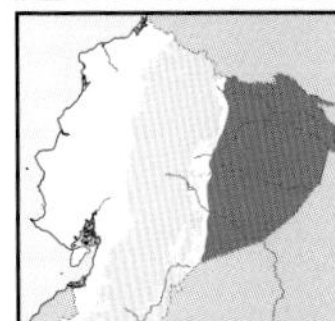

Black-bellied Cuckoo *Piaya melanogaster* 38–41cm

E lowlands (mostly to 400m). Upper strata of *terra firme* forest. ***Bright red bill, bluish orbital skin with yellow loral spot. Grey crown***, rufous upperparts, cinnamon-rufous towards breast, ***black belly***. Monotypic. Mostly forest-dweller, rarely in open areas; hops and bounds like Squirrel Cuckoo. **Voice** Not very vocal; only a loud, repeated *jeek-jéw..jeek-jéw..jeek-jéw...*, harsh *jeek-jerr..jeek-jerr..., jík-jew?, tjík-ji-tjík....* **SS** Commoner Squirrel Cuckoo (yellow bill, red orbital skin, grey belly). **Status** Rare, local.

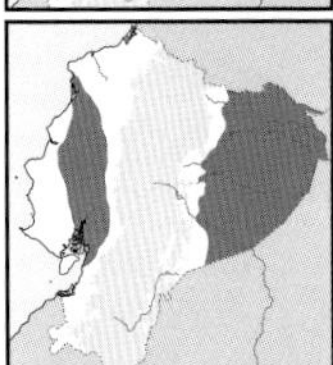

Greater Ani *Crotophaga major* 46–48cm

E lowlands; local in W lowlands (to 1,000m; locally higher in E). River and lagoons (E), swampy vegetation and river edges (W). ***Heavy arched bill, yellow iris, long, broad tail***. Black ***glossed greenish-blue***, more purplish on tail. Monotypic. Noisy groups (up to 20) cautiously forage low above water; breeds communally. **Voice** Guttural *ko-ko-ko-go'go'go'go-gog'gg'gg'gg...*; various ringing growls, croaks, hisses; nasal *jengggg!* **SS** Other anis much smaller, dark-eyed; cf. Great-tailed Grackle (habitat entirely different). **Status** Fairly common (E), rare, local (W).

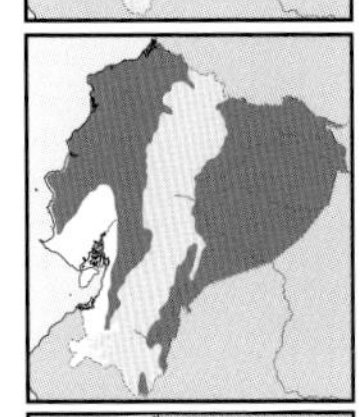

Smooth-billed Ani *Crotophaga ani* 33–36cm

E & W lowlands to foothills (to 1,800m). Shrubby borders, clearings, pastures, crops. ***Bill black, laterally compressed, culmen notably arched and smooth. Black*** with little gloss. Monotypic. Groups active in low vegetation; tame but nervous when approached, tends to follow cattle and horses; laborious flight. **Voice** Harsh, nasal, moaning *uuui, uuiik? uuui-uuiik?, uuuuik-uuuik-uuuik...*, and hissing notes. **SS** Groove-billed has longitudinal channels in culmen (not easy to see), bill less arched; prefers drier habitats. **Status** Common, widespread.

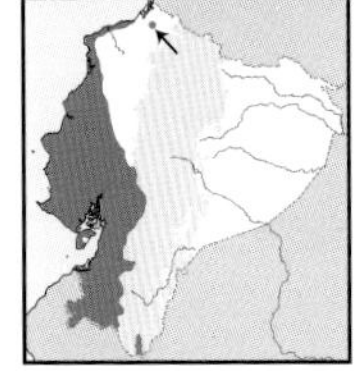

Groove-billed Ani *Crotophaga sulcirostris* 28–30cm

W lowlands, to foothills and subtropics in SW; local in extreme SE in Zumba region (to 2,300m; mostly below 1,300m). Shrubby borders, clearings, open areas. Very similar to Smooth-billed. Bill ***less arched, with longitudinal grooves***. Monotypic. Behaves like Smooth-billed, but prefers drier habitats; laborious flight, nervous flocks. **Voice** Dry series: *k'wiiik-k'wiiik-k'wiiik..., k'wiiik?..k'wiiik?...* plus growls and hisses. **SS** See Smooth-billed; voice less sharp and querulous. **Status** Fairly common.

adult
juvenile
adult
adult
juvenile
gracilis
minuta
Little Cuckoo
Dwarf Cuckoo
mesura
nigricrissa
Squirrel Cuckoo
Black-bellied Cuckoo
Greater Ani
Smooth-billed Ani
Groove-billed Ani

Rather poorly known in Ecuador, *Coccyzus* are slender and long-tailed, two are difficult to identify unless primaries are seen well. Their distributions are not well understood. Two northern migrants; one partially resident – but apparently engages in seasonal migration – another resident, partially migratory and austral migrant; and one of uncertain status. Striped is fairly common in semi-open and agricultural lands in W.

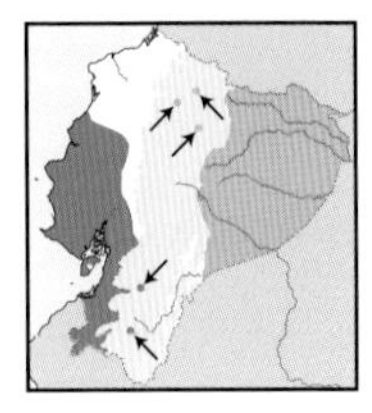

Dark-billed Cuckoo *Coccyzus melacoryphus* 27–28cm

E lowlands, locally to highlands, SW lowlands to foothills (to 1,900m, locally higher). Humid to deciduous forest borders, woodland, scrub, adjacent clearings. Black bill, narrow yellowish eye-ring. Grey crown, ***blackish mask***, olive-brown above, ***tawny-buff below, more vivid on breast, greyish patches on chest-sides***. Broad white tail feather tips. Monotypic. Fairly elusive dweller of dense understorey and midstorey, sometimes on exposed perches; singly. **Voice** Descending guttural *ku-ku...kolp-kolp-kolp* (RG). **SS** Grey-capped Cuckoo in W (rich tawny-rufous below, dark grey cap, no mask). **Status** Uncommon austral migrant to E lowlands and highlands (mainly Apr–Dec), apparently breeds in SW lowlands, where uncommon.

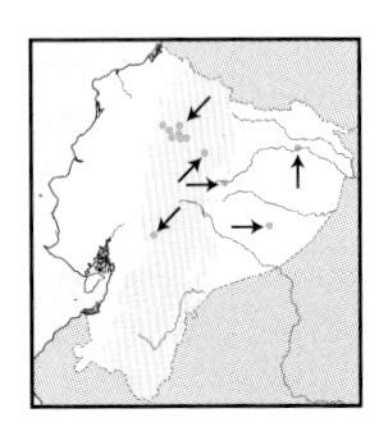

Yellow-billed Cuckoo *Coccyzus americanus* 29–31cm

Boreal transient, few records (to 2,800m). Apparently woodland and scrub. ***Yellow mandible base***, black towards tip. Narrow yellow eye-ring, brown eyes. Greyish-brown above, whitish below, whiter throat, creamier belly. Olive-brown wings, ***rufous fringes in primaries striking in flight***. Tail black below, broad white tips. Possibly nominate and *occidentalis* races. Habits unknown in Ecuador; a sharp decline has seemingly occurred, given number of early specimens (RG); elsewhere, similar to Black-billed Cuckoo. **Voice** Not heard in Ecuador. **SS** Similar but smaller Pearly-breasted Cuckoo (no rufous in wings); no other *Coccyzus* in range has yellow bill. **Status** Very few records (Sep–May).

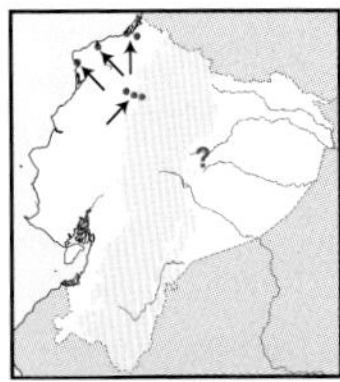

Pearly-breasted Cuckoo *Coccyzus euleri* 25–27cm

Uncertain, few records from N (to 1,300m). Habitat not certainly known, collected once in disturbed forest. Black-and-yellow bill like Yellow-billed Cuckoo. ***Plain olive-brown above, including wings***. Whitish below, ***pearly breast*** (but not very evident in field). Tail has broad white feather tips below. Monotypic. Apparently elusive or maybe casual visitor, poorly known. **Voice** Slow, guttural series of hollow *towlp, towlp, towlp* (R *et al.*). **SS** Very similar to Yellow-billed Cuckoo but ***lacks*** rufous in wings. **Status** Uncertain, probably irregular austral migrant (Mar–Apr, Sep).

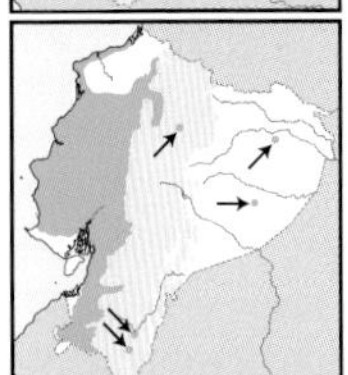

Black-billed Cuckoo *Coccyzus erythropthalmus* 28–29cm

Boreal transient mostly in W lowlands, few in highlands and E (regular to 1,000m, locally 4,000m). Open deciduous forest, woodland, scrub, clearings. ***Bill black, broad yellow eye-ring*** (breeders with red unlikely in Ecuador). Dull brown above, whitish below, creamier vent. ***Tail grey below, narrow whitish tips***. Sexes alike; juv. more ***tawny***-washed below. Monotypic. Fairly inconspicuous while foraging for invertebrates in understorey; might be less elusive on passage. **Voice** Not known in Ecuador. **SS** From other *Coccyzus* by tail pattern, whitish underparts and black bill; cf. Yellow-billed and Pearly-breasted. **Status** Regular in W (Mar–Apr, Nov, Feb), sporadic elsewhere.

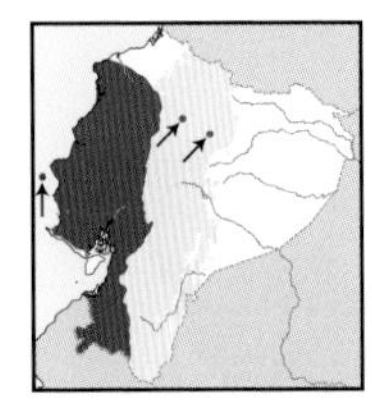

Grey-capped Cuckoo *Coccyzus lansbergi* 27–28cm

W lowlands to foothills, occasionally elsewhere (mostly to 1,300m). Deciduous and semi-deciduous forest, borders, adjacent clearings. Black bill, yellow eye-ring. ***Slate-grey cap, chestnut upperparts, rich buff underparts***. Tail has broad white feather tips below. Monotypic. Understorey and prefer lower strata; breeds during wet season (Jan–Jun) when highly vocal; then migrates. **Voice** Low-pitched, rather fast series of six hollow *coo* notes; guttural *t'uujj* and loud *cháj-cháj-cháj...* calls. **SS** Confusion unlikely but see Dark-billed Cuckoo (black mask, paler below, no cap). **Status** Uncommon breeder, migrates (N?) to more humid foothills.

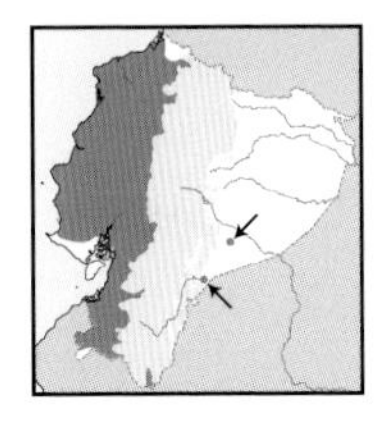

Striped Cuckoo *Tapera naevia* 28–30cm

W lowlands and foothills (to 1,500m). Shrubby clearings, gardens, crops. ***Ferruginous, erectile crest with black streaks, long buffy-white supercilium***, black malar. ***Dense streaking above, dark edgings in wings***. Whitish below, throat and breast tawnier. Nominate. Shy when foraging in dense vegetation or ground but conspicuous when vocalising (crest erect and alula spread); brood parasite. **Voice** Heard more than seen; rising, clear *wuu-weé, wuu-wee-weé, wuu-wuu-wuu-weedee-weedeé* whistles; *duu-deë!* Is common call. **SS** Confusion unlikely in range and habitat; cf. voice of Pale-browed Tinamou. **Status** Fairly common.

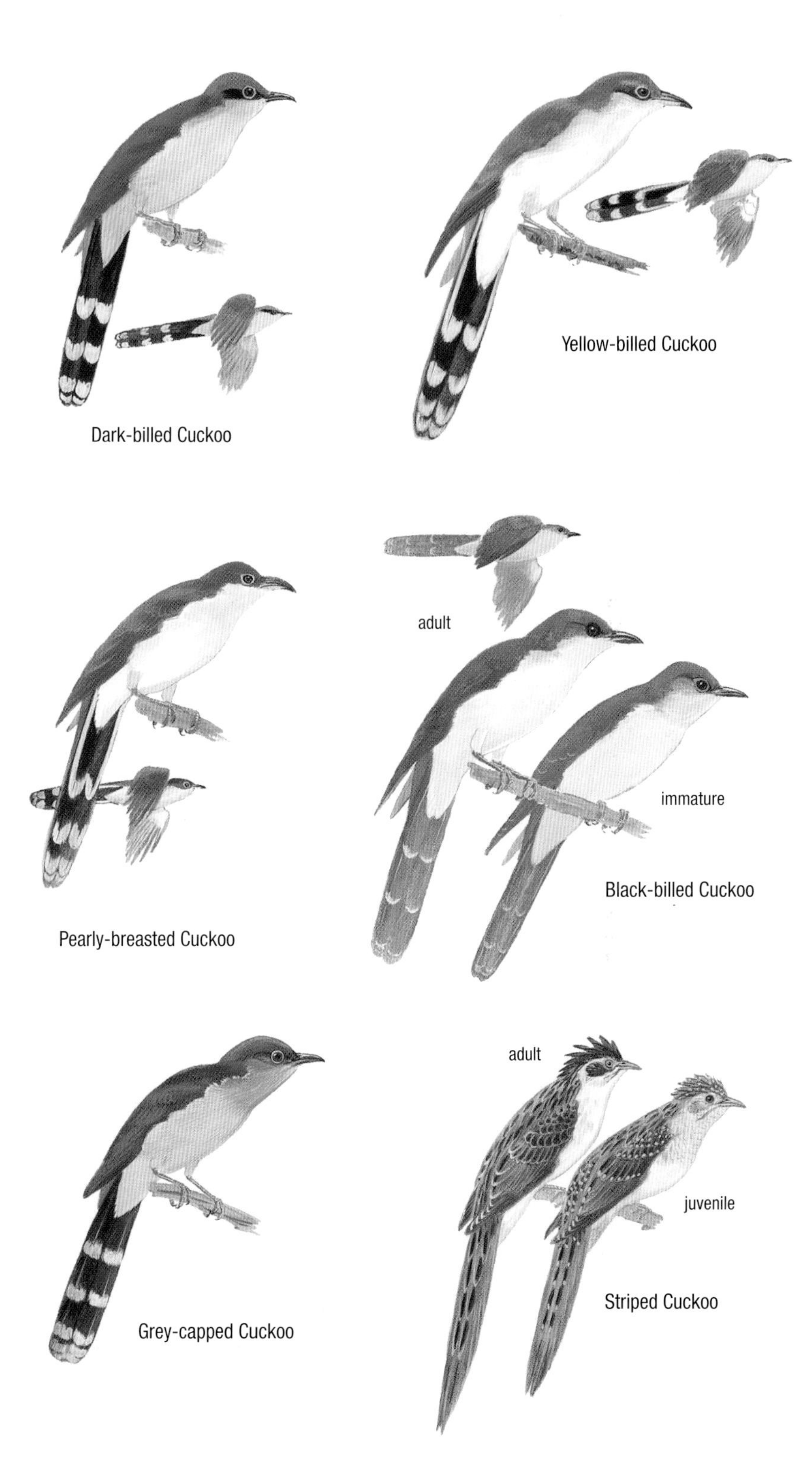
Dark-billed Cuckoo
Yellow-billed Cuckoo
adult
Pearly-breasted Cuckoo
immature
Black-billed Cuckoo
Grey-capped Cuckoo
adult
juvenile
Striped Cuckoo

PLATE 95: HOATZIN, CUCKOOS AND GROUND-CUCKOOS

A group of mostly rare birds. Hoatzin is rather common mostly around undisturbed rivers and lakes, forming weirdly noisy flocks in waterside vegetation. *Dromococcyx* are poorly known, seldom-recorded – even by voice – forest birds whose distributions are still poorly understood. Ground-cuckoos are large terrestrial birds hardly ever observed owing to their shy behaviour. One species endemic to Chocó lowlands is threatened by ongoing deforestation.

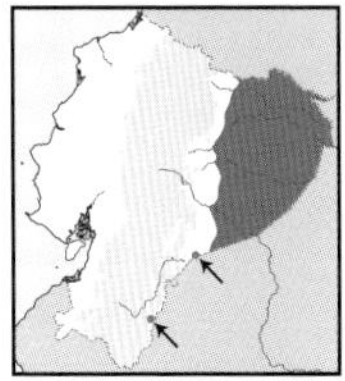

Hoatzin *Opisthocomus hoazin* 60–70cm

E lowlands (to 600m, to 1,000m in Nangaritza Valley). Lagoons and large-river edges. Odd, large, with *long, curly rufous crest, large bluish facial skin. Dusky wings with broad whitish covert edges and rufous remiges, long dusky tail*, broad buff tips. Monotypic. Gregarious and wary, feeds on leaves and buds; heavy hops and bounds, poor flight; alarmed young drop to water, swim and climb through vegetation using unique wing-claws. **Voice** Constant wheezing and moaning; nasal *wruaáj-waj-waj... jlájlájlájlá... kjrrrr!* and other sounds. **SS** Unmistakable. **Status** Locally common.

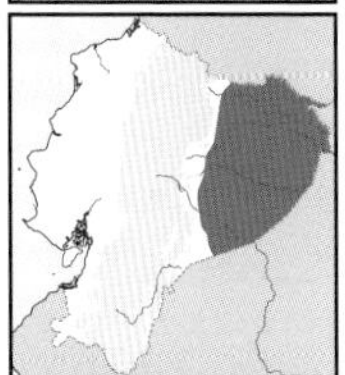

Pheasant Cuckoo *Dromococcyx phasianellus* 37–41cm

E lowlands (to 700m). Mid and low strata of *várzea* and swampy forests. Short, erectile brown-and-rufous crest, *long buffy-white postocular stripe*. Brown above, fine scalloping on mantle, white edging on wing-coverts. *White below, tawny breast, dusky streaking on throat and breast. Long, fan-like graduated tail*, narrow whitish tips below, *very long uppertail-coverts*. Monotypic. Elusive in dense growth, rarely seen (overlooked?); tail spread in flight; brood parasite. **Voice** Heard more often than seen; whistled, trilled *poo.. peh-pëë?, foo, fee-fuëë..., foo-fee-fee'dee*; nasal *dák* in alarm. **SS** Pavonine Cuckoo (less streaking / spotting on breast), voice fairly similar. **Status** Rare, local.

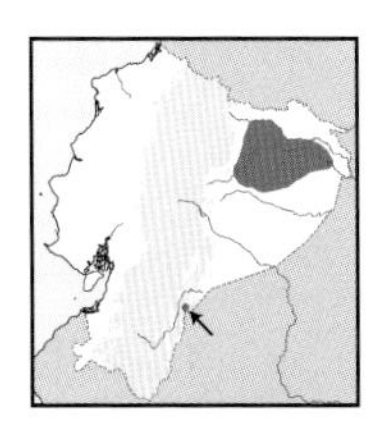

Pavonine Cuckoo *Dromococcyx pavoninus* 28–30cm

E lowlands (400m). Similar to Pheasant Cuckoo, but *smaller*, with short ferruginous erectile crest, *buff postocular stripe*, dark brown above, whitish scalloping and edging on mantle and wing-coverts. *Plain tawny to fulvous throat and breast*, white below. Long graduated tail, fan-shaped, but *uppertail-coverts shorter* than previous species. Monotypic. Skulking; habits and habitat in Ecuador virtually unknown; elsewhere prefers dense thickets, including bamboo; brood parasite. **Voice** Whistled, not trilled *puu-peé..püh-peë'pee* at long intervals. **SS** *Neomorphus* cuckoos terrestrial. **Status** Very few certain records.

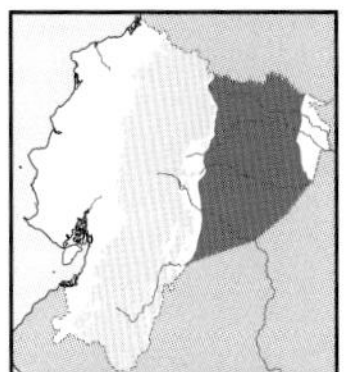

Rufous-vented Ground-Cuckoo *Neomorphus geoffroyi* 46–51cm

E lowlands (to 400m). *Terra firme* forest. Large, long-tailed, with *pale yellow bill, bluish orbital skin*, pointed crest. *Olive-brown head*, paler throat, more bronzy upperparts, purplish rump. *Narrow black band separates throat from buffy-brown breast*, bronzy flight feathers and tail. Race *aequatorialis*. Elusive, terrestrial in primary forest; singles, pairs or small groups at army ant swarms or follow peccary herds or monkey troops; swift runs, then stops, raising and lowering crest. **Voice** Low-pitched, moaning, long *wooooooOOp*; plus bill snaps. **SS** Unique. **Status** Rare and local; VU globally.

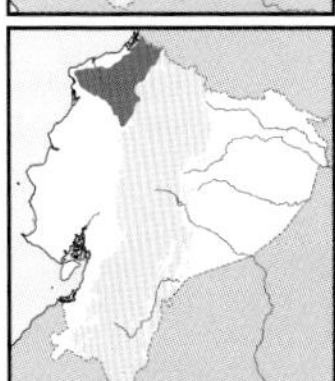

Banded Ground-Cuckoo *Neomorphus radiolosus* 46–51cm

Lowlands and foothills in NW (to 1,000m). Humid primary forest, rarely at borders. Black bill, *bluish orbital skin. Glossy black head, black mantle and underparts with whitish to buffy scalloping*. Chestnut wings and rump, greenish-black tail. Monotypic. Poorly known; singles, pairs or trios probably follow ant swarms and peccary herds; recently discovered nest placed 4–5 m above ground, a large, open cup of leaves; both ad. attend nest. **Voice** Bill snaps, *ták!*, also very deep *boooooo* every 3–5 sec or more. **SS** Unmistakable in range. **Status** Rare and local; EN.

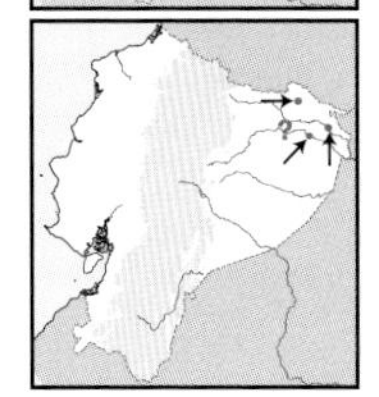

* Red-billed Ground-Cuckoo *Neomorphus pucheranii* 46–51cm

NE lowlands (200m). Few uncorroborated records from *terra firme* forest. *Red bill, red-and-blue orbital skin, glossy blue crest*. Bronzy-olive to brown above, purplish rump; *ashy-grey below with broad black breast-band* and rufous vent. Race uncertain, possibly nominate. Virtually unknown in Ecuador, but habits probably similar to other *Neomorphus*. **Voice** Short, deep hums, *wooOp.. wooOp..wooOp* at short intervals; *tek!* bill snaps. **SS** From Rufous-vented by bill, crest and throat. **Status** Uncertain, but possibly very rare and local; NT in Ecuador.

Hoatzin
Pheasant Cuckoo
adult
juvenile
Pavonine Cuckoo
adult
juvenile
Rufous-vented Ground-Cuckoo
juvenile
immature
immature
adult
adult
Red-billed Ground-Cuckoo
juvenile
Banded Ground-Cuckoo

Some distinctive swifts, along with three *Cypseloides*, all rare and very difficult to tell apart. Several swift species tend to flock, larger species often in higher airspace. Compare to bull-headed, shorter-tailed *Chaetura* (wings are butter-knife shaped).

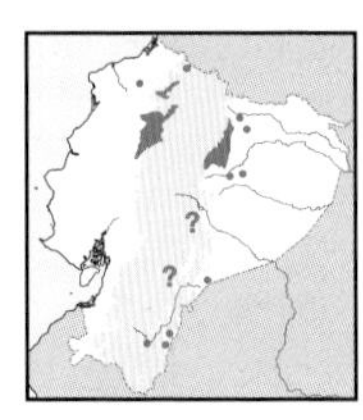

Spot-fronted Swift *Cypseloides cherriei* 14cm

Locally on N Andean slopes (500–2,050m). Above wet montane forest. ***White loral and postocular spots***; juv. has ***whitish vent scaling***. Monotypic. Nests on cliffs behind waterfalls. Flight: veering glides on wings below horizontal, swift banks and fast flapping, high in mixed flocks. **Voice** High-pitched, nasal chipping *pihk*, *pi-pihk*, accelerating into more rhythmic *chirr-pik-ti-ti-ti-ti-chirr*.... **SS** White-chinned Swift lacks facial spots; otherwise very similar (tail proportionately shorter, bat-like flight); juv. not safely identified; Chestnut-collared Swift (longer, notched tail). **Status** Rare, local, overlooked? DD globally.

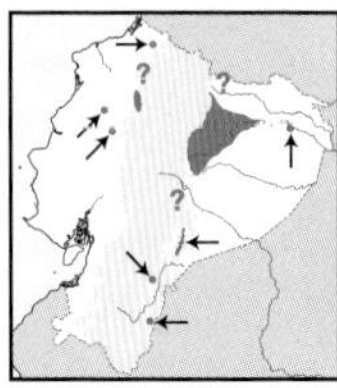

White-chinned Swift *Cypseloides cryptus* 15cm

Local in N lowlands and foothills (below 2,000m). Above humid forest. Variable ***small white chin patch***. Juv. has whitish vent scaling. ***Square tail proportionately shorter*** than previous species. Monotypic. Apparently breeds in foothills and disperses into lowlands to feed; tends to keep higher when flocking with other swifts; direct, bat-like flight, continuous flapping. **Voice** Possibly fast, musical chirps and clicks. **SS** cf. Spot-fronted (tail might look longer, shorter wingspan, more banking flight); also Chestnut-collared (longer, notched tail). **Status** Rare, local, overlooked?

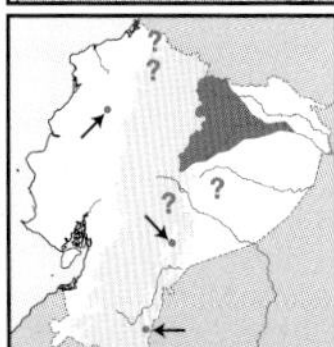

White-chested Swift *Cypseloides lemosi* 14cm

Local in E lowlands to subtropics (250–2,250m). Mainly above forest. ***White triangle on chest; tail long, obviously forked***. ♀ and juv. have fainter chest patch, ***fish-scaled underparts***. Monotypic. Apparently disperses to lowlands to feed; often with other swifts. **Voice** Series of steady-paced *chip-chip-chip*...; also faster, mellower *chi-chi-chi-chi*.... **SS** White-collared Swift (larger white collar, broader-winged), Chestnut-collared (darker, notched tail), silhouette, voice and flight recall Black Swift (not yet in Ecuador). **Status** Rare, overlooked?

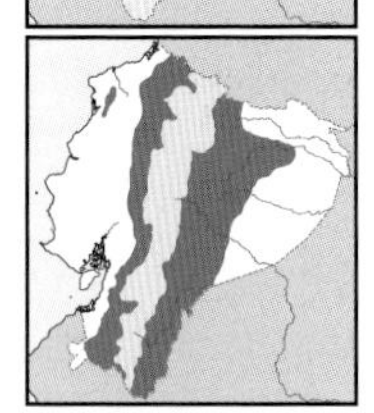

Chestnut-collared Swift *Streptoprocne rutila* 13cm

Andean slopes, inter-Andean valleys, locally to lowlands (500–2,700m). Above montane forest, clearings. Blackish-brown, ***chestnut collar down to chest patch; tail long, notched***. Chestnut less extensive in ♀ and juv, some even lacking. Race *brunnitorques*. High-flying flocks, often with White-collared and others; disperses widely. Flight fast, with permanent banking and veering, fast flapping, wings below horizontal. **Voice** High-pitched *bzzt* chatters, almost insect-like; dry *t-t-ts-ts-ts*... **SS** Chestnut collar not always visible; cf. shorter, less notched tail of blacker *Cypseloides*. **Status** Uncommon.

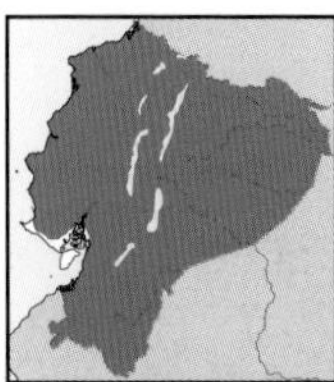

White-collared Swift *Streptoprocne zonaris* 20–22cm

Widespread in highlands and lowlands (to 4,000–4,200m). Above forested and open areas. Large, ***long-winged, conspicuous white collar***, broader on chest; tail long, slightly forked. Less extensive, scalier white in juv. Race *altissima* (highlands), *subtropicalis* (lower elevations), contact zone unknown. Small to large flocks, noisy, often very high overhead. Flight fast, flaps deep and floppy, often soaring. Disperses long distances when feeding, outnumbers other swifts in flocks. **Voice** Noisy clicks, squeaks, trills, buzzes and chatters. **SS** Largest swift, but cf. White-chested. **Status** Common.

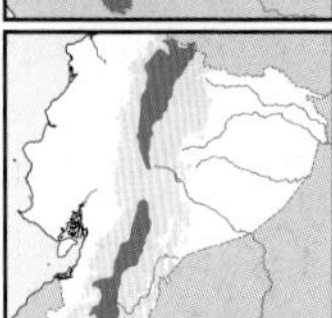

White-tipped Swift *Aeronautes montivagus* 13cm

Andean highlands to subtropics, inter-Andean arid valleys (1,300–2,700m, locally higher). Mostly over open dry areas. ***Contrasting white bib, whitish band on lower belly***. Tail long, ***forked, tipped white***. ♀ and juv. have less extensive white on throat, dark tail tips. Nominate. Very fast-flying flocks, not often with other swifts, but tend to occupy lower airspace when in flocks. **Voice** Fast buzzing trills and squeaks. **SS** Overlaps with fairly similar Lesser Swallow-tailed Swift unlikely. **Status** Uncommon.

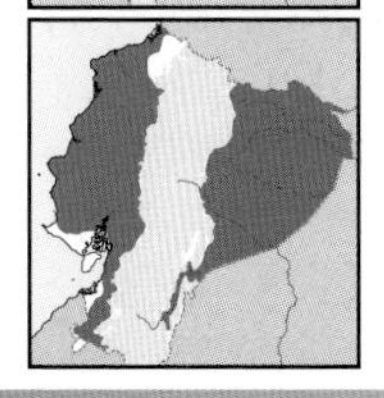

Lesser Swallow-tailed Swift *Panyptila cayennensis* 13cm

E & W lowlands and foothills (to 900m). Mainly above open areas. ***Contrasting white collar and throat, white flank tufts***. ***Tail long, deeply forked*** (open) or pointed (closed). ***Wings long, narrow, swept back***. Nominate. Mostly in pairs, loosely associated with other swifts, tending to occupy higher airspace. Flight fast and acrobatic, shallow wingbeats, much ***gliding***, fluttering wingbeats. **Voice** Thin, high-pitched chatters, *chit'tititit*. **SS** Fork-tailed Palm-Swift less bold, duller below. **Status** Uncommon.

Spot-fronted Swift
White-chinned Swift
adult
juvenile
White-chested Swift
adult
juvenile
Chestnut-collared
Swift
♂
♀
juvenile
adult
juvenile
White-collared Swift
adult
juvenile
White-tipped Swift
Lesser Swallow-tailed Swift

One of the most difficult genera in the Neotropics; *Chaetura* are not always safely identified to species. Subtle differences, fast flight, and light conditions all complicate identification. Note tail length, proportions, throat and rump patterns; carefully study every *Chaetura* seen. Fork-tailed Palm-Swift is slender and distinctive. When flocking, larger species tend to occupy higher airspace.

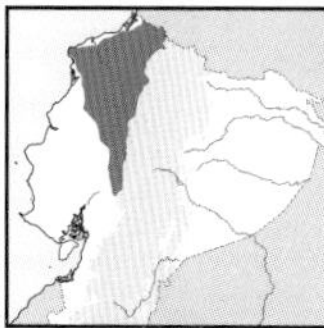

Band-rumped Swift *Chaetura spinicaudus* 11cm

NW lowlands and foothills (50–1,500m). Above humid forest. ***Narrow white band on rump***, darker uppertail-coverts. ***Dark*** underparts, ***whitish throat to upper chest***. Race *aetherodromus*. Flocks of 20–30 birds high overhead, lower when associated with other swifts, sometimes very low over water. Flight fluttery, little gliding. **Voice** Varied squeaks and twitters, *tsoo-si-si-si...* (ChD). **SS** Grey-rumped Swift (little contrast on rump, glossy upperparts, more uniform underparts). **Status** Uncommon.

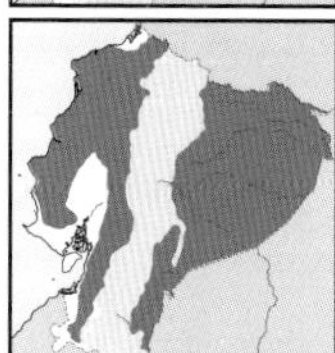

Grey-rumped Swift *Chaetura cinereiventris* 11cm

E & W lowlands to subtropics (to 1,700m, lower in W). Above humid forest, adjacent clearings. ***Glossy*** black above, ***pale grey rump and uppertail-coverts*** (***darker*** in W). Mainly brownish grey below. Races *occidentalis* (W), *sclateri* (E). Often in large flocks with other swifts and swallows; occupy lower airspace in mixed flocks. **Voice** Rapid, high-pitched, insect-like twitter: *cherr-cher-cheeeer....* **SS** In W Band-rumped Swift (narrow white rump); in E Pale-rumped (whiter rump); chunkier Short-tailed (dark rump, fastest flight, quickest wingbeats); see Chapman's. **Status** Uncommon.

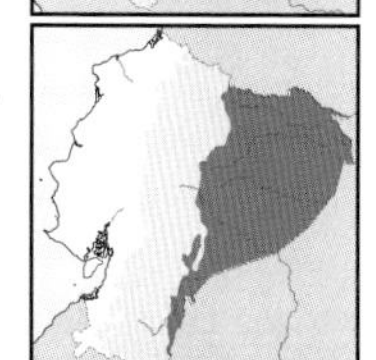

Pale-rumped Swift *Chaetura egregia* 11cm

Local in E foothills and adjacent lowlands (200–1,000m, once 1,800m). Above humid forest and clearings. ***Bronzy***-black above, ***whitish-grey rump and uppertail-coverts***. Sooty below, ***slightly paler throat***. Monotypic. Often with other swifts, especially Grey-rumped, lower in mixed flocks. Tends to glide more, wingbeats deeper. **Voice** Fast chippering. **SS** Appears longer-bodied, ***longer-winged*** than Grey-rumped, more contrasting rump; from Chimney, Chapman's and Short-tailed by pale rump. **Status** Rare, overlooked?

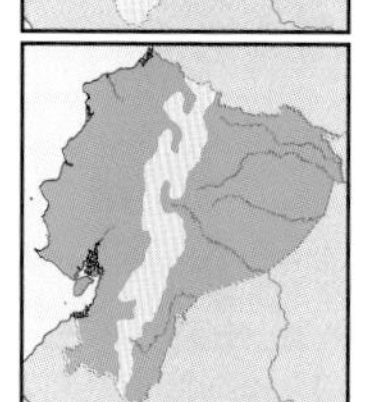

Chimney Swift *Chaetura pelagica* 13–14cm

Scattered records in lowlands and highlands (below 1,000m, once 3,200m). Open areas. Blackish-brown above, ***little contrast on rump***. Underparts sooty-grey, ***paler throat***. Monotypic. Poorly known in Ecuador, few records might reflect identification difficulties rather than rarity; flight more fluttering than *Cypseloides*; often low in mixed flocks. **Voice** Chatter of loose, chippering twitters reported (S *et al.*), but barely vocal. **SS** Difficult to separate from Chapman's (more contrasting rump, darker underparts, glossier above); Short-tailed (more compact, 'tail-less'); also White-chinned (broader wings). **Status** Rare boreal winter transient (Oct–Mar); NT globally.

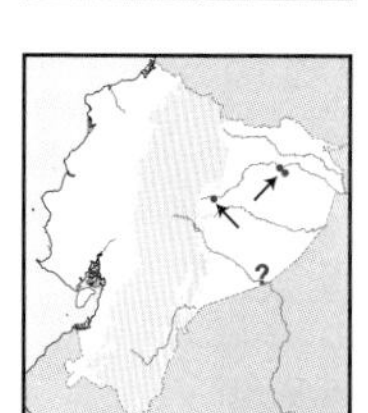

* Chapman's Swift *Chaetura chapmani* 13–14cm

Very few uncorroborated records in E lowlands at *c.*400m. Mostly sooty ***glossy*** black, few field marks; ***slightly paler rump*** and ***uniform*** underparts, ***little to no contrast on throat***. Monotypic. Poorly known in Ecuador; elsewhere often with other swifts; reportedly glides more than other *Chaetura*. **Voice** Almost ticking *che-e-e-e-e-ed* (ChD) **SS** Not safely separated from Chimney Swift without careful study (looks more ***capped*** due to paler throat); Short-tailed (darker below except pale undertail-coverts, much shorter tail); Grey-rumped (paler underparts and rump). **Status** Uncertain, wanderer?

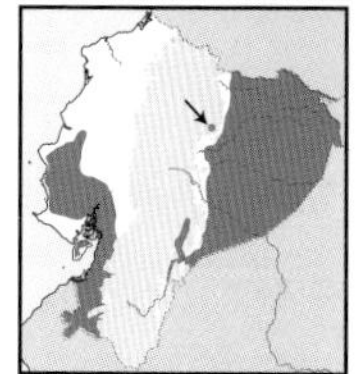

Short-tailed Swift *Chaetura brachyura* 10–11cm

E & SW lowlands and foothills (to 1,000m, locally higher). Above clearings, rivers, forest edge (E); deciduous forest (SW). Small, ***stubby tail***, broad wings, ***bulging inner primaries***. ***Pale grey rump and uppertail-coverts***. Dark underparts, ***pale undertail-coverts*** (E) or ***paler throat and undertail-coverts*** (W). Possibly race *cinereocauda* (E); *ocypetes* (SW) might be full species. Often with other swifts; flight more flappy, slower and unstable than other *Chaetura*. **Voice** Fast, high-pitched *chip-chip-chip-chirirp-chirp...*; very fast, insect-like *weez-zi-zi-zi....* **SS** cf. larger Chapman's and Chimney. **Status** Uncommon.

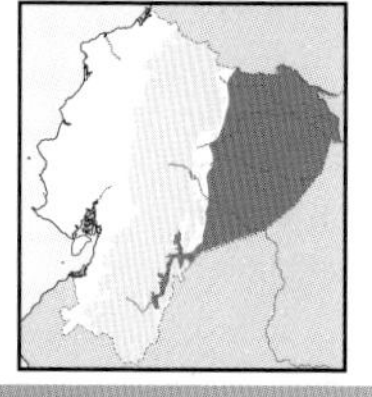

Fork-tailed Palm-Swift *Tachornis squamata* 13cm

E lowlands, locally into foothills (to 1,000m). Above forested areas, palms and waterside clearings. Slender with ***very long, deeply forked tail*** (pointed when closed). Blackish above, paler feather edging ***gives scaled appearance***. ***Whitish below***, sides and flanks ***scaled grey and white***. Race *semota*. Flocks often near water, rarely with other swifts. Flight fast with shallow, very fast wingbeats. **Voice** Buzzy, *dzzii...dzzii-zziit-zzii....* **SS** Confusion unlikely. **Status** Fairly common.

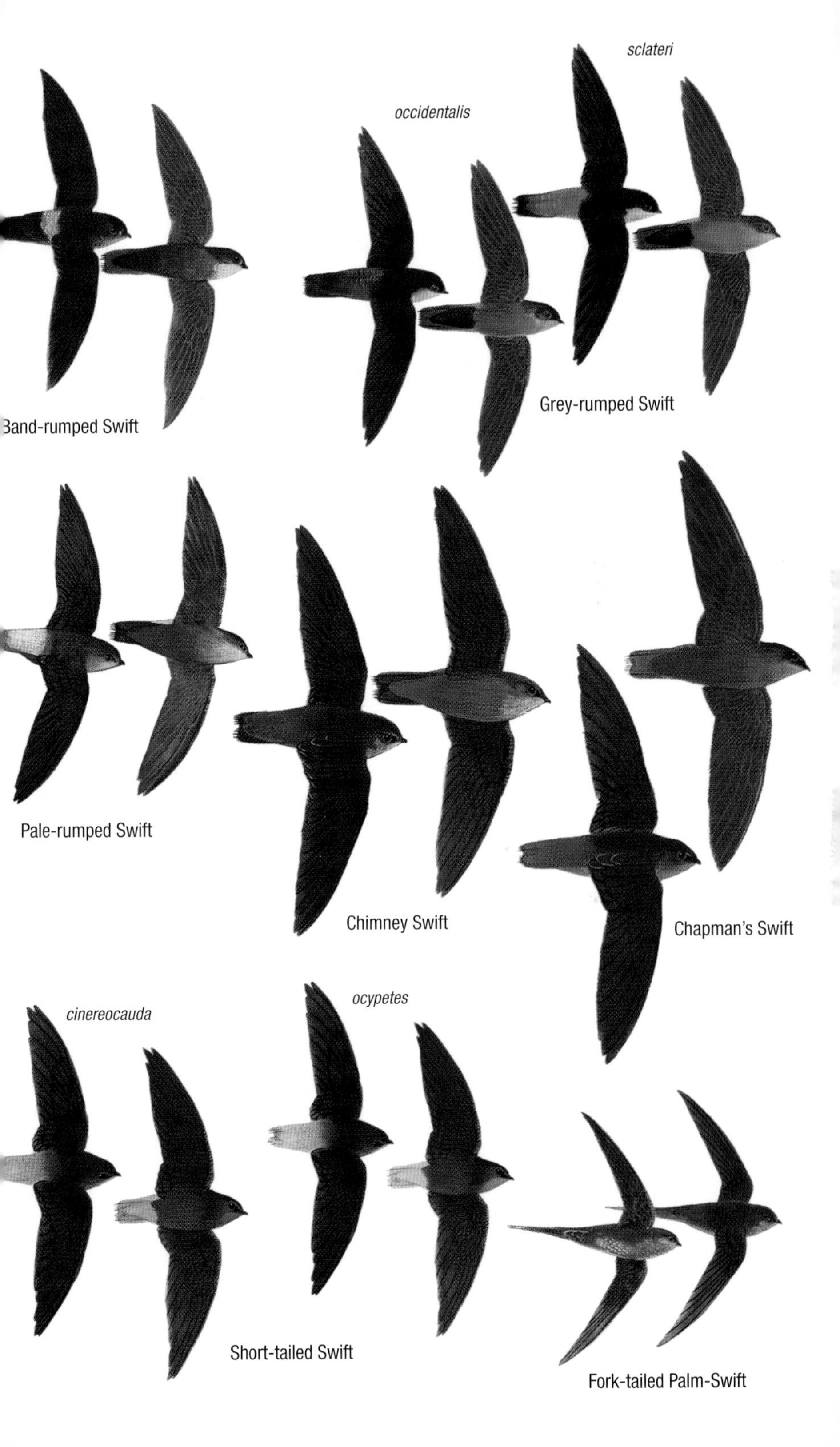
sclateri
occidentalis
Grey-rumped Swift
Band-rumped Swift
Pale-rumped Swift
Chimney Swift
Chapman's Swift
cinereocauda
ocypetes
Short-tailed Swift
Fork-tailed Palm-Swift

Glaucis and *Threnetes* recall *Phaethornis* hermits, but bills shorter, tail shorter and rounded. Sicklebills unique among hummers for their long and striking bill; note some plumage similarities with Tooth-billed Hummingbird. All are confined to forests and mature woodland.

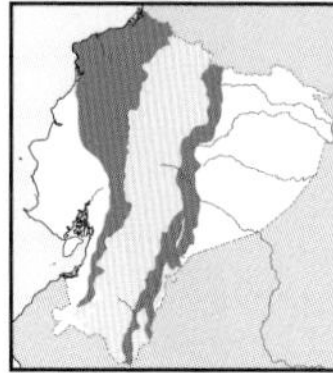

White-tipped Sicklebill *Eutoxeres aquila* — 12–14cm

W lowlands to subtropics, E Andean slope (100–1,900m; wanders to 2,500m). Undergrowth of wet montane forest, borders and woodland. Long (3.5cm) ***strongly decurved bill***. Dark metallic green above, ***whitish below densely streaked black. Greenish-black graduated tail tipped white***. Races *heterurus* (W), nominate (E). Singly in dense shady undergrowth, near *Heliconia* stands, rather territorial and elusive. **Voice** Piercing *tist* flight calls; song complex squeaky chatter: *tsi-see'ee'ee'eek-tsi-eek-tsi'ee'ee...*, simpler in W: *tsi-sri-sík*. **SS** See Buff-tailed Sicklebill in E. **Status** Uncommon.

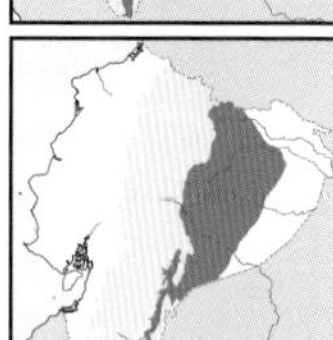

Buff-tailed Sicklebill *Eutoxeres condamini* — 12–14cm

E lowlands to foothills (300–900m, locally higher). Mature forest undergrowth, borders, older woodland. Long (3.5cm) ***strongly decurved bill***. Dark metallic green above, ***shining blue patch*** on neck-sides. ***Faint whitish superciliary and malar stripes. Whitish-buff below densely streaked black***. Tail mostly ***buff narrowly tipped white***. Nominate. Shy, often near shady streams, associated with *Heliconia* stands, clings to feed; trap-liner. **Voice** Loud, ringing *swee'p-ee-seé* in endless rhythmic series. **SS** From White-tipped by tail colour. **Status** Uncommon.

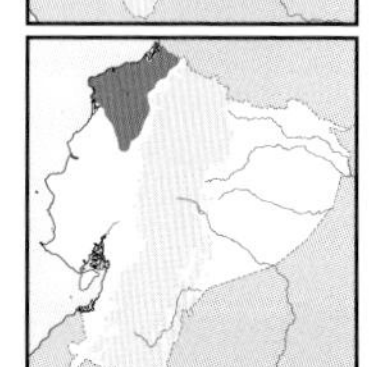

Bronzy Hermit *Glaucis aeneus* — 8–9cm

NW lowlands (to 600m). Primary and secondary forest undergrowth and borders. Bill long (3–3.3cm), decurved. Bronzy-green above, ***cinnamon-buff below***, short buff supercilium, ***whitish malar, dusky sub-malar***. Tail short, ***rounded, chestnut basally, black distally, tipped white***. Race *columbianus*. Solitary trap-liner near ground, not easy to see, shier than *Phaethornis*. **Voice** Sharp *tzeét-tzeét* in flight, fast trill *tzee-zee'ee'ee...* when excited. **SS** No hermits in range with similar tail shape and underparts. **Status** Uncommon.

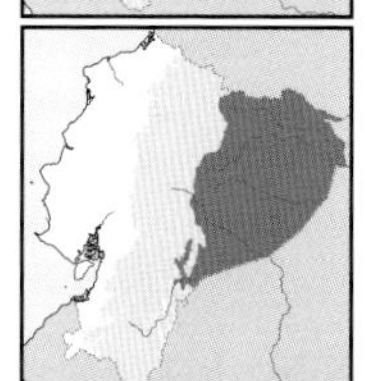

Rufous-breasted Hermit *Glaucis hirsutus* — 9–10cm

E lowlands to foothills (to 1,100m). *Terra firme* and *várzea* forest, borders, woodland, adjacent clearings. Bill long (3–3.3cm), decurved, mandible mostly yellow. Bronzy-green above, ***cinnamon-buff below***. Short buff supercilium, ***whitish malar stripe. Rounded tail, chestnut basally, black distally, tipped white***. ♀: richer below. Nominate. Prefers vicinity of water, also undergrowth and *Heliconia* stands. **Voice** Very fast, high-pitched, ascending *chéet'éet'éet'éet'eET'EET...*; sharp *sweep!-sweep!* in flight. **SS** Reddish and Black-throated Hermit (see face and tail pattern). **Status** Uncommon.

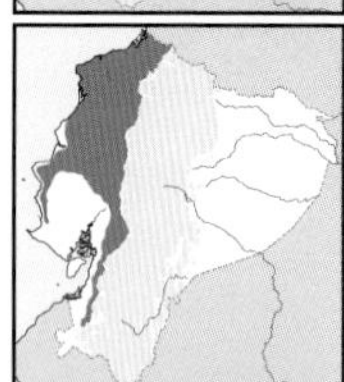

Band-tailed Barbthroat *Threnetes ruckeri* — 10–11cm

W lowlands and foothills (to 900m). Humid forest, borders, woodland. Long decurved bill (3cm), mandible mostly yellow. Bronzy-green above. Blackish throat, ***cinnamon breast-band, greenish-grey below***. Buffy-white postocular stripe, broad whitish malar. Tail ***rounded, white base, bluish-black narrowly tipped white***. Nominate. Forest undergrowth, more retiring than other hermits, often trap-lines *Heliconia* stands. **Voice** High-pitched, fast, repeated *tzi-tzi-tzi...tzi'tzi'tzi'zizizizi...*. **SS** Confusion unlikely given breast-band and tail pattern; cf. White-whiskered Hermit. **Status** Uncommon.

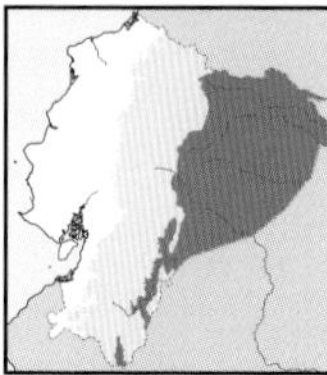

Pale-tailed Barbthroat *Threnetes leucurus* — 10–11cm

E lowlands to foothills (to 1,600m). *Terra firme* and *várzea* forest, borders and woodland. Long decurved bill (3cm), ***mandible mostly grey***. Bronzy-green above. Blackish throat, ***broad cinnamon breast-band, grey belly***. Buffy postocular stripe, whitish malar. Tail ***rounded, buffy-white, dusky subterminal band, narrow white tip***. ♀: duller. Race *cervinicauda*. Trap-lines in dense undergrowth, difficult to see; ♂ displays on low perch. **Voice** Thin, rhythmic *zee-zee-zee'ee-ee'ee, zee-zee'ee-ee...* in display; very thin *tsíp* in flight. **SS** Confusion unlikely. **Status** Uncommon.

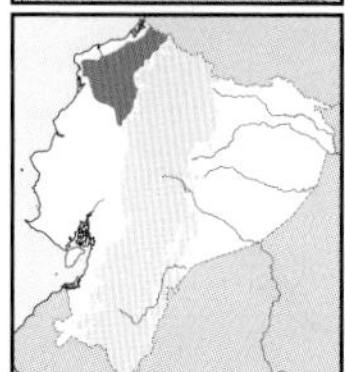

Tooth-billed Hummingbird *Androdon aequatorialis* — 10–11cm

NW lowlands to foothills (300–800m). Humid forest interior. Long *very straight bill* (4–4.2cm), yellow mandible base. Coppery-brown crown, ***white uppertail-coverts***. Greyish-white below, ***boldly streaked black to upper belly***. Long rounded tail, grey basally, ***bluish-black distally, broadly tipped white***. ♀: lacks coppery crown. Monotypic. Low to midstorey inside forest where trap-lines, also probes for spiders and small insects in dead leaves; displaying ♂ incessantly calls and wags tail. **Voice** High-pitched, rhythmic *tslE-tsét!, tslE-tsii-sút* in display. **SS** Unmistakable. **Status** Rare, NT in Ecuador.

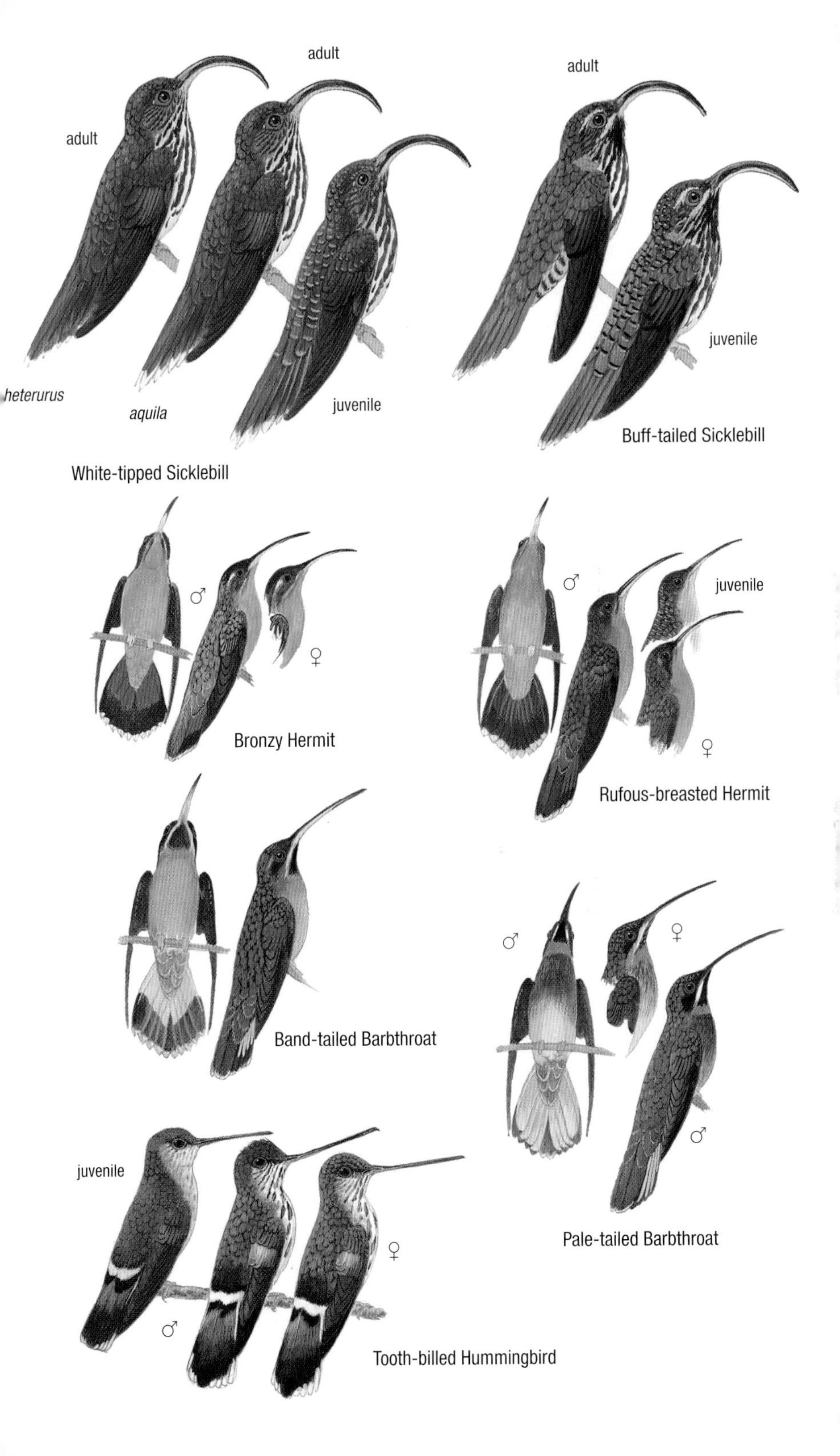
adult
adult
adult
heterurus
aquila
juvenile
juvenile
Buff-tailed Sicklebill
White-tipped Sicklebill
♂
♀
Bronzy Hermit
♂
juvenile
♀
Rufous-breasted Hermit
Band-tailed Barbthroat
♂
♀
♂
Pale-tailed Barbthroat
juvenile
♂
♀
Tooth-billed Hummingbird

Topazes are large and astonishing Amazonian hummingbirds of sluggish-flowing river edges, one definitely *not* found in Ecuador. The small *Phaethornis* hermits are shy understorey dwellers of forested habitats; less curious than their large congeners, yet also approach observers. Though plumages might cause confusion, they tend to replace each other geographically.

†Crimson Topaz *Topaza pella* 18cm

Formerly included in the Ecuadorian avifauna based on old specimens presumably collected more than 100 years ago from Río Suno and described as race *pamprepta*. A recent taxonomic study of *Topaza* concluded that *pamprepta* is the same taxon as *smaragdulus* from French Guiana and that the supposed Ecuadorian specimens were wrongly labelled. *T. pella* is thus removed from the Ecuadorian bird list.

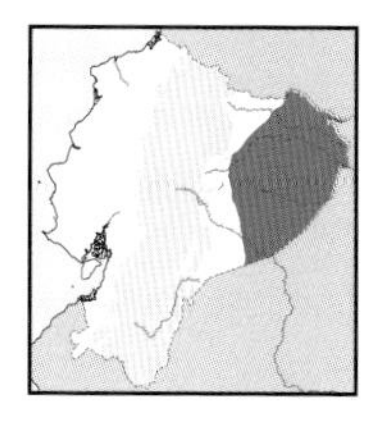

Fiery Topaz *Topaza pyra* 15–16cm

E lowlands (to 400m). Riverside vegetation in *várzea* and *igapó* forest, and forested streams. Spectacular ♂ has ***velvet black hood***, sparkling ***emerald-and-gold throat patch, fiery body, blackish thighs***, green vent, golden-green rump, purplish-black tail with ***two elongated and distally crossed feathers***. ♀: smaller, mostly bright green, ***fiery-orange throat patch***, long ***blackish tail, outer feathers half buff***. Buff underwing-coverts. Race *amaruni*. Often seen fly-catching insects or robbing spider webs, rarely at flowers, in riparian habitats, where territorial. **Voice** Sharp, almost ringing *zreép-zreép-zer'zer'zer'er..zreép-zreép'zer'er'errr....* **SS** Unmistakable. ♀ with Gould's Jewelfront. **Status** Rare, local, NT in Ecuador.

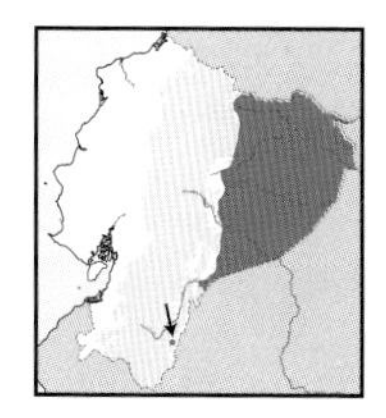

Black-throated Hermit *Phaethornis atrimentalis* 8–9cm

E lowlands to lower foothills (to 900m). Forest undergrowth, borders. Bill 2–2.5cm, slightly decurved, ***yellow mandible base***. Coppery-green upperparts, rufescent rump. Two distinctive whitish stripes border blackish mask. ***Dusky throat, dusky-rufous below***, paler vent. Blackish tail tipped white, with elongated central feathers. Sexes almost alike. Nominate. Singly, mostly at edges, trap-liner, also takes insects or spiders from leaves and twigs; small loose leks. **Voice** High-pitched *tss-tssi-tssi-tseeu, tssi-tssi-tse'seeu'seU* in display; thin *tsl!* in flight. **SS** Reddish and Grey-chinned Hermits lack dusky throat but otherwise similar. **Status** Rare.

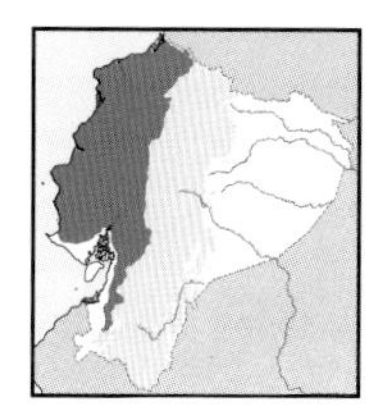

Stripe-throated Hermit *Phaethornis striigularis* 8–9cm

W lowlands to lower foothills (locally to 1,570m). Humid and deciduous forest interior and borders, also woodland. ***Small***. Bill 2–2.5cm, slightly decurved, ***yellow mandible basal half***. Dark coppery-green above, rufescent rump. Black mask, buffy-white superciliary and malar stripes. Pale greyish-cinnamon below, ***greyer throat and breast, throat finely streaked dusky***. Blackish tail tipped white, elongated central feathers broadly tipped white. Race *subrufescens*. Trap-liner in lower growth, often near ground, shy, often chased away by larger hermits; loose leks. **Voice** Long, jumbled, high-pitched *tsiit-sii-sii'sii'sii...* in squeaky series. **SS** Only small *Phaethornis* in most of range; see Grey-chinned Hermit in W Loja (no known overlap). **Status** Uncommon.

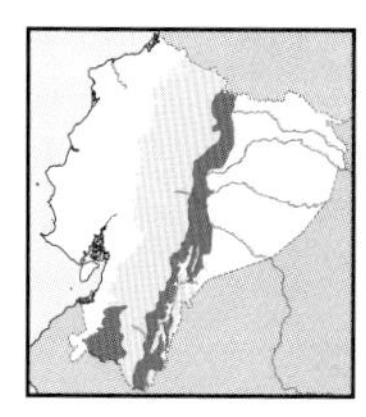

Grey-chinned Hermit *Phaethornis griseogularis* 8–10cm

E Andean slope, S Andes (600–1,700m in E; 900–2,000m in S). Humid forest undergrowth, borders. Bill 2–2.5cm, slightly decurved, ***yellow mandible base***. Coppery-green above, rufescent rump. Buff supercilium, ***narrow whitish malar stripe***. Rich cinnamon-buff below, ***narrow grey throat stripe***. Bronzy-green tail ***narrowly tipped buff***, elongated central feathers broadly tipped white. Nominate (E), *porcullae* (W); ***paler*** below, ***lateral*** tail feathers tipped white, ***black collar***, probably different species. Trap-liner in lower growth, elusive, often near ground; loose leks. **Voice** Gives *suee-swit-swit-sweeít...* squeaks in display; thin *tsui* in flight. **SS** Lower-elevation Black-throated Hermit (dusky-striped throat) and more richly coloured Reddish Hermit. **Status** Locally uncommon.

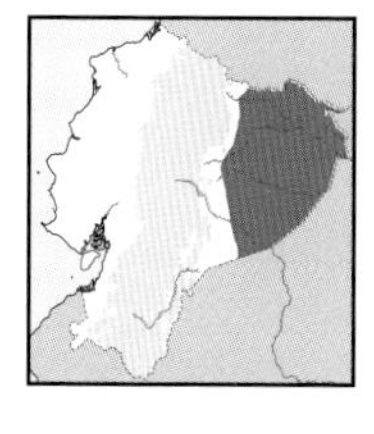

Reddish Hermit *Phaethornis ruber* 7–8cm

E lowlands (to 400m). Mostly *terra firme* undergrowth, borders, woodland. Bronzy-green upperparts, rufescent rump. Bill 2–2.5cm, slightly decurved, ***yellow mandible base***. Narrow buffy-white superciliary and malar. ***Cinnamon-buff below, narrow black band on lower breast***. Coppery tail ***narrowly tipped buff***, central feathers slightly longer, tipped whitish, lateral feathers tipped rufescent. ♀: ***paler below, no black breast-band***. Race *nigricinctus*. Trap-liner in forest undergrowth, with weaving fast flight; loose leks; flight recalls a coquette. **Voice** Descending, thin *zeé-zeé-zeé-zee'zeeuu'zeeuu'zeeu....* **SS** Richer coloured than sympatric Black-throated Hermit. **Status** Locally uncommon.

♂
♀
juvenile
Crimson Topaz
♂
♂
♀
♀
Black-throated Hermit
Stripe-throated Hermit
Fiery Topaz
griseogularis
porcullae
Grey-chinned Hermit
♂
♀
Reddish Hermit

PLATE 100: LARGE HERMITS

All large hermits, except Straight-billed, have long curved bills, colourful mandibles, striped faces, and long central tail feathers. Though mainly forest-based, some might inhabit woodland and adjacent wooded clearings. Fond of *Heliconia* and other flowers with curved corollas. All curious around observers. Straight-billed has similar plumage but a unique – in Ecuador – needle-like bill.

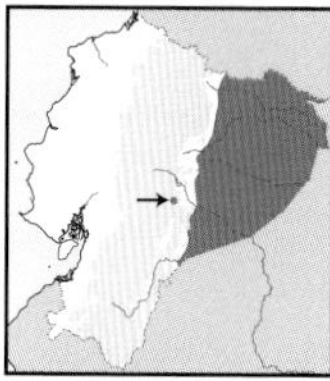

White-bearded Hermit *Phaethornis hispidus* 12–13cm

E lowlands (locally to 1,000m). *Várzea* undergrowth, riparian woodland, river islands. Long (3–3.3cm), decurved bill, ***yellow mandible tipped black***. Bronzy-green above, ***greyer rump***. Blackish mask, ***short white supercilium*** and malar, ***greenish-grey below, white mid throat***. Monotypic. Trap-liner near ground; small loose leks in dense thickets. **Voice** Evenly-pitched, fast, endless *tsíp-tsíp-tsíp...* in display; thin *ziip-ziip* in flight. **SS** Greyer and duller than Great-billed Hermit (more rufescent rump) and Straight-billed Hermit (no white beard, straight bill). **Status** Uncommon.

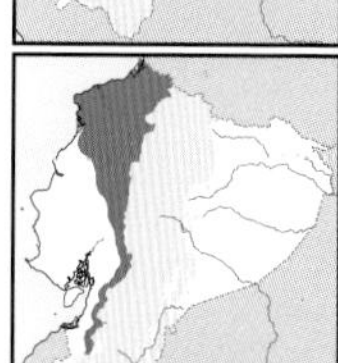

White-whiskered Hermit *Phaethornis yaruqui* 12–13cm

W lowlands and foothills (to 1,400m). Undergrowth of humid forest, woodland, dense gardens, borders. Long (4–4.5cm) bill, decurved, ***red mandible tipped black. Dark metallic green above, buff supercilium*** and ***whitish-and-buff malar stripe. Metallic green below***, greyer belly, ***white vent***, grey mid-throat stripe. Monotypic. Singly, trap-liner near ground, often approaches observers; dispersed leks. **Voice** Endless, nasal *shee-u, shee-u...*; sharp *diiik-dik-dik* in flight. **SS** Only green hermit in W. **Status** Common.

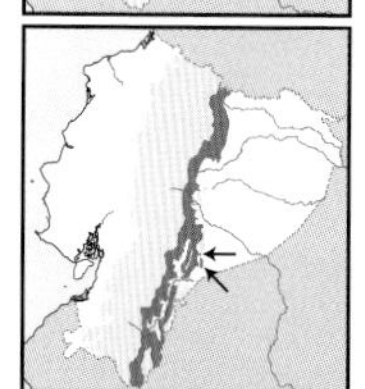

Green Hermit *Phaethornis guy* 11–12cm

E Andean foothills to subtropics (900–1,800m). Montane forest undergrowth, borders. Long (4–4.2cm) bill, decurved, ***red mandible tipped black. Dark metallic green above***, bluer rump; ***buff superciliary and malar stripes***. Greenish-***grey*** below, buffier belly, ***buff mid throat***. ♀: ***greyer below***, whiter tail. Race *apicalis*. Trap-lines low in forest; leks often large and compact. **Voice** Nasal, metallic, endless *whee-t*, *c.*1 every second; *sweep!* in flight. **SS** Only green hermit in range; cf. lower-elevation Great-billed Hermit (duller overall) and Pale-throated Barbthroat (shorter-billed, rounded tail). **Status** Uncommon.

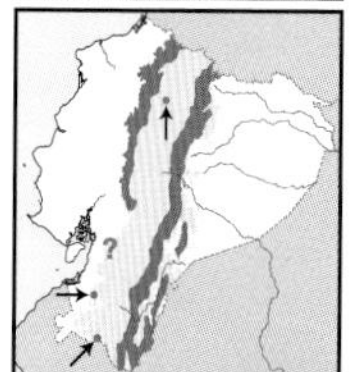

Tawny-bellied Hermit *Phaethornis syrmatophorus* 12–13cm

Subtropical to low temperate zones on both Andean slopes (900–2,500m, locally to 3,000m). Montane forest undergrowth, borders, woodland. Long (4–4.2cm) decurved bill, ***red mandible tipped black***. Bronzy-green above, ***ochraceous rump. Buff postocular stripe***, dusky sub-malar. ***Rich tawny-buff below***, whitish mid throat. Long tail ***tipped buff***. Nominate (W), *columbianus* (E). Curious trap-liner in lower growth, usually small leks. **Voice** High-pitched, thin *tsi-tseep, tsi-tseep..t'si-t'si-t'si...* every 1–2 seconds. **SS** Occurs above other hermits; see Green Hermit (E). **Status** Uncommon.

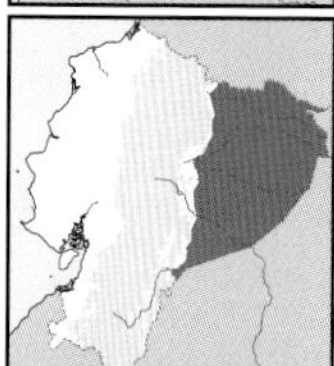

Straight-billed Hermit *Phaethornis bourcieri* 12–13cm

E lowlands to lower foothills (to 900m). *Terra firme* undergrowth, rarely borders. Long (3–3.3cm), ***straight bill, mandible mostly yellow***. Bronzy-green above, ***short whitish supercilium***, faint white malar. ***Greyish-buff below***, whiter throat and belly. Nominate. Solitary trap-liner near ground, less curious than other hermits; small loose leks. **Voice** Quavering, weak, often doubled *tsisisisi'seé, tsisisisi'seé..., tsi'i'i'i-tsiiii'siii...tsi-si-si-si-si'i'i'i'*, very thin *síp-síp-síp-sisisi'síp* in flight. **SS** From Great-billed and White-bearded Hermits by bill shape (red mandible in Great-billed). **Status** Uncommon.

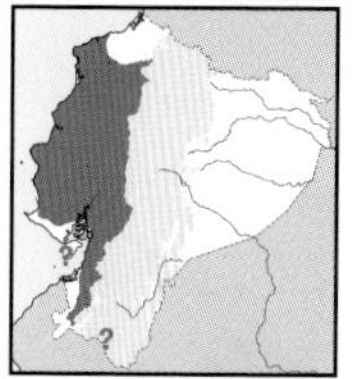

Long-billed Hermit *Phaethornis longirostris* 12–13cm

W lowlands to foothills (to 1,300m; locally higher). Humid to deciduous forest, borders, woodland. Long (4–4.2cm) decurved bill, ***mandible mostly red***. Dull green above, ***rufescent rump; conspicuous whitish supercilium and malar. Dull grey below***, throat and belly whiter. Race *baroni* (W) possibly valid species. Habits much like other hermits, small leks. **Voice** Repeated *tcheé-tcheé-tcheé-tcheé... c.*2 notes per second. **SS** Higher-elevation Tawny-bellied (richer below), and Green Hermits (mostly bright green); only large hermit in most of range. **Status** Fairly common.

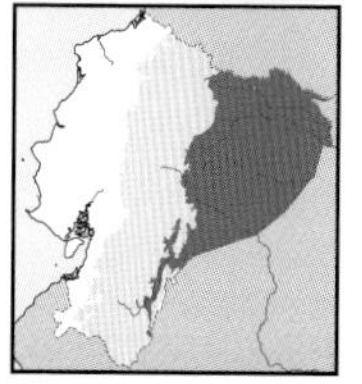

Great-billed Hermit *Phaethornis malaris* 12–13cm

E lowlands to foothills (to 1000m). Humid forest, borders, woodland. Long (4–4.2cm) decurved bill, ***mandible mostly red***. Coppery-green above, ***buff rump. Bold whitish supercilium and malar. Dull buff below***. Blackish tail narrowly tipped white, elongated central feathers broadly so. Race *moorei*; possibly *ochraceiventris* in extreme E. Curious trap-liner, fairly large leks. **Voice** Clear, quite nasal *sweép-sweép-sweép-sweép...* every 0.5 sec. **SS** Tawny-bellied (rich tawny below; higher elevation) and White-bearded (greyer below and rump; *várzea*). **Status** Common.

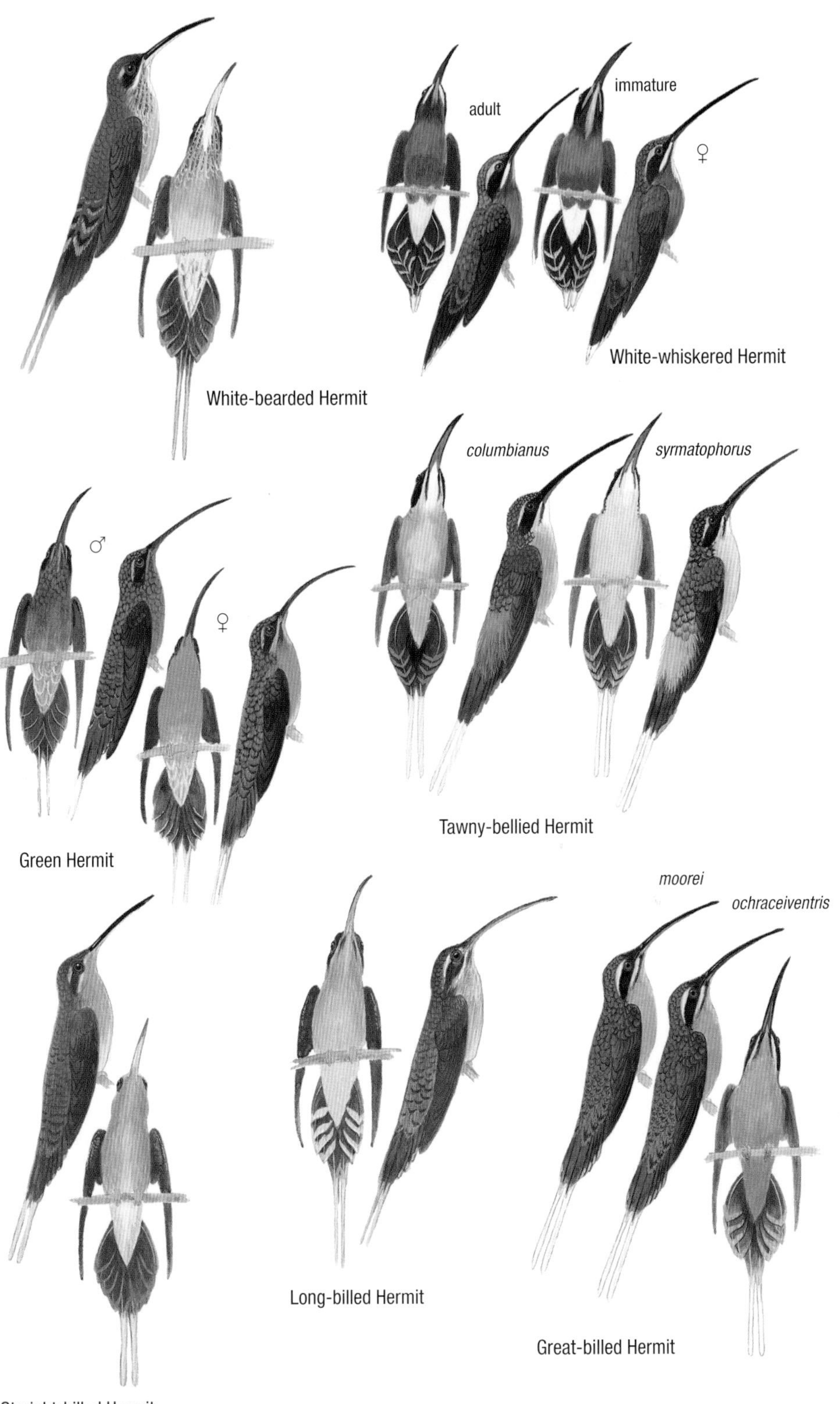
adult
immature
♀
White-whiskered Hermit
White-bearded Hermit
columbianus
syrmatophorus
♂
♀
Tawny-bellied Hermit
Green Hermit
moorei
ochraceiventris
Long-billed Hermit
Great-billed Hermit
Straight-billed Hermit

PLATE 101: JACOBIN, LANCEBILLS AND VIOLETEARS

Lancebills have very long and slender bills – even curved upwards – and might be difficult to separate where they meet (E Andean foothills). Violetears are fairly large and bold; Sparkling is the most familiar Andean hummingbird. Some may engage in seasonal movements, also move up and down slopes to feeders at regularly birded areas.

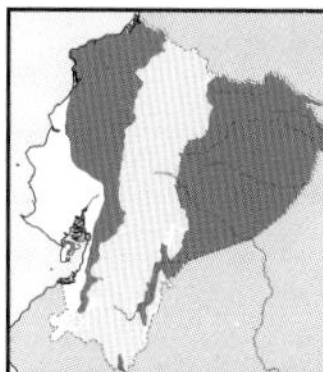

White-necked Jacobin *Florisuga mellivora* 9–10cm

E & W lowlands to foothills (to 1,600m, locally higher). Forest borders, woodland, adjacent clearings. Rather short (1.8–2cm), slightly curved bill. *Bright blue hood, white nape line; white belly to lower tail.* ♀: white underparts *boldly spotted green, white mid belly*, scaled vent, *blackish subterminal tail-band, broad white tips*. Nominate. Conspicuous, congregates with others at flowering trees and feeders; often to canopy. **Voice** Very high-pitched *tsíít-tsíít-tsíít...* or *shít-shít-shít....* **SS** ♀ with ♀ Violet-fronted and Green-crowned Brilliants; also Many-spotted Hummingbird. **Status** Common.

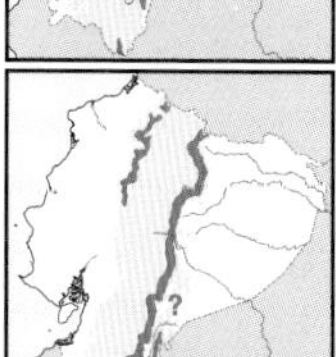

Green-fronted Lancebill *Doryfera ludovicae* 9–10cm

NW & E subtropics in Andes (1,100–2,100m, sometimes lower or higher). Undergrowth in forest interior, borders. *Long* (3.5–3.6cm) *straight bill. Bright green front, coppery crown and nape.* Otherwise mostly bronzy-green, brighter above, uppertail-coverts washed blue. Rounded black tail tipped grey. ♀: *smaller frontal patch*. Nominate. Regularly at shady borders near streams, trap-lines long flowers; territorial when breeding. **Voice** Thin *cht'shir'rp, chí!* calls. **SS** Blue-fronted Lancebill has shorter bill, dark blue underparts, violet front, but similar in dim light. **Status** Uncommon.

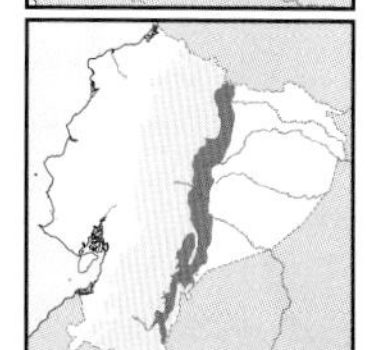

Blue-fronted Lancebill *Doryfera johannae* 9cm

Foothills of E Andes (400–1,500m). Undergrowth in montane forest, borders, locally to adjacent lowlands. *Bill shorter, straighter* than Green-fronted (2.7–3cm). *Violet-blue front* (*turquoise-blue* in ♀), *pale blue uppertail-coverts; dark blue below.* Rounded tail black. ♀: greyer below. Nominate. Mostly near streams; singles trap-line long flowers. **Voice** Not very vocal, occasional thin *tsir* in flight. **SS** Upper-elevation Green-fronted Lancebill (longer bill, paler below). **Status** Uncommon.

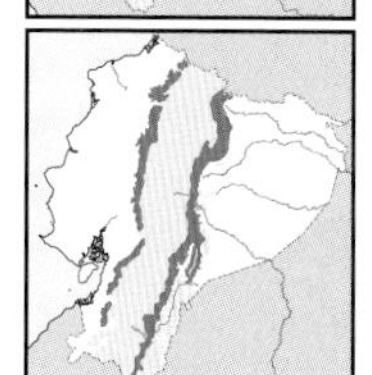

Wedge-billed Hummingbird *Schistes geoffroyi* 8–9cm

E & W Andean slopes, more local in W (800–2,000m, locally lower). Undergrowth of forest interior, borders, adjacent clearings. Short (1.5–1.8cm) *wedge-shaped bill.* Rounded tail, *broad blue subterminal band.* Bright green front, *violet patches on chest-sides, narrow white chest line*; ♀ has *smaller violet spots, white throat* (race *albogularis*, W). Nominate (E) *coppery above, blue on chest-sides, narrower, broken* white chest-band; ♀: *green throat spots.* Inconspicuous trap-liner, tends to remain in cover, often piercing corollas at base. **Voice** Fast simple *tsii-sii-sii-sii...* series (E); more jumbled *tsii-tsii-sueee-sii-sii...* in W. **SS** Confusion unlikely. **Status** Uncommon.

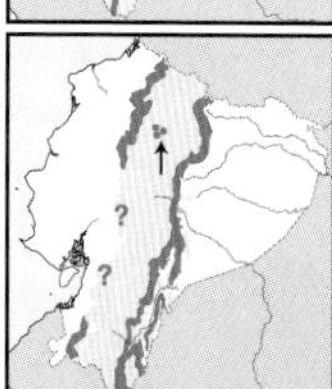

Brown Violetear *Colibri delphinae* 9–10cm

Subtropics on E & NW Andean slopes, also Cordillera de la Costa (1,000–2,200m, locally to 500m). Forest canopy, borders, adjacent clearings. Rather short bill (1.5–1.7cm), slightly curved. *Mainly dull brown, glittering violet ear-tuft*, glittering green throat, *pale buff malar*, rufous in vent and *rump. Tail coppery-greenish, dusky subterminal band.* Monotypic. Solitary, sometimes at feeders, chases other violetears. **Voice** Endless series of 8–10 *tcheep* notes, with short pauses, often jumbled at end. **SS** ♀ Lazuline Sabrewing (greener above, blue throat, no ear-tufts). **Status** Uncommon, local.

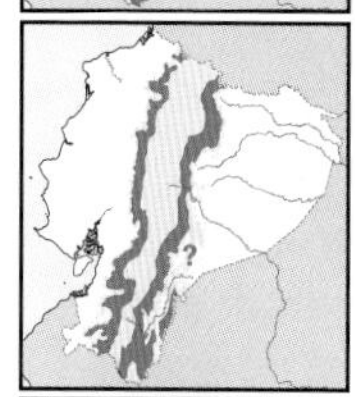

Green Violetear *Colibri thalassinus* 9–10cm

Subtropics on both Andean slopes (mostly 1,200–2,300m). Forest canopy, borders, woodland, shrubby clearings. Rather short bill (1.8–2cm). *Shiny green throat and breast, long violet-blue ear tufts.* Tail metallic blue-green, dark subterminal band. *Shorter ear-tufts* in ♀. Race *cyanotus*. Regular in semi-open areas, several may gather at flowering trees, aggressively attack other hummingbirds; ♂ vocalises incessantly on exposed perches. **Voice** Sharp, dry, metallic, long series of *tzu-zueek....* **SS** Higher-elevation Sparkling Violetear (larger with violet-blue belly); overlaps locally. **Status** Uncommon.

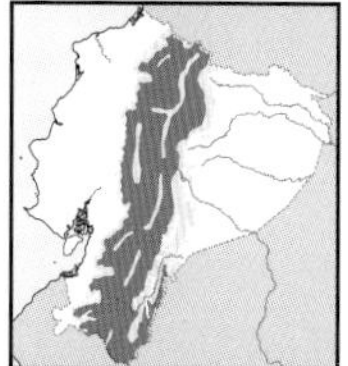

Sparkling Violetear *Colibri coruscans* 12–13cm

Andes (mostly 1,000–3,500m). Open and semi-open areas. Bill longer than previous species (2–2.2cm), slightly curved. Mainly bright green, *violet-blue throat to ear-tufts, violet-blue central belly.* Tail metallic blue-green, dark subterminal band. *Shorter ear-tufts* in ♀. Nominate. Widespread, aggressively defends territory against conspecifics, other hummingbirds, even passerines and small raptors; incessant singer, remarkable aerial display. **Voice** Endless, metallic *tik-tik...*series, often jumbled at end. **SS** Lower-elevation Green Violetear. **Status** Common.

♀
juvenile
♂
immature
♀
♂
♂
Green-fronted Lancebill
♀
White-necked Jacobin
♂
♀
immature
♀
♂
geoffroyi
♀
Blue-fronted Lancebill
♂
immature
♀
immature
♀
♂
adult
juvenile
♂
immature
albigularis
Wedge-billed Hummingbird
Brown Violetear
♂
♀
♂
♂
♂
Green Violetear
♀
dark morph
Sparkling Violetear

PLATE 102: FAIRIES, GOLDENTHROAT, AWLBILL AND MANGO

Few hummingbirds combine pure white and glittering green, like fairies, mangos and the small, mango-like, awlbill. The two fairies and the mango are regular in open habitats, borders and hedgerows, whilst the awlbill is apparently very rare – or erratic? –and the goldenthroat has yet to be definitely recorded in Ecuador.

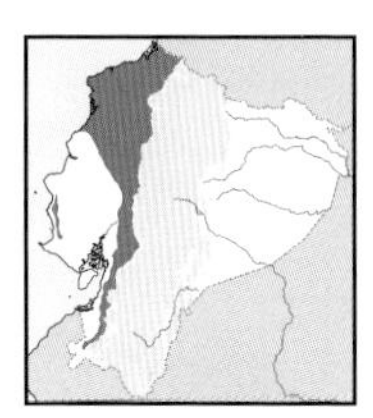

Purple-crowned Fairy *Heliothryx barroti* — 9–10cm

W lowlands and foothills, S in Andean foothills (to 1,300m). Humid forest, borders, woodland and adjacent clearings. Bill straight, rather short (1.5–1.8cm). Shining green above, ***purple crown, black mask, contrasting pure white underparts including long pointed, graduated tail***; blue-black central tail feathers. ♀: similar but no purple in crown and face. Monotypic. Solitary, regularly low in borders and shrubby clearings; fast darting flight, tail often fanned and lifted. **Voice** Very thin, jumbled *tsi-si'si'sisisisi'sir'r'r'r* in display. **SS** Only W hummingbird with all-white underparts. Black-eared Fairy only in E. **Status** Uncommon.

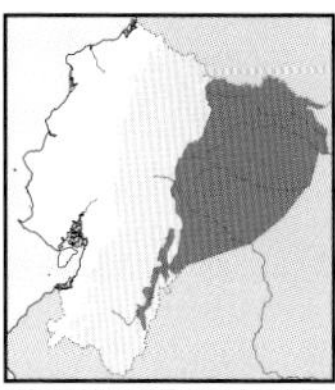

Black-eared Fairy *Heliothryx auritus* — 9–10cm

E lowlands to foothills (to 1,200m). Upper forest strata, borders, woodland, adjacent clearings. Bill straight, rather short (1.5–1.8cm). ***Shining green above***, narrow black mask, ***pure white underparts including long, pointed, graduated tail***. Central tail feathers blue-black. ♀: similar but ***faint greyish streaking on breast***. Nominate. Solitary, fast darting flight, mostly in forest canopy but lower in borders and shrubby clearings; often seen hawking insects. **Voice** Very thin *tsíp-síp...* more spaced than Purple-crowned Fairy. **SS** Confusion unlikely; Purple-crowned only in W. **Status** Uncommon.

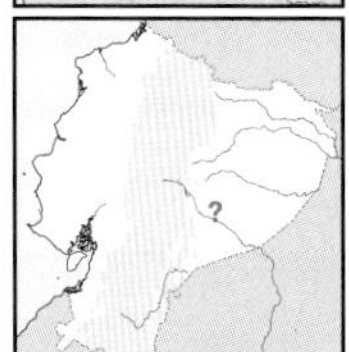

* Green-tailed Goldenthroat *Polytmus theresiae* — 8–9cm

One uncorroborated record in SE lowlands (*c.*200m). Habitat unknown in Ecuador, elsewhere in forest borders and shrubby clearings. Bill short (2cm) ***slightly decurved, reddish mandible. White spots in front of and behind eyes***. Mostly shining emerald-green, ***brighter underparts***, white vent with some green edges. ***Short, rounded tail, emerald-green***. ♀: similar, ***densely fringed white below***, tail narrowly tipped white. Race *leucorrhous*. Known in Ecuador from single old record at unknown locality; specimens probably wrongly labelled, but suspected records from Kapawi area in SE Pastaza exist. **Status** Uncertain, DD in Ecuador.

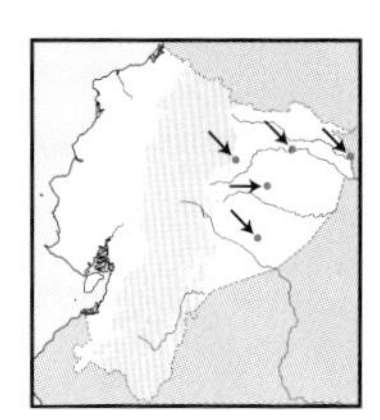

Fiery-tailed Awlbill *Avocettula recurvirostris* — 7–9cm

Few records in E lowlands (to 950m). Canopy and borders of *terra firme* and *várzea*. Bill short, black, ***upturned tip***. Bronzy-green above, ***glittering emerald throat to breast, black mid belly. Tail coppery-orange below***, central tail feathers green. ♀: ***white from throat- to belly-sides, black stripe on breast to belly centre***. Tail dusky-green, ***coppery-orange base*** below, central tail feathers green, outer ones tipped white. Monotypic. Singles or pairs trap-line high at forest edge; poorly known. Possibly engages in altitudinal, seasonal or erratic movements. **Voice** Not very vocal; thin *wip-wi-wip* notes reported. **SS** ♀ recalls ♀ Black-throated Mango but smaller, green throat, coppery tail; bill shape unique but hard to see. **Status** Uncertain, apparently very rare, DD in Ecuador.

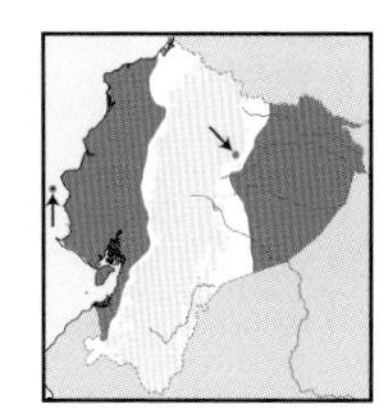

Black-throated Mango *Anthracothorax nigricollis* — 11–12cm

Lowlands of E & W (to 900m, locally higher). Forest borders, woodland, shrubby clearings; in W semi-deciduous to deciduous forest and woodland. Bill rather long (2.3–2.5cm) and ***slightly decurved***. Mainly bright green, coppery crown, brighter throat-sides. ***Solid black median stripe from chin to belly***. Undertail ***purplish-red tipped black***, central tail feathers green. ♀: ***white below contrasting with black central stripe***, tail tipped white. Nominate (E), *iridescens* (W), possibly separate species. Conspicuous, may congregate at flowering trees or feeders, quite aggressive; often hawks insects in open. **Voice** Rather quiet, sometimes thin *tsik* calls. **SS** Confusion unlikely; see smaller Fiery-throated Awlbill. **Status** Uncommon.

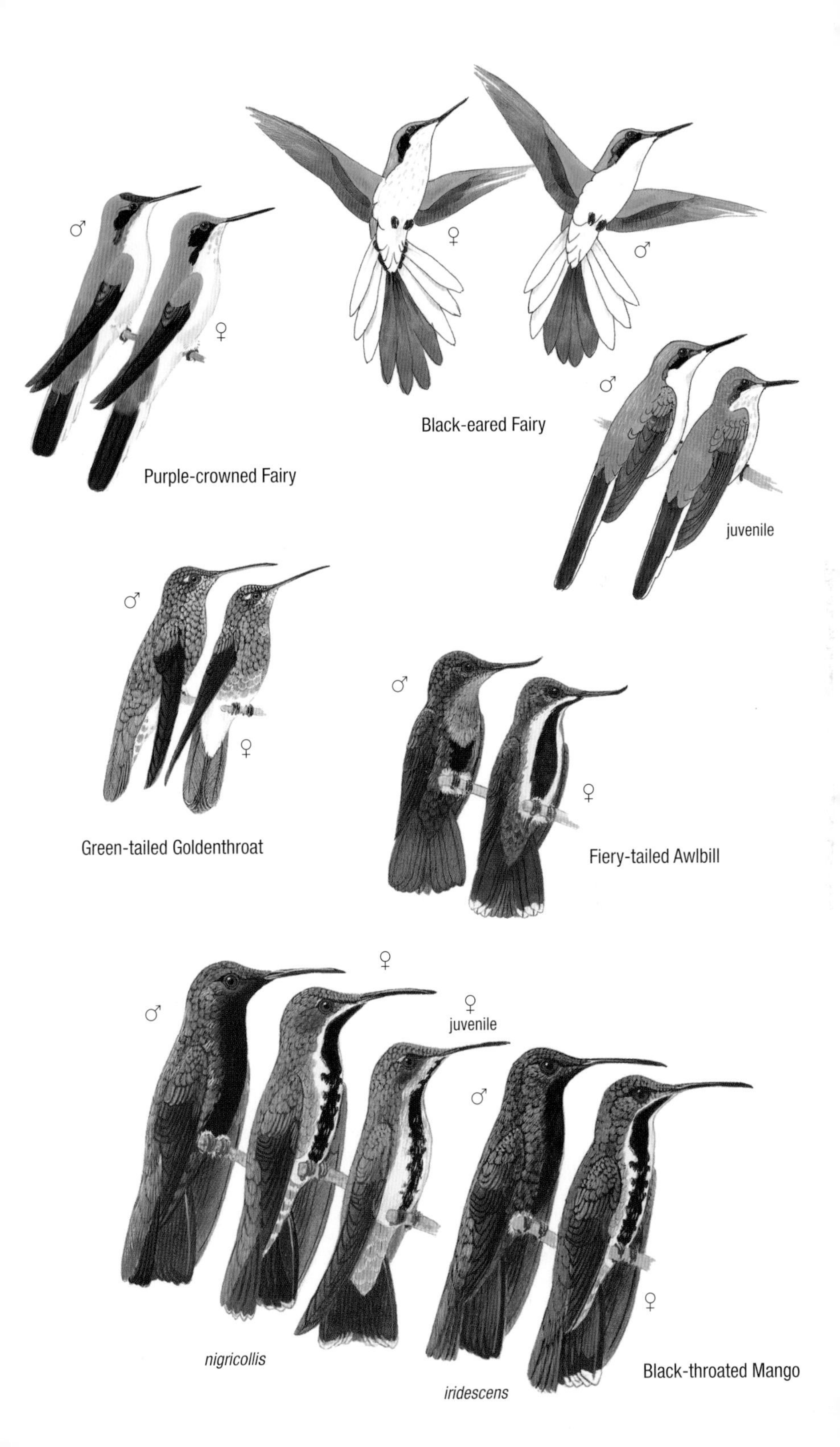
♂
♀
Purple-crowned Fairy
♀
♂
Black-eared Fairy
♂
juvenile
♂
♀
Green-tailed Goldenthroat
♂
♀
Fiery-tailed Awlbill
♀
♂
♀
juvenile
♂
♀
nigricollis
iridescens
Black-throated Mango

Forest hummingbirds primarily characterised by white chest crescents and / or glittering pink to purple throat patches; ♀♀ show variable glitter on throat. Purple-throated is the only sunangel in open and semi-open habitats, while Royal is strictly confined to forest and shrubland in Cordillera del Cóndor, and was only recently found in Ecuador. Bills short, delicate and straight.

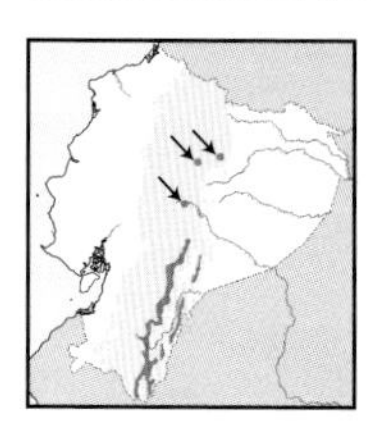

Amethyst-throated Sunangel *Heliangelus amethysticollis* 9–10cm

SE subtropics to temperate Andean slopes, possibly Sumaco massif (1,900–2,700m, locally higher). Montane forest canopy, borders, adjacent clearings. Bill short (1.5–1.6cm) straight, white postocular spot. Metallic green above, brighter, bluer frontlet. *Glistening fuchsia-amethyst throat, white crescent on chest.* Blue-black tail, central tail feathers green. ♀: *buff throat,* some with glistening amethyst spots. Race *laticlavius.* Singly, quite conspicuous and territorial, also at feeders; clings to flowers with wings outstretched for few seconds, also after perching. **Voice** Gravelly *tsir'sri'sri...*, but not very vocal. **SS** Tourmaline Sunangel lacks white breast-band (no known overlap). Little Sunangel has orange throat. **Status** Uncommon.

Gorgeted Sunangel *Heliangelus strophianus* 9–10cm

Subtropics on W Andean slope (1,700–3,000m, locally higher; 700–1,100m in SW). Montane forest, borders, woodland. Bill short (1.4–1.6cm) straight, white postocular spot. *Dark metallic green above,* brighter frontlet. *Glistening rosy to fuchsia throat, narrow white chest crescent.* Blue-black tail. ♀: similar but *smaller fuchsia throat patch,* white stripes on chin. Monotypic. More retiring than previous species, also regular at feeders, territorial at flowers; holds wings outstretched briefly. **Voice** Soft, fast *tsi'si'si'si..tsip-tsip-tsip,* sometimes starting with trill or jumble of notes. **SS** Confusion unlikely, but see Wedge-billed Hummingbird. **Status** Locally uncommon.

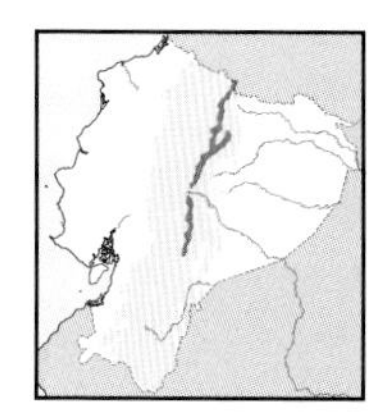

Tourmaline Sunangel *Heliangelus exortis* 9–10cm

Temperate areas of NE Andes (1,900–3,300m). Montane forest borders, dense shrubs, adjacent clearings. Bill short (1.5–1.6cm) straight, white postocular spot. Bright bluish-green frontlet and breast, violet-blue chin, *purplish-pink throat, white vent.* Tail longer than most sunangels, *forked,* central rectrices green. ♀: *variable white throat patch,* some with glistening pink spots. Monotypic. Singly, quite conspicuous at shrubby patches, territorial; also outstretches wings while clinging to flower or after perching. **Voice** Not very vocal; simple buzzy *jzip-jzip-jzip...* like other sunangels. **SS** Little Sunangel (orange throat) replaces it S (no known overlap). **Status** Uncommon.

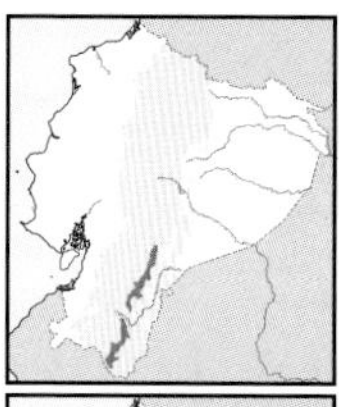

Little Sunangel *Heliangelus micraster* 8–9cm

Temperate areas on SE Andean slope (2,400–3,100m). Montane forest undergrowth, borders, woodland. Bill short (1.4–1.5cm) straight, white postocular spot. Mostly dark metallic green, *bright fiery-orange throat, white vent.* Tail long, forked, mostly green. ♀: *white throat* with some *green and glittering orange spots.* Nominate. Somewhat more retiring than Tourmaline Sunangel; territorial; outstretches wings briefly on perching. **Voice** Harsh, thin *ch'r-ch'r-ch'r....* **SS** See Tourmaline (orange throat; ♀ less variable); no overlap. **Status** Uncommon.

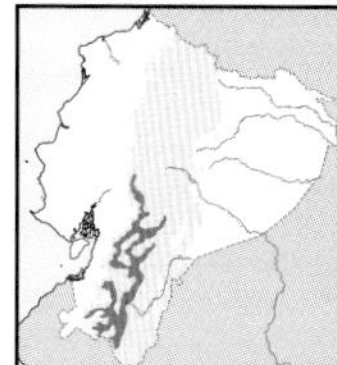

Purple-throated Sunangel *Heliangelus viola* 11–12cm

Andean highlands in S, N to Bolívar, possibly further N? (1,800–3,300m, locally higher). Montane forest, borders, shrubby clearings, woodland. Bill short (1.4–1.5cm) straight, white postocular spot. *Bright blue front, glistening violet to purple throat,* bluish breast, green below, *buffy-white vent.* Long black tail, *deeply forked.* ♀: throat *duller and streaked.* Monotypic, but possibly different subspecies to Peruvian populations (*pyropus*). Conspicuous in semi-open, often aggressively chasing other hummingbirds at flowering trees; regular in gardens. **Voice** Thin, trilled *t'rrp-t'rrp-t'rrp, tr't-tr't....* **SS** Confusion unlikely in range and habitat. **Status** Fairly common.

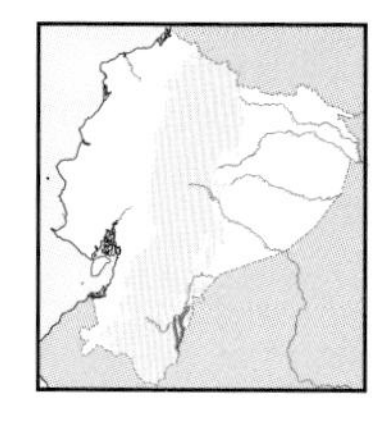

Royal Sunangel *Heliangelus regalis* 9–10cm

Few localities in extreme SE, in Cordillera del Cóndor and Nangaritza Valley (1,230–2,350m). Dense undergrowth of montane forest and shrubland on flat-topped hills. Bill short (1.8cm). *Deep violet-blue;* ♀ green above, green and *tawny below, paler crescent on chest,* streaked throat. Both have long, deeply forked, deep blue tail. Nominate. Territorial ♂ defends feeding area and performs flight display in low-stature shrubland. **Voice** High *teep, tchép, che'chúp-tchup..., chép-chép-chép...* **SS** ♂ unmistakable; ♀ recalls higher-elevation Amethyst-throated Sunangel but buffier below, tail more forked and blue. **Status** Exceedingly local but fairly common; EN.

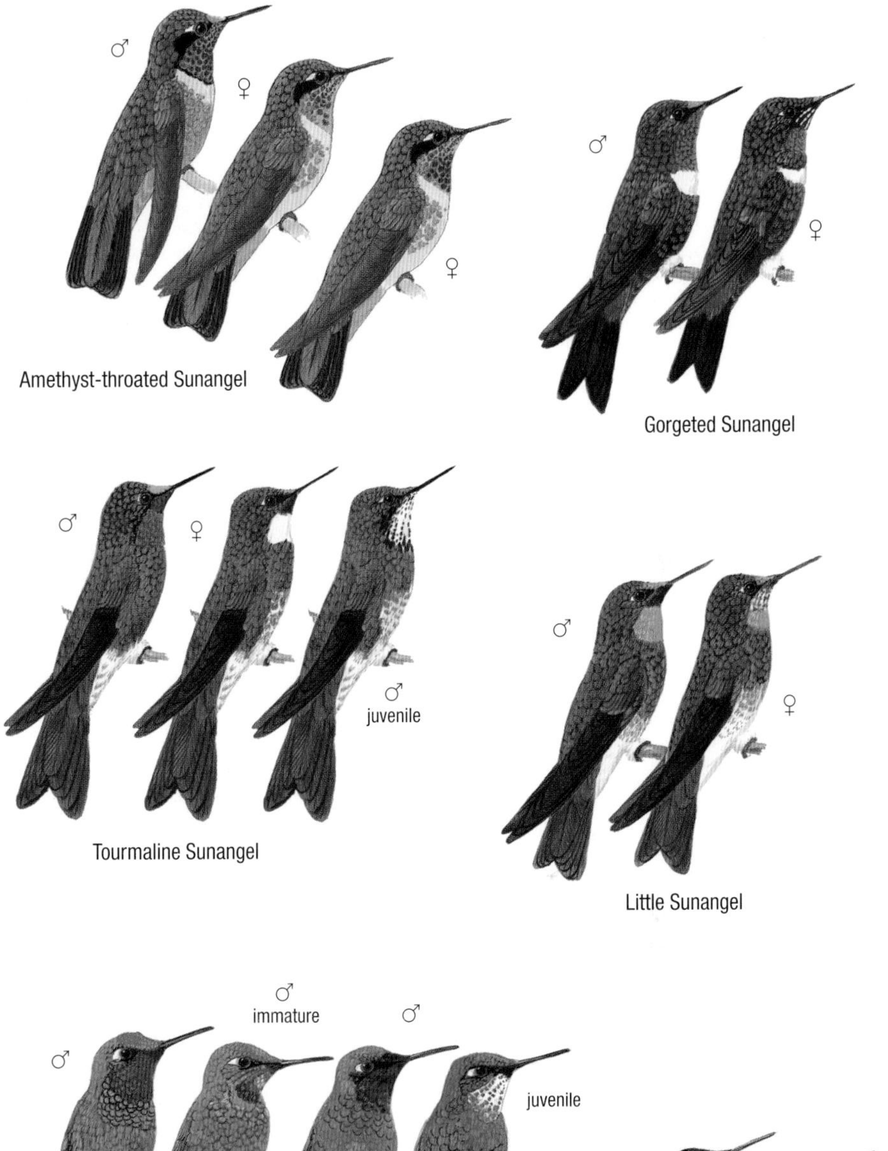
♂
♀
♀
Amethyst-throated Sunangel
♂
♀
Gorgeted Sunangel
♂
♀
♂
juvenile
Tourmaline Sunangel
♂
♀
Little Sunangel

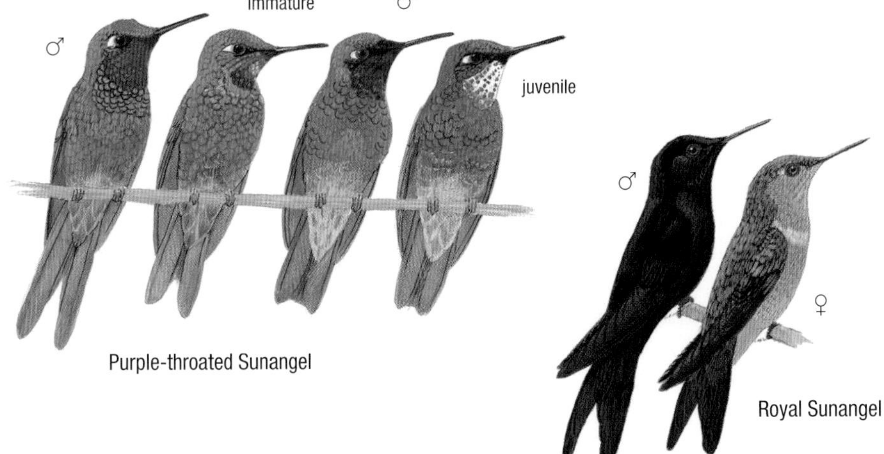
♂
♂
immature
♂
juvenile
Purple-throated Sunangel
♂
♀
Royal Sunangel

PLATE 104: THORNTAILS, COQUETTES AND PIEDTAIL

Tiny, handsome hummingbirds of humid forest, mostly in E lowlands and foothills. Sexual dimorphism well marked in thorntails and coquettes, but absent in the piedtail. ♂♂, which appear less numerous than ♀-plumaged birds, have fantastic plumages with glamorous adornments. Coquettes apparently engage in seasonal or erratic movements.

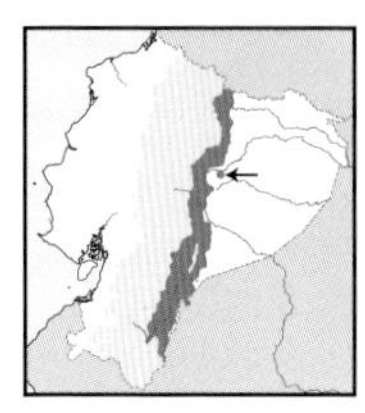

Wire-crested Thorntail *Discosura popelairii* — 7–9cm

Foothills and lower subtropics in E Andean slope (600–1,600m). Forest canopy, borders, adjacent clearings. Very short bill (1.2–1.3cm). Handsome, ***bright green crown and long coronal wires; narrow white rump***. Long, tapered, blue-black tail, ***white rachises from below***, long thistle-like outer feathers. ♀: coppery-green, blacker below, ***white whiskers, tufts on flanks and rump band***. Tail shorter, tipped white. Monotypic. Often congregates with other species at flowering trees, where inconspicuous; wasp-like flight, often cocks tail. **Voice** Rather silent; a liquid *tew* reported (S *et al.*). **SS** Black-bellied Thorntail lacks crown wires; ♀ longer-tailed, greener, with narrower moustachial. **Status** Uncommon; NT globally.

Green Thorntail *Discosura conversii* — 7–9cm

W lowlands to foothills, S in Andean foothills (300–1,300m, locally higher). Humid forest canopy, borders, adjacent clearings. Very short bill (1.2–1.3cm). ***Glittering green crown and throat, narrow white rump***. Long, tapered, blue-black tail, ***longer, thistle-like outer tail feathers with white shafts***. ♀: smaller, with blacker throat, ***white whiskers, flank tufts and rump band***. Tail much shorter, blue-black, tipped white. Monotypic. Flight and behaviour like other thorntails, regular at borders and feeders. **Voice** Piercing *chít* calls. **SS** Confusion unlikely. **Status** Uncommon.

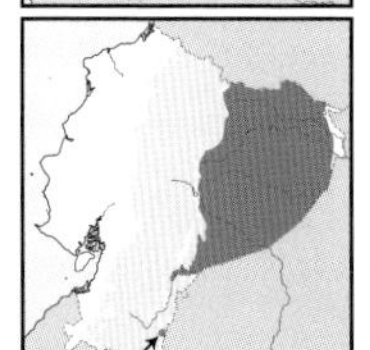

Black-bellied Thorntail *Discosura langsdorffi* — 7–9cm

Local in E lowlands (locally to 1,200m). *Terra firme* canopy, borders. Very short bill (1.2–1.3cm). ***Bright green crown and throat, white rump band, black belly***. Long, tapered, blue-black tail, ***long, grey outer tail feathers, white shafts***. ♀: ***less prominent white whiskers, flank tufts and rump band***, tail longer than Wire-crested, also tipped white. Race *melanosternon*. Canopy-dweller that may join other hummingbirds at flowering trees. **Voice** Loud, rapid *ti-ti-ti-ti* whistles in darting displays (HB). **SS** Higher-elevation Wire-crested Thorntail and rarer coquettes. **Status** Rare.

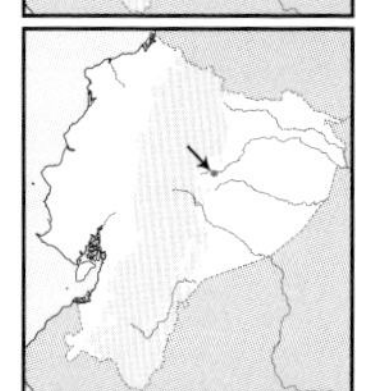

* Rufous-crested Coquette *Lophornis delattrei* — 7–8cm

Single uncorroborated record in E lowlands (400m). Forest borders. Spectacular ♂ has ***spikey orange bill***, tipped black. ***Long spikey orange-rufous crest, glittering green face and throat***, white outline to throat. Metallic green above, ***narrow white rump band***. Coppery tail, black subterminal band. ♀: ***extensive orange-rufous in face***, throat spotted orange-rufous. Race *lessoni*? Poorly known, possibly erratic; elsewhere behaves like other coquettes. **Voice** Soft *irrt* (Panama). **SS** ♂ Spangled Coquette (bold black spots in long crest; ♀ less orange-rufous in face). **Status** Uncertain, possibly wanderer.

Spangled Coquette *Lophornis stictolophus* — 7–8cm

Local in E Andean lowlands to foothills (to 1,200m). Forest borders, adjacent clearings, gardens. ***Spikey, short bill*** (1cm), mostly orange. Lovely ♂ has ***long orange-rufous bushy crest, each feather tipped black. Glittering green face and throat***, small orange-rufous ear-tufts; ***white band on coppery rump***. ♀: ***orange-rufous crown, throat whiter, small orange-rufous spots***, tail tipped rufous. Monotypic. Slow, weaving, wasp-like flight, often cocks tail; can congregate at flowering trees. **Voice** Ringing *chúip* calls. **SS** Rarer Rufous-crested Coquette. **Status** Rare.

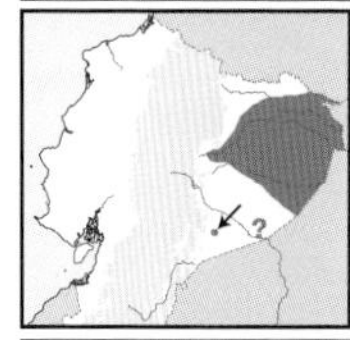

Festive Coquette *Lophornis chalybeus* — 7–8cm

Local in E lowlands (to 500m). Forest canopy, borders. Bill short (1.2–1.3cm), black. Lovely ♂ has ***bright green front***, blackish-green crown, ***long green facial plumes with white tips, white rump***, coppery uppertail-coverts and tail. ♀: dark metallic green, ***whitish malar, white rump***, coppery tail tipped whitish. Race *verreauxii*. Mainly alone in canopy; flight slow, wasp-like. **Voice** Soft *chui* calls (Peru). **SS** ♂ unmistakable; ♀ recalls thorntails but lacks flank tufts. **Status** Rare.

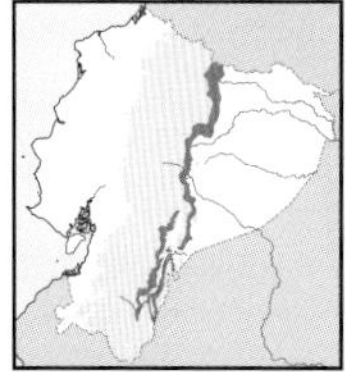

Ecuadorian Piedtail *Phlogophilus hemileucurus* — 7–8cm

Local in E Andean foothills to subtropics (900–1,400m, locally higher). Forest undergrowth. Short black bill (1.7–1.8cm), white postocular spot. Generally ***dull***, blue-green rump and uppertail-coverts. ***White breast-band, throat spotted green; mottled breast-sides and flanks***. Short graduated, ***fan-like, black and white tail***, blue-green central rectrices. Monotypic. Inconspicuous, near ground, mostly alone but displaying ♂♂ close together. **Voice** Very high-pitched *tzeee-tzeee-st'st'st'st...*, almost sparrow-like. **SS** Confusion unlikely due to tail pattern. **Status** Rare, NT in Ecuador; VU globally.

♂
♀
Wire-crested Thorntail
♂
♀
Green Thorntail
♂
♀
Black-bellied Thorntail
♂
♀
juvenile
Rufous-crested Coquette
♂
♂
juvenile
♀
Spangled Coquette
♂
♀
juvenile
Festive Coquette
Ecuadorian Piedtail

PLATE 105: LONG-TAILED HUMMINGBIRDS

All species on this plate boast long glittering tails, ranging from striking turquoise-blue / violet in sylphs, to green-and-black in trainbearers. The racket-tail is unique among Ecuadorian hummingbirds. Peruvian Sheartail, apparently only a visitor to Ecuador, resembles woodstars but its tail is distinctive.

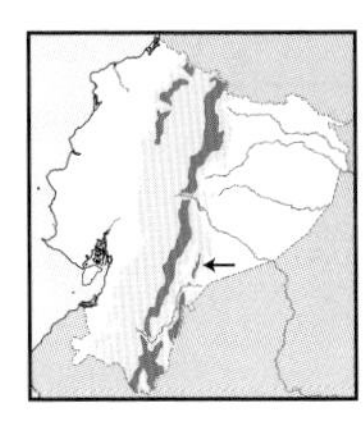

Long-tailed Sylph *Aglaiocercus kingii* 10–11cm

E & NW subtropics to temperate Andes (1,600–3,000m, locally higher; 2,300–2,800m in W). Montane forest, borders, woodland. Short bill (1.2–1.3cm). ♂: *strikingly long*, deeply forked tail (9–12cm), *turquoise to aquamarine* (E), *green* (W). *Glistening green crown, violet-blue throat* (E), *absent* in W. ♀: contrasting *cinnamon-buff underparts*, throat white, spotted green, short tail tipped white. Races *mocoa* (E), *emmae* (W). Sometimes with other hummingbirds at flowering trees; trap-liner, also fiercely defends feeding grounds. **Voice** Sharp, buzzy *dz-dz-jiit-jiit-jiit-dz-dz...* **SS** In NW Violet-tailed Sylph (violet-blue tail, blue throat patch; ♀ has white breast-band, turquoise crown). **Status** Uncommon.

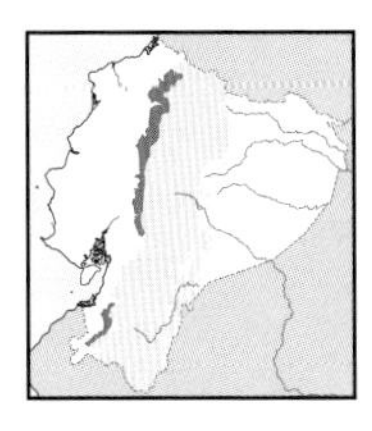

Violet-tailed Sylph *Aglaiocercus coelestis* 10–11cm

Foothills to subtropics on W Andean slope (800–2,000m). Montane forest, borders, woodland. Short bill (1.2–1.3cm). ♂ has *strikingly long*, deeply forked tail (10–12cm), *violet-blue* (turquoise at certain angles). Metallic green crown, *blue throat* (*lacking* in SW). ♀: *cinnamon-buff below*, throat spotted green, *white breast-band, turquoise-blue crown*. Nominate (most range), *aethereus* (SW). More forest-based than previous species, often at borders and feeders. **Voice** Fast, quavering *t'zi-zi-z'rrrrr, t'zit-t'zit-t'zit...* while foraging. **SS** Higher-elevation Long-tailed Sylph; races *emmae* of Long-tailed and *aethereus* of Violet-tailed more similar but no overlap. **Status** Uncommon.

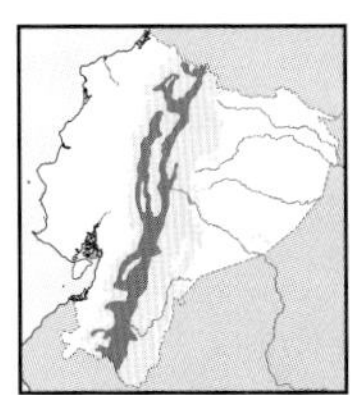

Green-tailed Trainbearer *Lesbia nuna* 9–10cm

Andean subtropics to highlands (1,900–3,200m, locally higher). Forest borders, shrubby clearings, woodland. Bill short (1cm), *straight*. ♂ has *long tail* but *shorter* than previous species (8–10cm), deeply forked, *mostly bright green* (except longest feathers). Mainly bright green, *shinier throat*. ♀: *shorter tail* (2–3cm), mostly bright green. Whitish below boldly spotted green. Race *gracilis*; possibly different in S. Prefers more humid situations; quite conspicuous and aggressive defender of feeding areas. **Voice** High-pitched, buzzy *bi-zizizizizi, b'zzzz-b'zzzz-b'zzzz...*; *tk'k'k'k* tail sounds. **SS** Black-tailed (longer-tailed, darker, longer-billed); slower and lighter flight in Green-tailed. **Status** Uncommon.

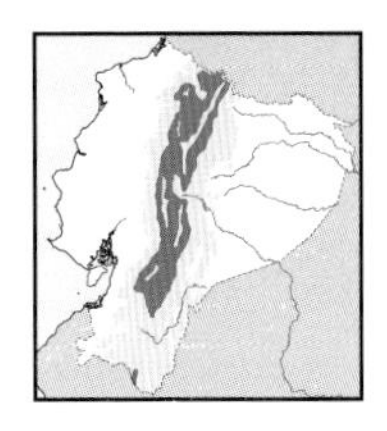

Black-tailed Trainbearer *Lesbia victoriae* 10–11cm

Widespread in Andes (2,500–3,800m). Shrubby clearings, borders, gardens. Bill short (1.3–1.5cm), *slightly decurved*. ♂ has spectacularly *long* (15–17cm), *deeply forked tail, mainly black with green edges above*. Metallic green, *shinier throat*. ♀: *shorter tail* (4–5cm), *green tips more conspicuous on upperside*, whitish below boldly spotted green. Nominate (most of range), *juliae* (S). Conspicuous in highland cities and towns, territorial; spectacular aerial display with mechanical sounds (*trik-trik-tr'rrrk*). **Voice** Accelerating, descending, high-pitched *tsee-tsee-tsee-titititi-tik-tik*, chattering, jumbled. **SS** Green-tailed Trainbearer tends to favour more humid habitats; greener overall, throat patch rounded not pointed, bill shorter, straighter. **Status** Common.

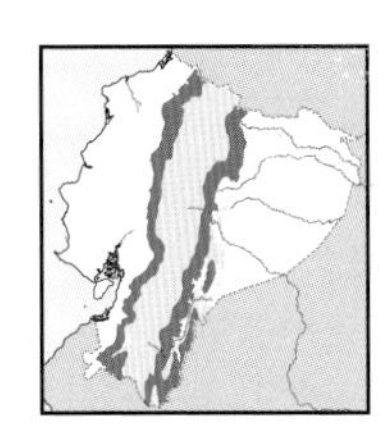

Booted Racket-tail *Ocreatus underwoodii* 8–9cm

Foothills and subtropics on both Andean slopes (900–2,200m, locally lower). Forest borders, adjacent clearings. Short straight bill (1.1–1.3cm). Spectacular ♂ has long *outer tail rackets* and *fluffy white thighs* (*rufous-buff* in E). ♀: metallic green above, *white below*, white postocular spot and *moustachial, mottled green* throat to belly-sides (*more spotty* in E), smaller *white puffs* (*buff* in E), tail shorter, *outer feathers tipped white*. Races *peruanus* (E), *melanantherus* (W). May gather with other species at flowering trees; flight slow, wavy, often cocks and spreads tail. **Voice** Constant humming / trilling *ti'r'rr'rr'rr...* while feeding. **SS** ♀ recalls ♀ *Urosticte*, even Ecuadorian Piedtail, but see thigh puffs. **Status** Uncommon.

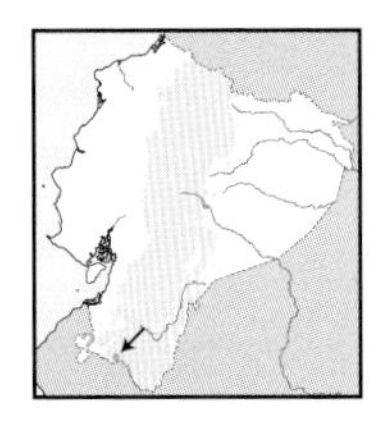

* Peruvian Sheartail *Thaumastura cora* 6–7cm

One record in extreme SW highlands (2,350m). Montane scrub edges. *Short* bill (1.2cm). ♂: bright green above, *greyish-white below, glittering fuchsia throat. Long tail* (7–8cm), *two feathers longer and whitish*. ♀ *buffy-white below*, tail much *shorter* (2–3cm), *greenish-black tipped white*, central tail feathers green. Monotypic. Poorly known, apparently seasonal or erratic in Ecuador (erratic in Peru); may gather with others at flowering trees, but territorial. **Voice** Fast jumbled series of chipping notes (S *et al.*). **SS** ♂ unique, ♀ resembles woodstars (especially Short-tailed and Purple-collared) but tail longer, bill shorter; see Tumbes Hummingbird. **Status** Uncertain.

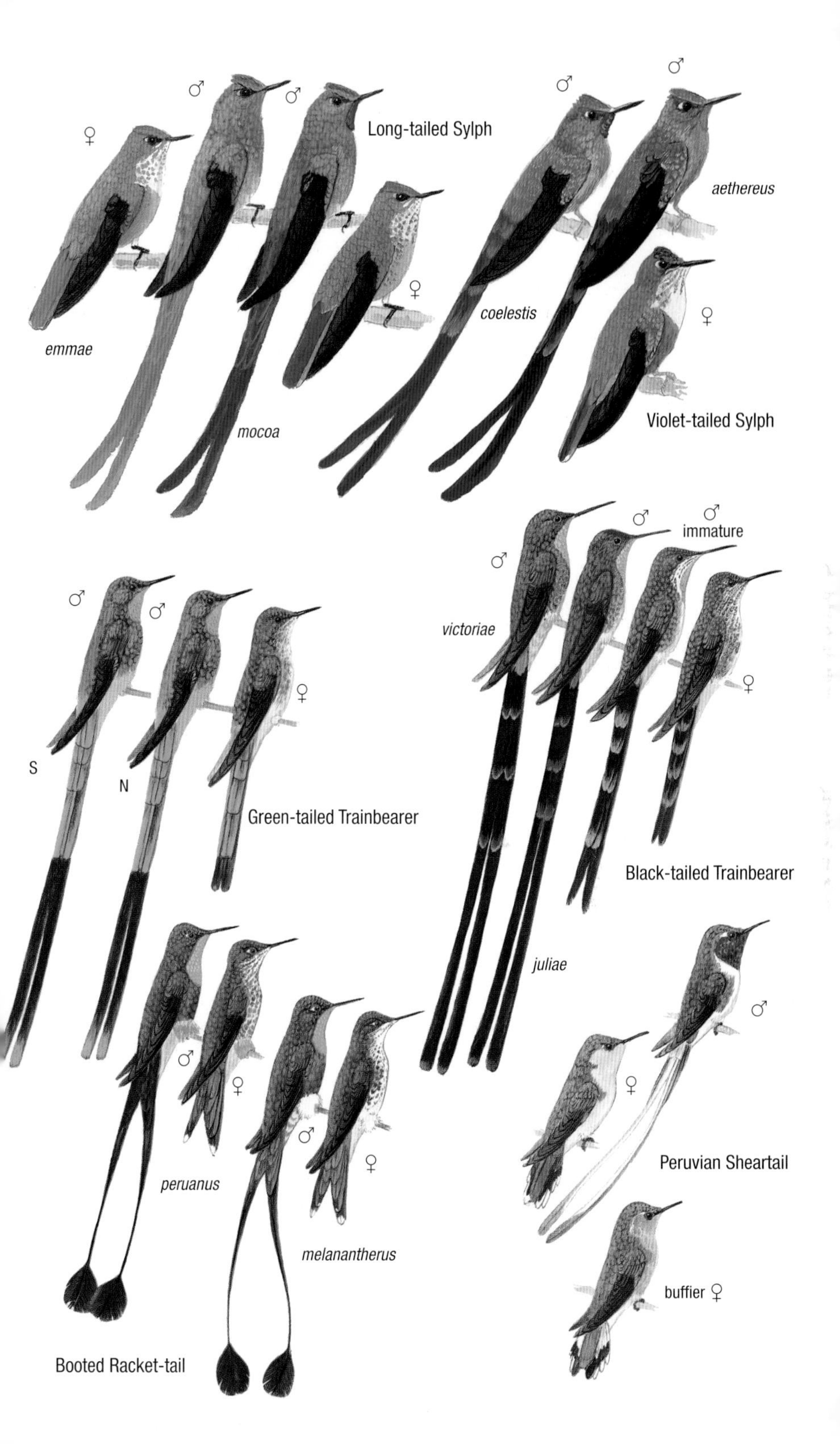
Long-tailed Sylph
♀
♂
♂
♀
emmae
mocoa
♂
♂
aethereus
coelestis
♀
Violet-tailed Sylph
♂
♂
♂ immature
victoriae
♀
Black-tailed Trainbearer
juliae
♂
♂
♀
S
N
Green-tailed Trainbearer
♂
♀
♂
♀
peruanus
melanantherus
Booted Racket-tail
♂
♀
Peruvian Sheartail
buffier ♀

PLATE 106: METALTAILS AND THORNBILLS

Small Andean hummingbirds. Metaltails have metallic glittering tails, but bodies tend to be more subdued. Thornbills are festooned by colourful glittering 'ties' and plain rufous crests. All feed by clinging to flowers, mainly low down, but apparently rob nectar occasionally.

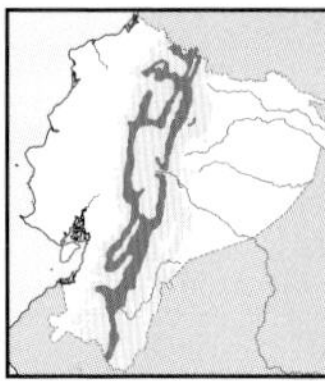

Tyrian Metaltail *Metallura tyrianthina* 8–9cm

Andes (mostly 2,300–3,800m). Montane forest, borders, woodland, adjacent clearings. Short straight bill (1–1.2cm). Metallic green above, ***glittering green throat patch***, bronzy-green below, paler and speckled belly to vent. ***Coppery-red tail*** slightly forked. ♀: ***buff throat to breast***, some green spots on sides. Race *quitensis* (W and inter-Andean valleys), *tyrianthina* (E, S). Often clings to flowers at edges, near ground, darting chases of others from territories. **Voice** Thin, fast jumble *dizidzdzdzdi...*, raspy *dzit'dzit'dzit...tiz-tiz-tiz....* **SS** Coppery tail diagnostic in range. **Status** Common.

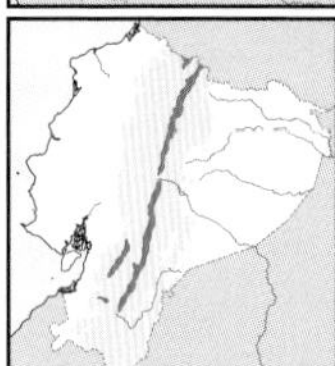

Viridian Metaltail *Metallura williami* 8–9cm

E Andes, local in W (2,700–3,700m). Treeline, shrubby *páramo*, *Polylepis* woodland. Short straight bill (1.2–1.3cm). Metallic green, ***glittering green*** (N) or ***black*** (S) ***throat***, vent speckled green-and-buff. ***Steel-green tail from below***. ♀: ***buff below densely speckled / streaked green***; buffier vent. Races *primolina* (N), *atrigularis* (S), might be separate species. Singles forage near ground, hover, cling or perch to feed; territorial. **Voice** Fast, short, descending *zee-zee-zee-...zrr*, often jumbled. **SS** Other metaltails less green, cf. Mountain Avocetbill (blue tail tipped white). **Status** Uncommon.

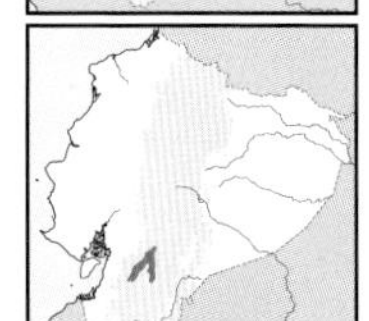

Violet-throated Metaltail *Metallura baroni* 8–9cm

Local in Azuay and Cañar highlands (3,100–4,000m). Treeline, shrubby *páramo*, *Polylepis* woodland. Short straight bill (1.2–1.3cm). ***Dull bronzy-green, dark violet throat patch***. Tail bronzy above, greener below. ♀: ***fainter throat patch***, mottled white and dusky below. Monotypic. Singles forage at various heights, rapid chases when defending feeding or display grounds. **Voice** Piercing, short *dzé-dzé-dzé*, followed by jumbled notes. **SS** Unmistakable in tiny range. **Status** Uncommon, range-restricted, EN.

Neblina Metaltail *Metallura odomae* 8–9cm

Local in highlands of extreme SE (2,900–3,650m). Treeline, shrubby *páramo*. Short straight bill (1.3–1.5cm). Dark bronzy-green, ***coppery-red throat patch***, more rufescent vent. ***Steel-green tail below***, greenish-blue above. ♀: duller, ***fainter throat patch***, mottled white on median underparts. Monotypic. Behaviour as other metaltails, some altitudinal movements possible. **Voice** Descending, very fast *dzee-dzee-dzee...drr* with jumbled end. **SS** From Tyrian Metaltail by undertail and throat, see shinier Little Sunangel with glowing throat. **Status** Uncommon, range-restricted, NT globally.

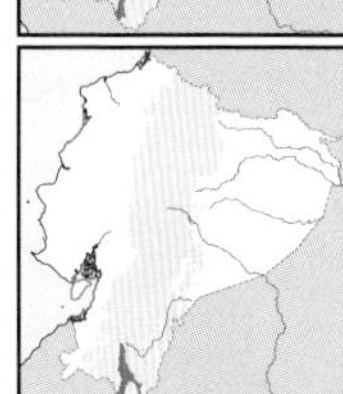

Rufous-capped Thornbill *Chalcostigma ruficeps* 9cm

Temperate zone in SE Andes (2,100–2,700m). Montane cloud forest, borders, adjacent shrubby clearings. Short spine-like bill (1–1.2cm). ***Rufous crown***, dull green above, ***buffy-green below, glittering green throat***, long forked tail, bronzy below. ♀: dull metallic green above ***including crown, buffy-green below***, with green spots mainly at sides. Monotypic. Singles forage near ground, regularly at landslides, borders and road embankments; retiring, clings to feed. **Voice** High *tzee* in flight, often chattered. **SS** ♂ with ♀ higher-elevation Rainbow-bearded (blue-black tail tipped white); ♀ with ♀ Tyrian Metaltail (smaller, coppery-tailed). **Status** Rare, local.

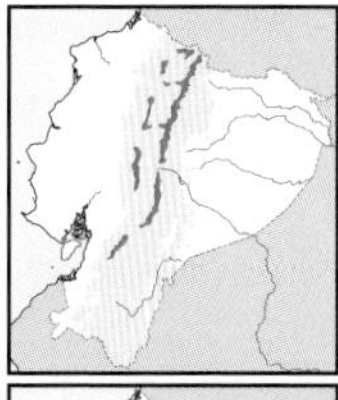

Blue-mantled Thornbill *Chalcostigma stanleyi* 10–12cm

Both Andean cordilleras, more local in W (3,600–4,500m). Shrubby *páramo*, *Polylepis* woodland. Short spine-like bill (1.1–1.2cm). ***Very dark overall, narrow and pointed throat mostly glittering green becoming glittering rosy-violet below. Blue-black long forked tail***. ♀: even darker as ***lacks throat patch***. Nominate. Singles forage from bush tops to near ground, territorial, mostly clinging to or perching next to small flowers. **Voice** Weak, fast *dizizizi'tzi* in flight. **SS** Confusion unlikely in habitat. In Azuay see Violet-throated Metaltail (dark but large violet throat, greener tail). **Status** Rare.

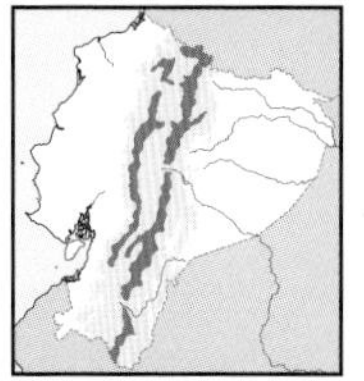

Rainbow-bearded Thornbill *Chalcostigma herrani* 10–11cm

Temperate zone on both Andean slopes (2,800–3,700m). Borders of montane forest and woodland, shrubby *páramo*, treeline. Short spine-like bill (1.1–1.2cm). ♂: ***rather long rufous crest, beautiful rainbow*** in ***narrow beard***. Otherwise dark metallic green, ***rufescent rump***, long forked ***deep blue*** tail, ***outer feathers broadly tipped white***. ♀ and imm. similar but throat has ***white speckles, no beard***. Nominate. Singly, mostly low, clinging to or perching next to flowers to feed; territorial and aggressive to other species including flowerpiercers. **Voice** Repeated chattering *cheet'dee'dee'dee'dee'cheet... dee-dee-cheet....* **SS** Lower-elevation Rufous-capped Thornbill (in SE). **Status** Uncommon.

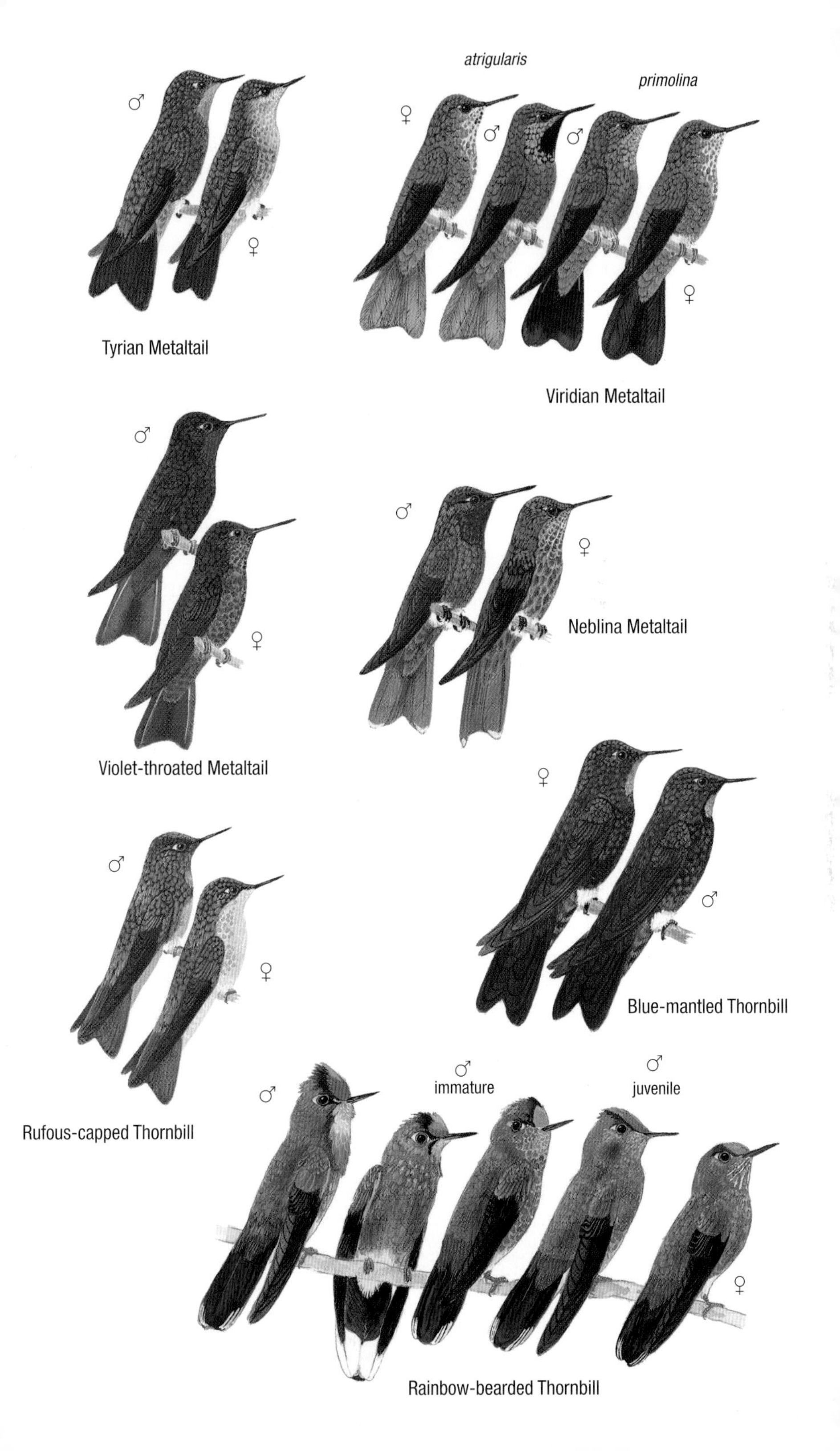

♂
♀
Tyrian Metaltail
atrigularis
primolina
♀
♂
♂
♀
Viridian Metaltail
♂
♀
Violet-throated Metaltail
♂
♀
Neblina Metaltail
♀
♂
Blue-mantled Thornbill
♂
♀
Rufous-capped Thornbill
♂
♂
immature
♂
juvenile
♀
Rainbow-bearded Thornbill

PLATE 107: MISCELLANEOUS MONTANE HUMMINGBIRDS

Speckled Hummingbird is among the most numerous Andean hummingbirds but might go undetected if not vocalising. Avocetbill and the thornbill apparently specialise in nectar-robbing. The two *Haplophaedia* pufflegs are also drab, their whitish puffs less distinct than in *Eriocnemis* pufflegs.

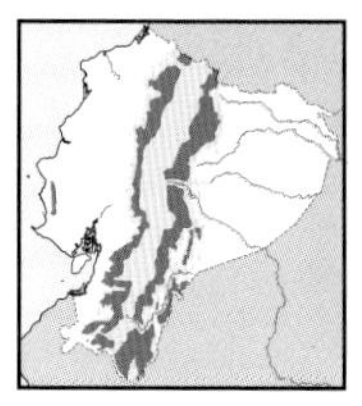

Speckled Hummingbird *Adelomyia melanogenys* — 8cm

Subtropics to temperate zone in Andes, also Cordillera de la Costa (mostly 1,400–3,200m, 600–800m in coastal mountains). Montane forest, borders, woodland, adjacent clearings. *Short bill* (1.2–1.3cm). Metallic green above, ***dusky cheeks, whitish supercilium. Whitish-buff below*** profusely spotted green on throat and sides. Bronzy tail *tipped buff*. Sexes alike. Nominate (E), *maculata* (W). Solitary, noisy flight, feeds low by clinging to or perching next to flowers. **Voice** Loud, distinctive *trrrrt* trill in flight; also piercing, descending *t't't'tsi-tsi-tsit-tsit-tseee-tseee-tseee….* **SS** Confusion unlikely despite dull pattern. **SS** Common.

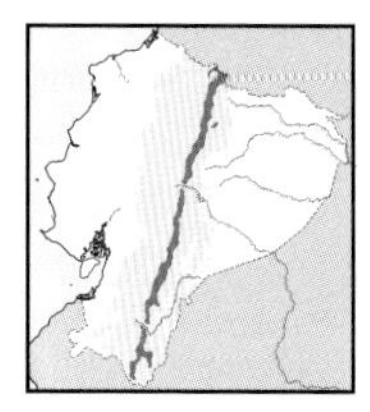

Mountain Avocetbill *Opisthoprora euryptera* — 10cm

Temperate zone on E Andean slope (2,400–3,300m). Montane forest undergrowth, borders. Short bill (1.3cm), ***upturned at tip; bold white postocular spot***. Bronzy-green head, ***whitish below boldly spotted / streaked green on central parts***, sides more solid, and ochraceous vent. Steel-blue tail, slightly forked, narrowly tipped white. Sexes alike. Monotypic. Shy singles in dense undergrowth, clings to flowers and often pierces them. **Voice** High-pitched, clear, inflected *wií-wií-wií-wií…*, sometimes more spaced. **SS** Streaks below separate from metaltails; see Tyrian Metaltail. **Status** Rare, local.

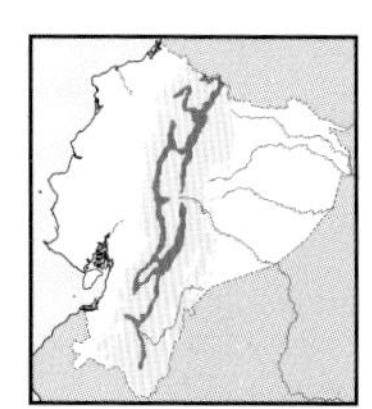

Purple-backed Thornbill *Ramphomicron microrhynchum* — 8–9cm

Temperate zone on both Andean slopes, mostly N (2,500–3,400m). Montane forest borders. ***Tiny spine-like bill*** (0.5cm), white postocular spot. ***Deep purple above***, glittering green throat, underparts densely speckled dusky. ***Long blackish tail***, somewhat forked. ♀: ***uniform green above***, white below, ***sides densely spotted green***, tail narrowly tipped white. Imm. purple above, green below. Nominate. Flies slow, trap-lining and often clings to or perches on open flowers, also uses holes pierced by flowerpiercers. **Voice** Long, raspy *tritritritrit*; more trilled *krrt* calls. **SS** Nearly unmistakable but see metaltails. **Status** Rare.

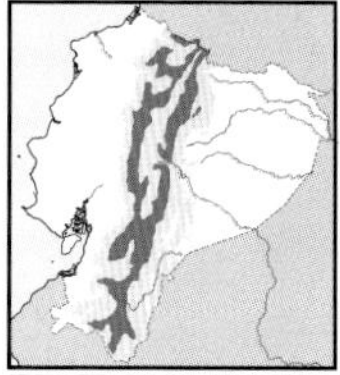

Mountain Velvetbreast *Lafresnaya lafresnayi* — 10–11cm

Andes (2,400–3,500m, locally higher or lower). Montane forest borders, shrubby clearings, treeline. Fairly long (2.2–2.5cm), notably ***decurved bill. Velvet black belly, long forked tail mostly white, broadly tipped black***, central tail feathers dull green. ♀: velvety black replaced by ***pure white on belly, speckled whitish-buff on median breast to throat***, tail as ♂. Race *saul*. Singly, at various heights, flashing white tail while hovering; ♂ territorial, ♀ more of trap-liner. **Voice** Clear *zeép!* well-spaced whistles. **SS** Unmistakable bill shape and tail pattern. **Status** Uncommon.

Greenish Puffleg *Haplophaedia aureliae* — 9–10cm

Local in E Andean subtropics (1,500–2,100m, locally lower). Wet montane forest. Mid-sized, straight bill (2cm), white postocular spot. ***Bronzy head***, otherwise mostly bronzy-green, more ***coppery uppertail-coverts***. ♀: dusky feather edges below give scalloped appearance. ***Small whitish thigh puffs. Deep blue forked tail***. Races *russata* (Andes), darker *cutucuensis* (outlying ridges) more scaled below. Forest interior, inconspicuous; singles forage at variety of flowers; ♂♂ may form loose singing assemblages. **Voice** Endless *trt'seé-trt'seé-trt'seé…* by displaying ♂; weak *zhír-zrt-zrt…* in flight. **SS** Nondescript but quite unique owing to tail and whitish puffs. **Status** Rare.

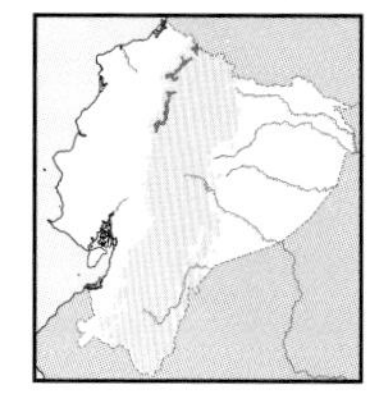

Hoary Puffleg *Haplophaedia lugens* — 9cm

Subtropics on NW Andean slope (1,700–2,100m, locally lower and higher). Inside wet montane forest, rarely at borders. Mid-sized, straight bill (2cm), white postocular spot. Mainly ***dull green with hoary wash overall***, throat somewhat scalloped white, greyer belly to vent, ***small white thigh puffs. Forked tail***, blackish-blue. Sexes alike. Monotypic. Mostly inside forest, rarely ventures to semi-open areas or feeders; shy and not well known. **Voice** Abrupt *tzík* or *djík!* when foraging. **SS** Few field marks, but this species is nearly unmistakeable; *Eriocnemis* pufflegs are larger, brighter and occur at higher elevations; drab violetears, brilliants, etc. differ markedly. **Status** Rare, local, nearly endemic, NT globally and in Ecuador.

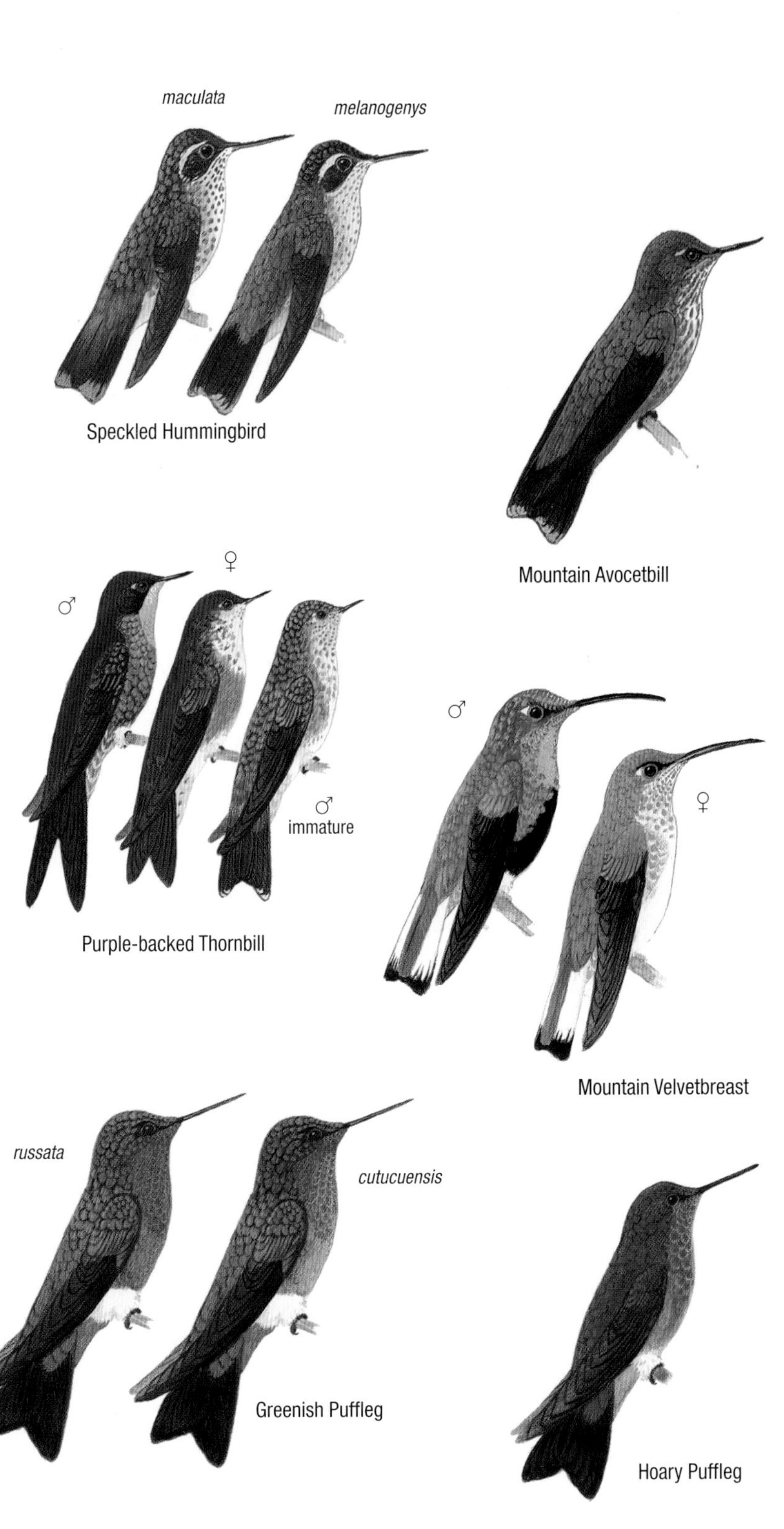

maculata
melanogenys
Speckled Hummingbird
Mountain Avocetbill
♂
♀
♂
immature
Purple-backed Thornbill
♂
♀
Mountain Velvetbreast
russata
cutucuensis
Greenish Puffleg
Hoary Puffleg

All have fluffy 'pants', black in one species. Confined to forested habitats in the Andes; three very restricted geographically are currently facing high risk of extinction. Up to four puffleg species may occur sympatrically in some areas, even feeding on same flowering shrubs. Emerald-bellied is the most dissimilar species, even in habitat choice (cloud forest interior).

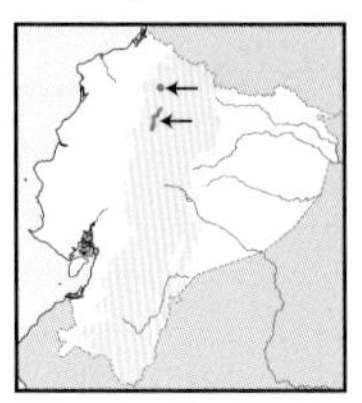

Black-breasted Puffleg *Eriocnemis nigrivestis* 8–9cm

Highlands in Pichincha and Imbabura (2,700–3,500m, historically higher, seasonally lower?). Montane and ridgetop forest, borders. Short, straight bill (1.5–1.6cm). Mainly dark green, glittering purple throat, ***breast velvety black, purple-blue vent, large*** white thigh puffs, blue-black forked tail. ♀: lacks black breast, ***small glittering turquoise throat patch***. Monotypic. Singles forage low at shrubby borders, fond of Ericaceae but wider diet, quite territorial, seemingly engages in seasonal movements. **Voice** Weak *tzeet-tzeet* perched; piercing *tzeet!* in flight. **SS** Smaller, shorter-tailed than sympatric pufflegs but cf. enigmatic Turquoise-throated Puffleg. **Status** Rare, tiny range, CR, endemic.

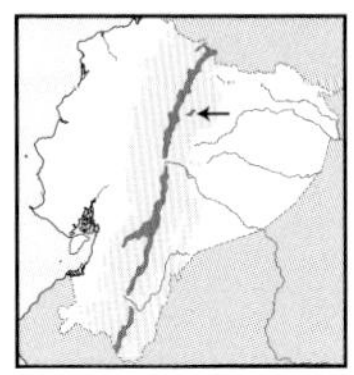

Glowing Puffleg *Eriocnemis vestita* 8–9cm

Highlands in E Andes (2,500–3,500m, locally lower). Montane forest, borders, dense shrubbery. Short, straight bill (1.5–1.7cm). Metallic green above, ***glittering purple throat, glittering green breast belly and rump***, white thigh puffs, blue-black forked tail. ♀: ***bronzy-buff throat and breast***, small turquoise throat patch, purple vent. Race *smaragdinipectus* (N), *arcosae* (S). Feeds by hovering and clinging low down, often several can gather, but territorial. **Voice** Not very vocal, thin *dzíp* calls in flight. **SS** Smaller, shorter-tailed than sympatric eastern pufflegs. **Status** Uncommon.

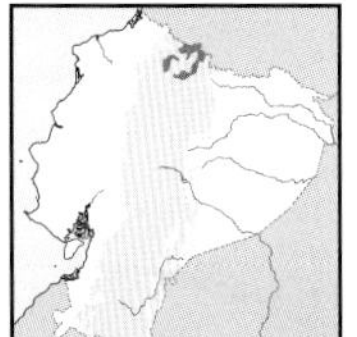

Black-thighed Puffleg *Eriocnemis derbyi* 8–9cm

Local in N highlands, S to Imbabura (3,000–3,500m, locally lower). Montane forest, borders, shrubbery. Short, straight bill (1.6–1.8cm). Metallic green above, ***glittering green uppertail-coverts***. Golden-green below, ***small black thigh puffs, glittering green vent, short, black, forked tail***. ♀: mottled green and dusky below, ***glittering green vent, dusky-whitish puffs***. Nominate. Mostly stays low at shrubby edges, may gather with other pufflegs but also territorial. **Voice** Weak *zrr-zrz-zrz....* **SS** Confusion unlikely though black puffs hard to see. **Status** Locally uncommon, NT.

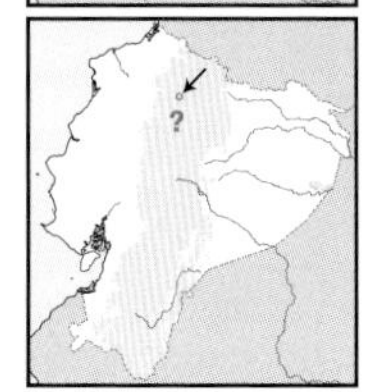

Turquoise-throated Puffleg *Eriocnemis godini* 8–9cm

Seemingly local in upper Río Guayllabamba valley (around 2,100–2,300m; formerly higher?). Habitat uncertain, perhaps semi-humid montane forest. Resembles Glowing Puffleg but ***throat patch turquoise not purple, golden throat and breast-sides***. ♀: similar but ***lacks*** golden-buff throat and breast. Monotypic. Unknown in life; few specimens with imprecise data collected in 1850s. Taxonomy uncertain, suggested to be subspecies of Glowing or hybrid. Recent fieldwork failed to locate it, but should be searched for in wooded gorges in Guayllabamba, Tumbaco, Perucho and similar valleys. Deemed CR, possibly EX.

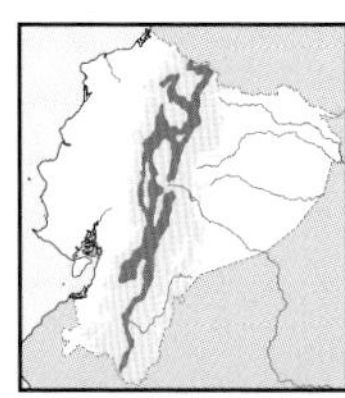

Sapphire-vented Puffleg *Eriocnemis luciani* 11–12cm

Andes S to Azuay (2,700–3,700m). Montane forest, borders, treeline, *Polylepis* woodland, shrubby clearings. Bill straight, ***long*** for a puffleg (2–2.1cm). Bronzy-green above, ***faint blue front. Glittering green below, purple vent***, large white thigh puffs. ***Blue-black long, deeply forked tail***. ♀: duller. Races *luciani* (N), *baptistae* (S). Bolder than other pufflegs in edges and semi-open, mostly alone and aggressive resource defender. **Voice** Thin *dzir'r-dzirt-zirt-zirt...* in short series. **SS** Golden-breasted (bronzy tail, dusky vent, golden breast); other pufflegs distinctively smaller. **Status** Fairly common.

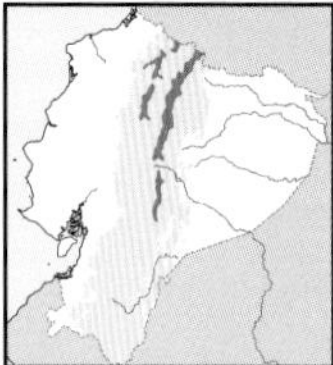

Golden-breasted Puffleg *Eriocnemis mosquera* 11–12cm

Andes in E (S to Morona Santiago) and NW (S to Cotopaxi) (3,000–3,600m, locally lower). Montane and ridgetop forest, borders, shrubby clearings. Bill straight, ***long*** for a puffleg (2–2.1cm). Bronzy-green above, metallic green below, ***breast to neck-sides golden-buff, dusky vent***, large white thigh puffs. ***Long bronzy-green, deeply forked tail***. Sexes nearly alike. Monotypic. Aggressive territory defender, more inside forest than other pufflegs, feeds by hovering or clinging. **Voice** Short, descending *tcheeu*. **SS** Similar-sized Sapphire-vented Puffleg. **Status** Uncommon.

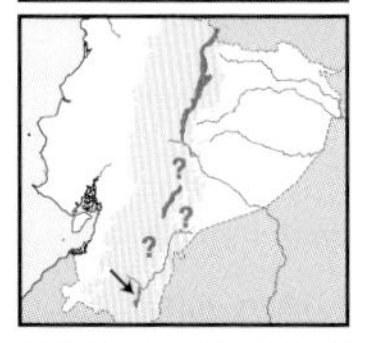

Emerald-bellied Puffleg *Eriocnemis aline* 7–8cm

Local along E Andean slope (1,800–2,300m). Cloud forest, borders. Short straight bill (1.5cm). Bronzy-green above, glittering green below, ***white irregular patch on mid breast***, glittering vent, ***large*** white thigh puffs; short, ***forked green tail***. ♀: less white in breast. Nominate. Singly inside forest, in treefall gaps, groves, small creeks; poorly known. **Voice** Very thin *zít* calls. **SS** Size, habitat and elevation separate from other pufflegs; puffs smaller in sympatric Greenish Puffleg. **Status** Rare, DD in Ecuador.

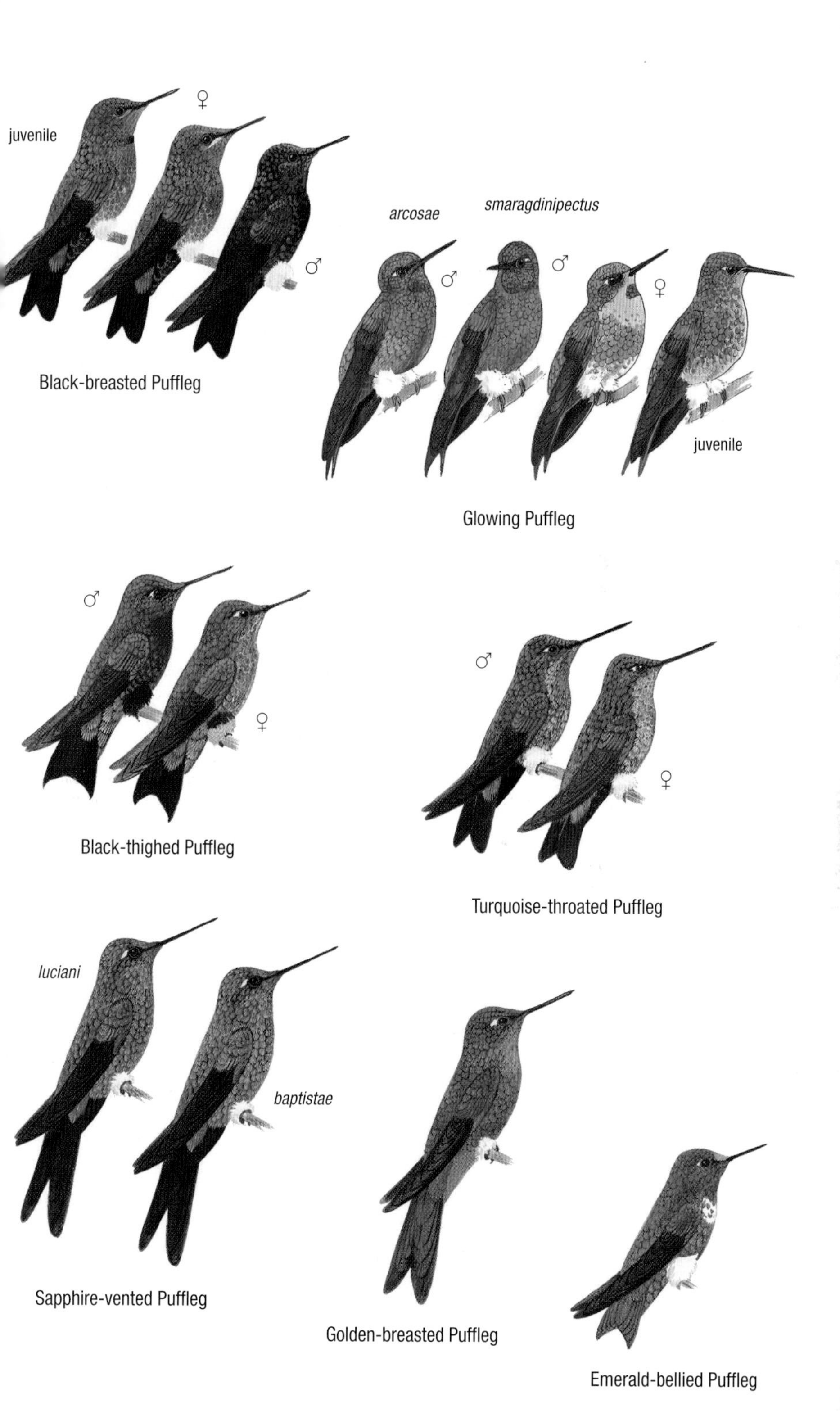
juvenile
♀
♂
Black-breasted Puffleg
arcosae
smaragdinipectus
♂
♂
♀
juvenile
Glowing Puffleg
♂
♀
Black-thighed Puffleg
♂
♀
Turquoise-throated Puffleg
luciani
baptistae
Sapphire-vented Puffleg
Golden-breasted Puffleg
Emerald-bellied Puffleg

PLATE 109: LARGE ANDEAN HUMMINGBIRDS

Fairly large hummingbirds, including the largest. All range high in the Andes, many confined to true *páramo*. High specialisation in at least two species on this plate: Ecuadorian Hillstar, nearly confined to *Chuquiraga* stands, and Sword-billed, which is necessarily specialised in flowers with very long corollas. Sapphirewing and sunbeam are among the commonest *páramo* hummingbirds.

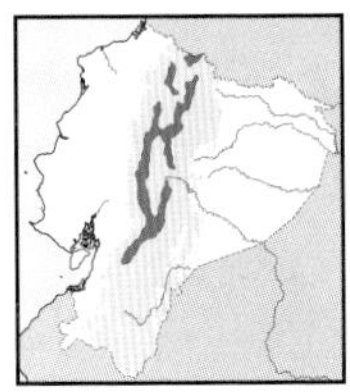

Ecuadorian Hillstar *Oreotrochilus chimborazo* — 11–12cm

Andes S to Azuay (3,600–4,600m). Dry *páramo*, mostly *Chuquiraga* stands. Medium-sized (2–2.2cm), slightly curved bill. ***Deep purple hood*** contrasting with ***immaculate white underparts and black median belly stripe. Most tail white***, central feathers blue-green. ♀: ***dull green above, greenish-buff below***, striped throat, tail mostly green, ***outer feathers broadly tipped white***. Races *jamesoni* (widespread), *chimborazo* (Volcán Chimborazo) has green throat. Fiercely territorial, with darting chases above *Chuquiraga* stands; often clings to flowers or perches to feed. **Voice** Fast jumbled *djidjidjidjidji...*, more abrupt *djít!* calls. **SS** Unmistakable. **Status** Locally uncommon.

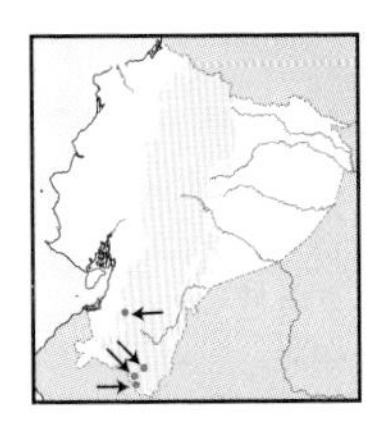

Andean Hillstar *Oreotrochilus estella* — 11–12cm

Local in extreme SE (3,250–3,500m). Wet rocky *páramo* and treeline. Medium-sized (2cm), slightly curved bill. ***Olive-green upperparts, glittering green throat*** bordered by black line, ***immaculate white underparts***, black median stripe on belly. ***White square tail***, central feathers green. ♀: resembles Ecuadorian Hillstar but greener above, greyer below, white equally extensive in tail. Race *stolzmanni*, possibly valid species. Aggressive, but not tied to *Chuquiraga* stands; ♀ less conspicuous than ♂. **Voice** Rising *spit* (S *et al.*). **SS** Unique in range. **Status** Uncommon but very local **Note** A new, undescribed form related to Andean Hillstar occurs in N Loja and adjacent El Oro highlands.

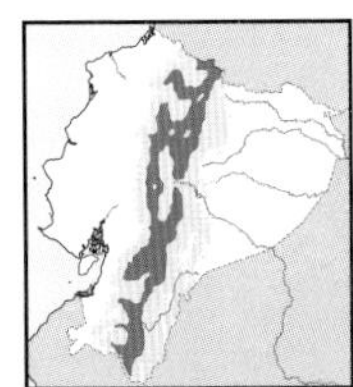

Shining Sunbeam *Aglaeactis cupripennis* — 11–12cm

Andes (2,800–3,900m, locally higher or lower). Grassy and shrubby *páramo*, treeline, woodland, shrubby clearings. Short straight bill (1.4–2cm). Dusky upperparts, ***glittering gold, purple and green rump to uppertail-coverts, cinnamon-rufous below***. ♀: ***lacks glittering rump***. Race *cupripennis* (most of range), *parvulus* (S). Conspicuous at edges and in scrub, where aggressive and territorial; holds wings open long after perching. **Voice** Weak *t'síp-t'síp-t'síp...*, faster *t'sípsípsíp....* **SS** See longer-billed Rainbow Starfrontlet in S. **Status** Common.

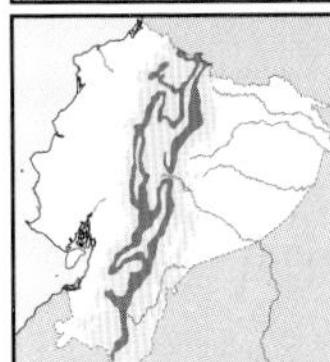

Sword-billed Hummingbird *Ensifera ensifera* — 13–14cm

Andes (2,500–3,400m, locally lower). Montane forest, borders, adjacent clearings. ***Bill incredibly long, slightly up-curved*** (9–10cm!). Bronzy-green head, tail long and deeply forked. ♀: white below, streaked bronzy-green on throat, speckled green below. Monotypic. Mostly trap-lines at long flowers by hovering but also perches beside them; not easy to find, but waiting by *Datura* or *Brugmansia* trees or *Passiflora* vines might be rewarded; often perches atop twigs. **Voice** Distinctive but infrequent *deet!* while feeding. **SS** Impossible to mistake. **Status** Rare.

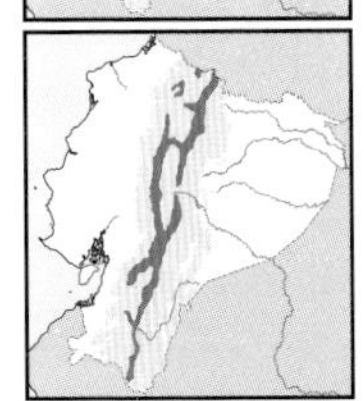

Great Sapphirewing *Pterophanes cyanopterus* — 15–17cm

Andes (2,900–3,600m). Forest borders, treeline, shrubby *páramo*. ***Long straight bill*** (2.8–3.2cm). Large, mostly bright green, ***long, deep blue wings including wing-coverts, long and forked tail***. ♀: ***cinnamon below, bluish-green wing-coverts, dusky flight feathers***. Race *peruvianus*. Slow wingbeats, even some glides, make flight strong; feeds by hovering, also perches with briefly open wings, territorial and trap-liner. **Voice** Thin, jumbled *djiír..djiír..djiír....* **SS** Unmistakable by size, wing length and colour, and flight. **Status** Uncommon.

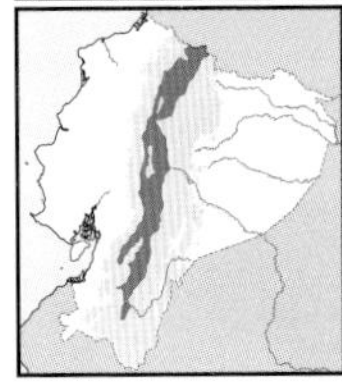

Giant Hummingbird *Patagona gigas* — 17–19cm

Inter-Andean valleys (1,800–3,300m, locally or seasonally higher). Arid scrub, agricultural areas, gardens, city parks. ***Huge*** for a hummingbird. Long straight bill (3.8–4.2cm). ***Dusky above***, contrasting ***irregular whitish rump, cinnamon-buff below***. ♀: similar but whitish throat to upper breast ***streaked and mottled dusky***. Race *peruviana*. Strong flight with slow wingbeats and short glides; aggressive defender of feeding areas, especially *Agave* flowers, but not tied to them. **Voice** Clear high-pitched *kueét!* **SS** Unique. **Status** Uncommon, seemingly engages in seasonal movements.

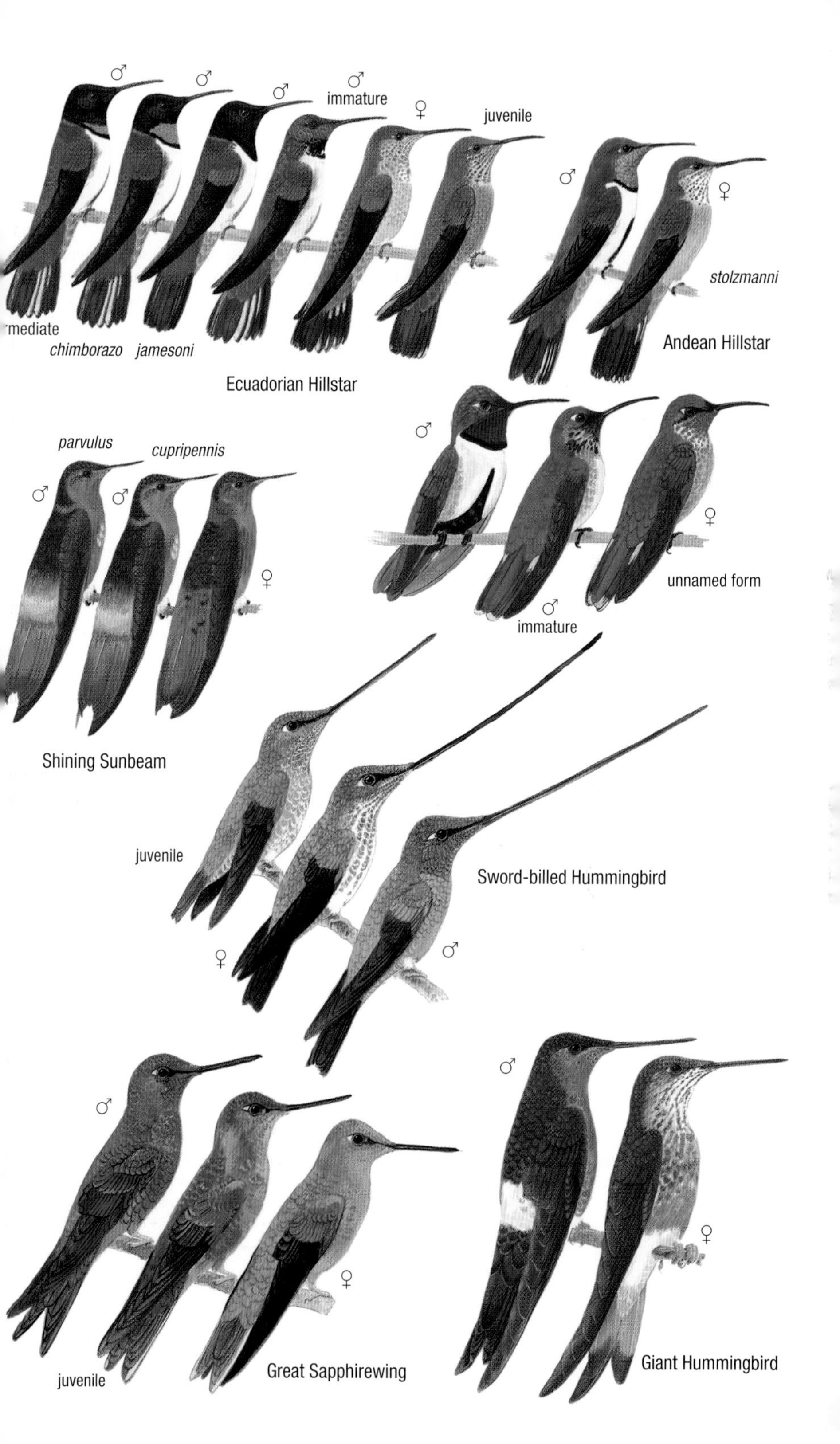

♂
♂
♂
♂
immature
♀
juvenile
♂
♀
stolzmanni
rmediate
chimborazo
jamesoni
Andean Hillstar
Ecuadorian Hillstar
♂
parvulus
cupripennis
♂
♂
♀
♀
unnamed form
♂
immature
Shining Sunbeam
juvenile
Sword-billed Hummingbird
♀
♂
♂
♂
♀
♀
Giant Hummingbird
juvenile
Great Sapphirewing

Spectacular hummingbirds of cloud forest to treeline, with long straight bills, tilted upwards when perched. Two brownish species lack sexual dimorphism, while the remaining three are more strikingly patterned and sexually dimorphic. The two brownish incas inhabit cloud forests either side of the Andes, starfrontlets tend to occur at higher elevations.

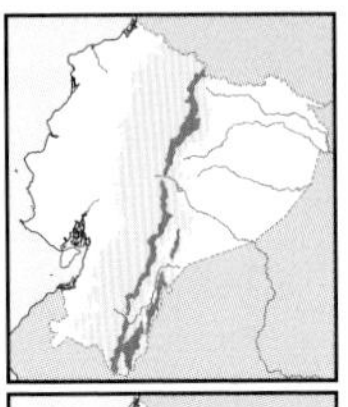

Bronzy Inca *Coeligena coeligena* 11–12cm

Subtropics on E Andean slope (1,400–2,300m). Montane forest, borders, adjacent clearings. *Long* (3.2–3.5cm) *straight bill.* Mostly *bronzy-brown, shinier above,* greener back and rump, throat and breast streaked white and dusky, somewhat forked tail is bronzy. Race *obscura.* Mostly inside forest, less often in semi-open areas; feeds by hovering in front of long colourful flowers; trap-liner, often chased away by other hummingbirds; briefly at feeders. **Voice** High-pitched, short *tzi-tzi-tzi*; also emphatic *sweet* or *szeet* calls. **SS** Confusion unlikely. Similar Brown Inca only in W. **Status** Uncommon.

Brown Inca *Coeligena wilsoni* 10–11cm

W Andean foothills to subtropics (800–2,000m, locally lower). Mid to lower growth of montane forest, borders. *Long* (3.2–3.3cm) *straight bill. Bronzy-brown above,* greener lower back and rump, dull brown below, small *violet patch on throat-sides, white patch on neck-sides,* bronzy tail. Monotypic. Trap-liner mostly inside forest, less often in semi-open areas; forages at various heights on long colourful flowers; brief visits to feeders. **Voice** Fast descending series of *tseee-tzeeé..dzeé* calls in flight; also squeaky *tsík-sík-sík-ik-ik....* **SS** Confusion unlikely; similar Bronzy Inca only in E; see Brown Violetear. **Status** Uncommon.

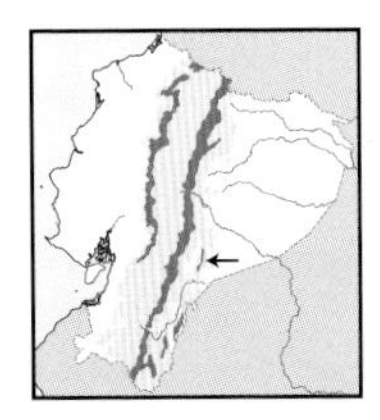

Collared Inca *Coeligena torquata* 11–12cm

Subtropics to temperate zone on both Andean slopes, not in SW (2,100–3,000m, locally to 1,800m). Montane forest undergrowth, borders, adjacent clearings. *Long* (3.2–3.5cm) *straight bill.* Mostly *velvet black,* greener throat (*brighter green* in W), *glittering purple patch on mid crown, contrasting white chest; white tail broadly tipped black.* ♀: similar but *greener* overall. Nominate (E), *fulgidigula* (W) has *blue crown patch, greener sheen* below. Conspicuously forages at various heights at long colourful flowers; regular trap-liner at borders and feeders; sometimes briefly joins mixed-species flocks. **Voice** Very fast twitters, *tee-dee-dee'ditditdit...* or longer *pip-pip-pip...* while foraging. **SS** Unmistakable. **Status** Fairly common.

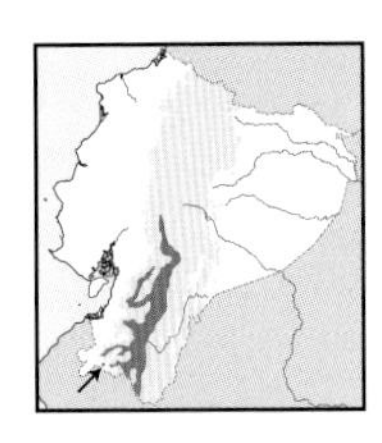

Rainbow Starfrontlet *Coeligena iris* 11–12cm

S Andes, N to Bolívar (2,000–3,300m, locally lower). Forest borders, woodland, shrubby clearings, gardens. *Long* (3–3.3cm) *straight bill.* Lovely, *rainbow-like crown* with glittering gold, orange, coppery and blue. Bright green throat to chest, dusky upperparts. *Rich rufous rear parts,* including tail. ♀: *crown mostly bright green.* Nominate (S), *hesperus* (N of Azuay) has *rufous limited to lower belly, coppery and purple crown.* More tolerant of semi-open and dry situations than congeners; conspicuously forages at various heights, trap-lining at variety of flowers, not necessarily with long corollas. **Voice** Very fast, high-pitched *dzidzidzi'trrr* trills in flight. **SS** Unmistakable. **Status** Fairly common.

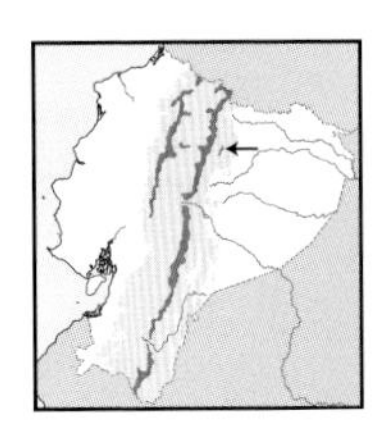

Buff-winged Starfrontlet *Coeligena lutetiae* 11–12cm

Temperate forest to treeline on both Andean cordilleras, S to Bolívar in W (mostly 2,700–3,700m). Montane forest undergrowth, borders, woodland, adjacent clearings, treeline. *Long* (3.2–3.5cm) *straight bill.* ♂ has bright green front and underparts, *small violet-blue patch on throat. Velvet black above, square buff patch on secondaries and tertials.* ♀: *metallic green* overall, somewhat scalloped below, *buff throat.* Bronzy-green tail slightly forked. Monotypic. Conspicuous trap-liner at various heights, often flying between distant forest patches; regular at feeders where aggressive. **Voice** Short squeaky, nasal *kwék* in flight, highly distinctive; jumbled *kwék-zitzitzit* series. **SS** Confusion unlikely owing to buff wing patch and squeaky calls. **Status** Fairly common.

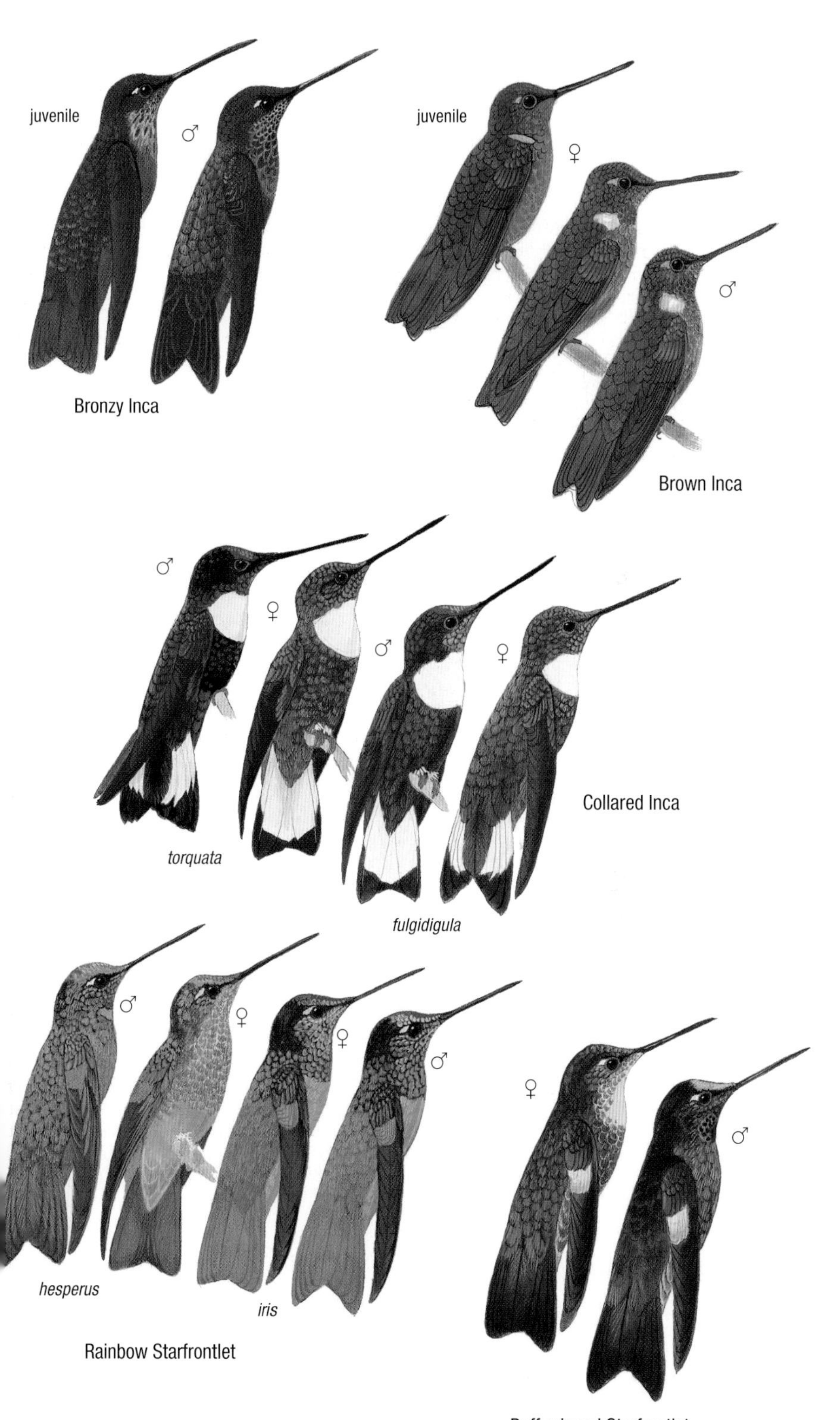

juvenile
♂
Bronzy Inca
juvenile
♀
♂
Brown Inca
♂
♀
♂
♀
Collared Inca
torquata
fulgidigula
♂
♀
♀
♂
hesperus
iris
Rainbow Starfrontlet
♀
♂
Buff-winged Starfrontlet

PLATE 111: CORONETS, HILLSTAR AND WHITETIPS

Coronets are smallish hummingbirds of Andean cloud forests, somewhat resembling incas but shorter-billed, with contrastingly bold tails; Velvet-purple Coronet is among the most striking hummingbirds in Ecuador. White-tailed Hillstar also resembles incas but plumage more subdued. Whitetips are smaller, shier, more uniformly glittering green and thrive in cloud forest undergrowth.

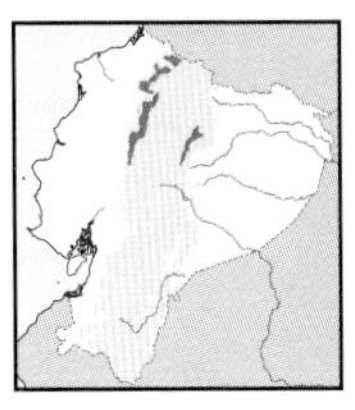

Buff-tailed Coronet *Boissonneaua flavescens* 10–11cm

Subtropics on NW Andean slope, also locally in NE (1,500–2,400m, locally higher). Montane forest, borders, adjacent clearings. Bill short (1.8–2cm) and straight, small white postocular spot. Mainly bright green, buffy-green median belly. *Wing bend and underwing-coverts rich buff.* Bronzy square tail, *outer feathers mostly buff*, tipped brown. ♀: buffier below. Nominate (E), *tinochlora* (W). Solitary, sometimes gathers at flowering trees, numerous at feeders, but territorial; holds wing aloft after perching. **Voice** Fast, rather buzzy *chip-chip-chip...*series. **SS** Rich buff underwing and pale buff tail conspicuous in flight, and confusion unlikely. **Status** Fairly common, rarer in NE.

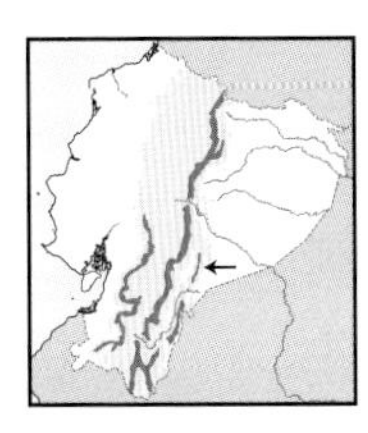

Chestnut-breasted Coronet *Boissonneaua matthewsii* 10–11cm

Subtropics on E & SW Andean slopes (1,600–2,800m, locally higher or lower). Montane forest, borders, adjacent clearings. Bill short (1.8–2cm) and straight, small whitish postocular spot. Metallic green above, throat buff spotted green, rest of *underparts rich chestnut including most of tail. Wing bend and underwing-coverts rufous-buff.* ♀: duller. Monotypic. Aggressively defends territory, but can gather with others at flowering trees and feeders; holds wings aloft after perching. **Voice** High *chíp-chíp-chi'chi'chí-chíp....* **SS** General pattern similar to Rainbow Starfrontlet but bill much shorter; underwing and tail conspicuous in flight; see Fawn-breasted Brilliant. **Status** Uncommon.

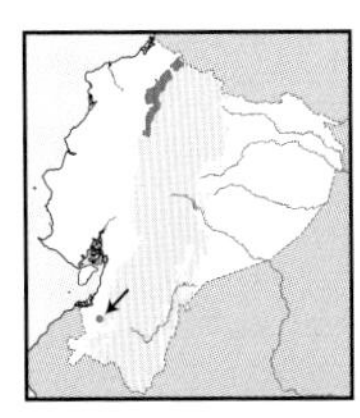

Velvet-purple Coronet *Boissonneaua jardini* 10–11cm

Foothills to subtropics in NW Andean slope (800–1,800m). Montane forest, borders, adjacent clearings. Bill short (2–2.2cm) and straight, small white postocular spot. Strikingly *dark glittering purple on front, throat and most underparts*; rest of head and underparts black. *Upperparts glittering bluish-green. Wing bend and underwing-coverts rufous*, tail *mostly white, edged and tipped black.* ♀: duller. Monotypic. Territorial and aggressive but may gather at flowering trees or feeders; holds wings aloft after perching. **Voice** A *si-siir-si-siir-si-siir*, first soft, second harsh. **SS** Looks dark but white tail and rufous underwing conspicuous in flight; cf. heavier White-tailed Hillstar. **Status** Rare.

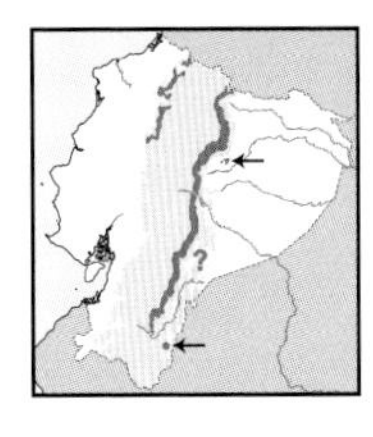

White-tailed Hillstar *Urochroa bougueri* 11–12cm

Subtropics on NE & NW Andean slopes (1,100–2,000m). Montane forest undergrowth, borders, near rushing water. Bill long (3–3.2cm), straight. Bronzy-green upperparts, coppery rump. Striking *orange-rufous malar* (W), glittering blue-green throat to breast, rest of underparts bronzy-green. Rounded tail *mostly white edged black.* ♀: duller. Nominate (W), *leucura* (E) *lacks* orange malar, *whiter tail*, possibly separate species. Aggressive and dominant but rather inconspicuous; regularly visits long tubular flowers. **Voice** Simple, weak *tsít, tsiit* in slow series; faster in W. **SS** Juv. brilliants (Green-crowned, Violet-fronted) show rufous moustachial but bill and tail longer, no white tail; see Napo Sabrewing. **Status** Rare.

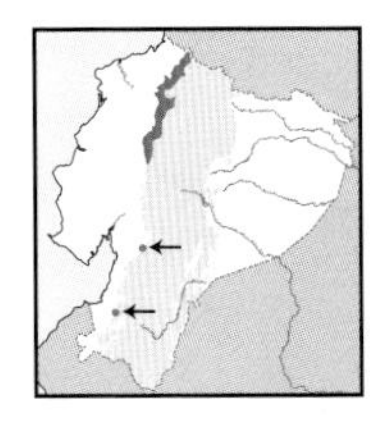

Purple-bibbed Whitetip *Urosticte benjamini* 8–9cm

W Andean foothills to subtropics, more local S of Cotopaxi (900–1,800m, locally lower). Montane forest undergrowth, borders. Short (1.8–2cm) straight bill, *white postocular stripe.* Metallic green sheen on throat and rump, *glittering fuchsia patch on throat* underlined by thin white breast-band. Blackish tail deeply forked, *central feathers with broad white tips.* ♀: similar above, white below *thickly spotted shiny green, white malar*, lacks white central tail tips but *outer feathers narrowly tipped white.* Monotypic. Mostly inside forest foraging at various heights, feeds by hovering; rather shy, but increasingly seen at feeders. **Voice** Buzzy *ch'rrr-ch'rrr-chi'chi'chí* in flight, *chí'chiri'rririri.* **SS** ♀ Booted Racket-tail has longer and less forked tail. **Status** Rare.

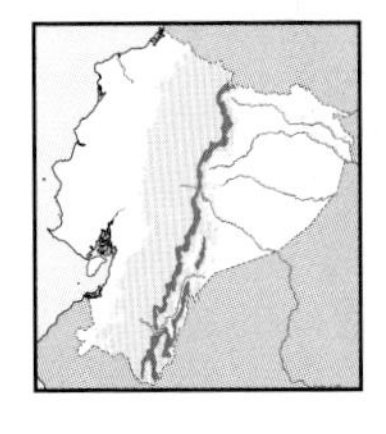

Rufous-vented Whitetip *Urosticte ruficrissa* 8–9cm

Subtropics on E Andean slope (1,200–2,300m). Montane forest undergrowth, borders. Short (1.8–2cm) straight bill, white postocular spot. Metallic *green sheen on throat to chest underlined by narrow white breast-band, buff vent.* Blackish tail deeply forked, *central feathers broadly tipped white.* ♀: resembles Purple-bibbed Whitetip but green spotting mostly to breast, *shorter* postocular stripe, *buff vent.* Monotypic. Rarely seen as skulks in forest undergrowth, mostly solitary. **Voice** Dry *jék!-jék!...* in chatter or singly. **SS** Purple-bibbed only in W; ♀ Booted Racket-tail also spotted below but smaller, shorter bill, buff thighs; also Ecuadorian Piedtail (very different tail pattern). **Status** Rare.

tinochlora
♂
Buff-tailed Coronet
flavescens
♂
♂
juvenile
♂
Chestnut-breasted Coronet
♀
♂
bougueri
♂
leucura
White-tailed Hillstar
♀
♂
Velvet-purple Coronet
♂
♀
♂ juvenile
♀ juvenile
Purple-bibbed Whitetip
♂
♀
♂ juvenile
Rufous-vented Whitetip

Brilliants are characterised by strong, longish, slightly curved bills with feathered base to maxilla. Tails rather long and forked, especially in the impressive Empress. Sexual dimorphism very marked in most species, ♀ being white dotted green below. Some ♂♂ are stunningly glittering. Two species in E lowlands, the rest in Andean foothills to cloud forests.

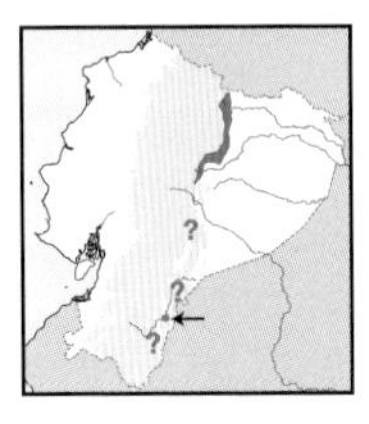

Pink-throated Brilliant *Heliodoxa gularis* — 9–10cm

Local in E Andean foothills (350–1,000m). Montane forest undergrowth, borders. Rather long, straight bill (2.5cm). *Glittering pink throat, white vent; forked tail, bronzy green.* ♀: ***smaller*** throat patch, ***short malar streak, mottled*** underparts, ***white vent***. Monotypic. Solitary and territorial in shady areas; fond of epiphytes. **Voice** Thin, nasal, squeaking *chéep-chéep*... or *chiúp-chiúp*... in repeated series; *cheep-chirriíp*, accelerating *chirp-prrriu* and jumbled *cheerríp-chichichichí...prriu-trtrtít'*; harsher *chrp-chrp-chrp* when feeding. **SS** Darker and smaller than other brilliants, white vent diagnostic. **Status** Rare, VU.

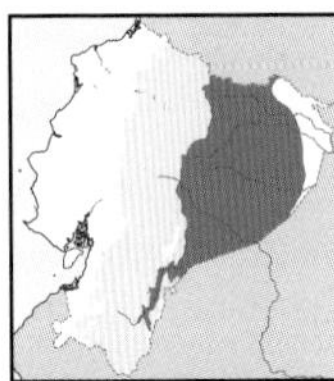

Black-throated Brilliant *Heliodoxa schreibersii* — 11–12cm

E lowlands to foothills (300–1,250 m, locally higher). *Terra firme* undergrowth, borders. Rather long decurved bill (2.5–2.6cm). Glittering frontlet, ***velvet black throat, glittering purple patch below, shining green breast, black belly***. Long, forked tail, blue-black. ♀: ***smaller*** purple patch, ***buffy-white malar streak***. Race *schreibersii*. Mostly low, rarely in open or higher strata, often near water. **Voice** Very high-pitched, thin, descending trill, *ch'rrrrrrrrrrr*, not hummingbird-like. **SS** Nearly unmistakable. **Status** Rare.

Gould's Jewelfront *Heliodoxa aurescens* — 9–10cm

E lowlands (to 900m, locally higher). *Terra firme* forest undergrowth, borders. Rather short, straight bill (1.8–2cm). *Glittering purple forecrown, velvety black chin, contrasting rufous chest-band. Chestnut tail* tipped green. ♀: similar but *buff* not black throat, smaller purple patch on forecrown. Monotypic. Singles in shady forest, often near streams; shy and seldom seen. **Voice** Piercing, thin *títh!* every *c.*1 sec. **SS** Rufous chest and chestnut tail unique. **Status** Rare, probably overlooked.

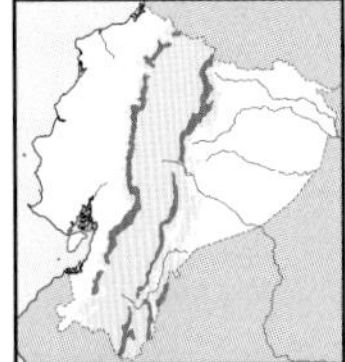

Fawn-breasted Brilliant *Heliodoxa rubinoides* — 10–11cm

Subtropics on both Andean slopes (1,100–2,400m, locally higher or lower). Montane forest undergrowth, borders, adjacent clearings. Fairly long, slightly decurved bill (2.3–2.6cm). Metallic green above, ***tawny below***, green speckling at sides, ***more ochraceous belly, pink patch on mid throat***. Bronzy tail long, forked. ♀: ***duller, mottled green on tawny throat***, no pink. Races *aequatorialis* (W), *cervinigularis* (E). Prone to emerge in open and congregates at trees or feeders, but rather shy. **Voice** Continuous *tchí-tchí-tchí*.... **SS** Chestnut-breasted Coronet and ♀ sylphs. **Status** Uncommon.

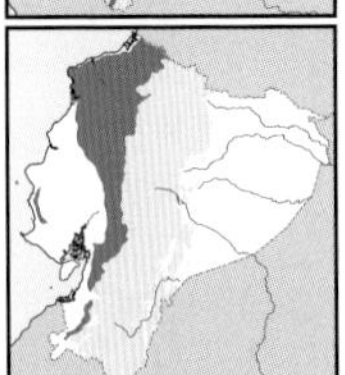

Green-crowned Brilliant *Heliodoxa jacula* — 10–12cm

W lowlands to subtropics (300–1,550 m, locally lower). Humid forest undergrowth, borders, adjacent clearings. Rather long straight bill (2.5–2.7cm). ***Glittering green crown, throat and breast***, small ***glittering purple throat patch***, fairly ***long forked steel-blue tail***. ♀: ***white malar streak***, below ***thickly spotted green, outer tail feathers tipped white***. Imm. ***buff moustache***. Race *jamesoni*. Conspicuous, mostly low inside forest, dominant at flowering trees and feeders, perching to feed. **Voice** Emphatic *chík! chík!* **SS** ♀ Empress Brilliant similar, but larger and longer-tailed. **Status** Uncommon.

Empress Brilliant *Heliodoxa imperatrix* — 12–13cm

Local on NW Andean slope (1,500–2,200m). Montane forest undergrowth, borders. Fairly long straight bill (2.5–2.7cm). Glittering ***golden-green belly, small purple throat patch. Long deeply forked tail, bronzy-green***. ♀ resembles Green-crowned Brilliant but ***golden-green*** belly, long ***bronzy-green tail*** shorter than ♂. Monotypic. Mainly inside forest, flight rather slow; hovers and clings to feed, regularly visits feeders. **Voice** Short *chi* in slow series. **SS** Smaller Green-crowned Brilliant has shorter, less forked, blue tail (tipped white in ♀). **Status** Rare, VU in Ecuador.

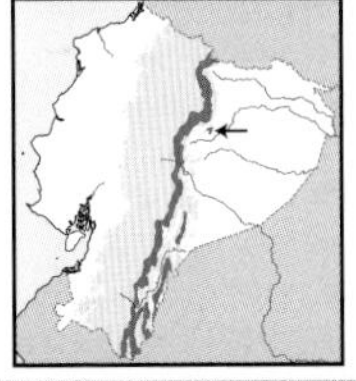

Violet-fronted Brilliant *Heliodoxa leadbeateri* — 11–12cm

E Andean subtropics (1,300–2,100m, locally lower). Montane forest, borders. Rather short, straight bill (2.4–2.5cm). ***Glistening violet front, glittering green throat to breast. Tail long, forked, blue-black.*** ♀: ***bluish forecrown, white malar streak***, mostly white below, ***densely spotted green, buffy-white mid belly***. Race *sagitta*. Solitary in forest cover though sometimes ventures into adjacent clearings; rather aggressive. **Voice** Slow, emphatic *cheép* series. **SS** Duller Many-spotted Hummingbird (solid white median underparts). From other brilliants by tail and forecrown. **Status** Uncommon.

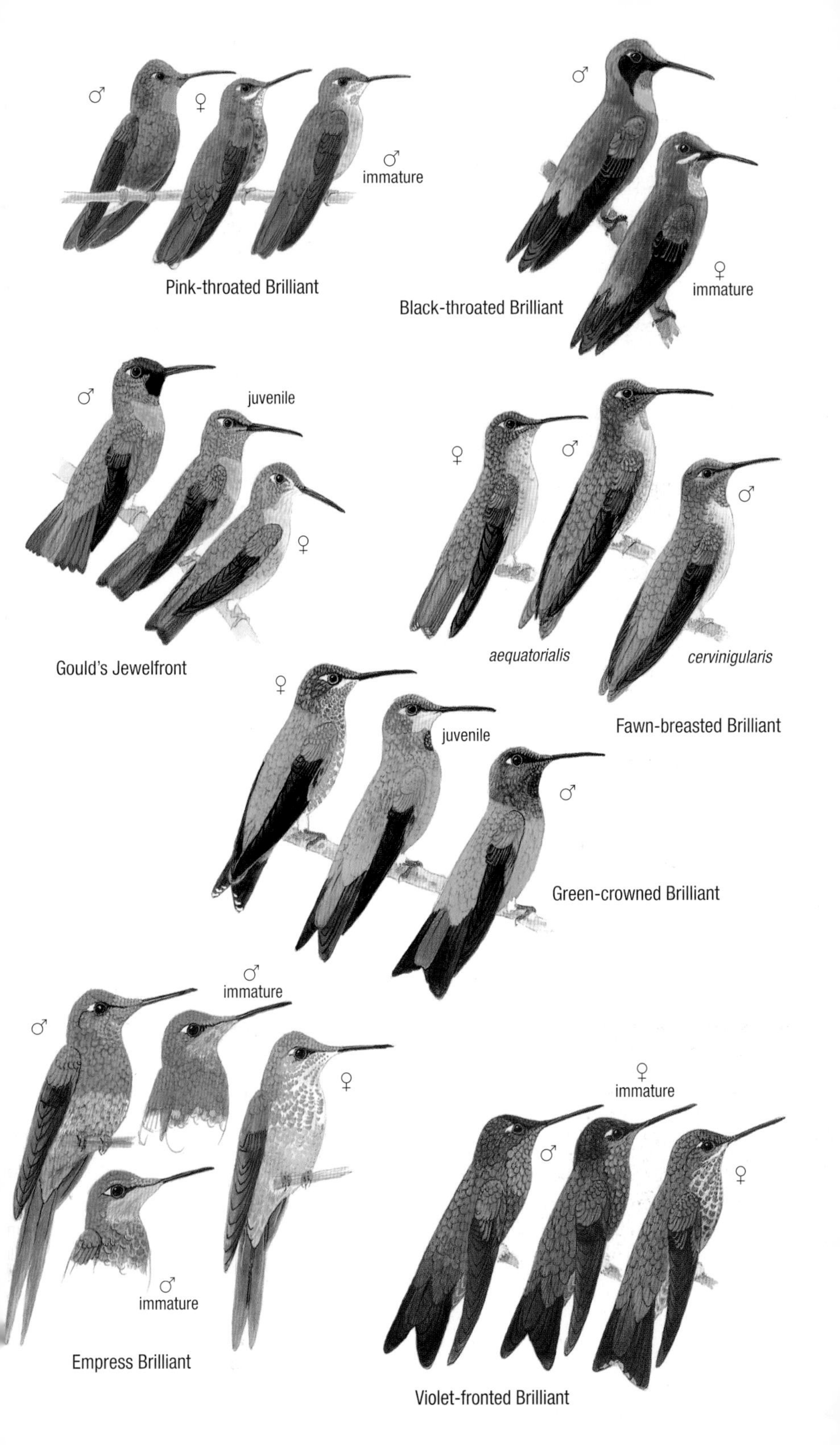
♂
♀
♂
immature
Pink-throated Brilliant
♂
♀
immature
Black-throated Brilliant
♂
juvenile
♀
Gould's Jewelfront
♀
♂
♂
aequatorialis
cervinigularis
Fawn-breasted Brilliant
♀
juvenile
♂
Green-crowned Brilliant
♂
immature
♂
♀
♂
immature
Empress Brilliant
♀
immature
♂
♀
Violet-fronted Brilliant

Tiny hummingbirds of forest, woodland and clearings, so small that they resemble bumblebees in flight. ♂♂ have very short tails, sharply pointed and forked, and stunning fuchsia-purplish bibs; ♀♀ more subdued and particularly difficult to identify in some. Note that ♂♂ in ♀ plumages (eclipse) are common. Complex plumage variation and seasonal movements poorly understood.

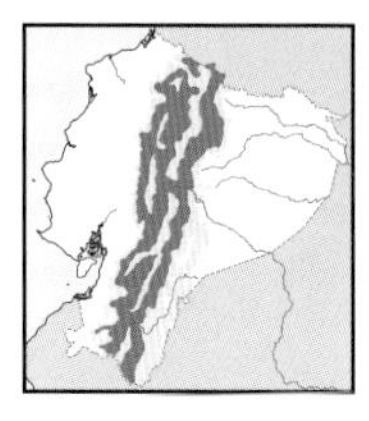

White-bellied Woodstar *Chaetocercus mulsant* — 6–7cm

Andes (1,100–3,500m). Montane forest, borders, shrubby clearings, gardens. Bill (1.8–2cm) straight. Glittering fuchsia throat, *white pectoral collar*, green breast-sides and flanks, *white median belly, white flank tufts. Short, deeply forked tail.* ♀: *whitish below* extending into *bold postocular stripe*, white to buffy throat, *rufescent sides and flanks, white flank tufts*; square, buff-tipped tail. Monotypic. Wasp-like flight, wags tail; sometimes several at flowering trees or feeders. Likely engages in altitudinal / seasonal movements **Voice** Metallic, weak *jirt* barely discernible from wing whirr. **SS** Only woodstar with *sharp white* median underparts (♂) or *extensive whitish* below (♀); large flank tufts, bold supercilium (both sexes). **Status** Uncommon.

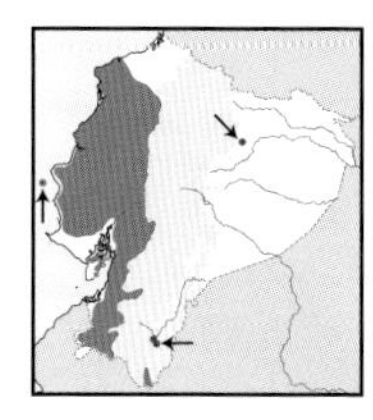

Little Woodstar *Chaetocercus bombus* — 6–7cm

W lowlands to foothills, also locally (seasonally?) in E foothills (to 1,200m, locally or seasonally higher). Humid to deciduous forest borders, adjacent clearings. Straight bill (1.3–1.5cm). *Buffy-white postocular stripe connected to whitish pectoral collar*, glittering violet-pink throat, dusky-green below; *fairly long, forked tail.* ♀: *uniform cinnamon-buff below* including *uniform* postocular stripe, short square tail, *broad cinnamon-buff tips.* Monotypic. Flight slow, wasp-like; probably engages in seasonal movements (Oct–Jan? on W Andean slope). **Voice** Rich *chúp* reported. **SS** ♀ Gorgeted Woodstar (rufescent uppertail-coverts to rump); Esmeraldas (contrasting supercilium, green central rectrices); from imm. Esmeraldas by tail pattern; see ♀ and imm. of larger Purple-throated (*broader* cheek patch, *paler collar*, longer bill, *dark* central rectrices). **Status** Rare, VU.

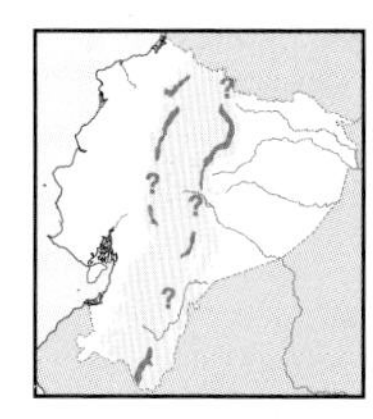

Gorgeted Woodstar *Chaetocercus heliodor* — 6–7cm

Local in E & NW subtropics (1,100–1,800m, locally higher). Montane forest canopy, borders. Straight bill (1.3–1.4cm), *small* white postocular spot. *Long and flared fuchsia throat, inconspicuous* whitish chest-band, *bluish-green below; short forked tail.* ♀: rich *ochraceous-buff underparts*, richest on flanks, *narrow* whitish-buff postocular stripe, *rufous uppertail-coverts*, black subterminal tail band. Race *cleavesi* (E), possibly *heliodor* (W). Similar to other woodstars, may congregate at flowering trees and disperse seasonally / altitudinally. **Voice** Sweet twitter *jiri'ri'ri'rí-ji'ri-ji'ri'ri'rí...* **SS** Some (young ♀?) are paler below, but lack white median underparts of White-bellied Woodstar; very similar ♀ Little Woodstar (rather longer-tailed, normally cinnamon-buff below). **Status** Rare, local.

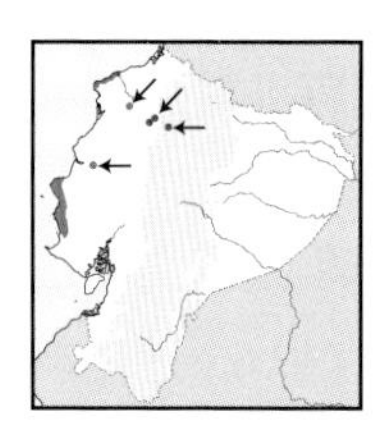

Esmeraldas Woodstar *Chaetocercus berlepschi* — 6–7cm

Local in coastal mountains (to 675m; seasonally? to 1,800–2,000m; M. Fogden). Humid forest borders, riparian woodland. *Long white postocular stripe*, violet-pink throat patch; mostly *white below, no* green pectoral collar; *fairly long tail, deeply forked.* ♀: *pale cinnamon-buff below, contrasting whitish* postocular stripe; broad black subterminal tail-band, *green central rectrices* with *buff tips*; imm. ♂: *lacks tips to central rectrices*, outer feathers *tipped whitish*, mostly *whitish below, broad* supercilium. Monotypic. Probably migrates seasonally or altitudinally; recorded late Oct–mid May in Ayampe region, Sep in W Esmeraldas, Aug and Dec in Tandayapa Valley. Wasp-like flight, strongly territorial, feeds at various heights; recently visiting gardens near forest. **Voice** Mellow *tsidi-tsidi-tdi-tdi-t't'tdi...* **SS** ♂ Little Woodstar (no white below), ♀ (*no green* in central tail, *no contrasting* postocular stripe , more *uniform cinnamon-buff* below), imm. ♂ (different tail pattern). **Status** Rare, local, EN, endemic.

♂
♂
immature
♀
White-bellied Woodstar
juvenile
♀
♂
♂
immature
Little Woodstar
♂
♂
immature
♀
juvenile
Gorgeted Woodstar
♂
♂
immature
♀
Esmeraldas Woodstar

Another assortment of tiny, bumblebee-like woodstars, three rather longer-tailed than *Chaetocercus* on previous plate; tails deeply forked. ♂ *Myrmia* has tiny square tail. *Calliphlox* are forest-based, the other species prefer arid to very arid habitats.

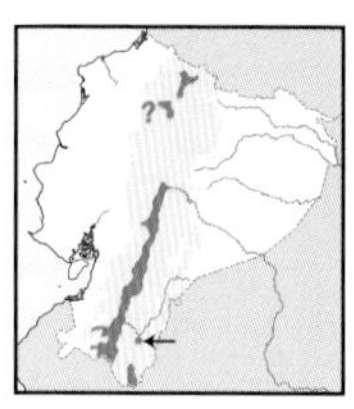

Purple-collared Woodstar *Myrtis fanny* — 6–7cm

Inter-Andean valleys (1,400–2,700m, locally or seasonally lower). Scrub, wooded borders, shrubby clearings, gardens. ***Decurved*** bill (1.8–2cm). Metallic green above, ***long white postocular stripe connected to white pectoral collar. Glistening bluish throat bordered by glistening purple band below.*** Underparts mostly whitish, white flank tufts. Bronzy-green tail, long, ***deeply forked.*** ♀ ***buffy-white below,*** narrow whitish postocular stripe, black square tail with ***broad white tips.*** Nominate. Flight faster than other woodstars, does not wag tail; regular trap-liner in gardens, at cacti and other food-rich resources; aggressive and often dominant over larger hummingbirds; spectacular aerial display by ♂. **Voice** Nasal *nhé-nhé-nhu!..nhé-nhé-ne-nhu!* in flight. **SS** Locally sympatric White-bellied Woodstar (bolder, greener below); Little Woodstar may overlap locally (also bolder). **Status** Uncommon.

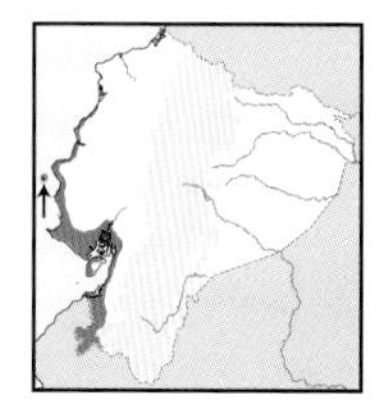

Short-tailed Woodstar *Myrmia micrura* — 6cm

SW lowlands (to 800m, locally higher). Xeric scrub, shrubby clearings, gardens. Bill nearly straight (1.3cm). Tiny, ***dusky***-green above with ***long white postocular and malar stripes connected to neck collar.*** Violet-pink throat, ***white below,*** white flank tufts; ***very short, rounded,*** dark-tipped tail. ♀: similar above, buffy-white postocular stripe, ***pale whitish-buff below, very short tail tipped white.*** Monotypic. Flight like other woodstars, regular in semi-open, feeding low, even near ground. **Voice** Thin *fi'iiii..fi-fi-fi...*; very fast *tí-tí-tí....* **SS** Tiny tail unique; ♀ Esmeraldas Woodstar (longer, buff-tipped tail with black basal half, bolder cheek patch, different habitat); Little Woodstar (larger, longer-tailed, dusky-green below). **Status** Common.

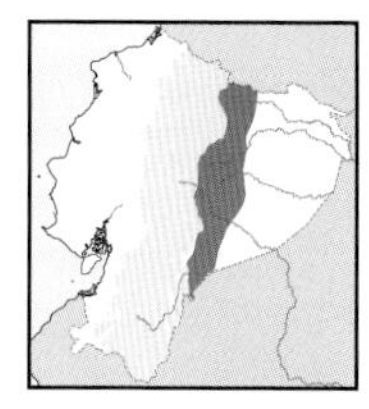

Amethyst Woodstar *Calliphlox amethystina* — 6–7cm

Local in E foothills to adjacent lowlands (300–1,200m, locally higher). Forest canopy, borders, adjacent clearings. Straight bill (1.3cm), white postocular ***stripe.*** Bronzy-green upperparts, ***large*** fuchsia-violet gorget outlined by ***white breast-band;*** greyish-green below. ***Long, deeply forked blackish tail.*** ♀: ***buffy-whitish throat and breast-band,*** dusky throat spots, ***rufous-buff below,*** fairly ***long tail, broadly tipped buff.*** Monotypic. Several may congregate at flowering trees, regular in semi-open, quite territorial; likely nomadic or performs seasonal altitudinal movements. **Voice** Undescribed? **SS** Confusion unlikely in lowlands; in foothills see White-bellied Woodstar (much white below in both sexes); Little Woodstar (shorter tail, ♀ paler below, ♂ bluish-green on belly). **Status** rare.

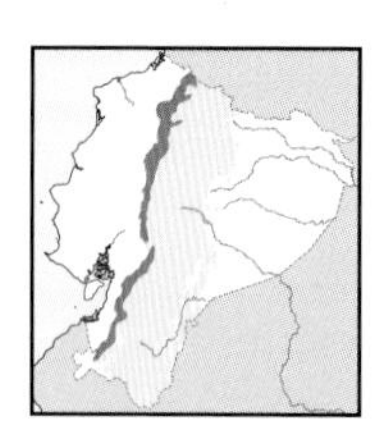

Purple-throated Woodstar *Calliphlox mitchellii* — 6–7cm

W Andean slope (mostly 800–1,800m). Forest canopy, borders, adjacent clearings. Straight bill (1.3–1.4cm), white postocular ***stripe. Large,*** violet-purple gorget, ***white pectoral collar to neck-sides. Dusky-green below, long, deeply forked black tail.*** ♀: ***buffy-white throat and pectoral collar,*** dusky-green chest-sides, ***buff to orange-rufous below,*** but variable; ***rather long, square tail, broadly tipped rufous.*** Monotypic. As previous species, regular at feeders. **Voice** Soft *chit* with wing whirr. **SS** From ♀ of sympatric *Chaetocercus* woodstars by longer tail, mostly rufous underparts, but see paler ♀ (imm.?) and imm. ♂; Little and Esmeraldas Woodstars (no collar, shorter tails, shorter bills, narrower cheek patch, bolder supercilium). **Status** Uncommon.

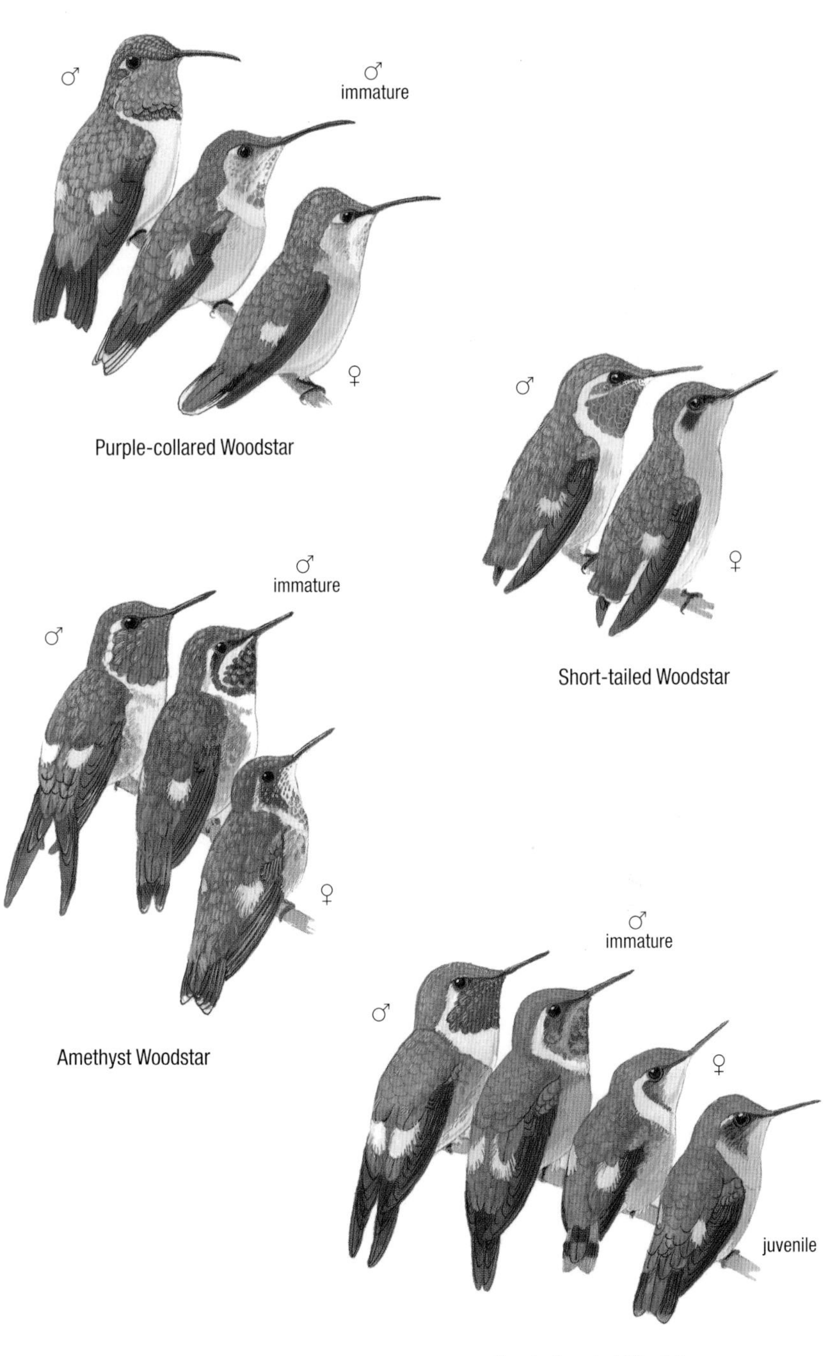
♂
♂ immature
♀
Purple-collared Woodstar
♂
♀
Short-tailed Woodstar
♂ immature
♂
♀
Amethyst Woodstar
♂ immature
♂
♀
juvenile
Purple-throated Woodstar

A mixture of species of open areas and forest edges, some small (emeralds) and others large (starthroats), all markedly dimorphic. Starthroats have very long, slightly curved bills and beautiful head patterns; emeralds are tiny, mostly green, quite featureless, with short, thin bills; *Chlorestes* resemble *Amazilia* hummingbirds; *Klais* is a unique monotypic hummingbird of E foothills with noticeably long square tail.

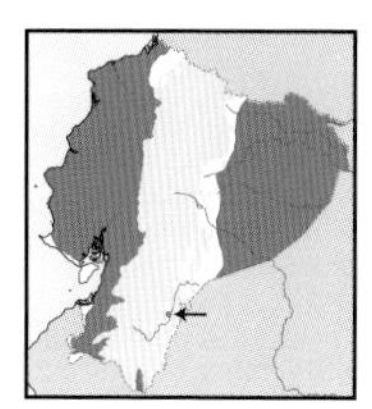

Long-billed Starthroat *Heliomaster longirostris* 9–10cm

E & W lowlands, locally to foothills (to 1,500m). Humid to deciduous forest, woodland, borders, adjacent clearings. *Very long* (3.5cm) *nearly straight bill. Bright bluish crown, glittering pink throat, white malar streak, white postocular stripe. Greyish below, white streak on rump. Long*, slightly forked tail, outer feathers tipped white. ♀: *lacks* bluish crown, smaller throat patch, *broader malar*. Nominate (E), *albicrissa* (W and Zumba region). Conspicuous in semi-open, often perching on exposed branches; may congregate, trap-line or defend flowering trees, but not aggressive. **Voice** Low *tsik* or *seik* notes. **SS** Unmistakable. **Status** Common.

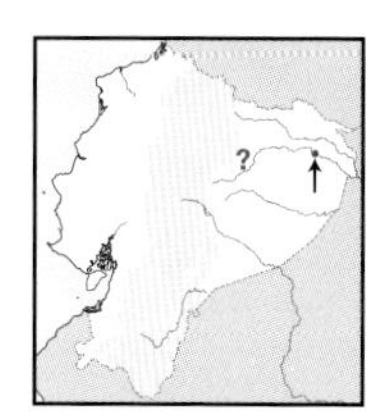

† Blue-tufted Starthroat *Heliomaster furcifer* 10–11cm

Two uncorroborated records at E lowland clearing (250m; 900m). Bill long, nearly straight (3cm). Striking ♂ has *bluish-green crown, glistening blue underparts, pink throat patch bordered by long bright blue tufts, metallic blue long forked tail*. ♀: coppery crown, *dull grey underparts, tail long, metallic blue basally, broad black subterminal band, white tips*. Monotypic. Unknown in Ecuador. Elsewhere regular in semi-open habitats, including gardens; might engage in seasonal movements. **SS** ♀ resembles Grey-breasted Sabrewing but has less white in tail. **Status** Uncertain; probably erroneous identifications.

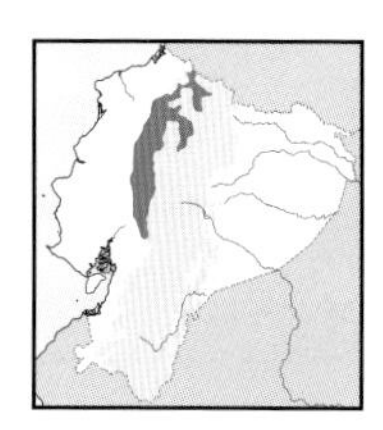

Western Emerald *Chlorostilbon melanorhynchus* 7–8cm

Inter-Andean valleys in N, also W Andean slope (600–1,800m in slopes, 1,500–2,700m in valleys). Montane forest, borders, woodland, gardens. Short straight bill (1.3–1.5cm). *Golden-green crown, emerald throat and breast*. Tail short, forked, *deep blue*. ♀: similar above, *pale grey below, whitish postocular stripe, dark ear-coverts patch*, tail *narrowly tipped white*. Nominate (Andean valleys), *pumillus* (W slope). Regular in open, often near ground, mostly a solitary, shy trap-liner. **Voice** Weak twittering *si-srr..si-srr..si-srr..si-srr...*, louder *shi'rrr* calls. **SS** Does not overlap with Blue-tailed Emerald! See ♀ Violet-bellied Hummingbird. **Status** Common.

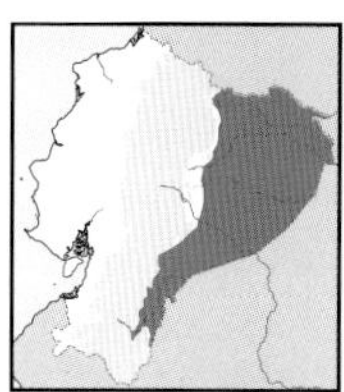

Blue-tailed Emerald *Chlorostilbon mellisugus* 7–8cm

NE lowlands, also SE foothills to subtropics (below 750m in NE, 900–2,600m SE; RG). Forest borders, shrubby clearings. Short straight bill (1.3–1.5cm). Mostly bright *emerald-green*, bluish on breast. *Deep blue, short forked tail*. ♀ closely recalls Western Emerald, likely inseparable in field. Race *napensis* (NE), possibly *phoeopygus* (SE). Solitary and elusive, often near ground; flight fast and direct. **Voice** Weak *chép-chép-chép...* feeding notes. **SS** Much smaller than Fork-tailed Woodnymph, no bicoloured underparts; see Blue-chinned Sapphire (♂ has rounded tail, red mandible). **Status** Rare.

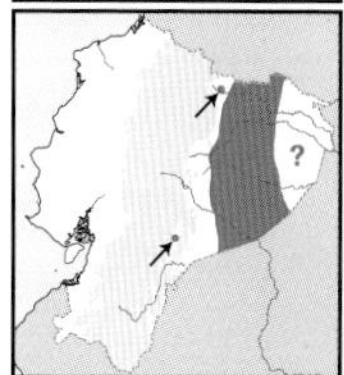

Blue-chinned Sapphire *Chlorestes notata* 8cm

Local in E lowlands (to 300m, once 900m). *Terra firme* and *várzea* borders, adjacent clearings. Straight bill (1.8cm), *mandible mostly red. Bright emerald below, chin slightly bluer*. Rather *long, rounded, metallic blue tail*. ♀: white below *thickly spotted glittering green on throat-sides*. Race *puruensis*. Mostly at borders; territorial and solitary, but rather shy, feeds at various heights. **Voice** High-pitched, fast *tss-tss-tss* series (RG). **SS** ♂ Glittering-throated Emerald has white median stripe below, square tail not blue; cf. greyer Fork-tailed Woodnymph (♀), other sapphires and Blue-tailed Emerald. **Status** Rare.

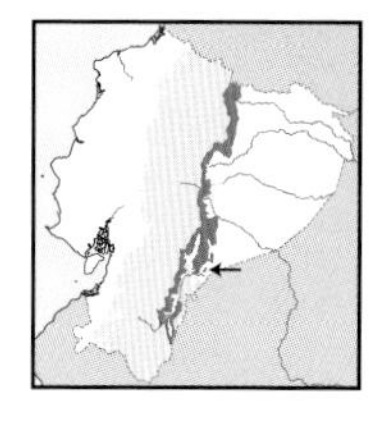

Violet-headed Hummingbird *Klais guimeti* 7–8cm

E Andean foothills to subtropics (800–1,900m, locally lower). Montane forest undergrowth, borders, adjacent clearings. Short straight bill (1.3cm), *large white postocular spot. Violet-blue head*, dark green above, *greenish-grey below, long square tail, mostly metallic green* narrowly tipped whitish. ♀: *blue limited to crown*, otherwise similar to ♂ with *broader tail tip*. Nominate. Usually singly, but may gather at flowering trees, often *Inga* trees and hedgerows; flight slow. **Voice** High-pitched *t'sí-sítsít...*or *tsi'tititití*. **SS** Violet-blue head and long square tail diagnostic. **Status** Uncommon.

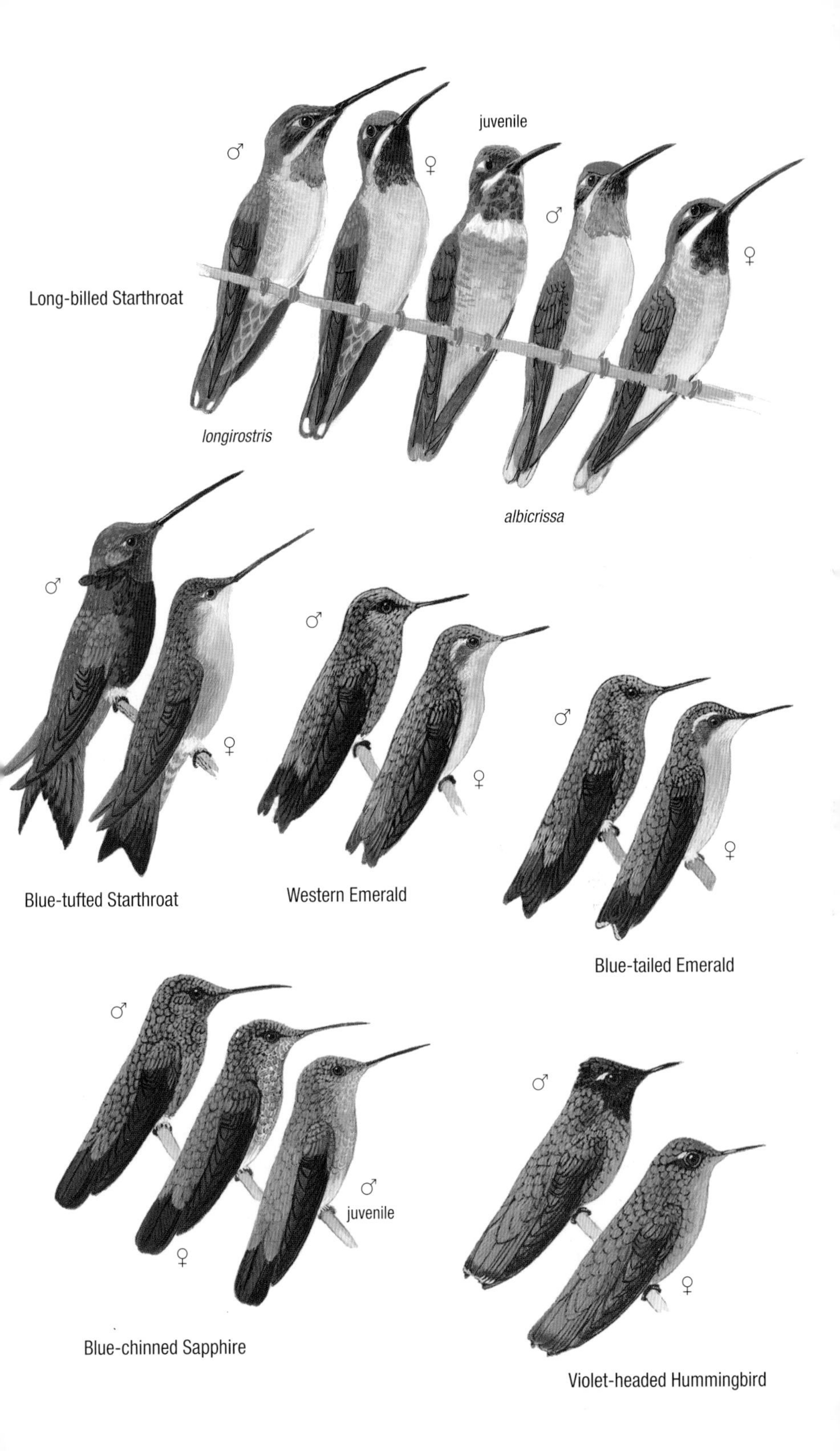
♂
♀
juvenile
♂
♀
Long-billed Starthroat
longirostris
albicrissa
♂
♀
Blue-tufted Starthroat
♂
♀
Western Emerald
♂
♀
Blue-tailed Emerald
♂
♀
♂
juvenile
Blue-chinned Sapphire
♂
♀
Violet-headed Hummingbird

PLATE 116: SABREWINGS AND PLUMELETEERS

Mid-sized hummingbirds of lower subtropics to lowlands, all with slightly long and curved bills, all but one (Grey-breasted Sabrewing) with well-marked sexual dimorphism. All sabrewings occur in E; plumeleteers in W. ♂ sabrewings have stiffened and curved outer primaries apparently used in display. Tails are particularly boldly patterned in two sabrewings, while vents are conspicuous in both plumeleteers.

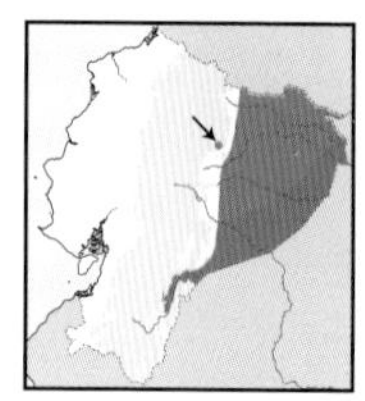

Grey-breasted Sabrewing *Campylopterus largipennis* 11–12cm

E lowlands, locally in foothills (mostly below 600m, to 1,000m in SE and locally in N). *Terra firme* and *várzea* forest, borders, woodland, adjacent clearings. Bill *long* (2.7–2.8cm) slightly decurved, white postocular spot. Dark green above, ***dull grey below, long forked tail broadly tipped white***. Sexes nearly alike. Race *aequatorialis*. Solitary and conspicuous, forages mostly in undergrowth, sometimes higher strata; regularly near water. **Voice** Sharp *tchép!* every 1–2 sec, faster when feeding. **SS** ♀ Fork-tailed Woodnymph smaller, lacks postocular spot, tail narrowly tipped white; see ♀ Napo Sabrewing (little overlap). **Status** Fairly common.

Lazuline Sabrewing *Campylopterus falcatus* 11cm

Local in subtropics on E Andean slope, mainly NE (1,400?–2,100m). Montane forest canopy, borders. Bill shorter than previous species (2.5cm) and ***decurved***, white postocular spot. ***Deep blue*** underparts becoming blue-green on belly. ***Vent and undertail chestnut***, central tail feathers tipped blue. ♀: ***smaller*** blue throat patch, ***grey below, grey malar stripe***, tail as ♂. Monotypic. Poorly known and seldom seen, should be looked for inside forest and sheltered borders with *Heliconia* and *Erythrina*. **Voice** Elsewhere (HB) varied *chic-it, chic-it...seet-chak*, but unknown in Ecuador. **SS** Only other chestnut-tailed hummingbird in range (Chestnut-breasted Coronet) is smaller, straight-billed, chestnut underparts. **Status** Very rare, local; NT in Ecuador.

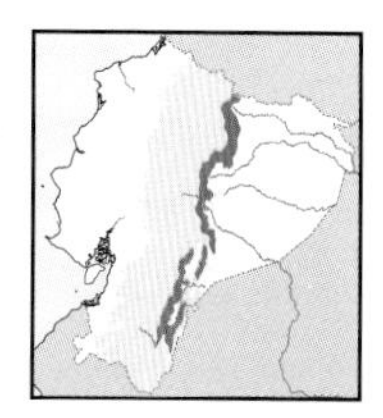

Napo Sabrewing *Campylopterus villaviscensio* 11–12cm

Local in foothills to subtropics on E slope (900–1,700m). Montane forest undergrowth, borders. Rather long (2.8cm) ***straight bill***, white postocular spot. ***Deep blue throat***, green below becoming dull grey on belly. Tail long, slightly forked, ***steel blue, paler at tip***. ♀: duller above, ***dark grey below***, tail as ♂ but ***outer feathers narrowly tipped white***. Monotypic. Mostly low inside forest and shady borders near rushing streams; poorly known. Might shyly show up at feeders. **Voice** Slow series *sut-chíp..sut-chíp.. sut-chíp....* **SS** Tail pattern diagnostic; ♀ resembles lower-elevation Grey-breasted Sabrewing and Fork-tailed Woodnymph; see White-tailed Hillstar (much white in tail). **Status** Rare, rather local, NT globally, DD in Ecuador.

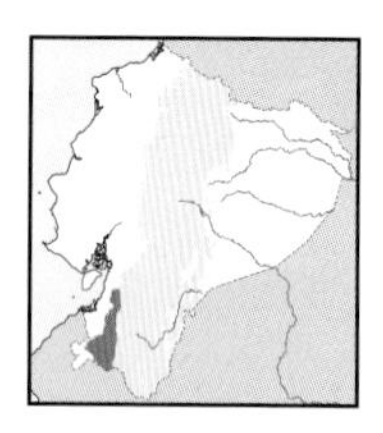

White-vented Plumeleteer *Chalybura buffonii* 10cm

SW Andean foothills (500–1,900m). Lower growth of montane forest, borders, adjacent woodland. Fairly long straight bill (2.5cm), ***rosy mandible, pink legs***. Bronzy-green above, more ***glittering breast. Vent distinctively white, deep blue tail*** rather long, slightly forked. ♀: ***pale grey below, white vent***, outer tail feathers broadly tipped white. Race *intermedia*, probably valid species. Prefers forest interior and shady borders, seldom seen, probably becoming rarer recently; often solitary and aggressive. Very locally at feeders. **Voice** Fast jumbled *chiripchiripchiri..., chirp* foraging calls. **SS** ♂ unique, ♀ with ♀ Crowned Woodnymph (deeply forked tail, grey only to breast) and Violet-bellied Hummingbird (smaller, whiter below, short tail, red mandible). **Status** Rare, NT in Ecuador.

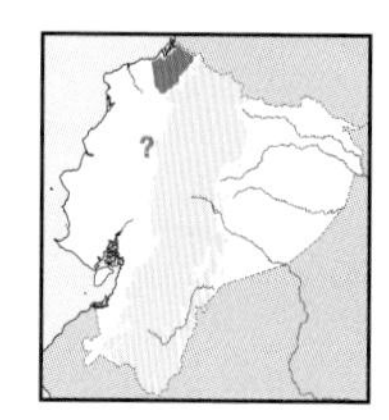

Bronze-tailed Plumeleteer *Chalybura urochrysia* 10cm

Lowlands to foothills in extreme NW (100–800m). Wet forest, less often at borders. Fairly long straight bill (2.5cm), ***rosy mandible, pink feet***. Bronzy-green above, bright green below, more ***glittering throat and breast***. Contrasting ***white vent***, rather long tail, slightly forked, ***bronzy*** in ♂ (***outer feathers tipped white*** in ♀). ♀ recalls previous species but ***tail bronzy***. Nominate. Mostly inside forest where fiercely defends feeding territories against conspecifics and others. **Voice** Loud *chú-checheche*; emphatic *chwép!* calls. **SS** ♀ Crowned Woodnymph (grey to breast, deeply forked blue tail with narrower whitish tips; no rosy mandible, no contrasting vent); Blue-chested Hummingbird (no white vent, duller underparts, red mandible); Purple-chested Hummingbird (blue tail, spotty below); Violet-bellied Hummingbird (smaller, short rounded blue tail). **Status** Uncommon, forest-based, VU in Ecuador.

Grey-breasted Sabrewing
♂
♀
Lazuline Sabrewing
♂
Napo Sabrewing
♀
♂
♀
Bronze-tailed Plumeleteer
♂
♀
White-vented Plumeleteer

PLATE 117: WOODNYMPHS AND OTHER HUMMINGBIRDS

Woodnymphs are forest species, among the commonest forest hummingbirds especially in Amazonian lowlands. Both sexually dimorphic, tails deeply forked, bills rather strong and curved. *Taphrospilus* and *Leucippus* also have curved, slightly strong bills, but plumages are subdued, with little sheen on upperparts; all inhabit open to fairly open areas, one on E river islands, two in arid scrub.

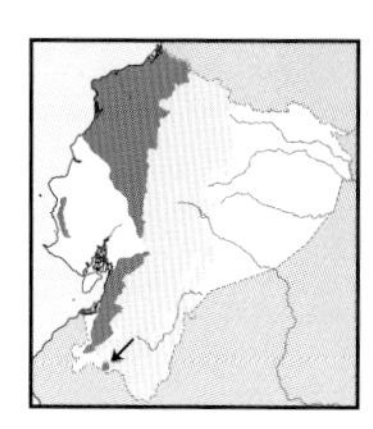

Crowned Woodnymph *Thalurania colombica* — 9–10cm

W lowlands, S in Andean foothills to El Oro (to 1,100m, locally higher). Humid forest undergrowth, borders, woodland. Short, nearly straight bill (1.8–1.9cm). Glittering green crown and ***throat to breast, glittering lavender-blue below*** (NW) or entire ***underparts glittering green*** (SW). ***Deeply forked steel-blue tail. Smaller*** ♀ (7cm) has ***grey throat to breast***, dusky-green below (NW) or underparts ***all grey*** (SW), ***outer tail feathers tipped white***. Races *verticeps* (N), *hypochlora* (S) perhaps separate species. Mostly solitary, frequents feeders; somewhat territorial, perches boldly on bare branches. **Voice** Weak *tsirp* trills, clearer in SW. **SS** Smaller Western Emerald; also ♀ plumeleteers and Violet-bellied Hummingbird. **Status** Uncommon.

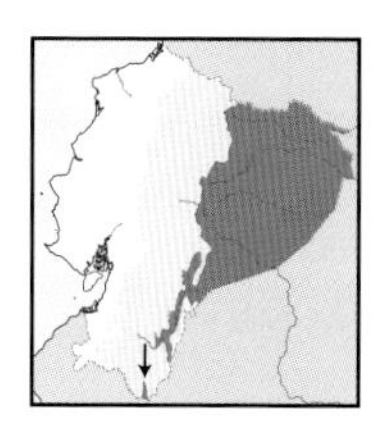

Fork-tailed Woodnymph *Thalurania furcata* — 9–10cm

E lowlands to foothills (to 1,200m, locally higher). Forest undergrowth, borders, adjacent clearings. Short, nearly straight bill (1.8–1.9cm). ***Glittering green throat to breast, lavender-blue below. Deeply forked tail, steel-blue. Smaller*** ♀: ***pale grey underparts***, tail less forked, ***outer tail narrowly tipped white***. Race *viridipectus*. Tends to remain inside forest, where one of the most numerous hummingbirds; solitary, fiercely defends territory, feeding at various heights. **Voice** Thin *tsít-tsít-tsít...*or *tsír-tsír...* while feeding; not very vocal. **SS** ♀ sabrewings similarly patterned but larger, longer-billed, different tails. See Blue-tailed Emerald (smaller, white supercilium). **Status** Common.

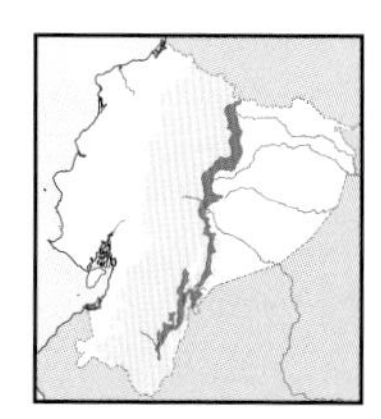

Many-spotted Hummingbird *Taphrospilus hypostictus* — 10cm

E Andean foothills (500–1,200m). Montane forest borders, adjacent clearings. ***Long***, slightly decurved bill (2.5cm), ***pinkish mandible base***, white postocular spot. Metallic green above, ***underparts thickly spotted metallic green, white median belly***. Rather ***long square tail, metallic green*** narrowly tipped buff. Sexes alike. Nominate. Mostly solitary but gathers with other species at flowering trees, notably at *Inga*. **Voice** Weak, buzzy *jirk-jirk-jirk-jirk...jir'r'r'r*. **SS** ♀ Violet-fronted Brilliant (no white belly, forked blackish tail, white malar); ♀ White-necked Jacobin (no postocular spot, more scaled below, tail tipped white), ♀ Blue-chinned Sapphire (few throat spots, short rounded tail). **Status** Rare.

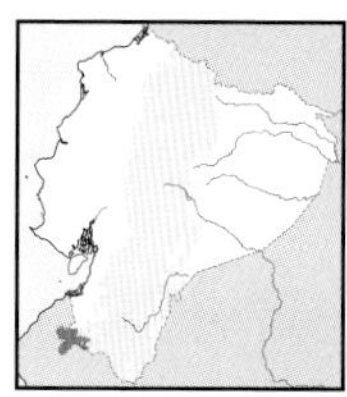

Tumbes Hummingbird *Leucippus baeri* — 8–9cm

Local in lowlands and foothills of extreme SW (to 1,000m). Arid scrub, borders, shrubby clearings. Short, nearly straight bill (2cm), ***conspicuous*** white postocular spot. ***All drab***: dull metallic green above, ***grey below, buffier*** throat and breast, ***rather long bronzy-green tail***, broad dusky subterminal band. Sexes alike. Monotypic. Inconspicuous, feeds mostly near ground; seasonally more numerous, regular near water. **Voice** Thin, buzzy *jirp..jirp..jirp...* running into warbling display. **SS** Drab plumage makes it easy to recognise in tiny range. **Status** Rare, locally or seasonally more numerous.

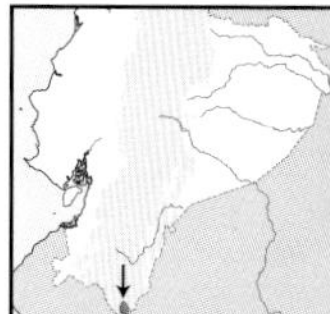

* Spot-throated Hummingbird *Leucippus taczanowskii* — 8–9cm

Single site in extreme SE Zamora Chinchipe, in Zumba (1,000–1,100m). Semi-deciduous woodland. Resembles Tumbes Hummingbird but ***mottled green throat***, sides to flanks more extensively dull green. Monotypic. Recently found in Ecuador feeding at *Inga* and *Erythrina* trees with other species. **Voice** Dry *tip* notes in chatter; complex series of chips and wheezing electric warbles (S *et al.*). **SS** No other very drab hummingbird in range. **Status** Uncertain; possibly rare local resident, or recent arrival.

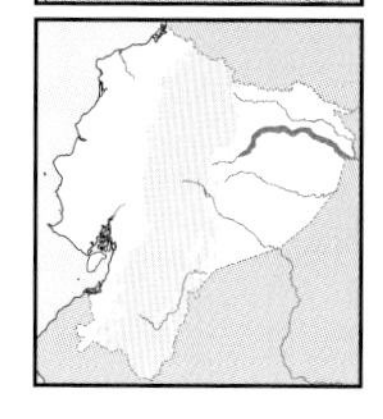

Olive-spotted Hummingbird *Leucippus chlorocercus* — 8–9cm

E lowlands (to 400m). Early second-growth scrub and low woodland on river islands. Short, nearly straight bill (1.7cm), ***conspicuous*** white postocular spot. ***Drab. Dull bronzy-green above, greyish-white below***, small green spots on throat, sides and flanks tinged dull green. Slightly forked tail, metallic green, outer tail tipped whitish. Sexes alike. Monotypic. Solitary in borders and open early successional vegetation, defends feeding territory around variety of pioneer plants; often perches exposed. **Voice** Accelerating *tsiyp-siyp-siyp-siyp'sipsipsipslp...*, briefly paused and restarted; slower *tsiyp* calls. **SS** Confusion unlikely in habitat, see Glittering-throated Emerald (not drab). **Status** Uncommon but very local.

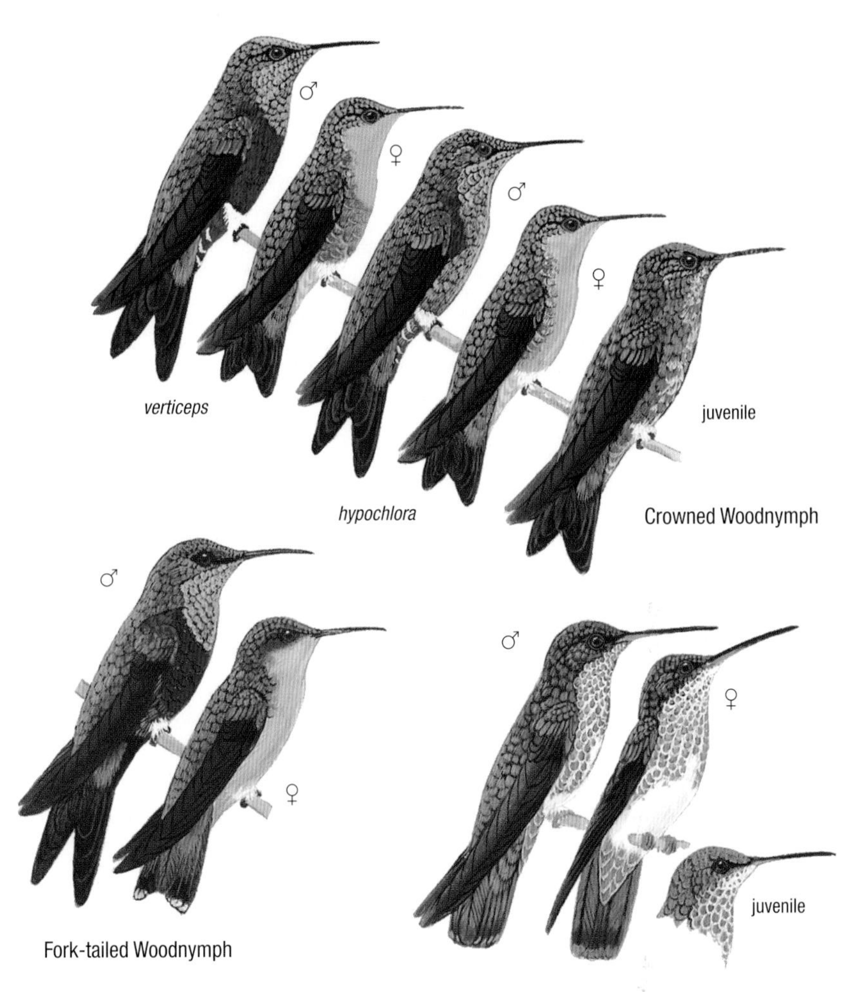
♂
♀
♂
♀
verticeps
juvenile
hypochlora
Crowned Woodnymph
♂
♂
♀
♀
juvenile
Fork-tailed Woodnymph
Many-spotted Hummingbird

juvenile
adult
Tumbes Hummingbird
Spot-throated Hummingbird
Olive-spotted Hummingbird

PLATE 118: AMAZILIAS

A varied and conspicuous genus of sexually dimorphic hummingbirds, some common in open and semi-open areas, others more forest-based. Most species in lowlands. Mandibles often pink, tails stunning coppery or bronzy in two species, but more inconspicuous in the other five. Identification can be difficult in ♀♀ of the two NW species (Blue-chested and Purple-chested).

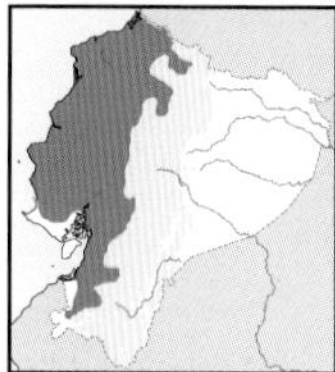

Rufous-tailed Hummingbird *Amazilia tzacatl* 9–10cm

W lowlands to subtropics, locally in inter-Andean valleys (to 2,500m). Forest borders, woodland, clearings, gardens. Rather long, faintly decurved bill (2.1–2.4cm), ***mandible mostly red***, sometimes the maxilla too. ***Glittering emerald throat and breast***, greyish-green belly, ***tail rich rufous***. ♀: duller below, throat somewhat spotted. Race *jucunda*. Numerous and conspicuous, constantly humming while hover-feeding, aggressively defends territory. **Voice** Weak *tseép-tsup-tsií..sií, tsi-tsi-si'sisisisí*.... **SS** Amazilia Hummingbird (white and rufous below). **Status** Common and widespread.

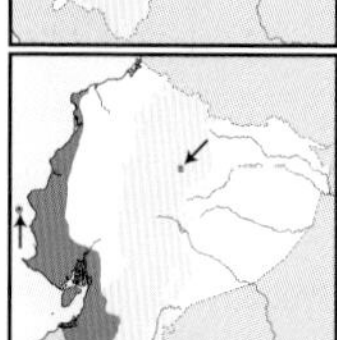

Amazilia Hummingbird *Amazilia amazilia* 9–10cm

W lowlands, to SW highlands (Azuay and Loja; to 2,500m). Xeric scrub, deciduous woodland, borders, gardens. Nearly straight bill (2.2–2.4cm), red tipped black. ***White below, throat densely spotted green, rufous flanks to belly. Coppery-green*** (lowlands) to ***rufous-chestnut tail*** (S highlands). Races *dumerilii* (lowlands), *alticola* (highlands) less rufous below, *azuay* (S Azuay-N Loja); might be separate species. Aggressive, mostly feeds alone, regularly perches in open. **Voice** Weak, rattled *tsee-tsee-tsee-tsee*..., sharper or faster when feeding. **SS** Unique in range. **Status** Fairly common.

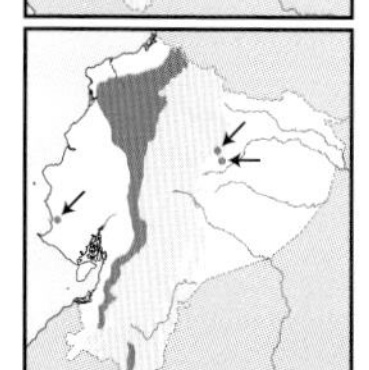

Andean Emerald *Amazilia franciae* 9–10cm

W foothills to subtropics, also SE foothills (300–1,800m in W, 900–1,600m in SE). Forest borders, adjacent clearings. Faintly decurved bill (2–2.2cm), mandible red tipped black, white postocular spot. ***Glistening bluish-green head*** (***bluer*** in SE), ***snowy-white below***, mottled to solid green sides. Slightly forked ***bronzy tail***. Sexes nearly alike. Races *viridiceps* (W), *cyanocollis* (SE). Trap-liner, sometimes with other species at flowering trees, rarer at feeders. **Voice** Fast, jumbled, thin *tsi'sitsí'tstststst*.... **SS** ♀ emeralds and sapphires less solid white below; see Violet-bellied Hummingbird. **Status** Uncommon.

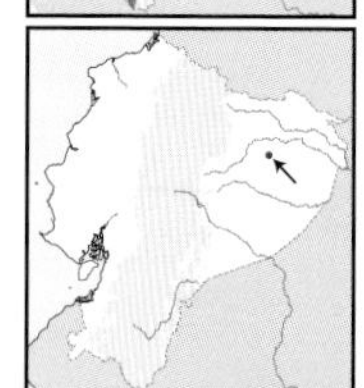

* Sapphire-spangled Emerald *Amazilia lactea* 8–9cm

One uncorroborated record from E lowlands (250 m). Borders of swampy forest. Nearly straight bill (2cm), mandible red tipped black. Similar to Glittering-throated, but ***throat to breast glittering blue***, fairly long blue-black tail, outer feathers faintly tipped greyish. Presumed race *bartletti*. Unknown in Ecuador; elsewhere solitary and territorial at various heights. **Voice** Foraging calls similar to other *Amazilia*, buzzy *j'zzzzt, di-jzijzijzi*... **SS** See Blue-chinned Sapphire (rounded tail, no white median stripe). **Status** uncertain; the only record might represent a misidentification (N Krabbe).

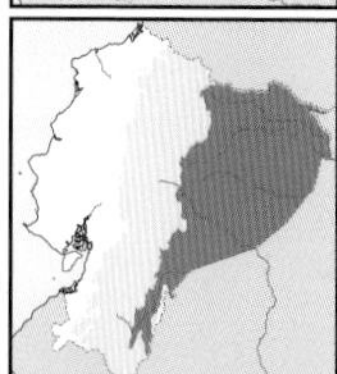

Glittering-throated Emerald *Amazilia fimbriata* 8–9cm

E lowlands to foothills (200–1,200m, locally higher). Second growth, shrubby clearings, borders, gardens. Vaguely decurved bill (2–2.2cm), mandible mostly red. ***Glistening throat and breast, white stripe on median underparts***, metallic blue tail. ♀: white stripe ***extends to throat***. Race *fluviatilis*. Numerous and widespread, even on river islands, often solitary trap-liner but may congregate at flowering spots. **Voice** Restless, buzzy *dz'írrt-dz'írrt-dz'írrt*..., *dz'iirr-dz'iirr-dz'iirr*..., raspy *tzrrr-tzrrr-tzrrr* calls. **SS** Blue-chinned Sapphire lacks white stripe below; see Sapphire-spangled Emerald. **Status** Common.

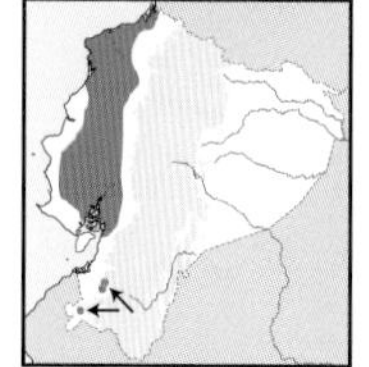

Blue-chested Hummingbird *Amazilia amabilis* 7–8cm

W lowlands (to 500m). Humid forest, woodland, borders, shrubby clearings. Nearly straight bill (1.8-2cm), mandible mostly red. ***Glistening green forecrown***, metallic green above, mottled below, ***upper chest glittering blue***. Rather long purplish tail. ♀ ***lacks glistening crown, less blue in chest***, more mottled below, greyish-buff belly. Nominate. Solitary, less often in semi-open; may gather at flowering trees where aggressive. **Voice** High-pitched, squeaky *pseét-pseét..pseét'pseét'pseét*..., often more rattled. **SS** Purple-chested Hummingbird (more solid blue pectoral patch, ***white vent***, no glittering crown); see Violet-bellied Hummingbird. **Status** Uncommon.

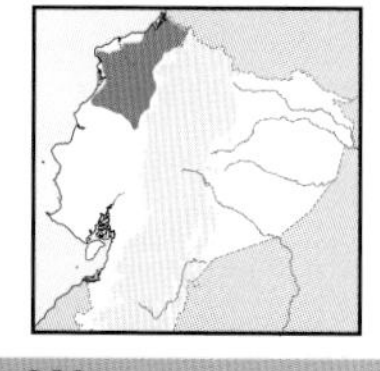

Purple-chested Hummingbird *Amazilia rosenbergi* 8–9cm

NW lowlands to foothills (to 600m). Humid forest undergrowth, borders. Straight bill (1.8–2cm), mandible mostly red. Metallic green above, ***glittering green throat, dark blue upper chest patch***. Greyish-green below, ***white vent***, blue-black tail. ♀: resembles Blue-chested Hummingbird but greyer, ***more mottled below, white vent***. Monotypic. More forest-based than Blue-chested; solitary trap-liner, apparently gathers at small leks; rare at feeders. **Voice** Piercing, fast *tsíp-tsíp-sí'sí'sí'síp-síp..tsítp*.... **SS** See Blue-chested; ♀ difficult to separate but see vent; also Violet-bellied Hummingbird. **Status** Fairly common.

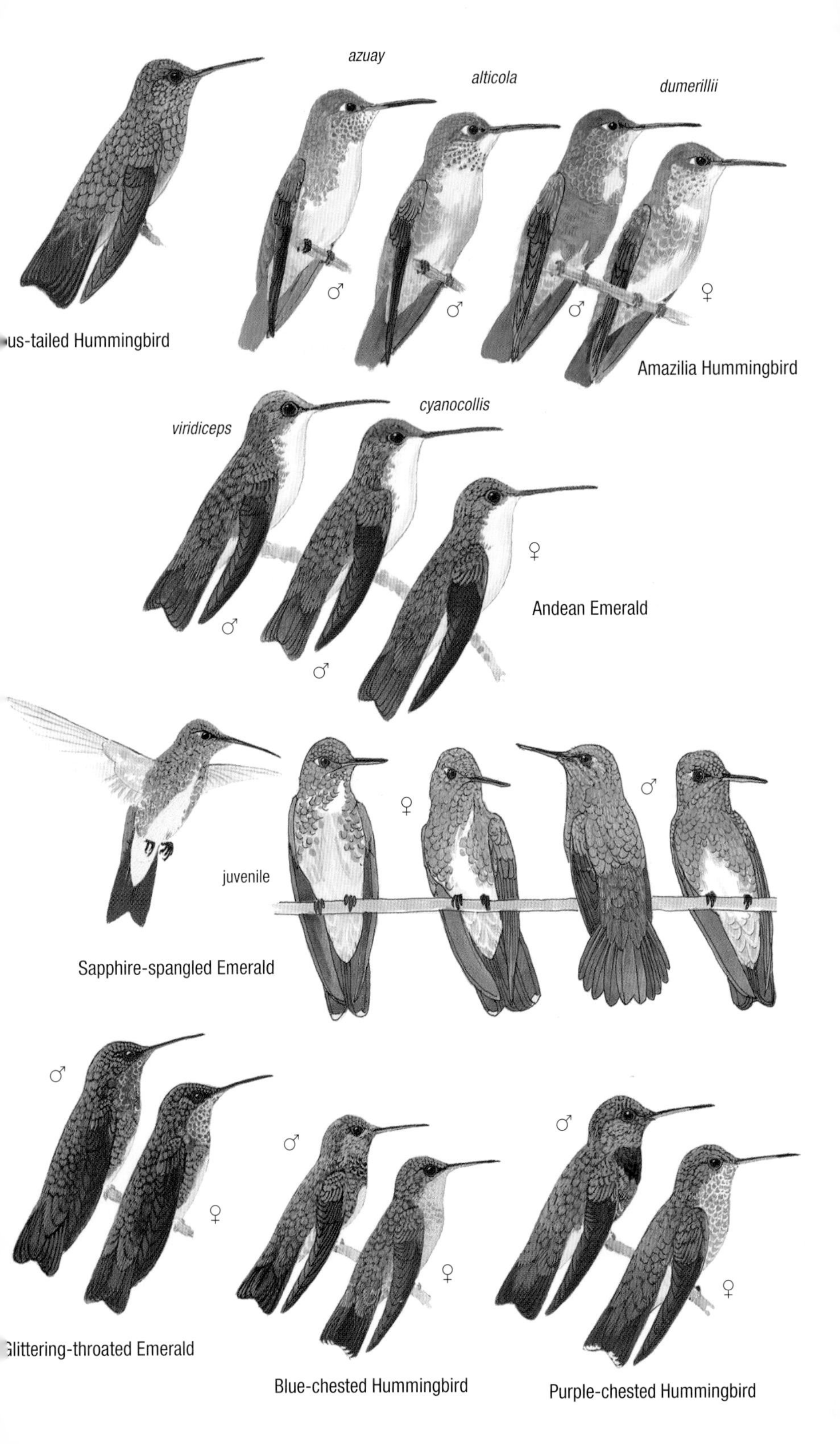
azuay
alticola
dumerillii
us-tailed Hummingbird
♂
♂
♂
♀
Amazilia Hummingbird
viridiceps
cyanocollis
♀
Andean Emerald
♂
♂
♀
♂
juvenile
Sapphire-spangled Emerald
♂
♀
Glittering-throated Emerald
♂
♀
Blue-chested Hummingbird
♂
♀
Purple-chested Hummingbird

Two *Hylocharis* are rare – possibly not resident – in Amazonian canopy, while one occurs in NW mangroves, and its closest relative in arid inter-Andean valleys. Golden-tailed, though placed in its own monotypic genus, resembles *Hylocharis* in its coral-red bill, but tail strikingly coloured. The small Violet-bellied Hummingbird recalls *Thalurania* woodnymphs but bill short, mandible red.

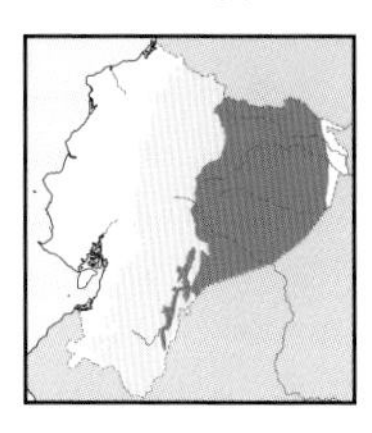

Golden-tailed Sapphire *Chrysuronia oenone* 9cm

E foothills and adjacent lowlands (400–1,200m, locally higher). Montane forest borders, woodland, adjacent clearings. Rather short (1.8–2cm) straight bill, *mandible red tipped black. Glittering purple-blue hood, bronze uppertail-coverts, golden-bronze tail.* ♀: emerald-green above, *greyish-white below*, some lateral spots, *tail as* ♂. Nominate. May gather at flowering trees like *Inga*; possible altitudinal movements. **Voice** Doubled, ringing *tsiu-tichí, tsiu-tichí...*, squeaky *t'chít..chew, t'ch'it...*; also more rattled *ch'rrrr* when feeding. **SS** White-chinned Sapphire (more extensive violet below, black tail); ♀ resembles *Hylocharis* sapphires but golden tail distinctive. **Status** Fairly common.

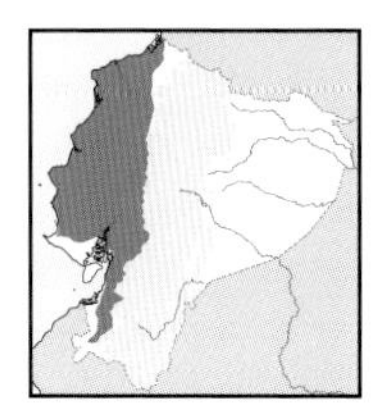

Violet-bellied Hummingbird *Damophila julie* 7–8cm

W lowlands to foothills (to 1,100m). Humid to semi-deciduous forest, borders, woodland, adjacent clearings. Short straight bill (1.3–1.4cm), *red mandible. Glittering emerald hood, bright violet-blue underparts*. Rather long *rounded tail*, blue-black. ♀: metallic green above, *whitish-grey below*, sides mottled, *blue-black tail, outer feathers narrowly tipped grey*. Race *feliciana*. Territorial, mostly low, may congregate at flowering trees. **Voice** Very high-pitched and thin trill: *v'iiirrr.* **SS** Similar to Crowned Woodnymph (N race) but much smaller, tail not forked, red mandible, no purple shoulders; ♀ resembles Western Emerald but no postocular stripe; see larger, rarer Humboldt's Sapphire (longer, forked green tail) and Blue-chested Emerald (see tail). **Status** Uncommon.

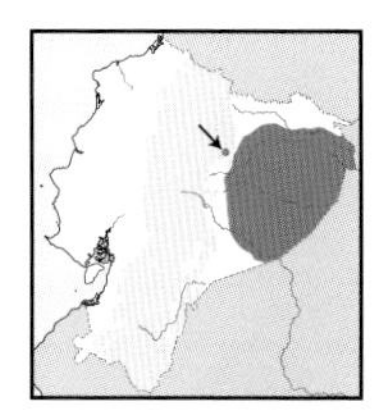

Rufous-throated Sapphire *Hylocharis sapphirina* 9cm

Local in E lowlands (to 400m). *Terra firme* canopy, borders, adjacent clearings. Straight *red* bill (1.8cm), *tipped black. Coppery upper- and undertail-coverts, rufous chin, glittering violet-blue throat. Coppery to rufous tail*, slightly forked. ♀: *only mandible red, rufous chin, throat white spotted glittering blue*, white below, tail narrowly tipped whitish. Monotypic? Poorly known in Ecuador; elsewhere solitary at various heights, fiercely defends feeding area, but will gather at flowering trees. **Voice** Very thin, piercing *sweé-sweé-sweé-sweé-ít...*, gravelly *ch'zee* reported. **SS** Tail colour differs from White-chinned; head and throat pattern separate from Golden-tailed. **Status** Rare.

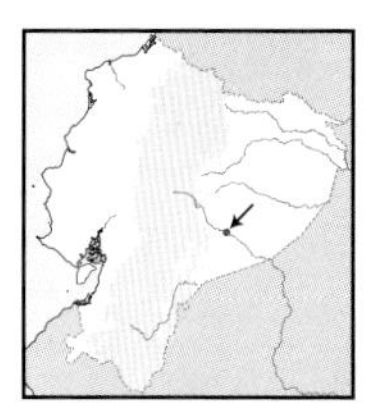

White-chinned Sapphire *Hylocharis cyanus* 9cm

Single record in SE lowlands (250 m). Probably forest canopy and borders. Straight *red* bill (1.8cm), *tipped black. Glittering violet-blue head to chest*, bright green upperparts, *rump and uppertail-coverts coppery*, bright green belly, *blue-black tail* fairly long and slightly forked. ♀ resembles Golden-tailed Sapphire, but *rump to uppertail-coverts coppery, blue-black tail*, outer feathers tipped whitish. Race *rostrata*. Unknown in Ecuador; elsewhere mainly a canopy species, solitary but may gather at flowering trees. **Voice** High-pitched, insistent *zweer-zweet-zweet-zweet-zweet...*, resembling Bananaquit. **SS** White chin almost imperceptible; from Golden-tailed and Rufous-throated Sapphire mostly by tail colour. **Status** Uncertain, perhaps not resident.

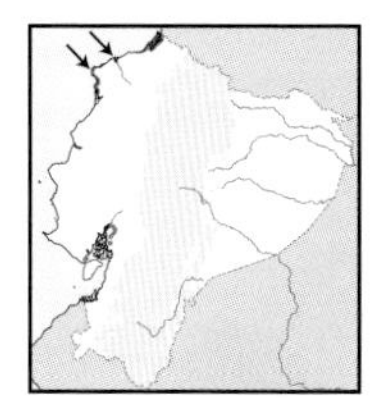

Humboldt's Sapphire *Hylocharis humboldtii* 9cm

NW lowlands (near sea level). Mangroves, adjacent humid forest, woodland and scrub. Straight *red* bill (1.8cm), *tipped black. Glittering blue face and chin, coppery rump*, whitish mid belly *extending as stripe onto breast, green forked tail.* ♀: *only mandible base red*, mostly bright green above, *greyish-white below*, green spotting on sides, *dark green tail*, outer feathers tipped white. Monotypic. Poorly known in Ecuador; elsewhere apparently territorial at mid to low heights, fond of epiphytes. **Voice** Undescribed? **SS** Nothing similar in small range; ♀ Violet-bellied Hummingbird (smaller, with short rounded tail). Blue-headed Sapphire only in highlands. **Status** Rare, local, VU in Ecuador.

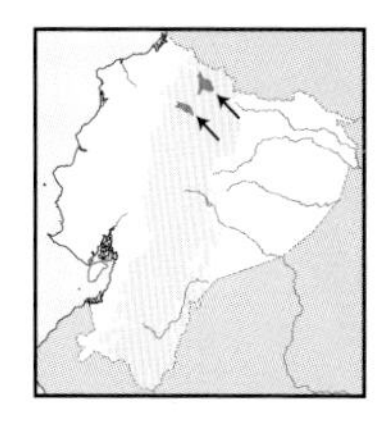

Blue-headed Sapphire *Hylocharis grayi* 9cm

Inter-Andean valleys in N (1,200–2,200m). Xeric scrub, semi-open areas, gardens. Straight bill (1.8cm), *mandible red tipped black. Glittering blue cap and face, glittering green chest*, green rump, slightly forked *steel-blue tail.* ♀: red mandible base, *greyish-white below*, mottled breast-sides, slightly forked *steel-blue tail*, outer feathers tipped whitish. Monotypic. Regular and conspicuous, but rather local, feeds at various heights, both territorial and gathers with other hummingbirds; fond of cacti. **Voice** Sharp, metallic *t'srrrr.* **SS** cf. Rufous-tailed Hummingbird with red bill; ♀ Western Emerald (pale supercilium, dark ear patch, shorter bill). **Status** Uncommon.

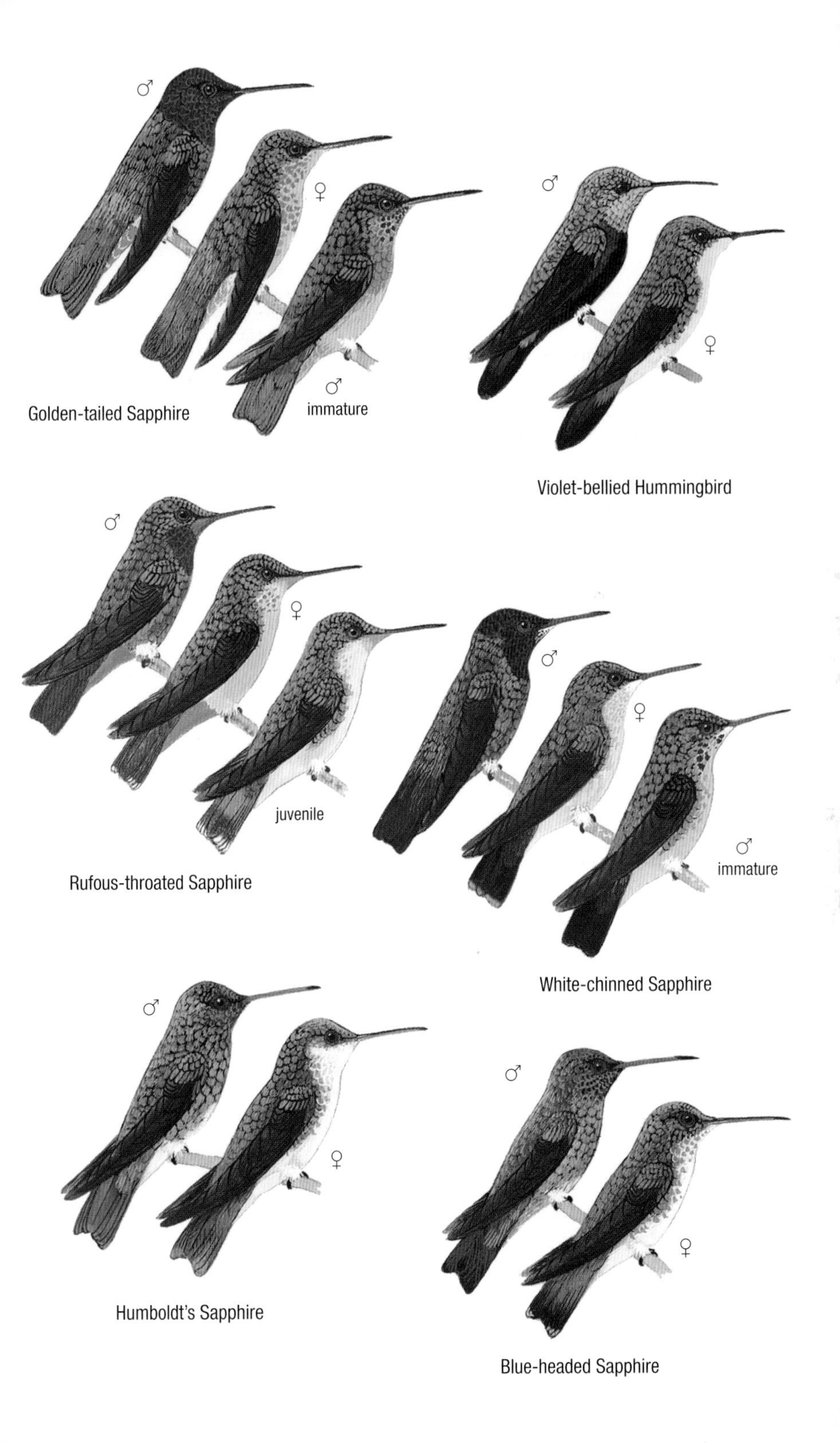
♂
♀
♂
immature
Golden-tailed Sapphire
♂
♀
Violet-bellied Hummingbird
♂
♀
juvenile
Rufous-throated Sapphire
♂
♀
♂
immature
White-chinned Sapphire
♂
♀
Humboldt's Sapphire
♂
♀
Blue-headed Sapphire

Quetzals are large, mostly frugivorous forest birds that tend to perch motionless for protracted periods, then swiftly 'sallying' after fruit taken in brief hovers; trogons included in this plate are more distinctive than those on next plates. Also forest-based and lethargic, typical of trogons, they tend to raise the tail rapidly and lower it down slowly while vocalising.

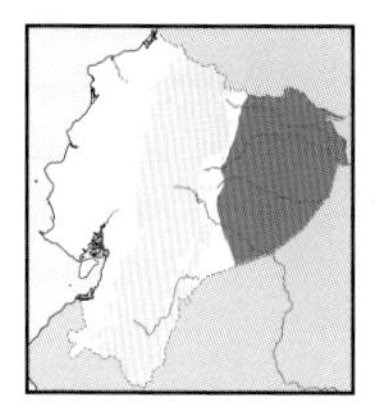

Pavonine Quetzal *Pharomachrus pavoninus* 33–34cm

E lowlands (to 600m, locally higher). Inside *terra firme* forest. ***Red bill*** (duskier in ♀) and iris. ***Bright golden-green head***, glittering emerald breast and upperparts, scarlet underparts. ***Elongated scapulars project over black wing feathers***, elongated uppertail-coverts. ♀: ***dusky head and breast***, upper breast blotched green; ***scapulars and uppertail-coverts shorter***. Nominate. Loose pairs perch high above ground, difficult to observe; undulating flight but often hard to locate after landing. **Voice** Whistled *khiiÜ'iiiiuuu..khiiÜ'iiiiuuu..khiiÜ'iiiiuuu...*, *kweeau..kweeau..kweeau....* **SS** Only quetzal in range; cf. Black-tailed Trogon (superficially similar plumage pattern). **Status** Uncommon.

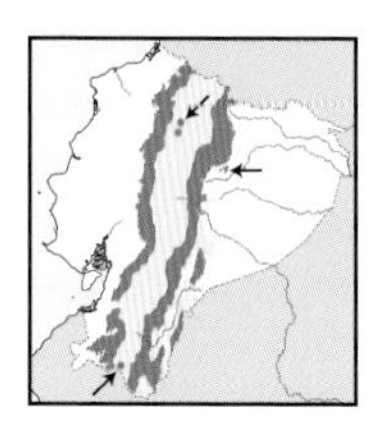

Golden-headed Quetzal *Pharomachrus auriceps* 33–35cm

E & W Andean slopes (1,000–2,800m, locally lower). Montane forest, borders, sometimes clearings. ***Yellow bill*** (dusky in ♀). ***Bright golden-green head***, emerald breast and upperparts. ***Elongated scapulars over black wings, elongated uppertail-coverts***. Scarlet below, ***black tail***. ♀: ***duskier head*** and pectoral band, ***black tail***. Nominate. Habits similar to other quetzals; several may congregate at fruiting trees, even with other frugivores; often heard before seen. **Voice** Noisy; far-carrying whistled *ka, kha-käooo, kha-käooo, kha-käooo...*; laughing *whe-de'de'de'de'dr...*, *whe-dee'dee'dee'rr.* **SS** ♀ Crested Quetzal has at least some white barring in tail. **Status** Fairly common.

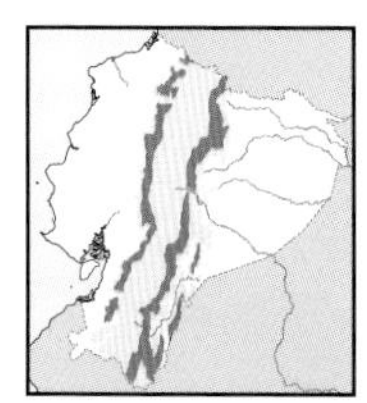

Crested Quetzal *Pharomachrus antisianus* 33–34cm

E & W Andean slopes (1,500–2,500m, locally to 1,100m). Montane forest, borders, sometimes clearings. ***Bushy emerald crest covering yellow bill***. Bright emerald foreparts and upperparts. Scapulars ***elongated over black wings, elongated uppertail-coverts***. Scarlet underparts, ***tail white below***. ♀: ***duskier***, lacks crest, red confined to lower belly, ***black tail with white edges and barring***. Monotypic. Singles or pairs perch high, remaining motionless for protracted periods; fruits plucked in aerial sallies and short hover. **Voice** Far-carrying, reiterated *küeép-kúao..küeép-kúao...*; noisy *kr-kr-kra-kára-kakára....* **SS** Golden-headed lacks crest, tail entirely black. **Status** Uncommon.

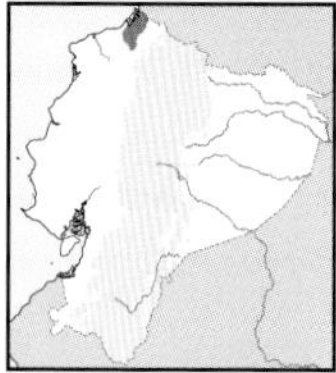

Slaty-tailed Trogon *Trogon massena* 32–33cm

Lowlands in extreme NW (below 200m). Humid forest, borders. ***Salmon-red bill*** (***dusky maxilla in*** ♀), ***dark eyes***. Shining green head, breast and upperparts, wing-coverts vermiculated black and white; red below, ***black tail***. ♀: ***grey instead of green***. Race *australis*. Pairs or singles in higher strata, hover-gleaning for prey and fruit. **Voice** Slow, nasal, uprising *kau*, in long series (30+ notes). **SS** Western red-and-green trogons have yellow bills and pallid eyes; voice lower-pitched, more nasal than Blue-tailed Trogon. **Status** Apparently very rare, EN in Ecuador.

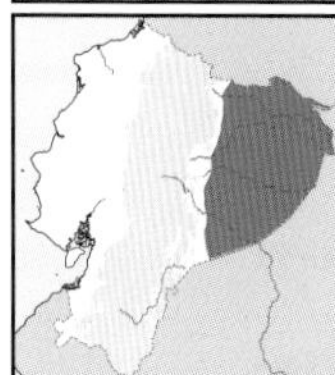

Black-tailed Trogon *Trogon melanurus* 30–32cm

E lowlands (below 400m). Humid forest, borders. ***Yellow bill*** (♀ dusky culmen), ***red eye-ring, dark eyes***. Green foreparts and upperparts, bluer rump and upper tail, ***thin white breast-band; tail black below***. ♀: ***grey and red***. Race *eumorphus*. Pairs in large sub-canopy branches, hovering for fruit and prey. **Voice** Slightly uprising 15–30, rather slow, *cuh-cuh-cuh...* or *kha-kha-kha-kha...*, often shorter; *ch'kja-kja-kj-kj-kj* calls. **SS** Collared (tail barred white) and Blue-crowned Trogons (green, blue and red; ♀ has broader eye-ring). **Status** Uncommon.

♂
immature
♀
Pavonine Quetzal
♂
immature
♂
♀
Golden-headed Quetzal
♀
♂
Crested Quetzal
♂
♀
♂
immature
♂
♀
Slaty-tailed Trogon
Black-tailed Trogon

Smaller red-and-green trogons, some with white in tail, all primarily forest-based and inconspicuous, despite bright and contrasting colours. ♀♀ are less stunning. Vocalisations are key identification features, as most species are more often heard than seen. One strictly montane species, but some ascend to lower subtropics.

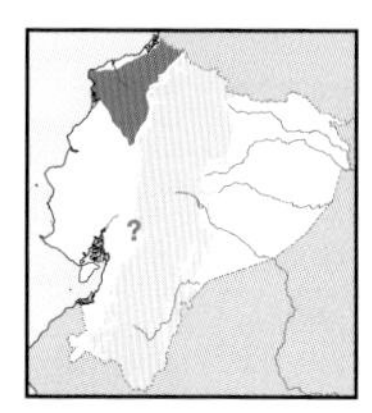

Blue-tailed Trogon *Trogon comptus* 30–32cm

NW lowlands and foothills (mostly 300–800m). Mature forest, woodland. ***Yellow bill*** **(*dusky culmen*** in ♀), ***grey eye-ring, white eyes***. Shiny green head, breast and upperparts. Red underparts, ***black undertail; bright green-blue rump***. ♀: grey above, tail blackish below. Monotypic. Pairs usually in large subcanopy branches; sudden flights to prey, fruit captured by hovering. **Voice** Slow, steady series of 10–30 *kow-kow-kow...* notes, higher-pitched, faster than Ecuadorian Trogon; *kr'kr'kra'kr* calls. **SS** Slaty-tailed (dark eyes, orange bill); Ecuadorian (red eye-ring, thin white breast-band). **Status** Uncommon, NT in Ecuador.

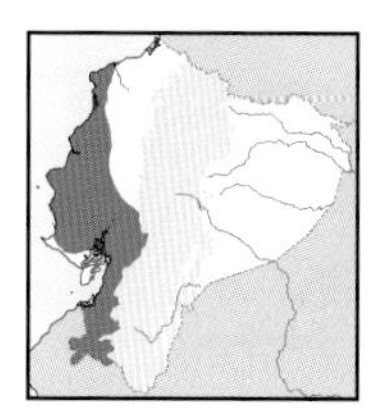

Ecuadorian Trogon *Trogon mesurus* 30–32cm

W lowlands, to SW subtropics (to 1,900m in SW). Semi-deciduous forest, borders. Yellow bill (dusky culmen in ♀), ***red eye-ring, whitish eyes***. *Green* foreparts and upperparts, red belly, ***narrow white pectoral line***, undertail black. ♀: grey and red, ***narrow white pectoral line***. Monotypic. Pairs in subcanopy, lazily perching for long periods. **Voice** Slightly uprising, slow *cu-cu-cu...*, often in +20-note series; higher-pitched than Black-tailed Trogon. **SS** Blue-tailed Trogon (slower voice, lacks red eye-ring and white pectoral line, including ♀); Collared Trogon (extensive white in tail, dark eyes); no overlap with Slaty-tailed. **Status** Uncommon.

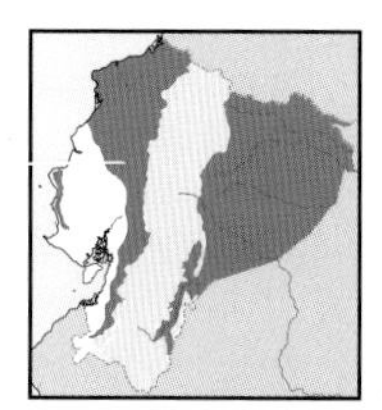

Collared Trogon *Trogon collaris* 24–26cm

E & W lowlands to lower subtropics (to 1,600m). Humid forest, woodland (in E mostly *várzea*). ***Metallic green upperparts and foreparts***, white breast-band, red belly. Wing-coverts ***vermiculated black and white***. Tail ***broadly and evenly barred white, broad white tips***. ***Brown*** ♀, white crescent-shaped eye-ring, ***tail flecked below***. Nominate (E), *virginalis* (W). Mainly forest-based, but habits as most trogons. **Voice** Short, somewhat high-pitched *caur-cáu-cáu-cáu-cáu, kwer-këw-këw*; soft *ch'urrrrr*. **SS** Higher-elevation Masked Trogon (thickly barred tail); Blue-crowned (blue foreparts, yellow eye-ring); voice rather ***simple***. **Status** Uncommon.

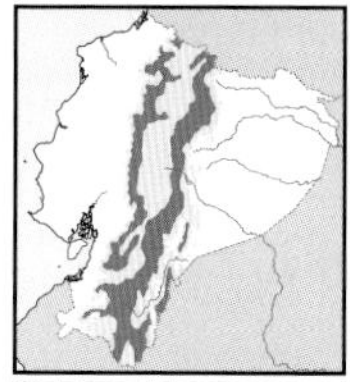

Masked Trogon *Trogon personatus* 25–26cm

E & W Andean slopes (1,500–3,400m; locally lower). Montane forest, borders. ***Metallic green foreparts and upperparts***, white breast-band, red belly. ***Tail finely barred white, broad white tips***. ♀: ***brown*** instead of green, tail as ♂. Races *assimilis* (W), *personatus* (E, below 2,500m), *temperatus* (E, above 2,500m); *temperatus* possibly separate species. Behaviour similar to other forest trogons. **Voice** Even, quite high-pitched *káu-káu-káu*, often longer (faster in E highlands); soft *chirrr* calls. **SS** Lower-elevation Collared Trogon (more evenly-barred tail). **Status** Uncommon.

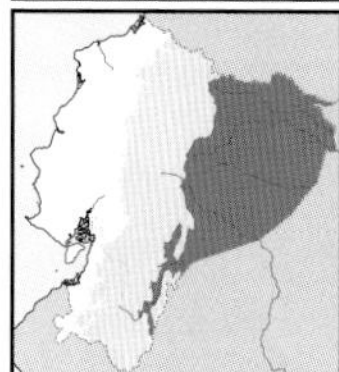

Blue-crowned Trogon *Trogon curucui* 24–26cm

E lowlands and foothills (to 1,100m). *Terra firme* and *várzea* forest, borders. Yellow bill, ***yellow eye-ring*** (***bluish, crescent-shaped*** in ♀). Bright ***greenish-blue foreparts, deeper crown***; blurry white breast-band, ***red belly. Tail barred white, broad white tips***. ♀: ***grey*** foreparts and upperparts, ***blacker tail***. Race *peruvianus*. Pairs frequent borders, where can be very phlegmatic; behaviour as other trogons. **Voice** Cadenced, fast *kau'kau'kau'kau...* series ending abruptly. **SS** Collared Trogon (green foreparts and uppertail, evenly-barred tail; ♀ brown); voice ***faster*** than Green-backed. **Status** Uncommon.

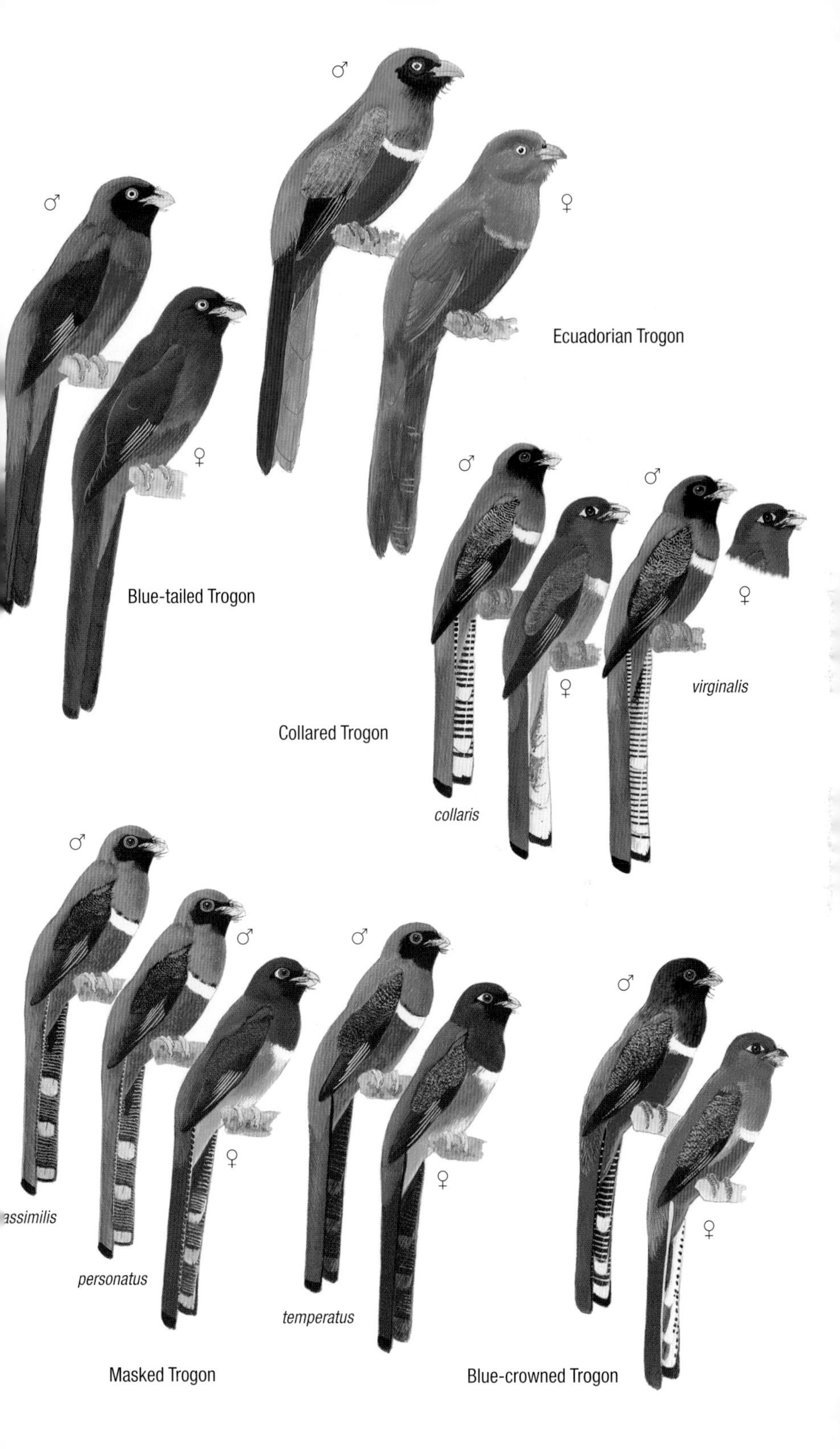
♂
♀
Ecuadorian Trogon
♂
♀
Blue-tailed Trogon
♂
♀
♂
♀
virginalis
Collared Trogon
collaris
♂
♂
♀
♂
♀
assimilis
personatus
temperatus
Masked Trogon
♂
♀
Blue-crowned Trogon

Yellow-bellied trogons of lowlands and foothills. Two species pairs, formerly treated as conspecific, occur each side of the Andes; very similar but differ in voice.

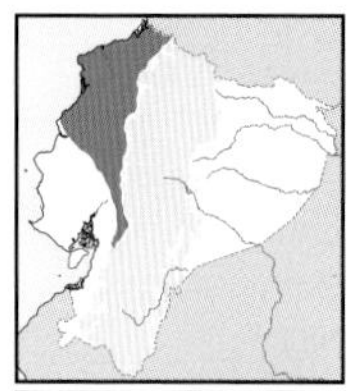

White-tailed Trogon *Trogon chionurus* 28–29cm

W lowlands (to 800m). Humid forest, borders. Greyish-blue bill, ***pale bluish eye-ring.*** Bright greenish-blue foreparts, greener back. Yellow underparts, ***white tail.*** ♀: ***grey above, black-and-white barring on white-tipped tail.*** Monotypic. Mostly pairs in upper strata, descending at borders; often tame. **Voice** Some 15–20 *caup* notes, fast, accelerating at end. **SS** Similar Gartered (less white in tail, yellow eye-ring in ♂); Black-throated voice ***slower***, more enunciated; Gartered's ***higher-pitched***, longer. **Status** Uncommon.

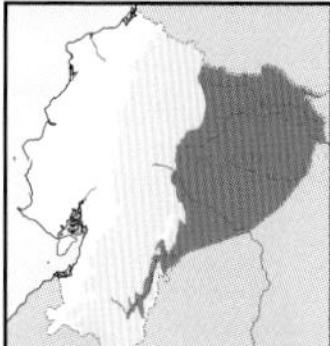

Green-backed Trogon *Trogon viridis* 28–29cm

E lowlands to foothills (to 1,200m). *Terra firme* and *várzea* forest, borders. Greyish-blue bill, ***pale bluish eye-ring.*** Bright greenish-blue foreparts, greener back; yellow underparts, ***tail mostly white below.*** ♀: ***grey*** replaces blue-green, ***black-and-white barring on white-tipped tail.*** Nominate. Mostly pairs that often tolerate close approach. **Voice** Fast, accelerating, uprising, slightly nasal series of 12–20 *caúp* notes; sometimes much slower; soft *chowk* calls. **SS** Similar Amazonian Trogon (less white in tail, ♂ has yellow eye-ring). **Status** Uncommon.

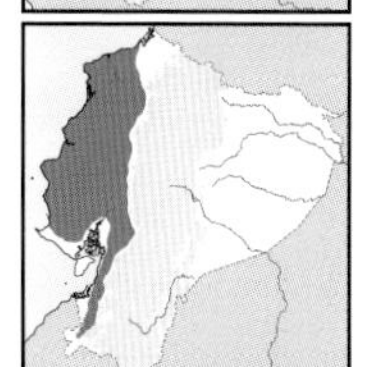

Gartered Trogon *Trogon caligatus* 22–23cm

W lowlands, locally foothills (to 900m). Humid to semi-deciduous forest, woodland. ***Bill greyish-blue, yellow eye-ring*** (***whitish*** in ♀). Bright greenish-blue foreparts, blue-green rump and uppertail; faint white pectoral line, yellow underparts. ***Tail black, barred white, broad white tips.*** ♀: grey and yellow, ***blacker tail.*** Race *concinnus*. Rather lethargic and confident, often in subcanopy and borders. **Voice** Long, fast, rather steady *ka-ka-ka-ka...* series, sharper than Amazonian Trogon. **SS** White-tailed Trogon has whiter tail, bluish eye-ring. **Status** Uncommon.

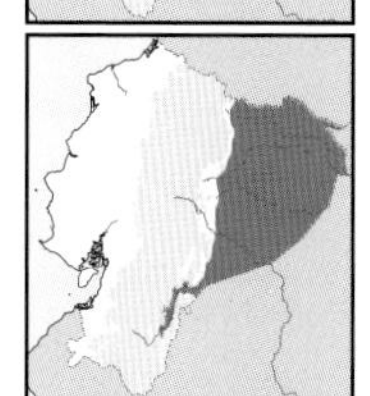

Amazonian Trogon *Trogon ramonianus* 22–23cm

E lowlands (to 500m, locally higher). Humid forest, borders. ***Bill greyish-blue, yellow eye-ring*** (***whitish*** in ♀). Bright greenish-blue foreparts, blue-green rump and uppertail-coverts. Faint white pectoral line, yellow underparts. ***Tail barred white, broad white tips.*** ♀: grey and yellow, ***more black in tail.*** Race *crissalis*? Lethargic, rather tame pairs in forest subcanopy, sometimes lower, may briefly join army ant followers. **Voice** slowly accelerating *caou-caou-caou-caou...* series. **SS** Green-backed Trogon (whiter tail); female has greyish eyering, broader white tail tips, no white pectoral band; Black-throated Trogon (green and yellow). **Status** Uncommon.

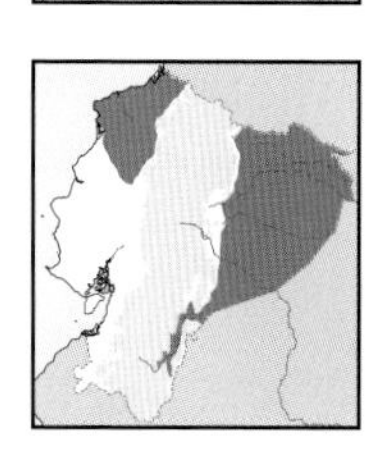

Black-throated Trogon *Trogon rufus* 25–26cm

E & W lowlands, to W foothills (to 800m). Midstorey, rarely borders. ***Pale yellow eye-ring. Bright green head, breast and back***, bronzy-green uppertail. ***Yellow belly***, tail with broad white barring, ***broad white tips.*** ♀: ***brown*** replaces green, tail whiter. Races *sulphureus* (E), *cupreicauda* (W). More forest-based than other small trogons, but behaviour similar. **Voice** Slow *kuá..kuá..kuá*, sometimes additional notes; longer in W. **SS** Only green-and-yellow or brown-and-yellow trogon; voice ***simple and brief.*** **Status** Uncommon.

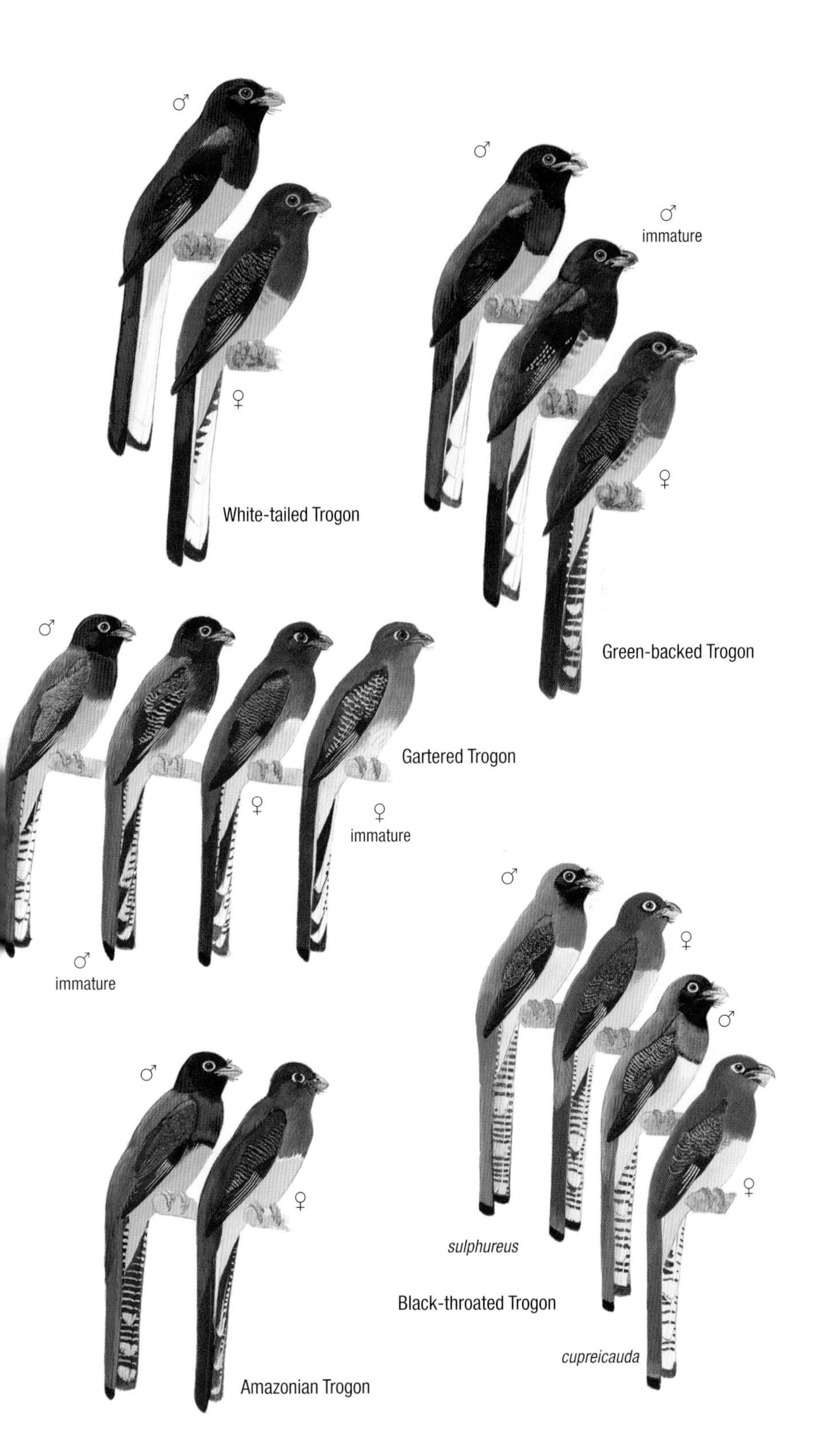
♂
♀
White-tailed Trogon
♂
♂
immature
♀
Green-backed Trogon
♂
♀
♀
immature
Gartered Trogon
♂
immature
♂
♀
sulphureus
♂
♀
Black-throated Trogon
cupreicauda
♂
♀
Amazonian Trogon

All kingfishers are closely tied to water but habitat differs, with Ringed occurring in a variety of clean to rather polluted waterbodies, but American Pygmy and Green-and-rufous prefer sluggish forested streams. Though size differs notably, structure and silhouettes are alike. Ringed and Amazon tend to perch higher above water than smaller species.

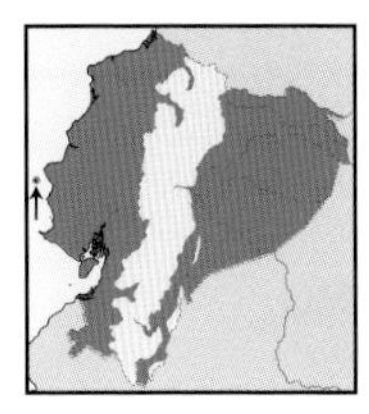

Ringed Kingfisher *Megaceryle torquata* 38–41cm

E & W lowlands to foothills (to 1,300m, occasionally to 2,600m). Rivers, large streams, lakes, estuaries, ponds, mangroves, artificial wetlands. *Massive bill. Bluish-grey above, broad white collar, rich rufous below.* White patches in primaries (in flight). ♀: *wide bluish-grey* and narrow white breast-bands. Nominate. Frequent in freshwater but also in saline environments; flies high and plunges for fish; perches high, usually on exposed branches, posts or wires. **Voice** Strident fast rattle, *krrEK-krrEK-krrEK-krrEK...*; harsh *ggreék!* and louder *rreEK!* in flight. **SS** Unmistakable, but cf. Belted Kingfisher (accidental in SW). **Status** Common.

* Belted Kingfisher *Megaceryle alcyon* 30 32cm

Boreal migrant, recorded twice near coast in SW, several records from Galápagos. Superficially similar to Ringed Kingfisher, but *underparts mainly white, with a single broad bluish-grey breast-band.* ♀: also has *narrow rufous band on chest* and rufous flanks. Monotypic. Recorded near sea in Ecuador; elsewhere behaves much like Ringed. **Voice** Strident rattle less harsh than Ringed. **SS** Commoner Ringed is larger, rich rufous below. **Status** Accidental (Sep, Dec).

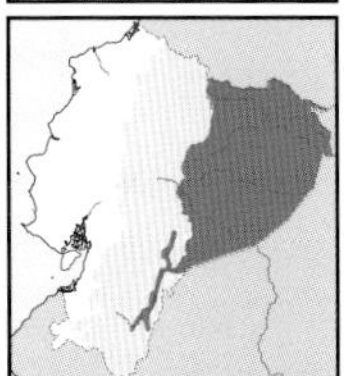

Amazon Kingfisher *Chloroceryle amazona* 28–29cm

Lowlands and foothills in E (to 1,000m). Rivers, large streams, ponds and lakes. Medium-sized with *long, strong, black bill. Dark shiny green head and upperparts,* white collar and belly, *broad rufous breast-band.* ♀: rufous band replaced by a *incomplete and narrower green band.* Monotypic. Flies fast and direct low over water; mainly observed perched at various heights above water, from where plunges for fish. **Voice** Descending, fast, squeaky *jí-jí-jik-jik-jik-jek-jek-jk*, sometimes shorter; loud harsh *chák* or *tják!* calls; *tják-tjak'jak....* **SS** Largest green kingfisher; cf. Green Kingfisher with obvious white spots in wings. **Status** Fairly common.

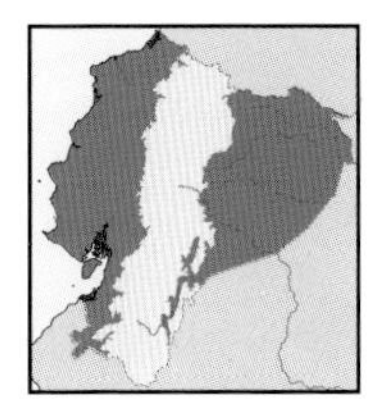

Green Kingfisher *Chloroceryle americana* 18–20cm

E & W lowlands to foothills (to 1,300m). Rivers, streams, lakes, ponds, mangroves, artificial wetlands. Smaller than Amazon Kingfisher, *dark shiny green above, wings have white spots.* Underparts mostly white, *rufous breast-band.* Outer tail feathers white (in flight). ♀: similar but *broken green breast-band.* Nominate (E), larger *cabanisii* (W). Flies fast, direct, low over water, plunges from low perches; more prone to occur in deforested habitat than other green kingfishers. **Voice** Raspy, sharp *dzrrt* often doubled or trebled: *dzrrt-zrrt, dzrrt-zrrt-zrrt, dzrr-zrr-zrr-zrrt.* **SS** Amazon Kingfisher larger, more pointed-headed, with no white in wings. **Status** Fairly common.

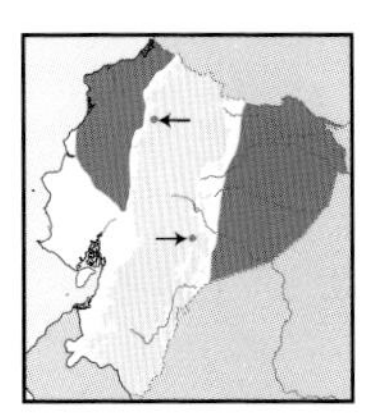

Green-and-rufous Kingfisher *Chloroceryle inda* 21–23cm

Lowlands in E & W (below 400m, locally higher). Slow-moving forest streams, shaded lake banks, flooded forest and *várzea. Dark shiny green head and upperparts, rich rufous collar, underparts and small facial streak,* minute white spots on wings. ♀: *green-and-white breast-band.* Monotypic. Difficult to observe as perches low, often under cover; dives directly from low perches in rapid and direct flight. **Voice** Gravelly, almost hummingbird-like, nasal, often trebled *zzhrrt-zhrrt-zhrr-zhrr...* or simpler *dzzirrt!*, drier than Green Kingfisher. **SS** American Pygmy Kingfisher has similar plumage and occupies similar habitat but much smaller. **Status** Uncommon, forest-based, rarer in W.

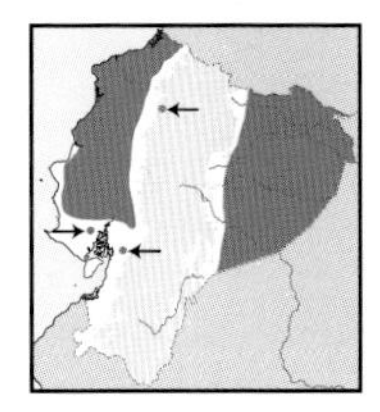

American Pygmy Kingfisher *Chloroceryle aenea* 13cm

Lowlands of E & W (mostly below 400m, locally higher). Slow-moving forest streams, shady lakesides, flooded forest, *várzea. Tiny* with dark *shiny green head and upperparts. Orange-rufous collar and underparts, white mid belly and vent.* ♀: *thin green breast-band.* Nominate. Difficult to observe as perches low above water mostly inside forest; very fast and direct flight, attacks prey from perch. **Voice** Shrill *t't't-tzt'zt'zt-tít* song; also feeble, hummingbird-like *dzeét* in flight and short *tik* notes. **SS** Larger Green-and-rufous Kingfisher has similar plumage pattern. **Status** Uncommon in E, forest-based, rarer in W.

♂
♀
juvenile
Ringed Kingfisher
♂
♀
♂
immature
Belted Kingfisher
♀
Amazon Kingfisher
♂
immature
♂
♂
♀
juvenile
♀
Green Kingfisher
♂
♀
Green-and-rufous Kingfisher
♂
immature
♂
♀
American Pygmy Kingfisher

Unique and unmistakable motmots characterised by long pendulum-like tails and strong bills. Two rufous species, very similar, best identified by voice. *Momotus* motmots were considered to represent a single, quite variable species, until recently. Currently, three species replace each other altitudinally or geographically.

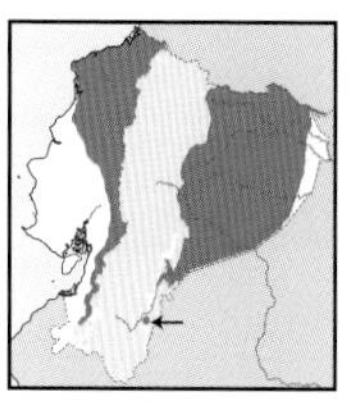

Broad-billed Motmot *Electron platyrhynchum* 33–36cm

E & W lowlands and foothills (mostly below 1,400m). Humid forest interior, borders. ***Broad***, slightly curved, ***keeled bill. Rich rufous foreparts***, black mask, ***green chin***, black mid-breast spots. Green upperparts, ***belly paler green***. Long green and turquoise-blue tail ending in square rackets (***not*** in E). Nominate (W), *pyrrholaemum* (E). Singles or pairs, often move tail sideways while perched; pre-dawn singer. **Voice** Nasal *cuuank* or *guaánh!* at long intervals; *ggua-ggua-ggua...* alarm calls. **SS** Larger Rufous Motmot (broader mask, no green chin, turquoise-blue primaries, green confined to *lower belly*). **Status** Fairly common.

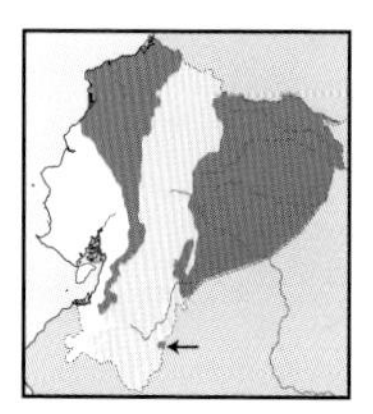

Rufous Motmot *Baryphthengus martii* 43–46cm

E & W lowlands and foothills (below 1,500m). Humid forest, borders. ***Heavy, serrated, longish bill. Rich rufous foreparts, broad*** black mask, black breast spots. Green upperparts and wings, ***turquoise-blue primaries***. Long green and turquoise tail, rackets only in W; pale green ***lower belly***. Nominate (E), *semirufus* (W). Behaves like Broad-billed but quieter, also moves tail like pendulum; more regular in borders and adjacent clearings. **Voice** Loud, resonant *hoop, hoop, hoop...hoo-doóp, hoo-doóp..hoo-hoo-hoo;* also trilled *hoo-oo'oo'oo'oo'oo....* **SS** From Broad-billed by mask, belly and vocalisations; voice more resonant than all *Momotus*. **Status** Fairly common.

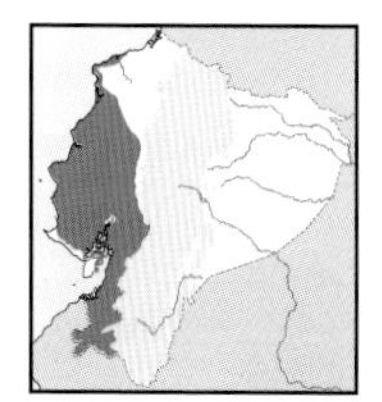

Whooping Motmot *Momotus subrufescens* 36–42cm

SW lowlands and foothills (to 1,500, higher in SW). Semi-deciduous and deciduous forest, borders, woodland. ***Bright green overall***, paler underparts, greener breast, more ochraceous belly. ***Silvery turquoise-blue crown*** extends over black central part. Black mask, striking red iris. ***Bicoloured*** tail racquets. Race *argenticinctus*, possibly deserves species status. Singles or pairs often in open and semi-open, including gardens and hedgerows; attacks prey from perch. **Voice** Single *whoOop!, whoo-wap-wap-wap* or guttural *guk-gk-gk-gk...* calls. **SS** Unmistakable in range, compare Rufous Motmot voice. **Status** Common.

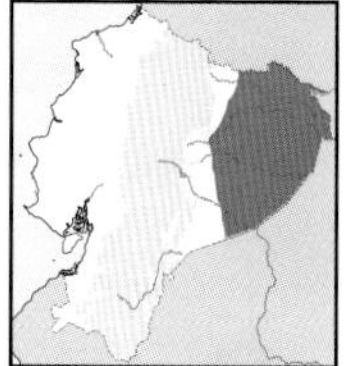

Amazonian Motmot *Momotus momota* 38–42cm

E lowlands (to 300m). Humid forest, borders. ***Bright green overall***, greener belly than breast (tinged tawny). ***Sky-blue crown***, black central part and often has ***thin rufescent posterior border. Black mask, striking red iris***. Race *microstephanus*. Singles or pairs in forest, sluggish but vocal, often among first birds to sing at dawn. **Voice** Hollow, fast *hoo-döop* every *c.*5 sec, first note higher and rising, second more abrupt; also *hoo-oop'oop'oop'oop*. **SS** Unmistakable, but compare more resonant and tremulous voice of Rufous Motmot. **Status** Uncommon.

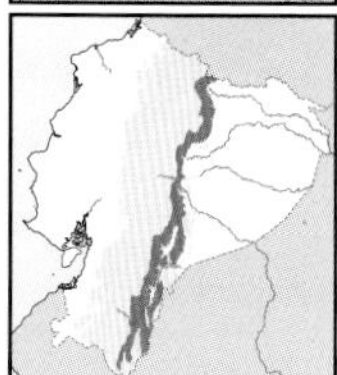

Andean Motmot *Momotus aequatorialis* 44–48cm

E Andean slopes (1,000–2,100m). Montane forest and borders. ***Bright green overall***, paler underparts, two ***well-marked*** black marks on mid breast. ***Turquoise-blue crown, black central part and posterior border. Black mask with narrow lower border, striking red iris***. Nominate. Sluggish pairs or singles mostly in forest, performing direct attacks after prey; more often heard than seen. **Voice** Faster, higher-pitched than Amazonian Motmot, *hoodoop*, both notes rapidly rising-falling. **SS** Unmistakable in montane range, no known overlap with Amazonian Motmot. **Status** Rare, possibly overlooked.

platyrhynchum
pyrrholaemum
Broad-billed Motmot
martii
semirufus
Rufous Motmot
Whooping Motmot
Amazonian Motmot
Andean Motmot

PLATE 125: TYPICAL JACAMARS

Characterised by their long slender bills and glistening upperparts, typical jacamars inhabit the forest interior, rarely venturing into edges and semi-open areas. Tend to perch lethargically for long periods, then swiftly sally after prey. Only one species in W lowlands, two in E foothills to subtropics, and others in E lowlands.

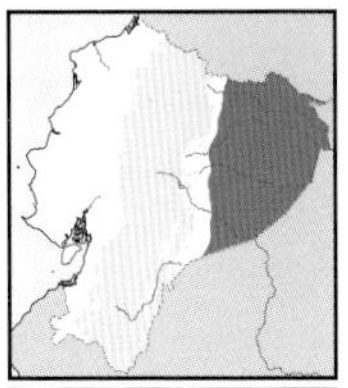

Yellow-billed Jacamar *Galbula albirostris* 19cm

E lowlands (to 400m). *Terra firme* forest, mature woodland. ***Bill yellow basally, yellow lores and eye-ring***. White throat, ***rufous underparts*** (throat and chin ***uniform rufous*** in ♀). Race *chalcocephala*, possibly separate species. Sluggish pairs at mid and low levels in forest, performing rapid and direct flights at flying insects; not with mixed flocks. **Voice** Ascending, whistled series *güüii..güüii..güüii* ending in *güi'ii'iiiii* trill; also slower *peeéf-peeéf-peeéf* series; *péw..péw..péw...* calls. **SS** Only forest jacamar with mostly uniform rufous underparts and yellow bill (but variable?). **Status** Uncommon.

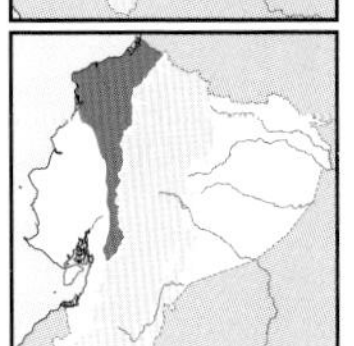

Rufous-tailed Jacamar *Galbula ruficauda* 24cm

W lowlands and foothills, mainly NW (to 700–800m). Humid forest, borders, adjacent clearings. Glittering coppery-green upperparts, chin and breast. ***White triangular throat patch, rich rufous below***, including tail. ♀: white replaced with ***buff on throat***. Race *melanogenia*. Lethargic pairs perch on fairly exposed branches, often near water. **Voice** Sharp *peeEEk!* often repeated rapidly; long, sharp, ascending, slowing *pii-pii-pll-pll-pll* series. **SS** Only *Galbula* in range. **Status** Uncommon, locally in Andean foothills to E Guayas.

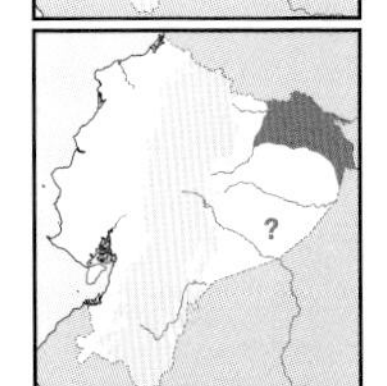

White-chinned Jacamar *Galbula tombacea* 23cm

NE lowlands, mostly N of Río Napo (to 400m). Low strata in *várzea* and stream-side forest. Golden-green above, ***greyish-brown forecrown, metallic green throat and breast, small whitish chin patch*** (fainter in ♀). ♀: most underparts ***orange-rufous***. Nominate. Tends to avoid *terra firme* forest; makes direct sallies for flying insects, not with mixed flocks. **Voice** Accelerating, ascending *piiíp..piiíp..piiíp.. píp-píp-píp-pi'ii'eee...*, descending at end, lower-pitched and shorter than Yellow-billed Jacamar; piercing *feeéyk* calls. **SS** Purplish Jacamar (no rufous below), higher-elevation Coppery-chested Jacamar (no whitish chin). **Status** Uncommon.

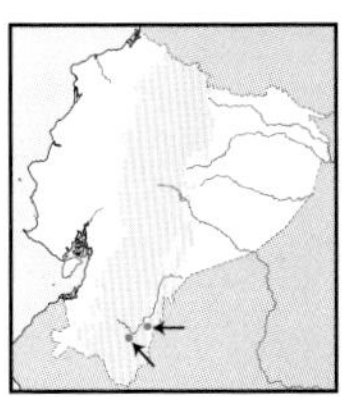

Bluish-fronted Jacamar *Galbula cyanescens* 23cm

Few records in SE foothills (880–1,000m). Montane forest. Much like White-chinned Jacamar, but forecrown ***bluish-green*** not brownish; chin to throat greener. ♀: ***washed-out*** lower underparts. Monotypic. Lethargic, fond of vine tangles, treefall gaps and forest edges. **Voice** Much as White-chinned, an accelerating, ascending *piíp, piíp, piíp...* (S *et al.*). **SS** Coppery-chested Jacamar (broad orange eye-ring, coppery chest; ♀ has buffy throat, more rufous belly). **Status** Uncertain; one historical plus one recent record.

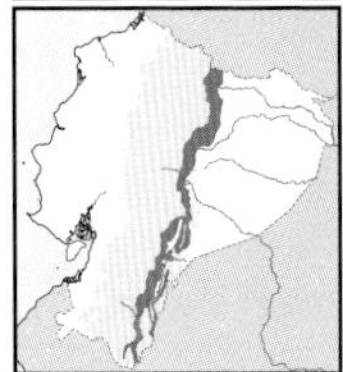

Coppery-chested Jacamar *Galbula pastazae* 24cm

E foothills to lower subtropics (700–1,500m). Montane forest, borders. ***Broad orange eye-ring***. Glittering green above, ***bluer foreparts, especially crown***; metallic green ***throat and breast***, rufous below. ♀: ***pale rufous throat***. Monotypic. Loose pairs on exposed branches, tends to prefer natural borders (gaps, ravines, landslides). **Voice** Fast, clear, ascending *peé-peé-peé-peé...* series, not trilled at end; clear *peép!* calls. **SS** Little overlap with Bluish-fronted Jacamar (no orange eye-ring, no coppery in chest); ♀ paler on belly. **Status** Uncommon, local; VU globally, NT in Ecuador.

Yellow-billed Jacamar
Rufous-tailed Jacamar
te-chinned Jacamar
Bluish-fronted Jacamar
Coppery-chested Jacamar

The species on this plate are all found in the E lowlands; Great also in NW. Paradise is rare and local in Ecuador. Great resembles typical *Galbula* jacamars but is larger and stronger-billed.

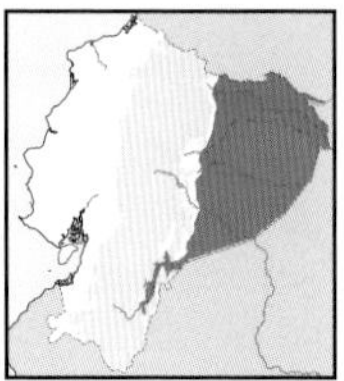

Purplish Jacamar *Galbula chalcothorax* 23–24cm

E lowlands to lower foothills (below 500m, locally to 1,000m). Humid forest, borders. ***Blue-green head, bronzy-purple upperparts, breast and upper belly; white throat, lower belly and vent, dusky tail below.*** ♀: ***buff*** instead of white below. Monotypic. Sluggish pairs sally for flying insects from low to mid-level perches; rather tame. **Voice** Ascending *güiii..güiii..güi..dii-dii'ii'ii'iiii'e, wuee-dii..wue-dii.. wuedii'wuedii-dudii'dii'ii'ii...* chatter ending in trill, notes often doubled; piercing *wued!* calls. **SS** Only Amazonian jacamar lacking green-and-rufous combination. **Status** Rare.

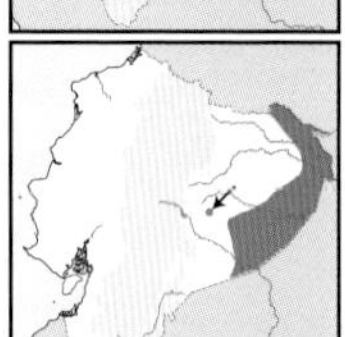

Paradise Jacamar *Galbula dea* 28–30cm

Local in E lowlands (below 250m). *Terra firme*, borders. Large, ***dark***. Mostly ***blue-black***, contrasting ***white triangular throat patch***. ***Long***, graduated tail with ***elongated central feathers***. Sexes very similar. Possibly race *brunneiceps*. Not well known, possibly similar to other *Galbula* but more in canopy, where fairly loose pairs are rather lethargic. **Voice** Long, descending *piiyr..piiyr-piyr-piyr-piur-piur-pur...* series; high-pitched *piúk!* calls. **SS** Unique. **Status** Rare, local.

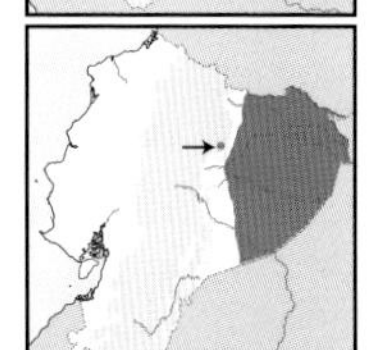

White-eared Jacamar *Galbalcyrhynchus leucotis* 19–20cm

E lowlands (below 400m, locally higher). *Várzea* forest, borders, lakesides, locally on river islands. Very ***long pointed pink bill***. Mostly ***chestnut, contrasting white ear patch***; darker wings, crown and tail. Small ***pink feet***, dark ***pink skin*** around eyes. Sexes alike. Monotypic. Pairs or small groups rather sluggish, often near ground; feeds mostly on flying insects caught in short, rapid sallies. **Voice** Rising-falling chatter, often starting with sharp *keeyp!.. keer-err'err'err'err'errerrerr...*; more spaced *keeyp!* calls, 1–2 per sec. **SS** Unmistakable. **Status** Fairly common, local.

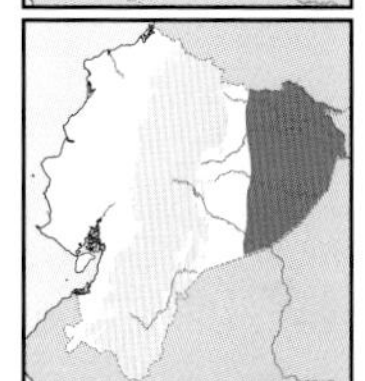

Brown Jacamar *Brachygalba lugubris* 16–17cm

E lowlands (below 300m). *Várzea* canopy, borders, riparian woodland and lakesides. ***Small, dusky-brown*** with ***thin, slender, long black bill***. Vague dusky streaking / mottling on foreparts, ***buffy mid belly***. ♀: ***duskier*** than ♂. Race *caquetae*. Small groups often forage at borders, making fairly long-distance sallies, perching sluggishly for long periods. **Voice** Descending, fast but slowing *pi-dididi'di'dii'dii'diw*; single piercing *peeey!* calls. **SS** Not likely confused due to small size and brown plumage. **Status** Uncommon and local.

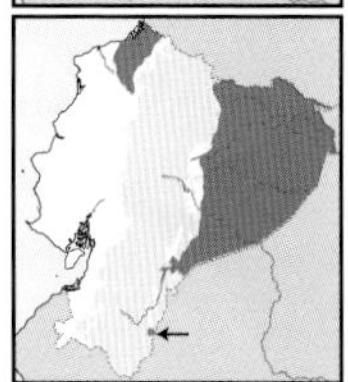

Great Jacamar *Jacamerops aureus* 29–31cm

E lowlands to foothills, locally in NW lowlands (below 750m, locally to 1,300m in E, below 450m in W). Humid forest. ***Largest*** jacamar. ***Stout, decurved bill. Glistening green upperparts and chin, white lower throat***, rich rufous below, ***long tail***. ♀: lacks white throat. Races *isidori* (E), presumably *penardi* (W). Shy pairs or singles; sudden attacks to branches or leaves for prey. **Voice** Mournful, far-carrying *weeee-ooouuuu* whistle, second part sometimes missing or shorter; sharp *gueeeaau!* mewing. **SS** Unmistakable; cf. Rufous-tailed Jacamar in W. **Status** Uncommon, rarer in W.

♂
♀
Purplish Jacamar
Paradise Jacamar
White-eared Jacamar
juvenile
adult
Brown Jacamar
♂
♂
♀
isidori
penardi
Great Jacamar

PLATE 127: PUFFBIRDS

Lethargic species that occupy canopy (*Notharchus*) or understorey (*Bucco*), vocalising little, usually before dawn, and attacking passing prey. All puffbirds on this plate, especially the *Bucco* species, can easily remain undetected owing to their sedate behaviour, lack of vocal activity and solitary lifestyle.

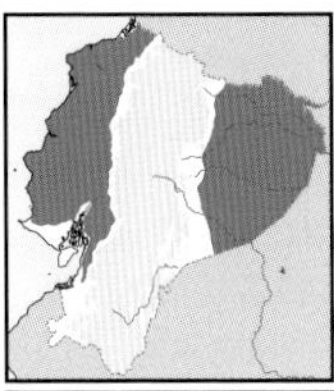

White-necked Puffbird *Notharchus hyperrhynchus* 25–26cm

E & W lowlands (to 600m). Canopy, borders of humid to semi-deciduous forests, secondary woodland, adjacent clearings. Large, with stout bill. Pied: ***black*** crown, upperparts and ***broad breast-band***; white ***front, throat, nuchal collar and belly***. Monotypic. Motionless and silently perch on exposed branches, performing sudden sallies to air or foliage, prey battered prior to ingestion. **Voice** Monotonous, rather soft trill *pt-tr'tr'tr'tr'tr* or *pt-t-t-t-t-t*. **SS** Black-breasted Puffbird (NW) has black front, less white on neck. **Status** Uncommon.

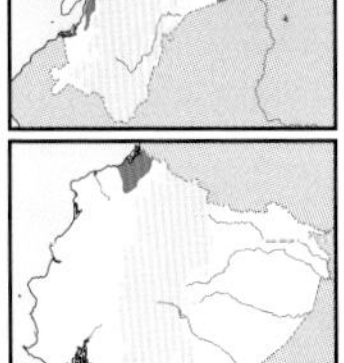

Black-breasted Puffbird *Notharchus pectoralis* 20–21cm

Humid lowlands in extreme NW (to 200m). Canopy, subcanopy and borders of wet forest. Pied, with ***black predominating; white confined to throat, ear-coverts, belly and vent***. Monotypic. Less prone to perch high and exposed than previous species, more often inside forest; also lethargic aerial sallier, but sometimes follows army ant swarms. **Voice** Whistled *kwee'kwee'kwee'kwee'kwee...*, slowing to *kwee'a-kwee'a..kwe'u..kwe'u*, descending at end. **SS** See whiter and larger White-necked Puffbird locally in NW. **Status** Rare, locally uncommon, confined to NW Esmeraldas.

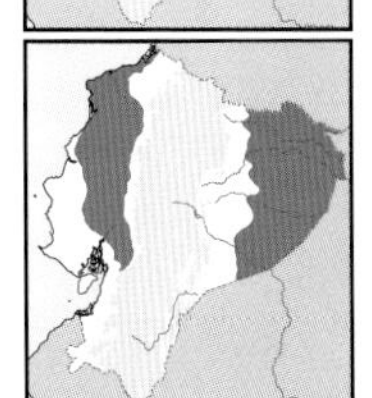

Pied Puffbird *Notharchus tectus* 14–17cm

E & W humid lowlands (to 500m). Subcanopy and borders of humid forest, secondary woodland, adjacent clearings. The ***smallest*** pied puffbird. ***Black crown with white spots*** (more prominent in E), ***broad white superciliary and throat patch***. Contrasting ***white patch in scapulars***; ***narrow*** black breast-band. ***White-tipped tail feathers*** from below. Races *subtectus* (W), larger *picatus* (E), possibly separate species. Pairs regular on exposed branches, sallying for flying insects; lethargic and not very vocal during day. **Voice** Piercing, whistled *güiii..güi-güii-güii-güii'güii..dii-dii-dii'dii'dii...*; *pii-pii, piidii-piidii-piidii-pii'dii'dii...*. **SS** Larger *Notharchus*. **Status** Uncommon.

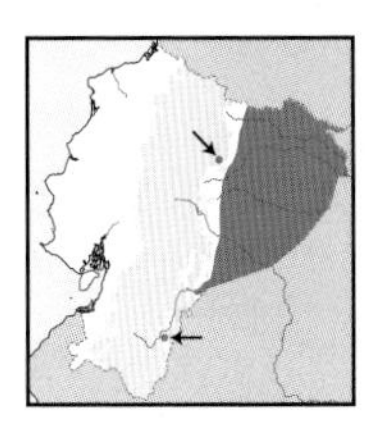

Chestnut-capped Puffbird *Bucco macrodactylus* 16–17cm

E lowlands (to 600m). Lower growth of humid *terra firme* and *várzea* forest, secondary woodland, adjacent clearings. Chestnut crown, ***black face bordered by white superciliary and malar stripes***, buffy-cinnamon throat and ***nuchal collar, contrasting black breast-band***. Nominate. Mostly inconspicuous, singles remain motionless for long periods, thus regularly overlooked; tends to prefer re-growth tangles; moves to upper strata at forest edges. **Voice** Mainly silent, but sometimes a plain, weak *fuiiooó*. **SS** Larger, rarer Spotted Puffbird (black-blotched breast to belly, less conspicuous head pattern). **Status** Uncommon.

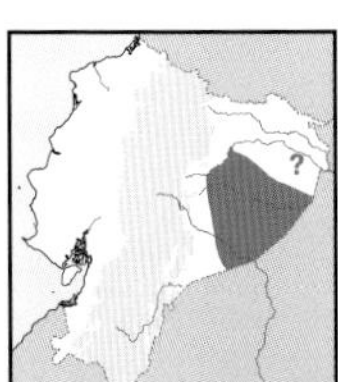

Spotted Puffbird *Bucco tamatia* 18–19cm

SE lowlands (below 300m). Lower strata of *várzea* forest. ***Rufous-orange front and throat***, duskier crown. ***Thin white malar stripe and nuchal collar***, striking ***black patch on neck-sides***. Dense ***black and white spotting on underparts***. Race *pulmentum*. Poorly known in Ecuador, but probably overlooked due to sluggish behaviour; mainly singly, quite regular at forest edges. **Voice** Monotonous, rising series of inflected *kuiiiíp?* notes, sometimes faster; weak *pyop* calls. **SS** Chestnut-capped Puffbird has black breast-band and browner barred belly. **Status** Rare and local.

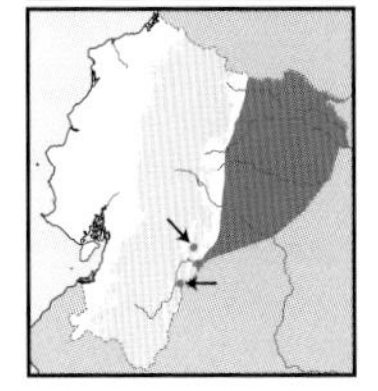

Collared Puffbird *Bucco capensis* 18–19cm

E lowlands (to 500m, locally or seasonally? higher). Mainly inside *terra firme* forest. ***Stout orange bill***, reddish-orange eyes. ***Striking orange head***, barred on crown; ***whitish throat, thin buffy-orange nuchal collar, black breast-band***. Rufous above, heavily barred. ***Mostly plain buff below***. Monotypic. Often in dense tangles, seldom at borders; perches motionless for protracted periods, and may briefly join mixed-species flocks. **Voice** Repeated *wua-weeu, wua-weeu..., wua-weeu-kleü, wua-weeu-kleü...* but seldom heard. **SS** Unmistakable. **Status** Rare, confined to forest.

White-necked Puffbird

Black-breasted Puffbird

picatus

subtectus

Pied Puffbird

Chestnut-capped Puffbird

Spotted Puffbird

Collared Puffbird

PLATE 128: PUFFBIRDS AND MONKLET

Two notably vocal canopy puffbirds (*Nystalus*) that might easily go undetected if silent. The three *Malacoptila* species dwell in the forest understorey, regularly join mixed-flocks, rarely venturing to forest borders and are usually difficult to detect. Lanceolated Monklet is the smallest puffbird, easily overlooked.

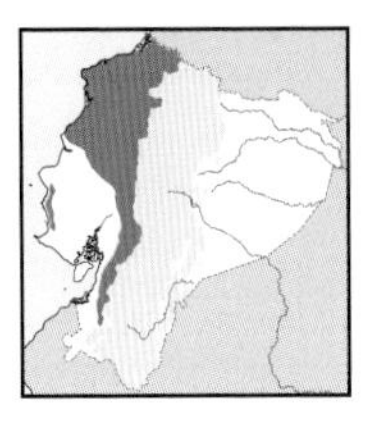

Barred Puffbird *Nystalus radiatus* — 20–22cm

W lowlands and foothills S in Andean foothills to El Oro and coastal ranges to N Santa Elena (to 1,400–1,500m). Humid forest, borders, adjacent clearings. Strong dark bill, yellow eyes. ***Rufous-chestnut crown, upperparts and tail, heavily barred black. Whitish-buff lores, buff nuchal collar*** and underparts, most underparts sparsely barred black. Monotypic. Singles or pairs on exposed branches for long periods, often moving tail sideways; may briefly join mixed-species flocks. **Voice** Whistled series of soft *wüeet?-wueéeu, wüip-wüuur*.... **SS** Striolated Puffbird occurs on E slope; confusion unlikely in range. **Status** Fairly common.

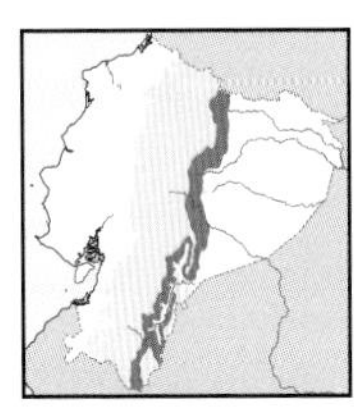

Striolated Puffbird *Nystalus striolatus* — 20–21cm

E foothills to subtropics (800–1,700m, occasionally to 400m). Montane forest and borders. Heavy dark bill, yellowish eyes. Dark chestnut above, including tail, with dense ***black bands. Buff nuchal collar and head-sides, blackish streaking on face to buffy breast***, whitish below. ***White lores, superciliary and throat***. Nominate. More confined to forest than Barred Puffbird, also perching motionless for long periods; tends to prefer higher forest strata, seldom with mixed flocks; hard to detect. **Voice** Melancholic, cadenced *wüp..wüu-wüup, wüuuuup*. **SS** No sympatry with superficially similar Barred Puffbird; cf. lower elevation White-chested Puffbird. **Status** Uncommon, overlooked.

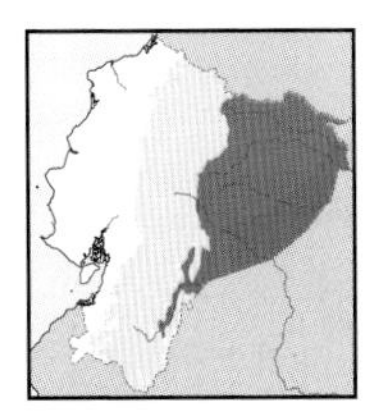

White-chested Puffbird *Malacoptila fusca* — 18–19cm

E lowlands and foothills (to 1,100–1,200m). *Terra firme* forest. Bill mostly ***orange***. Conspicuous ***white whiskers*** and crescentic breast-band. Dark brown head, throat and upperparts, ***with heavy buffy streaking***; paler underparts. Nominate. Shy and retiring, rarely observed; prefers forest undergrowth; sudden strikes at prey, often returning to same perch; regularly accompanies mixed flocks. **Voice** Rarely heard, high-pitched and descending *tsrrrrrrr* song; call a single, high-pitched, weak *pseeeú*. **SS** Higher-elevation Black-streaked Puffbird has black bill, less streaking above, bold buff throat; may overlap in lower foothills. **Status** Fairly common, confined to forest.

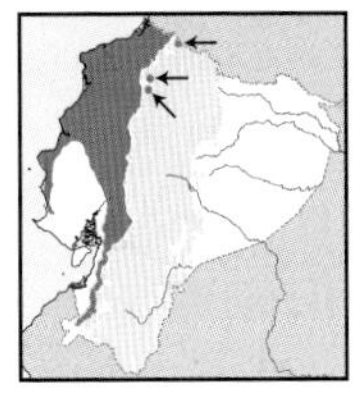

White-whiskered Puffbird *Malacoptila panamensis* — 18–19cm

W lowlands and foothills, S on Andean slopes to El Oro and in coastal ranges to N Santa Elena (to 900m). Humid forest, borders, second growth. Blackish bill, ***greenish-yellow mandible base***, red eyes. Rufescent-brown above, contrasting ***white whiskers, cinnamon to buffy-orange throat***. Dusky below, ***broad whitish streaking***. ♀: ***more greyish-brown*** overall, more subdued throat. Race *poliopis*. Pairs in forest undergrowth, often at edges, woodland and bamboo stands; less shy, tends to join mixed-species flocks. **Voice** High-pitched, descending trill, accented at end: *pseeeéuue*; more trilled *psee-see-see-suu*. **SS** No other *Malacoptila* in Ecuadorian range. **Status** Uncommon. **Note** Whiskered Puffbird (*Malacoptila mystacalis*) recently found in extreme NW Andes (plainer belly, richer breast, greyish base to mandible).

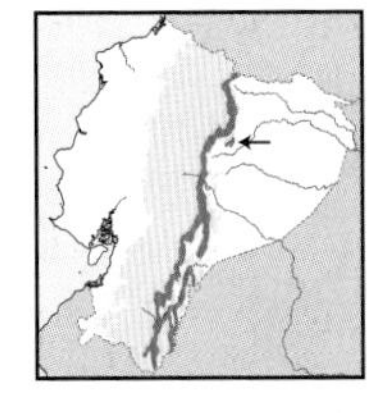

Black-streaked Puffbird *Malacoptila fulvogularis* — 18–19cm

E Andean foothills and subtropics (1,100–2,000m). Understorey of mature forest, sometimes borders. ***Black bill***, red eyes. ***Blackish-brown head finely streaked buffy-white***, brown upperparts with faintly streaked mantle. ***Conspicuous white whiskers and buffy-rufous throat patch***. Dusky below with broad whitish streaks. Nominate. As other *Malacoptila*, sluggish and easily overlooked; may join mixed-species flocks; probably at low densities. **Voice** Piercing and ascending *psiiiiuuií*, often *pseeuuueéE-seeé*. **SS** White-chested Puffbird at lower elevations (orange bill, streaked throat, white breast crescent). **Status** Rare, possibly overlooked; NT in Ecuador.

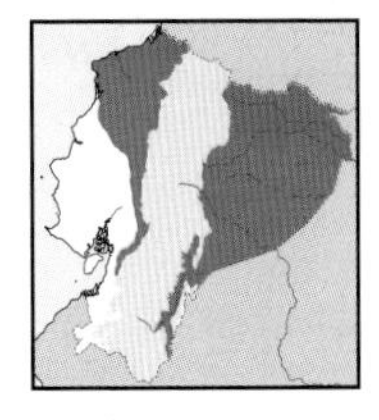

Lanceolated Monklet *Micromonacha lanceolata* — 13–14cm

E & NW lowlands and foothills, locally S to N Santa Elena and W Guayas (to 1,300m, locally higher). Humid forest, borders, secondary woodland. *Tiny* and plump, with black, proportionately strong bill. Plain brown above, ***small white whiskers outlined black***. Whitish below with ***broad and profuse blackish streaks***. Monotypic. Shy and very lethargic, may remain motionless for long periods, even permitting close approach; singles or pairs often on bare horizontal perches, suddenly attacking prey on branches, foliage or floor. **Voice** Short, accelerating series of inflected *sweeep?-swíp?-swíp?-swíp?-swíp?*.... **SS** Small size separates from other puffbirds. **Status** Rare and local, formerly ranked as NT.

Barred Puffbird

Striolated Puffbird

White-chested Puffbird

White-whiskered Puffbird

Black-streaked Puffbird

Lanceolated Monklet

PLATE 129: NUNLETS, NUNBIRDS AND SWALLOW-WING

Nunlets are inconspicuous, frequently overlooked even when vocalising. In contrast, lowland nunbirds are gregarious and noisy, often perching side by side. Unlike most bucconids, Swallow-winged prefers open areas. White-faced Nunbird is the most montane bucconid.

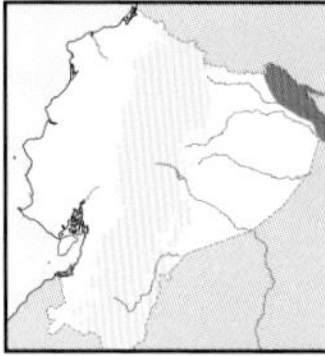

Rusty-breasted Nunlet *Nonnula rubecula* 14cm

NE lowlands, N of Río Napo (below 250m). *Terra firme* forest and borders. Small, with ***pale bluish-white eyering***. Greyish-brown head, brown upperparts and tail. Rusty loral spot, ***rusty throat and breast, whitish on belly***. Possibly race *cineracea*. Very poorly known in Ecuador; mainly confined to forest interior, tame as it perches lethargically for long periods; may follow mixed-species flocks. **Voice** Rising series of thin *wee-weep-weep...* whistles reported (S *et al.*); weak *jéeu* alarm calls. **SS** More widespread Brown Nunlet more uniform rufous below, lacks pale eye-ring. **Status** Rare, very local.

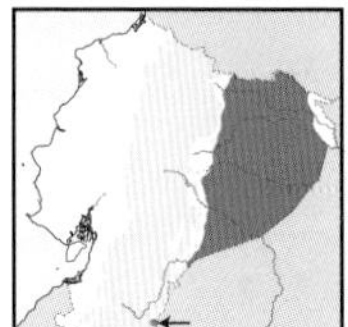

Brown Nunlet *Nonnula brunnea* 14cm

E lowlands (below 400m, locally to 700m). *Terra firme* forest interior and borders. Small, with ***pinkish eye-ring***. Rusty-brown above, ***more rufous below***, becoming ***paler on lower belly***, rufous loral spot. Monotypic. Lethargic, tame but difficult to observe; regular in dense tangles at forest borders and gaps; often in pairs that briefly join mixed-species flocks. **Voice** Repeated, long, uprising *preeú* series that fades at end. **SS** Rarer and local Rusty-breasted Nunlet has whitish eye-ring and rustier underparts. **Status** Uncommon, rather overlooked.

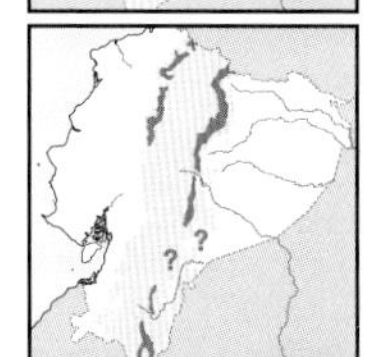

White-faced Nunbird *Hapaloptila castanea* 23–24cm

Subtropics on both Andean slopes (1,300–2,400m). Humid forest, borders, adjacent clearings. Strong black bill, ***white front, lores and throat***. Greyish-brown crown, black frontal stripe, dusky above, ***rich rufous-chestnut below***. Monotypic. Singles or pairs forage in mid to low strata, often motionless; takes small vertebrates and large invertebrates; rarely found, apparently occurs in low densities. **Voice** Slow series of inflected *puuoa?* notes, every 1–2 sec, but not very vocal. **SS** Confusion unlikely given plumage and range. **Status** Rare and local, might deserve threatened status.

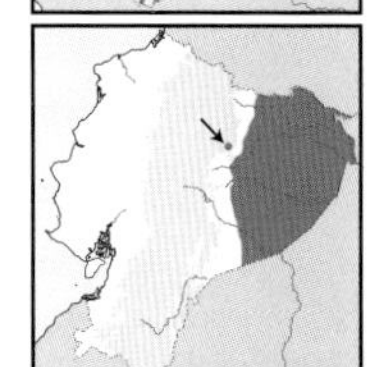

Black-fronted Nunbird *Monasa nigrifrons* 26–28cm

E lowlands (to 500m). *Várzea* and riparian forest, borders, secondary woodland, river islands. Conspicuous; ***black with striking scarlet bill***. Puffy feathers on front, lores and chin. Nominate. Groups of 4–6 birds actively move through midstorey; very noisy and rather nervous when vocalising; sometimes solitary or in pairs; follows monkeys or joins mixed-species flocks. **Voice** Sudden rollicking chorus of *cluurry-cluurry-kleeurr-kleeurr...* by groups. **SS** Other *Monasa* have white on front or yellow bill; segregates by habitat with more similar White-fronted Nunbird. **Status** Common.

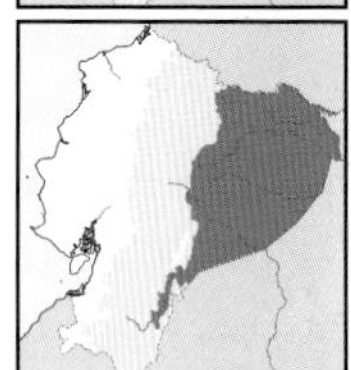

White-fronted Nunbird *Monasa morphoeus* 25–26cm

E lowlands and lower foothills (to 1,000–1,200m, rarely above). *Terra firme* forest interior, borders. Similar to Black-fronted Nunbird but contrasting ***white puffy front, lores and chin***. Race *peruana*. Replaces previous species in *terra firme* and more forested areas; also follows monkeys or joins mixed-species flocks, taking escaping invertebrates and small vertebrates; groups tend to be smaller, often only pairs; also very noisy, abruptly vocalising; usually pairs perch together when calling, facing in opposite directions. **Voice** Similar to Black-fronted but less chaotic, more rhythmic: *pleewuu-pleewuu... klëëo-klëëo-klëëo..., cluuyrr-cluuyrr....* **SS** Unlikely confused. **Status** Fairly common.

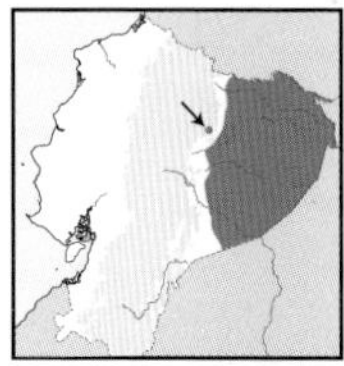

Yellow-billed Nunbird *Monasa flavirostris* 24–26cm

E lowlands (mostly below 400m, locally to 800m). Canopy, subcanopy, borders of humid forest and secondary woodland. ***Yellow bill. Slaty-black, white patch on scapulars and lesser wing-coverts.*** Monotypic. Regularly in pairs, rarely groups; less noisy than congeneric nunbirds; perches on exposed branches, often at borders; does not follow mixed-species flocks or monkeys; more lethargic than congeners. **Voice** Less lively chorus of reiterated *güiiyk-güiiyk-güiiyk... güii'kit-güiik, güii'kit-güiik....* **SS** Unmistakable by bill colour. **Status** Uncommon and local, confined to forested areas.

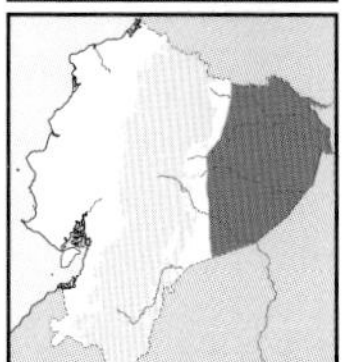

Swallow-winged Puffbird *Chelidoptera tenebrosa* 16–17cm

E lowlands (to 450m). Forest borders, woodland and clearings, especially near water. Small, robust, short-billed, conspicuous ***by behaviour and habitat***. Mostly slaty-black, ***contrasting white rump and cinnamon-rufous belly. White underwing-coverts*** in flight. Nominate. Pairs or small groups assume exposed perches; performs long aerial sallies at flying insects, usually returning to same perch; regular in treetops, exposed branches, posts and stumps on sandbanks. **Voice** Mainly silent; occasional *pit'ti'ti'ti'ti, uit-uit-uit-uit* twitters. **SS** Posture and habits recall martins, but different silhouette, behaviour, also white rump and underwing. **Status** Fairly common, eye-catching.

Rusty-breasted Nunlet
Brown Nunlet
White-faced Nunbird
juvenile
Black-fronted Nunbird
adult
adult
juvenile
White-fronted Nunbird
juvenile
adult
Yellow-billed Nunbird
adult
Swallow-winged Puffbird

PLATE 130: BARBETS

Chunky colourful birds of canopy and subcanopy. All are characterised by heavy stout bills, bright colours in stunning mosaics, and often memorable voices. *Capito* barbets are more vocal than *Eubucco*, which in contrast display more marked sexual dimorphism. Toucan Barbet lacks sexual dimorphism, but also has a deep voice; its distinctiveness from 'typical' barbets reflects a closer relation to toucans, it currently being classified in a family of its own.

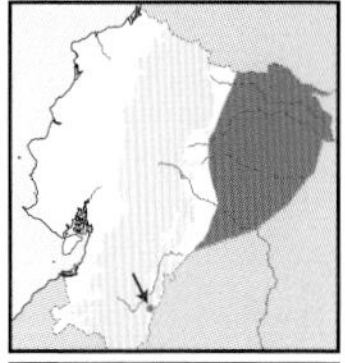

Scarlet-crowned Barbet *Capito aurovirens* 18–20cm

E lowlands (to 400m). Primary forest canopy, borders, secondary woodland in *várzea*, riparian and river island habitats. ♂ dusky-olive above, ***contrasting scarlet crown, orange throat and breast***. ♀: ***white crown***. Monotypic. Mainly in pairs foraging high above ground, perching impassively on semi-exposed branches; tends to shun mixed flocks. **Voice** Soft profound *crrouw*, 1–2 per sec, repeated up to 12 times. **SS** Sympatric Gilded Barbet also orange-throated, but plumage strikingly different: dotted (♀) or yellow (♂) below, blacker and bolder above (both). **Status** Common.

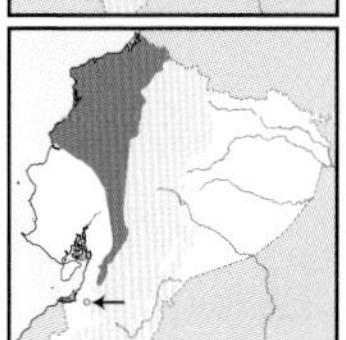

Orange-fronted Barbet *Capito squamatus* 17–18cm

W humid lowlands and foothills (to 1,000–1,200m) locally to NW El Oro. Forest canopy, borders, woodland, adjacent clearings. ***Lustrous black upperparts, white crown, red-orange front***. ♂: ***whitish below***, ♀: ***black throat to upper belly. White tertial tips***. Monotypic. Typically follows mixed flocks, frequents open areas with emergent trees, where descends lower. **Voice** Long rattle *trrrrrrrrrrr*, but not very vocal; *ggark* calls. **SS** ♂ of rarer Five-coloured Barbet has scarlet crown and yellow rear underparts, ♀ heavily dotted below. **Status** Fairly common but NT.

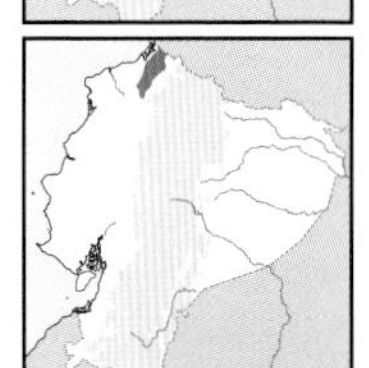

Five-coloured Barbet *Capito quinticolor* 17–18cm

Lowlands in extreme NW (to 560m). Primary and secondary forest canopy, borders. ♂: ***black above***, obvious ***scarlet crown***; slight ***V of yellow spotting on mantle***. White throat and breast, ***yellow belly***. ♀: thin ***tawny-yellow stripes on upperparts, heavily dotted below***. Monotypic. Joins mixed flocks, sometimes with Orange-fronted Barbet, possibly less often descends to lower strata at edges. **Voice** Ten or so resonant *ööp* notes at slow pace, also *gäak* calls. **SS** Orange-fronted has white crown in both sexes; no yellow below. **Status** Fairly common but very local, VU globally, EN in Ecuador.

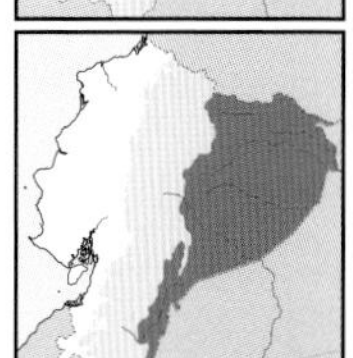

Gilded Barbet *Capito auratus* 17–18cm

E lowlands to foothills (to 1,200m). Forest and woodland canopy, borders. Attractive ♂ has ***golden-olive crown***, black upperparts, long ***yellow 'eyebrow'*** from front to mantle. ***Orange throat, yellow below***. ♀: ***black dotted streaks*** on underparts, thinner yellow stripe on sides of head and neck. Race *punctatus*. Pairs often accompany mixed flocks; moves heavily through vegetation cocking tail. **Voice** A resonant *joo-boop, joo-boop...* frequently heard even at long distance; throaty *jrrk* calls. **SS** Very different from sympatric Scarlet-crowned Barbet. **Status** Fairly common.

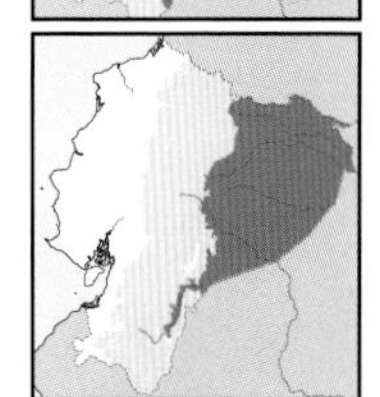

Lemon-throated Barbet *Eubucco richardsoni* 14–15cm

E lowlands (to 1,000m). *Terra firme* forest canopy, borders. Handsome ♂: ***red head, bluish nape, lemon-yellow underparts, pale red band on lower breast***. ♀: ***pale blue-grey front and malar***, dusky-olive crown, ***black mask***, thin and long ***pale orange superciliary*** extending to neck-sides. ***Pale orange chest-band***, dull below. Race *richardsoni*. Mostly in pairs vivaciously foraging, often inspecting clumps of dead leaves; regularly with mixed-species flocks. **Voice** Fairly fast series of 10–12 throaty *crrrruuc* notes. **SS** ♀ Red-headed Barbet has pale blue ear-coverts, yellower underparts. **Status** Uncommon, rarer in foothills.

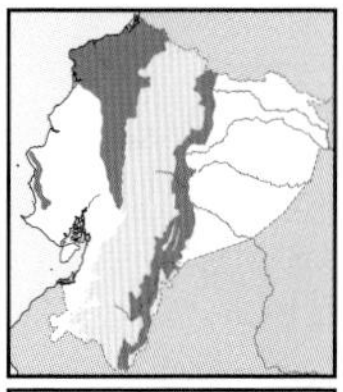

Red-headed Barbet *Eubucco bourcierii* 14–16cm

E & W foothills to subtropics, locally to W lowlands (800–1,900m, to 100m in W). Forest, borders, woodland. ***Scarlet-red hood***, small black mask. ♀: ***black front, lores and chin, bluish partial superciliary and ear-coverts***, orange neck-sides and upper breast-band. Races *aequatorialis* (W), *orientalis* (E). Nimble pairs often in canopy mixed flocks, descending lower at edges. **Voice** Brief, resonant, slow, *goo'o'o'o'o* (W); softer *torrrrrrr* (E), harsh *tjék*, *tjjt* calls. **SS** Lemon-throated Barbet (no red hood, ♀ has black ear-coverts). **Status** Fairly common.

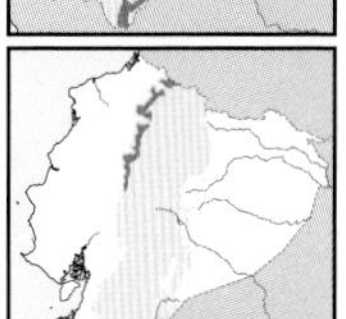

Toucan Barbet *Semnornis ramphastinus* 20–23cm

NW Andean slope (1,400–2,700m). Forest, woodland and borders. Gorgeous; at least ***eight different colours*** in plumage: ***black*** cap; ***grey*** throat; ***red, orange and yellow*** below; ***yellow rump, snowy superciliary***. Horn bill with ***black 'ring'***. Nominate. Heavy flight, so usually heard before seen; joins frugivorous flocks, great seed disperser; clumsily hops along branches. **Voice** Duets of far-carrying syncopated *kho-kha-kho-kha-kho-...kha-kho*, *kha-kho*; also bill claps. **SS** Unmistakable. **Status** Uncommon, local, NT.

♂
♀
Scarlet-crowned Barbet
♂
♀
Orange-fronted Barbet
♂
♀
Gilded Barbet
♂
♀
Five-coloured Barbet
♀
♂
immature
♂
Lemon-throated Barbet
♂
♀
aequatorialis
♂
♀
orientalis
♂
juvenile
♀
juvenile
Red-headed Barbet
Toucan Barbet

PLATE 131: TOUCANS AND TOUCANETS

Conspicuous and fairly well-known 'typical' toucans are classic birds of the New World tropics. Two white-bibbed species in E lowlands best identified by voice. This situation is mirrored in W lowlands by another species-pair (with yellow bibs). In both cases, the smaller counterpart croaks while the larger one yelps. *Selenidera* are small, forest-based, sexually dimorphic toucanets with striking golden ear patches and bright green backs.

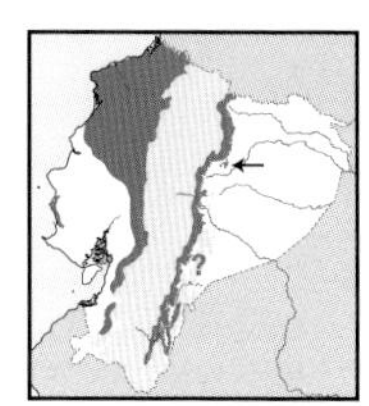

Black-mandibled Toucan *Ramphastos ambiguus* 52–56cm

W lowlands, E Andes slope (mostly below 1,400m in W, 1,000–1,600m in E). Humid primary, secondary forest, borders (W); foothill primary forest, borders (E). Black above, large contrasting ***lemon-yellow bib***. Maxilla mostly yellow with ***dark chestnut*** (W) or ***black*** (E) base and mandible. Yellow-green orbital skin. Two races formerly treated as full species (*swainsonii* W, *ambiguus* E). Mostly pairs in canopy, on exposed branches while vocalising; may congregate at fruiting trees. **Voice** Nasal *kyer-táy-táy, kya-tay-ta-kya...* transcribed as *Dios-te-dé*. **SS** Almost identical to Chocó Toucan (W), best told by voice. **Status** Uncommon (W), local and rare (E), VU in W Ecuador; NT globally.

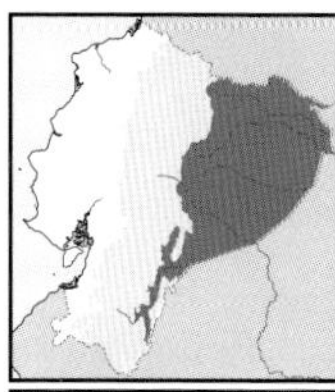

White-throated Toucan *Ramphastos tucanus* 53–57cm

E lowlands (up to 900m, locally higher). Humid primary forest, borders. ***Long*** bill mostly black, yellow tip, maxilla base and culmen, pale blue mandible base. Bright yellow rump, red vent; large ***white bib***. Race *cuvieri*. Pairs or small groups conspicuous in canopy, especially when vocalising from exposed perches; rarely seen inside forest. **Voice** Commonly heard *kyos-keí-keí, kyó-kay-kay-kay*, often omitting final syllables. **SS** Very similar to smaller Channel-billed Toucan, best separated by voice; White-throated bill often looks exceedingly large. **Status** Fairly common, confined to poorly disturbed areas.

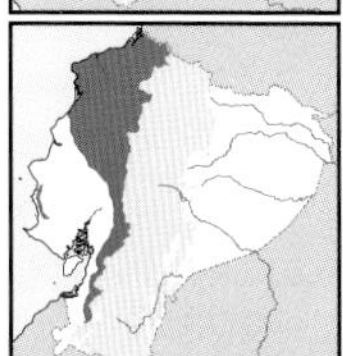

Chocó Toucan *Ramphastos brevis* 42–46cm

W humid lowlands and foothills (up to 1,500m, mostly below 1,000m). Humid primary and secondary forest, borders. Very similar to Black-mandibled Toucan but ***shorter bill, solid black*** mandible and lower maxilla. Smaller, but size difficult to judge in field. Monotypic. Actively forages in canopy and borders, but less conspicuous than Black-mandibled; also vocalises from exposed branches. **Voice** No barks as in Black-mandibled; a reiterated and less remarkable, croaking *criiik, griiuk*. **SS** Difficult to separate from Black-mandibled (W) but bill shorter, channelled, black replaces dark chestnut; focus on vocalisations. **Status** Uncommon, VU in Ecuador.

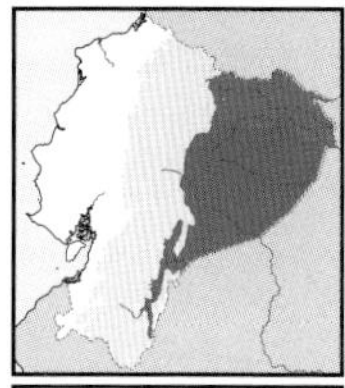

Channel-billed Toucan *Ramphastos vitellinus* 42–46cm

E lowlands to foothills (mostly below 1,100m, locally to 1,400m). Humid primary forest, borders. Plumage identical to White-throated Toucan. Bill ***shorter, channelled***. Race *culminatus*. Pairs, sometimes small groups forage mostly in canopy, but can move into upper understorey; regularly joins frugivores at fruiting trees; moves head up and down while calling. **Voice** A repeated croaking *crriik-crriik-crriik, crriik-criiuk, crriik-criiuk...* from exposed branches. **SS** Not safely identified from White-throated by plumage, best by voice. **Status** Fairly common in extensively forested areas.

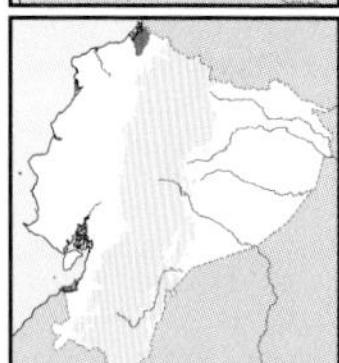

Yellow-eared Toucanet *Selenidera spectabilis* 37–38cm

Very few records from wet forest in NW (below 300m). Primary forest. ***Black head and underparts***, olive upperparts and wings. ***Bright yellow ear-coverts patch***. Bill as Black-mandibled Toucan but ***yellow parts greener***. ♀: lacks ear patches and has ***dark chestnut crown and nape***. Monotypic. Not well known in Ecuador; inconspicuous, rarely vocal, mostly in subcanopy, rarely in open and infrequently seen. **Voice** Not recorded in Ecuador; elsewhere, a weak *krrrik, krrrik* (HB). **SS** Confusion unlikely; Golden-collared Toucanet is confined to E lowlands. **Status** Uncertain; apparently very rare and local, seriously declining; DD in Ecuador.

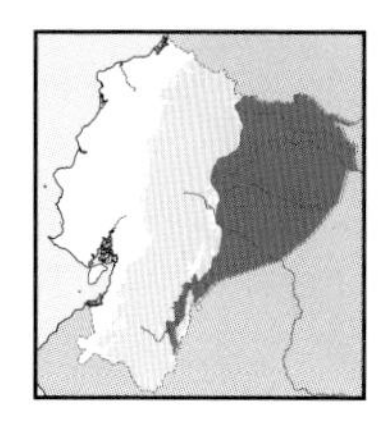

Golden-collared Toucanet *Selenidera reinwardtii* 30–32cm

E lowlands to foothills (to 1,100–1,200m). Mostly in primary forest, rarely at borders. ***Pale eyes with horizontal pupils***, bright blue orbital skin, ***reddish bill***, duller at tip. Black head and underparts, olive-green upperparts. Thin but striking ***golden nuchal collar. Golden ear-coverts patch***. ♀: similar, but ***chestnut*** replaces black, ***olive*** ear patch. Nominate. A skulking species of primary forest, where prefers subcanopy; mostly pairs and rarely in open. **Voice** Very guttural *ggrouw-ggrouw...* growls in slow series; rarely heard odd, loud *weeeaA!* **SS** Unmistakable in range; Yellow-eared Toucanet occurs in W lowlands. **Status** Uncommon, confined to forested areas.

swaisonii
ambiguus
Black-mandibled Toucan
White-throated Toucan
Chocó Toucan
Channel-billed Toucan
♀
♂
Yellow-eared Toucanet
♀
♂
Golden-collared Toucanet

PLATE 132: TOUCANETS AND MOUNTAIN-TOUCANS

All toucans on this plate are montane, some reaching close to the treeline, others approaching adjacent lowlands. Green toucanets easily overlooked if not calling or hopping about on branches. Bluish-grey mountain-toucans are bold and lovely, mostly confined to mature forest. All species may congregate at fruiting trees; some even joining mixed-species canopy flocks briefly. Voices are noteworthy.

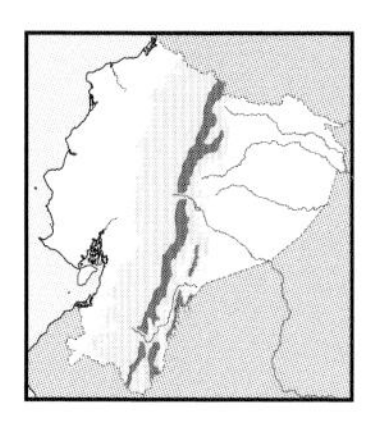

Emerald Toucanet *Aulacorhynchus prasinus* 32–34cm

E Andean slope, isolated eastern ranges (1,500–2,600m, to 3,200m locally). Montane to lower temperate forest, borders. ***Yellow maxilla and tip*** (N), ***only yellow tip*** (S). Emerald-green, ***whitish*** (N) or ***dark greyish-blue*** (S) ***throat***. Chestnut vent and tips to rectrices. Bill looks shorter and more curved than other green toucanets. Races *albivitta* (S to N of Pastaza River) and *cyanolaemus* (S), perhaps distinct species. Pairs to small groups vigilantly forage in canopy and borders, sometimes with other frugivores. **Voice** Fast, persistent *ryek-ryek-ryek* or *gyák-gyák*... barking, lasting *c.*20 sec; more nasal in S. **SS** Chestnut-tipped Toucanet (uniform bill, smaller throat patch, no chestnut vent). **Status** Common.

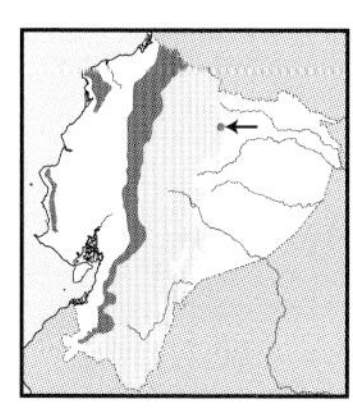

Crimson-rumped Toucanet *Aulacorhynchus haematopygus* 35–37cm

W & NE Andean foothills to subtropics, isolated coastal ranges (500–2,200m, to 300m in coastal ranges, to 2,700m elsewhere). Primary and secondary forest, borders, adjacent clearings. ***Dark red-and-blackish bill***. Green, contrasting ***red rump*** (often invisible when perched). Bluish on face and breast, chestnut tail tip. Races *sexnotatus* (W), *haematopygus* (NE). Pairs to small groups actively forage in forest, borders, gardens and crops, often joins mixed-species flocks or gathers at fruiting trees and feeding stations. **Voice** Nasal barking *gwá, gwá, gwá*... repeated for *c.*20 sec; more rasping *gguk* calls. **SS** Unmistakable in W; in E see Chestnut-tipped Toucanet. **Status** Common.

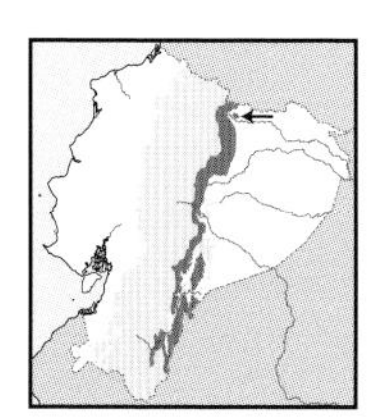

Chestnut-tipped Toucanet *Aulacorhynchus derbianus* 34–36cm

E Andean slope (800–1,800m, to 500m locally). Foothill and montane primary and secondary forest, borders. ***Blackish-and-chestnut bill***. Mainly emerald-green, ***small whitish throat patch***, bluish on face and nape, ***chestnut tail tip***. Nominate. Less numerous than Emerald Toucanet, often only in pairs; less conspicuous, sometimes with mixed-species flocks; tends to replace Emerald at lower elevations. **Voice** Not very vocal, frequently barks: *cruák-cruák-cruák...cruák-ku-kák-ku-kák..., guák-guák-guák...gák-gák-gák...*, more leisurely, harsher than Emerald. **SS** Crimson-rumped Toucanet lacks whitish throat, rump is red, more chestnut in bill; see Emerald (chestnut vent, yellow in bill, bolder throat patch). **Status** Rare to locally uncommon.

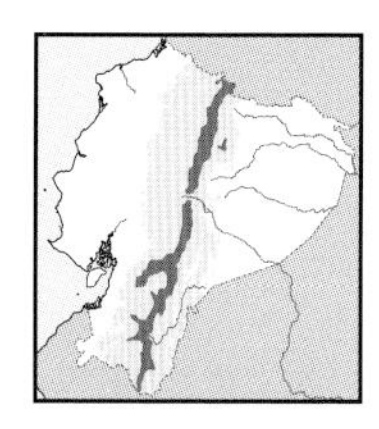

Grey-breasted Mountain-Toucan *Andigena hypoglauca* 42–45cm

Andean slopes in E, inter-Andean slopes locally, W slope in Azuay and N El Oro (2,200–3,500m). Temperate primary, old secondary forest and borders. ***Pale eyes, tricoloured bill*** (orange to pink, yellow and black). Black crown, ***bluish nape***, olive-brown mantle, ***dark bluish underparts***. Races *hypoglauca* (***orange maxilla, darker eyes***) in N, *lateralis* (***pink maxilla, yellow eyes***) in S. Pairs or small groups, rather shy in canopy and borders, often in cover. **Voice** Loud, nasal, querulous *guuaAAan.. guuaAan*, longer than Plate-billed; bill rattling and cackles. **SS** Lower-elevation Black-billed has *unicoloured* bill and whitish throat and neck. **Status** Fairly common but NT.

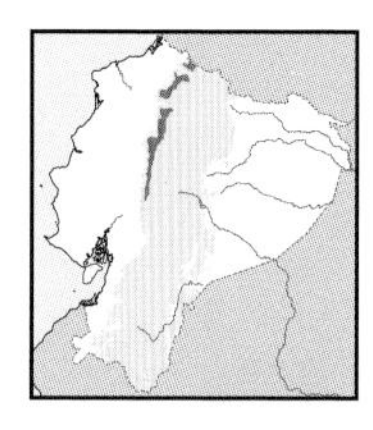

Plate-billed Mountain-Toucan *Andigena laminirostris* 42–44cm

NW Andean slope S to Cotopaxi (1,600–2,600m, locally higher). Primary, old secondary and montane forest borders. Black crown and nape, brown mantle, ***bluish underparts, bicoloured*** skin around eye, ***yellow plate near base of black-and-chestnut bill***. Monotypic. Pairs or small groups, very vocal, visit fruiting trees; may gather with other large frugivores; clumsy when leaping along branches. **Voice** Loud and nasal *k'wuaaa* repeated many times, bill rattling often interspersed. **SS** Unmistakable in range; cf. pale-eyed, more greyish Grey-breasted Mountain-Toucan (no known overlap); barking louder, 'cleaner' than Crimson-rumped Toucanet. **Status** Uncommon, near-endemic, NT.

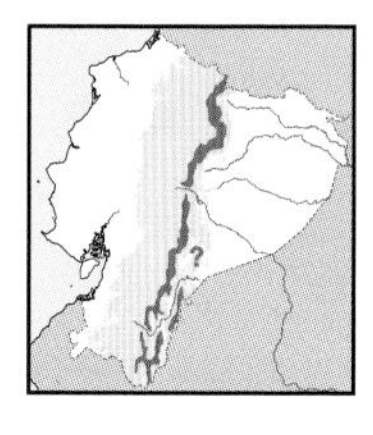

Black-billed Mountain-Toucan *Andigena nigrirostris* 42–44cm

E Andean slope (1,500–2,500m). Canopy and borders of primary and secondary montane forest. ***Bicoloured*** skin around eye, ***bill uniformly dark*** with dark red base. Black crown and ***nape***, brown mantle. ***Whitish throat and neck, pale bluish on belly***. Race *spilorhynchus*. Pairs to small groups at fruiting trees, not very vocal; often with other frugivores or briefly with mixed-species flocks. **Voice** Bill rattling often followed by *kwuuak* calls, repeated persistently. **SS** Grey-breasted occurs at higher elevations, but overlaps locally (lacks whitish throat, bill tricoloured). **Status** Rare to locally uncommon, less numerous than Grey-breasted; NT.

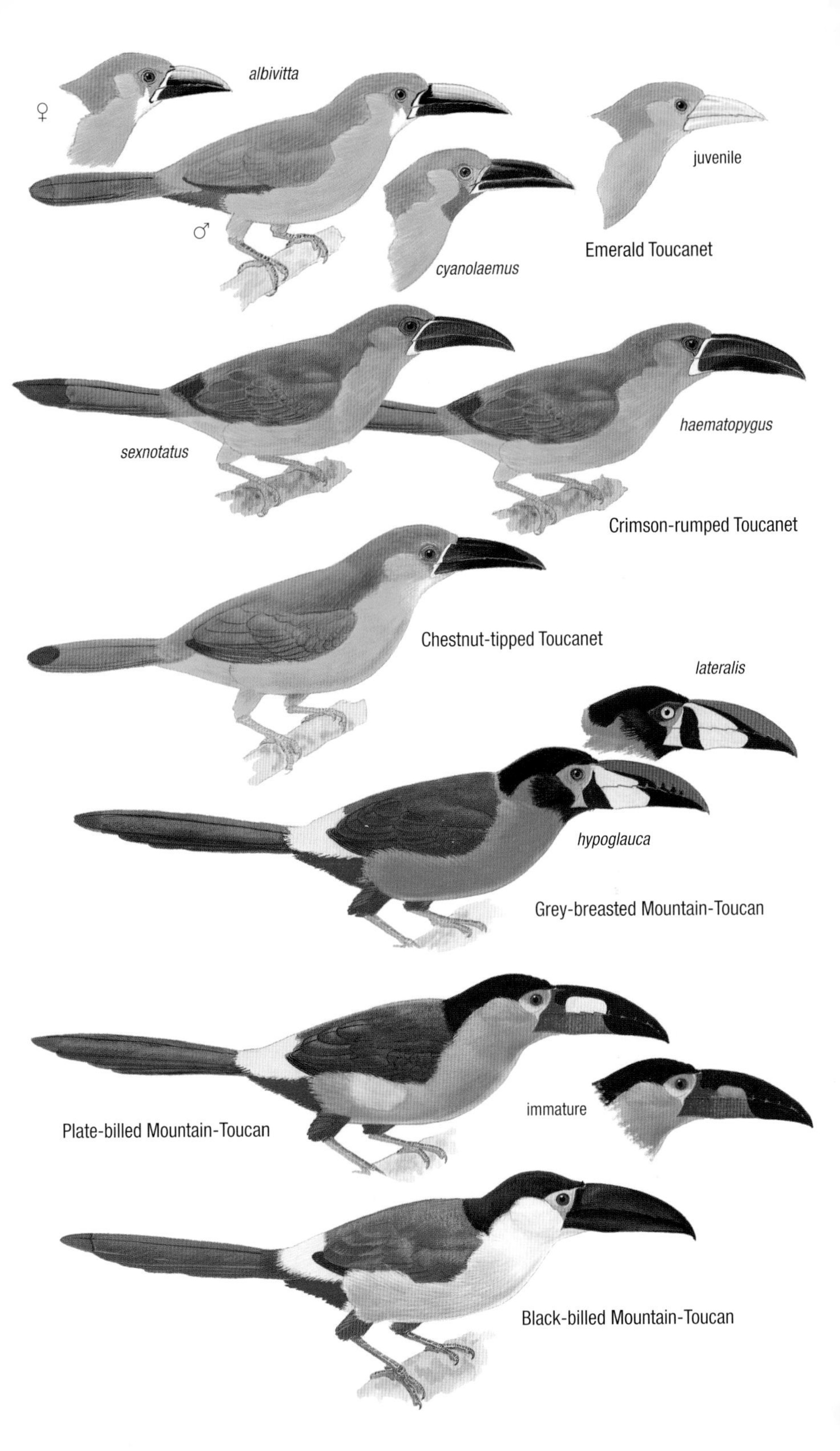
♀
albivitta
♂
cyanolaemus
juvenile
Emerald Toucanet
sexnotatus
haematopygus
Crimson-rumped Toucanet
Chestnut-tipped Toucanet
lateralis
hypoglauca
Grey-breasted Mountain-Toucan
Plate-billed Mountain-Toucan
immature
Black-billed Mountain-Toucan

PLATE 133: ARAÇARIS

More strikingly patterned – in bills and underparts – than toucans, araçaris are lowland forest species often seen in small compact groups, moving between canopy and midstorey. Less vocal than other toucans. Often congregate with other frugivores at fruiting trees, but are also ferocious when attacking chicks or robbing eggs.

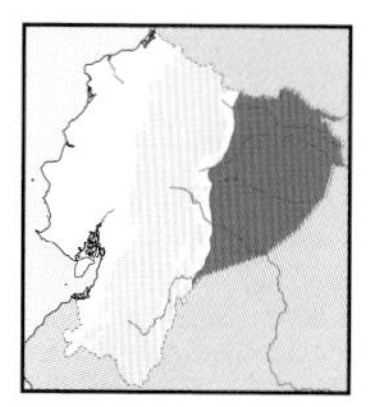

Lettered Araçari *Pteroglossus inscriptus* 33–35cm

E lowlands (mostly below 500m). Humid primary forest, borders, secondary woodland. ***Tricoloured*** orbital skin; unique bill, with black mandible, ***dull yellow maxilla*** with black culmen, tip and ***'letters' on tomia***. Black hood, ***uniform pale yellow below***. ♀: black confined to crown, rest dark chestnut. Race *humboldtii*. Small groups mainly in forest canopy; flock members often follow 'leader' in straight line with rapid wingbeats, rather clumsy when moving through vegetation. **Voice** Not very vocal; long series of guttural *scha'-scha'-scha...*; raspy *gg'ark* calls. **SS** Only araçari with plain yellow underparts. **Status** Fairly common in forested areas.

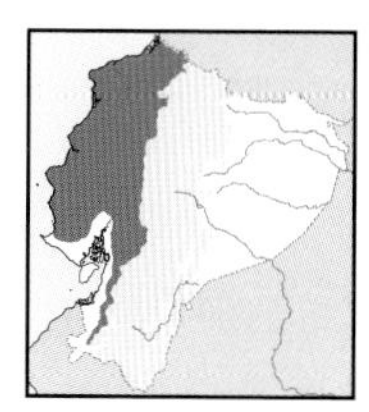

Collared Araçari *Pteroglossus torquatus* 40–43cm

W lowlands and foothills (to 1,200m, locally to 1,600m and higher). Primary and secondary forest, borders, woodland. Red orbital skin, pale yellow eyes. Bill ***mostly dull yellowish*** with black stripe on maxilla (*erythropygius*) or mostly black with yellowish limited to ***upper maxilla and base*** (*sanguineus*); possible intermediates. Black hood, yellow belly with ***large black spot on mid breast, single black band below*** and red stains elsewhere. Races *sanguineus* (extreme NW) and *erythropygius* (S), formerly separate species. Conspicuous and vocal, often visiting fruit farms and coffee or cacao plantations; mainly in small flocks. **Voice** Loud *psisiík-psiík-psiík...* sometimes transcribed as *pilís*. **SS** Only araçari in W. **Status** Uncommon, VU in Ecuador.

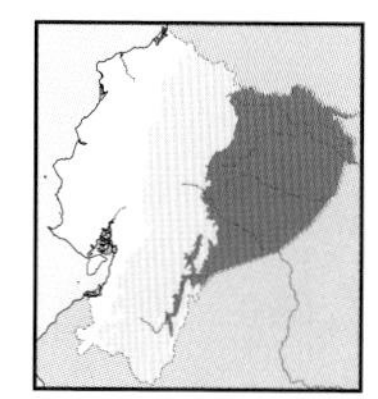

Chestnut-eared Araçari *Pteroglossus castanotis* 43–46cm

E lowlands to foothills (to 900–1,000m). Primary *várzea* and riverine forest, borders. Blue orbital skin, pale eyes, black and dull yellowish bill, blackish maxilla base. Black crown and throat, ***chestnut chin, sides of head to nape***. Yellow underparts, ***single red belly-band***. Nominate. Forages in canopy and subcanopy in small groups, favouring riverside habitats, including islands; rather inconspicuous from mid morning to noon. **Voice** A harsh, parrot-like *srrheeé-srrhee'yip-srrhee*; simpler *sreeyíp* calls. **SS** Many-banded has fairly similar head pattern (including bill), but has two black bands on yellow belly. **Status** Most frequent araçari in riparian habitats.

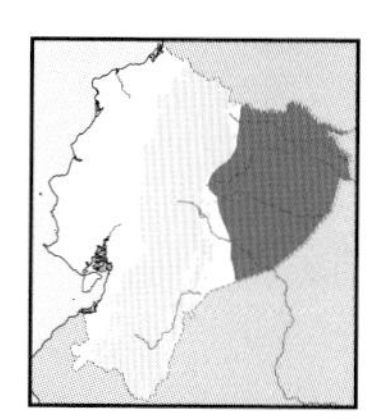

Many-banded Araçari *Pteroglossus pluricinctus* 40–43cm

E lowlands, locally into foothills (to 800m). Mainly primary *terra firme* forest, borders, locally into montane forest. Bicoloured bill with black mandible, culmen and base, and ***creamy-yellow maxilla***. Black head and throat, ***two black bands*** often with red irregular borders on ***yellow belly***. Monotypic. Flocks of up to 10–12 birds forage mainly in canopy, not regularly associating with other species; ungainly and lively when moving in foliage and on large branches. **Voice** Series of whiny *kyéek!* or *guéek!* notes, repeated quite tenaciously. **SS** Unique belly pattern separates from other Amazonian araçaris. **Status** Fairly common in forested areas.

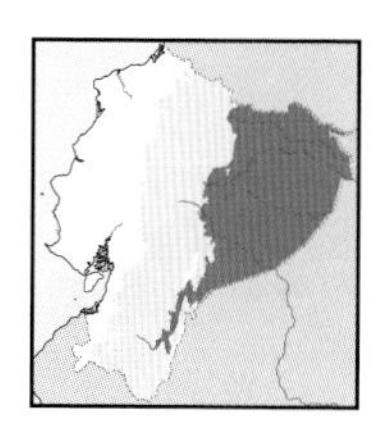

Ivory-billed Araçari *Pteroglossus azara* 33–35cm

Lowlands and foothills in E (up to 900–1,000m). Mostly primary *terra firme* forest. ***Ivory to yellowish bill***. Black crown, chestnut neck-sides to nape, but black extends to nape in ♀. ***Red breast-band***, black lower breast, ***yellow confined to belly***. Race *flavirostris*. Small lively groups in canopy and subcanopy, may join flocks of canopy frugivores; often lower at edges; flock members often follow a 'leader' in straight line with rapid wingbeats. **Voice** Far-carrying, high-pitched, querulous *kweéek-kweéek-kék'kék'kék-kweéek..., kiirr-kiir...* **SS** Pale bill and red breast-band separate from other Amazonian araçaris. **Status** Fairly common.

Lettered Araçari
sanguineus
intermediate
Collared Araçari
erythropygius
Chestnut-eared Araçari
Many-banded Araçari
Ivory-billed Araçari

Piculets are small, inconspicuous woodpeckers of the lowlands, regularly found with mixed flocks. As they lack a stiffened tail, piculets do not use this as a prop for climbing. Still, they behave much like woodpeckers, also clambering vertically and drumming. Plain-breasted does not occur in Ecuador.

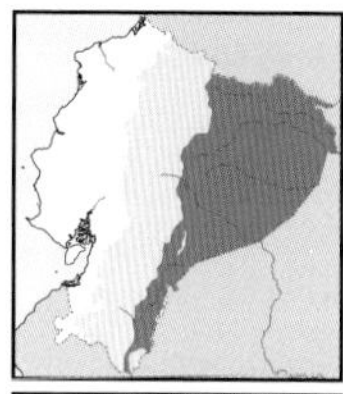

Lafresnaye's Piculet *Picumnus lafresnayi* 9cm

E lowlands and foothills (to 1,400m). Forest, borders, woodland, adjacent clearings in *terra firme* and *várzea*. Olive back, ***yellowish banding on mantle***. Black head ***densely spotted orange-red in front***, white at nape. Whitish below, ***contrasting black barring***. ♀: ***white dots*** in crown. Nominate. Frequently in mixed-species flocks, tends to remain in higher forest strata; climbs small, slender twigs and vines. **Voice** Not very vocal, but occasionally utters a sharp, thin *dsíít-síít*. **SS** Only barred piculet in Amazonian lowlands. **Status** Fairly common in lowlands, less frequent in foothills.

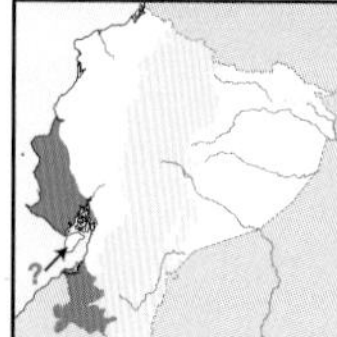

Ecuadorian Piculet *Picumnus sclateri* 9cm

SW dry lowlands (to 1,400–1,500m in El Oro and Loja). Semi-deciduous and deciduous forest, woodland, borders, adjacent clearings. ***Black head*** with ***golden*** dots in front and white dots on nape. Olive back, faint whitish bars. ***Densely barred chin to breast, streaked below***. ♀: white dots limited to ***crown***. Race *sclateri* (El Oro and Loja), more solidly barred below than *parvistriatus* (N). Regular with mixed-species flocks, lively creeping on small branches, twigs and vines in various forest and edge strata. **Voice** Not very vocal; thin, fast *tsiii'sii'sii'sii-sii-sii-iit* song; thin, piercing *psEEt*, *tsée* calls. **SS** Only barred piculet in range; Olivaceous (uniform below) prefers more humid habitats. **Status** Fairly common.

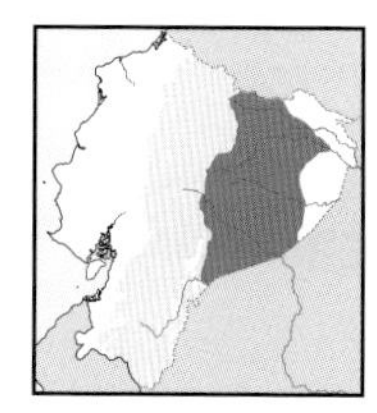

Rufous-breasted Piculet *Picumnus rufiventris* 9cm

Mostly E lowlands near Andes and foothills; locally to easternmost lowlands (to 1,000–1,100m). *Terra firme* forest, secondary woodland, borders, shrubby clearings. Unmistakable. Olive back, black crown with ***dense red spots almost forming stripes*** (♂) or ***white spots*** (♀). ***Uniform rufous underparts***. Nominate. Unobtrusive in lower forest strata and shrubby areas, favours tangled borders; difficult to observe as it moves through dense tangles, often with mixed-species flocks. **Voice** Not very vocal, a thin *tsiit-siit..sít*, or *tsíí-see-suu-suu* descending in pitch. **SS** Rufous underparts make this shy piculet unmistakable. **Status** Uncommon (overlooked?) and local, more frequent near Andes.

†Plain-breasted Piculet *Picumnus castelnau* 9cm

Not certainly recorded in Ecuador. Old specimens assigned to Ecuadorian Amazon were taken within current Peruvian territory, and no records exist for immediately adjacent Peru. ***Plain creamy-white underparts, uniform black crown*** (♀) or heavily ***dotted red*** (♂) make it unmistakable. Expected on river islands and in riparian woodland, if found. Although included here, it is no part of Ecuador's bird list (see Appendix).

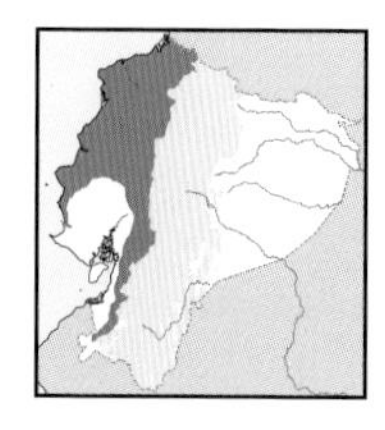

Olivaceous Piculet *Picumnus olivaceus* 9cm

Humid lowlands and foothills in W (to 1,400m). Forest borders, secondary woodland, adjacent shrubby clearings. ***Plain olive back***, black crown with yellow-orange dots in front, white behind (♀ ***only white dots***). Scaled chin and throat, ***plain olive breast, olive stripes on yellowish belly***. Race *harterti*. Fairly conspicuous in mixed flocks at forest borders, actively creeping on twigs and vines, even hanging upside-down; regular in mid to lower strata. **Voice** Not very vocal; soft, descending trill *ti'sssssssít*, *swi'i'i'i'i*. **SS** Only other piculet in W prefers drier habitats and is densely barred; might overlap locally, especially in coastal mountains. **Status** Fairly common.

Lafresnaye's Piculet

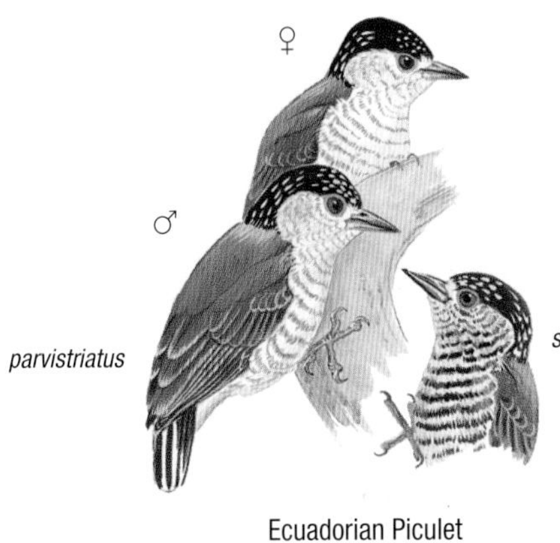

Ecuadorian Piculet

Rufous-breasted Piculet

Plain-breasted Piculet

Olivaceous Piculet

Small woodpeckers ranging from lowlands to near the treeline, sexual dimorphism not very marked, ♂♂ often having red crowns. Most prefer forest, but one is mostly found in non-forest habitats, coincidentally being bolder with scarlet back and whitish underparts. Regularly with mixed flocks.

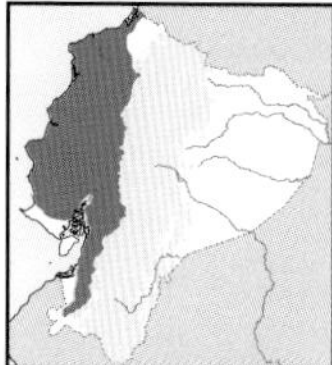

Red-rumped Woodpecker *Veniliornis kirkii* — 15–16cm

Lowlands, locally to foothills in W, S to coastal El Oro (to 1,200m). Humid to semi-deciduous forest, woodland. Red crown, *blond nape*, olive back, *contrasting red rump. Whitish chin, dusky and whitish barring below*. ♀: similar but *crown blackish*. Race *cecilii*. Regular in borders and clearings, very active in mid and lower strata, often with mixed-species flocks. **Voice** Slow series of 5–6 nasal *khléh* notes. **SS** Chocó Woodpecker (NW) lacks red rump, more heavily barred below, barred throat (more humid habitats). **Status** Uncommon.

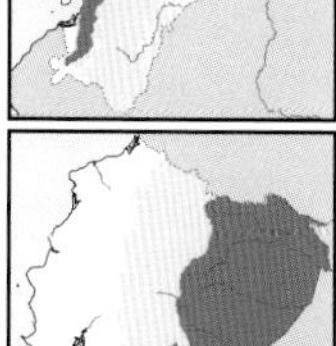

Little Woodpecker *Veniliornis passerinus* — 15–16cm

E lowlands, locally to foothills (to 1,200m). Riparian habitats, river islands, *várzea* forest, adjacent clearings. *Dusky front*, red crown and nape. Faint *whitish lines above and below ear-coverts*. Bright olive above, *dusky-olive below with faint whitish bars*. ♀: *olive-brown* crown and nape, facial lines more subdued. Race *agilis*. Often in borders and clearings, where forages in various strata. **Voice** Accelerating *wiík-wiík-wiík-wiík...*, single *wEEk!* calls. **SS** Red-stained Woodpecker larger, with red front, no facial lines, blond nape, sparser barring below. **Status** Uncommon.

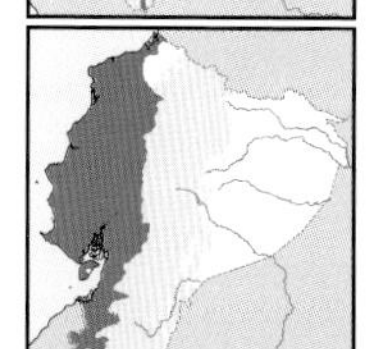

Scarlet-backed Woodpecker *Veniliornis callonotus* — 14cm

W lowlands to foothills (to 1,000m, to 1,800m in Loja). Deciduous forest, woodland, clearings, arid shrub. Small, *scarlet and white*, with fine barring below and some black in crown and nape, especially ♀. S race *major* (El Oro and Loja) more solidly barred, with whitish stripe behind ear-coverts; N race *callonotus* more pallid below. Tolerant of degraded habitats, including agricultural areas, often nesting in bamboo posts; regularly apart from mixed flocks. **Voice** Rapid, sharp rattle *kiiiiiirrr*, brief *kí-diík* calls. **SS** Unmistakable in range. **Status** Common, widespread.

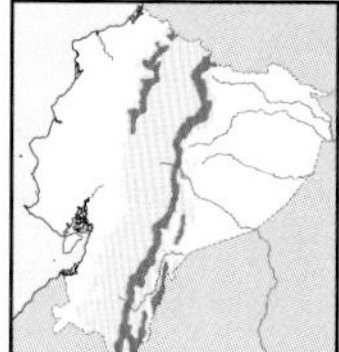

Yellow-vented Woodpecker *Veniliornis dignus* — 16cm

E & NW Andean slopes (1,400–2,700m). Montane forest, borders. Red head, *white line bordering ear-coverts*, olive back, *yellow below*, solid black bars to mid belly, *vent unmarked yellow* (♂). ♀: similar but *blackish crown and red nape*. Race *baezae* (E, more barred), nominate (W). Unobtrusive and not very vocal, often with mixed-species flocks in various forest strata. **Voice** Fast, rather long *kerrrrrrrrrrr* rattle. **SS** Higher-elevation Bar-bellied Woodpecker (larger, no yellow below, no red nape in ♀). **Status** Uncommon, forest-based.

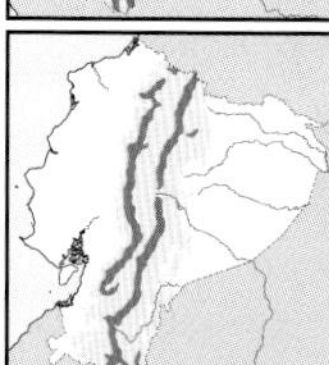

Bar-bellied Woodpecker *Veniliornis nigriceps* — 16–18cm

E & W Andean slopes, also above inter-Andean valleys (2,800–3,500m, locally lower). Forest, borders, treeline, *Polylepis* woodland. Red crown, *variable whitish superciliary and malar stripes*. Scaled chin, underparts *broadly barred tawny-white and black*. ♀: *blackish* crown and nape. Race *equifasciatus*. Shy dweller in various forest strata, even near forest floor; often with mixed-species flocks. **Voice** Long series of ascending, ringing *kyEE* notes, descending at end; ringing jumble of *kleé-kluyeé* notes. **SS** No other *Veniliornis* in range; cf. lower-elevation Yellow-vented. **Status** Rare, perhaps overlooked.

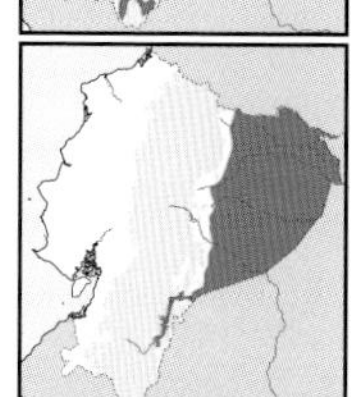

Red-stained Woodpecker *Veniliornis affinis* — 16–18cm

Amazonian lowlands (locally to 800m). Humid *terra firme* forest and borders. Red crown, *blond nape. Plain dusky face, scaled submalar*. Dusky-olive and whitish bars below. Some have *red staining* on back and wings, but not easy to see. ♀: *lacks red crown*. Race *hilaris*. Regular in mixed flocks in upper forest strata, less conspicuous than Little Woodpecker and more forest-based. **Voice** Not very vocal; fast, nasal series of up to 16–18 *kííj* notes; sometimes shorter. **SS** Little has obvious head-stripes and lacks yellow in nape. **Status** Uncommon.

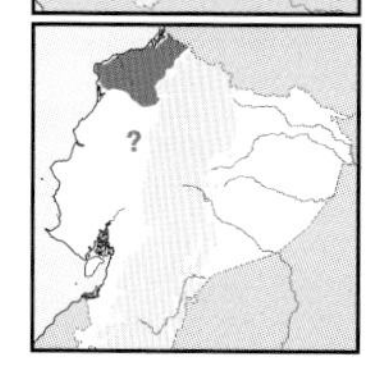

Chocó Woodpecker *Veniliornis chocoensis* — 15–16cm

Local in NW lowlands and foothills (200–700m). Wet forest and borders. Red crown, *blond nape, plain dusky face*, heavily *scaled submalar and chin*. Obvious *blackish-white barring below*. ♀: *dusky crown*. Monotypic. More confined to forest than other *Veniliornis*; poorly known, often with mixed-species flocks in upper strata. **Voice** Nasal, high-pitched *píík* notes in very fast, rattled series; more jumbled *week'a-week'a-week'a* calls. **SS** Red-rumped Woodpecker (less humid habitats) is less solidly barred below, whiter throat, red rump, paler face. **Status** Rare, NT globally, VU in Ecuador.

♂
♀
Red-rumped Woodpecker
♀
♂
Little Woodpecker
callonotus
major
♀
♀
♂
♂
Scarlet-backed Woodpecker
baezae
dignus
♀
♂
♂
Yellow-vented Woodpecker
juvenile
♀
♂
Bar-bellied Woodpecker
♀
♂
Red-stained Woodpecker
juvenile
♀
♂
Chocó Woodpecker

PLATE 136: MEDIUM-SZED WOODPECKERS

Mid-sized woodpeckers of forest and woodland, with bold head patterns, no marked sexual dimorphism, and barred to scalloped underparts. Most are confined to lowlands, but at least three reach lower subtropics. Regularly with mixed flocks. Spot-breasted recalls *Piculus* in head pattern, but rest of plumage and voice differ. Golden-olive recently separated from *Piculus* based on genetic differences.

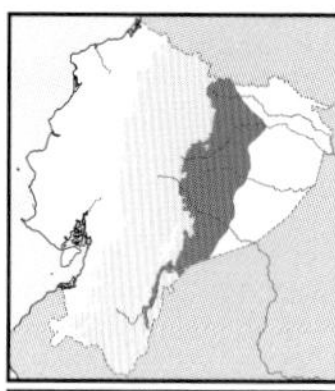

White-throated Woodpecker *Piculus leucolaemus* 18–19cm

E lowlands to foothills (to 1,000m). Mostly *terra firme* forest. Red crown, nape and ***broad malar stripe. Yellow and olive cheeks. White throat,*** broad blackish, olive and whitish barring below. ♀: ***red nape and lacks red malar.*** Monotypic. Singles or pairs in upper strata, on large limbs, mostly in forested areas, where loose pairs join mixed-species flocks. **Voice** Sudden, irritating, harsh *shirr...shirr-shirr-shirr*; *gígggg* hissing. **SS** Sympatric Yellow-throated Woodpecker (yellow throat), larger Golden-green (underparts entirely barred yellowish-olive, olive malar, pale eyes). **Status** Rare.

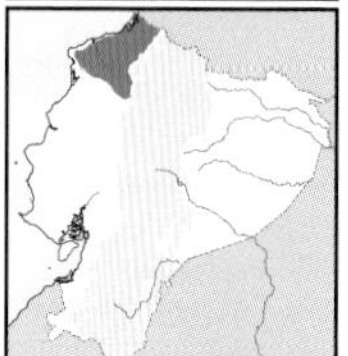

Lita Woodpecker *Piculus litae* 16–19cm

Humid lowlands and foothills in NW (to 800m). Forest, borders, secondary woodland. Fairly similar to White-throated Woodpecker (only in E), with bright red crown, nape and malar stripe (red limited to *thin nape patch* in ♀). ***Yellow stripe below eyes and ear-coverts.*** More scaled below than White-throated. Monotypic. Not well known; in upper strata mostly in forested areas, working on large limb; often with canopy flocks. **Voice** Speedy, abrupt, hissing *shiiggg*. **SS** Unlikely to be confused in range; Golden-olive Woodpecker has white face and solidly barred belly. **Status** Uncommon, fairly local, VU in Ecuador.

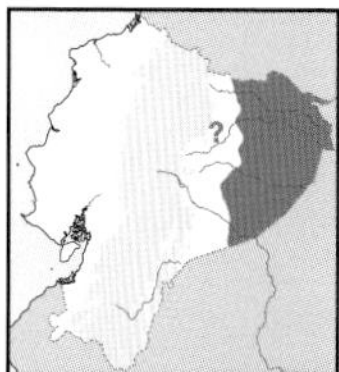

Yellow-throated Woodpecker *Piculus flavigula* 19–20cm

Amazonian lowlands (to 300m). Humid forest, mostly *terra firme*. Bright red crown and nape, ***yellow throat, head-sides, lores and eye-ring.*** Olive above, scaled olive-whitish below. ♀: ***crown golden-olive, red confined to nape.*** Race *magnus*. Mostly singles or loose pairs in mid to upper forest strata, rarely at borders, sometimes follows canopy flocks. **Voice** Far-carrying, hissing *sheeeeearr*, almost non-bird like. **SS** White-throated Woodpecker (white throat, ♀ with olive crown and little red in nape). **Status** Rare to uncommon, confined to forests.

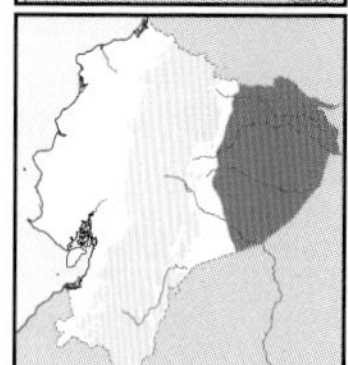

Golden-green Woodpecker *Piculus chrysochloros* 20–22cm

Amazonian lowlands (to 300m, locally to 600m). *Terra firme* and *várzea* forest, borders. Large *Piculus* with ***greyish eyes***; red crown and nape, ***long yellow stripe from lores to neck-sides, olive face. Solidly barred olive-yellowish below.*** Similar ♀ has ***green crown and nape.*** Unknown race, possibly ***undescribed.*** Also a forest species, but more frequently occupies borders, adjacent clearings and lower strata; fond of termites. **Voice** Doubled or tripled, drawn-out *sheeeygg, sheeegg-sheeegg-sheeegg*. **SS** Heavily barred underparts absent in other Amazonian woodpeckers; ♀ distinguished by dark olive head. **Status** Uncommon and somewhat local.

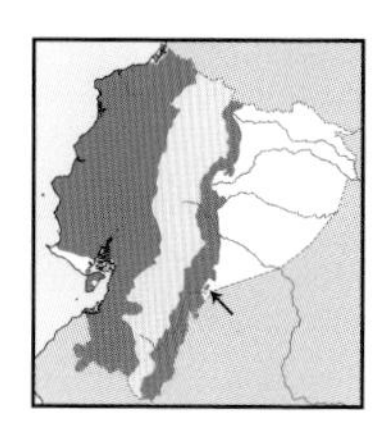

Golden-olive Woodpecker *Colaptes rubiginosus* 21–23cm

Lowlands to subtropics (W), Andean slope (E) (to 2,100m in W, 800–2,300m E). Primary, secondary forest, borders, adjacent clearings. Medium-sized, ***whitish face***, red nape and malar. ***Golden-olive above, yellow underparts and rump densely barred blackish and olive***; throat mostly black. ♀: ***dark grey crown, black malar***, red confined to ***nape.*** Races *rubripileus* (W), *buenavistae* (E), apparently *coloratus* (extreme SE). Easy to see; singles or loose pairs at borders, often with mixed flocks; bold, lively and vocal. **Voice** Fast *tree'ee'ee'ee* trill; also sharp, abrupt *kjEE* calls. **SS** Locally Crimson-mantled, Spot-breasted, or *Piculus* woodpeckers. **Status** Common.

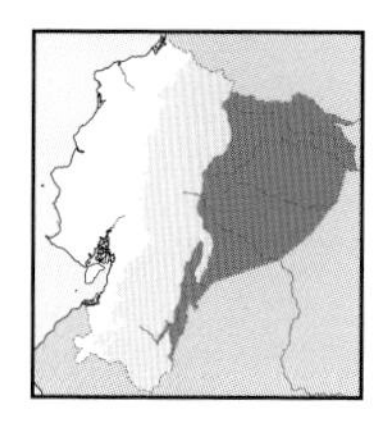

Spot-breasted Woodpecker *Colaptes punctigula* 20–21cm

E lowlands and foothills, locally higher (to 1,200m). Riparian habitats, river islands, *várzea*, secondary woodland, adjacent clearings. Handsome, with ***black front and crown***, red hindcrown and nape, ***extensive white in face, red malar.*** Olive above ***obviously barred blackish.*** Black throat heavily spotted white, yellow below with ***obvious blackish dots.*** ♀: ***lacks red malar.*** Race *guttatus*. Conspicuous, rarely descends to lower strata; apparently spreading towards Andes with deforestation; often in dead palms in riparian habitats. **Voice** Gives 10–12 loud, nasal, fast, pygmy-owl-like *khá-khá-khá...* notes. **SS** Solid barring above and dotting below make it unmistakable; cf. Golden-olive Woodpecker. **Status** Fairly common.

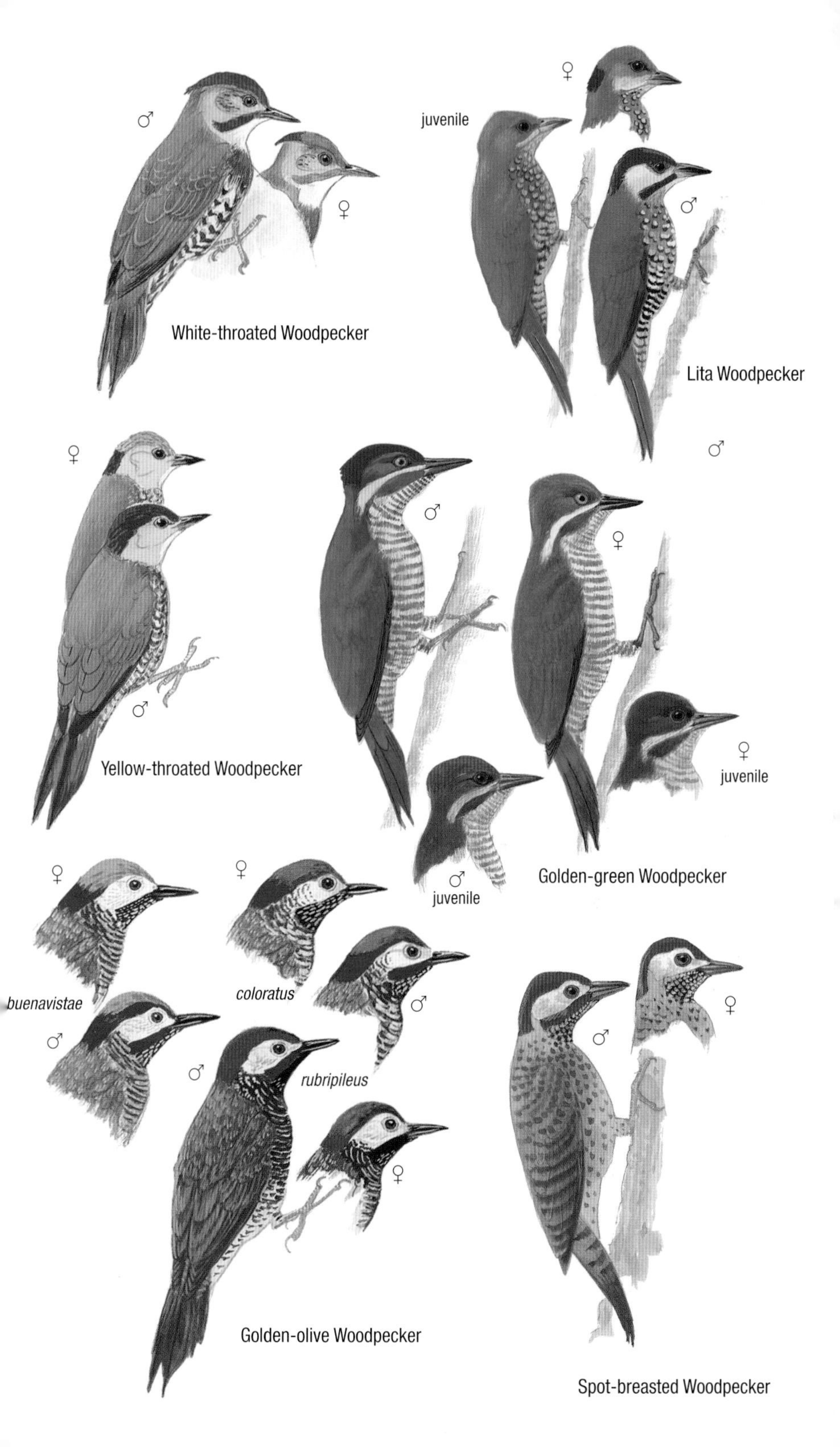

White-throated Woodpecker
juvenile
Lita Woodpecker
Yellow-throated Woodpecker
juvenile
juvenile
Golden-green Woodpecker
buenavistae
coloratus
rubripileus
Golden-olive Woodpecker
Spot-breasted Woodpecker

PLATE 137: MEDIUM-SZED WOODPECKERS AND FLICKER

Melanerpes are conspicuous at edges and in semi-open areas, and have loud voices. The other species are Andean in distribution, the flicker being the only true *páramo* species and feeds mainly on the ground. Crimson-mantled was formerly in the genus *Piculus*, but a genetic study revealed its true affinities. Smoky-brown was also recently removed from *Veniliornis* to *Picoides* based on genetics. All but the flicker may join mixed flocks, but *Melanerpes* are more often found away from flocks than with them.

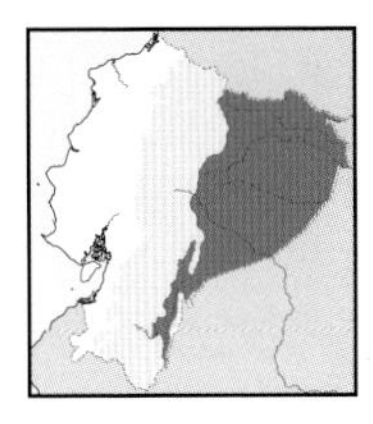

Yellow-tufted Woodpecker *Melanerpes cruentatus* 18–19cm

E lowlands and foothills (to 1,100–1,200m). Canopy and borders of humid *terra firme*, *várzea* forests, secondary woodland, adjacent clearings, cultivated land and towns. Mainly black, conspicuous ***yellow spectacles extend to nape, red mid crown, white rump***, reddish mid belly. ♀: ***white*** mid crown. Monotypic. Small groups but occasionally 10–12 birds perch high, often on exposed branches and dead snags; highly vocal and active even in hot midday; when landing holds wings briefly outstretched; feeds on invertebrates and fruits (can be seen 'attacking' papaya and banana plants). **Voice** Far-carrying *krrrr...keén-keén-keén-keén...*; also fast *kl-dídídí, krrrr-déh!-krrrr-déh!* often given by groups. **SS** Unmistakable. **Status** Common.

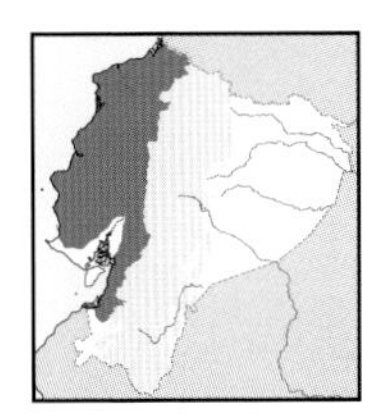

Black-cheeked Woodpecker *Melanerpes pucherani* 18–19cm

W lowlands and foothills (to 1,000m, locally higher). Humid forest canopy, borders, woodland, adjacent clearings, plantations, small towns. Conspicuous. ***Red crown and nape, yellowish front, black ear-coverts, buffy malar and throat***. Black above with fine white bars, ***white rump***. ♀: ***buffy forecrown, red nape***. Monotypic. Similar in habits to its Amazonian counterpart (Yellow-tufted), perches high on exposed branches and dead snags, but apparently less sociable; pecks at invertebrates, regularly sallies for flying insects and feeds on fruit; frequent in open. **Voice** Loud, ringing *kí'rrrr'rr*, 3–4 times. **SS** Unmistakable. **Status** Fairly common, apparently spreading with deforestation.

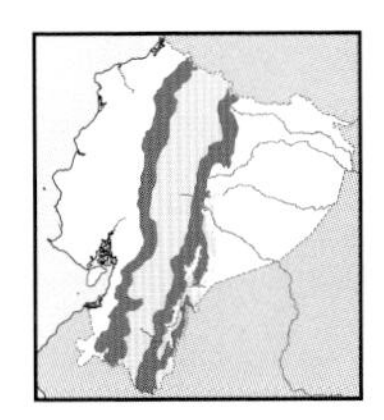

Smoky-brown Woodpecker *Picoides fumigatus* 18–19cm

E & W Andean foothills to subtropics (600–2,400m). Montane forest, woodland and borders. Small, ***mainly uniform brown, red crown and nape*** (♂). ♀: ***dusky-brown in crown***. Race *fumigatus* (most of range) and *obscuratus* (El Oro and Loja), darker overall. Inconspicuous, forages at borders and in forest interior, often with mixed-species flocks; in pairs or singly in lower strata, often on branches near ground. **Voice** Raspy, fast *kjá-kjá-kjá...* rattle; short, emphatic *pék!*, *chék-chék-chék...*; occasionally fast, loud *chék-chék...keewee-keewee-keewee-keewee-kwak'kwak'kwak* song. **SS** Sympatric *Veniliornis* are somewhat barred below, none as brown as Smoky-brown. **Status** Uncommon.

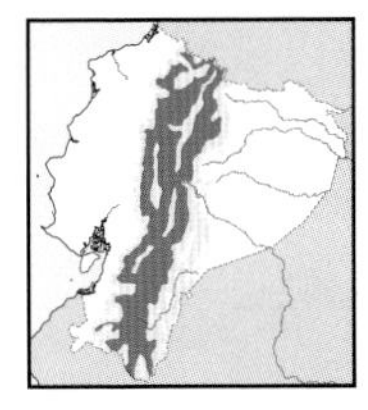

Crimson-mantled Woodpecker *Colaptes rivolii* 24–26cm

Andean slopes and valleys (1,800–3,300m). Forest canopy, secondary woodland, borders, adjacent clearings, sometimes *Eucalyptus* woodland. Handsome. ***Crimson above, yellowish-white face***, black throat, ***yellow below, densely scaled black and red on breast***. ♀: ***crown mostly black*** (black staining in ♂). Race *brevirostris*. Fairly easy to see, forages mostly in upper strata of forest and borders, regularly with mixed flocks, mostly singles or pairs. **Voice** Sharp, nasal, inflected *kyeep?-kyeep?-kyeep?-kyeep?*; rolling *tr'kkkkkk* rattle. **SS** Unlikely to be confused in range; cf. face pattern of lower-elevation Golden-olive Woodpecker. **Status** Fairly common, tolerates degraded habitat.

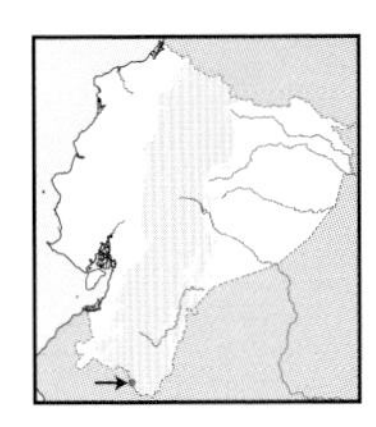

Andean Flicker *Colaptes rupicola* 32–33cm

Extreme SE (3,100–3,400m). Open rocky *páramo*, woodland patches. ***Long, slender, black bill. Grey crown, nape and malar stripe, red mark*** on face. Mantle ***banded black and whitish, contrasting yellow rump. Face, neck and underparts rich rufous-buff, breast finely scaled black***. ♀: ***lacks red on face***. Race *cinereicapillus*. Feeds mainly on ground probing long bill into muddy or loose terrain; perches on exposed rocks, roosts and probably nests in holes in banks and cliffs. **Voice** Several often vocalise together; ringing *kleéwk..kleéwk..kleéwk, khéé-khéé-khéé...*; fast *khE'khE'khE-khE...* ending abruptly. **SS** Unmistakable. **Status** Uncommon but very local (Cordillera Lagunillas).

Yellow-tufted Woodpecker

Black-cheeked Woodpecker

Smoky-brown Woodpecker

Crimson-mantled Woodpecker

Andean Flicker

PLATE 138: *CELEUS* WOODPECKERS

Six species of mid-sized to fairly large forest woodpeckers with triangular crests and often pale bills. Little sexual dimorphism. All can go undetected if not vocalising as they prefer higher strata and are inconspicuous. Two E species have strikingly patterned plumages, one eye-catchingly cream, and two predominantly chestnut. Only one species in W lowlands.

Cinnamon Woodpecker *Celeus loricatus* 19–21cm

Lowlands and foothills in W (to 800m). Humid forest and borders. Boldly patterned, ***cinnamon-rufous above*** with prominent ***black flecks on crown and mantle***. Wings darker cinnamon with black bars. ***Red chin to malar, yellowish-white*** underparts densely marked with ***blackish, U-shaped scales***. ♀: similar but head nearly ***unmarked***. Nominate. Mostly inside forest, where rather difficult to locate, prefers higher strata and tends to shun mixed-species flocks, but quite vocal. **Voice** Loud, far-carrying *wheeÉt-wheeÉt-wheeÉt-wheEE-wheEt-wheET*, resembling an antbird. **SS** Only *Celeus* in range. **Status** Uncommon, forest confined, NT in Ecuador.

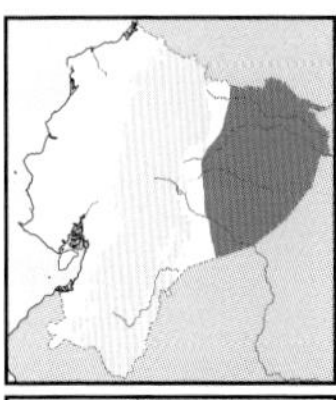

Scale-breasted Woodpecker *Celeus grammicus* 20–22cm

E lowlands (to 500m). Primarily *terra firme*, also *várzea* forest and borders. ***Chestnut overall, darker on foreparts***, brighter rump, ***sparse black flecks on upperparts***. Broad red malar. ***Underparts densely scaled black***. ♀: ***lacks*** red malar. Race *verreauxii*. Mainly inside forest where it occupies high strata and is often difficult to see; regularly with mixed flocks, climbing large branches; frequently feeds on fruit. **Voice** Nasal, strident *kuüw-kaa, küh..küh..küh*. **SS** Similar Chestnut Woodpecker larger and ***plainer*** (no markings below or on wings). **Status** Uncommon, confined to forest.

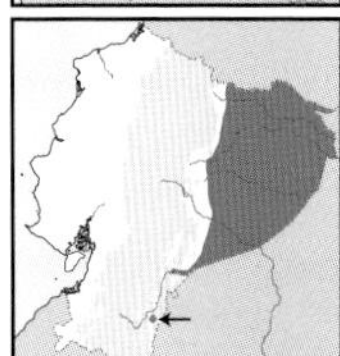

Chestnut Woodpecker *Celeus elegans* 26–28cm

E lowlands (to 600–700m). Primary *terra firme* and *várzea* forest, borders. Bill greenish-white. ***Unmarked chestnut***, darker foreparts, brighter, ***buffier rump***. Crimson malar streak (absent in ♀). Race *citreopygius*. Forages inside forest, prefers lower strata more than other *Celeus*; singles or pairs accompany mixed flocks, taking invertebrates and fruit. **Voice** Raspy, nasal *wýaa-waák-waák-waák...*; also loud, ringing *g'yeeE, gyeA-gyeA-gyeA* calls. **SS** Scale-breasted is also chestnut overall but smaller and obviously *scaled*, especially on chest; Chestnut more ***hammer-headed*** than Scale-breasted. **Status** Fairly common, the most frequent and widespread Amazonian *Celeus*.

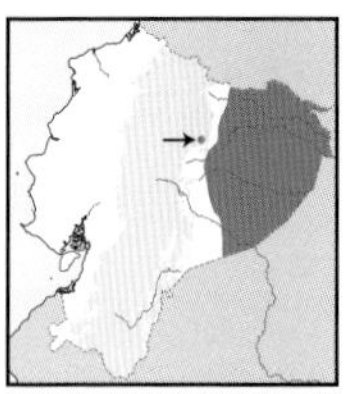

Cream-coloured Woodpecker *Celeus flavus* 25–27cm

Lowlands in E (to 600–650m). *Várzea* and riparian forests and borders, rarer in woodland and *terra firme* forest. Unique, ***creamy-yellow*** with red malar streak, contrasting ***dusky-brown wings*** and darker tail. ♀: ***lacks*** red malar streak. Possibly monotypic, but subspecies taxonomy complex. Explores termitaries and arboreal ant-nests where spends much time; tends to forage more in open than other *Celeus*, also joins mixed-species flocks. **Voice** Loud, ringing, descending, antbird-like *pEEer, péer, per, puh*. **SS** Confusion unlikely. **Status** Uncommon.

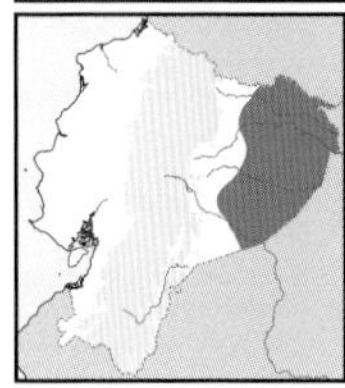

Rufous-headed Woodpecker *Celeus spectabilis* 25–27cm

Sparse in E lowlands (below 300m). Riparian forest and woodland where *Cecropia* trees dominate and *Gynerium* cane flourishes in understorey. Strikingly patterned. ***Rufous-chestnut*** head, ***reddish postocular streak***, crimson malar. Yellowish above with ***bold black scaling and barring***. Black chest, ***yellowish-buff below***, black scales on mid belly; ***rump buffy-yellow***. No crimson malar in ♀. Nominate. Singles or loose pairs in lower strata, inconspicuous, not with mixed-species flocks; often on *Cecropia* trunks and large branches. **Voice** Loud *kwEEÁ* followed by *glu-glu-glu-glu...*. **SS** Ringed Woodpecker (dark chestnut instead of yellowish above, barred belly and tail, paler head). **Status** Rare and local, likely overlooked.

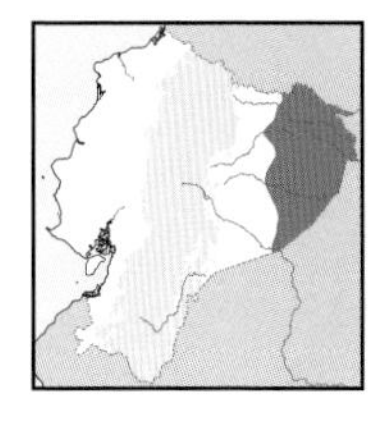

Ringed Woodpecker *Celeus torquatus* 26–28cm

E lowlands (below 300m). *Terra firme* and *várzea* forests and borders. Handsome, ***cinnamon head***, ♂ with red malar streak. ***Black chest and nuchal collar***. Dark chestnut upperparts, wings and tail with ***bold black barring. Buffy-ochraceous below with black banding***. Race *occidentalis*. Not with mixed-species flocks, difficult to see as it prefers canopy and subcanopy; singles or pairs often climb large branches and trunks, and visit termite nests. **Voice** Clear, ringing *kleé, kleé, kleé...* at low pace and even pitch. **SS** Head looks paler than any other *Celeus*; Rufous-headed is yellowish not chestnut. **Status** Rare, forest confined.

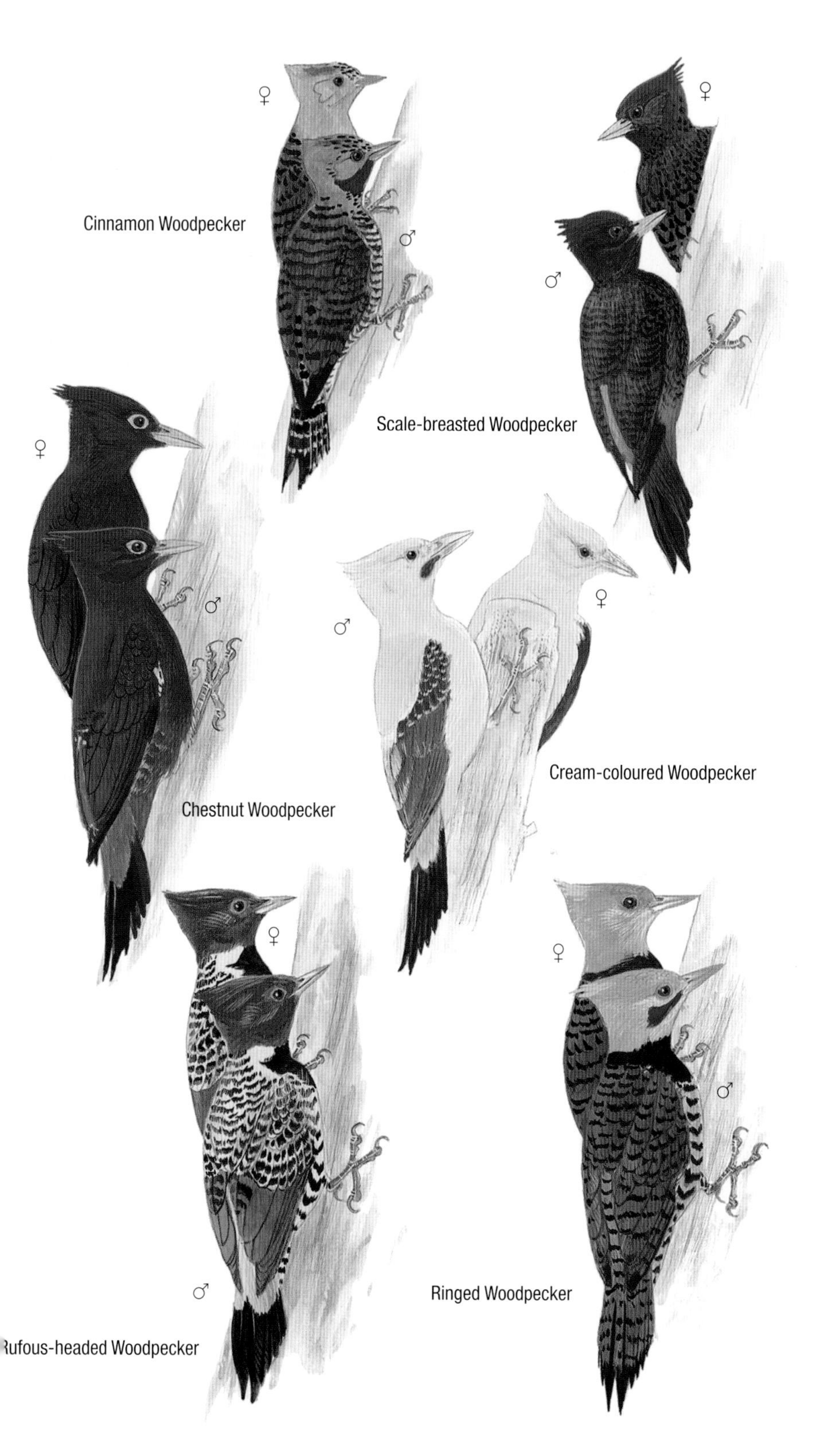
♀
Cinnamon Woodpecker
♂
♀
♂
Scale-breasted Woodpecker
♀
♂
Chestnut Woodpecker
♂
♀
Cream-coloured Woodpecker
♀
♀
♂
Ringed Woodpecker
♂
ufous-headed Woodpecker

PLATE 139: 'HAMMER-HEADED' WOODPECKERS

The largest woodpeckers, with bold triangular crests, red in all but one ♀. Bills strong, resulting in far-carrying drumming used for communication. Lineated closely resembles *Campephilus* but dorsal lines do not meet on back; voice also differs. One montane species, three predominantly in lowlands but reach lower subtropics, two confined to lowlands.

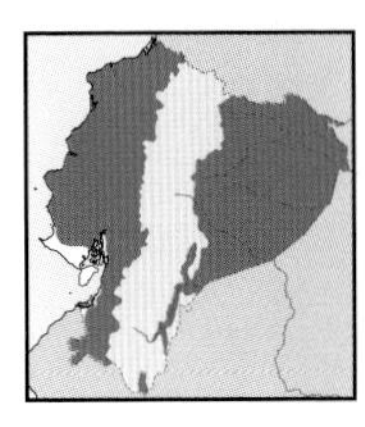

Lineated Woodpecker *Dryocopus lineatus* 33–35cm

E & W lowlands and foothills (to 1,300m; locally higher). Humid (not *terra firme*) to deciduous forests, borders, woodlands, clearings. Mainly black, red crest and ***malar streak, white line from lores to mantle***; lines ***do not meet*** on mantle. Black chest, whitish below, ***scaled black*** (brown and buff in W). ♀ has ***black front*** and malar. Nominate (E), *fuscipennis* (W); browner, smaller. Forages mostly on trunks and larger branches, often in isolated trees in clearings. **Voice** At least 20 loud, rising *kwEk* notes; abrupt *chÉK-kurrrr, chk-kurrrr* calls; 6–8 drums. **SS** White lines on back meet in similar *Campephilus*; cf. Guayaquil and Crimson-crested. **Status** Uncommon.

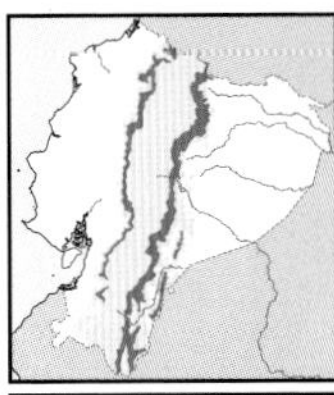

Powerful Woodpecker *Campephilus pollens* 33–36cm

Andean slopes in E & W (1,700–3,000m, to 3,500m locally). Montane forest, borders, treeline. Black above, bright red crest (***all black*** in ♀). Broad white line from bill base to mantle, ***where meet in V, white rump***. Lower breast and belly heavily ***banded black and chestnut***. Nominate. Forages on trunks and large limbs, mostly in pairs, rarely in clearings, sometimes with mixed-species flocks. **Voice** Loud, nasal whine, *Kyaáhn*, accelerating into *kyaaá-yaá'yaá'yaá'yaá...*; drums three times rapidly. **SS** White rump and ♀'s black head diagnostic. **Status** Rare, perhaps local.

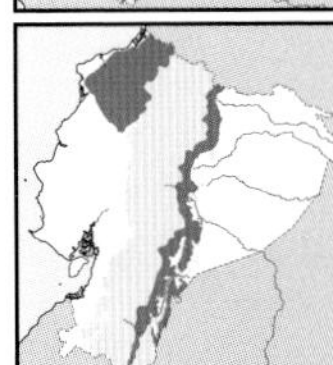

Crimson-bellied Woodpecker *Campephilus haematogaster* 33–35cm

NW lowlands and foothills (to 1,150m), E foothills and subtropics (1,000–2,000m). Forest interior. Red crest, ***broad buff line on lower face, buff postocular stripe***. Black above, ***contrasting red rump. Buffy-whitish underwing and flight feathers***, dark red underparts. ***Buff line extends to neck-sides*** in ♀. Nominate (E; redder below, black chest), *splendens* (W), perhaps separate species. Loose pairs, often near ground and briefly follow ant swarms; shier in W. **Voice** Strident *stk-ki'kkkk, stik-tík-tík-tikitík, stk-kikikiki*; 3–4+ drums, first louder, then bouncing (E); softer *stikt* and two loud drums in W. **SS** Buff facial lines and red rump diagnostic. **Status** Rare; NT in W.

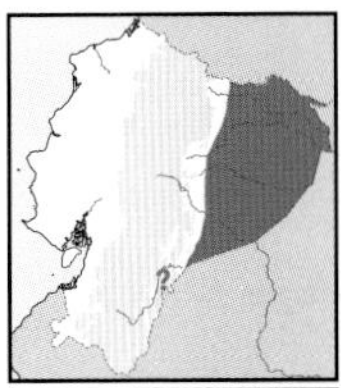

Red-necked Woodpecker *Campephilus rubricollis* 33–35cm

E lowlands (to 500m). Humid *terra firme* forest. ***Red head and neck***, small black-and-white patch in cheeks. ***Uniform black above, rufous-chestnut below; reddish patch in primaries***. ♀: ***broad whitish streak*** from bill base to cheeks. Nominate. Rarely seen in borders or clearings; forages on trunks and main branches in lower strata in *terra firme* forest, mostly in loose pairs. **Voice** Loud, very nasal, abrupt *tkyáah* calls; two-noted drumming, first louder. **SS** No other *Campephilus* has uniform rufous-chestnut underparts and very red crest and neck. **Status** Rare, probably local.

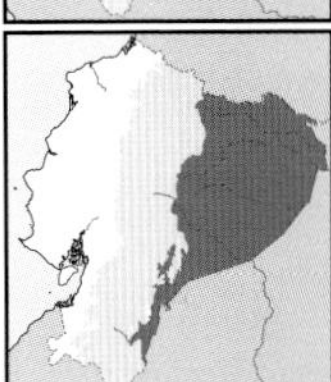

Crimson-crested Woodpecker *Campephilus melanoleucos* 33–35cm

E lowlands and foothills (mostly below 900m, locally to 1,300m). *Terra firme, várzea*, borders, adjacent clearings. ***Bright red head, whitish patch at bill base***, small black-and-white patch below ear-coverts. ***White V on neck to mantle***. Lower breast and belly ***buffy-white coarsely banded black***. ♀: ***black crest, broad white stripe*** from bill base to mantle. Nominate. Pairs, sometimes small groups, on trunks and large branches, less often in clearings. **Voice** Fast, strident, nasal rattle *t'kkkkkk, t'kkt'tkktk...*; rather weak *ték*; drums 3–4 times in swift succession. **SS** Lineated Woodpecker (head pattern, broken V in mantle). **Status** Uncommon.

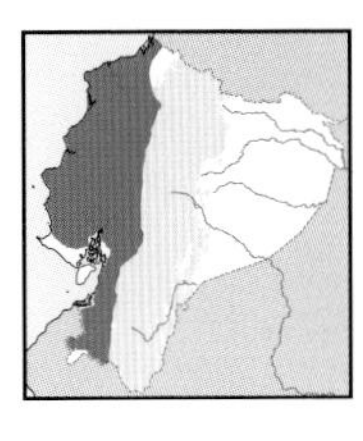

Guayaquil Woodpecker *Campephilus gayaquilensis* 33–35cm

W lowlands and foothills (to 1,500m). Humid to deciduous forest, borders, woodland. Bright red head, small black-and-white ear patch. ***White line on neck-sides, white dorsal V***. Blackish throat and chest, ***banded blackish-buffy below; rump banded as underparts***. ♀: ***broad white line from bill base to neck***. Monotypic. Loose pairs climb trunks and large limbs, rarely in clearings, often high on exposed branches. **Voice** Loud, ringing *tzék-tzék-tzék-tzék'zekekkk, stik-tík-tík-tikitík* similar to Crimson-crested, drumming also similar. **SS** Sympatric Lineated (blacker head in ♀, no banded rump, broken dorsal V). **Status** Uncommon, NT globally, VU in Ecuador.

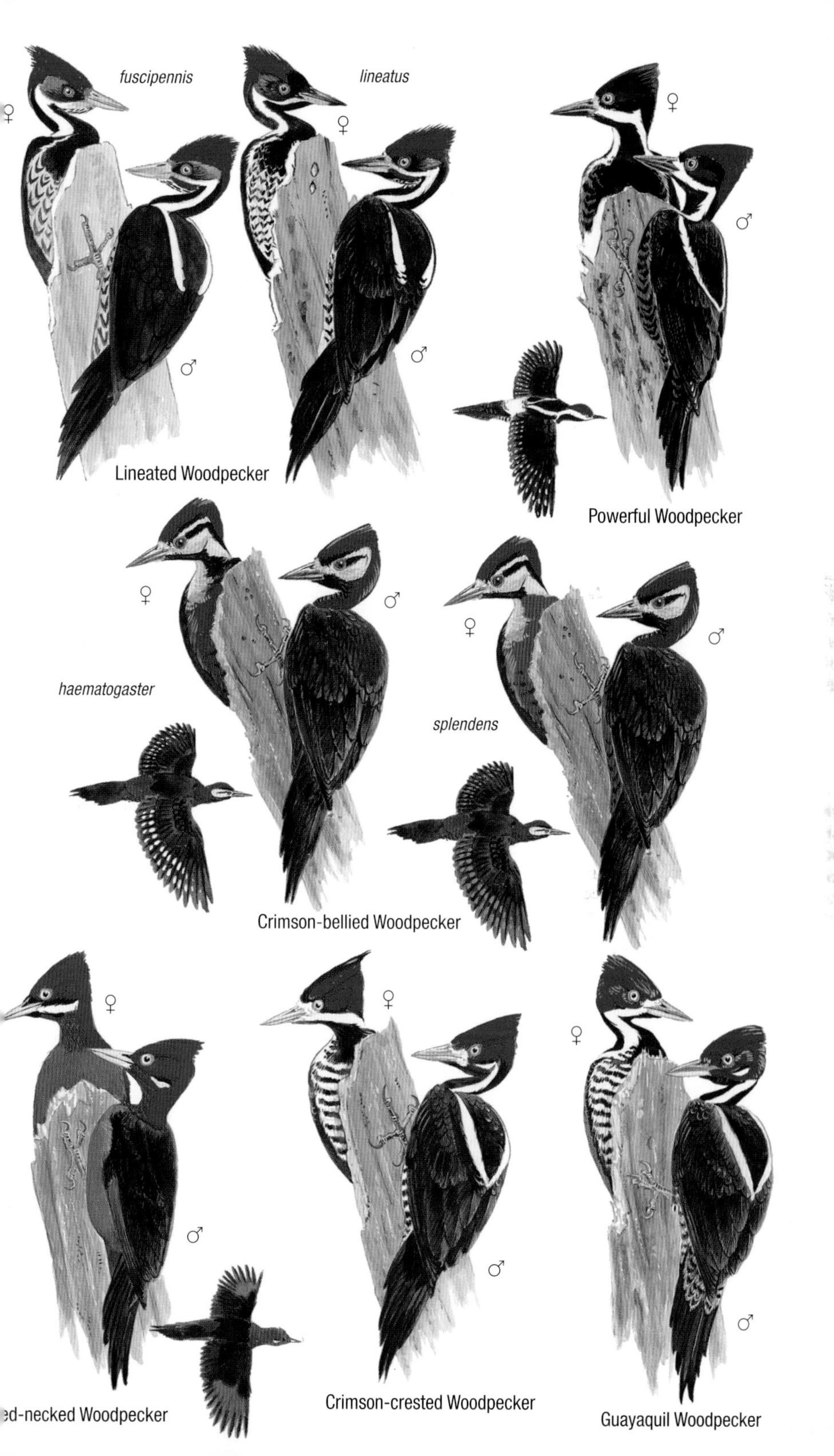
fuscipennis
♀
lineatus
♀
♂
♂
Lineated Woodpecker
♀
♂
Powerful Woodpecker
♀
♂
haematogaster
♀
♂
splendens
Crimson-bellied Woodpecker
♀
♂
ed-necked Woodpecker
♀
♂
Crimson-crested Woodpecker
♀
♂
Guayaquil Woodpecker

PLATE 140: WOODCREEPERS I

Dendrocincla are plain woodcreepers of humid forests, two in lowlands, one Andean. Often with mixed flocks, lowland species regularly follow ant swarms. *Deconychura* and *Certhiasomus* are faintly streaked or spotted, with long tails. Bar-bellied and Cinnamon-throated are relatively plain-bellied and pale-billed.

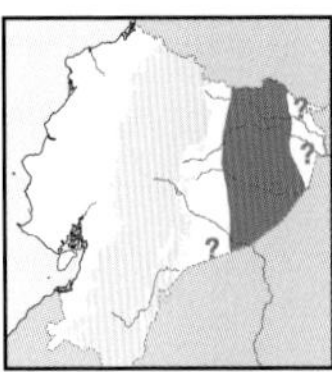

Spot-throated Woodcreeper *Certhiasomus stictolaemus* 16–17cm

E lowlands, only NE? (to 350 m). *Terra firme* forest undergrowth. ***Short*** slender bill. Mostly uniform brown, ***faint*** pale streaks on crown, ***buff spots in throat***, ***rufescent*** rump. Race *secundus*. Mostly singles with mixed-species flocks but not ant swarms, rather inconspicuous, always near ground; flicks wings. **Voice** Not very vocal; short, fast, descending-ascending trill *wr'reeerrEE* or *w'irrrrRRR*. **SS** Recalls Wedge-billed (tiny wedge-shaped bill, buff postocular stripe) and Long-tailed Woodcreepers (larger, longer-billed, longer-tailed, streakier throat to breast). **Status** Rare, local.

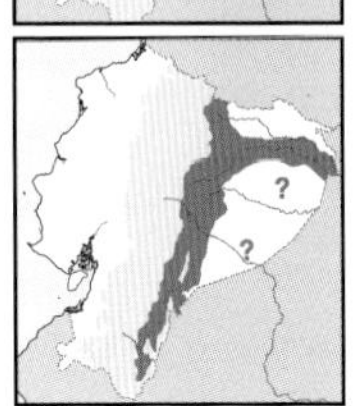

Long-tailed Woodcreeper *Deconychura longicauda* 19–20cm

Local in E Andean foothills to subtropics, locally in adjacent lowlands (to 1,700m). Montane forest undergrowth. ***Longish***, straight bill. Head finely streaked, ***indistinct postocular stripe, buff streaking to breast***, tail proportionately ***long***. Races *connectens* (Andes), *pallida*? (lowlands). Singles or pairs with mixed-species flocks, not ant swarms, mostly on trunks. **Voice** Short series of spaced, rising and falling, nasal *peer-peer-peEr-peEr-peer-peer-peur*, slightly slowing at end (Andes); slow, even-paced, descending *peé-pee-pee-peu-peu-puu-puu-puu* (lowlands). **SS** Smaller than all *Xiphorhynchus*, with eyebrow (see Elegant and Striped); Lineated (paler decurved bill, profuse streaking below, whitish throat); Spot-throated (smaller, more spotted below, shorter-tailed). **Status** Rare; NT globally.

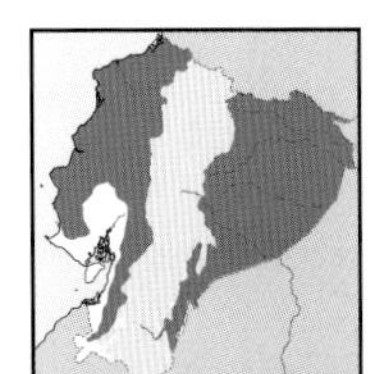

Plain-brown Woodcreeper *Dendrocincla fuliginosa* 20–21cm

E & W lowlands to foothills (mostly below 1,100m, to 1,700m locally). Humid forest, woodland, borders, treed clearings. Rather long-billed, ***brown eyes***. Mostly ***uniform brown, paler face, dusky malar stripe***. Races *neglecta* (E), *ridgwayi* (W). Often follows ant swarms, vigorous and noisy; also singles or pairs in mixed-species flocks. **Voice** Ascending rattle that slows at end: *kle-e'e'e'e'e'e'e'e'e'e-e-e..e..e*; also louder, shorter, whining *clé'clé'clé'cle-cle-cle..cle..cle...*; nasal *pyeek!* **SS** Only plain brown woodcreeper in W; in E see rarer White-chinned (bluish eyes, white chin). **Status** Fairly common.

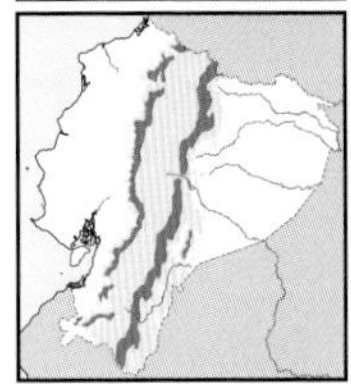

Tyrannine Woodcreeper *Dendrocincla tyrannina* 24–26cm

E & W Andean subtropics to lower temperate zone (1,400–3,100m). Montane forest lower growth, mature woodland. Large stout bill. Mostly ***uniform brown***, pale orbital area. Nominate. Singles regularly with mixed flocks, vigorously climbing, hitching, breaking bark, sallying; often on slim trunks. **Voice** Distinctive, loud, long ascending staccato series, rising in pitch but slowing at end: *dididididi'di'di'di-di-di-di..di..di..di*; calls abrupt *dididi-dík!* **SS** Larger, much heavier bill, plainer face than Plain-brown Woodcreeper, with little overlap. **Status** Uncommon.

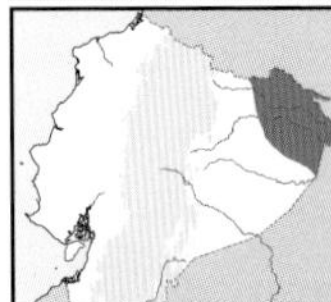

White-chinned Woodcreeper *Dendrocincla merula* 19–20cm

NE lowlands (below 300m). *Terra firme* forest undergrowth. ***Bluish eyes***. Mostly uniform brown, ***white chin***. Race *bartletti*. Obligate army ant follower, often in pairs near ground, less conspicuous than Plain-brown Woodcreeper, rarely high on trunks. **Voice** A rising-falling, slowing *pik-pík-peé-pee-pee-puu*; also harsher *pí'kik, pi'i'ik* calls. **SS** Larger Plain-brown (pale face and dusky malar streak); Spot-throated (dark eyes, no white chin, some pale spots on throat, streaks on crown). **Status** Rare.

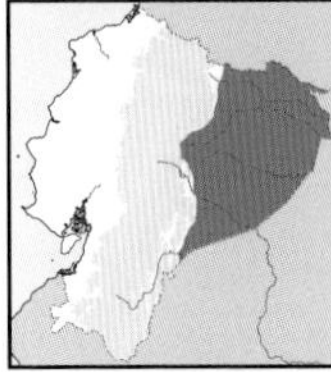

Cinnamon-throated Woodcreeper *Dendrexetastes rufigula* 25cm

E lowlands (to 500m). *Várzea* and riparian forest canopy, borders. ***Long heavy greenish bill***, reddish eyes, ***bold eye-ring***. Mostly rufescent, some ***white streaks on breast***. Race *devillei*. Mostly singles in higher strata, climbing and creeping on large branches and trunks, often searching palm fronds, large dead leaves, epiphyte clumps. **Voice** Fast, sudden rattle *chu'rr'rr'rr'rr'rr*, first ascending, then descending and fading, ending in lower-pitched *pchrrr*. **SS** Confusion unlikely; very local Bar-bellied Woodcreeper (red bill, streaky face and breast, barred belly). **Status** Uncommon.

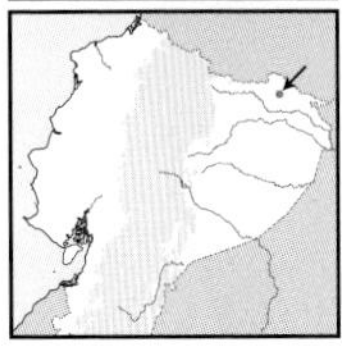

Bar-bellied Woodcreeper *Hylexetastes stresemanni* 25cm

One record in NE lowlands (250m). *Várzea* mid strata. ***Heavy reddish bill***. Uniform brown, ***whitish throat***, few narrow breast streaks, ***belly barred***. Presumed race *insignis*. Unknown in Ecuador; elsewhere, at various heights on large branches and trunks. **Voice** Loud, penetrating series of rising-falling whistles: *tchu'LEE-tchu'LEE-tchu'LEE--tchu'LEE-tchu'LEE*, resembling Strong-billed Woodcreeper but slower, shorter, less moaning; sneezing *tchEEz!* **SS** Amazonian Barred Woodcreeper (lighter bill, more barred underparts, barred back). **Status** Uncertain, possibly very rare.

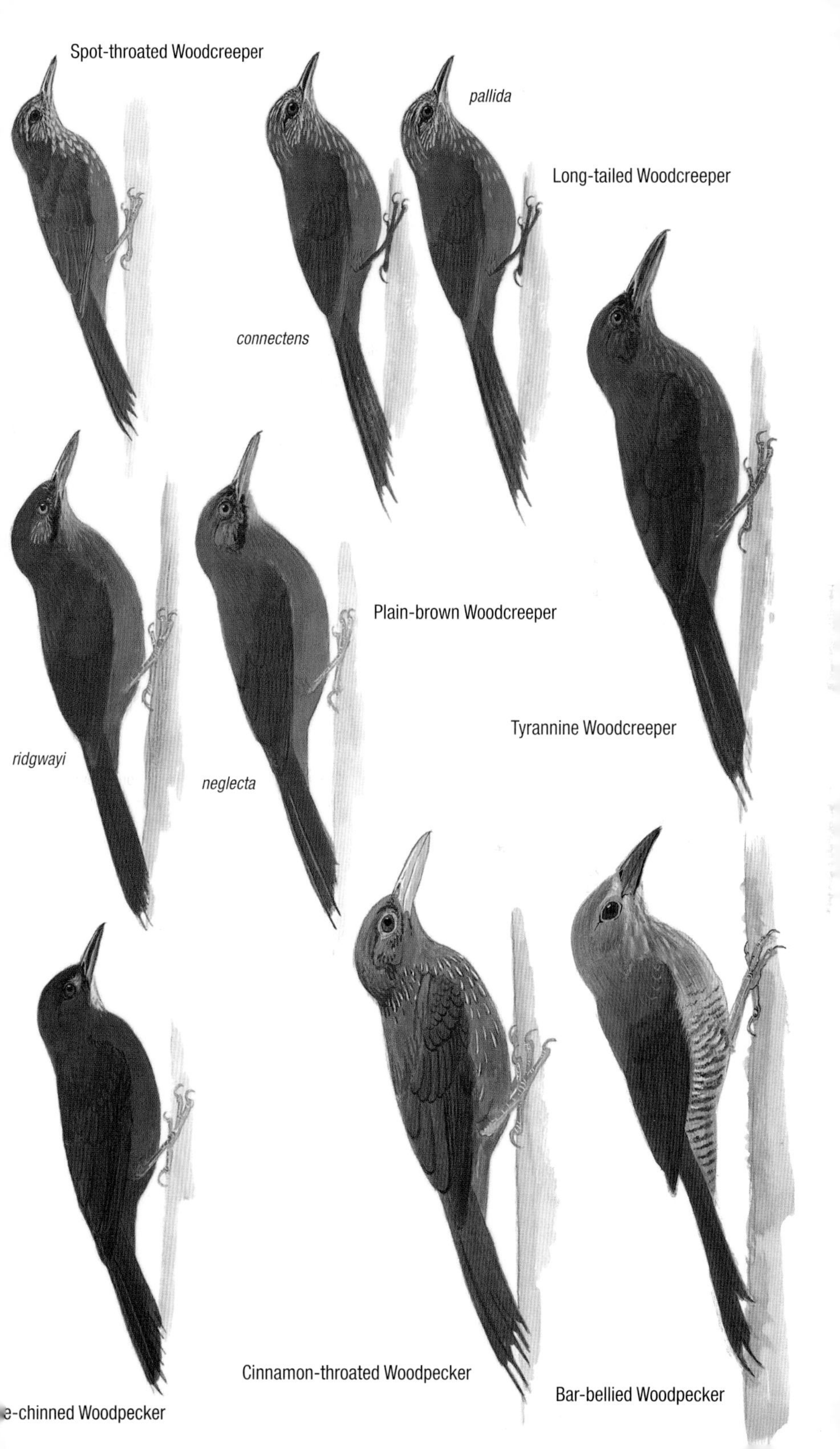
Spot-throated Woodcreeper
pallida
Long-tailed Woodcreeper
connectens
Plain-brown Woodcreeper
Tyrannine Woodcreeper
ridgwayi
neglecta
Cinnamon-throated Woodpecker
Bar-bellied Woodpecker
e-chinned Woodpecker

Large and distinctive woodcreepers, one confined to NW lowlands, and the Strong-billed complex also in Andes. Bills strong, prominently curved in Strong-billed. Note barred instead of streaked pattern below in *Dendrocolaptes*. Overall size and bill size separate these species from *Xiphorhynchus* (see Plates 142–143); therein cf. Buff-throated.

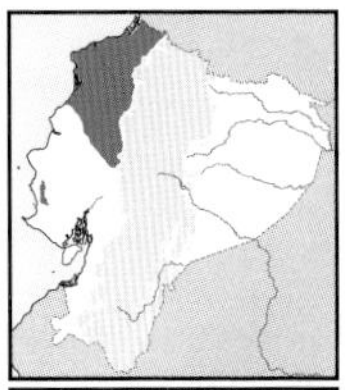

Northern Barred Woodcreeper *Dendrocolaptes sanctithomae* 27–28cm

NW lowlands to foothills (to 800m). Humid forest undergrowth. ***Strong blackish bill.*** Dark brown, ***heavily barred / scaled dusky***, darker lores, plain rufescent wings and tail. Nominate. Mainly at army ants but sometimes alone, rather inconspicuous as tends to remain low and hidden, but vigorous and vocal, especially just after dawn. **Voice** Far-carrying, rather nasal *ooíít-ooíít-ooíÍT-ooÍÍT-OOÍÍT-OOÍÍT*, gradually becoming louder, also loud *tch'clÉÉ!* **SS** Confusion unlikely. **Status** Uncommon, VU in Ecuador.

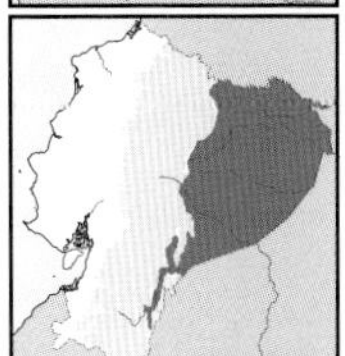

Amazonian Barred Woodcreeper *Dendrocolaptes certhia* 27–28cm

E lowlands (to 600m). *Terra firme* and *várzea* forest undergrowth. ***Strong reddish-brown bill.*** Recalls closely related Northern Barred Woodcreeper but bill colour differs, scaling ***more sparse and even, whiter throat.*** Race *radiolatus*. Most often seen following army ants, where sallies for prey; also alone on trunks and stumps. **Voice** Loud, fast, descending, slowing *tee-tee-tEE-tEE-tée-tee-tee-tee-teu-teu-uu*, slower, less staccato, more descending than Black-banded, loud, snarling *ch-cheff*, *chi-kuk* calls. **SS** Only profusely barred Amazonian woodcreeper with large bill; similar-sized Buff-throated and Black-banded show some streaking on foreparts. **Status** Uncommon.

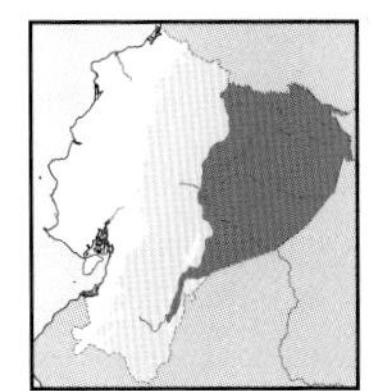

Black-banded Woodcreeper *Dendrocolaptes picumnus* 27–28cm

E lowlands to lower subtropics (to 1,500m). *Terra firme* and *várzea* undergrowth. ***Strong dark bill.*** Head heavily streaked / spotted buff, ***prominent postocular stripe***, nearly uniform rufous upperparts; ***broad*** buff throat stripes extend to breast, ***breast and belly thickly barred buff and dusky.*** Race *validus*. Often in pairs following army ants, dominant over smaller woodcreepers and dominated by larger ones. **Voice** Fast, briefly ascending then almost levelling series, *wee-wéé-wee-wee...wep*, clearer, more staccato, longer than previous species; also loud rattles and moans while feeding. **SS** Buff streaking below separates from Amazonian Barred Woodcreeper; Buff-throated has large buff throat patch, plain belly, broader streaks on breast and neck-sides; see larger Strong-billed Woodcreeper. **Status** Rare.

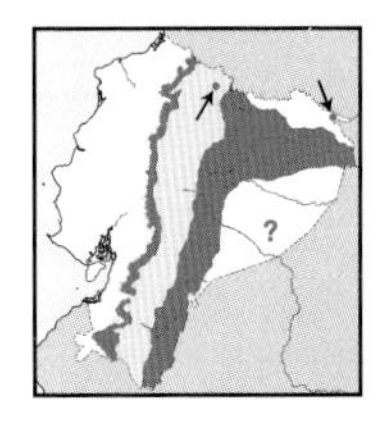

Strong-billed Woodcreeper *Xiphocolaptes promeropirhynchus* 28–30cm

E lowlands, E & W foothills to temperate zone (to 600m in E, 1,100–3,000m in Andes). Montane forest, borders, *várzea* forest. ***Massive dark bill.*** Variable; dark brown to blackish head, rufous brown upperparts, few mantle streaks. ***Buffy lores, face streaked dusky, blackish malar stripe***, whitish throat, whitish streaks on breast and flanks. Complex racial variation: nominate (Andes), *crassirostris* (SW), *orenocensis*? (E lowlands), *macarenae* or undescribed? (extreme SE); several species probably involved. Vigorous, aggressive pairs in mixed flocks, sometimes with ant-followers; vocalises at dusk. **Voice** Far-carrying, whining, descending series of paired *cht'tulEE-cht'tulEE-cht'tulEE-cht'tulEE...* (Andes); faster, first note higher in E lowlands; also nasal *www-nÁG*. **SS** Readily distinguished by size and massive bill; see Buff-throated and Black-banded Woodcreepers in E lowlands; voice of Bar-bellied weaker, shorter. **Status** Uncommon.

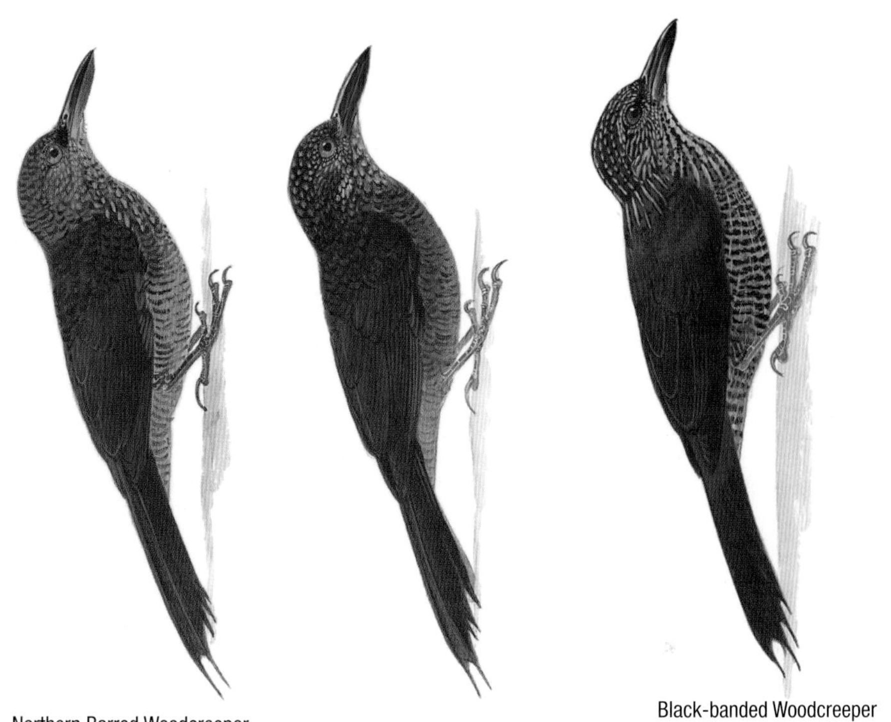

Northern Barred Woodcreeper

Amazonian Barred Woodcreeper

Black-banded Woodcreeper

orenocensis

promeropirhynchus

crassirostris

macarenae

Strong-billed Woodcreeper

PLATE 142: AMAZONIAN WOODCREEPERS

A complex group of woodcreepers confined to Amazonian lowlands. Plumage differences are subtle. Focus on extent of streaking on mantle and lower underparts, also shape of streaks. Voices are sometimes *the* identification clues. Bill more slender in Straight-billed and Duida (more curved bill). Buff-throated is larger and more reminiscent of species on previous plate.

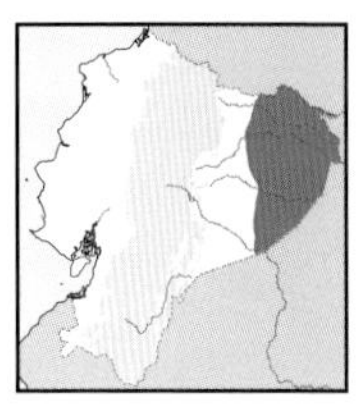

Striped Woodcreeper *Xiphorhynchus obsoletus* 20–21cm

E lowlands (to 300m). *Várzea* forest undergrowth, borders. ***Greyish-horn, slightly decurved bill.*** Brown above markedly ***streaked whitish on head to back, rosy-buff throat, bold whitish streaks to upper belly.*** Wings and tail more rufous-brown. Race *palliatus*. Singles or pairs on trunks and large branches at mid heights, regularly sallies, often with mixed flocks but seldom with ant-followers. **Voice** Abrupt, energetic, ascending staccato trill: *tr'e'e'a'a'ea'ea'ee'eé'EE!*; dry *tri-ti-ti* calls. **SS** Straight-billed (less streaked overall, straighter bill, less rufescent back), Ocellated and Elegant (less striped, darker-billed), and Duida Woodcreepers (lacks whitish streaking above, bill more slender). **Status** Uncommon.

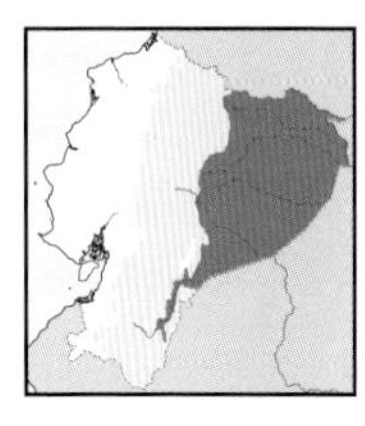

Ocellated Woodcreeper *Xiphorhynchus ocellatus* 21–22cm

E lowlands to lower foothills (to 800m). *Terra firme* forest undergrowth, locally in *várzea*. ***Straight dark bill.*** Blackish crown heavily spotted buff, ***little streaking on mantle.*** Buff chin, ***scaled / streaked throat to breast, unstreaked*** belly. Race *napensis*. Aggressive forest bird often with mixed flocks but not ant swarms, creeps on trunks and large branches, searches crevices, epiphyte clumps, dead leaves. **Voice** Fast, squeaky, descending *weé-wee-wee-wee-chee'ee'ee'ee'oo, weé-ee-ee-ee-ee-ee-chowchowchow*; calls shorter, simpler; less loud, shorter than Striped and Straight-billed. **SS** Elegant Woodcreeper more scalloped / spotted below, rosy-buff throat, more streaked back (primarily in *várzea*); Buff-throated (larger, more solid buff throat, heavier bill); Striped (paler bill, paler streaks, more streaky back, whitish throat). **Status** Fairly common.

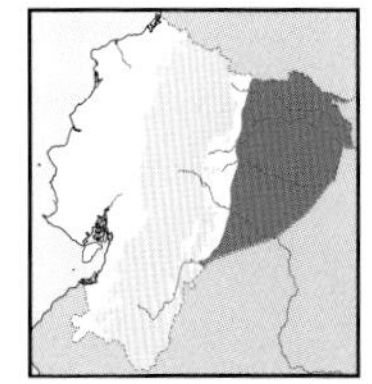

Elegant Woodcreeper *Xiphorhynchus elegans* 21–22cm

Local in E lowlands (to 400m). *Igapó* (and *várzea*?) forest undergrowth. ***Pale grey, slightly decurved bill.*** Dark brown upperparts, ***broad buff diamond-shaped dorsal streaks outlined black.*** Buffy throat, buff scales / streaks to upper belly. Race *ornatus*. Core species in mixed flocks, vigorously defends space; creeps and probes crevices, epiphyte clumps, etc. **Voice** Descending, squeaky, even-paced *tchE-tché-tche-tche-tche-chow-chow-chow-wher* with moaning end; more tremulous, musical, squeakier than Ocellated; calls shorter. **SS** Resembles Ocellated but back streaks are bolder (outlined black), throat paler, less ocellated; Ocellated mostly in *terra firme*. Note voices. **Status** Rare.

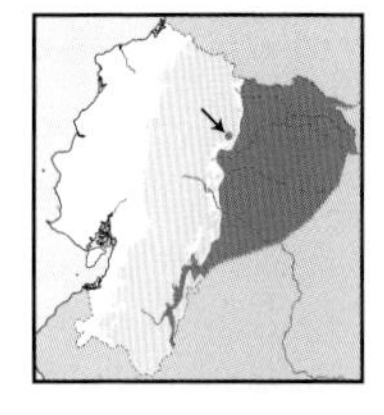

Buff-throated Woodcreeper *Xiphorhynchus guttatus* 25–26cm

E lowlands to lower foothills (to 700m). *Terra firme* and *várzea* forest. ***Long, strong, dusky bill.*** Blackish crown faintly dotted, few ***thin streaks on mantle. Large buff throat patch***, broad buff streaks to upper belly. Race *guttatoides*. Rather tame, vigorous, often in higher strata; usually follows mixed flocks, less often ant swarms. **Voice** Loud, explosive *du-du-dü-dú-dU-dU-dÜ-du..du..du*, louder with progress, slowing at end; also shorter *wüpEE-wüpEE-wup-wup-wüp*; ringing, descending *ky'yoow!, kýow!* **SS** Largest *Xiphorhynchus*, pattern recalls Ocellated (less buff on throat); Black-banded has barred belly. See Strong-billed (larger, much heavier bill). **Status** Fairly common.

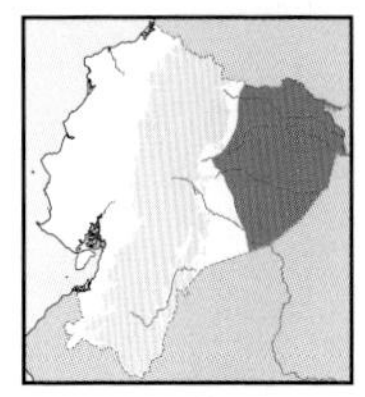

Straight-billed Woodcreeper *Dendroplex picus* 21–22cm

E lowlands (to 400m). *Várzea* forest borders, adjacent woodland. ***Whitish straight bill.*** Crown finely streaked whitish, ***uniform rufous-brown upperparts.*** Head heavily streaked whitish, ***whitish chin and scallop-like spots to upper belly.*** Race *peruvianus*? Rather tame, often conspicuous at edges; follows mixed flocks or found alone, rare at ant swarms. **Voice** Descending trill begins with stutter, then accelerates, ending upturned: *chirp-chip-dip-dip'dip'didididip'dep-dep-drrrrr* or shorter descending, accelerating trill, *chir'rrrrrp*. **SS** Striped Woodcreeper (streaked back, not scalloped on throat to breast, plain buff throat, shorter bill; see Zimmer's *D. keinerii* (potentially in far E). **Status** Fairly common.

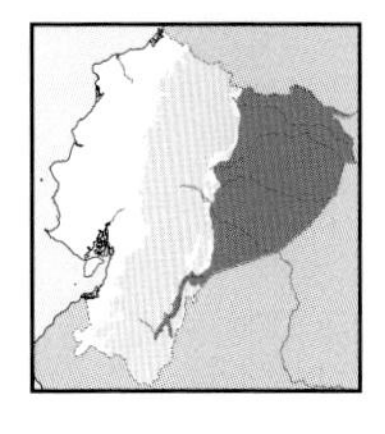

Duida Woodcreeper *Lepidocolaptes duidae* 19cm

E lowlands to foothills (to 900m; locally higher). *Terra firme* forest, old second growth, borders. ***Slender, brownish, decurved bill. Uniform brown upperparts***, more rufescent tail, indistinct ***pale supercilium.*** Whitish throat, brown underparts ***boldly streaked whitish.*** Monotypic. Often with mixed canopy flocks, where inconspicuous; hitches branches and higher trunks, might descend at borders. **Voice** Soft, accelerating and descending *pEE-pée-pee-pi-pi'pi'pi'pipipipupu*, resembling becard. **SS** Smaller than *Xiphorhynchus*, with narrow whitish postocular stripe, slender and decurved bill, and unstreaked back; no overlap with Montane Woodcreeper. **Status** Rare.

Ocellated Woodcreeper
Striped Woodcreeper
Elegant Woodcreeper
Buff-throated Woodcreeper
Straight-billed Woodcreeper
Duida Woodcreeper

PLATE 143: MONTANE AND WESTERN WOODCREEPERS

Three distinct woodcreepers of genus *Xiphorhynchus* occurring outside Amazonian lowlands. Two W species notably different. Two *Lepidocolaptes* included here in local contact as Streak-headed is spreading towards Andes with deforestation. Note thin, markedly curved bills (paler in Streak-headed), characteristic of genus; underparts, crown and postocular stripe useful for species identification.

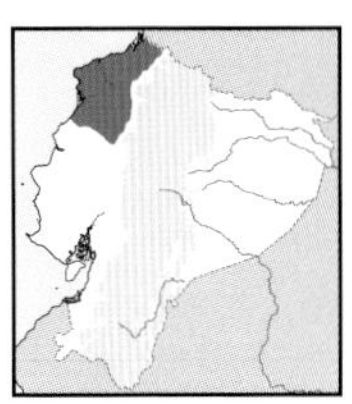

Black-striped Woodcreeper *Xiphorhynchus lachrymosus* 23–24cm

NW lowlands (to 450m). Humid forest canopy, subcanopy and borders. ***Strong dark bill. Mainly blackish, boldly streaked buffy-white***, white throat. Contrasting rufous wings and tail. Nominate. Singles or pairs on trunks, vine tangles, dead leaves, regularly with mixed flocks including ant-followers. **Voice** Rapid, descending, moaning *whee-whee-whee-whee-whea-whea-whea'*; also short loud laugh *whee-whee'deé* and loud, inflected *tchoo-ít?* **SS** Unmistakable in range; see duller Spotted and smaller Streaked-headed Woodcreepers, (more slender bill, less contrasting streaking above, buff throat, etc.; habitat also differs). **Status** Uncommon, VU in Ecuador.

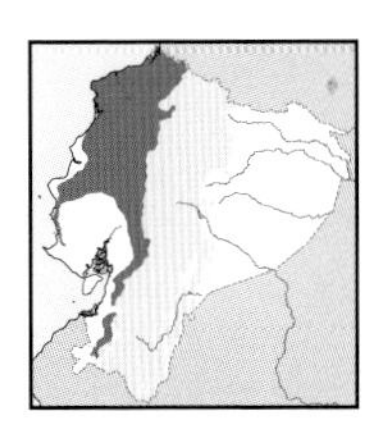

Spotted Woodcreeper *Xiphorhynchus erythropygius* 23–24cm

W lowlands locally to subtropics (mostly to 1,400m, locally to 2,000m). Humid forest, borders, secondary woodland. Long blackish bill. ***Olive-brown above, nearly unmarked. Buff eye-ring***, few spots on dark crown, cheeks and neck-sides. Buffy-white throat, ***buff spots on brown underparts***. Race *aequatorialis*. Active singles or pairs in mixed flocks, sometimes ant-following, hitches and probes crevices, mossy branches, dead leaves. **Voice** Short, quavering series *d'd'd'd'r-reeuw*, descending in pitch; calls are shorter moans *deewr*. **SS** Sympatric Black-striped Woodcreeper strikingly different; see smaller Montane in highlands and Streak-headed in lowlands, both boldly streaked below, whitish-buff throat, longer, more slender bills. **Status** Uncommon, NT in Ecuador.

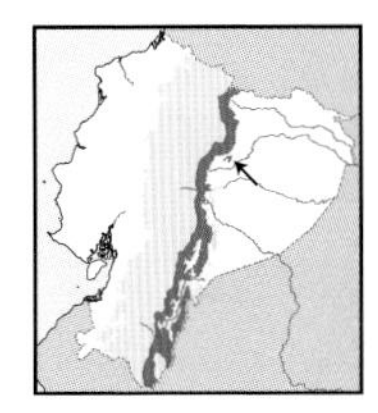

Olive-backed Woodcreeper *Xiphorhynchus triangularis* 23–24cm

E Andean foothills to subtropics (1,000–2,100m). Montane forest, woodland and borders. Long greyish bill. ***Dusky-olive head spotted whitish***, buff eye-ring, ***olive-brown upperparts***, very few streaks on upper mantle. ***Throat heavily scaled buff and dusky***, buff spots over rest of underparts. Nominate. Creeps on moss-covered trunks and large branches, often with mixed flocks; probes bark crevices. **Voice** sharp, accelerating series, slowing and fading at end: *we-we-we'we'we-we-we-wa..wa*; call piercing, downslurred *keeyuür!* **SS** Confusion unlikely with few sympatric woodcreepers; smaller Montane is boldly streaked below, has white postocular stripe; Long-tailed Woodcreeper has faint streaking below, smaller slender bill, buff throat. **Status** Uncommon.

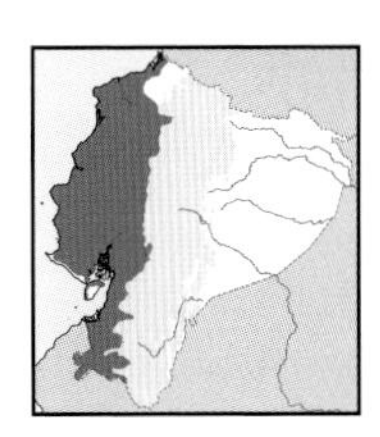

Streak-headed Woodcreeper *Lepidocolaptes souleyetii* 20–21cm

W lowlands, to subtropics in Loja (to 1,800m). Deciduous to humid forest and wooded undergrowth, borders, clearings. ***Long, slender, decurved pale bill.*** Dense buff streaking on crown, ***narrow buff postocular stripe***, few streaks on upper mantle, more rufescent wings. ***Buff throat, broad buff streaks on underparts***. Races *esmeraldae* (most range), *souleyetii* (extreme SW). Conspicuous and confiding, foraging in open, sometimes even on man-made structures, joins mixed flocks. **Voice** Ringing, descending trill: *pr'rrrrrrii*; call louder, short *trrrew!* resembling kingbird. **SS** Locally Montane Woodcreeper (especially with deforestation); Montane has whiter, drop-shaped streaks on underparts, less extensive buff on throat, spots on crown. **Status** Common.

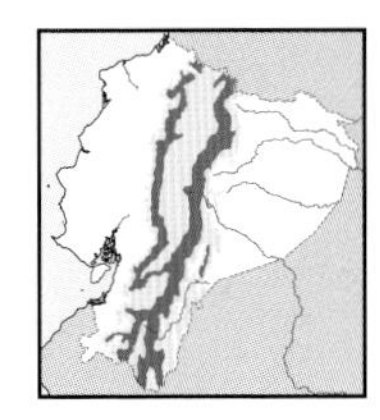

Montane Woodcreeper *Lepidocolaptes lacrymiger* 19–20cm

Andean subtropics to temperate zone (1,500–3,200m). Montane forest and borders, locally to treeline. ***Long, slender, decurved pale bill.*** Dark brown head with ***pale spots***, uniform rufous-brown upperparts, ***whitish eye-ring and postocular stripe***. White throat, underparts ***boldly drop-streaked white***. Races *aequatorialis* (most range), *warscewiczi* (extreme SE; buffier, less drop-shaped streaks below). Singles or pairs regularly with mixed flocks, agilely probing and creeping in moss clumps, epiphytes, frequently on underside of moss-covered branches. **Voice** Accelerating, high-pitched, rhythmic *ts-ts-ts'ts'tSS-tsi-tsi-tsi-tsií*, recalling a fruiteater or Pearled Treerunner (but slower, less ringing). Call a short, high-pitched *tss-tséé, tss-tséé*.... **SS** Confusion unlikely, overlaps with few other woodcreepers; see Streak-headed, Olive-backed and Spotted. **Status** Uncommon.

Black-striped Woodcreeper
Spotted Woodcreeper
Olive-backed Woodcreeper
esmeraldae
souleyetii
Streak-headed Woodcreeper
aequatorialis
warscewiczi
Montane Woodcreeper

PLATE 144: LONG-BILLED WOODCREEPER AND SCYTHEBILLS

Five species with remarkably long bills. Long-billed is distinctive at *várzea* forest edges. Scythebills occur primarily in humid forests, but the ubiquitous Red-billed also ranges into semi-deciduous habitats. The unique, montane Greater is readily separated from congeners. Identifying the other scythebills is difficult as bill colour and shape very similar. Note subtle plumage differences in throat and mantle, also voices. Red-billed and Brown-billed apparently do not overlap.

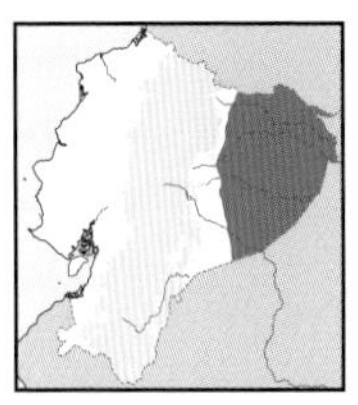

Long-billed Woodcreeper *Nasica longirostris* 35–36cm

E lowlands (to 400m). Upper strata in *várzea* and riparian forest, less numerous in *terra firme*. Unique, *very long ivory bill*. Blackish crown, *long white postocular stripe*. White throat and lower face, *bold white streaks on brown underparts*. Monotypic. Singles or pairs at various heights, lower at borders of lakes and streams with open vegetation, sometimes joins mixed flocks; probes vegetation clumps and bromeliads. Readily responds and approaches to whistled imitation of voice. **Voice** Gives 4–7 clear, mournful, rising-falling whistles, *wooOOooo, wooOOooo, wooOOooo...*, sometimes faster, longer, shorter; harsh *djág!* calls. **SS** Unmistakable. **Status** Uncommon.

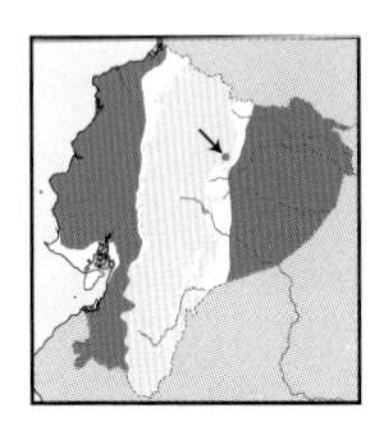

Red-billed Scythebill *Campylorhamphus trochilirostris* 23–24cm

E & W lowlands, to subtropics in Loja (to 400m in E, locally higher; to 800m in W, locally to 1,900m). Humid to semi-deciduous forest undergrowth, borders, woodland. *Very long, curved reddish bill*. Blackish crown, mostly dark brown *heavily streaked buff*, streaks broader below. More rufous wings and tail. Races *thoracicus* (W), *napensis* (E). Singles or pairs creep and probe trunks, large limbs, bark crevices, epiphyte clumps, may join mixed flocks. **Voice** Fast, descending series begins with trill *trrrt, tuee-tuee-tuee-tuee..tue..tue..tue!* or *tee'tee'tee-tuee-tuee-tuée-tuéé*; also simpler, slower *duee'duuee-duee-due-duu*, penetrating *twii-twii* calls. **SS** No known overlap with Brown-billed (darker overall, buff throat, browner bill); very similar Curve-billed (more curved bill, lacks back streaking, narrower streaks below). **Status** Uncommon.

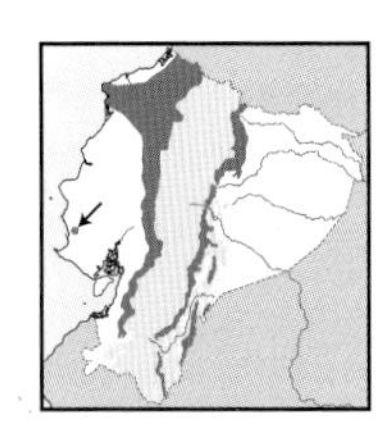

Brown-billed Scythebill *Campylorhamphus pusillus* 23–24cm

E & W foothills to subtropics, local in NW lowlands (600–2,100m, locally to 100m). Montane forest undergrowth, woodland, borders. *Long and curved dusky-brown bill*. Dark brown, more rufous flight feathers and tail. *Sparse thin buff streaking* on head, neck-sides, mantle and underparts to mid belly. Nominate (E), *guapiensis*? (W). Hitches on large mossy branches and trunks, regularly probing moss or epiphyte clumps, sometimes with mixed flocks. **Voice** Tremulous, mournful, rising-falling series *t'rrree-trr'wee-trreeu-teeurr-teeurr-tuee-tuee-tuee...*, sometimes shorter *tuee-tuee-tueea* version; also antbird-like calls. **SS** Red-billed Scythebill prefers less humid forests in W; bill colour differs only slightly, but Red-billed more profusely streaked, streaks broader, outlined blackish. **Status** Uncommon.

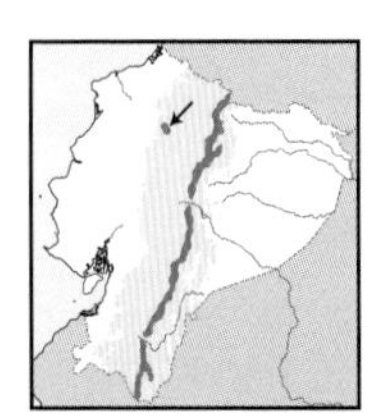

Greater Scythebill *Drymotoxeres pucherani* 28–29cm

Local in Andean subtropics to temperate zone (2,000–2,800m). Montane forest and borders. *Very long and curved horn bill*. Mostly dark rufescent-brown with faint pale streaks, more rufous flight feathers and tail. *Bold white stripes on face, dusky cheeks*. Monotypic. Poorly known, singles rare with mixed flocks, creeping on trunks and large mossy branches, active and acrobatic probing crevices or moss clumps. **Voice** Short, nasal, rather trembling and ascending *r'r'r-eEH!* or *ee'ee'ee'EH!*, sometimes nasal, loud but short *iík, i'ík, ik* calls. **SS** Confusion unlikely, lower-elevation scythebills markedly streaked, lack face pattern. **Status** Rare, local, NT; possibly overlooked?

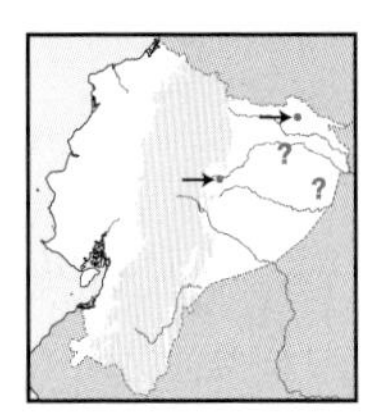

Curve-billed Scythebill *Campylorhamphus procurvoides* 23–24cm

Local in NE lowlands (250m). *Terra firme* forest interior. *Very long and very curved reddish-brown bill*. Closely recalls Red-billed Scythebill but *less rufescent*, nearly *unstreaked back*, streaks on underparts *very thin and not outlined blackish*, more rufescent rump. Race *sanus*. Poorly known in Ecuador, but presumably behaves much like Red-billed. **Voice** High, even or ascending series *whee-a, whee-a, whee-a, whee-whee-whe-wee'wee'wee'wee'h*, resembles Red-billed but slower, not upslurred, not trilled at start; *jtjtjít!* call. **SS** Not always safely separated from Red-billed (see streaking on back). **Status** Very rare and local.

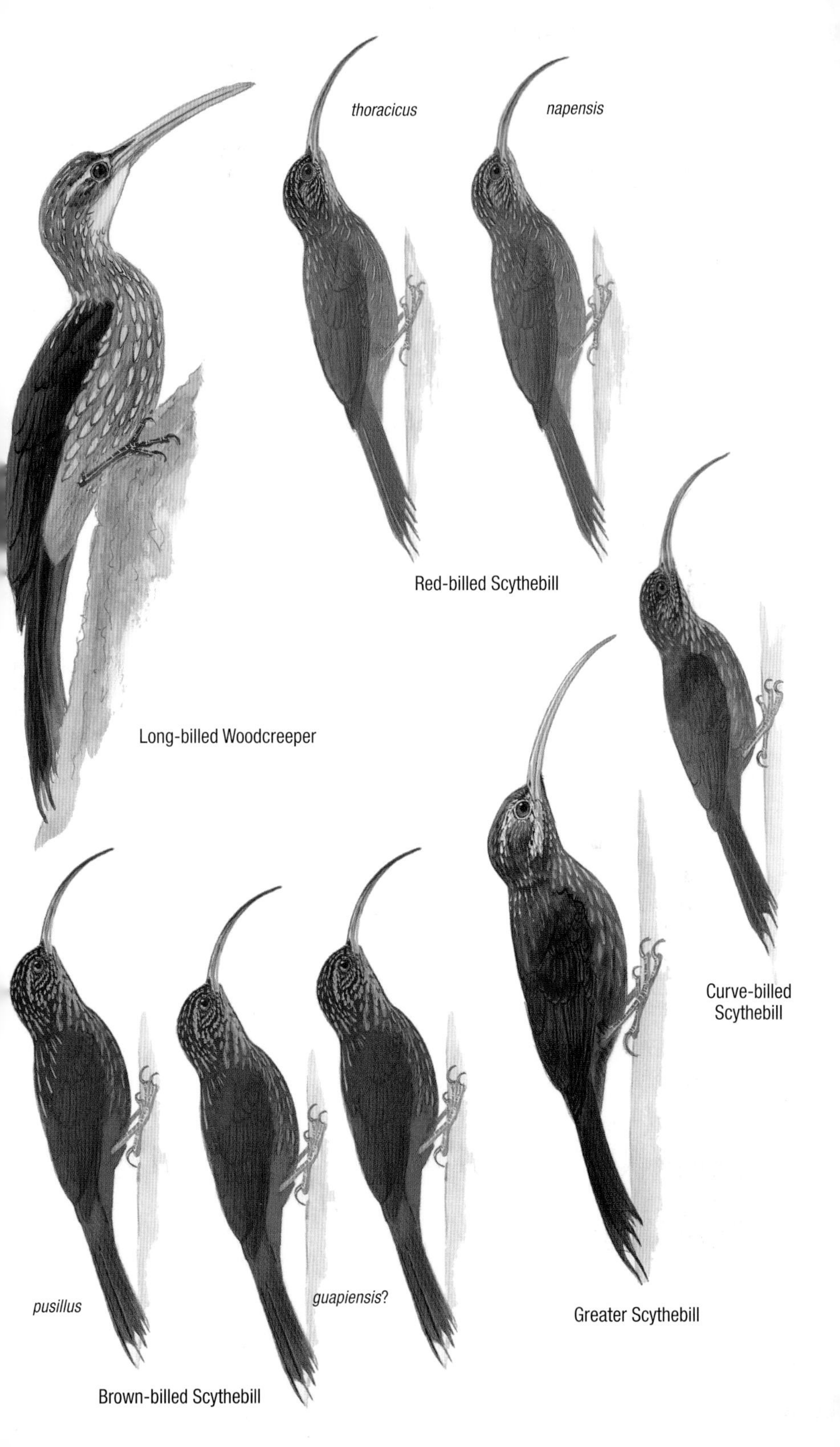
thoracicus
napensis
Red-billed Scythebill
Long-billed Woodcreeper
Curve-billed
Scythebill
pusillus
guapiensis?
Greater Scythebill
Brown-billed Scythebill

PLATE 145: SMALL WOODCREEPERS, XENOPS AND PRICKLETAIL

Xenops and prickletail are small and arboreal. They forage by clambering and hitching along branches. Identifying xenops not easy due to slight plumage differences. Note face pattern, extent of streaking below and on mantle. The prickletail is fairly similar but bill not upturned. The two smallest woodcreepers are distinctive despite size, behaviour and interior forest preferences.

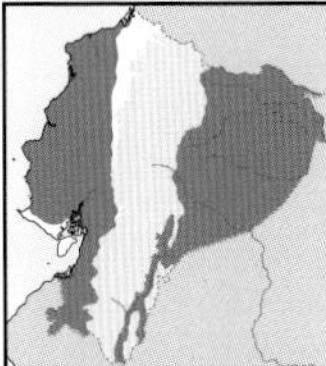

Olivaceous Woodcreeper *Sittasomus griseicapillus* 14–16cm

E & W lowlands to foothills, into SW subtropics (to 1,100m, to 2,000m in S). *Várzea* forest, woodland, borders (E); semi-deciduous to deciduous forest, borders (W). ***Plain grey head and underparts, rufescent back to tail*** (E); mostly ***plain greyish-olive, cinnamon-rufous rump to tail*** (W). Races *amazonus* (E), *aequatorialis* (W); probably separate species. Alone or in mixed flocks, rather inconspicuous, agile sallies. **Voice** Fast rolling, descending *prr-r-r-r-r-r-r'r'ru*, more trilled *prrrrrr...* (W); slower, rising *pu-pu-peu-peE-peÉ-pe-pehe* (E). **SS** Wedge-billed Woodcreeper browner, spotted. **Status** Uncommon.

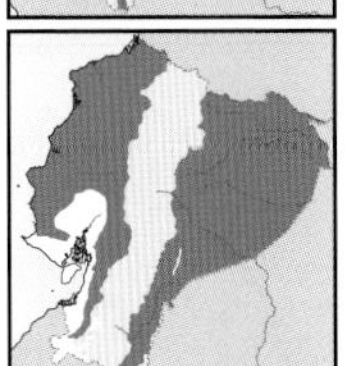

Wedge-billed Woodcreeper *Glyphorynchus spirurus* 13–14cm

E & W lowlands to lower subtropics (to 1,700m). Humid forest, borders, old second growth. ***Very short wedge-like bill.*** Mostly rufescent-brown, ***buff wing stripe visible in flight. Indistinct buff eyebrow,*** orange-buff throat, pallid breast spots. Races *castelnaudii* (E), *subrufescens* (W). Singles or pairs forage alone or in mixed-species flocks, rapidly hitching up vertical branches and trunks; numerous. **Voice** Fast, ascending, brief, rhythmic *twe-twe-tueé-tueé-tueep?*; sudden *djép!* call. **SS** Larger Spot-throated Woodcreeper has longer bill, no eyebrow. **Status** Common.

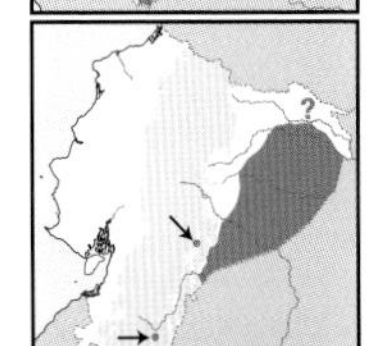

Slender-billed Xenops *Xenops tenuirostris* 11–12cm

Local in E lowlands (to 600m). *Terra firme* canopy, borders, locally in *várzea*. ***Slender, slightly upturned bill.*** Dusky-brown upperparts ***finely*** streaked buff, ***whitish postocular stripe, white malar streak.*** Olivaceous below ***faintly*** streaked whitish. Rufous wings, black primaries, long ***rufous-and-black tail.*** Race *acutirostris*. Singles or pairs join mixed-species flocks, climb thin twigs and vines. **Voice** High-pitched, short, steady *tsip-tsip-tsip-tsip*, slower than Streaked Xenops, not ascending-descending. **SS** Not safely separated from Streaked (see below), but little to no overlap. **Status** Rare, overlooked?

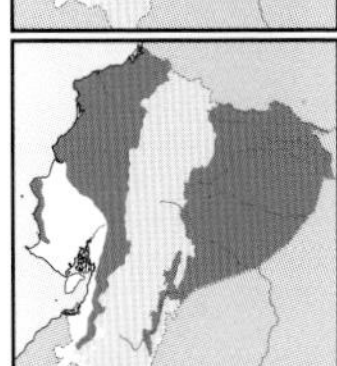

Plain Xenops *Xenops minutus* 11–12cm

E & W lowlands to foothills (to 900–1,000m). Humid forest canopy, borders, woodland. ***Upturned bill. Unmarked*** olive-brown upperparts, blacker head (E), ***buffy-white postocular stripe, prominent white malar***, few white streaks below, ***plain belly.*** Races *obsoletus* (E), *littoralis* (W); different species? Often lower than other xenops, with understorey flocks; agilely flakes open decaying bark. **Voice** Fast, ascending *tí-tí-títítítítí* (W); slower, ascending-descending *swé-swé-SWE-swé-swe, swe...* (E). **SS** Other xenops are streaked below; bill lighter, less upturned. **Status** Uncommon.

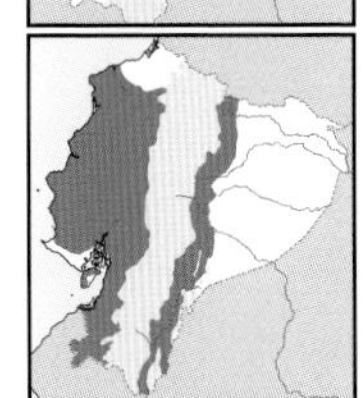

Streaked Xenops *Xenops rutilans* 12–13cm

W lowlands to subtropics, E Andean foothills to subtropics (to 2,000m in W, 800–2,000m in E, locally lower). Montane forest canopy, borders; to semi-deciduous and deciduous forest (SW). Closely recalls Slender-billed Xenops but bill ***more upturned and heavier, more rufous wings and tail.*** Streaks broader, better defined. Races *guayae* (W), *heterurus* (NE), *peruvianus* (SE). Singles or pairs with flocks, clambering on twigs and vines, flaking bark. **Voice** Short, ascending-descending, accelerating-decelerating *swee-swee-wEE-swEE-swee-swee-swea-swea...*; calls shorter. **SS** Lower-elevation Rufous-tailed Xenops (more profusely streaked, no white malar). **Status** Fairly common.

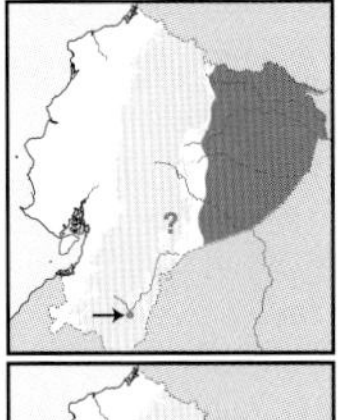

Rufous-tailed Xenops *Microxenops milleri* 11–12cm

E lowlands to adjacent foothills (to 1,000m). *Terra firme* canopy, borders. ***Straight bill.*** Black crown, brown upperparts, more olivaceous below; ***all streaked whitish-buff***, broader above. ***Long whitish postocular stripe, no white facial streak.*** Rufous wings and ***rather short tail.*** Monotypic. Joins flocks, clambers on thin branches, picking and probing in side-to-side manner. **Voice** Very fast, short, ascending but then fading trill: *chit-chi'chll'chll'chll'chll'chii'chii'chii....* **SS** Only xenops lacking white facial streak, also streakier upperparts, more rufous tail. **Status** Rare, local.

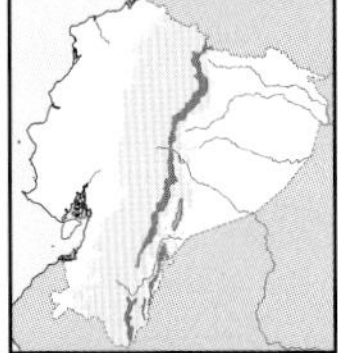

Spectacled Prickletail *Siptornis striaticollis* 12–13cm

Local in E Andean foothills to subtropics (1,200–2,300m). Montane forest canopy, borders. ***Decurved, slim bill.*** Plain rufous-brown above, ***chestnut crown. Whitish postocular stripe***, olivaceous-brown below, ***whitish streaks on throat to mid breast.*** Long, rufous, spiky tail. Race *nortoni*. Singles or pairs with mixed flocks search moss clumps, probe bark crevices, dead wood, hang upside-down. **Voice** Very thin, short, high-pitched, rising-falling quavering trill; also slower, slightly ascending, accelerating *chi'chu'chi'chi'chichichi*. **SS** Recalls xenops but longer, spiky tail, decurved bill, unmarked upperparts, no facial streak, are diagnostic. **Status** Rare.

amazonus
aequatorialis
ivaceous Woodcreeper
Wedge-billed Woodcreeper
obsoletus
littoralis
Slender-billed Xenops
Plain Xenops
Streaked Xenops
heterurus
guayae
peruvianus
Rufous-tailed Xenops
Spectacled Prickletail

PLATE 146: BARBTAILS, TREERUNNERS, PLUSHCROWN AND GREYTAILS

The distinctive, warbler-like plushcrown, plus a variety of small arboreal 'creepers' and gleaners. Greytails also recall warblers in shape and behaviour and often join mixed flocks. Barbtails and treerunners are superficially similar. Treerunners tend to creep more than the hopping manners of Spotted and clambering Rusty-winged Barbtails. Treerunners are primarily with mixed flocks.

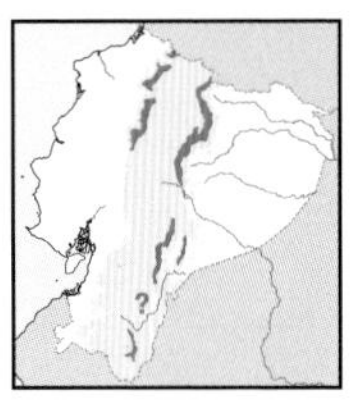

Rusty-winged Barbtail *Premnornis guttuliger* 14–15cm

Local in E & NW Andean subtropics (1,600–2,300m). Montane forest undergrowth. Brown above, ***long narrow buff supercilium, buff eye-ring***. Whitish throat, ***scaly buffy streaks to mid belly. Rufous-chestnut wings*** and long tail. Nominate. Mainly singles with understorey flocks, clambering through dense tangles, sometimes in curled dead leaves or moss clumps, not hitching on trunks. **Voice** Sharp *tseé, tseep-tseep-tseep* often ending in *si-si-si-si-si*, less trilled than Spotted. **SS** Spotted Barbtail (darker brown upperparts, shorter supercilium, densely spotted below); see Montane Foliage-gleaner (plainer underparts, more arboreal). **Status** Uncommon, local.

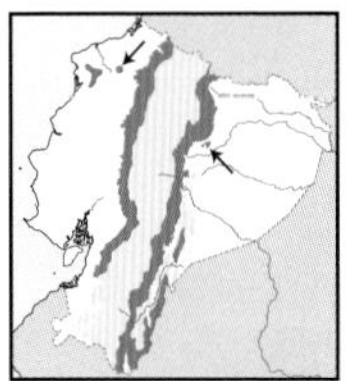

Spotted Barbtail *Premnoplex brunnescens* 13–14cm

E & W Andean foothills to subtropics. Local in N coastal ranges (900–2,500m, locally higher; to 500m in coastal mountains). Montane forest undergrowth. ***Dark brown above. Narrow buff eye-ring and short postocular stripe***. Cinnamon-buff throat, ***dense buff spotting below.*** Nominate. Unobtrusive, singles hop and hitch on branches and trunks, regularly with mixed flocks. **Voice** Short, explosive, fast *psserrrr'* trill; also high-pitched *tseé-tseé-tseé*. **SS** Rusty-winged Barbtail (rufescent wings and tail, streaky underparts); Pearled Treerunner (bolder spots, more arboreal). **Status** Fairly common.

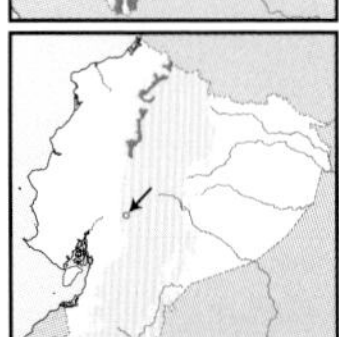

Fulvous-dotted Treerunner *Margarornis stellatus* 15cm

Local in NW Andean subtropics (1,200–1,900m). Upper strata of montane forest and borders. ***Rich orange-rufous above, brighter*** rufous below, ***snowy-white throat and teardrop spots on breast***. Long, graduated spiky tail. Monotypic. Seldom seen; hitches and clambers, sometimes using tail like woodcreepers, creeping upside-down on mossy branches. **Voice** Very thin, fast, quavering series of *tzít* notes; weak *zeet* or trilled *tt'zt*. **SS** Pearled Treerunner more dotted below with prominent postocular stripe, no known overlap. **Status** Rare, probably declining; NT globally, VU in Ecuador.

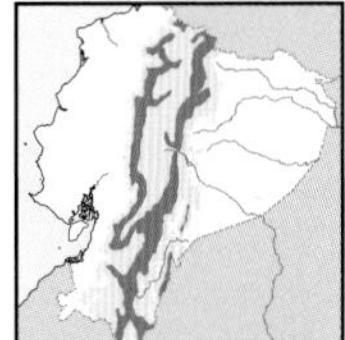

Pearled Treerunner *Margarornis squamiger* 15cm

Subtropics to temperate zone in Andes (1,800–3,500m). Montane forest canopy, borders, treeline, *Polylepis* woodland. Elegant. Chestnut-brown above, contrasting ***snowy-white supercilium, throat and pearl-shaped spots on underparts***. Long, graduated spiky tail. Race *perlatus*. Singles or pairs with mixed flocks, nimbly clambering on twigs, searching epiphytes, mossy clumps or dead leaves upside-down. **Voice** High-pitched, very fast series of *tsít-tsít-tsít-tsít*...; call a simpler *tsít*. **SS** Confusion unlikely; Montane Woodcreeper superficially similar but behaviour differs. **Status** Common.

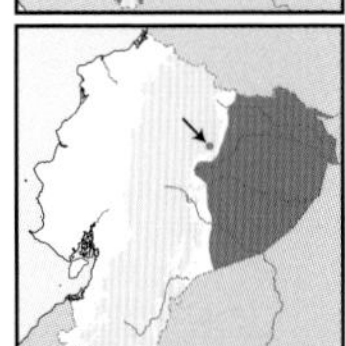

Orange-fronted Plushcrown *Metopothrix aurantiaca* 11–12cm

E lowlands to lower foothills (to 650m). Secondary forest, borders, treed clearings. Small, warbler-like with ***pale orange legs. Bright orange front***, narrow yellowish eye-ring. Olive-green above, ***yellowish-orange throat***, yellowish-olive underparts. Monotypic. Small groups, probably families, actively forage in semi-open, often near large stick nests; sometimes with mixed flocks. **Voice** High-pitched, rather variable *tseét-sweé-tseét, tsi-tsi-tsi-si-si-si*, piercing *psíí* calls. **SS** Resembles Tawny-crowned Greenlet or Yellow Warbler but note face, legs, behaviour and voice. **Status** Uncommon.

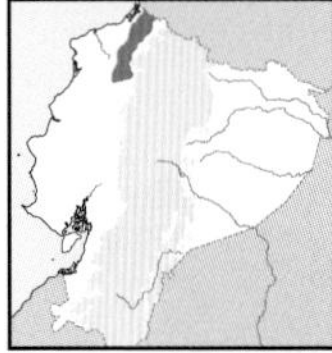

Double-banded Greytail *Xenerpestes minlosi* 11cm

Local in NW lowlands to lower foothills (150–500m). Humid montane, borders. Small, warbler-like with ***pale amber eyes. Grey upperparts, blacker crown, long white supercilium, two contrasting wingbars***. Pale yellowish-white below. Possibly race *umbraticus*. Pairs regular with mixed canopy flocks; gleans like warbler from undersides of leaves, twigs and in tangles. **Voice** High-pitched, dry, very fast trill, almost insect-like: *trrrrzzzzzzz'*; also dry *tzít*. **SS** Resembles some non-breeding migrant warblers, but plainer yellowish-buff below, pale eyes, long slender bill. **Status** Rare, VU in Ecuador.

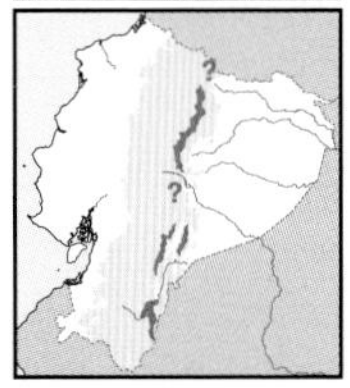

Equatorial Greytail *Xenerpestes singularis* 11–12cm

Local in foothills to lower subtropics in E Andes (1,000–1,600m). Montane forest canopy, borders. Small, warbler-like. ***Rufous front, grey upperparts, narrow white supercilium. Underparts coarsely streaked white and grey***. Monotypic. As Double-banded, Equatorial also joins canopy mixed flocks, acrobatically gleaning foliage; difficult to locate. **Voice** Fast, accelerating, sputtering, dry trill: *trrrzzzzrrzzzz*; dry *tzí, tzík* calls. **SS** Resembles Grey-mantled Wren, including behaviour, but wren lacks streaked underparts and whitish supercilium, and has longer, banded tail. **Status** Rare, NT globally.

Rusty-winged Barbtail

Spotted Barbtail

Fulvous-dotted Treerunner

Pearled Treerunner

Orange-fronted Plushcrown

Double-banded Greytail

Equatorial Greytail

PLATE 147: ARBOREAL AND FOREST FURNARIIDS

Tuftedcheeks are conspicuous Andean furnariids characterised by unique whitish 'scarves'. Often with mixed flocks, persistently skulking in epiphytes. The remaining species range in lowlands. The hookbill superficially recalls foliage-gleaners (see Plates 148–150) but has distinctive bill. The woodhaunters might be separated into W and E species as their voices differ strikingly; they also recall (and are closely related to) foliage-gleaners.

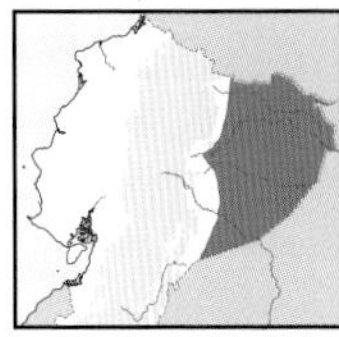

Point-tailed Palmcreeper *Berlepschia rikeri* 21–22cm

Local in E lowlands to lower foothills (below 650m). *Mauritia* palms. Long stout bill. Unique. ***Black*** with ***bold white streaks***, barred vent. ***Rufous-chestnut mantle, wings and long, graduated tail.*** Monotypic. Loose pairs (groups?) high in palm stands, where acrobatically searches curled palm frond blades, clumps or palm leaflets; very elusive. **Voice** Far-carrying, strident, ringing, slightly rising *tee-tee-tee...keé!-keé!-keé!-keé!...*. **SS** Unmistakable though very difficult to see. **Status** Rare, local.

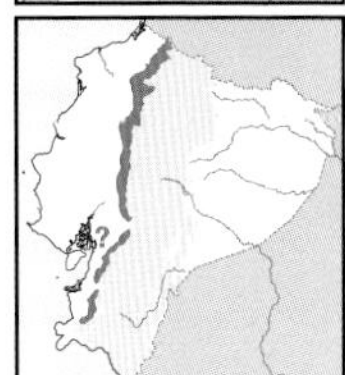

Buffy Tuftedcheek *Pseudocolaptes lawrencii* 19–20cm

Local in W Andes foothills to subtropics (700–1,700m). Montane forest canopy, borders. Blackish head streaked whitish, narrow buff supercilium, ***large buffy cheek patch. Plain rufous-brown above***, brown below with contrasting ***white streaks / spots to upper belly.*** Long rufescent tail. Race *johnsoni* might be valid species. Singles or pairs forage along horizontal branches, often probe noisily in epiphyte clumps, especially bromeliads; regular in mixed flocks. **Voice** Similar to Streaked Tuftedcheek, 1–2 loud *tchík* calls followed by fast trill *cht'trrrr*, loud, metallic *pínk* calls. **SS** Locally overlapping Streaked (more conspicuous white beard, paler underparts, streaked back). **Status** Uncommon, rather local, VU in Ecuador.

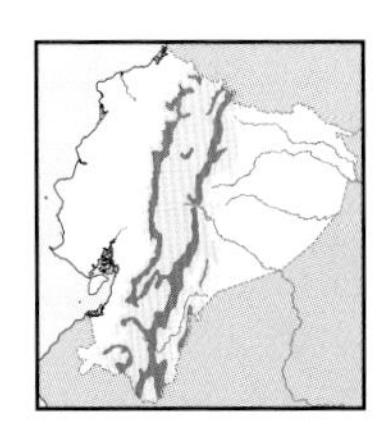

Streaked Tuftedcheek *Pseudocolaptes boissonneautii* 21–22cm

Subtropics to temperate zone on both Andean slopes (1,800–3,400m). Montane forest canopy, borders, treed clearings. Brown upperparts ***streaked whitish, long narrow buffy-white supercilium, snowy-white beard extends to cheeks.*** Breast boldly ***scalloped whitish***, cinnamon-fulvous belly. Long rufescent tail. Nominate (W), *orientalis* (E, extreme SW). Singles or pairs often with mixed flocks, vigorously probe epiphytes, especially bromeliads. **Voice** Loud *tcheék-tcheék* followed by more tinkling *teet* notes, then a fast, higher-pitched rattle *tcheék-tcheék-teet-teet-teet-teet, ptrtrtrtrtrt*, often rising. **SS** Buffy Tuftedcheek lacks puffy white throat, back unstreaked; voices very similar. **Status** Uncommon.

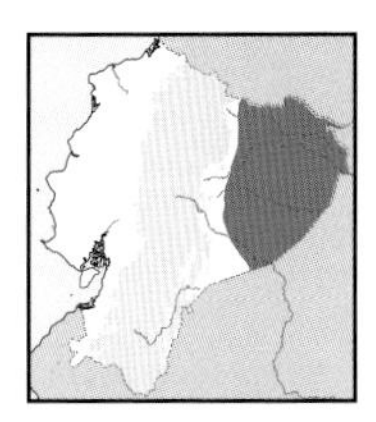

Chestnut-winged Hookbill *Ancistrops strigilatus* 18–19cm

E lowlands (to 400m). Mainly *terra firme* forest canopy. Blackish crown, dark olive-brown mantle, both ***finely streaked whitish-buff***, paler rump. ***Narrow yellowish-white supercilium***, whitish throat, pale buff below faintly streaked dusky. ***Rufous wings and long rounded tail.*** Monotypic. Singles or pairs with mixed flocks, forage on large horizontal branches and vines. **Voice** Some 30+ sec-trill, somewhat quavering *k'r'r'r'r'r'r...*; also shorter *gik'k'k'k'k'k* trill and raspy *guík*. **SS** Similar to Chestnut-winged Foliage-gleaner, although it lacks streaks on crown and back, breast streaking fainter; Rufous-tailed streakier but wings concolorous with upperparts. **Status** Rare.

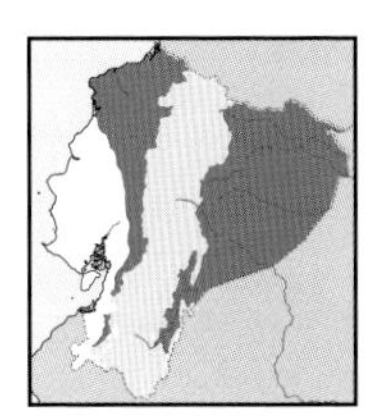

Striped Woodhaunter *Automolus subulatus* 17–18cm

E & W lowlands to foothills; S over W Andean foothills to El Oro (to 1,100m, locally higher). Humid forest, woodland, borders. Longish, straight, rather ***pointed bill.*** Brown upperparts ***finely streaked whitish, streaks confined to crown*** (W). Underparts olive-brown, ***vague but broad buffy streaks; plainer, more olivaceous below*** (W). Nominate (E), *assimilis* (W), probably separate species. Singles or pairs in lower growth with mixed flocks, glean foliage, clumps, epiphytes, palm fronds. **Voice** Short loud series, ringing, and even-paced *keeu-keeu-keeu-keeu!* (W) or loud, ringing *teuu-tEEw* or *teuu-téuu-téuu-téuu-téuu*, regularly followed by lower-pitched rattle *tr'r'r'r* (E). **SS** Darker, better-streaked Lineated Foliage-gleaner (plain buff throat, shorter bill); Uniform and Streak-capped Treehunters (darker, stouter bills, unstreaked back). See also Scaly-throated and Montane Foliage-gleaners. **Status** Uncommon.

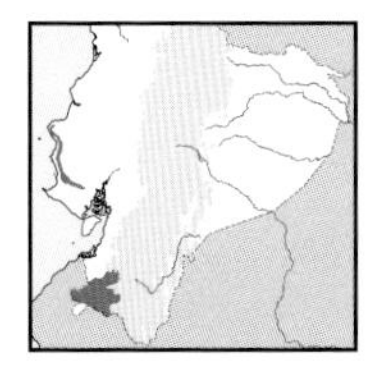

Henna-hooded Foliage-gleaner *Clibanornis erythrocephalus* 20–21cm

SW lowlands, to subtropics in Loja (300–1,800m, locally lower). Deciduous to semi-deciduous forest and wooded undergrowth, borders. Hazel eyes, grey mask. ***Bright orange-rufous hood***, paler on throat, contrasting with greyish-olive upper- and underparts. ***Rufous wings and tail.*** Nominate. Singles or pairs accompany mixed flocks, vigorous and noisy, on ground flicks leaves aside with bill, rummage in dense leaf litter. **Voice** Distinctive metallic, faltering *ch-khe'khe'khe'khe'khe* or *kruh-kruh-kruh-kruh...*. **SS** Unmistakable. **Status** Uncommon, local, VU.

Point-tailed Palmcreeper
Buffy Tuftedcheek
Streaked Tuftedcheek
Chestnut-winged Hookbill
assimilis
subulatus
Striped Woodhaunter
Henna-hooded Foliage-gleaner

PLATE 148: FOLIAGE-GLEANERS I

Mostly montane, arboreal furnariids. Buff-browed tends to prefer lower growth. Ranges do not overlap except Lineated and Buff-browed. These foliage-gleaners frequently join mixed-species flocks, are very acrobatic when inspecting mosses, dead leaves, clumps, bromeliads, and underside of foliage or branches. Identification can be complicated due to observation difficulties and resemblance to other furnariids (see Plates 149–150)

Scaly-throated Foliage-gleaner *Anabacerthia variegaticeps* 16–17cm

W foothills to subtropics, slopes of coastal mountains (700–1,700m, lower in coastal mountains). Montane forest canopy, borders. Blackish-brown crown finely streaked buff, ***long and broad ochraceous-buff eye-ring and postocular stripe***. Whitish ***scaly*** throat, olive-brown below ***streaked olive-buff on breast***, plain ***rufescent upperparts***, wings and long tail. Race *temporalis*. Singles or pairs with mixed flocks, forage on horizontal branches, inspecting their undersides, moving sideways. **Voice** Fast but steady and harsh *khee-khee-khee…* lasting *c.*5 sec; call harsh, simple *tjee!, tjee!…* **SS** Bold eyebrow, and streaked head and breast separate from other W foliage-gleaners; see Buff-fronted Foliage-gleaner; Rufous-necked prefers drier habitats, lacks whitish throat, etc. **Status** Uncommon.

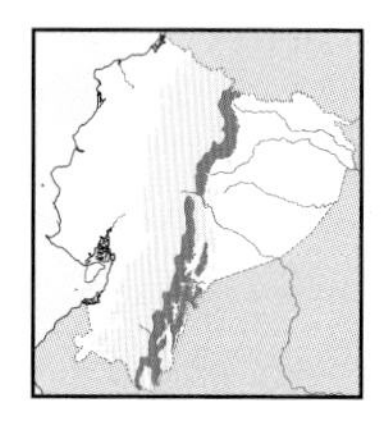

Montane Foliage-gleaner *Anabacerthia striaticollis* 16–17cm

E Andean foothills to subtropics (1,000–1,800m). Montane forest borders, mature woodland. Brown upperparts only ***vaguely*** streaked whitish. ***Whitish eye-ring and long narrow postocular stripe***; white throat, olive-brown below with ***faint pale streaks / spots on breast***, long rufescent tail. Race *montana*. Singles or pairs in mixed flocks, clamber and hop along horizontal branches, sometimes to terminal twigs, frequently inspect dead leaf clumps. **Voice** Squeaky notes in accelerating-decelerating, even-paced series, *tjeet-tjeet-tjeet…*; calls short *keet, keet…*. **SS** Streakier Lineated Foliage-gleaner has buff throat, narrower eyebrow. Other 'eye-browed' foliage-gleaners (Bamboo, Buff-fronted, Rufous-rumped) unmarked below. **Status** Fairly common.

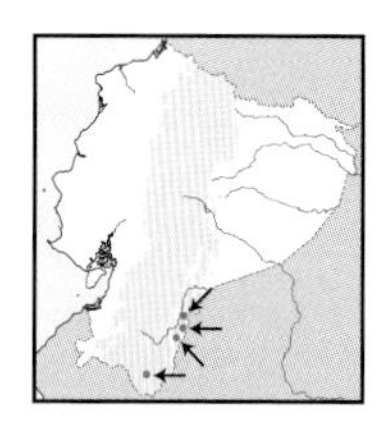

Buff-browed Foliage-gleaner *Syndactyla rufosuperciliata* 18–19cm

Local in extreme SE in Cordillera del Cóndor (1,700–1,900m; locally lower). Montane forest undergrowth. ***Unmarked dark olive-brown above, long narrow buff supercilium***, narrow buff eye-ring. ***White throat***, olive-brown below coarsely flammulated buff. Dark chestnut wings, more rufescent tail. Race *cabanisi*. Mostly in pairs with mixed understorey flocks; furtive and acrobatic. **Voice** Fast, slightly accelerating and ascending series or harsh *khu-khu-khu'ku'ku'ku'ku*; call a harsh and short *tjo-tjok*, spinetail-like in quality. **SS** Confusion unlikely in limited range; Lineated Foliage-gleaner streaked above, narrower and buffier supercilium, buff throat; louder, more nasal voice. **Status** Uncommon, local.

Lineated Foliage-gleaner *Syndactyla subalaris* 18–19cm

Andean foothills to subtropics on E & W slopes (1,000–2,300m). Montane forest, woodland, borders. Mostly brown to olive-brown ***profusely streaked buff***. Vague buff supercilium, ***plain yellowish-buff throat***. Dark chestnut wings, ***more rufescent tail***. Nominate (W), streakier *mentalis* (E). Mostly in pairs with mixed flocks, inspecting clumps, mosses, dead leaves, often upside-down; difficult to see. **Voice** Accelerating, nasal, almost toucanet-like *jan-aan-aan-aan'aan'aan'aan*; call a harsh *kr'rrk*. **SS** Little overlap with less streaked, longer-billed Striped Woodhaunter (longer billed, more rufescent, less profuse streaking); see also more elegant Montane Foliage-gleaner; darker Buff-browed (unstreaked above), and stouter Uniform, Streak-capped and Striped Treehunters. **Status** Fairly common.

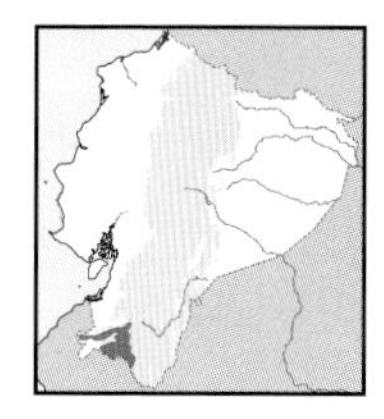

Rufous-necked Foliage-gleaner *Syndactyla ruficollis* 18–19cm

SW foothills to subtropics (1,300–2,300m, locally to 450m). Montane to semi-deciduous forest, woodland and borders. Mainly rufescent-brown, more chestnut wings and tail, pale lores, ***long and broad orange-rufous supercilium extends around dusky cheeks to neck-sides. Cinnamon-buff throat***, faint buff streaks on breast. Monotypic. Singles or pairs with mixed flocks hitch along large branches and limbs, inspect epiphytes, bromeliads, bamboo stems, debris. **Voice** Nasal, accelerating series ending in downwards rattle, *jhan-jhan-jan'jan'j'j'j'j*; harsh, loud, short *tjak, tjak*. **SS** Confusion unlikely in limited range; slightly similar Buff-fronted Foliage-gleaner in more humid areas (plainer overall). **Status** Uncommon, declining, VU globally, EN in Ecuador.

Scaly-throated Foliage-gleaner
Montane Foliage-gleaner
Buff-browed Foliage-gleaner
mentalis
subalaris
Lineated Foliage-gleaner
Rufous-necked Foliage-gleaner

Rather slender foliage-gleaners of humid forests. Some species in canopy, others in midstorey to understorey. Regularly with mixed flocks, acrobatically searching dead leaf clumps and bromeliads. Identification difficult, but note face patterns, wing, rump and tail coloration, and throat to breast pattern. Could be confused with bulkier, heavier-billed and more vocal *Automolus* foliage-gleaners.

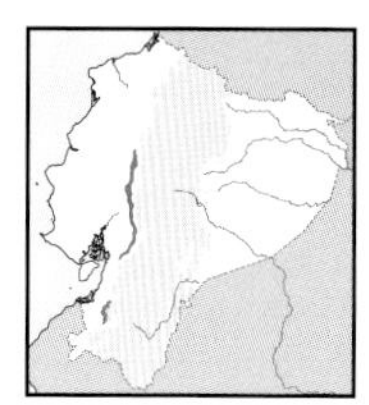

Slaty-winged Foliage-gleaner *Philydor fuscipenne* 17–18cm

Local in W foothills and adjacent lowlands (300–600m). Lower growth of very humid, premontane forest, old second growth. Rufous-brown above, duskier head, ***blackish wings. Cinnamon eye-ring and long eyebrow, dusky ocular line, cinnamon neck-sides. Ochraceous-buff below***, paler throat, orange-rufous tail. Possibly ***undescribed*** race. Mostly in pairs with mixed flocks, vigorously foraging on branches, dead leaf clumps, terminal twigs. **Voice** Harsh, metallic, fast *chef-chef-chef-chéf*; call similar but simpler *chéff!* **SS** Might occur locally with Buff-fronted Foliage-gleaner (rufous in wings, no dusky cheeks, duller rufous underparts, darker tail). **Status** Rare, VU in Ecuador.

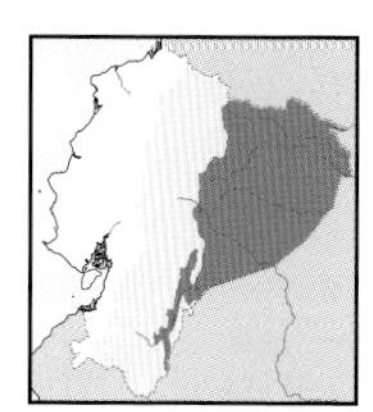

Rufous-rumped Foliage-gleaner *Philydor erythrocercum* 17–18cm

E lowlands to foothills (to 1,300m). *Terra firme* and montane forest. Plain olive-brown above, duskier wings, ***buff eye-ring and eyebrow,*** dusky cheeks. Buffy throat, more olivaceous below, ***no breast streaks. Rump*** and tail rufous-chestnut. Race *subfulvum*. Singles or pairs accompany mixed flocks, agile dead-leaf specialist in mid to lower growth. **Voice** Rising-falling, slightly accelerating series of rusty *tre-tré-trE-trE-trer-trer* notes; harsh shrill *wyeeeeg!* and *djék!-dják!* **SS** Rarer Rufous-tailed Foliage-gleaner (streaky neck sides and breast, duller underparts, darker rump difficult to see; possibly more arboreal, voice longer, more ringing). Bamboo Foliage-gleaner whiter below. **Status** Uncommon.

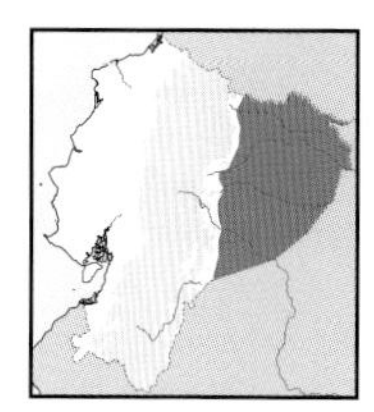

Chestnut-winged Foliage-gleaner *Philydor erythropterum* 18–19cm

E lowlands (to 400m). *Terra firme* forest mid strata. Olive-brown above, ***orange-buff lores and foreface***, paler eyebrow and narrow eye-ring. ***Ochraceous throat***, more yellowish-olive below. ***Rufous-chestnut wings and tail***. Nominate. Mostly singles with mixed flocks, often high on terminal twigs and inspecting dead leaves. **Voice** Short, low, slightly descending trill: *t'rrrrrrrrr* or *g'rriiiieeerr*; also shorter *k'ggg*. **SS** Chestnut-winged Hookbill has whiter throat, more flammulated underparts, narrower eyebrow, streaked back, heavier bill; no other foliage-gleaner has bright rufous-chestnut wings; see Rufous-tailed (dark wings, darker crown, duller face, streaky breast). **Status** Uncommon.

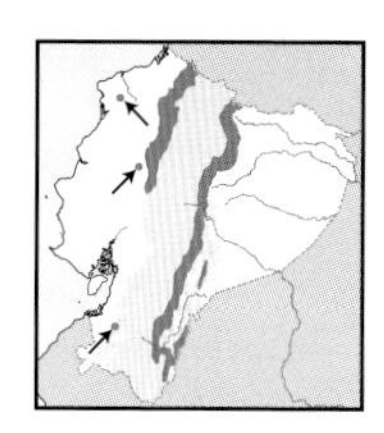

Buff-fronted Foliage-gleaner *Philydor rufum* 18–19cm

E & NW Andean foothills to subtropics (600–1,500m in W, 1,000–1,700m in E). Montane forest, borders. Mostly olive-brown above, duskier crown. ***Cinnamon-buff foreface, eye-ring, supercilium and throat*** (paler in E). Rest of underparts darker, more chestnut wings, tail slightly more rufescent than upperparts. Races *bolivianum* (E), *riveti* (W). Singles or pairs with canopy flocks, hopping along horizontal branches, inspecting terminal foliage, its undersides and dead leaves. **Voice** Fast, descending, but sometimes rising in volume *kí-kí-kikiki-ki-ki-kí-kí-kíí*; descending *tseu*. **SS** No known overlap with similar Slaty-winged and Cinnamon-rumped Foliage-gleaners (both dark-winged); Ruddy is darker, little contrast between upper- and underparts, no obvious supercilium. **Status** Fairly common.

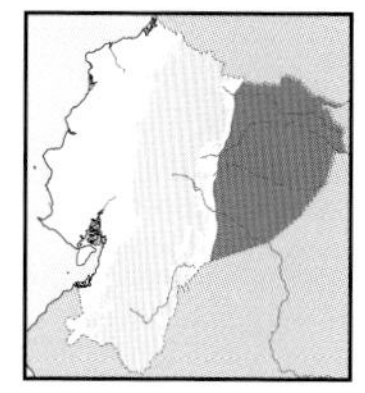

Cinnamon-rumped Foliage-gleaner *Philydor pyrrhodes* 17–18cm

E lowlands (to 400m). *Terra firme* and *várzea* forests. Rufescent-brown above, ***blackish wings, cinnamon-rufous rump and tail. Cinnamon eye-ring, eyebrow and lores, rich cinnamon underparts***. Monotypic. Singles or pairs in dense undergrowth, inspecting dead leaves, palm fronds and other large leaves; rarely follows mixed flocks. **Voice** Short, rising trill, *ku'rrrrrRRH!*, becoming louder and more vibrating; longer and louder than Chestnut-winged Foliage-gleaner; more complex *khu-khu-khu'krrrRRR-rruuu*; sometimes longer, ringing trill; also harsh *chajá* or *chaja-chák*. **SS** Recalls Slaty-winged Foliage-gleaner from W; in E unique combination of dusky wings and rich cinnamon. **Status** Rare.

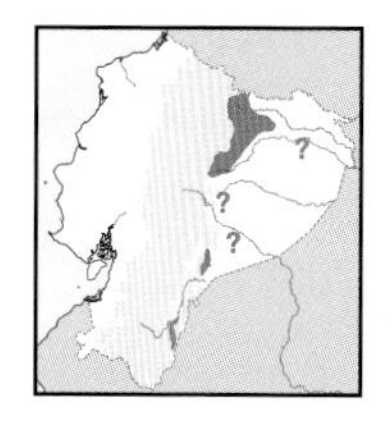

Rufous-tailed Foliage-gleaner *Anabacerthia ruficaudata* 17–18cm

Local in NE lowlands and adjacent foothills (to 1,000m, not below 200m?). *Terra firme* forest. Resembles Rufous-rumped Foliage-gleaner, differences rather subtle. Olive-brown above vaguely streaked on crown, long buff supercilium, dusky streaky cheeks. Yellowish-buff throat, ***flammulated breast***, rufescent tail. Race *subflavescens*. Mostly in upper strata with mixed flocks; inspects dead leaves. **Voice** Loud, staccato, slightly accelerating, ringing *khe-khe-khe'khe'khe'khe*, ending in descending trill. **SS** Similar Rufous-rumped (plainer breast, more yellowish below, rufous rump; rustier voice); Chestnut-winged (obvious rufous-chestnut wings, more orange throat). **Status** Rare.

Slaty-winged Foliage-gleaner
Rufous-rumped Foliage-gleaner
Chestnut-winged Foliage-gleaner
riveti
bolivianum
Buff-fronted Foliage-gleaner
Cinnamon-rumped Foliage-gleaner
Rufous-tailed Foliage-gleaner

Heavier than those on previous plate; two species in E & W. More skulking but also more vocal than *Philydor* foliage-gleaners, seldom seen, and if seen, visual identification not always straightforward. Voice necessary for 100% confident identification. Note face and throat patterns. Dusky-cheeked recalls *Automolus* but voice notably different.

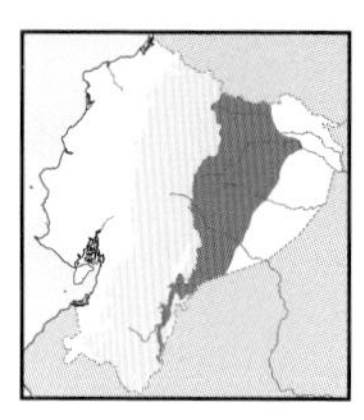

Dusky-cheeked Foliage-gleaner *Anabazenops dorsalis* 18–19cm

E foothills to adjacent lowlands (to 1,300m, not below 250m?). Second growth, borders, bamboo. Rufescent-brown above including wings, rufous uppertail-coverts and long tail. *White eye-ring and bold postocular stripe, dusky cheeks, white throat* grading into greyish-brown underparts. Monotypic. Singles or pairs inspect dead leaf clumps and bamboo nodes, often with mixed flocks. **Voice** Loud, steady, long *tcho'tcho'tcho'tcho'tcho…*, notes more separated at end; recalls pygmy-owl. **SS** Buff-throated Foliage-gleaner (buff eyebrow and orbital ring, no white throat in E, browner underparts); Olive-backed (greyer below, no supercilium). **Status** Rare.

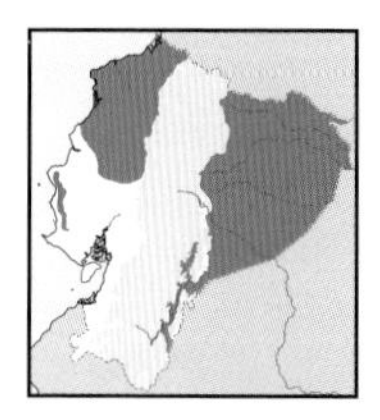

Buff-throated Foliage-gleaner *Automolus ochrolaemus* 18–19cm

E & W lowlands to lower foothills (to 800m). Swampy and riparian forest (E), humid forest, borders (W). Olive-brown above, rufous tail. *Bold buff eye-ring and postocular stripe; whitish-buff throat* (*whiter* in W), ochraceous below *faintly flammulated buff on breast* (more in E). Races *turdinus* (E), *pallidigularis* (W). Singles or pairs accompany mixed understorey flocks; acrobatic, often hangs upside-down; dead-leaf specialist. **Voice** Whiny, short, descending *khÉ-khé-khe-khe..khu-krr*; harsh *kh'ggg*. **SS** Bold orbital ring and postocular stripe separate from most foliage-gleaners; see Olive-backed, Dusky-cheeked, Rufous-rumped and Rufous-tailed. **Status** Fairly common.

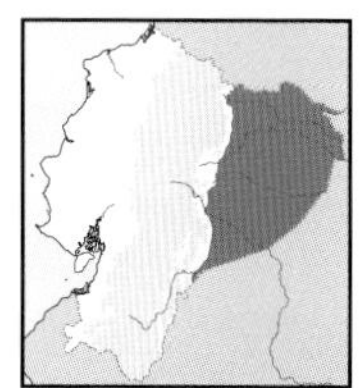

Olive-backed Foliage-gleaner *Automolus infuscatus* 19cm

E lowlands to lower foothills (to 700m). *Terra firme* forest. Plain *brownish-olive* upperparts, rufous uppertail-coverts and tail, *indistinct eye-ring and supercilium. White throat*, greyish-buff below. Race *infuscatus*. Singles or pairs with mixed understorey flocks, in dense tangles, inspects hanging dead leaves. **Voice** Fast, loud, staccato, slightly descending rattle: *ch-r-r-r-r-r-r*, similar to Chestnut-crowned but lower-pitched, less descending; also sharp *thyk-thák* or *pee-tAk*. **SS** Drabber face pattern than sympatric foliage-gleaners with white or pale throats. **Status** Fairly common.

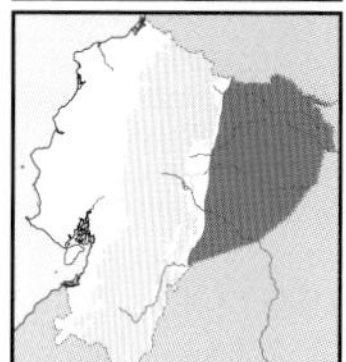

Brown-rumped Foliage-gleaner *Automolus melanopezus* 18–19cm

E lowlands, local in SE (to 400m). *Terra firme* forest, mostly near streams and swamps. *Orange eyes*. Dark brown above, rufescent tail. *Ochraceous-buff throat, brighter malar*, brown underparts, paler belly. Monotypic. Skulking singles or pairs often with mixed understorey flocks, specialising in dead-leaf foraging. **Voice** Unlike other Amazonian *Automolus*: fast, cadenced *whít-whít-whiid-didididi*, ending in *d'rrrrr* rattle. **SS** Chestnut-crowned Foliage-gleaner more rufescent overall, no ochraceous throat, habitat differs; Buff-throated has buff supercilium, eye-ring and throat, breast flammulated; Ruddy is darker overall. **Status** Rare.

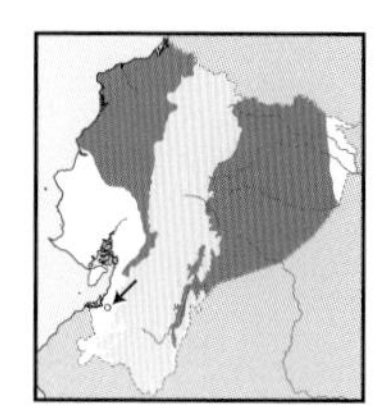

Ruddy Foliage-gleaner *Clibanornis rubiginosus* 18–19cm

E & NW lowlands to foothills (to 1,000m in E, to 1,300m in W). Humid forest undergrowth, borders. *Dark rufescent-brown upperparts* and cheeks, rufous supercilium (E); *rich rufous throat and breast* (buffier in E), *dark tail*. In W *little contrast* between upper- and underparts, only throat is ruddier. Races *brunnescens* (E), *caquetae* (extreme NE), *nigricauda* (W); different species? Rarely with mixed flocks, inspects dense and tangled dead leaves. **Voice** Nasal, querulous, woodcreeper-like *k'kweeeeyh*; little E–W variation. **SS** Darkest and plainest foliage-gleaner; see Brown-rumped (orange eyes, ochraceous throat); Chestnut-crowned (paler, richer rufous above). **Status** Uncommon.

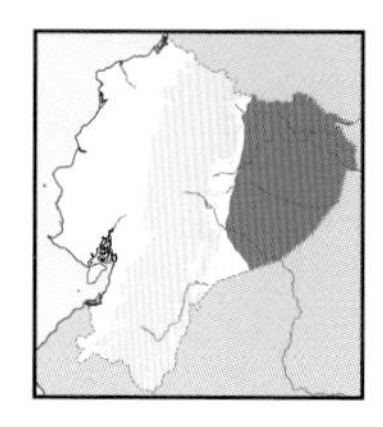

Chestnut-crowned Foliage-gleaner *Automolus rufipileatus* 19cm

NE lowlands (to 500m). *Várzea* and riparian forest undergrowth, swampy forest, river islands. *Pale orange eyes*. Mainly rufous-chestnut, *brighter on crown*, orange-rufous uppertail-coverts and tail. Very faint pale supercilium, buff throat. Race *consobrinus*. Singles or pairs only rarely, if ever, with mixed flocks; very skulking, dead-leaf specialist. **Voice** Short, staccato trill *t'rrrrrr* that descends and ends abruptly; lower-pitched, slower than Olive-backed Foliage-gleaner, lacking first hiccup; nasal, low-pitched *chá!* **SS** Darker Brown-rumped (brighter throat, darker crown) and Ruddy Foliage-gleaner. **Status** Uncommon.

adult
juvenile
Dusky-cheeked Foliage-gleaner
pallidigularis
turdinus
Buff-throated Foliage-gleaner
Olive-backed Foliage-gleaner
Brown-rumped Foliage-gleaner
caquetae
brunnescens
nigricauda
Ruddy Foliage-gleaner
juvenile
adult
Chestnut-crowned Foliage-gleaner

PLATE 151: TREEHUNTERS

Notably difficult to see and identify due to their skulking behaviour, the denseness of their preferred habitats, and subtle differences among some species. More than one species can occur at a single site, often replacing each other altitudinally, but also overlapping to some extent. May or may not join mixed flocks. Voices are better identification clues.

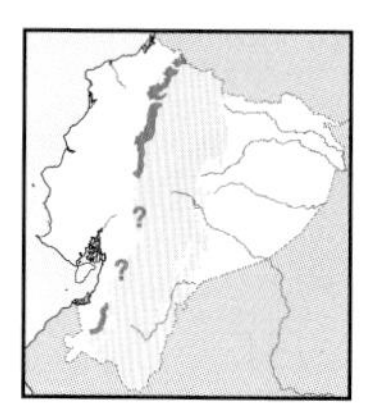

Uniform Treehunter *Thripadectes ignobilis* 19–20cm

W Andean foothills to subtropics (1,000–1,900m). Montane forest undergrowth and borders. Mostly ***uniform*** dark brown, faintly paler lores and ***narrow postocular stripe, few vague buffy streaks on throat and breast***. Monotypic. Singles in dense understorey but less secretive than other treehunters; often joins mixed flocks, moving low, gleaning dead leaves and foliage. **Voice** Short series of ringing *kyeep-kyeep-kyeep-kyeep*, louder, more staccato and slightly faster than Streak-capped; fast, simple *tueep*. **SS** Much less streaked than congeners; see Lineated Foliage-gleaner and Striped Woodhaunter (more slender bill, flammulated breast, paler underparts). **Status** Uncommon.

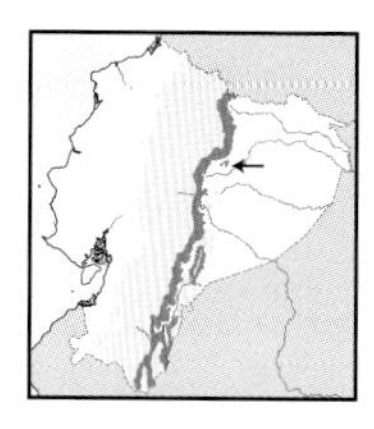

Black-billed Treehunter *Thripadectes melanorhynchus* 20–21cm

E Andean foothills to subtropics (1,000–1,700m). Montane forest undergrowth and borders. ***Blackish crown and browner back streaked buff***, rufescent-brown rump and tail. ***Ochraceous-buff scalloped throat, vague streaking below***. Nominate. Singly, very retiring and difficult to see as skulks in dense undergrowth, regularly in bamboo; not with mixed flocks. **Voice** Slow, slightly descending series of ringing and sharp, often doubled *KEEYP-kéeyp, keeyp-keeyp, keeyp-keeyp...* notes; fast, short *kyrr* or *k'rrt*. **SS** Not streaked below as Striped Treehunter; resembles Streak-capped (but back is unstreaked, throat lacks scalloped pattern); see slimmer Lineated Foliage-gleaner with plain buff throat. **Status** Uncommon.

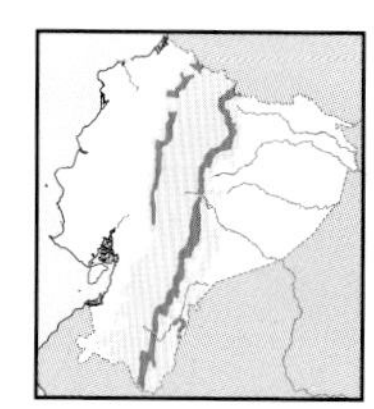

Striped Treehunter *Thripadectes holostictus* 21cm

E & NW Andean subtropics (1,500–2,500m). Montane forest undergrowth, borders. Dark brown upperparts ***heavily streaked buff***, more rufescent wings, rufous-chestnut rump to tail. Brown underparts ***streaked buff***, broad throat streaks nearly forming buff patch, weaker streaks on belly. Races *holostictus* (E), *striatidorsus* (W). Like other treehunters, furtive, stays well concealed in dense undergrowth, where gleans dead leaf clumps and debris. **Voice** Fast staccato, rather explosive trill: *k'rrRRRrrrrr*, higher-pitched than Flammulated, descending at end; short loud rattles *kwi-rik'rik, k'k'k*. **SS** More streaked than other treehunters, above and below. See Black-billed and Streak-capped; Flammulated more strikingly patterned; cf. more slender Lineated Foliage-gleaner (behaviour and habitat differ). **Status** Uncommon.

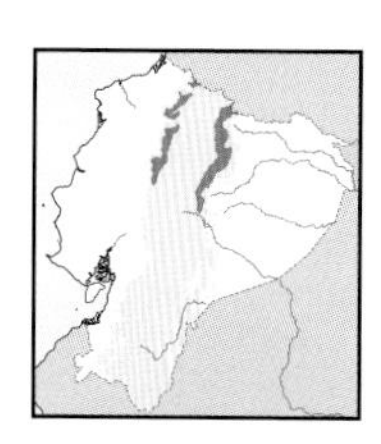

Streak-capped Treehunter *Thripadectes virgaticeps* 21–22cm

N foothills to subtropics (1,300–2,100m). Montane forest and borders. Resembles Black-billed but ***no streaks on back, more rufescent wings, throat to upper chest streaked buff***. Nominate (W), *sumaco* (E). Singles, sometimes pairs, rarely with mixed understorey flocks; inspects dense tangles and dead-leaf clumps; difficult to see. **Voice** Short series of 3–5 ringing, steady *cheep-cheep-cheep-cheep* notes; fast *chi'k*. **SS** In E cf. Striped Treehunter (streakier overall); in W, see darker Uniform Treehunter (lacks streaks above and only faint streaks below); see slimmer-billed Lineated Foliage-gleaner. **Status** Uncommon.

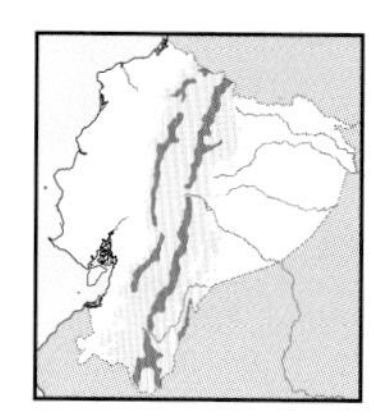

Flammulated Treehunter *Thripadectes flammulatus* 23–24cm

E & NW subtropics to temperate zone (2,200–3,500m). Montane forest undergrowth and borders, especially bamboo. ***Blackish-brown strikingly flammulated buff throughout***, rufous-chestnut wings and tail. Nominate. Mainly singles in very dense understorey; very difficult to see, usually providing only brief glimpses of an energetically moving bird. **Voice** Short, fast, staccato *t't'chechechechechecheche* or *pt'tchchchchchchch*, even in pitch and volume (not descending like Striped), loud *tchek* calls. **SS** Confusion unlikely; lower-elevation Striped Treehunter (little overlap) is streaked, not so boldly flammulated. **Status** Uncommon.

Uniform Treehunter
Black-billed Treehunter
holostictus
striatidorsus
Striped Treehunter
Streak-capped Treehunter
sumaco
virgaticeps
Flammulated Treehunter

PLATE 152: LEAFTOSSERS AND STREAMCREEPER

Dark terrestrial species of humid forest interior, difficult to find unless vocalising. Tails short, bills characteristically long and slender, used for teasing leaves apart and probing leaf litter. Identification complicated if not well seen (or if voices unknown); note throat patterns. Streamcreeper is similar in structure but its tail has spiky points. It dwells around fast-flowing forested streams.

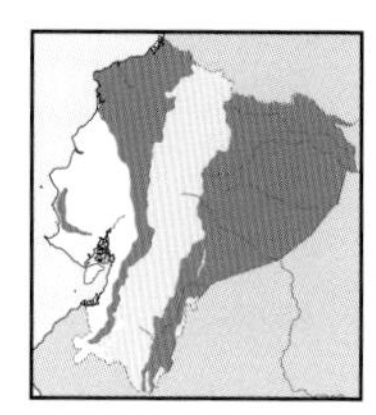

Tawny-throated Leaftosser *Sclerurus mexicanus* 16–17cm

E & W lowlands to subtropics (mostly below 1,500m, locally to 2,000m). Humid and montane forest interior. Long, slightly curved bill, narrow pale eye-ring. Mostly brown, ***rich tawny throat to upper breast***, rufescent rump, blackish tail. Races *obscurior* (W), *peruvianus* (E). Inconspicuous, easily missed as forages alone on ground, flicking leaves with bill. **Voice** Descending, slow, piping series, *péé-péé-pee-pee-peur* often ending in bouncing *pt't't't't*; slightly slower in W; sharp *phyeé* calls. **SS** Confusion unlikely in W (Scaly-throated has whitish throat, voice faster and longer); in E, see Short-tailed (bill noticeably shorter, less tawny throat; voice much faster). **Status** Uncommon.

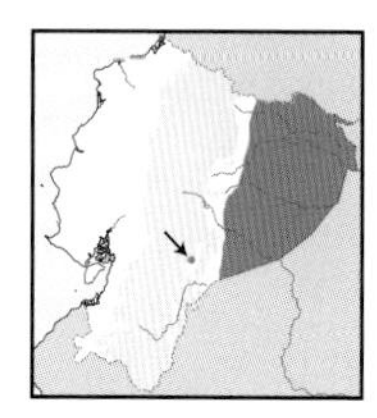

Short-billed Leaftosser *Sclerurus rufigularis* 16cm

E lowlands, locally to foothills (mostly below 300m, locally to 700m and 1,400m). *Terra firme* forest interior. ***Short***, slightly curved bill. Resembles Tawny-throated Leaftosser but duller, ***faint pale eyebrow***, buffier throat, more ***whitish chin***. Race *brunnescens*. Singles on ground, flicking leaf litter and probing leaf or moss clumps, not with mixed flocks. **Voice** High-pitched, chattering *téé-teé-tee-tu-tee, tEE-tEe-tu-tu-te*..., accelerating and ascending first, then decelerating; also sharp *tsuíp!* **SS** Whiter chin, less tawny throat, shorter bill separate from similar Tawny-throated; voice faster than other leaftossers; cf. especially Black-tailed. **Status** Rare.

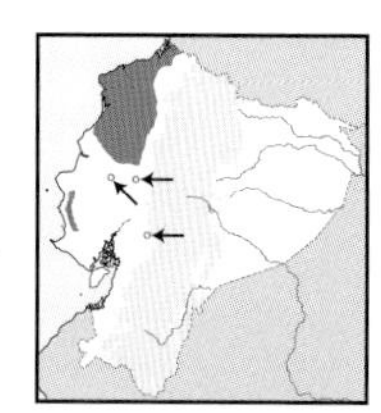

Scaly-throated Leaftosser *Sclerurus guatemalensis* 17–18cm

NW lowlands to foothills (to 800m). Humid forest interior. ***Long straight bill***. Mostly dull brown, paler face, ***white throat grading into whitish scaling on chest***, blackish tail. Race *salvini*. Like other leaftossers, hops on ground tossing leaves and debris aside, also inconspicuous but not very wary. **Voice** Fast series of two parts; first of falling, piping notes *peée-peeé-pee-peu*, followed by accelerating, rising, inflected *pee'pee'pee...peepepepe*; also accelerating *peé-peé-peé-peé-peé* and sharp *pueét* calls. **SS** Only sympatric with Tawny-throated, which lacks whitish scaling on throat, bill decurved at tip, rump more rufescent. **Status** Uncommon, VU in Ecuador.

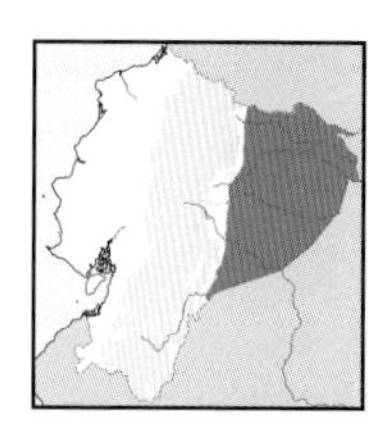

Black-tailed Leaftosser *Sclerurus caudacutus* 18–19cm

E lowlands (to 500m, locally higher). *Terra firme* forest interior. ***Long straight bill***. Mostly blackish-brown, ***faint pale supercilium, scaly whitish throat***, more rufescent rump, blackish tail. Race *brunneus*. Terrestrial and inconspicuous, not with mixed flocks, noisy when tossing leaves aside. **Voice** Loud, ringing, piping whistles, *pueé-pueé-pueé-pee...pee-pee-puu*, descending, slowing at end; also sharp *stýew!* **SS** Recalls western Scaly-throated Leaftosser (no overlap). Voice louder, more ringing than other Amazonian leaftossers; higher-elevation Grey-throated has larger whitish throat patch, more chestnut breast. **Status** Uncommon.

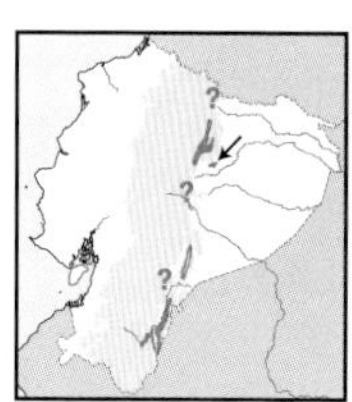

Grey-throated Leaftosser *Sclerurus albigularis* 17–18cm

Local in E Andean foothills and subtropics (1,000–1,700m, locally lower). Montane forest interior. ***Long straight bill***. Mostly dull brown, faint pale lores and supercilium, ***greyish-white throat, chestnut breast***, tawnier rump, blackish tail. Race *zamorae*. Behaves much like other leaftossers, hopping and gleaning on ground and rotten logs. **Voice** Short series of querulous *kwit-kwit-kwit-kwit?* notes, rising in pitch, sometimes ending in trill; also accelerating *kwi-kwi-kwi*... recalling Short-billed; calls a sharp, short, squeaky *kwít*. **SS** Occurs at higher elevations than most leaftossers. Tawny-throated has uniform tawny throat to chest, bill not straight; see also Short-billed. **Status** Rare, local, possibly overlooked; NT globally.

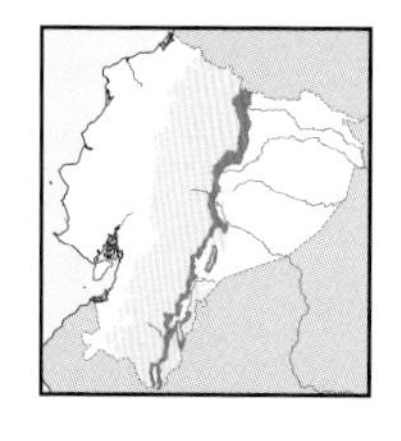

Sharp-tailed Streamcreeper *Lochmias nematura* 15cm

Local in E Andean foothills to lower subtropics (700–1,400m). Swift-flowing rocky streams and creeks in montane forest. Long, slightly decurved bill. Mainly dark brown, ***bold white spots over underparts***. Short, blackish, spiny tail. Race *sororius*. Inconspicuous, on or near ground probing debris, flicking leaf litter aside with bill in damp places, or on rocks and boulders. **Voice** Fast, dry, rising-falling *tit-tít-tititit't't't*, sometimes trilled at end; short, buzzy *see, see-k'r'r'r* trills. **SS** Confusion unlikely despite its resemblance to leaftossers. **Status** Uncommon.

peruvianus
obscurior
Tawny-throated Leaftosser
Short-billed Leaftosser
Scaly-throated Leaftosser
Black-tailed Leaftosser
Grey-throated Leaftosser
Sharp-tailed Streamcreeper

PLATE 153: GROUND-DWELLING FURNARIIDS

Terrestrial or primarily terrestrial furnariids, Miner and cinclodes occur in highlands, all horneros in lowlands. Cinclodes are conspicuous *páramo* birds, differing in overall size and bill shape. The miner – recently discovered and an endemic subspecies – occurs in dry sandy *páramo*. Horneros are particularly conspicuous in W lowlands, in E all horneros are rare or local. They walk sluggishly, bobbing heads and depressing their tails.

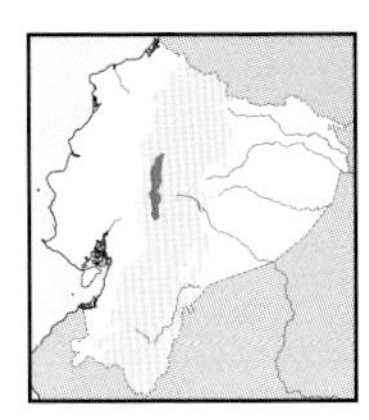

Slender-billed Miner *Geositta tenuirostris* 18cm

C Andes (3,350–4,000m). Open dry *páramo* and adjacent agricultural fields. ***Long, decurved, pale-based bill***. Greyish-buff upperparts, streaked crown, ***long whitish supercilium***; ***white throat, dusky-streaked breast***, greyish-buff below. Dusky wings, contrasting ***chestnut flight feathers***, chestnut outer tail feathers. Race *kalimayae*, possibly separate species. Singles to small groups entirely terrestrial, gleaning and probing; capable of strong and sudden flight; performs flight display. **Voice** Melodious, fast series of *kip-kip-kip-kip...kyie!...kikiki*. **SS** Unmistakable by bill shape and length; see both cinclodes. **Status** Rare, local.

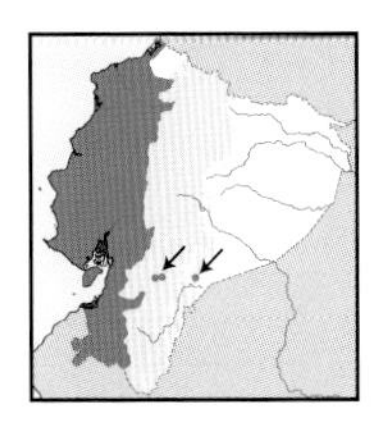

Pale-legged Hornero *Furnarius leucopus* 19–20cm

W lowlands to subtropics, locally to inter-Andean valleys; also Cordillera de Kutukú in SE (to 2,500m, locally higher in W; 520m in SE). Open areas, arid scrub (W), small-river edges (SE). ***Long, slightly curved, pale bill, whitish eyes*** (W). Greyish crown, ***long whitish eyebrow, orange-rufous upperparts***, whitish underparts. ***Pink legs***. SE birds ***darker*** rufescent above, ***brown eyes, dull buff breast*** to belly. Race *cinnamomeus* (W) probably a species; *tricolor* (SE). Conspicuous and familiar in W, terrestrial in variety of habitats, builds large, mud, oven nests. More shy in SE. **Voice** Explosive, strident, decelerating *klEklE-klé-klé..kle...kle* (W), faster, more ringing in SE; clear *kyéék* calls. **SS** Unmistakable. **Status** Common in W, rare in SE.

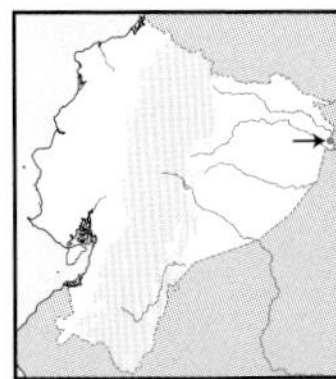

* Pale-billed Hornero *Furnarius torridus* 18–19cm

One record in extreme E lowlands at 200m. River islands. ***Long, decurved pink bill, pale legs***. Mainly ***dark chestnut above***, darker on crown, ***long whitish supercilium***, white throat, ***paler chestnut below***. Monotypic. Apparently only accidental in Ecuador; elsewhere, terrestrial, mainly in *várzea* and riparian forest borders. **Voice** Loud, decelerating series resembles Pale-legged Hornero, but rather shorter, higher-pitched, harsher. **SS** Commoner Lesser Hornero is smaller and paler, with short dark bill, blackish legs. **Status** Rare, possibly vagrant.

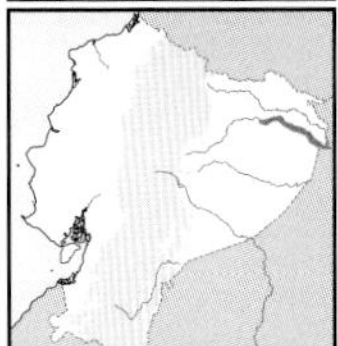

Lesser Hornero *Furnarius minor* 15cm

E lowlands (below 300m). Light woodland and regenerating scrub on Napo river islands. ***Short, slender, dark bill, blackish legs***. Greyish crown, ***long buffy supercilium, pale rufous above, pale buff below***, white throat. Monotypic. Rather skulking in dense undergrowth, may emerge at island edges if vegetated, gleans from ground and muddy riverbanks. **Voice** Accelerating, descending series, *krép-krép-krep..krep, kikikikiki*, similar to other horneros but drier, often with sputtering first part. **SS** Smaller than very locally sympatric Pale-billed Hornero; voices similar. **Status** Rare, locally uncommon.

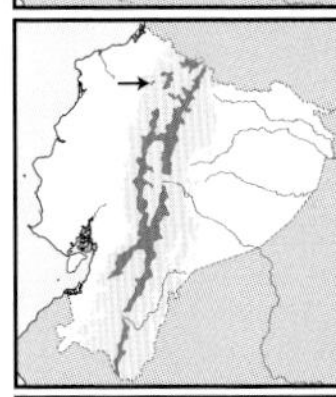

Chestnut-winged Cinclodes *Cinclodes albidiventris* 17–18cm

Andes (3,200–4,300m). Open and shrubby *páramo*, adjacent agriculture. ***Short, straight black bill***. Resembles Stout-billed Cinclodes but ***no markings*** on ***darker*** breast, more rufescent above. Race *albidiventris* (most of range), possibly *paramo* (extreme N). Tame pairs mostly terrestrial and fond of roadsides, stream edges, fields; runs and hops, probes in soft ground, mud, vegetation. **Voice** Short, rapid trills, rather high-pitched *t-ddddddd'*; call a short, sharp *pif*, sometimes short trills. **SS** See larger, heavier-billed Stout-billed; sympatric canasteros are all streaky. **Status** Fairly common.

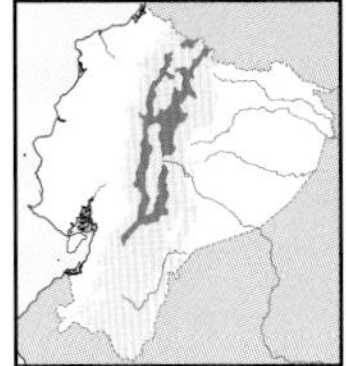

Stout-billed Cinclodes *Cinclodes excelsior* 20–21cm

Andes S to Azuay (3,300–4,500m). Shrubby and grassy *páramo*, *Polylepis* woodland, adjacent agriculture. ***Stout black bill***. Brown upperparts, long whitish eyebrow, whitish throat, buff-brown below, ***pale spots on breast to upper belly***. Duskier wings with rufous-chestnut in secondaries (mostly visible in flight). Nominate. Terrestrial pairs or singles sometimes perch atop shrubs, rather tame. **Voice** Loud rapid, rising trill, starting with a few slower notes, *tr-r-rrrrrr-iít*, louder than Chestnut-winged Cinclodes; also short, soft trills and low, nasal *duut, khut* calls. **SS** Chestnut-winged is smaller, with bill noticeably shorter and straighter; somewhat similar voice of White-chinned Thistletail is longer, more accelerating. **Status** Fairly common.

juvenile
adult
Slender-billed Miner
cinnamomeus
tricolor
Pale-legged Hornero
Pale-billed Hornero
Lesser Hornero
Chestnut-winged Cinclodes
Stout-billed Cinclodes

PLATE 154: SPINETAILS I

Relative drab spinetails that inhabit varied habitats and altitudinal zones. Spinetails occur in dense vegetation. Azara's, Dark-breasted and Slaty prefer regenerating scrub and shrubby clearings; Plain-crowned occurs in similar habitats but on river islands, whereas Dusky and Marañón occupy forest and woodland edges. Voices are key identification features; note also tail colour and throat pattern.

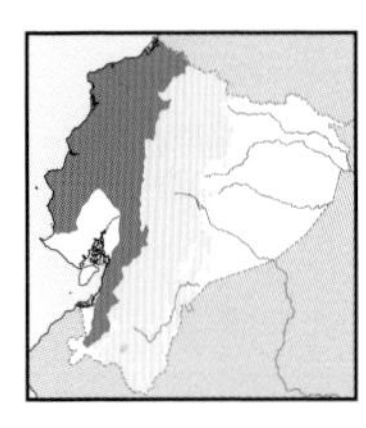

Slaty Spinetail *Synallaxis brachyura* 15–16cm

W lowlands to foothills (to 1,400–1,500m, locally higher). Shrubby clearings, scrub, woodland edges, hedgerows. Dusky-brown front, ***dark grey face and most underparts***, whitish scales on chin, rufous crown, rufous wing-coverts, ***dark brown tail***. Races *chapmani* (N), *griseonucha* (S). Less skulking than most *Synallaxis*, singles or pairs may emerge on exposed branches but rapidly seek cover. **Voice** Throaty, descending *chi'jrrrrr* or *tjit-tjit-ch'rrrrr* resembling scolds of wren; loud *chík* foraging calls. **SS** No overlap with similar Dusky Spinetail; higher-elevation Azara's paler overall, tail contrastingly rufous, squeakier voice, not rattled. **Status** Common.

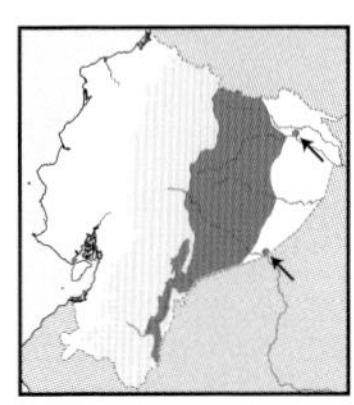

Dusky Spinetail *Synallaxis moesta* 15–16cm

E foothills and adjacent lowlands (200–1,350m). Dense undergrowth in primary and secondary forest, borders. ***Dusky-grey front, face and most underparts***, whitish scaled chin, ***rufous crown, wing-coverts and rather long tail***, rufous flight feathers. Race *brunneicaudalis*. Mostly in pairs, near the ground well hidden in dense tangles, sometimes in bamboo. **Voice** Low-pitched, nasal *rha-a-a-a-a-a-á...* **SS** Closely recalls Slaty Spinetail of W Andes. Dark-breasted has extensive white throat, less rufous in wings and tail, habitat differs; no known overlap with Azara's (paler overall, whitish throat, longer and richer rufous tail). **Status** Uncommon; NT globally.

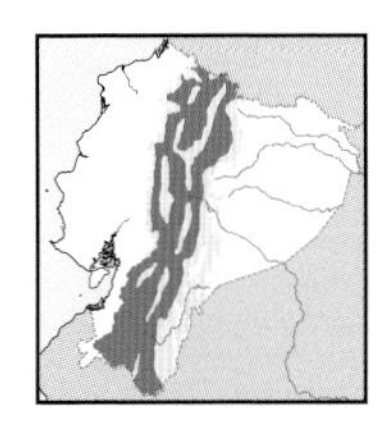

Azara's Spinetail *Synallaxis azarae* 17–18cm

Andes (1,500–3,200m, to 900m in SE). Forest and wooded borders, shrubby clearings, regenerating scrub. Greyish front, ***rufous crown***, faint pale supercilium, white sub-malar, ***scaly blackish throat, rufous wing-coverts, long rufous tail***. Race *media* (most of range), paler *ochracea* (SW). Singles to trios skulk in dense undergrowth, rarely with mixed flocks; rarely in open. **Voice** Loud, squeaky *kïp-kuiíp* or *quee-qweeék*; also accelerating *kukukukuku...*, short *krrt* calls while foraging; higher-pitched in SW. **SS** Normally at higher elevations than most spinetails; cf. locally Dark-breasted in E (similar face pattern but dark tail, darker breast); in W see much darker Slaty (no rufous tail). **Status** Common.

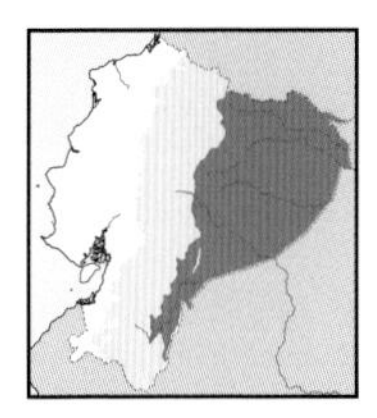

Dark-breasted Spinetail *Synallaxis albigularis* 15–16cm

E lowlands to lower subtropics (to 1,500m, locally higher). Dense regenerating scrub, shrubby clearings, woodland edges, undergrowth on river islands. Dusky front, ***grey face and most underparts***, rufous crown, ***white throat with black median patch***, rufous wing-coverts, ***dull brown, rather short tail***. Races *rodolphei* (N), nominate (S). Skulking but noisy, tends to prefer overgrown areas, thus probably increasing with deforestation. **Voice** Fast *téet-ti'di'di'di* or *téet-ti'di'di'di'drrrt*, louder, nasal *whé-whé-whé...* calls. **SS** More forest-based, darker Dusky Spinetail (no whitish throat, rufous tail, more rufous in wings); no overlap with Marañón Spinetail (dusky crown, rufous tail); see Azara's. **Status** Fairly common.

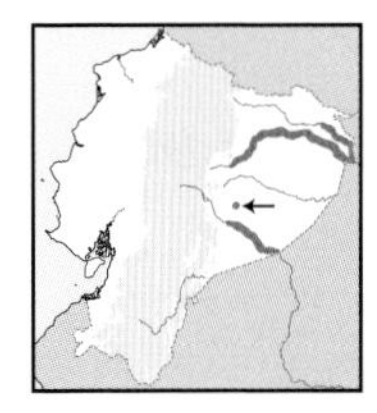

White-bellied Spinetail *Synallaxis propinqua* 16cm

E lowlands (below 400m). Early-successional scrub on river islands. ***Greyish-brown above***, paler below, ***whiter chin and belly, scaly blackish throat patch***; rufous wing-coverts, ***long orange-rufous tail***. Monotypic. Pairs in very dense tangles of *Gynerium* cane and regenerating shrub, very skulking but vocal. **Voice** Low-pitched, nasal, crake-like *k-rr'rr'rr'rr'rr...*, *khree-kják...kraah-kraah*, *krekrekre'krrrrrr...*, simpler, scratching *gk'gk'gk-gkgk* or *gk-gk*. **SS** Sympatric Plain-crowned Spinetail (darker, greyish-buff underparts, darker belly, no blackish throat, more rufous in wings; double-noted voice); see Dark-breasted (darker underparts, rufous crown). **Status** Uncommon.

Plain-crowned Spinetail *Synallaxis gujanensis* 16–17cm

E lowlands (below 400m). Dense regenerating scrub and riparian woodland undergrowth, especially *Gynerium* cane and *Cecropia* woodland. ***Dusky-brown upperparts***, paler and greyer underparts, ***white throat***, whitish lores, ***rufous wings, short, darker rufous tail***. Race *huallagae*. Skulking pairs furtively glean in tangles; very vocal. **Voice** Mellow, long repeated *kEE, küh*. **SS** Sympatric White-bellied Spinetail has dark patch on throat, longer and richer rufous tail, no pale lores, white belly; crake-like chattering voice; Dark-breasted (darker below, crown rufous). **Status** Fairly common.

griseonucha
adult
chapmani
adult
juvenile
Slaty Spinetail
Dusky Spinetail
rodolphei
juvenile
albigularis
media
adult
Dark-breasted Spinetail
adult
ochracea
Azara's Spinetail
juvenile
adult
Plain-crowned Spinetail
White-bellied Spinetail

Spinetails of varied habitats, mostly in lowlands. Four species confined to forest interior: Blackish-headed in deciduous and semi-deciduous forest, the rest in humid or semi-humid forests. Rufous is the most montane species on this plate. Blackish-headed and Necklaced are only spinetails in dry forests. Necklaced perhaps better classified in separate genus. Spinetails rarely join mixed flocks.

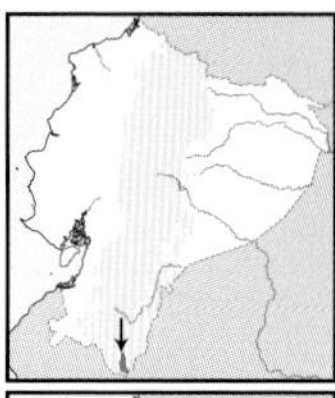

Marañón Spinetail *Synallaxis maranonica* 15–16cm

Extreme SE in Zumba region (650–1,200m). Understorey of deciduous to semi-deciduous forest and woodland. ***Dull brown above including crown***, greyish face and below, whitish chin, ***rufous-chestnut wings and rounded tail***. Monotypic. Skulking pairs can briefly join mixed flocks, remaining in dense cover; might briefly visit open. **Voice** Slow, nasal *keeeeuuw, keeeuw, keeéu*, also in duets *keéu...kiiu*; mellow *keeup* calls. **SS** No known overlap with other spinetails in limited range; Dark-breasted and Dusky occur in foothills to N (both have rufous crowns). **Status** Uncommon; CR globally.

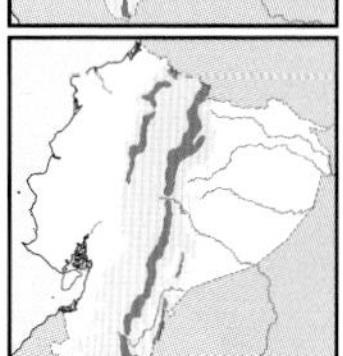

Rufous Spinetail *Synallaxis unirufa* 17–18cm

Andean subtropics to temperate zone in NW & E (2,000–3,200m). Undergrowth of montane forest, woodland, bamboo, borders. ***Entirely rich rufous*** with inconspicuous black mask, slightly paler throat, long tail. Juv. darker. Nominate. Pairs briefly join mixed flocks, in dense cover, gleaning foliage, small branches, bamboo. **Voice** Single loud, squeaky *kueeíh!*, often doubled, quite querulous *kui-kéeh?* resembling Azara's Spinetail but higher-pitched, more querulous. **SS** No other spinetail as rich uniform rufous; White-browed has shorter tail, white in face; see similar Rufous Wren (behaviour and voice totally differ). **Status** Uncommon.

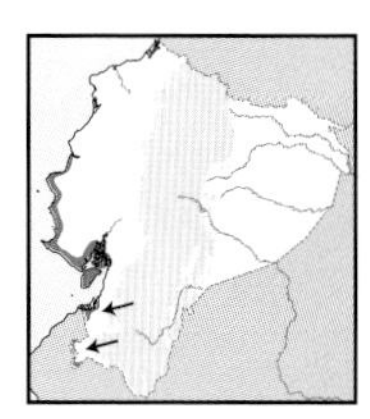

Necklaced Spinetail *Synallaxis stictothorax* 12–13cm

SW lowlands, possibly extreme SE in Zumba region (to 200m). Deciduous to semi-deciduous forest, woodland, borders, arid scrub. ***Long whitish supercilium***, blackish-streaked white forehead; ***white throat, fine necklace of brown streaks on breast***, whitish below. Greyish-brown above, ***long, orange-rufous tail***. Nominate (Guayas / Santa Elena / Manabí), *maculata* (El Oro / Loja). Mostly pairs, more arboreal and less skulking than other *Synallaxis*; explores branches, twigs, clumps in open understorey and borders. **Voice** Loud, squeaky, slowing series of sputtering *chchchch-chí, chchchch'chi-chi-chi-chi-chí, chchch-chí-chí-chí-chchchch*; also *chí, chí, chí* calls, sometimes ending in *chchchc, ch'rrrr* trills. **SS** Confusion unlikely. **Status** Fairly common.

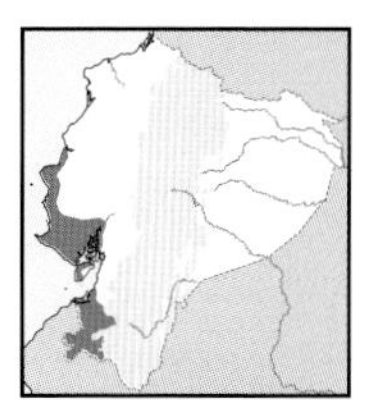

Blackish-headed Spinetail *Synallaxis tithys* 14–15cm

SW lowlands, to foothills in Loja (to 1,100m). Dense undergrowth of deciduous to semi-deciduous forest and woodland. ***Slaty-black head*** grading into greyish-brown upperparts and ***ash-grey underparts, cinnamon-chestnut wing-coverts***, dark brown short tail. Monotypic. Pairs briefly join mixed flocks, prefer dense, often spiny tangles where nests; sometimes on ground. **Voice** Short, fast, ascending trill *t-t-t-t-trrt!*, also loud *tee-dií, tee-dií...* or *dé-dé-dé*. **SS** Confusion unlikely in small range shared only with different Necklaced Spinetail (behaviour also differs, different voice is longer, squeakier, not trilled). **Status** Rare, EN.

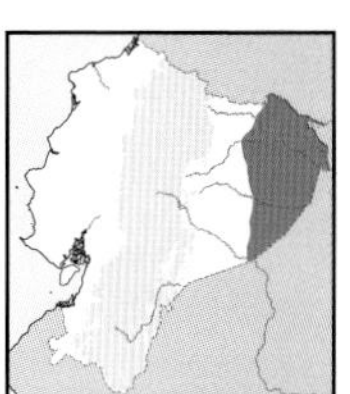

Ruddy Spinetail *Synallaxis rutilans* 14–15cm

E lowlands (below 300m). Dense undergrowth in *terra firme* forest. Mainly ***rufous-chestnut***, darker rear parts, ***black lores and throat patch***, short brown tail. Race *caquetensis*. Pairs low above ground, inspect dense tangles and dead leaves, sometimes briefly with mixed flocks. **Voice** Loud, squeaky *ket-ków?* or *keét-khew?*, long repeated; single *kík* calls similar in quality. **SS** Though no known overlap, cf. very similar Chestnut-throated Spinetail (darker above and rear parts, lacks black throat; prefers bamboo, voice has rattled first note). **Status** Rare.

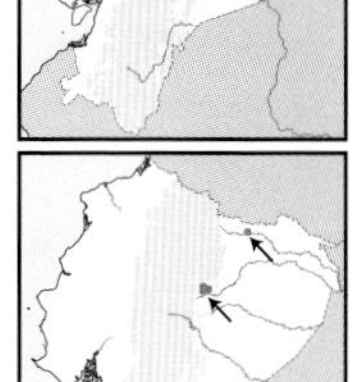

Chestnut-throated Spinetail *Synallaxis cherriei* 14cm

Local in NE foothills (300–900m). Dense undergrowth in secondary forest and borders. Mostly ***rufous-chestnut***, darker on crown and ***rear parts, richer throat to mid belly, short blackish tail***. Race *napoensis*. Poorly known and seldom seen; prefers dense second growth, apparently at low density; fond of bamboo. **Voice** Long repeated series of squeaky *trrrt-tuuí?* or *trrrt-tuuí-tuuí-tuuí?* **SS** Very similar Ruddy Spinetail has black throat; habitat and voice differ. Rufous Spinetail at higher elevations. **Status** Very rare, NT globally, VU in Ecuador.

Marañón Spinetail
juvenile
adult
Rufous Spinetail
stictothorax
maculata
Necklaced Spinetail
Blackish-headed Spinetail
Ruddy Spinetail
Chestnut-throated Spinetail

PLATE 156: SPINETAILS III

More arboreal spinetails, also more prone to join mixed-species flocks. Curiously, the three Andean species are more arboreal than Parker's and Speckled in the Amazon, which dwell in midstorey to lower growth. The small and shorter-tailed White-browed of Andean forests was formerly included in the genus *Synallaxis*.

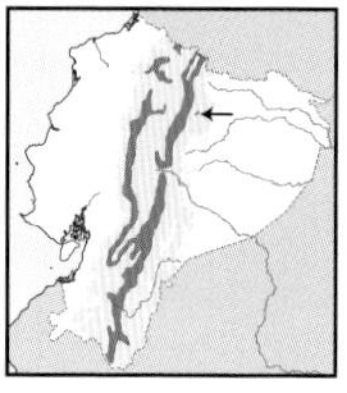

White-browed Spinetail *Hellmayrea gularis* 13–14cm

Highlands on E & W Andean slopes (2,500–3,700m). Montane forest and woodland undergrowth, borders. Mostly ***bright rufous*** with conspicuous ***white lores, supercilium and throat***, short spiny tail. Juv. mottled below. Nominate. Pairs in dense undergrowth probe and search moss clumps, dead leaves, crevices, bamboo; not very wary. **Voice** High-pitched, accelerating, louder first note, ending in trill: *chiip-chip-chip-chip-chichichichirr*; single *chep!* **SS** Confusion unlikely if well seen; longer-tailed Rufous Spinetail lacks white eyebrow, Line-cheeked has longer tail, no rufous underparts, richer rufous wings and tail, no white tail. **Status** Uncommon.

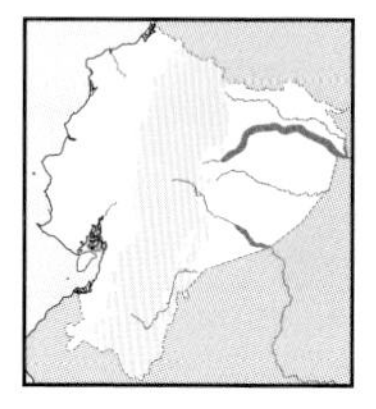

Parker's Spinetail *Cranioleuca vulpecula* 14cm

E lowlands (to 400m). Riparian woodland and regenerating scrub on river islands, especially *Cecropia* woodland. ***Uniform rufous above***, richer on crown, wings and tail, ***narrow whitish supercilium***, freckled face. ***Whitish throat, faint pectoral streaks***, buffier flanks. Monotypic. Less arboreal than other *Cranioleuca*, tends to remain in dense lower growth, especially *Gynerium* cane. **Voice** Accelerating, descending *tdéw-tdéw-tdéw-ch'rrrrr*; also squeaky, nasal *tew-tu, tew-tu, tew-tu* ending in fast *t'rrrrr*; fast, doubled *chú-chút* calls. **SS** Confusion unlikely in habitat, other spinetails (White-bellied and Plain-crowned) on river islands not rich rufous above, lack whitish supercilium, darker below. **Status** Uncommon.

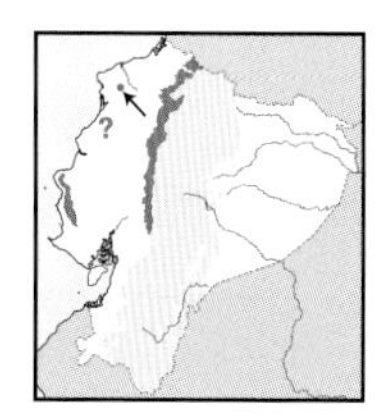

Red-faced Spinetail *Cranioleuca erythrops* 14cm

W foothills to subtropics, S to Cañar, cloud forest in coastal ranges (700–2,000m; 100–500m in coastal ranges). Montane forest and wooded borders. ***Rich rufous head, wings and outer tail feathers***. Otherwise dull olive-brown. Juv. has greyish-brown crown, ***narrow buff supercilium, more ochraceous below***. Nominate. Mostly in pairs nimbly searching, probing, hanging upside-down in moss or epiphyte tangles. **Voice** High-pitched, fast series of *pit-pit-pit-pit-pit* ending in short shrill, also more warbled *psit-psit-psit…chitik-chitik…tititititititi*; louder *chitik-tik-tik…* calls. **SS** Juv. with Line-cheeked Spinetail, especially juv., which also lacks rufous crown. Line-cheeked has paler, ***more contrasting*** supercilium. **Status** Fairly common.

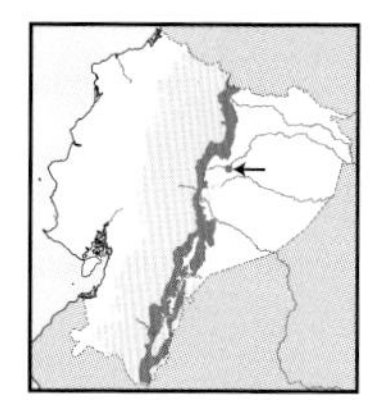

Ash-browed Spinetail *Cranioleuca curtata* 14–15cm

E Andean foothills to subtropics (900–1,700m). Canopy and borders of montane forest and woodland. ***Rufous crown, grey supercilium***, olive-brown above, ***rufous wings and long tail***. Pale throat, greyish-buff below. Juv. more ochraceous, with ***tawny eyebrow, dusky crown***. Race *cisandina*. As previous species, forages in tangles and moss, regularly with mixed flocks. **Voice** High-pitched, short, accelerating *chi-chi-chi-chi…* ending in short shrill (similar to Red-faced); also faster, more warbled *chiki-tiki, chiki-chiki…*. **SS** Confusion unlikely (no overlap with Line-cheeked Spinetail); *Synallaxis* spinetails not arboreal. Drab juv. recalls a foliage-gleaner but smaller, slimmer. **Status** Uncommon; VU globally.

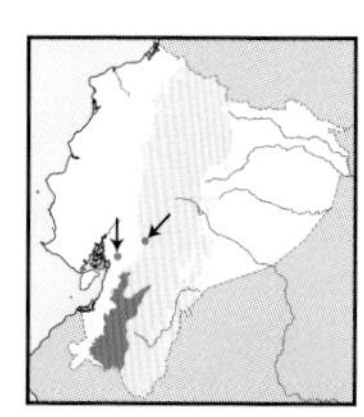

Line-cheeked Spinetail *Cranioleuca antisiensis* 14–15cm

SW subtropics to temperate zone (900–2,500m, locally higher and lower). Montane forest and wooded borders, montane scrub, adjacent clearings. ***Rich rufous crown, long white supercilium***, olive-brown above, rufous wings and long tail. Whitish throat, greyish-buff below. Nominate. Behaves like Red-faced Spinetail, also tame, in shrubby gardens and hedgerows, nimbly probing, not gleaning foliage. **Voice** High-pitched, fast, squeaky series ending in shrill *chí-chí-chí-chí-ch'rrrr*, generally shorter, lower-pitched and squeakier than Red-faced; also louder *chi'krr, chi'krr* calls. **SS** Whiter and bolder supercilium than other montane *Cranioleuca*. **Status** Fairly common.

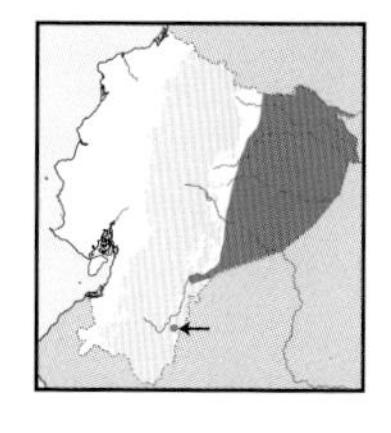

Speckled Spinetail *Cranioleuca gutturata* 14–15cm

E lowlands (to 400m, locally to 600m). Undergrowth of *terra firme* forest near natural clearings and streams. ***Orange-rufous crown, streaky face***, long buff supercilium, greyish-brown above, rufous wings and tail. Yellow chin, ***buffy-white underparts coarsely speckled brown***. Juv. duller, more ***mottled*** below, ***dusky crown***. Monotypic. Pairs often join mixed understorey flocks, hopping on branches, large twigs and vines; dead-leaf specialist. **Voice** Short, very high-pitched, slow, hummingbird-like *tsee-tsít-tsít-tsít…*; a quavering, descending trill *tch-t-t-t-t-t-t* reported (HBW). **SS** Confusion unlikely due to heavily speckled underparts plus rufous crown, wings and tail. **Status** Rare.

White-browed Spinetail

Parker's Spinetail

Red-faced Spinetail

Ash-browed Spinetail

Line-cheeked Spinetail

Speckled Spinetail

PLATE 157: CANASTEROS, THISTLETAILS AND OTHER FURNARIIDS

Distinctive furnariids of various habitats and elevations. Tit-spinetail and thistletails are highland species of shrubs, woodland and the treeline. All three species are acrobatic but skulking. Thistletails replace each other geographically. Canasteros also occur in highlands but are primarily terrestrial, though singing birds climb stumps or bushes. Also, the drab softtail of Amazonian forest gaps and thornbird of open areas in the extreme SE.

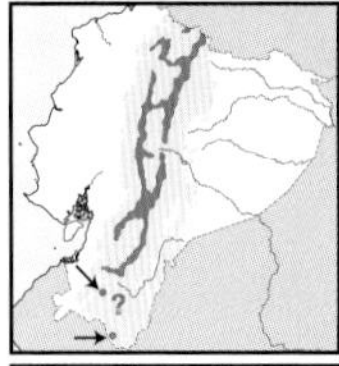

Andean Tit-Spinetail *Leptasthenura andicola* 16–17cm

Andes S to Azuay, also locally in extreme S (3,200–4,500m). Shrubby *páramo*, *Polylepis* woodland, treeline. Handsomely ***streaked rufous on crown, whitish on upper- and underparts; long white supercilium, very long and spiky***, graduated tail. Race *andicola*, *undescribed* in SE? Pairs to small groups very active in dense undergrowth, grass, tangles; creeps agilely through foliage. **Voice** Fast, accelerating, descending series of tinkling *tit-tit-ti-titiriu-titiriu...*; also short fast trills *trrrr...*, *tzik, tzi-zik*. **SS** Confusion unlikely; canasteros are more rufous, different behaviour, etc. **Status** Uncommon.

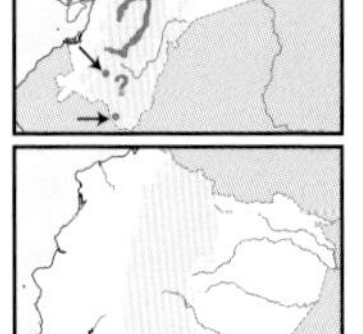

Rufous-fronted Thornbird *Phacellodomus rufifrons* 16–17cm

Extreme SE in Zumba region (650–1,500m). Deciduous to semi-deciduous woodland, treed clearings. ***Rufous-chestnut forecrown***, narrow supercilium. ***Plain olive-brown above, dirty white below***. Race *peruvianus*. Pairs to small groups noisily forage at various heights, including ground, often near large, colonial, stick nests. **Voice** Loud, hornero-like, abrupt, accelerating but slowing towards end: *chou-chóú-chou-chou...choo*; also sharp *tchék* calls. **SS** Confusion unlikely; see Marañón Spinetail (plain crown, longer tail, rufous-chestnut wings; behaviour and voice differ strikingly). **Status** Fairly common.

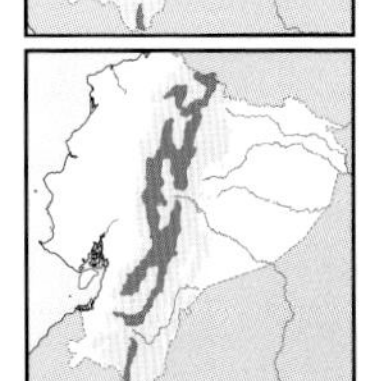

Many-striped Canastero *Asthenes flammulata* 16–17cm

Andes (3,200–4,500m). Grassy and bushy *páramo*, adjacent crops. Blackish above beautifully ***streaked rufous on crown, whitish on upper- and underparts***; buffy eyebrow, ***plain orange-buff chin. Rufous-chestnut fringes to wings, long spiny tail***, rufous outer rectrices. Nominate. Skulking, walks through dense grass but singing birds perch atop bushes or grasses. **Voice** Fast, long, accelerating, ending in trill *trreé-treé-treé-trr-trr-tr'tr'tr'tr'tr'tr'rrrrrr*; louder *tree* and buzzy *zrew* calls. **SS** Streak-backed Canastero (paler overall, plain belly, less streaky upperparts); see Andean Tit-spinetail. **Status** Fairly common.

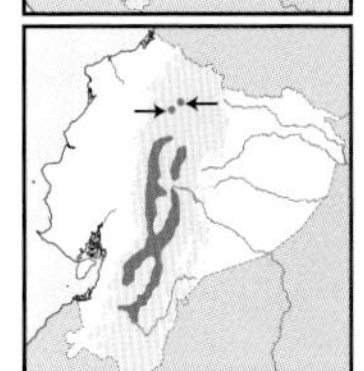

Streak-backed Canastero *Asthenes wyatti* 16–18cm

Andes from Cotopaxi to N Loja (2,800–4,100m, not above 3,100m in S). Dry grassy and shrubby *páramo*, adjacent scrub. Greyish-brown upperparts ***boldly streaked blackish, buff throat, plain*** or nearly plain ***greyish-buff below. Rufous flight feathers, long tail, rufous*** outer rectrices (N) or most tail feathers (S). Races *aequatorialis* (N), *azuay* (S). Skulking, easily missed unless vocalising atop bushes or grasses. **Voice** Very fast, ascending trill, *chi-p'rrrrrr*; calls shorter, doubled or tripled trills: *p'rrr, p'rrr....* **SS** Many-striped Canastero darker, streakier overall; chin more orange. **Status** Uncommon.

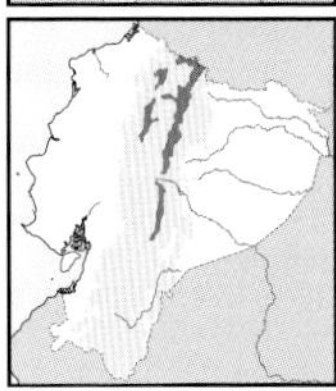

White-chinned Thistletail *Asthenes fuliginosa* 18–19cm

Both Andean slopes S to Chimborazo (2,800–3,500m). Shrubby treeline, *Polylepis* woodland, bushy *páramo*. ***Pale*** amber eyes. Brown above, greyish-buff below, ***narrow whitish eye-ring and short postocular stripe, whitish chin, long thistle-like chestnut-brown tail***. Nominate. Pairs in dense lower growth, irregularly with mixed flocks; acrobatic gleaner in foliage, tends to remain hidden. **Voice** High-pitched, slightly accelerating *tidididi...dit!*, also louder *ti-dlt-dlt, tididi, dididit...*; sharp *chynk*. **SS** No overlap with similar Mouse-coloured Thistletail (no obvious supercilium and white chin). **Status** Uncommon.

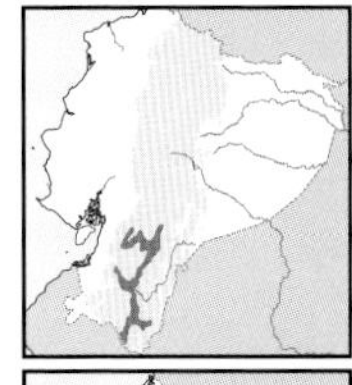

Mouse-coloured Thistletail *Asthenes griseomurina* 18–19cm

S Andes N to Azuay (2,800–4,000m, locally to 2,500m). Shrubby treeline, *Polylepis* woodland, bushy *páramo*. ***Brown*** eyes. ***Olive-brown above***, greyish below, ***white eye-ring, very faint greyish postocular stripe***, pallid lores. Tail darker but similar to previous species. Monotypic. Skulking, gleaning in low cover; pairs might briefly emerge in open. **Voice** Short, slightly accelerating, trilled at end: *swi-di-di-did'd'd'd'd*; calls loud, high-pitched, descending *pyeet!* or *pseew!* **SS** Unique in range, no overlap with White-chinned Thistletail. **Status** Uncommon.

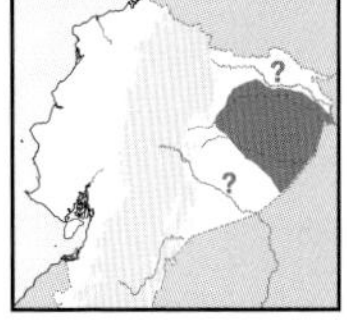

Plain Softtail *Thripophaga fusciceps* 16–17cm

Local in E lowlands (to 400m). *Várzea* forest, old secondary woodland, borders. ***Drab***. Olive-brown above, ***more rufescent-chestnut wings, rump and long, rounded tail***, greyish-buff below, faint buff supercilium. Race *dimorpha*. Pairs or small groups at natural forest gaps in dense vines, not with mixed flocks; possibly dead-leaf specialist. **Voice** Sharp, loud, long, chattered *tche-tche-ch'rrrrrr-tch'rrrr-r'tch'chrrrr*. **SS** Superficially resembles larger Olive-backed and Bamboo Foliage-gleaners, but no white throat, plainer overall, different behaviour. **Status** Rare.

adult
juvenile
Andean Tit-Spinetail
juvenile
adult
Rufous-fronted Thornbird
azuay
aequatorialis
Streak-backed Canastero
Many-striped Canastero
adult
juvenile
/hite-chinned Thistletail
Mouse-coloured Thistletail
Plain Softtail

PLATE 158: ANTSHRIKES I

The three largest antshrikes, all very vocal and bold, plus three obviously barred black-and-white (♂) species of dense scrub; these latter species replace each other geographically. Antshrikes are large to mid-sized antbirds with heavy and hooked bills, especially so in Fasciated, Fulvous and Great.

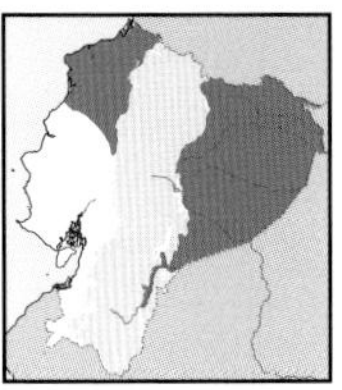

Fasciated Antshrike *Cymbilaimus lineatus* 17–18cm

E & NW lowlands and foothills (to 1,000m). Humid forest understorey, borders. ***Heavy hooked bill, red eyes. Entirely banded black and white, crown plain black*** (E). ♀: barred ***ochre and dusky, crown plain rufous***. Races *fasciatus* (W), *intermedius* (E). Pairs often with mixed flocks, rather deliberate but fast after prey. **Voice** Utters 3–10 slow, steady but downslurred, plaintive *cüw* notes in rising-falling series; complaining *déeuu* mewing call, also *déeuuu-tk'tk'tk'tk...*. **SS** Barred *Thamnophilus* antshrikes (less heavy bills, obviously crested, smaller, unbarred ♀); Fulvous Antshrike (larger, no bars below, plain throat, heavier bill). **Status** Uncommon.

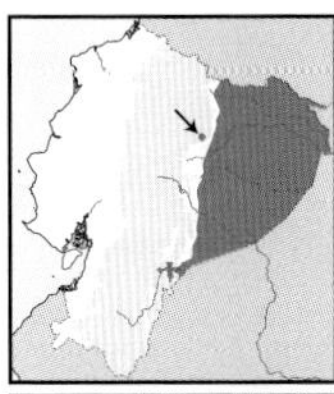

Fulvous Antshrike *Frederickena fulva* 22–23cm

E lowlands (to 700m). *Terra firme* forest undergrowth. ***Massive hooked bill.*** Entirely black with ***narrow whitish vermiculations***, but ***throat to breast solid black***. ♀: ***chestnut above, buffier below, sparser dusky vermiculations***. Monotypic. Skulking pairs rather sluggish in dense tangles, seldom with flocks. **Voice** Rather fast series of 10+ plaintive *üeé* whistles, rising and falling; nasal, high-pitched *kEEeyuur* alarm, and descending snarl *mEw'urrr*. **SS** Smaller Fasciated Antshrike more evenly barred, paler, crest solid black (♂) or rufous (♀), slower and shorter voice. **Status** Uncommon.

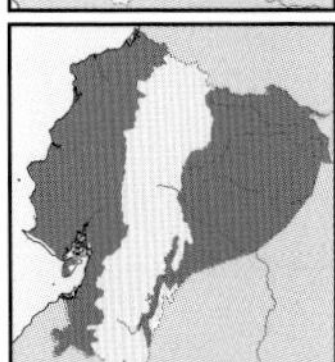

Great Antshrike *Taraba major* 19–20cm

E & W lowlands to foothills (to 1,000m, locally higher). Forest borders, dense second growth, shrubby clearings. Heavy bill, ***red eyes. Black above, white below***, white wingbars. ♀: ***rufous*** above, no wingbars. Races *transandeanus* (W), *melanurus* (E). Sluggish pairs skulk in very dense undergrowth, gleaning and even sallying for prey, only briefly with mixed flocks. **Voice** Loud, slowly accelerating *cü-cü-cü-cü-cü-cücücücü...* ending in snarl *gggaa*, faster in W; throaty *g'g'g'g'r*, also shorter *cÜ-cü-cü-cü..cü..cü*. **SS** Unmistakable if seen; trogon-like voice with snarl. **Status** Fairly common.

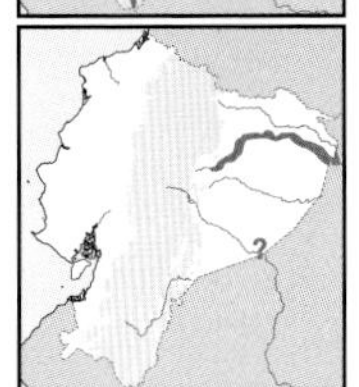

Barred Antshrike *Thamnophilus doliatus* 16–17cm

E lowlands along major rivers (below 250m). Dense riparian thickets on river islands. Yellow eyes. Boldly ***barred black and white***, plain bushy crest. ♀: ***plain cinnamon-rufous crown and upperparts***, underparts cinnamon-buff, ***face to neck-sides and nape streaked dusky and buff***. Race *radiatus*. Shy pairs rarely emerge in borders and open, only occasionally with flocks, very skulking. **Voice** Rapidly accelerating, nasal, staccato *khán-khán-khán'hánhánhanhanhun*, descending at end, terminating in emphasised *hÁn!*; call nasal *khá* or *nhá*. **SS** Confusion unlikely in range and habitat. **Status** Rare.

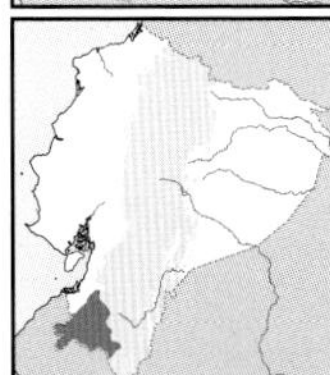

Chapman's Antshrike *Thamnophilus zarumae* 15–16cm

SW lowlands to subtropics (to 2,000m). Dense second growth, borders, regenerating scrub. ***Streaky bushy crest. Upperparts coarsely barred black and white, throat to upper belly only narrowly barred, belly buff.*** ♀: plain rufous crown and upperparts, ***dirty buff below, streaky face to nape***. Nominate. Active, vocal pairs deliberate at mid–low heights, seldom with flocks. **Voice** Fast, rising-falling series of 8–12 *cup* notes, increasing in middle, ending in rattle; short, nasal, mewing *khüw* calls. **SS** ♀ Collared Antshrike (boldly edged wings, unstreaked face to nape, cleaner underparts). **Status** Fairly common.

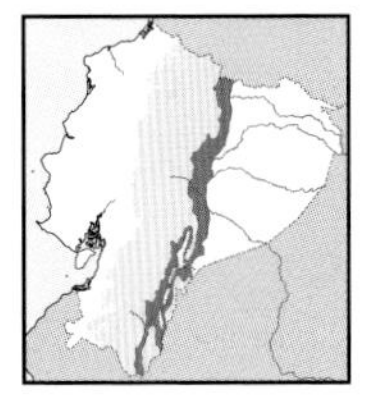

Lined Antshrike *Thamnophilus tenuepunctatus* 16–17cm

E Andean foothills to subtropics (to 1,750m). Montane forest borders, dense second growth. Yellow eyes. Recalls Barred Antshrike but ***blacker, white barring narrower above, unbarred crest***. ♀: ***plain rufous-chestnut crown and upperparts***, underparts and face to nape as ♂. Races *tenuifasciatus* (N), *berlepschi* (S). Vocal and responsive pairs, skulking but sometimes conspicuous, rarely with flocks, might briefly emerge into open. **Voice** Loud, bouncing, nasal *khü-khü-khü...khüa*, accelerating in middle, then falling, last note lower-pitched; nasal *khü!*, also weaker *ke'u*. **SS** Much blacker than Barred and Chapman's Antshrikes (neither is sympatric). **Status** Common; VU globally.

fasciatus
intermedius
♂
♀
♂
♀
Fasciated Antshrike
♀
immature
♂
Fulvous Antshrike
♀
Great Antshrike
♂
♂
immature
♂
♀
Barred Antshrike
♂
♀
Chapman's Antshrike
tenuifasciatus
♂
♀
♂
berlepschi
Lined Antshrike

Five Amazonian antshrikes, mostly characterised by subdued wing patterns, plus the only Andean antshrike. All are more often heard than seen, some notably difficult to observe; duetting voices notable and helpful for identification. Also a monotypic Amazonian antshrike resembling *Thamnophilus* but heavier bill and shorter tail.

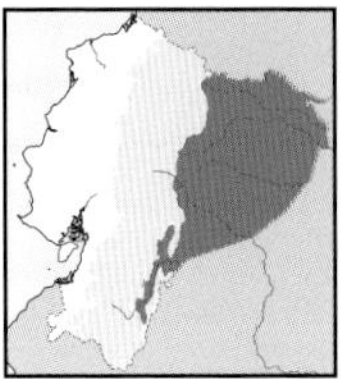

Plain-winged Antshrike *Thamnophilus schistaceus* — 14–15cm

E lowlands to foothills (below 1,000m). Mostly *terra firme* forest undergrowth. ***Red eyes***. Mostly ***plain grey, black crown*** (***concolorous*** in SE). ♀: ***rufous crown, olive-brown upperparts***, paler below, ***dusky cheeks***. Races *capitalis* (most range), *dubius* (extreme SE). Pairs skulk in dense tangles, rather lethargic, sometimes with mixed flocks, vocal daylong. **Voice** Fast, nasal *nha-nha-nha-nha-nha-nha'nha...nooh-nah*, last part lower-pitched, ♀'s shorter, higher-pitched; nasal *nyeeh*, *nyaahrr* and slower *narr..narr..narr*. **SS** Mouse-coloured Antshrike (dark eyes, paler, pale wing spots, uniform crown); see Cinereous, Dusky-throated and Spot-winged. **Status** Common.

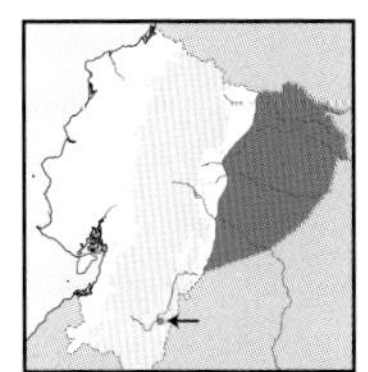

Mouse-coloured Antshrike *Thamnophilus murinus* — 13–14cm

E lowlands, N to Río Napo (below 450m). Hilly *terra firme* forest. ***Grey eyes***. ***Plain light grey, paler underparts. Small whitish spots in wing-coverts***, faint white tail tip. ♀: mostly ***olivaceous-brown***, more rufescent crown, ***paler below, pale buff*** wing dots. Race *canipennis*. Mostly pairs in dense to fairly open undergrowth, seldom with flocks. **Voice** Similar to Plain-winged, but slower-paced, shorter, higher-pitched *nya-nya-nya-nya-nhü'nAw*; abrupt *aw!* and whiny *awr*. **SS** Plain-winged (reddish eyes, unmarked wings); also Cinereous, Dusky-throated and Spot-winged Antshrikes. **Status** Uncommon.

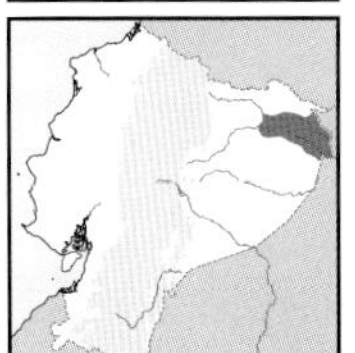

Cocha Antshrike *Thamnophilus praecox* — 16–17cm

Local in NE lowlands (below 300m). *Igapó* forest undergrowth, thickets, forested streams. ♂: ***all black***, white underwing. ♀: ***velvet black hood, deep cinnamon-rufous body***, more cinnamon below. Monotypic. Pairs skulk near water, seldom with flocks, little vocal activity at dawn. **Voice** Utters 8–18 hollow, evenly paced *cöh-cöh-cöh-cöh...* notes, shorter in ♀; mellow *pöw-pö*, *cuäh* and trilled *p'rrrrr*. **SS** Recalls White-shouldered Antbird but smaller, lacks bare bluish orbital skin and white line on wing bend; see also Castelnau's (no habitat overlap). **Status** Locally uncommon, NT.

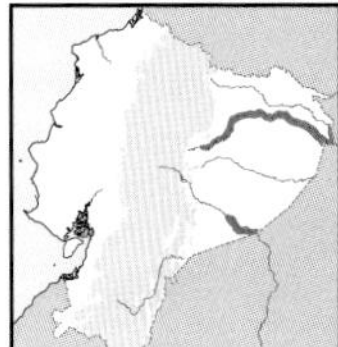

Castelnau's Antshrike *Thamnophilus cryptoleucus* — 17–18cm

Local in E lowlands along major rivers (below 300m). Mature regenerating woodland on river islands. ***Heavy bill***. ♂: ***all black***, with bold ***white-edged wing-coverts, semi-concealed white dorsal patch***, white underwing. ♀: ***completely*** black, ***no white*** in wing-coverts. Monotypic. Deliberate pairs, rather shy, seldom with flocks; lazily wags tail. **Voice** Short, rapidly accelerating, bouncing, nasal *keü, keü.. khü- khü-khü'khü'khü'khü*; nasal *khäw-khäw* calls, often ending in rattle *khäw-prrrrr*. **SS** ♀ like ♂ White-shouldered and Cocha Antshrikes; see White-shouldered Antbird. **Status** Fairly common; NT.

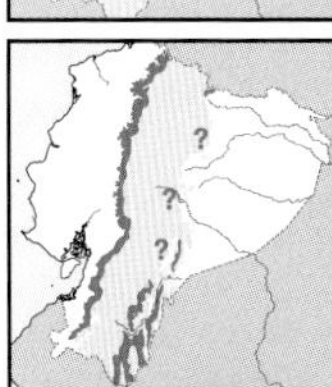

Uniform Antshrike *Thamnophilus unicolor* — 15–16cm

Foothills to subtropics on both Andean slopes (700–2,000m). Montane forest. ***Pale grey eyes***. ♂: ***plain slate-grey***, paler throat. ♀: mostly ***rufous***, more cinnamon below, with ***grey face***. Nominate (W), *grandior* (E). Mostly in vocal pairs skulking in dense undergrowth, difficult to see; briefly with mixed flocks. **Voice** Slow, plaintive, descending series of 3–6 *onh-anh-anh-anh-an* notes; slurred, mellow *kyö'rr* or *kä'rrr* calls. **SS** Confusion unlikely in most of range; in lower foothills cf. White-shouldered Antshrike (red eyes, black crown in ♂, no grey face in ♀). **Status** Uncommon.

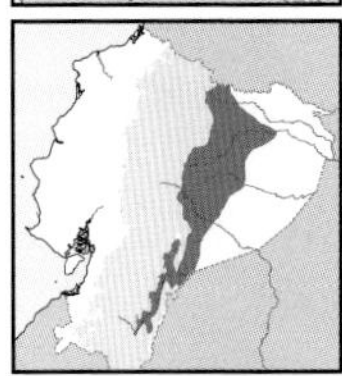

White-shouldered Antshrike *Thamnophilus aethiops* — 15–16cm

E foothills to adjacent subtropics and lowlands (to 1,000m, locally higher). *Terra firme* forest undergrowth. ***Reddish eyes***. ♂: ***entirely velvet black, tiny white dots in wing-coverts, white wing bend*** and underwing. ♀: ***deep rufous-chestnut***. Nominate. Pairs in dense thickets and tangles, seldom with flocks. **Voice** Slow, evenly paced, nasal, trogon-like *anh..anh..anh..anh...*; slurred *keyúrr*, *aaawr* calls. **SS** Few obvious field marks might cause confusion with other antbirds and antshrikes (♂ White-shouldered Antbird; ♀ Plain-winged, Cinereous and Dusky-throated Antshrikes). **Status** Rare.

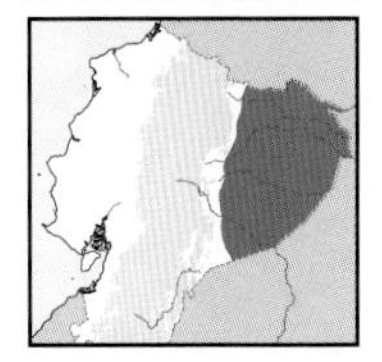

Spot-winged Antshrike *Pygiptila stellaris* — 13–14cm

E lowlands (to 400m, locally higher). *Terra firme* forest, adjacent woodland. ***Heavy bill. Black crown, bands of small white dots on coverts. Short*** blackish tail. ♀: grey above, ***unspotted browner wings, plain ochraceous-buff face to underparts***, short grey tail. Race *maculipennis*. Pairs in mid strata, often with flocks in dense vine tangles. **Voice** Sharp, short, semi-musical stutter *t't't't-téëuw*; sharp *chét!*, *chét-teéuw*, *téëuww-dját-dját* calls. **SS** ♀ Plain Antvireo, no dusky face; more arboreal than all *Thamnophilus* (see Plain-winged and Mouse-coloured), chunkier, short-tailed. **Status** Uncommon.

♀
capitalis
♂
♂
dubius
Plain-winged Antshrike
♀
Mouse-coloured Antshrike
♂
juvenile
♂
♀
Cocha Antshrike
♂
♀
Castelnau's Antshrike
♂
♀
Uniform Antshrike
♂
♀
White-shouldered Antshrike
♂
♀
Spot-winged Antshrike

PLATE 160: ANTSHRIKES III

Three *Thamnophilus* antshrikes with conspicuous wing patterns, ♂♂ predominantly grey, ♀♀ brown. Pearly is a bold, distinctive, monotypic antshrike of Amazonian hilly forests. The distinctive Russet resembles foliage-gleaners in plumage, but bill heavy and hooked. The bold Collared of xeric habitats was formerly placed in *Sakesphorus*.

Collared Antshrike *Thamnophilus bernardi* 16–17cm

SW lowlands, to subtropics in Loja (to 1,500m). Deciduous forest, borders, woodland, hedgerows. Unique. ***Black hood, bushy crest, white nuchal collar and lower underparts***; boldly marked wings. ♀: ***rufescent crown***, speckled face, ***ochraceous-buff below***. Nominate (N), *piurae* (S). Mostly in pairs, boldly foraging at mid–low heights, constantly wags tail, raises crest. **Voice** Bouncing, slowly accelerating *cüa-cüa-cüa-cüacücü-cúr'*; sharp *khü-k'rrr* calls and short mewing series (S); slower, more paced in N. **SS** ♀ Chapman's has streaky face to nape, unmarked wings. **Status** Common.

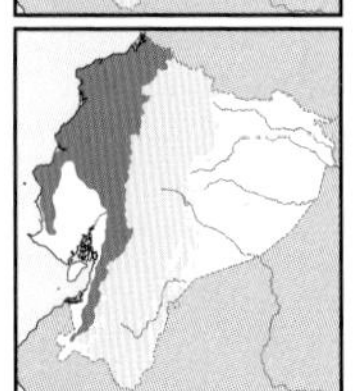

Black-crowned Antshrike *Thamnophilus atrinucha* 14–15cm

W lowlands to lower subtropics (to 1,100m, locally higher). Forest, borders, adjacent woodland. ***Blackish crown***; mainly ***slate-grey***, paler below, semi-concealed white dorsal patch. ***Boldly white-spotted wing-coverts***. ♀ similarly patterned but in ***brown and buff tones***, crown more rufescent than upperparts. Nominate. Pairs in dense to fairly open undergrowth, calmly foraging, often with mixed flocks. **Voice** Nasal *kün* or *khän* notes in slowly accelerating, rapid series, ending with inflected, higher-pitched *khüa?*; doubled *kün-küa!* **SS** Confusion unlikely in range; *Cercomacra* and *Cercomacroides* antbirds are more slender, with slenderer bills, longer tails, etc. **Status** Fairly common.

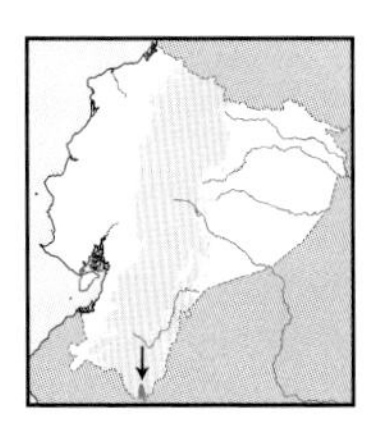

Northern Slaty Antshrike *Thamnophilus punctatus* 14–15cm

Extreme SE in Zumba region (600–650m). Deciduous to semi-deciduous woodland and borders. Recalls Black-crowned Antshrike, with which formerly lumped. Blackish crown, grey upperparts washed black, grey underparts, ***whiter belly***. Wings and tail ***boldly marked***. ♀: brown, crown more rufescent, ***whitish mid belly, white wing spots and edges***. Race *leucogaster* (might merit species status). Pairs apparently more skulking than Black-crowned, though emerge briefly at edges; deliberate hops, constantly wags tail. **Voice** Slow but accelerating series of clear, nasal *khün* notes ending in shorter *kü'kü'kü*; simple, hollow *khü* calls, and quick *chi'rrrrr* rattle. **SS** Unmistakable in limited range. **Status** Rare, local.

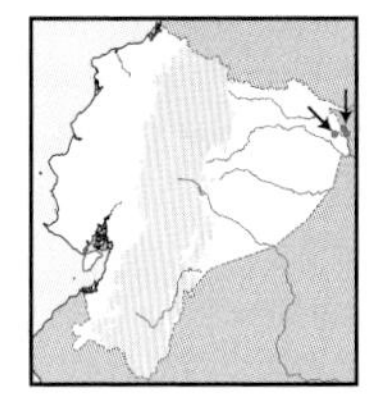

Amazonian Antshrike *Thamnophilus amazonicus* 14–15cm

Lowlands of extreme NE, in Río Lagarto region (200m). *Igapó* forest undergrowth, around streams. Mostly ***dark slate-grey, darker mid crown***, white semi-concealed dorsal patch, blacker wings, ***two white bars and white edges to flight feathers***. ♀: cinnamon to ***orange-rufous head and underparts***, upperparts brownish-olive, ***blacker wings and tail with white bars, edges and tips***. Race *amazonicus*? Poorly known in Ecuador; elsewhere, pairs forage near water level or higher, sometimes with mixed flocks. **Voice** Fast, ascending-descending, nasal *kuh..kuh-kuh-kuh-kuh'kuh'kuhkuhkuh'kuh-kuhn..kuhn*; whining *keeah!* **SS** Both sexes distinctive, especially ♀; no other grey antshrike with white wing markings share habitat. **Status** Rare, local.

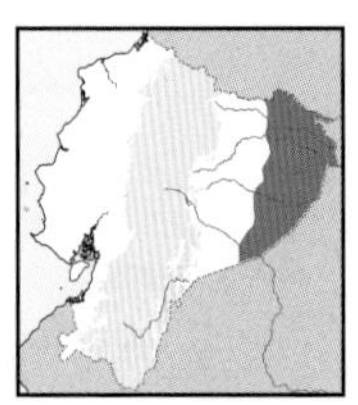

Pearly Antshrike *Megastictus margaritatus* 13cm

E lowlands (below 300m). *Terra firme* forest undergrowth. ***Pale grey eyes***. Mostly ***pearl-grey***, paler underparts. Blackish wings, uppertail-coverts and fairly short tail with ***bold white spots and tips***. ♀: similarly patterned but in ***brown and buff***. Monotypic. Usually in pairs away from flocks in fairly open understorey, calmly hops on branches and gleans upper side of foliage, often flicking tail. **Voice** Slow, querulous, nasal *wheep, wheep, wheep, wheep* followed by fast, raspy *jrr-jrr-jrr-jrr-jrr*; rising *kweep?* **SS** White (♂) or buff (♀) wing spots larger and bolder than any other antshrike or antbird; note eye colour; see tiny Plain-throated Antwren. **Status** Rare, local.

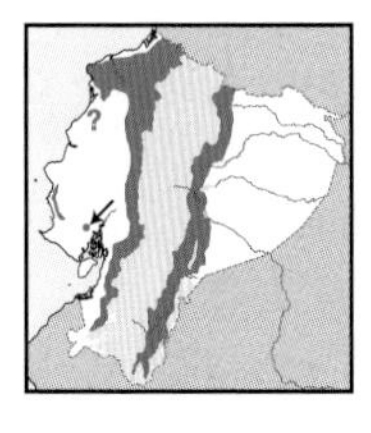

Russet Antshrike *Thamnistes anabatinus* 14–15cm

NW lowlands, S in Andes to N El Oro, locally in Cordillera de la Costa; E foothills to subtropics (400–1,700m, locally lower). Humid forest and woodland, borders. ***Heavy bill***. Brown above (more olivaceous in E), ***unmarked rufescent wings and tail; buffy supercilium, dusky ocular stripe***, buff to greyish-buff below. Races *intermedius* (W), *aequatorialis* (E). Active pairs regularly with mixed upper-strata flocks, inspects hanging dead leaves, clumps and undersides to branches and leaves. **Voice** Loud, ringing *peeü, pfé-pfé-pfé-pfé-pfé...*, faster, shorter in E; sharp *seew-psít* calls. **SS** Distinctive antshrike (no other has buffy supercilium), but cf. *Philydor* foliage-gleaners (Buff-fronted, Rufous-rumped, Slaty-winged); slenderer foliage-gleaners have finer, longer bills and longer tails. **Status** Uncommon.

♂
piurae
♂
♂
♀
bernardi
Collared Antshrike
♂
immature
♀
♂
♂
♀
Black-crowned Antshrike
♂
immature
Northern Slaty Antshrike
♂
♂
♀
♀
♂
Amazonian Antshrike
Pearly Antshrike
♂
intermedius
♀
aequatorialis
Russet Antshrike

PLATE 161: ANTVIREOS AND ANTSHRIKES

Chunky antbirds, short-tailed and heavy-billed. Antvireos occur in varied habitats, from deciduous woodland to very humid montane forest, none in Amazonia. Plain Antvireo is the most versatile in habitat choice. *Thamnomanes* antshrikes typically lead mixed-species flocks; they perch quite upright and have loud voices.

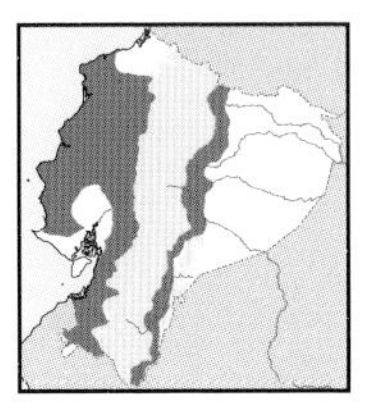

Plain Antvireo *Dysithamnus mentalis* 11–12cm

W lowlands to subtropics, not in wet NW; E foothills to subtropics (to 1,800m in W, 700–1,700m in E). Humid to deciduous forest (W), montane forest, borders (E). *Chunky.* Olive-grey above, grey below, whiter throat and belly (*yellowish belly*, W), *dark cheeks.* ♀: similarly patterned but in brown, yellowish and olive, more *rufescent crown, narrow white eye-ring.* Races *aequatorialis* (W), *napensis* (E), *tambillanus* (extreme SE). Deliberate pairs with mixed understorey flocks, glean foliage, flick wings and tail. **Voice** Accelerating, descending *püi-pÜI-püi-püi...püe-pu'pu'*; short, rising *pu'ee* and more nasal *jhá*; slower in E, first part more paced. **SS** Rather nondescript but confusion unlikely. **Status** Common.

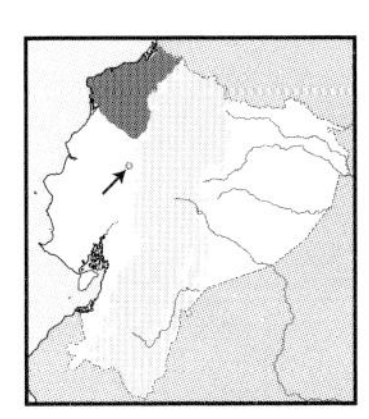

Spot-crowned Antvireo *Dysithamnus puncticeps* 11–12cm

NW lowlands to lower foothills (to 800m). Humid to wet forest. *Pale eyes. Crown densely streaked blackish and grey*, brownish-grey upperparts, blacker wings, white dotted wingbars. *Whitish underparts, throat streaked grey*, belly washed buffy. ♀: similarly patterned but in rufous and buff, *dusky throat and crown streaking.* Monotypic. Pairs in tangled undergrowth, deliberately following mixed flocks, may follow ant swarms. **Voice** Softly whistled, cadenced *fet-fet-fet...* accelerating at end; descending *f'rrr* trill. **SS** Plain Antvireo lacks streaks on head and throat, eyes are dark; no known overlap. **Status** Uncommon, NT in Ecuador.

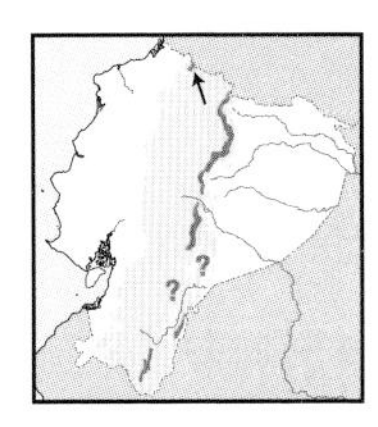

Bicoloured Antvireo *Dysithamnus occidentalis* 13–14cm

Local in E Andean subtropics and extreme NW (1,500–2,200m). Montane forest undergrowth, adjacent woodland. *Entirely blackish* with *small but conspicuous white dots on wing-coverts.* ♀: *chestnut crown*, dark brown upperparts, *bold whitish dots on wing-coverts. Grey face finely streaked whitish*, grey breast, buffy lower belly. Races *punctitectus* (E), possibly nominate (extreme NW). Mostly quiet, pairs forage apart from flocks, often near ground; flicks wings. **Voice** Fast, throaty, falling *je'eer-de'eer, je'eer-de'eer, je'eer-de'eer-du'ur*; clearer, querulous *peeu* call. **SS** Locally sympatric White-streaked Antvireo (greyer with blacker, more contrasting bib; ♀ heavily streaked whitish below). **Status** Rare, VU.

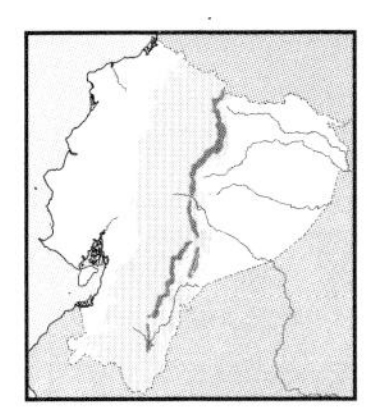

White-streaked Antvireo *Dysithamnus leucostictus* 12–13cm

E Andean foothills to subtropics (1,300–1,800m). Montane forest undergrowth. *Sooty-grey, blacker throat to breast*, faint white wing dots. ♀: chestnut crown, brown upperparts, *grey below obviously streaked whitish.* Nominate. Mostly pairs following mixed flocks, deliberately gleaning foliage; rather quiet. **Voice** Descending series of 6–8 soft *pu-pU-pu-pfu-pfu...*, last two notes sometimes accelerating; loud, clear *peü* calls. **SS** Bicoloured Antvireo (blacker, more dot-shaped wing markings; ♀ has more contrasting crown, whitish streaks *less extensive*); paler Plain Antvireo (no streaks below, ♀), (dark mask, ♂); Uniform Antshrike (pale eyes, uniform grey). **Status** Uncommon; VU.

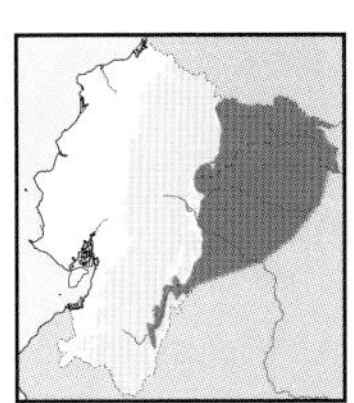

Dusky-throated Antshrike *Thamnomanes ardesiacus* 14–15cm

E lowlands (to 500m). Mainly *terra firme* undergrowth. *Entirely grey, paler belly, throat variably blackish.* Rather *short tail.* ♀: *olive-brown above, whitish throat*, dull *olive-buff chest, ochraceous below*, unmarked olive-brown wings. Nominate. Mostly in pairs with mixed flocks, perch low, leaning forwards; a gleaner that often leads flocks. **Voice** Raspy, accelerating, rising *grr..grr-grr-grr-gergergergirgirgírgír...jrrert*, sometimes last growl missing; sharp, squeaky *geért* and *chíft!* **SS** Very similar to Cinereous (looks longer-tailed, tends to perch more upright; underparts almost plain, belly not obviously paler; ♀ richer cinnamon-rufous below). **Status** Fairly common.

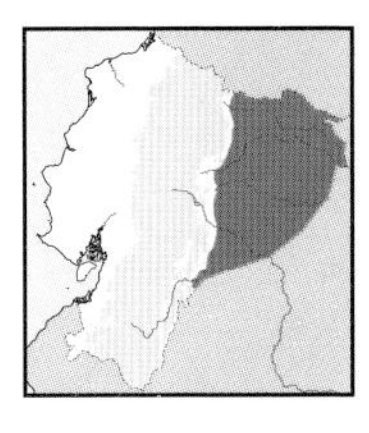

Cinereous Antshrike *Thamnomanes caesius* 14–15cm

E lowlands (to 600m). *Terra firme* and, less often, *várzea* undergrowth. Closely recalls previous species but *underparts plain, tail longer.* ♀: also similar but darker, browner above, *deeper cinnamon-rufous below, longer-tailed.* Race *glaucus.* Pairs to small parties in mixed flocks, often leading them, regularly with congener; perch more upright, more exposed and sally after prey. **Voice** Accelerating, rising-falling whistles: *wheer?-wheer?, wheer'wheerwhipwhip'w'p'p'p*, sometimes shorter; staccato *whér-whér-dEw-dEw-dEw...*, and *whërr-cheedididik* calls. **SS** See Dusky-throated with whiter belly (♂), duller ochraceous underparts (♀); narrow whitish tail tip in Dusky-throated hardly visible; differ in foraging behaviour, posture and voice. **Status** Fairly common.

♀
napensis, tambillanus
♂
♀
aequatorialis
♂
Plain Antvireo
♂
♂
Spot-crowned Antvireo
♀
immature
♂
♀
White-streaked Antvireo
♂
♀
'itectus
♂
♀
Bicoloured Antvireo
occidentalis?
♀
Dusky-throated Antshrike
♂
♂
♂
immature
♂
♂
immature
♀
Cinereous Antshrike

Small 'checker-throated' antwrens formerly included in diverse genus *Myrmotherula*, but split on basis of vocal and genetic characters (as well as general plumage); all but one occur in E lowlands and foothills. Three similar, closely related species are allopatric: Stipple-throated and Brown-backed separated by Napo River, and Foothill segregated by elevation.

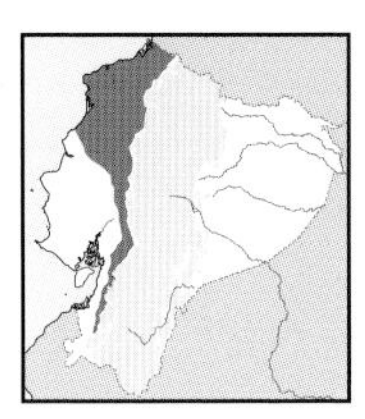

Checker-throated Antwren *Epinecrophylla fulviventris* 10–11cm

W lowlands to foothills, S along base of Andes to W Loja (to 900m). Humid forest undergrowth, borders, adjacent woodland. *Pale eyes*. Brown above, ***two buff wingbars; checkered throat***, greyish-buff underparts. ♀: ***little streaking on throat, fulvous underparts***. Monotypic. Pairs in mixed flocks inspect dry curled leaves and hanging litter. **Voice** Soft, descending series of 4–11 *chií* notes, last often more separated *chií-chií-chií-chií..chiu*; sibilant, quick *chíík* calls. **SS** ♀ with White-flanked (white flank tufts, duller above, no white streaking, fainter wing dots) and Slaty Antwrens (plainer wings, dark eyes, no throat streaking). **Status** Uncommon.

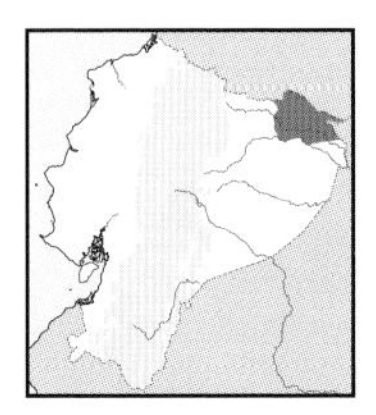

Stipple-throated Antwren *Epinecrophylla haematonota* 10–11cm

Local in extreme NE lowlands (250m). *Terra firme* forest. ***Pale eyes***. Fulvous-brown crown, ***rufous-chestnut mantle to rump***; grey face to belly, ***checkered throat***. Blackish wings with ***whitish dotted wingbars***. ♀: ***ochraceous-buff below, faint*** dusky streaks on throat, ***buff wingbars***. Race *pyrrhonota*. Acrobatic pairs with mixed flocks in tangled understorey, specialised on dead leaves. **Voice** Thin, high-pitched, sibilant *zee-zee-zEE-zee-zee*, short rattles. **SS** No sympatry with similar Brown-backed (no chestnut on upperparts); see Ornate (bolder overall, ♂ no stippled throat) and Rufous-tailed (longer rufous tail, no throat marks). **Status** Locally uncommon.

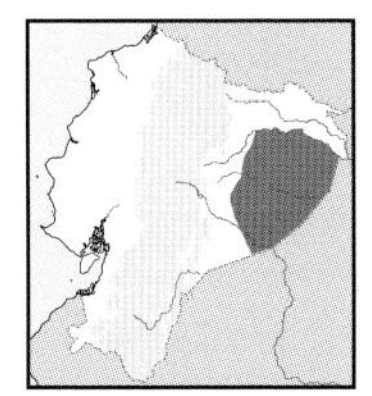

Brown-backed Antwren *Epinecrophylla fjeldsaai* 10–11cm

E lowlands S of Río Napo (below 250m). *Terra firme* forest undergrowth. ***Pale brown eyes. Uniform brown upperparts***, blackish wings, ***buff-dotted wingbars***, greyish face to lower breast, browner belly, ***checkered throat***. ♀: buff below, ***throat streaked blackish***. Monotypic, possibly not full species. Pairs in tangles near ground, fond of dead curled-up leaves and hanging debris in vines. **Voice** High-pitched, thin *zee'ee'ee'ee'ee*; *z'r'r'r* rattle, harsh *shee-shee-shee-she*. **SS** No overlap with similar but reddish-mantled Stipple-throated; Rufous-tailed lacks throat streaks, has longer rufous tail; ♀ Ornate has bolder wings and throat (not known overlap). **Status** Locally uncommon.

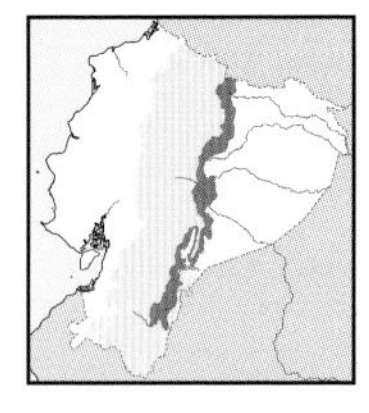

Foothill Antwren *Epinecrophylla spodionota* 10–11cm

E Andean foothills to lower subtropics (600–1,450m). Montane forest undergrowth. ***Brownish-grey foreparts, checkered throat***. Browner rear, blackish wings with ***white dotted wingbars***. ♀: ***uniform brown above***, ochraceous-buff below, ***throat vaguely streaked dusky, buff dotted wingbars***. Nominate. Pairs with mixed flocks, regularly with Slaty Antwren, foraging near ground mostly in curled-up dead leaves. **Voice** Very fast, thin, rising-falling series of ringing *chi-ri-RI-RI-ri-ri-ri* or *zee-ree-REE-REE-ree-ree*; thin *chit-chit-chit* and sharp *chét!* **SS** No overlap with fairly similar Brown-backed and Stipple-throated Antwrens; cf. ♀ Slaty (plainer wings, greyer upperparts, no throat marks); locally sympatric Ornate ♀ much bolder. **Status** Uncommon.

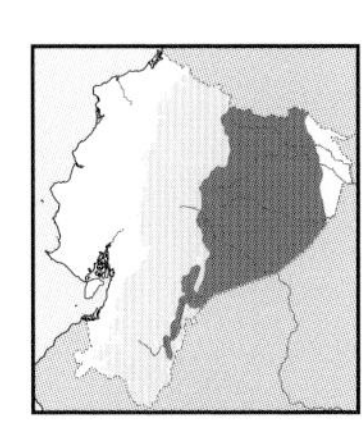

Ornate Antwren *Epinecrophylla ornata* 10–11cm

E lowlands to foothills, not in extreme E (250–1,200m). Tangled understorey of hilly *terra firme* forest. ***Grey head to belly, rufous-chestnut mantle to rump, bold black throat bib***, blackish wings, and ***two conspicuous white bars***. ♀: olive-brown crown, ***rufous-chestnut lower upperparts***, ochraceous-buff below, ***conspicuous black throat finely streaked white***. Race *saturata*. Acrobatic pairs with mixed flocks near ground, inspecting dead leaves and debris. **Voice** Thin, high-pitched, fading series of ringing *tseé-tsee-tsee-tsi'tsisisusu...*, sometimes shorter; short squeaky *pt-pt-see'see'see'see*. **SS** Although confusion unlikely, cf. Brown-backed, Stipple-throated and Foothill Antwrens. **Status** Rare.

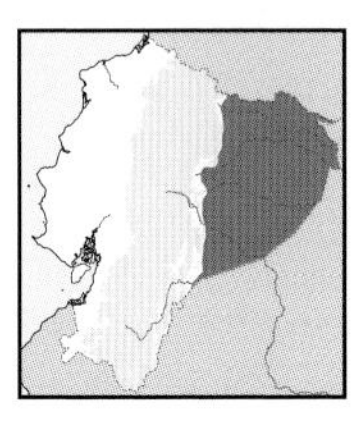

Rufous-tailed Antwren *Epinecrophylla erythrura* 10–11cm

E lowlands (to 700m, locally higher). *Terra firme* forest undergrowth. Olive-brown crown, ***rufous upperparts*** including ***longish tail. Pale grey face to lower breast***, olive-buff belly. Brown wings, faint buff bars. ♀: ***underparts mostly ochraceous-buff***. Nominate. Acrobatic pairs with mixed flocks, often inspects hanging dead leaves at low to mid heights. **Voice** Thin, high-pitched, somewhat squeaky *sweé-sweé-sweé-sweé-sweé...*; high *séép-séép* and others. **SS** Long (for an antwren) rufous tail and rufous-and-grey combination readily separate from any congener; see ♀ Slaty. **Status** Uncommon.

Checker-throated Antwren

Brown-backed Antwren

Ornate Antwren

Stipple-throated Antwren

Foothill Antwren

Rufous-tailed Antwren

The 'streaked group' of typical *Myrmotherula* antwrens, two species with black streaks above and yellow underparts; the rest being black and white (females orangey). Messy taxonomy in the blackish-and-yellow pair, with races switching from one species to the other. Vocal characters essential to resolve matters. One black-and-white species in W lowlands. Its 'sibling' lives in edge habitats adjacent to Amazonian streams and lakes.

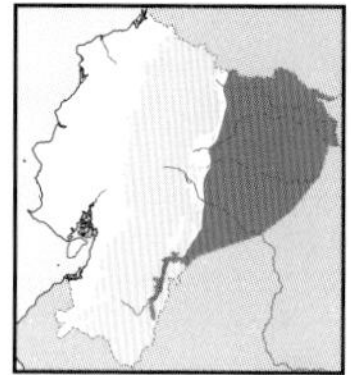

Pygmy Antwren *Myrmotherula brachyura* 8–9cm

E lowlands (to 600m, locally higher). *Terra firme* and *várzea* forest, mature woodland. ***Upperparts coarsely streaked black and white***, white throat, ***narrow black malar***, white cheeks, and pale yellow below. ♀: head to mantle and breast washed ***buff***. Monotypic. Lively pairs in subcanopy and mid strata, fond of vine tangles around treefall gaps, often joins mixed flocks briefly. **Voice** Fast, accelerating, descending at end *shree-shree'shee'shee'seuseuseuseurr*; *sh'rrr* trills. **SS** Closely recalls sympatric Moustached Antwren in E, with which often consorts; see malar (narrow vs. broad), streaky back (dense vs. faint), buffier ♀, voice (faster, accelerated finish). **Status** Uncommon.

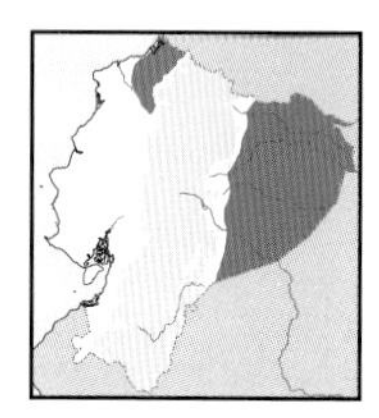

Moustached Antwren *Myrmotherula ignota* 7–8cm

E & NW lowlands (to 600m). Humid to wet forest, borders. Closely recalls Pygmy Antwren but looks ***blacker above*** owing to ***vague*** white streaking (***more streaked*** in W), ***black ocular stripe bolder, malar broader***. ♀: ***buffy*** head to throat and chest. Nominate (W), *obscura* (E); *obscura* formerly valid species, *ignota* formerly race of Pygmy. Lively pairs in mid strata, seldom join mixed flocks. **Voice** Accelerating, descending, semi-musical *tyee-tyee-tyee-tyee...*, last notes quite faster; call short *di'irr, tchéw* (W); in E slower than Pygmy. **SS** In E, cf. Pygmy Antwren voice (faster, accelerating at end). **Status** Uncommon, rarer in W.

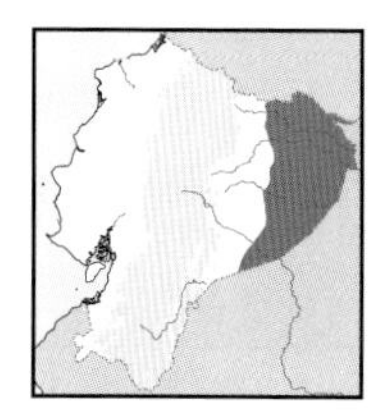

Amazonian Streaked Antwren *Myrmotherula multostriata* 9–10cm

E lowlands (to 300m). Dense shrubbery at lake and stream edges. ***Densely streaked black and white***, blackish wings with two white bars. ♀: ***orange-buff head, more ochraceous below***, black upperparts ***with narrower streaks***, wings as ♂. Monotypic. Pairs to small parties only briefly join mixed flocks, actively glean at low heights near water; brief sudden sallies. **Voice** Dry, rather high-pitched, evenly-paced *dree-er-er-er-er-er-r*; rhythmic, inflected *pur-pur-peé?-peé?-peé?-pee-pee-pur-pur* typically by ♀; *djeé-pu, djeé-pu-pu* calls. **SS** Stripe-chested has streaking limited to chest and upperparts, ♀ plainer and buffier or yellower (no known overlap and, if so, habitat also differs). **Status** Uncommon.

Pacific Antwren *Myrmotherula pacifica* 9–10cm

W lowlands and adjacent foothills, S to E Guayas (to 900m). Humid forest borders, second growth, shrubby clearings, gardens, hedgerows. ***Mostly streaked black and white***. Black wings with two white bars. ♀: ***orange-buff head to chest***, white belly, upperparts black streaked white, wings as ♂. Monotypic. Active pairs in tangles and clumps, only briefly with mixed flocks, glean from surface; easy to see. **Voice** Clear, inflected *dep?* or *chép?* in spritely series of 10–12, ascending and accelerating last 3–5 notes; *chir-chupchupchup*. **SS** Only black-and-white streaked antwren in W; see Moustached of more humid habitats (yellow belly, white cheeks, etc). **Status** Fairly common.

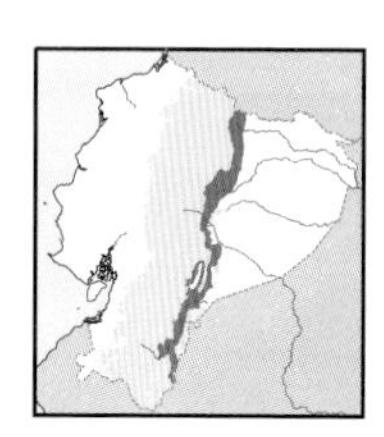

Stripe-chested Antwren *Myrmotherula longicauda* 9–10cm

E Andean foothills to adjacent lowlands (400–1,000m). Forest borders, secondary woodland, woody clearings. Black above densely streaked white, ***white below, chest streaked black***, narrow black malar. Black wings with two white bars. ♀: ***orange-buff above densely streaked black***, whitish throat and belly, ***ochraceous-buff breast*** with ***very faint blackish streaks***; wingbars ***orange-buff***. Races *soderstromi* (N), *pseudoaustralis* (S). Nimble pairs in mid to upper strata, glean from surface; may briefly join mixed flocks. **Voice** Slow, even series of 5–12 *cheétup-cheétup-cheétup...* or *chi'rp-ch'rp-ch'rp-ch'rp...*; *chu'rrip* or doubled *chiru'rrrrip* calls. **SS** Lower-elevation Amazonian Streaked Antwren (both sexes more streaked below, shorter tail); habitat differs. **Status** Fairly common.

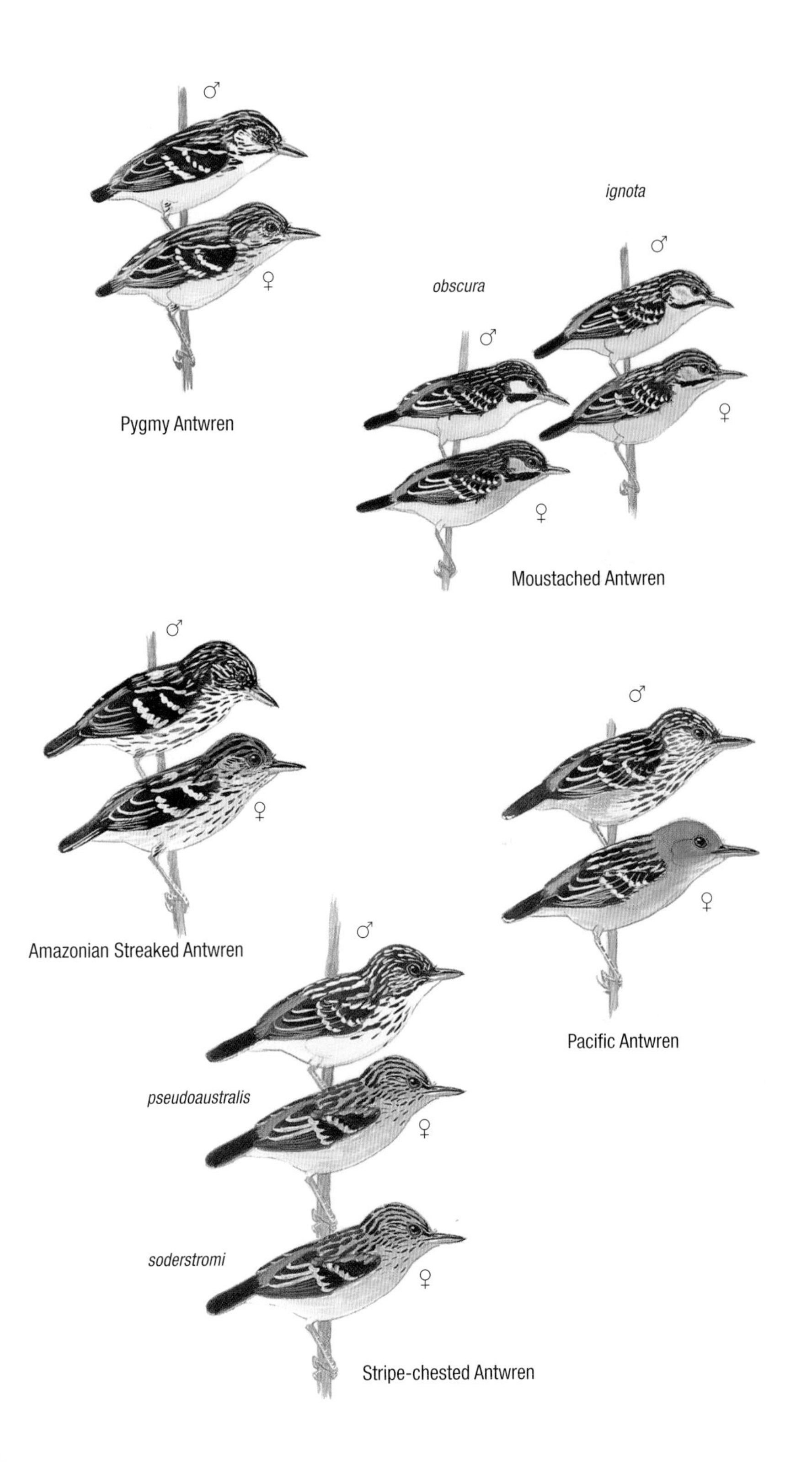
♂
♀
Pygmy Antwren
ignota
♂
obscura
♂
♀
♀
Moustached Antwren
♂
♀
Amazonian Streaked Antwren
♂
♀
Pacific Antwren
♂
pseudoaustralis
♀
soderstromi
♀
Stripe-chested Antwren

Mainly grey antwrens (olive-brown ♀♀), most with bold wing patterns. All species on this plate – excluding White-flanked – are confusing, especially ♀♀. Voices, bib size and shape, and wing patterns help. Plain-winged and Río Suno are rare and local. Slaty shares its montane range with the rare Plain-winged, but is much commoner.

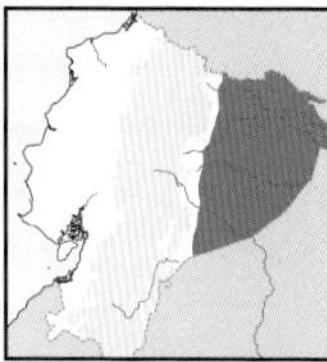

Plain-throated Antwren *Isleria hauxwelli* 10–11cm

E lowlands (to 400m). Mainly *terra firme* forest near streams. Mostly ***grey***, black wings, ***two bold white bars, white spots in tertials. Very short tail white-tipped and dotted.*** ♀: olive-brown above, ***ochraceous-cinnamon below***, paler throat, blackish wings, ***two buff bars and tertial spots***. Nominate. Pairs forage near ground, often hop on it, mostly apart from flocks. **Voice** Gives 6–10+ penetrating, accelerating, rising *chweér* notes; harsh *shígg!*, descending *ch'ch'ch'ch'ch*. **SS** Grey Antwren lacks white tertial and tail spots, wings and tail not blackish, duller ♀ lacks wingbars. **Status** Uncommon.

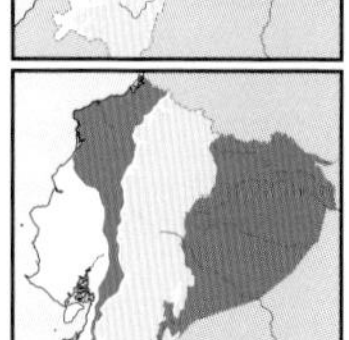

White-flanked Antwren *Myrmotherula axillaris* 10–11cm

E & W lowlands to foothills, S to W Loja (to 900m). Humid forest interior, borders, secondary woodland. Blackish, ***white flank tufts***, wings banded and edged white. ♀: brown above, buff below, paler throat, ***white flank tufts, faint buff wingbars***. Races *albigula* (W), *melaena* (E). Pairs or small groups in mixed understorey flocks, often flicks wings. **Voice** Slow, descending, whiny *pee-pee-peeu-peuh-peuh'pu'pu'pu*, last 2–3 notes faster; complaining *tchee'to-to, tchee?-tou, pew-pew-pew-pew* calls, rattled *ch'rrrrr*. **SS** ♀ recalls many congeners but flank tufts readily seen. **Status** Common.

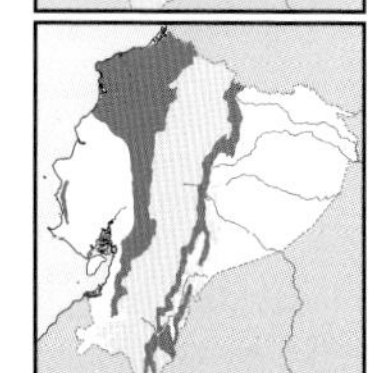

Slaty Antwren *Myrmotherula schisticolor* 10–11cm

W lowlands to foothills S to NW Loja, E foothills to subtropics (400–1,450m in W; 900–1,700m in E). Montane forest interior. Slate-grey, ***large black bib, faint white wing dots***. ♀: ***olive-brown*** (W) or ***greyish*** (E) ***above***, ochraceous-buff below, ***vague buff wingbars***. Nominate (W), *interior* (E). Active pairs in tangles and clumps, regularly with mixed flocks. **Voice** Inflected, doubled *jee-jeeít*; fast, complaining *chee-reewit, cheeu-chuir* scold. **SS** No overlap with paler Long-winged Antwren; cf. smaller, shorter-tailed Río Suno; Plain-winged (smaller bib, plain wings); ♀ Checker-throated (W). **Status** Uncommon.

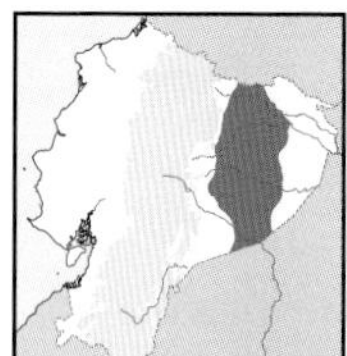

Río Suno Antwren *Myrmotherula sunensis* 9–10cm

E lowlands, not extreme E (to 250–500m, locally higher). *Terra firme* forest, adjacent woodland. Slate-grey, ***large black bib, white dotted wingbars, tiny tail***. ♀: dull brown above, buff below, whiter throat, faint pale supercilium, ***very faint wingbars***. Nominate. Poorly known, pairs near ground with mixed flocks. **Voice** Short, clear, rhythmic *sweé-suee'ee-suee*; *t'chéu* or *mhéu?* **SS** Slaty Antwren (longer-tailed, bib smaller); Long-winged (notably paler, longer tail tipped white, bolder wingbars); ♀ Plain-winged (less drab overall). **Status** Rare.

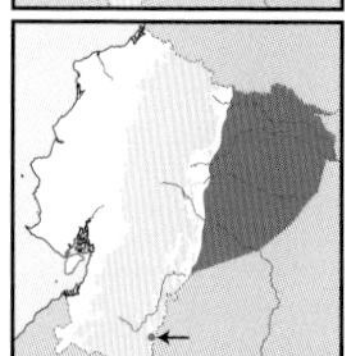

Long-winged Antwren *Myrmotherula longipennis* 10–11cm

E lowlands (to 500m, locally higher). *Terra firme* forest. ***Light grey, black bib***, darker wings, ***bold white bars, pale-tipped tail***. ♀: ***greyish-olive above, buff below; brown above, whitish belly*** (NE), faint wing markings, ***buff-tipped tail***. Races *zimmeri* (most range), nominate (extreme NE); separate species? Nimble pairs within mixed flocks, flicks wings constantly. **Voice** Harsh *chu'weer-chu'weer-chu'wee-chu'wee-chu'wee* ending abruptly; *djt-djt-djt* complaining calls; harsher *djú'heer-djú'heer...* song in NE. **SS** Darker Río Suno Antwren (larger bib, dotted and fainter wingbars and shorter tail); ♀ Grey (shorter, not buff-tipped tail, greyer upperparts). **Status** Rare.

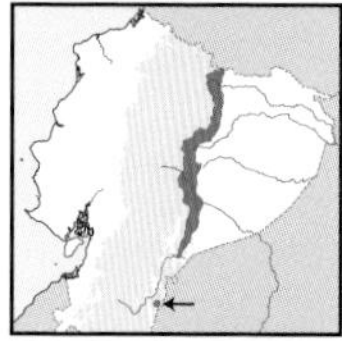

Plain-winged Antwren *Myrmotherula behni* 9–10cm

Local in E Andean foothills (800–1,600m). Montane forest. Dark grey, ***small black bib, plain wings, very short tail***. ♀: brown above, ***plain wings, whitish throat***, olive-buff below, ***tiny tail***. Nominate. Active pairs consort with Slaty Antwren in understorey flocks; acrobatic gleaner. **Voice** Steady series of loud 3–9 *cheu-cheu-cheu*; call short, emphatic *chuít-ek!* or *kweu-ék!* **SS** Plain wings and tiny tail distinctive; cf. ♀ Slaty Antwren (longer tail, greyer above, visible wingbars). **Status** Rare.

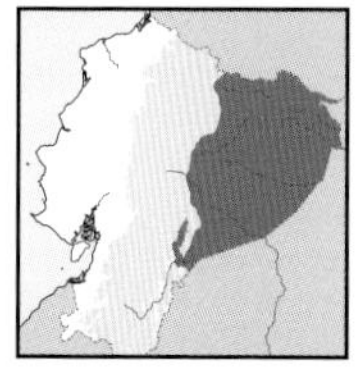

Grey Antwren *Myrmotherula menetriesii* 10–11cm

E lowlands (to 600m). *Terra firme* forest interior. ***Grey***, paler below, darker wings, ***three narrow white bars***, and short tail ***narrowly tipped white***. ♀: grey to greyish-brown above, faint buff wingbars, ***ochraceous-buff below***. Race *pallida*. Restless pairs to small parties ***twitch tail sideways***, often forage higher than most congeners. **Voice** Thin, quavering series of 8–14 *peeu*, rising and slightly accelerating; spritely *chii'wii'wii-wik, chee'week!* and other calls. **SS** ♀ Long-winged very similar but deeper below (note behaviour). **Status** Fairly common.

♂
melaena
♀
♂
♀
Plain-throated Antwren
♂
albigula
♀
White-flanked Antwren
♂
interior
♀
♂
schisticolor
♀
♂
juvenile
Slaty Antwren
♂
♀
Río Suno Antwren
longipennis
♂
immature
♂
juvenile
♀
zimmeri
♀
longipennis
Long-winged Antwren
♂
♂
immature
♀
juvenile
Plain-winged Antwren
♂
♀
Grey Antwren

Herpsilochmus inhabit forest canopy, where they join mixed flocks but remain inconspicuous unless vocalising. Two species in lowlands, two in foothills and subtropics. Among the boldest thamnophilids, with conspicuous face, tail and wing patterns. Also, monotypic Dot-winged of forest undergrowth, characteristic of both canopy and understorey flocks.

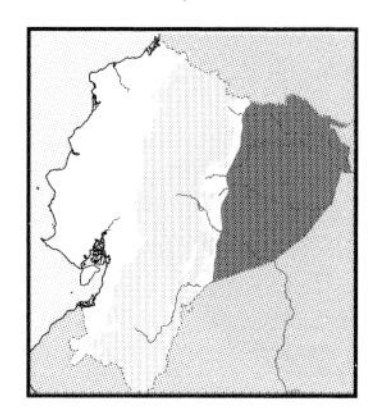

Dugand's Antwren *Herpsilochmus dugandi* 11–12cm

E lowlands (to 450m). *Terra firme* forest canopy. ***Black crown, grey upperparts, long white supercilium***, blackish ocular stripe, ***greyish-white underparts***. Blackish wings with white scapulars, bars and edges to flight feathers, long tail with ***white dorsal spots and broad tips below***. ♀: ***rufous crown***, face, upperparts and tail as ♂, ***underparts pale buffy***. Monotypic. Close pairs with mixed flocks, inspect outer foliage, quivers tail constantly. **Voice** Accelerating, fast, cadenced *ch..ch..ch-ch-ch'ch'ch'chchchch*; calls rich *chee-chut, chee-chee-chu'chut* and softer *cheew*. **SS** Similar Rufous-winged Antwren occurs at higher elevations (bold rufous flight feathers, yellowish belly); very local Ancient Antwren has yellowish underparts and eyebrow. **Status** Uncommon.

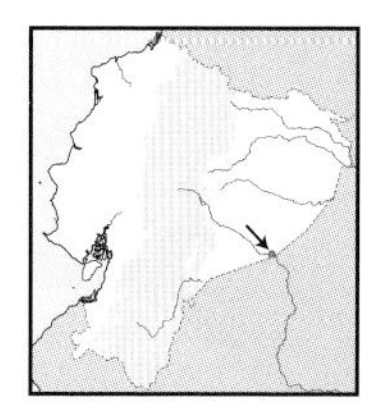

Ancient Antwren *Herpsilochmus gentryi* 11–12cm

Local in extreme SE lowlands (200m). *Terra firme* canopy in white-sand forests. Black crown, blackish upperparts. ***Long yellowish supercilium***, black ocular stripe, ***yellowish underparts***. Blackish wings, ***bold white edges and bars***; blackish tail edged and broadly tipped white. ♀: similar but ***crown spotted whitish, ochraceous breast wash***. Monotypic. Agilely hops and flutters in outer branches and tree crowns, often in pairs with mixed flocks, regularly flicks wings. **Voice** Fast, cheerful, descending series *chedEEdeedee-dee-deu-duh-duh-duh*; rich *chwép* calls. **SS** Confusion unlikely; Dugand's pale grey above (♂), buffy below (♀); much smaller Pygmy and Moustached Antwrens have short tails, streaky upperparts, dwell at lower heights. **Status** Rare, local, NT globally.

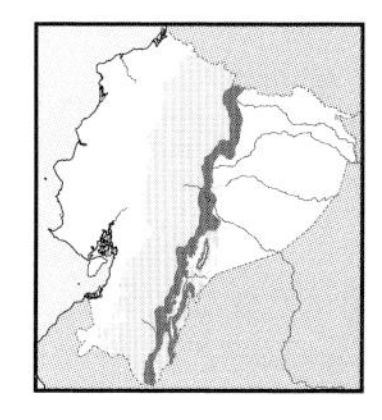

Yellow-breasted Antwren *Herpsilochmus axillaris* 11–12cm

E Andean foothills to subtropics (800–1,700m). Montane forest canopy, borders. ***Black crown spotted white, white freckled supercilium***, black ocular stripe. Whitish throat, ***yellow below***, blackish wing-coverts with ***two white bars***, broad white tail tip. ♀: similar but ***crown rufous***. Race *aequatorialis*. Active pairs with mixed flocks, inspect tree crowns and outer foliage. **Voice** Accelerating, rather sharp, rising-falling *chui-reeREErerep, chui-reeREErirep*; short descending *chí-reereu* calls. **SS** Rufous-winged Antwren lacks crown spots, underparts less yellow, bold rufous flight feathers; ♀ has more olive upperparts, buffier breast, rufous flight feathers, paler below; Rufous-rumped (plain crown, little yellow below, bold yellow wingbars). **Status** Rare, local; VU globally.

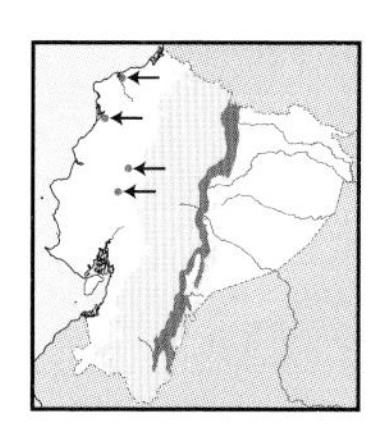

Rufous-winged Antwren *Herpsilochmus rufimarginatus* 11–12cm

E Andean foothills, local in NW lowlands (600–1,300m in E, to 200m in W). Montane forest canopy, borders, woodland. ***Black crown, long white supercilium***, black ocular stripe. ***Greyish-olive upperparts, yellowish-white underparts***. White scapulars and wingbars, ***bold rufous flight feathers***, tail narrowly edged white. ♀: similar but ***crown rufous***. Race *frater* (E), unknown race in NW. Active pairs in outer foliage follow mixed flocks, often descending at edges. **Voice** Loud, nasal, descending *pë-pë-pë-prrr'üp, pë-pë-pë-prrr'üp*; nasal and short *pë-peereereep* calls. **SS** Readily distinguished from other montane antwrens by rufous wings; cf. differently shaped Rufous-winged Tyrannulet (bold wingbars, short bill, bluish-grey crown, olive upperparts). **Status** Uncommon.

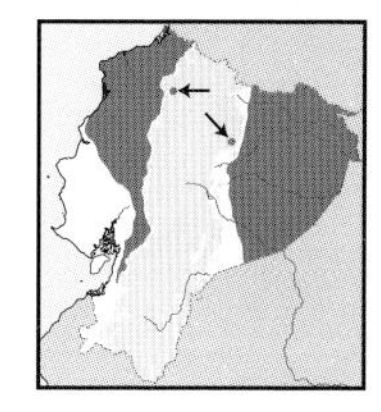

Dot-winged Antwren *Microrhopias quixensis* 12–13cm

E & W lowlands, S to S Guayas in W (to 500m, locally higher). Humid forest borders, secondary woodland. ♂: black, ***very bold white-dotted wing markings***, long graduated tail ***broadly tipped white***. ♀: ***chestnut*** below (W), ***black throat*** (E). Races *consobrinus* (W), nominate (E). Very active pairs to small groups in mixed flocks in various strata, mostly gleans; constantly spreads and cocks tail; easy to see. **Voice** Short, fast, descending whistles *whip-whip-whip-wi'i'i'i'u*, fading at end; clear *kee'yu* or *cheeu* (W), harsher *dyí-duuk* (E) calls, also harsh *r'r'rj*, softer *téew* and others. **SS** Unmistakable; White-flanked Antwren has much shorter tail and white flank tufts. **Status** Fairly common.

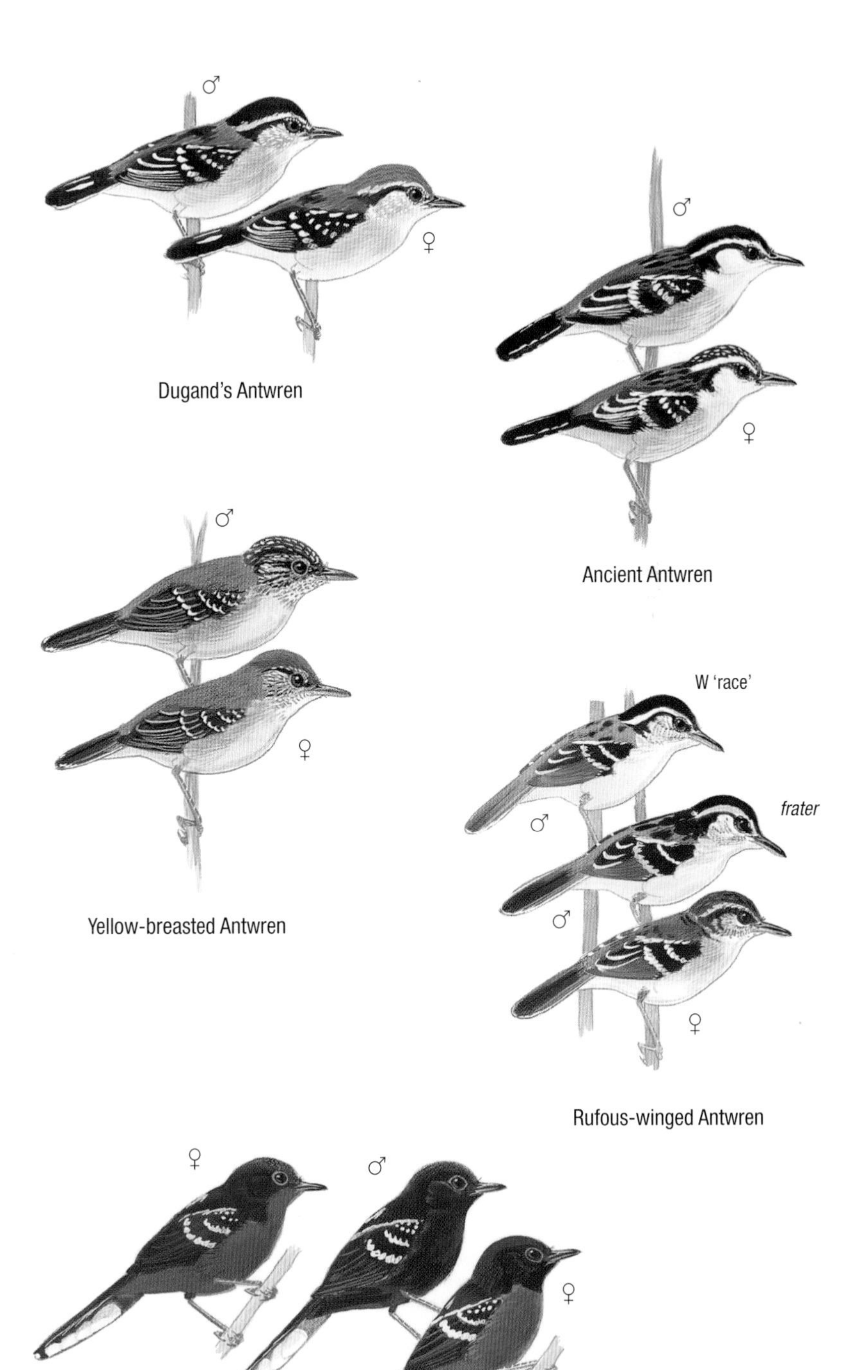
♂
♀
Dugand's Antwren
♂
♀
Ancient Antwren
♂
♀
Yellow-breasted Antwren
W 'race'
frater
♂
♂
♀
Rufous-winged Antwren
♀
♂
♀
consobrinus
quixensis
Dot-winged Antwren

Euchrepomis antwrens are sometimes bestowed the odd name 'ant-warblers' owing to their shape and behaviour. Small canopy antwrens, very inconspicuous unless vocalising. Also boldly patterned, especially their wings. Additionally, two very bold and notably vocal antbirds with long, graduated tails. Both associated with bamboo, with Striated only recently rediscovered in Ecuador.

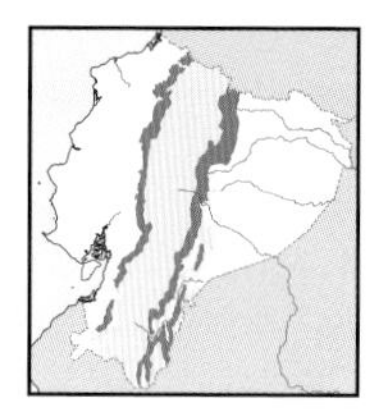

Rufous-rumped Antwren *Euchrepomis callinota* 11–12cm

Foothills to subtropics on both Andean slopes (900–1,800m). Montane forest canopy, borders. ***Black crown***, long greyish-white supercilium, greyish-white cheeks to breast, pale yellow below. Olive back, ***rufous rump***, long olive tail. Duskier wings, ***yellow shoulders and two bold bars***. ♀: crown ***olive, paler below***. Nominate. Restless pairs with mixed flocks inspect outer branches, foliage and undersides of leaves. **Voice** Thin *ssip..si-si-si-si-sisiririp* running into chipper; calls thin *sseep* or *siip, siirip*. **SS** Lower-elevation Chestnut-shouldered Antwren has distinctive, though not always discernible, chestnut shoulder patch; ♀♀ difficult to separate but Chestnut-shouldered has browner crown. **Status** Uncommon.

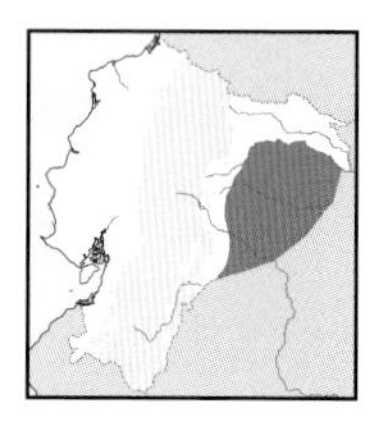

Chestnut-shouldered Antwren *Euchrepomis humeralis* 11–12cm

E lowlands N to Río Napo (below 600m). *Terra firme* forest canopy. Resembles previous species but underparts greyer, paler yellow flanks and vent, ***chestnut shoulders***. ♀: ***crown browner***, upperparts duller. Nominate. Acrobatic pairs or small compact groups nearly always with mixed flocks, inspect outer foliage and its underside. **Voice** Thin, accelerating *see-si-si-sisisisiririr'r'r'r*, soft *sép* calls. **SS** Much rarer Ash-winged Antwren shares rufous rump but is grey overall (♂) or brownish-olive (♀), lacking shoulder patch; no known overlap as separated by Napo; higher-elevation Rufous-rumped very similar (♀), but crown duller, slightly paler underparts. **Status** Rare.

Ash-winged Antwren *Euchrepomis spodioptila* 10–11cm

Very local in lowlands and foothills of extreme NE (to 900m). *Terra firme* forest canopy. Mostly ***ash-grey***, blacker crown, contrasting ***rufous lower back and rump***, vague whitish wingbars, ***pale grey supercilium***. ♀: ***dull brown above, rufous lower back and rump, faint supercilium***, greyish-buff below. Possibly race *signata*. Poorly known in Ecuador, but presumably behaves much like congeners. **Voice** Similar to congeners, a thin, high-pitched fast *see-see-see-sisisisiririri*. **SS** Rather nondescript ♀ might resemble greenlets, vireos or even conebills but note wingbars and contrasting rufous rump. No known overlap with other *Euchrepomis*, from which told by ash-grey (♂) or brownish (♀) body, no shoulder patch, vague wingbars. **Status** Rare, local.

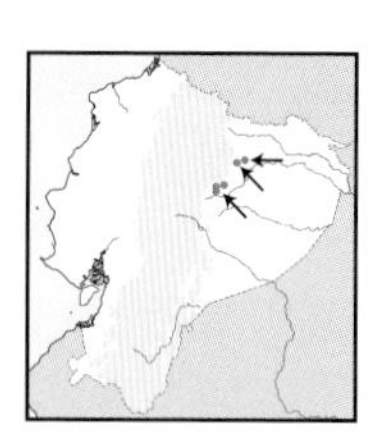

Striated Antbird *Drymophila devillei* 13–14cm

Very local in NE foothills to adjacent lowlands (300–750m). Dense second growth, bamboo. ***Boldly streaked blackish and white from crown to back***, lower back and rump cinnamon-rufous. White below, ***streaked breast-sides***, more rufescent flanks to lower belly; two whitish wingbars. Long graduated tail with ***bold white tip***. Nominate. Pairs to small groups inspect bamboo nodes and tangles; often with mixed flocks, but also away. **Voice** Raspy *jt-jt-JITITI'titi, jt-jt-jt-jt-jí-titititi, jít-jijijí*, similar in quality to commoner congener; metallic, paired *cht'jit, cht'jit, jít, ji-jít* calls. **SS** Recalls commoner Streak-headed (no overlap); see short-tailed, chunkier, very streaked Peruvian Warbling Antbird. **Status** Very rare.

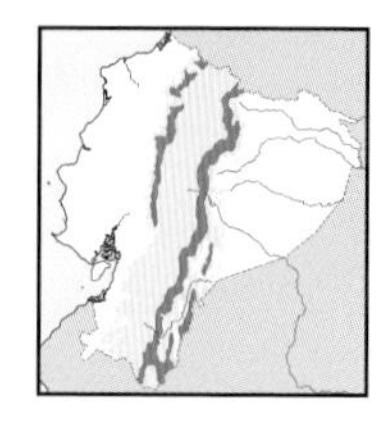

Streak-headed Antbird *Drymophila striaticeps* 14–15cm

Subtropics on both Andean slopes, more local in W (1,500–2,600m). Montane forest undergrowth, borders, bamboo. ***Boldly streaked blackish and white*** (♂) or ***buff*** (♀) ***from crown to back and breast***, lower back and rump rufescent. White below, bright cinnamon-rufous lower belly, flanks and vent. Black wings, white dots on coverts. ***Long graduated tail with white tip***. Race *occidentalis*. Active and vocal pairs away from flocks or only briefly with them, inspect dense foliage and bamboo tangles. **Voice** Raspy *cheek-cheek...whiít-whiít, cheek-cheek...kuiít-kuiít-kuiít...*; call harsh *chee'kík, chee'kík*. **SS** Unmistakable in most of range; no overlap with much rarer Striated Antbird. **Status** Common.

♂
♀
Rufous-rumped Antwren
♂
♀
Chestnut-shouldered Antwren
juvenile
♂
♀
Ash-winged Antwren
juvenile
♂
♀
Striated Antbird
♂
♀
Streak-headed Antbird

Long-tailed antbirds, from ash-grey to jet black (♂♂). Bills long and rather slender for an antbird. They dwell in lower growth of forest and edges, where pairs duet; two species in W, four in E. The fire-eye is a bold, large antbird of forest undergrowth, each subspecies once considered a valid species. Very vocal and often follows ant swarms.

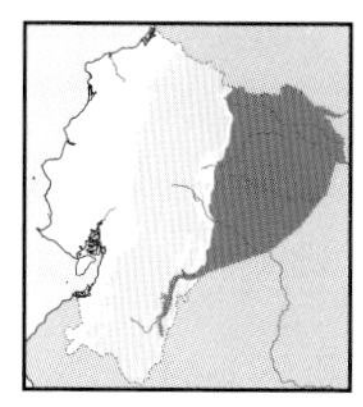

Grey Antbird *Cercomacra cinerascens* 14–15cm

E lowlands to lower foothills (to 900m). *Terra firme* interior, borders, woodland. ***Grey***, wings ***nearly unmarked, long blackish tail, broad white tip***. ♀: dull greyish-olive above, greyer rump, nearly ***unmarked wings, broad white tail tip***. Nominate. Pairs regularly higher above ground than many antbirds, fond of vine tangles, with or apart from mixed flocks; vocalises most of day. **Voice** Utters 5–9 doubled, raspy *tjj'kúrr-tjj'kúrr-tjj'kúrr...*, second note of each pair higher; nasal *keéyur* calls, fast *kjkjkjkjk*, mewing *kheew*, simpler *kéw*. **SS** White tail tip distinctive; paler than sympatric congeners. **Status** Common.

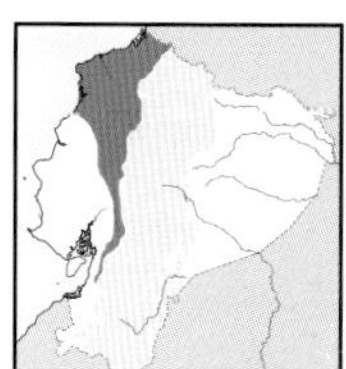

Dusky Antbird *Cercomacroides tyrannina* 14–15cm

W lowlands, S in foothills to N El Oro (to 1,400m). Humid forest borders, woodland. ***Smoky-grey, narrow white wingbars, long tail*** with ***narrow whitish tip***. ♀: dull brown above, ***ochraceous-buff below, vague buff wingbars and tip to tail***. Nominate. Skulking pairs in dense tangles near ground, briefly join mixed flocks. **Voice** Fast, slowly accelerating, whistled series, inflected at end: *kyü-kyü-kee'kee'kikiki?*; nasal *kyuú* calls, scratchy *w'iirrr*. **SS** Locally overlaps with Jet Antbird (noticeably blacker, broader tail tip; ♀ blackish-grey, streaky throat). **Status** Fairly common.

Blackish Antbird *Cercomacroides nigrescens* 14–15cm

E Andean foothills (1,000–1,800m). Montane forest borders. ***Sooty-grey, narrow white fringes to wing-coverts***, long ***plain tail***. ♀: dull ***olive-brown*** above, ***rich orange-rufous underparts***, long ***dusky-brown*** tail, ***unmarked*** wings. Race *aequatorialis*. Rather shy pairs in dense tangles near ground, seldom with mixed flocks. **Voice** Antiphonal duets *dür-CHI'ti'tititi*, sometimes longer *dü-dü-dü-chi'ti'tití*, followed by more enunciated *pür..pü-pi'pi'pi?* by ♀; also harsh *ji'g'g'g* calls. **SS** Not safely separated from Black Antbird but greyer; ♀ Black has greyer tail (voices and habitat differ). **Status** Uncommon.

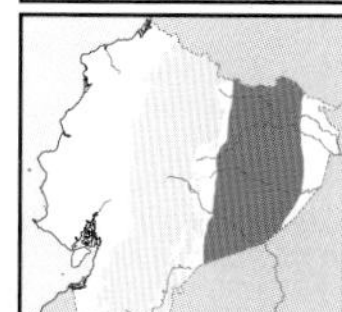

Riparian Antbird *Cercomacroides fuscicauda* 14–15cm

Local in E lowlands (to 500m). Riparian second growth, shrubby clearings. Very similar to Blackish Antbird but wings ***plainer*** in ♂; ♀ has ***greyer*** tail. Monotypic. Shy but vocal pairs in dense tangles near ground, seldom with mixed flocks. **Voice** Antiphonal duets simpler but louder than Blackish, more drawn-out *dür-chEyrr*; *dür-chEyrr...dür-dür-dür-dür*; also harsh *ji'g'g'g* calls. **SS** Paler than Black, darker than Grey Antbird; no habitat overlap; voices differ. **Status** Rare.

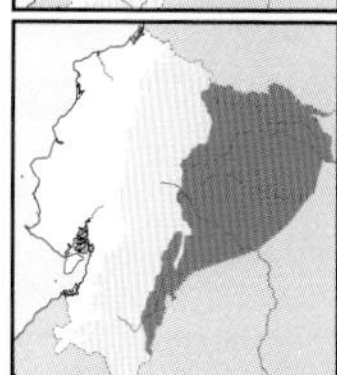

Black Antbird *Cercomacroides serva* 14–15cm

E lowlands to foothills (to 1,400m). Shrubby forest borders, adjacent woodland. Recalls Blackish Antbird, but ***blacker***, white fringes to wing-coverts, ***long unmarked tail***. ♀: more ***olivaceous-brown*** above, ***orange-buff*** below, but less rich on face and forecrown. Nominate. Pairs in dense tangles near ground, briefly with mixed flocks but mostly apart; not in riverside vegetation like Riparian. **Voice** Loud, harsh, ascending, slightly accelerating *dör-dür-dee-dee-deedeedididi?* ♀ higher-pitched, slower; call a fast *dEr-drdr'dr'dr'dr*. **SS** Grey Antbird paler, greyer, tail boldly tipped white; ♀ paler and duller, well-marked wings and tail. **Status** Uncommon.

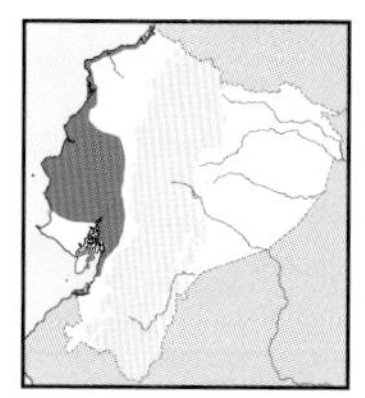

Jet Antbird *Cercomacra nigricans* 14–15cm

SW lowlands, N along coast to W Esmeraldas (below 500m). Semi-deciduous forest borders, secondary woodland, shrubby clearings. ***Black***, bold ***white wing markings, long tail with broad white tip***. ♀: ***blackish-grey*** with ***narrow whitish streaks on throat to upper chest***, narrow white wingbars, tail as ♂. Monotypic. Rather shy pairs in tangled undergrowth mostly apart from flocks. **Voice** Loud, short, fast series of 3–5 *chi'cö, chi'cö, chi'cö...*; emphatic, fast *chakrr-chakrr-krr...* calls. **SS** Confusion unlikely; locally sympatric Dusky Antbird is grey not black, wings and tail less patterned in white; Jet Antbird prefers drier habitats. **Status** Uncommon.

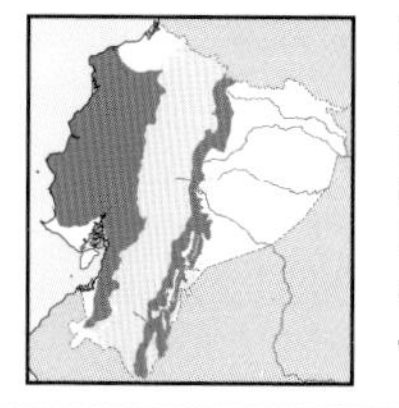

White-backed Fire-eye *Pyriglena leuconota* 17–18cm

W lowlands to foothills; E Andean foothills to subtropics (to 1,450m in W, 1,000–2,000m in E). Montane (E), deciduous to humid forest (W), borders. ***Glossy black, bright red eyes, large*** white dorsal patch, ***long tail***. E ♀ ***black and chestnut-brown***; W ♀ ***dull olive-brown above, paler below***. Races *pacifica* (W), *castanoptera* (E). Pairs to small groups skulk in dense tangles, regularly flock at ant swarms. **Voice** Rather fast, steady series of loud *kír* whistles; nasal, loud *chí, chïkíp* calls. **SS** In W, Jet Antbird (white-patterned wings and tail); in E Blackish and Black Antbirds. **Status** Uncommon.

♂
♀
Grey Antbird
♂
♀
Dusky Antbird
♀
♀
♀
♂
♂
♂
♀
Black Antbird
Blackish Antbird
Riparian Antbird
♂
castanoptera
♂
♀
♀
Jet Antbird
♀
pacifica
White-backed Fire-eye

Small, chunky Amazonian antbirds, all are short-tailed, but two are boldly streaked on richly coloured backgrounds. The other three are predominantly grey and black (♂♂) or brown and white (♀♀). One species is confined to river islands. All may join mixed flocks but move separately in family groups too.

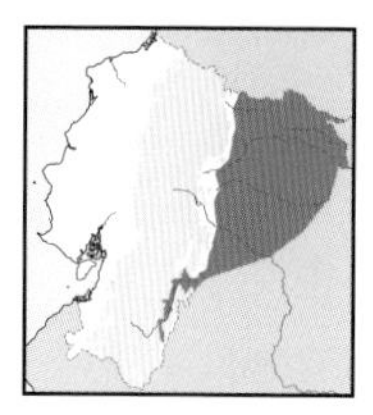

Peruvian Warbling Antbird *Hypocnemis peruviana* 12–13cm

E lowlands to lower foothills (to 900m). Mostly dense *terra firme* forest. ***Black crown to mantle coarsely streaked whitish, white supercilium, face to lower breast streaked blackish***, brown rump, ***orange-rufous flanks***. Black wing-coverts with ***bold white bars***, short tail ***narrowly tipped whitish***. ♀: ***buff replaces white*** in crown to mantle and wing-coverts; ***fainter streaking***. Race *saturata*. Vocal pairs in dense tangles, sometimes away from mixed flocks, gleans, hops and sallies after prey. **Voice** Raspy, growling *heer-heer-heer-heer...hurh-hurhh*, often ending in hoarser *djjrr*; ♀ higher-pitched; nasal *wöoor*, raspy *djérrr, djee-jerr-jerr* calls. **SS** Very local Striated Antbird (long, graduated tail tipped white, bolder wingbars and edges). **Status** Fairly common.

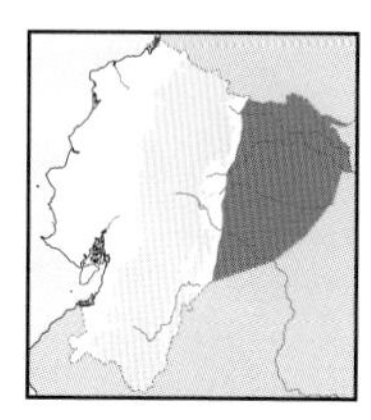

Yellow-browed Antbird *Hypocnemis hypoxantha* 12–13cm

E lowlands (below 400m). *Terra firme* forest interior. ***Black crown, white central crown-stripe*** (***yellow*** in ♀). Olive above, ***rich yellow face, including supercilium, and entire underparts***. Black ocular and malar stripes, ***black streaks continue from malar onto flanks; three bold white wingbars***. Nominate. Pairs regularly with mixed flocks in fairly open understorey, deliberate hops. **Voice** Recalls Peruvian Warbling Antbird but slower, more measured and hesitant, a descending *péer..peer..peer-peer-peer*, lacking hoarse end; fast, raspy *wree-pür, wree-pür-pü, preeéuk*. **SS** Unmistakable; Peruvian Warbling Antbird streaked black and white, rich rufous flanks, etc. **Status** Uncommon.

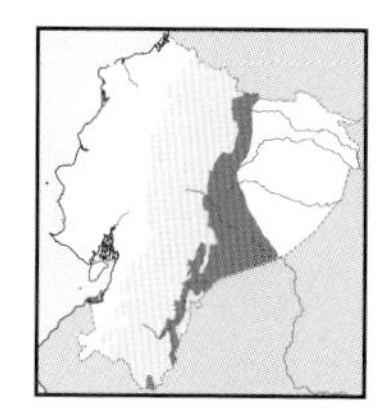

White-browed Antbird *Myrmoborus leucophrys* 13–14cm

E foothills to adjacent lowlands (400–1,100m, locally lower). *Terra firme* forest borders, shrubby second growth. Mostly slate-grey, ***broad white supercilium, black face to throat***. ♀: rufescent above, ***broad buff supercilium, small black mask, white underparts***, faint buff-dotted wingbars. Nominate. Mostly pairs, sometimes small parties, in dense tangles near ground, often with mixed flocks, sallies and hops from vertical perches. **Voice** Fast, descending, ringing *EE-féé-fififififififi'fr'fr'fr'fr'fr* slowly fading at end; weak, harsh *jerrg* calls. **SS** Little overlap with Black-faced Antbird, which has narrower supercilium, reddish eyes, white (♂) or buff (♀) fringes to wing-coverts, whitish-grey (♂) or buffy (♀) breast to belly. **Status** Uncommon.

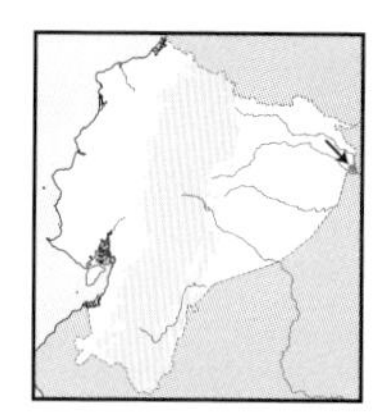

Ash-breasted Antbird *Myrmoborus lugubris* 13–14cm

Known from river islands in extreme E of Río Napo (150m). Riparian shrubbery and woodland. ***Red eyes***. Grey above, ***black mask to throat, whitish-grey underparts***. ♀: chestnut crown, brown upperparts, ***black mask, whitish throat, pale grey below***, rufescent wing-coverts, vague buff fringes. Race *berlepschi*. Mostly pairs in dense lower growth, often *Heliconia* stands, perches vertically, sallying and hopping. **Voice** Fast, descending series, similar to White-browed, *FEEEfééfifi'fi'fi'fi-fi-fi-fu-fu* but shorter, slowly fading at end; complaining *peer?-pur?, peer?-puur?-peer?*, descending *peuw* calls. **SS** Confusion unlikely in limited habitat and range; from Black-faced (no habitat overlap) by lack of supercilium, unmarked wings, brighter eyes. **Status** Uncommon, very local; VU globally.

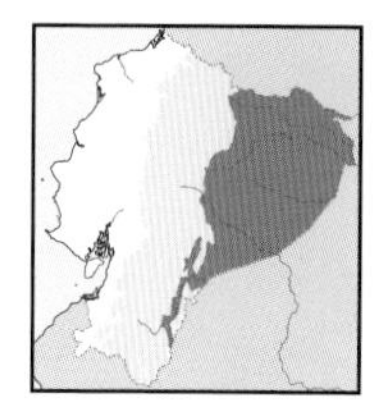

Black-faced Antbird *Myrmoborus myotherinus* 13–14cm

E lowlands to foothills (to 1,300m). *Terra firme* forest interior, adjacent woodland. Grey above, ***diffuse pale grey eyebrow bordering black mask***. Whitish-grey underparts, ***blacker wing-coverts, 2–3 white bars***. ♀: brown above, ***black mask, white throat, ochraceous-buff eyebrow and underparts, buff wingbars***. Race *elegans*. Mostly pairs, perch vertically, sally and hop for prey, fond of fairly open understorey. **Voice** Loud, descending, harsh series of 6–13 *djéé-djee-djee...*; harsh *jeert!, chízt!, jée'urr*, descending *cheeu*; *j'tttt* rattle. **SS** Confusion unlikely by long pale eyebrow, obvious black mask and marked wings. **Status** Uncommon.

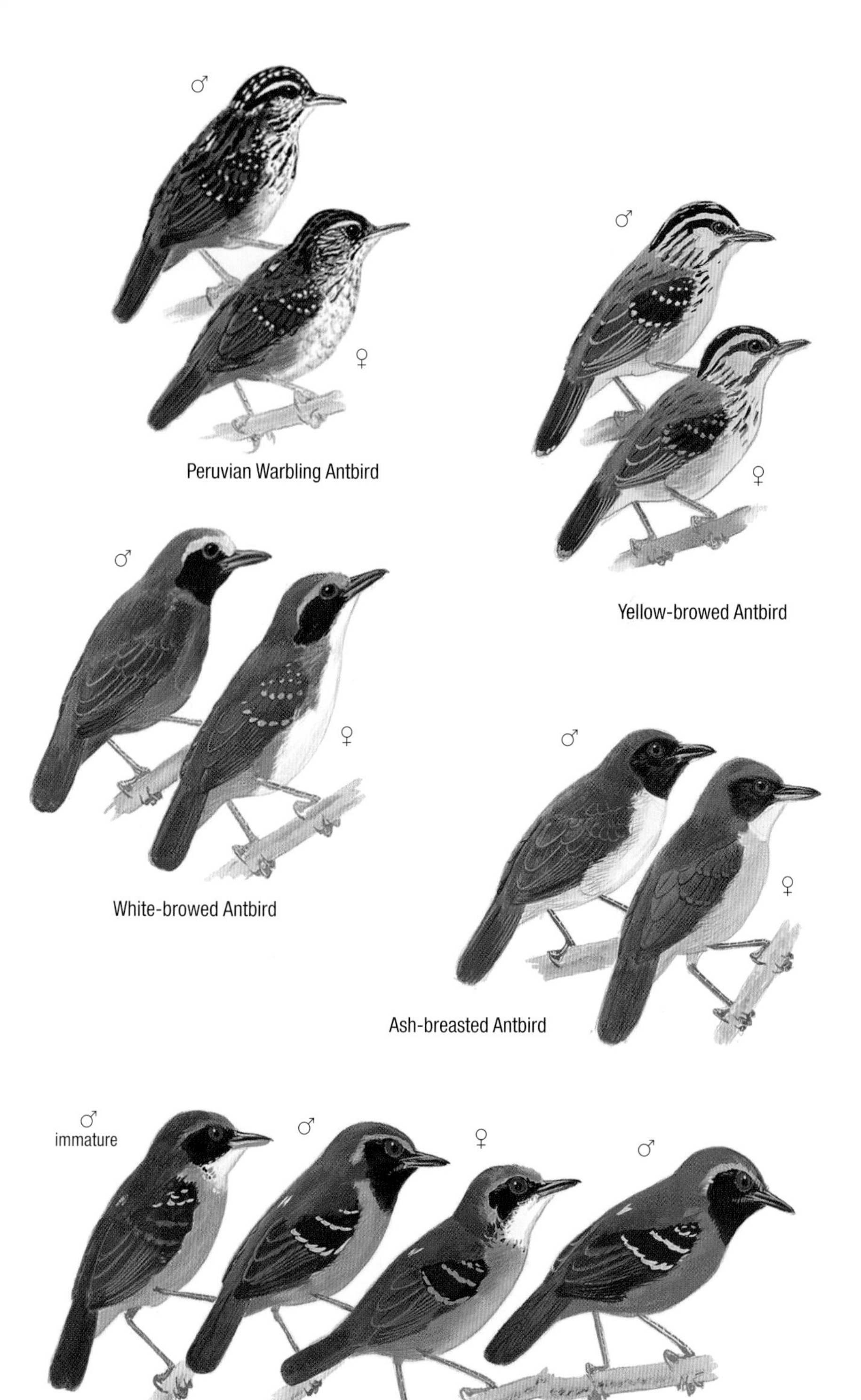
♂
♀
Peruvian Warbling Antbird
♂
♀
Yellow-browed Antbird
♂
♀
White-browed Antbird
♂
♀
Ash-breasted Antbird
♂
immature
♂
♀
♂
Black-faced Antbird

Four very distinctive Amazonian antbirds; Black-chinned and Silvered occur in flooded forests; Black-and-white on river islands; and Black Bushbird in dense secondary scrub. Most are poorly known. Also, the two rare species of *Schistocichla*, which might not be safely identified except by voice – although their habitat preferences differ.

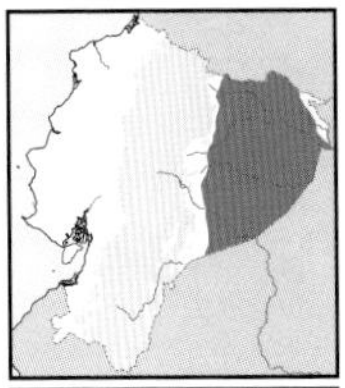

Black Bushbird *Neoctantes niger* 15–16cm

E lowlands (to 600m). *Terra firme* forest, dense secondary woodland. ***Upturned bill*. ♂: *all black*, ♀: *chestnut pectoral patch*.** Monotypic. Skulking pairs in very tangled undergrowth, often apart from flocks, inspect putrid logs and litter or glean foliage and vines. **Voice** Long, steady, cadenced *weop-weop-weop...*; nasal *ki'yoo* calls, slow *bi'r'r'r'r* rattle. **SS** Unique bill shape separates sympatric black antbirds, as well as large chestnut patch in ♀; cf. Black-throated, Sooty and Black Antbirds. **Status** Rare.

Black-chinned Antbird *Hypocnemoides melanopogon* 11–12cm

Lowlands of extreme NE (to 200m). *Várzea* and *igapó* undergrowth. ***Pale grey eyes***. Mostly grey, ***contrasting black throat***. Blacker wings, ***three narrow white bars, very short tail tipped whitish***. ♀: ***throat whitish freckled grey***, breast to flanks ***mottled grey***. Race *occidentalis*. Pairs to small noisy and active groups forage near ground, sometimes on it, and low above seasonally flooded areas. **Voice** Loud, accelerating *djeé-djeé-djeé'djee'djeedjeedjee...*, descending, with raspier end; hoarse *djéuu* and emphatic *chék*. **SS** Confusion unlikely in habitat; cf. Long-winged Antwren (no habitat overlap) and larger, pink-legged, grey-and-white Silvered Antbird. **Status** Fairly common but local.

Black-and-white Antbird *Myrmochanes hemileucus* 11–12cm

E lowlands on Río Napo islands (to 300m). Early successional riparian second growth. ***Long slender bill. Black above, large*** white semi-concealed dorsal patch, grey rump. ***Whitish underparts, white spotty wingbars***, black tail tipped white. ♀: ***short white loral stripe***, yellowish rear parts. Monotypic. Active and not inconspicuous pairs in low dense growth but sometimes in subcanopy, fond of *Tessaria* shrubs, young trees and *Gynerium* cane. **Voice** Fast series of *tu* notes rising in pitch, slightly accelerating *tu-tu-tu'u'u'i'i'i*, followed by inflected *tuut?, tuu-tuu-tuut?*; short *tu'rrt* rattle, hollow *pu-pu-pu-purr*, descending *peew*. **SS** Confusion unlikely. **Status** Uncommon.

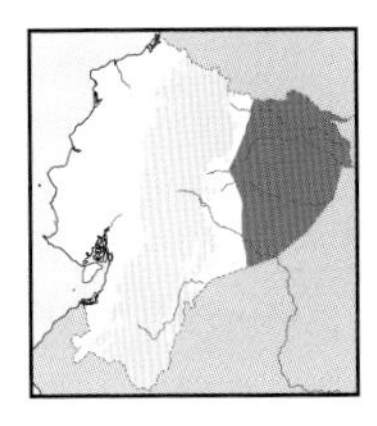

Silvered Antbird *Sclateria naevia* 14–15cm

E lowlands (to 450m). *Várzea* and *igapó* forest undergrowth, palm swamps. ***Long pink legs, long bill***. Grey above, ***silvery-white underparts; three white-spotted wingbars***. ♀: rufescent-brown above, ***richer orange supercilium and sides to flanks, faint buff*** wing spots. Race *argentata*. Active pairs low above black-water streams and in seasonally flooded forests, deliberately foraging, sometimes even on wet ground. **Voice** Ringing, accelerating, rising-and-falling series: *DJEÉ, ti-ti-ti-titiTITITIti-ti'trr*; harsh *djuip*, loud *jit-jit-chík!*, rattled *ch'trrr*. **SS** Long pink legs and grey-and-white (♂) or brown-and-white (♀) patterns distinctive; see shorter-legged, short-tailed, grey-eyed Black-chinned Antbird. **Status** Uncommon.

Slate-coloured Antbird *Schistocichla schistacea* 14–15cm

Local in extreme NE lowlands (below 250m). *Terra firme* forest. ***Grey eyes***. Slate-grey, blacker wings, with ***very faint whitish spots***. ♀: ***rufescent-brown above***, some ***dark crown streaks, cinnamon to orange-rufous below, buff wing spots***. Monotypic. Pairs or small parties in dense undergrowth, often on ground, shy, avoids mixed flocks. **Voice** Slow, piercing, slightly accelerating *eeyr-eeyr-eeyr-eeyr-eeyr...*; dry *tjít!*, whistled *éeu* calls. **SS** ♂ not safely separated from Spot-winged except by voice; cf. grey vs. brown eyes, blackish vs. bluish mandible, faint vs. clearer wing spots. ♀ Spot-winged has grey head, dark eyes; see ♀ Black Antbird (unmarked wings, olivaceous-brown upperparts, dark eyes) and ♀ antshrikes (heavier hooked bills). **Status** Rare.

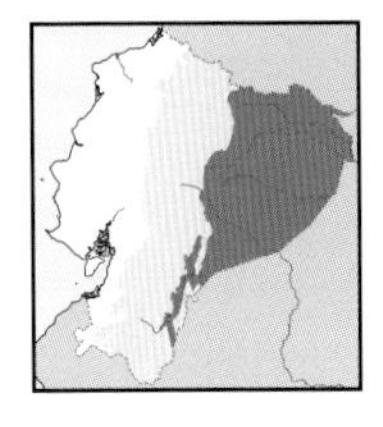

Spot-winged Antbird *Schistocichla leucostigma* 14–15cm

E lowlands, locally to foothills (mostly below 600m, locally to 1,100m). Swampy areas near forested streams. ***Brown eyes, bluish-grey mandible***. Slate-grey, blacker wings, ***three rows of white spots***. ♀: ***grey-headed***, brown upperparts, ***rich rufous below, buff spotty bars***. Race *subplumbea*. Retiring pairs or small groups near ground, seldom joins mixed flocks. **Voice** Short, fast, ringing *pii'pii'piillliuuu...*, descending and fading; sputtering *cht't't't't-t*, with whining *keéyu!* interspersed; dry *chi'tít!* **SS** Rarer Slate-coloured (see above); ♀ Plumbeous (larger, bolder wing spots, bluish orbital skin, grey back), Zimmer's (bold wing pattern, orange-buff breast, cinnamon tail). **Status** Rare.

♂
♀
Black Bushbird
♂
♀
Black-chinned Antbird
♂
♀
Black-and-white Antbird
♂
♀
Silvered Antbird
♂
♀
Slate-coloured Antbird
♂
immature
♂
♀
Spot-winged Antbird

PLATE 170: WESTERN ANTBIRDS

All western '*Myrmeciza*' antbirds are distinctive but difficult to observe but their voices are loud. Complex taxonomy; major recent splits. Esmeraldas and Stub-tailed moved back to genus *Sipia*; Chestnut-backed now in monotypic *Poliocrania*; and Zeledon's moved to *Hafferia*. The only truly montane *Myrmeciza* (Grey-headed) currently separated in *Ampelornis*, as its voice and plumage are notably different. It is one of the most threatened antbirds in Ecuador.

Chestnut-backed Antbird *Myrmeciza exsul* — 13–14cm

W lowlands to foothills, S along base of Andes to N El Oro (to 1,500m). Humid forest undergrowth, adjacent secondary woodland. ***Bluish orbital skin. Slate-grey head to belly. Dark chestnut back***, duskier wings with small white dots. ♀: ***grey crown, dark chestnut back***, wings as ♂ but spots buffier, ***rich orange-rufous below***. Race *maculifer*. Noisy pairs in overgrown vegetation at treefall gaps; often joins mixed flocks and attends ant swarms. **Voice** Ringing, doubled or tripled *tcheé-chéea, tchep-chéea, chee-chee-chéea*; harsh *jreéur!* and *t'ipipipip*. **SS** ♀ Esmeraldas has red eyes, stippled throat, larger wing spots, no orbital skin; otherwise confusion unlikely. **Status** Common.

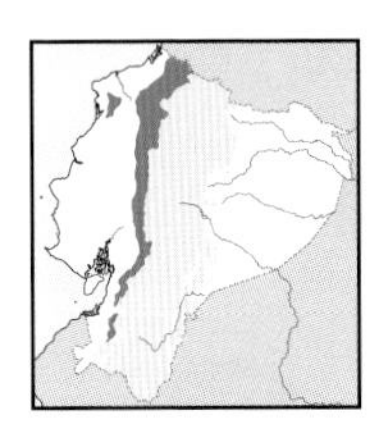

Esmeraldas Antbird *Myrmeciza nigricauda* — 13–14cm

W lowlands to foothills, S along Andes to N El Oro (to 1,300m). Humid to wet forest undergrowth, adjacent woodland. ***Red eyes. Entirely sooty-grey, three bold rows of white dots*** on wing-coverts. ♀: ***slate-grey head to belly, throat stippled whitish and black***, chestnut-brown above, ***wing dots buffier***. Monotypic. Mostly pairs in dense tangles, sometimes joins mixed flocks and attends ant swarms. **Voice** Slow, high-pitched, emphatic *je'jEE'je'je'je'je..jEE, je'jEE'jejeje'jijEE* or *che'chEE'checheche...*; throaty *d'jezz-d'jeyr-tjtj* calls. **SS** ♂ Chestnut-backed (dark eyes, plain throat, blue orbital skin); blacker ♂ Stub-tailed (red eyes, bold white wing dots); White-backed Fire-eye (much larger, long tail, large white dorsal patch). **Status** Uncommon.

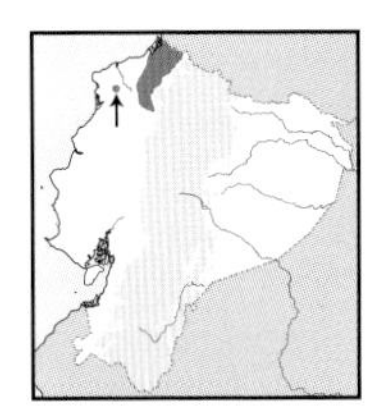

Stub-tailed Antbird *Myrmeciza berlepschi* — 13–14cm

NW lowlands (to 400m, locally higher). Humid forest undergrowth, adjacent woodland. Rather ***long*** blackish bill, ***brownish-red eyes***. ***Black***, with large white semi-concealed dorsal patch. ♀: ***throat to lower breast spotted whitish***, vague white dots form wingbars. Monotypic. Pairs in dense tangles, often at treefall gaps, shy but vocal, sometimes joins mixed flocks. **Voice** Ringing, single-noted *teeu-teeu-teeu-teéu-teéu...* or *cheeu-cheeu...* slowing, but rising in pitch, each note rather downslurred; sharp *chit, ch'dít* calls. **SS** Eye colour and plain wings separate from greyer Esmeraldas Antbird; note brown upperparts, stippled throat, buff wing dots in ♀. **Status** Uncommon, NT in Ecuador.

Zeledon's Antbird *Myrmeciza zeledoni* — 17–18cm

W lowlands to lower subtropics, not in dry SW (to 1,400m, locally to 2,000m). Humid forest understory, borders, adjacent woodland. ***Large bluish orbital skin. Mostly blackish, white wing bend***. ♀: mostly rufescent-brown above, duskier tail, unmarked wings, ***dusky face and throat, rufous-chestnut below***. Race *macrorhyncha*. Pairs or small groups regularly at ant swarms; clings to vertical stems. **Voice** Gives 10–12 whistled, ringing *peeü* notes in steady, fast series; call more emphatic *teé-teé-teé*, also simpler *theép* or *cheep!* **SS** Confusion unlikely in range, larger than most W antbirds, with bold bluish orbital skin; see Chestnut-backed. **Status** Fairly common but NT in Ecuador.

Grey-headed Antbird *Myrmeciza griseiceps* — 13–14cm

Local in foothills to subtropics of extreme SW (600–2,500m). Dense montane forest borders, secondary woodland, bamboo. ***Slate-grey head, throat, neck and most of underparts, black breast patch***. Rufous upperparts, duskier wings with ***bold white bars, long graduated tail tipped white***. ♀: ***paler*** overall, breast often mottled blackish and whitish. Monotypic. Rare, poorly known; mostly shy pairs, difficult to spot and not very vocal; often join mixed flocks in very dense vegetation; acrobatic. **Voice** Fast, brief, rather explosive and descending trill *ch'rrrrrr*, sometimes last note more enunciated *chi'rrrrrT*; sharp, whining *nyeee-nyé* or *gyeerr-geért* calls. **SS** Unmistakable in limited habitat and range. **Status** Rare, local, declining, EN in Ecuador, VU globally.

♂
♀
Chestnut-backed Antbird
♂
♀
Esmeraldas Antbird
♂
♀
Stub-tailed Antbird
♂
♀
Zeledon's Antbird
♂
♀
Grey-headed Antbird

PLATE 171: AMAZONIAN ANTBIRDS III

Five Amazonian antbirds, all characterised by rather long bills and legs. Three species are notable for their bare facial skins. No species in current genus *Myrmeciza*. Zimmer's now classified in *Sciaphylax*; Black-throated in *Myrmophylax*; White-shouldered in *Akletos*; Plumbeous in *Myrmelastes*; and Sooty in *Hafferia*. Curiously, the bare-skinned species follow ant swarms at least momentarily, whilst the other two species tend to avoid even mixed flocks. The handsome White-plumed is unique, but chances to find its congener White-masked are good in extreme SE.

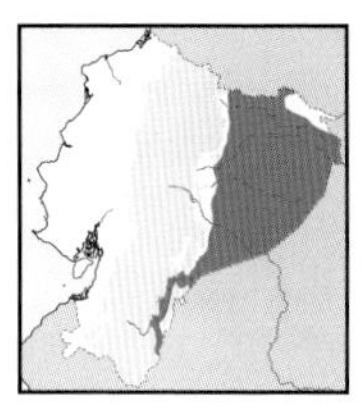

Zimmer's Antbird *Myrmeciza castanea* 12–13cm

Local in E lowlands to foothills (200–1,450m). *Terra firme* forest interior. ***Dark grey crown to breast-sides***, mantle browner, ***rufous lower back to short tail. Black throat to lower breast, white central belly***. Bold buff wing spots. ♀: ***rich cinnamon-buff throat to lower breast, white central belly***. Race *centunculorum* (most of range), *castanea* (extreme SE). Pairs in dense tangles often at treefall gaps, seldom with mixed flocks or ant swarms. **Voice** Slowly accelerating, high-pitched, enunciated *tee-tee-tee-tit?* or *téé-téé-teu-ti-titititirt?*; short whining *chi'irrt* calls. **SS** Confusion unlikely; ♀ somewhat resembles ♀ Black-throated but greyer crown, whiter throat, short rufous tail. **Status** Rare.

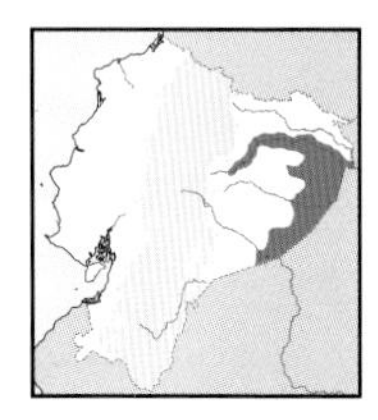

Black-throated Antbird *Myrmeciza atrothorax* 14–15cm

E lowlands along major rivers (to 400m). Tangled forest borders, second growth, riparian woodland. ***Dark greyish-brown above, blacker underparts***, blackish-grey sides to flanks; very ***faint whitish wing spots***. ♀: olive-brown above, ***small*** buff wing spots, ***greyish face, white throat*** and median belly, ***rich orange-buff breast, long dark tail***. Race *tenebrosa*. Retiring pairs difficult to see but vocal, regular in swampy tangled areas, rarely with mixed flocks. **Voice** Piercing, fast, ascending but decelerating *tee'tee, cheew-cheew-chee-chEE-chEE*; loud *chi'chrrrt-phEew, jéeer* calls. **SS** Few antbirds syntopic; ♀ somewhat recalls ♀ Zimmer's and Spot-winged but has plain rufous underparts, greyish head; see Riparian Antbird. **Status** Rare.

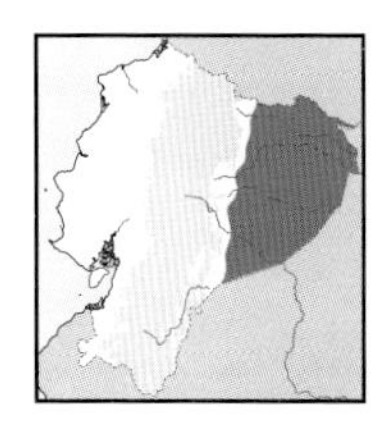

White-shouldered Antbird *Myrmeciza melanoceps* 17–18cm

E lowlands (to 500m). *Terra firme* forest borders, secondary woodland, *várzea* and riparian forest. ***Small bluish orbital skin. Black, small white patch on shoulders***. Elegant ♀ has ***black hood, rich rufous upperparts and cinnamon-orange below***. Monotypic. Pairs to small groups in dense tangles near ground, may attend ant swarms or mixed flocks. **Voice** Loud, ringing, far-carrying, slow *er-er, EEur, EE'er-EE'er-EE'er-EE'er...*; loud *kléup!*, mewing *kyeAA!*, abrupt *keeh'ur'ur'ur* when disturbed. **SS** ♂ very similar to Sooty but latter's orbital skin larger, plumage less lustrous; ♀ somewhat recalls smaller ♀ Cocha Antshrike, which lacks bare orbital skin. **Status** Uncommon.

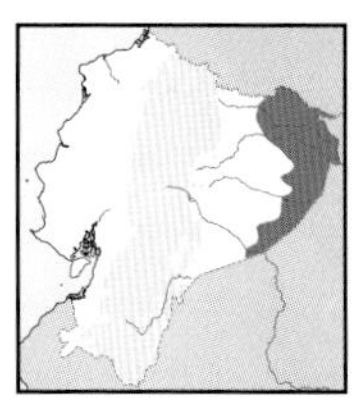

Plumbeous Antbird *Myrmeciza hyperythra* 17–18cm

NE lowlands (to 300m). *Várzea* and *igapó* forest undergrowth. ***Bluish orbital skin. Grey, with small white wing spots***. ♀: resembles ♂ above, but ***underparts rich orange-rufous***. Monotypic. Pairs to small groups in tangles, often in swampy areas, rarely with mixed flocks or ant swarms, quite tame. **Voice** Ringing, tittering *do-do-do-dududrrrr* or *do-doo-doo'durrrr'r'r'p* ending in rattle; also fast *chrr-chrr-chrr-chrr* chattering, clear *woo-klúr* and low churrs. **SS** ♂ Sooty (blacker, lacks white wing spots; ♀ differs notably); smaller Spot-winged and Slate-coloured somewhat similar but wing spots less conspicuous, no orbital skin. **Status** Fairly common.

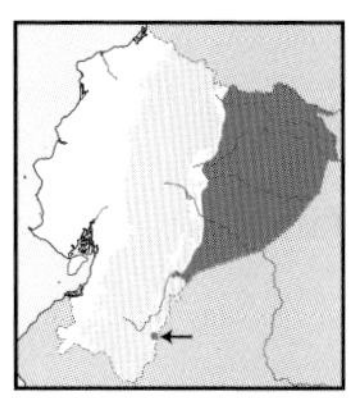

Sooty Antbird *Myrmeciza fortis* 17–18cm

E lowlands (to 600m, locally higher). *Terra firme* forest interior. ***Large bluish orbital skin***. Entirely ***sooty-blackish***, white wing bend. ♀: rufescent-brown crown, brown upperparts and tail, unmarked rustier wings. ***Face, neck-sides and most of underparts grey***. Nominate. Pairs with or apart from mixed flocks but regular at ant swarms, perches vertically near ground, might descend to it. **Voice** Ascending series of 8–12 penetrating *tééu-tééu-tééu-tééu...*, each note descending; low-pitched *tu'rtrtrtrt* rattle, mewing *haau*. **SS** ♂ a drabber version of White-shouldered with larger orbital skin; ♀'s rufescent crown and contrasting underparts distinctive. **Status** Fairly common.

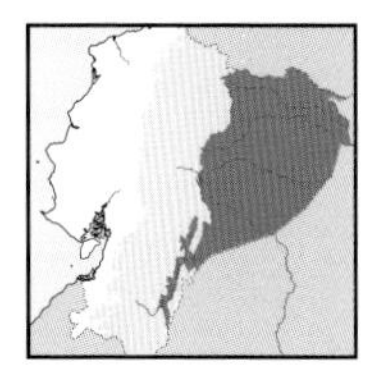

White-plumed Antbird *Pithys albifrons* 12–13cm

E lowlands to adjacent foothills (mostly below 600m, locally higher). *Terra firme* forest. Gorgeous! ***Long white face plumes, grey above, rich chestnut nape, underparts and tail***. Juv. lacks facial plumes, head greyer. Race *peruvianus*. Obligate ant-follower; several pairs at swarms, often perches vertically. **Voice** Thin *pséeeu*, often *pseeeu-chíp-pseeeu*; descending *chiurr*, throaty *tjtjjj* calls. **SS** White-masked Antbird *P. castaneus* might occur in extreme SE (***white limited to face***, black head, ***entire body rich chestnut***; long, complaining *huuuuue?* song; S *et al.*). **Status** Fairly common.

ntunculorum
castanea
Zimmer's Antbird
Black-throated Antbird
White-shouldered Antbird
Plumbeous Antbird
Sooty Antbird
juvenile
White-plumed Antbird

PLATE 172: 'PATTERNED' ANTBIRDS

Antbirds with fairly short to very short tails and bold adornments including dots, streaks and bands on dark or pale backgrounds. Also, the highly distinctive and enigmatic Wing-banded Antbird of uncertain taxonomic affinity, but once considered the 'link' between antbirds and antthrushes. Some usually avoid mixed flocks, but others attend ant swarms. Scale-backed was recently separated from *Hylophylax*, following vocal and molecular studies.

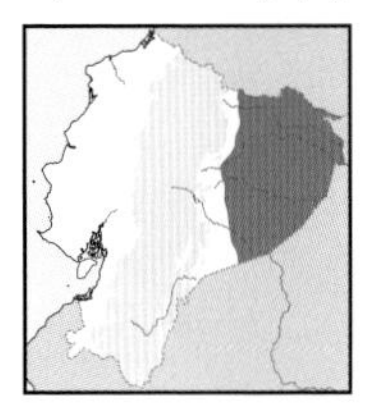

Banded Antbird *Dichrozona cincta* 10–11cm

E lowlands (to 450m). *Terra firme* forest. ***Long slender bill.*** Rufescent-brown above, ***blackish lower back, white rump band. White underparts, black spots form breast-band. Bold buff wingbars, tiny tail.*** ♀: ***buff*** shoulder spots and rump band. Monotypic. Terrestrial, singles or pairs sometimes in fairly open understorey, not with mixed flocks; constantly bobs head and fans tail. **Voice** Long, rather slow, steady series of 9–16 ringing *pueeEE* notes; also descending *chee'rrrr.* **SS** Breast pattern recalls Spot-backed and Dot-backed Antbirds, but rump band, bill and tiny tail unique; voice longer than Scale-backed, each note longer. **Status** Rare.

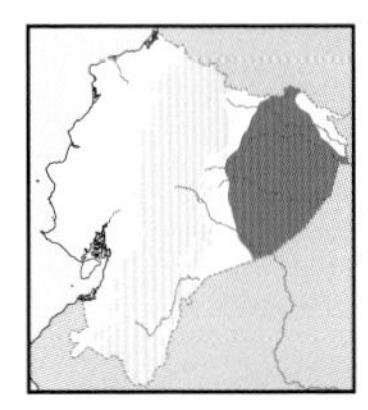

Wing-banded Antbird *Myrmornis torquata* 15–16cm

Local in E lowlands (to 400m). *Terra firme* forest. Long slender bill, ***small bluish orbital skin.*** Mainly rufescent-brown above, ***very short rusty tail. Black foreface to bib, neck-sides to semi-collar stippled black and white,*** rest of underparts grey. Blackish wings, ***bold buff bars*** and ***rusty edges*** to flight feathers. ♀: ***rich orange-rufous throat.*** Nominate. Shy singles or pairs hop on forest floor, rarely with flocks, flick leaf litter and probe ground. **Voice** Emphatic, slow, ascending, whistled *tuee-tuee-tueé-tueE-tueE-tuEE-tuEE!-tuEÉ!*; nasal, modulated *wuirrr!* **SS** Unmistakable by plumage and, especially, silhouette and terrestrial habits. **Status** Rare, local.

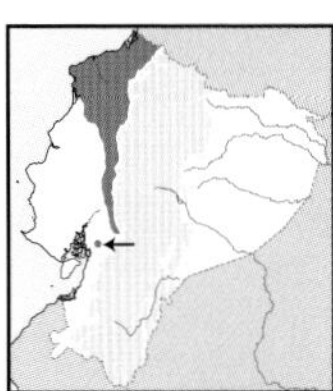

Spotted Antbird *Hylophylax naevioides* 11–12cm

NW lowlands, locally in E Guayas (to 300m, locally higher). Humid to wet forest interior, adjacent woodland. ***Grey head,*** rufescent-brown upperparts, very short brown tail tipped buff. ***Black throat,*** white rest of underparts, ***band of large black spots on breast; two bold rufous wingbars.*** ♀: dirty brown crown, faint dusky breast spots. Nominate. Singles to small groups regularly follow ant swarms, perch vertically and directly attack prey on ground or foliage. **Voice** Loud, high-pitched series of paired *teedeé-teedeé-tee-deé...* or *whedeé-whedeé-whedeé...,* somewhat descending; sharp *téwk* calls. **SS** Unmistakable in range. **Status** Uncommon, NT in Ecuador.

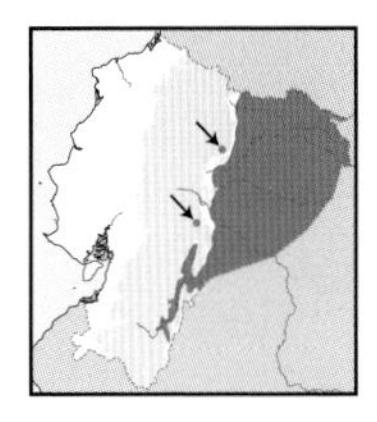

Spot-backed Antbird *Hylophylax naevius* 11–12cm

E lowlands to lower foothills (to 1,000m, locally higher). *Terra firme* and *várzea* forest. Greyish head, ***bold buff dorsal spots. Black throat,*** white below, ***black stripes form breast-band. Bold whitish wingbars.*** ♀: ***tawny-ochraceous below, white throat.*** Race *theresae*, complex vocal variation suggests intricate taxonomy. Singles or pairs near ground, often attend ant swarms, perch vertically on stems, directly attacking prey. **Voice** Fast, high-pitched, rising-falling, rhythmic series of doubled notes: *duu'dee-duu'dee-duu'dEE-duu'déé-duu'dee... dur..deé'ur-deé'ur-deé'ur...* (*várzea*); higher-pitched, slower-paced, hesitant, less rhythmic *tuu'EET-tuu'EET-tuu'EEt...* (*terra firme*); sharp *diít!, dee-diít!*; in *terra firme dee'et'et'et'et* calls. **SS** Dot-backed Antbird (smaller white dots above, whitish face, brown crown); Banded Antbird (white rump band). **Status** Uncommon.

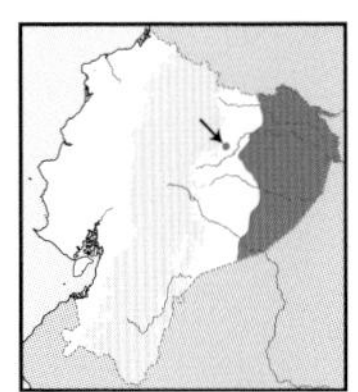

Dot-backed Antbird *Hylophylax punctulatus* 11–12cm

E lowlands (to 300m, locally to 900m). *Várzea* and *igapó* undergrowth. Recalls Spot-backed Antbird, but back to ***rump heavily dotted white,*** spots ***smaller*** than Spot-backed. ***Whitish face, greyish legs,*** black tail. ♀: ***black confined to malar, underparts mainly whitish,*** belly tinged yellowish-buff. Monotypic. Pairs, away from mixed flocks and army ants, rather elusive. **Voice** Well-spaced, enunciated series of doubled *wheé-HEur, wheé-HEur, wheé-HEur...*; downslurred *péeyr* and *péer-peur, péer-peur* calls. **SS** Commoner Spot-backed has larger buff, less extensive dorsal spots, grey head, pink legs, tawnier underparts, grey head. **Status** Rare.

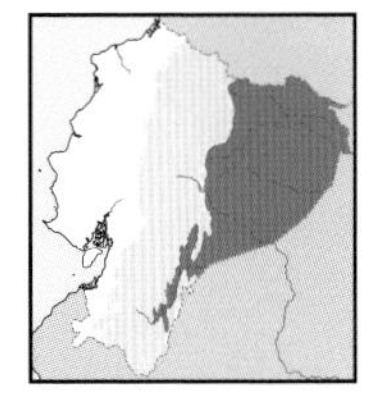

Scale-backed Antbird *Willisornis poecilinotus* 13cm

E lowlands to adjacent foothills (mostly below 700m, locally to 1,100m). *Terra firme* interior. Mostly grey, ***back, wings and rump blacker scaled white;*** longish tail ***dotted and tipped white.*** ♀: ***rufous-chestnut to cinnamon-rufous,*** no grey. Race *lepidonotus*. Pairs regularly at army ant swarms, where several pairs attend; perch vertically, sallies to ground, air or foliage. **Voice** Gives 3–10 piercing, inflected whistles in slowly ascending series: *tuue?-tuué?-tuEE?-tUEE?-tUÉÉ?*, each note higher-pitched and more emphatic than previous; loud *cheéft!* sneezing calls, often doubled; thin *t'prprprr* rattle. **SS** Unmistakable, but see ♀ Lunulated Antbird. **Status** Fairly common.

♀
♂
Banded Antbird
♂
♂
immature
♀
Wing-banded Antbird
♂
♀
Spotted Antbird
♀
♂
juvenile
Spot-backed Antbird
♀
♂
♂
juvenile
Dot-backed Antbird
♀
Scale-backed Antbird

Obligate ant-followers, all with conspicuous plumage or bare facial skin. All but two restricted to Amazonian lowlands. Bare-eyes are large and very conspicuous 'professional' ant-followers – among the most conspicuous of all antbirds. Despite their names – which are used for all thamnophilids – no antbird actually feeds on ants.

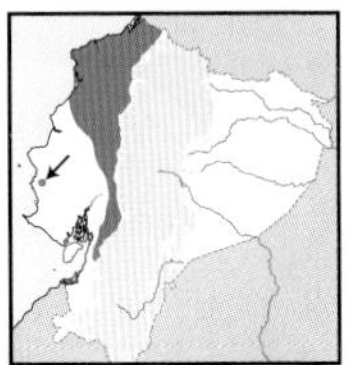

Bicoloured Antbird *Gymnopithys bicolor* 14–15cm

NW lowlands to foothills (to 1000m). Humid forest. ***Bluish orbital skin***. Brown above, ***white below, blackish and grey cheeks***. Race *aequatorialis*. Obligate army ant-follower; perches vertically, directly attacks prey. **Voice** Fast, rising-falling series starting with 2–4 whistled *cheé* notes, then accelerating *cheecheecheechee'chërr'chërr'chërr-chërr-chërr*, last notes snarled; raspy *kirrr'r*, short *churt* calls. **SS** Only bicoloured antbird with bluish orbital skin in range. **Status** Fairly common.

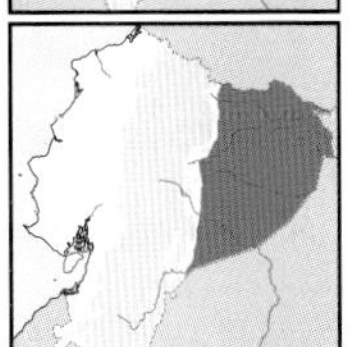

White-cheeked Antbird *Gymnopithys leucaspis* 14–15cm

E lowlands (to 750m) *Terra firme* forest, adjacent woodland. ***Bluish orbital skin. Rusty brown*** above, ***white underparts to cheeks***. Race *castaneus*. Several congregate at army ants; perches vertically, directly attacks prey. **Voice** Rising-falling series starting with 2–4 *cheé* whistles, then accelerating *cheecheecheechee'chërr'chërr'chërr*, last notes snarled; slower than Bicoloured; raspy *kirrr'u*, short *churt* calls. **SS** Only bicoloured antbird with bluish orbital skin in range. **Status** Fairly common.

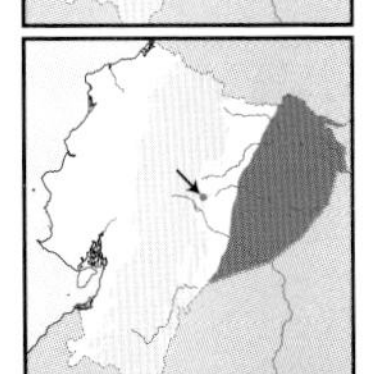

Lunulated Antbird *Gymnopithys lunulatus* 14–15cm

Local in E lowlands (mostly below 300m, once at 950m). Mainly *várzea* forest, locally in *terra firme*. Mostly grey, ***narrow white supercilium, white throat***. ♀: olive-brown above, ***whitish supercilium and throat, black crescent-shaped spots on back***, buff wing markings, ***whitish tail spots***. Monotypic. Pairs attend army ant swarms, where few pairs may gather, shier than other obligate ant-followers. **Voice** Fast, decelerating *chEE'chEE-cheecheechee-chee-chee..chuu*, last notes slower; *chirr, d'juurrr* calls. **SS** Voice of White-cheeked Antbird rising-falling, higher-pitched, ending in snarl. **Status** Rare, local.

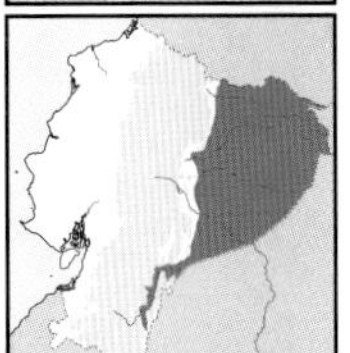

Hairy-crested Antbird *Rhegmatorhina melanosticta* 16cm

E lowlands (to 750m, locally higher). *Terra firme* forest, adjacent woodland. ***Large round pale orbital skin. Greyish to buffy, hairy crest, black mask and throat***. ♀: obviously ***dotted blackish on back***. Nominate (most of range), *brunneiceps* (extreme SE). Pairs attend ant swarms, perch vertically, directly attack to ground for escaping prey. **Voice** Slow, rising-falling, wheezy, whistled *déer-déer-deer-derr-derrz-derzz-drrz-drrz*, becoming harsher at end; descending harsh *djerrzl*, richer *chip!* **SS** Confusion unlikely, other bare-eyed antbirds are mostly grey to blackish, without hairy crests. **Status** Uncommon.

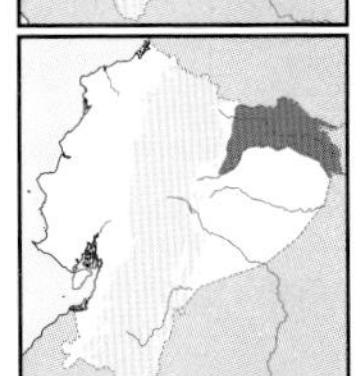

Black-spotted Bare-eye *Phlegopsis nigromaculata* 18–19cm

NE lowlands (to 400m). *Várzea* undergrowth, locally in swampy areas. ***Large red orbital skin***. Black head and underparts, ***olive-brown above, wing-coverts boldly spotted black***. Nominate. Pairs at ant swarms, where aggressively defends feeding territories, dominant over other ant-followers; performs sudden attacks or tosses litter. **Voice** Slow, raspy *dEEEu-djuú* or *djEEu..dju-djú*; coarse *djeeuu!*, clearer *pew* calls. **SS** Reddish-winged Bare-eye (black upperparts scaled white, reddish wingbars, black tail, shorter voice). **Status** Uncommon.

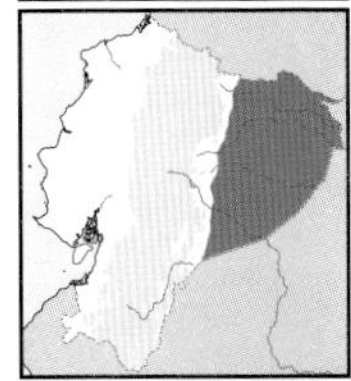

Reddish-winged Bare-eye *Phlegopsis erythroptera* 18–19cm

E lowlands (to 750m). *Terra firme* forest interior. ***Large red orbital skin***. Black with ***narrow white scaling on upperparts***, wings ***boldly marked reddish and black***. ♀: brown above, more cinnamon below, ***bold whitish-buff wingbars and edges***. Nominate. Shy pairs attend ant swarms, rather aggressive but elusive, also tosses litter; dominant over other passerines. **Voice** Short descending series of piercing, harsh *díg-dgEEu-dgeeu-dgeur-dgeurz*; harsh *d'jéugg*, downslurred *chigg* calls. **SS** Unmistakable; longer and faster voice than Black-spotted Bare-eye. **Status** Uncommon.

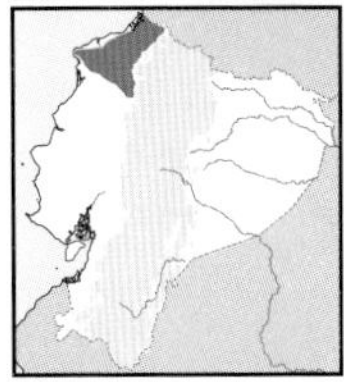

Ocellated Antbird *Phaenostictus mcleannani* 18–19cm

NW lowlands to adjacent foothills (to 700m). Humid to wet forest undergrowth. ***Large blue orbital skin***, pink legs. ***Black bib***, rufous nape to breast-band. ***Upperparts and belly ocellated blackish with buff edges***. Race *pacificus*. Pairs to small groups, obligate ant-followers, aggressive and dominant over others, dashes between vertical branches and ground, often flicking tail. **Voice** High-pitched, first accelerating, then slowing series of *dzée-dzée-dzédzédzé-dzé-dzé-dzé-ee-ee-ee-eer-rr*, brief stutter near slower and lower end; higher-pitched *chirr, chip* calls. **SS** Unmistakable. **Status** Rare, VU in Ecuador.

juvenile
♂
:oloured Antbird
♀
♂
White-cheeked Antbird
juvenile
♂
♀
Lunulated Antbird
♂
immature
brunneiceps
♂
♀
melanosticta
Hairy-crested Antbird
adult
juvenile
:k-spotted Bare-eye
adult
juvenile
Occellated Antbird
♀
♂
♂
immature
Reddish-winged Bare-eye

PLATE 174: CRESCENTCHESTS, GNATEATERS AND *PITTASOMA*

Two small, distinctive families distinguished on genetic, morphological and vocal grounds. Gnateaters are confined to mature Amazonian forests in lowlands and foothills. The Chocó endemic Rufous-crowned Antpitta was recently transferred to the Conopophagidae based on molecular evidence. Crescentchests occur in arid to semi-deciduous forest and scrub, in two separate but biogeographically related regions.

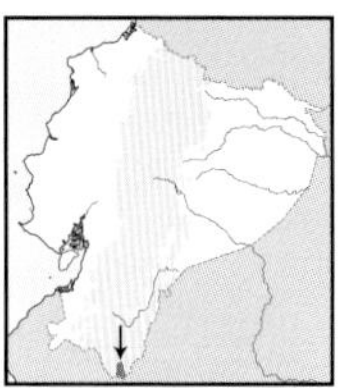

Marañón Crescentchest *Melanopareia maranonica* 15–16cm

Local in extreme SE in Zumba region (600–1,100m). Deciduous to semi-deciduous scrub and borders. ***Black crown, mask and chest, creamy supercilium and throat.*** Olive-brown upperparts, rich chestnut breast, paler below. Blackish wings ***edged white, flight feathers edged buff.*** Long tail. ♀: similar, but has narrower chest-band, ***tawnier*** or ochraceous below. Monotypic. Solitary, often on ground but rarely emerges from cover; frequently cocks tail; not easy to spot. **Voice** Loud, clear, steady paced, slightly accelerating *chu-chu-chu-chu'chu'chu¡chu...*; also short, yelping *whep*, more metallic *cherr* and soft *chit.* **SS** Unique in range. **Status** Rare, local, NT.

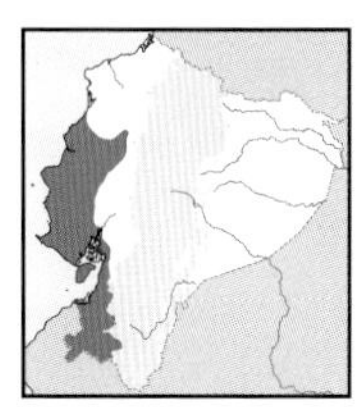

Elegant Crescentchest *Melanopareia elegans* 14–15cm

SW lowlands to subtropics in Loja (to 2,300m). Dense xeric scrub, deciduous to semi-deciduous woodland. ***Black crown, mask and breast-band, creamy supercilium and throat.*** Olive-brown upperparts, rich chestnut breast, ***ochraceous-buff belly to vent. Wing-coverts and inner flight feathers broadly edged rufous.*** Long tail. ♀: similar but ***narrower breast-band,*** more ochraceous ***buff*** below. Nominate. Singles in dense to fairly open understory, often on ground; skulking and difficult to see; frequently cocks tail. **Voice** Clear, steady, 12+ whistles: *chek-chek-chek-chek* or *cho-cho-cho-cho...*; also *trrrrk* trill and clicking *tick* calls. **SS** Unique in range. **Status** Uncommon.

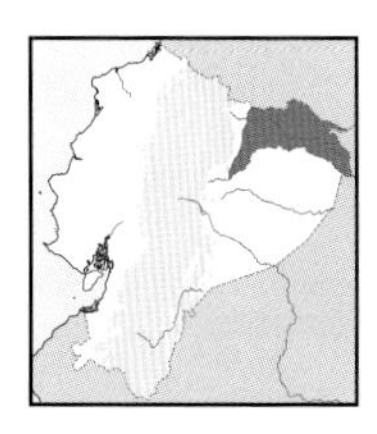

Chestnut-belted Gnateater *Conopophaga aurita* 12–13cm

E lowlands N of Río Napo (to 300m). *Terra firme* undergrowth. ***Black mask and throat, chestnut crown and breast, long white tuft behind eyes.*** Olive-brown upperparts, ***rufous-chestnut belt,*** whitish mid belly. ♀: lacks black mask, ***white postocular stripe as ♂; rich rufous neck-sides and breast.*** Race *occidentalis.* Pairs or singles in tangled understorey, e.g. treefall gaps and vine tangles; very hard to see. **Voice** Short, harsh, slightly rising rattle: *sch'frrrr* or *pt'rrrrr,* sneezing *shéff!* or *chogh!* **SS** ♀ Ash-throated Gnateater (scaly upperparts, dotted wing-coverts), but not sympatric. **Status** Rare to uncommon.

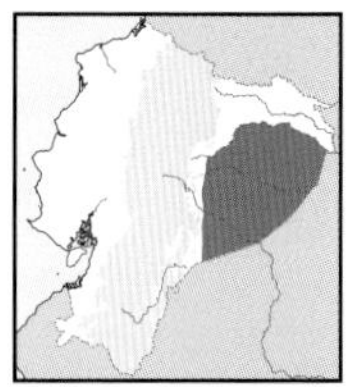

Ash-throated Gnateater *Conopophaga peruviana* 11–12cm

E lowlands S of Río Napo (to 600m). *Terra firme* undergrowth. Brown crown, ***white superciliary tuft,*** ash-***grey neck-sides, mantle and breast,*** white throat and belly. Upperparts ***finely scalloped dusky, small buff dots*** in wings. ♀: grey replaced by ***orange-rufous,*** more rufous crown. Monotypic. Pairs or singles in tangled understorey, very wary and difficult to follow. **Voice** Harsh *sweeEEk!* or sneezing *sweEEk-éét,* at 4–5-sec-intervals, similar to Chestnut-belted. **SS** Scaled upperparts and dotted wings distinctive, but no known sympatry with other gnateaters. **Status** Uncommon.

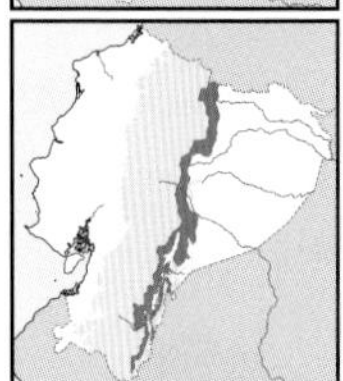

Chestnut-crowned Gnateater *Conopophaga castaneiceps* 11–12cm

Local in E Andean foothills to subtropics (800–2,000m). Montane forest and woodland undergrowth, borders. ***Rufous-chestnut crown,*** brighter forecrown, ***long white postocular tuft.*** Olive-brown upperparts faintly scaled dusky; ***sooty-olive underparts,*** whitish lower belly. ♀: ***rich orange-rufous head and throat to breast,*** olive-brown upperparts, white belly. Nominate (N), *chapmani* (S). Singles or pairs hop in tangled vegetation, rarely on ground, perches on vertical saplings, seldom with mixed flocks. **Voice** Harsh 8–10-note rattle, starting with brief stutter: *shrk'shrk-shreeik'shreik'shreik'shreik...,* slowing at end; calls a loud, harsh *chuink!* and sharp *tsEEW.* **SS** At higher elevations than other gnateaters, but might overlap locally with Ash-throated (scalloped back). **Status** Uncommon.

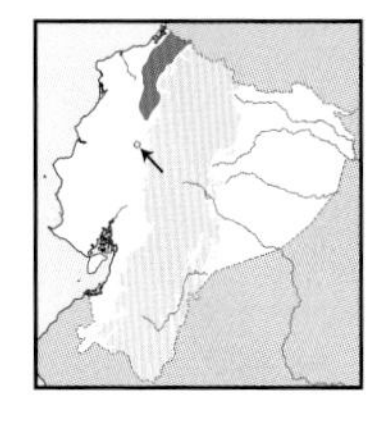

Rufous-crowned Antpitta *Pittasoma rufopileatum* 16–18cm

NW lowlands to foothills (to 950m). Humid forest interior. ***Chestnut crown, broad black facial line.*** Olive-brown above, ***black stripes on back,*** small whitish dots on wings. Ochraceous face and throat, ***underparts densely barred blackish.*** ♀: ***whitish spots on face, tawny to ochraceous underparts*** faintly spotted or streaked. Nominate. Singles or pairs hop and freeze in forest undergrowth, often following army ants. **Voice** Loud *kEY-eeup* whistles at rather long intervals; also piercing *ch-ch-ch-chiekk-kk* and more guttural *kuk-kuk-kuk....* **SS** Unmistakable; cf. short, louder calls of Black-headed Antthrush. **Status** Rare, VU in Ecuador; NT globally.

Marañón Crescentchest
♂
♂
♀
Elegant Crescentchest
♂
♀
Chestnut-belted Gnateater
♂
castaneiceps
♀
♂
chapmani
♀
Chestnut-crowned Gnateater
♂
juvenile
♀
♂
Rufous-crowned Antpitta
♂
♀
♀
Ash-throated Gnateater

PLATE 175: SCALED AND BANDED ANTPITTAS

Five 'typical' antpittas of primarily montane distribution. Undulated and Giant, which are both large with similar tremulous voices, replace each other altitudinally. Other elevational replacements include Moustached / Scaled in W Andes, and Moustached / Plain-backed / Scaled in E Andes. Their voices are also similar. Recently, seeing antpittas has been revolutionised by worm-feeding.

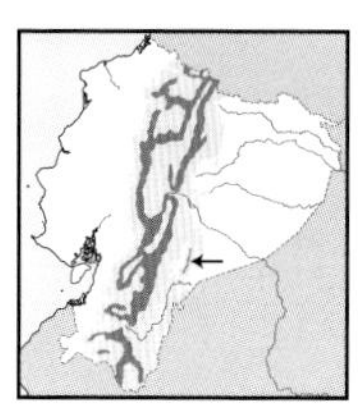

Undulated Antpitta *Grallaria squamigera* 21–24cm

Subtropics to temperate zone on both Andean slopes (2,000–3,700m). Montane forest undergrowth, mature woodland. Heavy blackish bill. Greyish-brown above, greyer on head, *buffy-white lores, blackish submalar stripe, whitish throat. Face and underparts ochraceous to buff obviously barred dusky*. Nominate (most of range), *canicauda* (SE) greyer above. Mostly confined to cover but might emerge in open during overcast conditions; difficult to see; brief vocal season (Nov–Jan). **Voice** Short, slightly rising, low-pitched, tremulous trill: *hohohohohohoho*... lasting 4–6 sec. **SS** Larger Giant Antpitta at lower elevations (more rufescent below, no white throat or blackish malar); voice longer, higher-pitched. **Status** Uncommon

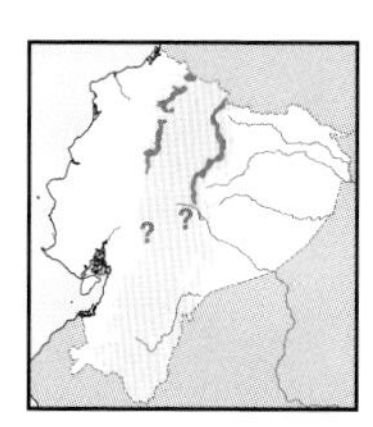

Giant Antpitta *Grallaria gigantea* 24–25cm

Local in subtropics on both Andean slopes of N (1,400–2,400m, not below 2,000m in E). Montane forest, often in damp areas. Heavy blackish bill. Greyish-olive crown, *rufescent front*, rufous-buff lores. Olive-brown upperparts, *face and underparts rufous-chestnut* (W) to *ochraceous* (E) *coarsely waved dusky*. Nominate (E), *hylodroma* (W) possibly valid species. Mainly restricted to wet forest interior; occasionally emerges at borders, trails and adjacent clearings. **Voice** Trill longer than previous species, lasting 7+ sec; quavering and slightly rising; faster in W. **SS** On E slope, smaller Undulated Antpitta (tawnier below); rather similar voices. **Status** Rare, VU; rarer in E.

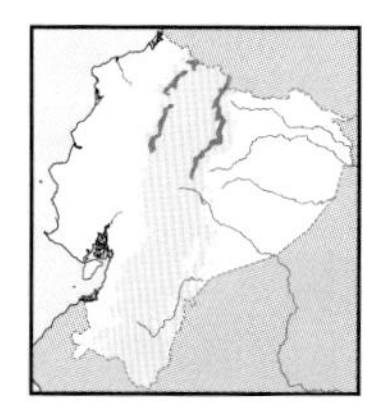

Moustached Antpitta *Grallaria alleni* 16–17cm

Local in subtropics of N Andes (1,800–2,200m). Montane forest, mature woodland. Grey crown, buffy-white lores, *whitish malar streak* ('moustachial') *with black freckles*. Olive-brown upperparts, ochraceous to buff below, *white streaks on lower throat, creamy to buff-white belly*. Race *andaquiensis*. Elusive ground dweller, often in wet areas; might emerge in adjacent clearings during overcast weather. **Voice** Short, ascending *dugugugugugu*.... **SS** Lower-elevation Scaled Antpitta very similar but more ochraceous belly, whitish malar less obvious; lower-pitched, faster voice, bouncing at end; cf. more trilled, longer voice of Giant. **Status** Rare, EN in Ecuador, VU globally.

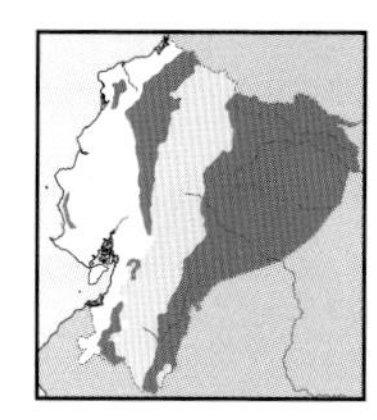

Scaled Antpitta *Grallaria guatimalensis* 16–17cm

E & W lowlands to lower subtropics, more local in W lowlands (to 2,000m). Primary and mature secondary forest undergrowth. Grey crown to nape, olive-brown upperparts with *faint dusky scaling*; *whitish malar streak*, brown throat and chest, *semi-concealed white crescent on lower throat, rufous to ochraceous-buff below*. Race *regulus*. Retiring, difficult to see, often near streams but also on slopes; rarely emerges at borders. **Voice** Low-pitched, quavering, accelerating-decelerating, rolling, slightly bouncing *duguru'ru'ru'ru'rURURUru'ruru*. **SS** Closely recalls Moustached Antpitta of higher elevations, cf. voices; Plain-backed lacks white moustachial and throat crescent, slower and more hollow voice. **Status** Uncommon.

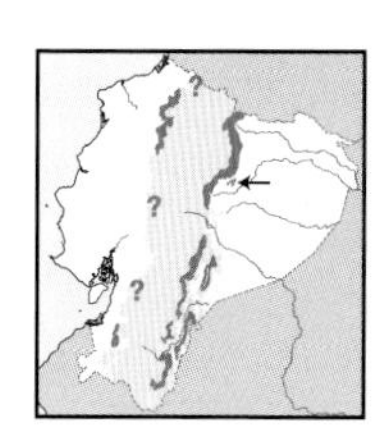

Plain-backed Antpitta *Grallaria haplonota* 16–17cm

Local in Andean foothills to lower subtropics, in W separated in N & S (700–1,300m in W, 1,100–1,700m in E). Mature forest undergrowth. *Olive-brown upperparts*, buff lores and *broad malar streak, dusky sub-malar. Whitish throat, rufous breast grading into ochraceous belly*. Races *parambae* (W), *chaplinae* (E) duller below, somewhat scaled above. Elusive, prefers tangled undergrowth, very rarely emerges in open understorey. **Voice** Slow, hollow, slightly rising, rolling and falling at end: *dub-dub-dub-DUB-DUB-dub-dub*.... **SS** Plainer than Scaled and Moustached Antpittas, slower-paced voice, more hollow; cf. shorter voice of Thrush-like Antpitta. **Status** Rare.

Giant Antpitta
Undulated Antpitta
juvenile
hylodroma
gigantea
squamigera
canicauda
mmature
adult
Moustached Antpitta
juvenile
immature
young
adult?
adult
chaplinae
parambae
Scaled Antpitta
Plain-backed Antpitta

PLATE 176: STREAKED AND STRIPED ANTPITTAS

Striped antpittas of three different genera, mostly confined to lowland rainforests. The common Chestnut-crowned occurs in Andean scrub and Watkins's – its close relative – in dry scrub of the Tumbesian region. As typical of all antpittas, most species on this plate are more often heard than seen.

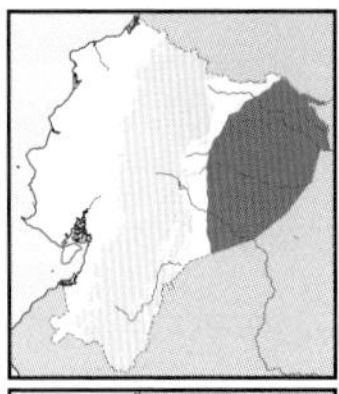

Ochre-striped Antpitta *Grallaria dignissima* 18–19cm

E lowlands (to 450m). *Terra firme* forest undergrowth. Olive-brown above, whitish loral spot, darker rump ***streaked white. Ochraceous throat and breast***, faint white streaks on breast, whitish belly, ***sides and flanks heavily streaked, elongated flank feathers***. Monotypic. Singles or pairs regularly in vicinity of forest streams, skulking and difficult to see. **Voice** Far-carrying, melancholic *düü-dÜUup* or *düü-dÜUep*; sadder, mewing *whöarrr* call. **SS** Sympatric White-lored and Thrush-like Antpittas less boldly streaked below, pale ocular spot, smaller. **Status** Rare, confined to forest.

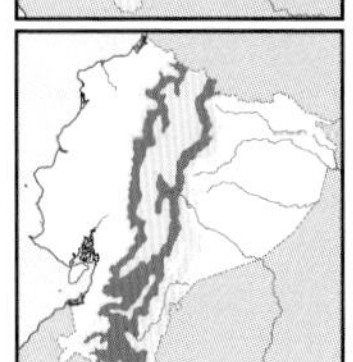

Chestnut-crowned Antpitta *Grallaria ruficapilla* 19–20cm

Subtropics to temperate zones on both Andean slopes (1,900–3,200m). Forest borders, second growth, bamboo, woodland. ***Rich rufous-chestnut crown***, olive-brown upperparts, ***white throat; sides, flanks and belly white coarsely streaked dusky-brown***. Nominate (most of range), *connectens* (SW). More tolerant of degraded habitats than other forest antpittas, occasionally but briefly emerges in open. **Voice** Loud, whistled *wee-püü-wEE*; also loud *kyeeeu* calls. **SS** In SW *connectens* is paler overall resembling even paler Watkins's Antpitta, but their voices differ; cf. Yellow-breasted and White-bellied Antpitta voices (last two notes higher). **Status** Common.

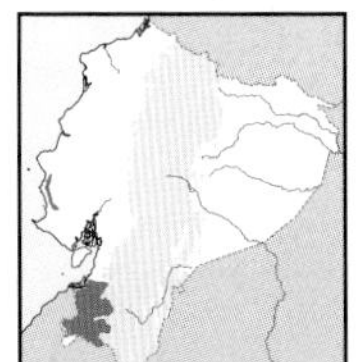

Watkins's Antpitta *Grallaria watkinsi* 18–19cm

SW lowlands to subtropics in Loja (to 1,800m). Deciduous to semi-humid forest, woodland, borders, regenerating scrub. ***Pale chestnut crown*** faintly streaked, ***whitish lores***, olive-brown upperparts. ***Whitish below*** with ***pallid dusky-brown streaking*** mainly on sides and flanks. ***Pinkish legs***. Monotypic. Tolerates certain degree of habitat alteration, but never in non-forested areas; difficult to see. **Voice** Emphatic, whistled *ti, tow-tow-tow, t-toowii* (locally called *tun-tu-ruway*); also *tee, tow-t-toowii*; *t-t-toowi* and others. **SS** Sympatric Chestnut-crowned (darker, greyish legs). **Status** Uncommon, NT.

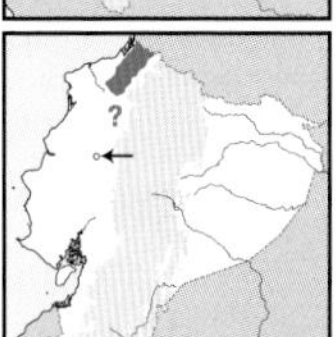

Streak-chested Antpitta *Hylopezus perspicillatus* 14–15cm

Local in NW lowlands to foothills (to 800m). Humid forest undergrowth. Ash-grey crown, buff lores, ***bold ochraceous orbital ring***, blackish malar, back faintly streaked whitish, buff-spotted wing-coverts, ***breast to flanks coarsely streaked blackish***. Race *periophthalmicus*. Confined to forest, shy, solitary; sometimes follows ant swarms. **Voice** Melancholic, whistled *pO-po-pee-pee-peu..peu..peu.. pew* slowing at end; sharp rolling *chup* and others. **SS** Thicket Antpitta lacks bold eye-ring, wing spotting and heavy streaking below; song shorter, rising first. **Status** Rare, VU in Ecuador.

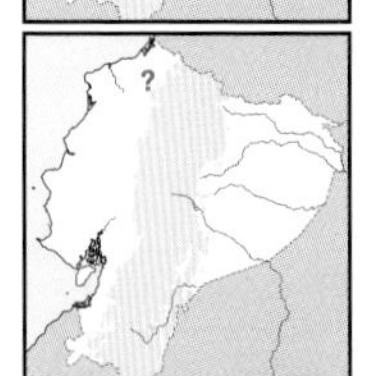

†Thicket Antpitta *Hylopezus dives* 14cm

One uncorroborated record in NW lowlands at 150m. Wet forest undergrowth in ravines. Grey crown to nape, ***buff lores*** and cheeks, narrow blackish malar; olive-brown above. White throat and belly, ***ochraceous breast-sides to flanks faintly streaked blackish***. Presumed race *barbacoae*. Poorly known in Ecuador; recorded at single site where not numerous (B. Palacios); elsewhere skulks in dense undergrowth. **Voice** Short, mellow, whistled *oh-oh-ou-oü-oü-oü-oü*, rising first, then levelling; rolling *t'cho'trrrr* calls. **SS** Streak-chested Antpitta boldly streaked below, broad pale eye-ring. **Status** Rare, local.

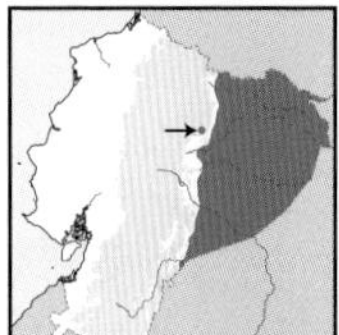

White-lored Antpitta *Hylopezus fulviventris* 14–15cm

E lowlands (to 700m). Forest borders, woodland, regenerating scrub. Pinkish legs. ***Sooty-grey head, contrasting white lores and postocular patch***. Olive-brown upperparts, white throat and belly, ochraceous-buff breast-sides to ***flanks and vent, dusky breast streaks*** extend to flanks. Nominate. Ground-dweller in very dense and tangled undergrowth, difficult to spot. **Voice** Abrupt, low-pitched, short, slow and sad *kwoh, kwoh, kwoh, kwoh*, sometimes faster. **SS** Voice, white lores and breast pattern separate from Ochre-striped and Thrush-like Antpittas. **Status** Uncommon.

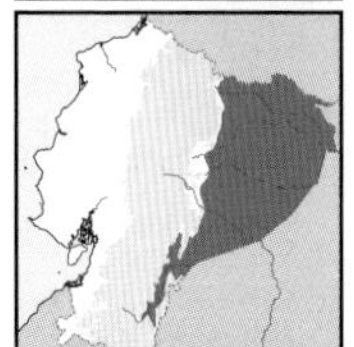

Thrush-like Antpitta *Myrmothera campanisona* 14–15cm

E lowlands to lower foothills (to 700m. Whitish below, ***breast finely streaked dusky***, greyish-brown flanks. Race *signata*. Singles or pairs skulk in dense thickets, often at treefall gaps or forested streams, where difficult to see. **Voice** Gives 3–7 hollow whistles, briefly rising then levelling *kow-KOW-kow-kow-kow*, ending abruptly; sometimes a rattled *kow'ororororrr*. **SS** Duller than other sympatric antpittas; voices faster, less hollow than White-lored. **Status** Fairly common.

Ochre-striped Antpitta
adult
ruficapilla
immature
Chestnut-crowned Antpitta
adult
immature
connectens
juvenile
chick
Watkins's Antpitta
Streak-chested Antpitta
Thicket Antpitta
White-lored Antpitta
Thrush-like Antpitta

PLATE 177: PLAIN ANTPITTAS

Seven Andean antpittas without stripes, scales or bars. Some species are rare, local and seemingly declining in numbers because of habitat loss. Like whole family, loud and onomatopoeic voices are better identification clues. In recent years, some of these jewels, including the breathtaking Jocotoco, are being attracted to feeding stations using earthworms.

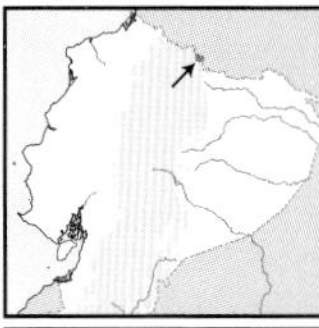

Bicoloured Antpitta *Grallaria rufocinerea* 15–16cm

Andes of extreme NE (2,550m). Montane forest and borders. ***Rufous-chestnut hood***, browner upperparts, ***underparts grey***, pale eye-ring. Race *romeroana*. Elusive, difficult to see, but responds to playback and whistled imitations. **Voice** Loud whistle *wyeeeea* or *treeeea*; also tremulous *tree-tree-tree*.... **SS** Larger Chestnut-naped Antpitta lacks rufous throat and has pale postocular patch; voice slightly reminiscent of Chestnut-crowned but longer and clearer. **Status** Rare, local, VU.

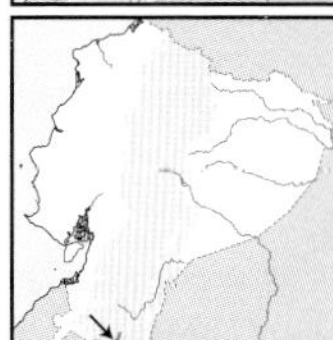

Jocotoco Antpitta *Grallaria ridgelyi* 22–23cm

Andes of extreme SE (2,300–2,650m). Montane forest, mainly bamboo. Handsome! ***Black crown and cheeks, contrasting white facial patch***, red eyes. Slate-grey neck to mantle, olive-brown rest of upperparts. ***White throat***, greyer underparts. Monotypic. Fond of tangled and steep habitats, probably occurs at very low density and occupies large territories; vocalises mostly at dawn and dusk; a curious pair now often seen at a worm-feeder. **Voice** Hollow, owl-like whistles: *hoo-hoo-hoo*... steadily repeated; also *hoo-joo, hoo-joo...hoo-hoo-hoo-hoo*.... **SS** Unmistakable. **Status** Rare, very local, EN.

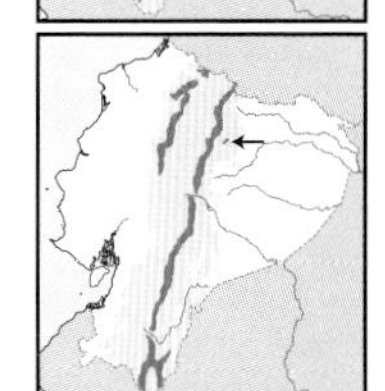

Chestnut-naped Antpitta *Grallaria nuchalis* 19–20cm

E & NW Andean slopes (2,000–3,000m). Montane forest, bamboo. ***Chestnut crown to nape***, greyish lores, ***white postocular patch***, grey eyes. Olive- to rufous-brown upperparts, ***blackish throat, sooty-grey below. Chestnut confined to nape, olive crown***, sootier underparts in NW. Nominate (E), *obsoleta* (NW) might deserve species status. Prefers tangled understorey, but can emerge on trails and borders when overcast, and visits worm-feeders in SE, where shy. **Voice** Clear, high-pitched *tr-te-te, trew-trew-teu-teu-te-te-te* starting with soft trill, not rising at end (W); or clear, metallic, bouncing, accelerating *teu, teu, te-te-tep-tep-tep-tep-tep-tep'tititi?*, last note higher (E). **SS** Bicoloured Antpitta in extreme NE. **Status** Uncommon.

Yellow-breasted Antpitta *Grallaria flavotincta* 17cm

Subtropics on NW Andean slope (1,500–2,300m). Montane forest and borders. ***Rufous-brown upperparts, pale yellow throat to upper belly***, white belly. Monotypic, formerly a subspecies of White-bellied Antpitta. Prefers tangled ravines and steep slopes where difficult to see even after playback; visits worm-feeders. **Voice** Three loud whistles, *puu-pïï-pEEp*, first note softer, inaudible at distance; also piercing *eeeeeeek*. **SS** White-bellied Antpitta occurs only in E; cf. two whistles of Rufous-breasted Antthrush. **Status** Rare, confined to forest, VU in Ecuador.

White-bellied Antpitta *Grallaria hypoleuca* 17cm

Subtropics in E Andean slope (1,400–2,200m). Montane forest, secondary woodland, bamboo, borders. ***Rufous-brown upperparts***, darker crown, brighter neck-sides. ***White throat grading into greyish-white belly***, rufous breast-sides to flanks. Race *castanea*. Persists in degraded areas provided some dense vegetation remains, may briefly emerge in adjacent clearings; difficult to see even with playback. **Voice** Resembles Yellow-breasted Antpitta; short, loud *pu-pïï-pEE* at long intervals; also owl-like *to-to-to-to*... (RT) **SS** Slower, more pronounced voice of Chestnut-crowned. **Status** fairly common.

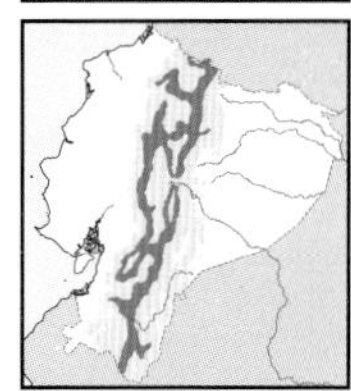

Rufous Antpitta *Grallaria rufula* 14–15cm

E & W Andean slopes, inter-Andean slopes (2,200–3,300m, locally higher). Montane forest, borders, bamboo. ***Uniform rich rufous***, somewhat paler on lores, throat and belly, narrow buffy orbital ring. Nominate. Pairs or singles in dense tangles but also fairly open understorey in wet forest, including *Polylepis* woodland; easier to see than most congeners. **Voice** Long, almost trilled *tititititi*... song; loud, rather high-pitched, bouncing *píp-pip'pi'tr* or *péép-pipipee* repeated every 2–3 sec; also short version of *tititi* song and single-noted *phew*. **SS** Unmistakable, other uniform rufous Andean birds have different shapes. **Status** Fairly common.

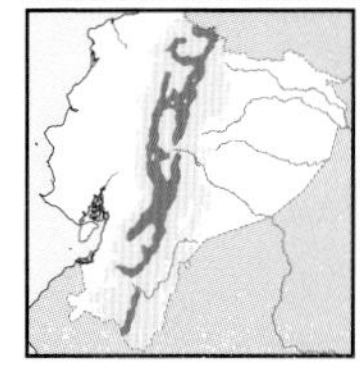

Tawny Antpitta *Grallaria quitensis* 16–17cm

Andes (3,000–4,500m). Open and shrubby *páramo*, *Polylepis* woodland, borders. ***Dull olive above***, buff-white lores. ***Whitish throat, buff to tawny below***, duller breast-sides and flanks. Race *quitensis*. Easy to see for an antpitta, hops and bounds usually within vegetation, but also on roads or in fields; often perches on stumps, bunchgrass or bushes. **Voice** Loud, musical, clear whistles, *tee-tüü-tüü*, last two notes inflected, may be louder; also sharp, far-carrying *kyuup!* **SS** Unmistakable. **Status** Common.

Bicoloured Antpitta
Jocotoco Antpitta
obsoleta
nuchalis
Chestnut-naped Antpitta
Yellow-breasted Antpitta
White-bellied Antpitta
Rufous Antpitta
Tawny Antpitta

These beautifully patterned, tiny antpittas of the Andes are primarily confined to E slopes and foothills. Long-legged like other antpittas, but not terrestrial, feeding in dense undergrowth up to 2–3m above ground. Very elusive, voices quite weak. Still, vocalisations are often only indication of their presence. Rusty-breasted only recently discovered in Ecuador.

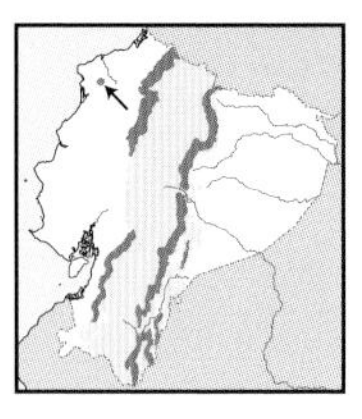

Ochre-breasted Antpitta *Grallaricula flavirostris* 10cm

Local in foothills and subtropics on both Andean slopes, NW coastal range (600–2,000m, lower in W). Montane forest undergrowth, mature woodland, borders. ***Blackish, dusky and yellowish*** or ***yellowish bill***. Greyish-olive upperparts, ***ochraceous-buff lores and broad orbital ring. Throat, breast and neck-sides rich buff to ochraceous, plain*** to almost plain (SW), ***streaked*** (NW) or ***coarsely streaked*** (E). Whitish belly. Races *flavirostris* (E), *mindoensis* (NW), *zarumae* (SW); also differ in bill colour. Singles or occasionally in pairs lower growth, less often on ground. **Voice** Short, simple *veet* at 7–10-sec intervals. **SS** In E, see Slate-crowned (plain and richer underparts; voices differ); Peruvian Antpitta's voice more plaintive, sharper; in W, see plainer Rusty-breasted. **Status** Uncommon; NT globally.

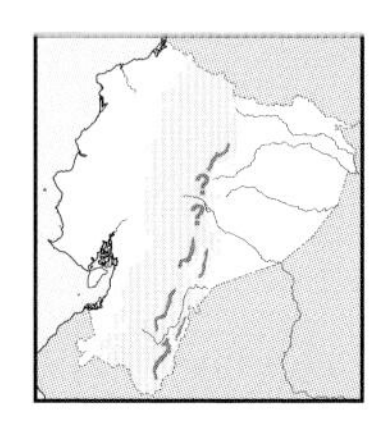

Peruvian Antpitta *Grallaricula peruviana* 10cm

Local in E Andean subtropics (1,700–2,100m). Montane forest undergrowth. ***Rufous crown, buff lores and orbital ring***, dusky crescent on lores, ***white malar, dusky sub-malar stripe***. Olive-brown above, ***white throat, whitish below coarsely scalloped dusky on breast and flanks***. ♀: olive-brown crown. Monotypic. Shy and rather silent, in lower growth, less often on ground; once attracted by worm-feeding, but shy. **Voice** Simple, plaintive *seeep* whistles at long intervals, similar to Ochre-breasted, but sharper. **SS** Confusion unlikely if seen well; sympatric Ochre-breasted has ochraceous throat, bold eye-ring, no rufous crown, dusky breast streaking, no scalloping. **Status** Rare but somewhat overlooked; NT.

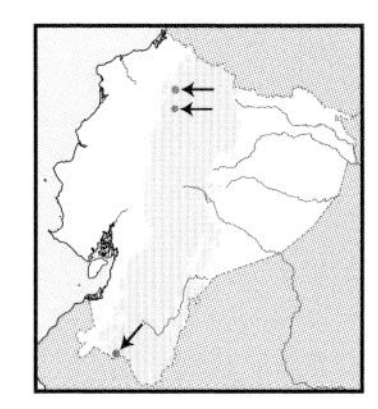

Rusty-breasted Antpitta *Grallaricula ferrugineipectus* 10–11cm

Local in W Andes, few localities (2,000–2,500m). Montane forest, woodland understorey, bamboo. Olive-brown above, ***faint buff loral spots***, bold ***whitish eye-ring. Rich ferruginous breast to flanks***, white throat and belly. Possibly race *leymebambae*, might deserve species status. Shy and difficult to see, but may approach after playback, occasionally flicks wings. **Voice** Gives 3+ *chep-chep-chep...* loud, evenly high-pitched. **SS** Recalls SW race of Ochre-breasted but orbital ring is white, rustier underparts, mid belly white, more olive upperparts; Ochre-breasted voice is simpler. **Status** Rare, possibly overlooked.

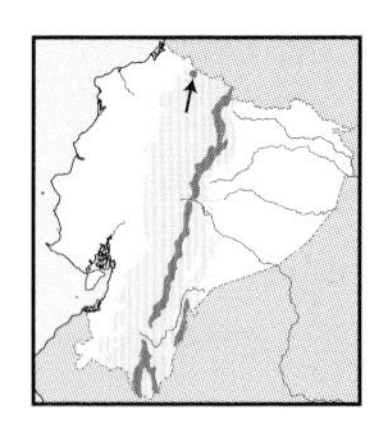

Slate-crowned Antpitta *Grallaricula nana* 10–11cm

E and extreme NW Andean slopes (2,000–2,900m). Montane forest undergrowth, bamboo, borders. ***Slate-grey crown to nape***, ochraceous lores and orbital ring, olive-brown upperparts. ***Rich ochraceous to rufous below***, buffy-white mid belly. Nominate. Singles or pairs in dense lower growth, sometimes emerging at borders; more vocal than congeners, but might not respond well to playback. **Voice** High-pitched, short, slightly descending and fading *we-didididi...*; also sharp *jiip* calls quite similar to Peruvian but less plaintive. **SS** Lower-elevation Ochre-breasted (no grey crown, more ochraceous below, dusky breast streaking); Rusty-breasted lacks grey crown, has whitish eyering and buffy lores; Peruvian notably differs, calls rather similar but sharper; Crescent-faced voice longer, higher-pitched, more ascending. **Status** Uncommon.

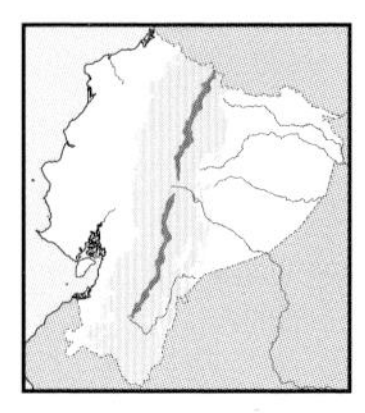

Crescent-faced Antpitta *Grallaricula lineifrons* 11–12cm

Local in temperate zone of E Andes, S to N Loja (2,900–3,400m). Montane forest undergrowth, bamboo. Beautiful and ***striking!*** Slate crown to nape, ***long white crescent in front of eye, long white malar streak grading into buff neck-sides***, black sub-malar. Brown upperparts, whitish underparts, more tawny and ochraceous on breast, sides and flanks, ***obviously streaked black***. Monotypic. Might briefly emerges at borders after playback, but otherwise shy, skulking and not very vocal. Fond of very dense and tangled undergrowth. **Voice** High-pitched, fast, ascending *du-du-di-di-di-di-di...di?* **SS** Unmistakable; shorter, simpler, fading, more bouncing and slower-paced voice of Slate-crowned. **Status** Rare, NT.

Slate-crowned Antpitta

Crescent-faced Antpitta

PLATE 179: TAPACULOS I

Three monotypic tapaculos: Rusty-belted, restricted to Amazonian *terra firme* forest; Ash-coloured and Ocellated occur in Andean forests. The spectacular Ocellated, the largest tapaculo, is among the most astonishing and hard-to-see understorey birds in the Neotropics. *Scytalopus* are actually *the* most difficult Neotropical genus to identify visually. The three species on this plate occur above other *Scytalopus* species. Use voice for correct identification.

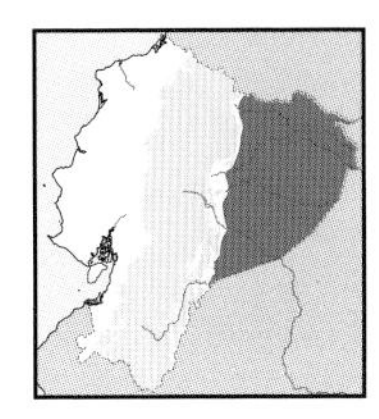

Rusty-belted Tapaculo *Liosceles thoracicus* — 19–20cm

E lowlands (to 600m). *Terra firme* forest. Grey head, brown upperparts, ***long white supercilium. White throat to breast, faint orange breast-band***, scaled breast-sides, and flanks to vent barred buff and dusky. ***Long tail***. Race *erithacus*. Elusive, mostly solitary, hops on floor, especially where there are large fallen logs. **Voice** Mellow, slightly fading, slowly accelerating *poö, poö-poö-poö...* or *poö, poö-poö-poö... pupupupupu*; also loud, *dö...dö-dö-dö*, harsh *creé-creé, tchót!* **SS** Confusion unlikely; superficially resembles Coraya Wren or Lunulated Antbird; voice slightly similar to Thrush-like and White-lored Antpittas. **Status** Uncommon.

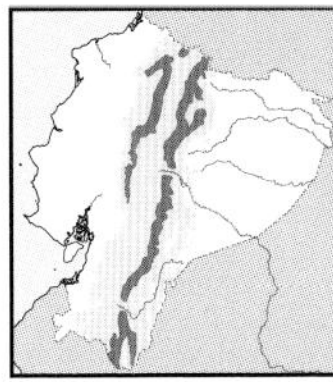

Ocellated Tapaculo *Acropternis orthonyx* — 21–22cm

E & NW Andean slopes (2,500–3,500m, locally lower). Montane forest undergrowth, bamboo. Spectacular! Large bill, ***face to chest and rearparts rich rufous***. Otherwise mainly ***blackish, boldly dotted white*** including wing-coverts and inner flight feathers. Race *infuscatus*. Hops and bounds in undergrowth, but terrestrial when foraging; difficult to see as skulks, foraging birds scratch ground apparently with long claw. **Voice** Loud, far-carrying *kiyeéw, kiyeéw...* whistles; also steady *pyiu-pyiu-pyiu-pyiu....* **SS** Unmistakable; its voice resembles a jay or hawk. **Status** Uncommon.

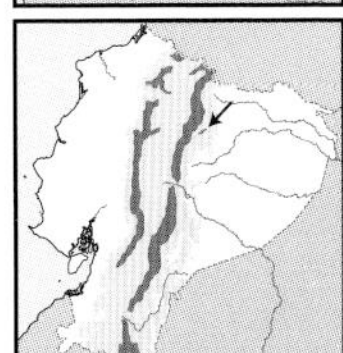

Ash-coloured Tapaculo *Myornis senilis* — 14–15cm

E & NW Andean slopes (2,000–3,500m). Montane forest undergrowth and borders. Mostly ***mouse-grey. Long tail***, similar in length to body. Juv. ***brown, pale below***, whitish throat and belly, ***dusky barring on wings and tail***. Monotypic. Non-terrestrial, skulks in dense undergrowth, including bamboo thickets, with tail cocked. **Voice** Piercing, accelerating *chá-chá-chá...chadá-chadá...* ending in rapid, often repeated, long trill. **SS** Longer-tailed than similar *Scytalopus* tapaculos; juv. somewhat resembles darker Sharpe's Wren, but behaviour different; voice recalls trills of Spillmann's (longer, faster), Chusquea (longer, lower-pitched) and Nariño Tapaculos (more paced, slower). **Status** Uncommon.

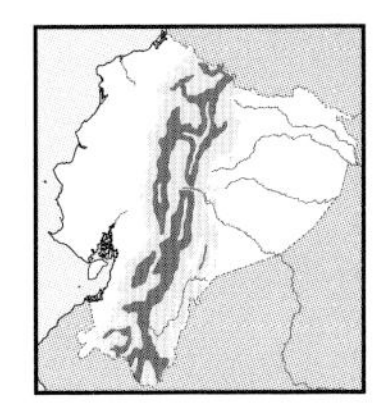

Blackish Tapaculo *Scytalopus latrans* — 12–13cm

E & W Andean slopes (1,700–4,000m; 1,900–2,500m in E). Montane forest undergrowth, bamboo, shrubbery. ***Blackish-grey*** (W, NE) to ***sooty-black*** (E, SW), ad. ***lack any brown barring below***. E & SW ♀ ***grey***, brown lower flanks and vent. Races *latrans* (W, extreme NE), *subcinereus* (SW), possibly undescribed race in most of E; separate species probably involved. Solitary, skulks like a mouse, always low, nearly impossible to see. **Voice** Repeated, rather steady *poop-poop-poop-poop-poop...* (*latrans*); faster, shorter, sharper *peup-peup-peupeu-peupeupeu...* (*subcinereus*); also faster, higher-pitched *pepepepep...* (unnamed race) beginning with a stutter; call a loud *we-wer!*, sharper in E & SW. **SS** Locally sympatric tapaculos have faster, more trilled (less 'whistled') songs (Spillmann's, Páramo, Chusquea); Long-tailed utters longer, accelerating series. **Status** Common.

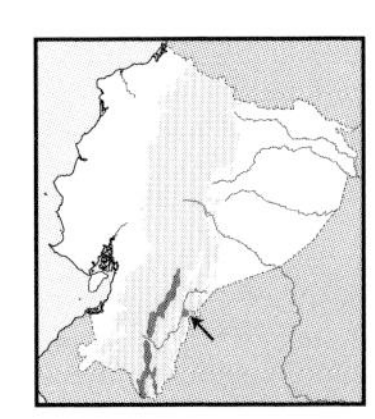

Chusquea Tapaculo *Scytalopus parkeri* — 12–13cm

SE Andean slope (2,250–3,150m). Dense bamboo, forest borders. Closely recalls more widespread Spillmann's, mouse-grey, browner nape, ***rump to tail, lower belly slightly tinged silvery***. Monotypic. Habits like other *Scytalopus*. **Voice** Long, quite musical rolling trill, stuttering and descending first, then levelling off: *cht-tiirrrrrrrrrr...* lasting up to 8–10 sec; ♀ response higher-pitched; call a loud *chiep!* **SS** Locally sympatric Blackish Tapaculo (*subcinereus* or unnamed race) has whistled, not trilled voice; higher-elevation Páramo has drier, faster, insect- or swift-like voice. Plumage and song resemble Spillmann's but no known overlap. **Status** Uncommon.

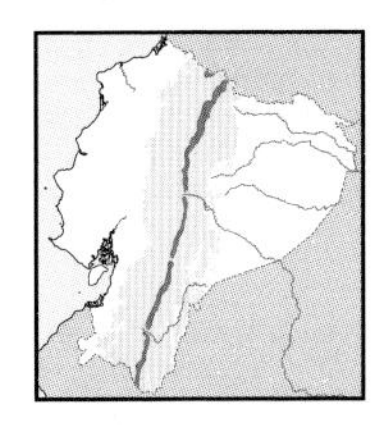

Páramo Tapaculo *Scytalopus opacus* — 10–11cm

Highlands in E & extreme NW Andes (3,000–4,000m). Treeline, *Polylepis* woodland, *páramo* shrubbery. Small, dark grey, dark brown flanks to vent; ♀ ***browner above***, pale grey below, ***paler brown flanks to vent***. Nominate (most of range), *androstictus* (SE), with ***white wing spot***. Habits as previous species. **Voice** Fast, dry, 5–10 sec, insect-like trill, stuttering at first: *cht-prrrrrrr...*; calls 5–9 *ki-ki-ki-ki...*; in SE slightly different, rising-falling song; calls shorter (*c*.1 sec) churring phrases. **SS** Almost no range or habitat overlap, cf. locally Spillmann's and Chusquea whose songs are rather slower, louder, longer, less insect-like. **Status** Uncommon.

adult
juvenile
Rusty-belted Tapaculo
Ocellated Tapaculo
immature
Ash-coloured Tapaculo
juvenile
adult
♂
♂
♀
latrans
subcinereus
♀
♀
latrans
immature
E slope
Blackish Tapaculo
adult
juvenile
opacus
androstictus
♂
juvenile
♀
♂
Chusquea Tapaculo
Páramo Tapaculo

Six nondescript tapaculos of Andean slopes, all extremely difficult to identify by plumage, even in the hand. Use voice for correct identification. Tapaculos tend to replace each other altitudinally, but some overlap exists between species pairs (Long-tailed / White-crowned in E, Nariño / Spillmann's in W). Two rare, local, recently described and closely related species (Chocó and Ecuadorian) replace each other geographically in W foothills.

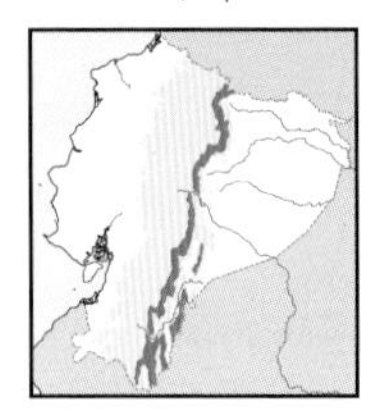

Long-tailed Tapaculo ***Scytalopus micropterus*** 13–14cm

E Andean slope (1,250–2,300m). Montane forest undergrowth, second growth, dense shrubbery, borders. *Large, sooty-blackish*, rufescent-brown rump to flanks, barred black, *fairly long tail*. ♀: duller, *browner*. Monotypic. Habits as other *Scytalopus*. **Voice** Repeated, resonant, coupled *chuu-rík, chuu-rrík...*, slow at first then accelerating; ♀ replies in more descending bursts *tchu-rík-chu-chu, rík-chu-chu, rík-chu-chu...*; call a single *kick*; also *ki-kikik* in distress. **SS** Blacker White-crowned Tapaculo replaces at lower elevations; its song is not two-noted, more rolling. Locally overlaps with Spillmann's (long trill) and Blackish T (steadily repeated 'barks'); cf. Ash-coloured Tapaculo's voice, which ends in trill **Status** Fairly common.

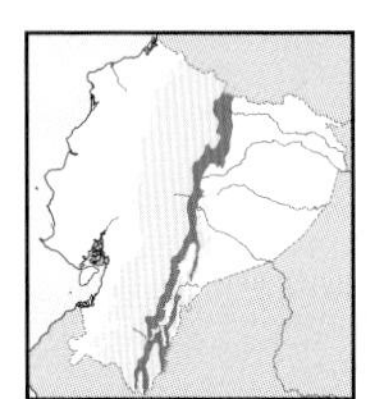

White-crowned Tapaculo ***Scytalopus atratus*** 12–13cm

Foothills to subtropics on E Andean slope (850–1,650m). Montane forest undergrowth, borders. *Blackish, small white spot on crown, flanks distinctly barred* brown and dusky. ♀: smaller crown spot, sometimes *whitish throat*, brown wash above. Nominate. Habits like other *Scytalopus*. **Voice** Even-paced series of 5–8 *wk-wk-wk-wk...wr-wr-wr* or *wir-wir'we-wir-wir'we...* notes; first part of each note louder, ending in stutter, with somewhat rolling quality. **SS** Locally sympatric with Long-tailed Tapaculo, which is slightly paler, longer-tailed, lacks white crown spot (which is difficult to see); voice two-noted repetitions, slightly higher-pitched. **Status** Uncommon.

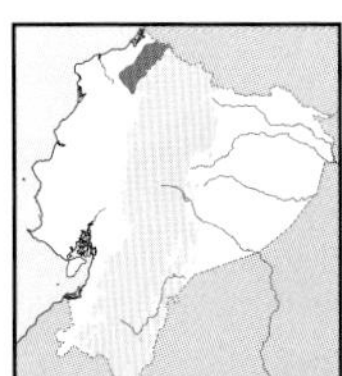

Chocó Tapaculo ***Scytalopus chocoensis*** 11–12cm

Local in NW Andean foothills (350–1,050m). Humid undergrowth, mostly primary forest. *Small*. Dark mouse-grey, *slightly paler below*, flanks barred dark brown. Sexes nearly alike. Monotypic. Habits like other *Scytalopus*. **Voice** Long series, starting with 2–4 rapid, lower-pitched notes *d'p-p-p, chúp-chúp-chúp-chúp...* lasting up to *c*.1 min; calls 4–8 *chú-pú-pú-pú...* and buzzy *zzik* by ♀. **SS** Only tapaculo in NW foothills, might locally overlap with (or at least occur in proximity to) larger but similar Nariño (shorter, more rolling, higher-pitched song; single-noted calls). **Status** Uncommon; EN in Ecuador.

Ecuadorian Tapaculo ***Scytalopus robbinsi*** 11–12cm

Local in SW Andean foothills (700–1,250m). Wet forest undergrowth. Closely recalls Chocó Tapaculo, but no overlap; Ecuadorian somewhat *darker*, flanks to vent more *cinnamon-brown*. Sexes nearly alike. Monotypic. Habits as other *Scytalopus*. **Voice** Long *chú-chú, chú-chú, chú-chú...* lasting up to 30+ sec, first part of each note louder; ♀ gives a shorter, descending series of falling *chu-chu-chu...* notes; call a single loud *quick!* or *chuip!* **SS** Only tapaculo in SW foothills, no known overlap with higher-elevation, blacker Blackish Tapaculo (*subcinereus*); voice shorter, more 'barked', louder. **Status** Uncommon, EN.

Nariño Tapaculo ***Scytalopus vicinior*** 12–13cm

NW Andean foothills to subtropics (1,250–2000m, locally higher). Montane forest undergrowth. *Dark* mouse-grey, slightly paler below, rufescent rump, *dusky cinnamon-brown barring* on flanks to vent, *rather long tail*. Sexes nearly alike. Monotypic. Habits like other *Scytalopus*. **Voice** Fast series of clear, ringing, decelerating *k-kikiki-kï-kï-kï...*; also single-noted, loud *keé* calls, or short *keekeekeekee*. **SS** Closely recalls locally overlapping Spillmann's Tapaculo, but usually replace each other elevationally; Spillmann's song and calls are louder, faster, more trilled; see also lower-elevation Chocó. **Status** Fairly common.

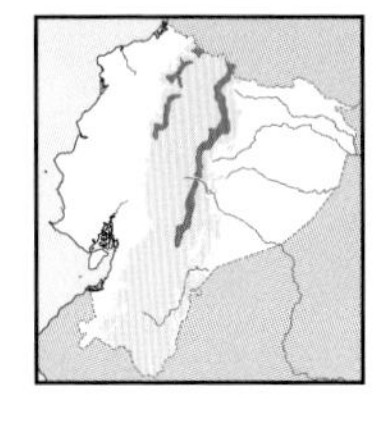

Spillmann's Tapaculo ***Scytalopus spillmanni*** 12–13cm

Andean slopes S to Azuay (1,900–3,200m). Montane forest undergrowth, bamboo. *Sooty*, paler rear, *dusky-brown and dusky barring* on flanks and vent; ♀: paler, *tinged brown above*. Monotypic. Habits as other *Scytalopus*. **Voice** Long, very fast trill, starting with slower stutter: *w'w'ch-prrrrrrrrrrrr*, slightly rising at end; call a short trill. **SS** Closely recalls Nariño Tapaculo, shorter-tailed than Long-tailed, darker than Páramo. In W compare slower, less-trilled voice of Nariño, and rapid barks, not trills of Blackish. Compare slower trills of Ash-coloured that begin with loud *chá-chá...*; ♀ Spillmann's song also includes a loud, descending *ché-ché-ché-ché...*; also Blackish (song not trilled); Long-tailed's two resonant notes *cu-rrik, cu-rrik...*; Páramo's harsher, shorter trill, stuttering at first; no known overlap with Chusquea. **Status** Common.

Long-tailed Tapaculo
♀
♂
juvenile
White-crowned Tapaculo
♂
juvenile
♀
Chocó Tapaculo
♂
♀
Ecuadorian Tapaculo
♂
♀
juvenile
Nariño Tapaculo
♀
♂
juvenile
Spillmann's Tapaculo

PLATE 181: ANTTHRUSHES

Antthrushes are ground-dwelling birds, confined to mature forest interior and heard more often than seen. They were recently grouped in a separate family from antpittas based on molecular and vocal evidence. *Formicarius* are smaller than *Chamaeza*, and walk with tail cocked. Four species in lowlands, three on Andean slopes.

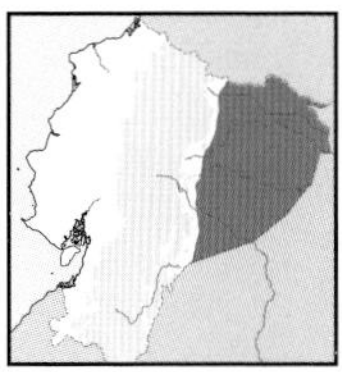

Rufous-capped Antthrush *Formicarius colma* — 17–18cm

E lowlands (to 500m). *Terra firme* forest undergrowth. ***Rufous crown to nape, black forecrown, cheeks and underparts***, grading into slaty belly and vent. Olive-brown upperparts, short black tail. Juv. has some ***white*** on face, throat and breast. Race *nigrifrons*. Solitary or pairs on forest floor flicking leaves, walk with tail cocked. **Voice** Fast, first descending but ascending at end: *tEE-ti-ti-te-te-tu-tu...ti-tí* lasting 5–6 sec; call a sharp, descending *tchéu*. **SS** Sympatric Black-faced lacks rufous crown, vent is ***rufescent***, bare orbital skin; voices differ. **Status** Uncommon to rare.

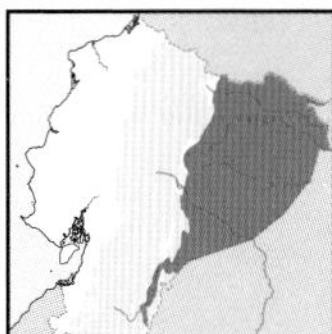

Black-faced Antthrush *Formicarius analis* — 17–18cm

E lowlands to foothills (to 1,000m). *Terra firme* forest, mature woodland undergrowth. ***Pale bare orbital skin. Olive-brown above*** including crown. ***Black face and throat***, sooty-grey below, ***rufescent vent***, black tail. Race *zamorae*. Probably more numerous than previous species, mainly terrestrial, sometimes in *várzea* and damp forests. **Voice** Starts with loud *pÚ*, then a fast, descending and fading *di'didididi...de-du-du-du..du*; also *t'wéep!*, *twep-dididi....* **SS** Rufous-capped Antthrush (rufous crown, dark vent, black breast) has shorter, more trilled voice. **Status** Fairly common.

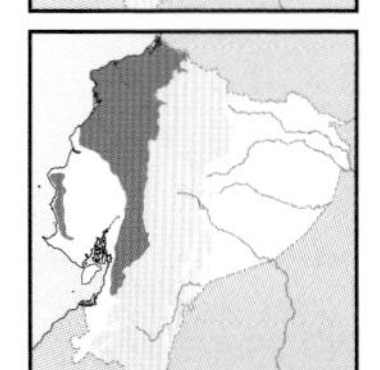

Black-headed Antthrush *Formicarius nigricapillus* — 17–18cm

W lowlands to foothills (to 900m, locally higher). Humid forest undergrowth. ***Pale orbital skin. Sooty-black head and most underparts***, rufescent vent. Olive-brown back, more rufescent rump, black tail. ♀: belly tinged brown. Race *destructus*. Solitary forest-dweller, shy, not easy to see, rarely joins army ant followers. **Voice** Loud series of *dup* whistles, accelerating first then slowing, ending abruptly; also *tchwep* and others. **SS** Locally Rufous-breasted Antthrush (no black head, rich rufous breast); voice lower-pitched and louder than sympatric antbirds. **Status** Uncommon.

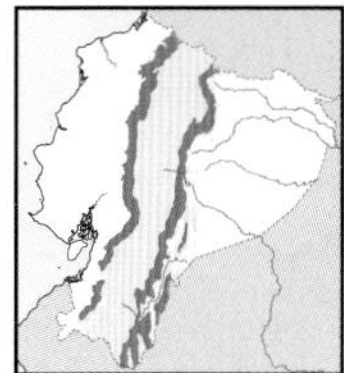

Rufous-breasted Antthrush *Formicarius rufipectus* — 18–19cm

E & W Andean slopes (1,100–2,200m). Montane forest and mature woodland. Chestnut crown to nape, browner upperparts, black face and throat, ***rich rufous breast to belly***, rufescent vent. Races *carrikeri* (W), *thoracicus* (E) has dark crown. Mainly terrestrial, alone or in pairs in steep, dense forest. **Voice** Two clear, loud *pu-pü* whistles; also *twep* calls. **SS** Unique in range, where note louder, three-whistled voice of Yellow-breasted Antpitta (NW), whose first note is often not audible. **Status** Uncommon.

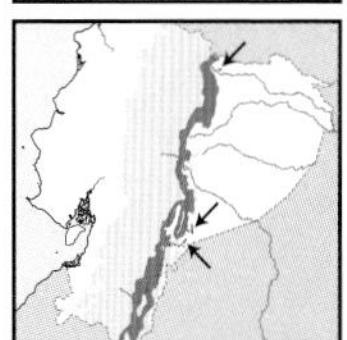

Short-tailed Antthrush *Chamaeza campanisona* — 19–20cm

E Andean foothills to subtropics (900–1,700m). Montane forest. Bicoloured bill. Olive-brown above, contrasting ***white lores, long eyebrow and malar stripe***. Whitish throat, underparts ***buffy***-white ***streaked dusky-brown***. Short blackish tail ***narrowly tipped whitish, buff*** vent finely speckled dusky. Race *punctigula*. Shy and difficult to see; often in pairs, silently walking through dense undergrowth; tail cocked. **Voice** Accelerating series of 15–20 *dub-dub-dub...* notes ending in 7–9 bouncing, fading *wuo-wuo-wuo-wuo*; also sharp *whip*, *wi-wi...wip*. **SS** Lower-elevation Striated Antthrush (whiter below, more scalloped), higher-elevation Barred (darker, barred below). **Status** Rare.

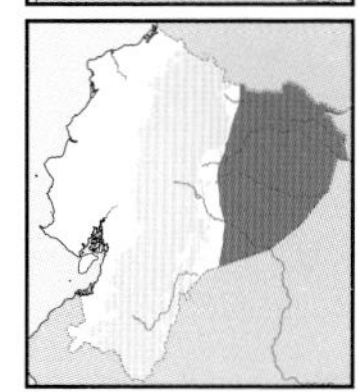

Striated Antthrush *Chamaeza nobilis* — 22–23cm

E lowlands (to 500m). *Terra firme* and *várzea* undergrowth. Brown above, ***pale lores, bold white postocular stripe and streak on neck-sides***. White throat, ***white below coarsely scalloped brown***, whiter belly. Tail brown with ***narrow white tip***. Race *rubida*. Very skulking and difficult to see as walks away. **Voice** Long, resonant, accelerating, bouncing *woo-woo-woo-wup'wup'wup...wuwuwuwu...* ending in 6–10 bouncing, fading *wöop-wÖOp-wÖop-wöop-wöop* notes; also loud *whöp! ... whöp!-whöp!-whöp!* and *ko-woop?* **SS** Upper-elevation Short-tailed is streaked below, not scalloped, and buffy not white; voice higher-pitched, shorter and slower (no known overlap). **Status** Uncommon.

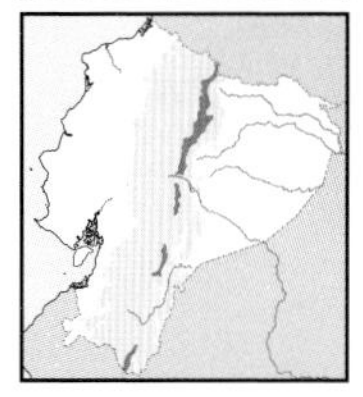

Barred Antthrush *Chamaeza mollissima* — 20–21cm

Local in temperate E Andes (2,000–2,600m). Montane forest undergrowth. Chestnut-brown upperparts, ***pale lores, white and black freckled stripes on face. Underparts white, heavily barred blackish-brown***, including vent. Nominate. Very secretive and hard to see as skulks in dense undergrowth on steep slopes. **Voice** Long, accelerating and ascending series of *cud-cud-cud...cu-cu-cu-cu-cu...* notes lasting *c.*15 sec.; also *whit, whi, whip* calls. **SS** Unique in montane range, voice not bouncing at end like Short-tailed. **Status** Rare.

♂
juvenile
♂
Black-faced Antthrush
juvenile
♀
juvenile
Rufous-capped Antthrush
adult
thoracicus
Rufous-breasted Antthrush
Black-headed Antthrush
carrikeri
Short-tailed Antthrush
Barred Antthrush
Striated Antthrush

Small forest tyrannulets, often with mixed flocks in upper strata. Short, stubby bill and bold wingbars (except in Sooty-headed) differentiates from other tyrannulets. All are acrobatic gleaners.

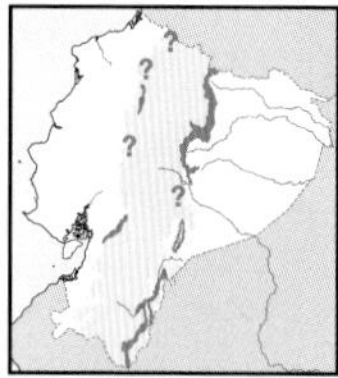

Rough-legged Tyrannulet *Phyllomyias burmeisteri* 11–12cm

Scattered records in E & W foothills to subtropics (600–1,500m). Montane forest, borders. *Pale mandible. White forecrown, face, throat and narrow supercilium, ear-coverts frosted grey-white.* Two bold yellowish wingbars and edges to flight feathers. Race *leucogonys*. Singles or pairs with mixed flocks scan foliage horizontally, flag one wing up; less nimble than congeners. **Voice** High-pitched, brief *suee*, *psee*, sometimes repeated *psee-psee-psee*. **SS** Extensive white in face and lack of facial crescent distinctive but not easy to see; cf. Ecuadorian Tyrannulet. **Status** Rare.

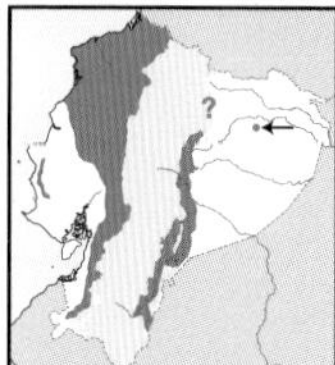

Sooty-headed Tyrannulet *Phyllomyias griseiceps* 10–11cm

W lowlands to foothills; local in E foothills (mostly to 1,300m). Humid to semi-deciduous forest canopy, borders, woodland, clearings. *Slate brown crown, narrow white supercilium and eye-ring*, dusky ear-coverts crescent, white throat. Olive upperparts, pale yellow underparts, breast tinged olive. Dusky wings *lacking bars*. Monotypic. Singles or pairs rapidly scan foliage and dart to underside, tend to avoid mixed flocks. **Voice** Loud, rhythmic, rollicking *wee-wiip-widi-wiip-widi* or *wiip-wii-diriip*, faster in E. **SS** Lack of wingbars distinctive, but see other tyrannulets and *Myiopagis* elaenias. **Status** Uncommon.

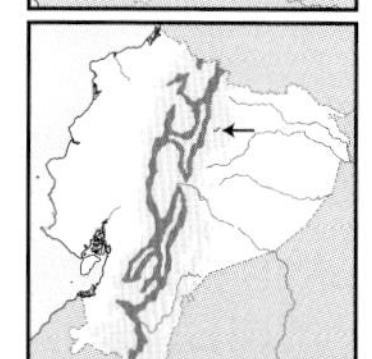

Black-capped Tyrannulet *Phyllomyias nigrocapillus* 11–12cm

E & W Andean subtropics to temperate zone (2,300–3,300m). Montane forest borders, secondary woodland, clearings. *Black crown, long white supercilium*, black ocular stripe. Dull olive above; *two bold white wingbars*, ochraceous edges to secondaries. Nominate. Singles or pairs with mixed flocks, dash agilely through dense foliage and mossy branches, gleans nervously. **Voice** High-pitched *tii-tr'tiii* or *tz-tz-tzrii*, often repeated insistently. **SS** Overlaps with Tawny-rumped (browner above, distinct rump); White-tailed and White-winged lack black crown, have longer bills. **Status** Uncommon.

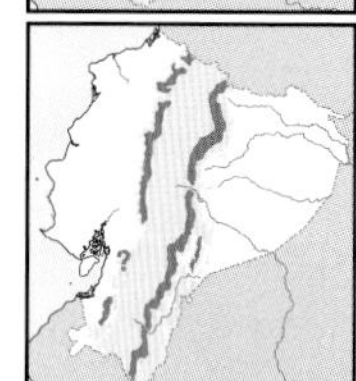

Ashy-headed Tyrannulet *Phyllomyias cinereiceps* 11–12cm

E & W foothills to subtropics, local in SW (1,350–2,500m, lower in SW). Montane forest borders, woodland. *Bluish-grey crown*, faint eyebrow and eye-ring, bold *black ear-coverts crescent*, *yellow underparts*, breast faintly streaked dusky; two narrow yellowish bars and edges to flight feathers. Monotypic. Singles or pairs with mixed flocks, perch more vertically than congeners, often flick one wing up. **Voice** Thin, high-pitched, sibilant *psee-see'ee'ee'ee*; ending trilled. **SS** Black ear-coverts crescent, bluish-grey crown and breast streaking distinctive. **Status** Uncommon.

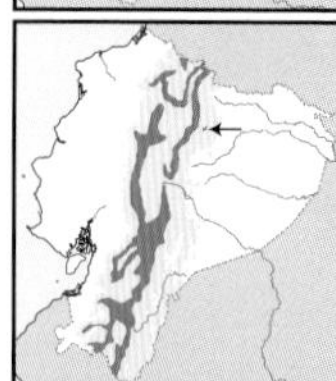

Tawny-rumped Tyrannulet *Phyllomyias uropygialis* 11–12cm

E & W Andean subtropics to temperate zone (2,100–3,100m). Montane forest borders, woodland, shrubby edges, clearings. *Blackish-brown crown*, white supercilium, narrow black ocular stripe, white throat. Olive-brown back, *tawny-ochraceous rump*, olive-grey breast, yellow belly. Black wings, two *buffy bars*. Monotypic. Mostly in pairs with mixed flocks, hovers agilely below foliage. **Voice** Thin, high-pitched, almost buzzy *tzz-tzzeéu*; high-pitched *psi-psr'sr* foraging calls. **SS** Distinctive rump precludes confusion with Black-capped, White-tailed and White-banded Tyrannulets. **Status** Uncommon.

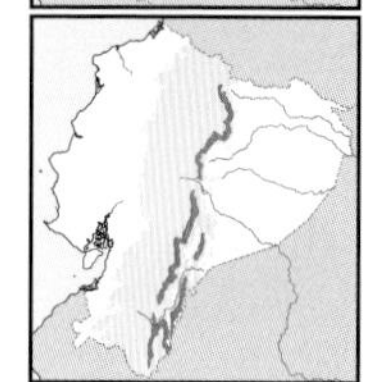

Plumbeous-crowned Tyrannulet *Phyllomyias plumbeiceps* 11–12cm

Local in E Andean subtropics (1,200–2,200m). Montane forest canopy, borders. *Plumbeous crown*, narrow white eyebrow, *blackish ear-coverts crescent bordered white*. Dusky wings, *two bold yellowish bars*, yellow edges to flight feathers. Monotypic. Singles or pairs with mixed flocks; sallies and hover-gleans. **Voice** Unlike congeners, a 'bright' series of 5–8 high-pitched *pík-pík-pík*, sometimes ending in brief trill. **SS** Recalls Ecuadorian Tyrannulet (longer bill, weaker face pattern); also Ashy-headed (bluer crown) and Rough-legged Tyrannulets (whiter face), Marble-faced and Variegated Bristle-Tyrants, and Slaty-capped Flycatcher (behaviour and posture differ). **Status** Rare.

Rough-legged Tyrannulet

Sooty-headed Tyrannulet

Black-capped Tyrannulet

Ashy-headed Tyrannulet

Tawny-rumped Tyrannulet

Plumbeous-crowned Tyrannulet

Small elaenias more similar to *Phyllomyias*, *Tyrannulus* and *Mecocerculus* tyrannulets than to typical elaenias, even in foraging behaviour; they often cock their tails. A complex genus characterised by semi-concealed coronal patches and bold wingbars. Identification tricky unless voices are learnt. One species notable for its sexual dimorphism and plumage variation. Excluding the recently described Foothill, *Myiopagis* are typically found in lowlands.

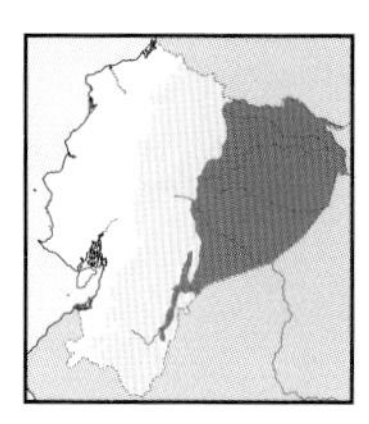

Forest Elaenia *Myiopagis gaimardii* 12–13cm

E lowlands to foothills (to 1,000m). Mainly *várzea* and *igapó* canopy, borders. ***Greyish crown*** grading into olive upperparts, whitish throat and ***breast clouded olive***. Narrow white eyebrow and eye-ring, ***whitish semi-concealed coronal patch. Two pale yellowish wingbars***. Nominate. Pairs with mixed flocks, glean horizontally, often cock tail slightly. **Voice** Emphatic *chweE-wiid'wíd* or simpler *chweED?* **SS** Resembles Yellow-crowned Elaenia (yellower wingbars, plainer breast, yellow crown patch); ♀ Grey (three bolder wingbars, greyer face, little breast flammulations); Foothill has three wingbars, greyer crown; Greenish (duller crown, plainer wings). **Status** Uncommon.

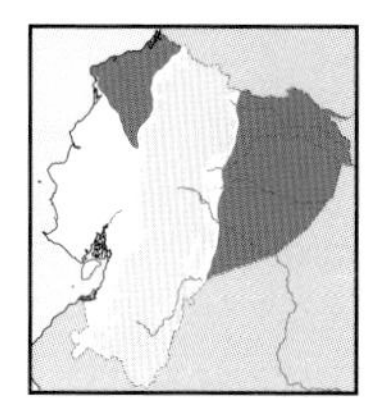

Grey Elaenia *Myiopagis caniceps* 12–13cm

E & NW lowlands to foothills (to 1,050m). *Terra firme* forest canopy, borders. ***Ash-grey crown, grey to greyish-olive above; pale grey below***, short white eyebrow, white coronal patch. Blackish wings, ***three white to yellowish wingbars (two obvious)***, pale edges to flight feathers. ♀: more ***olive above***, yellower below. Races *parambae* (W), *cinerea* (E); possibly separate species. Pairs with mixed flocks, glean rather horizontally in treetops. **Voice** Fast, descending shrill *tsee-see'ee'ee'ee'ee* (W); faster, often doubled *ts'see-seuw-seé-seeuw, ts'see-seuw-seé-seeuw...tsEE-slsisisi* (E). **SS** Bolder wingbars, blacker wings, grey underparts (♂) than other greenish *Myiopagis*. **Status** Uncommon.

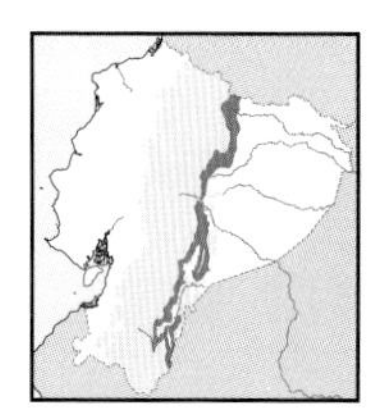

Foothill Elaenia *Myiopagis olallai* 12–13cm

Local in E Andean foothills to lower subtropics (1,000–1,500m). Montane forest canopy, borders. ***Grey crown***, short white eyebrow, white coronal patch, face somewhat grizzled. Olive upperparts, whitish throat, ***breast clouded olive***, yellowish below. ***Three pale yellowish wingbars*** and edges to flight feathers. Nominate. Pairs follow mixed flocks, perch rather horizontally, make short sallies to underside of foliage. **Voice** Harsh ascending trill *trr'see'ee'ee'ee'ee'*, lasting 2 sec, short falling introductory notes. **SS** Little, if any, overlap with congeners; cf. Forest (bolder wingbars, darker crown, darker upperparts); ♀ Grey (grey crown, yellow crown patch). **Status** Rare, VU globally.

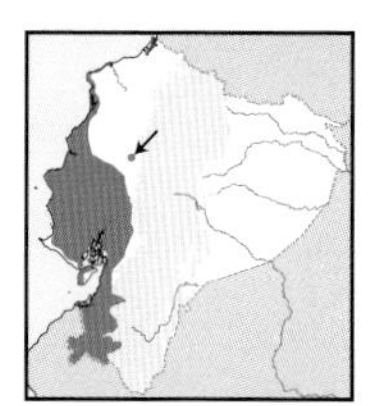

Pacific Elaenia *Myiopagis subplacens* 13–14cm

SW lowlands to subtropics, N along coast to Esmeraldas (to 1,700m). Semi-deciduous to deciduous forest undergrowth, borders, woodland. Olive upperparts, ***browner crown***, yellow semi-concealed coronal patch, ***long whitish supercilium, dusky grizzled cheeks***. Throat to lower breast has vague whitish-grey streaking. ***Two weak wingbars, whitish edges to inner flight feathers***. Monotypic. Singles or pairs perch upright, briefly join mixed flocks. **Voice** Loud, endless *chrrt-chrrt-chi'weér*, also *chiirt-cho'wiit, pích-cho'wiit....* **SS** From Greenish Elaenia by longer and grizzled eyebrow, brownish-olive crown, duller underparts; stronger voice. **Status** Fairly common.

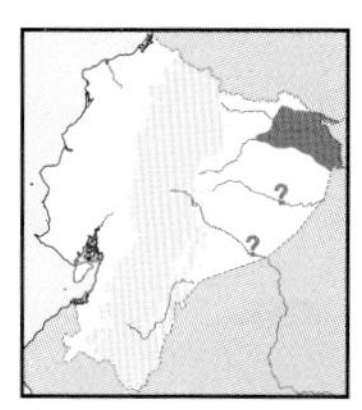

Yellow-crowned Elaenia *Myiopagis flavivertex* 13–14cm

NE lowlands (below 300m). *Várzea* forest, borders. ***Olive crown*** to upperparts, short whitish eyebrow, ***yellow semi-concealed coronal patch***. Whitish throat grading into ***olive clouded breast. Two obvious yellow wingbars***. Monotypic. Singles or pairs perch fairly upright in lower growth, seldom with mixed flocks, easy to miss unless vocalising. **Voice** Burry, rather descending *WEE-djee-djee'reew*, sometimes *WEE-djee'jee'jee'jee'djeew, WEEh-djeu-djeu-dju, djE-jijijijewjrr.* **SS** ♀ Grey (three bolder, whiter wingbars, blacker wings); Forest (paler wingbars, streaky breast, bolder face pattern; habitat and voice differ notably). **Status** Uncommon.

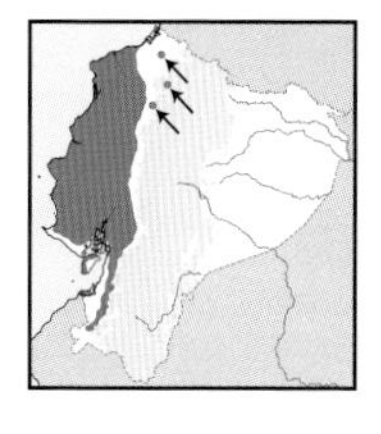

Greenish Elaenia *Myiopagis viridicata* 13–14cm

W lowlands to foothills, not in wet NW (to 1,000m). Deciduous to semi-deciduous forest, borders, woodland. Greyish crown, olive upperparts, yellow coronal patch, ***short whitish eyebrow, grizzled face to cheeks***. Whitish throat, ***olive mottling on breast; weak wingbars and narrow pale edges*** to flight feathers. Race *implacens*. Singles perch fairly upright in lower growth, rather unobtrusive, easily overlooked unless vocalising. **Voice** Thin, high-pitched *shee'ri..shee* at dawn; emphatic, slurred *sheeeer* or *shee-ryee, shee, sheer-yee.* **SS** Recalls Pacific Elaenia but weaker face pattern, crown greyer; voice weaker; Grey (bold wingbars, blacker wings). **Status** Uncommon.

immature
juvenile
adult
Forest Elaenia
adult
grey morph
parambae
nerea
juvenile
adult
green morph
adult
yellow morph
Grey Elaenia
Foothill Elaenia
Pacific Elaenia
adult
juvenile
Yellow-crowned Elaenia
adult
juvenile
Greenish Elaenia

All species on this plate, except Yellow-crowned, occur in shrubby vegetation or semi-open habitats. Yellow-crowned is superficially similar to *Phyllomyias* tyrannulets, but its coronal patch is richly coloured. The small and attractive Grey-and-white Tyrannulet and Tawny-crowned Pygmy-Tyrant are often found on slopes facing the sea. Both very vocal, as is the beardless tyrannulet. Mouse-coloured is variable and complex, and might comprise a multi-species group.

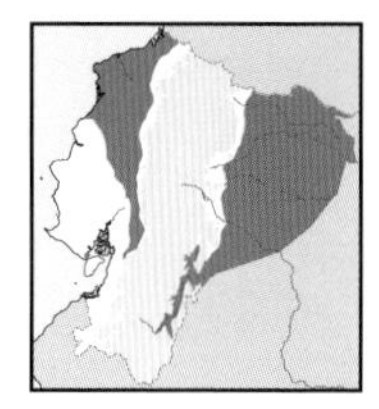

Yellow-crowned Tyrannulet *Tyrannulus elatus* 10–11cm

E & NW lowlands (to 600m). Humid *terra firme, várzea* forest canopy, borders, woodland. Blackish crown, ***semi-concealed yellow crown patch, greyish-white supercilium and face to lower throat***. Olive upperparts, yellow underparts. ***Two whitish wingbars*** and white edges to inner flight feathers. Monotypic. Singles or pairs apart from flocks, vocal and nimble, perch more upright than *Phyllomyias*. **Voice** Simple, clear *fee-teéw, fee-te-heéw*. **SS** Greyish supercilium, face and crown patch (if seen) distinctive; cf. Sooty-headed Tyrannulet, and Greyish, Forest and Greenish Elaenias. **Status** Uncommon.

Grey-and-white Tyrannulet *Pseudelaenia leucospodia* 12–13cm

SW lowlands, Puná and La Plata islands (below 100m). Xeric scrub, light woodland. ***Short bushy crest, white crown patch***. White supraloral stripe. ***Grey above, whitish-grey throat to breast***, pale yellow belly. ***Weak*** pale wingbars and edges. Race *cinereifrons*. Nimble pairs glean horizontally from outer foliage to larger limbs, sometimes with mixed flocks; cocks tail. **Voice** Sharp, emphatic *cht-trrrt*, more nasal *chee-vík!, chevík-chi, chevík-chi*. **SS** Smaller Southern Beardless Tyrannulet has larger but ***not forked*** crest, no crown patch and bolder wingbars. **Status** Uncommon.

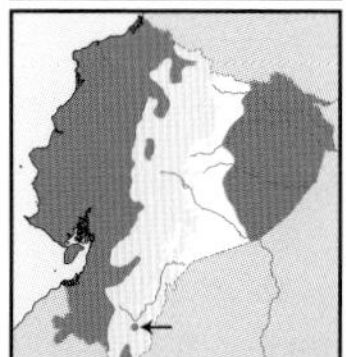

Southern Beardless Tyrannulet *Camptostoma obsoletum* 9–10cm

W lowlands to subtropics, inter-Andean valleys, NE lowlands, Zumba region (to 2,800m, below 300m in E, below 1,500m in Zumba). Humid to xeric woodland, scrub, clearings, gardens. ***Triangular bushy crest. Greyish to pale olive upperparts, paler rump***, whitish throat, olive breast, yellowish belly, ***two buffy to whitish wingbars***. Races *sclateri* (W), *olivaceum* (E), possibly *maranonicum* (SE). Restless, vocal, ubiquitous; in E confined to canopy. **Voice** Rather gravelly *kleé-klee-gligli, kleé-klee-gli-glirt!, glee-glee-klíklíklí*; simpler *gleeé, kleé'u, chrr'glee'u*; in E fast, descending *fleeé-gligligligli*. **SS** Recalls a tiny *Elaenia*, bushy crest diagnostic; see Grey-and-white Tyrannulet (drabber, weaker wing pattern). **Status** Common.

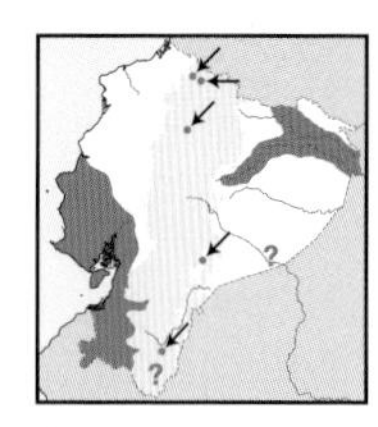

Mouse-coloured Tyrannulet *Phaeomyias murina* 12–13cm

E & SW lowlands, locally to slopes and N inter-Andean valleys (to 2,300–2,400m in W, mostly below 600m in E). Humid (E) to arid (W) light woodland, shrubby clearings. Flesh mandible. ***Dusky***-brown above, ***long*** but ***weak whitish supercilium***, greyish breast, ***whitish-yellow*** (W) to ***yellow*** (E) ***belly***. ***Two whitish*** (E) ***or ochraceous*** (W, Andes) ***wingbars***. Races *wagae*? (E), *incomta* (Andean valleys), *tumbezana* (W), possibly separate species. Usually in cover, gleans and sallies. **Voice** Dry, jumbled *puh'peh'pEh'peh'pegz, puh'peeg'g; peu-peé* calls (E); *drt-wizzzí, kd-zzeee't*, also squeaky *skeeit!, skee-kit!* (W). **SS** Little resemblance to other drab lowland tyrannids; see Fuscous Flycatcher (longer-billed, browner). **Status** Uncommon.

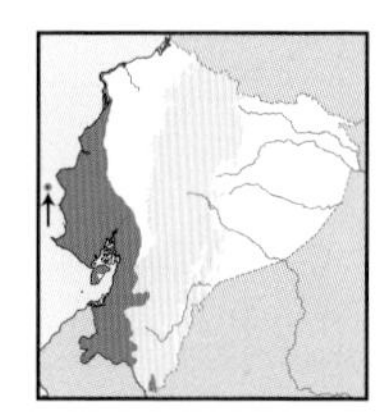

Tawny-crowned Pygmy-Tyrant *Euscarthmus meloryphus* 10–11cm

SW lowlands, to subtropics in Loja, extreme SE (Zumba region) (to 1,500m, locally 2,000m in S). Arid scrub, shrubbery, woodland. ***Tawny-orange forecrown and face***, greyish-olive above, duskier wings, ***two narrow buff wingbars***. White throat to breast, whitish below. Race *fulviceps*. Singles or pairs near ground, skulking but vocal; gleans foliage and searches debris. **Voice** Loud, fast, mechanical *t't't't-chirrr-tr-tr-tr-trr...*, call explosive *tiri-dík, tiri-di'dík, tiridí-tiridí*. **SS** Confusion unlikely in range and habitat; Scale-crested Pygmy-Tyrant (pale eyes, streaked breast) prefers more forested habitats; Grey-and-white Tyrannulet and Southern Beardless Tyrannulet share habitat but noticeably different. **Status** Fairly common.

Grey-and-white Tyrannulet
Yellow-crowned Tyrannulet
maranonicum
juvenile
moulting
into 1st
adult
juvenile
olivaceum
sclateri
Southern Beardless
Tyrannulet
fresh
worn
adult
juvenile
tumbezana
incomta
fresh
worn
wagae
Tawny-crowned Pygmy-Tyrant
Mouse-coloured Tyrannulet

PLATE 185: ELAENIAS

One of the most complex genera in the Neotropics. Some species are nearly impossible to identify by sight, especially White-crested, Sierran and Lesser. Voices are *the* identification clue, although austral migrants (Large, Small-billed, and possibly some White-crested) seldom call. Note wing pattern, lower belly colour and distribution (however, migrants complicate identification by range).

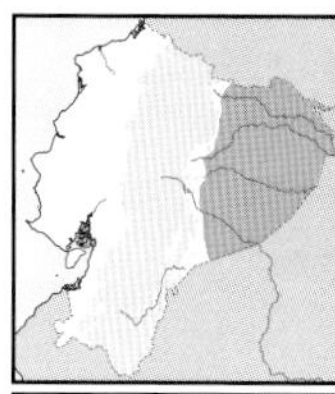

Large Elaenia *Elaenia spectabilis* — 17–18cm

Austral migrant to E lowlands (to 600m). Forest borders, riparian woodland, shrubby clearings. Flesh base to mandible, faint pale eye-ring, ***short crest, no white crown patch***. Brownish-olive above, ***pale grey*** throat to breast, yellow belly. Blackish wings, ***three whitish bars***, white edges to flight feathers. Monotypic. Mostly singly, perches exposed, gleans and plucks from perch. **Voice** Rarely utters a soft *t'cléuw!* **SS** Only large elaenia with *slight* crest and three wingbars; see *Myiarchus* flycatchers (sootier above, longer bills, no wingbars). **Status** Rare visitor (Mar–Sep).

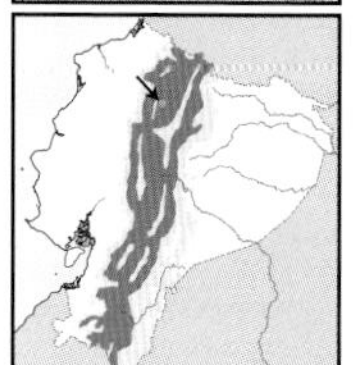

White-crested Elaenia *Elaenia albiceps* — 14–15cm

Andes and inter-Andean valleys (1,900–3,400m). Woodland, borders, shrubby clearings, *Eucalyptus* woodland. ***Prominent crest***, white crown patch, faint eye-ring. ***Brownish-olive above, dingy*** greyish underparts, breast-sides tinged dull olive, grading into ***whitish to yellowish-white mid belly***. Dusky wings, two pale bars. Races *griseigularis*, possibly austral migrant *chilensis*. Singles to loose groups perch quietly, sally briefly, hover-glean or pluck fruit. **Voice** Burry *béeurr*, buzzy *féeyrr!* every 3–5 sec. **SS** Lesser, Sierran and Small-billed Elaenias have yellowish not whitish mid belly, but easily confused (all prefer more humid habitats); voice of Sierran similar too (which see). **Status** Common.

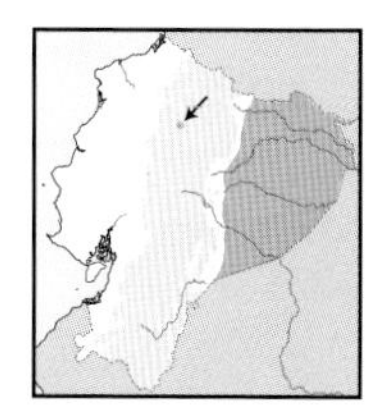

Small-billed Elaenia *Elaenia parvirostris* — 14–15cm

Austral migrant to E lowlands (mostly below 400m; once 2,800m). Forest borders, riparian woodland, clearings. Almost ***no crest, complete whitish eye-ring. Pale olive above***, whitish throat to breast, ***pale yellow belly***. Blackish wings, often ***three whitish bars***. Monotypic. Singles perch low, sometimes gather at fruiting trees and join mixed flocks. **Voice** At dawn *dweer'dwe'jit?*; clear *tcheeu* calls. **SS** White-crested, Lesser and Sierran are all dingier, tend to lack third wingbar (upper bar might be lost in worn plumage); incomplete eye-rings; larger Large Elaenia has yellower belly, more crested. **Status** Uncommon visitor (Apr–Oct).

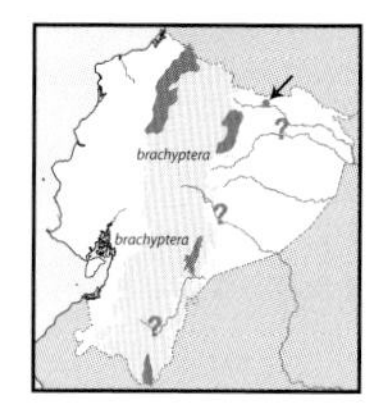

Lesser Elaenia *Elaenia chiriquensis* — 13–14cm

NW foothills to subtropics, SE foothills in Zumba region (700–2,800m in NW, 300–1,300m in E). Wooded borders, shrubby clearings. Similar to White-crested Elaenia but ***belly yellowish***. Some have ***three wingbars***. Crest somewhat more prominent than White-crested and Sierran. Races *brachyptera* (NW, NE), *albivertex* (extreme SE), currently separate species. Behaviour recalls White-crested and Sierran; Lesser tends to occur at humid edges. **Voice** Burry *chee'vir* or *jeé-bir*, longer in SE; also plaintive whistle *wiíb*. **SS** Not safely separated from White-crested (whitish mid belly) and Sierran (crest less squared), except by more whistled, less burry voice. **Status** Rare, seasonal movements?

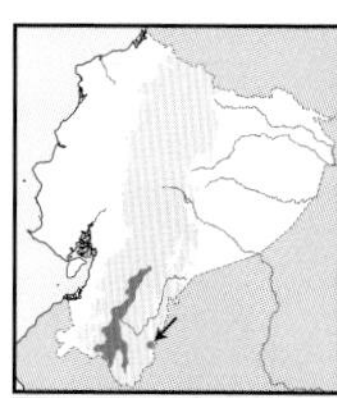

Highland Elaenia *Elaenia obscura* — 17–18cm

Local in S Andean subtropics to temperate zone (2,100–3,000m; once 1,700m). Montane forest borders, woodland. ***Short bill. Rounded head***, no white crown patch, ***yellowish eyebrow and narrow eye-ring***. Dull olive above, ***dull yellow below***, breast to flanks clouded olive. Two yellowish wingbars and edges to flight feathers. Nominate. Unlike congeners, remains in undergrowth; perches leaning forward, hover-gleans for fruit. **Voice** Burry, rolled *burrr, buuerrt, brrieet*. **SS** All other elaenias are crested, greyer below, lack eyebrow. **Status** Rare.

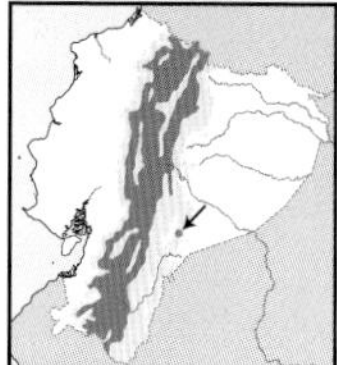

Sierran Elaenia *Elaenia pallatangae* — 14–15cm

E & W Andean subtropics to temperate zone, local in inter-Andean valleys (1,500–2,800m, locally higher). Montane forest borders, light woodland, shrubbery. Recalls White-crested Elaenia and not safely identified by plumage alone. Mid belly is ***yellowish***, whitish eye-ring sometimes more prominent; might show a third wingbar. Nominate. Behaves like White-crested but prefers more humid areas. **Voice** Burry, sharp *beer-ey, béerr, béeyrr, béeyr'r*, recalling White-crested. **SS** Lesser (squarer crest, dingier below, less burry voice); White-crested (whiter belly, often in drier areas, voices are similar, Sierran somewhat burrier; some silent birds must remain unidentified). **Status** Uncommon.

Large Elaenia
juvenile
griseigularis
griseigularis
White-crested Elaenia
chilensis
immature
fresh
worn
Lesser Elaenia
Small-billed Elaenia
juvenile
brachyptera
worn
Highland Elaenia
fresh
albivertex
adult
adult
juvenile
Sierran Elaenia

PLATE 186: ELAENIAS AND SCRUB FLYCATCHERS

Two large elaenias with distinctive large crests, often erected, plus a drab, vagrant elaenia. Scrub flycatchers recall elaenias but lack crests, heads are more rounded and bills are very short.

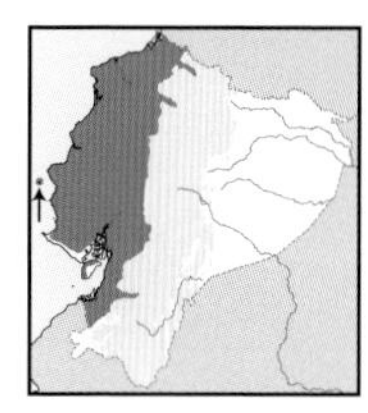

Yellow-bellied Elaenia *Elaenia flavogaster* 16–17cm

W lowlands to foothills, extreme SE in Zumba region, local to Andean valleys (to 2,300m). Forest borders, light woodland, shrubby clearings. ***Long erect crest, white semi-concealed crown patch.*** Greyish-olive upperparts, short whitish supraloral, narrow eye-ring. Whitish throat, greyish breast, yellowish belly, two whitish wingbars. Race *semipagana*. Singles to small parties, 'cheerful' and vocal; sallies, hover-gleans or picks fruit. **Voice** Loud, hoarse *chree-déyeeur* or *wree-wréeyuu, wreee-yeeeuú*; hoarse whistle *dreeeyr* sometimes in fast succession. **SS** Only large, crested elaenia in range; cf. Large (more rounded head). **Status** Fairly common.

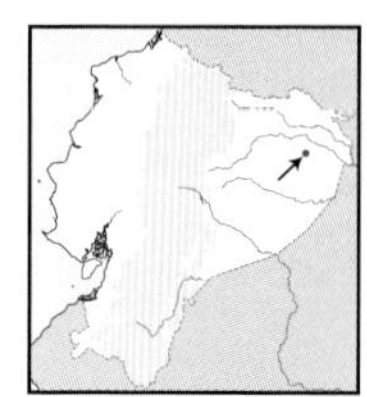

Slaty Elaenia *Elaenia strepera* 15–16cm

One record in E lowlands (J. Blake). *Terra firme* forest; expected in riparian woodland, forest edges. Very ***dark***, mostly ***slate-grey, very faint wingbars***, narrow pale eye-ring. Juv. greyish above, ***dirty olive breast to flanks***, yellowish belly, ***two bolder rufescent wingbars***, narrow white eye-ring and supraloral stripe. Monotypic. Inconspicuous as perches quietly, though often sallies. **Voice** Dry growling *t'grrrrr* (S *et al.*). **SS** Darker than other elaenias, nearly unmarked wings; cf. Crowned Slaty Flycatcher. **Status** Seasonal or accidental visitor? (Mar).

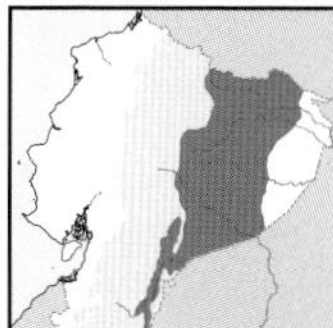

Mottle-backed Elaenia *Elaenia gigas* 18–19cm

E lowlands to foothills, not extreme E (to 1,250m). Riparian clearings, *Tessaria* scrub on river islands. ***Long, erect, forked crest***, white semi-concealed crown patch. Greyish-olive upperparts, ***back mottled dusky***. Whitish throat, greyish breast, yellowish belly, dusky wings, two prominent white wingbars. Monotypic. Pairs or small groups sometimes perch atop bushes or stumps. **Voice** Dry *drírt*; more complex, nasal, wheezy *whuet, chEt-tjit-tjet-jt* at dawn; *whü'eet, wüy-wüy'eet*. **SS** Confusion unlikely given mottling and long crest; Large Elaenia lacks long crest, upperparts are plain. **Status** Uncommon.

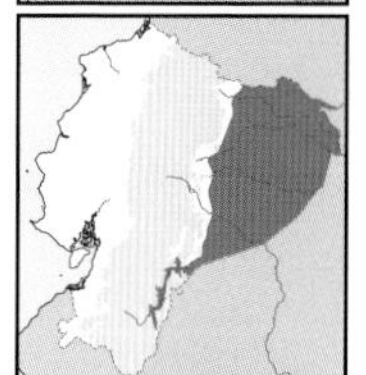

Amazonian Scrub Flycatcher *Sublegatus obscurior* 14–15cm

Local in E foothills and adjacent lowlands (to 900m). *Terra firme* forest canopy, borders. ***Drab. Short black bill.*** Sooty-grey upperparts, ***pale lores, faint eye-ring and supercilium. Ash-grey throat to breast***, pale yellow belly. ***Two or three faint wingbars.*** Monotypic. Sedentary pairs, do not join mixed flocks, sallies, sometimes picks fruit in flight. **Voice** Sweet, repeated *towhee?'chéudee...*; high-pitched *dzéeuw*, descending *dreeu* calls. **SS** Drab plumage and short supercilium distinctive; see Mouse-coloured Tyrannulet (dull yellowish below) and accidental Southern Scrub Flycatcher. **Status** Rare.

Southern Scrub Flycatcher *Sublegatus modestus* 14cm

Single record from shrubby clearing in Sumaco area at 1,120m (J. Nilsson). ***Short stubby bill, faint eye-ring and narrow supercilium.*** Olive-brown above, duskier wings, ***two or three bold whitish bars***; pale grey throat to breast, ***clear-cut contrast*** with pale yellow belly. Nominate? Quiet, unobtrusive, mainly in canopy where sallies to foliage or hover-gleans; austral migrant to Amazonia. **Voice** Weak, descending, high-pitched *pseeu* or *di-di-di-dzeéé* (RT, S *et al.*). **SS** Recalls Amazonian but smaller-billed, yellower below, bolder wingbars; see drabber Mouse-coloured Tyrannulet. **Status** Accidental visitor (Jun).

fresh
worn
Yellow-bellied Elaenia
♂
♀
Slaty Elaenia
♂
juvenile
Amazonian Scrub Flycatcher
-backed Elaenia
juvenile
Southern Scrub Flycatcher
fresh
worn

Small *Ornithion* tyrannulets have short tails and few markings. *Mecocerculus* are fairly small, arboreal tyrannulets characterised by clear-cut face patterns, bold wingbars and small bills. Active and agile, most glean horizontally, often with mixed flocks; cf. *Phylloscartes* and *Phyllomyias* tyrannulets. Careful identification of White-banded and White-tailed.

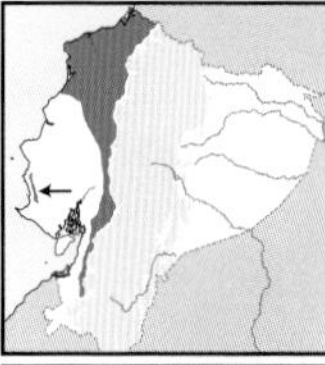

Brown-capped Tyrannulet *Ornithion brunneicapillus* 8–9cm

NW lowlands, S along base of Andes to N El Oro (below 400m). Humid forest canopy, borders, woody clearings. ***Short stubby bill. Sooty-brown crown, white supercilium***, dusky cheeks. Olive above, yellow below, ***unmarked dusky wings. Short tail***. Nominate. Mostly pairs with mixed flocks, active but very inconspicuous, glean foliage for prey. **Voice** High-pitched, whistled, descending *deeé-pi-pi-pi-piy*, sometimes only introductory *deeé* or *pleeé*. **SS** Short tail and unmarked wings distinctive; Sooty-headed shares unmarked wings but tail longer, longer supercilium, whitish throat. **Status** Uncommon.

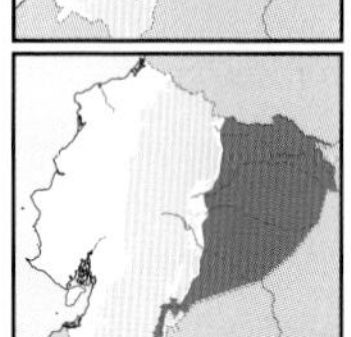

White-lored Tyrannulet *Ornithion inerme* 8–9cm

E lowlands (mostly below 600m). Mainly *várzea* and *igapó* forest canopy, borders, light woodland. ***Short stubby bill. Grey crown, short white supercilium and narrow eye-ring***. Greyish-olive cheeks and nape grading into olive upperparts. ***Two white, spotted wingbars***. Monotypic. Mostly pairs, energetic and fast but rather inconspicuous, might descend near water at edges. **Voice** High-pitched, rising *wee-dídídít?, wee-dídídít?, wee-dee-deet?*, more chortled when alarmed. **SS** Spotted wingbars, short tail and well-defined eyebrow and eye-ring diagnostic. **Status** Uncommon.

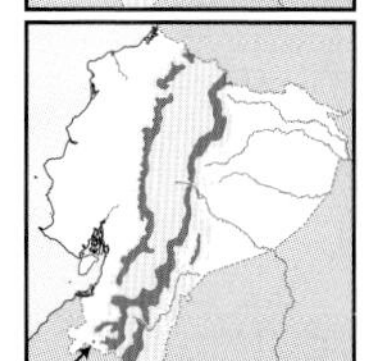

White-tailed Tyrannulet *Mecocerculus poecilocercus* 11–12cm

E & W subtropics (1,500–2,500m, locally higher). Montane forest canopy, borders. Bluish-grey crown, ***white supercilium, yellowish rump. Two buffy yellow wingbars***, whitish edgings to inner remiges. ***White outer tail feathers below***. Monotypic. Singles or pairs with mixed flocks; sallies, warbler-like hover-gleans or gleans; perches horizontally, often in epiphytes, moss clumps, tree ferns. **Voice** Thin, high-pitched *tsí-tsí-tse, tsí-tsí-tsi-tse*; simpler *wsí-wsí, tsí-tsi-wik*. **SS** Higher-elevation White-banded has ***sharper*** face pattern, bolder ***white*** wingbars, no yellowish rump, no white in tail; Rufous-winged (rufous wing panel and edgings). **Status** Fairly common.

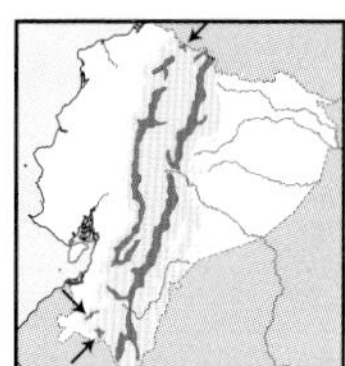

White-banded Tyrannulet *Mecocerculus stictopterus* 12–13cm

E & W subtropics to temperate zone (2,400–3,500m). Montane forest canopy, borders, woodland, clearings. Like previous species but ***better-demarcated supercilium (bolder head pattern), broader white wingbars, no white*** in tail, rump ***concolorous*** with upperparts. Nominate. Singles or pairs with mixed flocks, nimble, warbler-like, horizontally gleans outer foliage, twigs. **Voice** Thin, long, slurred *wiizzzz?* or *skuizzz?*, often becoming a descending trill *skuizzz-dr'r'r'r*; thin, squeaky *wsss-wss-t't't't-wis* at dawn. **SS** Rufous-winged (locally overlap) has distinct rufous in flight feathers. **Status** Fairly common.

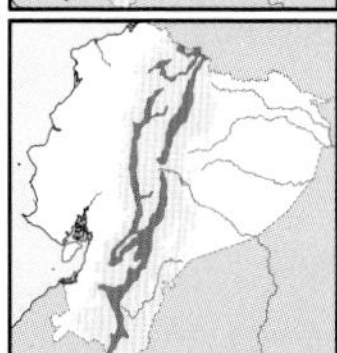

White-throated Tyrannulet *Mecocerculus leucophrys* 14cm

E & W temperate zone to *páramo* (2,800–3,500m). Montane forest borders, woodland, bushy 'islands' in *páramo*. ***Brown*** upperparts, white supercilium, ***puffy white throat***. Olive-brown chest, ***two rufous wingbars, long tail***. Race *rufomarginatus*. Pairs to small groups with mixed flocks, perches vertically, sometimes exposed, sallies to air or foliage. **Voice** Warbled *tidi-k'k, tidi-k'k..., tidi'k-tidi'k-tidi'k'k'k*; shorter *píf-píf-píf* calls. **SS** Other *Mecocerculus* lack puffy white throat and perch horizontally; Brown-backed Chat-Tyrant (chocolate-brown upperparts, orangey underparts, bolder supercilium). **Status** Uncommon.

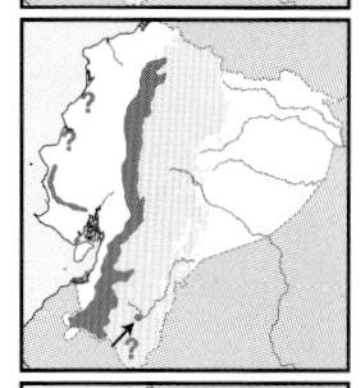

Rufous-winged Tyrannulet *Mecocerculus calopterus* 11–12cm

W Andean foothills to subtropics, local in coastal mountains, once in SE (700–2,000m, locally to 300m in SW). Humid to semi-deciduous forest, borders, woodland. Grey crown, ***long white supercilium, blackish ocular stripe***. Two yellowish wingbars, ***rufous panel in secondaries and tertials***. Monotypic. Singles or pairs with mixed flocks, often low, mostly gleans from foliage; possibly makes seasonal movements. **Voice** Gives *pü-weé, pü-wee'ki, pü-wee-ki'ki'wi*. **SS** White-tailed Tyrannulet (no rufous in wings); also Rufous-winged Antwren (different shape, behaviour, etc.). **Status** Uncommon.

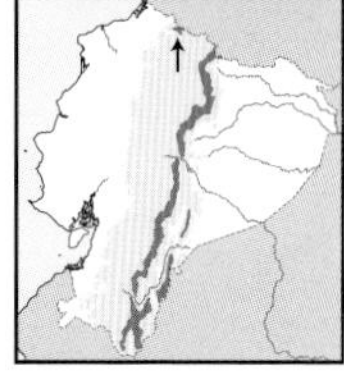

Sulphur-bellied Tyrannulet *Mecocerculus minor* 11–12cm

E and extreme NW subtropics to lower temperate zone (1,600–2,800m). Montane forest canopy, borders. Greyish crown, olive above, ***long white supercilium***, dusky ocular stripe. ***Sulphur-yellow underparts***, whiter throat. ***Two bold buff wingbars***, long tail. Monotypic. Singles or pairs with mixed canopy flocks, gleans horizontally, nimbly inspecting foliage and twigs. **Voice** Squeaky *chew-chew-chew-chew*, sometimes with additional emphasised *chéú* at end; more nasal *skwi-skwi-skwe-skwu* (RG), shorter *che-we-ku*. **SS** Smaller Variegated Bristle-Tyrant, Plumbeous-crowned and Ecuadorian Tyrannulets have dark cheek crescent and perch more vertically. **Status** Uncommon.

Brown-capped Tyrannulet
juvenile
adult
White-lored Tyrannulet
juvenile
adult variant
adult
White-tailed Tyrannulet
juvenile
adult
White-banded Tyrannulet
juvenile
adult
Rufous-winged Tyrannulet
White-throated Tyrannulet
Sulphur-bellied Tyrannulet

PLATE 188: TIT-TYRANTS AND PYGMY-TYRANTS

Small Andean tyrannids mostly found away from flocks, except Agile Tit-Tyrant. Tit-tyrants have bold streaky bodies and erect crests, and are less elusive than both pygmy-tyrants illustrated here. Pygmy-tyrants of forest interior are rarely seen owing to their behaviour and habitats; their trilled voices and bill snaps are better indications of their presence and identity.

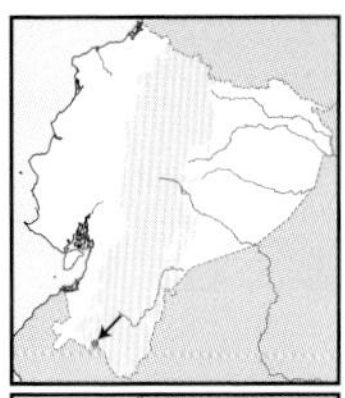

Black-crested Tit-Tyrant *Anairetes nigrocristatus* 12–13cm

Temperate zone in extreme SW Andes (2,400–2,500m). Humid woodland and scrub, borders. ***Red mandible***. Handsome, mostly black above, ***very long erect black crest, white crown to nape***, streaked ***black and white*** above and on breast, two white wingbars. Undescribed race? Singles or pairs at edges, glean foliage or sally, but more skulking than Tufted; sometimes perch on exposed branches to sing. **Voice** Explosive, fast *whi'r-ti-ti-ti-ti-ti'r* or *whí-ti'ti'ti'ti'ti*. **SS** Confusion unlikely, smaller Tufted has shorter crest, pale eyes, yellow belly, etc. **Status** Rare, very local.

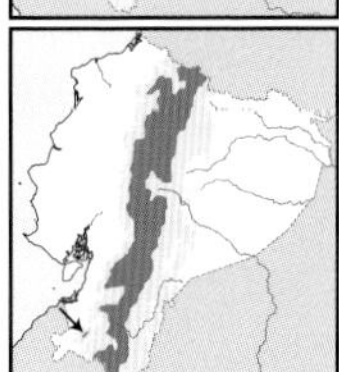

Tufted Tit-Tyrant *Anairetes parulus* 10–11cm

Andean temperate zone in E & W, inter-Andean slopes (2,500–3,500m). Montane forest borders, woodland, dense shrubbery. ***Pale eyes***. Olive-grey above, ***long curly blackish crest, whitish eyebrow***. Whitish cheeks to lower breast ***streaked dusky***, yellowish belly, two whitish wingbars. Race *aequatorialis*. Agile and nervous, singles or pairs in undergrowth and edges, gleans, hover-gleans or briefly sallies; seldom joins mixed flocks. **Voice** Fast and brief harsh trill *wi'rrr't* or *chuirr't*, more explosive, faster and high-pitched buzzy *chrrrt-chuirt-chi'di-chi'di'di'di...* in interactions. **SS** Larger Agile has similar pattern but back streaked, lacks erect crest, eyes darker, tail longer. **Status** Common.

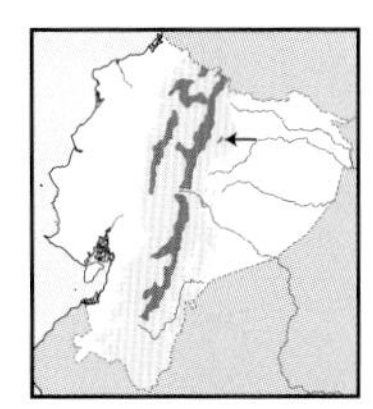

Agile Tit-Tyrant *Uromyias agilis* 13–14cm

E & W temperate zone, S to N Loja in E, to Cotopaxi in W (2,600–3,500m). Montane forest undergrowth, borders, bamboo. ***Dark eyes***. Blackish crown, ***long flat crest, long whitish supercilium***. Olive-brown ***streaky*** upperparts, more rufescent rump. Whitish underparts ***coarsely streaked dusky***, plainer yellowish belly, ***no wingbars***. Monotypic. Acrobatic pairs or small groups rapidly glean outer foliage and twigs, mainly with mixed flocks. **Voice** Brief soft trill *pi'rrrrr't* or *pi'iiii'irt*, also louder, more jumbled *te'ee'rrr, treerrr* interspersed with clearer *tsidí* or *tsee-dí* notes. **SS** Smaller Tufted has curly crest, pale eyes, wingbars and shorter tail. Voice recalls Cinnamon Flycatcher's trills. **Status** Uncommon.

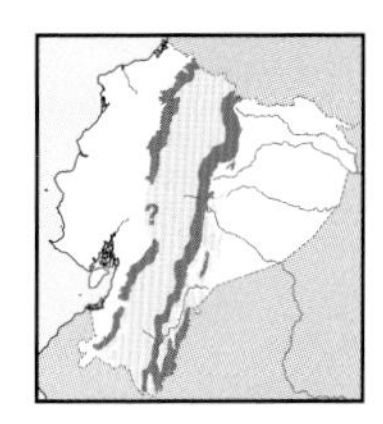

Bronze-olive Pygmy-Tyrant *Pseudotriccus pelzelni* 11–12cm

E & W foothills to subtropics, sparser in W (600–2,000m, locally lower). Montane forest, borders. ***Reddish eyes. Dark to greenish-olive***, whiter throat, yellower belly. Nominate (E), *annectens* (W). Mainly solitary in shady understorey, seldom with mixed flocks; short sallies to underside of leaves. **Voice** High-pitched, weak series of 3–6 *piít* or *pií* notes, sometimes one separated at end, slightly accelerating; weak, high-pitched shrill *pr'eeeeee* or *shr'eeeee*; also short *triii* trill and wing snaps. **SS** Duller and drabber than most sympatric tyrannids; Rufous-headed has contrasting head and wings, Buff-throated and Cinnamon-breasted Tody-Tyrants have distinct head and breast patterns; ♀ manakins in E (Green, Jet, Golden-winged) are distinctively plumper, with different behaviour, leg colours, etc. **Status** Uncommon.

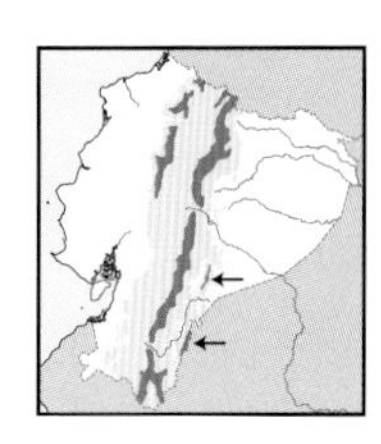

Rufous-headed Pygmy-Tyrant *Pseudotriccus ruficeps* 11–12cm

E & W subtropics to temperate zone, S to Chimborazo in W (2,000–3,300m). Montane forest undergrowth, borders. ***Rufous-orange hood, rufous wings and outer tail feathers***; olive above, yellowish-olive below, yellower belly. Monotypic. Unobtrusive in shady understorey, usually not with mixed flocks, performs short strikes to underside of leaves. **Voice** Very fast, long, dry trill *z'rrrrreeeu?*, first ascending then descending, varying as brighter *teeeee'iii-ip*; calls include shorter trills *tz'rrr* or *t'rrrr*. **SS** Imm. lacks complete rufous head and might resemble Bronze-olive, but wings already rufous; see Rufous-crowned Tody-Flycatcher with rufous head but bolder pattern overall. **Status** Uncommon.

Black-crested Tit-Tyrant

Tufted Tit-Tyrant

Agile Tit-Tyrant

Bronze-olive Pygmy-Tyrant

Rufous-headed Pygmy-Tyrant

PLATE 189: TYRANNULETS AND ROYAL FLYCATCHER

A combination of forest and forest edge tyrannulets. The small-billed *Zimmerius*, being primarily frugivorous, occur in forest canopy, borders and adjacent clearings. They are characterised by yellow edges to their wing-coverts and flight feathers. Complex taxonomy in Golden-faced, from which Chocó was recently separated. The Royal – which almost certainly comprises two species in Ecuador – has a unique, spectacular, spreadable and multi-coloured crest.

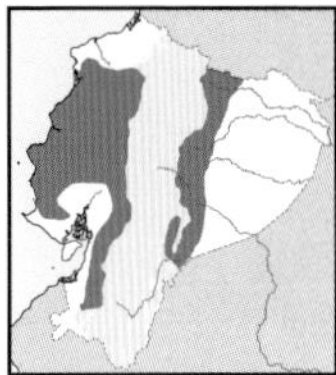

Yellow Tyrannulet *Capsiempis flaveola* 11–12cm

W lowlands to foothills, NE foothills, not in wet NW (to 1,500m). Forest borders, secondary woodland, bamboo. Bright olive above, ***lemon-yellow supercilium and lower face to underparts***. Blackish wings with ***two yellow bars and edges to flight feathers***. Races *cerula* (E), *magnirostris* (W). Mostly restless pairs, horizontally gleans foliage, briefly sally and hover-glean; seldom with mixed flocks. **Voice** Short nasal trill *ph'rrrr*; more pleasing song *pee-ee, pee-tee, pee-peetee, pee-peetee...prr-peetee....* **SS** Confusion unlikely, the yellowest tyrannulet in Ecuador. **Status** Fairly common.

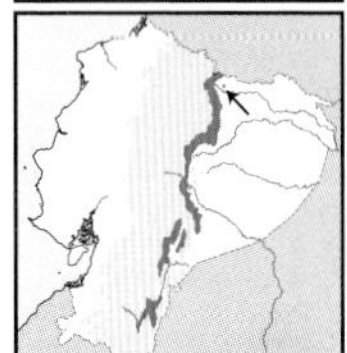

Red-billed Tyrannulet *Zimmerius cinereicapilla* 11–12cm

Scattered records in E Andean foothills (900–1,350m). Montane forest canopy and borders, secondary woodland. ***Reddish mandible, pale amber eyes. Bluish-grey crown to nape***, bright olive upperparts, ***narrow whitish supercilium***. Whitish throat, yellow below. Darker wings with ***sharp yellow edges***. Monotypic. Mostly singly, hops and gleans in outer foliage, often with mixed flocks; easy to miss unless vocalising. **Voice** Gives *chédew-chédew-chédew, chew-chechecheche*, also slightly descending *chew-dedededu*. **SS** No overlap with similar Slender-footed; commoner Golden-faced has yellow eyebrow, dark eyes, olive crown. **Status** Rare; VU globally.

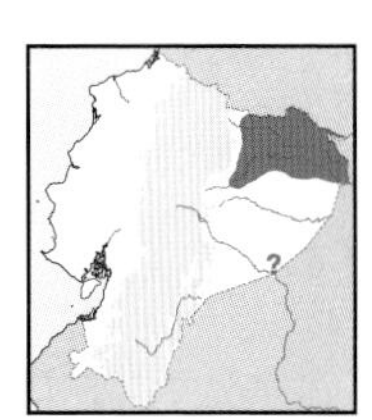

Slender-footed Tyrannulet *Zimmerius gracilipes* 10–11cm

NE lowlands (below 300m). *Terra firme* and *várzea* canopy, borders. ***Grey eyes. Greyish-olive crown, narrow whitish supercilium***. Bright olive above, ***yellow below***, throat whitish. Blackish wings, ***prominent yellow edges***. Nominate. Pairs with mixed flocks; gleans horizontally and rapidly hops in outer foliage. **Voice** Thin, *deeu-di'riri, deeu-di'riri...*, lower-pitched than Golden-faced, shorter introductory note, slightly descending; also simpler *tuee?* **SS** Golden-faced (similar wing pattern) has yellow face, dark eyes. Wing pattern separates from other small tyrannids; Sooty-headed has browner crown, bolder supercilium. **Status** Uncommon.

Chocó Tyrannulet *Zimmerius albigularis* 11–12cm

W lowlands to subtropics (to 1,900m?). Humid to montane forest canopy, borders, adjacent clearings. Closely recalls Golden-faced but ***chin to throat whiter***, greyer breast, little yellow on belly. Monotypic. Mainly singly, nimbly hover- and perch-gleans using horizontal movements, briefly perches upright in open, rapidly hops through outer foliage. **Voice** Querulous, slightly ascending *dreeu-de-dididi?*, simpler *dri-dri-dirí...* calls. **SS** Replaced in SW by similar *flavidifrons* race of Golden-faced, which is ***greyer*** below; on extreme NW slopes cf. Golden-faced, which replaces it above 1,200–1,400 m. **Status** Fairly common.

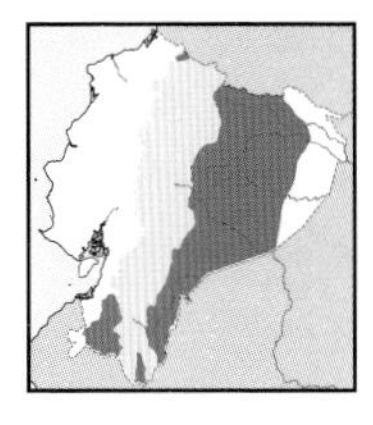

Golden-faced Tyrannulet *Zimmerius chrysops* 11–12cm

E lowlands to subtropics, S foothills to subtropics, NW subtropics (to 2,200m in E, 900–2,400m in S, 1400?–2600m in NW). Humid forest, borders, woodland, adjacent clearings. ***Brown eyes. Golden-yellow eyebrow to foreface***. Wings as congeners. Nominate (E, extreme NW?), *flavidifrons* (SW–SE) possibly valid species or race of Peruvian Tyrannulet *Z. viridiflavus*. Active singles or pairs in outer foliage and canopy, with or apart from mixed flocks. **Voice** Thin, cheerful *teeeu-dididi* song, plaintive *deeuú* calls (E and NW); loud, ascending *treeu-truuu-eé*, querulous *deeeuu'IT?* (SW). **SS** Locally overlaps with Chocó Tyrannulet from which better told by voice; in E lowlands see Slender-footed (grey eyes, greyer crown, no yellow in face, yellower belly), in foothills Red-billed (whitish eyes, no yellow in face, grey crown). **Status** Common.

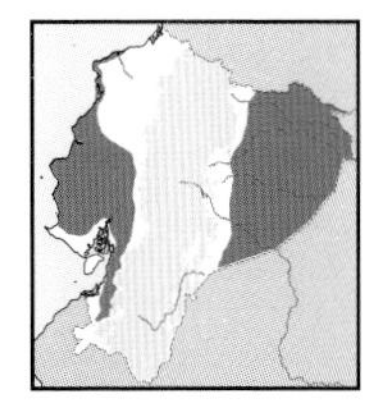

Royal Flycatcher *Onychorhynchus coronatus* 15–16cm

E & SW lowlands (to 400m in E, 600m in W). *Várzea* forest (E), deciduous to semi-deciduous forest, borders (W). ***Long bill. Chocolate-brown*** (E) to ***ruddy*** (W) ***upperparts***, whitish wing dots, ***long rufescent tail. Tawny-cinnamon below***, breast mottled brown (E). ***Long, fan-shaped, extendable, red and deep blue crest***. Races *castelnaui* (E), *occidentalis* (W), possibly separate species. Mostly in pairs, unobtrusive, perch upright and alert, makes darting sallies; often near water. **Voice** Loud, short, squeaky *kEE'oóp!* **SS** Note hammer-shaped head. **Status** Uncommon, W race VU.

cerula
magnirostris
Yellow Tyrannulet
Red-billed Tyrannulet
nder-footed Tyrannulet
Chocó Tyrannulet
chrysops
flavidifrons
Golden-faced Tyrannulet
♂
juvenile
castelnaui
♂
Royal Flycatcher
occidentalis

PLATE 190: BRISTLE-TYRANTS AND TYRANNULETS

The genus depicted here was formerly separated into *Pogonotriccus* and *Phylloscartes*, and two groups can be distinguished. The bristle-tyrants (former *Pogonotriccus*) have slender bills, bold wingbars, grizzled faces and ear-covert crescents. They perch somewhat upright, long tails held almost vertical and regularly flick one wing upward. The other two tyrannulets move more horizontally, cock their tails and also flick one wing up, foraging in upper strata.

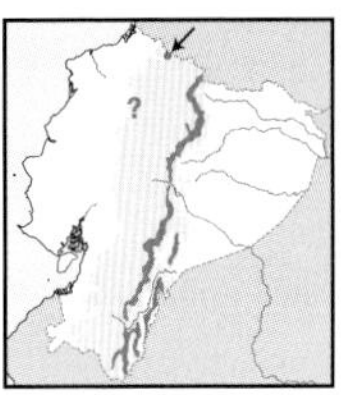

Variegated Bristle-Tyrant *Phylloscartes poecilotis* 11–12cm

E and extreme NW subtropics (1,500–2,000m). Montane forest undergrowth. ***Pink to yellow mandible.*** Grey cap, white eye-ring, ***grizzled grey-and-white face, white-and-black ear crescent.*** Olive above, yellow below. Dusky wings, ***two bold cinnamon bars***, olive edges to flight feathers. Nominate. Singles to small groups often with mixed flocks, perch vertically, make short sallies to foliage or twigs; flick wings up alternately. **Voice** Thin *chít* calls, often running into longer *chít-chít-chít...chs-chs-chs-chsít.* **SS** Marble-faced (black bill, fainter yellow wingbars); Ecuadorian (no grizzled face, yellow wingbars) and Sulphur-bellied Tyrannulets (long white supercilium, no grizzling, buffier wingbars); larger Slaty-capped Flycatcher (blackish bill, breast washed olive). **Status** Rare.

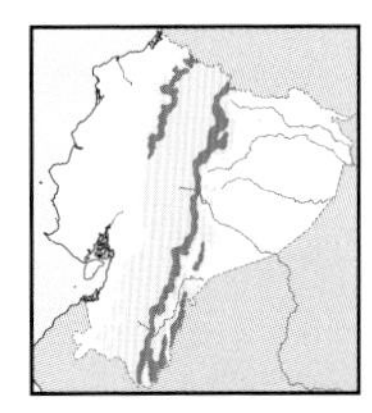

Marble-faced Bristle-Tyrant *Phylloscartes ophthalmicus* 11–12cm

E & NW Andean foothills to subtropics (1,100–2,100m). Montane forest undergrowth and borders. ***Uniform blackish bill.*** Grey cap, ***grizzled white-and-grey face, and black-and-white ear-covert crescent.*** Olive above, yellow below, dusky wings, ***two yellow bars***, olive edges. Nominate. Singles or pairs with mixed flocks, perch vertically, hover-glean and make short sallies. Regularly flicks one wing up. **Voice** High-pitched *kéek-t't't't*, fast *psee-ee-ee-ee...sí-sí-sí*; call a doubled *ts-trt.* **SS** Variegated (cinnamon-buff bars, bicoloured bill); Ecuadorian (no face grizzling, less conspicuous ear crescent, different behaviour), smaller Plumbeous-crowned (no facial grizzling, shorter bill) and Ashy-headed Tyrannulets (more bluish crown, streaky breast, shorter bill, no facial grizzling); also Slaty-capped Flycatcher. **Status** Uncommon.

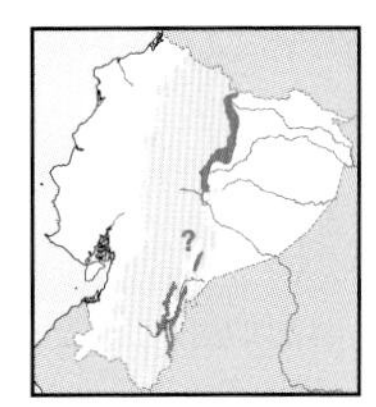

Spectacled Bristle-Tyrant *Phylloscartes orbitalis* 11–12cm

Local in E Andean foothills (700–1,400m). Montane forest undergrowth. ***Flesh mandible.*** Grey cap, ***pale lores and eye-ring, yellowish cheeks, greyish-and-white weak ear-covert crescent.*** Bright olive above, yellow below, black wings with two yellowish bars, whitish edges to flight feathers. Monotypic. Perches vertically and sallies to pick prey from foliage or twigs, regular in mixed flocks; flicks one wing up, then opposite wing. **Voice** Fast, thin, descending trills followed by emphatic notes: *t'rrr..pít-pít-pít.* **SS** Confusion unlikely but see Ecuadorian Tyrannulet (long white supercilium, faint eye-ring but bolder face pattern, uniform blackish bill, and perches horizontally, cocking tail). **Status** Rare, overlooked? NT in Ecuador.

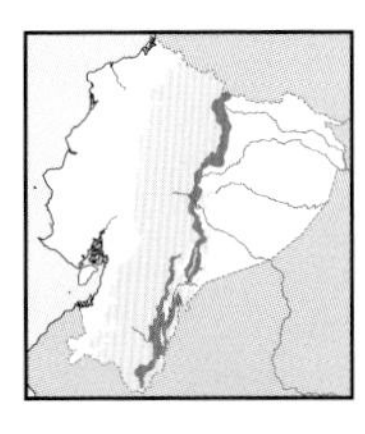

Ecuadorian Tyrannulet *Phylloscartes gualaquizae* 11–12cm

Local in E Andean foothills (700–1,400m). Montane forest canopy, borders. ***Greyish bill.*** Grey cap, ***narrow white supercilium, yellowish-white cheeks, faint dusky ear-covert crescent and dusky malar.*** Bright olive above, yellow below, breast clouded dusky. Black wings, two yellow bars, greenish yellow edges. Monotypic. Singles or pairs with mixed flocks, gleans foliage horizontally and sallies briefly, often cocks tail and flags one wing up. **Voice** Thin *freeee'i'i* at dawn, more trilled *friiiiii* and thin *feee.* **SS** Posture differs from bristle-tyrants but cf. Marble-faced; Plumbeous-crowned (bolder face pattern, with more contrasting ear crescent, stubbier bill) and Rough-legged Tyrannulets (whiter face, no ear crescent and flesh mandible). **Status** Uncommon; NT.

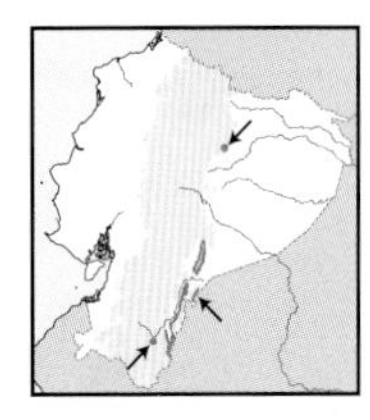

Rufous-browed Tyrannulet *Phylloscartes superciliaris* 11–12cm

Local in isolated SE mountain ranges (Kutukú, Cóndor; 1,300–1,700m). Montane forest canopy. Grey cap, ***narrow*** but distinctive ***rufous-buff supercilium. White cheeks narrowly outlined black.*** Olive upperparts, ***greyish-white underparts***, breast clouded grey. Dusky wings with yellowish edges to coverts and inner flight feathers. Race *griseocapillus.* Active pairs to small groups with mixed flocks, also glean foliage and move through terminal twigs in high strata; body held horizontal, tail rather cocked. **Voice** 'Joyful' *pree-ee-ee-ee-i* or *tri-didididi.* **SS** Confusion unlikely in limited range. **Status** Uncommon.

Variegated Bristle-Tyrant
Marble-faced Bristle-Tyrant
adult
Spectacled Bristle-Tyrant
juvenile
Ecuadorian Tyrannulet
juvenile
Rufous-browed Tyrannulet
immature
adult

PLATE 191: FLYCATCHERS I

Three primarily frugivorous, flycatchers of forest understorey, with short bills and little pattern to wings; often join mixed flocks, but otherwise inconspicuous. Streak-necked and Olive-striped are easily confused. Note mandible colour, underparts and amount of grey on head. *Leptopogon* recall bristle-tyrants in their grizzled faces, ear crescent and single-wing flicks, but bills longer and wingbars less conspicuous. Often with mixed flocks.

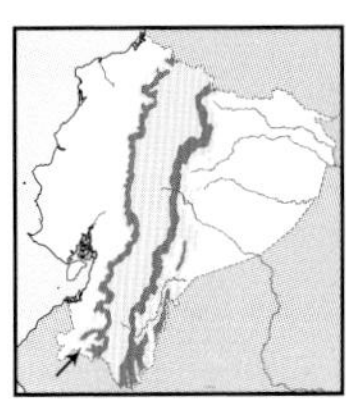

Streak-necked Flycatcher *Mionectes striaticollis* 13–14cm

E & W upper foothills to subtropics (1,500–2,500m, locally higher). Montane forest interior, borders, secondary woodland. Fleshy mandible base, white postocular spot. ***Grey head to neck***, olive upperparts, underparts greyish-olive coarsely streaked pale yellowish but lower underparts ***plainer yellow***. Dark wings with two faint buff bars. Races *colombianus* (E), *viridiceps* (W). Singles or loose pairs with mixed flocks, gleans and hover-gleans, perches upright but leans forward; ♂♂ display at fairly loose leks. **Voice** Long-repeated, thin, hummingbird-like *tsi-tsiri, tsi-tsiri*... or *sli-sle, sli-sle*.... **SS** Olive-striped Flycatcher lacks grey on head, streaking more extensive on underparts, bill often entirely dark. **Status** Fairly common.

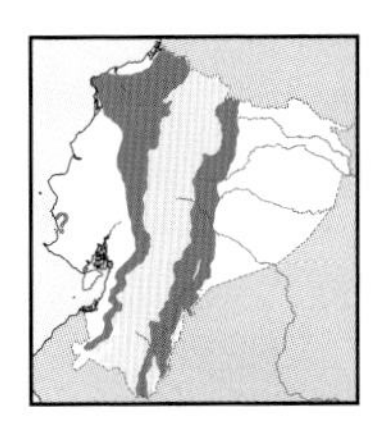

Olive-striped Flycatcher *Mionectes olivaceus* 13–14cm

E & W subtropics to adjacent lowlands, in NW extends to lowlands (to 2,000m). Montane forest interior, borders, woodland. Flesh to ***dark*** mandible base, white postocular spot. Dark olive above, dull olive below ***coarsely streaked pale yellow except yellowish lower belly***. Races *fasciaticollis* (E), *hederaceus* (W). Singles or loose pairs with mixed flocks, glean and hover-glean for invertebrates and fruit; ♂♂ display at exploded leks. **Voice** Long-repeated, thin, high-pitched, hummingbird-like *seeee-seee-see-sé, seeee-seee-see-sé*...; also thin, dry *dzeeeee* and descending *dseei* calls. **SS** Closely recalls higher-elevation Streak-necked Flycatcher (yellower belly, grey hood – but less evident in W). **Status** Fairly common.

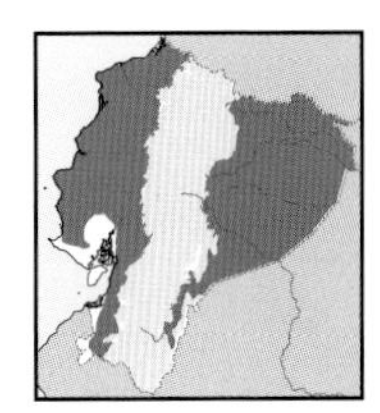

Ochre-bellied Flycatcher *Mionectes oleagineus* 12–13cm

E & W lowlands to foothills (to 1,000m). Humid to deciduous forest interior, secondary woodland, borders. Olive head to upperparts, ***throat to chest ochraceous clouded dull olive, plain ochre belly***. Two faint ochraceous wingbars. Races *hauxwelli* (E), *pacificus* (W). Mostly singles, sometimes with mixed flocks; gleans foliage and makes rapid darts; also congregates at fruiting trees; when perched fans wings alternately. ♂♂ display at dispersed leks. **Voice** At leks loud sneezing *pchá-pchá-pchá-pchá*... interspersed by *weéb, widee, wíp* and other calls; otherwise rather silent. **SS** Smaller Ruddy-tailed Flycatcher (distinct ruddy tail and wings). **Status** Uncommon.

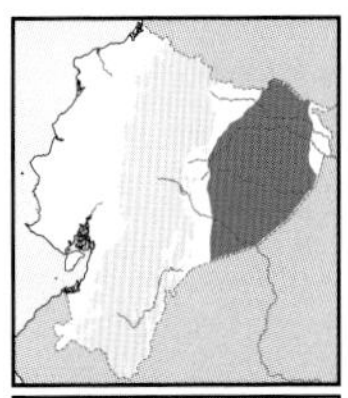

Sepia-capped Flycatcher *Leptopogon amaurocephalus* 13–14cm

E lowlands (to 450m). *Terra firme* and *várzea* forest interior. ***Maroon-brown crown, whitish face grizzled grey, dusky ear-coverts***. Olive above, whitish throat grading into greyish-olive chest, yellow belly. Dark wings, ***two buff bars***, pale edges to inner flight feathers. Race *peruvianus*. Mostly in pairs with understorey flocks, sally to foliage, make upward strikes and pick fruit; often raises one wing up while perched. **Voice** Accelerating but falling rattle *drE-e'e'e'e'eu'u'u*; low *puk* calls. **SS** Confusion unlikely in range. **Status** Uncommon.

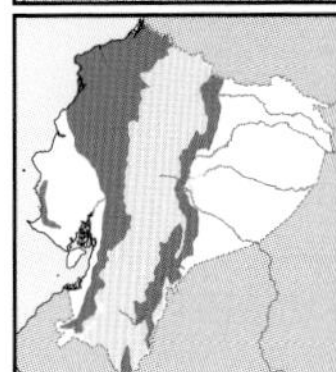

Slaty-capped Flycatcher *Leptopogon superciliaris* 13–14cm

W lowlands to upper foothills, SW coastal ranges, E foothills (to 1,500m in W, not below 600m in E). Montane to semi-humid forest interior, woodland, borders. ***Grey crown, frosty white eyebrow and most of face, black ear-coverts***. Olive above, greyish-olive breast. ***Two bold ochraceous to yellowish wingbars***. Nominate race. Singles or pairs with mixed flocks, perch erect, sally and hover-glean; lift wings up. **Voice** Nasal, squeaky *teyeér-teyeér* or *k'yeér-tyerrt*, loud *wyeek* (E); less nasal, high-pitched *tss-tsriit* (W). **SS** In E Variegated (uniform yellow below, pink mandible, cleaner face pattern) and Marble-faced Bristle-Tyrants (fainter yellow wingbars); smaller Plumbeous-crowned Tyrannulet (bold white supercilium, yellower underparts, no ochraceous wingbars). **Status** Common.

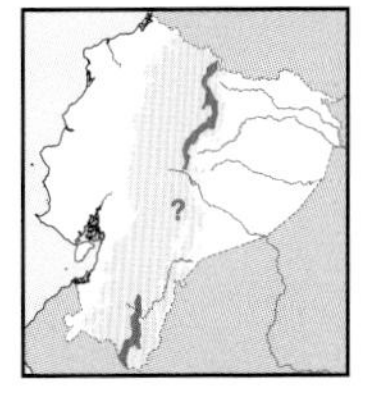

Rufous-breasted Flycatcher *Leptopogon rufipectus* 13–14cm

E subtropics (1,600–2,500m). Montane forest interior, borders, secondary woodland. ***Rufous-buff forecrown and eyebrow***, grey crown, dark ear patch, olive upperparts. ***Rufous-buff throat to breast***, yellow below, two buff wing bars, yellowish edges to inner flight feathers. Nominate. Singles or pairs with mixed flocks, perch erect, perform hover-gleans and upward sallies to leaves and foliage. **Voice** Squeaky *kyeét* or *squeéyt*, often in fast succession; also brief *kwik*. **SS** Handsome Flycatcher (bright buff breast, no rufous in face, no dark grey crown and heavier Fulvous-breasted Flatbill). **Status** Uncommon.

columbianus
juvenile
Olive-striped Flycatcher
fasciaticollis
viridiceps
Streak-necked Flycatcher
hederaceus
hauxwelli
pacificus
juvenile
Ochre-bellied Flycatcher
adult
variants
Sepia-capped Flycatcher
Rufous-breasted Flycatcher
Slaty-capped Flycatcher

Small to very small flycatchers with the appropriate name of pygmy-tyrants. *Myiornis* are exceptionally small, with tiny tails. Very inconspicuous by size and retiring behaviour. *Lophotriccus* also retiring but vocal. Also a drab tody-flycatcher that might resemble sympatric Double-banded Pygmy-Tyrant.

White-bellied Pygmy-Tyrant *Myiornis albiventris* 6–7cm

Single locality in extreme SE in Nangaritza Valley (980m). Foothill forest borders. ***Pink legs***, narrow eye-ring, ***short white supraloral stripe***. Olive above, darker wings with two yellowish bars and edges. Face, neck-sides and most underparts ***whitish with coarse dull grey streaks***; lower belly pale yellowish. Monotypic. Small groups or pairs within mixed flocks; nimble and vocal. **Voice** Mellow, tinkling, rising-falling trill: *triiiieeuw*; sometimes stuttered notes interspersed (S *et al.*); call a quieter *trrrrl*. **SS** Tail length and size diagnostic; see Scale-crested. **Status** Rare resident.

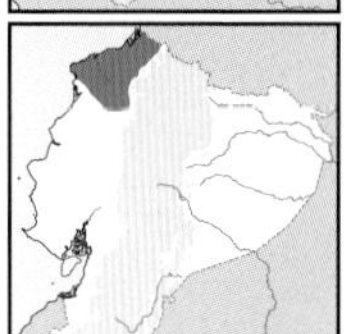

Black-capped Pygmy-Tyrant *Myiornis atricapillus* 6–7cm

NW lowlands to lower foothills (to 800m). Humid forest borders, woodland. ***Blackish crown (grey in ♀), white supraloral and narrow eye-ring***. Olive above, mostly white below; conspicuous yellowish wingbars and edges to flight feathers. ***Tiny tail***. Monotypic. Rather inactive, perches motionless for long periods, suddenly darts to under- or upperside of leaves for prey; difficult to spot, moves fast between perches. **Voice** Brief, explosive, inflected *brrip* or *greek*, high-pitched; shorter cricket-like *giik* or *grrk* and hummingbird-like wing buzz. **SS** Unmistakable in range. **Status** Uncommon.

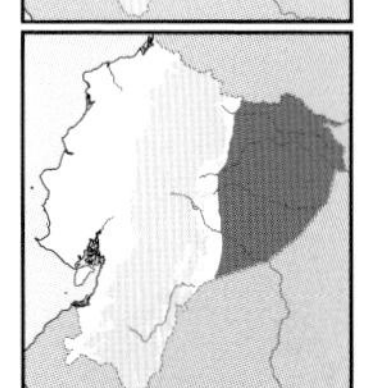

Short-tailed Pygmy-Tyrant *Myiornis ecaudatus* 6–7cm

E lowlands (to 400m). *Terra firme* forest interior and borders, *várzea* borders. ***Head mostly grey, white supraloral and narrow eye-ring***, dark lores. Olive upperparts, mostly white underparts, wings as previous species (formerly considered conspecific). ***Tiny tail***. Nominate. Regular in vine tangles, darting flight produces insect-like sound and motion effect; attacks prey on under- and upperside of leaves. **Voice** An insect-like short trill *birrrp* or *grrrip-rrrip..rrrip*; shorter *birrp*, *bírp*, *trérp* calls. **SS** Unmistakable by size and tail length; see plumper Dwarf Tyrant-Manakin. **Status** Uncommon.

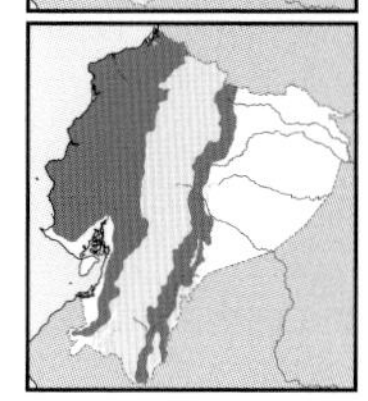

Scale-crested Pygmy-Tyrant *Lophotriccus pileatus* 10–11cm

W lowlands to subtropics, E foothills to subtropics (to 1,700m in W, 700–1,700m in E). Humid to semi-deciduous forest, borders, woodland. Pale eyes. Long ***orange-black crest*** often held flat (sometimes ***fanned***). Olive above, yellowish-white below ***coarsely streaked olive***. Nominate (E), *squamaecrista* (W). Mostly alone in shady understorey, rarely with mixed flocks; suddenly darts to foliage; ♂ vocalises day long in lek-like associations. **Voice** Fast, rolling, ascending-descending, whistled *tr-tr-tr-tr-tr…*, more inflected *trrreí?* (W); 2–3 short consecutive trills *trree-trreu*, *trree-trree-trreu*; rising and inflected *prrruí?*, longer *trrrrruí* (E). **SS** Lower-elevation Double-banded Pygmy-Tyrant has bluish-grey crest. **Status** Common.

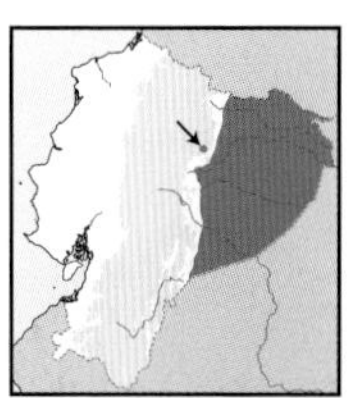

Double-banded Pygmy-Tyrant *Lophotriccus vitiosus* 10–11cm

E lowlands (to 600m, locally higher). *Terra firme* and *várzea* undergrowth, borders. Resembles Scale-crested but ***crest bluish-grey and black, faint streaks on throat to lower breast***. Race *affinis*. Performs short sallies for prey mainly to underside of leaves; shy. **Voice** Short, harsh, descending trill *t'rrrrreuur* less vigorous than congener; short *trruep* call. **SS** No known sympatry with Scale-crested (orange-rufous in crown); White-eyed Tody-Tyrant lacks dotted crest; though crest not always easy to see, plain grey vs. dotted effect noticeable. **Status** Uncommon.

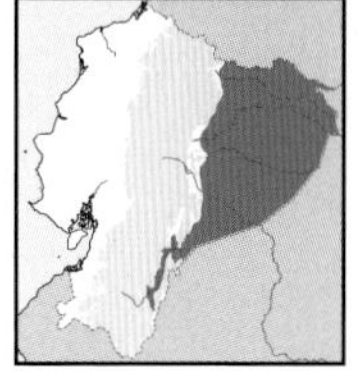

Rusty-fronted Tody-Flycatcher *Poecilotriccus latirostris* 9–10cm

E lowlands to lower foothills (to 700m). Dense shrubbery, borders, riparian woodland, river islands. Greyish crown, ***orange-ochre lores to orbital area***; olive above, ***pearly below, two ochraceous wingbars***. Race *caniceps*. Singles or pairs, rarely with mixed flocks, shyly inspects tangled understorey, makes darting sallies to underside of leaves, also gleans dense foliage. **Voice** Sharp, short, rather slow, low-pitched trill *túk-turrr*, *túk-turrrt-turrrr*; rich *túk!* call. **SS** ♀ Black-and-White Tody-Flycatcher has rufous crown, grey face and breast-sides, yellowish bar in flight feathers, no wingbars. **Status** Uncommon.

White-bellied Pygmy-Tyrant
Black-capped Pygmy-Tyrant
Short-tailed Pygmy-Tyrant
♂
pileatus
♂
juvenile
squamaecrista
♀
Scale-crested Pygmy-Tyrant
Double-banded Pygmy-Tyrant
Rusty-fronted Tody-Flycatcher

Small, inconspicuous and quite confusing tyrannids of humid forests, all but one subspecies of Black-throated being confined to the E. White-eyed, Johannes's and Zimmer's are best recognised by voices. Note head, amount of grey or olive in crown and sharpness of breast streaks; habitat also differs. The three montane species are more distinctive, but seldom seen. Voices are better indication of their presence and identity.

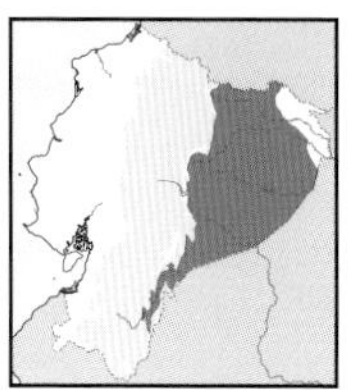

White-eyed Tody-Tyrant *Hemitriccus zosterops* 11cm

E lowlands to foothills, not N of Río Napo? (to 1,000m). Hilly *terra firme* forest. ***Pale iris***. Greyish-olive above, ***short white supraloral stripe, white eye-ring***, whitish throat, breast finely streaked olive, pale yellow belly. ***Two olive-yellow wingbars***, whitish edges to tertials. Nominate? Inconspicuous singles or pairs in midstorey, not with mixed flocks; makes upward strikes to foliage. **Voice** Staccato *píp'íp'íp'íp'ip*, often accelerating into *pí'i'i'i'i* trill, others with higher *pEEp* introductory note; *píp!* call. **SS** Zimmer's (duller crown, browner cheeks, whiter underparts, less yellow in flight feathers, shorter tail); see also Johannes's (habitat differs); Double-banded Pygmy-Tyrant has elongated blue-scaled crest. **Status** Uncommon.

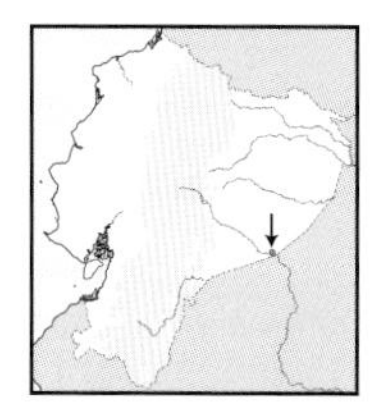

Johannes's Tody-Tyrant *Hemitriccus iohannis* 11cm

One record from lowlands of extreme SE (200m). Riparian woodland and oxbow lake edges. ***Yellow-creamy eyes. Greyish head***, greyish-olive upperparts, ***pale lores and orbital area***. Whitish throat to breast, ***streaked dull olive, yellow belly. Faint yellowish wingbars*** and edges. Monotypic. Singles or pairs skulk in dense tangles and thickets near water level, not with mixed flocks, gleans prey from underside of leaves. **Voice** Short, fast, fading trill with introductory *tew-tew, tri'r'r'r'r?*, rising staccato *tew-tew-tew....* **SS** White-eyed and Zimmer's (less grey in crown, more prominent wingbars; both in *terra firme* forest upper strata). **Status** Rare.

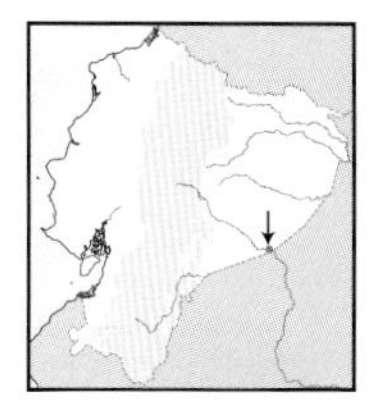

Zimmer's Tody-Tyrant *Hemitriccus minimus* 11cm

One record from lowlands of extreme SE (200m). *Terra firme* forest subcanopy. Recalls White-eyed but crown looks duller, browner cheeks and lores ***buffier, throat streaks more sharply defined, less marked eye-ring, no contrasting pale edges to outermost primaries*** (only to ***inner flight feathers***). Johannes's is ***yellower below***, with less obvious wingbars. Monotypic. Little known in Ecuador; elsewhere, easily overlooked except by voice, apparently prefers white-sand forests. **Voice** Slightly rising-and-falling trill *tre'e'E'E'e'e*; also faster, rising-and-falling *tututulT'tututu*. **SS** Double-banded Pygmy-Tyrant has characteristic crest and wingbars; no habitat overlap with Johannes's. **Status** Rare.

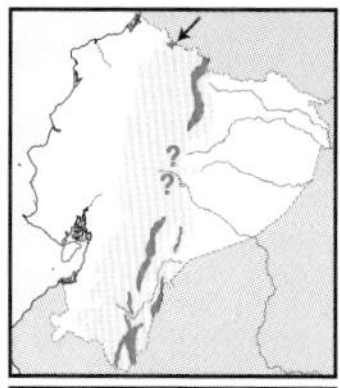

Black-throated Tody-Tyrant *Hemitriccus granadensis* 10–11cm

Andean subtropics to temperate zone in E and extreme NW (1,700–3,000m). Montane forest borders, adjacent woodland. Bold ***white*** (N) or ***buff*** (SE) ***face, black throat to cheeks***, grey chest, ***whitish below***, olive above. Races *pyrrhops* (SE), nominate (NW, NE); possibly two species. Singles or pairs with or apart from mixed flocks, rapidly switches perches and sallies to underside of foliage; ♂ hovers in front of ♀ in display, making a loud wing whirr. **Voice** Short series of 4–6 emphatic *wík-wík-wík-wík, wi-wi-wík-wík-wí-wí* (S); *we-kík-krrrr* or short *ke-kirrr* trill (N). **SS** Unmistakable. **Status** Uncommon.

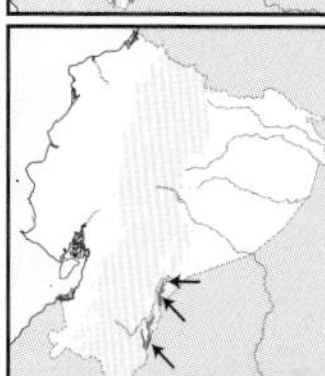

Cinnamon-breasted Tody-Tyrant *Hemitriccus cinnamomeipectus* 10cm

Local in subtropics of Cordillera del Cóndor (1,700–1,900m). Montane forest and woodland undergrowth. Brownish-olive cap, greyish-olive upperparts, ***cinnamon face and throat to lower breast***, belly pale yellowish. Dark wings with ***pale yellow edges to tertials***. Monotypic. Seldom seen and poorly known; skulks in dense thickets, with or apart from mixed flocks; makes short strikes or sallies to foliage, changing perches. **Voice** Short, dry, falling trill *pi'rrrrr*. **SS** Confusion unlikely in small range; Buff-throated has dull buff throat, never so rich, eyes are pale. **Status** Rare, local, VU.

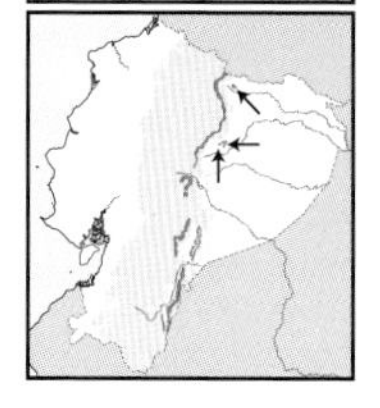

Buff-throated Tody-Tyrant *Hemitriccus rufigularis* 11–12cm

Local in E Andean lower subtropics (1,300–1,500m). Montane forest undergrowth. ***Pale amber eyes***. Greyish-olive upperparts, duskier wings. ***Dull buff orbital area and neck-sides to throat***, whitish below, faint pale streaking to upper belly. ***No wingbars***. Monotypic. Singles or pairs with or apart from mixed flocks, apparently sallies to underside of foliage and occurs in more open situations than previous species. **Voice** Gives 4–10 bright, somewhat nasal, inflected *wyeek, gyeek* or *yeek*, sometimes doubly or singly *gyeek-gyeek, gyeek, gyeek-gyeek, gyeek....* **SS** Unlikely to be confused in small range; Cinnamon-breasted has bolder head to breast. **Status** Rare, local, NT globally.

White-eyed Tody-Tyrant

Johannes's Tody-Tyrant

Zimmer's Tody-Tyrant

Black-throated Tody-Tyrant

Cinnamon-breasted Tody-Tyrant

Buff-throated Tody-Tyrant

Poecilotriccus are mostly confined to forest interior and are best located by their distinctive voices. These three species have little in common. In fact, Golden-winged resembles *Todirostrum*, which resulted in it being long classified in the latter genus. *Todirostrum* have longer bills that are noticeably flat. They often cock their tails when moving through vegetation, and all but Yellow-browed prefer edges and clearings.

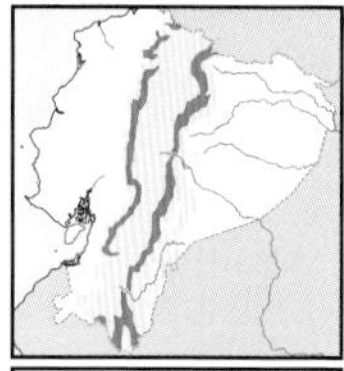

Rufous-crowned Tody-Flycatcher *Poecilotriccus ruficeps* 9–10cm

E & W Andean subtropics, S to Azuay in W (1,500–2,500m). Montane forest, dense woodland, shrubby edges. ***Rufous crown, black outline to buffy cheeks***. Grey nape, olive upperparts, ***white throat, broken blackish breast-band, rich yellow belly***, two buff wingbars. Nominate (E), *rufigenis* (W), *peruvianus* (extreme S). Singles or pairs in dense undergrowth, infrequently with mixed flocks, sallies upwards or forwards. **Voice** Brief shrill *t'rrreu, pít-trrrew, pi't-trrrew.* **SS** Unmistakable if seen, no overlap with Black-and-white (rufous-crowned ♀ has grey in face and sides). **Status** Uncommon.

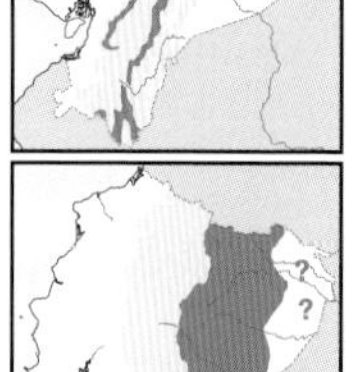

Black-and-white Tody-Flycatcher *Poecilotriccus capitalis* 9–10cm

E lowlands to foothills, not in extreme E (300–1,350m). *Terra firme* forest, borders, bamboo. ***Black above***, mainly ***white below***. Whitish loral spot, ***yellow panel in flight feathers, yellow wing bend***. ♀: ***rufous crown, grey face to breast-sides***, olive upperparts, ***white underparts***, wings as ♂ but dusky-olive. Nominate. Singles or pairs in dense thickets, dart forwards or upwards to underside of foliage, rarely with flocks. **Voice** Explosive, rapidly accelerating *tik-tik-t'weey-weey-weey…, tr'eeew*; also sharp *tk-t'r'r'r'ew*, shorter *tk-t'rrrr.* **SS** Locally Rusty-fronted Tody-Flycatcher and *Lophotriccus* pygmy-tyrants. **Status** Uncommon.

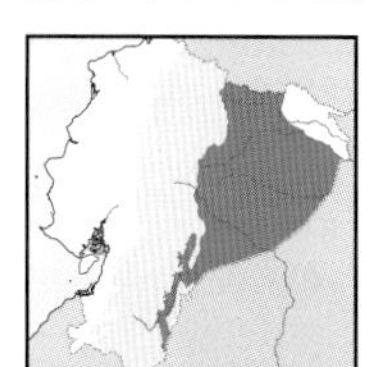

Golden-winged Tody-Flycatcher *Poecilotriccus calopterus* 9–10cm

E lowlands to foothills, not in extreme E (300–1,300m). *Terra firme* forest undergrowth and borders. ***Black head***, olive upperparts, ***white throat***, rich yellow below. Black wings with conspicuous ***chestnut shoulders, golden-yellow greater coverts***, and yellow edges to inner flight feathers. Monotypic. Sally-gleans in dense understory, often venturing to higher strata; does not join mixed flocks. **Voice** Sprightly, dry *drrreee-duu'rrr-d'deuur*, or simpler, descending *drrrew.* **SS** Wing pattern distinctive; recalls *Todirostrum*, e.g. Common (pale eyes, yellow throat, simpler wing pattern). **Status** Uncommon.

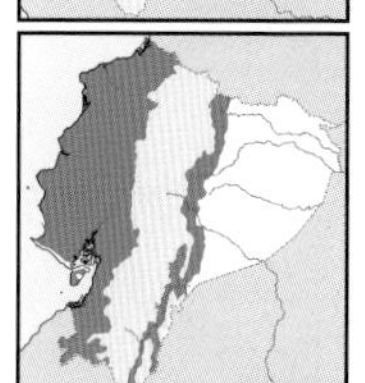

Common Tody-Flycatcher *Todirostrum cinereum* 9–10cm

W lowlands to foothills, E subtropics to adjacent lowlands (to 1,500m in W, 400–1,900m in E). Light woodland, borders, shrubby clearings, gardens. ***Pale eyes. Black crown, slate-grey upperparts*** (more ***olive*** in W). ***Rich yellow below*** (***whitish throat*** in W). Races *peruanum* (E), *sclateri* (W). Conspicuous in semi-open, nimble and noisy, regularly flicks tail sideways or cocks it, fond of gleaning undersides of leaves. **Voice** Fast, trilled *tik'tk'tk'tk'tk*, more inflected *te'eeee'eek*, shorter *tic'tic'tic* calls. **SS** Golden-winged and Black-headed (dark eyes, more contrasting throat, bolder wings). **Status** Common.

Spotted Tody-Flycatcher *Todirostrum maculatum* 10–11cm

E lowlands (to 250m). Riparian woodland, shrubbery, river islands. ***Orange eyes. Grey cap***, whitish throat, yellow below, ***throat to upper belly densely streaked dusky***. Race *signatum*. Singles or pairs near water, more arboreal than congeners. **Voice** Clear series of *fík* notes followed by *peék* in duets; also briefer *shrip* and excited *fee'fík-fee'fík…*. **SS** White-eyed, Johannes's and Zimmer's Tody-Tyrants are streaked below, but duller overall; bills shorter, caps not pure grey. **Status** Uncommon.

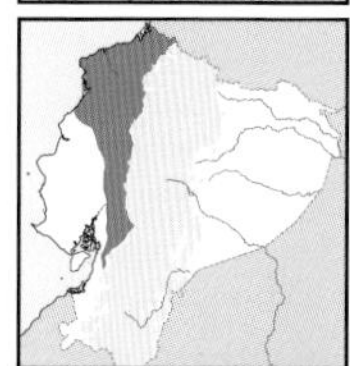

Black-headed Tody-Flycatcher *Todirostrum nigriceps* 9–10cm

W lowlands to foothills, S to N El Oro (to 900m). Humid forest canopy, borders, adjacent woodland, clearings. ***Black cap*** contrasting with ***bright olive upperparts, white throat***, rich yellow below. Monotypic. Singles or pairs remain high above ground, attack prey in short upward or forward sallies; less active than previous species; might go unnoticed unless vocalising. **Voice** Far-carrying, slow series of up to ten bright *peep* or *tsép* notes. **SS** Common Tody-Flycatcher has pale eyes, grey back and duller throat; does not range into forest. **Status** Fairly common.

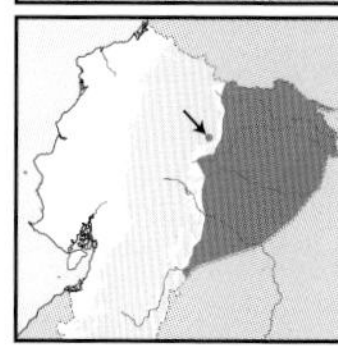

Yellow-browed Tody-Flycatcher *Todirostrum chrysocrotaphum* 9–10cm

E lowlands (to 600m). *Terra firme* and *várzea* forest canopy, borders, adjacent woodland. ***Yellow eyes. Black cap***, whitish supraloral spot, ***broad yellow postocular stripe. Bright olive above***, yellow below, ***spotted malar to breast sides***. Race *guttatum*. Singles or pairs high above ground, rarely with mixed flocks; not very active, performs upward sallies to foliage. **Voice** Long, accelerating series of 12+ emphatic *tíc!* or *chíp!* notes. **SS** Yellow eyebrow diagnostic; Spotted Tody-Flycatcher does not venture to forest canopy (orange eyes, no eyebrow, white throat, grey cap). **Status** Uncommon.

Rufous-crowned Tody-Flycatcher
peruvianus
ruficeps
rufigenys
♂
♀
juvenile
Black-and-white Tody-Flycatcher
Golden-winged Tody-Flycatcher
adult
sclateri
peruanum
adult
mmon Tody-Flycatcher
immature
juvenile
adult
juvenile
Spotted Tody-Flycatcher
juvenile
adult
Black-headed Tody-Flycatcher
Yellow-browed Tody-Flycatcher

PLATE 195: TWISTWING AND FLATBILLS

Rather inconspicuous, large-headed flycatchers of forest interior. The twistwing is distinctively drab. The two flatbill groups, despite their superficial similarities, are not closely related. *Rhynchocyclus* have broader and flatter bills, quite bold eye-rings and few wing markings. In contrast, *Ramphotrigon* have less flat bills, bold wingbars and, often, narrower broken, eye-rings.

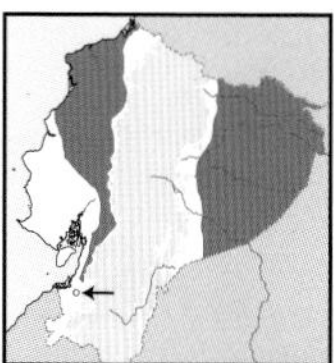

Brownish Twistwing *Cnipodectes subbrunneus* — 16–18cm

E & W lowlands, not wet NW or arid SW (to 600m). Humid to deciduous forest undergrowth, woodland. ***Orange eyes, stout bicoloured bill. Brown above to breast, rufescent rump and long tail; rufescent-buff wingbars and edges.*** Nominate (W), *minor* (E). Singles in tangles, motionless for long periods, often stretching one wing up; ♂'s twisted primaries possibly for display. **Voice** Emphatic, doubled or tripled *kuwéeh-kuwéeh*, often preceded by *tk'kik*; nasal *éeewt?* **SS** Flatbills olive above, yellow below, pale eye-ring; Royal Flycatcher has distinctive head shape. **Status** Rare.

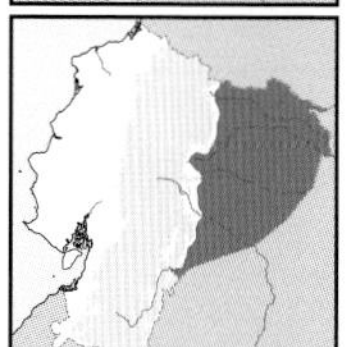

Olivaceous Flatbill *Rhynchocyclus olivaceus* — 15–16cm

E lowlands to lower foothills (to 700m). Mainly *terra firme* forest interior. ***Wide, stout, bicoloured bill, pale eye-ring.*** Olive head to upperparts, pale yellow underparts, ***breast to flanks flammulated olive.*** Duskier wings, ***faint buff bars***, yellowish edges to flight feathers. Race *aequinoctialis*. Singles or pairs perch erect, suddenly dart to underside of leaves; often with mixed flocks. **Voice** Strident *shriit*, *griit*, possibly ascending *tuee tee tee ti ti?* song (RG). **SS** Sympatric but rarer, smaller and larger-headed *Ramphotrigon* flatbills have prominent wingbars and supraloral stripes. **Status** Rare.

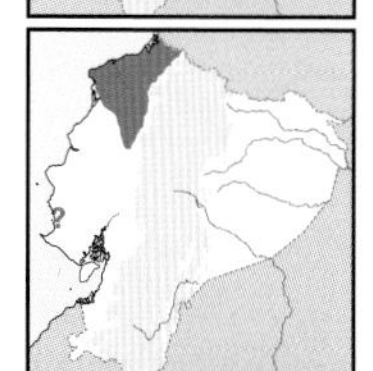

Pacific Flatbill *Rhynchocyclus pacificus* — 15–16cm

NW lowlands to lower foothills (to 800m). Wet to humid forest interior. ***Broad, stout, bicoloured bill, pale eye-ring. Olive upperparts and breast***, paler throat, yellowish belly, faint streaking on breast. Dark wings, ***faint rufescent bars*** and edges. Monotypic. Singles or pairs with mixed flocks, foraging in similar fashion to Olivaceous Flatbill, often changing perches. **Voice** Burry series of descending *tsheee-she-she-she-she*; hissing *shreeiw* calls. **SS** Higher-elevation (overlap?) Fulvous-breasted has evident fulvous-tawny throat to breast; see Sapayoa. **Status** Rare, rather local, NT in Ecuador.

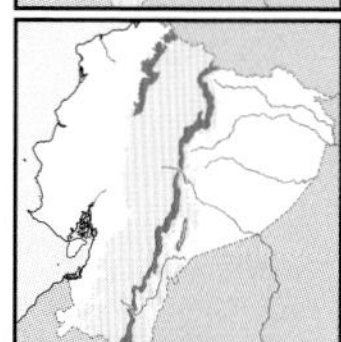

Fulvous-breasted Flatbill *Rhynchocyclus fulvipectus* — 15–16cm

E & NW Andean foothills to subtropics (900–1,800m). Montane forest undergrowth. ***Broad, stout, bicoloured bill, pale bluish eye-ring.*** Olive head to upperparts, ***fulvous-ochre throat to breast***, pale yellow belly. Dark wings with rufous edges. Monotypic. Singles or pairs with mixed flocks, perch erectly and suddenly dash to underside of foliage, then change perch. **Voice** Buzzy, rather weak *jiiipi* or *zhiip'r.* **SS** Breast coloration distinctive even under poor light; see Pacific Flatbill locally in NW, Olivaceous Flatbill in E, Brownish Twistwing, and smaller Rufous-breasted Flycatcher. **Status** Uncommon.

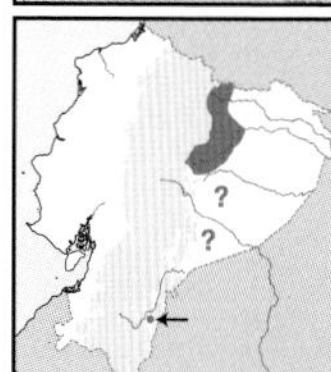

Large-headed Flatbill *Ramphotrigon megacephalum* — 13–14cm

Local in E lowlands and foothills (300–1,300m). Humid forest borders, bamboo. ***Broad flat bill***, fleshy mandible base. Dull olive above, browner crown, ***short pale yellow eyebrow and eye-ring, dusky lores***, vaguely streaked breast. Dusky wings, ***two prominent buff bars***. Race *pectorale*. Mostly alone in dense undergrowth, perches erect for long periods, darts after prey. **Voice** Slow, sad, whistled *fuü-feë-fuü* at dawn; sad, loud *baa-buü*, last note lower-pitched. **SS** Tail shorter than congeners, yellowish supercilium and eye-ring distinctive; cf. rarer Dusky-tailed. **Status** Uncommon.

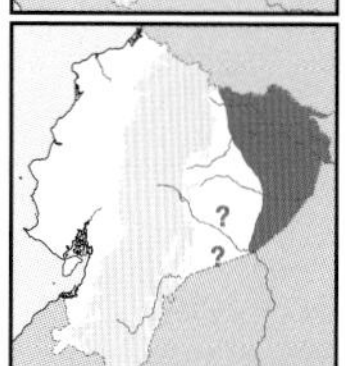

Rufous-tailed Flatbill *Ramphotrigon ruficauda* — 15–16cm

Scattered records in E lowlands (mostly below 300m). *Terra firme* undergrowth. ***Broad flat bill***, fleshy mandible base, ***narrow broken white eye-ring.*** Olive above, ***uppertail-coverts to tail rich rufous.*** Yellowish below finely streaked olive, plainer belly. ***Prominent rufous bars and edges to flight feathers.*** Monotypic. Singles or pairs rarely with mixed flocks, upright for long periods, often on exposed branches. **Voice** Long, mournful, rather trembling *wuuueeéuuU* whistle followed by shorter, more emphatic *e'yuú.* **SS** Rufous wings and tail separate from similar Dusky-tailed Flatbill. **Status** Rare.

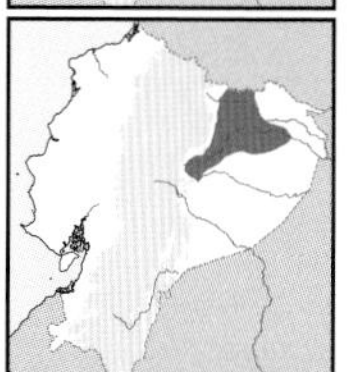

Dusky-tailed Flatbill *Ramphotrigon fuscicauda* — 15–16cm

Scattered records in NE lowlands to foothills (250–800m). *Terra firme* undergrowth, secondary woodland, bamboo. ***Broad flat bill***, fleshy mandible base, ***broken white eye-ring.*** Olive above, yellowish below, breast flammulated olive, plainer belly. Blackish wings, two ***prominent cinnamon bars, yellowish-buff edges to flight feathers. Dusky tail.*** Monotypic. Apparently at low density; behaves much like congeners. **Voice** Clear, ascending, mellow *feuuuu, p'rr-fey-fey-feeey*, last note higher; mournful *peéeu-wuoo.* **SS** Tail separates from Rufous-tailed; Large-headed has prominent yellowish eyebrow and eye-ring; Olivaceous has fainter wingbars. **Status** Very rare, local.

subbrunneus
adult
adult
minor
adult
Brownish Twistwing
juvenile
adult
Olivaceous Flatbill
Pacific Flatbill
Fulvous-breasted Flatbill
juvenile
adult
Large-headed Flatbill
adult
juvenile
Rufous-tailed Flatbill
Dusky-tailed Flatbill

Tolmomyias species are difficult to identify, being mostly yellow and olive, some with bold yellow wing edges and bars, pale eye-rings and supraloral stripes, and pale eyes. Best identified by their voices. Complex taxonomy; many current 'species' certainly comprise multiple species.

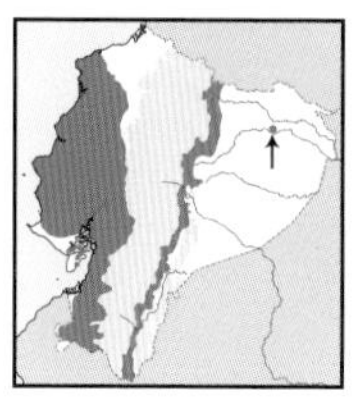

Yellow-olive Flycatcher *Tolmomyias sulphurescens* 13–14cm

W lowlands to subtropics, not wet NW; E foothills to subtropics (to 1,800m in W, 900–1,700m in E). Deciduous to semi-deciduous forest, borders, woodland. Grey crown, greyish-olive upperparts, ***short white supraloral stripe, white eye-ring, dusky auriculars***, yellow underparts; ***bold yellow wingbars and edges***. Races *aequatorialis* (W), *confusus* (NE foothills), *peruvianus* (SE foothills) and *insignis* (E lowlands). Understorey, singles or pairs deliberately sally or chase prey, sometimes with flocks. **Voice** Short, sharp *tsiíp* in brief series, less penetrating than Yellow-breasted (E); shorter, more piercing quick notes (W). **SS** Yellow-margined (drabber face pattern, more contrasting crown, narrower wingbars, white wing spot; more humid forest-based). **Status** Common.

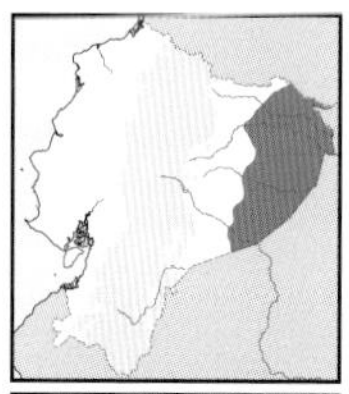

Orange-eyed Flycatcher *Tolmomyias traylori* 13–14cm

Scattered records in E lowlands (below 300m). *Várzea* forest canopy, subcanopy, borders, riparian woodland. ***Pale orange eyes***. Grey crown grading into olive upperparts; ***pale tawny lores, whitish eye-ring***. Yellow below, ***throat to upper belly clouded tawny-olive***. Yellow wingbars and edges to flight feathers. Monotypic. Poorly known, singles or pairs follow mixed flocks, often in gaps. **Voice** Slowly accelerating series of wheezy *zzreep*, last note higher; also buzzy *wyEEeez-zriii, wyeeez-zrii*…. **SS** Orange eyes and tawnier breast distinctive; see Grey-crowned. **Status** Rare.

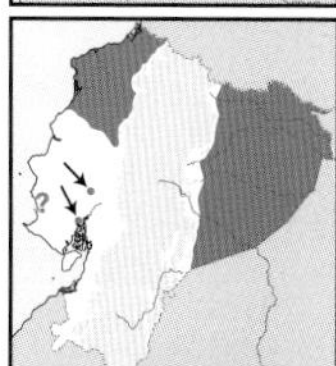

Yellow-margined Flycatcher *Tolmomyias assimilis* 13–14cm

E & NW lowlands, to E lower foothills (to 900m, lower in W). *Terra firme, várzea* forest, borders. ***Dark eyes. Narrow whitish supraloral***, white eye-ring. ***Grey crown***, olive upperparts. Whitish throat, yellow below, ***white wing spot***. Races *obscuriceps* (E), *flavotectus* (W), could be separate species. Singles or pairs in canopy, sally and chase prey, sometimes with flocks. **Voice** Utters 2–3 plaintive whistles, each longer and higher-pitched than previous, last inflected: *whee-wheeé-wheeEE* or *wee-weeO-weEE?*, descending *wEE-tzeu* calls (E); short shrill series *zhrií-zhrip-zhrip-zhrip* (W). **SS** Brown eyes and white wing spot separate from Grey-crowned and Yellow-olive. **Status** Uncommon.

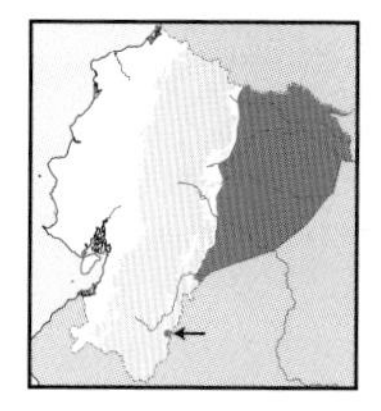

Grey-crowned Flycatcher *Tolmomyias poliocephalus* 12–13cm

E lowlands (to 600m). *Terra firme* and *várzea* subcanopy, borders. ***Mostly dark bill, orangey eyes***. Grey crown, olive upperparts. Whitish supraloral, white eye-ring. Yellow below, whiter throat, breast clouded olive; ***narrow yellow wingbars and edges***. Nominate. Mostly in pairs, following mixed flocks, sometimes perch upright and cock tail. **Voice** Short inflected whistles *fuee?-fuee?-fuee?-feeUuEet?* Repeated at 10+ sec intervals; single *tuee?* call. **SS** Yellow-margined (darker grey crown, darker eyes, white wing spot, bolder wing feather edges) and Orange-eyed Flycatchers (brighter orange eyes, tawny lores and throat-breast). **Status** Uncommon.

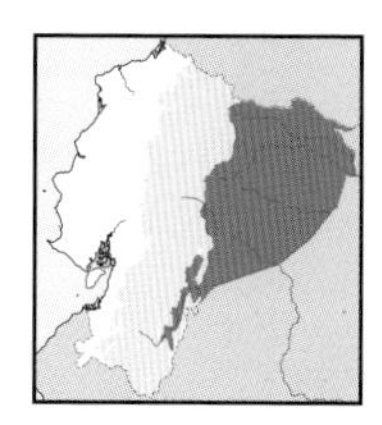

Yellow-breasted Flycatcher *Tolmomyias flaviventris* 12–13cm

E lowlands to lower foothills (to 800m). Light riparian woodland, shrubbery, *várzea* borders. ***Dark eyes. Olive crown to upperparts***, narrow pale eye-ring. ***Olive-yellow below***, duskier breast. Two yellow wingbars and edges to flight feathers. Race *viridiceps* possibly a valid species. Singles or pairs sometimes with mixed flocks, regularly at gaps. **Voice** Short, inflected, emphatic *tsuép?* in series 3–7, rising in volume; rising *tseET!* calls. **SS** Note uniform crown to back (no contrasting grey crown), dark eyes, no supercilium or eye-ring. **Status** Common.

aequatorialis
Yellow-olive Flycatcher
confusus
insignis
peruvianus
Orange-eyed Flycatcher
flavotectus
Yellow-margined Flycatcher
adult
juvenile
obscuriceps
adult
Grey-crowned Flycatcher
immature
juvenile
Yellow-breasted Flycatcher

PLATE 197: ANTPIPIT AND SPADEBILLS

Small, bull-headed flycatchers of forest undergrowth, mostly in lowlands although two species also venture into subtropics. Bills flattened and wide, eyes look disproportionately large, legs and tail short. Semi-concealed crown patches are characteristic but rarely seen. All are shy and sluggish. Plus, the notably distinctive, terrestrial antpipit of forest interior that recalls an antbird or gnateater.

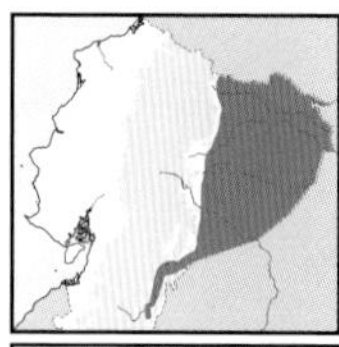

Ringed Antpipit *Corythopis torquatus* 13–14cm

E lowlands (to 600m). *Terra firme* forest. Pinkish legs. Olive-brown above, duskier wings and tail. *White throat, collar of broad blackish streaks*, whitish below. Race *sarayacuensis*. Mostly alone, walks on ground, bobs head and pumps tail; sallies to take prey from undersides of leaves. **Voice** Clear, rhythmic double whistle *pEEeeeu-peeuyd*, second note lower; also *pEéeud-peeu-dit*. **SS** Confusion unlikely; recalls an antbird in behaviour, but none truly similar. **Status** Uncommon.

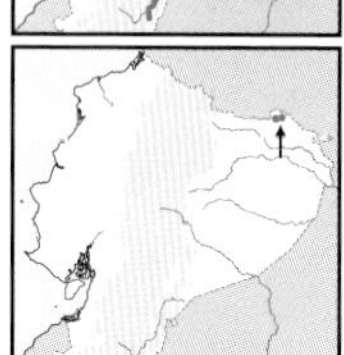

Cinnamon-crested Spadebill *Platyrinchus saturatus* 9–10cm

Few records from lowlands of NE in Sucumbíos (200m). *Terra firme* forest. *Broad flat bill*. Brown above, *orange semi-concealed crown patch, faint pale supraloral stripe and eye-ring. White throat*, dingy whitish below, browner flanks. Nominate. Poorly known; tends to remain in understorey, where might loosely follow mixed flocks. **Voice** Unknown in Ecuador; elsewhere sharp *ka-wíp-díp-díp*; song possibly short *sh'rrr* trill. **SS** Might only occur with Golden-crowned (more complex face pattern, more conspicuous crown patch) and White-crested Spadebills (dark grey head, yellower below, white crown patch). **Status** Very rare.

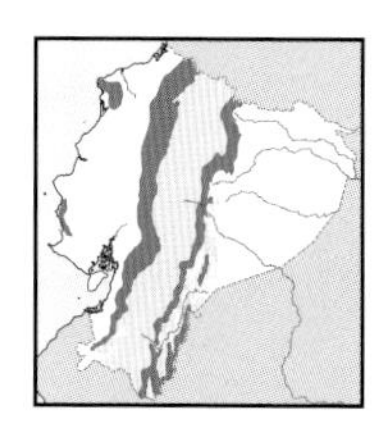

White-throated Spadebill *Platyrinchus mystaceus* 9–10cm

E & W foothills to subtropics, adjacent lowlands in NW, coastal ranges (600–2,000m in W, 1,000–2,000m in E). Montane forest, woodland. *Broad flat bill*. Olive-brown above, darker crown, *golden-yellow semi-concealed crown patch*; narrow *yellowish-buff line outlining dusky face, buff eye-ring and cheeks. White throat*, greyish-buff breast to flanks. Races *zamorae* (E), *albogularis* (W). Singles unobtrusive in dense undergrowth, seldom with mixed flocks; remains motionless for long periods, making sudden sallies to underside of foliage. **Voice** Short song, rather slow and ascending rattle *tir'r'r'r'rí* or *tui'r'r'r'r'rít!*, sometimes inflected at end; calls loud, sharp *quit* or *skít-skít*, *si-suik-squit!* **SS** Confusion unlikely; might locally overlap with Golden-crowned with bolder crown patch, no white throat. **Status** Uncommon.

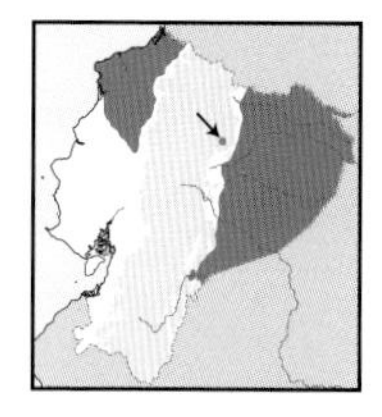

Golden-crowned Spadebill *Platyrinchus coronatus* 9–10cm

E & NW lowlands to lower foothills (to 700m). *Terra firme* and humid forest. *Broad flat bill. Golden-orange crown outlined blackish; buff lores, eye-ring, postocular stripe and cheeks, dusky outline to cheeks*. Olive-brown above, dingy yellowish-white below, breast-sides more olive-brown. Nominate (E), *superciliaris* (W). Singles or pairs in fairly open understorey, unobtrusive and lethargic, makes sudden upward strikes to underside of foliage, rarely joins mixed flocks. **Voice** Ascending-descending weak trill, rather long and high-pitched: *eeeeeeee'eez?* or *re'e'e'e'e'e'eea?* **SS** Crown patch more visible and conspicuous than congeners, head pattern resembles White-throated but throat not white. **Status** Uncommon.

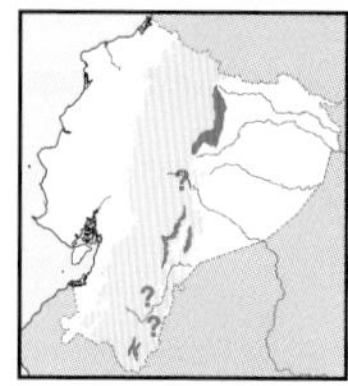

Yellow-throated Spadebill *Platyrinchus flavigularis* 9–10cm

Local in E Andean foothills to subtropics (750–1,900m). Montane forest undergrowth. *Broad flat bill. Rusty head* with *white semi-concealed crown patch*, buffy-yellow lores and narrow eye-ring. Olive upperparts, *yellow underparts*, breast-sides tinged dull olive-yellow. Nominate. Like congeners, lethargic in undergrowth but more prone to perch in open understorey or to emerge at borders, landslides and ridges. **Voice** Loud, sharp *tyuu!* or *pyeeu!* every 4–5 sec; also short, rising *trrrrRR* followed by *pyeeu!* **SS** Confusion unlikely in limited range where overlaps with White-throated (bold head pattern, white throat, faded underparts). **Status** Rare to uncommon, NT in Ecuador.

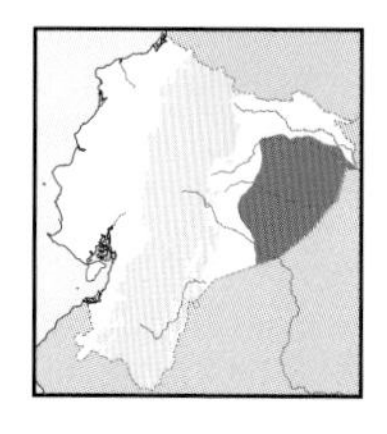

White-crested Spadebill *Platyrinchus platyrhynchos* 10–11cm

Scattered records in E lowlands (below 300m). *Terra firme* forest undergrowth. *Broad flat bill. Blackish crown* with *white semi-concealed crown patch*; buff lores, *white throat*. Olive-brown upperparts, *ochraceous-yellow underparts*, breast to flanks browner-olive. Race *senex*. Poorly known in Ecuador but possibly more numerous than few records indicate; higher in forest strata than congeners. **Voice** Song a burry, rising-falling trill, *bi'rrrRRrrrii*; commonly heard call is loud, explosive, squeaky *keeut!* every 2–5 sec, resembling Golden-crowned. **SS** Confusion unlikely; other E lowland spadebills lack combination of blackish head, white throat and ochraceous underparts. **Status** Rare, local.

juvenile
Ringed Antpipit
adult
♂
♀
juvenile
Cinnamon-crested Spadebill
albogularis
juvenile
zamorae
White-throated Spadebill
♀
♂
coronatus
superciliaris
♂
juvenile
Golden-crowned Spadebill
adult
juvenile
Yellow-throated Spadebill
juvenile
adult
juvenile
White-crested Spadebill

PLATE 198: FLYCATCHERS II

Rather small, primarily Andean flycatchers. Five forest-based species are clad in yellows and olives, two edge and shrub species are drabber and paler. Semi-concealed crown patches and bold wingbars are characteristic; coronal patches reduced or absent in ♀♀. All perch upright, some join mixed flocks. Handsome and Orange-banded are more arboreal than congeners.

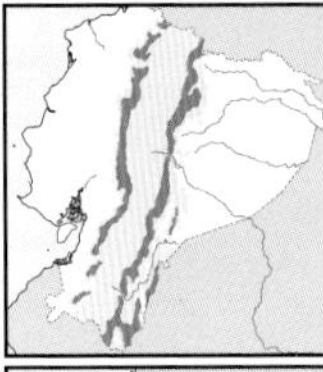

Flavescent Flycatcher *Myiophobus flavicans* 12–13cm

E & W upper foothills to subtropics (1,300–2,500m, locally higher). Montane forest, borders. *Yellow loral spot and broken eye-ring*, yellow semi-concealed crown patch. Bright olive above, *two bold ochraceous-buff bars*, buff edges to flight feathers. Nominate. Singles or pairs in shady areas, perch erect and sally, often with mixed flocks. **Voice** Song a fast series *khadí-khadí-khadí…*; emphatic *chíip*. **SS** Yellow lores, broken eye-ring and ochraceous wingbars separate from Orange-crested and Handsome; Roraiman is browner above, wings pattern more contrasting. **Status** Uncommon.

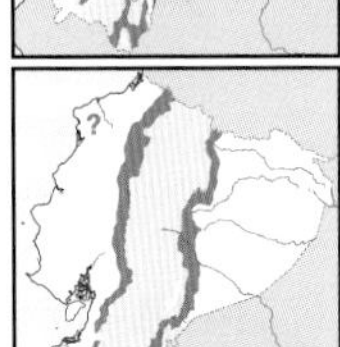

Orange-crested Flycatcher *Myiophobus phoenicomitra* 11–12cm

Local in E & W foothills (600–1,550m). Montane forest undergrowth. *Fleshy mandible base*. Faint but *complete yellowish eye-ring*, orange crown patch. Olive above, *two weak ochraceous wingbars* and *narrow* flight feather edges. Nominate (E), *litae* (W). As Flavescent Flycatcher, sedentary singles or pairs, only briefly with mixed flocks. **Voice** High-pitched, weak, buzzy, ascending-descending *chi-dzíp-zip*. **SS** Recalls higher-elevation Flavescent (prominent yellow lores and broken eye-ring, black bill); see also browner, bolder-winged Roraiman. **Status** Uncommon.

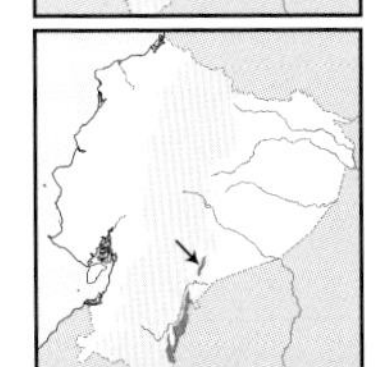

Roraiman Flycatcher *Myiophobus roraimae* 13–14cm

Locally Cordillera del Cóndor and Kutukú subtropics (1,400–1,700m). Montane forest interior. *Yellowish mandible. Faint pale supraloral stripe, narrow eye-ring. Brown above*, orange-rufous crown patch, *two bold cinnamon-rufous bars and broad flight feather edges*. Race *rufipennis*. Like congeners, perches low in shady areas, sometimes joins mixed flocks. **Voice** Explosive, buzzy *tché, tchéw-tché'e'e'e'e*, rising and descending, often buzzier introductory notes; also simpler *tché-chichichí*. **SS** Darker above than Flavescent (black bill) and Orange-crested, both with narrower wingbars; see also Euler's Flycatcher. **Status** Rare.

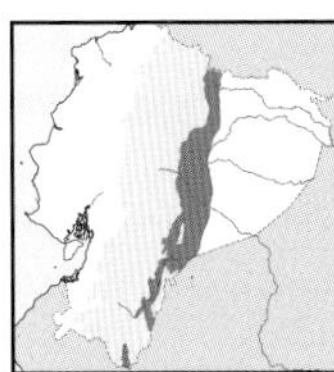

Olive-chested Flycatcher *Myiophobus cryptoxanthus* 12–13cm

E foothills to adjacent lowlands, extreme SE in Zumba region (400–1,400m, locally higher). Shrubby edges, clearings. *Weak whitish supercilium, vague broken eye-ring. Greyish-brown above*, yellow crown patch. *Two whitish wingbars and flight feather edges, dull greyish-olive breast streaks*. Monotypic. Singles or pairs perch erect near ground, seldom with flocks. **Voice** Incessant *chuee* at dawn; *wue-d'd'd'd'd* and harsh *chi'gi-chi-gi…* call. **SS** Duller than Bran-coloured (streakier breast, bolder wings); larger Euler's Flycatcher (whitish belly, buff wingbars, browner). **Status** Uncommon.

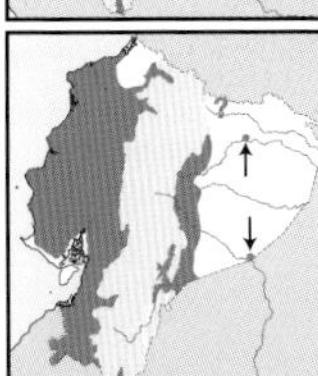

Bran-coloured Flycatcher *Myiophobus fasciatus* 12–13cm

W lowlands to upper foothills, E foothills to adjacent lowlands (to 1,500m in W, 300–1,100m in E). Light woodland, shrubby edges, clearings. *Short whitish supercilium*, narrow eye-ring. *Dull olive-brown* (W) or *rufescent-brown* (E) *above*, yellow-orange crown patch. *Two bold buff wingbars*. Whitish below, *breast coarsely streaked dusky to tan-brown*. Nominate (E), *crypterythrus* (W). Singles or pairs rarely with mixed flocks, sally to air or foliage; rather inconspicuous. **Voice** Utters *wee-we...wü-dü* at dawn; also fast *wee..di-di-di-di-di*. **SS** Confusion unlikely; in E see Olive-chested. **Status** Common.

Handsome Flycatcher *Nephelomyias pulcher* 10–11cm

E & W upper foothills to subtropics, more local in E (1,500–2,400m). Montane forest, borders, woodland. *Fleshy mandible. Narrow pale loral stripe, narrow eye-ring. Greyish crown*, orange crown patch; two *cinnamon-buff bars* and *prominent flight feather edges*. Bright yellow below, *chest tinged tawny*. Races *bellus* (E), nominate (W). Arboreal pairs to small parties with mixed flocks, perches horizontally. **Voice** Rhythmic, fast, bright *pri-pri-ri, pri-ri-ri-ri*. **SS** Grey crown and ochraceous-tawny breast diagnostic; see Orange-banded (no overlap?). **Status** Uncommon.

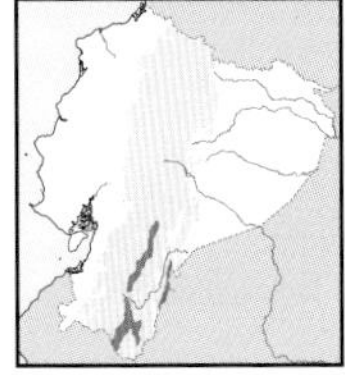

Orange-banded Flycatcher *Nephelomyias lintoni* 12–13cm

SE Andean subtropics to temperate zone (2,250–3,200m). Montane forest, borders, elfin woodland atop isolated cordilleras. *Fleshy mandible, pale eyes*, narrow eye-ring. Olive-brown head and upperparts, orange crown patch, breast tinged olive; *two bold spotty orange-cinnamon bars and narrow flight feathers edges*. Monotypic. Arboreal pairs to small groups often with mixed flocks. **Voice** Clear, sharp *bieek* or *peeyk* calls; incessant *beet-beet-beet…*. **SS** Handsome (tawnier breast, grey crown, lower elevations), yellower Flavescent (yellow eye-ring) and browner Roraiman Flycatchers (more rufescent wingbars). **Status** Rare, NT.

Orange-crested Flycatcher
Flavescent Flycatcher
litae
phoenicomitra
♂
juvenile
Roraiman Flycatcher
Olive-chested Flycatcher
juvenile
crypterythrus
fasciatus
Bran-coloured Flycatcher
bellus
pulcher
juvenile
Handsome Flycatcher
adult
juvenile
Orange-banded Flycatcher

PLATE 199: FLYCATCHERS AND MANAKIN-TYRANT

Ornate and Cinnamon are small, vocal and attractive flycatchers of montane forest. Ruddy-tailed and the tyrant-manakin are rather similar, but the former is more flycatcher-like, whereas the latter is more manakin-like. *Myiobius* are also small forest flycatchers that typically lean forwards, wings slightly drooped and tail fanned. Appear oddly large-eyed and long-whiskered.

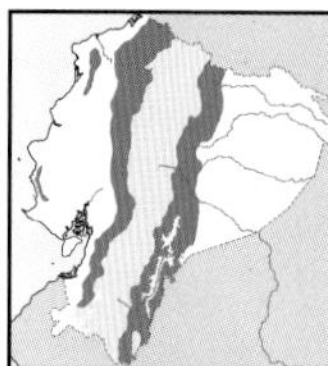

Ornate Flycatcher *Myiotriccus ornatus* 12–13cm

E & W foothills to subtropics, coastal ranges (800–2,000m, lower in NW, 200–400m in coastal ranges). Montane forest, borders, woodland. *Lovely! Slate-grey head, white loral spot*, olive back and chest, *bright yellow belly and rump, rufous-and-brown* (W) or *rufous tail* (E). Races *phoenicurus* (E), *stellatus* (W). Singles or pairs sally quickly and return to nearby perch, highly sedentary; briefly with mixed flocks. **Voice** Emphatic, ringing *pyép!*, often with additional *pyép-píp!*; faster 3–5-note series at dawn. **SS** Confusion unlikely; see much duller *Myiobius*. **Status** Common.

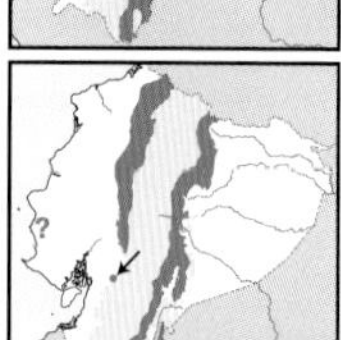

Tawny-breasted Flycatcher *Myiobius villosus* 13–14cm

E & NW foothills to subtropics (1,000–1,700m). Montane forest and mature woodland interior. Olive brown crown and upperparts, concealed yellow crown patch; whitish throat, *tawny-brown breast to flanks and vent, yellow lower belly and rump*, rounded black tail. Races *clarus* (E), nominate (W). Active pairs in undergrowth, often with mixed flocks, pursues prey; constantly fans tail and flicks wings. **Voice** Brief, sharp *tiúp!* or *chiúp!* **SS** Underparts richer tawny-brown than congeners, with yellow confined to central-lower belly. **Status** Uncommon.

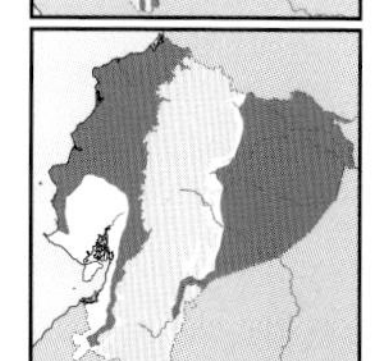

Sulphur-rumped Flycatcher *Myiobius barbatus* 12–13cm

E lowlands, W lowlands to foothills (mostly to 600m in E, to 1,000m in W; higher in S). *Terra firme* to wet forest, adjacent woodland. Olive head and upperparts. Whitish throat, *greyish-olive* (E) to *tawny-cinnamon* (W) *breast*, pale *yellow belly, sulphur-yellow rump*. Nominate (E), *sulphureipygius* (W); possibly different species. Singles or pairs with understorey flocks, sallies or hover-gleans, constantly fans tail and droops wings. **Voice** Thin *tzíp-tzíp-tziíp-tziíp*, rising and falling; sharp, short calls *jíp*. **SS** Closely recalls Black-tailed; in E note buffier tinged breast, dingier throat; Black-tailed tends to prefer secondary habitats. **Status** Uncommon.

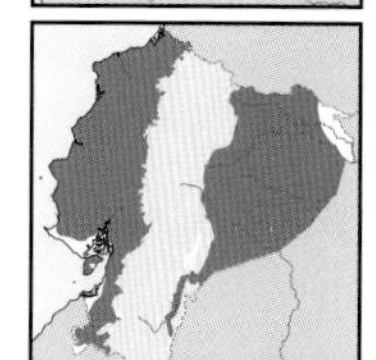

Black-tailed Flycatcher *Myiobius atricaudus* 12–13cm

E & W lowlands to foothills, not extreme E (to 1,000m). Humid forest, adjacent woodland. Olive head and upperparts. *Dingy* whitish throat, *dull olive-buff breast*, yellow belly, *sulphur rump, blackish wings*. Races *adjacens* (E), *portovelae* (W). Behaves much like congeners, but more fond of secondary habitats. **Voice** Weak *twik, twik-wek*. **SS** Tail not blacker than congeners; in E breast buffier and wings blacker than Sulphur-rumped; higher-elevation Tawny-breasted is browner below. **Status** Uncommon.

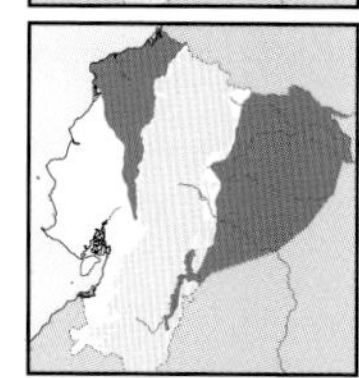

Ruddy-tailed Flycatcher *Terenotriccus erythrurus* 10cm

E & NW lowlands to foothills, S to NE Guayas (to 1,000m). Humid forest undergrowth. Short bill, *pinkish legs*. Olive-grey above, *orange-cinnamon underparts, rump and tail*. Wings *extensively edged cinnamon-rufous*. Races *signatus* (E), *fulvigularis* (W). Singles, seldom with mixed flocks, perch erect and dart after prey; sometimes flicks wings over back. **Voice** Weak *pew-hEE* (E) or more rhythmic *see-peeeu'sue-peeeu'su-peeeu'su...* (W); thin doubled *peeeu-tít, teeu-tirrt* calls. **SS** Cinnamon Manakin-Tyrant chunkier, tail shorter, greyish legs, bill smaller, more rounded crown. **Status** Uncommon.

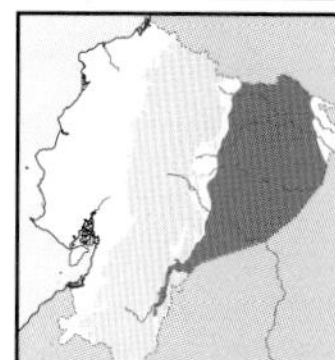

Cinnamon Manakin-Tyrant *Neopipo cinnamomea* 9–10cm

E lowlands (mainly below 400m). *Terra firme* undergrowth. *Chunky*, manakin-like, short bill, *greyish legs*. Greyish-olive crown, semi-concealed yellow crown patch. *Cinnamon underparts. Wings extensively ruddy-cinnamon; rump and tail same*. Nominate. Perches hunchbacked inside forest; pairs apparently follow mixed flocks, but not very active. **Voice** Ascending-descending then slowing and fading *pee-pee-peé-peE-pEE-peé-pee-pee-pu-puh*; thin *éeeu* at long intervals. **SS** Though chunkier and shorter-billed, recalls Ruddy-tailed Flycatcher (longer tail, pinkish legs, no crown patch, less rounded head). **Status** Rare, local.

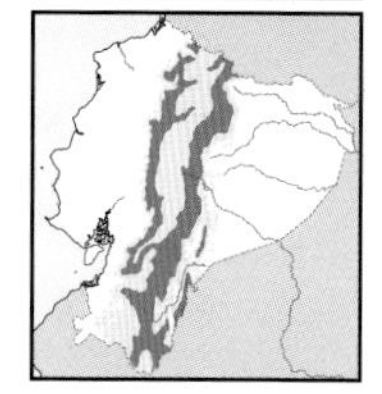

Cinnamon Flycatcher *Pyrrhomyias cinnamomeus* 12–13cm

E & W upper foothills to temperate zone, inter-Andean slopes (1,200–3,000m). Montane forest borders, woodland, woody clearings. Olive-brown above, pale *cinnamon rump. Rich cinnamon-rufous below* extends to neck-sides. Blackish wings, *two rufous bars* and conspicuous *rufous panel in secondaries*. Race *pyrrhopterus*. Mostly in pairs, often at edges or on prominent perches, repeatedly makes short, sudden sallies, returns to same or nearby perch; constantly vocalising. **Voice** Dry, abrupt *t'trrrrrt* trill. **SS** Confusion unlikely in montane habitat. **Status** Common.

Ornate Flycatcher
phoenicurus
stellatus
juvenile
♀
♂
♂
clarus
villosus
Tawny-breasted Flycatcher
juvenile
sulphureipygius
adult
portovelae
♂
adjacens
adult
adult
juvenile
juvenile
barbatus
Black-tailed Flycatcher
Sulphur-rumped Flycatcher
juvenile
ulvigularis
juvenile / ♀
signatus
♂
adult
Cinnamon Flycatcher
Cinnamon Manakin-Tyrant
Ruddy-tailed Flycatcher

Drab flycatchers with well-marked wingbars but little overall pattern. Visual identification risky in the migratory *Empidonax*, particularly between Alder and Willow, which are best identified by voice. *Lathrotriccus* are understorey dwellers, one either side of Andes. All perform short sallies and often change perches afterwards.

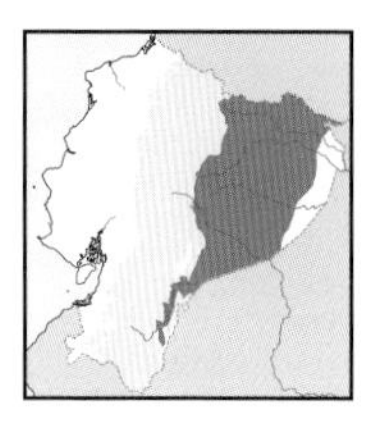

Euler's Flycatcher *Lathrotriccus euleri* 13–14cm

E lowlands to foothills, not in extreme E (300–1,300m). *Terra firme* and *várzea* forest undergrowth, adjacent woodland, bamboo. ***Bicoloured bill***, narrow buff eye-ring, vague ***pale supraloral stripe***. Olive-brown above, whitish below, ***olive-brown chest. Two dull buff wingbars***. Race *bolivianus*. Mostly singles, rather unobtrusive and inactive, sally to foliage. **Voice** Short, shrill *phréeew, phréeew... phreu'ét* at dawn; simpler *phrEeu'ur*; descending shrill *pheeyw-pheeyw-pheeyw...*; dry *chéeur* calls. **SS** See Fuscous (long supercilium, richer wingbars, unicoloured longer bill; habitat differs), Willow and Alder Flycatchers (olive breast-band, more olive upperparts, yellower belly). **Status** Rare, overlooked?

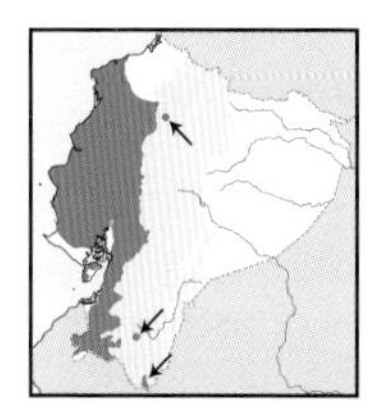

Grey-breasted Flycatcher *Lathrotriccus griseipectus* 13–14cm

W lowlands, to subtropics in Loja, not in wet NW (to 1,700m). Humid to deciduous forest, borders. Bicoloured bill, ***broken white eye-ring, white supraloral stripe. Sooty-grey head and breast-band***, whitish rest of underparts; ***two whitish wingbars***, white edges to flight feathers. Monotypic. Singles or loose pairs in lower growth, perch upright, briefly sally to foliage. **Voice** Short, burry *pheeu-phee-phu'r'r* at dawn; *pee-urr, peer-peer-peer, peeu-peeu-peer-peer'rr* song, often more trilled at end. **SS** Alder, Acadian and Willow Flycatchers lack obvious eye-ring and supraloral, breast-band and sooty-grey crown; see Tropical Pewee (no grey, no eye-ring and supraloral, different posture, behaviour and silhouette). **Status** Rare, VU.

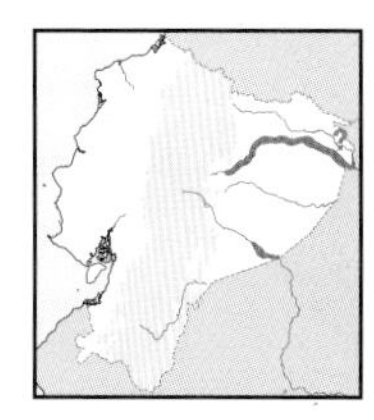

Fuscous Flycatcher *Cnemotriccus fuscatus* 14cm

E lowlands along Ríos Napo and Pastaza (below 400m). Riparian woodland, shrubby river islands. ***Black bill***, narrow broken buff eye-ring, blackish lores, ***bold buff supercilium. Fuscous-brown above***, whitish-buff below, ***buffy-brown breast-band. Two bold tawny-cinnamon wingbars***. Race *fuscatior*. Elusive, not very active, perches upright in dense undergrowth, briefly sallies or picks prey from underside of leaves. **Voice** Loud, thin *heeéew* at long intervals; also lower-pitched, gravelly *ghö, jéy-jéy-jéy-jéy-jéy..., tjEE-víu* calls. **SS** Confusion unlikely in range and habitat; cf. Mouse-coloured Tyrannulet (no breast-band, shorter bicoloured bill, whitish supercilium, different behaviour and posture). **Status** Uncommon.

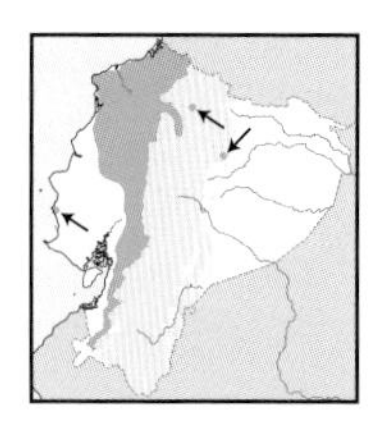

Acadian Flycatcher *Empidonax virescens* 13–14cm

Boreal migrant to W lowlands and foothills, local in Andes (to 1,200m; locally higher). Humid forest, woodland, borders. ***Fleshy mandible, white eye-ring***, pale lores. Olive above, whitish below, dull olive chest; ***two buffy-white wingbars*** and edges to inner flight feathers. Monotypic. Singles in lower growth, perform short sallies to foliage or air; vocal and territorial. **Voice** Sharp, brief, inflected *véek!* or *whýek!* every 8–10 sec, repeated often. **SS** From Alder and Willow by pale eye-ring, more olive upperparts, less dry, more inflected voice; tends to prefer more forested situations. **Status** Rare winter resident, probably overlooked (Oct–Mar).

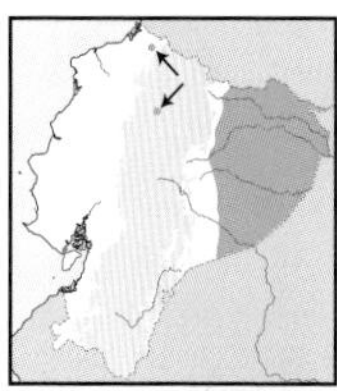

Willow Flycatcher *Empidonax traillii* 13–14cm

Boreal migrant to E lowlands (below 500m, once 2,800m). Light woodland, shrubby clearings, gardens. Virtually identical to Alder and unlikely to be identified by plumage alone. Possibly nominate and *campestris*. Like Alder, also prefers semi-open and edges, where perches erect and repeatedly sallies and calls. **Voice** Short, dry *whít* or *vít* calls similar to Alder but less 'piping'; seldom-heard song a short, sharp *fEEt'zbew*; with accent on first syllable (second in Alder). **SS** See also Acadian and Euler's Flycatchers. **Status** Rare winter resident (Oct–Apr).

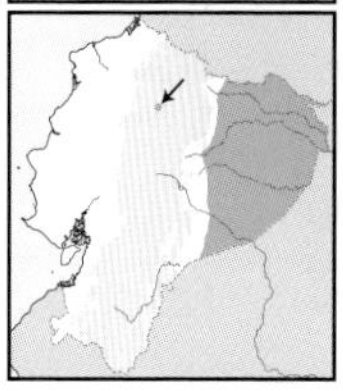

Alder Flycatcher *Empidonax alnorum* 13–14cm

Boreal migrant to E lowlands (below 500m; once 2800m). Light woodland, shrubby clearings, gardens. ***Fleshy mandible***, pale lores. Brownish-olive above, whitish throat, breast suffused olive, yellowish below; ***two whitish wingbars***, buffy to white edges on inner flight feathers. Monotypic. Mostly singly, perch upright on low exposed branches, repeatedly sallying and vocalising. **Voice** Short, flat *pép* or *típ*; song a burry *ffee-bEE'o*. **SS** Indistinguishable from Willow unless vocalising; cf. Acadian (bold eye-ring), migratory wood-pewees (size, head shape, no yellow belly), Fuscous and *Lathrotriccus* flycatchers. **Status** Uncommon transient and resident (Sep–Apr).

grey variant
olive variant
Euler's Flycatcher
juvenile
adult
Grey-breasted Flycatcher
juvenile
adult
Fuscous Flycatcher
juvenile
adult
Acadian Flycatcher
adult fresh
adult worn
juvenile
Willow Flycatcher
juvenile
adult worn
adult fresh
Alder Flycatcher

PLATE 201: PEWEES

Drab tyrannids with little overall pattern. Pewees perch on exposed perches (bare branches, snags, posts) and perform long aerial sallies. Two resident species are plain smoky-grey; resident Tropical is nondescript; wood-pewees are featureless and confusing. Identifying wood-pewees is complicated unless they are vocalising.

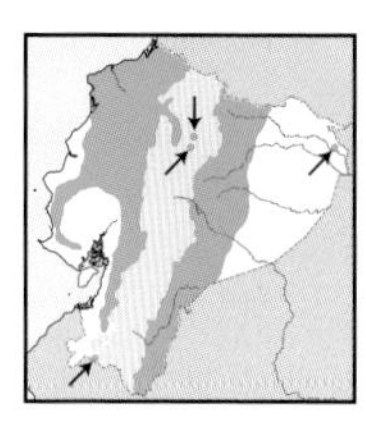

Olive-sided Flycatcher *Contopus cooperi* 17–18cm

Boreal migrant to E & W lowlands to upper foothills, not extreme E and dry SW (400–1,500m, locally lower and higher). Forest borders, woody clearings. Pink mandible. *Large triangular crest.* Dull greyish-olive above. *White tuft behind wings* sometimes visible. *White underparts, flanks densely mottled greyish-olive.* Monotypic. Singles atop stumps or prominent branches, perform long aerial sallies, also restlessly defend winter territory. **Voice** Tripled *tip-tip-tip* calls, sometimes with additional *tip*; less often *tic-thee-uu* song. **SS** Larger than congeners, white median underparts distinctive, tail proportionately shorter. **Status** Uncommon resident, NT globally (Sep–Apr).

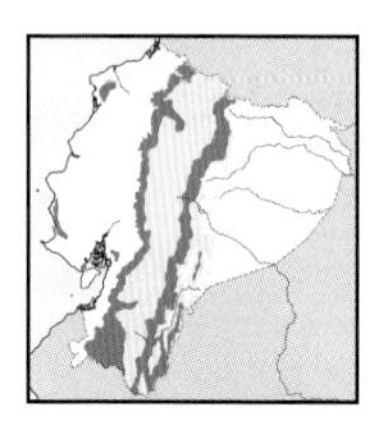

Smoke-coloured Pewee *Contopus fumigatus* 16–17cm

E & W foothills to upper subtropics, atop coastal mountains, locally in inter-Andean valleys (800–2,600m, to 500m in coastal ranges). Montane forest borders, shrubby clearings. *Long, pointed, triangular crest. Mostly smoky-brown to grey.* Races *ardosiacus* (E), *zarumae* (W). Singles or loose pairs atop stumps, exposed branches, wires or low branches, make long sallies, return to same spots. **Voice** Loud *píp-píp-píp*, additional *píp* sometimes added; also *píp-píp-píp..fee*; *we-dudít-wee'*, *we-dudít-wee'* at dawn. **SS** Occurs above similar Blackish Pewee (note size, less prominent crest; voices differ); Olive-sided Flycatcher is larger, not uniform smoky; see Smoky Bush-Tyrant. **Status** Fairly common.

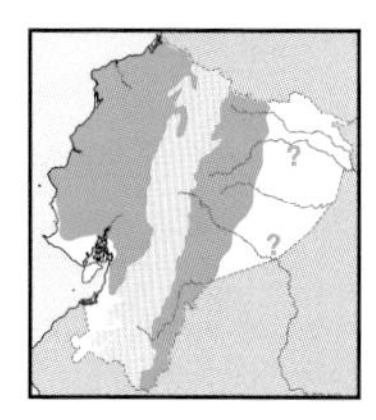

Western Wood-Pewee *Contopus sordidulus* 14–15cm

Boreal migrant to W, inter-Andean valleys, E foothills (to 2,800m, not below 400m in E). Humid to montane forest borders, shrubby clearings. *Pale mandible base or all-dark bill.* Dull greyish-olive above, two pale wingbars, the *upper bar fainter.* Whitish throat, pale greyish-olive breast, yellowish-white belly. Nominate. Mostly alone, perches on low to mid branches, makes long aerial sallies, returns to same perch. **Voice** Sad, burry *dryeee* or *pieeer*; less often heard song, a burry *free'e-free'e, fur-didi, free-e....* **SS** Not safely separated from Eastern except by voice. Subtle differences in bill and underparts obscured in juv. or moulting individuals. **Status** Uncommon transient and winter resident (Sep–Apr).

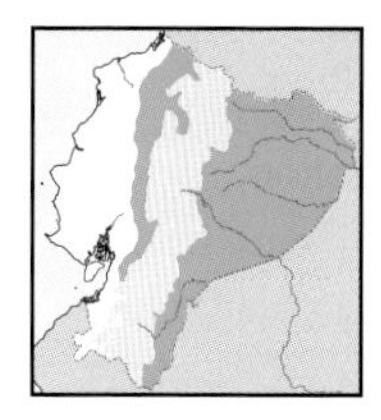

Eastern Wood-Pewee *Contopus virens* 14–15cm

Boreal migrant to E lowlands to subtropics, inter-Andean valleys to W foothills and adjacent lowlands (200–2,800m). Humid forest borders, shrubby clearings, woodland. Nearly identical to Western. Mandible often *entirely pale, underparts somewhat paler, less extensive greyish on chest.* Monotypic. Mainly alone, perches and sallies like Western. **Voice** Plaintive, somewhat inflected *fyéeeo?*, with *fee-ur* interspersed; longer and sadder than Western. **SS** Reported to hold tail *angled downwards*, Western *straighter in relation to back and wings*; see Acadian, Willow and Alder Flycatchers (smaller, less crested, more yellowish below, more patterned wings not projecting towards mid tail). **Status** Uncommon transient and winter resident (Sep–Apr).

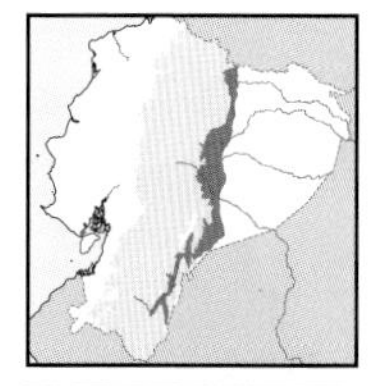

Blackish Pewee *Contopus nigrescens* 13–14cm

Local in E Andean foothills to adjacent lowlands (400–900m). Montane to *terra firme* forest canopy, borders. *Short bill, yellowish mandible. Short triangular crest. Mostly sooty-grey*, darker crown, wings and tail, slightly paler underparts. Nominate. Mainly in pairs atop high branches, perform long aerial sallies, return to same perch, often wag tail on perching; highly sedentary. **Voice** Song *pee-béew*, bright *píp-píp-píp-píp*, not tripled and less pronounced than Smoke-coloured. **SS** Larger Smoke-coloured has longer, bushier crest, more ringing voice. **Status** Rare.

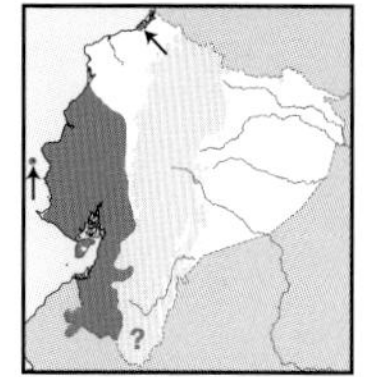

Tropical Pewee *Contopus cinereus* 13cm

SW lowlands, to Loja subtropics (to 1,500m). Light woodland, forest borders, adjacent clearings. *Fleshy mandible base.* Triangular crest, *pale lores.* Dull greyish-olive above, two *vague* wingbars. Whitish throat, greyish-olive breast, yellowish belly. Race *punensis* (may merit species status). Singles or loose pairs at low to mid heights, sometimes on exposed branches, sally long distances and return to same spot, wags tail on landing. **Voice** High-pitched, bright *peéé-pít*; simpler, upslurred *peép*; faster *peé-pi'dit.* **SS** Larger wood-pewees best separated by voice; both have less yellowish on belly, less contrasting throat, darker face; only regular pewee in arid habitat; see Grey-breasted Flycatcher. **Status** Uncommon.

Olive-sided Flycatcher
adult
non-breeding
first-winter
ardosiacus
zarumae
adult
adult
juvenile
Smoke-coloured Pewee
juvenile
adult
Western Wood-Pewee
juvenile
adult worn
adult fresh
Eastern Wood-Pewee
adult
juvenile
Blackish Pewee
adult
juvenile
Tropical Pewee

PLATE 202: WATERSIDE TYRANNIDS

A miscellaneous mix of tyrannids occurring by rivers and lakes including the unique – in Ecuador – Subtropical Doradito, the handsome wagtail-tyrant and the dull River Tyrannulet of river islands. Torrent Tyrannulet and, especially, Black Phoebe are widespread beside rushing Andean rivers and streams. The water-tyrant resembles forest-based chat-tyrants but is much drabber.

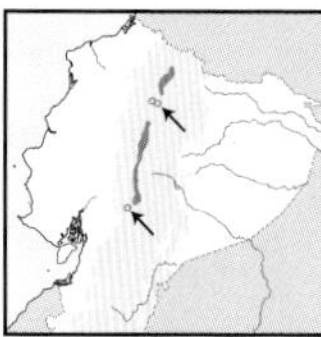

Subtropical Doradito *Pseudocolopteryx acutipennis* 11–12cm

Local in inter-Andean valleys (2,400–3,500m). Reedbeds, *Scirpus* marshes, damp grassland. ***Olive upperparts***, dusky cheeks, ***rich yellow underparts, unmarked wings***, fairly long tail. Monotypic. Singles or pairs glean and pick in cover, briefly perching atop a reed stem. **Voice** Nasal series of 3–5 *wüik* notes followed by wing whirr and snorting *wök*; simple *tzit* foraging call. **SS** Unmistakable in habitat and range. **Status** Rare, local, declining, VU in Ecuador.

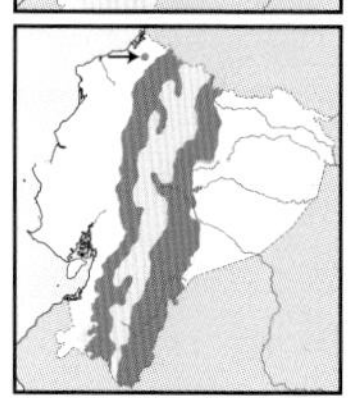

Torrent Tyrannulet *Serpophaga cinerea* 11–12cm

E & W foothills to temperate zone, more locally in inter-Andean valleys (700–3,100m). Fast-flowing rocky streams and rivers. ***Black cap and wings*** contrast with ***ash-grey upperparts, paler grey underparts***. Two narrow white bars on wing-coverts. Nominate. Pairs often perch on rocks and trunks above rushing water, regularly flick tail, make short sallies after flying prey or catch them in riparian vegetation. **Voice** Sharp *cheep* or *sheep* audible over rushing water, sometimes repeated several times, also longer *seep-seep-seep-see'pr-see'pr-see'pr*. **SS** See larger, mostly black, longer-tailed, bolder Black Phoebe; and much larger, chunkier, etc. White-capped Dipper. **Status** Common.

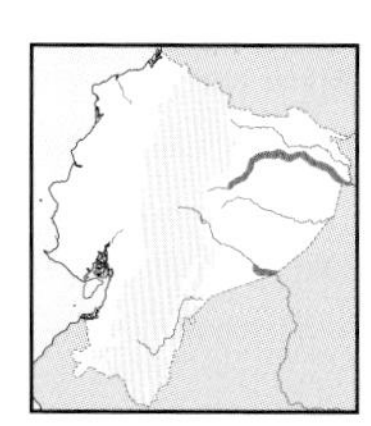

River Tyrannulet *Serpophaga hypoleuca* 11–12cm

Scattered records in E lowlands along Ríos Napo and Pastaza (below 400m). Young successional scrub on river islands. ***Dusky-grey above, blackish crest***, white supraloral stripe. ***Underparts dirty white***, breast-sides mottled olive. ***Unmarked dusky wings and long tail***. Nominate. Not on rocky streams like congener, perches upright in bushes, gleans from foliage or performs short sallies, not returning to same perch. **Voice** Rather slow, clear *tseEt-seee-seee'eu*, more explosive sputtering *teep-de'dir'r'r*; call dry, upslurred *du-eeerrrt*. **SS** Greyer above than commoner Drab Water-Tyrant, more contrast, wings and tail duskier, only short supraloral stripe; behaviour differs. **Status** Rare.

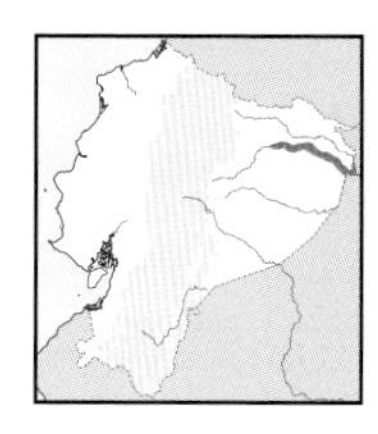

Lesser Wagtail-Tyrant *Stigmatura napensis* 13cm

E lowlands along lower Río Napo (below 300m). Young successional scrub on river islands. ***Long graduated tail***. Olive above, ***yellow below, yellow supercilium***, black ocular stripe. Wings and tail black, ***conspicuous whitish bands on wing-coverts*** and inner flight feathers, ***tail feathers broadly tipped white***. Nominate. Singles or lively pairs, fan and jerk tail, perch-glean or briefly sally. **Voice** Clear, querulous chatter *keeur?-kewee?-kéykéykéy*, sometimes jumbled at end; more rollicking syncopated duets *keewe-kuw-keewe-kuw'kuw-krr...*; *kEyuur* call. **SS** Long tail, wing and tail patterns, short bill readily distinguish it from other river island birds, including conebills, spinetails, etc. **Status** Uncommon.

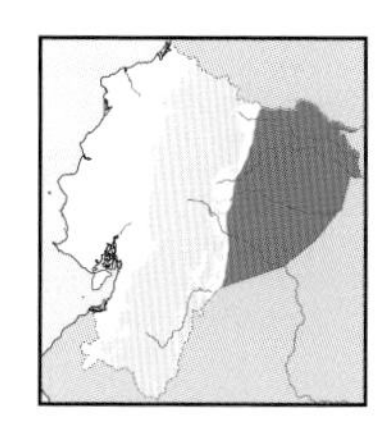

Drab Water-Tyrant *Ochthornis littoralis* 13–14cm

E lowlands (to 400m). Large river edges. ***Mostly sandy-brown***, browner crown, darker wings and tail, paler rump, throat and belly. ***Short whitish supercilium*** and narrow eye-ring, dusky ocular stripe. Monotypic. Singles or pairs on open branches, snags, roots in exposed banks, or sandbars, performs short sallies for flying insects; tends to repeatedly fly ahead of boats. **Voice** Excited *wee'p-wee-wee'p-diit- diit-diit*, also soft *dreeep* call. **SS** Rarer River Tyrannulet (smaller, different profile, shorter supercilium, greyer overall); Fuscous Flycatcher (whitish wingbars, white throat, browner upperparts, bolder supercilium); rarer Little Ground-Tyrant (slenderer bill, longer tail, different posture and profile). **Status** Fairly common.

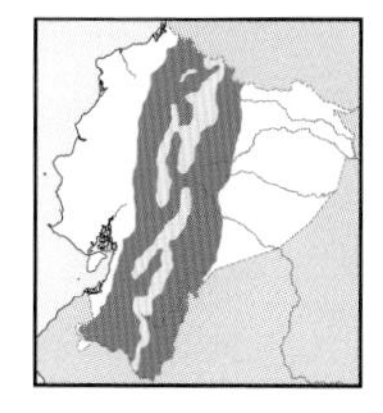

Black Phoebe *Sayornis nigricans* 17–18cm

E & W foothills to low temperate zone, inter-Andean valleys (500–2,800m). Open and semi-open areas, vicinity of streams. All ***sooty-black*** with conspicuous ***white lower belly, white bars and edges on wings***, outer tail feathers mostly white from below. Race *angustirostris*. Confiding pairs perch on or near ground, also higher on exposed branches, wires, fence posts; make aerial sallies, constantly flick tail upwards; can be seen in polluted rivers. **Voice** Simple shrill *zhrrrre* followed by clearer *feebee*, also short *pyeeert* reminiscent of Tropical Kingbird and more complex *t't't't-eest-kiut't't't...* in flight display. **SS** Unmistakable. **Status** Common.

Subtropical Doradito
adult
juvenile
Torrent Tyrannulet
River Tyrannulet
Lesser Wagtail-Tyrant
Drab Water-Tyrant
immature in
fresh plumage
juvenile
♂
typical adult
adult in worn
plumage
Black Phoebe

The Tufted-Flycatcher of NW forest borders is readily recognised by its short pointed crest and short bill. The gorgeous Vermilion Flycatcher is common and widespread in Andean valleys and dry SW (also austral migrant to E). The ♂ is one of the most striking of all tyrannids. All *Knipolegus* tyrants are inconspicuous; Rufous-tailed is the most distinctive, lacking the strong sexual dimorphism of congeners.

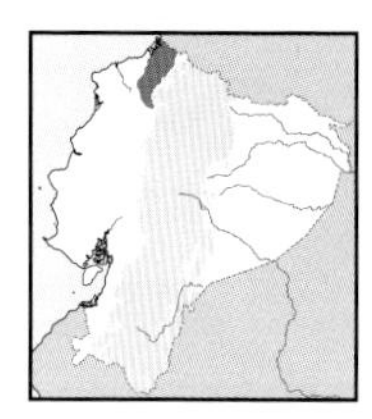

Tufted Flycatcher *Mitrephanes phaeocercus* 12–13cm

NW lowlands to lower foothills (100–600m). Humid forest borders, natural clearings. *Short, stubby, bicoloured bill. Pointed bushy crest*, narrow pale eye-ring, and pale lores. Dull olive head and upperparts, duskier wings and tail. *Tawny throat to chest*, yellow belly. Two very vague whitish wingbars. Race *berlepschi*. Singles or duos return to same exposed branch or stump after long sally, quivering tail on landing. **Voice** High-pitched, piercing and accelerating *peét, pee-pee-pee-pee-pee, pheé*.... **SS** Confusion unlikely given pointed crest, short bill and tawny-yellow underparts. **Status** uncommon.

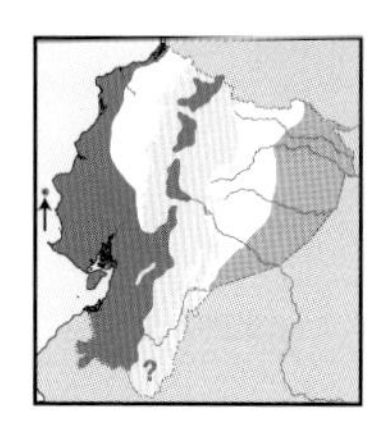

Vermilion Flycatcher *Pyrocephalus rubinus* 14–15cm

SW lowlands, to subtropics in Loja, inter-Andean valleys, austral migrant to E lowlands (to 3,000m, 200–300m in E). Open and semi-open arid to semi-humid areas. *Vermilion and black*. ♀: greyish-brown above, whitish below, *faint dusky breast streaks, flanks to lower belly washed pink to yellowish-pink*. Races *piurae* (W, Andes), nominate (austral migrant to E), possibly separate species. Mostly pairs on exposed perches, sally to air or ground. **Voice** Utters *tic-tic-tic, tiribí* at dawn, *tic-tic-t't-tiridíít* in flight display, thin *píc* calls. **SS** ♀: recalls Fuscous Flycatcher but breast always streaked dusky, wings nearly unmarked; see also Bran-coloured (short supercilium, well-marked wingbars). **Status** Common, rare migrant to E lowlands (Jul–Aug).

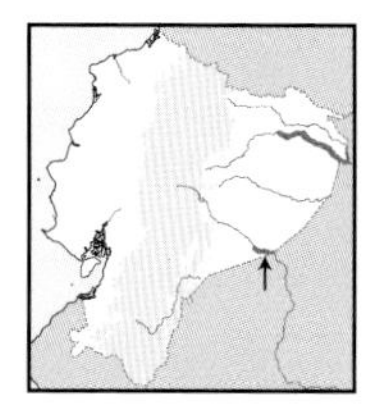

Riverside Tyrant *Knipolegus orenocensis* 15cm

Scattered records in E lowlands (below 250m). Young successional scrub on river islands, *Tessaria* stands. *Bluish-grey bill tipped black*, brownish eyes. *Entirely black*. ♀: dull *greyish above*, tail duskier; *faint pale eye-ring*, paler below, some dull greyish-olive streaks. Race *sclateri*. Singles or pairs at edges and semi-open, perch upright and make sallies to air or ground; also aerial displays. **Voice** Short, metallic or liquid *púk* followed by emphatic *pik'CHIt*, sometimes accelerating into sputter. **SS** Amazonian Black Tyrant of flooded forest; browner ♀ with rufescent uppertail-coverts, bold wingbars, etc. **Status** Rare.

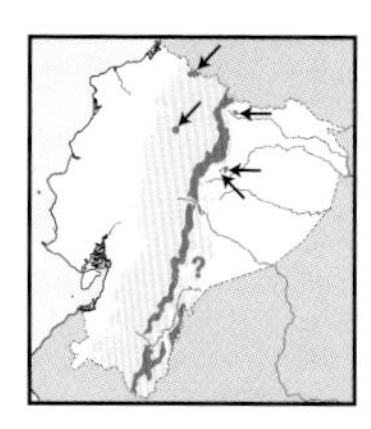

Rufous-tailed Tyrant *Knipolegus poecilurus* 14–15cm

Local in E Andean foothills to subtropics, extreme NW (1,000–2,000). Montane forest borders, adjacent clearings. *Reddish eyes*, narrow pale eye-ring. Dull brownish-grey upperparts, pale lores. Underparts whitish-buff grading into *cinnamon-buff belly. Rufescent inner webs to tail feathers*. Race *peruanus*. Singles or duos, upright on semi-open perches, perform aerial sallies, lift tail when alighting. **Voice** High-pitched, short, gravelly *tz-zr"zr'zrip* in flight display, often followed by fast jumble of notes. **SS** Red eyes striking and extensive rufous in tail diagnostic, but cf. ♀ of more local Andean Tyrant with dark brown underparts, rufescent rump. **Status** Rare.

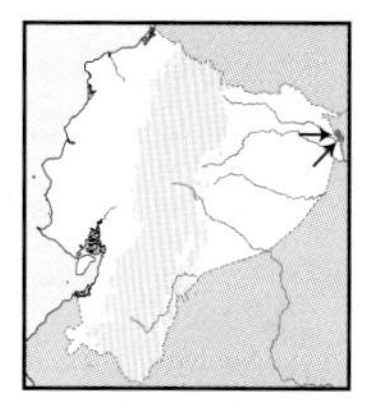

Amazonian Black Tyrant *Knipolegus poecilocercus* 13–14cm

Few records in extreme NE lowlands, Imuya region (200m). *Igapó* forest undergrowth and thickets. *Glossy blue-black, silvery bill tipped black*. ♀: *dusky-brown above, more rufescent lower rump. Bold whitish eye-ring, pale lores*. Whitish below, breast flammulated brownish-olive; *two clear buff wingbars*. Monotypic. Singles or pairs near water level make short sallies to ground, floating vegetation or water surface; aerial display. **Voice** High-pitched, very thin metallic *tík* or *tsík, pt-tsík* in flight display. **SS** Larger Riverside Tyrant sootier black, ♀ all sooty grey, lacks rufescent rump; habitat differs. **Status** Rare, local, overlooked?

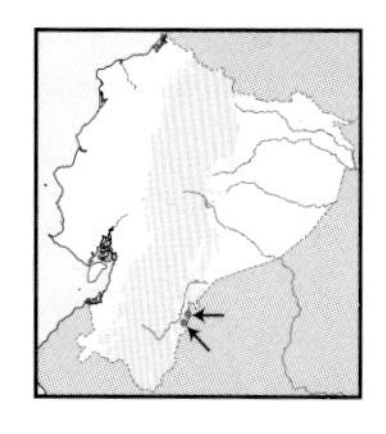

Andean Tyrant *Knipolegus signatus* 15–16cm

Local on Cordillera del Cóndor (1,900–1,950m). Stunted forest borders. *Entirely sooty-black, reddish eyes*. ♀: *dull greyish-brown above and breast*, more *rufescent rump*, two pale buff wingbars. Throat and belly whitish-olive streaked dull brown. Nominate. Poorly known in Ecuador, elsewhere mostly alone, perches upright in shady edges, often near water, constantly shivers tail sideways. **Voice** Not very vocal, a thin, nearly imperceptible, mechanical *tíc* by displaying ♂. **SS** Rufous-tailed Tyrant not known to be sympatric with Andean (no rufescent rump, has rufous inner webs to tail feathers, cinnamon-buff belly). **Status** Rare, very local.

Tufted Flycatcher
juvenile
rubinus
piurae
♀
♀
♂
Vermilion Flycatcher
♂
juvenile
♀
Riverside Tyrant
Rufous-tailed Tyrant
♂
♀
Amazonian Black Tyrant
♂
♀
juvenile
Andean Tyrant

PLATE 204: GROUND-TYRANTS AND FIELD-TYRANT

Terrestrial tyrants with sand-coloured plumages in open, often barren terrain. Two resident, locally common species in highlands. Two austral migrants (Dark-faced and White-browed) and a possible wanderer (Little) are rarer. They walk and run like thrushes, and perch upright. Use head pattern, underparts tone and size for identification. The field-tyrant is a small, virtually tailless bird in the arid SW.

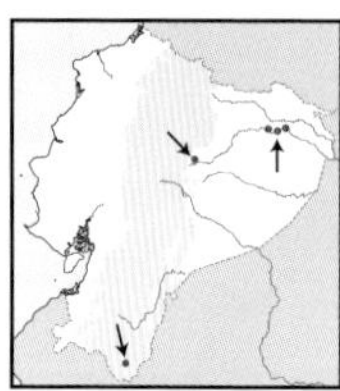

Little Ground-Tyrant *Muscisaxicola fluviatilis* — 13–14cm

Scattered records in E lowlands and foothills (200–1,150m). Sandbars, semi-open areas. ***Pale mandible base***, narrow pale eye-ring. Dull sandy-brown above, ***dusky lores***, vague pale supercilium, ***whitish*** throat and ***belly***, dull buff breast. Dusky wings, two ***brownish bars***. Monotypic. Singles or pairs run and hop on sandbars or agricultural fields. **Voice** A rising *peeép* (R *et al.*). **SS** Recalls Spot-billed (bolder supercilium, tawnier underparts and not sympatric); superficially similar to Drab Water-Tyrant but behaviour and overall shape differ. **Status** Rare, possibly wanderer or accidental (Jul–Aug).

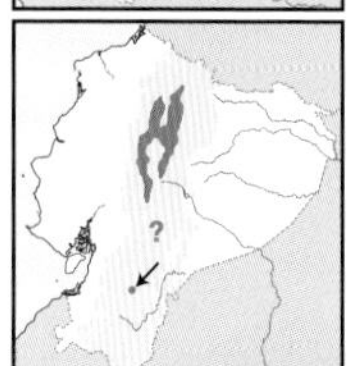

Spot-billed Ground-Tyrant *Muscisaxicola maculirostris* — 14–15cm

Andes and high valleys in C Ecuador (2,400–3,500m). Arid open and semi-open areas, *páramo*. ***Pale mandible base***, narrow broken whitish-buff eye-ring, ***short whitish supercilium***. Dull sandy-brown above, duskier wings and tail with buff edges, ***pale tawny below***. Race *rufescens*. Terrestrial, singles or pairs, occasionally perch on boulders or stumps, usually near cliffs and landslides. **Voice** Fast *clee-clee-clee-clee'lee'a* in display flight; *tleé* calls in display; single *tít* or *tík* calls. **SS** Smaller than sympatric Plain-crowned (longer supercilium, greyer below); no known overlap with similar Little Ground-Tyrant (fainter supercilium, white belly). **Status** Uncommon.

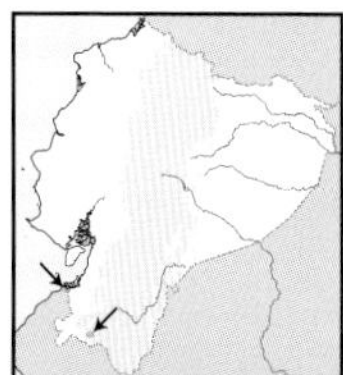

* Dark-faced Ground-Tyrant *Muscisaxicola maclovianus* — 16–17cm

Two isolated records in SW (0, 2,500m). Open grassy terrain. ***Blackish face***, dull brown upperparts, ***greyish-buff underparts***, whiter belly, duskier wings and tail, white outer tail. Race *mentalis*. Walks and hops on ground like congeners, might be expected near arid land and sandy beaches. **Voice** Not recorded in Ecuador, elsewhere a rapid series of emphatic *cheep!* notes (RG) **SS** No other ground-tyrant has blackish face without supercilium or eye-ring; see much bolder Short-tailed Field-Tyrant. **Status** Very rare austral visitor (Jan, Jun).

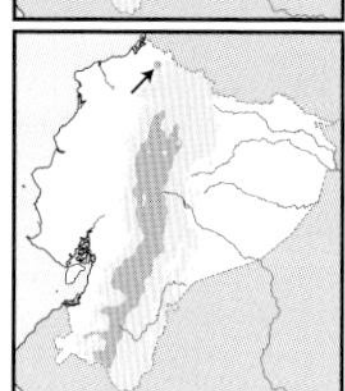

White-browed Ground-Tyrant *Muscisaxicola albilora* — 17cm

Austral migrant to *páramo* and inter-Andean valleys (2,400–3,700m). Open *páramo*, adjacent agriculture. ***Black bill. Long white supercilium***, narrow broken white eye-ring, dark lores. Dull buff-brown above, ***more rufescent rear crown***, duskier wings and tail, pale edges to wing-coverts. Dull whitish-buff below, white outer tail. Monotypic. Pairs or small, dispersed groups run, hop and stand erect on ground. **Voice** Not expected to vocalise much; elsewhere a soft *tuip* reported (FK). **SS** Seasonally sympatric with Plain-capped and Spot-billed Ground-Tyrants; Plain-capped has broader but shorter supercilium, no rufescent nape, paler underparts; Spot-billed has pale mandible base, shorter and fainter eyebrow, tawny underparts. **Status** Uncommon visitor (Jun–Aug, Nov).

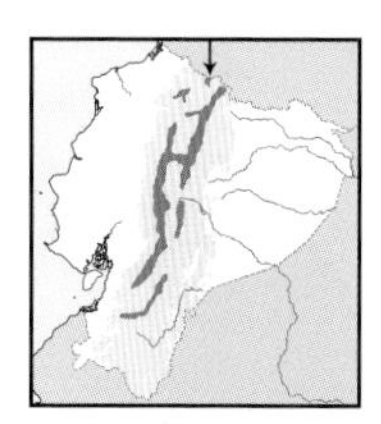

Plain-capped Ground-Tyrant *Muscisaxicola alpinus* — 18–19cm

Andes in E & W, S to Azuay (3,800–4,600m). Dry and grassy *páramo* and adjacent agriculture. ***Black bill, broad white supraloral stripe extends as narrower supercilium***, narrow broken whitish eye-ring. Dull buff-brown above, duskier wings and tail, dull greyish-buff below. Nominate. Pairs to small groups run and hop on ground, stands erect, sallies for prey. **Voice** Soft, 'bright' *pít-pít-pitít...pipít-pít...*, but not very vocal. **SS** Larger than sympatric Spot-billed and White-browed Ground-Tyrants, with broader but shorter supercilium, black bill, no tawny underparts, no rufous rear crown. **Status** Fairly common.

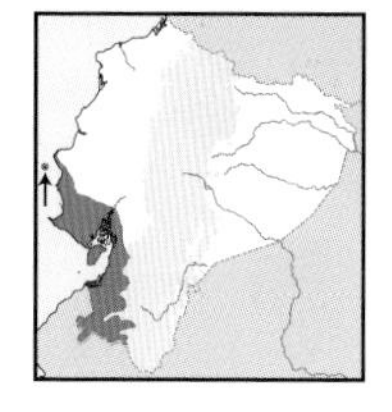

Short-tailed Field-Tyrant *Muscigralla brevicauda* — 11–12cm

SW lowlands, to lower subtropics in Loja (to 1,500m). Barren arid land, adjacent agriculture. ***Long pale legs***, short whitish eyebrow and narrow eye-ring. Dusky-brown above, ***yellowish rump, rufous band on uppertail-coverts***. Greyish-buff breast, whiter throat, yellowish belly. Dusky wings with whitish bars and edges. ***Very short tail tipped chestnut***. Monotypic. Singles or duos run and hop on ground, dart after prey or make brief sallies, not with flocks. **Voice** Sibilant, rattled *drrr-dzzzz'* or *tizzz-tízzzz*, often with several *tík* notes first; fast *tík-tík-tík, tr'zzzz* in flight display. **SS** Unmistakable in range by long legs and tiny tail, almost recalling a slim antpitta. **Status** Uncommon.

Little Ground-Tyrant

Spot-billed Ground-Tyrant

Dark-faced Ground-Tyrant

adult

juvenile

White-browed Ground-Tyrant

Plain-capped Ground-Tyrant

Short-tailed Field-Tyrant

PLATE 205: CLIFF FLYCATCHER, SHRIKE-TYRANTS AND BUSH-TYRANTS

Shrike-tyrants are large, dull-coloured inhabitants of open fields, with characteristic heavy and sharply hooked bills. The other species share extensive rufous to orange on flight feathers, but Cliff differs in structure, voice and behaviour, whereas white in wings is distinctive of Red-rumped. Bush-tyrants are uncommon at forest edges, but Streak-throated sometimes enters towns.

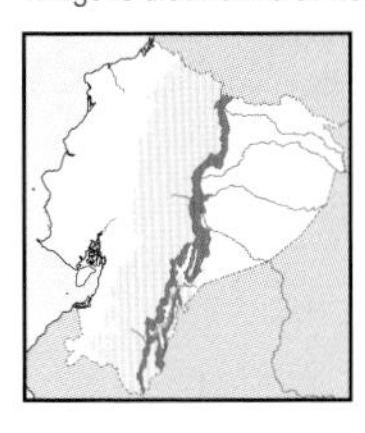

Cliff Flycatcher *Hirundinea ferruginea* 18–19cm

E foothills to subtropics (900–1,800m). Rocky cliffs. Silhouetted like a *large swallow*. Brown above, ***rich rufous-chestnut below, frosty whitish face***, buffy-white supercilium; ***most flight feathers orange-rufous, tail mostly rufous below***. Race *sclateri*. Pairs to small groups at natural and road-cut cliffs and rock faces make eye-catching, long, swallow-like sallies. **Voice** Chattering, often tireless, high-pitched *whér-dededede, whér-de-e'e'e'e, dwuep, dirr'wee, kli-kli-kli*…. **SS** Confusion unlikely; upper-elevation Streak-throated Bush-Tyrant has different silhouette, throat obviously streaked, less rufous in tail, etc. **Status** Uncommon.

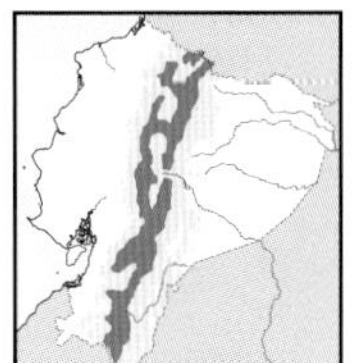

Black-billed Shrike-Tyrant *Agriornis montanus* 23–24cm

High inter-Andean valleys and *páramo* (3,000–4,000m, to 2,500m in S, locally lower). Dry and grassy *páramo*, adjacent agricultural and shrubby fields. ***Hooked black bill, white eyes***. Dull dusky-brown above, duskier wings, paler underparts. ***Short whitish supraloral stripe, whitish throat streaked dusky***, whitish belly. Most tail feathers white. Race *solitarius*. Mostly alone, runs on ground or perches atop grass clumps, stumps, rocks or posts; makes sudden attacks to ground prey, often hovers briefly. **Voice** Far-carrying *weee-di'uuu* with ringing quality. **SS** Bill and eye colour, plus extent of dusky throat streaking separate from similar but larger and stockier White-tailed; ground-tyrants are slimmer, with less white in tail, etc. **Status** Uncommon.

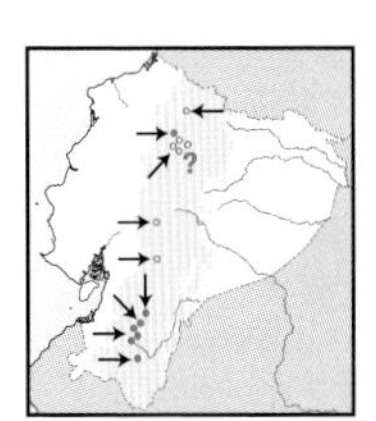

White-tailed Shrike-Tyrant *Agriornis albicauda* 27–28cm

Scattered records in high inter-Andean valleys and adjacent *páramo* (2,400–3,100m). Dry and grassy open areas, arid scrub. ***Heavy black bill with pale mandible base, dark eyes***. Dull dusky-brown above, paler below, whiter belly. ***Whitish throat streaked dusky, whitish supraloral stripe extends as narrow supercilium***. Most tail feathers white. Race *andicola*. Behaves like previous species but tends to prefer drier and more open situations, perches atop boulders, rocks, outcrops; apparently low density. **Voice** Melodic, far-carrying, ringing *wëëu-deeü-deeü-deeü, wëëÜ-diiu-diiu-diiu, teéü-teéü-teéü'chtr-chtr*…; also more chattering *ch'ch'ch'tchew-ch'ch'ch*…. **SS** See commoner Black-billed with pale eyes, smaller and darker bill. **Status** Rare, local, declining, EN in Ecuador, VU globally.

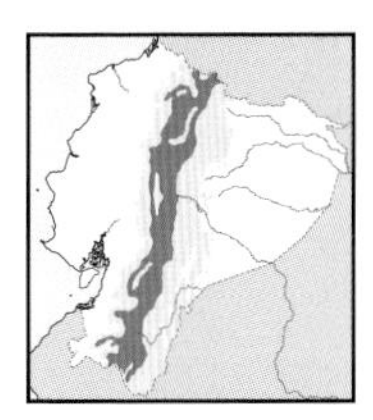

Streak-throated Bush-Tyrant *Myiotheretes striaticollis* 23–24cm

E & W temperate zone, inter-Andean valleys (2,400–3,200m). Montane forest borders, shrubby clearings, woodland, gardens. Brown above, long whitish supercilium, whitish throat ***streaked dusky*** to upper chest, ***pale orange rest of underparts. Orange-cinnamon underwing-coverts and panel in flight feathers***, tail mostly cinnamon-orange below. Nominate. Singles or pairs perch high above ground, perform long swooping sallies and return to same perch. **Voice** Far-carrying, long whistle *weeeeeu* or more inflected *weee'it?-weee'it?-weee'it?* and *tsi-si-rít* resembling Tropical Kingbird. **SS** Smaller, habitat-specific Cliff Flycatcher is richer rufous below, unstreaked throat, different shape; smaller Smoky Bush-Tyrant is darker overall. **Status** Uncommon.

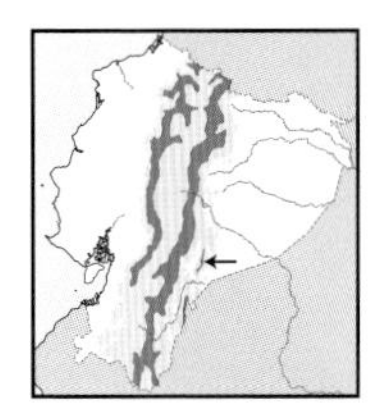

Smoky Bush-Tyrant *Myiotheretes fumigatus* 20–21cm

E & W subtropics to temperate zone, locally in inter-Andean valleys (2,000–3,200m). Montane forest, woodland, borders. Mainly ***dull brown. Faint pale supercilium, whitish throat streaked dusky. Orange-cinnamon underwing-coverts and rufous panel in flight feathers***. Nominate (most of range), *cajamarcae* (SE). Singles or pairs, shyer than previous species, often on canopy perches or edges, perform long sallies or attack foliage. **Voice** Gives *peea..peea..peea..peeu-peeu* at dawn; also steady *pë-pë-pë-pë-pë*…. **SS** Confusion unlikely; underwing similar to Streak-throated but overall pattern differs notably; Smoke-coloured Pewee is slaty-grey overall, crested head, no cinnamon underwing, etc. **Status** Uncommon.

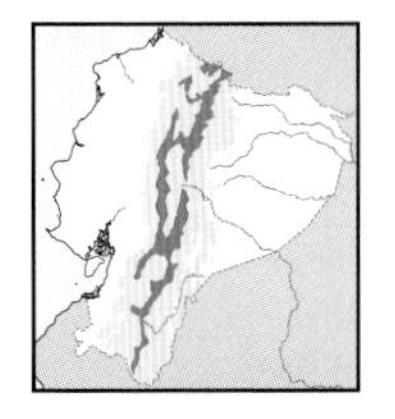

Red-rumped Bush-Tyrant *Cnemarchus erythropygius* 23–24cm

E & W temperate zone and *páramo* (2,850–4,100m). Grassy, semi-open shrubby areas, *Polylepis* woodland. ***Whitish-grey face to throat***, grey crown to ***back and breast. Orange-rufous rump, belly and basal half of outer tail feathers***; whitish panel in inner flight feathers, cinnamon-orange underwing-coverts. Nominate. Singles or loose pairs perch atop boulders, shrubs or posts, perform long aerial sallies to ground, return to same perch. **Voice** Not very vocal but utters a sharp, far-carrying *pyééér, shyeeér* or shrilled *kyeee*. **SS** Nearly unmistakable, but see Streak-throated. **Status** Rare.

Cliff Flycatcher
♂
juvenile
♀
Black-billed
Shrike-Tyrant
♂
White-tailed Shrike-Tyrant
Streak-throated Bush-Tyrant
fumigatus
cajamarcae
Smoky Bush-Tyrant
Red-rumped Bush-Tyrant

Another miscellany of flycatchers found in open and semi-open habitats. Pied Water-Tyrant and the marsh-tyrant wander to E lowlands. Cattle Tyrant is colonising as a result of deforestation. The beautiful Tumbes Tyrant was only recently discovered in the arid extreme SW. Masked Water-Tyrant and Long-tailed Tyrant are fairly common, but Long-tailed occurs in more forested situations than the widespread Masked.

Pied Water-Tyrant *Fluvicola pica* 13–14cm

Very few records in NE (to 350m; once 1900m). Riverbanks. ***Snowy-white head, underparts, scapulars and rump, black rear crown, most upperparts, wings and tail.*** Monotypic. Not resident in Ecuador and probably not spreading as records are scattered; elsewhere, conspicuous in open areas, walks on floating vegetation or marshy shores. **Voice** Not heard in Ecuador; elsewhere, a nasal *dreep* or *zhreep*, soft *pik* calls; buzzy *choo-wér* song (R *et al.*). **SS** Can only be confused with Masked but no overlap; Black-backed Water-Tyrant *F. albiventer* does not occur in Ecuador. **Status** Uncertain, probably vagrant (Jan, Jul).

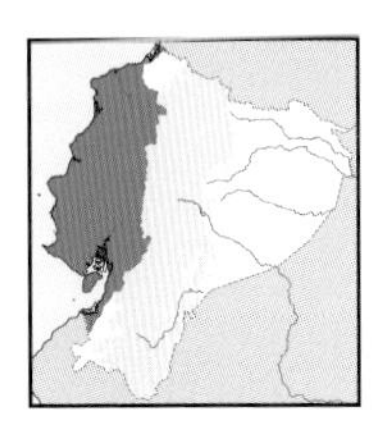

Masked Water-Tyrant *Fluvicola nengeta* 14–15cm

W lowlands to foothills (to 900–1,000m). Open and semi-open areas, damp grassland, rice fields, gardens. ***Mostly snow-white*** with dirtier back, ***narrow black ocular stripe, black wings and tail,*** latter broadly tipped white. Race *atripennis*, might deserve species status. Conspicuous and noisy, in pairs or small groups, walks on muddy shores, over floating vegetation, extensive pastures, etc.; mostly terrestrial, often fans and cocks tail, and lowers wings. **Voice** Loud sharp *py-kíft* or *ki'ít* repeated in excitement, often by pairs or group members, also simpler but brighter *kyt!* **SS** Unmistakable. **Status** Common.

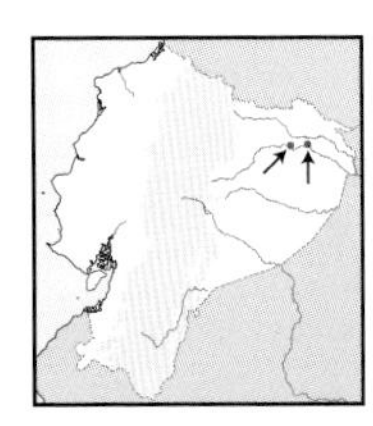

* White-headed Marsh-Tyrant *Arundinicola leucocephala* 13cm

Two records from E lowlands along Río Napo (300m). Grassy and young successional growth on river islands. ***Snowy-white hood contrasting with black body.*** ♀: duller, white face and underparts grade into ***greyish-white breast to flanks. Dull grey above,*** blacker wings and tail. ***Rosy-flesh mandible.*** Monotypic. Only wanders to Ecuador; elsewhere perches conspicuously atop grasses and stems, performs short aerial sallies. **Voice** Not heard in Ecuador; elsewhere, high, sharp *sedík!* (R *et al.*). **SS** ♀ diagnostic; no other Ecuadorian tyrant is patterned in grey and white, with fleshy mandible; see Pied Water-Tyrant; more arboreal, larger sirystes (blackish crown, white rump, etc.). **Status** Uncertain, possibly wanderer (Aug, Dec).

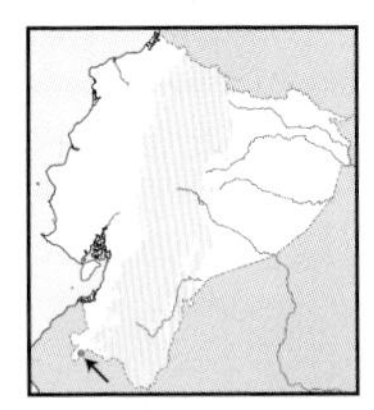

Tumbes Tyrant *Tumbezia salvini* 13–14cm

Two records from extreme SW lowlands near Zapotillo (N. Athanas, L. Ordóñez) (150 m). Dry forest and scrub. Unmistakable: dull above, ***long broad yellow-and-white supercilium, lemon-yellow below,*** obvious ***white wing patch***, tail white below. Monotypic. Singles, less often pairs, usually perch low, not fully exposed and quite inconspicuous; regularly wags and fans tail. **Voice** Fast and mellow *dyer'r'r'R, dyer'r-DEEw-de'r'r* (S *et al.*); calls more emphatic, clearer. **SS** Confusion unlikely in small range; chat-tyrants are superficially similar but occur in higher, more humid forests and scrub, and none is so boldly yellow. **Status** Rare, possibly spreading or only wandering from Peru; NT globally.

Cattle Tyrant *Machetornis rixosa* 19–20cm

Few recent records from E lowlands (to 300m). Open grassy areas. ***Uniform dull olive above***, yellow below, paler throat, ***long legs***. Probably race *obscurodorsalis*. Recent arrival to Ecuador following deforestation; elsewhere widespread in open habitats, often follows and perches atop cattle, may hit windows, rear-view mirrors, etc. **Voice** Recalls Tropical Kingbird *t'te'te'tree'e* but higher-pitched, squeakier, more strident (R *et al.*). **SS** Recalls kingbirds, especially Tropical, but readily distinguished by terrestrial habits and different profile when perched (no triangular crest, long legs, shorter tail, etc.). **Status** Possibly colonising.

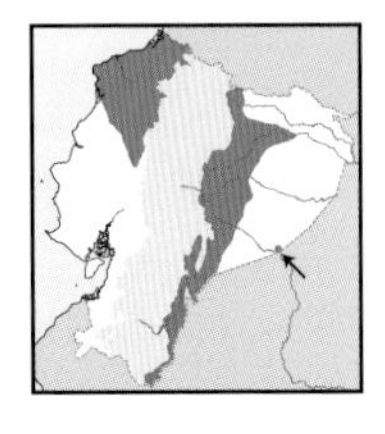

Long-tailed Tyrant *Colonia colonus* 20–23cm

NW lowlands to foothills, E foothills to adjacent lowlands (to 1,100m, not below 300m in E). Humid forest borders, woody clearings. Mostly ***slaty-black***, crown to upperparts ***variably glazed whitish***, white rump, ***two very long central tail feathers.*** ♀: somewhat paler below, ***tail shorter.*** Races *fuscicapillus* (E), *niveiceps* (SE), *leuconotus* (W). Pairs or small groups perch on exposed branches and snags, make long sallies for flying insects; sedentary, reuses same perches for long periods. **Voice** Soft, pure-toned *sweet* or *suee?*, sometimes paired or tripled, or runs into brief rattle, repeated every 2–5 sec; also smoother *swee* and *wee'e* calls. **SS** Unmistakable. **Status** Fairly common.

juvenile and ♀
♂
Pied Water-Tyrant
Masked Water-Tyrant
White-headed Marsh-Tyrant
immature
adult
Tumbes Tyrant
♂
♂
niveiceps
♀
immature
leuconotus
fuscicapillus
Cattle Tyrant
♂
niveiceps
juvenile
Long-tailed Tyrant

An attractive genus of Andean tyrants characterised by bold eyebrows and, in some, well-marked wingbars. Some species (three formerly in genus *Silvicultrix* and Slaty-backed) are more retiring inhabitants of forest understorey and borders, while the others are more conspicuous, preferring edges and, quite coincidently, having longer tails.

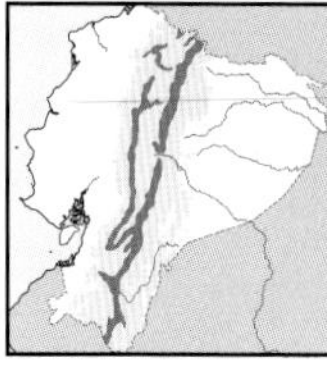

Crowned Chat-Tyrant *Ochthoeca frontalis* 12–13cm

E & W subtropics to temperate zone, sparser in CW (2,800–4,000m). Montane forest, elfin woodland, treeline. Dark brown above, brownish-grey below, *yellow frontlet, long white supercilium, dark brown wings*. Nominate. Mainly singly in dense undergrowth, rather lethargic, seldom with mixed flocks, performs rapid sallies to moss clumps. **Voice** Prolonged, descending, rather weak trill, *ts'rrrrrr*; short *tseeá* calls every 1–2 sec at dawn. **SS** No known overlap with lower-elevation Jelski's (browner back, buff wingbars). **Status** Uncommon.

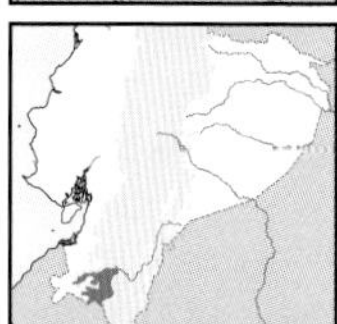

Jelski's Chat-Tyrant *Ochthoeca jelskii* 12–13cm

S subtropics to lower temperate zone (2,200–3,100m). Montane forest and woodland. Like previous species but upperparts more *rusty, two rufescent-buff wingbars*. Monotypic. Shy inhabitant of dense low undergrowth but apparently more tolerant than Crowned of degraded habitats. **Voice** Sharp *tseeeé*, more emphatic than Crowned, often varied as a more trilled *tss-trrr't, tsee-krrr*. **SS** Larger White-browed lacks yellow front, fainter wingbars; no known overlap. **Status** Rare.

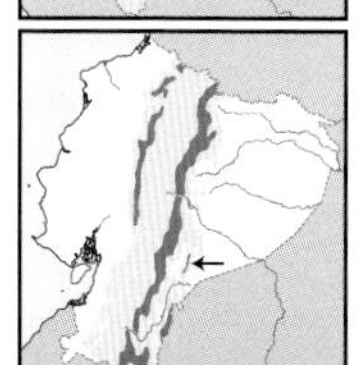

Yellow-bellied Chat-Tyrant *Ochthoeca diadema* 12–13cm

E & W subtropics to lower temperate zone, sparser in CW (2,200–3,100m). Montane forest, adjacent mature woodland. Brown above, *olive-yellow below*, more olive breast. *Long bold yellow supercilium*, two vague rufescent wingbars. Race *gratiosa*. Shy pairs tend to go unnoticed in dense, mossy undergrowth; may briefly join mixed flocks, but sedentary and territorial. **Voice** Repeated *tsíe'u* notes every 2–4 sec at dawn, also fairly long dry trill *tssrrtsé'rrrr* or shorter *ws'rrrr*. **SS** Confusion unlikely due to yellow supercilium and most underparts; see larger Lemon-browed Flycatcher (longer bill, underparts, wing pattern, behaviour, habitat, voice). **Status** Uncommon.

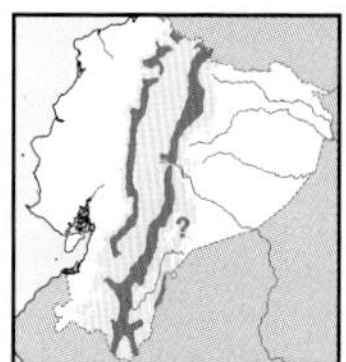

Slaty-backed Chat-Tyrant *Ochthoeca cinnamomeiventris* 12–13cm

E & W subtropics to lower temperate zone, S in W to Chimborazo (1,700–2,800m). Montane forest and mature woodland. *Mostly blackish* with conspicuous but *short white eyebrow* and *rich chestnut central breast to belly*. Nominate. Singles or pairs, not with mixed flocks, perch for long periods above or near forested streams, sometimes edges, perform short sallies. **Voice** Long, dry, high-pitched, far-carrying burry whistles *dzeéeee'uw* or *t'zeeéeuw*, sometimes followed by 2–4 clear *tseé* notes, mainly at dawn. **SS** Confusion unlikely in habitat. **Status** Uncommon.

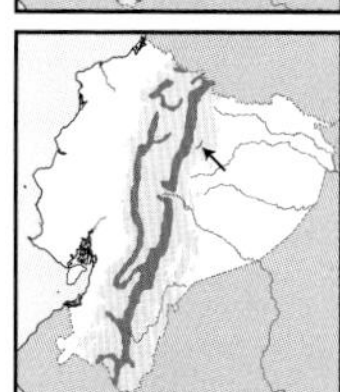

Rufous-breasted Chat-Tyrant *Ochthoeca rufipectoralis* 13–14cm

E & W Andean subtropics to temperate zone (2,500–3,300m). Montane forest borders, secondary woodland, adjacent clearings. Brown above, *long bold white supercilium, orange-rufous breast*, greyish-white belly. Dusky wings with *one bold rufous bar*. Race *obfuscata*. More arboreal and conspicuous than most congeners, perches on exposed branches, sometimes at edges or in open, performs long sallies; joins mixed flocks. **Voice** Gives *tirí-dééuw..tirí-dééuw* at dawn; loud, squeaky *trr'E, trr'E*... or *clerE, clerE* and *ch'brrre, ch'brrre*.... **SS** Overlaps with next species, which has uniform underparts, two wingbars, browner back, etc. **Status** Fairly common.

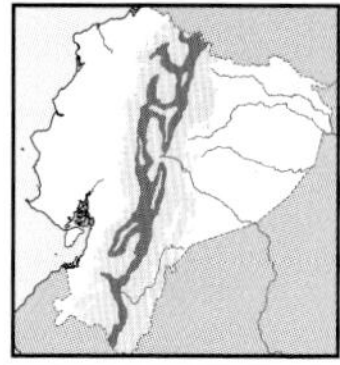

Brown-backed Chat-Tyrant *Ochthoeca fumicolor* 14–15cm

Andes (2,800–4,500m). *Polylepis* woodland, grassy and shrubby *páramo*. Rufescent-brown above, *dull orange-buff to cinnamon below*, grey throat, *long broad whitish-buff supercilium*; dusky-brown wings, *two rufescent bars*. Race *brunneifrons*. Singles or pairs perch atop stumps, bushes, grasses, perform short sallies to air or ground, often with mixed flocks. **Voice** Clear *clee-dé-clee-dé-clee-dé* or *cheé-dé-cheé-dé* in succession, sometimes slightly accelerating; simple *pseé, pséii* or *prii-prii*... calls. **SS** Confusion unlikely; only *páramo* tyrannid with long broad supercilium. **Status** Fairly common.

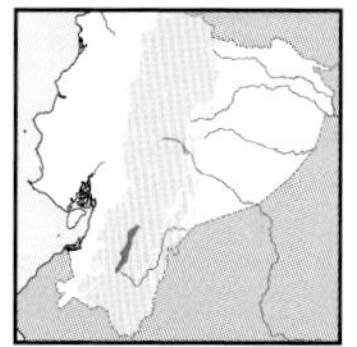

White-browed Chat-Tyrant *Ochthoeca leucophrys* 14–15cm

Local in S highlands (2,200–2,800m). Montane woodland and scrub. Dusky-brown above, *ash-grey below, long bold white supercilium, two faint rufescent wingbars*. Race *dissors*. Not common in Ecuador where poorly known; elsewhere rather conspicuous, perches on exposed branches, fence posts, atop bushes, makes short sallies to air or ground. **Voice** Clear, rather sharp and nasal *kueeu* or *küee* calls; also *keeu-kee-keeu-kee*.... **SS** Jelski's (no known overlap) has bold yellow supraloral stripe and two bold rufescent-buff wingbars. **Status** Rare.

Jelski's Chat-Tyrant
adult
adult
juvenile
juvenile
Crowned Chat-Tyrant
Slaty-backed Chat-Tyrant
juvenile
adult
Yellow-bellied Chat-Tyrant
Rufous-breasted Chat-Tyrant
adult
juvenile
Brown-backed Chat-Tyrant
White-browed Chat-Tyrant

Three common and widespread flycatchers of open lowland habitats, spreading to lower subtropics following deforestation. Although plumage of Social and Rusty-margined recalls larger kiskadees or similar-sized *Conopias* flycatchers, their short stubby bills are distinctive. Dusky-chested and Sulphury are rare canopy dwellers, the former in *terra firme* and the latter in *várzea*.

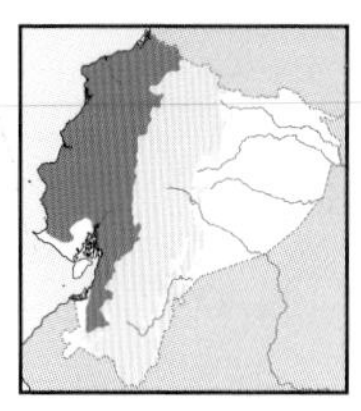

Rusty-margined Flycatcher *Myiozetetes cayanensis* 16–17cm

W lowlands to upper foothills (to 1,400m). Forest borders, woodland, clearings. ***Short bill. Black-and-white striped head.*** Brown above, white throat, yellow below; dusky wings, ***rufescent edges to flight feathers.*** Dusky tail ***edged rusty.*** Race *hellmayri*. Singles to small groups, often with other flycatchers, make long sallies to foliage, air or ground, regularly near water. **Voice** Long, whining *peeeea*, also noisier *feeee-chéi-chéi-cherechere*.... **SS** Closely recalls Social (no rusty wings, more olive back; best told by voice); White-ringed lacks rusty in wings, white encircles crown, longer bill; both kiskadees have larger bills, longer superciliary. **Status** Fairly common.

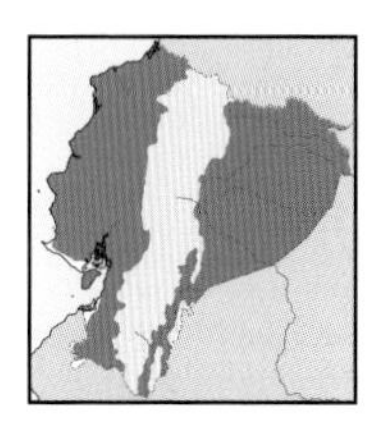

Social Flycatcher *Myiozetetes similis* 16–17cm

E & W lowlands to lower subtropics (to 1,700m). Agricultural land, clearings, forest borders. ***Short bill. Blackish mask, olive-brown crown*** to upperparts, long white supercilium. Blackish wings with ***pale edges.*** Nominate (E), *grandis* (W). Singles to small groups regularly consort with other flycatchers; noisy and conspicuous, makes long sallies to air, ground or foliage. **Voice** Noisy, including harsh *chreeew, chreei-pee-pee...; greeu-greeu-greeu, peei...*; also *chrii-chrii-chreu'chreu* (E). **SS** See Rusty-margined (wing edges and voice); White-ringed (longer bill, white diadem), Yellow-throated and Three-striped (yellow throats); kiskadees (larger, longer-billed, heavier); Grey-capped (lacks supercilium, has paler eyes). **Status** Common.

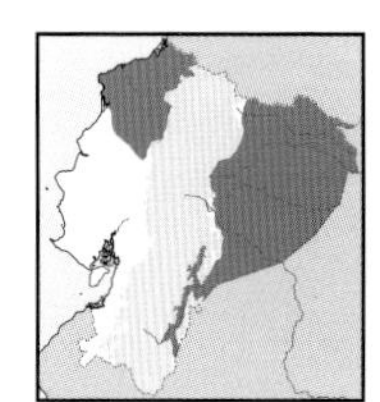

Grey-capped Flycatcher *Myiozetetes granadensis* 16–17cm

E & NW lowlands to foothills (to 1,000m). Humid forest borders, shrubby clearings. ***Short bill, pale eyes. Grey cap, whitish front and eyebrow,*** blackish mask. Brownish-olive above, white and yellow below. Races *obscurior* (E), *occidentalis* (W). Mostly singles or pairs at forest edges, less often in semi-open; sallies to air or foliage and takes much fruit. **Voice** Harsh, piercing *chík-chík-chík-keu, kép-kép-kép, jék-kEY..., kíp-kíp-jekEY..., ke'e'e'e'e-kéy-kéy-kéy...*; at dawn rather short *kip, kip, kip, keeuw-kreh* in W, or ends with *kE-kE-kee-yí* in E. **SS** Grey cap, pale eyes and almost plain wings distinctive. **Status** Fairly common.

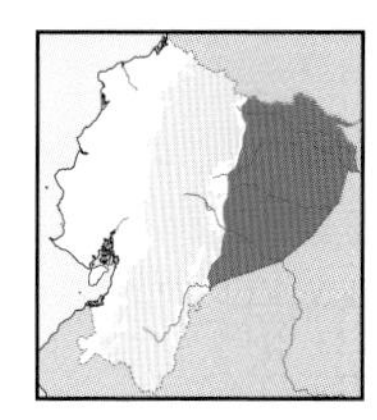

Dusky-chested Flycatcher *Myiozetetes luteiventris* 14–15cm

E lowlands (to 600m). *Terra firme* forest canopy, borders. ***Featureless. Very short bill. Dusky-brown above.*** Whitish throat, rest of underparts yellow, ***neck-sides to chest flammulated dusky. Unmarked wings.*** Nominate. Mostly pairs, perch upright on exposed branches, perform short sallies to air and hover for fruit. **Voice** Piercing, rather nasal, whining *meeew-meeew, deeepít, meeew-meeew-meeew, meew-chee-chaá-chaá...* reminiscent of Grey-capped; *eEw-eEw-EEw-chee'chu'chEEdee* at dawn. **SS** Smaller than Sulphury (greyer head, darker mask, more extensive white throat); see Grey-capped. **Status** Rare.

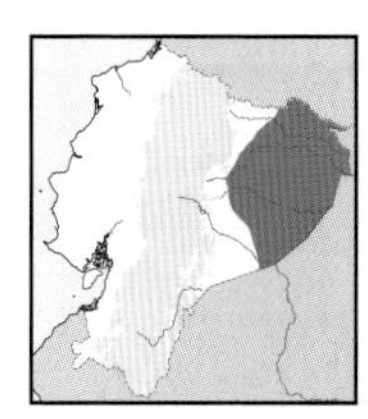

Sulphury Flycatcher *Tyrannopsis sulphurea* 18–19cm

E lowlands (to 400m). *Várzea* forest, *Mauritia* swamp forest. ***Short bill. Sooty-head*** grading into greyish-olive upperparts, ***blackish mask. Whitish throat, olive flammulations on chest,*** yellow below. Unmarked dusky wings, ***rather short, square tail.*** Monotypic. Singles or pairs perch erect atop palm fronds, perform long aerial sallies, seldom with mixed flocks, takes much fruit. **Voice** Noisy; gives a piercing, high-pitched *jeeét-jeeét-jeeét-jeeét-jee'éé..., tjeeeé-jeep-tjrrrp, djEE-djt'djt'djt..., jeEEZ-jeEEZ-jeEEZ..., dzirrt-zzirrt-zzirrt,* also louder *ghEE.* **SS** Dusky-chested is much smaller and entirely dusky-brown above, bill shorter and stubbier; see widespread Tropical Kingbird (no flammulations below, but duskier chest, longer and notched tail, longer bill, etc.). **Status** Rare.

adult
juvenile
Rusty-margined Flycatcher
adult
adult
juvenile
grandis
similis
Social Flycatcher
adult
adult
juvenile
bscurior
occidentalis
Grey-capped Flycatcher
adult
juvenile
Dusky-chested Flycatcher
Sulphury Flycatcher

Conopias recall the common Social and Rusty-margined, but their bills are larger and more slender. The status of Yellow-throated and Three-striped is not fully understood; more in forest canopy than *Myiozetetes*. Lemon-browed ranges in foothills to subtropics. Kiskadees and Boat-billed are fairly large and conspicuous flycatchers of open areas.

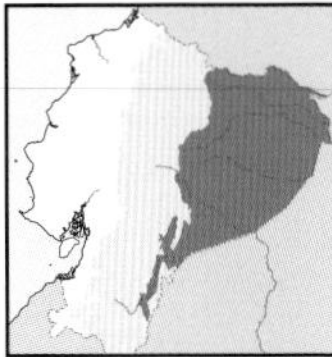

Great Kiskadee *Pitangus sulphuratus* 20–21cm

E lowlands to foothills (to 1,000m). Wooded lake and riverbanks, adjacent clearings. ***Heavy bill.*** Black crown and mask, ***long white supercilium nearly encircling crown.*** Dusky tail ***edged rusty, rusty edges*** to wing feathers. Nominate. Conspicuous, singles or pairs, very vocal, sally and attack wide array of prey. **Voice** Loud, distinctive *kEEs-kaa-deé!*, similar *kee-kEE-ku'awr* at dawn; loud, hawk-like *keeeér* calls. **SS** Bill less massive than Boat-billed Flycatcher, but longer, less slender than Lesser Kiskadee; Boat-billed has less rusty in wings; *Myiozetetes* are smaller, bills included. **Status** Common.

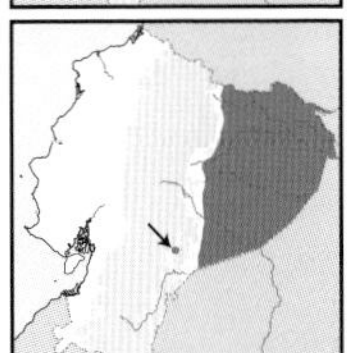

Lesser Kiskadee *Pitangus lictor* 17–18cm

E lowlands (to 500m, locally higher). Lakes and sluggish riverbanks, adjacent clearings. ***Long slender bill.*** Overall pattern recalls larger Great Kiskadee; mask somewhat less prominent, plainer wings and tail. Nominate. Singles or pairs perch upright a few metres above water, sally swiftly to water's surface, floating vegetation or ground. **Voice** Nasal, raspy, distinctive *dzzree-dzzrEE-dzzree-dzzree-zru*; fast *quéèz-see, quezz-meé!*; chattering *kzaa'deé-deé-deé....* **SS** Bill shape distinctive; cf. larger Greater Kiskadee and smaller Social and Rusty-margined Flycatchers. **Status** Uncommon.

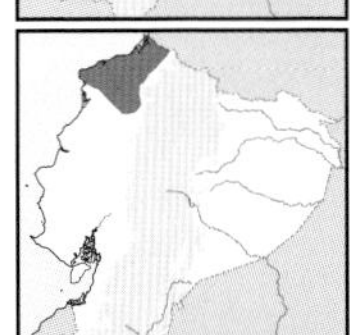

White-ringed Flycatcher *Conopias albovittatus* 16–17cm

NW lowlands (to 500m). Wet forest canopy and borders. ***Rather long bill. Black mask and crown, long white diadem,*** white throat. Brownish-olive above, yellow below, dusky ***unmarked*** wings and tail. Nominate. Pairs to small parties sometimes with mixed flocks, perch upright on exposed branches or atop large leaves, sally to upperside of foliage. **Voice** Fast, dry, abrupt rattle *tri'rrrr't, treé, tri'rrrr't, tree....* **SS** Recalls Social and Rusty-margined Flycatchers, but white encircles crown, bill longer, wings plainer. **Status** Rare to uncommon, NT in Ecuador.

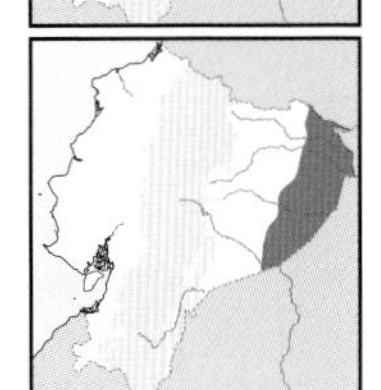

Yellow-throated Flycatcher *Conopias parvus* 16–17cm

Scattered records in far E lowlands (mostly below 200m). *Terra firme* forest canopy. Head as White-ringed but underparts ***all yellow. Dull olive upperparts,*** duskier wings. Monotypic. Poorly known, seldom seen; elsewhere perches atop crowns and exposed twigs, hover- or perch-gleans, seldom sallies to air. **Voice** Distinctive, far-carrying, ringing *klee-leé, klee-leé..., klu-yee'ee'ee'ee'ee*, oft-repeated; also rising *klu-kléé-krEE.* **SS** Much like Three-striped (best told by harsher voice); latter lacks crown patch, head paler, brighter olive back; habitat probably differs. **Status** Uncertain, possibly very rare or wanderer.

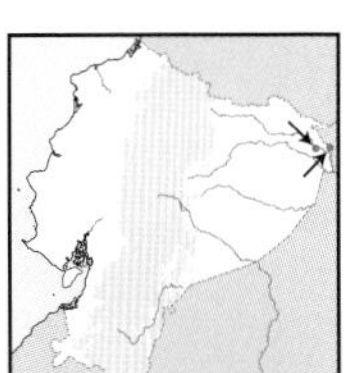

Three-striped Flycatcher *Conopias trivirgatus* 14–15cm

Very local in lowlands of extreme NE (below 200m). *Várzea* and *igapó* forest canopy. Blackish crown and mask, ***white superciliary encircling crown, no crown patch; olive upperparts,*** yellow underparts, duskier wings and tail. Possibly race *berlepschi.* Poorly known in Ecuador; elsewhere fond of high canopy perches, where perhaps behaves like previous species. **Voice** Harsh, grating *djeeuw*, often varied as *djeeu-djeeu-djeeu.* **SS** Yellow-throated has duller head, darker back, and yellow crown patch (not easy to see); voice is less harsh, more ringing; see also Social. **Status** Rare, wanderer?

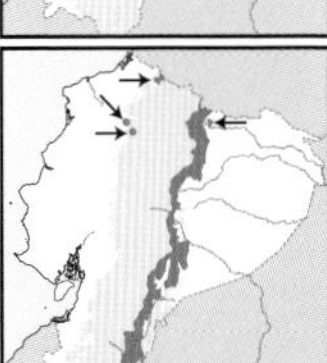

Lemon-browed Flycatcher *Conopias cinchoneti* 16–17cm

E & NW foothills to subtropics (1,000–2,000m, locally lower in SE). Montane forest borders, adjacent clearings. Olive upperparts and ***mask*** with ***long conspicuous yellow supercilium.*** Lemon-yellow underparts. Nominate (E), probably *icterophrys* (NW). Loose pairs or trios on exposed perches or atop large leaves, sally to foliage or pick fruit; nests in abandoned cacique baskets. **Voice** Quavering trill *pt'teerr-pt'teerr-pt'teerr* or *pre'e'e-terr, pre'e'e-terr....* **SS** Confusion unlikely, no other *Conopias* has yellow supercilium; see smaller Yellow-bellied Chat-Tyrant (different voice, habitat, behaviour, overall pattern, size, wing pattern, etc.). **Status** Uncommon; VU globally.

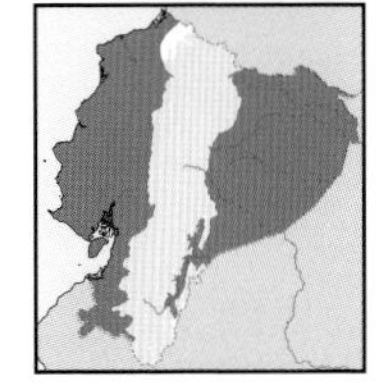

Boat-billed Flycatcher *Megarynchus pitangua* 22–23cm

E & W lowlands to upper foothills (to 1,300m). Humid to deciduous forest borders, woodland, adjacent clearings. ***Massive bill.*** Recalls smaller kiskadees but crown patch fiery orange, ***less rufescent*** wings and tail. Nominate (E), *chrysogaster* (W). Singles or pairs feed on large invertebrates, smashing them against branches; forages in higher strata but darts to ground. **Voice** Strident, fast *khreé-khreé-khreé-khreé-khreeen*; trilled nasal *khi'i'i'i'i* (W); nasal, complaining *nhyE-nhyE-nhyE-nhyE* or *ghEE-ghEE-ghEE*; softer *shree'u* at dawn; shrill *shrEE'ee!* (E). **SS** Unmistakable by bill size and loud voice. **Status** Common.

Great Kiskadee
adult
juvenile
Lesser Kiskadee
adult
juvenile
White-ringed Flycatcher
Yellow-throated Flycatcher
juvenile
adult
Three-striped Flycatcher
adult
adult
juvenile
cinchoneti
terophrys
chrysogaster
Lemon-browed Flycatcher
pitangua
Boat-billed Flycatcher

Mid-sized to large flycatchers of edges and open habitats, mostly in E lowlands. Two have short bills, one of them is an austral migrant (Variegated), the other a seasonal migrant (Piratic). *Myiodynastes* are heavy-billed but otherwise rather heterogeneous. Two heavily streaked species (one of them an austral migrant) and two predominantly yellow (one of them a boreal migrant). Most tend to use exposed high perches.

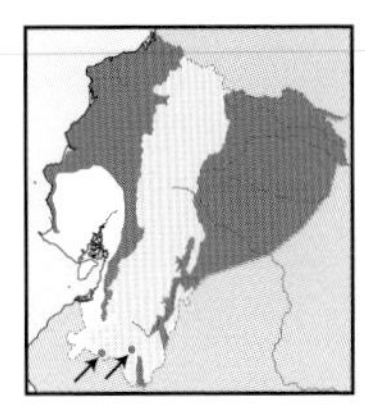

Piratic Flycatcher *Legatus leucophaius* — 14–15cm

E & W lowlands to foothills (to 800m). Humid forest borders, woodland, clearings. ***Short black bill.*** Greyish-brown upperparts, ***long white supercilium, broad whitish malar, thin dusky sub-malar, black mask.*** Whitish throat to breast, sides streaked dusky, yellowish belly. ***Nearly unmarked dusky wings and tail.*** Nominate, also migratory *variegatus*? Mostly alone, on exposed perches, very vocal; feeds mostly fruit, robs nests. **Voice** Loud, whining *freeE-yeé, piri'ri'ri'ri*, often only first part; also oft-repeated, ringing *peé-peé-peé-peé....* **SS** Larger Variegated Flycatcher (more mottled back, rufescent tail, longer bill, patterned wings). **Status** Uncommon.

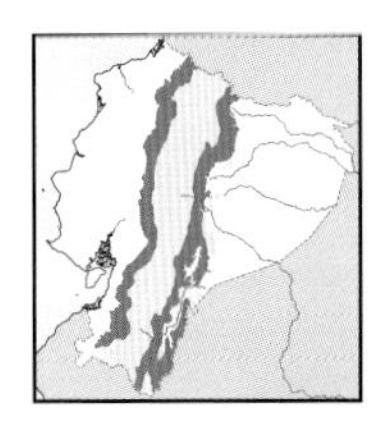

Golden-crowned Flycatcher *Myiodynastes chrysocephalus* — 20–21cm

E & W Andean foothills to subtropics, locally in N coastal ranges (800–2,200m, lower in coastal mountains). Montane forest canopy, borders, woody clearings. Greyish-olive head to upperparts, ***whitish superciliary and broad malar***, dusky sub-malar. Whitish throat, ***yellow below, breast flammulated olive. Dusky wings edged rufous.*** Race *minor.* Mostly pairs, on exposed perches; may briefly join mixed flocks. **Voice** Whining *skwaá-kiyaá-kiyaá-skuaa'yi...*; loud *kyaaa!-kyaaa!*; *skeee-ky'y'r'r* at dawn. **SS** Streaked (very streaked) and Baird's Flycatchers (rufescent wings, less patterned face). **Status** Fairly common.

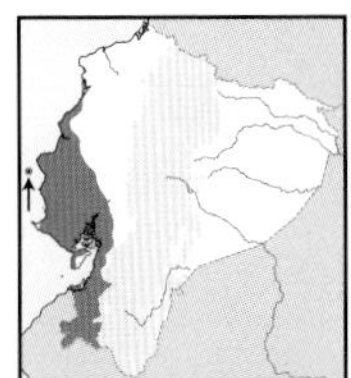

Baird's Flycatcher *Myiodynastes bairdii* — 22–23cm

SW lowlands, to subtropics in Loja (to 1,600m). Deciduous to semi-deciduous forest borders, woodland, scrub. Olive-brown above, ***broad whitish supercilium, black frontlet to mask.*** Whitish throat, ***ochraceous breast***, yellow belly, ***rufous rump.*** Wings and tail ***mostly rufescent.*** Monotypic. Singles to trios perch in open, long sallies to air, ground or foliage. **Voice** Rather shrill *jur'r'r'r-eét!* or *wrree-ít!* at dawn or dusk; also chattering *jimelorí-jimelorí....* **SS** Readily identified in range by rufescent wings and tail, face pattern differs from congeners. **Status** Uncommon, range-restricted.

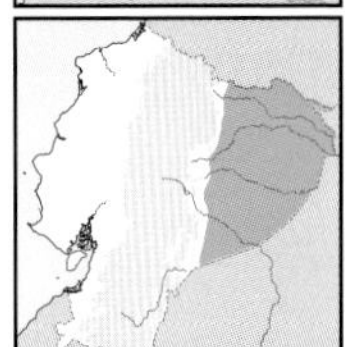

Sulphur-bellied Flycatcher *Myiodynastes luteiventris* — 20–21cm

Boreal migrant to E lowlands, once in SW foothills (mostly below 400m). *Terra firme, várzea* and *igapó* forest borders, adjacent woodland, clearings. Recalls Streaked Flycatcher but belly bright ***sulphur-yellow, longer and broader sub-malar reaches neck-sides.*** Monotypic. Singles or pairs sally in long bouts and hawk in air, but also take much fruit and sally-glean to foliage. **Voice** Mostly silent, occasionally gives a buzzy, but loud *quEEz-leek!, quéEEe!* **SS** Resident and migrant E races of Streaked Flycatcher; also Piratic Flycatcher. **Status** Uncommon transient (Dec–Feb).

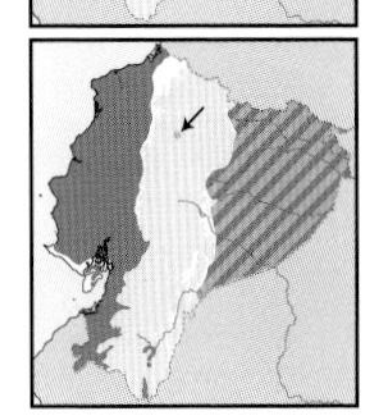

Streaked Flycatcher *Myiodynastes maculatus* — 20–22cm

E & W lowlands to foothills, migrant to E lowlands (to 1,150–1,200m, once 2,800m). Forest borders. Dusky above ***densely streaked whitish, whitish below streaked dusky. Broad whitish superciliary and malar***, blackish mask and sub-malar. ***Rufous rump and outer tail feathers.*** Nominate (E), *chapmani* (W), *solitarius* (austral migrant in E). Singles or pairs on exposed branches, sally and chase prey, also take fruit. **Voice** A *wueE-chede-deede* at dawn; squeaky *chép, khép!*, often a longer *chépechépechépe....* **SS** Larger, streakier, heavier-billed than Variegated and Piratic; Sulphur-bellied (longer and broader sub-malar, belly rich yellow, no rufous wings). **Status** Uncommon resident and austral migrant (Apr–Aug).

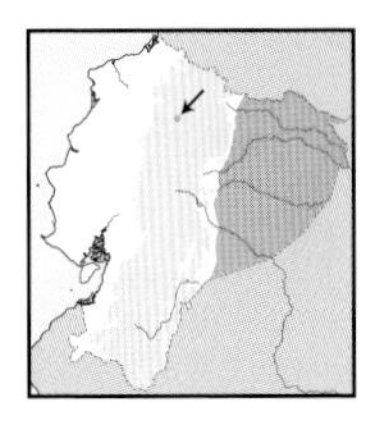

Variegated Flycatcher *Empidonomus varius* — 18–19cm

Austral migrant to E lowlands (to 500m, once 2,800m). *Terra firme* and *várzea* borders, adjacent clearings, riparian shrubbery. ***Pink mandible base.*** Blackish crown and mask, long whitish superciliary, ***broad malar.*** Greyish-brown above ***streaked dusky***, whitish-buff below, ***sides streaked dusky.*** Wings ***boldly edged whitish, rufous uppertail-coverts and edges to outer tail feathers.*** Nominate. Mostly singly on low perches, sometimes congregates at fruiting trees. **Voice** Occasionally a thin, quavering *psíí'i'i.* **SS** Smaller than bolder Streaked Flycatcher; Piratic (no rufous tail, shorter bill, plainer wings). **Status** Uncommon austral winter resident (Mar–Sep).

adult
juvenile
Piratic Flycatcher
adult
juvenile
Golden-crowned Flycatcher
Baird's Flycatcher
Sulphur-bellied Flycatcher
solitarius
chapmani
maculatus
Streaked Flycatcher
adult
juvenile
Variegated Flycatcher

Fairly large flycatchers of open areas, edges or canopy. Most kingbirds are migratory; White-throated and Grey are rarely seen. In contrast, Tropical Kingbird is one of the commonest species in tropical areas, some even reaching the subtropics and inter-Andean valleys. Pale-looking individuals might cause confusion with other kingbirds, especially White-throated.

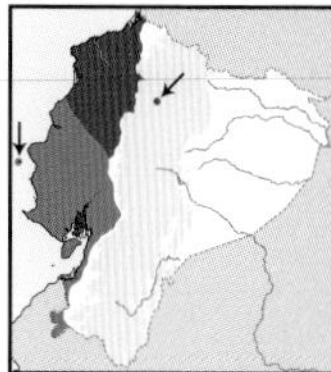

Snowy-throated Kingbird *Tyrannus niveigularis* 18–19cm

W lowlands to foothills (to 1,300m). Forest borders, woodland. ***Grey crown to mantle, clear-cut black mask. White throat, greyer chest***, yellow below. Blackish wings, ***narrow pale edges***. Monotypic. Singles atop trees or bushes, sometimes near ground, sallies and hawks. **Voice** Dry, upslurred *kí-kí-ki-kir're'reít* at dawn; dry, sharp *kít!, ki-tít!, ki-titit-kít-kit*. **SS** Some Tropical Kingbirds have whitish throat, but black mask fainter, wings plainer, more olive above, notched tail. **Status** Uncommon, migrates N seasonally (Jun–Nov).

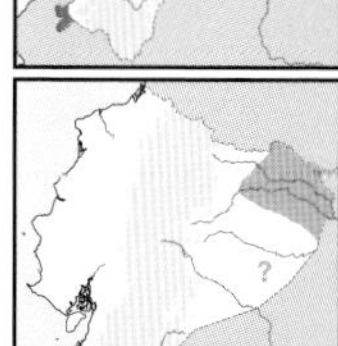

White-throated Kingbird *Tyrannus albogularis* 20–21cm

Austral migrant to E lowlands (below 300m). *Terra firme* forest borders, adjacent clearings. ***Pale grey crown to nape***, olive upperparts, ***black mask. White throat, yellow rest of underparts***, dusky wings narrowly edged pale. Monotypic. Singles perch atop bushes and trees, less conspicuous than Tropical. **Voice** apparently silent in Ecuador, elsewhere a trilled *tic-tri'i'i'i-tic-tic-tric-tri'i'i'i*, less shrill than Tropical. **SS** See widespread Tropical (duller grey head, less contrasting mask, no ***clear-cut throat***). **Status** Rare visitor (Jul–Aug).

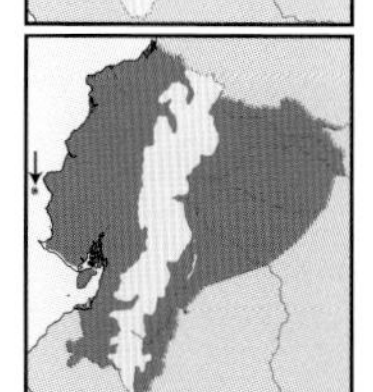

Tropical Kingbird *Tyrannus melancholicus* 21–22cm

E & W lowlands to subtropics, locally to inter-Andean valleys (to 2,500m; locally higher). Open and semi-open areas. ***Dull grey crown, faint blackish mask***, olive upperparts, ***dirty olive breast***, and yellow below, ***notched tail***. Nominate. Singles to small groups on wires, exposed branches, etc., conspicuously vocal and makes long aerial sallies; aggressive to larger birds. **Voice** Shrill *tree'ee'ee'ee'ee?* at dawn; various high-pitched chatters and twitters, *pree-ee'ee'ee'eer....* **SS** Snowy-throated and White-throated Kingbirds (white throats, more contrasting masks); see Sulphury Flycatcher. **Status** Common.

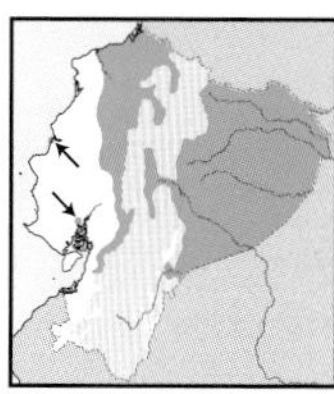

Eastern Kingbird *Tyrannus tyrannus* 20–21cm

Boreal migrant to E & NW lowlands, scattered Andean records (mainly to 800m, locally higher). Humid forest, wooded borders, clearings. ***Black cap, blackish upperparts***, white edges to wing feathers. ***Pure white below***, blackish tail ***tipped white***. Monotypic. Singles to large flocks often congregate at borders and with other kingbirds; eats much fruit. **Voice** Mostly silent, soft *tiriritic* chatter and others. **SS** Juv. Fork-tailed Flycatcher (longer tail, paler upperparts contrasting with crown); Grey Kingbird (grey above, obvious mask, dark tail tip). **Status** Uncommon transient (Sep–Nov, Mar–Apr).

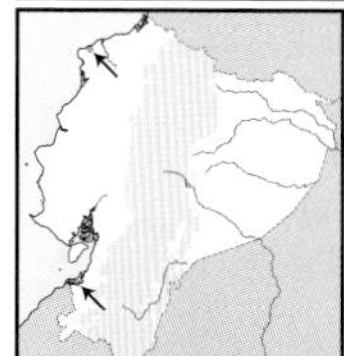

Grey Kingbird *Tyrannus dominicensis* 21–22cm

Two records from SW and NW coast (sea level). Mangroves, wetland edges. ***Grey crown to entire upperparts, blackish mask***, bold pale edges to wings. ***Greyish-white below; notched tail***. Nominate. Behaves like other kingbirds. **Voice** Apparently silent in Ecuador; elsewhere a three-noted, rolling chatter *pe-chiirr-ry* (RT). **SS** Recalls Tropical, Snowy-throated and White-throated Kingbirds but white below; Chocó Sirystes is more contrasting blackish-grey and white; Eastern Kingbird is obviously black and white, tail tipped white. **Status** Accidental (Dec-Jan, Apr).

Snowy-throated Kingbird
pale adult
White-throated Kingbird
adult
juvenile
Tropical Kingbird
juvenile
adult
Eastern Kingbird
immature
adult
Grey Kingbird

The two largest *Myiarchus* flycatchers might resemble kingbirds more than their congeners; see wing and tail patterns. Both sirystes are 'intermediate' in structure and shape between kingbirds and *Myiarchus*; the two species in Ecuador were formerly united as one. Additionally, two austral migrants.

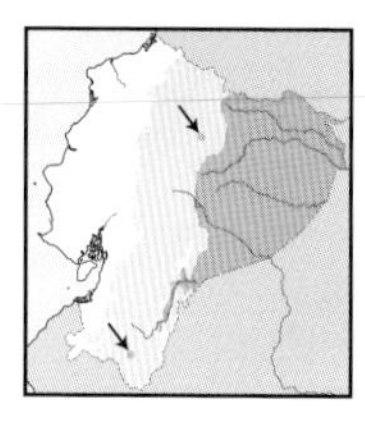

Crowned Slaty Flycatcher *Empidonomus aurantioatrocristatus* 18cm

Austral migrant to E lowlands and foothills (to 1,100m). *Terra firme*, *várzea* and *igapó* forest borders, adjacent clearings. ***Black crown***, semi-concealed yellow crown patch, vague blackish mask, ***long grey supercilium***. Brownish above, ***grey below***, duskier wings ***without markings***. Nominate. Singles atop trees, bushes and exposed branches, sally and return repeatedly in pewee-like fashion; often at fruiting trees. **Voice** Mostly silent, occasionally a thin, martin-like *pséu-pseE'eep*. **SS** Confusion unlikely, only 'stripe-headed' flycatcher that is all grey below; see sirystes, black tyrants and pewees. **Status** Uncommon austral winter resident (Mar–Sep).

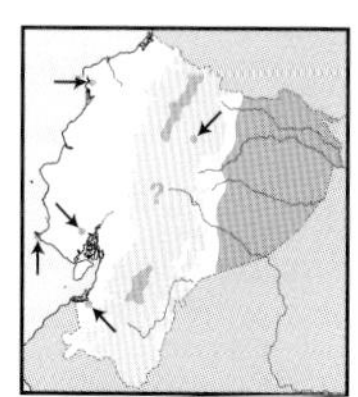

Fork-tailed Flycatcher *Tyrannus savana* 30–40cm

Austral migrant to E lowlands, scattered W & Andean records (mainly below 400m, higher in Andes). Forest borders, woodland, adjacent clearings. ***Black cap, grey upperparts, white underparts. Very long, deeply forked tail***. Nominate. Small groups (sometimes a few tens) or singles atop trees, bushes, fence posts, wires, etc.; eats much fruit. **Voice** Apparently silent in Ecuador; elsewhere a weak *tic-tic-tir'r'r'h*. **SS** Tail of juv. shorter, but pattern and length distinctive. **Status** Uncommon transient (Feb–Sep).

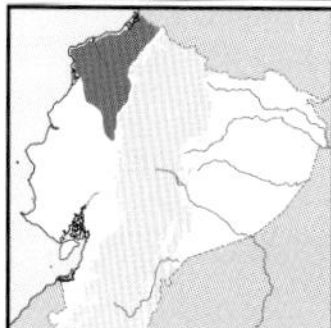

Chocó Sirystes *Sirystes albogriseus* 18–19cm

NW lowlands (to 500m, locally higher). Humid forest, borders. ***Black crown***, grey upperparts, ***whitish rump. Greyish-white underparts***, blackish wings ***conspicuously edged white***. Monotypic. Singles or pairs sally down from treetops, often with mixed flocks; leans forward when perching. **Voice** Gravelly *chep-chep-chepche-che*, often faster. **SS** Confusion unlikely; not yellow below like *Myiarchus* flycatchers; Eastern Kingbird more black and white; Grey Kingbird (dark rump, pure white below). **Status** Rare, possibly declining.

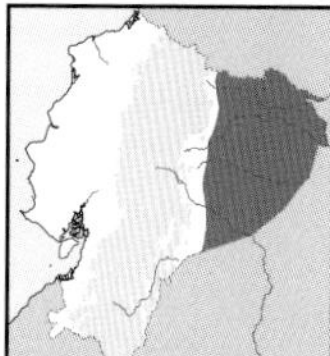

White-rumped Sirystes *Sirystes albocinereus* 18–19cm

E lowlands (to 500m). *Várzea*, *igapó*, forest borders, adjacent clearings. ***Black crown***, grey upperparts, ***whitish rump. Whitish underparts***, blackish wings only faintly ***edged white***. Monotypic. Behaviour as previous species. **Voice** Whining *güeeer-pEE'uu, güeer-peu-peu-peu-pEE'wu; güeer-pEw....* **SS** Confusion unlikely; recalls *Myiarchus* flycatchers but not yellow below; Eastern Kingbird obviously black and white; Grey Kingbird (dark rump, pure white below). **Status** Uncommon.

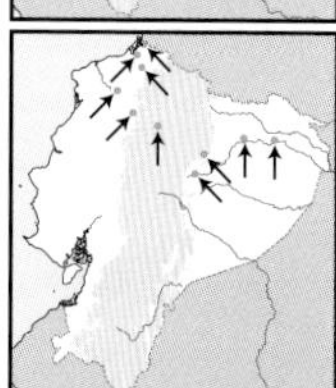

Great Crested Flycatcher *Myiarchus crinitus* 19–20cm

Boreal migrant to few sites in E lowlands, local in highlands (below 500m, once 2,800m). Forest canopy, borders. ***Fleshy mandible base, heavy bill***. Brownish-olive back, dusky-grey breast, ***large crest. Rufous edges to primaries*** and ***inner webs of tail feathers***. Monotypic. Mainly alone, sallies, plucks, hover-gleans; rather sluggish. **Voice** Loud, inflected whistle *wheeEp?* **SS** Rufous in wings and tail, and size separate from congeners, but see local Brown-crested (less contrast between grey and yellow below, browner crown). **Status** Rare or casual visitor (Feb–Mar, Nov).

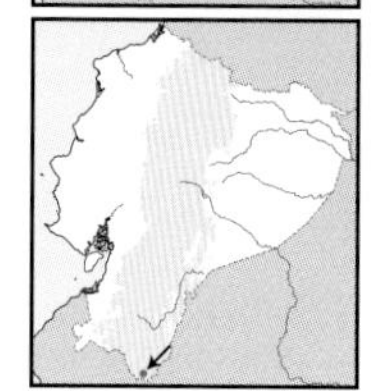

* Brown-crested Flycatcher *Myiarchus tyrannulus* 19–20cm

Extreme SE in Zumba region (600–1,000m). Semi-deciduous woodland, shrubby clearings. Recalls previous species, but ***crown browner, weak*** breast / belly contrast, browner back, ***less*** rufous in tail. Race *tyrannulus*? Singles to trios perch low, sally to foliage, air or ground. **Voice** Loud, rolling whistles *whéép-pi'rl'l* at dawn; clear *wírp!*, sometimes more chattered. **SS** Sooty-crowned and Dusky-capped (smaller, sootier crowns, black bills, no rufous in wings and tail); no known overlap with Great Crested. **Status** Locally uncommon.

adult
juvenile
ɔwned Slaty Flycatcher
adult
juvenile
immature
Fork-tailed Flycatcher
Chocó Sirystes
White-rumped Sirystes
Great Crested Flycatcher
adult
juvenile/
immature
Brown-crested Flycatcher

PLATE 213: *MYIARCHUS* FLYCATCHERS

Smaller than kingbirds, duller and crested, some forest-based. One of the trickiest tyrannid genera to identify to species, but note head to breast tones, wing and tail patterns, colour of mandible, voice and, to some extent, habitat and range. Learning Dusky-capped will enable comparison with less widespread species.

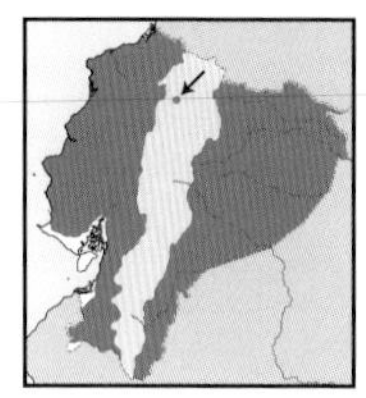

Dusky-capped Flycatcher *Myiarchus tuberculifer* 16–17cm

E & W lowlands to subtropics (to 1,500m in N, 2,500m in S). Forest borders, woodland, shrubby clearings. *Blackish-grey crown, sooty-olive upperparts.* Pale grey throat to breast, yellow belly, pale wingbars and edges. Nominate (E), *nigriceps* (W), *atriceps* (S). Singles or pairs, with or without flocks, lean forwards on exposed perches. **Voice** Whining, clear whistle *féeeeeu..fEEeu* at dawn; similar *féeeeu* during day; short, rattled *wee'eer-peeur-wee'eer-purr.* **SS** Crown always darker than congeners, but is sooty-brown in E, where cf. Swainson's (paler crown and underparts) and Short-crested (browner above); in W see Panama (paler crown and face, more olive back) and Sooty-crowned (paler-crowned, more olive back, whitish tail tips and edges). **Status** Common.

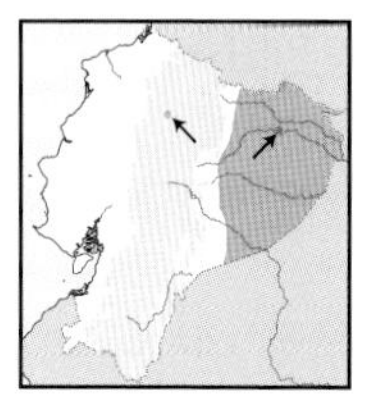

Swainson's Flycatcher *Myiarchus swainsoni* 18–19cm

Austral migrant to E lowlands (to 400m). Lake and river sides with scattered trees. *Fleshy mandible base. Brownish to greyish-olive above,* duskier wings and tail. Pale underparts, *little contrast.* Race *ferocior,* possibly *swainsoni; phaeonotus* breeds locally. Singles perch low, often with mixed flocks, lean forward on perch. **Voice** Plaintive *feeeuu,* more descending than Dusky-capped; often run into *fee'ee'ee'ee'eu* or *fEE'ee'ur'r'r.* **SS** Flesh mandible separates from sympatric Dusky-capped and Short-crested; paler overall, race *ferocior* with obvious dusky cheeks; larger Great Crested has rufous in wings and tail. **Status** Rare winter resident (Apr–Sep, local breeder).

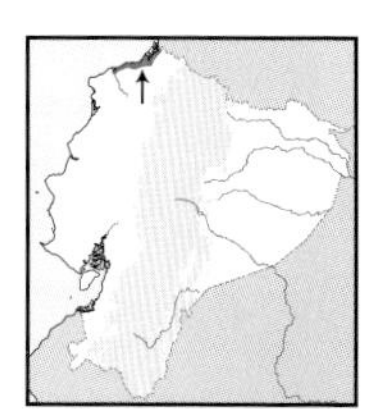

Panama Flycatcher *Myiarchus panamensis* 17–18cm

NW below 50 m. Mangroves and low scrub. Greyish-brown crown, almost ***concolorous*** with brownish-olive upperparts, ***pale grey face*** to chest; faint pale edges to tail feathers. Nominate? Pairs atop trees and scrub; typically sallies and hover-gleans. **Voice** A running, mellow series of whistles, *feu?-dee'dee'dee'dee'dee'déé,* also shorter whistles (HB). **SS** Sooty-crowned (darker crown, more olive back; faster voice); Dusky-capped (blacker crown). **Status** Possibly very local resident.

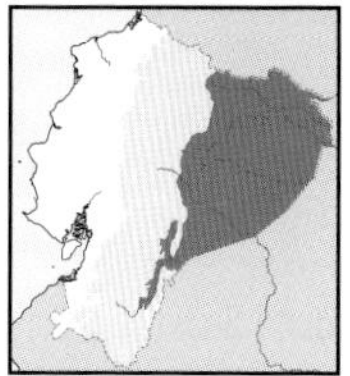

Short-crested Flycatcher *Myiarchus ferox* 17–18cm

E lowlands to foothills (to 1,000m). Forest borders, light woodland, adjacent clearings. ***Dusky-brownish crest to upperparts,*** duskier wings and tail. Grey throat to breast ***contrasting*** yellow belly. Nominate. Singles or pairs on low perches, sally to foliage where hover-gleans; perches leaning forward. **Voice** Short, sweet rolling *pui'rrrt;* shorter *qui'rrt* calls. **SS** Darker and ***browner*** than sympatric Dusky-capped, which has squarer crest, cleaner underparts; Swainson's is paler overall, fleshy mandible base distinctive, with marked mask. **Status** Fairly common.

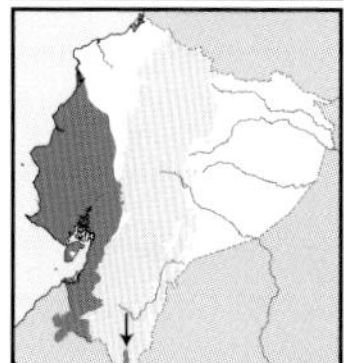

Sooty-crowned Flycatcher *Myiarchus phaeocephalus* 17–18cm

SW lowlands to foothills, along coast to S Esmeraldas, SE in Zumba region (to 1,100m). Deciduous forest, woodland, scrub. ***Sooty-grey crown, duskier rear crown,*** contrasting greyish-olive upperparts; ***pale tips and edges to outer tail feathers.*** Race *phaeonotus* (W), *interior* (SE). Singles or pairs sometimes with flocks, hover-gleans in foliage and plucks fruit. **Voice** A *fueeé, di-ri-ri-re,* second part descending; querulous, sharp, inflected *fueeee?,* shorter *freeu?;* rising-falling song in SE. **SS** Dusky-capped (blacker crown, duskier upperparts, no whitish tail tips and edges); in SE see larger Brown-crested (rufous in wings and tail). **Status** Common.

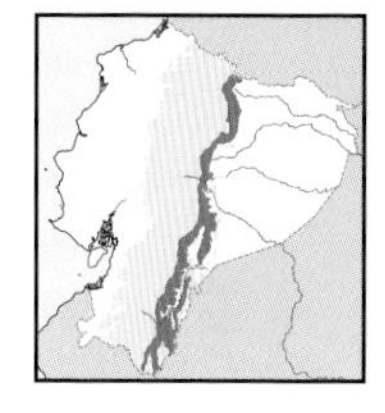

Pale-edged Flycatcher *Myiarchus cephalotes* 17–18cm

E Andean foothills to subtropics (1,000–2,200m). Montane forest borders, woodland, adjacent clearings. Overall pattern recalls congeners, especially Dusky-capped, Short-crested and Sooty-crowned, but nearly ***concolorous crown and upperparts,*** and ***outer webs to outer tail feathers obviously whitish.*** Nominate. Singles or pairs often with mixed flocks, perch low, lean forward, sally and hover-glean; shyer than Dusky-capped. **Voice** Descending, mellow *weép-peü..peep-peep-peep-peep-peep...;* loud, clear *peép!* call. **SS** Lacks Dusky-capped's contrast between crown and upperparts, tail pattern also distinctive; from Short-crested (little overlap) by tail pattern. **Status** Uncommon.

Dusky-capped Flycatcher
juvenile
nigriceps
atriceps
tuberculifer
swainsoni
ferocior
adult fresh
adult
juvenile
phaenotus
adult
Swainson's Flycatcher
adult
Panama Flycatcher
mmature
fresh
worn
juvenile
Short-crested Flycatcher
interior
phaeonotus
Sooty-crowned Flycatcher
adult
juvenile
Pale-edged Flycatcher

PLATE 214: MOURNERS AND ATTILAS

These mourners resemble other crested flycatchers but plainer coloured. Notably, they recall *Rhytipterna* mourners, now transferred to the Tityridae; also recall larger pihas (Rufous and Screaming). Attilas are also unusual tyrannids by plumage, heavy hooked bills and loud far-carrying voices. An attila no longer considered part of Ecuadorian avifauna is included.

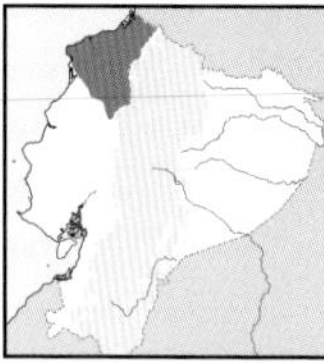

Rufous Mourner *Rhytipterna holerythra* 20–21cm

NW lowlands to foothills (to 1,150m). Wet forest interior, borders. ***Flesh mandible base, 'flat' crown.*** Mostly ***rufous***, duskier crown, rufous wing edgings. Race *rosenbergi*. Sluggish singles or pairs perch low down for long periods, suddenly sally to foliage to snatch prey; may join mixed flocks. **Voice** Mournful, whistled *whép-wheeur..whéeo*, first note up, second down; slow *wheeeip* whistle. **SS** Rufous Piha (larger, rounder crown, unmarked wings); Speckled Mourner (rounder head, obvious wingbars, pale eye-ring), voices differ; cf. becards. **Status** Rare, local, NT in Ecuador.

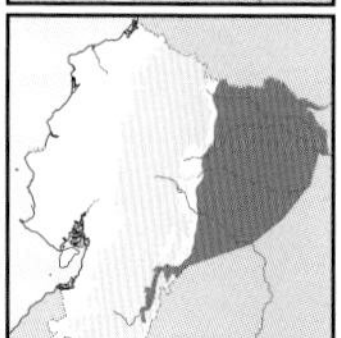

Greyish Mourner *Rhytipterna simplex* 20–21cm

E lowlands to foothills (to 1,100m). *Terra firme* and *várzea* interior. Orange-brown eyes, ***triangular crest. Grey above, paler below***, paler lores, ***belly tinged yellowish-buff***, duskier wings. Race *frederici*. Singles or pairs in subcanopy to understorey, suddenly sally or hover-glean for prey; may join mixed flocks. **Voice** Fast *t't't't't*, followed by explosive sneeze *t'ché!*; *tu-tu-tu-tu-te-te-te-tEEW-tEEW* slightly rising, ending abruptly. **SS** Recalls very vocal Screaming Piha but latter has rounded head, browner wings and tail, no yellowish on belly. **Status** Uncommon.

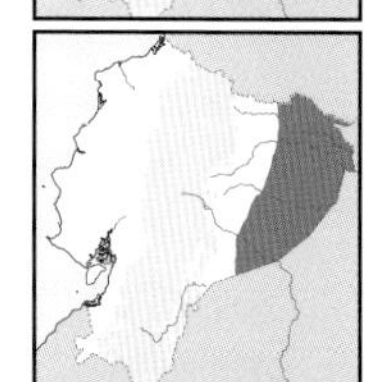

Cinnamon Attila *Attila cinnamomeus* 19–20cm

E lowlands (to 300m). *Várzea* and *igapó* canopy, borders. ***Black bill***, dark eyes. ***Mostly cinnamon-orange to cinnamon-rufous***, paler throat, belly and rump. Dusky wings, two faint cinnamon-rufous wingbars and edges to inner flight feathers. Monotypic. Singles or pairs often over water, rather calm, hops sluggishly and makes brief sallies. **Voice** Persistent, leisurely, whistled *tuoor, thue-thueeeer-her, thue-thueeer...*; ringing, plaintive *pü'éeeeuur!* **SS** Bright-rumped (contrasting yellowish rump) and Citron-bellied (grey head, pale yellowish belly and rump) do not share Cinnamon's habitat; cf. Chestnut-crowned Becard and Várzea Schiffornis (different size, shape, behaviour). **Status** Uncommon.

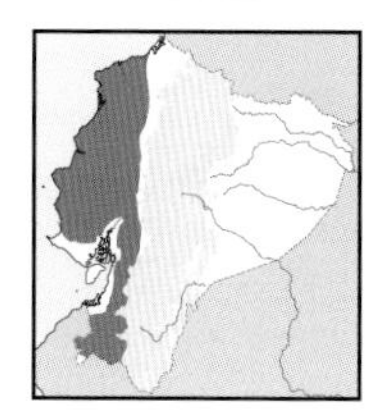

Ochraceous Attila *Attila torridus* 20–21cm

W lowlands, to subtropics in Loja (to 1,500m, locally higher). Humid to semi-deciduous forest, woodland, adjacent clearings. ***Cinnamon above, ochraceous below***, yellower throat, ***belly and rump***, rufescent tail. ***Blackish wings***, two cinnamon-rufous wingbars and flight feather edges. Monotypic. Lethargic, singles or pairs, raise tail when perched, briefly join mixed flocks. **Voice** Rising, far-carrying, cadenced *woo-wheép-wheép-wheép-wheép...*, or *wee'ü-weee'ü-wee-wee-wee*; hawk-like *wheeeiu* and emphatic *wheeík* calls. **SS** Confusion unlikely; sympatric with Bright-rumped (brighter rump, paler belly, darker upperparts); see ♀ Cinnamon, One-coloured and Slaty Becards (smaller, more rufous above, crested). **Status** Uncommon, VU.

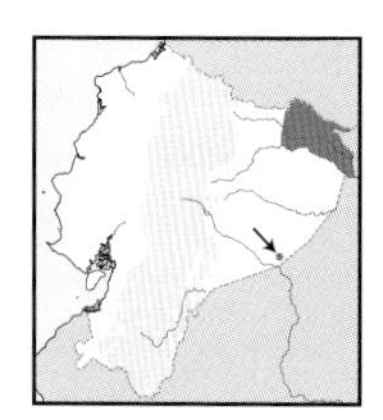

Citron-bellied Attila *Attila citriniventris* 18–19cm

E lowlands (to 300m). *Terra firme* canopy, borders. ***Grey head***, rufous-brown upperparts, ***cinnamon-yellow lower rump, ochraceous-yellow lower belly***, throat faintly streaked. Nearly ***unmarked duskier wings***. Monotypic. Singles or duos calmly forage at fruiting trees and sally for aerial prey. **Voice** Endless series of *cuu?* or *whoop?* notes ending in lower-pitched *wo*, less 'enthusiastic' than Bright-rumped, not doubled; fast *wïp-wïp-wïp...*, *wïp-wE* call. **SS** Bright-rumped can show grey head but also back, rump always brighter, yellower, throat has yellowish streaks, banded wings. **Status** Rare.

†Dull-capped Attila *Attila bolivianus* 19–21cm

Two records from *várzea* and riparian habitats in E lowlands now believed to pertain to white-eyed individuals of Bright-rumped Attila. From latter by duller crown, plainer ***dull cinnamon below, no yellow rump, no wingbars***. **Voice** Leisurely, descending *whéép-whéép-whéép... wheuu, wheeuu...* (S *et al.*). **SS** Cinnamon Attila (dark eyes, cinnamon crown). **Status** No longer part of Ecuador's avifauna.

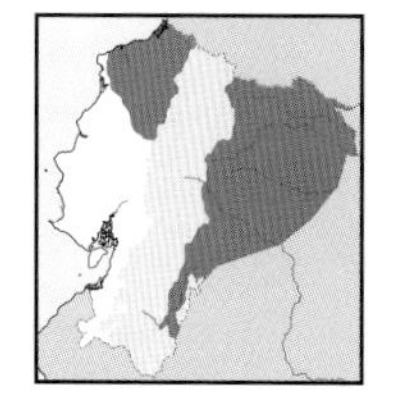

Bright-rumped Attila *Attila spadiceus* 18–19cm

E & NW lowlands to foothills (to 1,300m). *Terra firme* to humid forest, borders. ***Variable. Brownish to whitish-amber eyes***. Olive (more common?), greyish or rufescent above, ***vague pale supercilium***. Olive, grey or cinnamon-rufous throat to breast, ***streaked yellowish, paler belly; yellow rump, two wingbars***. Nominate (E), *parambae* (W). Like other attilas, lethargic, sallies or hover-gleans. **Voice** 'Enthusiastic', leisurely, doubled *woor, wue-beer..wue-béer..wue-bEEr..wue-BEER-wÉr?*; faster, rising-falling *wüü-weer'weer'wEEr'weer* and other calls. **SS** Rump yellower and wings more boldly barred than congeners, throat / breast streaking distinctive; plumage variation not well understood. **Status** Common.

Rufous Mourner
adult
juvenile
Greyish Mourner
Cinnamon Attila
Ochraceous Attila
Citron-bellied Attila
Dull-capped Attila
Bright-rumped Attila
non-rufous morph
pale-eyed morph
rufous morph
brown morph
olive morph
grey morph

PLATE 215: SHARPBILL AND FRUITEATERS

The attractive fruiteaters are plump, short-legged Andean cotingas, mostly in upper strata. Marked sexual dimorphism but even striking ♂♂ are difficult to observe because all are lethargic. Sharpbill is an enigmatic species recently reclassified in its own family. Superficially recalls fruiteaters, but bill and feeding behaviour strikingly different.

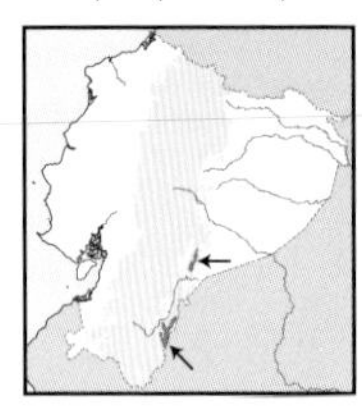

Sharpbill *Oxyruncus cristatus* 15–17cm

Few records in SE outlying Cordilleras del Cóndor and Kutukú (900–1,350m). Montane forest canopy, borders. *Sharp bill.* Head and neck *whitish boldly scaled black*, semi-concealed red coronal patch, olive upperparts, whitish yellow underparts *markedly spotted dusky.* ♀: orange coronal patch. Race *hypoglaucus?, tocantinsi?, cristatus?* or perhaps *undescribed.* Deliberate, picks fruits and searches for invertebrates in dead leaves, moss and foliage, often hanging upside-down, with mixed flocks. **Voice** Long, high-pitched, buzzy whistle ending in short trill, fading: *tdzeeeeeeu-dzz.* **SS** Confusion unlikely; cf. plumper fruiteaters and smaller Spotted Tanager. **Status** Rare, local.

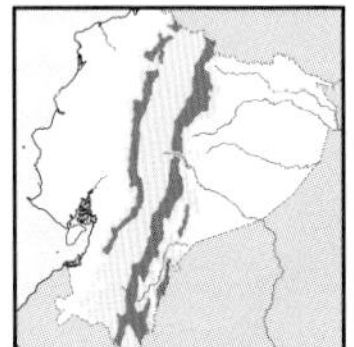

Green-and-black Fruiteater *Pipreola riefferii* 18–19cm

E & W Andean subtropics (1,700–3,200m, locally lower). Montane forest, borders. *Dark eyes. Black hood, narrow yellow collar, mottled flanks. Inner flight feathers tipped black, edged whitish.* ♀: *green breast, yellow belly spotted green.* Races *confusa* (E), *occidentalis* (W). Singles to small groups follow mixed flocks, hovers to pluck fruits. **Voice** Piercing trill *t-t-s-s-s-s-s-seee* fading at end; thin *seep* calls. **SS** Only dark-eyed fruiteater; Black-chested (plain belly, no white tertials tips), ♀ Orange-breasted (dark legs); Barred has single-noted song, no lilting. **Status** Fairly common.

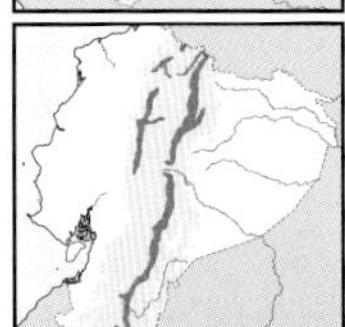

Barred Fruiteater *Pipreola arcuata* 22–23cm

Andes in E & NW (2,500–3,300m, locally lower). Montane forest, borders. *Pale grey eyes. Black hood*, olive-green upperparts, yellowish underparts *boldly barred black.* Wing-coverts and tertials with *yellowish dots* edged black, *broad black subterminal tail-band.* ♀: lacks black hood, *barred below.* Nominate. More lethargic than other fruiteaters, occasionally with mixed flocks; plucks fruit in flight. **Voice** Very high-pitched, descending *tsssssssiu*, lasting 2–3 sec. **SS** Lower-elevation Green-and black lacks ventral barring, greener above, dark eyes. **Status** Uncommon.

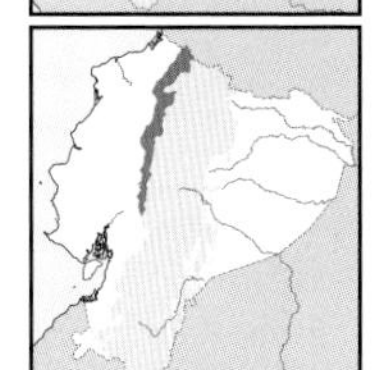

Orange-breasted Fruiteater *Pipreola jucunda* 17–18cm

W Andean foothills to subtropics, S to Chimborazo (600–1,700m). Montane forest, borders. *Yellow eyes, olive legs. Black head, orange breast* suffused yellow, yellow belly. ♀: underparts *streaked yellow and green.* Monotypic. Less often in mixed flocks, mostly in pairs calmly moving through dense vegetation, sometimes perches on exposed branches. **Voice** Thin, very high-pitched, ascending *tsssiiiií*; upslurred *sweet* calls. **SS** Green-and-black Fruiteater (black chest, dark eyes, red legs; ♀ green breast, mottled belly, red legs). **Status** Rare, NT in Ecuador.

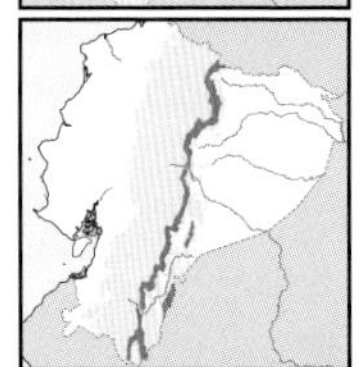

Black-chested Fruiteater *Pipreola lubomirskii* 17–18cm

E Andean subtropics (1,500–2,100m). Montane forest, borders. *Yellow eyes, olive legs. Black head to chest*, green upperparts, *yellow underparts.* ♀: black replaced by *green on head and chest, belly streaked green and yellow.* Monotypic. Retiring and stolid, mostly in pairs, often joining mixed flocks. **Voice** Piercing, ascending *psssiiíí* or *pseeeeéét*; also shorter *pseet.* **SS** From Green-and-black Fruiteater by pale eyes and olive legs (both sexes), plain yellow belly (♂), ♀ mottled below; voice more ascending than Green-and-black, longer than Scarlet-breasted. **Status** Uncommon.

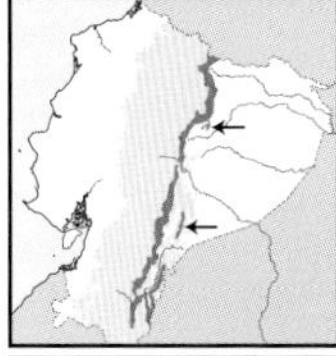

Scarlet-breasted Fruiteater *Pipreola frontalis* 15–16cm

E Andean foothills to subtropics (1,000–1,700m, locally lower). Montane forest. *Yellow eyes.* Green head and upperparts, *tertials tipped white. Fiery red lower throat*, yellow throat, breast and central belly. ♀: green above, *tertials tipped white*, yellow below *densely scaled green.* Race *squamipectus.* More arboreal than other fruiteaters, often joins mixed flocks in sluggish pairs to small groups. **Voice** Very high-pitched, short, sharp *tssiiit* or *tsít, tsít....* **SS** Larger than Fiery-throated (♂ green below, paler head; ♀ more barred than scaled below). **Status** Uncommon.

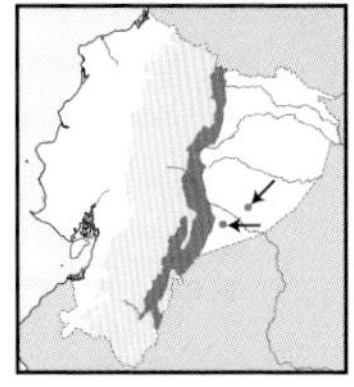

Fiery-throated Fruiteater *Pipreola chlorolepidota* 12–13cm

E Andean foothills, locally to adjacent lowlands (600–1,300m). Mid strata of forest interior. *Small, short orange bill, whitish eyes.* Mostly green, flight feathers tipped white, *fiery orange throat suffused yellow.* ♀: *flight feathers tipped whitish, heavily barred underparts.* Monotypic. Retiring and never numerous, singles or pairs follow mixed-species flocks. **Voice** Short, very high-pitched *tsiiiu*, similar to Scarlet-breasted but more descending; short *tsí* calls. **SS** Higher-elevation Scarlet-breasted (yellow below), ♀ more scaled below. **Status** Rare, possibly overlooked, NT.

Green-and-black Fruiteater
♂
♀
♂
occidentalis
confusa
Sharpbill
juvenile
♀
♂
♂
red-eyed variant (N?)
♀
immature
♀
Barred Fruiteater
♂
Black-chested
Fruiteater
♂
♀
♀
Orange-breasted
Fruiteater
♂
♀
♂
♀
Scarlet-breasted Fruiteater
Fiery-throated Fruiteater

Three Andean cotingas, highly frugivorous and sluggish, with little or no sexual dimorphism. *Ampelion* characterised by long crests held flat when relaxed but raised when excited. Chestnut-bellied is rare, recently described (1994), extremely elusive. Also, a monotypic, plump fruiteater with strikingly complex plumage and similarly sluggish. The fruitcrow and red-cotinga are remarkable species of Amazon canopy and subcanopy.

Scaled Fruiteater *Ampelioides tschudii* 19–21cm

Foothills and subtropics on both Andean slopes, local in coastal ranges (900–1,900m, 400–600m in coastal ranges). Montane forest upper strata, borders. Spectacular! *Orange eyes. Black crown, whitish lores and throat*, broad yellowish malar and narrow nuchal collar. *Blackish above scaled olive. Pale yellow below markedly scalloped olive.* ♀: *olive crown*, no nuchal collar, *bolder black scalloping below*. Monotypic. Mostly pairs, regularly with mixed flocks, hopping stolidly on branches. **Voice** Long, loud, fading *weeeeééúuuu* whistle. **SS** More striking than other fruiteaters; cf. local Shrike-like Cotinga (plain olive upperparts, dark eyes, yellower below). **Status** Rare.

Chestnut-bellied Cotinga *Doliornis remseni* 21–22cm

Local in highlands in E Andes (2,900–3,500m). Treeline, *páramo* woodland. *Blackish crown, concealed maroon crest*, unmarked *sooty*-grey upperparts to breast, *deep chestnut belly*. ♀: similar but *crown much duller*. Monotypic. Poorly known, pairs to small groups seldom seen, as apparently occurs at low densities; perches motionless atop treeline trees, primarily in *Escallonia*, regularly at landslides; rarely with mixed flocks. **Voice** Harsh, rising *krr'yé, krr'yéw* first note vibrating; harsh *yew* calls (J. Nilsson). **SS** Notably darker than locally sympatric Red-crested, lacks tail band, with darker bill. **Status** Rare, local, VU.

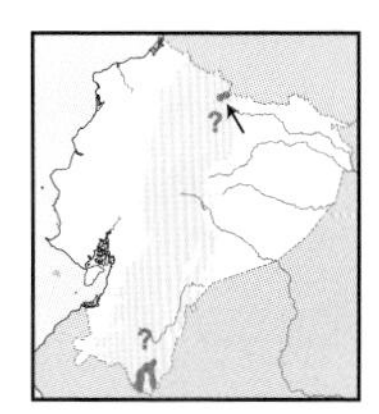

Chestnut-crested Cotinga *Ampelion rufaxilla* 20–21cm

Local in upper subtropics on E Andean slope (1,800–2,700m). Montane forest canopy, borders. *Grey bill tipped dark*. Handsome. Black crown, *long chestnut crest, rufous-chestnut cheeks, neck, nape and throat*. Olive-grey upperparts faintly streaked dusky, olive grey chest band, *yellowish belly streaked black, maroon-chestnut shoulders*. Race *antioquiae* (NE), *rufaxilla* (SE), race elsewhere unknown. Poorly known, singles or pairs perch motionless for long periods; not with mixed flocks. **Voice** Raspy, nasal *k, kre-rrreé; k, krrreé, kk*. **SS** More colourful than Red-crested (see juv. with yellowish lower belly, streaking to throat, white tail-band). **Status** Rare, local.

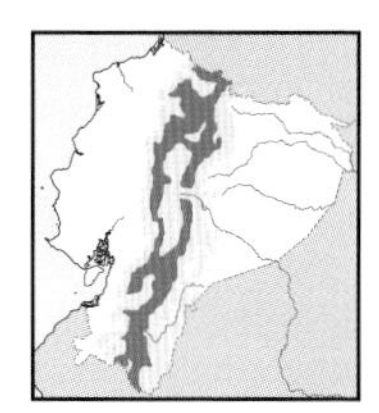

Red-crested Cotinga *Ampelion rubrocristatus* 21–23cm

Andes in both cordilleras (2,500–3,500m, locally lower). Montane forest, borders, treeline, *Polylepis* woodland, humid shrubbery. *Whitish bill tipped dark*, red eyes. Mostly *slaty-grey, white-streaked rump and vent. Long maroon crest* usually concealed. *Broad white tail-band*. Juv. profusely streaked dusky, *yellow belly*, tail as ad. Monotypic. Singles or pairs motionless atop bushes or exposed branches; fond of mistletoes. **Voice** Soft, frog-like chipping *k-k-k-k-rrrá*; loud, throaty *gur-grt*, softer *que-que-que* calls. **SS** Rarer Chestnut-bellied local in E Andes (darker, no pale bill, no white tail-band, chestnut belly); juv. somewhat recalls lower-elevation Chestnut-crested, but less yellow below, no chestnut on neck. **Status** Uncommon.

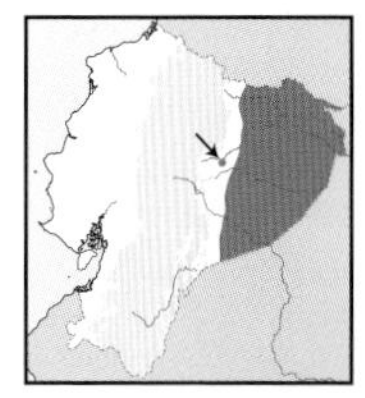

Black-necked Red-Cotinga *Phoenicircus nigricollis* 22–24cm

E lowlands (to 400m). *Terra firme* forest interior, rarely at borders. *Scarlet crown, underparts, rump and tail*, contrasting *black throat, cheeks, nape, upperparts and tail tip*, brownish-black wings. ♀: similarly patterned but *dull brown* replaces black, *no scarlet rump*, narrower tail tip. Monotypic. Elusive, mostly singles but may gather at fruiting trees; ♂ forms loose leks, producing mechanical noises with modified primaries. **Voice** Explosive, sharp *dyEk-dyEk-dyEk*... or *yOk-yOk-yOk*... at slow pace; also *ch-ch-cheE-cheE-cheEk*.... **SS** Unmistakable. Masked Crimson Tanager shares red-and-black pattern but otherwise totally dissimilar. **Status** Rare, local.

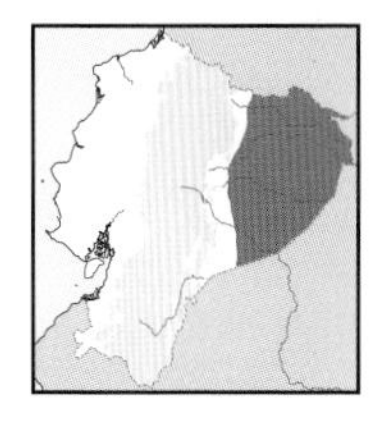

Bare-necked Fruitcrow *Gymnoderus foetidus* 32–38cm

E lowlands (to 400m). *Terra firme* and *várzea* forest, river edges, river islands. Mostly velvet black, *bluish-grey bill, neck, throat and nape wattles*, grey wing-coverts and flight feathers. ♂: *neck skin smaller*, no bluish-grey in wings. Juv. *whitish*, oddly *barred and blotched dusky*, lacks neck wattles. Monotypic. Singles to small wary groups move heavily through canopy; often fly over rivers, with deep wingbeats. **Voice** Mostly silent, but deep *ooooom* reported (HBW). **SS** Confusion unlikely; Amazonian Umbrellabird looks larger-headed in flight (voices similar); in flight can recall a woodpecker but grey in wings and / or neck distinctive. **Status** Uncommon.

♂
♀
Scaled Fruiteater
♂
♀
Chestnut-bellied Cotinga
Chestnut-crested Cotinga
antioquiae
rufaxilla
adult
juvenile
Red-crested Cotinga
ndertail
♀
♂
Black-necked Red-Cotinga
♂
juvenile
♀
Bare-necked Fruitcrow

PLATE 217: LARGE COTINGIDS

An assortment of large to very large cotingas, some spectacularly adorned, others brightly coloured. Diets predominantly fruit-based, but invertebrates and even small vertebrates are taken, especially when rearing young. Lekking behaviour notable in Andean Cock-of-the-rock and both umbrellabirds. Of the fruitcrows, Red-ruffed is very rare in Ecuador.

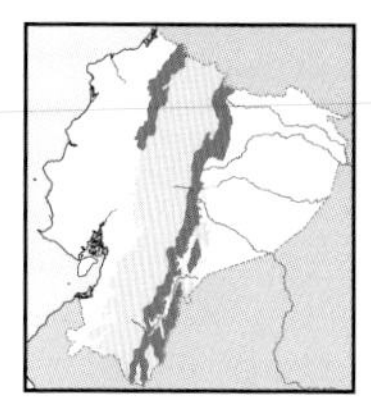

Andean Cock-of-the-rock *Rupicola peruvianus* 30–32cm

E & NW Andean foothills to subtropics (900–2,100m, locally higher and lower). Forested ravines, gorges, montane forest canopy, borders. Striking: ***orange*** (E) or ***red*** (NW), black wings and tail, ***broad silvery-grey square patches in wings***. Yellow bill concealed by ***large bushy-combed crest***. ♀: ***chocolate-brown*** (more orange in E), shorter crest. Nominate (E), *sanguinolentus* (NW). Shy, mostly solitary when not lekking. Leks spectacular: ♂♂ dance, jump, bow, flap wings, while vocalising. **Voice** Loud, pig-like grunting *yeen, yeaank, yeeeenk...* plus other chuckles and squeaks at leks; foraging birds utter loud *kwaank?* **SS** Unmistakable. **Status** Uncommon.

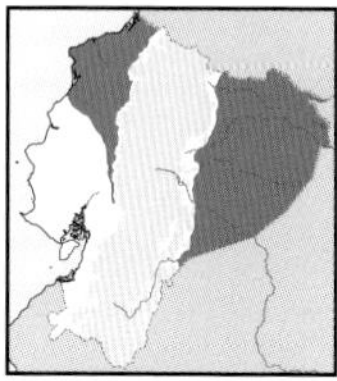

Purple-throated Fruitcrow *Querula purpurata* 25–28cm

E & NW lowlands (to 500m, locally higher). Humid forest, borders. Black with contrasting ***deep red-purple throat***. ♀ and juv. lack purple throat. Monotypic. Small noisy groups with few ♂♂, active in subcanopy, sometimes with other large passerines; bounding flight. **Voice** Loud *hówa-hówa-hówa, huoo-huoo, huoo-oo, hówa-kuoo...*; harsh *gwak-gwok-gooak* calls. **SS** ♀ with heavier ♀ umbrellabirds (stouter bill, larger head, tail proportionately shorter). Black caciques more slender, longer-billed, etc. **Status** Uncommon.

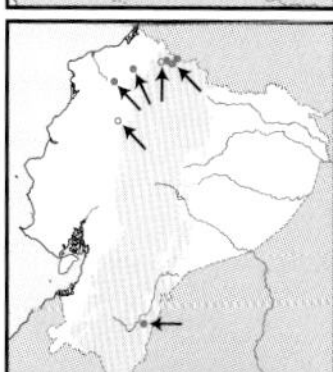

Red-ruffed Fruitcrow *Pyroderus scutatus* 36–41cm

Very local in NW & SE Andean foothills to subtropics (600–1,800m). Montane forest canopy, possibly borders. Heavy grey bill. Black upperparts, large ***flame orange throat and chest patch***, reddish-brown below (blacker in SE?), flanks and lower belly black. ♀: smaller. Races *occidentalis* (NW), *masoni*? (SE). Poorly known in Ecuador; elsewhere, mostly solitary at various heights, tame; ♂♂ gather in leks where each holds display territory. **Voice** Very deep booming *ooom-ooom* at leks; otherwise rather silent. **SS** Unmistakable; cf. umbrellabirds' deep booming calls. **Status** Rare, very local, EN in Ecuador.

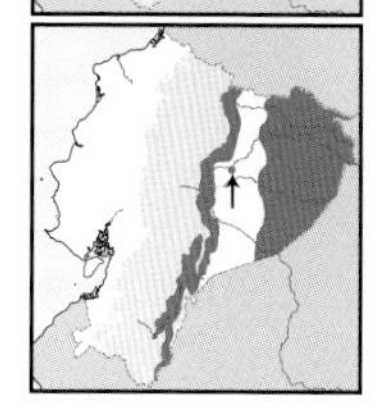

Amazonian Umbrellabird *Cephalopterus ornatus* 42–49cm

E lowlands and foothills (below 300m and at 900–1,400m). Canopy of *várzea* and riparian forest, river islands (lowlands); montane forest canopy, borders (foothills). Heavy black bill, ***white eyes. Glossy black***. Large ***'Elvis Presley' hairdo, white shafts in frontal feathers, long glossy wattle***. ♀: smaller, ***small wattle and crest, no*** white shafts. Monotypic. Singles to small groups forage in upper strata, moving heavily, often seen in undulating flight above rivers; ♂♂ gather at leks, where they lean forward, erect crest, inflate wattle and boom. **Voice** Very deep *hooooom*. **SS** Unmistakable, but cf. smaller Purple-throated (smaller head, weaker bill, more sociable). **Status** Rare to uncommon.

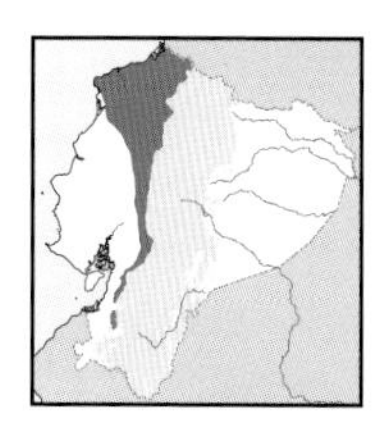

Long-wattled Umbrellabird *Cephalopterus penduliger* 36–42cm

NW lowlands, S in Andean foothills to El Oro (200–1,100m, local to 1,600m). Humid forest and borders. Heavy blackish bill, dark eyes. ***Glossy black***. Large glossy ***'umbrella'*** curving downwards in front, ***very long glossy wattle***. ♀: smaller, smaller bushy crest, much shorter wattle. Monotypic. Singles to small groups forage in upper forest strata, hop heavily between branches, undulating flight. Long-occupied leks where ♂♂ lean forwards, inflate wattle, erect crest and boom. **Voice** Low-pitched, deep *booooo* or *wooooo* in leks. **SS** Unmistakable; cf. ♀ with rarer Red-ruffed Fruitcrow (utters two consecutive booms). **Status** Rare to uncommon, VU.

♂
sanguinolentus
♂
peruvianus
♀
♀
juvenile
Andean Cock-of-the-rock
♂
♀
♂
♂
juvenile
Purple-throated Fruitcrow
masoni
occidentalis
Red-ruffed Fruitcrow
♀
♀
♂
♂
Amazonian Umbrellabird
Long-wattled Umbrellabird

Brightly coloured cotingas including the duller but elegant Purple-throated in Amazonia. *Cotinga*, *Xipholena* and *Carpodectes* are canopy dwellers, quite inconspicuous despite striking colours. All are plump, silhouettes like stocky doves. ♀♀ much duller. Loose leks, though not as noisy and spectacular as in other cotingas.

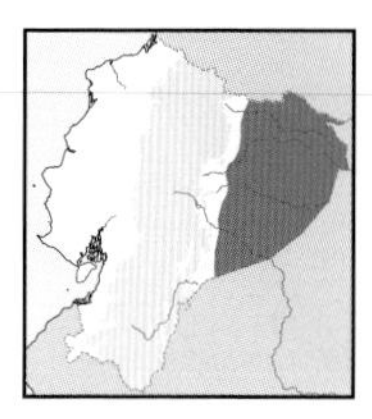

Purple-throated Cotinga *Porphyrolaema porphyrolaema* 17–19cm

E lowlands (to 400m). Upper to mid strata of *terra firme* and *várzea* forest, borders, adjacent clearings. *Black upperparts*, faint white scaling on mantle to rump, *more solid white in scapulars*, black wings with some white fringes. *Purple throat* grading into upper chest, *pure white below*. ♀: mostly dull brown *heavily barred buff*, more dotted crown, more scaled upperparts, *throat and vent plain cinnamon-buff*. Monotypic. Pairs to small groups perch atop trees, plucking fruit like *Ficus* and *Cecropia*. **Voice** Long, descending, whining *weeeeeuud*, reminiscent of some flycatchers but longer. **SS** ♀♀ of genus *Cotinga* are not barred below, lack contrasting plain throat. **Status** Rare.

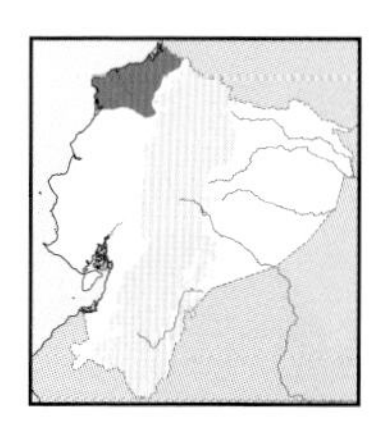

Blue Cotinga *Cotinga nattererii* 18–19cm

NW lowlands (to 300m, locally higher?). Humid forest canopy, borders, treed clearings. Handsome, mostly *shining deep turquoise-blue, deep violet throat and central belly*. ♀: mostly *dusky-brown heavily speckled buff above, mottled dusky below*. Monotypic. Singles or small groups at canopy level, often perch motionless for protracted periods atop trees or on exposed branches. **Voice** No vocal sounds reported, but wing whir in flight display. **SS** ♂ unique in range; cf. brown ♀ with grey ♀ White-tipped Cotinga (orange eyes, no speckling, white wing fringes). **Status** Rare, rather local, VU in Ecuador.

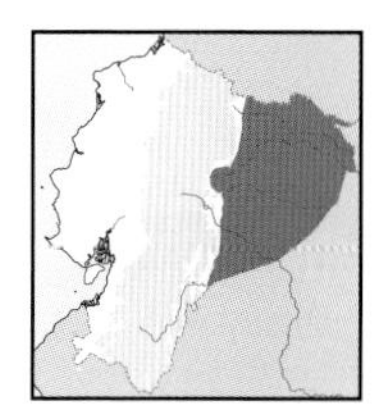

Plum-throated Cotinga *Cotinga maynana* 19–20cm

E lowlands (to 700m). *Terra firme* canopy, borders, *várzea* borders, adjacent clearings. *Yellow eyes*. Mostly *shining turquoise-blue, deep violet throat patch*. ♀: *greyish-brown* thickly scaled buff above, *whitish-buff throat*, dusky-mottled breast, *more ochraceous-buff belly to vent, cinnamon-buff underwing-coverts*. Monotypic. Singles or pairs atop trees and exposed branches, sometimes with Spangled Cotinga; ♂♂ display alone in canopy. **Voice** Very soft *hom* or *mmm*, ascending *te-deeEEA!* calls. **SS** ♂ plain blue and pale-eyed (cf. Spangled); ♀ closely recalls Spangled but less dusky, more ochraceous plain belly and cinnamon underwing-coverts. **Status** Uncommon.

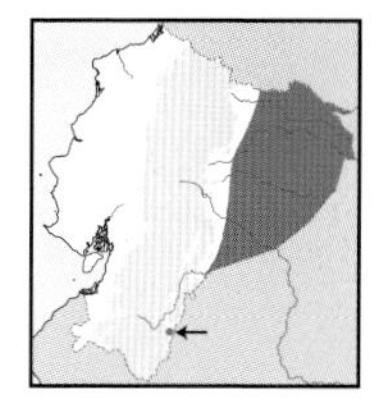

Spangled Cotinga *Cotinga cayana* 19–20cm

E lowlands (to 400m, locally higher). *Terra firme* forest canopy, river borders, adjacent clearings. *Dark eyes*. Mostly *pale shining turquoise-blue spangled black* on upperparts, *violet throat patch*, black flight feathers and tail. ♀: *dark brown heavily scaled whitish*, paler underparts *faintly scaled* dusky to belly, *pale buff underwing-coverts*. Monotypic. Singles to small groups, occasionally with Plum-throated, regularly perching quietly; ♂ displays black spangles while vocalising. **Voice** Gives 2–3 soft, mournful *hoo* notes similar to Plum-throated but hollower, shorter. **SS** Paler turquoise than ♂ Plum-throated; ♀ darker, less ochraceous belly; see ♀ Purple-throated. **Status** Uncommon.

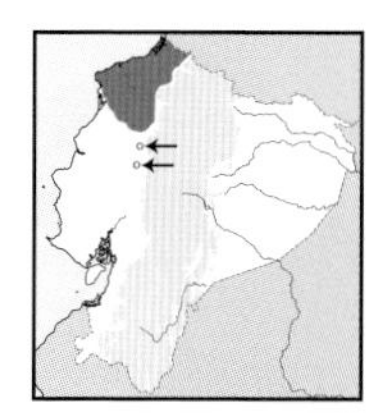

Black-tipped Cotinga *Carpodectes hopkei* 23–25cm

NW lowlands to lower foothills (to 500m, locally to 700m). Humid forest canopy, borders. Unique ♂ is *immaculate white*, with small black dots on outer flight feather tips and central rectrices, *orange-yellow eyes*. ♀: *ash-grey*, paler below, blacker wings with contrasting *white fringes to coverts and inner flight feathers, orange-yellow eyes*. Monotypic. Lethargic ♂ perches atop snags and treetops for long periods, often in groups. **Voice** Squeaky *quip* reported (KN). **SS** ♀ with darker and smaller, more slender White-winged and Black-and-white Becards (dark eyes, not uniform ash-grey, different tail pattern, different behaviour). **Status** Uncommon, VU in Ecuador.

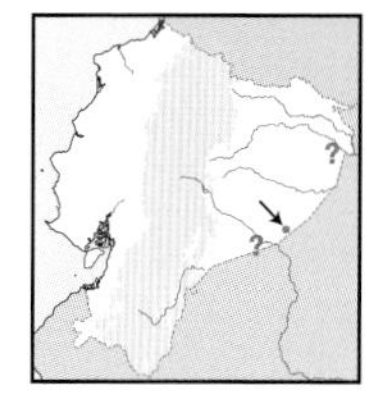

Pompadour Cotinga *Xipholena punicea* 19–20cm

Very local in SE lowlands, at 250 m. *Terra firme* forest canopy. Striking. *White eyes. Bright vinaceous-red, pure white flight feathers*, elongated scapulars. ♀: *uniform grey*, whiter belly, darker wings with *white edges, pale eyes*. Monotypic. Unknown in Ecuador; elsewhere might engage in long-distance dispersal following fruiting events, singles to small groups; displaying ♂♂ displace each other from perches. **Voice** Short *kórp!* barks. **SS** ♀ with smaller, less plump mourners (no white wing fringes, dark eyes) and becards (not uniform ash-grey, dark eyes, no rounded head). **Status** Uncertain, known from single specimen collected in 1960s and few uncorroborated sightings.

♂
♀
Purple-throated Cotinga
♂
♀
Blue Cotinga
♂
♀
Plum-throated Cotinga
♂
♀
juvenile
Spangled Cotinga
♀
♂
Black-tipped Cotinga
♂
♀
juvenile
Pompadour Cotinga

Inconspicuous cotingas of forested habitats. Sluggish and difficult to observe, but most having very loud voices audible from long distances. *Snowornis*, of Andean distribution, was recently split from *Lipaugus*; both species are less vocal, Olivaceous even appears to be silent.

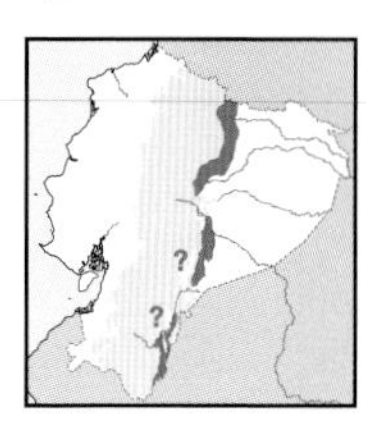

Grey-tailed Piha *Snowornis subalaris* 23–24cm

Foothills and subtropics on E Andean slope (500–1,400m). Mid strata of humid montane forest. Narrow ***pale grey eye-ring***. Mostly olive, black mid crown, faint yellowish stripes on throat and breast, ***grey lower belly, rump and tail***, yellow underwing. Black coronal stripe nearly absent in ♀. Monotypic. Rather lethargic, more active when briefly joining mixed flocks, mostly solitary or pairs; forages by hover-gleaning and sallying. **Voice** Loud, drawn-out, rising, ringing *cheeér-eEIN* at long intervals, also shorter *chueen?* **SS** Overlaps with Olivaceous Piha in E foothills; Olivaceous more silent, tail dusky-olive, yellower below, yellow eye-ring. **Status** Rare, local, NT.

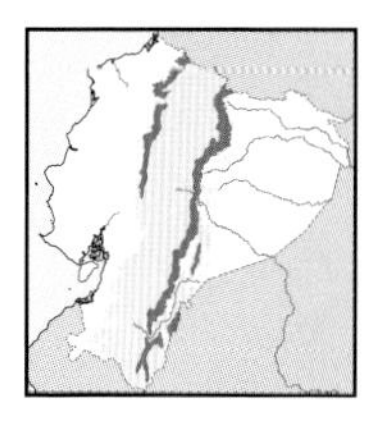

Olivaceous Piha *Snowornis cryptolophus* 23–25cm

Subtropics on E & NW Andean slopes (1,000–1,800m, locally to 2,000m). Mid strata of montane forest and borders. Narrow ***yellow*** eye-ring. Mostly olive, black coronal stripe, ***yellow belly***, yellow underwing-coverts. No coronal stripe in ♀. Nominate (E), *mindoensis* (NW). Rather lethargic and inconspicuous, mostly singles or pairs with mixed flocks, sallying and hover-gleaning. **Voice** Harsh, metallic, ascending rattle *t'trrrrr* lasting <1 sec., or shorter, slower *tk-k'k'k'k* (S. Olmstead & N. Athanas). **SS** In E see Grey-tailed Piha (grey rear parts, duller underparts, grey eye-ring); cf. some montane tanagers and bush-tanagers (Carmiol's, Ochre-breasted, Dusky); behaviour and overall shape differ notably. **Status** Uncommon, local.

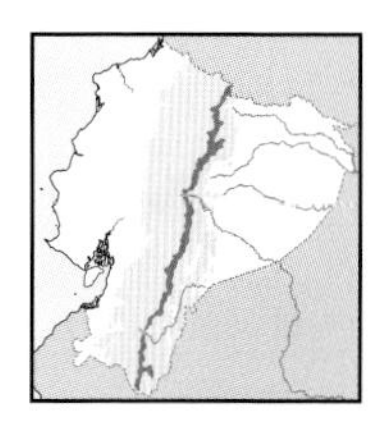

Dusky Piha *Lipaugus fuscocinereus* 32–33cm

Subtropics to temperate zone in E Andes (1,700–2,600m). Upper strata of montane forest, woodland and borders. ***Large***, long-tailed. Mostly ***mouse-grey***, duskier wings and tail, paler below, ***olivaceous-grey belly***. Sexes alike. Monotypic. Rather lethargic but may join mixed flocks, especially of large fruit-eating passerines; ♂♂ apparently gather at loose leks, but poorly known; wing whir in display. **Voice** Far-carrying, rising-falling, penetrating *dwéeeo-wuuEEY*, sometimes just one part, slightly fading near end but rising again. **SS** Confusion unlikely; only sympatric piha is Olivaceous; Great Thrush has notably different silhouette, behaviour, etc. **Status** Rare, rather local, possibly overlooked.

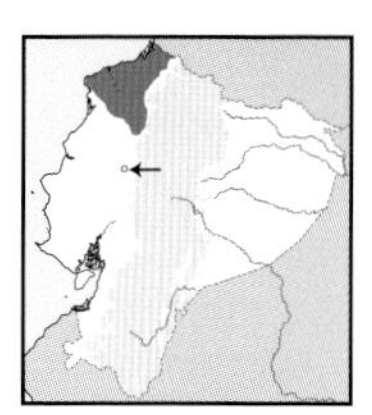

Rufous Piha *Lipaugus unirufus* 23–24cm

NW lowlands to lower foothills (to 700m). Mid strata of very humid forest. Mostly ***cinnamon-rufous***, brighter below, ***whitish throat***. Sexes alike. Race *castaneotinctus*. Rather lethargic, solitary or in small loose flocks, rarely joins mixed flocks; loose leks; sudden sallies after prey or to pluck fruit. **Voice** Short, explosive whistle *DYÉEEU* or ch-*WHYEEÜ*; also fading *chee-dudududu*, single *peer* whistles and metallic *dirgg*. **SS** Speckled and Rufous Mourners (smaller, with wingbars). Speckled more cinnamon below, mottled dusky; Rufous more slender, longer-tailed, slimmer bill, duller overall, has flatter and duskier crown; softer whistled voice. **Status** Uncommon but local, VU in Ecuador.

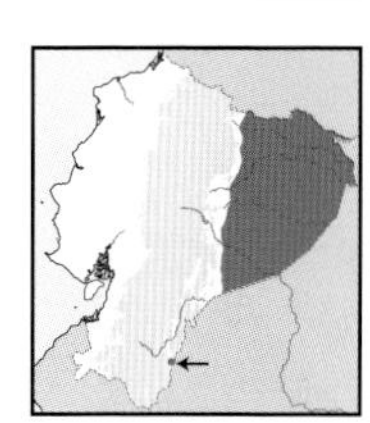

Screaming Piha *Lipaugus vociferans* 24–26cm

E lowlands (to 500m, locally higher). Mid strata of *terra firme* forest, locally in *várzea*. ***Grey above***, duskier wings and tail. ***Pale grey below***, lighter belly. ***Bright orange gape***. Sexes alike. Juv. tinged brown on wings. Monotypic. Difficult to see as perches silently. ♂♂ gather in loose flocks, otherwise lethargic. **Voice** One of most familiar songs in Amazonia, short mewing whistles *woA-woA*, followed by loud, powerful, explosive *SQWEÉÉ-QUEEE'Á!*, sometimes omits the mewing; querulous *kweeeuu?*, repeated *weeé?-yuu, weeé?-yuu…* calls. **SS** Smaller, slimmer Greyish Mourner (flatter head, paler belly, weaker bill, reddish eyes, proportionately longer tail; voices very different). **Status** Common.

♀
♂
Grey-tailed Piha
♀
♂
Olivaceous Piha
Rufous Piha
Dusky Piha
adult
juvenile
Screaming Piha

A mix of four lovely manakins, another tiny, drab and inconspicuous (Dwarf), and the distinctive *Piprites* of uncertain taxonomic affinities. Golden-winged, Club-winged, and White-bearded, all notably dimorphic, have remarkable display behaviours and mechanical sounds.

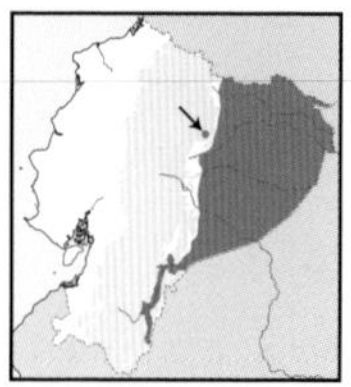

Dwarf Tyrant-Manakin *Tyranneutes stolzmanni* 7–9cm

E lowlands to foothills (to 900–1,000m). *Terra firme* and *várzea* forest undergrowth. *Tiny.* Bill proportionately 'strong', ***pale eyes***, grey legs. Very ***drab***. Mostly olive-green above, duskier wings and tail. Whitish throat faintly smudged olive, olive breast to flanks, ***yellowish mid belly***, short tail. Sexes alike. Monotypic. Vocalises from same branch for long periods, but very difficult to see; does not gather at leks, occasionally with mixed flocks. **Voice** Metallic, sneezing *tchu-eet*; also louder *duít-duít...duít-duít-duít...* calls. **SS** Confusion unlikely by size and voice, see Short-tailed Pygmy-tyrant; cf. softer, more metallic voice of Blue-crowned Manakin. **Status** Uncommon.

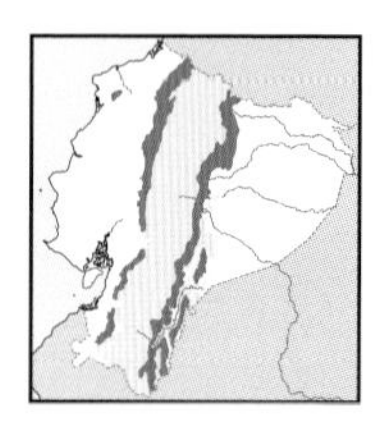

Golden-winged Manakin *Masius chrysopterus* 10–11cm

Foothills to subtropics in E & W Andes (800–2,000m, locally lower). Montane forest undergrowth, mature woodland. Flesh bill, ***orange-flesh legs***. ***Golden-yellow crown and throat, flame orange to reddish nape, bright yellow in wings and tail.*** Otherwise ***velvet black***, two black lateral 'horns'. ♀: mostly olive, ***yellowish throat and mid belly***. Races *coronatulus* (W), *pax* (most of E), *peruvianus* (extreme SE). Briefly with flocks, takes many berries; leks include metallic sounds and displays yellow ornaments. **Voice** Metallic grunt *nh'ar!*, also *tsseeeuuu-tssi-ttk....* **SS** ♀ Club-winged has whitish throat, buffy face, darker bill and legs; cf. also larger Green Manakin. **Status** Uncommon.

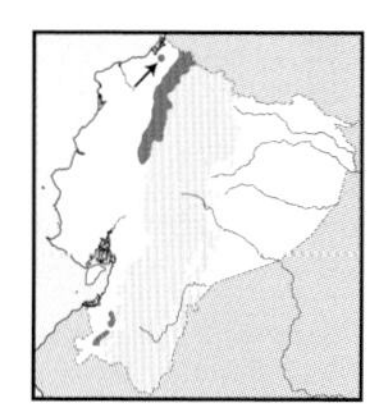

Club-winged Manakin *Machaeropterus deliciosus* 9–10cm

Foothills to subtropics in W Andes (600–1,600m, locally lower). Montane forest undergrowth, mature woodland, borders. Mostly ***chestnut, red forecrown***. Black wings, ***club-shaped inner flight feathers broadly edged white***, yellow wing bend, ***white underwing-coverts***. ♀: mostly olive, ***greyish-white throat, buff-tinged face***, yellowish belly, ***whitish inner secondary webs***, yellow wing bend. Monotypic. Displays include metallic sound produced by clubbed wing feathers and dances with wings open and held upwards; fond of melastome berries, briefly joins mixed flocks. **Voice** Electronic, ringing *tip-chnnk, tip-tip-chnnk* produced by vibrating wings; also high-pitched *beew* notes. **SS** Confusion unlikely; see ♀-plumaged Golden-winged. **Status** Uncommon.

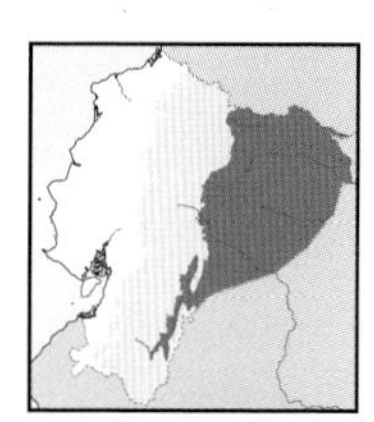

Striped Manakin *Machaeropterus regulus* 8–9cm

E lowlands to foothills (to 1,100m). *Terra firme* forest undergrowth, mature woodland. ***Tiny. Bright scarlet crown***, bright olive face to upperparts, white edges to inner flight feathers. White throat, ***yellowish-white below heavily streaked red.*** ♀: mostly olive above, ***yellowish underparts with faint reddish streaks on breast to belly***, white mid belly. Race *striolatus*, might deserve species status. Loose leks, lone ♂ displays within earshot of next; utters soft whistles, shows breast and gives metallic sounds using wings; otherwise seldom seen. **Voice** High *woe-cheéw* at long intervals; other buzzing and sneezing sounds at leks. **SS** Confusion unlikely. **Status** Uncommon.

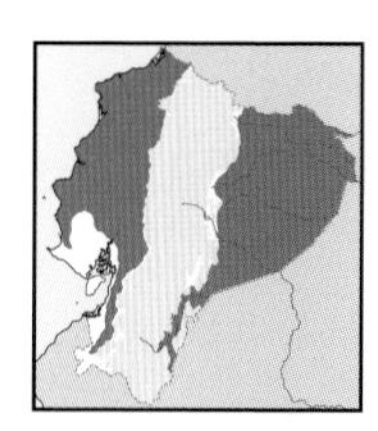

White-bearded Manakin *Manacus manacus* 10–11cm

E & W lowlands to foothills (to 800m). Second growth, borders. Handsome ***black-and-white*** ♂ has ***bright orange-red legs***, puffy white beard. ♀: mainly olive, yellowish mid belly, ***bright*** legs as ♂. Races *interior* (E), *bangsi* (NW), *leucochlamys* (CW), *maximus* (SW). Conspicuous leks, varied fire-cracking mechanical sounds and dances; snatches prey or fruit by sallying, may follow army ants. **Voice** Loud *peeurr... péeurr...* followed by 'twig-breaking' *trchk!*; inflected *chiurp?* **SS** In E larger ♀ Blue-backed (paler legs, tail proportionately shorter) and Yellow-crowned Manakins (pale bill, paler legs); in W Red-capped (duller overall, paler legs). **Status** Common.

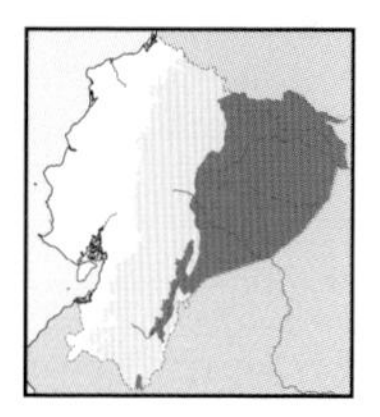

Wing-barred Piprites *Piprites chloris* 12–13cm

E lowlands to foothills (to 1,100m, locally higher). Mid strata of *terra firme* forest, woodland, borders. Blackish bill, ***large eyes, pink legs***. Olive above, ***yellow lores and orbital ring***, grey nape, ***yellow underparts***, greyer breast-sides. Dusky wings, ***two yellow wingbars, inner flight feathers broadly edged yellow***. Dusky-olive tail, pale tips. Sexes alike. Race *tschudii*. Rather lively singles or pairs in upper strata, often with mixed flocks, gleaning insects while perched. **Voice** Loud, cadenced, hesitant *whep, peep-peep, dee-deep, whep-whep?, whep-whep-whep-whe-whe*. **SS** Confusion unlikely despite resemblance to tyrannids or even vireos (cf. Yellow-throated). **Status** Rare.

♂
Golden-winged Manakin
Dwarf Tyrant-Manakin
peruvianus
coronulatus
peruvianus
♂
♂
immature
pax
♂
immature
♀
♂
♂
immature
Club-winged Manakin
♀
Striped Manakin
♀
leucochlamys
maximus
interior
♂
♂
♂
bangsi
♀
♂
♀
Wing-barred Piprites
White-bearded Manakin

Manakins with very marked sexual dimorphism; females – and young ♂♂ – much duller than ad. ♂♂. ♂ Blue-crowned and Blue-rumped have less conspicuous display behaviours at dispersed leks, while ♂ *Pipra* and *Ceratopipra* perform spectacular dances at more compact leks. Imm. ♂♂ also dance at lek peripheries. ♀ rears young alone. Typically understorey species.

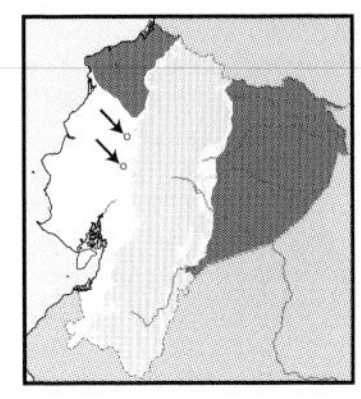

Blue-crowned Manakin *Lepidothrix coronata* 8–9cm

E & NW lowlands (to 900m, lower in W). *Terra firme* forest, mature woodland. ***Tiny. Dark*** bill and ***legs***. Mostly black, ***shining blue crown*** (deeper in W). ♀: ***grass-green upperparts and breast, yellowish belly***. Nominate (E), *minuscula* (W). ♂ displays in dispersed leks, performing simple dances; plucks berries and sallies to foliage. **Voice** Repeated, high-pitched *tsee-purrrí…tsee-purrrí…*, less enunciated than Dwarf Tyrant-Manakin; also high-pitched trill *treerrrr* and hesitant *sweEE?* **SS** ♀ brighter than other manakins, with reddish eyes; cf. Blue-rumped Manakin (foothills) with yellower crown, brighter green rump. **Status** Common.

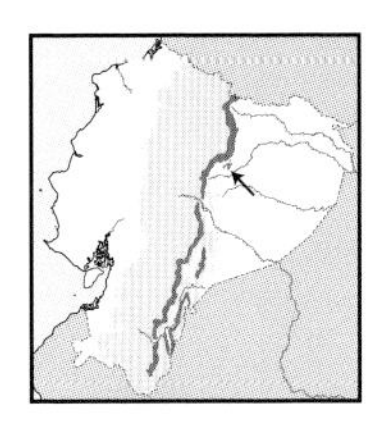

Blue-rumped Manakin *Lepidothrix isidorei* 7–8cm

E Andean foothills to subtropics (1,000–1,700m, locally lower). Montane forest, mature woodland. *Tiny*. Dark bill and legs, reddish eyes. ***Bright snowy-white crown***, velvet black overall, ***pale blue rump***. ♀: recalls Blue-crowned Manakin, but ***yellower cast to crown, brighter green rump***, and yellower belly. Nominate. ♂ displays in loose leks, apparently less dispersed than Blue-crowned; dances include back-and-forth flights. **Voice** A *dooit-dooit-doo-eet-doo-eet…*; other calls during display, including softer *shew*. **SS** ♀ smaller, greener than sympatric Golden-winged and White-crowned Manakins (bulkier, greyer crown, duller below); cf. Blue-crowned (more uniform green below). **Status** Uncommon; NT globally.

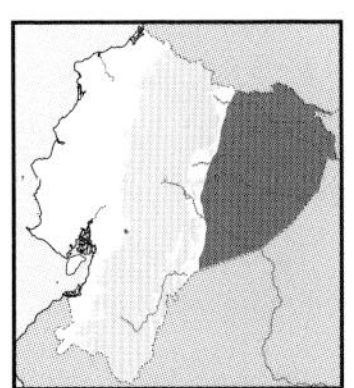

Wire-tailed Manakin *Pipra filicauda* 11–12cm

E lowlands (to 500m, locally higher). *Terra firme* and *várzea* forest. ***Dark bill and legs, white eyes***. Fantastic ♂ ***scarlet, black and yellow***, with ***long tail wires***. ♀: mostly olive, ***yellower***-green throat and belly, ***tail with shorter wires, white eyes***. Nominate. Mechanical sounds and complex displays at dispersed leks, often two birds dancing together. **Voice** Nasal, descending *neeeuw* or *nheeeeaa*; also upslurred *sweee?* longer than Blue-crowned Manakin; short *deeo* and others. **SS** White eyes and tail wires diagnostic, but cf. ♀ White-crowned and Golden-headed. **Status** Fairly common.

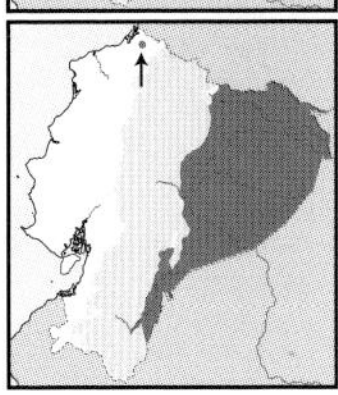

White-crowned Manakin *Dixiphia pipra* 9–10cm

E foothills to lowlands; once in NW (500–1,500m, locally to 250 m, 1,800m). Montane to *terra firme* forest. ***Red eyes***. Velvet black with ***white crown***. ♀: dusky-olive, ***bluish-grey crown***, yellowish to greyish below, paler throat and belly. Races *coracina* (Andean slopes), *occulta* (extreme SE), *discolor* (E lowlands), *minima?* (NW). ♂♂ gather at small dispersed leks and collective display sites; largely frugivorous, often in dense undergrowth. **Voice** Buzzy, descending *kk-cheeéw*, metallic, at rather long intervals; short *péew* calls; in Andes a hoarser *pgee..cheEEo!*; in SE simpler, ascending *chirr'ríp!* **SS** Crown distinctive in ♀; see Golden-headed and Blue-backed Manakins (pale legs), Green (greener, yellow belly, dark eyes), White-bearded (red legs). **Status** Uncommon.

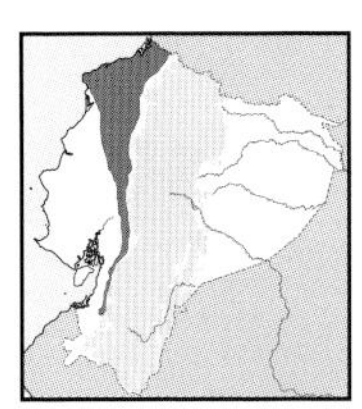

Red-capped Manakin *Ceratopipra mentalis* 10–11cm

NW lowlands, S in Andean foothills to E Guayas (below 500m). Humid forest, mature woodland. ***White eyes***. Black with ***bright red hood, yellow thighs***, yellow underwing-coverts. ♀: dark eyes, ***flesh legs***. Mostly olive, yellowish throat and belly. Race *minor*. Rather dispersed leks where ♂ dances, moving up and down thin branches; largely frugivorous. **Voice** High-pitched *tzip-tzip…tzweeeuu-tzip*; also sharp, short *psit-psit-psit…*; wing snaps like a typewriter. **SS** ♀'s flesh legs distinctive; duller than sympatric Green Manakin; duskier than Blue-crowned; White-bearded has redder legs. **Status** Uncommon.

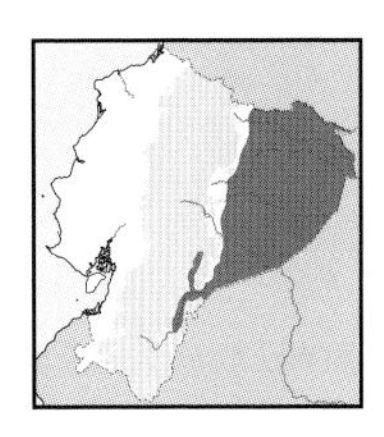

Golden-headed Manakin *Ceratopipra erythrocephala* 9–10cm

E lowlands, locally in foothills (to 1,100m). *Terra firme* forest, mature woodland. ***Flesh legs, white eyes***. Velvet black, with ***bright golden hood***, reddish thighs. ♀: ***pale bill, flesh legs, dark red eyes***. Mostly ***dull*** olive, yellowish-white throat and belly. Race *berlepschi*. Difficult to see away from leks or fruiting bushes; leks rather dispersed, dances and noises similar to Red-capped but no mechanical sounds known. **Voice** Sharp, buzzy *zeek-zeek..prirrr..zeek*; ringing *weeeu-weeeu-zirrk-zirrk…*, short *zit* calls. **SS** Paler bill and legs separate from sympatric ♀ manakins; see White-crowned (greyer crown), Blue-backed (larger, dark bill, pale eye-ring). **Status** Fairly common.

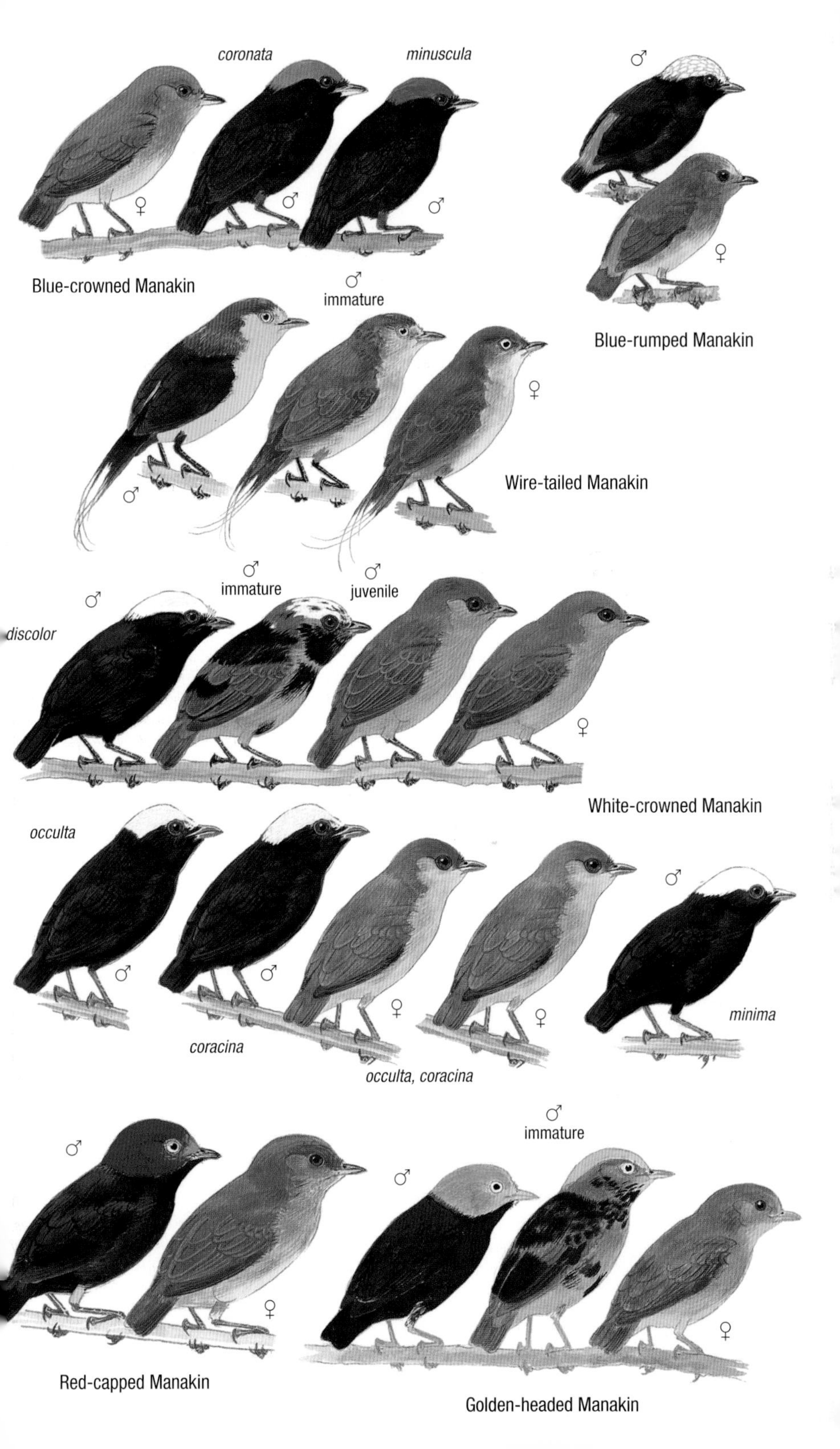

coronata
minuscula
♀
♂
♂
Blue-crowned Manakin
♂
♀
Blue-rumped Manakin
♂
immature
♀
♂
Wire-tailed Manakin
♂
immature
♂
juvenile
♂
discolor
♀
White-crowned Manakin
occulta
♂
♂
♀
♀
♂
minima
coracina
occulta, coracina
♂
immature
♂
♀
♂
♀
Red-capped Manakin
Golden-headed Manakin

Five 'large' manakins. Blue-backed lek behaviour is spectacular, but other species on this plate do not form leks. Green Manakins are mostly solitary, two Andean species (Jet and Yellow-headed) are rare, seldom seen and inconspicuous even by voice. Orange-crowned is distinctive and solitary. Sapayoa is the only Neotropical representative of Old World suboscines.

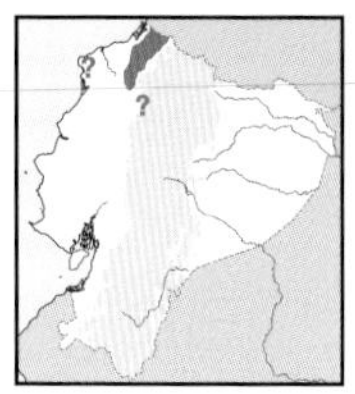

Sapayoa *Sapayoa aenigma* 15cm

NW lowlands (below 500m). Wet forest undergrowth. ***Broad, flat, bicoloured bill***, large brownish eyes. Mainly ***olive***, paler, ***yellower*** throat and central belly, darker wings and tail, ***semi-concealed yellow coronal patch; long tail***. ♀: lacks coronal patch. Monotypic. Singles or pairs rather inconspicuous as perches quietly, most easily seen after sudden sally; often joins mixed-species flocks. **Voice** Soft trill *chir'r'r'r*; also louder *ch-prprprrrr'*. **SS** Duller than Green Manakin, more olive than Northern Schiffornis, bill broader, flatter; behaviour also different from schiffornis. **Status** Rare, local, confined to forest, VU in Ecuador.

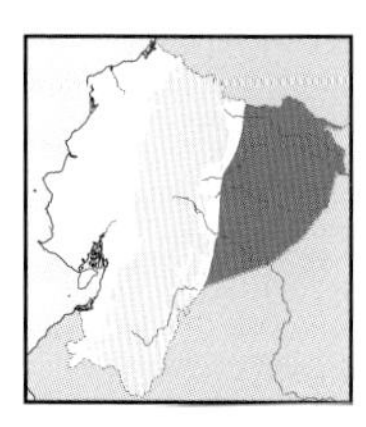

Blue-backed Manakin *Chiroxiphia pareola* 12–13cm

E lowlands (to 500m). *Terra firme* undergrowth. Glossy black with ***bright red crown, blue back, orange-red legs***. ♀: mainly ***greyish-olive***, slightly paler throat and belly, ***pale pink legs***, narrow greyish orbital ring. Race *napensis*. Elusive and inconspicuous; ♂♂ jump over each other at noisy leks led by one bird; fond of aroid fruits. **Voice** Various calls at leks, including ringing whistles: *treee-roop, treee-roop* or *chírr'up-chírr'up*; nasal, frog-like *whern-whern-whernn...*; schiffornis-like *tuu-eeét, tuuee-eet!, tueet-tueet-téét...*. **SS** ♀ larger, somewhat duskier than 'typical' manakins, legs paler than White-bearded; cf. Green Manakin (grass-green, black legs). **Status** Uncommon.

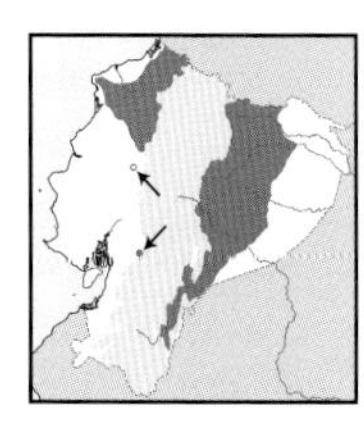

Green Manakin *Cryptopipo holochlora* 12–13cm

E & NW lowlands to foothills (300–1,200m in E, to 1,100m in W). Humid forest undergrowth, mainly in foothills. ***Grass-green upperparts***, duskier wings, yellower cast to throat and ***belly to vent***. Narrow pale eye-ring, ***dark legs, longish*** tail. Nominate (E), *litae* (W; duller overall, more contrasting yellowish belly), possibly different species. Mostly solitary, sometimes with mixed flocks; does not gather at leks. **Voice** Not very vocal; soft *fuiit* song, also rapid *chirrrr* calls. **SS** Recalls dark-legged but ***much smaller*** Blue-crowned and Blue-rumped Manakins; cf. ***pink- to red-legged*** ♀ Blue-backed, Golden-winged and White-bearded; brighter, ***yellower*** Yellow-headed, ***darker*** ♀; larger, ***heavier-billed*** Sapayoa (NW). **Status** Uncommon.

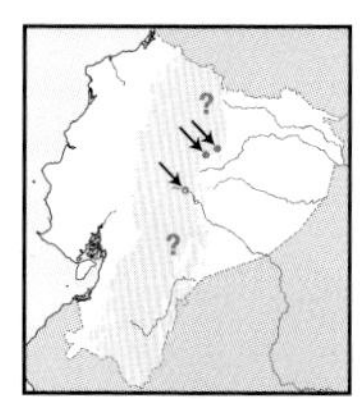

Yellow-headed Manakin *Chloropipo flavicapilla* 12–13cm

Local in foothills to subtropics on E Andean slope (1,500–2,100m). Montane forest undergrowth. ***Red eyes, grey legs. Bright yellow crown to nape***, bright olive upperparts, yellowish-olive cheeks, throat and breast, ***yellow belly to vent***, more olive flanks, ***white underwing-coverts***. ♀: duller, ***yellow belly***. Monotypic. Shy, difficult to find, possibly occurs at low densities but occasionally at borders. **Voice** Short, downslurred *hoeeet* or *wheep* whistle reported (KG). **SS** ♀ brighter and yellower than Green (especially crown and belly) and locally sympatric ♀ Jet Manakins. **Status** Rare, local, possibly overlooked. VU globally, EN in Ecuador. Few uncorroborated records.

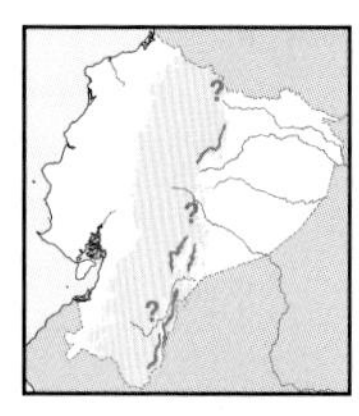

Jet Manakin *Chloropipo unicolor* 12cm

Local in foothills to subtropics on E Andean slope (1,300–1,700m). Montane forest undergrowth. Dark eyes, black legs. ***Glossy blue-black, white underwing-coverts***. ♀: ***dusky*** olive-grey, ***white underwing-coverts***. Monotypic. Rare, seldom seen so very poorly known; probably at low density, joins mixed-species flocks. **Voice** Soft, short downslurred *teeuur* or *que-urr* whistles; also longer *tuuuit* and higher-pitched *tuíít*. **SS** ♂ unmistakable; ♀ much darker than lower-elevation Green Manakin and (even more so) than locally sympatric Yellow-headed; see ♀ Golden-winged (pink legs, greener overall, yellow throat and belly). **Status** Rare, local, likely overlooked.

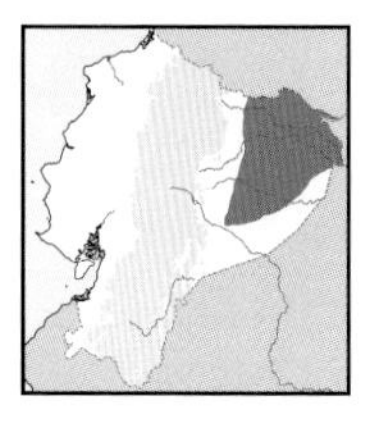

Orange-crowned Manakin *Heterocercus aurantiivertex* 14–15cm

E lowlands (below 300m). *Várzea* and *igapó* forest and woodland undergrowth. Dark eyes, ***black legs***. Olive upperparts, contrasting but somewhat concealed ***orange patch in crown***, greyer cheeks. ***Whitish throat*** extends to neck-sides, ***buff to pale cinnamon underparts***, more olive flanks. ♀: duller, ***lacks crown patch***. Monotypic. Solitary, often perches above or facing water using same perch for long periods; does not gather at leks, solitary aerial displays. **Voice** Weak ascending-descending-ascending *tErtirEtirtErtitEt* trill, rather flycatcher-like; louder, bouncing *we-te'te'te'te*, sibilant *seee!* and other calls. **SS** Confusion unlikely by plumage and habitat. **Status** Rare, rather local, nearly endemic.

♂
♀
Sapayoa
♂
juvenile
♂
♂
immature
♀
Blue-backed Manakin
holochlora
litae
Green Manakin
Yellow-headed Manakin
juvenile
♂
♀
♂
♀
Jet Manakin
♂
♀
Orange-crowned Manakin

Tityras are plump canopy dwellers with short tails and rather stout bills. Quite inconspicuous despite their stunning white plumages. The two mourners occur in humid forests, where heard more often than seen. Taxonomic affinities of all species on this plate long debated, being sometimes considered cotingids, others tyrannids. Recent genetic studies have grouped them in a separate family (Tityridae), which contains a 'miscellany' of long difficult-to-place species.

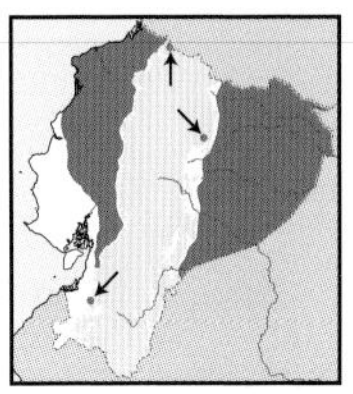

Black-crowned Tityra *Tityra inquisitor* 18–19cm

Lowlands in E & W (to 800m). Humid forest canopy and borders, adjacent clearings; semi-deciduous forest in SW. ***Black bill. Black cap***, pearl-grey upperparts, immaculate white underparts, black flight feathers, ***broad black subterminal band in white tail*** (W) or tail all black (E). ♀: whitish front, ***chestnut face***, some brown streaks above. Races *buckleyi* (E), *albitorques* (W). Pairs to small groups forage rather calmly in canopy, not with mixed flocks, aggressive towards other passerines. **Voice** Rather silent; doubled *zik, tzik*, weaker than other tityras. **SS** Although plumage similar, other tityras have rosy-red bills and orbital area. **Status** Uncommon.

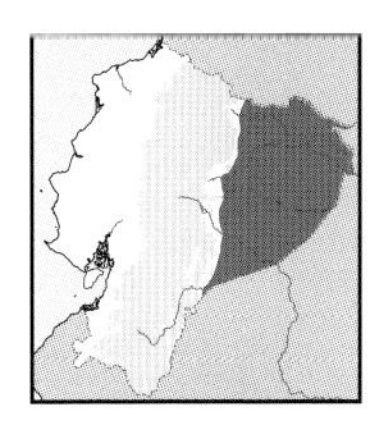

Black-tailed Tityra *Tityra cayana* 21–22cm

E lowlands (to 500m, locally higher). Canopy and borders of *terra firme* and *várzea* forests, woodland, adjacent clearings. ***Rosy-red bill tipped black, rosy-red orbital skin. Black head***, pale pearl-grey upperparts, immaculate white underparts, black flight feathers, ***black tail***. ♀: similar but obviously ***streaked dusky***, fainter on underparts. Nominate. Mainly pairs in canopy, rather calm, does not join mixed flocks, being aggressive to other species. **Voice** Somewhat dry, metallic grunting *ggik* or *erk*, often doubled *erk-erk*.... **SS** Mostly at lower elevations than Masked Tityra (incomplete black on head, unstreaked ♀); smaller Black-crowned has black bill. **Status** Fairly common.

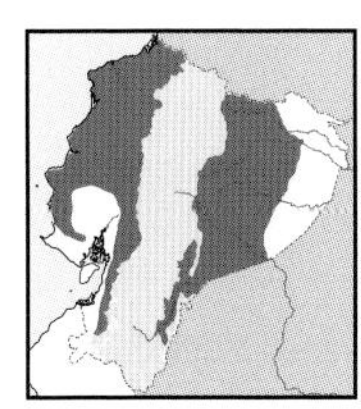

Masked Tityra *Tityra semifasciata* 21–22cm

W lowlands to foothills, E foothills locally to adjacent lowlands (to 1,100m, locally to 1,500–1,700m; not regular below 400m in E). Humid forest canopy, borders, woodland, adjacent clearings. ***Rosy-red bill tipped black. Black mask*** bordering ***rosy-red orbital skin***, very ***pale grey above***, white underparts, black flight feathers, ***greyish-white tail, broad dark subterminal band***. ♀: ***entire head greyish-brown***, greyer upperparts. Races *fortis* (E), *nigriceps* (W). Pairs to small groups fairly conspicuous, aggressive to other species. **Voice** Doubled, dry, nasal grunting *girk, girrk*; similar to Black-tailed but rather higher-pitched. **SS** Black-tailed has more extensive black on head (♂), coarse back streaking (♀). **Status** Fairly common.

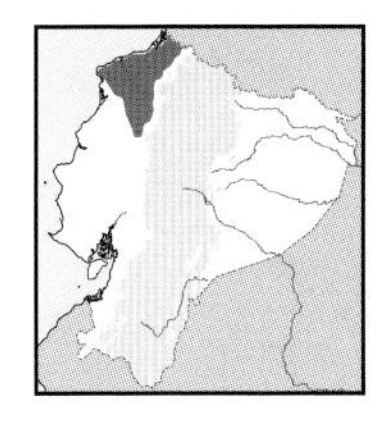

Speckled Mourner *Laniocera rufescens* 19–20cm

Local in NW lowlands (below 500m). Humid forest interior. Black bill, ***narrow pale buff eye-ring***. Mostly ***rufous***, with duskier-brown speckling on head, upperparts and ***breast; yellowish tufts on breast-sides, more rufescent rump***; dark wings with ***two rufous wingbars*** and ***edges to flight feathers***. Race *tertia*. Seldom seen, somewhat sluggish remaining well hidden in dense under- to midstorey, often with mixed flocks; ♂♂ gather in loose leks. **Voice** Far-carrying, ringing, repeated *tee-yeee, tee-yeee, tee-yeee*.... **SS** Darker Rufous Mourner lacks speckling and wingbars, voice slow and whistled, head flatter. Rufous Piha larger, uniform rich rufous, voice louder, whistled. **Status** Rare, confined to forest, VU in Ecuador.

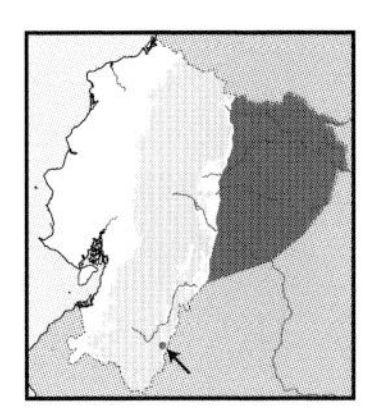

Cinereous Mourner *Laniocera hypopyrra* 19–20cm

E lowlands (to 400m, locally higher). Mid-strata to canopy of *terra firme* forest. ***Narrow buff eye-ring***. Mostly ***grey***, paler below, ***cinnamon-orange to yellow tufts on breast-sides***, sparse cinnamon and black dots on breast and belly, especially in juv. Blackish wings with conspicuous ***cinnamon-buff spotted wingbars and tertials tips***, dark grey tail ***tipped rufous***. Sexes alike. Monotypic. Seldom seen, more often heard, as remains immobile in dense tangles; occasionally with mixed flocks. **Voice** High-pitched, ringing, descending, squeaky *cheey-hoeeét, cheey-hoeeét, cheey-hoeeét…, peew-wEE'eet.. peee-wEE'eet*; also plaintive *pewéet*. **SS** Superficially recalls Greyish Mourner and Screaming Piha but note size, wingbars, belly spots and voice. **Status** Rare.

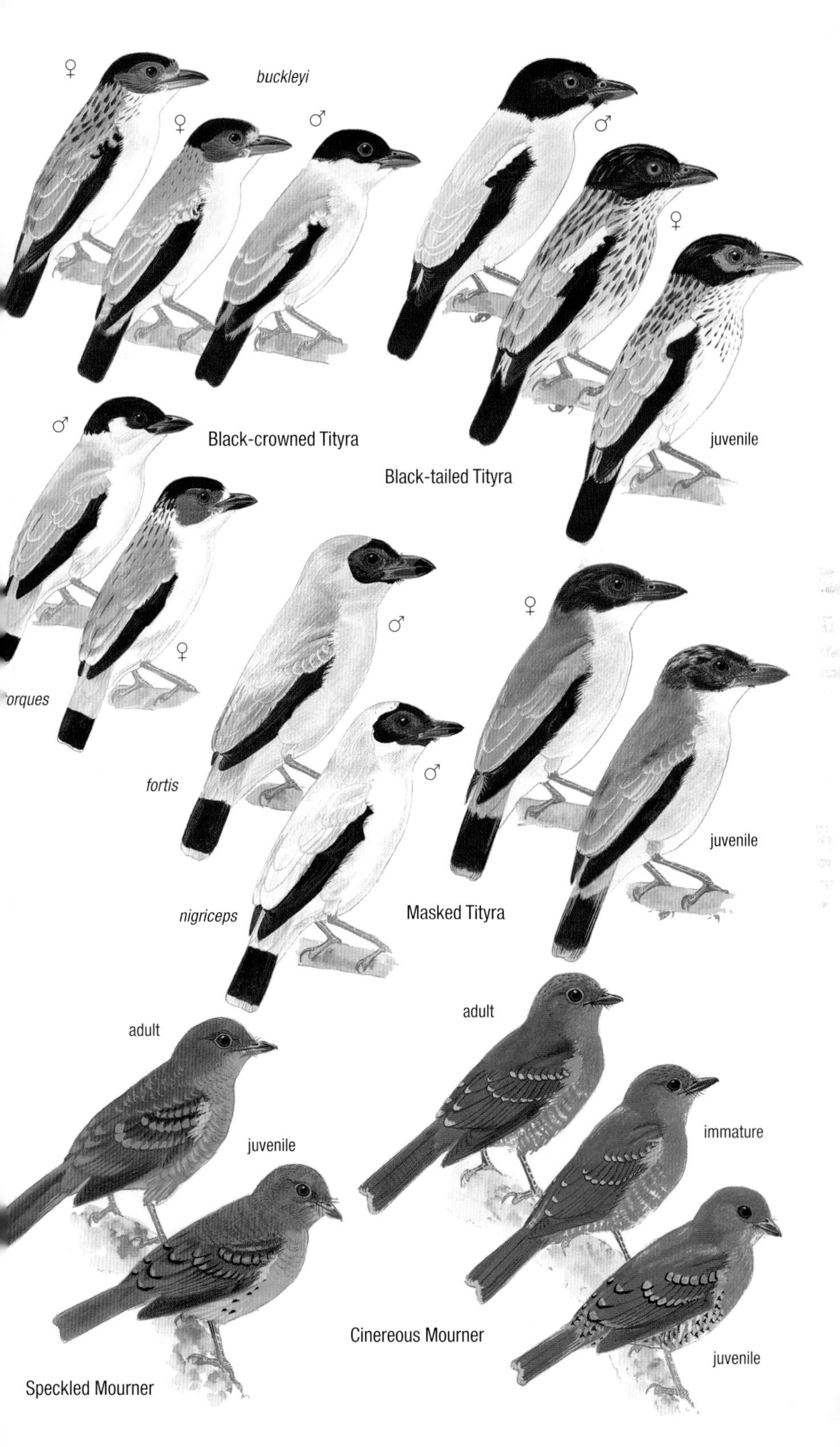
♀
buckleyi
♀
♂
♂
♀
♂
Black-crowned Tityra
Black-tailed Tityra
juvenile
♀
'orques
♂
♀
fortis
♂
juvenile
nigriceps
Masked Tityra
adult
adult
juvenile
immature
Cinereous Mourner
juvenile
Speckled Mourner

Several former manakins (*Schiffornis*) and cotingas (purpletuft and *Laniisoma*). Taxonomic affinities long debated and still quite controversial. The purpletuft inhabits Amazonian forest canopy, while the other species are mostly confined to lower forest growth, being more often heard than seen.

Várzea Schiffornis *Schiffornis major* 15cm

E lowlands (below 300m). *Várzea* forest undergrowth. Large 'manakin', mostly ***rufous to cinnamon-rufous***, darker wings, paler throat, central belly, vent and rump, ***grey cheeks***, sometimes extends to nape and crown. Some lack grey. Nominate. Elusive, stays well hidden in dense tangles; sluggish. **Voice** Leisurely series of slurred notes: *toee-dee-wEEt, doe-owEEt?... déo-owEEt?*; also *deo?-üwit, deo?-üwit... d-chrrrrr.* **SS** Duller, browner Brown-winged Schiffornis; voice slower, shorter; habitats also differ. Similarly coloured and canopy-dwelling Cinnamon Attila differ in size, shape, behaviour, habitat, etc. **Status** Locally uncommon.

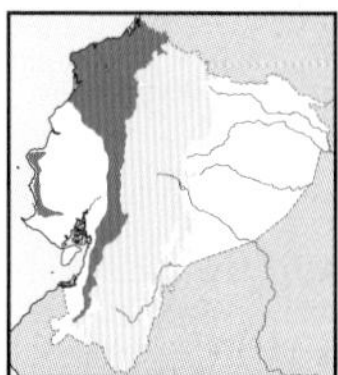

Northern Schiffornis *Schiffornis veraepacis* 16cm

W lowlands to foothills (to 1,300m). Forest undergrowth, mature second growth. ***Olive-brown*** above, ***dull*** brown wings, underparts ***paler, yellower*** throat, faint ***pale eye-ring***. Sexes alike. Race *rosenbergi*. Unobtrusive in lower growth where difficult to see; seldom joins mixed flocks, feeds by sallying to foliage, sometimes clinging to stems. **Voice** Slow, rising *deeeé...wue-wEE?*, first note long, followed by two very short ones. **SS** More olive Sapayoa has broader bill, underparts tinged yellow; voice not whistled. Green Manakin much greener, yellow-bellied, etc. **Status** Uncommon.

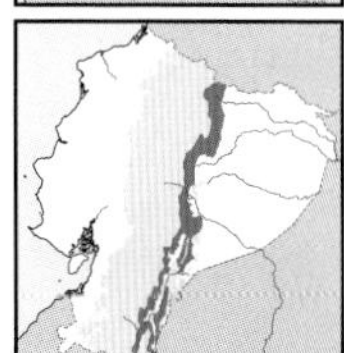

Foothill Schiffornis *Schiffornis aenea* 16cm

E foothills to lower subtropics (900–1,700m). Forest undergrowth, mature second growth. Recalls Brown-winged Schiffornis but darker, ***more olive-brown*** above, ***less greyish below, breast browner.*** Nominate. More often heard than seen as stays hidden and not very active; rarely joins mixed flocks, forages by sallying to air or foliage. **Voice** Mournful *deuu...teé'wOOee, tew-tew,* second note rising-falling; faster, more rhythmic than previous species. **SS** ♀ Jet Manakin much darker, with white underwing, tail shorter; no known overlap with paler Brown-winged (greyer below, more contrasting wings; voices differ). **Status** Uncommon.

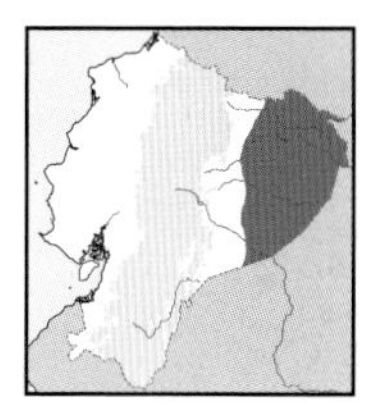

Brown-winged Schiffornis *Schiffornis turdina* 16–18cm

Local in E lowlands (below 300m). *Terra firme* forest undergrowth, mature second growth. Large 'manakin', mostly ***dull brown, more rufescent wings*** and tail, ***greyish-brown*** below, tawnier throat, central belly washed grey, ***narrow pale eye-ring***. Sexes alike. Race *amazonum*. Seldom seen as stays well hidden in dense understorey, rarely joins mixed flocks; feeds by sallying to foliage or air, sometimes perches upright on vertical stems. **Voice** Far-carrying, slow, 3–4-note whistled *weet-deeEE-wee?, we-eé-toeeeEE-weed?*, second note longer. **SS** Much brighter Green Manakin; ♀ Blue-backed Manakin smaller, pink-legged, greener; and rich cinnamon-rufous Várzea Schiffornis. **Status** Rare, locally uncommon.

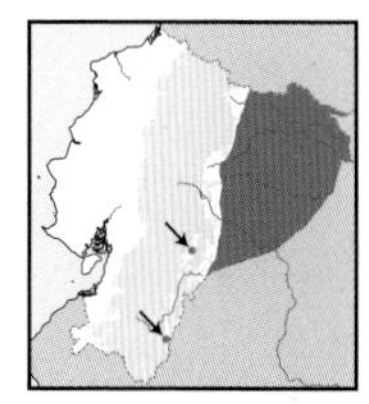

White-browed Purpletuft *Iodopleura isabellae* 11–12cm

E lowlands (to 500m). *Terra firme* and *várzea* forest canopy, borders, woodland, adjacent clearings. ***Small and chunky***, short bill, short tail. Blackish upperparts, contrasting ***white lores, malar and postocular streaks, white rump. White underparts***, sides and flanks mottled blackish-brown, small semi-concealed ***purple tufts on flanks***. ♀: tufts white. Nominate. Pairs to small groups mostly in canopy, lower at borders, often perches atop bare twigs and snags; apparently wanders widely for fruit. **Voice** Very thin, melancholic *wheee?*, faster *weer-eer'eer* in chases. **SS** Confusion unlikely. **Status** Uncommon.

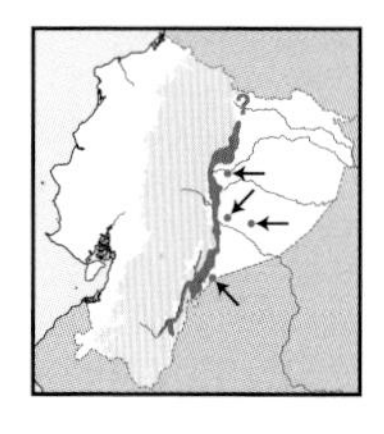

Shrike-like Cotinga *Laniisoma elegans* 17–18cm

E Andean foothills, locally in lowlands (400–1,350m). Montane forest, borders. Narrow eye-ring. ***Sooty-black crown***, olive upperparts. ***Bright yellow underparts, scaled blackish*** on flanks and vent. ♀: ***more profusely scaled below***, dull olive crown. Juv. has ***faint rufous-dotted wingbars.*** Race *buckleyi*. Singles or pairs in understorey to subcanopy; lethargic, inconspicuous, very difficult to see. **Voice** High-pitched, thin *psiiiiiiuueeee!* rising at end; recalls fruiteaters and Black-streaked Puffbird; also louder *chee-kik!* **SS** Confusion unlikely, bright-billed fruiteaters are greener, plumper, etc. **Status** Rare, local, NT in Ecuador.

grey
morph
rufous
morph
Várzea Schiffornis
adult
juvenile
Northern Schiffornis
adult
juvenile
Foothill Schiffornis
adult
Brown-winged Schiffornis
♂
♀
White-browed Purpletuft
♂
♀
Shrike-like Cotinga

PLATE 225: BECARDS

Becards are large-headed birds of forested habitats. Reminiscent of some tyrannid genera; in fact, until recently grouped within the Tyrannidae. Species on this plate have bold wing patterns and marked sexual dimorphism. Green-backed and Barred Becards are boldly patterned in both sexes. ♂♂ of the remaining three species are mainly grey and black, ♀♀ olive and yellow.

Green-backed Becard *Pachyramphus viridis* 14–15cm

E Andean foothills to subtropics (650–1,700m). Montane forest, treed clearings. ***Black crown, white lores, lemon-yellow cheeks and throat, whitish belly. Olive-green upperparts***, dusky wings, coverts edged pale greenish. ♀: whitish lores, ***grey crown, cheeks and throat***, greenish breast, ***olive upperparts, rufous shoulders***. Race *xanthogenys*, possibly valid species. Slow-moving singles or pairs glean at forest edges, often with mixed-species flocks. **Voice** Fairly loud, fast, slightly accelerating *we-pi-pi-pi-pipipipipi*; also *ti-wee?* **SS** Little overlap with Barred Becard, with yellow cheeks but profusely barred dusky below, white (♂) or chestnut (♀) in wings, etc. Voices differ. **Status** Locally uncommon.

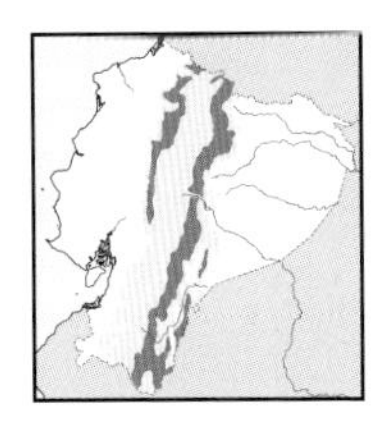

Barred Becard *Pachyramphus versicolor* 12–13cm

E & NW subtropics to temperate zones (1,500–2,600m). Montane forest canopy, borders, treed clearings. ***Black crown and above, yellow lores and eye-ring, yellowish cheeks and throat***, whitish belly; ***underparts profusely barred dusky***, contrasting ***white in scapulars and wing-coverts***. ♀: sooty-grey crown, yellow lores, ***dusky-yellow cheeks, yellowish underparts*** profusely ***barred dusky***. Olive upperparts, ***extensive chestnut in wings***. Nominate (widespread), *meridionalis* (extreme SE). Singles or pairs glean and sally with mixed-species flocks. **Voice** Soft but high-pitched, fast *we-di-di-di-dididi*; *dredredre* rattles or *dididi* while foraging. **SS** Occurs at higher elevations than other becards, confusion unlikely. **Status** Uncommon.

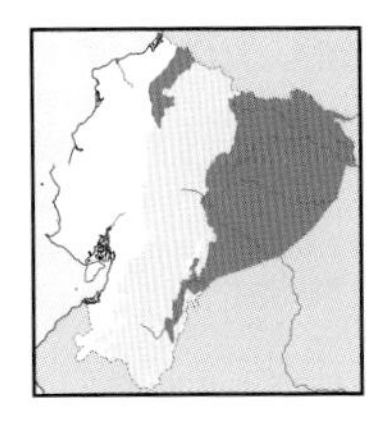

White-winged Becard *Pachyramphus polychopterus* 15cm

E lowlands to foothills, NW foothills to lower subtropics (to 1,200m in E, 600–1,900m in NW). Forest borders, woodland, river islands. ***Glossy black crown and upperparts***, sooty blackish-grey (E) or paler grey (W) from ***nape to underparts. White scapulars, wing-coverts and flight feather edges***. ♀: ***olive-brown above*** (darker crown in W), whitish supraloral stripe and broken eye-ring, yellowish underparts (duskier in E); ***buffy-chestnut*** in wings. Races *tenebrosus* (E), *dorsalis* (NW); possibly valid species. Lethargic pairs sally short distances at various heights. **Voice** Soft, musical *chyú-chu-chuchu-dudup-dup*; *teü-teü-tu-tu-deedu-deedu* (NW); faster, sweet *deü-deü-düdüdüdüdeü...*, number of *dü* notes vary (E). **SS** ♂ Black-and-white Becard in NW (grey back, whitish lores, shorter call; ♀ has chestnut crown). In E, ♀ Black-and-white (chestnut crown) and Black-capped (no cinnamon scapulars, rufous crown). **Status** Uncommon.

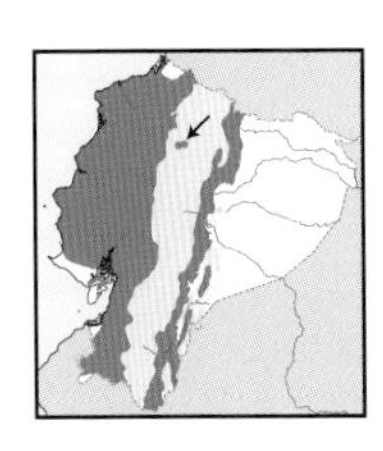

Black-and-white Becard *Pachyramphus albogriseus* 14–15cm

W lowlands to subtropics, E foothills to subtropics (to 2,000m in W, 900–1,800m in E; few records at 2,800–2,900m). Montane forest canopy, borders; semi-deciduous forest (SW). Black crown, ***whitish lores, cheeks and underparts. Grey upperparts***, black wings, ***white bars*** and edges to flight feathers. ♀: ***chestnut crown, black postocular. Olive upperparts***, yellow underparts, ***black wings edged buff***. Races *guayaquilensis* (W lowlands), *salvini* (E & possibly W slopes). Singles or pairs with mixed flocks sallying or hover-gleaning for prey. **Voice** Soft, musical *dwedwe-dwip?, dwedwe-dwip?, dwedwe-dwip?*; longer, more plaintive in E. **SS** White-winged Becard in NW (♀ has browner back, duskier cap); no known overlap with Black-capped in E. **Status** Uncommon.

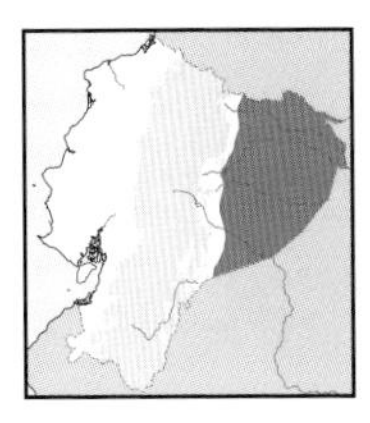

Black-capped Becard *Pachyramphus marginatus* 13–14cm

E lowlands (to 700m). *Terra firme* canopy, borders. Black crown, ***grey lores, cheeks and underparts. Black back***, paler rump, ***white scapulars***, black wings, coverts and inner flight feathers ***edged white***. ♀: recalls White-winged Becard but ***crown more rufous, no cinnamon scapulars***, upperparts more ***olive-green***. Race *nanus*. Mostly pairs, often with mixed flocks, actively peers around, sallies to foliage for prey. **Voice** Soft, melodic *teuu-wee-deeu?, teuu-wee-deeu?*, shorter than White-winged; also faster repetition of *teuu-tidi, teuu-ti-dididi*. **SS** Confusion unlikely; E race of White-winged blacker throughout, ♀ has olive crown, cinnamon scapulars. **Status** Uncommon.

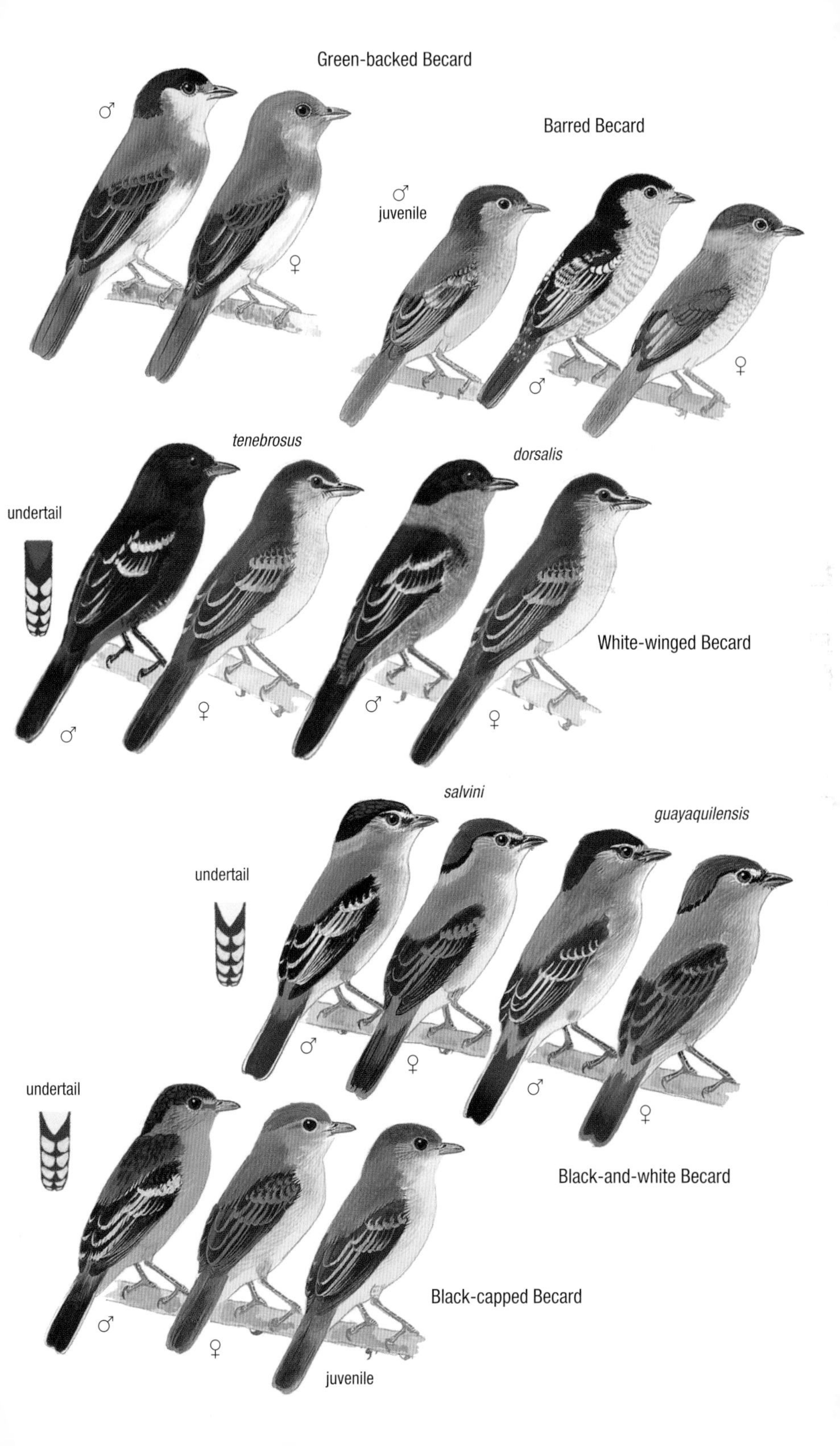
Green-backed Becard
♂
♀
Barred Becard
♂
juvenile
♂
♀
tenebrosus
dorsalis
undertail
White-winged Becard
♂
♀
♂
♀
salvini
guayaquilensis
undertail
♂
♀
♂
♀
undertail
Black-and-white Becard
Black-capped Becard
♂
♀
juvenile

PLATE 226: UNICOLOURED BECARDS

Becards that lack wing markings and are much plainer than those on preceding plate. Two species lack sexual dimorphism. One-coloured, Pink-throated and Crested Becards were formerly separated in *Platypsaris*. Tricky identification in western species, but note face patterns, voice and habitat. Crested and Cinereous erroneously assigned to Ecuador in previous literature.

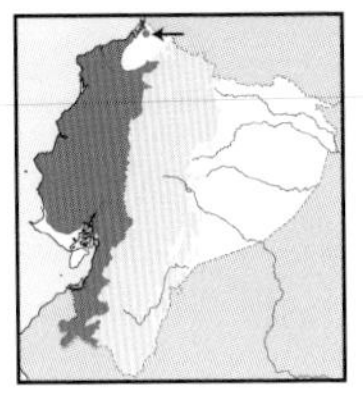

Slaty Becard *Pachyramphus spodiurus* 14cm

W lowlands to lower foothills (to 600m). Semi-deciduous to deciduous forest, borders, adjacent woodland. Blackish crown, ***slate-grey upperparts***, paler rump, blackish wings with contrasting ***white edgings***. Sooty-grey underparts, paler throat and belly, ***greyish lores***. ♀: chestnut-brown crown, ***greyish lores***, cinnamon-rufous upperparts, darker wings, buffier underparts, paler throat and belly. Monotypic. Pairs inconspicuous, tend to remain low, rarely with mixed flocks. **Voice** Fast, accelerating *whip-whip-wi-wi-dididiririi*, increasing in pitch at end. **SS** Larger ♂ One-coloured (slower, louder, more sputtering voice) lacks white in wings; ♀ has duskier lores; also Cinnamon Becard with whitish supraloral (more humid habitats). **Status** Rare, nearly endemic, EN.

†Cinereous Becard *Pachyramphus rufus* 13–14cm

Old records from SE Andes now known to be misidentified Chestnut-crowned Becard. ♂ differs from other becards in Ecuador by black cap, ash-grey upperparts, nearly ***plain wings, whitish frontlet***, all-dark tail. ♀: mostly rich cinnamon, ***whitish lores***; from Chestnut-crowned by plain cinnamon nape (***no grey***). **Status** No longer part of Ecuadorian avifauna (O. Janni & C. Pulcher).

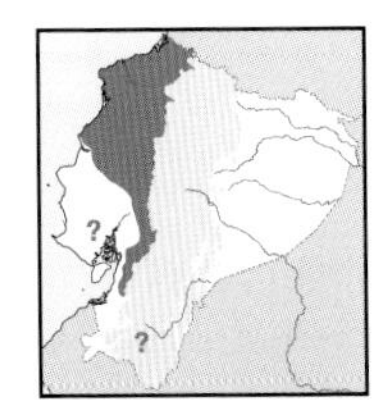

Cinnamon Becard *Pachyramphus cinnamomeus* 14–15cm

W lowlands to foothills (to 1,000m, locally higher). Humid forest borders, woodland. ***Rufous-brown crown, whitish supraloral stripe***, cinnamon-rufous upperparts, darker flight feathers. ***Pale*** cinnamon-buff below, ***paler*** throat and central belly. Sexes alike. Nominate. Pairs more conspicuous than Slaty Becard, sallying at borders and in clearings, rarely with mixed flocks. **Voice** Mellow *tuee, tou-tou-tou*; sometimes louder first note and more *tou* notes; short, reedy *swee-dwee-dwee* calls. **SS** Larger ♀ One-coloured Becard (dusky lores, no whitish supraloral); ♀ Slaty Becard (no whitish supraloral stripe). **Status** Fairly common.

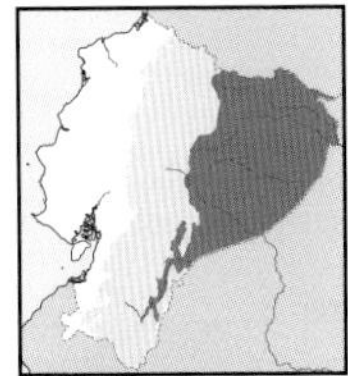

Chestnut-crowned Becard *Pachyramphus castaneus* 14–15cm

E lowlands to lower foothills (to 1,000m). *Terra firme* and *várzea* borders, adjacent clearings. ***Chestnut-brown crown***, grey lores, ***broad grey stripe behind eyes to nape***. Rufous-chestnut above, pale cinnamon-buff below. Sexes alike. Race *saturatus*. Sallies or gleans in canopy, lower at borders; irregularly with mixed flocks, mostly in pairs. **Voice** Mellow, descending *teéE, déuu-deuu-deuu*, number of *deuu* notes varies. **SS** Confusion unlikely, but cf. larger ♀ Pink-throated Becard (grey crown); see Cinnamon Attila (different size, behaviour, proportions, size). **Status** Uncommon.

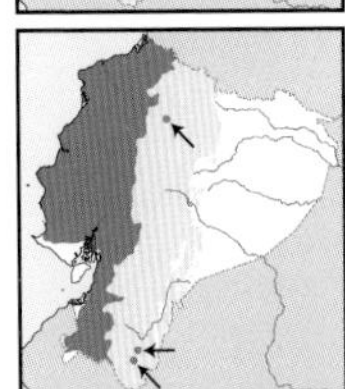

One-coloured Becard *Pachyramphus homochrous* 16–17cm

W lowlands to foothills (to 1,500m, locally higher). Humid to deciduous forest, borders, woodland, adjacent clearings. ***Slaty-black upperparts***, darker crown, mantle and tail, ***unmarked blackish wings; sooty***-grey underparts, paler throat and belly. ♀: rufous-chestnut above, slightly darker crown, ***dusky lores***, pale cinnamon-buff below, whiter throat and belly. Nominate. Mostly pairs with mixed-species flocks; active and conspicuous at various heights in edges; often nods head. **Voice** Loud, sputtering, quite variable but never musical *prr-tretret-tee-peeuw*; also soft *twéeeeuu*. **SS** Slaty Becard ♂ (white wing edgings; ♀ very similar); Cinnamon Becard (whitish supraloral). **Status** Fairly common.

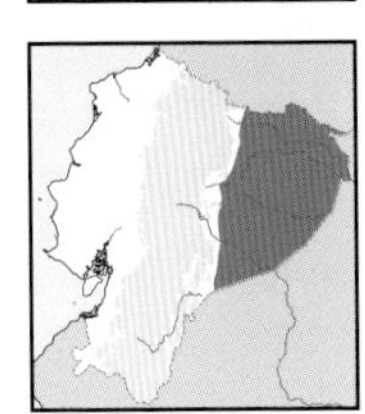

Pink-throated Becard *Pachyramphus minor* 16–17cm

E lowlands (to 600m). *Terra firme* forest, borders. Mostly black, ***unmarked wings, pink patch on lower throat***. ♀: brownish-grey above, ***grey crown, cinnamon-buff cheeks and underparts***. Monotypic. Mostly pairs high in canopy, often with mixed flocks; forages by sallying or gleaning. **Voice** Squeaky *dtueeét* followed by fast *di-di-di-di-we*, also fast, quavering *tiri-riririri* and soft *pik*. **SS** Only becard with pink throat; cf. ♀ with Chestnut-crowned Becard (chestnut crown); see also Citron-bellied Attila and Royal Flycatcher (different behaviour, voice, shape, etc.). **Status** Rare.

†Crested Becard *Pachyramphus validus* 18–19cm

Single, erroneous record in SE Andes (2,550–2,600m). Montane forest canopy, borders. Crown more ***glossy black*** than One-coloured Becard, underparts more ***olive-grey***. ♀: ***sooty-grey crown***, paler lores, ***cinnamon-rufous upperparts***. Possibly behaves like congeners but silent and inconspicuous. **Voice** Clear *sui-sui-sui*... (RG). **SS** One-coloured known to move altitudinally; black confined to crown (not down to eye). **Status** Record no longer valid; recently re-identified as One-coloured Becard.

♂
♀
Slaty Becard
♂
♀
♀
Cinereous Becard
immature
Cinnamon
Becard
Chestnut-crowned
Becard
♂
♂
immature
♀
One-coloured Becard
♂
♂
immature
♀
♂
♀
Pink-throated Becard
Crested Becard

Large and conspicuous forest birds, two species in lowlands (one either side of the Andes) and four in the Andes. Gregarious, noisy and sometime ferocious predators; *Cyanolyca* are characterised by black masks and brighter or paler crowns and throats, whereas *Cyanocorax* are heavier, and have black bibs and faces.

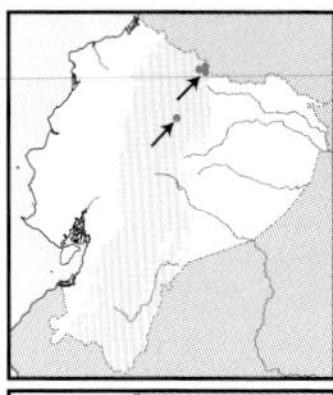

Black-collared Jay *Cyanolyca armillata* 32–34cm

Local on NE temperate Andean slope (2,100–3,150m). Montane forest, borders. Black mask and collar. ***Bright ultramarine-blue head and throat*** grading into deep greenish-blue back and breast to tail. Race *quindiuna*. Noisy pairs to small groups in canopy and edges, often in mixed flocks with other large omnivorous birds. **Voice** Loud *shrrriej* or *shrrwee*; *shaj*, *shá-shá* and others. **SS** Similar Turquoise Jay paler, more turquoise-blue overall, more contrast between whitish crown and throat and rest of body; voice somewhat harsher, whistles lower-pitched. **Status** Rare, very local, VU in Ecuador.

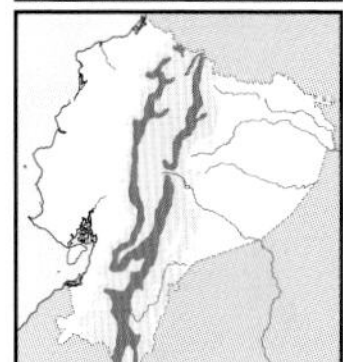

Turquoise Jay *Cyanolyca turcosa* 31–33cm

Subtropics to temperate zone on both Andean slopes (2,000–3,300m). Montane forest, borders, woodland, adjacent clearings. Black mask and narrow collar. Mostly ***turquoise-blue***, paler, ***washed silvery on crown to nape and throat***. Monotypic. Small noisy groups in mid strata to canopy, briefly join mixed flocks; may engage in seasonal movements following fruiting events. **Voice** High-pitched, metallic *jeey, jeey...* resembling a 'huge chicken'; also *pchee-pchee...*, *jee-jeey, jee-jeey...* and other whistles, hisses, cries. **SS** Paler crown and throat, and brighter turquoise-blue than possibly sympatric Black-collared Jay; voices, although variable, differ. **Status** Common.

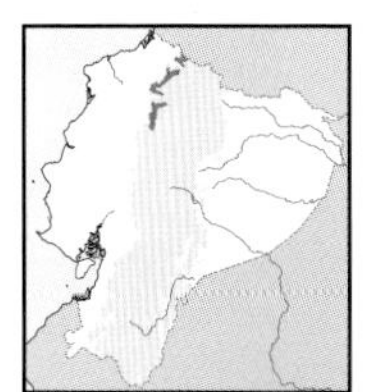

Beautiful Jay *Cyanolyca pulchra* 27–28cm

Subtropics on NW Andean slope (1,300–2,200m). Montane forest and borders. Black mask, ***silvery crown*** grading into bluish nape, ***deep purplish-blue*** overall, blacker on mantle and breast, ***ultramarine-blue throat***. Monotypic. Pairs to small groups forage in dense cover, rarely at borders; elusive, rarely with mixed flocks. **Voice** Whistled *chueep-chueep* or *tweep-tweep* and metallic, harsh *jash-jash-jash*; also raspy *tjj-tjj-tjj...*, *chwep-chwep* etc. **SS** Turquoise Jay much paler, bright turquoise, whiter throat outlined black; voice higher-pitched, less ringing. **Status** Rare, VU in Ecuador, NT globally.

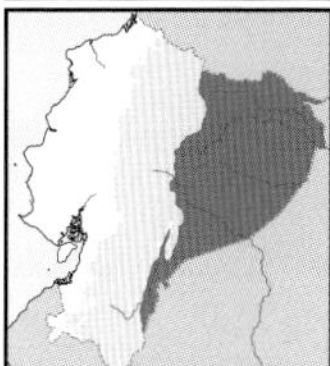

Violaceous Jay *Cyanocorax violaceus* 35–37cm

E lowlands locally to adjacent foothills (mostly below 500m, locally to 1,200m). *Terra firme*, *várzea* forest, borders, woodland, adjacent clearings, river islands. Large. ***Black head to breast, silvery-white rear crown, violaceous-blue*** upperparts, more ***purplish wings and tail***, paler underparts. Nominate. Noisy and conspicuous groups mostly at borders, often in mixed flocks with other large, canopy frugivores; move long distances. **Voice** Loud, metallic *shrrié, shrrié...* or *jéiiy, jéiiy, jéiiy...*, *jéeyk!, jeeýk!* **SS** Unmistakable in range. **Status** Common.

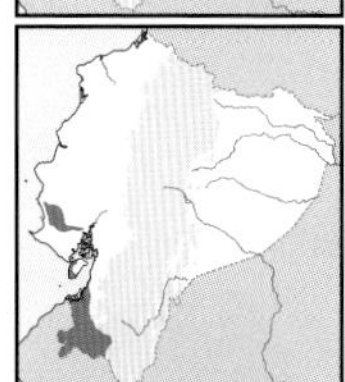

White-tailed Jay *Cyanocorax mystacalis* 31–33cm

SW lowlands, to foothills and subtropics in W Loja (mostly below 1,200m, to 2,000m locally). Deciduous and semi-deciduous forest, woodland, arid scrub, borders. Beautiful. Yellow eyes. Black face, throat and breast, contrasting with ***white moustachial and small supercilium. White rear crown to upper mantle***, deep blue upperparts, pure ***white below***, white outer tail feathers and tail tip. Monotypic. Small, rather elusive groups in dense cover, irregularly with mixed flocks, often near ground. **Voice** Metallic *cleek-cleek-cleek-cle-cle!* or *choek-choek-choek-chek-chek...*; loud *tchek-tchek-tchek...*, other rings and clicks. **SS** Only jay in range. **Status** Uncommon.

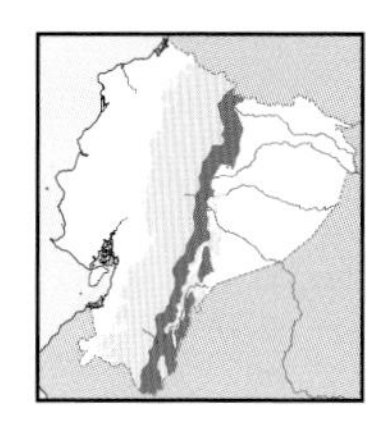

Green Jay *Cyanocorax yncas* 29–32cm

Subtropics on E Andean slope (1,300–2,200m, to 700m in extreme SE). Montane forest, woodland, borders. Unique. Black face and bill, ***blue moustachial***, eyebrow and bushy front. White crown, ***bright green upperparts***, bright ***yellow underparts***, tail yellow below, green above. Nominate (most of range), possibly *longirostris* in extreme SE. Small noisy groups, irregular with mixed canopy flocks, but often with icterids, usually near ground; permanently vocal with wide repertoire. **Voice** Loud, metallic *jeen-jeen-jeen-jeen, cleeng, cleeng... jish-jish...*; also *tjtjtjtj, t-gaga, t-gaga, nya-nya-nya...*; other trills, hisses, clicks, etc. **SS** Unmistakable plumage and voice. **Status** Fairly common.

juvenile
adult
adult
Turquoise Jay
Black-collared Jay
juvenile
Beautiful Jay
adult
Violaceous Jay
juvenile
adult
yncas
longirostris
White-tailed Jay
Green Jay

Small swallows, most are bicoloured, with fairly long, notched or forked tails. *Pygochelidon* and *Orochelidon* are glossed blue above; both were formerly in genus *Notiochelidon* and occur in Andes, up to 4,000+m elevation. White-thighed is smallest swallow in Ecuador, and the most subdued in colour. White-banded and Southern Rough-winged are common and widespread in lowlands.

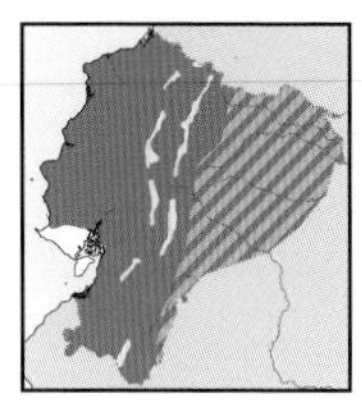

Blue-and-white Swallow *Pygochelidon cyanoleuca* 12–13cm

Andes, more local in W lowlands; migrant to E lowlands (to 3,200m). Open and semi-open areas, fields, waterbodies, towns. ***Glossy steel-blue above, pure white below, black vent***, dark underwing. Juv. browner above, 'dirtier' below. Nominate (most of range), *patagonica* (austral migrant to E; Mar–Sep), ***white vent***. Graceful flight from near ground to mid heights, flocks often perch on wires, walls and other man-made infrastructure. **Voice** Scratchy or buzzy *dzee-e'e'e'e'e'e'e'?* or *drrrzzzeeé*; thin *tzet* in flight. **SS** Locally sympatric, forest-based Pale-footed Swallow not pure white below (buff throat); Bank Swallow (brown above, dusky-brown breast-band). **Status** Common, rarer in E lowlands.

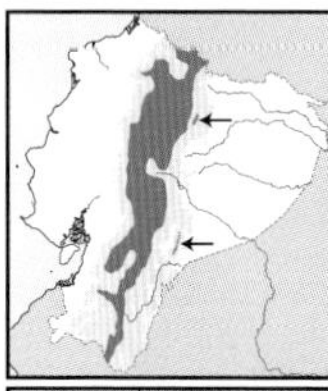

Brown-bellied Swallow *Orochelidon murina* 13–14cm

Andes (2,500–4,400m). Open *páramo*, semi-open and open fields. ***Dark steel blue upperparts, greyish-brown underparts***, black vent. Juv. duller. Nominate. Small flocks, graceful flight at various heights, below Blue-and-white Swallow when together, rarer in anthropogenic habitats and dry valleys. **Voice** Scratchy *tjj-tjj-tjjt-prrrr....* **SS** No known overlap with darker White-thighed Swallow; juv. Blue-and-white looks 'dirty' below, but never smoky-brown; Blue-and-white's voice longer, rising, less dry, and more pleasant. **Status** Fairly common.

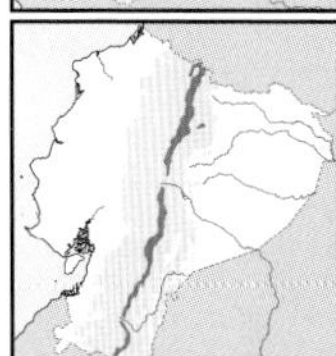

Pale-footed Swallow *Orochelidon flavipes* 11–12cm

Temperate zone in E Andes (2,650–3,300m). Montane forest and adjacent clearings. ***Dark steel-blue above, ochraceous-buff throat***, white below, ***dark-mottled sides***, dark vent. Monotypic. Graceful flocks, more forest-based than other swallows, not in anthropogenic habitats, seen mostly in flight. **Voice** Dry, scratchy *drrr-dre'e'e'e-dre'e'e'e...drr-drr... drrr-e'e'e'e, drr-e'e'e'e...* interspersed by warbles; flight call thin *tzi-dit*. **SS** Blue-and-white Swallow in more open areas; buff throat not always easy to see, but looks less extensively white below than Blue-and-white; voice recalls Brown-bellied but less buzzy, drier. **Status** Rare, rather local.

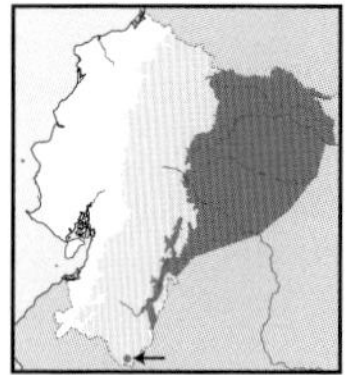

White-banded Swallow *Atticora fasciata* 14–15cm

E lowlands to foothills (to 1,100m). Slow-moving rivers, large streams, lakes. ***Entirely glossy blue-black, contrasting white breast-band***, long deeply forked tail. Juv. duller, some pale-edged feathers in wings and below. Monotypic. Always above water in rapid zigzagging flight, regularly low; often perches on rocks, snags, wires, driftwood, etc. **Voice** Fast, dry *t'rrrrdt, tit-ti-dirrt* or *zzzrr'éep-zzzrr'éep-zzzrr'éep....* **SS** Unmistakable. **Status** Fairly common.

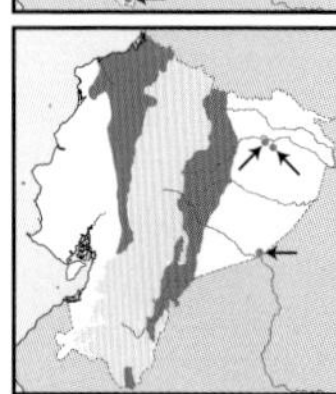

White-thighed Swallow *Atticora tibialis* 10–12cm

E & W foothills to adjacent lowlands (to 800m in W, 1,250 m in E). Montane forest borders, adjacent clearings. Small. Glossy ***blackish-brown above***, paler rump, ***greyish-brown below, small white thighs***. Darker race *minima* (W), *griseiventris* (E). Small flocks fly low above or through canopy, sometimes with Southern Rough-winged, seldom over open areas, fast and erratic flight. **Voice** Thin *cheep-cheep...cheep; td-td-chee-dreep, t-deet*, shorter, less buzzy than Blue-and-white. **SS** Larger Brown-bellied Swallow browner below, bluer gloss above (no known overlap); Southern Rough-winged larger, much paler below, more contrasting pale rump, etc. **Status** Uncommon to rare.

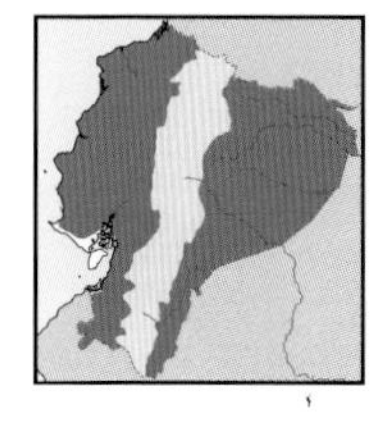

Southern Rough-winged Swallow *Stelgidopteryx ruficollis* 13–14cm

E & W lowlands to subtropics (to 2,050m). Open and semi-open areas. Greyish-brown above, contrasting ***whitish*** (W) or ***dull brown*** (E) ***rump, cinnamon-buff throat*** grading into yellowish-buff below. Nominate (E), *uropygialis* (W). Fast, rather strong and direct flight, little circling; often flies low over open areas or water, perches on branches and stems protruding from banks; sometimes above canopy. **Voice** High-pitched, fast *dzzzzeer-dzzz-zeer, d-zzzz, zzzzérr, djzerrrt-djert...drrt-drrt-drrt....* **SS** Pale rump and cinnamon throat distinctive; see darker White-thighed Swallow. **Status** Common.

Blue-and-White Swallow
patagonica
cyanoleuca
cyanoleuca
juvenile
patagonica
Pale-footed Swallow
adult
juvenile
Brown-bellied Swallow
adult
White-banded Swallow
juvenile
minima
griseiventris
White-thighed Swallow
uropygialis
juvenile
ruficollis
ruficollis
uropygialis
Southern Rough-winged Swallow

Migrant or vagrant swallows, plus a common resident (White-winged). Most species have showy patterns or ornaments. *Tachycineta* are glossed green above, two with white rumps, but Tree and Tumbes difficult to see. Cliff and Barn are regular boreal visitors that could appear anywhere in Ecuador.

* Tree Swallow *Tachycineta bicolor* — 13–14cm

One record in NW lowlands (50m). Semi-open areas. ***Glossy greenish-blue above***, pure white below, ***dark underwing***. Juv. browner above, dirty white below, some ***dirty mottling on breast***. Monotypic. Very poorly known in Ecuador, but might appear in fairly large, compact flocks; feeds over water. **Voice** Not heard in Ecuador, elsewhere various chips and twitters, *chirp-chirp-chirp, zjid, tick-tick-tick*... (RG). **SS** Smaller Blue-and-white (infrequent in NW lowlands) has black vent, darker upperparts; juv. resembles Bank Swallow but lacks breast-band; superficially similar Tumbes (white rump) only in extreme SW. **Status** Rare, accidental visitor (Oct).

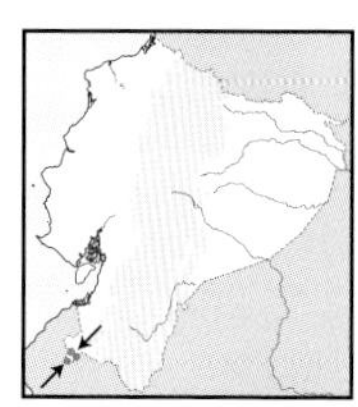

Tumbes Swallow *Tachycineta stolzmanni* — 11–12cm

Few records in SW lowlands (150m). Semi-open scrub and adjacent fields. ***Dark glossy bluish-green above, whitish rump***, dirty white below with ***dusky shaft-streaks, whitish underwing***. Juv. browner. Monotypic. Often in small flocks, not near water; apparently seasonal or erratic in Ecuador; nests in tree holes. **Voice** Dry and simple *br-brz-brz-zzet*.... **SS** Confusion unlikely, locally sympatric Blue-and-white Swallow lacks whitish rump, vent black, darker upperparts. **Status** Rare, local.

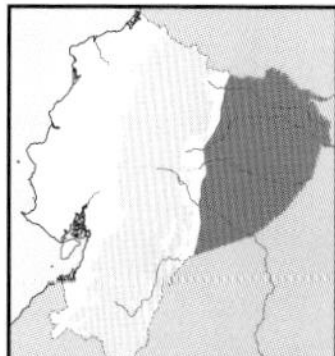

White-winged Swallow *Tachycineta albiventris* — 13–14cm

E lowlands (to 400m). Open areas near lakes and slow-moving rivers. Glossy greenish-blue above, ***solid white rump, large white patch in wings***, pure white below. Juv. 'dirtier' below, browner above, less white in wings. Monotypic. Small flocks to singles always near water, low over surface, often perch on snags, rocks, driftwood, bridges, etc. **Voice** Sweet, buzzy *dréeeet* or *wrréeet*. **SS** Confusion unlikely due to white rump and wings. **Status** Fairly common.

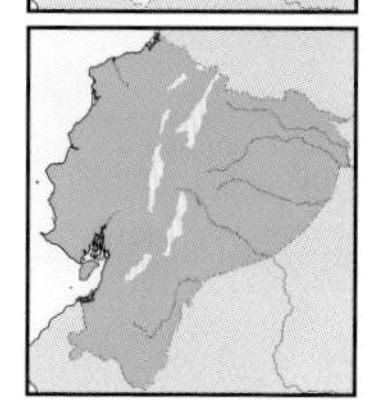

Barn Swallow *Hirundo rustica* — 14–17cm

Widespread boreal migrant (to 3,500m). Open and semi-open areas. ***Glossy blue-black upperparts, rufous-chestnut front and throat***, narrow blue-black throat-band, ***cinnamon-buff to buffy-white underparts, long deeply forked tail***, white undertail markings. Juv. has ***tawny front and throat***, buff underparts, shorter, less forked tail. Race *erythrogaster*. Swift, banking flight, low, often in large flocks with other swallows. **Voice** Migrants often silent; but gives high-pitched, semi-musical twittering *viit-viit-iit-ii-ii-ii-i'i'i'i'i'i'i'i', vii-viit-ii-viit-'i'i'i'i'*.... **SS** Tail not always long and forked; cf. heavier Cliff Swallow (tawny rump, greyish nuchal collar, short, notched tail, streaky back). **Status** Fairly common transient (Jul–Nov; Feb–Apr).

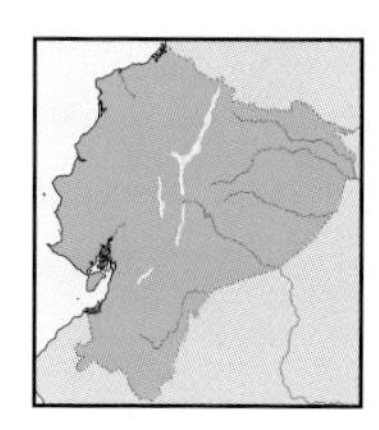

Cliff Swallow *Petrochelidon pyrrhonota* — 13–14cm

Widespread boreal migrant, mainly lowlands (to 3,300m, locally even higher). Open, semi-open areas. Glossy blackish above, some ***whitish streaks*** on back, ***buff front***, dark chestnut throat, ***greyish-buff underparts and nuchal collar, cinnamon rump, notched or square tail***. Juv. duller. Races *melanogaster* and *pyrrhonota*; others possible. Fast, low flight, regularly with Barn Swallow. **Voice** A thin *churr*, but mostly silent in winter. **SS** Chestnut-collared Swallow in SW (richer rufous underparts, rump and nuchal collar, white throat); Barn Swallow has longer, deeply forked tail, concolorous upperparts. **Status** Uncommon transient (Aug–Oct; Mar–Apr).

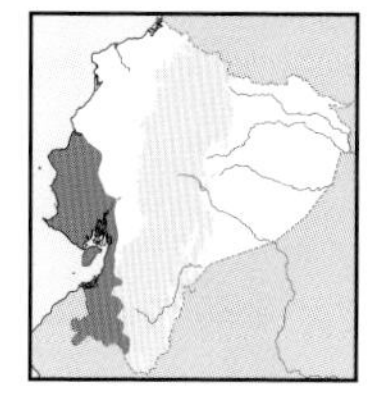

Chestnut-collared Swallow *Petrochelidon rufocollaris* — 12–13cm

SW lowlands, to subtropics in Loja (to 2,000m). Open and semi-open areas, fields, woodland, towns. Glossy blue-black crown and upperparts, some white streaks on back. Rufous-chestnut front, ***white throat and belly, orange-rufous nuchal collar, 'vest' and rump***. Race *aequatorialis*. Small flocks fly swiftly, low, over open, semi-open and water, may congregate at colonies under eaves, bridges, cliffs, etc. **Voice** Rather nasal *chrrrt-ch'ch'ch'ttt-ch-t't't-chrrrt-chrrrt; chrrrt-chrrrt* in flight. **SS** Confusion unlikely in range and habitat, see migrant Cliff Swallow with greyish-buff nuchal collar and underparts, paler rump and forehead, white throat. **Status** Uncommon.

Tumbes Swallow
White-winged Swallow
juvenile
Tree Swallow
♂
juvenile
♀
worn
juvenile
fresh
fresh
adult
non-breeding
Barn Swallow
dult
breeding
adult
breeding
juvenile
adult
pyrrhonota
juvenile
Chestnut-collared
Swallow
Cliff Swallow
adult
pyrrhonota
juvenile
melanogaster

Large 'swallows' with broader wings and fairly long, forked tails. Sexual dimorphism marked in the two migratory species; ♂♂ very difficult to identify. Brown-chested and Grey-breasted monomorphic, easily recognised by colours of upperparts and breast. Also Bank Swallow, a smaller boreal migrant – sometimes called Sand Martin.

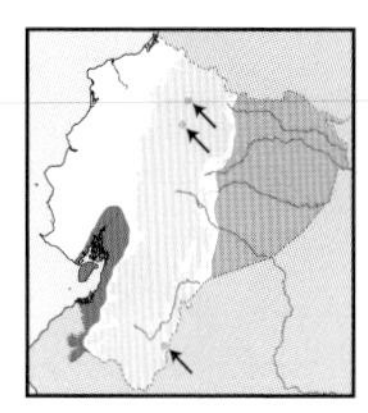

Brown-chested Martin *Progne tapera* 17–18cm

E & SW lowlands (mostly to 600m, migrant to 2,600m). Open and semi-open areas, waterbodies. ***Brown above, dusky-brown breast*** continuing as ***spotted line to central belly*** (E); dirty ***greyish-brown breast***, no spotted line, ***no contrasting throat*** (W). Whitish throat and belly to vent. Nominate (W), *fusca* (austral migrant to E). Rather strong, swooping flight, low, often gliding on bowed wings; several gather at regular perches. **Voice** Weak *chiur-chiur-chip*, buzzier, downslurred gurgling *j'jrt* or *d'jirt, dch-tirrt, dj-ji-rrrt*. **SS** Only brown martin; juv. and ♀ Grey-breasted and Purple (grey throat, darker above, latter with pale nuchal collar); Bank Swallow similar but much smaller. **Status** Fairly common.

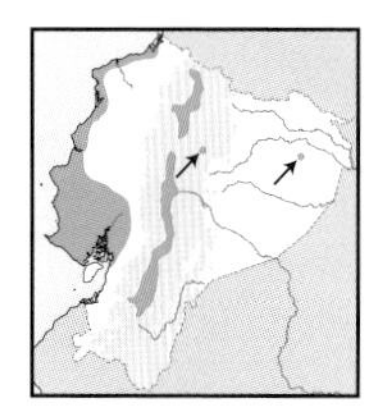

Purple Martin *Progne subis* 18–19cm

Few records in lowlands and inter-Andean valleys (to 2,600m). Open and semi-open areas. ***All glossy steel-blue***. ♀ and juv. blackish-blue above; dusky throat to breast, pale greyish below, variably ***mottled or streaked; pale forecrown and nuchal collar***, juv. duller. Race/s unknown. Glides, flapping briefly, describing circles; singles to small flocks. **Voice** Not heard in Ecuador; elsewhere low whistles; descending *cherr*, buzzy *geerrt* (R *et al.*). **SS** Ad. ♂ practically indistinguishable from Southern (which see); ♀ and juv. recall Grey-breasted but whitish front and ***nuchal collar*** distinctive; no white throat and brownish breast like Brown-chested; ♀ Southern much darker below. **Status** Rare boreal transient (Sep–Apr).

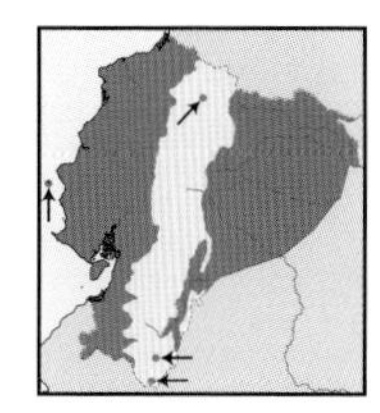

Grey-breasted Martin *Progne chalybea* 18–19cm

E & W lowlands to subtropics (to 2,000m). Mainly at waterbodies. ***Dark glossy blue-black upperparts, smoky or brownish-grey throat to 'vest'***, whitish below. ♀: less glossy above, juv. duller above, duskier below. Nominate. At considerable height, always higher than Brown-chested; long glides and short flapping bursts; thousands gather on wires and under eaves. **Voice** Rich, buzzy *ch'rrrrrt, chírr... chírr-chirrt..., chrreet-chrrert-churrr-churr....* **SS** Ad. ♂ not all glossy blue-black like Purple and Southern Martins; cf. ♀ Purple (grey extends to breast, paler front and nuchal collar) and Southern (underparts mostly blackish-brown); see ♀ Peruvian *P. murphyi* (all-grey below, grey rump; not yet recorded in Ecuador but possible). **Status** Common.

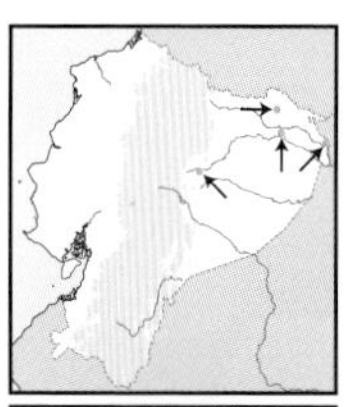

Southern Martin *Progne elegans* 19–21cm

Vagrant to E lowlands (below 400m). Semi-open areas. ***Glossy steel-blue above***. ♀ and juv. as ♂ above, ***underparts smoky to brownish-grey***, very ***densely scaled whitish***. Monotypic. Often with Grey-breasted Martin, perches on logs, stumps and dry wood above water. **Voice** Not heard in Ecuador; elsewhere a very buzzy, dry *j'rrr*. **SS** ♂ nearly identical to Purple, but tail longer and more deeply forked (very difficult to see in field); ♀ and juv. mottled and much darker below, no whitish belly like Purple and Grey-breasted. **Status** Rare, possibly casual austral transient (Apr, Jun–Aug).

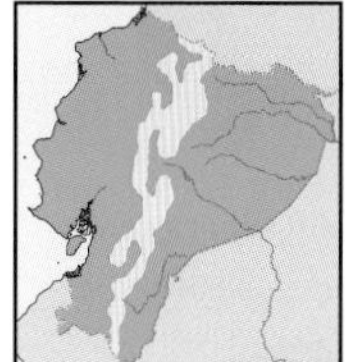

Bank Swallow *Riparia riparia* 12–13cm

Widespread boreal migrant, mostly lowlands, some in inter-Andean valleys (to 2,500m). Open and semi-open areas. ***Dull brown above***, whitish below, narrow ***dull brown breast-band***, short notched tail. Nominate. Flickering flight on rapid, shallow wingbeats, often low over water; associates with migrant swallows. **Voice** Not very vocal, sometimes a quick *b'rrt-b'rrt-b'rr....* **SS** White-thighed Swallow all dark below, Southern Rough-winged has whitish to brown rump, no breast-band, tawny throat; recalls Brown-chested Martin but smaller, different silhouette and flight. **Status** Uncommon transient (Aug–Nov, Mar–Apr). [Alt: Sand Martin]

Brown-chested Martin
fusca
tapera
♂
juvenile
♀
Purple Martin
juvenile
♀
♂
♂
Grey-breasted Martin
♀
♂
♀
adult
Bank Swallow
Southern Martin

PLATE 231: PLAIN WRENS

Four relatively plain Andean wrens of forests and edges, together with one of the commonest birds in Ecuador (and Neotropics): House Wren. Sedge Wren lives in grassy fields, mostly in *páramo*, whereas Mountain often moves with mixed flocks, and is not as tame as its congener. Rufous and Sharpe's move in noisy family groups; though not easy to observe, their 'cheery' voices are readily recognised.

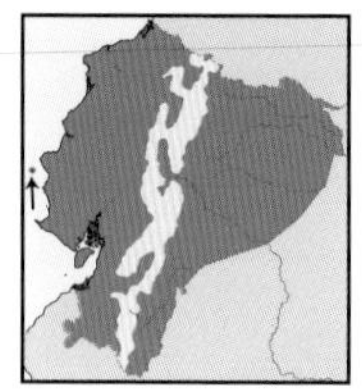

House Wren *Troglodytes aedon* 11–12cm

Widespread from lowlands to highlands (to 3,300m). Open habitats, scrub, fields and gardens. Small. Slender bill, ***dusky-brown upperparts, pale supercilium***, dusky eyestripe, wings and tail coarsely barred dusky. Buffy-white underparts, whiter throat and belly. Juv. browner and mottled. Race *albicans*. Common and conspicuous near houses and gardens, noisy when foraging; restless, mostly near ground. **Voice** Variable, fast, with melodies, wheezes, warbles, etc.: *cheet't't't, bsbsbsbs, whep-whep-whep…, tr'r'r'r-T, tch-yy, tch-yy….* **SS** Mountain Wren, more a 'wild' bird, is more rufescent, supercilium more contrasting; familiar and readily identified. **Status** Common.

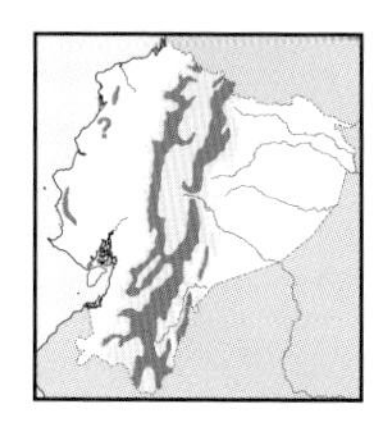

Mountain Wren *Troglodytes solstitialis* 11cm

Subtropics to temperate zone on both Andean slopes, local in coastal ranges (1,500–3,300m, 500m in coastal mountains). Montane forest, borders, woodland. ***Rufescent-brown above***, wings and ***short*** tail finely barred dusky. Contrasting ***long, buff supercilium***, rufescent-brown lores and ear-coverts. Cinnamon-buff underparts, whiter throat and belly. Nominate. Pairs forage in dense moss or vine tangles, regularly in semi-open; often with mixed flocks. **Voice** Short, fast, tinkling *tdsi-tdsi-tdsi… tdsi-í*; also *dzz-r'r'r* calls. **SS** House Wren in semi-open habitats; cf. Sharpe's and Rufous Wrens (longer tail, deeper colour, more complex voices). **Status** Fairly common.

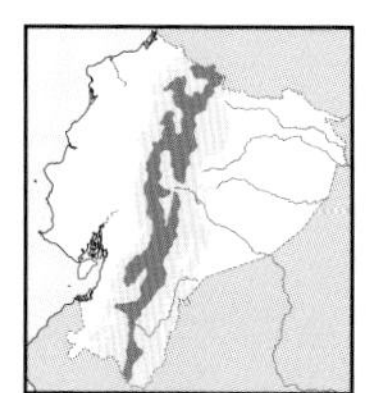

Sedge Wren *Cistothorus platensis* 10–11cm

Andes (2,800–4,000m, locally or seasonally? to 1,900m). Grassy *páramo*, adjacent fields and pastures. Small. Brown above, ***buff-streaked mantle***, wings and tail barred dusky. Whitish supercilium, ***buff underparts***, whiter central belly. Race *aequatorialis*. Skulking, remains near ground well hidden in grass, noisy; difficult to flush, but often sings from exposed perch. **Voice** High-pitched *tt-che'e'e't…* or *tw-titititititi-tw, tyer-tyer-tyer*; calls harsh *tyer-tyer-tyer….* **SS** Confusion unlikely in range; see House and Mountain Wrens, some *páramo* furnariids, and Paramo Pipit, which has a softer, more musical voice ending in short trill. **Status** Fairly common.

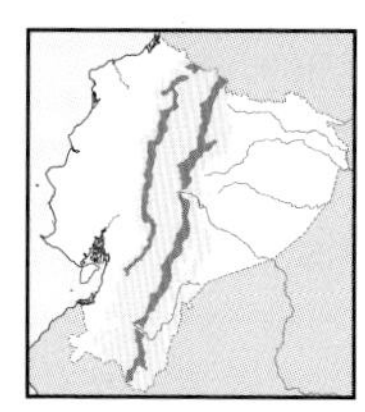

Rufous Wren *Cinnycerthia unirufa* 16–17cm

Temperate zone on E & W Andean slopes, S to Azuay in W (2,200–3,400m). Montane forest and borders. Slender black bill. ***Uniform rich rufous***, blackish lores, flight feathers and tail ***faintly*** barred dusky. Race *unibrunnea*. Groups forage in tangles, regularly in bamboo, noisily and nimbly moving through foliage, sometimes with mixed flocks. **Voice** Melodious, rapid *dldldl-tuudeé-tuudeé-tuudeé… dldldl-tuudeé…* duets or other birds join in chorus *tedee-tuudeé, tedee-tuudeé…*; also short *whik-whi-duk, whi-duk…*, sharp *tchik, t-tche-kee* calls. **SS** Lower-elevation Sharpe's Wren (darker brown, some with white faces, wings and tail more marked); Rufous Spinetail (very different behaviour, voice); cf. more musical voice of Grey-breasted Wood-Wren. **Status** Uncommon.

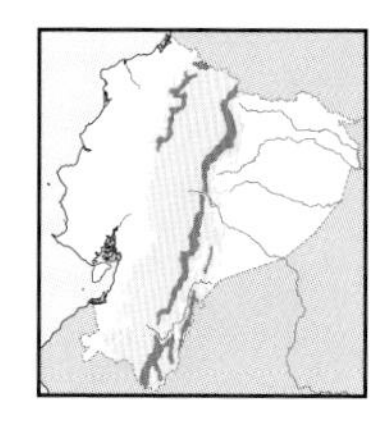

Sharpe's Wren *Cinnycerthia olivascens* 16–17cm

Subtropics on E & NW Andean slopes (1,500–2,500m). Montane forest, borders, bamboo. Slender black bill. ***Dark chocolate-brown***, flight feathers and wings ***markedly*** barred dusky, some whitish on throat. Older birds show ***white on forehead***. Nominate. Small groups forage in dense and tangled vegetation, often in bamboo and with mixed flocks; agile and restless. **Voice** Various musical phrases *deeu-deeu, bubububu, deeu, dewaap-dewaap-dewaap…* in duets or choruses; loud scolds and wheezes: *ch'ch-qrrr* or *tjtj-qrrr.* **SS** Little overlap with bolder Rufous Wren (more melodious voice, but shorter phrases); see Chestnut-breasted (different silhouette, behaviour, voice). **Status** Uncommon.

adult
juvenile
House Wren
juvenile
adult
Mountain Wren
adult
juvenile
Sedge Wren
juvenile
adult
Rufous Wren
head pattern
variations
Sharpe's Wren

Large *Campylorhynchus* are complexly patterned, rather furtive, and their voices are notably loud, variable and fascinating. The three *Pheugopedius* (formerly *Thryothorus*) have marked face patterns, often move in pairs and are very secretive and seldom seen. However, they have complex voices comprising antiphonal duets. Plain-tailed is mostly confined to bamboo.

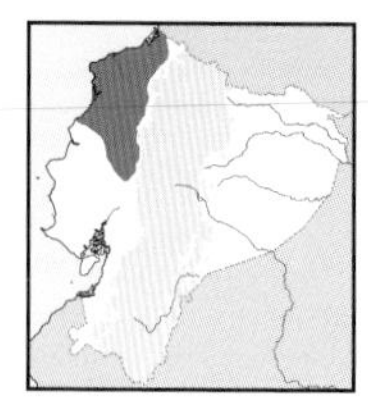

Band-backed Wren *Campylorhynchus zonatus* 18–19cm

NW lowlands to lower foothills (to 800m). Wet forest canopy and borders. Greyish-brown head speckled dusky, long buffy-white supercilium, dusky postocular stripe. ***Blackish back and wing-coverts coarsely banded buff; buffy-white below, dusky spots, ochraceous-buff*** flanks to lower belly; long barred tail. Juv. much duller, darker crown. Race *brevirostris*. Small noisy groups, often in palms, hanging upside-down, teasing bark and moss apart, etc. **Voice** Harsh jumbled *tj, tj, tj-k'k-kiki, tjak, tj-k'k-kiki*; calls loud *tjak*. **SS** Fasciated Wren occurs in drier habitats, no known overlap (see underparts and voice). **Status** Uncommon.

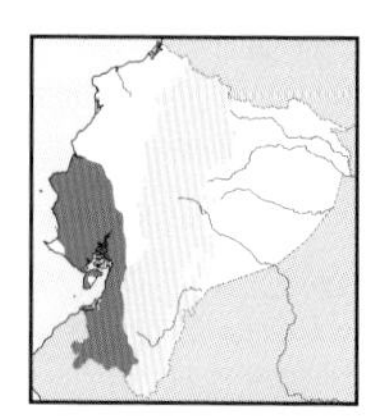

Fasciated Wren *Campylorhynchus fasciatus* 18–19cm

SW lowlands, to subtropics in Loja (to 2,500m). Arid scrub, deciduous woodland, borders, shrubby clearings. Slender bill, pale mandible. Attractive ***grey***, brown and whitish, complexly streaked and ***banded above, spotted and banded below***. Long whitish supercilium, blackish postocular stripe, and long barred tail. Race *pallescens*. Small noisy groups at various heights, conspicuous and rather tame. **Voice** Harsh, loud churring *tj-tjk-tjk-tjk...tj-tj, kikik'k'k-tijík*; calls loud *tjak, t-jak, t-jak*. **SS** No overlap with Band-backed Wren (less barred below, ochraceous-buff lower belly, less harsh voice). **Status** Fairly common.

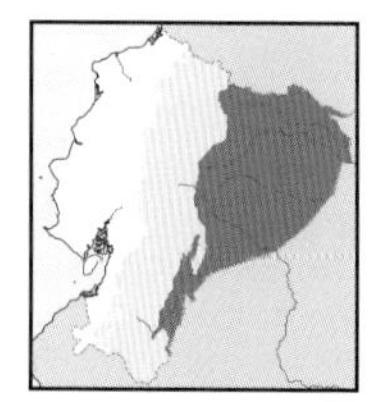

Thrush-like Wren *Campylorhynchus turdinus* 20–21cm

E lowlands to foothills (to 1,100m, locally higher). Forest canopy, borders, adjacent clearings. Dull greyish-brown upperparts lightly mottled dusky, ***long white supercilium***, dusky cheeks. ***Whitish underparts markedly spotted dusky***, plain throat, more ochraceous belly; long tail. Race *hypostictus*. Pairs to small noisy groups mostly in canopy, regular in vines and palms. **Voice** Variable, rhythmic, cheerful duets or choruses: *chóókadadOk-chóókadadOk-chóókadadOk..., chooka-choOk-chok-chok... chO-chO-chO..., chooka-choók...; chek-yooó-chok, chek-yooó-chok..., cheke-chó, cheke-chó...*; also *jiff* scolds. **SS** Unmistakable in range by plumage and voice. **Status** Fairly common.

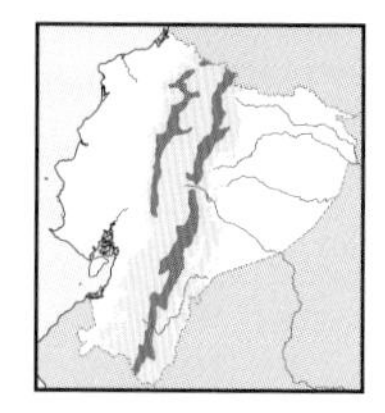

Plain-tailed Wren *Pheugopedius euophrys* 16–17cm

Temperate zone on both Andean slopes, S in W to Bolívar (2,200–3,200m). Montane forest, bamboo. Rufous-brown upperparts, ***darker crown***, long white supercilium, blackish ocular stripe, ***whitish cheeks and malar, blackish submalar***. White throat, ***some blackish breast spots***, plain below. ***Unmarked*** rufous wings and tail. Nominate (W), *longipes* (E), *atriceps* (extreme SE). Pairs in dense undergrowth, very difficult to see but readily respond to playback. **Voice** Fast, rollicking, antiphonal duets, *wo'odiri-wo'odiriü...*; calls loud *cho-kít, cho-kít...* and dry *tjktjktjk* alarm. **SS** Only *Pheugopedius* with plain tail, occurs above all other congeners. **Status** Fairly common.

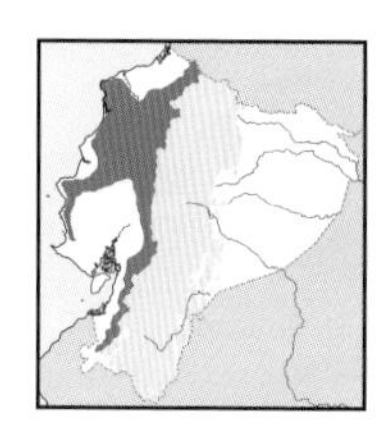

Whiskered Wren *Pheugopedius mystacalis* 16–17cm

W lowlands to lower subtropics (to 1,500m). Forest undergrowth, borders, dense secondary woodland. Greyish crown to ***upper mantle***, rufous upperparts, long white supercilium, ***heavily streaked cheeks, broad white malar, black whiskers***. White throat, greyish-buff below, browner vent. Long ***barred*** tail. Nominate. Pairs very difficult to see as skulk in very dense foliage, often high above ground. **Voice** Very melodious, liquid *wewodee-wedewo-cheewo'odu*; also 3–4 loud *kwee* calls. **SS** Confusion unlikely if seen; cf. less musical Bay and Stripe-throated Wrens, and more jumbled songs of wood-wrens. **Status** Uncommon.

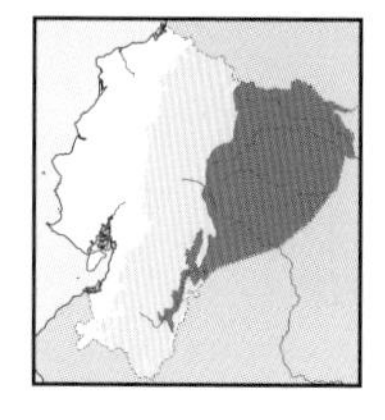

Coraya Wren *Pheugopedius coraya* 14–15cm

E lowlands to lower foothills (mostly to 700m, locally 1,200m). *Terra firme* and *várzea* forest undergrowth, borders. Brown crown to nape, rufous above. ***Black cheeks thickly streaked white, narrow white eyebrow, broad black malar***. White throat, greyish-buff breast, more ochraceous belly. Long barred tail. Race *griseipectus*. Pairs regularly with mixed flocks in vine tangles, but not especially skulking. **Voice** Musical, repeated, fast duets: *weoo?-chudchudchudchúd..., wuuo?-chuchuchú..., dee'Eeew-yów'yów'yów, deeEee'yow'yow'yow...*; shorter phrases *dj'eét-chukoow, chewEet-choo'w*; clear *chidip* calls. **SS** Confusion unlikely if seen (White-breasted Wood-Wren has tiny tail); Buff-breasted Wren's voice is louder, more staccato but less rich. **Status** Uncommon.

juvenile
adult
Band-backed Wren
juvenile
adult
Fasciated Wren
Thrush-like Wren
atriceps
adult
euophrys
adult
longipes
adult
juvenile
adult
Whiskered Wren
juvenile
Plain-tailed Wren
adult
juvenile
Coraya Wren

Five understorey wrens formerly in *Thryothorus* that are seldom seen but often heard; their voices are loud, variable, complex antiphonal duets. Small to mid-sized, with striking face patterns and banded tails. Grey-mantled is an atypical wren of upper strata; bears more resemblance to warblers, antwrens, or graytails (Furnariidae).

Grey-mantled Wren *Odontorchilus branickii* 12–13cm

E Andean foothills to subtropics, NW foothills to adjacent lowlands (1,100–1,900m in E, 100–500m in NW). Montane forest canopy, borders. Rufous forehead, ***grey crown to upperparts***, head finely streaked. ***Pearl-grey below***, whiter belly. ***Long barred tail***. Races *branickii* (E), *minor* (NW). Singles or pairs in canopy with mixed flocks, gleaning and peering into mosses, epiphytes, never green foliage; tail often held cocked. **Voice** Dry, short trill, *drdrdrdr*, also *t-swe, t-swe-sew...* (E), louder *t-swe-swe-swe-swe-swe...* in W. **SS** Recalls Equatorial Graytail even in behaviour, but see unstreaked underparts, no supercilium, cocked tail; voices differ; see Tropical and Slate-coloured Gnatcatchers. **Status** Uncommon, rarer in NW.

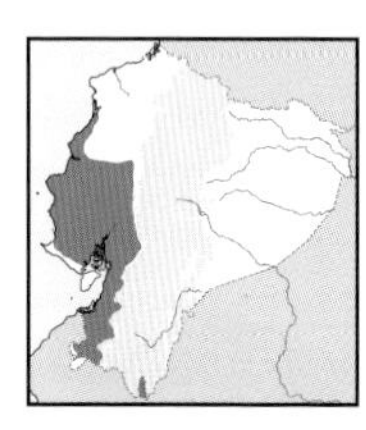

Speckle-breasted Wren *Pheugopedius sclateri* 13–14cm

SW lowlands, to subtropics in Loja; extreme SE in Zumba region (to 1,600m). Semi-deciduous to deciduous forest, borders. Rufous-brown upperparts, long white supercilium, ***face freckled black and white***; ***breast densely speckled black***. Long barred tail, ***unmarked wings***. Races *paucimaculatus* (W), *sclateri* (SE), possibly separate species. Pairs active in mixed flocks, more often near ground. **Voice** In W fast *d-ch, ch-d-weebléé, d-wlïï, d-wlïï, d-wlïï..., d-d-d, weebliblée, d-d-d, weebliblée...*; fast churring *br'r'r'reé*; in SE generally faster *d-d-ï, weeblee-ee-éü..., d-ií, weeblee-ú, weeblee-ú*. **SS** Confusion unlikely; cf. rather simpler voice of Superciliated Wren (SW). **Status** Uncommon.

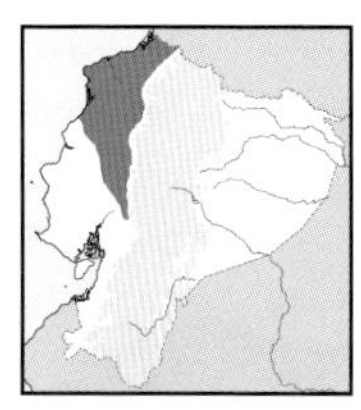

Stripe-throated Wren *Cantorchilus leucopogon* 12–13cm

NW lowlands to lower subtropics, S in Andean foothills to Azuay (to 750 m). Humid forest midstorey, borders. ***Small. Amber eyes***. Brown upperparts, wings narrowly barred black. Long white supercilium, ***whitish throat to cheeks heavily streaked blackish***, cinnamon-brown below. ***Short*** barred tail. Nominate. Pairs at various heights, often with mixed flocks, fond of dense vine tangles. **Voice** Loud, tuneless *cht-ee-weé'üü, cht-ee-weé'üü...*, steady *tcheee...tcheee...teee...teee...*; *tch'k'k'k'k* alarm. **SS** Speckle-breasted Wren (whiter, speckled below) does not reach NW; cf. more complex and melodious voice of Whiskered, more rapid and repetitive voice of Bay W. **Status** Rare, NT in Ecuador.

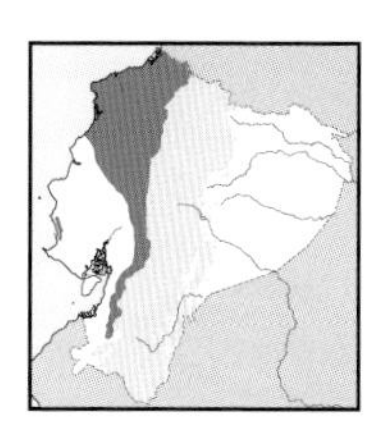

Bay Wren *Cantorchilus nigricapillus* 14–15cm

NW lowlands to foothills, S in Andean foothills to El Oro (to 1,300m). Wet forest borders, woodland. ***Black head with broad white moustachial***. Bright rufous upperparts, wings thickly barred blackish. White below barred blackish, rufescent flanks. NW birds ***uniformly barred below***. Long barred tail. Races nominate (most of range), *connectens* (NW). Pairs in undergrowth, rather curious and not difficult to see, regularly in gardens. **Voice** Loud, ringing, repetitive *dee-dee-waä-waä... t't't-tdó-tdó, t'r'r..., ch't't't't...* **SS** Unmistakable if observed; cf. more melodious voice of Whiskered; slower, more musical Stripe-throated; higher-pitched, more melodious and variable White-breasted Wood-Wren. **Status** Fairly common.

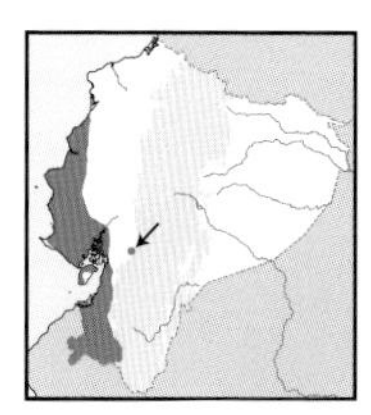

Superciliated Wren *Cantorchilus superciliaris* 14–15cm

SW lowlands to subtropics in Loja (to 1,500m). Arid scrub, deciduous and semi-deciduous forest borders, woodland, adjacent clearings. Rufous-brown above, wings heavily barred dusky, ***long white supercilium***, blackish ocular stripe. ***White throat, neck-sides to mid belly***, buffier flanks. Long barred tail. Races *baroni* (NW), nominate (SW). Pairs mostly near ground, sometimes but not always with mixed flocks, less skulking than other wrens. **Voice** Rather simple series of warbles and churrs: *chueéd-chueéd-chueéd...*; *ch'd'r'r'r-chuë-chuwee, chueë-chöw, chrrr-chue-deé, chue-deé...*; also *t-ch't't't't*. **SS** Confusion unlikely; cf. Speckle-breasted Wren's richer musical voice. **Status** Fairly common.

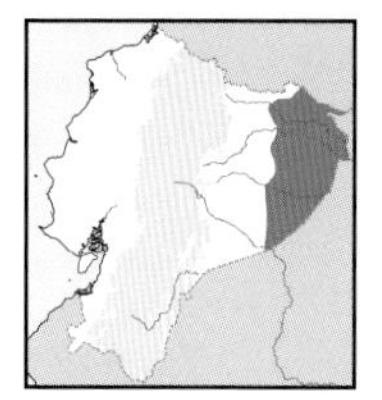

Buff-breasted Wren *Cantorchilus leucotis* 14cm

E lowlands (below 300m). *Várzea* and riparian forest undergrowth. Rufescent-brown upperparts, ***long white supercilium***, ***white freckled cheeks***. Duskier wings heavily barred blackish. White throat, ***buff breast***, more ***cinnamon*** below. Long barred tail. Race *peruanus*. Mostly in pairs foraging near ground but higher in vine tangles, not with mixed flocks. **Voice** Fast, staccato, long series *chOow'chOow-chOow'chOow..., cho'wreé-cho'wreé-cho'wreé-cho'wreé..., woroocheeree-woroocheeree...* often interspersed by clear whistles; harsh *chrr-chrr-chrr...*; calls loud staccato *chi'dok- chi'dok*. **SS** Sympatric Coraya Wren has different face pattern, plain wings; less strident voice is less staccato, but richer. **Status** Uncommon.

branickii
juvenile
minor
Grey-mantled Wren
sclateri
paucimaculatus
Speckle-breasted Wren
Stripe-throated Wren
nigricapillus
connectens
Bay Wren
iperciliaris
baroni
Superciliated Wren
adult
juvenile
Buff-breasted Wren

Microcerculus are forest-floor wrens with long legs and far-carrying, lovely songs that often represent the best indication of their presence, as both are very difficult to spot. Tiny-tailed wood-wrens also have variable melodious songs and are difficult to observe. Overlap / replacement altitudinal zones rather complex among wood-wrens, as is vocal variation throughout Grey-breasted range.

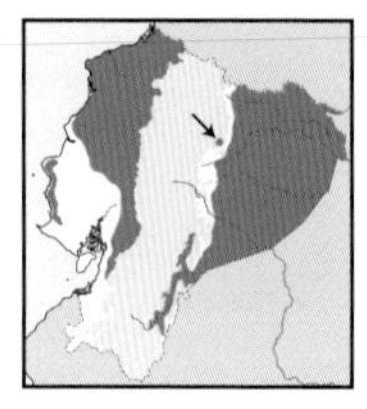

Scaly-breasted Wren *Microcerculus marginatus* 11–12cm

E & W lowlands to foothills (to 1,100m, locally higher). Wet and humid forest undergrowth, dense secondary woodland. Slender ***long*** bill. Dark brown above, ***whitish below, scaled dusky*** on sides to mid belly, brown flanks to vent. SW birds ***thickly scaled white and dusky below. Very short tail***. Nominate (E), *taeniatus* (most of W), *occidentalis* (NW). Solitary or loose pairs, not with mixed flocks, shy and inconspicuous except by voice; constantly teeters rear body. **Voice** Whistled *we-we-EE-ee.. ee...ee...ee....ee.....ee*, initially fast, then falling and at longer intervals. **SS** Confusion unlikely, voice very distinctive; Wing-banded has more complex voice. **Status** Fairly common.

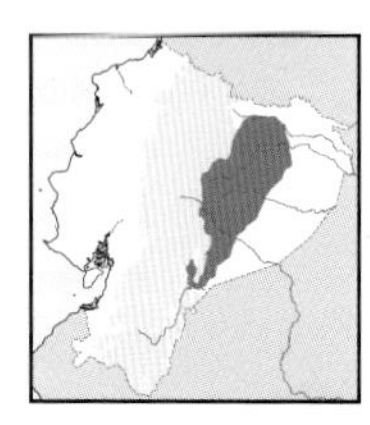

Wing-banded Wren *Microcerculus bambla* 11–12cm

E Andean foothills to adjacent lowlands (300–1,300m). Humid forest undergrowth. Brown above, duskier wings with contrasting ***white bars. Greyish-white below***, browner flanks to vent. ***Very short unmarked tail***, rather long legs. Race *albigularis*. Also shy, heard more than seen; terrestrial; visits fallen logs. **Voice** Descending and accelerating series of whistled *e-e-ee-ee-ee-ee'ee'ee'ee'ee...*, sometimes starting with 2–3+ *dué* notes; also decelerating *de-de-de-de...de...de....de...*; slowing *t-b'r'r'r-r-r...r*; metallic *chétk* calls. **SS** Scaly-breasted lacks wingbar; slower voice, notes more separated and slowing; see Banded Antbird (bolder pattern, different voice, etc.). **Status** Rare.

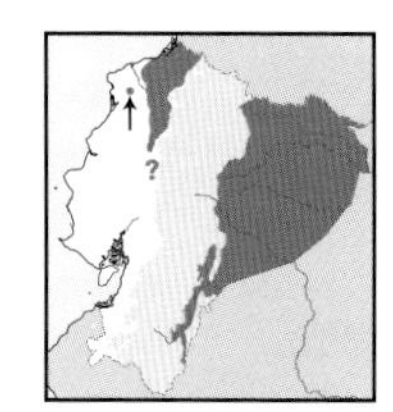

White-breasted Wood-Wren *Henicorhina leucosticta* 10–11cm

E & NW lowlands to foothills (to 1,000m). Humid forest undergrowth, borders. Black crown, ***narrow long white supercilium, cheeks speckled white and black***. Rufous-brown upperparts, faintly barred wings. ***White below***, rufous flanks. NW birds have ***white throat*** grading into grey below. ***Short barred tail***. Races *hauxwelli* (E), *inornata* (NW). Nimble pairs in dense undergrowth, mostly near ground, not very wary; holds tail cocked. **Voice** Variable series of 5+ notes: *chéé-wu'rrrr, yëër-teedyeé-teedyeé-teedyeé..., weeér-tye-tuu-eé...tye-tuu-tye-tuu..., wiir-witidi-yü-yü-yü-yü..., veeú-veeú..., veer-t'dididi...*; also *chi'rrrr* scolds, *ch'rrr-wEE-ch'rrrr-ch'rrrr* alarm calls. **SS** Higher-elevation Grey-breasted is greyer below, darker above, but differences less marked in NW; best told by voice (faster, more complex in Grey-breasted, but similar scold). **Status** Fairly common.

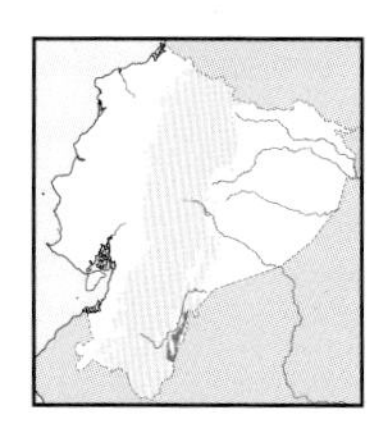

Bar-winged Wood-Wren *Henicorhina leucoptera* 10–11cm

Extreme SE in Cordillera del Cóndor (1,700–2,000m). Montane and ridgetop forest undergrowth, borders. Dark brown crown, rufous-brown upperparts, dusky wings with ***two white wingbars***. Long, narrow white supercilium, face speckled white. ***White throat to central belly***, breast-sides washed grey, brown flanks. ***Short dark tail***. Monotypic. Pairs near ground in dense tangles, difficult to see as creep rapidly; tail cocked. **Voice** Soft, ringing *ttd, weu-deé, t-we-deweé..., t-dyu-tdeé, td-dyú...*, with frequent trills interspersed. **SS** Might overlap locally with Grey-breasted, which lacks wingbars; voices similar, but rather louder, slower, less ringing in Grey-breasted. **Status** Very local but uncommon; NT globally.

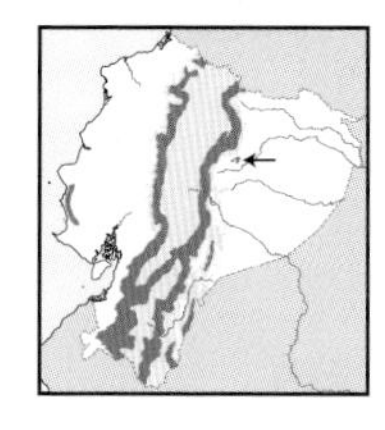

Grey-breasted Wood-Wren *Henicorhina leucophrys* 10–11cm

E & W foothills to subtropics, local in coastal ranges (1,500–3,000m, locally higher; 400–500m in coastal ranges, locally lower). Humid forest undergrowth, dense woodland, borders. Dark brown crown, rufescent upperparts. Long narrow white supercilium, face speckled white. ***Grey breast to mid belly***, rufous-brown below. ***Short barred tail***. Nominate (most of range), possibly *brunneiceps* (NW), *hilaris* (SW) has paler chest; separate species probably involved. Busy pairs near ground, difficult to see as rapidly creep through vegetation; cocks tail. **Voice** Fast loud phrases: *iiií-tee-wuü-ú, tee-wuü-ú..., cheechee-chee-woowaa, chee-chee-woowaa...*and others, more staccato than White-breasted; *k'rrrr* scolds; notable altitudinal and geographic variation. **SS** Little overlap with congeners, NW race of White-breasted has greyish chest. **Status** Common.

marginatus
juvenile
occidentalis
taeniatus
Scaly-breasted Wren
juvenile
hauxwelli
adult
Wing-banded Wren
inornata
Bar-winged Wood-Wren
juvenile
adult
juvenile
White-breasted Wood-Wren
leucophrys
hilaris
brunneiceps
juvenile
Grey-breasted Wood-Wren

PLATE 235: WRENS, DONACOBIUS AND DIPPER

Three almost terrestrial forest wrens not known to be sympatric. They have possibly the loveliest musical voices of all Ecuadorian birds. *Donacobius* is a monotypic genus whose taxonomic position has long been debated, now separated in a family of its own. It has unique social and vocal features. The dipper of rushing Andean streams is the sole representative of this remarkable family in Ecuador.

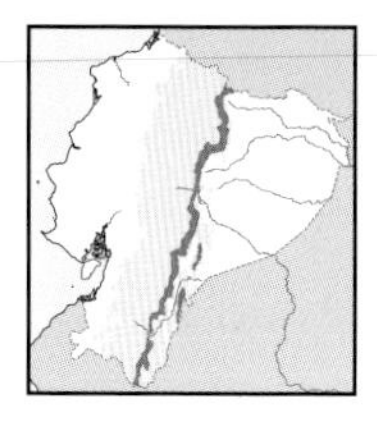

Chestnut-breasted Wren *Cyphorhinus thoracicus* 14–15cm

E Andean foothills to subtropics (1,100–2,000m). Montane forest undergrowth. Heavy blackish bill, rather long legs. Dark brown upperparts and lower belly to vent, ***orange-rufous face to mid belly, unmarked wings and tail***. Race *dichrous*. Pairs to small groups near ground, often noisy; not with mixed flocks. **Voice** Beautiful, but rather simple cadenced, clear whistles, *deeE-eeEÉ-uuuu*, last note descending, sometimes additional *deee* at end; also tremulous *deee-üüüü* and *kurrr* or *krrurr* calls. **SS** Smaller Musician Wren might overlap locally; see wing and tail barring, rufous supercilium; voices differ; see Sharpe's Wren and ♀ Uniform Antshrike (pale eyes, different bill shape, etc.). **Status** Uncommon

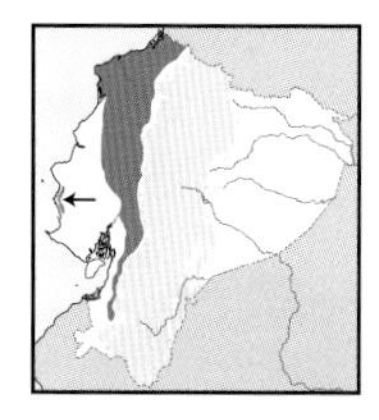

Song Wren *Cyphorhinus phaeocephalus* 13–14cm

W lowlands, S in Andean foothills to El Oro (to 900m). Wet forest undergrowth. ***Bare bluish facial skin***. Dark brown above and belly to vent, ***deep rufous throat, cheeks and breast***, wings and tail barred dusky. Nominate. Pairs to small groups near ground, often flick leaves aside or at fallen logs; not with mixed flocks. **Voice** Beautiful, complex series: *chg-chg, cuuo'ö'ö'ö'ö, chg-chg, cuu ö'ö'ö'ö, woo-wu-EE-ö, woo-wu-EE-ö..., chg-chg, cuuo'ö'ö'ö* with fluty notes interspersed; also *cheeg!* or *chuurr!* **SS** Unmistakable voice; cf. blue-faced antbirds (Chestnut-backed, Zeledon's, Ocellated), but different behaviour, voice, bill shape, plumage pattern. **Status** Uncommon, NT in Ecuador.

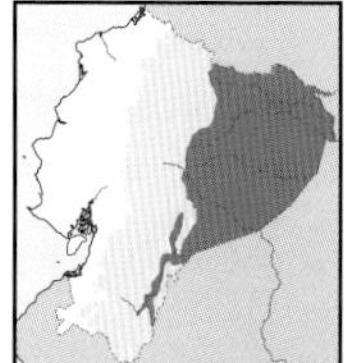

Musician Wren *Cyphorhinus arada* 12–13cm

E lowlands to lower foothills (to 1,000m). *Terra firme* forest undergrowth. Brown above and belly to vent, ***long orange-rufous supercilium, orange-rufous throat to lower breast***. Wings and tail barred black. Race *salvini*. Retiring pairs to small groups near ground, rarely joining ant-following flocks, often flicks leaf litter. **Voice** Very beautiful, memorable series of pure, musical whistles all over scale: *peeü-pü-EO-peeu-pü'ü-pu'Ü*, etc. interspersed by harsh *churr!* or *grrr!*; antiphonal duets. **SS** Might resemble some antbirds but behaviour and, especially, voice strikingly different. **Status** Rare.

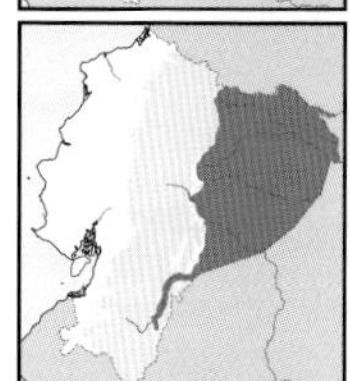

Black-capped Donacobius *Donacobius atricapilla* 21–22cm

E lowlands, local in foothills (to 600m, to 1,400m locally). Damp grasslands, lake edges, river islands. Unmistakable. ***Black cap, orange-yellow eyes***, brown back, rufescent rump, ***ochraceous-buff below***, white spot in primaries, long blackish tail, ***broad white corners***. Race *nigrodorsalis*. Pairs or small family groups perch atop bushes, fan and wag tail; noisy and conspicuous, inflates throat pouch when singing. **Voice** Large repertoire; duets utter loud whistles, bubbling and hissing notes, including *wooít-wooít-wooít..., kéew-kéew-kéew...*, harsh *geesh-geesh..., chwirr'r'rr, chwee-chwee-chwee....* **SS** Confusion unlikely. **Status** Fairly common.

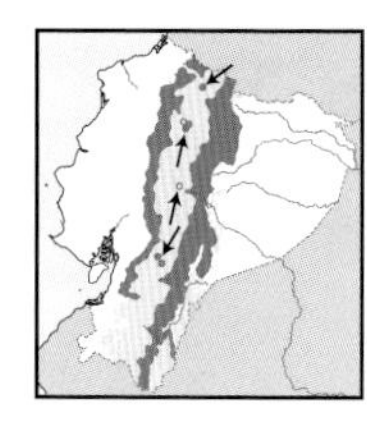

White-capped Dipper *Cinclus leucocephalus* 15–16cm

Foothills to temperate zone on both Andean slopes (700–3,800m, locally lower). Mostly forested, fast-flowing rocky rivers and streams. Unmistakable ***chunky black-and-white*** bird unique in habitat and range. Sexes alike. Race *leuconotus*. Singles or pairs perch on rocks, boulders, less often logs, next to rushing water; flies low over water while calling; feeds on aquatic invertebrates picked from surface. **Voice** Loud *tiik-tiik-tiik*, audible against noisy rivers. **SS** Unmistakable. **Status** Uncommon.

Chestnut-breasted Wren
Song Wren
Musician Wren
Black-capped Donacobius
juvenile
adult
juvenile
adult
White-capped Dipper

PLATE 236: GNATWRENS AND GNATCATCHERS

Beautiful small birds, mostly in forests, some with notably long tails, others with long bills and tiny tails; might resemble small antbirds and antwrens. Gnatcatchers range from canopy to midstorey, whereas gnatwrens occur in undergrowth. Voices are much simpler than in wrens.

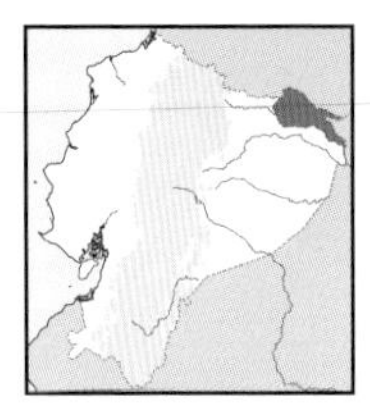

Collared Gnatwren *Microbates collaris* 10–11cm

Local in NE lowlands (below 250 m). *Terra firme* undergrowth. ***Long slender bill.*** Brown upperparts, ***long narrow white supercilium, black-and-white face pattern***, white throat, ***black breast-band***, greyish flanks, white belly, short tail. Sexes alike. Race *colombianus*. Retiring pairs regularly with ant-followers, fond of dense vine tangles. **Voice** Sad *peééeeo*, higher-pitched than Half-collared, ascending then descending at end; *p-trrrrrr'rr'rr* scolds weaker than Half-collared. **SS** Commoner Half-collared has different face pattern and lacks breast-band; voice louder, sadder also recalls Tawny-crowned Greenlet. **Status** Rare, local.

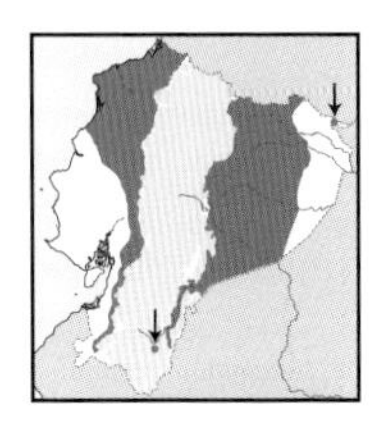

Half-collared Gnatwren *Microbates cinereiventris* 10–11cm

E & W lowlands to foothills (to 1,000m, not below 250 m in E?). Humid forest undergrowth. ***Long slender bill.*** Rufous-brown upperparts, browner crown, buff lores, ***cinnamon-tawny face***, white throat, ***black partial streaky collar***, grey belly, ochraceous rump to flanks, short tail. Nominate (W), *hormotus* (E). Pairs with mixed flocks, wags tail; fond of dense vine tangles. **Voice** Clear plaintive *eeeéuw*, *pr-t't't't't* scolds, nasal *nyeeea* calls, accelerating into *chít-chít-chít-chít…*, *chít-chít-chít-chít-nyeew*; song recalls Tawny-crowned Greenlet and Slaty-capped Shrike-Vireo but less plaintive. **SS** Long-billed Gnatwren buffier overall, bill and tail notably longer, trilled voice; also some antwrens (e.g., Checker-throated, Grey, Long-winged, Rufous-tailed). **Status** Uncommon.

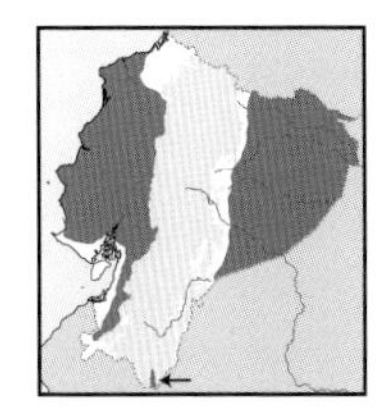

Long-billed Gnatwren *Ramphocaenus melanurus* 11–12cm

E & W lowlands to foothills (to 600m in E, locally higher; to 1,300m in W). Humid forest undergrowth, secondary woodland, borders, locally in semi-deciduous forest. ***Very long, straight bill.*** Brown above (greyer in W), ***long blackish tail tipped white. Buff face to breast***, white throat. Races *rufiventris* (W), *duidae* (NE), *badius* (SE); possibly different species in W & E. Active, not shy, pairs at vine tangles and treefall gaps, wag and flip tail; sometimes with flocks. **Voice** Rising short trill *cht'trrrriíl*, *pu-riririreéE* (W) or longer, vibrating, semi-musical *pur-ee'ee'ee'ee'ee'ee'ee*, often shorter and higher-pitched (E). **SS** Long bill distinguishes this tiny 'Cyrano' bird, also long tail. **Status** Fairly common.

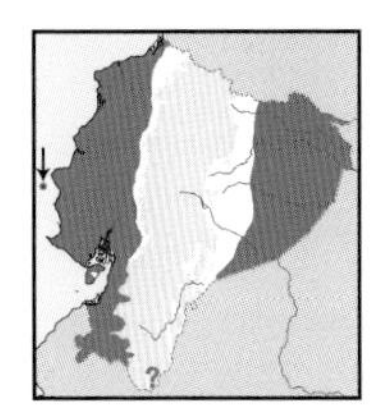

Tropical Gnatcatcher *Polioptila plumbea* 11cm

E lowlands, W lowlands to foothills (below 300m in E, to 1,500m in W). Humid to deciduous forest, woodland, borders, locally in arid scrub. ***Black crown***, grey upperparts, ***white lower face and underparts***, long ***black-and-white tail***. ♀: ***grey crown***. Races *bilineata* (W; ***largely white face***), *parvirostris*? (E); possibly different species. Pairs regularly with mixed flocks; active, flicks and wags tail constantly. **Voice** Simple, descending series of soft *sweep*, also nasal *wyeen!* notes (W); in E longer, faster, more vibrating, evenly-pitched series, *chee-chee-chee-chee*.... **SS** Confusion unlikely by striking pied plumage, agility and long 'flicking' tail; cf. Slate-throated Gnatcatcher in NW. **Status** Fairly common, rarer in E.

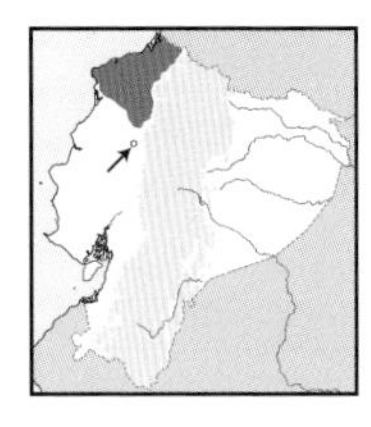

Slate-throated Gnatcatcher *Polioptila schistaceigula* 11cm

NW lowlands to lower foothills (to 750m). Wet forest canopy and borders. Mostly ***slaty-grey***, pale eye-ring, ***white belly to vent***, long blackish tail, with outer feathers narrowly tipped white. Slightly paler ♀. Juv. has ***faint pale supercilium***. Monotypic. Pairs remain high above ground, often with mixed canopy flocks; also cocks and wags tail. **Voice** Soft, short ascending trill, first note higher *trrrrrrr?*, also descending and longer *tsi-tsi-tsi-tsi-tsi*…. **SS** Darker than commoner Tropical Gnatcatcher, only belly is white, less white in tail; voices differ. **Status** Rare, rather local, VU in Ecuador.

adult
juvenile
cinereiventris
hormotus
immature
Collared Gnatwren
Half-collared Gnatwren
badius
duidae
rufiventris
Long-billed Gnatwren
♂
♂
parvirostris
♂
bilineata
♀
♀
Tropical Gnatcatcher
♂
♀
♂
Slate-throated Gnatcatcher

Forest thrushes only rarely observed due to secretive habits. Solitaires in understory, nightingale-thrushes more terrestrial, the migratory Grey-cheeked and Swainson's more arboreal. Songs of solitaires and nightingale-thrushes are beautifully musical.

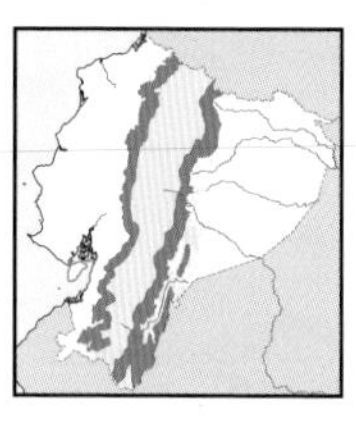

Andean Solitaire *Myadestes ralloides* 16–18cm

E & W Andes foothills to subtropics (1,000–2,500m, locally lower). Montane forest, borders. *Grey head and underparts, rufescent-brown upperparts. Silvery panel in flight feathers* and *outer tail feathers*. Juv. heavily *streak-spotted*. Race *plumbeiceps* (W), *venezuelensis* (E). Elusive, more often heard than seen; singles or pairs in undergrowth and inactive for long periods. **Voice** Lovely, slow, fluty, whistled *til'l'l, teew'ee, tee-l'l'l'l...twel'l'l...*, ventriloquial; throaty *grrou* calls; less musical, more shrill in E. **SS** Little overlap with Rufous Solitaire (lacks grey foreparts); melodious voice of Slaty-backed Nightingale-Thrush more phrased, less repetitive. **Status** Uncommon.

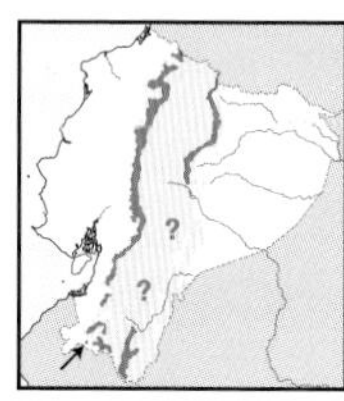

Slaty-backed Nightingale-Thrush *Catharus fuscater* 18–19cm

Subtropics on both Andean slopes, sparser in E (1,200–2,600m, locally lower). Montane forest. *Bright orange bill, legs and narrow eye-ring, white eyes*. Blackish head, *slaty-grey upperparts. Ash-grey underparts*, whitish central belly to vent. Juv. brownish, dark-eyed. Race *fuscater* (most of range), possibly *caniceps* (extreme SE). Singles or pairs in dense undergrowth hopping on ground; nervous, very difficult to see. **Voice** Musical, fluting *too-teedeë, too-teedeë, tiidiitoo!...*; calls raspy *meeeh*, high-pitched *teeee*. **SS** Confusion unlikely if seen; cf. Andean Solitaire's more musical, longer song; Spotted Nightingale-Thrush has shorter, lower-pitched phrases. **Status** Uncommon.

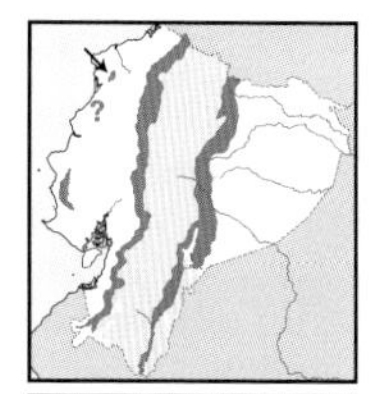

Spotted Nightingale-Thrush *Catharus dryas* 17–18cm

E & W Andean foothills to subtropics, also coastal ranges (650–1,800m, 400–500m in coastal ranges). Montane forest undergrowth. Bright *orange bill, legs and narrow eye-ring*, dark eyes. Black head, olive-grey upperparts, *yellow underparts spotted dusky*. Juv. *streaked above*. Race *maculatus*. Elusive, difficult to find; mostly in pairs hopping on ground, sometimes with mixed flocks. **Voice** Musical, liquid, short-phrased *treetlee, wuweedü...tleë...tloo-weé...tloo-wee-dü...tleeleé...*; calls dry *tcheee*. **SS** cf. voice of Slaty-backed Nightingale-Thrush (more complex, longer phrases). **Status** Uncommon.

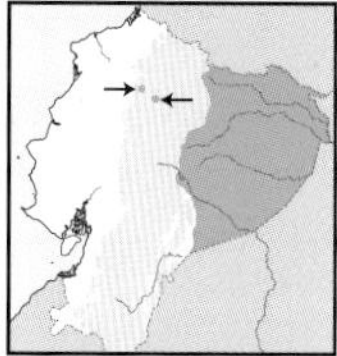

Grey-cheeked Thrush *Catharus minimus* 18–19cm

Boreal winter resident in E lowlands to foothills, occasionally highlands (mostly 200–1,300m, locally to 2,800m). Humid forest undergrowth, borders. Blackish bill, fleshy mandible base, narrow pale eye-ring. Dull olive above, *greyish lores and cheeks, whitish below*, throat-sides to lower breast and flanks *spotted dusky*. Juv. profusely streaked, more spotted below. Races *minimus*, *aliciae*. Rather elusive, mostly solitary in cover. **Voice** Seldom sings; calls nasal *véeur* or *wéaar*, and flycatcher-like *pweep*. **SS** Commoner Swainson's Thrush has brighter face pattern, buffier breast. **Status** Rare (Oct–Apr).

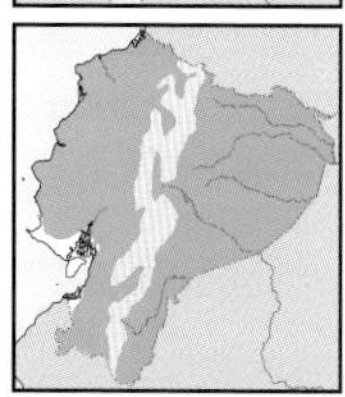

Swainson's Thrush *Catharus ustulatus* 18–19cm

E & W lowlands to subtropics, locally in highlands (to 3,000m). Forest undergrowth, woodland, borders, gardens. *Bold buff eye-ring. Olive-brown* above, *buff lores and cheeks. Whitish-buff below*, spotted dusky on throat to lower breast. Juv. *more* spotted below, streaked above. Nominate, possibly *almae*. Mostly alone, more prone to enter open areas than previous species; can be numerous. **Voice** A *quee-birr-rre, quee-birr-rre-rrerré-rrerré...* liquid song on spring passage; calls *weep-wip..., weep-weep-brrr...* and mellow *weep*. **SS** Grey-cheeked Thrush has greyer face, *no spectacles*; voices differ. **Status** Common boreal winter visitor (Oct–Apr).

Black Solitaire *Entomodestes coracinus* 22–23cm

NW foothills to lower subtropics (1,100–1,600m). Montane forest, borders. Bicoloured bill, *red eyes*. Mostly sooty-black, contrasting *white broad moustachial, white band in primaries and outer tail feathers*. Monotypic. Difficult to see, not very active, occasionally with mixed flocks; may curiously approach observers briefly. **Voice** Melancholic, far-carrying, simple, long *truuuuuü* or *treeeeeeë*, almost tinamou-like; also soft *srreeu*. **SS** Confusion unlikely. **Status** Rare, local, NT in Ecuador.

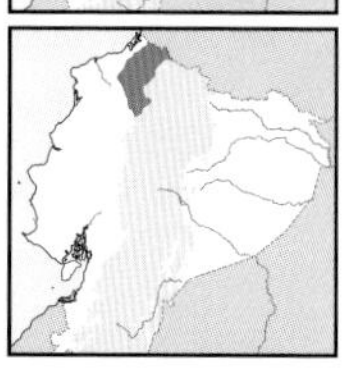

Rufous-brown Solitaire *Cichlopsis leucogenys* 20–21cm

Locally in NW foothills (400–1,200m, locally or seasonally lower and higher). Montane forest, borders. Narrow pale eye-ring, *short legs* and *short bill*. Mostly *reddish-brown, richer* head and throat, *orange-cinnamon vent*. Race *chubbi*, probably separate species. Shy; singles or pairs rarely with mixed flocks, perches more horizontally. **Voice** Rapid, variable phrases: *t'l'deuu...tl'dií... tl'deuueeé... tliwoodií, tsss-tsss...tliiidii-í...*, less flute-like than Andean Solitaire. **SS** Andean S (greyer below); Pale-vented and female Pale-eyed Thrushes (duller brown, longer-legged, longer-billed). **Status** Rare, NT.

nezuelensis
plumbeiceps
adult
juvenile
adult
Andean Solitaire
juvenile
Slaty-backed
Nightingale-Thrush
adult
fuscater
caniceps
adult
juvenile
adult
Spotted
Nightingale-Thrush
minimus
aliciae
swainsoni
almae
y-cheeked Thrush
Swainson's Thrush
juvenile
adult
Black Solitaire
juvenile
adult
Rufous-brown Solitaire

Four dark-plumaged Andean thrushes, two in open country (Great and Chiguanco) and two in humid forests and borders (Glossy-black and Pale-eyed). Interestingly, the two forest-based species are more sexually dimorphic than their open-country counterparts. Not difficult to see; they have loud musical voices. Chestnut-bellied is more forest-based, with a simpler voice. Also, the intra-tropical migrant Slaty Thrush.

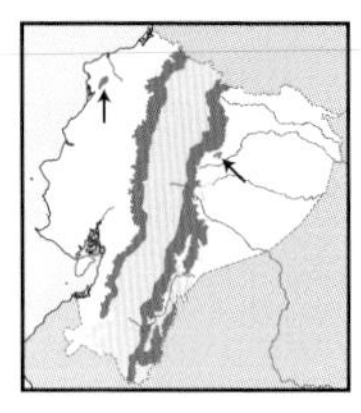

Pale-eyed Thrush *Turdus leucops* 20–22cm

Foothills and subtropics on both Andean slopes, local in NW coastal ranges (1,000–2,000m, to 500m in coastal ranges). Montane forest, borders. ***Yellow-orange bill and legs, white eyes, no eye-ring.*** ♂: ***glossy black***, ♀: mostly ***brown***, paler below, faint throat streaking, ***brown eyes, brownish bill, greenish-yellow legs.*** Monotypic. Shy, difficult to see; forages at various heights, sings from treetops. **Voice** Squeaky, musical, short phrases: *dwe-wee, dew-chedeuu, sh'sh'deeweu, cheu-cheu-cheu… ezz-ezz, chedeuu*, lower-pitched, more varied than Glossy-black Thrush, less melodic than Andean Solitaire and nightingale-thrushes. **SS** Larger Glossy-black (♂ with bright orange bill and legs, dark eyes, yellow eye-ring, ♀ darker brown, yellow eye-ring); Pale-vented and Black-billed Thrushes have whitish bellies. **Status** Rare.

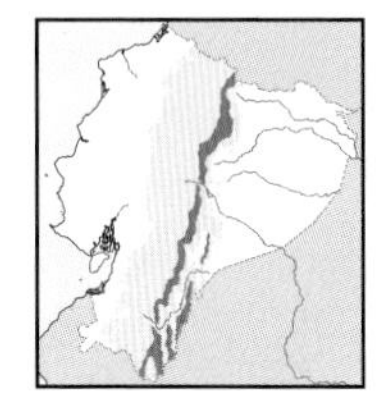

Chestnut-bellied Thrush *Turdus fulviventris* 24–25cm

E Andean subtropics (1,500–2,800m). Montane forest, borders. Greenish legs, narrow orange eye-ring. Blackish head, brownish-grey upperparts, ***rufous-chestnut belly.*** ♀: similar but ***head duskier.*** Juv. mottled dusky below, streaked whitish above. Monotypic. Retiring and wary, usually in pairs foraging low, sometimes emerges in adjacent clearings or briefly joins mixed-species flocks. **Voice** Fast, hesitant *che'e'e'e'e-cherr, cheert-chee'e'e'e, chuwirt, eet-eet*; less melodious than sympatric Glossy-black and Great Thrushes, Spotted Nightingale-Thrush and Andean Solitaire; clear *peet* calls. **SS** Only highland thrush with chestnut belly; cf. other species' voices. **Status** Uncommon.

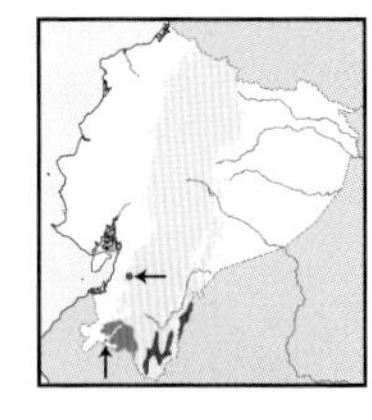

Slaty Thrush *Turdus nigriceps* 21–22cm

S subtropics (1,400–1,800m, locally to 1,000m). Montane forest canopy, borders. ***Bright yellow bill and legs, narrow yellow eye-ring.*** Blackish head, ***slaty-grey upperparts. Grey underparts***, white throat coarsely streaked black, white central belly to vent. ♀: dark brown, brown bill, greenish legs, ***yellow eye-ring***, paler underparts, ***white throat streaked dusky.*** Nominate. Retiring, intra-tropical migrant post-breeding (Jan–May). **Voice** High-pitched *too-tlï-hele-tlïï-ii-i…tlïjeelele-tlï-zz…tle-zz-leelë…tleë-le-le*; calls dry *kle-ë*, harsher *kaa*. **SS** ♀ with Pale-eyed (no eye-ring, paler overall), Glossy-black (darker), Ecuadorian (paler, broad eye-ring), White-necked and Black-billed Thrushes (white pectoral crescents). **Status** Rare, local, seasonal migrant; NT in Ecuador.

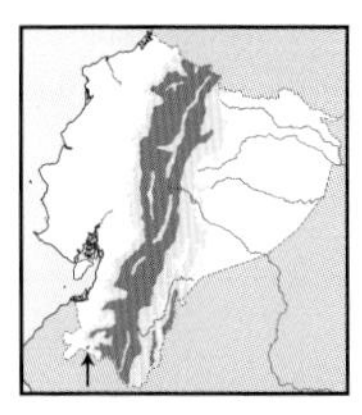

Great Thrush *Turdus fuscater* 30–33cm

Andes (2,500–4,000m, locally higher). Montane forest borders, woodland, clearings, open areas. ***Bright orange bill and legs, orangey eye-ring. Sooty to brownish-black.*** ♀: ***lacks yellow eye-ring.*** Juv. duller overall. Races *quindio* (S to Azuay), *gigantodes* (N to Azuay). Widespread in anthropogenic habitats, walks and runs on ground; perches high to sing. **Voice** Rapid, variable, musical *pwyee-pyee, p't't, wyeer-wyeer, tutüdüt, whee-tyduw*; more frequent calls: *chuup-chuup-chuup, wyeert-wyeert* or *keeyee-keeyee*; more varied than Glossy-black, more complex than Chiguanco Thrush. **SS** Smaller, glossier Glossy-black Thrush (♀ brown); Chiguanco duskier, paler, dark eye-ring. **Status** Common.

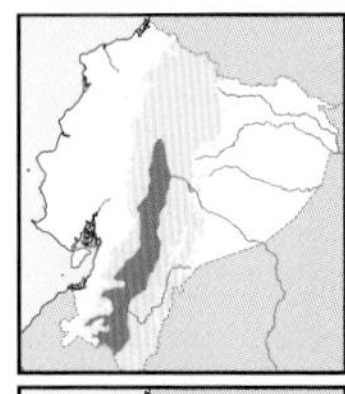

Chiguanco Thrush *Turdus chiguanco* 27–29cm

Andes N to Cotopaxi (1,500–3,200m). Arid scrub, agriculture, towns. Orange bill and legs, ***narrow grey eye-ring. Dusky brownish-grey***, darker wings and tail. Sexes alike. Juv. duller brown and mottled. Race *conradi.* Conspicuous, rather tame near houses, runs and hops; very rarely with Great Thrush. **Voice** Rather simple but pleasant, musical *d'wed-d'wed, se-se…* or *wee-dee, wee-dee, seu-seu…*often jumbled at end; 4–5 loud *chee-chee-chee*, also *tzep-tzep-tzep….* **SS** Larger Great Thrush (*gigantodes* race) is darker than Chiguanco, has yellow eye-ring; voice more complex. **Status** Common.

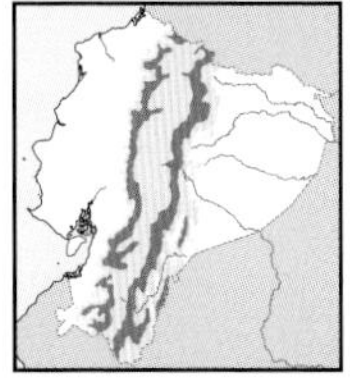

Glossy-black Thrush *Turdus serranus* 24–26cm

Subtropics to temperate zone on both Andean slopes (1,500–2,800m, locally higher). Montane forest, borders. ***Bright orange bill and legs, narrow yellow eye-ring, dark eyes.*** ♂: ***glossy black***, ♀: ***dark brown, yellowish-brown bill and legs, narrow yellow eye-ring***, faint throat streaks. Race *fuscobrunneus.* More wary and arboreal than Great Thrush, rarely walks on ground or breaks cover. **Voice** Rather weak, short, melodic *dee-dee-do-dee'et, dee-dee-doo'it*, simpler than Great Thrush; rapid *kee-kee-kye-kye….* **SS** Smaller and glossier than Great; Pale-eyed Thrush lacks eye-ring in both sexes. **Status** Fairly common.

immature
♀
♂
e-eyed Thrush
♀
juvenile
♂
Chestnut-bellied Thrush
uvenile
♀
♂
Slaty Thrush
♂
quindio
♀
♂
gigantodes
quindio
immature
juvenile
Great Thrush
juvenile
adult
immature
Chiguanco Thrush
♀
♂
juvenile
Glossy-black Thrush

Typical thrushes of lowlands, three species in W, four in E. All drab and rather difficult to recognise by plumage alone. Voices, luckily, differ, some notably so. Lawrence's is among the most prolific mimics in the Neotropics. Most species are forest dependent, but at least Black-billed and Ecuadorian tolerate moderate degrees of habitat disturbance.

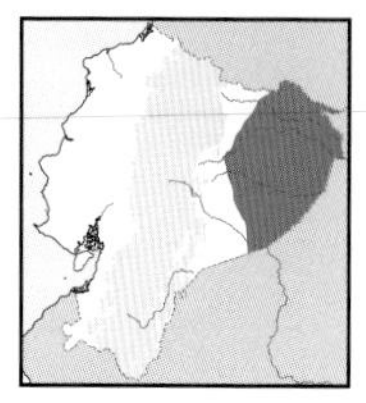

Hauxwell's Thrush *Turdus hauxwelli* 22–23cm

E lowlands (below 300m). *Várzea* and riparian forest, woodland, river islands. ***Dark bill, grey eye-ring. Mainly dark rufescent-brown***, richer below, ***buff streaks on paler throat***, white central belly to vent. Monotypic. Shy forest dweller at various heights, sometimes with mixed flocks. **Voice** Rising-falling *wlo-clee-wlo-dew... wlo-wloo-woít-woít...t-woiít*, regularly repeated; *vyéek-vyéek!...* and more musical *würll-jek-jek*; apparently also imitations. **SS** Darker brown than sympatric thrushes, no yellow eye-ring and bill as Lawrence's and Várzea Thrushes *T. sanchezorum* (not yet recorded in Ecuador) ; no white throat crescent as White-necked and Black-billed; cf. White-necked (monotonous voice). **Status** Rare.

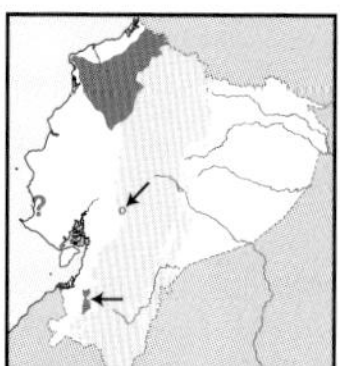

Pale-vented Thrush *Turdus obsoletus* 22–23cm

NW foothills, locally S in Andes to El Oro (550–1,100m). Humid forest interior, borders. ***Dark bill and eye-ring***. Dark brown above, ***paler below***, throat streaked brown, ***white belly to vent***. Juv. ***heavily*** spotted. Race *parambanus*. Arboreal and shy, sometimes joins mixed-species flocks. **Voice** Melodious, far-carrying, rather fast *w'l, clee-clëo-clëo-eet, widï...weeoweët, clëo-weet...*; calls simple *wep*, loud *byeek-byeek*. **SS** White-throated Thrush (yellow bill, white throat crescent, yellow eye-ring, slower, less quavering voice); Ecuadorian (paler, yellow bill, broad eye-ring). **Status** Rare, somewhat local.

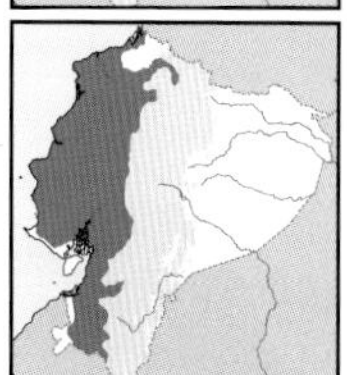

Ecuadorian Thrush *Turdus maculirostris* 22–23cm

W lowlands to subtropics, locally to inter-Andean valleys (to 2,000m, locally higher). Deciduous to humid forest borders, woodland, adjacent clearings. ***Greenish-yellow bill, broad yellowish-buff eye-ring***. Dull olive-brown above, ***buffier below, white throat streaked dusky***. Monotypic. Wary but regular in semi-open habitats, mainly arboreal, sometimes join mixed flocks. **Voice** Short melodic *w'l-wo-ö, clyë...wleeö-clyë, wo-ö, clyë...*; cat-like querulous *nyeew?*, harsh *chag* or *chuk, cleo, chag-chag-chag...* calls; also short version of song: *cloö, weelö...cloö, cleë, wee-ë...*. **SS** Broadly overlaps with notably different Plumbeous-backed Thrush; also White-throated Thrush. **Status** Common.

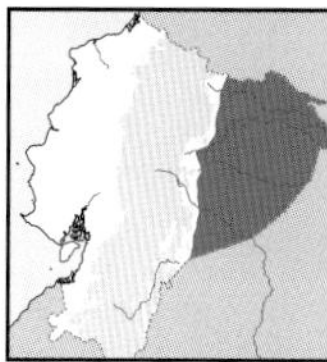

Lawrence's Thrush *Turdus lawrencii* 22–23cm

E lowlands (to 600m). Higher strata in *terra firme* and *várzea*. ***Yellow bill and eye-ring***. Brown upperparts, ***whitish-buff throat narrowly streaked dusky, warmer brown below***. ♀: ***faint*** eye-ring. Monotypic. Arboreal, unobtrusive even when singing; may join mixed flocks. **Voice** Extraordinary imitator with wide repertoire, rarely repeating same phrase twice in song bout, mimicry sometimes begins with far-carrying *klo-klo-tlee-lí?*; calls *kup-kup-kít* or *kup-kup*. **SS** Only Amazonian thrush with yellow bill and eye-ring; ♀'s duller soft parts might resemble Hauxwell's. **Status** Uncommon.

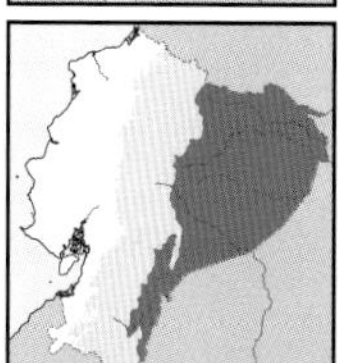

Black-billed Thrush *Turdus ignobilis* 22–23cm

E lowlands to lower subtropics (to 1,200m, locally 1,700m). Forest borders, woodland, clearings. ***Black bill, dark eye-ring***. Dull ***olive-brown above. White throat streaked brown, faint white throat crescent. Greyish-buff underparts***. Juv. paler, ***densely*** spotted. Race *debilis*. Rather tame, only thrush in open areas of E. **Voice** Soft, pleasant *clee-ootoot, cle-ee-clee-ee...clee-oweet, clee-oweet, clee-clee...*, often repeating phrases in sequences; calls loud *wee-E!*, short *kwet!, kwet-keeyu*. **SS** Bill, eye-ring and throat crescent separate from sympatric, more forest-based thrushes. **Status** Fairly common.

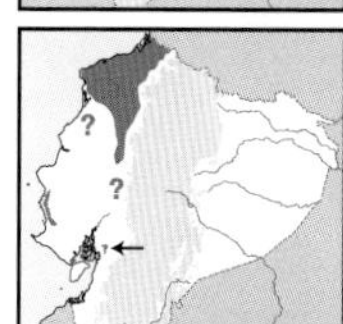

White-throated Thrush *Turdus assimilis* 22–23cm

Local in W lowlands, mainly NW (to 600m). Humid forest. ***Dull yellowish bill, yellow eye-ring, white throat densely streaked brown, white throat crescent, ochraceous-buff below***. Race *daguae*, perhaps valid species. Arboreal, sometimes join mixed flocks; elusive. **Voice** Musical, somewhat monotonous *clee-clë, woodoocleë, woodoocleë, woö-clee-clë...*; simple *yep!, weik*. **SS** Pale-vented Thrush (warmer underparts, no white in throat, dark bare parts). **Status** Uncommon, NT in Ecuador.

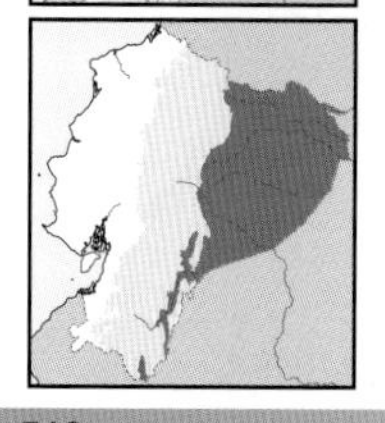

White-necked Thrush *Turdus albicollis* 21–22cm

E lowlands to foothills (to 1,000–1,100m). *Terra firme* forest, borders. ***Dusky bill, narrow orange-yellow eye-ring***, greenish legs, ***white throat densely streaked brown, white throat crescent; greyish-buff below***. Juv. ***profusely streaked above***. Race *spodiolaemus*. Mostly in undergrowth, wary, often joins mixed flocks at ant swarms. **Voice** Melancholic, far-carrying, monotonous *d'yee-clo-woo, dlye-clo-woo, dlye-glo-öö...*; harsh *guk* calls. **SS** Habitat differs from Black-billed Thrush, which has dark eye-ring, is duskier and has fainter throat crescent and streaks; voice slower than Hauxwell's. **Status** Uncommon.

Pale-vented Thrush
adult
juvenile
Hauxwell's Thrush
adult
juvenile
juvenile
Ecuadorian Thrush
adult
♀
♂
Lawrence's Thrush
juvenile
adult
juvenile
Black-billed Thrush
adult
juvenile
White-throated Thrush
adult
White-necked Thrush

Two distinctive thrushes with limited ranges in deciduous and semi-deciduous forest, woodland and edges. Also, two large, dull-coloured mockingbirds, with neat face patterns. In open and semi-open areas, Tropical has recently become established and is still spreading. Páramo Pipit is the only breeding representative of its widespread family.

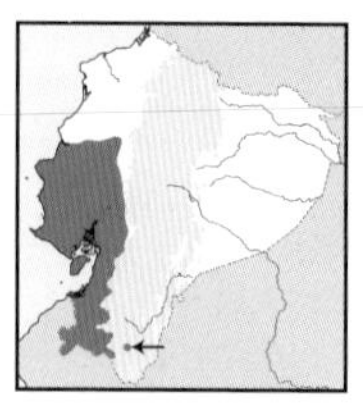

Plumbeous-backed Thrush *Turdus reevei* 22–23cm

SW lowlands to subtropics (to 1,600m). Deciduous to semi-deciduous forest, woodland, scrub. Pale yellow bill, grey eye-ring, ***white eyes, yellow legs. Bluish-grey upperparts.*** Buffy-white throat to lower breast, dusky streaks on throat, ***buff belly.*** Sexes nearly alike. Juv. ***heavily streaked buff,*** greyish-brown above, buff below. Monotypic. Rather tame and conspicuous, often in groups and joins mixed flocks; seasonal movements. **Voice** Fast, clear warbling *cleë-cleeo-leoleelee-cleëo-lelele... cleo-lelele...cleeowi-cle-ë...*; calls piercing, descending *pzeeeü* or *pi-tzeeü.* **SS** No known overlap with darker Slaty Thrush (grey underparts, dark eyes, brown ♀); sympatric Ecuadorian Thrush differs strikingly. **Status** Uncommon.

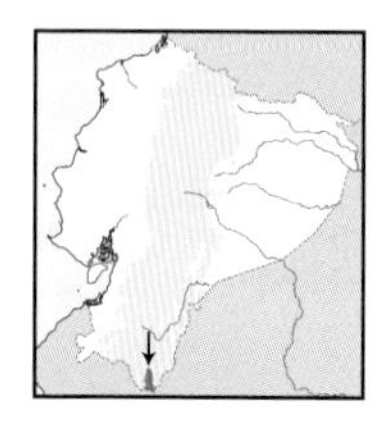

Marañón Thrush *Turdus maranonicus* 21–22cm

Extreme SE in Zumba region (650–1,900m). Forest borders, woodland, adjacent clearings and gardens. Black bill, narrow pale eye-ring, greenish legs. Brown upperparts, ***whitish underparts markedly spotted brown,*** especially on breast-sides, solid white central belly to vent. Sexes alike. Juv. ***streaked*** above. Monotypic. Rather tame and conspicuous, often feeds in recently cleared areas and shady woodland alike. **Voice** Melodious but rather simple and monotonous *blë-err, blü-bleë-rr, woü-blë-blërr...wlöö-bliu, wlöö-bliu...*; calls simple *chüp.* **SS** Confusion unlikely in limited range. **Status** Fairly common but very local.

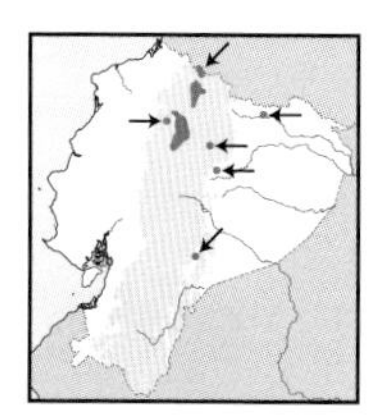

Tropical Mockingbird *Mimus gilvus* 24–26cm

N inter-Andean valleys, locally on E & W slopes (scattered records at 600m, 1,900–2,600, and 3,100m; locally higher and lower). Non-forested habitats. ***Yellow eyes. Long white supercilium, black ocular stripe. Grey upperparts,*** blackish wings and tail, two white wingbars. ***Pearl-grey underparts,*** washed pale yellowish on breast-sides. ***Broad white tail tip.*** Juv. browner, ***spotted chest, dark eyes.*** Race *tolimensis.* Singles or small conspicuous, tame groups; apparently a recent, and still-spreading arrival in Ecuador. **Voice** Rambling musical series of clucks, chucks and wheezes: *trüdeoo-drëë-drütdee..., trü-duu-deet..trü-dëëd, dreeeow-drrow-tched-tchou...* and others; harsh *jép, jeép'jeép.* **SS** Long-tailed Mockingbird has complex face pattern, streaked upperparts, etc. **Status** Probably increasing, currently rare.

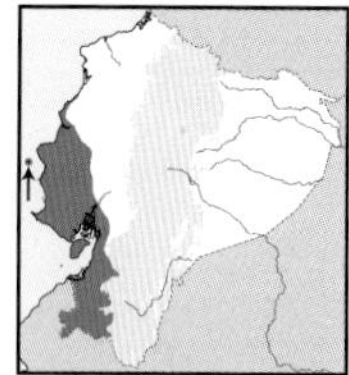

Long-tailed Mockingbird *Mimus longicaudatus* 28–30cm

SW lowlands, to subtropics in S (to 1,500m, locally higher). Arid scrub, borders, fields, gardens. Dusky-brown above, ***dull streaking;*** complex face pattern with long whitish supercilium. Whitish underparts, ***duskier breast and flanks;*** dusky wings with ***large white patch,*** very long brown tail broadly tipped white. Races *albogriseus* (most of range), *platensis* (La Plata Island). Conspicuous, active and noisy, mostly in pairs or groups; long glides. **Voice** Various chucks, wheezes, whistles and gurgles: *ghouk-ghouk, clák-clák, gzz-gzz, go-gk, go-gk...tzi-pú, tzi-pú....* **SS** Unmistakable in range. **Status** Common.

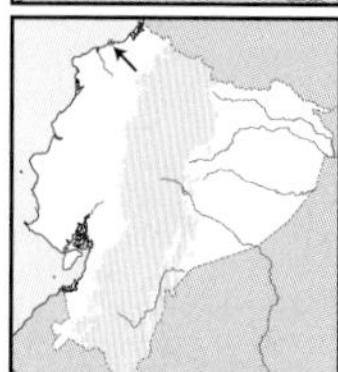

Red-throated Pipit *Anthus cervinus* (not illustrated) 14cm

Single record from NW coast (Río Verde; D. Brinkhuizen). Coastal scrub. Resembles Páramo Pipit but has more ***buffy face and eyebrow,*** supercilium ***nearly reaches nape, prominent dark malar,*** underparts rich buff, ***breast to flanks heavily streaked blackish,*** heavily streaked rump, ***two white wingbars.*** Pale yellow mandible base. Monotypic. On ground at high-water mark next to coastal shrub, shy. **Voice** Gives 1–3 sharp *psssssii* in flight; elsewhere very high, slightly falling *speeeeeh* or *pseeeeu* (LJ). **SS** Resident Paramo Pipit never in lowlands; other migrant pipits might eventually be recorded. **Status** Unexpected boreal vagrant (Mar).

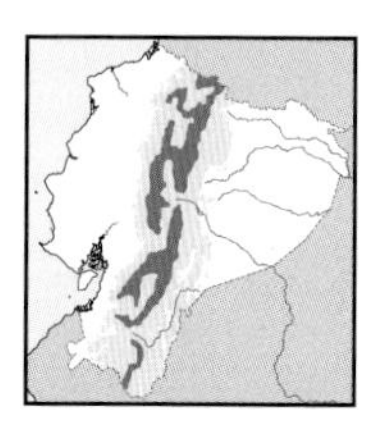

Páramo Pipit *Anthus bogotensis* 15cm

Andes (3,000–4,300m). Grassy, dry and shrubby *páramo,* adjacent fields. ***Slender*** bicoloured bill, ***pinkish legs.*** Upperparts ***streaked*** black and buff, ***long buffy-white supercilium; buffy-white below,*** some ***dusky spotting on breast,*** central belly white. Fairly long ***notched tail,*** outer feathers whitish. Nominate. Singles or pairs on ground, not easy to see; often seen when suddenly escapes at one's feet. **Voice** Perched birds utter a warbling *tss-tleedleelee, tss-tleedleelee...*; complex *dss-dleedlee-dsds, ñzzzzzz, chi'i'i'i'i'...t'e'e'e'e...* buzzy warbling and wheezing in display flight. **SS** ♀ Plumbeous Sierra-finch has heavier conical bill, coarse streaking below. **Status** Uncommon.

juvenile
adult
Plumbeous-backed Thrush
juvenile
adult
Marañón Thrush
adult
juvenile
Tropical Mockingbird
albogriseus
adult
platensis
adult
juvenile
Long-tailed Mockingbird
juvenile
adult
Páramo Pipit

PLATE 241: PEPPERSHRIKES AND VIREOS

Peppershrikes and shrike-vireo are fairly large vireos with heavy, hooked bills and patterned heads; all are quite difficult to see. Fortunately, their voices are distinctive. While peppershrikes are melodious, the shrike-vireo endlessly repeats a simple whistle. The two vireos on this plate were only recently discovered in Ecuador; both are quite distinct from 'typical' vireos on the next plate.

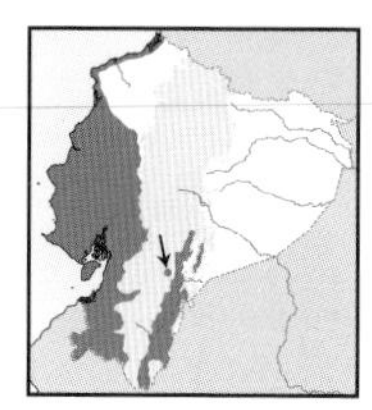

Rufous-browed Peppershrike *Cyclarhis gujanensis* 15 cm

W lowlands to foothills, to subtropics in Loja; subtropics in SE Andes (to 2,000 m in W, locally higher; 900–1,900 m in SE). Deciduous to semi-deciduous forest, borders (W), montane forest canopy, borders (SE). ***Pale horn bill, pink legs.*** Broad ***rufous-chestnut supercilium***, olive upperparts, ***lemon-yellow cheeks to breast***, whitish below. In SE darker, ***more chestnut on head, olive yellow and grey below.*** Races *virenticeps* (W), *contrerasi* (SE). Mostly in pairs, skulking in cover, mainly in higher strata; often with mixed flocks. **Voice** Variable, melodious *dre'e-dre'e-dre'ü*; *chew-dreu-dreu-chew-dreu-dreu-dre'üü*; *cheedu-e-e-e'uü* and others. **SS** See Black-billed Peppershrike in SE and NW (black bill, grey legs, little rufous on head). **Status** Fairly common.

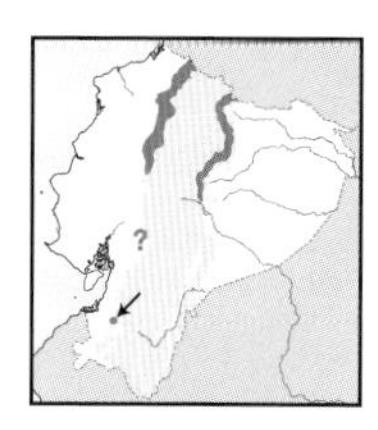

Black-billed Peppershrike *Cyclarhis nigrirostris* 15 cm

Foothills to subtropics in E & W Andes, mostly N (650–2,300 m). Montane forest and borders. ***Bill mostly black, pale eyes, grey legs. Short chestnut supercilium***, olive above, ***olive-yellow cheeks to breast-sides***, greyish throat to vent. Races *nigrirostris* (E), *atrirostris* (W). Pairs mostly in canopy, heard much more frequently than seen, often with mixed flocks. **Voice** Resembles previous species but more melodious and cadenced: *deeree-deeree-de'ü*; *deeree-deeree-deeree-e'ü*; *derere-derere-teewö*, and others. **SS** Might prove sympatric with Rufous-browed in SE; see soft parts, crown and voice. **Status** Uncommon.

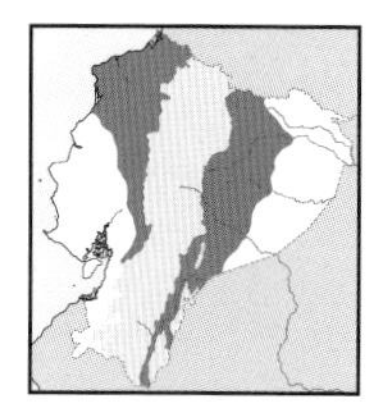

Slaty-capped Shrike-Vireo *Vireolanius leucotis* 14 cm

E & W lowlands to foothills (to 1,100 m). Canopy and borders of humid forest, secondary woodland. ***Pale grey eyes. Bluish-grey head***, with ***long and broad lemon-yellow eyebrow***, yellow teardrop, ***white streak on cheeks*** (E); olive above, ***lemon-yellow below.*** Nominate (E), *mikettae* (W), possibly separate species. Pairs often with mixed flocks, rather sluggish and not easy to spot as remain in higher strata. **Voice** Loud, tireless, slightly descending *teeeeeuu'* every *c.*1 sec, become more penetrating after several repetitions; in E shorter *teeeu'*, with more descending effect. **SS** Unmistakable if seen; voice recalls Tawny-crowned Greenlet but shorter, less plaintive. **Status** Uncommon.

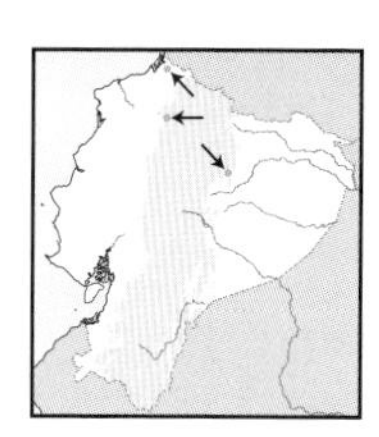

Yellow-throated Vireo *Vireo flavifrons* 14–15 cm

Three records from NW & NE lowlands and foothills (100 m, 1,495 m). Treed clearings. Olive above, grey rump, ***yellow supraloral and spectacles***, dark lores, ***lemon-yellow throat and breast***, contrasting ***white*** belly to vent. Dark wings, with ***two whitish bars.*** Monotypic. Singles search for insects low down, loosely associated with other insectivores; elsewhere, singles with mixed flocks in canopy. **Voice** Pleasing, rather buzzy whistles, well-separated *rreeyoo-rreeyoo-rreeyoo* (R *et al.*, HBW). **SS** See boreal migrant warblers, especially Blackburnian, Blackpoll, ♀ Cerulean, all with white wingbars and yellowish breasts. **Status** Rare vagrant (Nov, Jan, Feb).

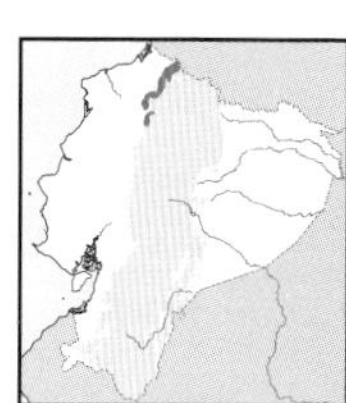

Chocó Vireo *Vireo masteri* 14 cm

NW foothills to adjacent lowlands (800–1,500 m). Canopy and borders of steep, wet montane forest. Dark olive above, ***long yellowish supercilium, olive ocular stripe***, whitish throat, yellow rest of underparts. Dusky wings with ***two conspicuous whitish wingbars.*** Monotypic. Singles or pairs in outer foliage, gleaning undersides of leaves, somewhat like a *Setophaga* warbler, sometimes with mixed flocks. **Voice** Short, musical, high-pitched warbling *chi'chipít-chi'piiít-chi'ii'ii'ií...*, downslurred, almost trilled at end; also sharp, brief *chip* calls. **SS** See locally sympatric Brown-capped and Red-eyed Vireos (no wingbars, voices differ markedly); also boreal *Setophaga* warblers (Blackpoll, ♀ Cerulean, ♀ Bay-breasted). **Status** Rare, local, EN.

virenticeps
contrerasi
Rufous-browed Peppershrike
nigrirostris
atrirostris
Black-billed Peppershrike
mikettae
♂
leucotis
♂
♀
juvenile
Slaty-capped Shrike-Vireo
Yellow-throated Vireo
Chocó Vireo

Plain, unadorned vireos; one montane species readily distinguished by plumage and voice, but Red-eyed and Yellow-green are particularly difficult to identify by sight, as only subtle differences in facial pattern exist. As Yellow-green is a boreal migrant, vagrants might appear within the range of the most similar races of Red-eyed.

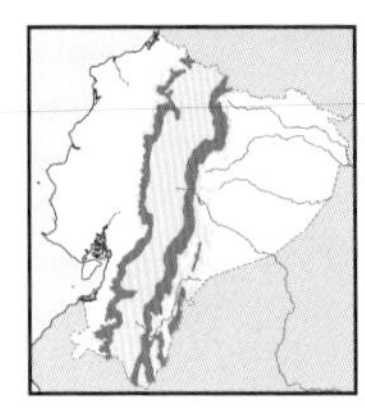

Brown-capped Vireo *Vireo leucophrys* 12–13 cm

E & W Andean subtropics, local in foothills and Andean valleys (1,300–2,700 m, locally lower). Forest canopy, borders, shrubby clearings. ***Brown crown, long white supercilium, pale yellowish underparts***, whiter throat. Nominate (E), *josephae* (W). Mostly pairs, commonly in mixed flocks at various heights; active, deliberate. **Voice** Short, fast, hesitant warbling *diurlee-diuryee, diurlee-dee-diuryiyi…*; also buzzy, descending *shrieeu* or *zreeeu*. **SS** Regularly at higher elevations than other resident vireos. Larger Red-eyed and Yellow-green have different face patterns, grey crowns, Chocó has wingbars; also some hemispinguses (richer underparts) and tyrannulets (most with wingbars), especially Sooty-crowned Tyrannulet. **Status** Common.

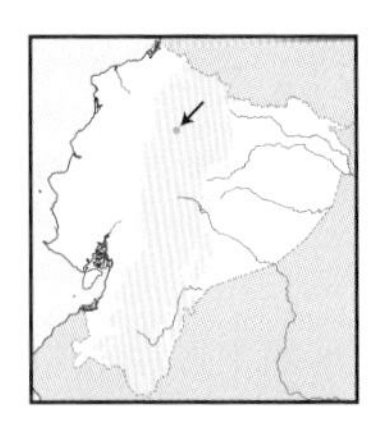

Philadelphia Vireo *Vireo philadelphicus* 11–13 cm

Single record (R. Ahlman) in Quito Botanical Garden (2,800 m). Woodland. Recalls larger Red-eyed and Yellow-green Vireos but supercilium ***more diffuse, broader and shorter***, ending right ***behind eyes, whitish chin***; dark lores and ocular line that ends behind eyes. Eyes ***dark***, bill short. Monotypic. Winters in Central America; forages in upper strata, gleaning foliage, sometimes fluttering; often with mixed flocks. **Voice** Not expected to vocalise; elsewhere utters *zeet* calls. **SS** Red-eyed and Yellow-green Vireos have crispier face patterns, longer, outlined supercilia, longer ocular stripes; see also noticeably different Brown-capped V. **Status** Accidental boreal visitor (Apr).

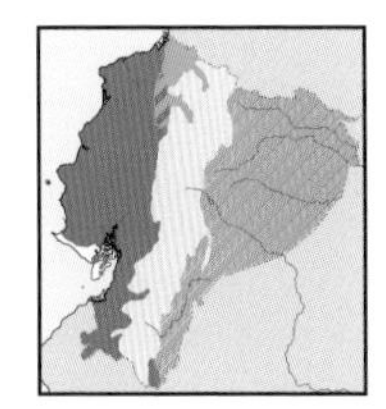

Red-eyed Vireo *Vireo olivaceus* 14–15 cm

E & W lowlands to foothills, locally to subtropics, migrants higher, extreme SE in Zumba region (mostly below 1,300 m, some to 2,800 m). Forested and semi-open habitats. ***Grey crown, white supercilium outlined black*** (especially W, SE), olive upperparts, whitish underparts, sides and flanks tinged greenish. Races *olivaceus* (boreal migrant in E, Andes; Sep–Apr) mostly whitish below, ***red eyes***; *chivi* (austral migrant to E; May–Aug) duller white below, ***dark reddish eyes***; *griseobarbatus* (resident W) bright ***greenish-yellow breast-sides to vent***, greenish back, darker eyes; *pectoralis* (extreme SE) ***throat to breast tinged buffy-greyish***, flanks to vent yellowish, dull back; *solimoensis*? (extreme E lowlands) ***brighter*** yellow below, greyish crown extends to hindneck. Active, nimble pairs or groups often with mixed flocks. **Voice** Short, fast, repeated *yeo'wíi-yeo'wíi-yeo'wíi...yoo'dyée-yoo'dyéu-yoo'dyée...* (*solimoensis*); short, fast *cheeyr-cheewí-chee...* song, nasal *jyee* calls (*pectoralis*); fast *cheeye-cheeu-cheewí…* song, nasal *jeee!* calls (*griseobarbatus*); migrants rarely vocalise. **SS** Very similar to Yellow-green Vireo, with ***weaker face pattern*** (no narrow black outline to supercilium, less contrast between crown and supercilium, more olive crown). E *solimoensis* race cleaner yellow below, more vocal; W *griseobarbatus* and SE *pectoralis* very similar to Yellow-green (stronger face patterns); vagrant Yellow-green might appear almost anywhere; see Philadelphia Vireo (weaker face pattern, broader supercilium); cf. Tennessee Warbler. **Status** Common.

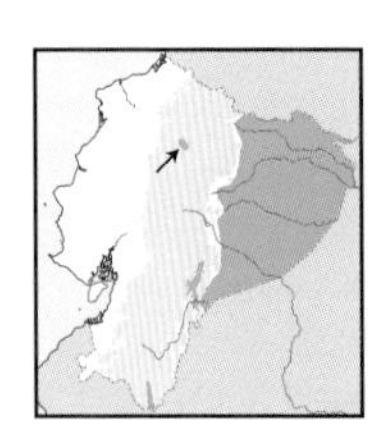

Yellow-green Vireo *Vireo flavoviridis* 14–15 cm

Boreal migrant to E, locally in highlands (to 1,000 m, once 2,800 m). Canopy, borders, woodland, adjacent clearings. ***Greyish-olive crown***, long whitish supercilium, ***not outlined dark; greenish-yellow sides, flanks to vent***. Races *flavoviridis*, *forreri*, possibly *perplexus*. Active, deliberate forager, lower in borders and semi-open; not very vocal, regularly with mixed flocks. **Voice** Short, musical *ch'reé-ch'rëu-ch'ree-ch'reéreé…*; bouts longer, more monotonous than sympatric Red-eyed. **SS** Recalls *griseobarbatus* and *pectoralis* races of Red-eyed but normally sympatric with others (blackish eyestripe and borders to supercilium, plus darker grey crown provide ***stronger facial pattern***, less yellow on sides to flanks); migrant Red-eyed not very vocal; Philadelphia (broader and bolder supercilium); Black-whiskered (not in Ecuador but could occur) has black moustachial; Warbling Vireo (very recently found in Quito Botanical Garden; R. Ahlman) is greyer below, weaker and shorter supercilium. **Status** Uncommon winter visitor (Sep–Apr).

juvenile
leucophrys
josephae
Brown-capped Vireo
Philadelphia Vireo
juvenile
solimoensis
pectoralis
chivi
olivaceus
Red-eyed Vireo
griseobarbatus
flavoviridis
perplexus
forreri
Yellow-green Vireo

Small, inconspicuous canopy dwellers, often overlooked unless vocalising (fortunately, they vocalise quite frequently). Little overall pattern and resemblance to unrelated passerines (antwrens, warblers, tyrannulets) make identification somewhat complicated. Recent taxonomic revolution has separated formerly larger *Hylophilus* into several genera.

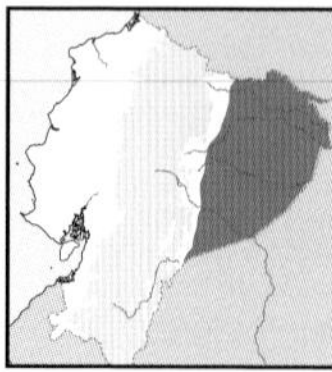

Lemon-chested Greenlet *Hylophilus thoracicus* — 12–13 cm

Local in E lowlands (below 400 m). *Terra firme* forest canopy and borders. ***Pinkish bill, white eyes.*** Olive upperparts, greyish-white underparts, ***broad lemon-yellow breast-band***. Race *aemulus*. Singles or pairs often very active and restless in mixed flocks, otherwise hard to detect. **Voice** Fast, simple, repetitive *déedit-déedit-déedit…*, *déeri-déeri-déeri…* or *dëë'r-dëë'r-dëë'r…*. **SS** Sympatric Dusky-capped Greenlet is solid yellow below, has dark eyes, brown upperparts; voice shorter; see also ♀ Guira Tanager (solid yellow below, dark eyes). **Status** Rare, local, possibly overlooked.

Olivaceous Greenlet *Hylophilus olivaceus* — 12–13 cm

E Andean foothills to lower subtropics (600–1,600 m). Montane forest borders, woodlands, regenerating scrub. ***Pinkish bill, whitish eyes***, pink legs. Drab. Olive upperparts, ***yellowish forecrown and lores***, olive-yellow underparts. Monotypic. Mainly in pairs remaining in dense cover, acrobatic and lively, sometimes follows mixed flocks. **Voice** Long, fast repetition of ascending *seeweé* or *deeweé* notes, sometimes slower, notes more separated. **SS** Lower-elevation Dusky-capped Greenlet has brown upperparts, dark eyes; Tawny-crowned has tawny-orange forecrown; voice somewhat recalls Slate-throated Whitestart. **Status** Uncommon; NT globally.

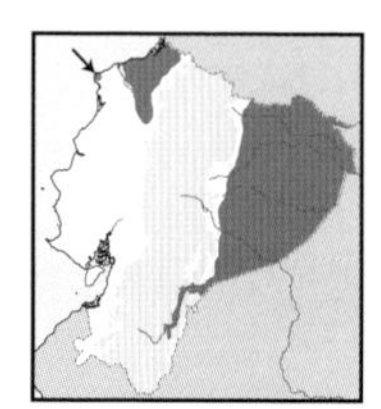

Tawny-crowned Greenlet *Tunchiornis ochraceiceps* — 11–12 cm

E & NW lowlands, locally in lower foothills (to 700 m). Wet forest undergrowth. ***Pinkish bill, pale grey eyes, flesh-coloured legs. Tawny-orange forecrown***, dusky-olive above, greyish-white throat, dull olive-yellowish below, yellower vent. Races *ferrugineifrons* (E), *bulunensis* (NW). Restless and agile pairs with mixed understorey flocks, rarely in borders or higher strata. **Voice** Descending, long *pEeeeeu*, softer and longer than Slaty-capped Shrike-Vireo, with plaintive cadence, sometimes doubled *peeeu-yeeee*; in W shorter and higher-pitched; loud *nýah-nýaah…* calls. **SS** Tawny forecrown and white eyes separate this drab greenlet from sympatric congeners; voice also resembles Half-collared Gnatwren. **Status** Rare.

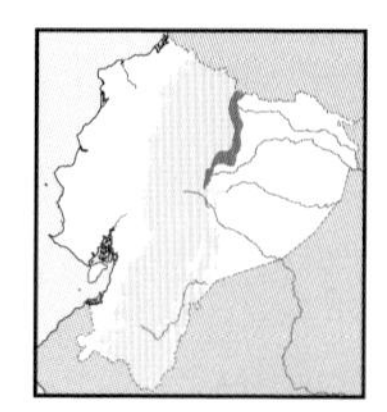

Rufous-napped Greenlet *Pachysylvia semibrunnea* — 12–13 cm

Local in NE foothills (900–1,400 m). Montane forest midstorey, borders. Dark eyes. ***Orange-rufous crown to nape***, olive upperparts; whitish lores, ***short white supercilium, white face to upper breast***, orangey breast-sides, greyish-white belly. Monotypic. Singles or pairs restless in mixed flocks, in outer foliage, not easy to spot. **Voice** Fast, rising but soft *cheedowiít-cheedowiít…*repeated at 6–8 sec-intervals. **SS** Sympatric greenlets lack rufous crown and white throat combo; Olivaceous Greenlet voice a longer series of fast notes; see Brown-capped Vireo, with fairly similar face pattern but longer supercilium, brown limited to crown. **Status** Rare, local, probably overlooked.

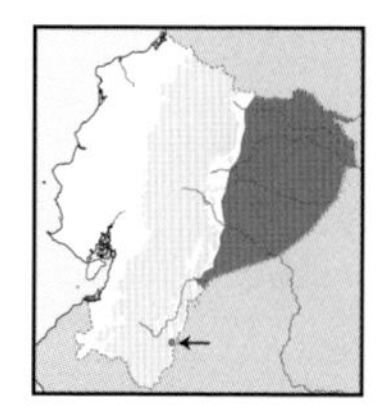

Dusky-capped Greenlet *Pachysylvia hypoxantha* — 11–12 cm

E lowlands (to 400 m). Upper strata in *terra firme* forest and borders. Pinkish mandible, ***dark eyes***. Quite nondescript. ***Brown to bronzy-olive upperparts***, somewhat duskier crown, whitish lores, ***eye-ring*** and ***faint supercilium;*** greyish-white throat, ***yellow breast to belly and vent***. Race *fuscicapilla*. Restless, acrobatic and noisy pairs with canopy mixed flocks, rarely at borders. **Voice** Fast, musical *peetchoocheewee-cheewo'í-wo'i…chocheewëë-chocheewee'weew-cheewee'weew…choo-weew-choo-weew…*, last notes often repeated. **SS** Tawny-crowned Greenlet has pale eyes, dusky-olive upperparts, tawny-orange forecrown (more often in low strata); sympatric Lemon-chested not solid yellow below. **Status** Fairly common.

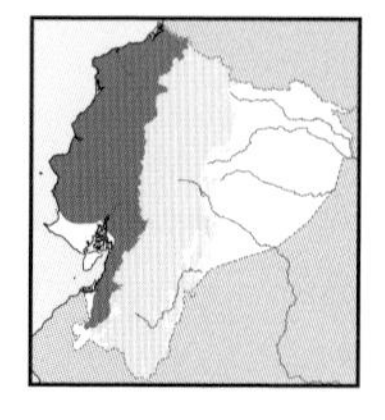

Lesser Greenlet *Pachysylvia decurtata* — 10–11 cm

W lowlands to foothills (to 1,100 m). Humid to semi-deciduous forest and borders, secondary woodland. ***Bicoloured bill***. Olive upperparts, white in face including narrow ***white spectacles***, white underparts, ***lemon-yellow breast-sides to flanks, short tail***. Race *minor*. Pairs to small groups, very active and acrobatic gleaners often in mixed-species flocks in various strata, including close to ground. **Voice** Short, musical, long-repeated *wichee-wee, wíchee, wichee-wee-wíchee…*; also louder, nasal *kyee-kyee…* and drier *tree-ree-ree-er* calls. **SS** Lacks whitish supercilium of sympatric vireos (Red-eyed, locally Brown-capped); ♀ Guira Tanager larger, solid yellow below; migratory, larger Tennessee Warbler has finer, pointed bill, faint supercilium. **Status** Fairly common.

juvenile
adult
Lemon-chested Greenlet
Olivaceous Greenlet
bulunensis
ferrugineifrons
Tawny-crowned Greenlet
Rufous-napped Greenlet
Dusky-capped Greenlet
Lesser Greenlet

Small boreal migrants of forest and edge upper strata, all lacking the dusky lateral streaks typical of many *Setophaga*. Yellow Warbler has one distinct resident subspecies in mangroves and adjacent habitats, and others that migrate to forests, borders and adjacent fields. Two species are probably involved, as regularly suggested.

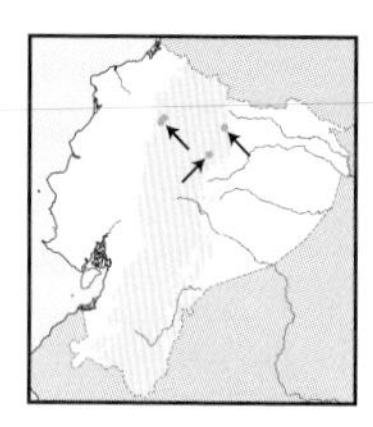

Golden-winged Warbler *Vermivora chrysoptera* 12–13cm

Few records in N subtropics (1,300–2,100m). Montane forest borders, woodland. Slender ***pointed*** bill. ***Yellow forecrown, grey face and throat*** (black in breeders), ***whitish superciliary and malar stripes***. Grey upperparts, paler below, ***yellow wing-coverts*** (golden-yellow in breeders). ♀: similar but duller. Tail shows much ***white***. Monotypic. Singles in upper strata with mixed flocks, nimbly searching dead leaves and gleaning foliage. **Voice** Sharp, dry *chik* calls (R *et al.*). **SS** Confusion unlikely given yellowish forecrown, yellow wings and face pattern. **Status** Very rare boreal winter visitor (Jan–Mar); NT globally.

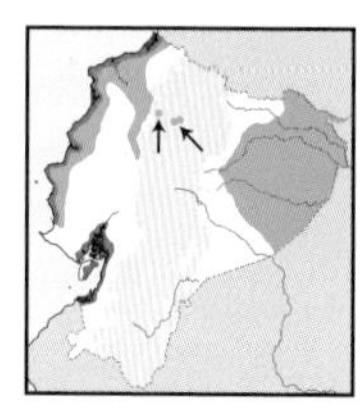

Yellow Warbler *Setophaga petechia* 12–13cm

Resident on coast, migrant to lowlands and foothills, locally in highlands (migrants up to 2,800m). Forest borders, woodland; resident in mangroves. Resident *peruviana* race: ***reddish crown***, olive-yellow above, ***bright yellow below, streaked breast to flanks; yellow*** undertail. ♀: olive-yellow above, ***bright yellow below***, darker cheeks. Juv. ***greyer, olive-yellow*** in wings. Migrant races ***lack*** red crown, ***red streaks fainter*** or absent. Migrant races *aestiva, amnicola, morcomi, sonorana*. Singles to small groups (resident) active in various strata. **Voice** Rapid *swe-swe-swe'swee'sweswesweswe*... ending in warble (*peruviana*); strong *chíp* calls. **SS** ♀ recalls Prothonotary, Tennessee, Citrine and Wilson's Warblers, ♀ yellowthroats and Orange-fronted Plushcrown, but note yellow wing edgings and yellow undertail. **Status** Rare boreal winter resident (Sep–Apr), common resident.

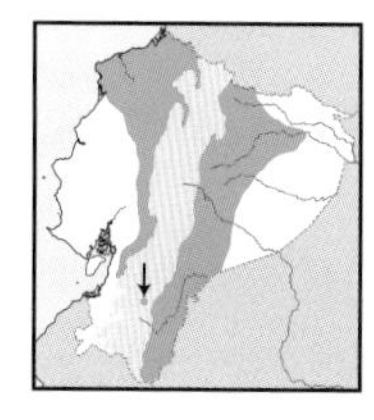

Canada Warbler *Cardellina canadensis* 13–14cm

Migrant to NW & E subtropics, foothills to adjacent lowlands, local in inter-Andean valleys (500–2,800m). Montane forest borders, woodland, shrubby clearings. ***Bluish*** to olive-***grey above***, bold ***yellow spectacles***, yellow underparts, ***grey-streaked breast-band*** (more solid black in breeding ♂), ***white vent***, pink legs. Monotypic. Singles or pairs often with mixed flocks, active and nimble forager, regularly cocks and fans tail. **Voice** Utters short versions of jumbled warbling song: *teedeeweedee-teedee*, also sharp *tchík* and soft *tsi* calls. **SS** Yellow spectacles and greyish 'necklace' distinctive even in dullest plumages; see Wilson's, Mourning and Connecticut Warblers. **Status** Fairly common boreal winter resident (mainly Oct–Apr).

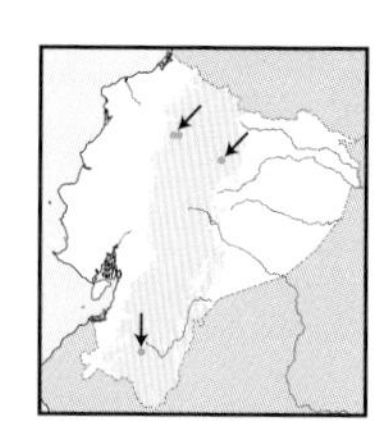

Wilson's Warbler *Cardellina pusilla* 13–14cm

Few sites in foothills to subtropics (1,250–2,000m). Montane forest borders, woodland. Olive above, ***all yellow below, yellowish lores, supercilium and narrow eye-ring, duskier cheeks***, flesh-pink legs. Breeders have ***black rear crown***. Race unknown. Singles observed following mixed flocks or alone, foraging in mid to low strata. **Voice** Dry, emphatic *chet* (C *et al.*). **SS** Although quite drab, yellowish face pattern distinctive; Yellow Warbler (chubbier, yellower face), Canada (yellow spectacles, greyish necklace), ♀ yellowthroats (less rounded head, longer bills, darker above), Prothonotary (yellower, bluish-grey wings), greenlets (less yellow below, no supercilium). **Status** Very rare boreal winter visitor (Oct–Mar).

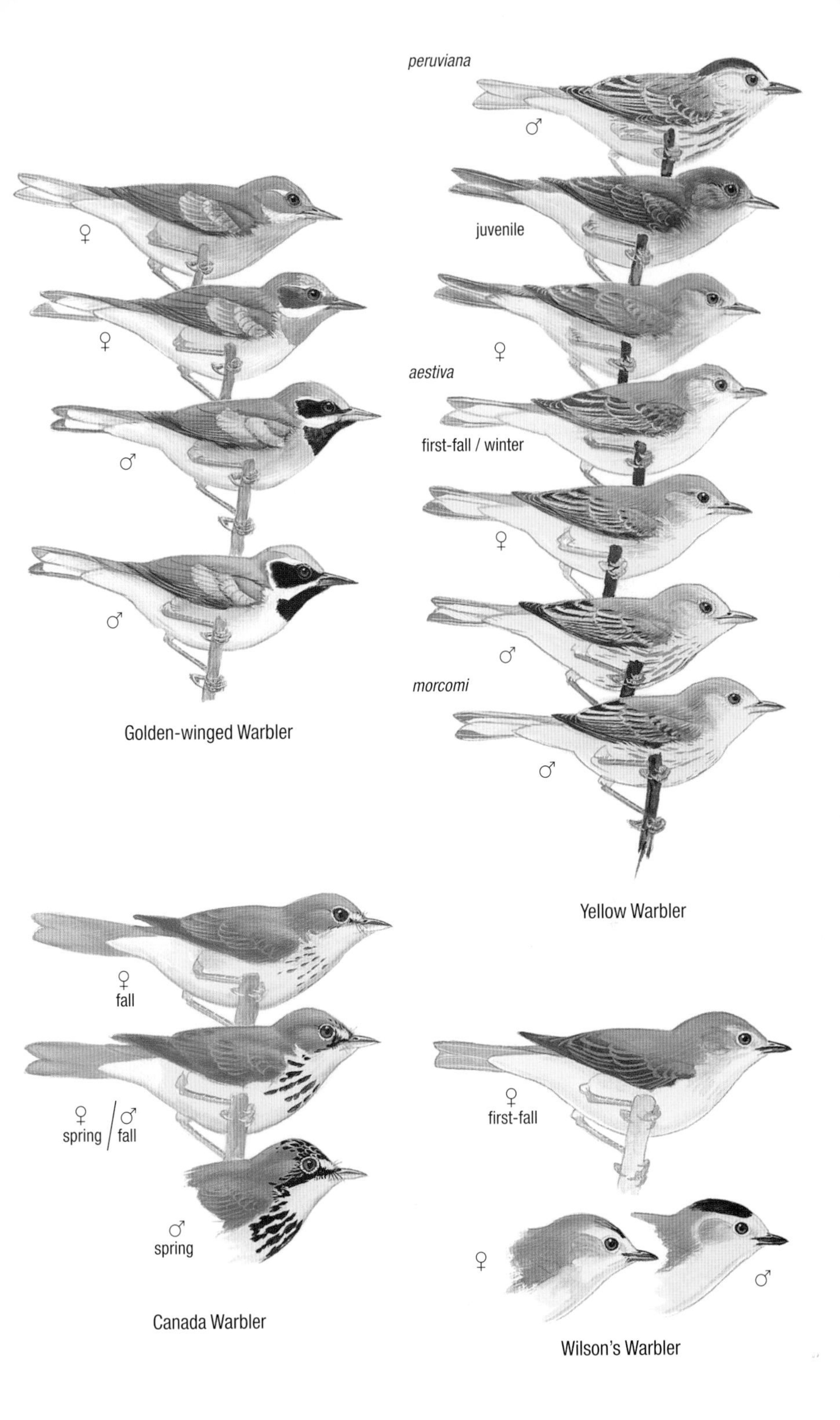

♀
♀
♂
♂
Golden-winged Warbler
peruviana
♂
juvenile
♀
aestiva
first-fall / winter
♀
♂
morcomi
♂
Yellow Warbler
♀
fall
♀ spring / ♂ fall
♂
spring
Canada Warbler
♀
first-fall
♀
♂
Wilson's Warbler

PLATE 245: MIGRANT WARBLERS I

Four rare to accidental *Setophaga* warblers and slim Tennessee of forest borders and scrub. Compare common Blackpoll and Blackburnian (next plate) for identification; also consult specialist literature (Curson *et al.*). These migrant species have so far been recorded in Andean foothills and slopes, but might show up as stragglers elsewhere. Cerulean is of global conservation concern.

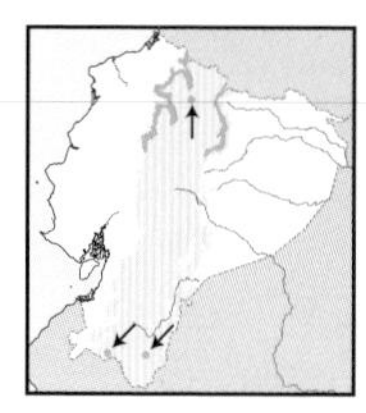

Tennessee Warbler *Leiothlypis peregrina* 12–13cm

Scattered records in Andes (1,300–2,800m). Montane forest borders, humid scrub. Slender *pointed* bill. *Plain olive above, narrow yellowish-white supercilium*, black ocular stripe, *slight wingbar*. Whitish to yellowish below, yellower sides to flanks, *white vent, white subterminal tail-band*. Monotypic. Mainly singles with mixed flocks in various strata, may congregate at flowering trees. **Voice** Thin, clear *see* calls. **SS** Yellow Warbler may look superficially similar but plumper, yellower, bill heavier. Larger Red-eyed and Yellow-green Vireos have thicker bills, greyer crowns, broader superciliary and ocular stripes, longer tails, and no white tail-band; see greenlets and ♀ Chestnut-vented Conebill (buff below, no eyebrow). **Status** Rare boreal winter visitor (Oct–Mar).

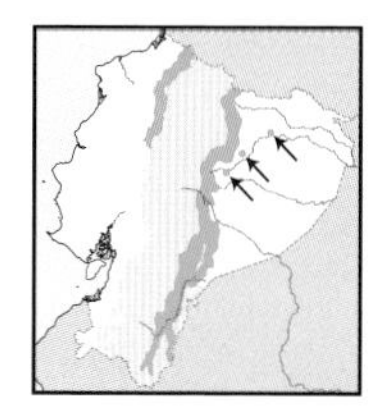

Cerulean Warbler *Setophaga cerulea* 12–13cm

Local in E Andean foothills to adjacent lowlands, scattered records in NW (400–1,500m; once 2,200m). Montane forest, woodland, borders. *Short* tail. *Pale bluish above* (richer in breeders), *two white wingbars*. White below, *faint dusky necklace* and streaked sides to flanks. ♀ *lime-green above, two white wingbars, long yellowish supercilium, pale yellow below*, few lateral streaks. Monotypic. Singles, less often pairs, accompany mixed canopy flocks. **Voice** Emphatic, musical *chip* calls (C *et al.*). **SS** Drabber Tennessee (spikey bill, olive above, narrower supercilium), non-breeding Blackpoll (greyer olive, streaked upperparts, fainter supercilium), ♀ Bay-breasted (duller above, paler below, fainter supercilium). **Status** Rare boreal winter visitor/transient (Oct–Apr), VU globally.

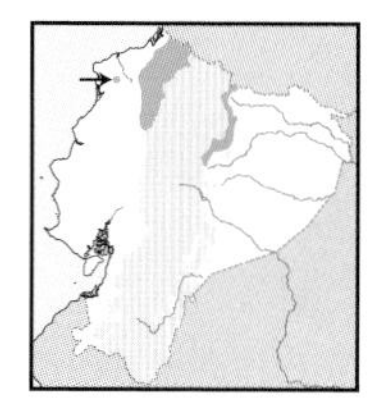

Bay-breasted Warbler *Setophaga castanea* 12–13cm

Scattered records in N lowlands to subtropics (100–1,500m). Montane forest canopy, borders. Greyish-olive upperparts *finely streaked dusky* (much *vaguer* in ♀); duskier wings, *two bold whitish wingbars. Narrow buffy supercilium*, buffy-white below, *sides to flanks washed pale cinnamon*. Rufous crown and throat to sides in breeders. Monotypic. Singles with mixed flocks, agile and deliberate, cock tail. **Voice** Thin *sip* (C *et al.*). **SS** Chestnut-sided (plain upperparts, prominent eye-ring); ♀ Cerulean (greener above, yellower below); Blackpoll (streaks below more prominent, white vent, greater contrast in eyebrow/ocular stripe gives severer expression); larger Chocó Vireo has similar pattern but duller, unstreaked above, stouter-billed. **Status** Rare boreal winter visitor (Jan–Feb).

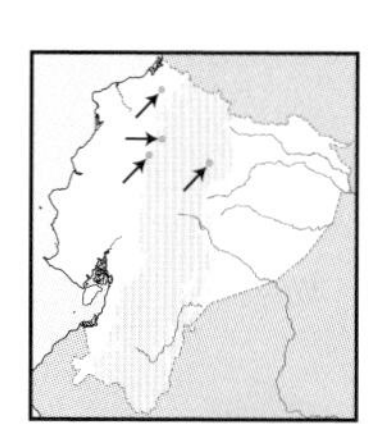

* Chestnut-sided Warbler *Setophaga pensylvanica* 12–13cm

Scattered records in NW lowlands and foothills (70–700m). Montane forest borders, woodland. *Bright* olive above, darker wings with *two yellowish-white wingbars. Prominent pale eye-ring. Whitish-grey below, including vent and undertail, sides to flanks faintly washed chestnut* (bolder in breeders, also extensive yellow in crown, bold black malar and ocular stripes). ♀: duller, *less chestnut* on sides, *no* eyebrow. Monotypic. Active, constantly cocks and flicks tail and droops wings. **Voice** Sharp *tchíp* (C *et al.*). **SS** See Cerulean and Bay-breasted (white, not yellowish wingbars; pale eyebrow, no eye-ring); Blackpoll (dusky sides to flanks, faint supercilium, streaky back). **Status** Very rare boreal winter visitor (Apr–May).

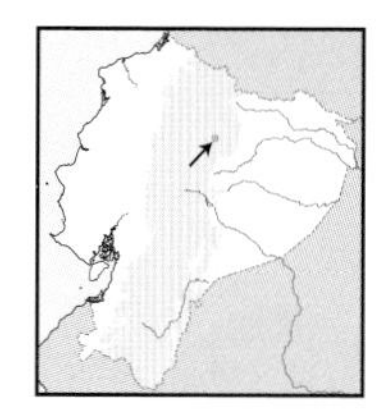

Black-throated Blue Warbler *Setophaga caerulescens* 12–13cm

Two records from NE subtropics in Cosanga / Baeza region (1,900–2,170 m). Montane forest borders, woodland, regenerating scrub. *Greyish-blue above, black face and throat*, blackish-mottled sides. *White below, white wing patch*. ♀: *dull olive above*, greyish-white below, *short, narrow whitish supercilium, smaller, fainter white wing patch*. Nominate? Observed with mixed flock, gleaning foliage from 3 m to canopy. **Voice** Gives 3–4 metallic *twik* and flat *stip* calls (C *et al.*). **SS** ♀ drab and confusing; greener Tennessee W (no pale wing patch, white vent, prominent tail-band, spiky bill); ♀ Cerulean W (brighter, greener, long eyebrow, bold wingbars). **Status** Uncertain, possibly only vagrant (Jan–Jun).

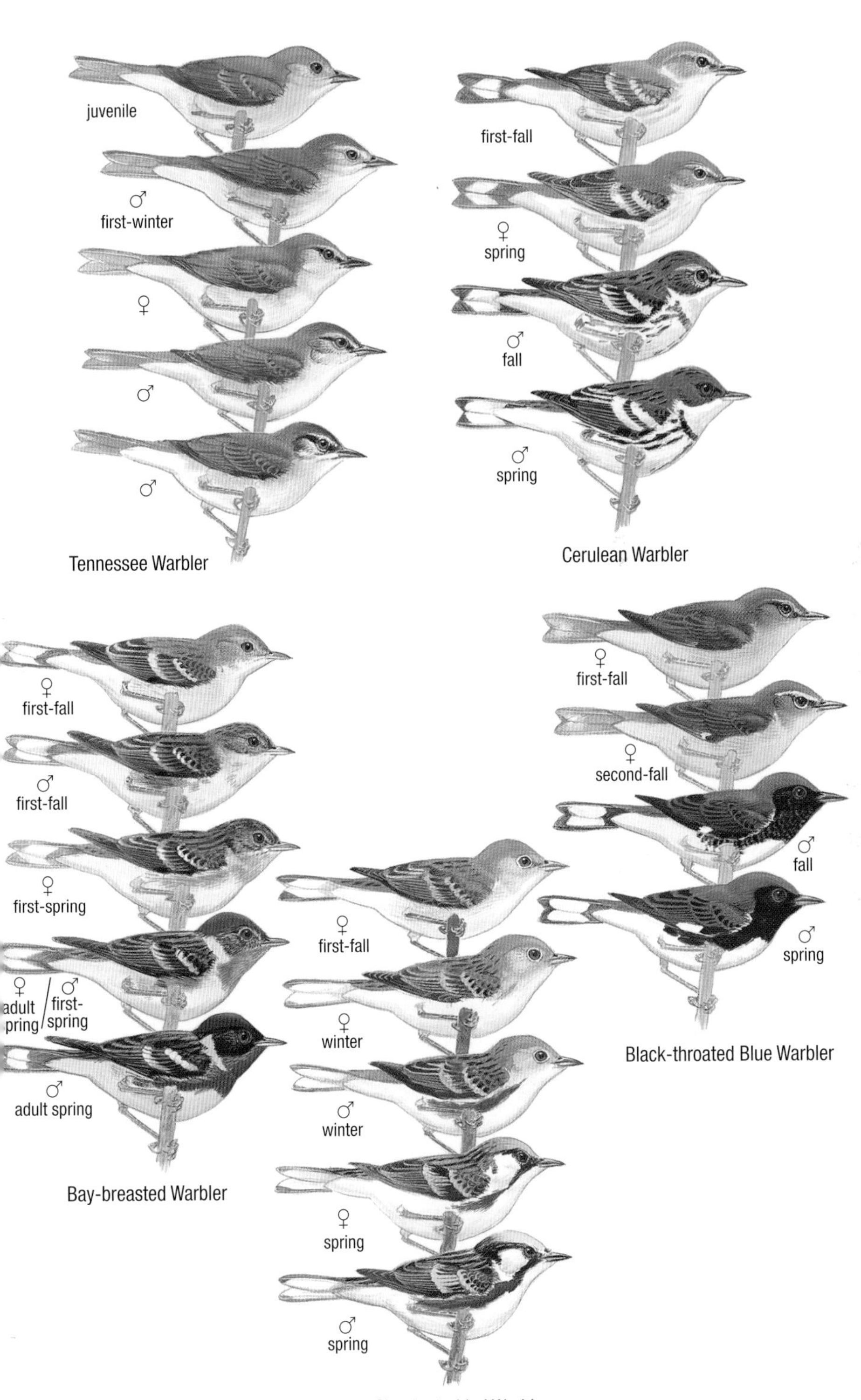
juvenile
♂
first-winter
♀
♂
♂
Tennessee Warbler
first-fall
♀
spring
♂
fall
♂
spring
Cerulean Warbler
♀
first-fall
♂
first-fall
♀
first-spring
♀
adult
spring
♂
first-
spring
♂
adult spring
Bay-breasted Warbler
♀
first-fall
♀
winter
♂
winter
♀
spring
♂
spring
Chestnut-sided Warbler
♀
first-fall
♀
second-fall
♂
fall
♂
spring
Black-throated Blue Warbler

Three common warblers plus a rare visitor (Black-throated Green). Blackpoll and Blackburnian are the only common and widespread migratory *Setophaga* in the country. Monotypic Black-and-white superficially resembles *Setophaga* but behaviour very different.

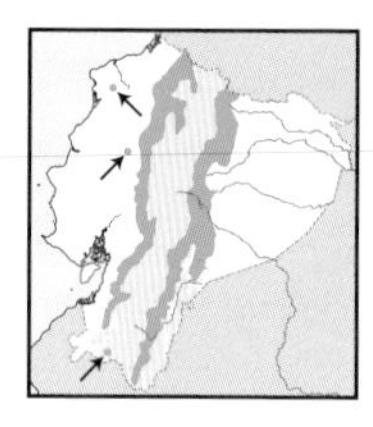

Black-and-white Warbler *Mniotilta varia* 12–13cm

Migrant mainly to N Andean subtropics and foothills, local in highlands and W lowlands (700–1,900m, locally higher). Montane forest borders, woodland. *Zebra*-patterned. ***Blackish-and-white striped head, streaked upperparts, two white wingbars***, whitish below, sides to flanks show ***blackish streaks***, whitish subterminal tail-band. Breeders bolder. Monotypic. Singles or pairs creep on large limbs and trunks, often with mixed flocks, in mid strata. **Voice** Sharp *tík* and softer *tzeet* (C *et al.*), but mostly silent. **SS** Breeding ♂ Blackpoll has solid black crown, white cheeks, greyer upperparts, no stripes on head; otherwise unmistakable. **Status** Uncommon boreal winter resident (Sep–Apr).

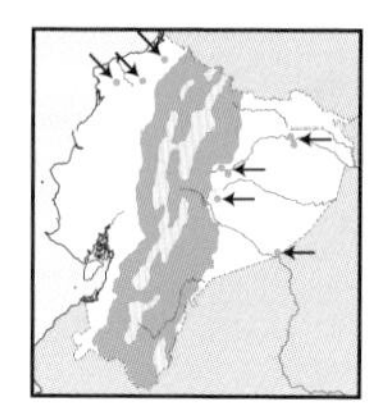

Blackburnian Warbler *Setophaga fusca* 12–13cm

E & W foothills to temperate zone, inter-Andean valleys, local in lowlands (mainly 900–3,000m). Montane forest borders, woodland, shrubby clearings, drier scrub, gardens. Dusky above ***finely streaked blackish; broad buffy supercilium, dusky cheeks outlined buffy or yellowish***; blacker wings, two ***bold white wingbars. Yellowish throat to breast***, grey streaks on sides to flanks. Breeders bolder, ***black upperparts, bright orange face to breast, black cheeks*** and streaks on sides. Birds moulting from breeding to non-breeding often seen. Monotypic. Singles to small groups in various strata and forest types, with mixed flocks; vigorous and deliberate. **Voice** Sharp *chíp* or buzzier *tseet*. **SS** Strong face pattern distinctive; Blackpoll (fainter, narrower supercilium, paler cheeks). **Status** Common boreal winter resident (Oct–Apr).

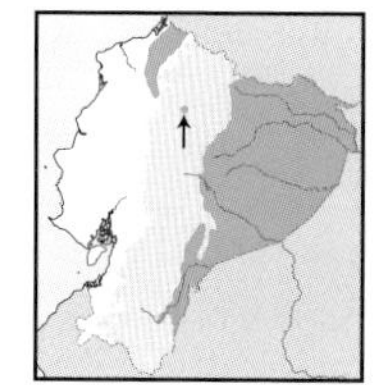

Blackpoll Warbler *Setophaga striata* 12–13cm

E lowlands to foothills, local in NW lowlands and highlands (mostly to 1,400m, locally higher). Humid forest, borders, woodland. Dull olive above ***finely streaked dusky, faint pale supercilium***, blackish orbital stripe; ***two bold white wingbars***. Whitish-yellow underparts, ***faint dark streaks on sides to flanks, white belly***. Breeder has ***black crown, white cheeks, black malar***, and white below. Monotypic. Singles to small groups deliberately forage at various heights, often in terminal canopy, with mixed flocks. **Voice** Emphatic, dry *chít*; short version of song *swi-si'si'si'si...*. **SS** Rarer Bay-breasted (buffier below, unstreaked sides, duller vent, 'softer' face pattern), Blackburnian (stronger face pattern, yellower below), Chestnut-sided (no supercilium, yellower wingbars), ♀ Cerulean (greener, longer supercilium). **Status** Uncommon boreal winter resident (Oct–Apr).

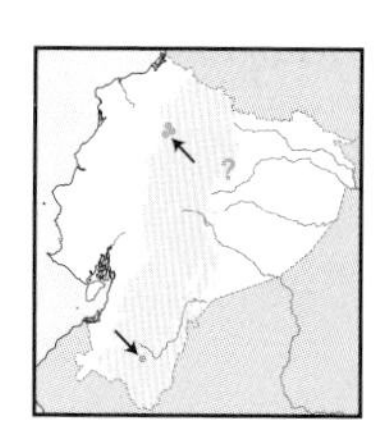

* Black-throated Green Warbler *Setophaga virens* 12–13cm

Scattered records in Andean subtropics (1,400–2,500m). Montane forest borders, woody clearings. ***Bright olive above***, black wings with two white wingbars (***upper bolder***), ***yellowish face*** (***bright*** yellow in breeders), ***olive cheeks***, whitish below, ***blackish streaking on breast to flanks*** (***solid black*** throat to breast in breeders and moulting birds). ♀: duller, only faintly streaked. Monotypic. Singles or, less often, pairs, expected with mixed canopy flocks, active, regularly droops wings. **Voice** High, metallic *tick* (C *et al.*). **SS** Smoother face pattern, paler cheeks than Blackburnian, which is darker and streaked above; see Blackpoll. **Status** Very rare boreal winter visitor, accidental (Dec–Mar).

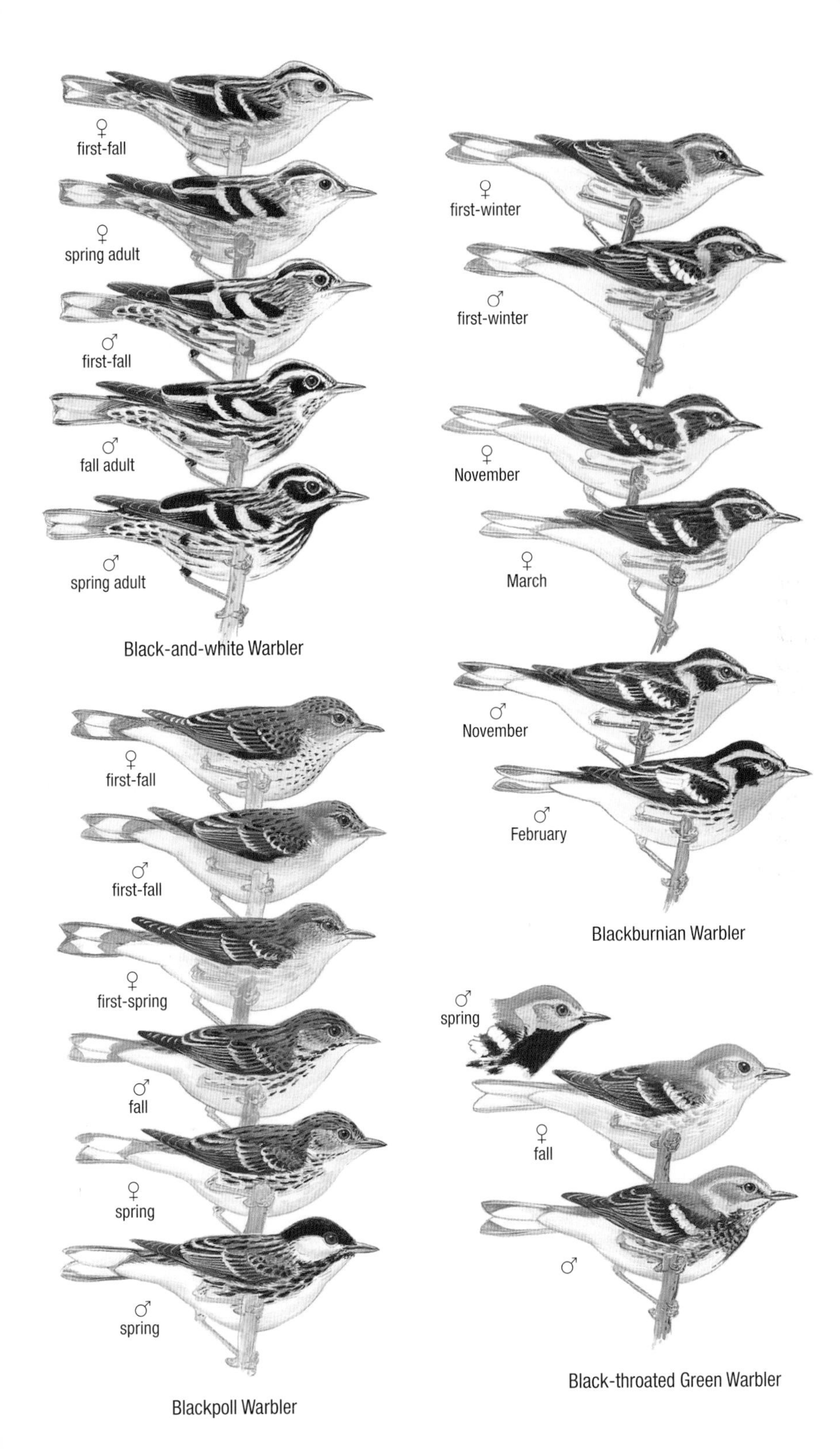
♀
first-fall
♀
spring adult
♂
first-fall
♂
fall adult
♂
spring adult
Black-and-white Warbler
♀
first-fall
♂
first-fall
♀
first-spring
♂
fall
♀
spring
♂
spring
Blackpoll Warbler
♀
first-winter
♂
first-winter
♀
November
♀
March
♂
November
♂
February
Blackburnian Warbler
♂
spring
♀
fall
♂
Black-throated Green Warbler

American Redstart is similar in shape and behaviour to resident *Myioborus* redstarts, all flashing tail. The resident redstarts replace each other altitudinally, but overlap broadly and often occur in same mixed flocks. ♀ yellowthroats can be difficult to separate so be cautious in area of overlap in N Manabí, Los Ríos and Guayas. Olive-crowned prefers wetter areas.

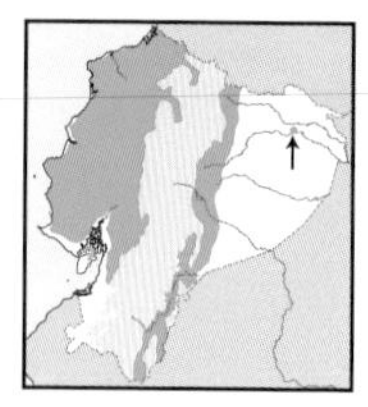

American Redstart *Setophaga ruticilla* — 12–13cm

Migrant to E & W lowlands to foothills (to 1,300m, locally higher). Humid to deciduous forest borders, woody clearings, woodland. Grey head, greyish-olive above, ***bright yellow panel in flight feathers, yellow breast-sides and tail base***. Narrow white broken eye-ring, whitish underparts. Breeders black, yellow replaced by ***flame orange***, white belly to vent. Monotypic. Singles very active, gleaning foliage, sallying and jumping, tail often fanned, wings drooped; regularly territorial. **Voice** Thin, sibilant *ship* or *tsip*. **SS** Confusion unlikely; *Myioborus* redstarts similar but tail white, not yellow. **Status** Uncommon boreal winter resident (Oct–Mar).

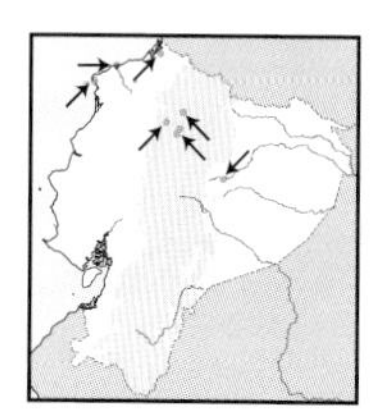

Prothonotary Warbler *Protonotaria citrea* — 13–14cm

Scattered records in N lowlands to foothills (to 1,300m, locally higher). Secondary woodland, humid forest borders, woody clearings. Mostly ***bright yellow***, olive upperparts, ***bluish-grey wings and tail, white vent*** and outer tail feathers. ♀ / imm. duller, more olive crown. Monotypic. Singles or pairs regularly with mixed flocks at low heights, sometimes near water or on ground, around fallen logs and rotten wood. **Voice** Ringing *chink*, softer, sibilant *psit* (C *et al.*). **SS** Yellow Warbler also bright yellow but different wing pattern, tail yellow below, shorter bill, slimmer; more arboreal. **Status** Very rare or casual boreal winter visitor (Oct–Mar).

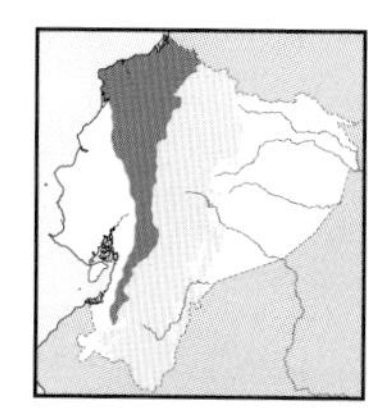

Olive-crowned Yellowthroat *Geothlypis semiflava* — 13–14cm

W lowlands to foothills, S along base of Andes to N El Oro (to 1,200m, locally to 1,700m; after deforestation?). Humid shrubbery, grassland. ***Black forecrown and large mask***, olive upperparts, yellow below. ♀: ***lacks*** mask, ***head olive, narrow yellowish supercilium***, fading behind eyes. Nominate. Skulking singles or pairs in dense cover. **Voice** Sweet warble, *sew-sew-swe'swe'swe-seewuu-seewuu...* often ending in jumble; recalls resident Yellow Warbler but slower, more melodious; calls harsh *cheewr*. **SS** Masked Yellowthroat (smaller black mask, greyer crown; ♀ has more prominent eyebrow, greyer crown); Masked prefers drier habitats. **Status** Uncommon.

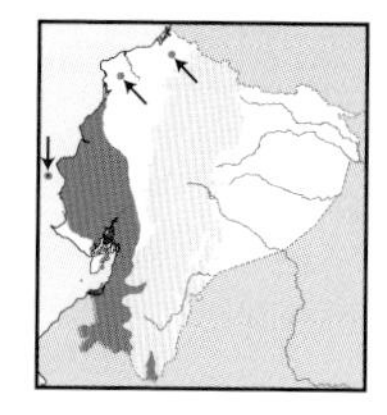

Masked Yellowthroat *Geothlypis aequinoctialis* — 13–14cm

SW lowlands to foothills, extreme SE in Zumba region (to 1,100m in W, 850–1,650 m in SE). Semi-humid shrubbery, woodland, clearings. ***Grey crown, small black mask***, olive upperparts and yellow underparts. ♀: ***greyish crown, yellow supercilium and narrow eye-ring***. Races *peruviana* (SE), *auricularis* (W), might be valid species. Singles or pairs, rather skulking except when ♂ sings from exposed perch. **Voice** Short, clear, cheerful, ending in sweet warble: *swe-swe-swe-t'wo-wo-t'wlee-wlee-wlee...* (W), calls harsh *werr*; in SE faster warble: *we-we'chee-we'chee-wiichii-wiwiwiwi* or *we-we'clii-we'clii-wii-tilililili*. **SS** ♂ Olive-crowned (large black mask, grey crown; ♀ supercilium fainter, crown olive; habitat differs); see ♀ Yellow Warbler. **Status** Uncommon.

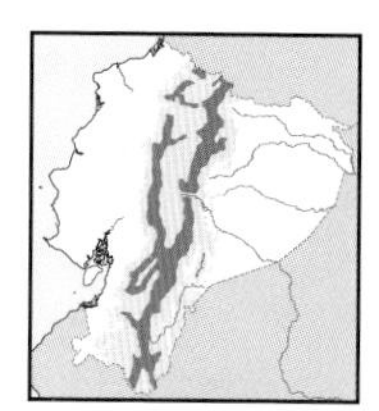

Spectacled Redstart *Myioborus melanocephalus* — 13–14cm

Andean subtropics to *páramo* woodland in E & W (2,200–4,000m, locally lower). Montane forest, borders, woodland, shrubby clearings, treeline. Black head, chestnut coronal patch, ***yellow spectacles***. Slate-grey upperparts, ***bright yellow underparts, white vent and most tail feathers***. Populations in NE, near Colombia have extensive ***yellow in face***. Race *ruficoronatus*. Behaves much like previous species, also tends to lead mixed flocks, characteristic for flashing tail; may join Slate-throated in mixed flocks. **Voice** Melodious series of high-pitched *swe-vi-vi-vi, swii-swii-vi'vi'vi...* that varies and continues for long periods; clear *chip*. **SS** See previous species. **Status** Fairly common.

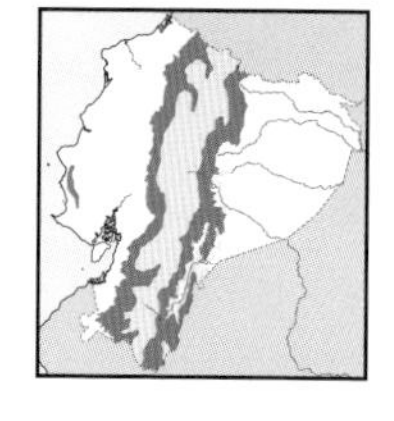

Slate-throated Redstart *Myioborus miniatus* — 13–14cm

Andean foothills to subtropics in E & W, atop coastal ranges, locally to inter-Andean valleys (800–2,400m, 400–600m in coastal ranges). Montane forest borders, woodland, clearings. ***Slate-grey hood and upperparts***, chestnut coronal patch, bright yellow below, ***white vent and outer tail feathers***. Races *ballux* (N), *verticalis* (SE), *subsimilis* (SW). Singles to small parties, very nimble, glean and sally at various heights, regular with mixed flocks, apparently leading them; permanently flicks and fans tail, and droops wings. **Voice** Short, slightly accelerating, sometimes rising, series *see-seew-seew-see'see'see-ss'ss*; calls various *see* and *tic*. **SS** Spectacled Redstart (distinctive face pattern, all yellow below). **Status** Common.

juvenile
♀
spring
♂
spring
♂
♂
American Redstart
juvenile
♀
♂
fall
♀
♂
Prothonotary Warbler
♀
♂
♀
first-year
♂
first-year
Olive-crowned Yellowthroat
juvenile
♂
immature
♀
♂
Masked Yellowthroat
juvenile
ballux
verticalis
subsimilis
Slate-throated Redstart
juvenile
adult
northern variant
Spectacled Redstart

Resident and widespread Tropical Parula of forests and woodland is common in mixed canopy flocks. Additionally, four other species are rare migrants, mostly terrestrial and skulking. Ovenbird, although bearing a name suggestive of a furnariid, is closely related to waterthrushes of N America. Buff-rumped Warbler is distinctive and conspicuous in waterside habitats, even along small forested streams.

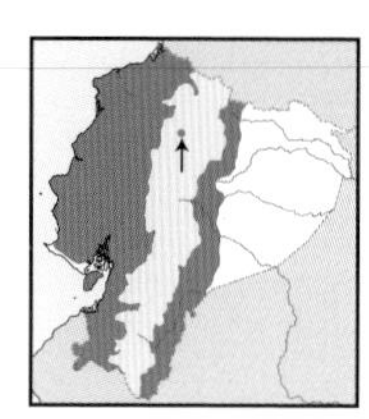

Tropical Parula *Setophaga pitiayumi* 10–11cm

W lowlands to subtropics, E foothills to subtropics (to 2,000m in W, 900–2,000m in E, twice in Quito). Humid to semi-deciduous forest canopy, borders, woodland, shrubby clearings. ***Greyish-blue upperparts, olive mantle, two white wingbars***. Bright yellow below, ***orange breast, white lower belly to vent***. ♀: duller, ***no*** orange breast. Races *pacifica* (W), *alarum* (E). Mostly in pairs with mixed flocks; agilely works on terminal twigs, *Cecropia* catkins, foliage. **Voice** Thin notes followed by brief, buzzy trill: *tsi-tsi-tsi-tsi-t'sssssp*, sometimes more twittering *tse-tse-tse-ts's's's-tse-tsee...*; thin, clear *típ*. **SS** Unmistakable; ♀ Cerulean and Bay-breasted Warblers, with white wingbars, differ in overall appearance, size and pattern. **Status** Fairly common.

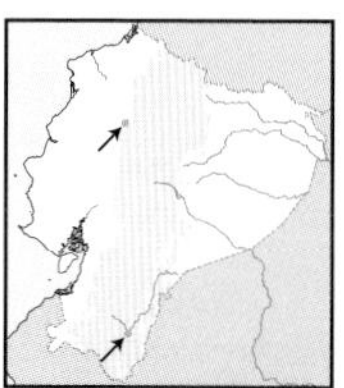

Ovenbird *Seiurus aurocapilla* 14–15cm

Accidental, few sites in NW & SE foothills (700–1,000m). Humid forest, woodland. ***Tawny to orange crown outlined black, narrow whitish eye-ring, large eyes***. Greyish-olive above, white below, ***black malar, prominent streaks to sides***. Juv. duller. Nominate. Singles difficult to see as terrestrial and silent, but rather deliberate, regularly teeters rear while walking, not necessarily near water. **Voice** Sharp, dry *chip* and higher-pitched *seeé* (C *et al.*). **SS** Northern Waterthrush lacks coronal patch, has long buff supercilium, more yellowish underparts. **Status** Very rare boreal migrant (Mar, Sep).

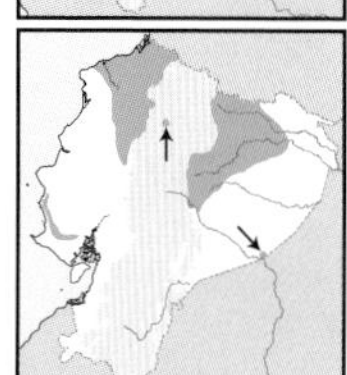

Northern Waterthrush *Parkesia noveboracensis* 14–15cm

Scattered records in W & E lowlands to subtropics, local in highlands (to 2,000m, once 2,800m). Forest and woodland around streams and lakes, damp shrubby clearings, mangroves. Flesh legs. Olive-brown above, ***long buffy supercilium***, dusky ocular stripe. ***Buffy-white below coarsely streaked dusky***, unstreaked belly. Monotypic. As previous species, singles often territorial near water or damp areas, teetering rear body, tossing leaves aside. **Voice** Loud, sharp, metallic *chínk* and buzzier *tseep* (C *et al.*). **SS** Rarer Ovenbird has distinct head pattern, whiter underparts; recalls a small, streaky thrush. **Status** Rare boreal winter resident (Oct–Apr).

Connecticut Warbler *Oporornis agilis* 13–14cm

One record in NW lowlands (50m). Secondary forest near water. Slender bill, pale mandible base, pink legs, ***white eye-ring***. Mostly ***greyish-olive to brownish head, breast and upperparts***, paler throat, more olive breast, yellow belly to vent. ♀ / juv. duller, especially hood. Monotypic. Accidental; single bird mist-netted; elsewhere very skulking and shy, regularly ***walks*** on or near ground, slightly bobbing tail. **Voice** Sharp metallic *plink* and buzzier *zee* (C *et al.*). **SS** Similar Mourning Warbler has narrower, incomplete, vaguer eye-ring; darker, greyer hood and upperparts, long undertail-coverts give short-tailed appearance. See similarly patterned Grey-hooded Bush-Tanager. **Status** Vagrant (Nov).

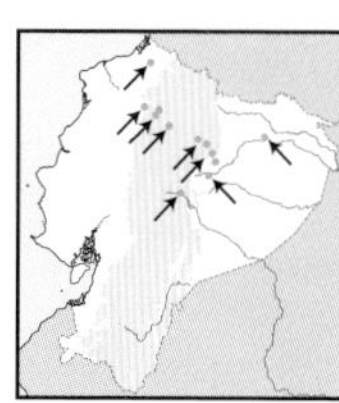

Mourning Warbler *Geothlypis philadelphia* 13–14cm

Scattered records in N lowlands to subtropics (250–2,500m). Shrubby clearings, woodland and borders near water. ***Slender*** bill, ***very*** narrow, ***broken*** pale eye-ring (sometimes nearly ***absent***). ***Grey to brownish head to chest***, paler throat. Olive upperparts, yellow lower underparts. Breeders have ***black breast patch***. Monotypic. Singles skulk on or near ground, mostly ***hopping*** and gleaning prey. **Voice** Rough *jik* and buzzy *zee* calls (C *et al.*). **SS** Much rarer Connecticut (***complete*** eye-ring, less grey hood, 'shorter' tail); ♀ more confusing with less defined hood (***broken, narrow*** eye-ring distinctive); ♀ yellowthroats lack hood, show supercilium, weaker bills. Grey-hooded Bush-Tanager similar but bill pink, red eyes, different behaviour. **Status** Very rare boreal winter resident (Oct–Apr).

Buff-rumped Warbler *Myiothlypis fulvicauda* 13–14cm

E & W lowlands to lower subtropics, S to El Oro and C Zamora Chinchipe (mainly below 1,000m, locally to 1,600m). Small rushing to mid-sized and more sluggish forested streams and damp areas. Greyish-olive upperparts, ***buff supercilium, most underparts, rump and basal half of tail***. Juv. brownish ***to breast***. Nominate (E), *semicervina* (W). Singles or pairs rather nervous near water; hops on or near ground fanning tail and switching it sideways. **Voice** Clear, strong, rising series of *teew-teew-teww.. tchEW-tchEW-tchEW...*, last part faster, more emphatic; *tchép* calls. **SS** Unmistakable. **Status** Fairly common.

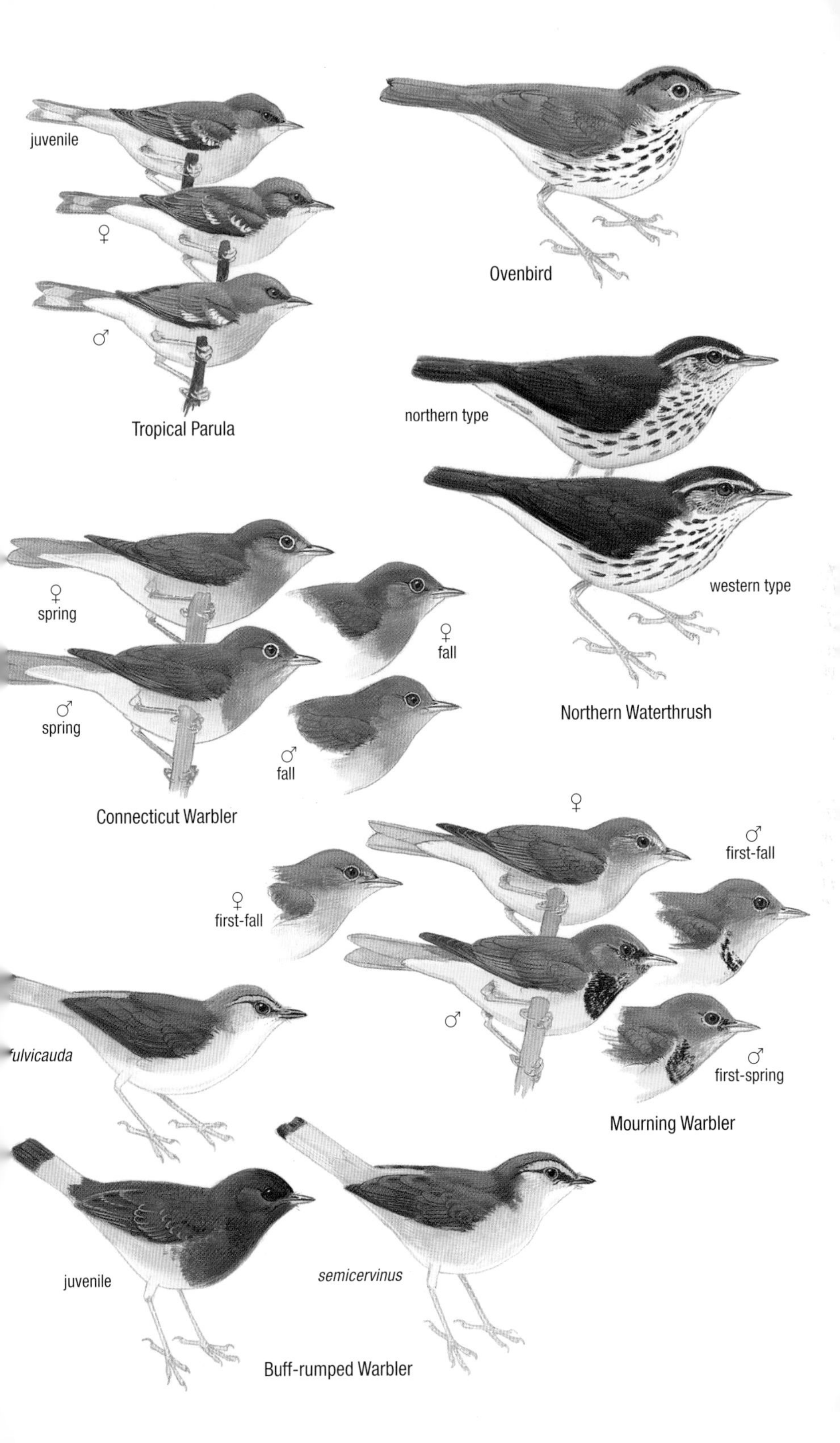

juvenile
♀
♂
Tropical Parula
Ovenbird
northern type
western type
Northern Waterthrush
♀
spring
♂
spring
♀
fall
♂
fall
Connecticut Warbler
♀
♂
first-fall
♀
first-fall
♂
♂
first-spring
Mourning Warbler
fulvicauda
juvenile
semicervinus
Buff-rumped Warbler

Resident forest warblers with duller plumages than migrant species on their breeding grounds, but lovely voices. In forest understorey, where they move in family groups and mixed flocks. Interesting distribution patterns, with altitudinal replacements and overlaps.

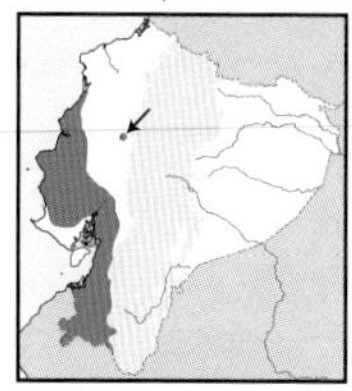

Grey-and-gold Warbler *Myiothlypis fraseri* 13–14cm

SW lowlands, to subtropics in Loja (below 550m, to 1,700m in Loja). Humid to deciduous forest undergrowth, woodland. ***White supraloral stripe***, black crown, ***bright orange*** (N) or ***yellow*** (S) ***semi-concealed patch, grey head to upperparts***. Lemon-yellow underparts. Races *ochraceicrista* (N), *fraseri* (S). Mainly in small groups, active and noisy, rarely in open, briefly with mixed flocks. **Voice** Melodious series of clear notes *tij-tij-tij-tlowee'wee-dlee, tiju-tiju-tlee-wee'clee-teechee'tee...* in duets, reminiscent of Russet-crowned Warbler; brief *tip, cheet* calls. **SS** Locally sympatric with Three-banded Warbler (drabber overall, head paler and striped). **Status** Uncommon.

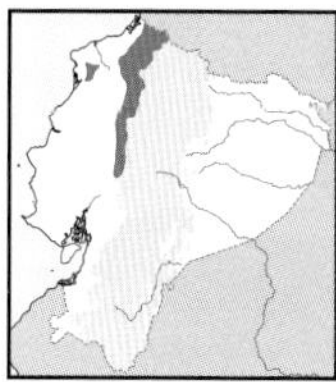

Golden-bellied Warbler *Myiothlypis chrysogaster* 12–13cm

NW Andean foothills to adjacent lowlands, Mache-Chindul hills in S Esmeraldas (400–1,200m). Montane forest, borders, woodland. ***Orange-red coronal patch outlined by broad black stripes***, dull yellowish eyebrow, ***yellower supraloral***. Olive above, ***dull olive-yellow below***, brighter throat and belly. Race *chlorophrys* possibly valid species. Pairs to small family groups in lower to mid strata, active and nervous; often with mixed flocks. **Voice** Fast, buzzy series of thin *t-t-t-t-t-t-tit'tzzzzzzi*. **SS** See Three-striped Warbler with bolder head pattern, paler yellow below. **Status** Fairly common.

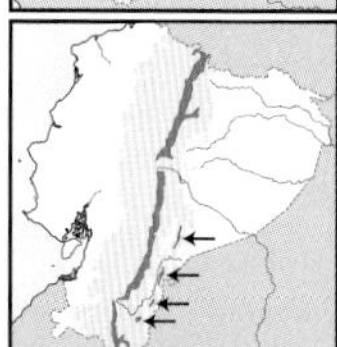

Citrine Warbler *Myiothlypis luteoviridis* 14–15cm

Subtropics to temperate zone in E Andes (2,500–3,200m). Montane forest, borders, mature woodland. Olive above, ***short yellow supercilium***, dusky-olive cheeks, dull yellow below. Nominate. Pairs to small groups in lower growth, very active and agile, mostly with mixed flocks. **Voice** Undulating, very fast, long series of clear *teewee-te'e'e-chet'te'te'te-seewit...*, often run into a trill; high *tit* calls. **SS** Sympatric Black-crested Warbler has more contrasting head pattern, brighter yellow underparts; see duller Oleaginous Hemispingus (grey legs, fainter supercilium, more olive wash below). **Status** Uncommon.

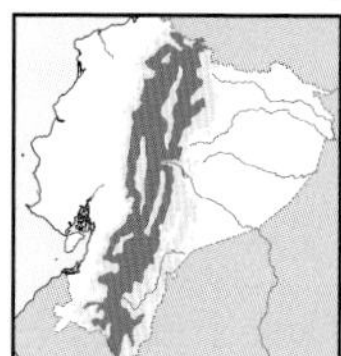

Black-crested Warbler *Myiothlypis nigrocristata* 13–14cm

E & W subtropics to temperate zone, inter-Andean highlands (2,000–3,500m). Montane forest, borders, woodland, hedgerows. ***Black crest bordered by broad yellow supercilium***, narrow black ocular stripe. Olive upperparts, ***bright yellow underparts***. Monotypic. Pairs to small groups active and nervous in dense lower growth, including bamboo; often joins mixed flocks. **Voice** Accelerating *chut-chut-tit-tit-tit-t'ch'chii'iiir*, first part clearer, ending in more emphatic *tchéw*; dry *tik* calls. **SS** In E see Citrine Warbler (shorter eyebrow, no black crown, duller ocular area). **Status** Fairly common.

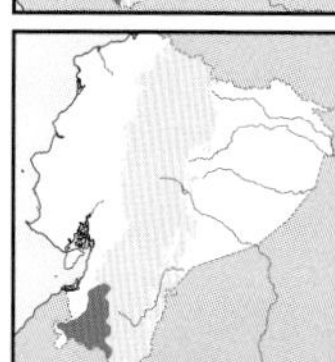

Three-banded Warbler *Basileuterus trifasciatus* 12–13cm

SW Andean foothills to subtropics (800–2,400m). Montane forest, woodland, shrubby borders. Blackish lateral crown-stripes, *olive coronal patch, pale grey supercilium and face to throat, narrow* blackish ocular stripe. Race *nitidior*. Pairs to small groups with mixed understorey flocks, nervous, often in fairly open edges. **Voice** Very fast, accelerating, rising series of high-pitched *sit-sit-sit-tsee-tsee..ts'ttttt* running into trill; high *tit* and buzzy *zeer*. **SS** Three-striped Warbler (overlaps locally?) has bolder facial pattern, no pale grey throat to chest (sharper voice). **Status** Uncommon.

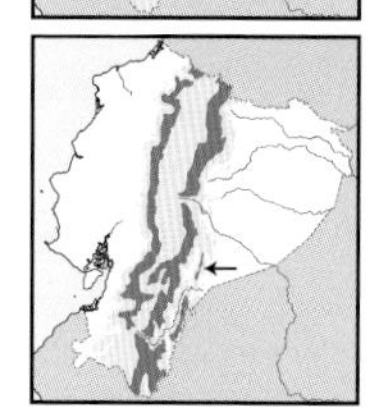

Russet-crowned Warbler *Myiothlypis coronata* 13–14cm

E & W Andean subtropics to temperate zone (1,500–3,300m). Montane forest, woodland, borders. ***Head grey, orange crown***, black coronal and ocular stripes; whiter throat, ***underparts yellow*** (W) or ***mainly grey*** (E, SE, SW). Races *elata* (NW), *castaneiceps* (SW), *orientalis* (E), *chapmani*? (extreme SE). Pairs to small noisy groups regular with mixed understorey flocks, sometimes higher at borders. **Voice** Melodious, whistled, accelerating antiphonal duets *tli-tli-tli-gloo'loo-lii?*, response often shorter and descending; buzzy shorter version: *tiglii-z'z'z...* and other calls. **SS** Confusion unlikely; see smaller Three-banded and Golden-bellied Warblers (no grey head). **Status** Common.

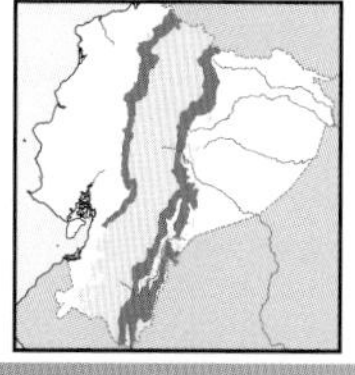

Three-striped Warbler *Basileuterus tristriatus* 12–13cm

E & W foothills to subtropics, S in W to NW Azuay (mostly 1,000–2,300m, locally lower). Montane forest, woodland, borders. Bold ***blackish crown-sides and cheeks, olive-yellow coronal-stripe, long whitish supercilium***. Races *daedalus* (W), *baezae* (NE), *tristriatus* (SE). Restless vocal pairs to small groups with mixed understorey flocks, possibly briefly leading them. **Voice** Very fast, ringing, jumbled *tttt-tsit-tsit-tse-tse-se'se'sesesese...* ascending and accelerating at end; clear *chep* calls. **SS** Locally Golden-bellied and Three-banded Warblers (weaker head patterns). **Status** Fairly common.

juvenile
fraseri
ochraceicrista
Grey-and-gold Warbler
Golden-bellied Warbler
juvenile
adult
Black-crested Warbler
juvenile
adult
Citrine Warbler
Three-banded Warbler
juvenile
elatus
orientalis
castaneiceps
chapmani
Russet-crowned Warbler
juvenile
tristriatus
daedalus
baezae
Three-striped Warbler

PLATE 250: RED-CAPPED CARDINAL AND TANAGERS

Several distinctive tanagers. Magpie and White-capped are the largest tanagers in Ecuador, resembling jays, especially in habits. Black-faced is locally conspicuous in semi-open habitats, Rufous-crested joins montane canopy flocks, while Black-and-white is seasonal in dry forest. The cardinal was, until recently, considered to be allied to grosbeaks and relatives, but is separated on the basis of molecular evidence.

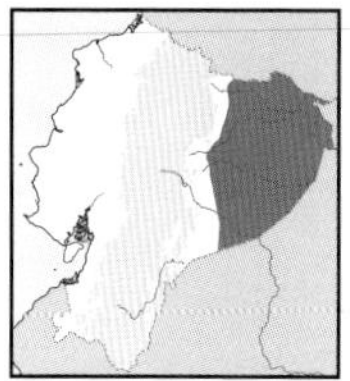

Red-capped Cardinal *Paroaria gularis* 16–17cm

E lowlands (to 300m). Shrubbery beside sluggish rivers, streams and lakes, flooded forest. ***Bright red head***, blackish orbital area, black chest and ***upperparts, pure white underparts***. Juv. has red replaced by ***buff-brown, upperparts brownish***. Nominate. Pairs to small groups conspicuous, often on exposed branches; rather tame and noisy. **Voice** Sweet brief series of clear, repetitive, hesitant *vui-vee'wee?* or *twë-wee-weeú?*; call harsher *chew!* **SS** Unmistakable. **Status** Fairly common.

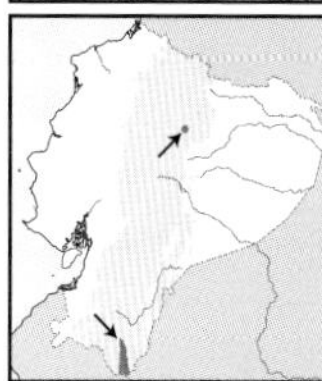

Black-faced Tanager *Schistochlamys melanopis* 18cm

Extreme SE in Zumba region (600–1,600m, locally higher). Shrubby clearings, woodland, borders. ***Silvery bill. Velvet black face to breast, dark grey upperparts, paler*** grey below; juv. ***olivaceous***, yellower below, ***yellowish orbital ring***. Race *grisea*. Singles or pairs, rarely with other species, fairly conspicuous. **Voice** Rich, melodic, whistled *tu-whit-WHEER-tu-whit-WHEER...* recalling a grosbeak; sharp *twíík-twíík....* **SS** Ad. unmistakable, juv. might resemble a bush-tanager, saltator, *Piranga* tanager or Black-and-white Tanager, but yellowish spectacles, long tail, plain underparts distinctive. **Status** Locally fairly common.

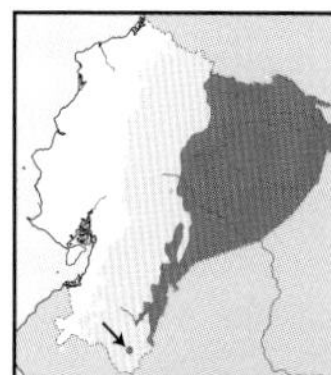

Magpie Tanager *Cissopis leverianus* 25–27cm

E lowlands to foothills (to 1,200m, locally 1,800m). *Terra firme* forest borders, woodland, woody clearings. Unique, large striking tanager with ***pied plumage, bright yellow eyes*** and ***long tail***. Nominate. Conspicuous and rather noisy, mostly in groups but rarely with mixed flocks. **Voice** Loud, metallic *tchÍT, tchÍT, tchÍT...t-cht'cht'cht'cht*, sometimes jumbled and complex; also loud *tééeeowí...*, ringing *chlt!* **SS** Might recall a jay, but confusion unlikely. **Status** Fairly common.

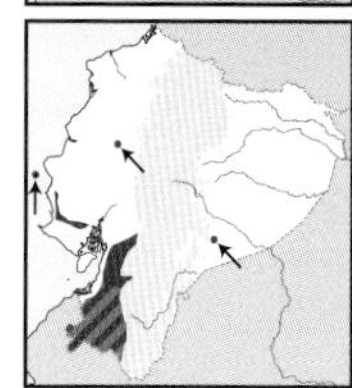

Black-and-white Tanager *Conothraupis speculigera* 16–17cm

SW subtropics, local elsewhere in S & W (500–1,700m, once near sea level). Open woodland, scrub, borders. Bright bluish-grey bill, red eyes. ***Glossy blue-black head to breast***, greyer rump, contrasting ***white belly***, small ***white wing patch***. ♀: ***bright olive above, bold yellow eye-ring***, streaky face, yellower below, ***breast to flanks coarsely flammulated olive***. Juv. duller, duskier cheeks, dark eyes. Monotypic. Only present during wet breeding season (Feb–May), when quite common and conspicuous, especially ♂. **Voice** Loud, ringing *chee-yoón, chee-yoón, chee-yoón...* repeated up to 6–8 times, recalling Scrub Blackbird; call simpler *cheé, cheé....* **SS** ♀ with more local Black-faced and *Piranga* tanagers (all lack breast flammulations and bold eye-ring). **Status** Fairly common during rainy season; NT globally.

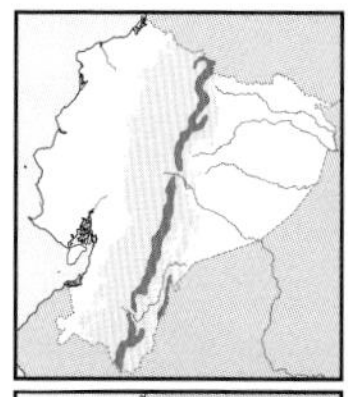

White-capped Tanager *Sericossypha albocristata* 24–25cm

Subtropics to temperate zone in E Andes (1,900–2,700m, locally to 3,300m). Montane forest, borders. Handsome, large, with ***striking white cap*** and ***red throat to chest***, ♀: bib duller, absent in juv. Monotypic. Noisy flocks move heavily through upper strata in jay-like fashion, bound along branches, briefly perch in open; irregularly with mixed flocks. **Voice** Loud, piercing calls while on move, *CHEWP-CHEEYP-CHËP-CHËP-CHEEYÉP...*, simpler *cheé-cheyp* when perched. **SS** Voice recalls a jay (particularly Turquoise Jay), but when seen is unmistakable. **Status** Uncommon; VU.

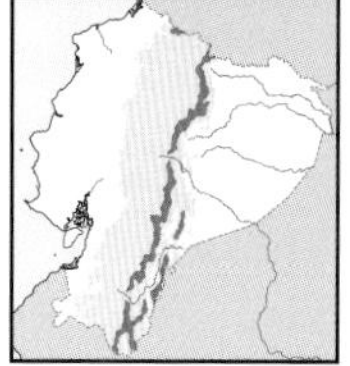

Rufous-crested Tanager *Creurgops verticalis* 15–16cm

Subtropics in E Andes and extreme NW (1,500–2,500m). Montane forest and borders, sometimes *Chusquea* stands. ***Reddish eyes. Slate-grey upperparts*** with ***semi-concealed orange-rufous crown patch*** outlined black; contrasting ***orange-rufous underparts***. ♀: ***lacks*** crown patch. Monotypic. Mostly pairs with mixed flocks from canopy to near floor at borders, deliberately seeking invertebrates. **Voice** Not very vocal; very sharp *tss* or *tsit*, often a fast warbling *tss'ss't-t-t-tss't-t-t-tss'ss'ss....* **SS** Smaller Black-eared Hemispingus has rather similar pattern but black mask, no coronal patch, duller below. **Status** Uncommon.

adult
immature
juvenile
Red-capped Cardinal
adult
immature
juvenile
Black-faced Tanager
adult
juvenile
Magpie Tanager
♂
♀
♂
juvenile
Black-and-white Tanager
♂
immature
♂
♀
White-capped Tanager
♀
♂
juvenile
Rufous-crested Tanager

PLATE 251: HEMISPINGUSES

Rather nondescript Andean tanagers that often join mixed-species flocks. Nimble and acrobatic, seen more often than superficially similar *Basileuterus* and *Myiothlypis* warblers. Several species might occur at some sites but altitudinal replacements also common. Complex taxonomy in Black-eared, possibly comprising three species. Recent major taxonomic shuffle classified most species in separate genera.

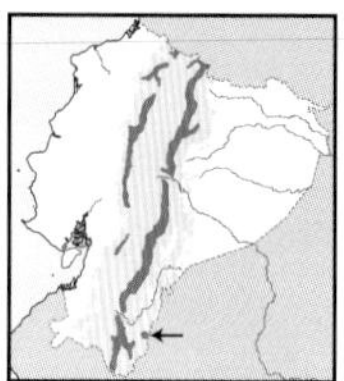

Black-capped Hemispingus *Hemispingus atropileus* 17–18cm

E & W temperate zone in Andes (2,300–3,200m, locally higher or lower). Montane forest undergrowth, secondary woodland, borders. ***Blackish head, very long white supercilium.*** Olive above, dull olive-yellow below. Nominate. Small groups, sometimes pairs, with mixed-species flocks, often in bamboo. **Voice** At dawn a high-pitched, simple, repeated *teew-dseé, teew-dseé…*, sometimes repeated *teew*; song a fast, sputtering warble, *tss-tss-tss-t't'TSEEU….* **SS** Only hemispingus with black-and-white head, Superciliaried has shorter supercilium, olive head, bright yellow underparts; see Black-winged Saltator and Tanager Finch, with long white supercilium but otherwise very different. **Status** Uncommon.

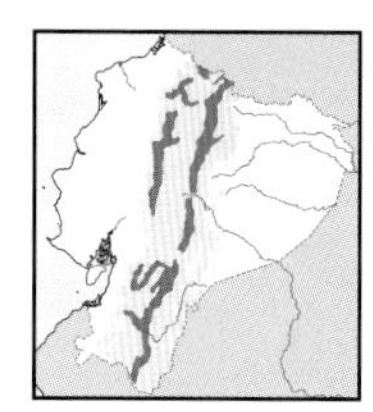

Superciliaried Hemispingus *Hemispingus superciliaris* 14–15cm

Temperate zone in E & W Andes (2,400–3,200m). Montane forest undergrowth, woodland, borders. Mostly dull olive above, duskier crown, ***white supercilium***, duskier cheeks. ***Bright lemon-yellow below.*** Races *nigrifrons* (most of range), *maculifrons* (S). Pairs to small groups with mixed-species flocks, forage in warbler-like fashion in various strata, often terminal foliage. **Voice** Rather explosive, jumbled and very fast series of warbles and chatters *T't'ss's's's't't't…*; simple *tss-tss* foraging calls. **SS** Black-capped (more prominent eyebrow) and Oleaginous Hemispingus (fainter supercilium); Black-crested Warbler has pink legs and black on crown, Citrine Warbler duller below, yellow supercilium, pale legs; see Brown-capped Vireo. **Status** Uncommon.

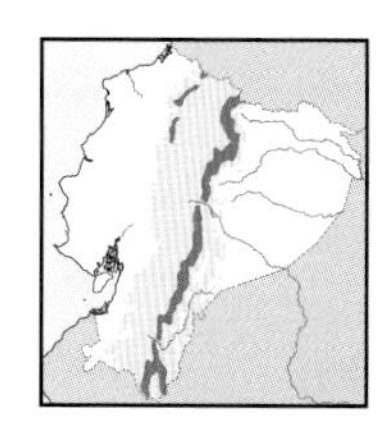

Oleaginous Hemispingus *Hemispingus frontalis* 14–15cm

Subtropics in E and extreme NW Andes (1,500–2,500m). Montane forest, borders. Mostly olive above, ***faint yellowish supercilium***, dusky cheeks. ***Dull olive-yellow underparts.*** Nominate. Rather shy pairs or trios in lower strata, consort with similar-looking *Basileuterus* and *Myiothlypis* warblers. **Voice** Explosive, fast, jumbled warbles and chatters *T'T't's'ser'ss'ts'ser'ts'ser'ss'ss…*; simple, sharp, buzzy *dsíip* and brief *chip* calls. **SS** Locally Citrine Warbler (pinkish legs, more marked supercilium; behaviour differs); somewhat reminiscent of a *Chlorospingus* bush-tanager, especially the local Yellow-green (no overlap?), but less sturdy, with more yellow throat, bolder eyebrow. **Status** Uncommon.

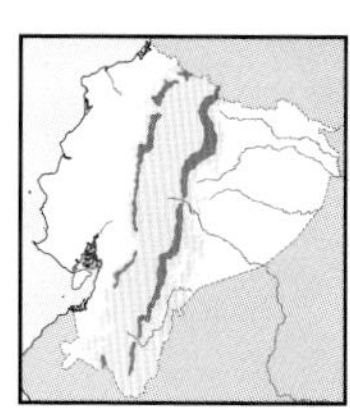

Black-eared Hemispingus *Hemispingus melanotis* 14–15cm

Subtropics in E & W Andes (1,800–2,700m in E, 1,600–2,200m in NW, 2,000–2,500m in SW). Montane forest undergrowth, borders, woodland. ***Olive-grey upperparts, blackish face*** (*vaguely darker* in NW), very faint pale eyebrow. ***Ochraceous-buff underparts*** (*duller* in NW). SW race strikingly different, ***head black, long white supercilium, rich cinnamon below.*** Races *melanotis* (E), *ochraceus* (NW), *piurae* (extreme SW); likely separate species. Pairs or small groups with understorey flocks, noisy and active but not easy to see; E & SW races favour bamboo. **Voice** Explosive, very fast, high-pitched sputtering *T'R'R'SS's's's's'st't't...*; calls a sharp *tss's* (E); fast *chit'chit'chit…* or *tsit'tsit'tsit…* in SW. **SS** In E more boldly patterned Rufous-crested and Fawn-breasted Tanagers; in NW White-sided Flowerpiercer (♀), larger Fawn-breasted Tanager, bolder Dusky Bush-Tanager; unmistakable in SW. **Status** Rare to locally uncommon.

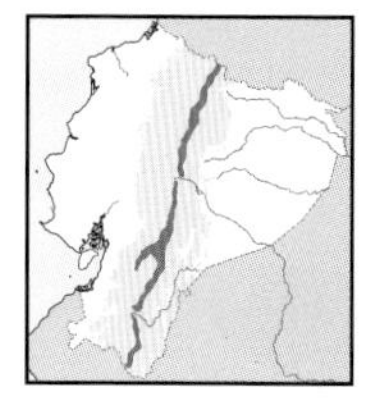

Black-headed Hemispingus *Hemispingus verticalis* 14–15cm

Temperate zone in E Andes (2,700–3,400m). Montane forest undergrowth, woodland, shrubby clearings, borders. ***Creamy-white eyes*** contrast with ***black head***, buff mid crown. ***Grey*** above, whitish-grey below. Monotypic. Pairs or small groups in mixed-species flocks, agile and rather conspicuous at edges and atop bushes and small trees. **Voice** High-pitched jumbled chatters, less explosive than other hemispingus; piercing *tdít*, sometimes repeated and followed by jumbled song. **SS** Confusion unlikely. **Status** Rare to locally uncommon.

Black-capped Hemispingus
nigrifrons
maculifrons
Superciliaried Hemispingus
adult
juvenile
Oleaginous Hemispingus
ochraceus
melanotis
adult
juvenile
piurae
Black-eared Hemispingus
Black-headed Hemispingus

The lovely *Thlypopsis* are small arboreal tanagers, one species in E lowlands, the others on Andean slopes and valleys; all three mostly insectivorous members of mixed canopy flocks. The monotypic bush-tanager resembles *Chlorospingus* bush-tanagers (Plate 269) but differs in bare-parts coloration and behaviour. Also a distinct, monotypic, understorey Amazonian tanager (Grey-headed).

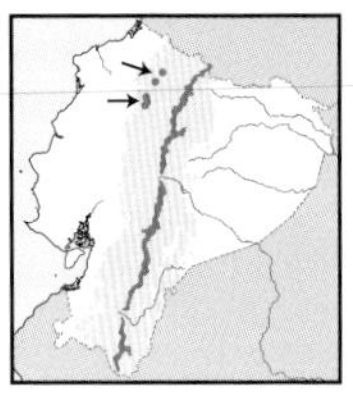

Grey-hooded Bush-Tanager *Cnemoscopus rubrirostris* 15cm

Subtropics to temperate zone in E Andes, very local in NW (2,200–3,100m). Montane forest canopy, borders, shrubby clearings. ***Pale pink bill***, reddish eyes, ***pink legs. Grey head to breast***, olive upperparts, bright yellow belly. Nominate. Mainly pairs in outer foliage of upper strata with mixed-species flocks; constantly wags tail. **Voice** Fast, high-pitched, squeaky *tswee-tss', swee-swee'ur*; piercing *chí'típ-ch'típ-ch'típ*, sometimes more jumbled. **SS** Recalls *Chlorospingus* bush-tanagers, but tail wagging, pink bill and legs distinctive; ♀ Capped Conebill has similar pattern and tail-wagging behaviour but slimmer, lacking pink bare parts. **Status** Uncommon.

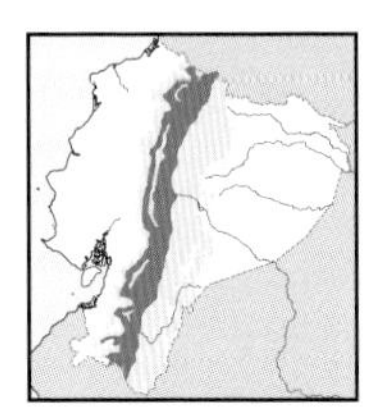

Rufous-chested Tanager *Thlypopsis ornata* 12–13cm

Andean highlands to subtropics (1,800–3,000m). Montane forest borders, secondary woodland, woody clearings. ***Bright rufous-orange head and most underparts***, contrasting ***white mid belly***, olive-grey upperparts. Sexes alike. Juv. duller, duskier above. Nominate (most of range), *media* (S). Mostly pairs, sometimes small groups, conspicuously foraging in outer limbs of tall trees and bushes, often loosely associated with mixed flocks. **Voice** High-pitched, piercing, slightly accelerating or jumbled *tser-tser-tser...*; louder group calls *tse-sje-tse-je...*, also simpler and lower calls. **SS** No overlap with other *Thlypopsis*; see Cinereous Conebill (browner, with long supercilium, etc.) and ♀ Rusty Flowerpiercer. **Status** Uncommon.

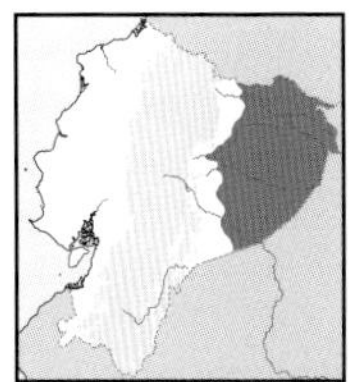

Orange-headed Tanager *Thlypopsis sordida* 13–14cm

NE lowlands, also SE? (to 500m). Riparian woodland, second growth on river islands. ***Bright orange hood***, contrasting ***grey upperparts and pale grey underparts***, whiter mid belly. Olive juv. has rich ***orangey-yellow face and bib***. Race *chrysopis*. Pairs or trios in borders and semi-open, forage in foliage and dead leaves in various strata, often with mixed flocks. **Voice** Thin repeated *séét* in fast series, followed by high-pitched stuttering *td'td'td'td'td'dididit*. **SS** No overlap with congeners; juv. Bicoloured Conebill plainer yellowish below, with ***more pointed bill***, more olive above. **Status** Uncommon.

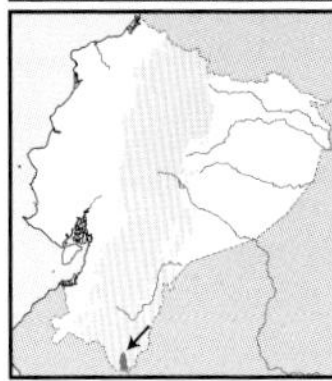

Buff-bellied Tanager *Thlypopsis inornata* 12–13cm

Extreme SE in Zumba region (600–1,200m). Semi-deciduous forest borders, shrubby clearings. ***Orange-rufous crown to nape, orange-buff face and underparts***, contrasting grey upperparts and rather long tail. ♀ and juv. have ***dusky crown, buff front and supercilium***, paler below. Monotypic. Pairs in various strata, regular near ground; joins mixed-species flocks, gleans outer foliage, probes flowers and buds. **Voice** At dawn a sweet, musical *sweep-sweep-sweep-sweep, seeu'l-seeu'l-seeu'l, sweep-see'l-see'l-see'l...*; sharp *tsíip-tsíip...* in flight. **SS** Confusion unlikely in limited range; duller ♀ might be challenging but usually with ♂; no overlap with similar (but more rufescent below) Rufous-chested Tanager. **Status** Uncommon locally.

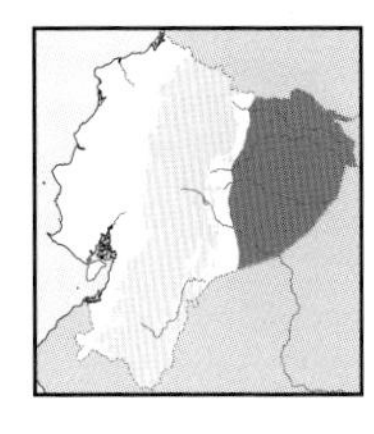

Grey-headed Tanager *Eucometis penicillata* 17–18cm

E lowlands, mostly in NE (below 400m). *Várzea* forest. ***Grey hood, whiter on throat and long shaggy crest***. Bright olive upperparts, ***bright olive-yellow below***, richer on belly. Duller juv. has olive head and lacks crest. Nominate. Pairs noisy in undergrowth, often in swampy areas; may follow mixed flocks; usually flicks wings and tail when excited. **Voice** Long-repeated, loud, jerky *chíp* or *ztít*; jumbled series of buzzy, sputtering *pzzt-pzzt-fzzt-fzzt-feezzee-feezzeet-zziit-fzeet....* **SS** Confusion unlikely in *várzea*; ♀ White-shouldered Tanager similar but habitat, strata and behaviour very different. **Status** Rare to locally uncommon.

Grey-hooded Bush-Tanager
adult
juvenile
Rufous-chested Tanager
adult
juvenile
Orange-headed Tanager
adult
juvenile
Buff-bellied Tanager
adult
juvenile
Grey-headed Tanager

PLATE 253: LOWLAND TANAGERS

Usually with mixed flocks, one species (the shrike-tanager) leading them in the canopy. Marked sexual dimorphism in *Tachyphonus*; drab ♀♀ can be confusing, see underparts. Not especially vocal. Shrike-tanager distinctive, often seen sallying for prey.

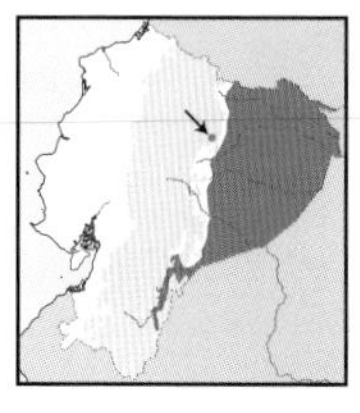

Flame-crested Tanager *Tachyphonus cristatus* 15–16cm

E lowlands (to 600m, locally to 1,000m). *Terra firme* forest, borders. Silvery mandible base. Mostly black, ***long, flame orange crest, golden-orange central throat*** and rump, white shoulders. Drab ♀: olive-brown head, more ***rufescent mid crown and upperparts, whitish-buff throat, ochraceous below***. Race *fallax*. Mainly insectivorous in mixed canopy flocks, gleaning outer foliage and large limbs. **Voice** Thin hummingbird-like *zéép* foraging calls; soft *tzét-tzét-tzét-zítzít*. **SS** Long crest and throat patch distinctive (♂); ♀ resembles Fulvous-crested (greyer crown, duller upperparts) and larger and stockier Fulvous Shrike-Tanager (duller and plainer below). **Status** Uncommon.

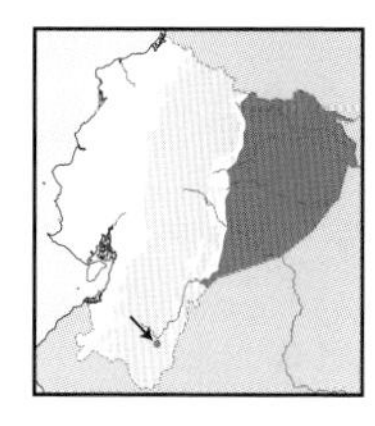

Fulvous-crested Tanager *Tachyphonus surinamus* 15–16cm

E lowlands (to 900m). *Terra firme* forest and borders. Silvery mandible base. Mostly black, ***small semi-concealed rufous-orange crown patch***, yellow-orange rump, ***reddish lower flanks,*** white shoulders. ***Bicoloured*** ♀ has ***greyish-olive crown***, olive upperparts, whitish-buff throat, ***and pale buff underparts***. Race *brevipes*. Pairs or small groups forage below canopy, often with mixed flocks; insectivorous. **Voice** Weak *zeét* foraging calls; also fast, emphatic *tzeé'zít* followed by buzzy *ssít-ssít-ssít...*. **SS** ♂ lacks long crest and throat patch of Flame-crested; ♀ Flame-crested has rufescent crown and upperparts, more ochraceous underparts. **Status** Uncommon.

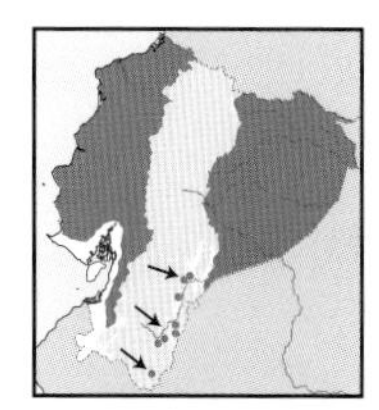

White-shouldered Tanager *Tachyphonus luctuosus* 13–14cm

E & W lowlands, to foothills in W (to 1,000m). Canopy and borders of humid to deciduous forest and woodland. Silvery mandible base. Mostly glossy black with ***contrasting white wing-coverts***. ♀: ***grey head***, whiter throat, bright olive above, ***rich yellow below***. Races *panamensis* (W), nominate (E). Pairs or small groups forage for invertebrates in outer twigs and foliage, often with mixed flocks. **Voice** Thin, high-pitched *zit-zit-zit...zit-zit* or *t'chirt-t'chirt-zit-zit-t'chirt-t'chirt-zit....* **SS** ♂ unmistakable; drab ♀ resembles Grey-headed Tanager but habitat and behaviour differ. In W foothills, see Ashy-throated Bush-Tanager (similar pattern but not rich yellow below). **Status** Fairly common.

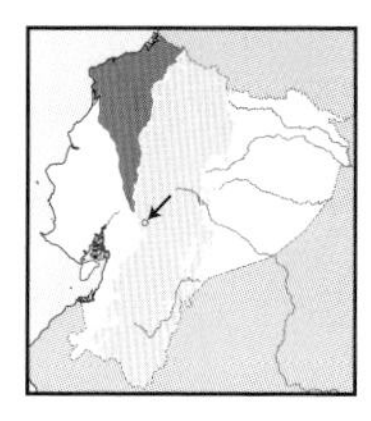

Tawny-crested Tanager *Tachyphonus delatrii* 14–15cm

NW lowlands (to 800m). Very humid forest, borders. Silvery mandible base. ***Glossy black*** with striking ***tawny-orange crown patch***. ♀: drab and ***nondescript***, mostly ***dull olive-brown***, slightly paler on throat, darker wings and tail. Monotypic. Mostly in fairly large monospecific flocks moving through subcanopy and edges, sometimes joining mixed flocks. **Voice** Thin, high-pitched *chít-chít-chít...* sometimes rattled at end, constantly repeated while foraging. **SS** ♂ unmistakable in range; ♀ is darkest tanager in range; might be misidentified with brighter White-lined Tanager, large-billed ♀ Blue-black Grosbeak and even dull brown solitaires, mourners and foliage-gleaners; behaviour of latter species always differs. **Status** Rare.

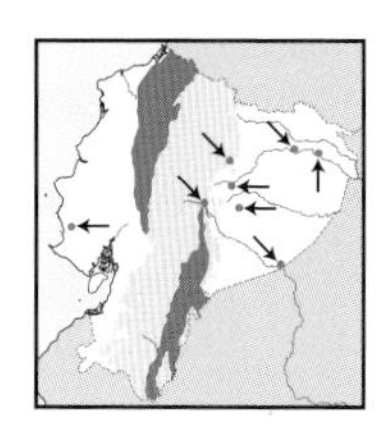

White-lined Tanager *Tachyphonus rufus* 18–19cm

NW lowlands to foothills, E Andean foothills to subtropics (to 2,000m). Forest borders, adjacent clearings, woodland. ***Silvery mandible base. Entirely black*** with ***white wing bend and underwing-coverts***, conspicuous in flight. ♀: mostly ***rich rufous***, paler below. Monotypic. Mainly pairs away from mixed flocks, foraging in edges and semi-open, rather inconspicuous; may flick wings before flying. **Voice** More musical than congeners, song a simple, repetitive *cheü-chooï-cheü-chooï-cheü-chooï ...*, first note stronger; thin *shíp* call. **SS** ♂ resembles a blackbird but underwing distinctive; ♀ might recall rufous mourners, attilas and becards, but behaviour, voices and silhouette differ. Silvery bill distinctive. **Status** Uncommon.

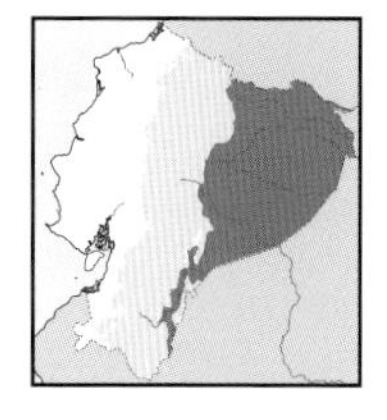

Fulvous Shrike-Tanager *Lanio fulvus* 17–19cm

E lowlands to foothills (to 1,100m). Mostly *terra firme* forest, borders. ***Heavy, hooked bill. Black hood, wings and tail***, contrasting ***fulvous body***, darker chestnut breast and rump. ♀: olive-brown above, paler below, ***more rufescent wings and rear***, duskier crown. Race *peruvianus*. Mainly in pairs leading mixed flocks, foraging in higher strata, often sallying for insects. **Voice** Loud, sharp *t'CHEW*, doubled *tswe-SÉEU* or tripled *tsee-tséu-tséu...t'sss*. **SS** ♂ unmistakable; cf. ♀ with smaller ♀ Fulvous-crested and Flame-crested Tanager, both paler, with lighter bills, greyer head (Fulvous-crested), more rufescent crown (Flame-crested). **Status** Uncommon.

♀
♂
Flame-crested Tanager
♂
Fulvous-crested
Tanager
♀
♂
immature
luctuosus
♀
♂
♀
panamensis
♂
♀
juvenile
♂
immature
White-shouldered Tanager
♂
♂
♂
♀
Tawny-crested Tanager
♂
♂
immature
juvenile
♀
White-lined Tanager
♂
♂
immature
♀
Fulvous Shrike-Tanager

Common and widespread species of open and semi-open habitats, often in pairs or groups, only briefly with flocks. Blue-capped is more forest-based than the rest. *Ramphocelus* characterised by swollen mandible base. Silver-beaked, Flame-rumped and Blue-grey are among the most numerous birds of open habitats. Blue-and-yellow is confined to Andean valleys.

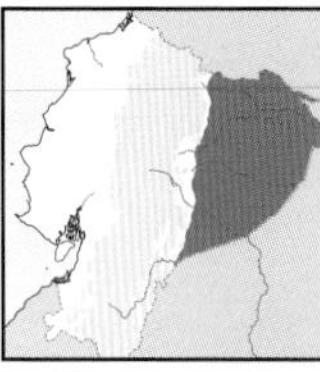

Masked Crimson Tanager *Ramphocelus nigrogularis* 18–19cm

E lowlands (to 600m). Forest borders, rivers, secondary woodland. ***Blood red*** with ***velvet black mask, back, wings, central belly and tail***. ♂: bill has ***swollen silvery mandible base***. Monotypic. Pairs or groups forage low down, often near water, sometimes joining mixed flocks. **Voice** Simple *ti'pip-wheeu, ti'pip-wheeu, ti'pip-wheeu...* or *wit'weet-weér, wit'weet-weér..., wheét-doowee-wheét...*; loud, ringing *tíínk*. **SS** Unmistakable, only other red-and-black Amazonian species (Black-necked Red-Cotinga) differs in too many respects to be serious confusion risk. **Status** Uncommon.

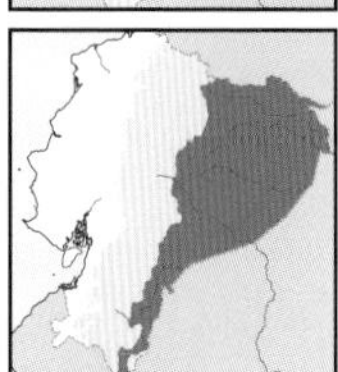

Silver-beaked Tanager *Ramphocelus carbo* 17–18cm

E lowlands to foothills (to 1,100m, locally to 1,700m). Forest borders, woodland, adjacent clearings, parks. ***Bright silvery swollen mandible base***. Entirely ***wine red***, redder front. ***Redder-brown*** ♀ lacks silvery beak. Nominate. Noisy and energetic groups of ad. and juv. low down in open and semi-open areas, often in riparian scrub; sometimes confiding. **Voice** Simple *teechuuwee, tchuíít-wee-tchuíít, teet'chuuveet...* song; commonly heard loud, ringing *tchínk*. **SS** Brighter red *Piranga* tanagers (pale bills) and Red-crowned Ant-Tanager of forest interior. **Status** Common, spreading with deforestation.

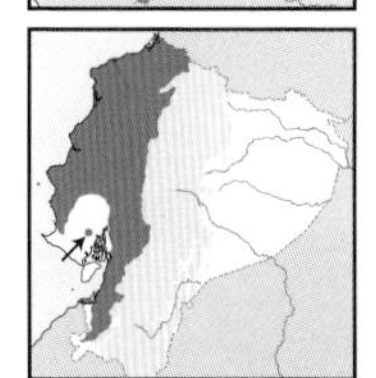

Flame-rumped Tanager *Ramphocelus flammigerus* 18–19cm

W lowlands to subtropics (to 2,000m). Forest borders, woodland, adjacent clearings, gardens. ***Silvery-blue bill tipped black***. ♂: ***velvet black*** with ***striking lemon-yellow rump***. ♀: dusky above, ***lemon-yellow rump and most underparts***. Race *icteronotus*. Noisy pairs to small groups conspicuously forage at roadsides and other open areas, often seen flying between bushes, flashing yellow rump. **Voice** Loud, raspy *t'choou, t'che-tcheoo-tche...*, more rattled *t'chchchch't't*. **SS** ♂ looks like small cacique, but size, behaviour, etc. strikingly different. **Status** Common, spreading with deforestation.

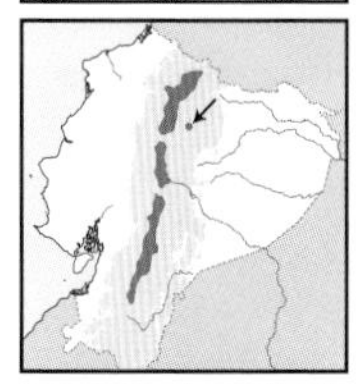

Blue-and-yellow Tanager *Pipraeidea bonariensis* 16–18cm

Dry inter-Andean valleys (1,800–3,000m). Dry woodland, scrub, gardens, hedges. ***Sky-blue hood***, olive mantle, rich ***golden-yellow rump and most underparts, bluish wings*** and tail. ♀: duller, ***bluish wash on crown***, dull olive-brown above, ***dull yellowish-buff below***. Juv. drabber. Race *darwinii*. Mostly pairs to small groups, sometimes surprisingly silent and easy to miss. **Voice** Somewhat musical, not squeaky *swee-teét, swee-teét, swee-sew-teét* or *s'duseé-duseé-duseé...*; high *típ* calls. **SS** ♂ unmistakable; in range cf. ♀ with imm. Scrub and Fawn-breasted Tanagers. **Status** Fairly common.

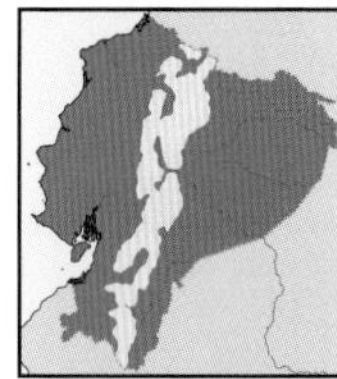

Blue-grey Tanager *Thraupis episcopus* 15–17cm

E & W lowlands to subtropics, inter-Andean valleys (to 1,500–1,700m on Andean slopes, to 2,800m in Andean valleys). Forest borders, woodland, scrub, hedgerows, gardens, parks. Mostly ***pale bluish-grey***, bluer wings and tail (extensive ***white on wing-coverts*** in E). Sexes alike. Races *quaesita* (W), *coelestis* (E), *caerulea* (SE). Noisy pairs or small groups at various heights, sometimes quite tame at feeders, etc. **Voice** High-pitched, thin, squeaky *s'weé-s'weé-swswsweé', sweé-sweé'ee'ee-s'wwwwee....* **SS** Unmistakable, but see darker Palm Tanager. **Status** Common and widespread.

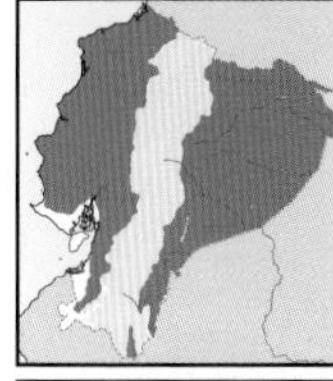

Palm Tanager *Thraupis palmarum* 16–17cm

E & W lowlands to foothills (to 1,300m, locally to 1,800m). Forest borders, woodland, adjacent clearings, gardens. Mostly ***greyish-green***, darker back, ***brighter wing-coverts and primary 'window', black flight feathers and tail***. Sexes alike. Races *melanoptera* (E), *violilavata* (W). Pairs or small groups regularly with Blue-grey Tanager at fruiting trees, feeders, crops, etc; fond of palms, agilely searches for prey in fronds. **Voice** Like Blue-grey, somewhat less squeaky, more ringing; *see-you, see-you, see-you...s's's's, ss'wee-ss'wee....* **SS** Darker than sympatric Blue-grey. **Status** Common.

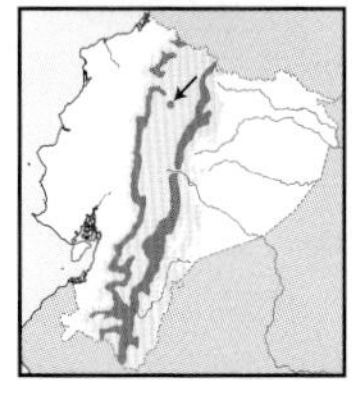

Blue-capped Tanager *Thraupis cyanocephala* 17–18cm

Subtropics to temperate zone on both Andean slopes (1,800–3,000m). Montane forest, borders, woodland. ***Bright blue crown to nape***, blackish mask, olive upperparts, ***bluish-grey below***, yellow vent to ***thighs and underwing-coverts***. Nominate. Pairs or small flocks in upper strata, regularly with mixed flocks, also monospecific groups or large aggregations of frugivores at fruiting trees. **Voice** High-pitched, squeaky, very fast twittering *tsuu-tséétseetsee's's's...*, sometimes preceded by sharp *tsiip* calls. **SS** ♀ Blue-and-yellow Tanager is duller, paler blue in crown; see smaller, more agile, less bold ♀ Capped Conebill. **Status** Uncommon.

♂
♀
Masked Crimson Tanager
♂
♀
juvenile
Silver-beaked Tanager
♂
♂
immature
♀
♀
juvenile
Flame-rumped Tanager
♂
♀
Blue-and-yellow Tanager
coelestis
lea
uaesita
juvenile
Blue-grey Tanager
violilavata
melanoptera
juvenile
Palm Tanager
adult
juvenile
Blue-capped Tanager

PLATE 255: ANDEAN TANAGERS AND MOUNTAIN-TANAGERS

Colourful tanagers, mostly Andean, regularly with mixed flocks. Mountain-tanagers are large and heavy-built treeline dwellers, Hooded being more conspicuous than the other two. *Bangsia* species are rare, rather local and apparently restricted to moss-laden forests. Orange-throated has a very limited range and specific habitat requirements. Vermilion superficially recalls a *Piranga*.

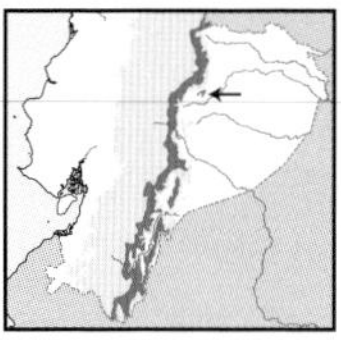

Vermilion Tanager *Calochaetes coccineus* 17–18cm

Subtropics on E Andean slope (1,100–1,800m). Montane forest canopy and borders. Striking monomorphic tanager in ***black and bright red***. Monotypic. Small groups usually with mixed canopy flocks; searches for prey in foliage, mosses, branches and limbs. **Voice** Not very vocal, sharp *chip-chi'ip-chip...chititit*, descending *tset*. **SS** White-winged Tanager has obvious white wingbars, Scarlet lacks black mask and bib, and bill is pale. **Status** Uncommon.

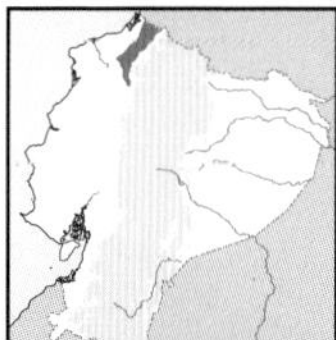

Golden-chested Tanager *Bangsia rothschildi* 16cm

NW Andean foothills to adjacent lowlands (to 700m). Upper strata in wet forest and borders. ***Dark bill, glossy blue-black*** with conspicuous ***golden-yellow breast patch***, creamy lower belly to vent. Monotypic. Mainly pairs loosely associated with mixed flocks, often behind them; slowly hops along larger branches. **Voice** Sharp, high-pitched, buzzy, descending trills: *tzz-tzz-tzz-tzz, tzz-tzz-tzz*. **SS** Moss-backed is mostly green (but can look dark in dim light), with smaller breast patch, pale mandible, pale blue face, etc.; see Purplish-mantled. **Status** Rare, local, possibly declining, VU in Ecuador.

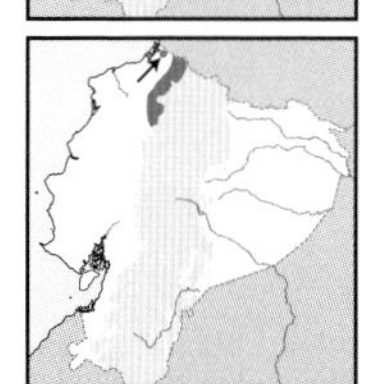

Moss-backed Tanager *Bangsia edwardsi* 16cm

NW Andean foothills (400–1,600m). Upper strata in montane forest, secondary woodland and borders. ***Rosy mandible***. Black head, ***blue face*** and wing-coverts. Otherwise mostly ***moss-green, bright yellow breast patch***. Monotypic. Mainly in pairs with mixed flocks but not very active, tends to remain still for long periods, allowing close approach. **Voice** Ascending and descending chippered trill, *tr'e'e'E'E'e'e-tr'e'E'E'e'e'e...* not tanager-like. **SS** Black-chinned Mountain-Tanager has similar-coloured back but no blue in face, yellow underparts, etc. **Status** Uncommon, quite local.

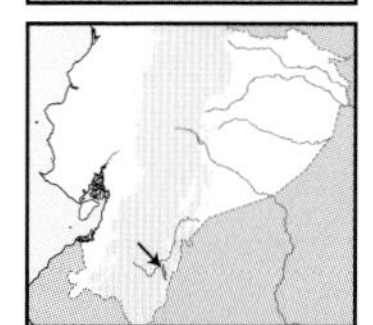

Orange-throated Tanager *Wetmorethraupis sterrhopteron* 18cm

Foothills of Cordillera del Cóndor in Nangaritza Valley (900–1,000m). Montane forest, tall second growth and borders. Beautifully clad in ***glossy black, bright blue, deep orange and yellow***. Monotypic. Conspicuous monospecific flocks hop along larger limbs, searching for prey on twigs, mosses and epiphytes. **Voice** At dawn, soft *tsoo-swi̇t, tsoo-swi̇t, tsoo-swi̇t...*, sometimes three-noted *wit'soo-swi̇t, wit'soo-swi̇t...*; sharp *chi̇t, chít, chi̇t...*when foraging. **SS** Unmistakable. **Status** Rare, very local, VU.

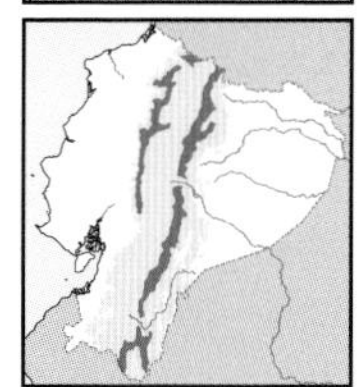

Hooded Mountain-Tanager *Buthraupis montana* 22–23cm

Subtropics to highlands in E & W Andes, S to Chimborazo in W (2,000–3,200m). Montane forest, tall secondary woodland, borders, woody clearings. ***Blood-red eyes. Black hood, blue upperparts, rich yellow underparts***. Race *cucullata*. Noisy and conspicuous groups move more like jays, hopping along large branches, sometimes with flocks of other large passerines. **Voice** Loud *weee'ëk-weee'ëk...* or *to-weee'ek, to-weee'ek...* at dawn; in flocks a *to-twee'twee'tee-to'twee-to'twee-to'twee*, and loud *tséét-tséét-tséét-tséét* in flight. **SS** Smaller Black-chested Mountain-Tanager lacks red eyes and all-black hood, upperparts green; Buff-breasted also dark-eyed, has bold bluish facial line and buffy breast-band. **Status** Common.

Masked Mountain-Tanager *Buthraupis wetmorei* 20–21cm

Local in E Andean temperate zone (2,950–3600m). Treeline and adjacent montane forest. Dark eyes. ***Black mask, bright yellow crown-sides*** ('eyebrows'), ***rump and most underparts***. Bright olive-green above, ***black wings, blue wingbars***. Monotypic. Mainly pairs, follows mixed flocks but discreet and difficult to see as usually in cover. **Voice** Song a sharp series of steadily repeated *tzeet-zeet-zeet...* at dawn, sometimes starts with short rolling *t't't-tzeet-tzeet-tzeet...*; short *zíp* foraging calls. **SS** Confusion unlikely; smaller Blue-winged Mountain-Tanager at lower elevations (all yellow below, yellow crown, black upperparts, etc.). **Status** Rare, likely overlooked, VU.

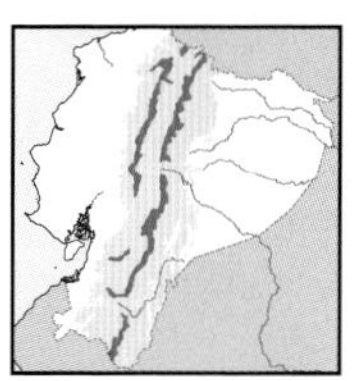

Black-chested Mountain-Tanager *Cnemathraupis eximia* 21–22cm

E & W Andes temperate zone, S to Chimborazo in W (2,750–3,300m). Montane forest, borders, old second growth, treeline, *páramo* woodland. ***Dark eyes. Black face to breast, blue crown to nape, green upperparts***, bright yellow belly. Race *chloronota* (most of range), *cyanocalyptra* (SE). Pairs to small groups often alone; tend to remain under cover and to forage quietly. **Voice** At dawn simple, two-noted *cheëp-t'cheëw, cheëp-t'cheëw...*; sharp *tzee-tzee-tzee* or *cheép*. **SS** Buff-breasted lacks solid blue crown, upperparts not green, conspicuous blue 'supercilium'. **Status** Uncommon to rare.

Vermilion Tanager
Golden-chested Tanager
Moss-backed Tanager
Orange-throated Tanager
Hooded Mountain-Tanager
Masked Mountain-Tanager
Black-chested Mountain-Tanager

Bold and colourful tanagers of montane forests to treeline. These mountain-tanagers are smaller than those on preceding plate, but more conspicuous and noisy, more frequently with mixed flocks. Grass-green and Buff-breasted are monotypic tanagers that recall mountain-tanagers; the voice of Grass-green is distinctive.

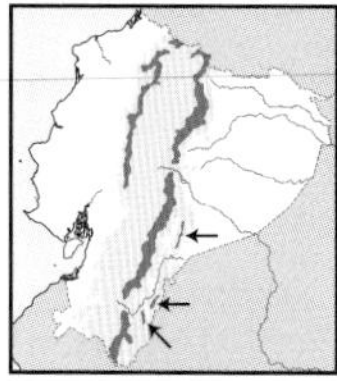

Grass-green Tanager *Chlorornis riefferii* 20–21cm

Andean subtropics to temperate zone in E & NW (2,000–3,200m). Canopy of montane forest and borders. Clad in ***bright green and red***, with ***bright red bill and legs***. Nominate. Small parties regular with mixed flocks, sometimes rather low at borders, very bold as quite stolidly hops and peers about. **Voice** Loud, nasal *queek-queek-quee'k'k'k, quik-quik-ke'ke'ke'ke-kiik*; dawn song pretty, faster and less nasal *qi'qi'qi-quee'u-quee'u-quee'u*; nasal *hink*. **SS** Simply unmistakable! **Status** Uncommon.

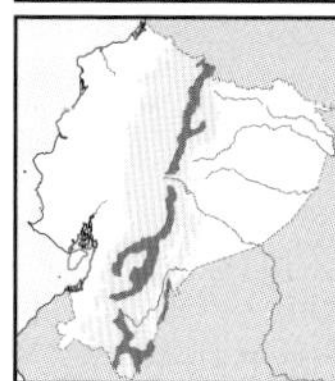

Lacrimose Mountain-Tanager *Anisognathus lacrymosus* 18–19cm

Temperate zone in E & S Andes (2,300–3,200m). Montane forest, borders, secondary woodland, treeline. ***Sooty-black head*** to breast-sides, ***yellow 'teardrop' and ear patch. Dull ochraceous-orange underparts***, blue-and-black wings and tail. Races *palpebrosus* (most of range), *caerulescens* (S). Noisy groups often apparently lead mixed flocks; forage in various strata, often near ground. **Voice** Accelerating and decelerating, jumbled and sputtering series of high-pitched, mechanical *zeek-tzee-tzee-dii-dii-widi-widi*...; call simple *ziit* notes. **SS** Lower-elevation Blue-winged Mountain-Tanager. **Status** Fairly common.

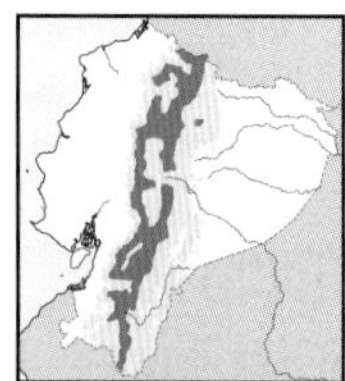

Scarlet-bellied Mountain-Tanager *Anisognathus igniventris* 18–19cm

Andes and inter-Andean valleys (mostly 2,500–3,500m). Montane forest, woodland, borders, hedgerows. Handsome in ***velvet black and scarlet***, with ***bluish shoulders and rump***. Race *erythrotus*. Pairs to small flocks often noisy but not always easy to see as prefer to stay hidden; forage in various strata, lower at edges but usually in upper strata. **Voice** Loud, semi-musical and tinkling *tik'tik-tlee'lee-tlee'lee-tlee'lee-tlee'lee-tEE*, sometimes simpler and faster, with mechanical quality; call sharp *t'zeet, t'zeet, sít-sít-sít*.... **SS** Unmistakable. **Status** Common.

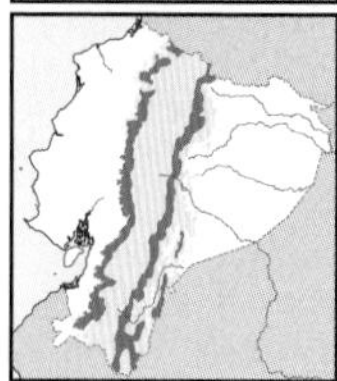

Blue-winged Mountain-Tanager *Anisognathus somptuosus* 18–19cm

E & W Andean subtropics to temperate zone (1,200–3,200m). Montane forest, secondary woodland, borders, woody clearings. ***Bright lemon-yellow crown to nape and underparts***, black forecrown, ***face*** to upperparts (mossy-green in NE), ***blue wing bend***, blue-and-black wings and tail. Nominate (SE), *baezae* (NE), *cyanopterus* (NW), *alamoris* (SW). Noisy groups of up to a dozen apparently lead mixed flocks, regularly visiting feeders. **Voice** Loud, high-pitched, jumbled *ti-ti-ti-wít-ti-ti-ti-ti-wiít-TI-TI-TI'ti'ti'ti*..., simpler rhythmic *seet-seet-seet-seet*... at dawn; fast *zit'zit'zit* when foraging. **SS** In W Black-chinned Mountain-Tanager (orange cast below, smaller and duller crown patch, greener back and shoulders). **Status** Common.

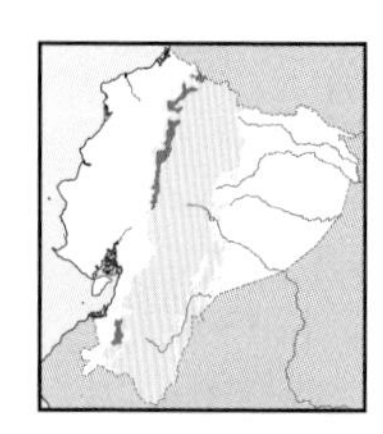

Black-chinned Mountain-Tanager *Anisognathus notabilis* 18–19cm

W Andean foothills to subtropics (1,400–2,200m, to 900 in SW). Montane forest and borders. ***Black head with contrasting yellow mid crown. Mossy-green upperparts and wing bend***, bright blue-and-black wings and tail, ***orangey-yellow underparts***. Monotypic. Pairs to small flocks often with Blue-winged Mountain-Tanager or mixed canopy flocks; with other frugivores at fruiting trees, occasionally at feeders. **Voice** Recalls previous species but higher-pitched, long-repeated *t't-tsit-tsit-tsee-tseeu-tsit-tsit-sit-sit*, call a louder *tic-tic-tic*.... **SS** Blue-winged Mountain-Tanager and Moss-backed Tanager (different head pattern). **Status** Uncommon.

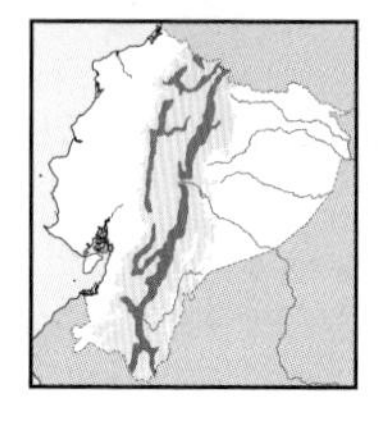

Buff-breasted Mountain-Tanager *Dubusia taeniata* 19–20cm

Andean subtropics to highlands (2,250–3,500m). Montane forest borders, humid scrub, secondary woodland, hedgerows. ***Black hood with freckled blue face outline***. Bluish-black upperparts, ***creamy breast-band***, yellow below. Nominate. Small groups with or apart from mixed flocks, quite noisy but not easy to see as stolid in vegetation. **Voice** Sweet, semi-musical *feeu-feee-deuu*, sometimes two-noted, resembling Rufous-collared Sparrow. **SS** Black-chested lacks creamy breast-band and blue 'supercilium'; see Masked and Hooded Mountain-Tanagers. **Status** Uncommon.

Grass-green Tanager
Lacrimose Mountain-Tanager
adult
juvenile
Scarlet-bellied Mountain-Tanager
juvenile
baezae
cyanopterus
somptuosus, alamoris
Blue-winged Mountain-Tanager
Black-chinned Mountain-Tanager
Buff-breasted Mountain-Tanager

Iridosornis are lovely Andean tanagers, regularly with mixed flocks but shyer than mountain-tanagers. Purplish-mantled is puzzlingly rare in Ecuador. Fawn-breasted is a distinctive, primarily frugivorous tanager of quite uncertain affinities within the family. Also two gems of cloud forests, one either side of the Andes, which recall typical *Tangara* but are shorter-tailed, smaller and plumper.

Purplish-mantled Tanager *Iridosornis porphyrocephalus* 16cm

Local in NW Andean subtropics (1,500–2,200m). Montane forest, borders. ***Bright purplish-blue head and breast*** grading into ultramarine and turquoise-blue rear; black mask, contrasting ***bright yellow throat***, buff median belly. Monotypic. Poorly known, sometimes with mixed flocks, moves unobtrusively through undergrowth, quiet and inconspicuous. **Voice** Buzzy *trr-zee-zee, ts-ts-zee-zeéuiit…*, call short *zít*. **SS** Recalls Yellow-throated but overlap unlikely (despite fairly recent, but probably mistaken, records of Purplish-mantled within latter's range). Purplish-mantled deep blue below; Golden-chested is blacker-blue overall, golden patch in breast, not throat. **Status** Very rare, local, NT globally, VU in Ecuador.

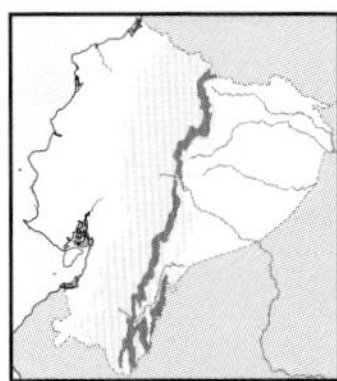

Yellow-throated Tanager *Iridosornis analis* 16–17cm

E Andean subtropics (1,400–2,300m). Montane forest, borders, adjacent woodland. ***Deep purplish-blue head*** grading to turquoise-blue back; black mask, ***bright yellow throat, buff below*** with pale bluish wash on sides to flanks. Monotypic. Pairs or small groups follow mixed flocks; more active and conspicuous than previous species, and more prone to emerge at edges. **Voice** Not very vocal, at dawn a simple, down-inflected *ts-sééuur, ts-sééuur…*, sometimes *t'ts-seeuur*; call a simple *tseuu*. **SS** Unique in range; see previous species. **Status** Uncommon.

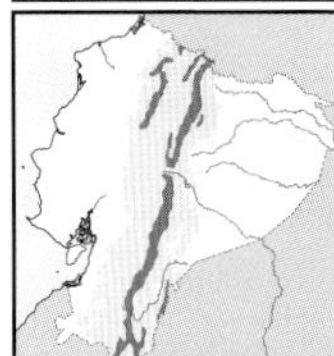

Golden-crowned Tanager *Iridosornis rufivertex* 16–17cm

E & NW Andean temperate zone (2,500–3,500m, locally lower). Montane forest, woodland and scrub, borders, treeline. ***Velvet black hood*** with contrasting ***golden mid crown***. Otherwise deep ***purplish-blue***, more turquoise wings, rufous vent. Nominate (E), *subsimilis* (NW). Pairs to small groups with mixed flocks, often well inside tangles, but briefly emerges at edges. **Voice** Not very vocal; a thin, high-pitched, buzzy *tziit* or longer, similar-quality *tsiít-tsít-tsít*. **SS** Unmistakable; other *Iridosornis* occurs at lower elevations and has bright yellow throat, no black head, etc. **Status** Fairly common.

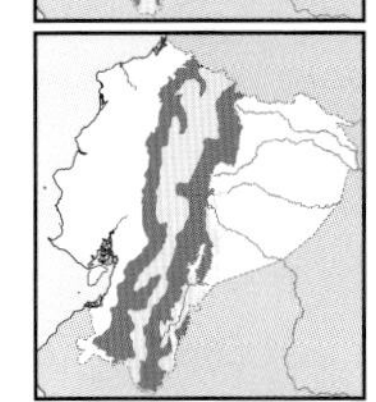

Fawn-breasted Tanager *Pipraeidea melanonota* 14cm

E & W Andean subtropics, inter-Andean valleys (1,000–2,800m, locally lower). Forest borders, woodland, humid scrub, hedgerows, gardens. ***Bright red eyes*** (dark in juv.). ***Bright sky-bluish crown and rump, black mask***, dark upperparts, blue-and-black wings and tail, ***buffy-yellow to beige underparts***. ♀: duller above; juv. even more so. Race *venezuelensis*. Pairs or trios with or apart from mixed flocks, forage conspicuously at edges and perch on exposed branches. **Voice** Accelerating but simple high-pitched *see-see-see…*, sometimes faster, more trilled *see'see'seeseesee* or warble. **SS** Ad. readily distinguished; juv. might resemble a hemispingus (Black-eared), bush-tanager (Yellow-green) or other tanagers (Blue-capped, Blue-and-yellow, Rufous-crested). **Status** Uncommon.

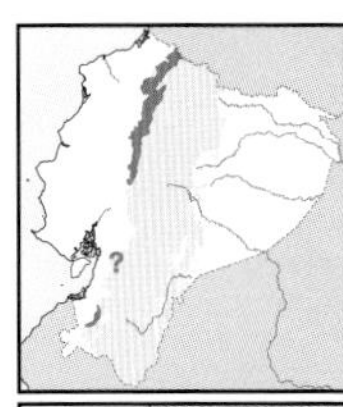

Glistening-green Tanager *Chlorochrysa phoenicotis* 12–13cm

Foothills to subtropics in W Andes, S to N El Oro (600–1,900m). Montane forest, old secondary woodland, borders. Mainly ***brilliant green, orange and bluish on ear-coverts***, bluish patches on face-sides and shoulders. Monotypic. Pairs to small groups often with mixed flocks, agilely hop along twigs, cling upside-down, search mosses and sally. **Voice** High-pitched, somewhat nasal *tseet-teét-teét-teét…ts-ts-ts-teét-teét-tseet*, accelerating, sometimes descending and trilled. **SS** Confusion unlikely; similar-shaped *Tangara* less stocky, bills more conical and none is so glistening. **Status** Uncommon.

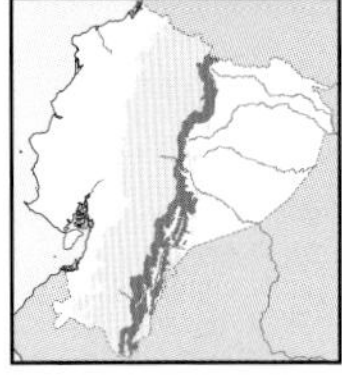

Orange-eared Tanager *Chlorochrysa calliparaea* 12–13cm

Foothills to subtropics in E Andes (1,000–1,700m). Montane forest canopy, borders, adjacent woodland. Spectacular ♂ mostly ***brilliant green, velvet black throat, red cheeks, bright orange rump and patch on forecrown***, bluish-washed central belly, green-and-black wings. ♀: ***paler*** green, ***smaller orange cheek patch, dirty buff throat***. Race *bourcieri*. Pairs to small groups with mixed flocks, search mosses, foliage, small branches, move rapidly tree to tree. **Voice** High-pitched *zeep-zeep-zeep-zeep* at dawn, similar but simpler *zeep* or *síip*. **SS** Confusion unlikely as no overlap with Glistening-green; no sympatric tanager is so shiny; see Golden-eared (names more confusing than plumages!). **Status** Uncommon.

♂
juvenile
Purplish-mantled Tanager
Yellow-throated Tanager
adult
juvenile
Golden-crowned Tanager
♂
♀
juvenile
Fawn-breasted Tanager
♂
juvenile
Glistening-green Tanager
♂
♀
Orange-eared Tanager

A mix of montane tanagers in which blue to green tones prevail; most are somewhat 'plain' (at least among *Tangara*). Scrub is the only *Tangara* that occurs in dry inter-Andean valleys. Beryl-spangled is astonishing when seen in direct sunlight.

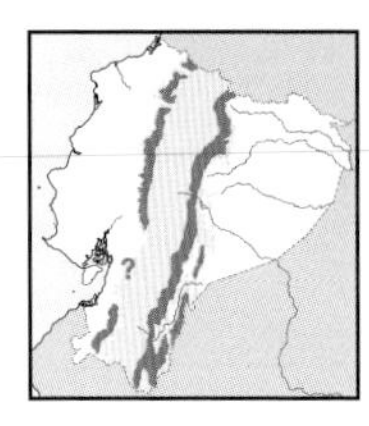

Beryl-spangled Tanager *Tangara nigroviridis* 13–14cm

E & W subtropics, in W only local S of Cotopaxi (1,400–2,500m, locally lower; to 1,000m in SW). Montane forest, borders, woodland, adjacent clearings. Handsome! ***Metallic greenish-blue spangles over black background. Brighter and yellower forecrown***, black mask and mantle, ***solid opalescent-green rump***, whiter central belly. Nominate (E), *cyanescens* (W). Pairs to small groups forage with mixed flocks, glean outer foliage and bare twigs. **Voice** Thin, high-pitched, buzzy *see* or *sit* often in bursts. **SS** Unmistakable, but cf. Metallic-green and Black-capped Tanagers; duller juv. might be more confusing. **Status** Fairly common.

Golden-naped Tanager *Tangara ruficervix* 13–14cm

E & W Andean subtropics (1,500–2,100m, to 700–800m in SW). Montane forest, borders, secondary woodland, adjacent clearings. Mainly ***bright turquoise, black crown*** with striking ***golden nape band***, upperparts finely spotted black, ***buff central belly to vent***. Races *leucotis* (W), *taylori* (E). Pairs to small groups regular with mixed flocks, on large limbs, gleaning outer foliage and sallying for prey. **Voice** Not very vocal; a thin, insect-like *dzip* in chattering series recorded. **SS** Metallic-green and Blue-browed Tanagers (black wing, blue shoulders, no golden nape; Blue-browed black above). **Status** Fairly common.

Scrub Tanager *Tangara vitriolina* 14–15cm

Inter-Andean valleys in N (800–2,800m). Dry woodland, hedgerows, gardens, arid scrub. ***Rufous-orange crown*** (***duller*** in ♀ and juv.); ***broad black mask***, mostly ***pale green***, paler below, black wings edged green. Monotypic. Pairs to small monospecific groups in semi-open areas; eats much fruit, also seeks invertebrates in bushes, small trees, etc. **Voice** Chattering *tss-tss-tss, tsstss'chd'chd'chd...*, high-pitched and fast; also thin *dzip, dz-zip...*calls. **SS** Though not conspicuous, confusion unlikely in restricted habitat and range; see juv. Fawn-breasted Tanager. **Status** Uncommon.

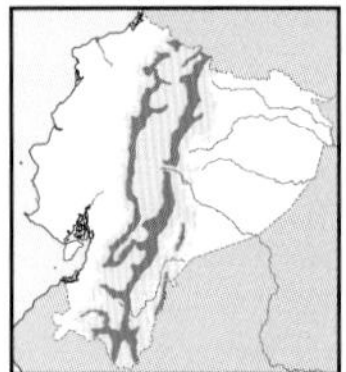

Blue-and-black Tanager *Tangara vassorii* 13–14cm

Andes in temperate zone (2,000–3,300m, locally or seasonally lower). Montane forest, borders, secondary woodland. Mostly ***deep blue with black mask, most of wings and tail***. Juv. duller, ***greyer, sparsely spotted*** below. Nominate. Pairs to small groups with mixed flocks, nimbly gleaning dense outer foliage, with other tanagers at berry-laden trees and bushes. **Voice** Thin, high-pitched *shít* notes, sometimes *shi-shit, shi-shit...*, also in rapid series. **SS** Other blue passerines (Masked Flowerpiercer or Tit-like Dacnis) have different bill shape and behaviour; 'spotted' juv. duller than Beryl-spangled Tanager. **Status** Uncommon.

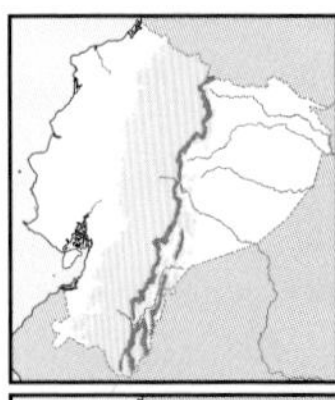

Blue-browed Tanager *Tangara cyanotis* 12–13cm

Subtropics in E Andes (1,400–1,900m). Montane forest, borders, secondary woodland. Bicoloured, ***turquoise-and-black*** tanager with bold ***blue supercilium, shoulders and wingbar***, pale buff central belly. Race *lutleyi*. Pairs to small groups with mixed canopy flocks, work terminal foliage and twigs. **Voice** Clear, high, chattering *zhep-zhep-zp-zp-zhep-zp'zp'zp...*, also simpler *tip*. **SS** Similarly coloured to Metallic-green and Golden-naped Tanagers but pattern is different, especially upperparts and head. **Status** Uncommon.

Metallic-green Tanager *Tangara labradorides* 13–14cm

NW subtropics, local in SE (mostly 1,300–2,000m, to 800m in SE). Montane forest, borders, secondary woodland. Mainly ***metallic opalescent-green, bright silvery-green cast to forecrown and 'supercilium', black crown to nape***, most wings and tail, pale central belly to vent. Nominate (W), *chaupensis* (SE). Pairs or small groups with or apart from mixed flocks, on small twigs and in terminal foliage, sometimes hanging upside-down. **Voice** Clear, squeaky *jeet-jeet, jee-jeet...* or *eék-eék...*, also *jee-jeet, jt-jt-jt...*. **SS** Similarly coloured Golden-eared, Blue-browed and Golden-naped Tanagers all have distinctive head patterns. **Status** Uncommon, rarer in SE.

nigroviridis
cyanescens
juvenile
Beryl-spangled Tanager
♂
leucotis
♂
♂
taylori
juvenile
Golden-naped Tanager
♂
♀
juvenile
Blue-and-black Tanager
♂
juvenile
Scrub Tanager
♂
juvenile
Blue-browed Tanager
chaupensis
labradorides
Metallic-green Tanager

Miscellaneous *Tangara*, including the hooded trio (Masked, Blue-necked and Hooded). Silvery, Black-capped, and Green-throated are well adorned and strikingly patterned, with marked sexual dimorphism (unusual in the genus). Silvery / Green-throated have remarkably similar patterns but bright colours replaced on back and throat.

Silvery Tanager *Tangara viridicollis* 13–14cm

Subtropics in S Andes (mostly 1,300–2,300m, once 900m in Zumba region). Montane forest, borders, woodland, adjacent clearings. ***Black head and most underparts, silvery-bluish upperparts and flanks, orange-rufous throat.*** ♀: ***brown crown***, paler throat, dull green upperparts, greyish-green below. Race *fulvigula*. Pairs to small groups with other tanagers, inspect lower branches and twigs. **Voice** Song a high-pitched, falling *dz-zíp-z'z'z'z'z'z'z'z'z...*, also *dz-dz-dzzzzzz*, ending in trill. **SS** Confusion unlikely, note similar pattern but different colours of Green-throated Tanager. **Status** Uncommon.

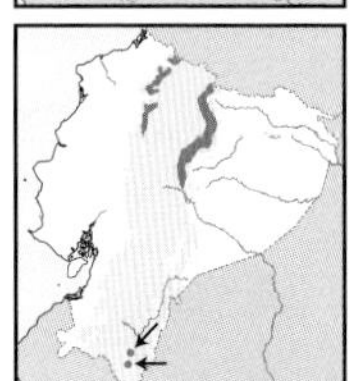

Black-capped Tanager *Tangara heinei* 13–14cm

Upper foothills to subtropics in NE, also locally in SE & NW (1,100–1,900m, local to 2,000–2,100m). Montane forest, borders, woodland, adjacent clearings. ***Black crown to nape, metallic silvery-greenish, speckled throat*** fading into ***'reef' blue underparts***. ♀: ***dusky scaling on cap, yellowish-green instead of blue body***. Monotypic. Mostly pairs with or apart from mixed flocks, nimbly works dense foliage, sometimes peers head down. **Voice** High, rather metallic *zheet-zheet-zheet...*, clearer *she-sheet....* **SS** Beryl-spangled more noticeably spangled; ♀ might resemble Metallic-green, but speckled below. **Status** Uncommon to rare; spreading?

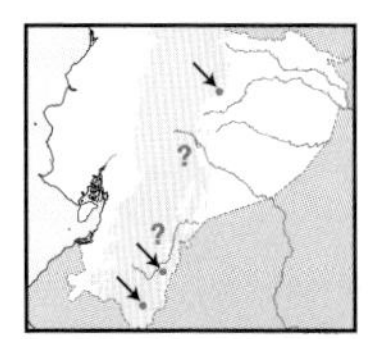

Green-throated Tanager *Tangara argyrofenges* 13–14cm

Very local in SE subtropics (1,350–1,600m). Montane forest canopy, edges. Similar to Silvery Tanager but orange throat replaced by ***metallic greenish***, silver upperparts and rump by ***straw-yellow***. ♀: ***dusky-greenish above, duller throat***, greyer underparts. Race *caeruleigularis*. Small groups often with other tanagers agilely foraging low down. **Voice** Not very vocal, long nasal *weee*, descending *tsew* and dry *tip* (S *et al.*). **SS** ♀ similar in pattern to Silvery but throat is green, not orange. **Status** Rare; VU globally.

Grey-and-gold Tanager *Tangara palmeri* 14–15cm

NW humid lowlands to foothills (400–1,000m). Forest, borders, woodland, adjacent clearings. Mostly ***silvery with black mask, scapulars, flight feathers and tail***; black and ***gold speckled breast-band and neck-sides***. Monotypic. Small groups with mixed canopy flocks or in monospecific groups; conspicuous and noisy, fly long distances, often the first species to move from fruiting trees; perches on high exposed branches. **Voice** High-pitched, semi-musical, pretty *chiiíp, tsi-tsi-tsi-chiiíp-chi'dipdip, sweeé....* **SS** Confusion unlikely. **Status** Uncommon, NT in Ecuador.

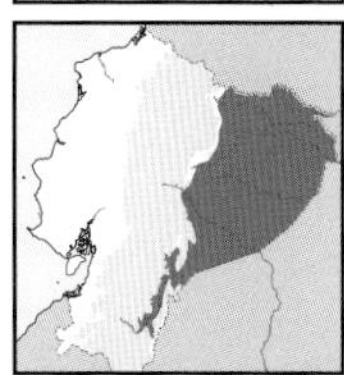

Masked Tanager *Tangara nigrocincta* 13–14cm

E lowlands (to 600m, locally to 1,000m). Canopy of *terra firme* and *várzea* forest and borders. ***Pale blue hood***, black mask, ***mantle and chest***, bluish ***rump*** and flanks, ***bluish-white central belly***. Monotypic. Pairs to small groups often with mixed canopy flocks, especially at fruiting trees with small berries. **Voice** Dry, almost buzzy chattering *tzeuu* notes, recalling Golden-hooded Tanager, sometimes ending in accelerating trill or wheezy series; high *sik* calls. **SS** Blue-necked Tanager local in foothills, has distinctive golden-green rump and golden shoulders; see White-bellied Dacnis. **Status** Uncommon.

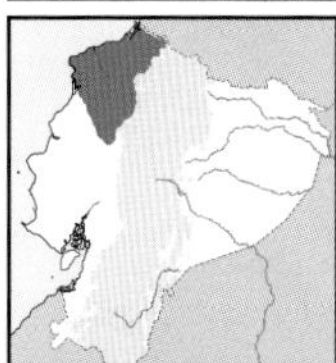

Golden-hooded Tanager *Tangara larvata* 13–14cm

NW lowlands to foothills (to 800m, locally higher). Humid forest, borders, secondary woodland, adjacent clearings. ***Golden hood*** with variable ***blue foreface***, black mask. Otherwise resembles Masked Tanager but rump more opalescent. Race *fanny*. Pairs to small groups, independent of mixed flocks, but sometimes joins them at fruiting trees; regular at edges. **Voice** Dry, almost buzzy *tic* or *zic* notes, sometimes with trill near end and ending in clear, metallic *teek*. **SS** More richly coloured Blue-necked Tanager, with golden shoulders. **Status** Fairly common.

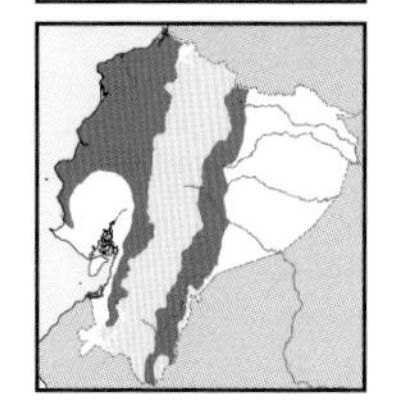

Blue-necked Tanager *Tangara cyanicollis* 13–14cm

W humid lowlands to lower subtropics, E Andean foothills to subtropics (to 1,700m in W, 500–1,800m in E). Humid to deciduous secondary forest, borders, clearings, gardens. ***Bright blue head***, black mask and ***most of body; golden-green wing-coverts*** (less obvious in W), ***metallic yellow-green rump*** (E) or turquoise-blue (W). Races *caeruleocephala* (E), *cyanopygia* (W). Small flocks with or apart from mixed flocks, in various strata; fond of small berries and buds, often also sallies for prey. **Voice** High *chíp* or *síít* notes, sometimes followed by louder *cheep*. **SS** Golden-hooded (W) and Masked Tanagers (E), have greenish-blue in wings, white central belly, paler blue in hood. **Status** Common.

♂
Silvery Tanager
♀
♂
second-year
♂
first-year
♂
Green-throated Tanager
♀
♀
Black-capped Tanager
Grey-and-gold Tanager
adult
juvenile
♂
Masked Tanager
♂
juvenile
cyanopygia
♂
caeruleocephala
♂
juvenile
immature
Blue-necked Tanager
adult
juvenile
Golden-hooded Tanager

Two colourful and rather similar tanagers that overlap locally in NW foothills (Bay-headed and Rufous-winged), along with four striking Amazonian species. Opal-crowned and Opal-rumped share general blue colour, whereas Paradise is the most striking *Tangara* in Ecuador. These species (except Rufous-winged) often flock together in the Amazon, but Bay-headed is more often found in clearings.

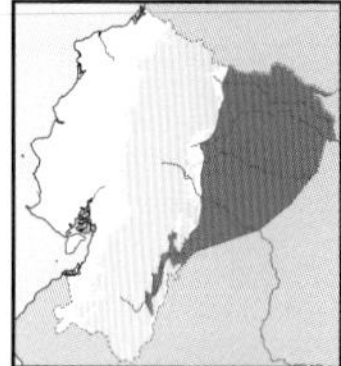

Opal-rumped Tanager *Tangara velia* — 14–15cm

E lowlands (mostly to 600m). *Terra firme* and *várzea* canopy, woodland and borders. Mainly ***deep blue and black***, with ***silvery rump*** and ***chestnut lower belly to vent***. Race *iridina*. Pairs to small groups invariably with mixed flocks; eats fruit but also searches terminal branches, epiphytes and underside of larger branches for invertebrates. **Voice** Thin, weak, buzzy *tititití*... or *tsi-tsi-tslt-tslt-tsit*..., often rising and falling. **SS** Opal-crowned Tanager has similar opal rump but conspicuous silver supercilium, no chestnut belly. **Status** Uncommon.

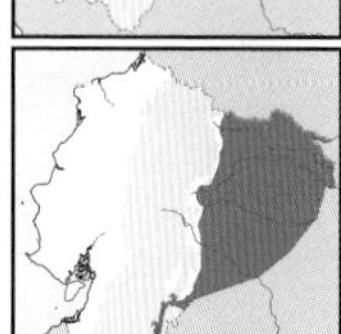

Opal-crowned Tanager *Tangara callophrys* — 14–15cm

E lowlands (mostly to 600m). *Terra firme* and *várzea* canopy, woodland and borders. Upperparts mainly black with ***broad silvery eyebrow*** and ***silvery rump***; ultramarine-blue underparts, ***dark*** lower belly. Monotypic. Pairs to small groups forage with mixed canopy flocks, eats fruit but also searches undersides of large branches and debris for insects. **Voice** High-pitched, thin, often paired *tzet'zét-tzet'zét*..., lower than Opal-rumped. **SS** Recalls Opal-rumped Tanager but crown distinctive. **Status** Uncommon.

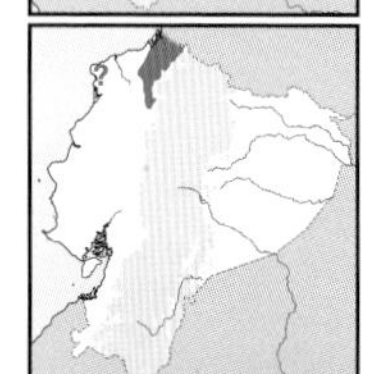

Rufous-winged Tanager *Tangara lavinia* — 13–14cm

NW wet foothills to adjacent lowlands (to 800m). Wet forest canopy and borders. Resembles Bay-headed Tanager but ***golden-yellow mantle, green rump, greener underparts, rufous wings***. Nominate. Small restless groups work large bare branches, more nimble but less numerous than Bay-headed. **Voice** Buzzy, slightly metallic *tzzzt, tzzzt, tzt-tzt*, more twittering *ti-ee-it, ti-ee-it*.... **SS** Juv. lacks reddish hood and has little rufous in wings; see Green Honeycreeper. **Status** Rare, declining? VU in Ecuador.

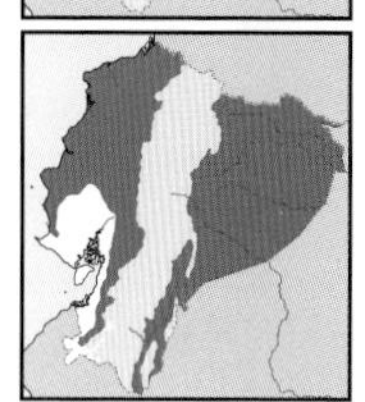

Bay-headed Tanager *Tangara gyrola* — 13–14cm

E & W lowlands to lower subtropics (to 1,500–1,700m, higher in E). *Terra firme* and *várzea* forest, borders, secondary forest, adjacent clearings. ***Brick-red head***, narrow yellow nape, ***green upperparts, wings and tail***, bright blue rump and ***most underparts***. Juv. lacks reddish head. Races *nupera* (W), *catharinae* (E), *parva* (extreme SE). Small groups eat much fruit, also work large bare branches and hang head down. **Voice** Nasal *dzéeEt*, more twittering *seew*. **SS** In wet NW, Rufous-winged Tanager (golden-yellow back, green underparts, extensive rufous in wings); juv. with ♀ Blue Dacnis and Green Honeycreeper. **Status** Common.

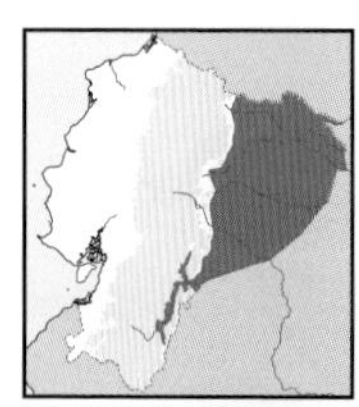

Turquoise Tanager *Tangara mexicana* — 13–14cm

E lowlands (mostly below 600m, locally to 1,000m). *Terra firme* and *várzea* forest, woodland, borders, adjacent clearings. Not turquoise! Mainly ***metallic blue and black***, with conspicuous ***yellow central belly to vent, and shoulders***. Race *boliviana*. Small monospecific flocks in upper strata inspect small bare branches, dead twigs and mosses, often only briefly with mixed flocks. **Voice** Fairly continuous twittering, often fast, very thin *tíc* or *tsíp* notes. **SS** Readily identified by yellow belly; cf. Opal-crowned and Opal-rumped Tanagers. **Status** Fairly common.

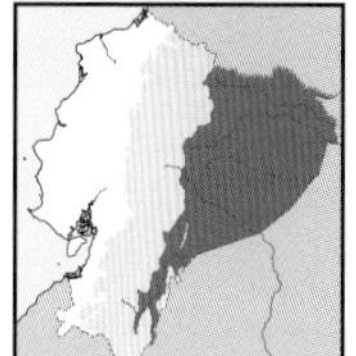

Paradise Tanager *Tangara chilensis* — 14–15cm

E lowlands to foothills (to 1,200m). *Terra firme* and *várzea* canopy, woodland, borders, adjacent clearings. Unbelievably adorned, this five-coloured tanager has a glistening ***lime-green head***, rich purple throat, ***turquoise-blue underparts*** and ***scarlet rump***. Nominate. Groups of 10–12 forage in upper strata, lower at edges, inspecting bromeliads, foliage; briefly with mixed flocks. **Voice** High-pitched, twittering, often upslurred repetition of rich *chip-zeét* or *chit-sweéE* notes, sometimes with thin *zeeé* and lower-pitched *tsut* interspersed. **SS** Unmistakable. **Status** Fairly common.

♂
juvenile
♀
Opal-rumped Tanager
Opal-crowned Tanager
♂
♀
Rufous-winged Tanager
catharinae
nupera
parva
juvenile
Bay-headed Tanager
adult
juvenile
Turquoise Tanager
Paradise Tanager
adult
juvenile

PLATE 261: CLOUD-FOREST *TANGARA* TANAGERS

Beautiful montane tanagers often found foraging together. Golden and Rufous-throated are very distinctive, while Silver-throated looks like a dingy Golden. The other three share greenish underparts and orange to fiery heads. The co-occurrence of several *Tangara* species is ecologically fascinating.

Rufous-throated Tanager *Tangara rufigula* 12–13cm

Foothills to subtropics in W Andes, spottily S to N El Oro (600–1,400m). Montane forest, secondary woodland, borders. ***Velvet black head*** with contrasting ***rufous throat***. Black back feathers ***edged rufous***, greenish rump, greenish-white below with ***bold black spots*** on breast, solid whitish central belly. Monotypic. Mainly in pairs gleaning leaves in outer foliage and twigs; active and quick. **Voice** Chattering *chitichi'chi'chichichi....* **SS** Confusion unlikely due to head and breast patterns. **Status** Uncommon.

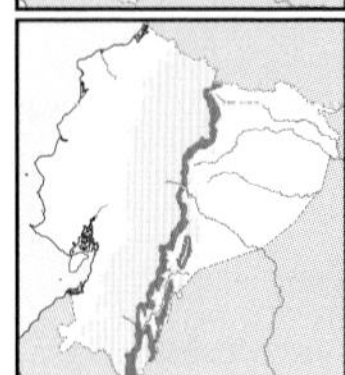

Golden-eared Tanager *Tangara chrysotis* 13–14cm

Upper foothills to subtropics in E Andes (1,000–1,700m). Montane forest, borders, secondary woodland, adjacent clearings. ***Metallic golden-yellow face***, especially ***cheeks***, with ***black crown and malar***. Rich turquoise-green with black-streaked upperparts and ***orange-rufous central belly to vent***. Monotypic. Pairs join mixed flocks inspecting undersides of mossy limbs and branches, sometimes near their tips. **Voice** A piercing *tzík-ziík-tz-tz-ziík-tz...* in slow series, sometimes accelerating and more chattering, also low-pitched *tzuk!* **SS** Face pattern and central belly coloration distinctive; cf. Metallic-green, Saffron-crowned and Orange-eared Tanagers. **Status** Uncommon.

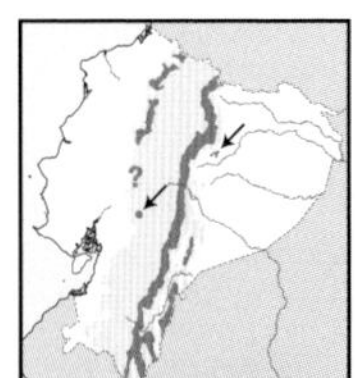

Saffron-crowned Tanager *Tangara xanthocephala* 13–14cm

E & NW Andean subtropics (1,500–2,300m). Montane forest, borders, adjacent clearings. ***Bright saffron-yellow head***, black mask and chin, ***rich turquoise***-green body, profuse black streaking on back, black wings edged green, ***yellowish-buff central belly to vent***. Race *venusta*. Small groups with mixed flocks; eats much fruit but also gleans outer and small twigs. **Voice** High, clear *chit!-chi-tít!*, often becoming a wheezy, fast chattering. **SS** In E, see Golden-eared Tanager (similar body pattern but no saffron crown, more rufescent central belly to vent; also Flame-faced). **Status** Fairly common.

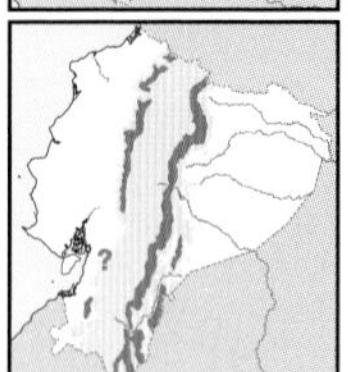

Flame-faced Tanager *Tangara parzudakii* 14–15cm

Subtropics in E & W Andes, S to N Cotopaxi in W, more spottily S (1,500–2,400m, to 900–1,000m in SW). Montane forest, borders, secondary woodland, adjacent clearings. ***Bold flame red face*** grading into ***golden-yellow crown and nape; black mask, ear-coverts, throat and back***. Greenish ***opalescent*** shoulders, rump and underparts washed buff towards belly. Nominate (E), *lunigera* (W; less fiery face). Methodically inspects hanging moss and lichen clumps, hanging upside-down; regularly with mixed flocks. **Voice** High-pitched, sharp *chit!*, sometimes becoming an accelerating, almost trilled, dry chatter. **SS** Confusion unlikely, but see Saffron-crowned and Golden-eared Tanagers. **Status** Fairly common.

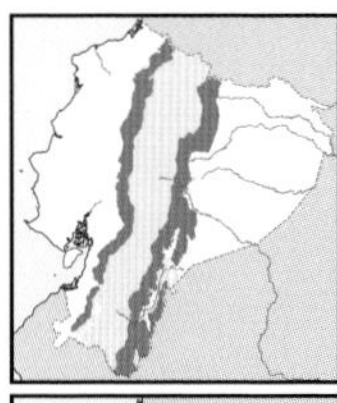

Golden Tanager *Tangara arthus* 13–14cm

E & W foothills to subtropics (700–2,000m). Montane forest, borders, woodland, adjacent clearings, hedgerows. Rich ***golden-orange*** (deeper in E) with contrasting ***black ear patch***, back streaking, wings and tail. Races *goodsoni* (W), *aequatorialis* (E). Small groups mostly in mixed flocks, often the most numerous *Tangara*; inspects underside of mossy and bare branches and limbs, also epiphytes. **Voice** Short, penetrating, clear *cheét* or *chuúp* while foraging, also fast *chip-di-dip....* **SS** Confusion unlikely; Silver-throated is yellower, with whitish throat, no black ear patch. **Status** Common.

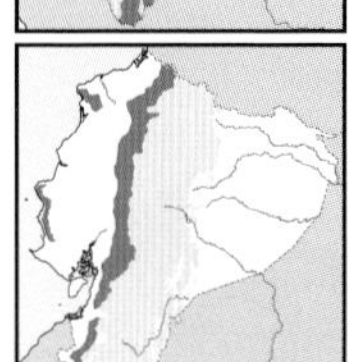

Silver-throated Tanager *Tangara icterocephala* 13–14cm

Foothills to subtropics in W Andes (500–1,750m, lower in coastal mountains). Montane forest, borders, secondary woodland. ***Lemon-yellow*** with bright ***silvery-white throat extending to neck-sides***; back profusely streaked black, black wings and tail edged yellow and greenish. ♀: duller. Nominate. Small groups with canopy flocks, peering head down on mossy branches and epiphytes. **Voice** Harsh, buzzy, often paired *cheez!* or *dzee!-dzee!* **SS** Confusion unlikely, but Golden Tanager in W. **Status** Uncommon.

Golden-eared Tanager
♂
Rufous-throated Tanager
♀
juvenile
adult
juvenile
Saffron-crowned Tanager
lunigera
parzudakii
juvenile
Flame-faced Tanager
goodsoni
aequatorialis
juvenile
Golden Tanager
♂
juvenile
Silver-throated Tanager

Six mostly bright green *Tangara*, four in lowlands, two in foothills. Sympatric species pairs in lowlands readily separated by face pattern (Blue-whiskered vs. Emerald in W) and presence of spots (Green-and-gold vs. Yellow-bellied). The other two heavily spotted species are separated by the Andes; Spotted in E and Speckled in SW.

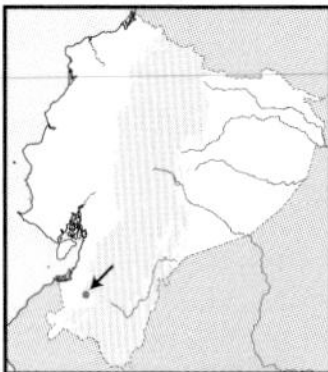

Speckled Tanager *Tangara guttata* 13–14cm

Few records in SW foothills in El Oro prov. (900m). Humid forest borders. Recalls Spotted Tanager but brighter, whiter background to underparts, ***yellower face, more turquoise-blue in wings***. Race unknown, possibly ***undescribed***. Elsewhere, behaves as other *Tangara*. **Voice** Not heard in Ecuador, elsewhere a high, clear twittering *tit-tit-chit'a-chit*..., with ringing quality, often accelerating into a trill (S *et al.*). **SS** No overlap with Spotted; confusion unlikely in limited range; see juv. Bay-headed Tanager and ♀ Green Honeycreeper and Blue Dacnis. **Status** Uncertain, very few observations.

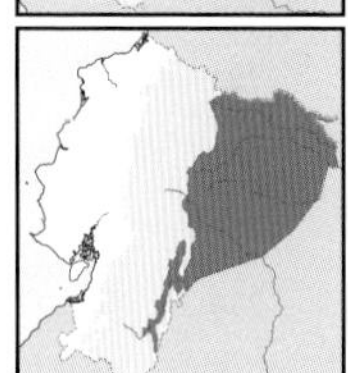

Yellow-bellied Tanager *Tangara xanthogastra* 12–13cm

E lowlands to foothills (to 1,100m). *Terra firme* and *várzea* canopy, secondary woodland, borders and adjacent clearings. Mostly ***brilliant green heavily spotted black***, solid green rump, ***lemon-yellow belly***. Nominate. Pairs to small groups join mixed canopy flocks, especially other *Tangara*, rarely at edges or openings, inspects slender branches and twigs. **Voice** Weak, high-pitched, buzzy *tit-tst-dzeééet* or *tit-tit-dz-dzeet*, high *tic* calls. **SS** In foothills see Spotted Tanager with white central belly and most underparts, spotting more prominent; Green-and-gold has distinctive head pattern, lacks spots. **Status** Uncommon.

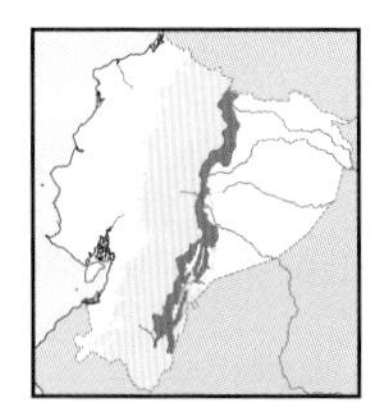

Spotted Tanager *Tangara punctata* 13–14cm

E Andean foothills to lower subtropics (900–1,500m). Montane forest, secondary woodland, borders. Metallic greenish-blue head, green upperparts, ***whitish underparts, entirely spotted black*** except green rump and ***whitish lower belly***. Juv. duller, little spotting. Race *zamorae*. Mainly small groups with mixed canopy flocks on mossy and terminal branches, even perching on large leaves. **Voice** High, accelerating chattering *chit-chit't't't't...chit-chit't't't...*, sometimes ending in a trill; thin *zzeep*. **SS** Yellow-bellied Tanager of primarily lower elevations; brighter green, spotting less striking, belly yellow; see bluer, darker Beryl-spangled Tanager. **Status** Uncommon.

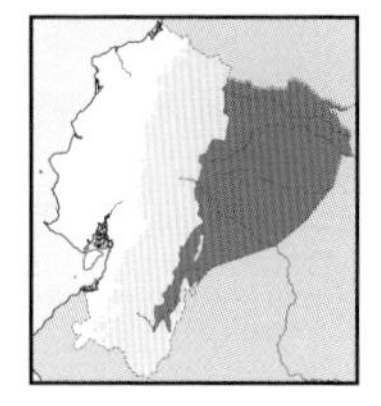

Green-and-gold Tanager *Tangara schrankii* 13–14cm

E lowlands to foothills (to 1,100m, locally higher). *Terra firme* and *várzea* canopy, secondary woodland, borders, woody clearings. Mostly ***brilliant emerald-green***, black front and ***cheeks***, ***bright yellow crown, central breast and lower belly to rump***. Back profusely streaked black. Nominate. Pairs to small groups with other tanagers in canopy but also join understorey flocks, mainly peering into dense outer foliage and slender branches. **Voice** Piercing, very high-pitched *chít* series, sometimes twittering or even trilled at end. **SS** Only other green *Tangara* (Yellow-bellied) in range is profusely spotted black. **Status** Uncommon.

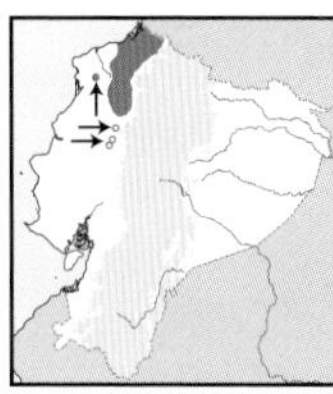

Blue-whiskered Tanager *Tangara johannae* 13–14cm

NW lowlands to adjacent foothills (to 500m). Wet forest canopy, second growth and borders. Mainly ***brilliant emerald-green***, contrasting ***black face and throat***, black spotting on bluer forecrown, ***blue whisker***. Back profusely streaked black, yellow rump, ***blue in wing-coverts*** and tail. Monotypic. Pairs to small flocks often with other tanagers in mid to upper forest strata, but low at edges; hangs upside-down. **Voice** Very high, insect-like *tzzeét-tzzeét-tz-zeét*. **SS** Emerald Tanager lacks extensive black in face, central belly is yellower, lacks blue whisker. **Status** Rare, declining? NT globally, VU in Ecuador.

Emerald Tanager *Tangara florida* 13–14cm

Local in NW humid lowlands to foothills (400–1,200m). Montane forest, borders, secondary woodland. Mainly ***brilliant emerald-green***, brighter and yellower crown and rump, conspicuous ***black ear patch***. Back profusely streaked black, black wings and tail edged green. Monotypic. Pairs to small flocks often with mixed canopy flocks, search moss clumps and undersides of mossy branches. **Voice** High, repeated *zeeee*, a sharp and raspy *dzreee*, sometimes accelerating and twittering. **SS** Upper-elevation Metallic-green Tanager more opalescent, metallic blue-green, with more black in crown and wings; Blue-whiskered similarly coloured but face pattern very different. **Status** Uncommon.

Speckled Tanager
Yellow-bellied Tanager
Spotted Tanager
♀
♂
juvenile
Green-and-gold Tanager
juvenile
adult
Blue-whiskered Tanager
Emerald Tanager

Dacnises are small lovely tanagers of lowland forests, with small and slender bills and bright or pale eyes. Marked sexual dimorphism. ♂♂ among the most beautiful canopy birds; ♀♀ drab and rather confusing. Swallow Tanager very distinct in plumage, voice, posture and behaviour. Its relationships were long debated, even being separated by some authors in a family of its own.

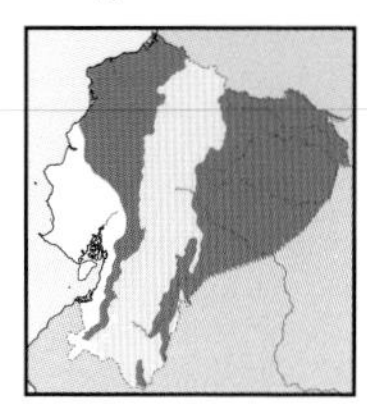

Swallow Tanager *Tersina viridis* 15–16cm

E & W lowlands to foothills (to 1,400m). Forest canopy, borders, secondary woodland. Bright ***turquoise*** with ***bold black mask*** and ***white central belly*** to vent. ♀: ***bright green***, dusky scaling on throat, ***yellow central belly*** to vent. Race *occidentalis*. Perches on exposed branches, sallies like a swallow, briefly with flocks; nests in burrows in banks. **Voice** Resembles Blue-grey Tanager; accelerating, squeaky *dzeét-zzeép-dzeé-zeét'z'z'z-zz'chee-zz...*; clear, rising *tsweé?* **SS** Unmistakable by shape, posture and overall pattern; see Emerald, Green-and-gold and Yellow-bellied Tanagers, and Spangled Cotinga. **Status** Uncommon and seasonal.

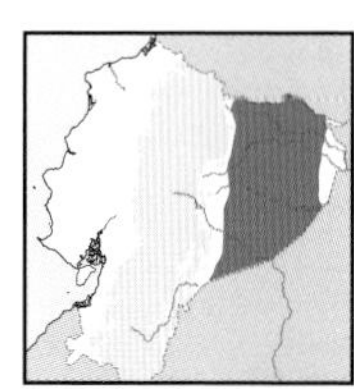

White-bellied Dacnis *Dacnis albiventris* 11–12cm

Local in E lowlands (below 300m, once 900m). *Terra firme* and *várzea* canopy, borders, woodland. ***Bright deep blue*** with ***black mask***, most upperparts and tail, ***white belly to vent***. ♀: yellow eyes, ***olive-green*** above, ***yellower-green below, greyish throat***, paler belly. Monotypic. Singles or pairs with mixed canopy flocks; apparently at very low density. **Voice** Thin, lisping *see-suuu-see-le'e'e'e, see-su-se'e'e'e*; thin *seéE* calls. **SS** Black-faced Dacnis (brighter blue, more black above); Turquoise and Masked Tanagers; ♀ Black-faced Dacnis (more olive above, duller below, longer bill). **Status** Rare.

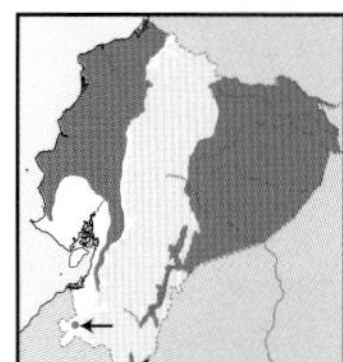

Black-faced Dacnis *Dacnis lineata* 11–12cm

E & W lowlands to foothills (to 1,200m, locally higher). Humid forest canopy, borders, woodland, adjacent clearings. ***Turquoise-blue and black, white central belly to vent*** (***bright yellow*** in W). ♀: ***olive above, greyish-buff below*** (E) or has ***yellow pectoral tuft and central belly to vent*** (W). Nominate (E), *aequatorialis* (W), might be separate species. Pairs to small groups with mixed flocks; agile and acrobatic. **Voice** Thin *tsrrip, tsrrip-srrip*, often with louder *sherrp!* interspersed. **SS** In E, rarer White-bellied Dacnis (deeper blue, smaller mask; ♀ greener above, yellower below). **Status** Uncommon.

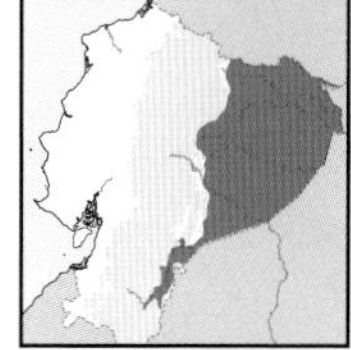

Yellow-bellied Dacnis *Dacnis flaviventer* 12–13cm

E lowlands to foothills (to 1,100m). *Terra firme* and *várzea* canopy, borders, adjacent clearings. ***Green crown, black throat, mask to back, wings and tail. Yellow lower face to underparts, scapulars and rump.*** ♀: ***red eyes***, dull olive above, yellower below, ***faint dusky streaking on breast***. Monotypic. Pairs with or apart from mixed flocks, foraging for berries and gleaning invertebrates from foliage. **Voice** Thin, high, quite buzzy *zrreet*. **SS** Confusion unlikely; other ♀ dacnises lack red eyes and dusky breast. **Status** Fairly common.

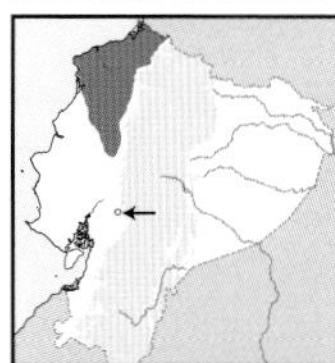

Scarlet-thighed Dacnis *Dacnis venusta* 12–13cm

NW lowlands to foothills (to 800m). Wet forest canopy, borders, secondary woodland, adjacent clearings. Red eyes. ***Bright turquoise-blue hood, scapulars and upperparts, black mask to throat***, underparts and wings, ***scarlet thighs***. ♀: ***dull turquoise hood***, upperparts suffused dull bluish, ***greyish underparts***. Race *fuliginata*. Mainly small groups with or apart from mixed flocks; gleans in upper strata. **Voice** Buzzy *zirp* or *wezt*. **SS** ♀ Blue (greener overall, pink legs, longer, pale-based bill) and Black-faced Dacnises (golden eyes, yellow belly, no hood). **Status** Rare, NT in Ecuador.

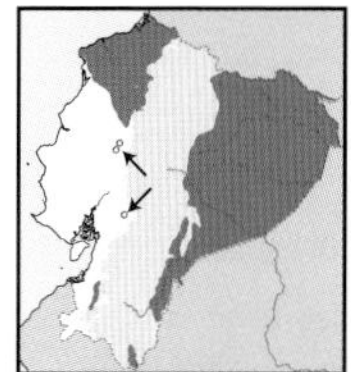

Blue Dacnis *Dacnis cayana* 12–13cm

E & NW lowlands to foothills (to 1,000m). Forest canopy, borders, woodland, adjacent clearings. Red eyes, ***rosy bill base and legs. Bright turquoise-blue with small black mask, throat, back and tail.*** Green ♀ has similar bare parts, ***turquoise head***, and whitish throat. Darker race *baudoana* (W), *glaucogularis* (E). Pairs to small groups glean foliage and pick small berries, often with mixed flocks. **Voice** A thin wheezy *tsírt, tseé-swee-tsírt-see-swee....* **SS** Bluest dacnis; ♀'s bluish head distinctive (Scarlet-thighed greyer below); see honeycreepers. **Status** Common.

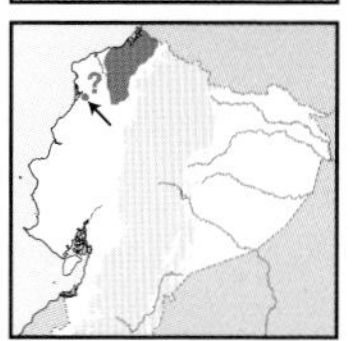

Scarlet-breasted Dacnis *Dacnis berlepschi* 12cm

NW lowlands to foothills (to 600m). Wet forest, borders. Yellow eyes. ***Brilliant blue head to chest***, paler blue upperparts, ***whitish-streaked mantle; flame red lower breast*** grading into pale orange belly and vent. ♀: ***mostly brown***, breast to belly ***paler orange*** than ♂. Monotypic. Mostly pairs, often with mixed canopy flocks, lower in edges, agile; poorly known. **Voice** Thin, high-pitched, slightly rapid *t'tsitsitsitsitsitsitsit*, also thin *ts-sit* calls. **SS** Confusion unlikely; ♀ very distinctive. **Status** Rare, possibly declining, VU.

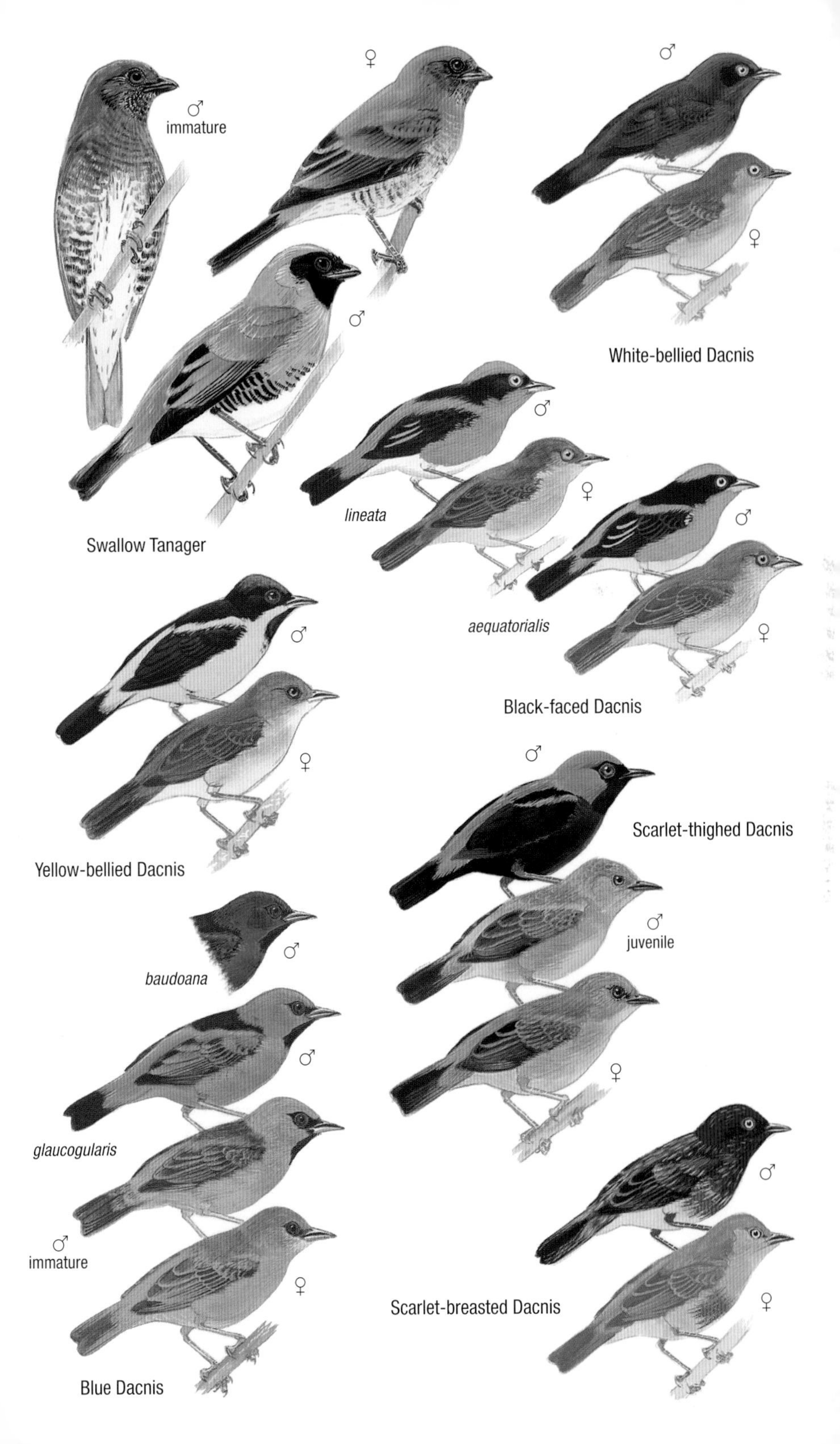

♂
immature
♀
♂
♂
White-bellied Dacnis
♀
♂
lineata
♀
Swallow Tanager
♂
aequatorialis
♀
♂
Black-faced Dacnis
♀
Yellow-bellied Dacnis
♂
Scarlet-thighed Dacnis
♂
juvenile
♂
baudoana
♂
♀
glaucogularis
♂
♂
immature
♀
Scarlet-breasted Dacnis
♀
Blue Dacnis

Small and slender like dacnises, but with longer and finer bills, especially in *Cyanerpes*, which also have brightly coloured legs. Green and Golden-collared are quite similar in overall structure, but Golden-collared was sometimes placed within *Tangara*. Sexual dimorphism marked in all species; ♀ *Cyanerpes* might be confusing.

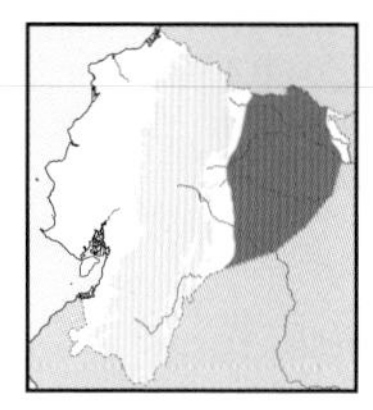

Short-billed Honeycreeper *Cyanerpes nitidus* 9–10cm

Local in E lowlands (to 400m). *Terra firme* forest, borders. ***Short bill, pink legs. Deep blue*** with small black mask, ***large black bib***. ♀: green above, ***dusky lores, short*** blue malar, yellowish throat grading into ***streaky yellowish-white and green below. Short tail.*** Monotypic. Pairs join mixed-species flocks in upper strata, rarely descending even at borders; eats much fruit, nectar, but also gleans for prey. **Voice** Thin, rising *swéé, swéé, swéé*… at slow pace. **SS** Bill length and leg colour separate from other honeycreepers, no black bib in Red-legged; ♀ Purple has large blue malar, buffy face, longer bill, yellowish legs; drabber ♀ Red-legged has yellowish eyebrow, bill is longer. **Status** Rare and local.

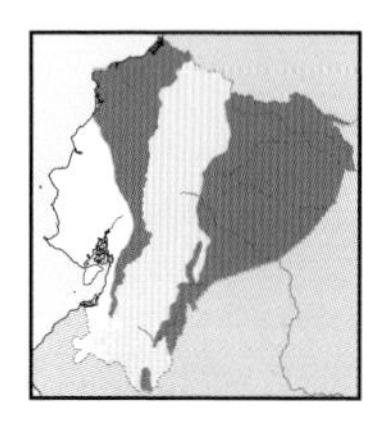

Purple Honeycreeper *Cyanerpes caeruleus* 10–11cm

E & W lowlands to foothills, S in Andes to N El Oro in W; extreme SE in Zumba region (to 1,200m). Humid forest canopy, borders, woodland, adjacent clearings. ***Long*** decurved bill, ***yellow legs. Rich purple-blue***, black mask, wings and ***small bib***. ♀: green above, ***blue malar, buffy foreface and throat, streaky face***, whitish and green streaking below. Races *chocoanus* (W), *microrhynchus* (E). Pairs to small groups with mixed flocks, lower at borders, feed on berries, nectar and gleans for invertebrates. **Voice** Thin, short *zik-zik-zi-zik-zik*; buzzy *whééz-whééz*…. **SS** Leg colour and black bib separate from Red-legged; Short-billed has larger bib, shorter bill, pink legs; ♀ has dusky lores, greener head, shorter malar. **Status** Common.

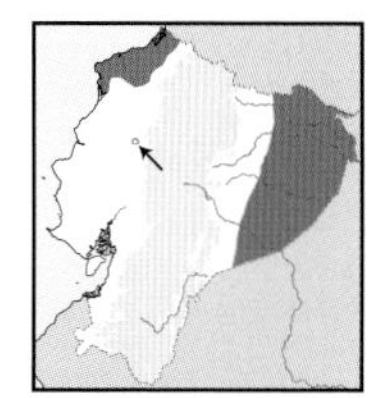

Red-legged Honeycreeper *Cyanerpes cyaneus* 11–12cm

NW lowlands near coastline, local in W interior and E lowlands (to 300m). Humid forest borders, woodland, riparian scrub. ***Long*** decurved bill, ***red legs. Bright purple-blue, brighter turquoise crown***, black mask and ***mantle, yellow underwing-coverts***. ♀: ***reddish legs***, green upperparts, ***pale supercilium***, underparts ***streaked whitish***. Races *pacificus* (W), *dispar* (E). Pairs to small groups mostly in canopy but low on W coast, sometimes with mixed flocks. **Voice** Thin inflected *tsiírp?*, harsher *zzhrree*. **SS** Leg colour, bill shape, no bib, yellow underwing and longer black tail separate from other honeycreepers; ♀ drabber overall, lacks blue malar. **Status** Rare, local in E.

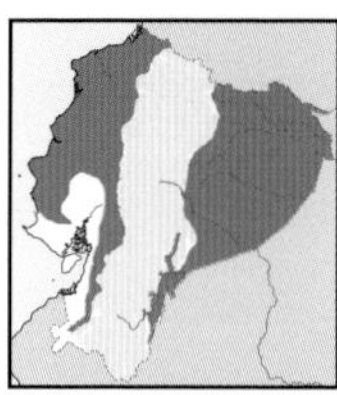

Green Honeycreeper *Chlorophanes spiza* 13–14cm

E & W lowlands to lower foothills (to 800m). Humid forest canopy, borders, secondary woodland, adjacent clearings. Curved bill, ***pale yellow mandible, grey legs. Black head, bright turquoise-green*** body. ♀: mostly ***bright green***, paler throat and median belly. Races *exsul* (W), *caerulescens* (E). Mainly pairs, may congregate at flowering trees or feeders, joins mixed tanager flocks but sometimes alone. **Voice** Thin, high *chíp* or *tsíp!* **SS** ♂ unmistakable; ♀ recalls more slender-billed *Cyanerpes* honeycreepers and Blue Dacnis, but bill heavier and pale yellow below. **Status** Fairly common.

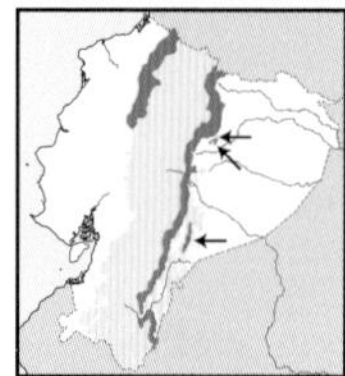

Golden-collared Honeycreeper *Iridophanes pulcherrimus* 12–13cm

Subtropics to foothills on E and NW Andean slopes, few records in Pichincha (1,100–2,000m, locally lower in W). Montane forest canopy, borders, adjacent woody clearings. Pale yellow mandible. ***Black hood and back, golden nuchal collar, blue wings***, opalescent-yellow underparts and rump. ♀: ***olive-grey hood and upperparts, duller collar***, paler underparts, ***bluish-green wings***. Races *pulcherrimus* (E), *aureinucha* (NW). Pairs with mixed flocks, often at edges, sometimes several at fruiting or flowering trees. **Voice** High, thin, lisping *seep*, sometimes in fast bursts. **SS** Confusion unlikely, golden collar distinctive even in duller ♀, bill thinner, more curved than tanagers. **Status** Uncommon.

♂
microrhynchus
♀
♂
juvenile
♀
juvenile
♂
Short-billed Honeycreeper
♀
chocoanus
♂
♂
♂
moulting
:ificus
dispar
♂
immature
♀
♂
♂
Purple Honeycreeper
Red-legged Honeycreeper
caerulescens
exsul
♀
juvenile
♂
♂
ureinucha
pulcherrimus
♀
Green Honeycreeper
Golden-collared Honeycreeper

PLATE 265: TANAGERS AND BANANAQUIT

Three small, warbler-like tanagers of lowland rainforest canopy, all sexually dimorphic. *Hemithraupis* are short-tailed, acrobatic tanagers of mixed flocks, ♀♀ quite confusing. Scarlet-browed is a heavier, distinctive species with little sexual dimorphism. Also, the unique Bananaquit, whose taxonomic affinities are controversial, having shifted from Coerebidae to Thraupidae.

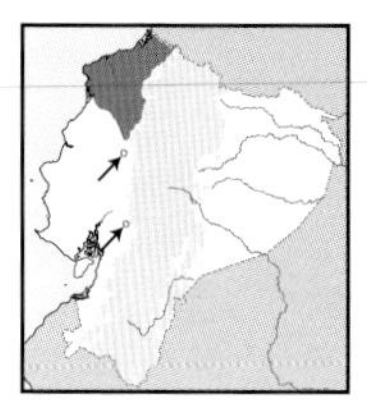

Scarlet-browed Tanager *Heterospingus xanthopygius* 17–18cm

NW lowlands to lower foothills (to 800m). Humid forest canopy, old second growth, borders. *Black* with ***red-and-white brow, bright yellow shoulders and rump, white tuft under wings***. ♀: ***greyer*** overall, ***no red supercilium***, no yellow shoulders. Race *berliozi*. Mainly in pairs in upper strata, sometimes perches unobtrusively for fairly long periods; regularly with mixed canopy flocks. **Voice** High, clear *dzeep, d'zeep* in fast bursts, also squeakier *weero-chiri, weero-chií, weero-chií....* **SS** Smaller and more boldly marked (♂) than commoner and widespread Flame-rumped Tanager; white tufts distinctive in both sexes. **Status** Uncommon; NT in Ecuador.

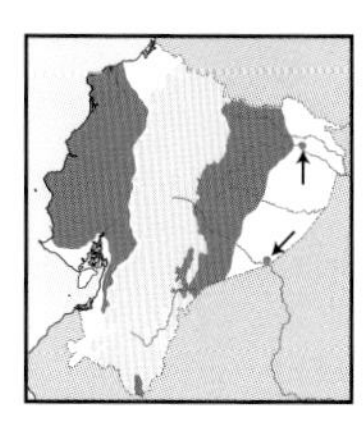

Guira Tanager *Hemithraupis guira* 12–13cm

E & W lowlands to foothills; not in wet NW or extreme E; also extreme SE in Zumba region (to 1,100m, locally higher). Humid forest canopy, borders, woodland, adjacent clearings. Olive above, ***black mask outlined yellow. Orange rump, rufous-orange breast***, yellow below, greyish flanks. ♀: olive above, yellower rump, ***faint yellowish supercilium, eye-ring and cheeks***; yellow below, greyer flanks. Races *guirina* (W), *huambina* (E). Mostly pairs join mixed flocks, very agile in outer foliage. **Voice** High, squeaky *teep-teep-teep*, often accelerating suddenly *td'td'td'tddd....* **SS** ♀'s yellowish eyebrow and face, and ***plainer wings*** separate from Yellow-backed Tanager; see smaller, chunkier, shorter-billed euphonias. **Status** Fairly common.

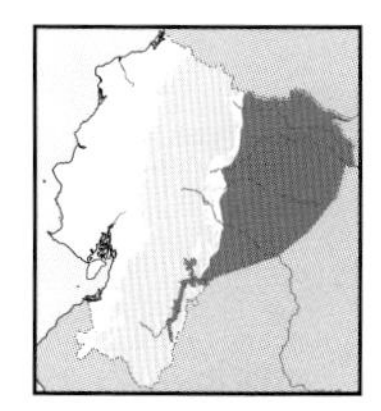

Yellow-backed Tanager *Hemithraupis flavicollis* 12–13cm

E lowlands to lower foothills (to 700m). *Terra firme* forest canopy, borders. Mostly ***black above, rump bright yellow***. Yellow lores and ***throat*** contrast with ***greyish-white underparts***, breast mottled dusky. ♀: olive above, yellow below, duskier wings ***edged yellowish***; ***dusky cheeks***, very vague eyebrow. Race *peruana*. Pairs often with mixed flocks, glean in outer foliage, sometimes sallying. **Voice** High, fast bursts of *seek* or *tseet* notes, nearly hummingbird-like. **SS** ♀ can be confusing but luckily, always accompanied by ♂! See ♀ Guira (yellower face); also ♀ dacnises, larger *Tachyphonus* and *Piranga* tanagers, and smaller, chunkier euphonias. **Status** Uncommon.

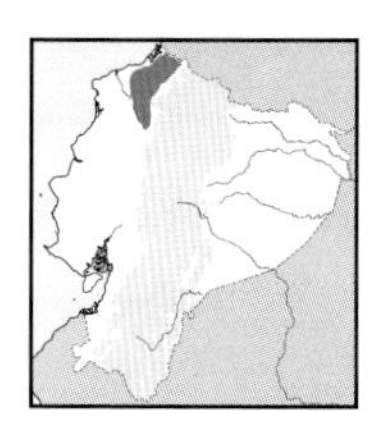

Scarlet-and-white Tanager *Chrysothlypis salmoni* 12–13cm

NW lowlands to lower foothills (to 800m). Humid forest canopy, borders, secondary woodland, woody clearings. Lovely ♂ strikingly clad in ***orange-red and white***. ♀: similar but red replaced by ***dull olive-brown***, paler on throat. Monotypic. Mainly in small groups, often with mixed canopy flocks, nimbly gleans for invertebrates in outer foliage. **Voice** Fast, short, descending and rather weak *tss-tsi-tsi-tsi-tse..tse*, squeakier *tseé* calls. **SS** ♂ unmistakable; ♀ can be confusing, but extensive white in underparts distinctive; see Black-faced Dacnis and Slate-throated Gnatcatcher. **Status** Locally uncommon; NT in Ecuador.

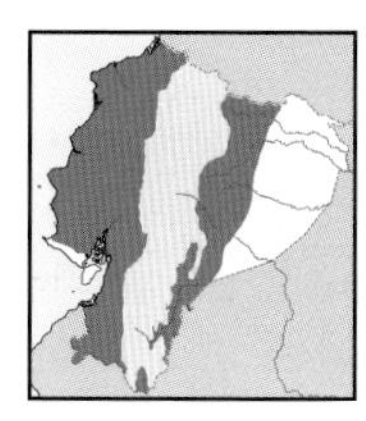

Bananaquit *Coereba flaveola* 10-11cm

E & W lowlands to lower subtropics, in E never away from Andes, also SE in Zumba region (to 1,100m, locally to 1,800m). Forest borders, woodland, clearings, gardens, crops. Slender, ***short decurved bill. Black head*** with ***bold white supercilium***, grey throat. Dusky-olive above, ***yellow rump, yellow below***, greyish flanks. Dusky wings with ***small white speculum***. Races *intermedia* (most range), *magnirostris* (extreme SE). Very active and nervous, non-forest species that moves singly, in pairs or congregates at flowering or fruiting trees, even hummingbird feeders. **Voice** Very high, short, insect-like *bzzzz-zzzy-zzyr*. **SS** Confusion unlikely due to bill shape, supercilium, underparts; but see small tanagers and warblers. **Status** Common, widespread in W.

♂
Scarlet-browed Tanager
♀
♂
♂
guirina
huambina
♀
♀
immature
Guira Tanager
♂
♂
immature
♀
Yellow-backed Tanager
juvenile
intermedia
♂
♂
magnirostris
Bananaquit
♂
♀
Scarlet-and-white Tanager

PLATE 266: CONEBILLS

Small, warbler-like tanagers with short, sharply pointed bills. All *Conirostrum* are rather drab and show little sexual dimorphism, except Capped. Two species in highlands, Blue-backed up to treeline; one species on E river islands. Giant is a very distinctive conebill with longer and very pointed bill used to inspect loose *Polylepis* bark.

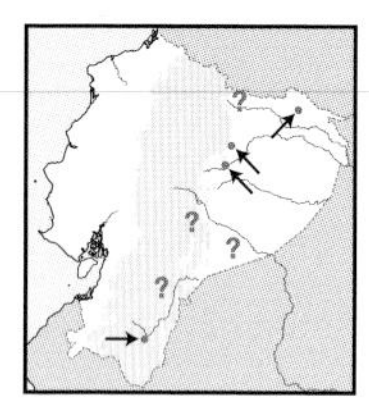

Chestnut-vented Conebill *Conirostrum speciosum* 10–11cm

Local in E Andean foothills, more local in NE lowlands (400–1,200m, locally at 220m). Humid forest canopy, borders, woodland. ***Bluish-grey***, whitish mid belly, ***chestnut vent, small white speculum in wings***. ♀: duller, ***pale bluish-grey crown***, olive upperparts, ***whitish lores and short supercilium***, buffy-white underparts. Race *amazonum*. Occasionally at flowering *Inga* trees; seen near ground in Cuyabeno. **Voice** High, thin *ti-ti...ti-ti-ti...ti-widi, ti-widi, ti-widi...*; repeated, hummingbird-like *wi'd-dí, wi'd-dí* song at dawn (Peru). **SS** Although some ♂♂ lack wing spot, confusion unlikely; ♀ more confusing, resembles rarer Bicoloured Conebill, smaller greenlets and Tennessee Warbler (clearer supercilium, dark legs). **Status** Rare, local, erratic?

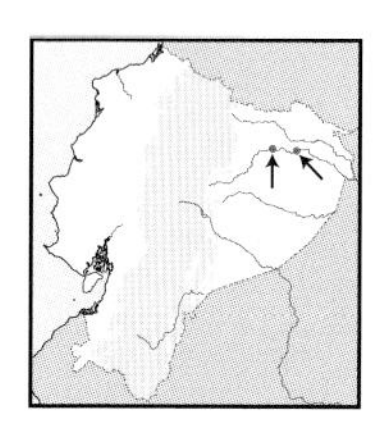

Bicoloured Conebill *Conirostrum bicolor* 12–13cm

Few records on lower Río Napo (below 300m). Regenerating woodland on river islands. ***Flesh legs. Pale bluish-grey upperparts, whitish-buff underparts***. ♀: duller above, less bluish, yellowish-tinged below. Race *minus*? Poorly known in Ecuador, elsewhere in *Cecropia*-dominated woodland; gleans invertebrates and picks seeds mostly in canopy, with mixed flocks. **Voice** High, sibilant *tsik, pit-tsik...*, running into fast, squeaky *t'wit-t'wit-t'wit...* **SS** Confusion unlikely in limited range and habitat; see Red-eyed Vireo; juv. Orange-headed Tanager (more orangey face and throat); also greyer, duller Pearly-breasted Conebill *C. margaritae* (not unlikely in Ecuador). **Status** Only wanderer or rare resident; NT globally.

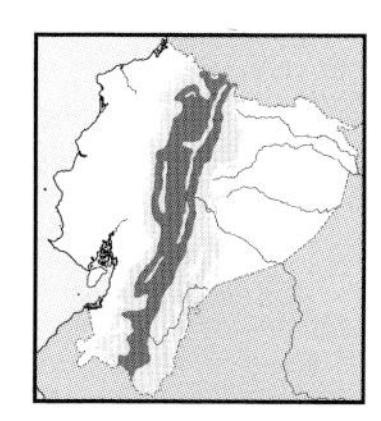

Cinereous Conebill *Conirostrum cinereum* 12–13cm

Andes and inter-Andean valleys (2,500–4,500m, mostly below 4,000m). Humid to xeric scrub, woodland, treeline, hedgerows, gardens. Olive-brown upperparts, ***long buff supercilium, buff underparts***, black wings with ***large white patch***. Race *fraseri*. Singles, pairs or small groups inspect foliage, undersides of leaves, debris and probe flowers, may join mixed flocks. **Voice** Fast, semi-musical *ti-wid'd'd'd* at dawn, also faster twittering *t'd'd'd'd'd'w'w* day song; call a thin *zee, zeet...*. **SS** Confusion unlikely, see superficially similar Rusty Flowerpiercer (no supercilium, upturned bill, bluish-grey upperparts; ♀ with flammulated breast). **Status** Fairly common.

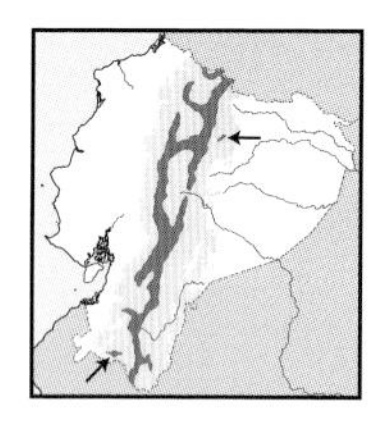

Blue-backed Conebill *Conirostrum sitticolor* 12–13cm

Andes in E & W (2,500–3,500m, locally higher). Montane forest borders, secondary woodland, treeline. ***Black hood, blue upperparts***, blacker flight feathers and tail, ***reddish underparts***. Sexes alike. Nominate. Pairs to small groups with mixed flocks; inspect foliage, terminal twigs, undersides of leaves, debris or probe in flowers. **Voice** Pretty, sibilant twittering *chi-wee'dee, chi-chi-chi'wee-dii...*, also faster, more jumbled *t'chi-wee'd'd...*and *zeet, zzít* calls. **SS** Confusion unlikely; Scarlet-bellied Mountain-Tanager has similar overall pattern but different size, bill shape, behaviour, voice, etc. **Status** Fairly common.

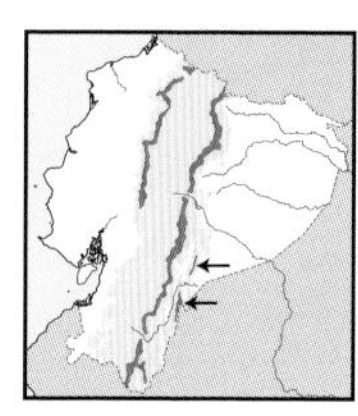

Capped Conebill *Conirostrum albifrons* 12–13cm

Subtropics to low temperate zone on E & NW Andean slopes, S in W to Chimborazo (2,000–2,800m). Montane forest canopy, borders, woodland. All black with ***dark blue forecrown*** (***white*** in NE), ***scapulars and rump***. ♀: ***bluish-grey forecrown, grey head to breast***, yellowish-olive upperparts and belly. Races *atrocyaneum* (most of range), *centralandium* (extreme NE). Pairs to small groups with mixed flocks, gleans permanently wagging tail. **Voice** Short, rapid *dsuduiit..dsee-sau, dsee-sau, dsee-sau...* at dawn, with jumbled start; call thin *sueep* or *ssít*. **SS** Tail wagging in both sexes distinctive; cf. ♀ with larger but rather similar Blue-capped and Blue-and-yellow Tanagers, and Grey-hooded Bush-Tanager. **Status** Uncommon.

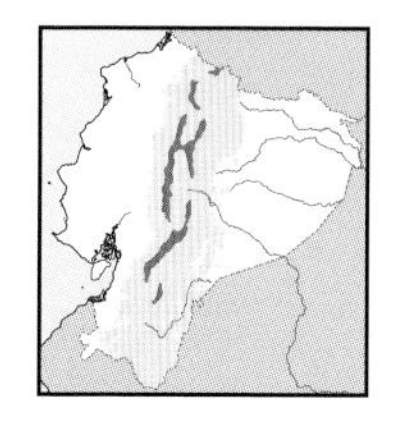

Giant Conebill *Oreomanes fraseri* 15–16cm

Andean *páramo* S to N Loja (3,500–4,200m). *Polylepis* woodland. ***Long pointed bill. Grey upperparts*** contrasting with ***chestnut underparts. Bold white moustache***, freckled forecrown, chestnut supercilium. Juv. somewhat streaked dusky below. Nominate. Loose pairs rather shy but noisy when flicking pieces of *Polylepis* bark, often clambering on trunks and large branches; may accompany mixed flocks. **Voice** Gives *chee-wee, chee-wee'd, cheeveét..., chee-cheevee-chee*; also simpler *chíí*. **SS** Unmistakable. **Status** Rare, NT globally, VU in Ecuador.

♀
♂
Chestnut-vented Conebill
♂
♀
Bicoloured Conebill
Cinereous Conebill
Blue-backed Conebill
centralandium
♂
♂
yaneum
♀
Capped Conebill
immature
juvenile
adult
Giant Conebill

Nectar-robbing flowerpiercers with upturned bills and sharply hooked tips useful for piercing corolla bases. Rusty and White-sided show marked sexual dimorphism. All are Andean; Glossy and Black to treeline. Plushcap is a small, attractive, finch-like bird that bears little resemblance to other tanagers. Its small, deep and stubby bill suggests a granivorous diet. Its taxonomic affinities are also intriguing.

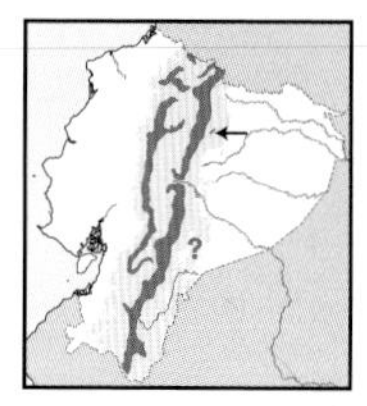

Glossy Flowerpiercer *Diglossa lafresnayii* — 14–15cm

Temperate zone in E & W Andes, S in W to Azuay (2,700–3,500m, locally lower and higher). Montane forest borders, woodland, treeline. Entirely ***glossy black*** with conspicuous ***pale bluish shoulders***; rather long bill. Duller, ***greyer*** juv. Monotypic. Singles or pairs mainly apart from mixed flocks; though active, not always easy to see as clings to flowers or gleans for invertebrates. **Voice** Semi-musical twittering *chiff-we'd'd'd'd, chi-chi-d'd'd, chiff-we'd'd'd…, chiff-chee-wi-di'di'd…*; call a short clear *cheé* or *chint*. **SS** Black Flowerpiercer lacks bluish wing patch, is smaller, shorter-billed and occurs in more open situations. **Status** Fairly common.

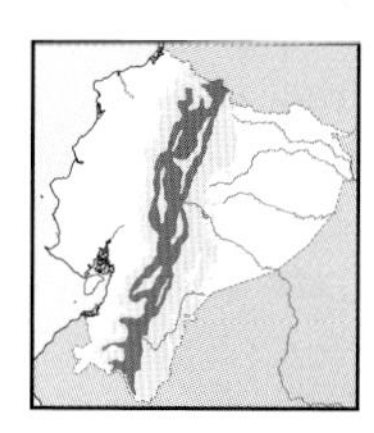

Black Flowerpiercer *Diglossa humeralis* — 13–14cm

Andes, inter-Andean valleys (2,500–4,550m, mostly below 4,000m). Montane forest borders, woodland, shrubby clearings, *Polylepis* woodland, gardens. Short bill. ***Entirely glossy black***. Juv. ***slate-grey to blackish***, some with streaked underparts. Race *aterrima*. Singles or pairs not with mixed flocks, nimbly cling to flowers to pierce corollas but also probe open ones; can be seen carrying pollen on forecrown. **Voice** Very fast twittering *t'd'd'dd'd'ddd-dd-di*, sometimes rather trilled; resembles Cinereous Conebill but more musical, longer, different pace. **SS** Larger Glossy (bluish shoulders) and smaller White-sided Flowerpiercers (white sides to flanks). **Status** Common.

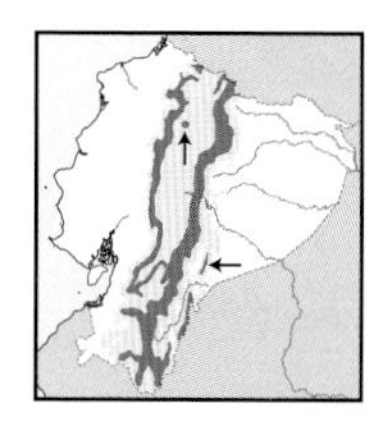

White-sided Flowerpiercer *Diglossa albilatera* — 12cm

Subtropics to temperate zone in E & W Andes (1,900–3,200m, locally to 1,600m). Montane forest borders, secondary woodland. Mostly ***blackish*** with contrasting ***white sides to flanks*** (not always visible but flash when wings spread). ♀: ***brown above, buffier below***, also has ***white sides***. Juv. as ♀ but somewhat streaked blackish below. Nominate (most of range), *schistacea* (Loja). Restless pairs move through lower growth and canopy alike, clings to flowers and gleans for insects, often with mixed flocks. **Voice** Short, very fast, dry trill *T-p'rrrrr* or *SSE-ti'ri'ri'ri'ri…*. **SS** Confusion unlikely as white tufts easy to see; cf. Black, Glossy and Rusty (♀) Flowerpiercers. **Status** Fairly common.

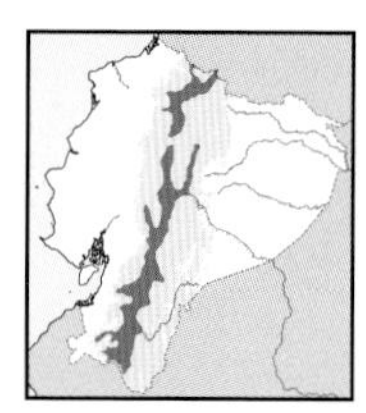

Rusty Flowerpiercer *Diglossa sittoides* — 11–12cm

Subtropics to lower temperate zone on both Andean slopes, inter-Andean valleys (1,700–2,800m). Montane forest borders, light woodland, shrubby clearings, hedges. ***Flesh mandible and legs. Bluish-grey upperparts***, blackish mask, ***cinnamon-buff underparts***. ♀: olive-brown above, yellowish-olive below faintly ***flammulated olive on throat and breast***. Race *decorata*. Mostly in pairs, rarely with other passerines, even flowerpiercers; briskly clings from flowers and moves from one to next. **Voice** Short twittering *ch'd'd'd'd* and faster, more trilled *d'rrrrrr'r*, sometimes shorter and ends in lower-pitched note. **SS** ♀ with White-sided Flowerpiercer, which has white tufts in sides to flanks; juv. Black Flowerpiercer always greyer. **Status** Uncommon.

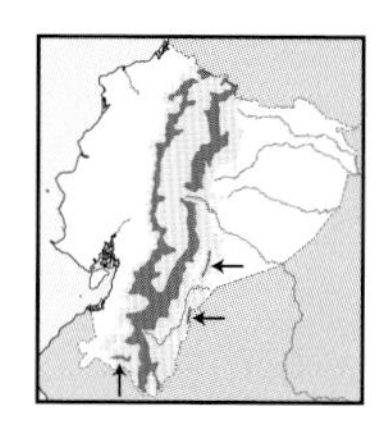

Plushcap *Catamblyrhynchus diadema* — 13–14cm

Subtropics to temperate zone on both Andean slopes (2,000–3,500m). Montane forest undergrowth, secondary woodland, bamboo. Small, seedeater-like bill. ***Bright yellow 'brush' on forecrown***, black crown to nape. Grey upperparts, including rather long tail, contrasting ***rufous-chestnut underparts***. Juv. brownish above, duller chestnut below. Nominate. Singles or pairs often with mixed flocks, rather shy but active in undergrowth, probing at leaf whorls, bamboo nodes, stems, debris. **Voice** Long, unmusical, weaving *t'chip-chip-wee-dip-chip*, rather 'random' and hummingbird-like. **SS** Bill shape resembles several seedeaters and even grosbeaks but plumage is distinctive. **Status** Uncommon.

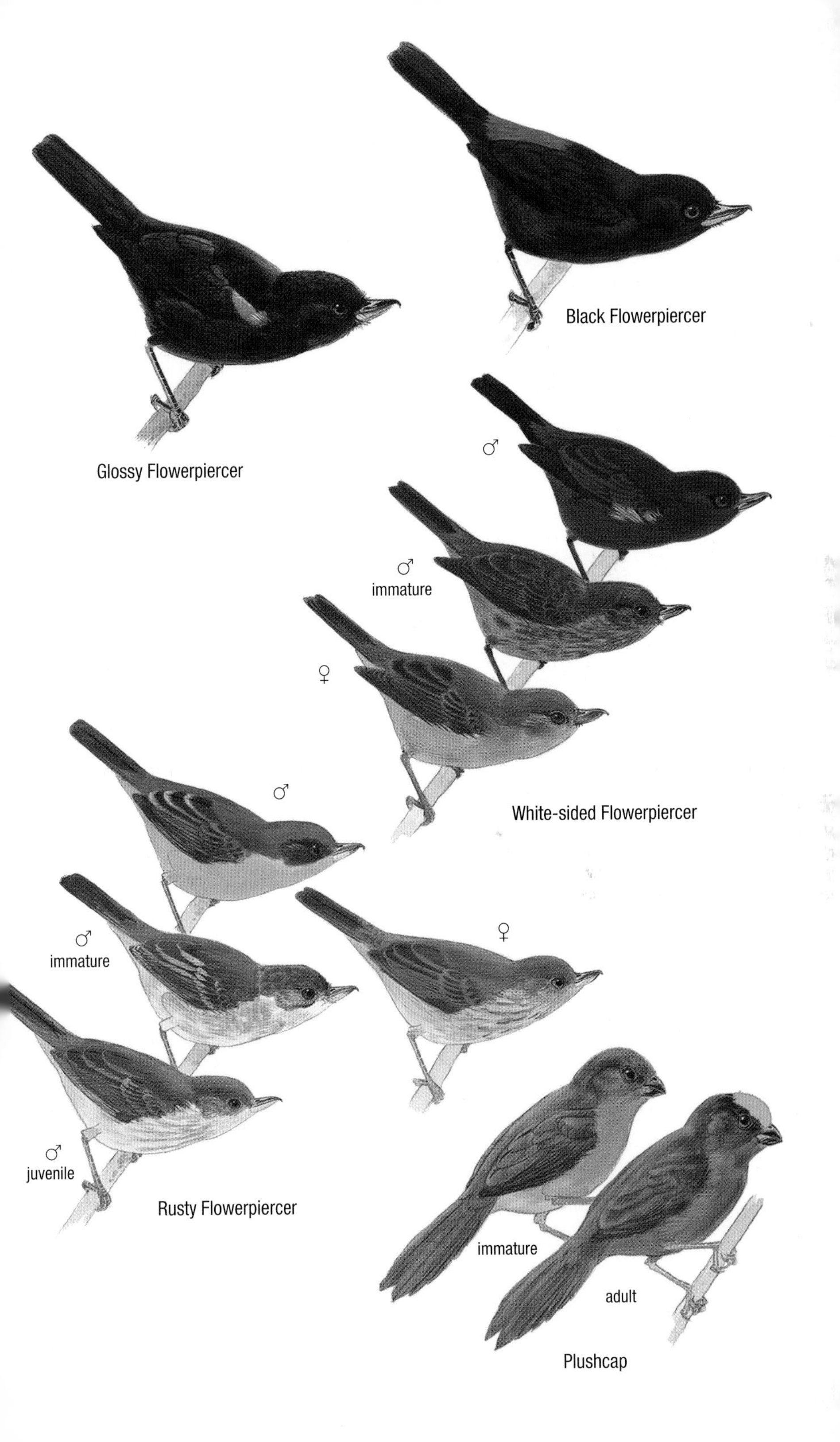
Black Flowerpiercer
Glossy Flowerpiercer
♂
♂
immature
♀
White-sided Flowerpiercer
♂
♂
immature
♀
♂
juvenile
Rusty Flowerpiercer
immature
adult
Plushcap

Flowerpiercers with less hooked and upturned bills; their diets are apparently more frugivorous. No sexual dimorphism. Rarer Indigo and Deep-blue are exquisite deep purplish-blue, differing markedly from Bluish and the common Masked. The small, intriguing Tit-like Dacnis is also deep blue but has a tiny bill, marked sexual dimorphism and distinctive behaviour.

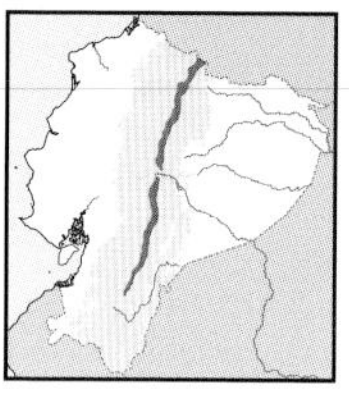

Black-backed Bush-Tanager *Urothraupis stolzmanni* 15cm

Temperate zone to *páramo* in E Andes S to E Azuay (3,200–4,000m). Treeline, elfin forest, *Polylepis* woodland. ***Black upperparts, snowy-white throat*** grading into ***greyish-mottled upper belly, blackish lower belly to vent***. Monotypic. Mainly small groups with mixed flocks, forage in dense foliage and edges. **Voice** Thin song, high-pitched, rather jumbled twittering *chip-chip-chi'ri'r'r'rip*; similar fast *tsi-ri-rip* foraging calls; simpler, thinner *chi-chirp-chir*. **SS** Recalls a brush-finch but none has black-and-white pattern; smaller Black-headed Hemispingus (white eyes, black hood, buffy crown, grey upperparts). **Status** Uncommon.

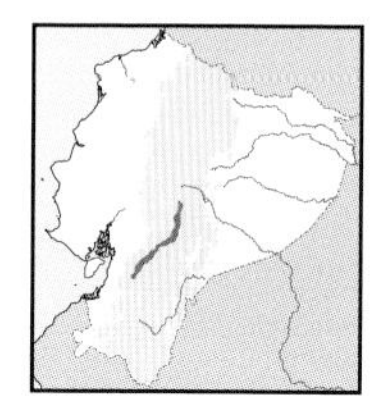

Tit-like Dacnis *Xenodacnis parina* 13–14cm

Very local in S Andes; Azuay N to Volcán Altar region (3,700–4,000m). *Polylepis* and *Gynoxys* woodland and edges. ***Short pointed bill***. Entirely ***deep blue*** finely ***streaked silvery on foreparts***. ♀: ***blue forecrown and orbital area***, brown upperparts, cinnamon-buff underparts, ***blue edges to wings and tail***. Race undescribed. Pairs to small noisy groups work dense foliage of *Gynoxys*, probing underside of leaves for aphids and sugary secretions. **Voice** Loud, fast, bubbling *wueep-wueep-wueep-zeet-zeet-zeet*, squeaky quality with some harsher notes interspersed; also *dyeet-dyeet-dyeet..zzuee-zzuee-zzuee'*. **SS** Readily distinguished from flowerpiercers (locally sympatric Masked, Glossy, Rusty?) by bill shape. **Status** Uncommon but very local, EN in Ecuador.

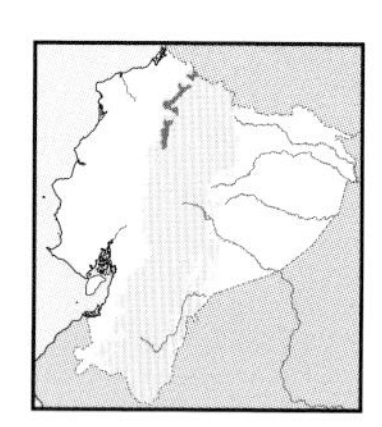

Indigo Flowerpiercer *Diglossa indigotica* 11–12cm

Local in W Andean subtropics, also in El Oro? (1,600–2,000m, lower in El Oro?). Wet montane forest, borders, and ericaceous shrubs. ***Bright red eyes***. Rich ***indigo-blue*** with more turquoise flight feathers and tail. Monotypic. Mostly pairs or trios, often with mixed flocks; pierces flowers, gleans invertebrates and pecks berries; often with mixed flocks. **Voice** Fast twittering, similar to Bluish but higher-pitched; also high-pitched, clear *teep, teep, teep* or *chep, chep*.... **SS** Bluish Flowerpiercer decidedly paler, bill less upturned; higher-elevation Masked is less vivid, with large black mask. **Status** Very rare and local, VU in Ecuador.

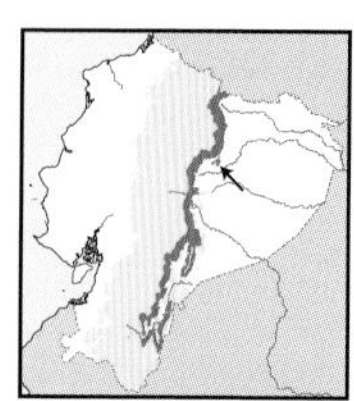

Deep-blue Flowerpiercer *Diglossa glauca* 12–13cm

E Andean foothills to subtropics (1,000–1,800m). Montane forest borders, old secondary woodland. ***Golden eyes. Deep purplish-blue***, blackish lores. Race *tyrianthina*. Pairs to small groups often with mixed flocks, gleans for invertebrate on mossy branches and bromeliads; less often pierces flowers; tends to be in upper strata, regularly at natural gaps. **Voice** High-pitched, fast, descending *ti-ti-ti-wiii'dii-di'di'di'de*..., higher-pitched, less musical than other flowerpiercers; call simple, clear *ti', ti', ti'*.... **SS** Bluish is largely paler and has red eyes. **Status** Rare to uncommon.

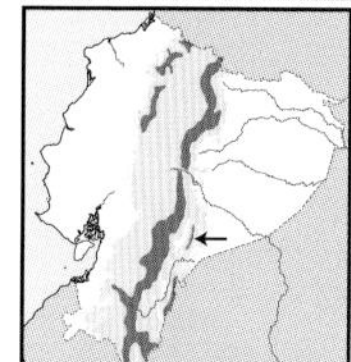

Bluish Flowerpiercer *Diglossa caerulescens* 13–14cm

E & NW subtropics to lower temperate zone, also W & C Azuay (1,700–2,700m, locally higher). Montane forest borders, secondary woodland. ***Reddish eyes, nearly straight, long bill***. Mostly ***pale bluish, greyer below***, slightly paler throat and belly, ***blackish lores***. Races *media* (E), *saturata*? (W). Singles to small groups often with mixed flocks, probe flowers, glean invertebrates and pick small berries. **Voice** Fast, high, descending, whistled *eet-eet-eet-tee-tee-teea-teeaa-welee-welee*..., recalling rarer Deep-blue and Indigo Flowerpiercers but rather lower-pitched; high, metallic *tsik* calls. **SS** Higher-elevation Masked (deeper blue) has black mask (though mask in juv. is not well defined); bill obviously upturned and hooked. **Status** Uncommon.

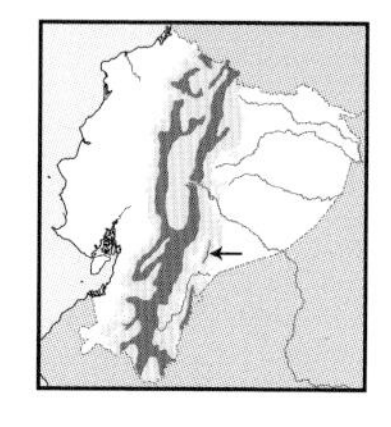

Masked Flowerpiercer *Diglossa cyanea* 14–15cm

E & W subtropics to temperate zone (2,400–3,500m, locally to 2,000m and lower). Montane forest borders, woodland, shrubby clearings. ***Red eyes. Rich blue*** with ***black forecrown and mask***. Sexes nearly alike. Juv. duller. Nominate (most of range), *dispar* (SW). Pairs to small noisy groups with mixed flocks, sometimes appears to lead them; probes and pierces flowers, and picks small berries. **Voice** Accelerating thin *seet-seet-see-see-see'ss'ss'ss'ss'sll'sll'sll*..., ending in fast twitter; call a thin *zíp* or *zzip*. **SS** Bluish (duller, tinged grey, straighter bill, no mask); see lower-elevation Deep-blue and Indigo Flowerpiercers, and Blue-and-black Tanager (straight conical bill, black wings, etc.). **Status** Common.

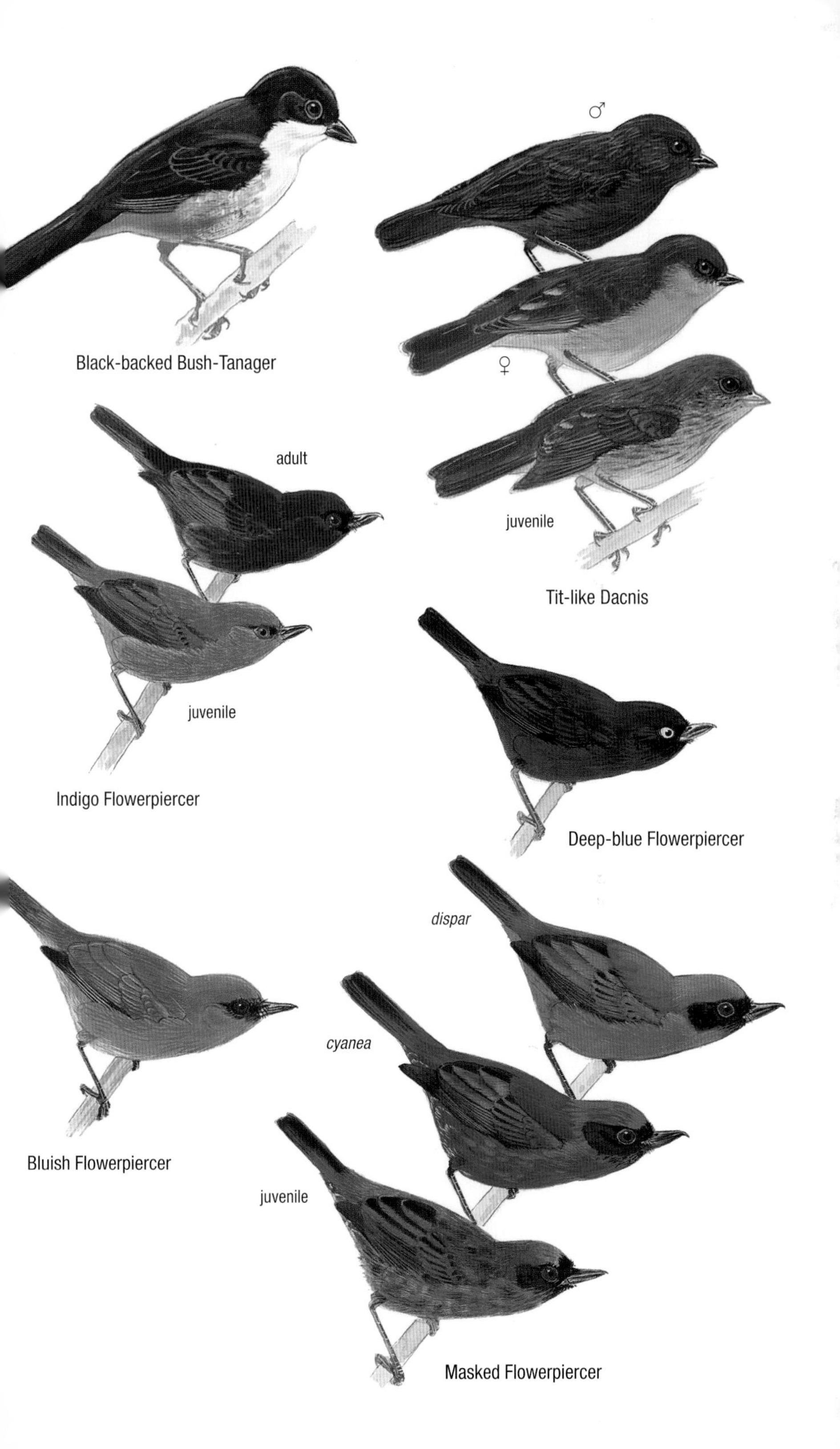

♂
Black-backed Bush-Tanager
♀
adult
juvenile
Tit-like Dacnis
juvenile
Indigo Flowerpiercer
Deep-blue Flowerpiercer
dispar
cyanea
Bluish Flowerpiercer
juvenile
Masked Flowerpiercer

PLATE 269: BUSH-TANAGERS

Drab Andean 'tanagers' that move in small flocks, less often in pairs, and frequently with mixed-species flocks. Conspicuous and noisy. Particularly challenging is the rare, seldom seen and notably difficult to recognise Yellow-green, whose taxonomic affinities remain unsolved. Recent molecular studies show that *Chlorospingus* bush-tanagers are not true tanagers, but finches.

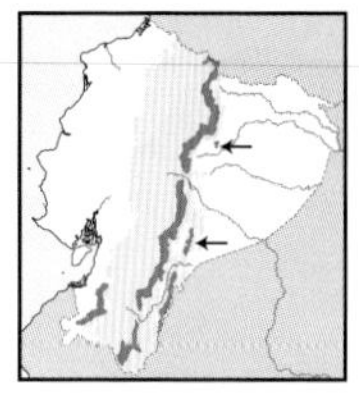

Common Bush-Tanager *Chlorospingus flavopectus* 14–15cm

E & SW foothills to subtropics (mostly 1,500–2,500m; 700–1,500m in SW). Montane forest borders, adjacent clearings. ***Pale eyes. Dusky-grey head, greyish-white throat*** and belly, ***greenish-yellow breast-band***. Race *phaeocephalus* (E); undescribed? (SW). Mostly in groups, noisy and conspicuous following midstorey mixed flocks; regularly moves tail sideways. **Voice** Rhythmic series of *chíp* notes lasting up to 25 sec, becoming a sputter; high *tip* calls. **SS** Pale eyes separate from Ashy-throated Bush-Tanager, which is paler and brighter, E birds with white postocular; other *Chlorospingus* lack yellow breast-band. **Status** Fairly common.

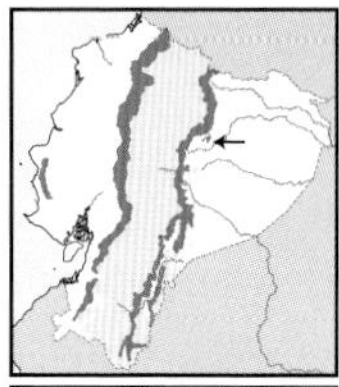

Ashy-throated Bush-Tanager *Chlorospingus canigularis* 13–14cm

E & W foothills to subtropics, SW coastal mountains (1,000–1,900m in E, 700–1,300m in W, 400–500m in coastal ranges). Montane forest, borders. ***Grey head, greyish-white throat and belly, yellow breast-band. White postocular stripe, darker cheeks*** (E). Races *signatus* (E), *conspicillatus*? (NW), *paulus* (SW). Small groups with mixed flocks, noisy and conspicuous, mostly in upper strata. **Voice** Sputtering *chít, tsít, zít*. **SS** Common (darker, pale-eyed, drabber and dirtier underparts); see ♀ White-shouldered Tanager (plain yellow underparts except throat). **Status** Uncommon.

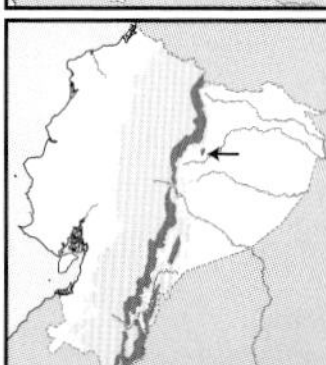

Short-billed Bush-Tanager *Chlorospingus parvirostris* 14–15cm

E Andean foothills to subtropics (1,200–2,250m). Montane forest and borders. ***Pale eyes***. Dark olive above, greyish below, ***bold yellow whisker***. Race *huallagae*. Pairs to small groups, sometimes with mixed flocks but also alone, seemingly forages lower than Yellow-throated. **Voice** Harsh, vibrating *t'sss'rrr* or *tsrreet* while foraging. **SS** Yellow-throated has darker eyes, more solid yellow throat with clear-cut separation from breast, paler underparts. Common has prominent yellow breast-band. **Status** Uncommon.

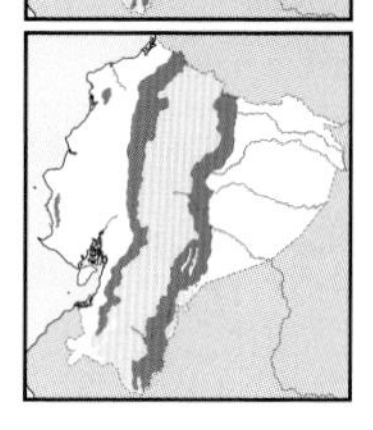

Yellow-throated Bush-Tanager *Chlorospingus flavigularis* 14–15cm

E & W foothills to subtropics, coastal mountains (700–1,800m, locally lower; 400–500m in coastal cordilleras). Montane forest, borders. ***Brown to hazel eyes***. Greyish underparts, ***grey lores, yellow throat*** (E) or ***throat-sides*** (W). Nominate (E), *marginatus* (W). Noisy groups (up to 10–12) with mixed flocks or alone, mostly in mid to upper strata. **Voice** Alternating *tuué-tseeu...tseeu-tuué...* (E); monotonous, long-repeated, high-pitched series of *tzét* notes (W); also simpler *tssét, chút, tzeet*. **SS** Short-billed Bush-Tanager (pale eyes, yellow only in whiskers); in NW see Dusky (darker head and underparts, no yellow throat, orange to reddish eyes). **Status** Uncommon.

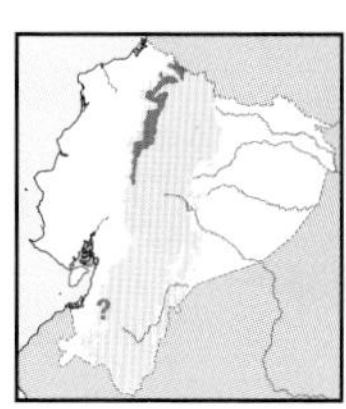

Dusky Bush-Tanager *Chlorospingus semifuscus* 14–15cm

NW foothills to subtropics (1,200–2,200m, locally higher). Montane forest borders, shrubby clearings. ***Orange to red eyes. Dusky-grey head, greyish underparts***, dark olive upperparts. Nominate. Pairs to groups (up to 10–12) with mixed understorey flocks, sometimes higher, apparently forms displaying lek-like aggregations on ridges. **Voice** High-pitched *tzít* notes, accelerating and ending in dry sputter. **SS** Greyer and darker than lower-elevation, rarer Yellow-green Bush-Tanager, both unmarked (see also eyes). Yellow-throated is brighter above, obvious yellow throat. See also heavier, plainer and darker olive Ochre-breasted Tanager. **Status** Fairly common.

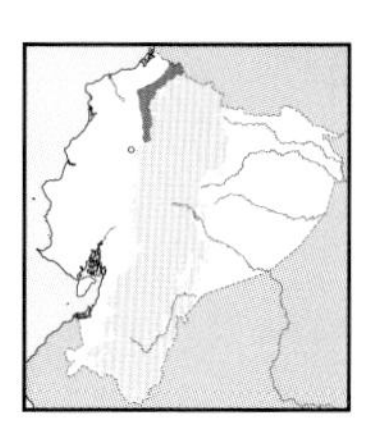

Yellow-green Bush-Tanager *Chlorospingus flavovirens* 14–15cm

Local in NW foothills (400–1,100m). Wet forest, borders. Dark olive above, ***greenish-yellow below, dusky cheeks***. Monotypic. Very poorly known in Ecuador; pairs to small groups with upper strata mixed flocks, work moss-laden branches and epiphytes. **Voice** Ringing *díp* and *chép* notes, becoming a chatter *díp-díp-di-cht'chrrrrr* and thin *ti, ti-ti-titi*. **SS** Drabber than most bush-tanagers. Dusky has dull grey underparts, darker head (no known overlap); see larger, heavier and noisier Ochre-breasted Tanager; plainer brown ♀ Tawny-crested Tanager and Olive Finch (grey face, chestnut crown, etc.). **Status** Very rare, few confirmed records, VU.

Common Bush-Tanager

Ashy-throated Bush-Tanager

Short-billed Bush-Tanager

Yellow-throated Bush-Tanager

Dusky Bush-Tanager

Yellow-green Bush-Tanager

PLATE 270: GREEN TANAGERS AND ANT-TANAGER

A group of plump, thick-billed 'tanagers' recently demonstrated to be related to cardinals (except Dusky-faced, of uncertain affinities). Three W species are drab in plumage but rather conspicuous in behaviour and voice. *Chlorothraupis* mostly in mixed flocks – even leading them. Dusky-faced is mostly found away from flocks. The ant-tanager is a very distinct ant-follower, also drab despite reddish plumage (♂).

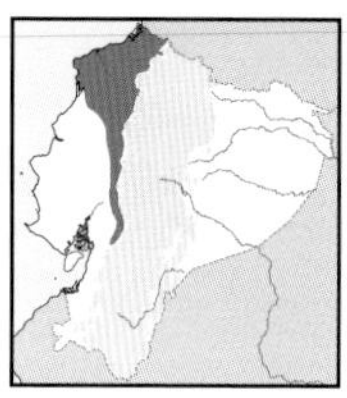

Dusky-faced Tanager *Mitrospingus cassinii* 17–18cm

W lowlands to foothills, S at base of Andes to Guayas / Cañar boundary (to 800m). Humid forest borders, second growth near streams. Fairly long bill, ***whitish eyes. Dark grey face, throat and upperparts***, darker wings and tail. ***Dull ochraceous-yellow crown to nape and rest of underparts***. Nominate. Noisy flocks of up to 10–12 skulk in dense undergrowth; usually not with mixed flocks. Twitches wings and tail constantly. **Voice** Song a harsh, scratchy, jumbled *wss-wss-wss'wrr, wss-wss-wss'wrr…*; also incessant harsh *ch'trrr-ch'trrr-ch'trrr…* flock calls. **SS** Confusion unlikely; see ♀ White-shouldered Tanager. **Status** Fairly common.

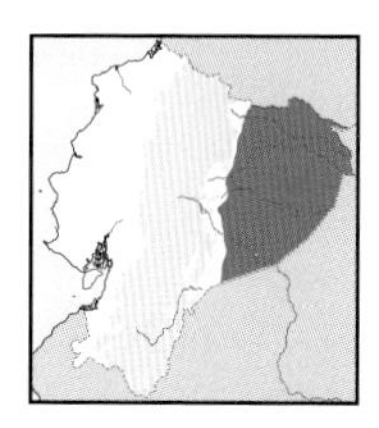

Red-crowned Ant-Tanager *Habia rubica* 17–18cm

E lowlands to foothills (to 700m). *Terra firme* forest undergrowth. ***Dull red, bright red semi-concealed crown patch***, paler belly, ***flanks washed grey***. ♀: red replaced by ***brownish-olive above, yellowish-olive below, tawny-orange coronal patch***. Race *rhodinolaema*. Active but shy pairs with mixed flocks, often follow army ants. **Voice** Melodic song *pü-déé-poö-déé-poö, plü-déé-plü-déé…*, recalls a thrush; rapid, rough *chá-chá-chá-chá…*. **SS** Confusion unlikely; few tanagers in forest undergrowth; see ♀ Silver-beaked; might locally overlap with Carmiol's (greener than ♀, heavier-billed, shorter-tailed); see also understorey foliage-gleaners. **Status** Rare to locally uncommon.

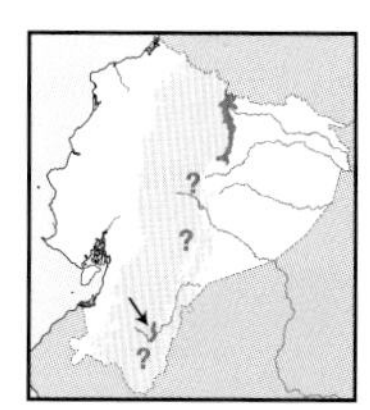

Carmiol's Tanager *Chlorothraupis carmioli* 17–18cm

Local in E Andean foothills (600–1,100m). Montane forest undergrowth. ***Heavy black bill***. Mostly ***dark olive, yellowish throat*** with some ***olive streaking***, paler central belly. Generally paler ♀ has ***yellow lores and plainer throat***, underparts more yellow. Race *frenata*, sometimes treated as valid species. Small groups, very active and noisy, mostly with mixed flocks in shady creeks and ravines. **Voice** Sweet, melodious, liquid and variable *swee-swee-swee-swu-swu-t'veeu-t'veeu, d'teew-d'teew-d'teew…*at dawn; also harsh, rather metallic *jyé-jyé* or *jhé-jhé* calls. **SS** Only *Chlorothraupis* in range. Yellow-throated and Yellow-whiskered Bush-Tanagers have brighter yellow throat, slighter bill; see also more slender ♀ *Tachyphonus* and *Lanio*. **Status** Uncommon.

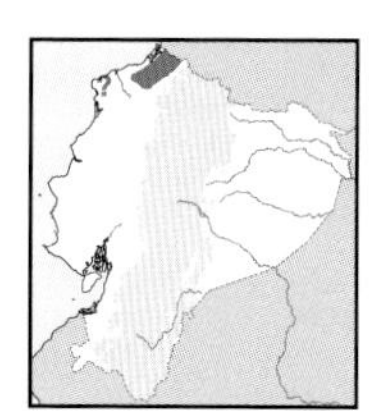

Lemon-spectacled Tanager *Chlorothraupis olivacea* 17–18cm

Lowlands in extreme NW (below 450m). Wet forest undergrowth and borders. Fairly heavy blackish bill. Mostly olive-green, ***bold yellow lores and spectacles, yellow throat*** becoming fine streaks on upper chest. Sexes nearly alike. Monotypic. As previous species, small noisy groups with understorey flocks. **Voice** Musical, pleasant dawn song recalls that of Carmiol's Tanager, a variable *swee-swee-swu-swu, suwee-suwee-s'wuu-wuu-wuu, ss'ss'ss…*, also nasal, annoying *nya!-nya!-nya!* or *trya!-trya!* **SS** No overlap with darker Ochre-breasted Tanager, which lacks yellow facial marks. Equally noisy Dusky-faced Tanager has white eyes, grey face and upperparts, etc. **Status** Uncommon.

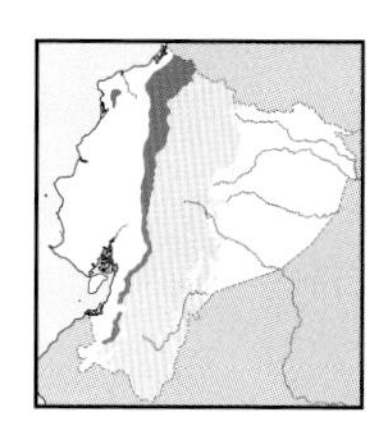

Ochre-breasted Tanager *Chlorothraupis stolzmanni* 17–18cm

W Andean foothills to subtropics, S to N El Oro, also coastal mountains in S Esmeraldas (400–1,500m). Montane forest undergrowth and borders. ***Heavy black bill***, greyish eyes. ***Drab***. Mainly ***dark olive above***, paler, ***duller ochraceous below***. Sexes alike. Nominate (most of range), *dugandi* (N Esmeraldas, to Carchi?). Groups of up to a dozen noisily forage in lower strata, sometimes higher; can join mixed flocks, but apparently does not lead them. **Voice** Long-repeated, harsh *weeu-wee-wee-wee-wee… ws'ws'ws'ws'ws* at dawn; also harsher, monotonous *w-s's's's's, w'ss'ss'ss…, whee-ss'ss'ss'ss…; jeé*, harsh *jé-jék'k'k'k…*. **SS** Heavier than bush-tanagers, lacks obvious characters; Yellow-green Bush-Tanager has dusky lores, yellower underparts. **Status** Fairly common.

Dusky-faced Tanager

Red-crowned Ant-Tanager

Carmiol's Tanager

Lemon-spectacled Tanager

Ochre-breasted Tanager

Saltators recall grosbeaks (Plate 282) but their bills are less massive. Recent studies have shown that saltators are probably more related to finches or tanagers. Some species have a well-patterned head, and most have melodious, far-carrying songs.

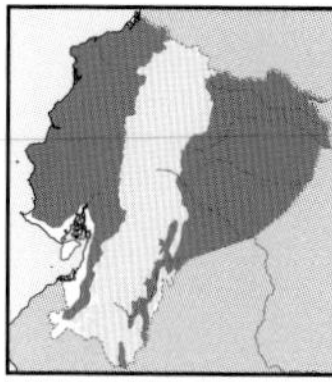

Buff-throated Saltator *Saltator maximus* 20–21cm

E & W lowlands to lower subtropics (to 1,600m). Humid forest borders, woodland, adjacent clearings. ***Bright olive upperparts***, short white eyebrow, grey face, ***black malar, orange-buff throat***. Nominate. Pairs regularly join mixed flocks, tend to remain hidden, but may perch on exposed branches when singing; regular at feeders. **Voice** Short, mellow, repetitive *cheëreeuu...cheë-tleeu, cheëreeuu...cheë-tleeu...; cheëtleeu-chetlëelee, cheëtleeu-chetlëelee...*; also faster *tlee-teleeutlee'chwee-chee....* **SS** Greyish Saltator lacks buff throat, upperparts grey; voice slower, simpler whistles. **Status** Common.

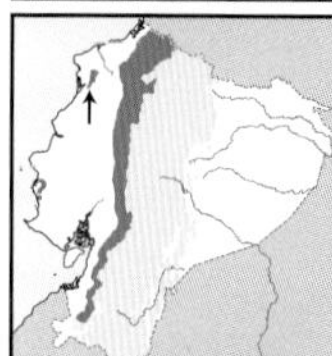

Black-winged Saltator *Saltator atripennis* 20–21cm

W Andean foothills to lower subtropics, local in N coastal ranges (to 1,500m, locally higher). Montane forest canopy, borders, woody clearings. ***Black head, long white eyebrow, white ear patch, black wings and tail***. Monotypic. Singles to small groups join mixed flocks, tend to remain high and hidden; often with Buff-throated Saltator. **Voice** Loud, melodious *tooow-tlee-tlee, tooow-tleetlee-teeow...*, sometimes ending in fast trill and high whistle; also leisurely *tloooö?-tleeeé...*; high-pitched *gleéc*. **SS** Only saltator with black wings and strong facial pattern. **Status** Uncommon.

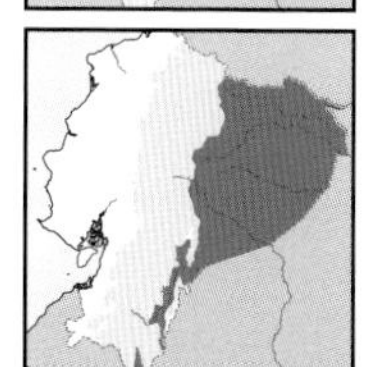

Greyish Saltator *Saltator coerulescens* 20–21cm

E lowlands to foothills (to 1,300m). *Várzea* and riparian forest borders, woodland, shrubby clearings. Mostly ***ash-grey***, short white eyebrow, ***black malar bordering white throat***. Race *azarae*. Pairs alone or with mixed flocks, less wary than other saltators, often in semi-open. **Voice** Mellow, clear *teewee-teewee-teewee, tee-chooow...; teew-teew, chóooow; teewee-chuoow, teweë-teweë...* (SE); *toow-toow-tdEéeu, tcheuu-tcheuu-tcheuu-tdeeew*, nasal *twink-twink-twink-twink...* (lowlands); squeaky *chuin* call. **SS** Buff-throated Saltator (olive upperparts, buff throat, more complex voice). **Status** Uncommon.

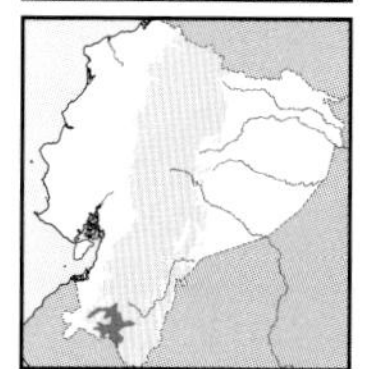

Black-cowled Saltator *Saltator nigriceps* 22cm

S Andean subtropics to highlands (1,700–2,900m). Montane forest borders, woodland, dense shrubbery. ***Bright red bill. Black hood***, dark grey upperparts, paler below, cinnamon-buff belly to vent. Longish tail. Monotypic. Skulking and difficult to see, regularly joins mixed species flocks in mid strata. **Voice** Loud, fast *wet-chloowii-iít* at dawn; ringing, short *klut-swee-lít...swee-lít*; loud *clee-gLÍl* calls. **SS** Only other saltator with red bill (Slate-coloured Grosbeak) lacks black hood and cinnamon-buff vent; no known overlap. **Status** Uncommon to rare, NT in Ecuador.

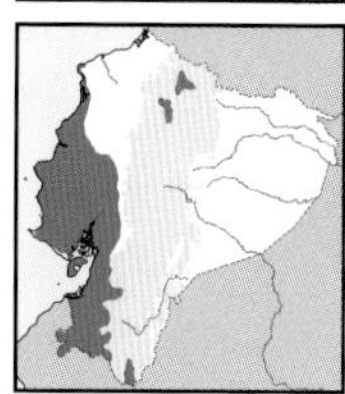

Streaked Saltator *Saltator striatipectus* 19–20cm

W lowlands to SW subtropics, N inter-Andean valleys S to Pichincha, extreme SE (300–2,500m). Arid woodland, scrub, gardens. ***Black bill, short white eyebrow, greyish-olive above, coarsely streaked breast to flanks*** (Andean nominate) or ***black bill tipped yellow, long*** white eyebrow, ***olive upperparts, plain whitish below*** (W *flavidicollis*), juv. resembles nominate; bill as W, ***short*** whitish eyebrow, ***olive above, streaked dusky below*** (SE *peruvianus*); possibly separate species. Rather tame, conspicuous, arboreal pairs to small groups. **Voice** Loud, with up-slurred ending: *t'chee-t'cheew-t'cherr'r; t'chew-t'cheerrw...; t'cheerrw-t'chooou...; t'chew-t'che-chow, t'cheew-chow...*; differs between races. **SS** Long white eyebrow and streaky underparts distinctive. **Status** Fairly common.

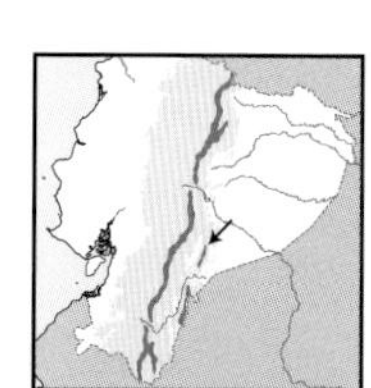

Masked Saltator *Saltator cinctus* 21–22cm

Local in E Andean subtropics (2,000–2,700m). Lower growth and borders of montane forest. ***Black-and-reddish bill, reddish eyes***. Grey upperparts, ***striking black mask to collar, white throat*** and lower underparts, flanks washed grey. ***Long graduated, blackish tail tipped white***. Monotypic. Skulking pairs sometimes with mixed flocks; seemingly engages in non-seasonal movements following *Podocarpus* fruiting. **Voice** Rather simple, melodious *cheucheucheu-chuwit?*; also piercing *chiip-t'chiip-chiip't* calls. **SS** Unmistakable. **Status** Rare, nomadic? NT.

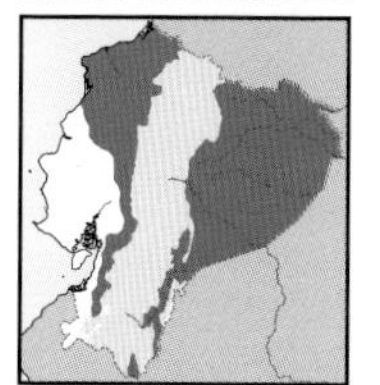

Slate-coloured Grosbeak *Saltator grossus* 20cm

E & W lowlands to foothills, S in Andean foothills to El Oro (mostly below 1,200m, locally to 1,500m). Humid forest mid strata, borders, woodland. ***Bright red bill. Bluish-grey***, blacker face and chest, ***white throat***. ♀: ***lacks black on face and chest***. Races *saturatus* (W), nominate (E). Pairs, seldom with mixed flocks, not easy to spot unless vocalising, fond of vine tangles. **Voice** Rich, melodious, quite monotonous song: *dleeo-dee-o'deeu, dleeo-dee-ouu'douu...; three-dlee-woodlee..., three-dlee-woodlee-glou'glou...*; nasal *weeah!* **SS** Unmistakable by striking bill colour. **Status** Uncommon.

adult
Buff-throated Saltator
juvenile
adults
Black-winged Saltator
adult
juvenile
Greyish Saltator
adult
juvenile
Black-cowled Saltator
flavidicollis
adult
peruvianus
adult
juvenile
juvenile
striatipectus
adult
juvenile
Streaked Saltator
adult
juvenile
Masked Saltator
♀
♂
juvenile
Slate-coloured Saltator

Sparrows are small, streaked birds of open and semi-open habitats that often form small flocks to feed. Some species (e.g., Rufous-collared, Yellow-browed) are opportunistic around man. Grasshopper is apparently extirpated in Ecuador. Sierra-finches are slightly larger birds of open areas, often barren terrain in the high Andes and arid SW.

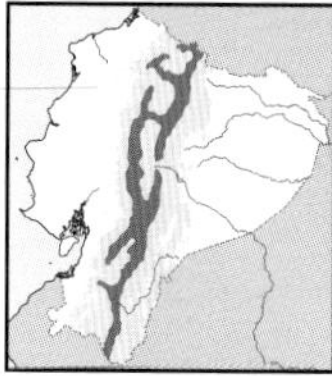

Plumbeous Sierra-Finch ***Phrygilus unicolor*** **15–16cm**

Andes (3,000–4,800m). Grassy and shrubby *páramo* and adjacent fields. Blackish bill, pink legs. ***Mainly dark grey***, darker wings and tail. ♀: ***densely streaked brown and dusky above***, whitish and dusky below, ***plain grey rump***. Race *geospizopsis*. Pairs to small groups feed on ground, often at roadsides, also perch and hop on boulders, rocks, stumps. **Voice** Brief buzzy trill *brzzz'e* or *zhrrre*. **SS** Larger, dark-billed than *Catamenia* seedeaters; other sierra-finches smaller, ♀♀ less streaky, whiter below (♂ Ash-breasted paler grey); cf. crested Rufous-collared Sparrow. **Status** Fairly common.

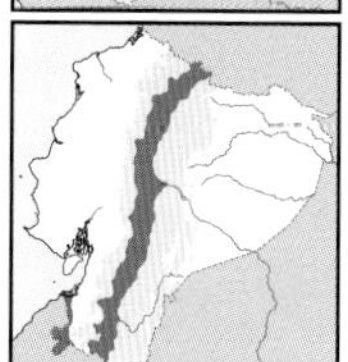

Ash-breasted Sierra-Finch ***Phrygilus plebejus*** **12–13cm**

Inter-Andean valleys, highlands, locally to coasts (mainly 1,500–3,500m, but to sea level). Grassy areas, fields, arid shrubbery. ***Mostly grey, thin white eyebrow. Upperparts striped blackish except rump***, underparts paler. ♀ and juv. brownish-grey above, ***whitish below*** all ***streaked dusky***. Race *ocularis*. Flocks feed on ground, rather tame often approaching habitation. **Voice** Short, buzzy trill *zrrrrri*, *zrrrrriii-tsu-tsu-tsu*; also more rattled *tzi'eeee*; call a sharp *chí*. **SS** ♀ Plain-coloured Seedeater (pale stubby bill, buffier); ♀ Band-tailed Sierra-Finch (buffier, paler-billed). **Status** Fairly common.

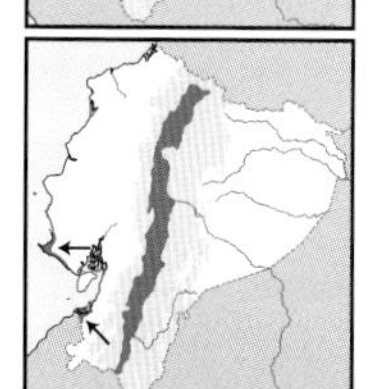

Band-tailed Sierra-Finch ***Phrygilus alaudinus*** **14–15cm**

Inter-Andean valleys and arid SW lowlands (1,200–3,200m, locally higher, also sea level). Very arid semi-open grassland, sandy areas. ***Bright yellow, long bill and legs***. Black foreface, ***grey head to breast***, browner upperparts streaked blackish, ***pure white belly, white median tail-band***. ♀: ***horn bill, buff brown streaked dusky, belly white***. Races *bipartitus* (Andes), *humboldti* (SW). Singles or pairs, tame and conspicuous, seldom with mixed flocks. **Voice** Gives *dzzi-zzi-dii-ziiuw*, *d'zzz-d'zzzi-zzzzzhhhh*; call a thin *dip*. **SS** Smaller Plain-coloured and Band-tailed Seedeaters (stubbier bills, no white below); in SW see rarer Cinereous Finch (all grey). **Status** Uncommon, declining?

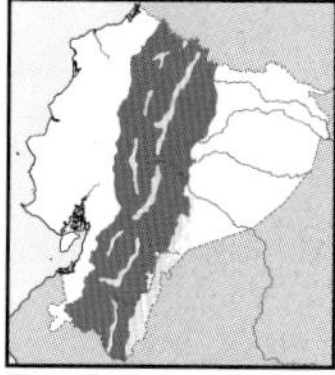

Rufous-collared Sparrow ***Zonotrichia capensis*** **13–14cm**

Widespread in Andes (1,500–3,500m, locally lower and higher). Open and semi-open areas. ***Triangular crest. Black and grey striped head, rufous nuchal collar***, streaked upperparts, ***two white wingbars***. Juv. drabber brown, ***densely streaked dusky***. Race *costaricensis*. Singles to small groups, tame and conspicuous, commensal; probably spreading downslope following land conversion. **Voice** Familiar but variable whistled *zee-zeee'oo-zeee*, call sharp *chíp*. **SS** Confusion unlikely but cf. locally Tumbes and Grasshopper Sparrows; also sierra-finches and *Catamenia* seedeaters. **Status** Common.

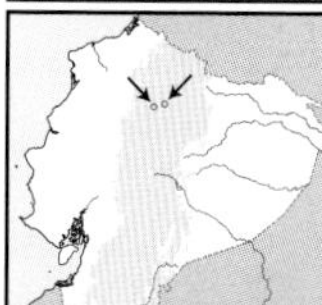

Grasshopper Sparrow ***Ammodramus savannarum*** **11–12cm**

Inter-Andean valleys in N (2,800–2,900m). Tall grass and pastures. Horn bill, ***short spiny tail***. Coarsely streaked above, ***buff crown-stripe and supercilium***; ***yellow bend of wing, buff to tawny below***. Race *caucae*. Unknown in Ecuador, where collected in undisturbed grassland; elsewhere shy and skulking. **Voice** Unknown in Ecuador, in Colombia a hissing high-pitched *pit-tzzzzzzzz* trill (HB). **SS** Shorter tail than all sympatric finches and sparrows, no crest; sierra-finches lack striped head. **Status** Possibly extinct, both museum records more than 100 years ago.

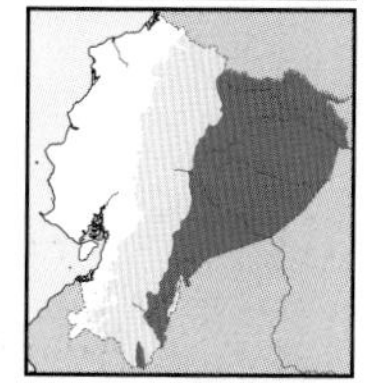

Yellow-browed Sparrow ***Ammodramus aurifrons*** **12–13cm**

E lowlands to foothills, extreme SE in Zumba region (mainly to 1,300–1,400m, locally higher). Open grassy and shrubby areas. ***Yellow foreface, broad supercilium, crown to lower back streaked blackish***. Greyish to ***whitish-buff below***. Nominate. Singles to small groups feed on ground, tame and conspicuous, seldom with other seed-eating birds. **Voice** High-pitched, insect-like *tzzz-zzzzz*, often with short *tíc* first; piercing *tík* call. **SS** Confusion unlikely in range. **Status** Common, likely increasing.

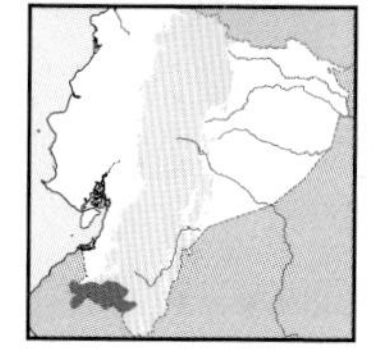

Tumbes Sparrow ***Rynchospiza stolzmanni*** **14–15cm**

Lowlands to foothills in SW (to 1,300m). Xeric scrub, dense shrubby. ***Bicoloured heavy bill. Greyish head*** with ***bold chestnut stripes, blackish sub-malar***. Dull brown upperparts streaked dusky, ***rufous-chestnut shoulders***, whitish to buffy below. Monotypic. Singles or pairs on ground, curious but rather skulking, often apart from mixed flocks. **Voice** Short, sweet, somewhat metallic and variable *tre'tre-dzzeé, tre'tre-dzzeé*...at dawn, *t'zzz-de-de-de-de* breeding song; call a short *tsí, tzz*. **SS** Head-stripes distinctive, no crest like Rufous-collared Sparrow. **Status** Uncommon.

♀
♂
Plumbeous Sierra-Finch
juvenile
adult
Ash-breasted Sierra-Finch
♀
♂
Band-tailed Sierra-Finch
juvenile
adult
Rufous-collared Sparrow
juvenile
adult
Grasshopper Sparrow
juvenile
immature
adult
adult
Yellow-browed Sparrow
♀
♂
Tumbes Sparrow

Various finches of Andean grasslands (Grassland Yellow), montane forest (Slaty), xeric open areas (Sulphur-throated) and modified habitats (Saffron). Status of Wedge-tailed and Cinereous in Ecuador poorly understood. Sulphur-throated apparently engages in seasonal or erratic movements, whereas Saffron is spreading N, E and into Andean valleys.

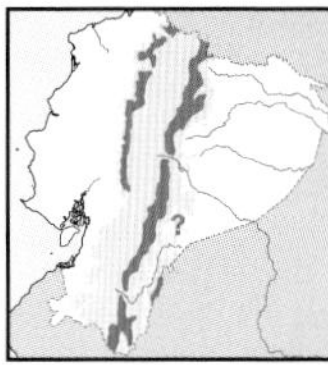

Slaty Finch *Haplospiza rustica* 12–13cm

Local in E & NW Andean subtropics to temperate zone, S in W to Chimborazo-Bolívar (1,500–3,400m). Montane forest, borders, bamboo. ***Conical pointed bill. Grey,*** darker wings. ♀: dull olive-brown above, ***buffy below coarsely streaked dusky***. Nominate. Singles or pairs rather shy but more conspicuous when joins flocks; locally numerous when extensive bamboo seeding. **Voice** Fast series of buzzy notes: *d'r-zzzz, d'r-zzzz'rr, dr'zzz-dr'zzzziir....* **SS** ♀ recalls other streaky finches and seedeaters, but nearly plain upperparts and short tail distinctive. **Status** Uncommon.

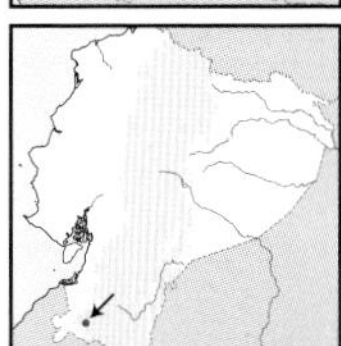

*** Cinereous Finch** *Piezorhina cinerea* 16–17cm

One record from SW foothills (1,100m). Desert-like shrubby areas. ***Thick yellow bill, pale yellow legs***. Black foreface, ***mostly pale grey, paler below***, whiter throat and belly. Monotypic. Unknown in Ecuador, possibly only wanderer; in Peru conspicuous and tame, feeds in small groups on ground. **Voice** Not heard in Ecuador, in Peru a pleasant, clear, whistled *chew-che-weeít-che'cheweeít-che'wir-cheweét...*; call an emphatic *chúp*. **SS** Confusion unlikely; Band-tailed Sierra-Finch has more slender bill, immaculate white belly, streaky back. **Status** Uncertain, possibly vagrant.

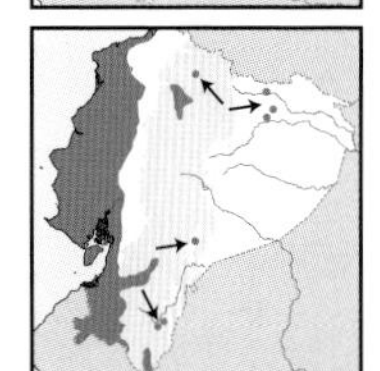

Saffron Finch *Sicalis flaveola* 13–14cm

SW lowlands to Andes, extreme SE in Zumba region, spreading in W lowlands, E lowlands, and N inter-Andean valleys (to 2,500m). Open areas. Mostly ***bright yellow, forecrown tinged orange*** in breeding ♂; bright olive-yellow upperparts. Juv. ***greyish-buff, yellower breast, neck-sides and rump***; streaked above. Race *valida*. Pairs to small family groups feed on ground, rather bold perching atop bushes or trees to sing. **Voice** Clear, semi-musical *dreé-dre'eé, tseet-dsee-it, dsee-it..dseetii-dseetii*; call a clear *tsíp, dz-chíp*. **SS** In Andes cf. streakier, smaller Grassland Yellow-Finch. **Status** Common.

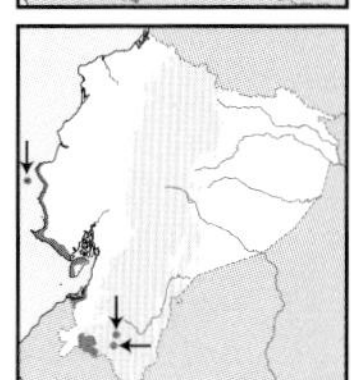

Collared Warbling-Finch *Poospiza hispaniolensis* 12–13cm

SW lowlands, La Plata Island, to subtropics in Loja (mostly below 300m, to 1,450m in Loja). Xeric scrub, woodland. Grey or yellowish bill. ***Grey above, bold white supercilium, black cheeks***, white throat. ***Incomplete black chest-band***, grey sides to flanks, white belly, ***white outer tail feathers***. Duller ♀ similar but ***greyish-brown streaked dusky***. Monotypic. Tame pairs to small groups on ground or in low bushes. **Voice** Far-carrying, semi-musical *swee-swee-s'keeuw* or *swee-swee-seeuu*, sometimes four-noted or faster; various emphatic ringing calls. **SS** See heavier Black-capped Sparrow (black head, darker grey upperparts, no white in tail). **Status** Fairly common.

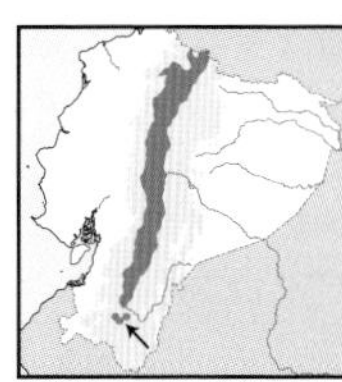

Grassland Yellow-Finch *Sicalis luteola* 12–13cm

Inter-Andean valleys and highlands S to N Loja (2,200–3,200m, locally to 1,300m). Pastures, fields, around lakes. ***Yellow supercilium and most underparts, brighter throat and belly***. Upperparts densely streaked blackish, ***plain dull yellow rump***. ♀ and juv. duller, especially on face. Race *bogotensis*. Small to fairly large groups on ground and in low bushes, apparently seasonal. **Voice** Fast buzzy series, somewhat ringing and trilled *tzz-tzz-tzzz-zz-zz-zzn..z'z'z'z, t-z'z'z'z'z'z*; call an emphatic *tzít-t'zít....* **SS** No other streaky finch or sparrow is so yellow; Saffron Finch larger and brighter. **Status** Uncommon.

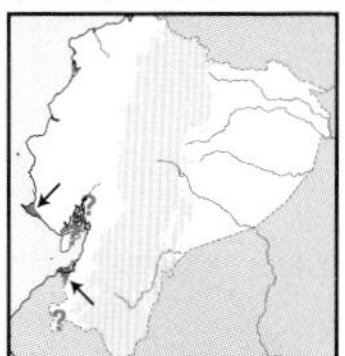

Sulphur-throated Finch *Sicalis taczanowskii* 12–13cm

Local along SW coast (to 50m). Bare sandy areas with scattered xeric shrubs. ***Massive pinkish bill.*** Mostly buffy-grey, some ***yellow in foreface and especially throat***, crown and upperparts striped dusky-brown, ***yellowish in tail base.*** Monotypic. Fairly large and compact flocks feed on bare ground; erratic or seasonal movements, might appear in fairly large numbers but disappears for protracted periods. **Voice** Very fast buzzy trill *t'wee'zzzz-dzzz-de-de-d'zzz....* **SS** Unmistakable by massive bill; see Parrot-billed Seedeater. **Status** Rare, possibly seasonal but declining, NT in Ecuador.

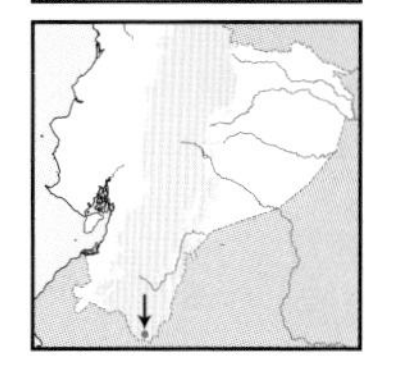

Wedge-tailed Grass-Finch *Emberizoides herbicola* 17–18cm

Extreme SE in Zumba region (1,150m). Open grassy fields near semi-deciduous woodland. ***Long, pointed, mostly yellow bill. Short whitish supercilium*** and narrow eye-ring. ***Greyish above densely streaked blackish. Very long pointed tail***. Race *apurensis*? Very poorly known in Ecuador where only recently found; elsewhere skulks in tall grass, not easy to see unless flushed or vocalising atop vegetation. **Voice** Semi-musical, clear *che-yeeu, che-wee-che-yeë, che'd-yeuw, chu-deëü....* **SS** Confusion unlikely in small range; tail unique among Ecuadorian emberizids. **Status** Rare, local.

♂
♀
Slaty Finch
Cinereous Finch
♂
♀
♀
juvenile
♂
juvenile
Saffron Finch
♀
♂
immature
♂
brown-backed
♂
grey-backed
Collared Warbling-Finch
juvenile
♀
♂
Grassland Yellow-Finch
Sulphur-throated Finch
Wedge-tailed Grass-Finch

Grassquits and *Sporophila* seedeaters often form multi-species feeding flocks. Bills of grassquits are slighter and more pointed than those of seedeaters, which are usually stubby. Marked sexual dimorphism in seedeaters. Bolder adult ♂♂ often easy to recognise (but cf. Grey and Slate-coloured), but ♀♀ and young ♂♂ are extremely difficult. Grassquits probably not closely related to seedeaters, despite similarity.

Slate-coloured Seedeater *Sporophila schistacea* 10–11cm

Local in N lowlands to subtropics (150–1,850m in NW, 200–400m in NE). Humid forest borders, adjacent shrubby clearings. ***Yellow bill. Mostly grey***, darker above, blacker wings with ***white patch,*** often ***white bar in coverts. White moustachial streak***, whiter central underparts. ♀: dull brownish-olive, paler below, ***dusky-horn bill***, some have white ***wing patch***. Races *incertae* (NW), *longipennis* (NE). Poorly known, apparently performs seasonal or erratic migrations associated with bamboo seeding; not with other seedeaters. **Voice** High-pitched, fast, unmusical bursts: *ch'ziit-zii'it-cheew-cheew..zíí-zíí-zíí..z-z-z...*, with sibilant whistles and short trills. **SS** Grey Seedeater lacks neck streak and wingbar (as do some Slate-coloured!), bill more swollen; ♀'s dusky-horn bill not easy to see; more forest-based and arboreal than most congeners. **Status** Rare, local, erratic?

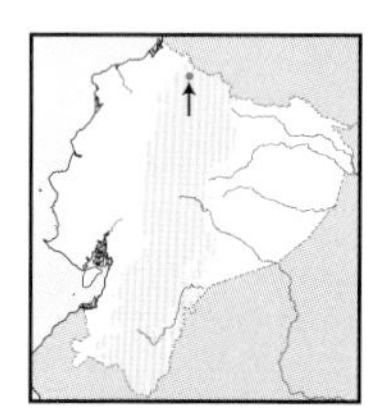

* Grey Seedeater *Sporophila intermedia* 10–11cm

Single record from dry Mira Valley (1000m). Roadside grass. Closely recalls rare Slate-coloured Seedeater but duller yellow bill ***more swollen, especially culmen, no*** white moustachial, ***paler*** grey above, ***no*** wingbar (but often has white speculum). ♀ not safely identified. Possibly race *bogotensis*. Recorded just once (J. Nilsson) with mixed seedeater flock, possibly an accidental visitor. **Voice** Not heard in Ecuador; in Colombia a fast series of twitters and trills, more musical than Slate-coloured Seedeater (HB). **SS** Slate-coloured Seedeater; ♀ indistinguishable from other *Sporophila* but note bill shape; see Ruddy-breasted and Yellow-bellied (both somewhat paler-billed, rustier). **Status** Uncertain.

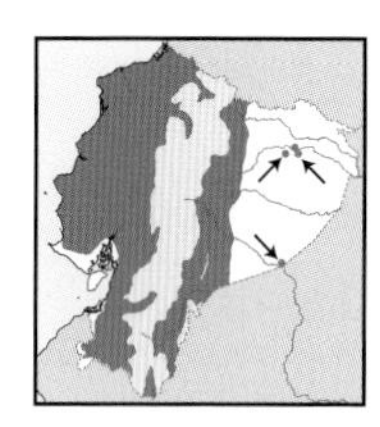

Yellow-bellied Seedeater *Sporophila nigricollis* 10–11cm

W lowlands to subtropics, inter-Andean valleys, E foothills to adjacent lowlands, extreme SE in Zumba region (to 2,400m, locally higher). Open pasture, fields, hedgerows, shrubby clearings. Dull black head to lower breast, ***olive-brown upperparts, pale yellow below***. ♀: much as other *Sporophila*. Races *vivida* (W), nominate (E), *inconspicua*? (Zumba region). Pairs to small flocks often with other seedeaters, sometimes in fair numbers where food abundant, probably moving up- and downslope seasonally. **Voice** Short musical twitters *tsee-tsee-chee-cheeoo-chee-pitiichii...*, often ending in buzzier notes; harsh, buzzy *j'iii*. **SS** Black-and-white Seedeater; ♀ rather rustier, best told by accompanying ♂; ♀ Variable yellower. **Status** Fairly common.

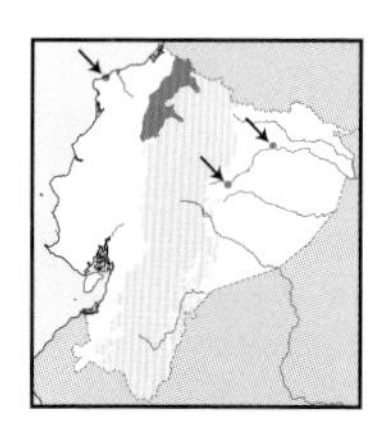

Yellow-faced Grassquit *Tiaris olivaceus* 9–10cm

NW foothills to subtropics, few records in NE and inter-Andean valleys, once on NW coast (mainly 600–1,800m, locally higher; 300–400m in E). Roadsides, pastures, shrubby clearings. Olive upperparts, blacker median underparts, ***bold yellow face and throat***. ♀ and juv. duller below, with less yellow in face. Race *pusillus*. Singles to small flocks forage low in grass, often with other seed-eaters, sometimes in low canopy of bushes; hangs from grassy stems, bending them down. **Voice** Weak, short *tir'r'rr'rr* or *te'eeeee* trill. **SS** Confusion unlikely; face pattern evident even in dull birds. **Status** Uncommon, but increasing?

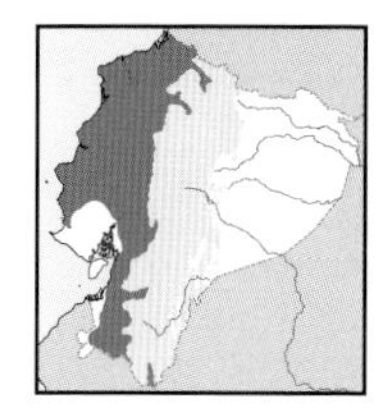

Dull-coloured Grassquit *Tiaris obscurus* 9–10cm

W lowlands to lower subtropics, locally in inter-Andean valleys, extreme SE in Zumba region (to 1,400–1,500m, locally to 2,400m). Shrubby clearings, bamboo, hedgerows, wooded borders. ***Short, bicoloured bill. Dull*** greyish-olive to dusky-olive above, paler below, whiter central belly. Sexes alike. Race *pauper*. Singles to small parties, sometimes with other seed-eaters though more skulking, tends to remain in cover. **Voice** Dry, explosive buzzy *zzee'iig-zzee'iig...zee'diig...zee'yeee'e*. **SS** Closely recalls several ♀ *Sporophila* (Slate-coloured, Black-and-white, Yellow-bellied, Lesson's, Lined), but shorter, less stubby, dusky-and-yellowish bill distinctive; easily told by buzzy, unmusical voice. **Status** Uncommon.

♀
juvenile
♂
intermediate
normal (grey)
morph
♂
old adult
Slate-coloured Seedeater
♀
♂
immature
♂
Grey Seedeater
♂
♂
♂
vivida
nigricollis
inconspicua
♂
immature
♀
Yellow-bellied Seedeater
♂
♂
♀
Yellow-faced Grassquit
Dull-coloured Grassquit

Four lowland seedeaters, one ranging to subtropics and Andean valleys, and the distinctive Blue-black Grassquit. ♂ seedeaters are readily identified but are outnumbered by ♀♀ and young ♂♂, which can be trickier. Some ♀-plumaged seedeaters on this plate are more easily recognised than those on next and previous plates, by bill size (Parrot-billed), banded wings (Drab) or streaky back (Chestnut-throated).

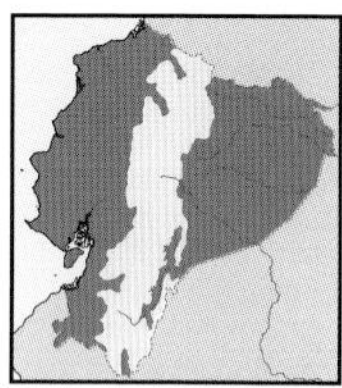

Blue-black Grassquit *Volatinia jacarina* 10–11cm

E & W lowlands to subtropics, in E not in lowest areas; extreme SE in Zumba region (to 1,800m, locally higher). Pastures, crops, shrubby clearings, gardens. Sharp *pointed bill. Glossy blue-black.* ♀: dull brown above, whitish below *coarsely streaked dull brown over chest to flanks.* Juv. ♂ splotched blue-black. Race *splendens* (E), *peruviensis* (W). Singles to small groups flock with other seed-eaters, inspecting grass stems. ♂ engages in characteristic 'jumping' display. **Voice** Dry, insect-like *d'zzzzir*; dry *jít*. **SS** Confusion unlikely by bill shape, streaky underparts (♀). **Status** Common.

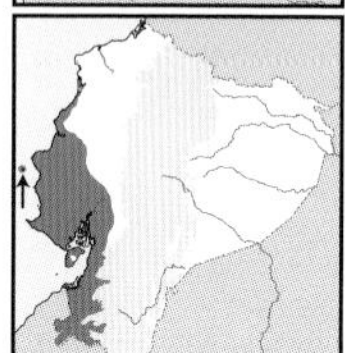

Parrot-billed Seedeater *Sporophila peruviana* 11–12cm

SW lowlands to foothills (to 800m, locally to 1,200m in Loja). Shrubby pastures, roadsides, arid scrub. *Pale yellow swollen bill*, duskier in ♀. Dull grey above, blacker face, *black throat, white moustachial.* Wing has *white speculum*, often *two white wingbars.* ♀ whitish-buff below, whitish speculum and *two wingbars.* Race *devronis.* Small to fairly large flocks with other seedeaters, apparently seasonal. **Voice** Harsh series of spaced *jueét* notes followed by 2–3 faster, shorter *je-je-je* or simpler *jué*; also shorter *juee-jee-jee-juú.* **SS** Bill unique among seedeaters; cf. smaller-billed ♀ Chestnut-throated and Drab Seedeaters. **Status** Uncommon.

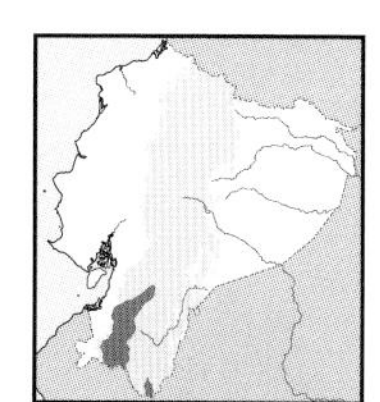

Drab Seedeater *Sporophila simplex* 10–11cm

SW subtropics to inter-Andean valleys, locally in Zumba region (mostly 650–1,900m). Xeric scrub, adjacent shrubby areas, fields. Small, dusky-horn bill. Dull greyish-brown above, buffier below, paler belly, *whitish speculum* and *two wingbars* (*buffier* in ♀). Monotypic. Pairs to small parties often with other seedeaters, remain low in cover, also climb to scrub canopy to search foliage. **Voice** Short, harsh rolling *d'd-djee-djee-djee-dzee-djee-dzee-dzee-djee...*, sometimes with sweeter *zee* notes first. **SS** Note bill shape in Parrot-billed; Chestnut-throated is streaked above (no overlap?); in Zumba region see Black-and-white and Yellow-bellied (no wingbars). **Status** Rare, possibly erratic.

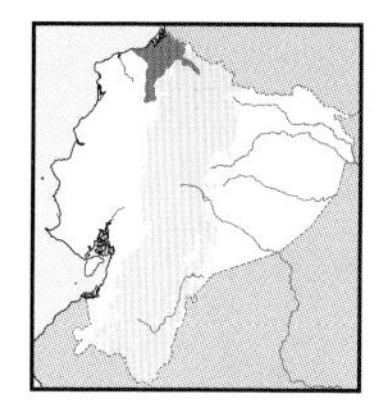

Ruddy-breasted Seedeater *Sporophila minuta* 10–11cm

NW lowlands to inter-Andean valleys (to 1,000m). Young second growth, shrubby clearings. Dull *greyish above, rusty below and on rump*, duskier wings edged buff, *white speculum.* ♀: pale buffy-brown, duller above, *duskier wings edged buff, flesh to pale horn bill.* Nominate. Small flocks with other seedeaters at roadsides and other open and semi-open grassy areas, fast-moving and nervous. **Voice** Sweet warbled *pt-zee'ee'eet, pt-zee'ee'eet, p'tzeer...*often starting with 2–4 clear *chét-chét.* **SS** ♂ unmistakable; cf. ♀ Yellow-bellied and Variable (both lack buff wing edges and white speculum, bills somewhat darker; Variable yellower overall). **Status** Rare, local.

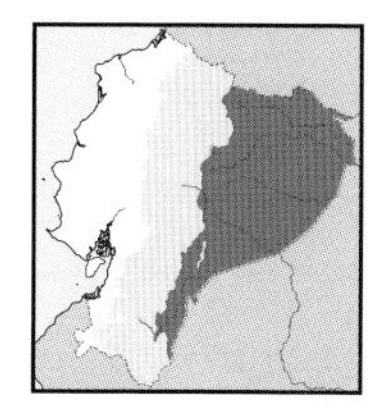

Chestnut-bellied Seedeater *Sporophila castaneiventris* 10–11cm

E lowlands to upper foothills (to 1,300m, locally higher). Roadsides, shrubby clearings, fields, river and lake edges. *Slate-grey above*, darker wings and tail, *chestnut below.* ♀: dull greyish-brown, tawnier below, paler throat and central belly, *bill dusky-horn.* Monotypic. Like other seedeaters, often in groups with congeners; rather tame and eye-catching. **Voice** Sweet warbling, rather slow *cheew-cheew-t'cheeoo-t'chee-t't't'trrr...cheeo-ched-ched...*; call clear *déeu.* **SS** No overlap with Ruddy-breasted; ♀ resembles other *Sporophila* but browner (less dusky); ♀ Caqueta yellower overall, with blacker bill; see also Lined and Lesson's ♀♀. **Status** Common.

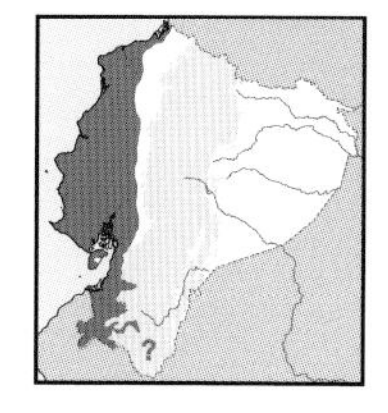

Chestnut-throated Seedeater *Sporophila telasco* 10–11cm

W lowlands, local to subtropics in Loja and El Oro (mainly below 500m, locally to 1,400m). Grassy areas, shrubby clearings, roadsides, fields, sugarcane crops. *Grey and streaky above, white in rump* and *underparts, small chestnut throat* (lacking in non-breeders). ♀: *horn bill*, buffy-brown above, whiter-buff below, *upperparts coarsely streaked dusky, fainter streaks on breast to flanks.* Monotypic. Small to fairly large groups with other seedeaters; conspicuous and noisy when feeding. **Voice** Sweet staccato warble, *t'chew-cheey-chee-weed-chew...*, sometimes shorter, or ends in brief sputter. **SS** Similar ♀ Parrot-billed has heavier, swollen bill. Drab also unstreaked. **Status** Common.

♂
breeding
♂
after
postnuptial
moult
♀
♂
juvenile
Blue-black Grassquit
♂
♀
juvenile
Parrot-billed Seedeater
♀
♂
Drab Seedeater
♂
brown type
♂
grey type
♂
first-year
♀
juvenile
Ruddy-breasted Seedeater
♂
♂
second-year
♂
first-year
♀
juvenile
Chestnut-bellied Seedeater
♂
♂
first-year
♀
juvenile
Chestnut-throated Seedeater

Black-and-white ♂♂ are readily identified but be careful with the migratory Lesson's and Lined of E lowlands. ♀-plumaged birds – which outnumber ♂♂ – are much more difficult. Often, the only way to recognise them is by their accompanying adult ♂. ♂♂ apparently take several years to acquire adult plumage. Variable is among the commonest grassland birds in W.

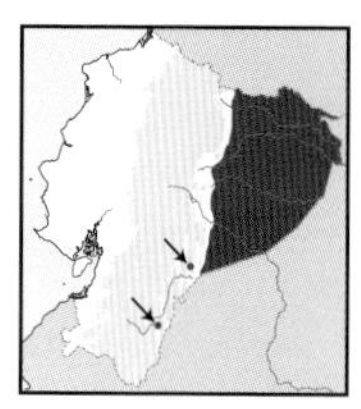

Lesson's Seedeater *Sporophila bouvronides* — 11–12cm

Intra-tropical migrant to E lowlands (mostly below 400m). Pastures, roadsides, grassy river and lake edges. *Uniform black above*, whitish rump. *Black throat, white malar,* whitish underparts, white speculum. ♀: *bicoloured bill*, olive-brown above, more yellowish below, whiter belly. Race *restricta*. Small flocks within larger aggregations of seedeaters, including similar Lined; migrates from breeding grounds further E in Amazonia. **Voice** Not recorded in Ecuador; elsewhere a harsh, fast, almost trilled *chaaw-cheee-chea-chea*...or *chee-didididi*.... **SS** Lacks white coronal stripe of otherwise similar Lined Seedeater; ♀ not safely recognised unless accompanying ♂ seen. **Status** Rare visitor (mostly Nov–Apr, several anomalous Jun and Aug records).

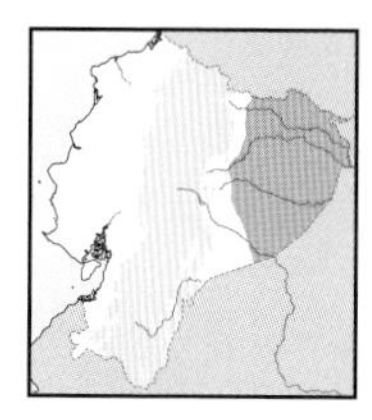

Lined Seedeater *Sporophila lineola* — 11–12cm

Local austral visitor to E lowlands (below 400m). Pastures, roadsides, grassy river and lake edges. Resembles previous species but ♂ has *white coronal stripe* and is *whiter below*. ♀: essentially identical. Monotypic. Singles to small parties with other seedeaters in tall grass; apparently sympatric with Lesson's only for a few months each year. **Voice** Not recorded in Ecuador; in Brazil a fast, metallic, variable warbling. **SS** White coronal stripe separates from very similar Lesson's, though not always easy to see. See also ♀ of rarer (in range) Black-and-white and commoner Caqueta Seedeaters. **Status** Uncertain, possibly casual transient or austral migrant resident (Aug–Dec, Feb).

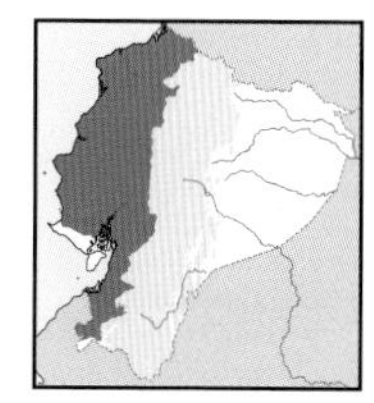

Variable Seedeater *Sporophila corvina* — 11–12cm

W lowlands to subtropics, not in arid lowlands (to 1,500m). Pastures, agricultural fields, parks, gardens, shrubby and grassy roadsides. *Black upperparts, breast-band* and tail, whitish rump, *white throat, partial collar, lower underparts and speculum*. ♀: *yellowish-olive to buffy-brown*, paler belly, dusky bill. Race *ophthalmica*. Small to fairly large flocks, outnumbering other seedeaters in mixed groups. **Voice** Long, rather variable, musical twittering *t'cheew-t'chee-t'chee-t'bzzz..cheew-cheerr-cheert*...; call clear *cheeu*. **SS** ♀ more yellowish-olive than Yellow-bellied and Slate-coloured; see bicoloured bill of Dull-coloured Grassquit. Other ♀ seedeaters in range show buff wing edges (Ruddy-breasted) or streaky upperparts (Chestnut-throated). **Status** Common.

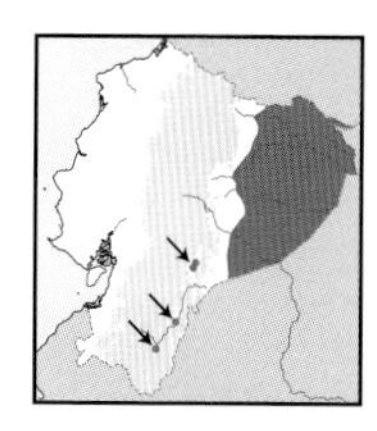

Caqueta Seedeater *Sporophila murallae* — 11–12cm

E lowlands, local in foothills (mainly below 400m, locally to 900m). Grassy roadsides, damp pastures, shrubby fields at river and lake edges. *Black upperparts*, whitish rump. *White below to partial collar, broken black breast-band*. Black wings, vague *white wingbars and speculum*. ♀: indistinguishable from Variable Seedeater. Monotypic. Small flocks wherein ♀-plumaged birds outnumber ♂♂; feeds on grasses but also in flowering trees. **Voice** Rapid, melodic, long jumbled series, with somewhat staccato quality, often starting with buzzy *tzziir-tzii-tzii-tzziir*..., phrases often repeated. **SS** ♂ Lesson's and Lined lack breast-band and show white malar streak, Black-and-white has simpler pattern. ♀ is more yellowish and darker-billed than sympatric seedeaters. **Status** Uncommon.

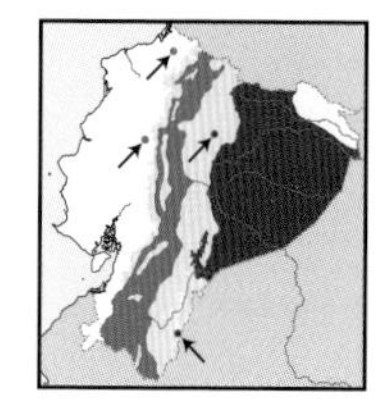

Black-and-white Seedeater *Sporophila luctuosa* — 10–11cm

Subtropics to inter-Andean valleys, also locally / seasonally in NW & E lowlands to foothills (to 1,200m in lowlands; to 2,400m in Andes). Roadsides, shrubby clearings, fields. *Black upperparts and large hood, white belly to vent and speculum; bluish bill*. ♀ olive-brown, paler throat and belly. Monotypic. Pairs to largish flocks with other seedeaters in tall grass, ♀-plumaged birds more numerous; apparently moves seasonally to E lowlands. **Voice** Fast, harsh *td't't't't-tseeu-tseed'pitirrt-pitirrt*, often ending in brief twitter, others starting with euphonia-like ringing *theé, theé-to*. **SS** ♀ not safely identified from Yellow-bellied (♂ yellower below, more olive above, no speculum), and locally sympatric Lesson's, Lined and others; ♀ Variable and Caqueta rather yellower. See Dull-coloured Grassquit. **Status** Uncommon, only visitor to lowlands.

Lesson's Seedeater
♂
♂
immature
♀
juvenile
♀
♂
♂
immature
juvenile
Lined Seedeater
♂
♀
Variable Seedeater
♂
♂
second-year
♀
juvenile
♂
♀
Caqueta Seedeater
Black-and-white Seedeater

PLATE 277: SEED-FINCHES AND SEEDEATERS

Seed-finches are larger than other seedeaters, forming smaller flocks. Bills huge and square-looking in two species and rather heavy in the other two. Identification complex between the larger species pair. *Catamenia* are small, attractive Andean seedeaters that form small flocks. Attractive ♂♂ have coloured bills; ♀♀ are streaky and often quite confusing. Broad range and habitat overlap.

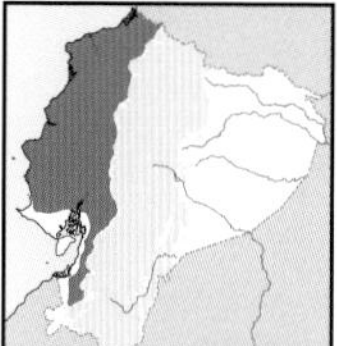

Thick-billed Seed-Finch *Sporophila funerea* 12–13cm

W lowlands to foothills (to 1,200–1,300m). Humid shrubby clearings, wooded borders. ***Heavy blackish bill. Glossy black***, white wing spot (not always visible), white underwing-coverts. ♀: ***brown above, fulvous-brown below***, bill and underwing as ♂. Monotypic. Singles or pairs often with other seedeaters, but not tied to grassy areas. **Voice** Sweet, jumbled, deliberate series: *t'dee-cheeuu-t'cheew-t'dee-cheew-tititi*.... **SS** ♀ smaller seedeaters, most being duller, all with smaller bills. See also ♀ Blue Seedeater (smaller bill) and larger seed-finches with much heavier bills. **Status** Uncommon.

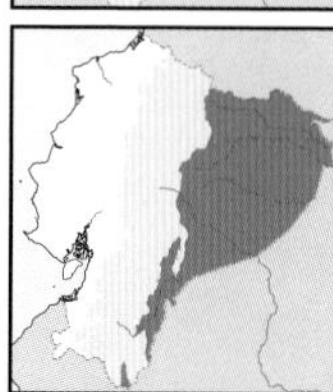

Chestnut-bellied Seed-Finch *Sporophila angolensis* 12–13cm

E lowlands to foothills, extreme SE in Zumba region (to 1,200–1,300m). Humid shrubby clearings, wooded borders. ***Heavy blackish bill. Glossy black upperparts to breast, chestnut belly***, white wing spot, white underwing-coverts. ♀: like previous species (formerly conspecific) but somewhat darker. Race *torridus*. Singles to small loose flocks near ground, sometimes in open with other seedeaters. **Voice** Sweet jumbled series of twitters, interspersed by trills: *t'deeu-cheew-cheew-chee-chee-ti-ti-t'deeu-chee-cheew*.... **SS** ♀ smaller seedeaters are duller, smaller-billed. Black-billed larger, bill much thicker, uniform black below; see Large-billed. **Status** Uncommon.

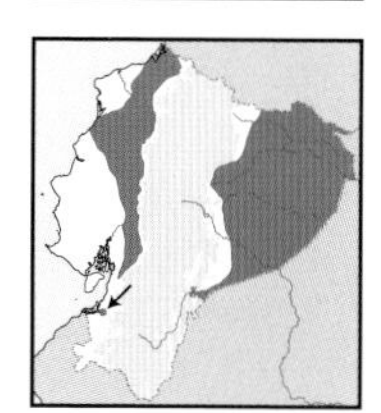

Large-billed Seed-Finch *Sporophila crassirostris* 14–15cm

Sparse in E & W lowlands to foothills (to 900–1,000m). Damp grasslands, lake edges, roadsides. Massive ***whitish bill. Black, with bold white wing spot***, white underwing-coverts. ♀: ***larger*** than Thick-billed / Chestnut-bellied Seed-Finches, ***massive dark bill***. Races *occidentalis* (W), nominate (E). Singles or pairs sometimes with other seedeaters; apparently seasonal movements. **Voice** Jumbled, running series of twitters and short trills: *td-ch'ch'ch-deewee-deewee-d'd'd'd*.... **SS** In E see Black-billed; somewhat paler ♀ Large-billed has smaller bill; best told by accompanying ♂! See more forest-based ♀ Blue-black Grosbeak. **Status** Rare, local; NT globally.

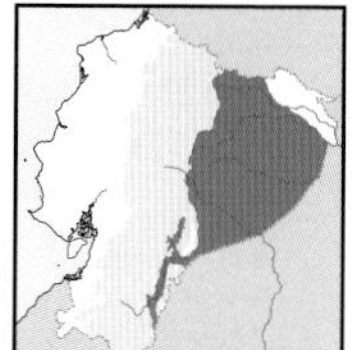

Black-billed Seed-Finch *Sporophila atrirostris* 16–17cm

E lowlands to lower foothills (to 1,350m). Damp grass, lake edges. ***Massive blackish bill. Black with white wing spot***, white underwing-coverts. ♀: ***darker, heavier-billed*** than previous species. Nominate. Singles or pairs near ground, sometimes with other seedeaters; apparently moves seasonally. **Voice** Rich accelerating warble: *terr-twee-twee-terrr..dü-twee-twee..dü-twee-twee-wheet-wheet*...; call a clear *cheúd*. **SS** Bill size difference from Large-billed (♀) not always easy to appreciate; ♀ Blue-black Grosbeak (forest-based); smaller Chestnut-bellied Seed-Finch (smaller bill). **Status** Rare, local.

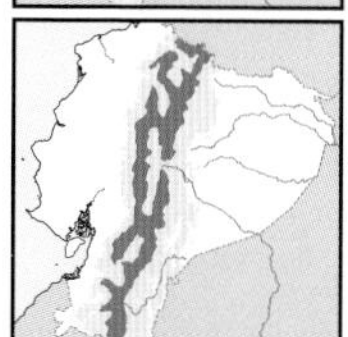

Plain-coloured Seedeater *Catamenia inornata* 13–14cm

Andes, inter-Andean valleys (2,600–4,500m). Open grassy fields and *páramo*. ***Pink bill. Mostly grey, mantle faintly streaked blackish***, belly washed pale buff, ***unmarked*** grey tail. ♀: brown above, ***yellowish-buff*** below, ***coarsely streaked dusky*** except belly, pale bill, pale rufescent vent. Race *minor*. Pairs to small groups with other seedeaters, on grass stems or on ground. **Voice** Variable, long buzzy trill starting with fast rattle: *chi'titi-trriii-ziiiiii-bzzz*. **SS** ♂ paler than Páramo Seedeater, ♀ paler and streakier; paler Band-tailed ♀ has white tail-band. **Status** Common.

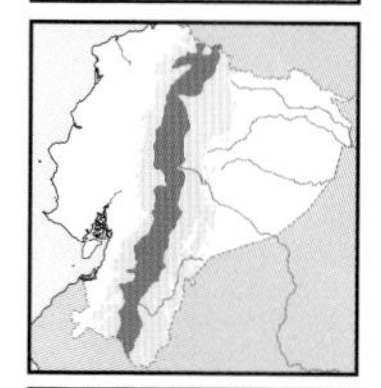

Band-tailed Seedeater *Catamenia analis* 13–14cm

Andes, inter-Andean valleys (1,500–3,000m). Open grassy fields, crops, parks. ***Yellow bill***. Mostly ***slate-grey, blacker face***, whiter belly, ***white band on median tail***. ♀: ***darker bill***, greyish-brown above, buffier below, streaked dusky, ***whitish belly, buff vent***. Race *soderstromi*. Pairs to small groups sometimes with other seedeaters, not skulking, regularly on ground, also perch atop bushes. **Voice** Short buzzy trill *pr'sisisisi*. **SS** From congeners by white tail-band, ♀ is paler. **Status** Fairly common.

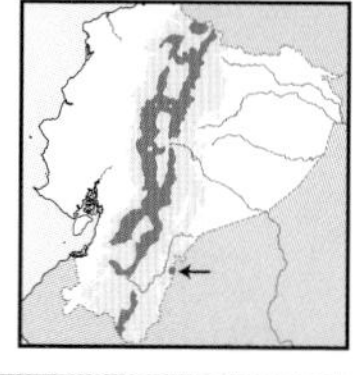

Páramo Seedeater *Catamenia homochroa* 13–14cm

Andes (2,500–3,500m, locally higher). Humid shrubbery, *páramo*, treeline. ***Pale yellow to pinkish bill. Entirely coal-grey***, blacker face. ♀: ***pinkish bill***; dark brown, paler belly, pale rufescent vent, dusky streaks above, ***faintly streaked to unstreaked below***. Nominate. Pairs to small flocks rather skulking, often in bamboo. **Voice** Prolonged whistle *theeeeeeëë*, followed by shorter, harsher, higher-pitched *dzeeé*, at well-spaced intervals. **SS** Darker than congeners, bill very pale in ♂, brownish-pink in ♀ (more conical than congeners); ♀ darkest *Catamenia*. **Status** Uncommon, rather local.

♂
♂
Chestnut-bellied Seed-Finch
Thick-billed Seed-Finch
♀
♀
♂
juvenile
Large-billed Seed-Finch
juvenile
♀
♂
Black-billed Seed-Finch
♂
♀
juvenile
Plain-coloured Seedeater
♂
fresh
♂
worn
♀
juvenile
Band-tailed Seedeater
♂
breeding
♀
juvenile
Páramo Seedeater

Attractive sparrows found in forests, woodland, borders and adjacent clearings, where they forage on or near ground, rarely with mixed flocks. Also, three Andean species that were only recently lumped into genus *Arremon*, also confined to dense forest cover.

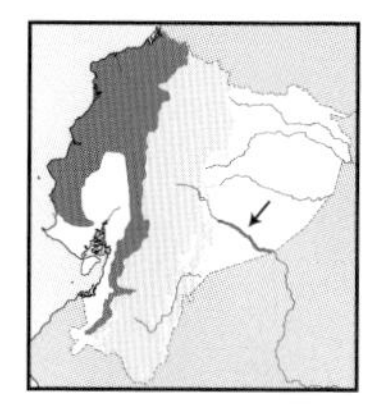

Black-striped Sparrow *Arremonops conirostris* 16–17cm

W lowlands to foothills, also along Pastaza and Palora Rivers (to 1,400m, 180–850m in SE). Woodland, borders, shrubby clearings, *Tessaria* riverine scrub. *Grey head* (*darker* in SE), *bold black crown- and face-stripes*, olive upperparts (*greyer* in SE), *pale grey underparts*, yellow wing bend. Races *striaticeps* (W), *pastazae* (SE), possibly valid species. Singles or pairs on ground or in low bushes, locally numerous in semi-open. **Voice** Clear, whistled *tuuee-chö, tuee-chö, whee-chö, whee-chö*; running *cho.. cho.cho-cho-chochocho*...(W); higher-pitched *tuuee-cheey, tueee-cheey, tueey-chee'chee'chee*... in SE. **SS** Black-capped Sparrow (white throat and supercilium, black breast-band); see Grey-browed Brush-Finch. **Status** Uncommon.

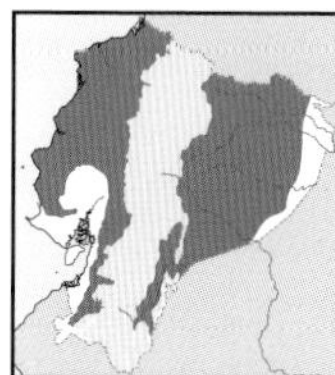

Orange-billed Sparrow *Arremon aurantiirostris* 15–16cm

E & W lowlands to foothills; in E not far from Andes (to 1,100–1,200m). Humid forest, borders, mature woodland. *Bright orange bill.* Black head, *bold whitish supercilium*, white throat, *black breast-band.* Olive above, whitish below, greyish sides, yellow (W) or orange (E) wing bend. Races *spectabilis* (E), *occidentalis* (most of W), *santarosae* (SW). Mostly in pairs near ground, sometimes in open. **Voice** Very high-pitched, jumbled fast *ts-tsu-tsi-tsitsitsisi* (W); buzzy, slower *t'tzeeee-zeee-t'tzeeee* (E); calls sharp *zít.* **SS** Unmistakable, but see Black-striped Sparrow. **Status** Uncommon.

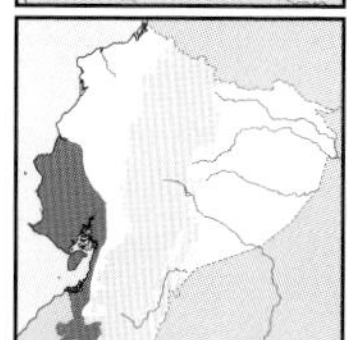

Black-capped Sparrow *Arremon abeillei* 15–16cm

SW lowlands, local in SW subtropics; extreme SE in Zumba region (below 800m, locally to 1,600m). Deciduous to semi-deciduous forest, woodland, shrubby clearings. *Black bill.* Black head, *bold white supercilium*, white throat, *black breast-band. Grey* (SW) or *olive* (SE) above, white below. Nominate (SW), *nigriceps* (SE). Singles or pairs skulk in dense undergrowth. **Voice** High-pitched, fast, short *tssu-tsu-ti-ti, tsu-tsu-tipi, tsu-tsee-tseét-ti* (SW); in SE slower *tsee-tsee-ts'ts'ts'ts, tse-tse-sisisi*; call harsh *tsíí-tsí-tsíp*...**SS** Grey-browed Brush-Finch (no breast-band), Orange-billed Sparrow (orange bill), Collared Warbling-Finch (grey head). **Status** Uncommon.

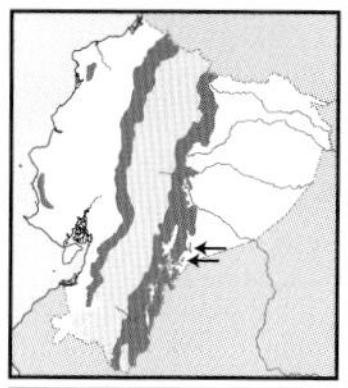

Chestnut-capped Brush-Finch *Arremon brunneinucha* 18–19cm

E & W Andean foothills to subtropics, S in W to N El Oro; atop coastal cordilleras (mainly 700–2,500m, locally higher; 400–500m in coastal ranges). Montane forest, borders, woodland. *Black face, chestnut crown*, white supraloral, *white throat, narrow black breast-band* (lacking in SW). Olive above, whitish below. Races *frontalis* (most of range), *inornatus* (SW). Singles or pairs in forest cover, skulking, forages mostly on ground. **Voice** Very high-pitched *tseé-tse-sweé-suéép, dseé-du-zéé-zéép, didí-di-típ-típ*, calls thin *tzzz* or *ziii.* **SS** Confusion unlikely. **Status** Fairly common.

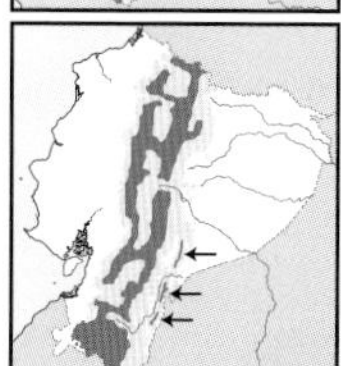

Grey-browed Brush-Finch *Arremon assimilis* 18–19cm

Andean subtropics to temperate zone (N: 1,900–3,500m; S: 900–3,000m). Montane forest, borders, mature woodland. Head *broadly striped dark grey and black*, olive upperparts, white below, greyish-olive sides to flanks, yellow wing bend. Nominate (E, NW), *nigrifrons* (SW). Singles or pairs in forest cover, shy, partially terrestrial, flicks leaves aside; seldom with flocks. **Voice** High-pitched short *tsue-tseé-tsue..tw'zeeu*...; also slower *tsuee-su-seék* or *tsueé-seék, tzuué-zeé-zuweé*...; call thin *tsí.* **SS** Lacks chestnut crown of Chestnut-capped Brush-Finch, see Black-striped and Black-capped Sparrows. **Status** Uncommon.

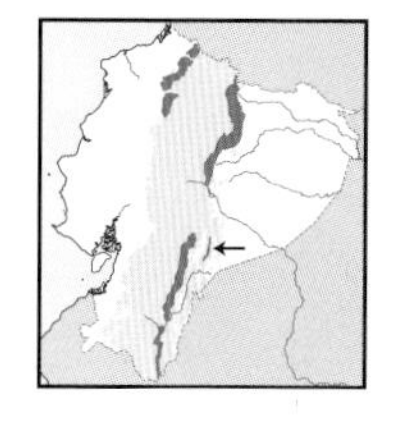

Olive Finch *Arremon castaneiceps* 15–16cm

Local in E & NW Andean foothills to subtropics (800–1,800m). Montane forest undergrowth. *Rufous-chestnut crown, freckled grey face; dark olive*, slightly paler below. Monotypic. Skulking, singles or pairs in dense tangles, often near streams and in ravines, rarely joins mixed flocks; sometimes rather tame. **Voice** Very high-pitched, thin, unmusical *tsee-tsee-ts-ts-tsee-see-sisisi-tsee*...; calls shorter *ts-ts-sisi-ts-ts* or *t-tssss* trill. **SS** Pattern resembles White-rimmed Brush-Finch but lacks white around eyes; no white underparts like other chestnut-capped finches. **Status** Rare, overlooked? NT globally.

juvenile
immature
adult
pastazae
adult
striaticeps
Black-striped Sparrow
juvenile
♂
spectabilis
♂
santarosae
♂
occidentalis
Orange-billed Sparrow
abeillei
nigriceps
Black-capped Sparrow
inornatus
adult
frontalis
adult
juvenile
Chestnut-capped Brush-Finch
immature
adult
adult
nigrifrons
assimilis
rey-browed Brush-Finch
juvenile
juvenile
adult
Olive Finch

PLATE 279: BRUSH-FINCHES

Brush-finches occur in Andean forests and montane scrub. They move in small family parties, often also with flocks. The yellow-bellied species replace each other altitudinally (Yellow-breasted and Pale-naped in E; Yellow-breasted and Tricoloured in W), whereas the grey-bellied do not overlap.

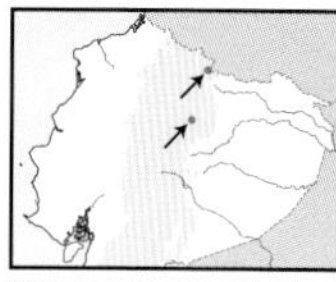

White-naped Brush-Finch *Atlapetes albinucha* 17–18cm

Few records from shrubby areas in extreme NE (2,100–2,450 m). Crown to nape ***entirely white***, underparts mostly ***dirty white*** except ***yellow throat***. Race *gutturalis*, sometimes treated as separate species. Forages low, rarely on ground, in pairs or family groups; nervously pumping tail. **Voice** Weak, thin *o see mee, o see...* **SS** Confusion unlikely. **Status** Uncertain, possibly recent arrival, locally uncommon.

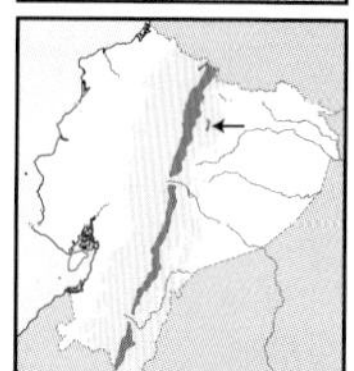

Pale-naped Brush-Finch *Atlapetes pallidinucha* 17–18cm

Temperate zone to *páramo* in E Andes (2,700–3,700m). Montane forest borders, secondary woodland, treeline, *páramo* 'islands'. Pale cinnamon front, ***white mid crown to nape***. Black face, blackish upperparts, yellow below tinged greyish-olive on flanks and sides. Race *papallactae*. Noisy pairs to small parties at various heights, including canopy, briefly in open, often with mixed flocks. **Voice** Clear, slightly accelerating *tsee-tsee-teet-teet-swee*, also simpler *swee-swee-ts'ts'ts*, *swee-swee-wededede...*, *whee-ee-tew-tew-tew*; calls sharp *zíí-zíí*, thin *sip-sip....* **SS** Locally sympatric Yellow-breasted Brush-Finch (no white in crown, brighter underparts). **Status** Fairly common.

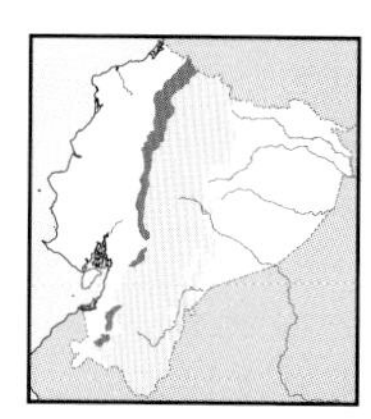

Tricoloured Brush-Finch *Atlapetes tricolor* 17–18cm

W Andean foothills to subtropics, spottily S (600–2,300m, locally at 3,100m). Montane forest, secondary woodland, borders. ***Fulvous-brown crown to nape***, black mask, dark ***olive-brown upperparts***, dull yellow underparts, sides washed greyish-olive. Race *crassus*, possibly separate species. Pairs to small groups in lower to midstorey, sometimes briefly in open, often with mixed flocks. **Voice** Clear, high-pitched, wheezy *teeu-teé-tee-tee*, *teéu-tsi-tititi...*; first note downslurred, slightly accelerating at end; call thin *zít-zít*. **SS** Might locally overlap with Yellow-breasted Brush-Finch (richer yellow underparts, richer rufous crown, blacker upperparts). **Status** Uncommon.

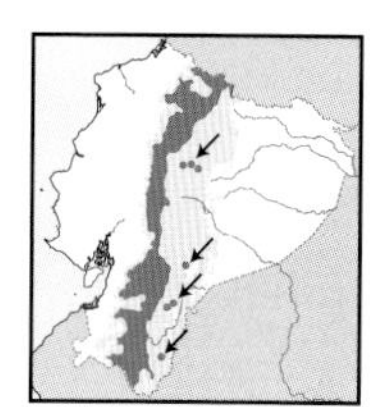

Yellow-breasted Brush-Finch *Atlapetes latinuchus* 17–18cm

Andes, slopes, inter-Andean valleys (1,500–3,200m). Montane forest borders, woodland, shrubland, hedgerows. ***Rufous crown, blackish face and upperparts, yellow below***. S races have ***black to faint black submalar stripe***; conspicuous ***white speculum*** in SE; yellow loral spot in SW. Race *spodionotus* (inter-Andean valleys and W Andes S to Chimborazo), *comptus* (W Andes from Chimborazo S), nominate (E Andes from Azuay S). Pairs to small groups mostly apart from mixed flocks, noisy, easy to see. **Voice** Fast, pretty *twee-tuee-fee-fee-fefefé...*, *twee-tuee-fefefe*; also more complex *twee-tee-tee-f'f'f'f-teew-teew...* starting with short trill; call thin *tít*. **SS** Pale-naped (E) and Tricoloured (W). **Status** Common.

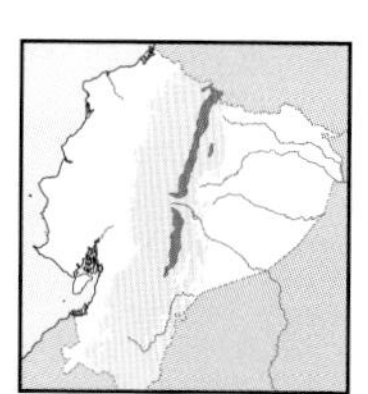

Slaty Brush-Finch *Atlapetes schistaceus* 17–18cm

E temperate zone to *páramo*, S to Río Paute (2,500–3,400m, locally lower). Montane forest borders, shrubbery, treeline, *páramo* 'islands'. ***Chestnut crown, black moustachial, white loral spot, malar stripe and throat. Dark grey above***, white speculum. Nominate. Pairs to small groups join mixed flocks, prefers to remain in cover, emerges briefly. **Voice** Sweet *weeye-chuwít*, or *wee..cheyee-cheyee*, *chewee-chewee-chey'chey'chey*; call thin *tzít*. **SS** No overlap with White-winged and Bay-crowned Brush-Finches; from sympatric Yellow-breasted and Pale-naped by underparts. **Status** Uncommon.

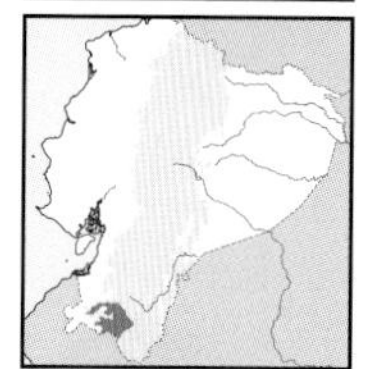

Bay-crowned Brush-Finch *Atlapetes seebohmi* 16–17cm

S subtropics (1,300–2,300m). Montane forest borders, woodland, shrubby clearings. ***Rufous-chestnut crown to nape, black moustachial***, whitish loral spot. ***Whitish underparts***, grey upperparts. Race *simonsi*. Pairs to small groups mostly in cover, sometimes with mixed flocks; skulking. **Voice** Accelerating series of thin *sít-sít..tsya-tsya-tsy-tsy'tsy'tsy*, *tseet-tseet-see-see'see'seeseesee*, ending in brief trill; call fast *sít-sít-si-si....* **SS** No overlap with Slaty Brush-Finch (white speculum); White-winged (speculum, more cinnamon crown); Chestnut-capped (black collar, no malar). **Status** Rare.

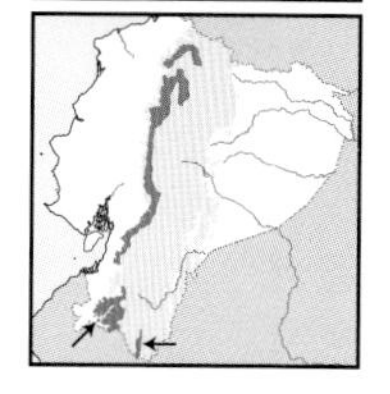

White-winged Brush-Finch *Atlapetes leucopterus* 15–16cm

Inter-Andean valleys, W Andean subtropics, extreme SE (1,000–2,600, locally to 3,200m). Arid scrub, woodland, likely spreading. ***Cinnamon-rufous crown***, whitish loral spot. Blackish above, ***dirty white below, white speculum. Variable white in face*** (S). Nominate (S to Azuay), *dresseri* (Loja and El Oro), *paynteri* (extreme SE). Pairs to small parties often with mixed flocks in various strata. **Voice** Jumbled *td-weé-td-dee-dee-weededede*, *te-te-te-wee'dededede*; call thin *tseé*. **SS** Bay-crowned (richer crown, no speculum, darker); White-headed (white face, black rear crown). **Status** Uncommon.

adult
Pale-naped Brush-Finch
juvenile
juvenile
adult
White-naped Brush-Finch
Tricoloured Brush-Finch
juvenile
adult
juvenile
latinuchus
adult
comptus
adult
spodionotus
adult
Yellow-breasted Brush-Finch
adult
juvenile
Slaty Brush-Finch
Bay-crowned Brush-Finch
juvenile
paynteri
dresseri
dresseri
leucopterus
White-winged Brush-Finch

Three brush-finches distinctive by plumage and behaviour (White-rimmed) or habitat (dry forests, woodland and scrub). The endemic Pale-headed is a conservation icon in mainland Ecuador. Tanager Finch is monotypic, shy but very distinctive. Also, two distinctive red-and-black finches of deciduous and semi-deciduous habitats in Tumbesian lowlands and Marañón Valley. Crimson-breasted performs local migrations.

Tanager Finch *Oreothraupis arremonops* 20–21cm

NW foothills to subtropics (1,300–2,300m). Montane forest undergrowth and borders. ***Black head with bold greyish-white superciliary and coronal stripes. Rufous-chestnut upperparts, richer below***, greyish central belly. Monotypic. Pairs to small groups methodically skulk in dense vegetation, seldom joins mixed flocks; apparently at low density, often in or near bamboo. **Voice** Accelerating series of thin buzzy notes starting with sparrow-like *tzzz-tzzz…tze-tzé-tze-zeë-zeë-zee*; sharp *zét* call and frog-like *wäk* (not heard in Ecuador?). **SS** Unmistakable. **Status** Very rare and local, likely declining, VU.

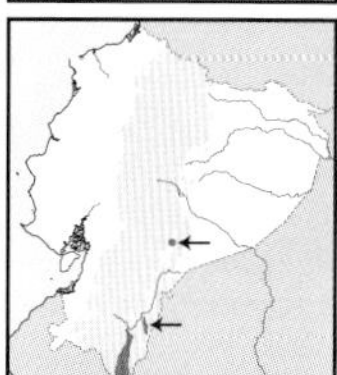

Red-crested Finch *Coryphospingus cucullatus* 13–14cm

SE, mainly in Zumba region (1,100–1,400m). Semi-deciduous secondary woodland, scrub, gardens. ***Red crest outlined black***, narrow whitish eye-ring. ***Dusky-reddish above***, browner wings and tail, ***redder underparts***. ♀: ***lacks crest***, mostly reddish-brown, ***rosier underparts, crimson rump***, whitish at bill base, ***narrow white eye-ring***. Race *fargoi*. Singles, pairs or small parties feed on or near ground; may remain in cover but briefly emerge, even perching atop bushes. **Voice** Gives *chéeu-wee'wít, chéeu-wee'wít, chéeu-wee'wít…*, or simpler *weet-chee'weét, weet-chee'weét*. **SS** Confusion unlikely in limited range. **Status** Locally uncommon, spreading N following deforestation.

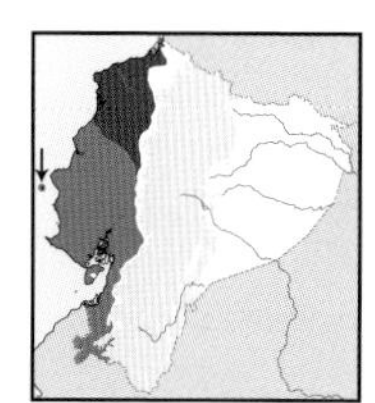

Crimson-breasted Finch *Rhodospingus cruentus* 11cm

Mostly W lowlands, including Isla de la Plata (to 500m). Deciduous to semi-humid forest borders, light woodland, scrub. Fairly long, ***pointed bill. Black forecrown and face to upperparts, red mid crown and underparts***. Non-breeder duller red overall, black replaced by scaly brown. ♀: ***drab***, mostly pale brown above, ***pale buff below***, buff lores. Monotypic. Gregarious, larger groups when breeding; joins mixed arboreal flocks or seedeater parties near ground; apparently moves N post-breeding (Jun–Dec). **Voice** Insect-like, high-pitched *b'zzzz-tzzt, t'zzzzí* or *tzzz-zzzt* recalling Blue-black Grassquit but longer and more buzzy; call short *tzít*. **SS** ♂ unmistakable; ♀ resembles some ♀ seedeaters and small tanagers, but bill shape distinctive; see accompanying ♂. **Status** Uncommon.

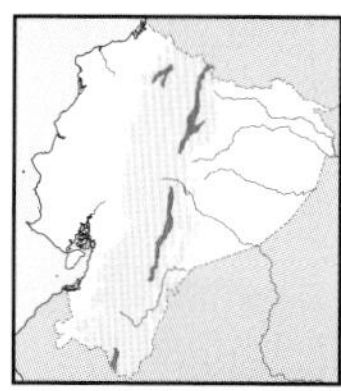

White-rimmed Brush-Finch *Atlapetes leucopis* 17–18cm

Local in subtropics to temperate zone in E & NW (2,200–3,100m). Montane forest. ***Dark chestnut crown***, blackish-olive face and upperparts, ***bold white orbital rim***, olive below. Monotypic. Mainly pairs skulking in dense understorey, shy and meticulous, sometimes joins mixed flocks, poorly known. **Voice** Musical series of clear, whistled *teeuw-twee-twee-deü-deü-deydeydey, twee-dee-dee-dedede…* often repeating phrases; call thin *síip* reminiscent of other brush-finches. **SS** No other brush-finch has white orbital area; Olive Finch has similar crown but grey face. **Status** Rare, local, rarer in NW.

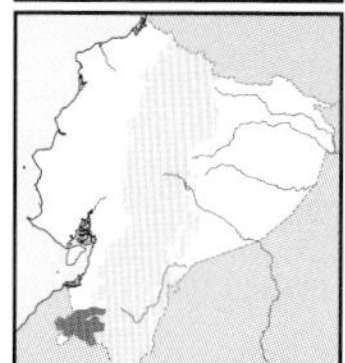

White-headed Brush-Finch *Atlapetes albiceps* 16–17cm

SW lowlands to lower subtropics (200–1,100m). Deciduous forest, woodland, scrub, borders. ***White forecrown, face and throat, black rear crown to nape***. Grey above, greyish-buff below, ***bold white speculum***. Monotypic. Pairs or small groups in lower growth, sometimes higher at edges, shy but easier to see than previous species; may join mixed understorey flocks. **Voice** Jumbled *tsee-tsee-see-swee-swee-se'se'se'se-swee…*, sometimes simpler, ending in trill; recalls other brush-finches but slower than sympatric White-winged. **SS** Confusion unlikely; though some White-winged can show much white in face, crown is always chestnut. **Status** Uncommon.

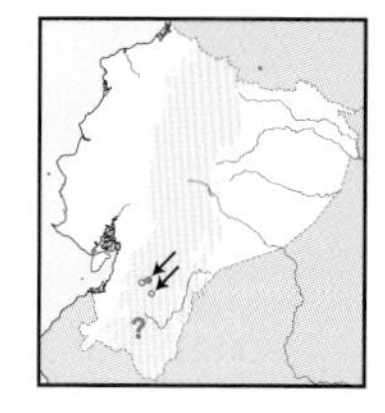

Pale-headed Brush-Finch *Atlapetes pallidiceps* 16–17cm

Very local in inter-Andean valleys of Azuay (1,500–2,100m, now 1,630–1,970m). Xeric woodland and scrub. ***Pale buff head, faint duskier stripes*** on crown and cheeks. ***Dusky upperparts***, white wing spot, ***whitish-buff underparts***. Monotypic. Pairs to small groups independent of mixed flocks but may join them briefly; tiny population (*c.*100 breeding territories), heavily parasitised by Shiny Cowbirds, now controlled, and numbers apparently increasing. **Voice** Fast, jumbled, high-pitched *zét-zét-zét-ts-ts-ts't't't…* at dawn; *zít-zi-t'sisisisisi* sparrow-like day song, lacking whistles of other *Atlapetes*; cascading calls weaker than congeners, call thin *zít* or *tsíp*. **SS** Unmistakable in tiny range; no overlap with White-headed Brush-Finch. **Status** Very localised population under careful management; CR in Ecuador, EN globally.

Tanager Finch
♂
♀
Red-crested Finch
♂
♂
immature
♀
juvenile
Crimson-breasted Finch
White-rimmed Brush-Finch
White-headed Brush-Finch
juvenile
adult
immature
Pale-headed Brush-Finch

PLATE 281: *PIRANGA* 'TANAGERS'

Arboreal 'tanagers' recently shown not to be true tanagers but related to cardinals. Often join mixed flocks but also found in monospecific pairs or small parties. Marked sexual dimorphism in most. Imm. ♂♂, especially of migratory species, mostly olive but varied red blotching throughout.

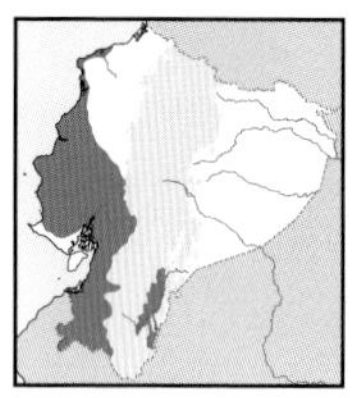

Hepatic Tanager *Piranga flava* — 17–18cm

W lowlands to subtropics in Loja, also local in SE foothills (to 1,900m in SW, 1,000–1,400m in SE). Deciduous to semi-deciduous forest, borders, shrubby clearings. ***All red*** with ***dusky bill, dark lores,*** duskier wings. ♀: mostly ***greenish-olive***, yellower below, ***dusky lores, blackish bill.*** Imm. shows some red splotches. Race *lutea* (possibly valid species). Mainly pairs with mixed canopy flocks, descending at edges; deliberately gleans foliage and picks berries. **Voice** Sweet, musical song recalls a saltator or thrush: *che'wëë, dö'wee, dö, chë...che'wëë, dö'wee*; calls louder *chup-chu'tip, chip-chu'tip....* **SS** Recalls Summer Tanager (paler bill, rosier overall, no dusky lores; ♀ yellower belly, no dusky lores. Scarlet Tanager (blacker wings); also smaller, duller ♀ *Tachyphonus*. **Status** Uncommon.

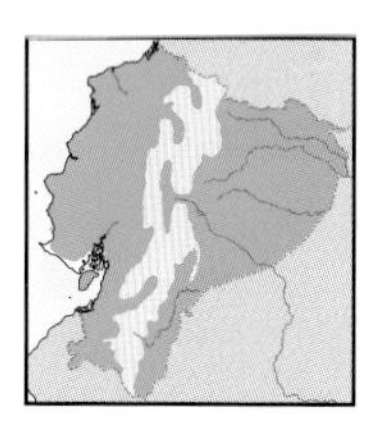

Summer Tanager *Piranga rubra* — 17–18cm

Boreal migrant to E & W lowlands, subtropics and inter-Andean valleys (to 2,800m, but mostly below 1,500m). Humid forest borders, woodland, adjacent clearings, hedgerows. ***Entirely rosy-red***, paler underparts, ***pale horn to horn bill***, wings slightly darker. ♀: greenish-olive above, yellow underparts with some ***orange to pink wash, pale lores***, bill as ♂. Imm. recalls ♀ but shows some red splotches. Nominate. Pairs tend to join mixed flocks, in upper forest strata or low at edges, crops and gardens. **Voice** Short, vibrating *ti-ki-ti-kup, ti-ki-ti-ku'p, ti-ki-ku'up*. **SS** Closely recalls resident Hepatic Tanager, with darker bill and lores. **Status** Uncommon (Oct–Apr).

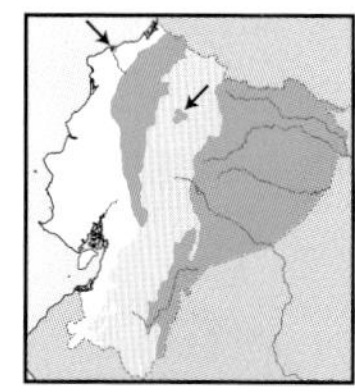

Scarlet Tanager *Piranga olivacea* — 17–18cm

Boreal migrant to E lowlands and subtropics, local in W lowlands to subtropics (to 1,500m). Humid forest borders, woodland, adjacent clearings. ***Entirely scarlet*** with ***black wings and tail, horn bill.*** ♀: bright olive above, yellower below, ***blackish to dusky wings***, bill as ♂. Non-breeding ♂ resembles ♀ but ***wings and tail blacker***. Moulting imm. shows some red splotches, wings ***dusky***, not blackish. Monotypic. Mainly in pairs with mixed flocks, more often in canopy, where meticulously gleans. **Voice** High *chlP-churr*. **SS** Wings always blacker than congeners, bill smaller; see Masked Crimson, Vermilion and White-winged Tanagers. **Status** Uncommon transient (Oct–Apr).

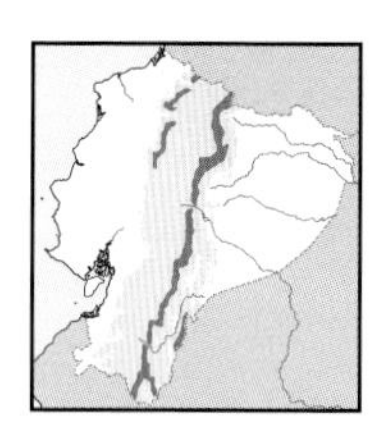

Red-hooded Tanager *Piranga rubriceps* — 17–18cm

Upper subtropics to temperate zone in E & NW Andes (2,200–3,000m, locally to 1,800m in E). Montane forest canopy and borders. ***Scarlet hood to breast***, bright olive mantle, yellower rump, ***yellow shoulders and rest of underparts***. Blackish wings with yellow edges. ♀: similar but ***red only to throat***. Monotypic. Pairs to small groups deliberately forage in upper strata, often with mixed flocks, regularly perches high and exposed. **Voice** Loud, sweet *ti-ti-ti-tí, tsi-wee-tsí*, first part trilled, second musical; also sweet *tit'swee, tit'swee-tit'de-de-de-de...*, last trill more piercing. **SS** Nearly unmistakable; see moulting Scarlet and Summer Tanagers with some red blotches over olive background. **Status** Rare, even rarer in NW.

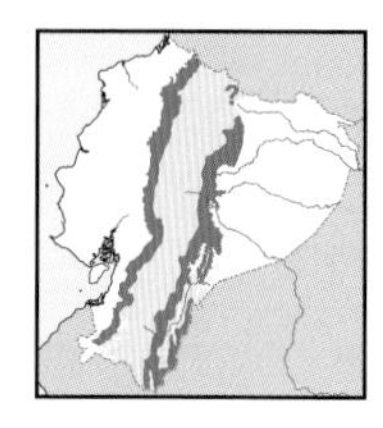

White-winged Tanager *Piranga leucoptera* — 14–15cm

E & W Andean foothills to subtropics (800–1,800m, locally to 400m in SW). Montane forest canopy, borders, adjacent clearings. Entirely ***bright scarlet*** with small black mask, ***black wings with two bold white wingbars***, black tail. ♀: similar but red replaced by ***olive above, yellowish below***. Race *ardens*. Active pairs with mixed flocks, deliberately gleaning outer foliage and terminal twigs, more often in upper strata. **Voice** Loud, semi-musical *tsee-tsee-tsee-tsuu-tsee, tsee-tsee-tsee-tsuu-tsuu...* song; calls sharper *tsee'pu, tsee-su*, faster *tsu-ts'ts'ts'u-tsee'suu...*, also buzzy *swee* or *schee*. **SS** Unmistakable; similarly patterned Scarlet Tanager lacks wingbars. **Status** Uncommon.

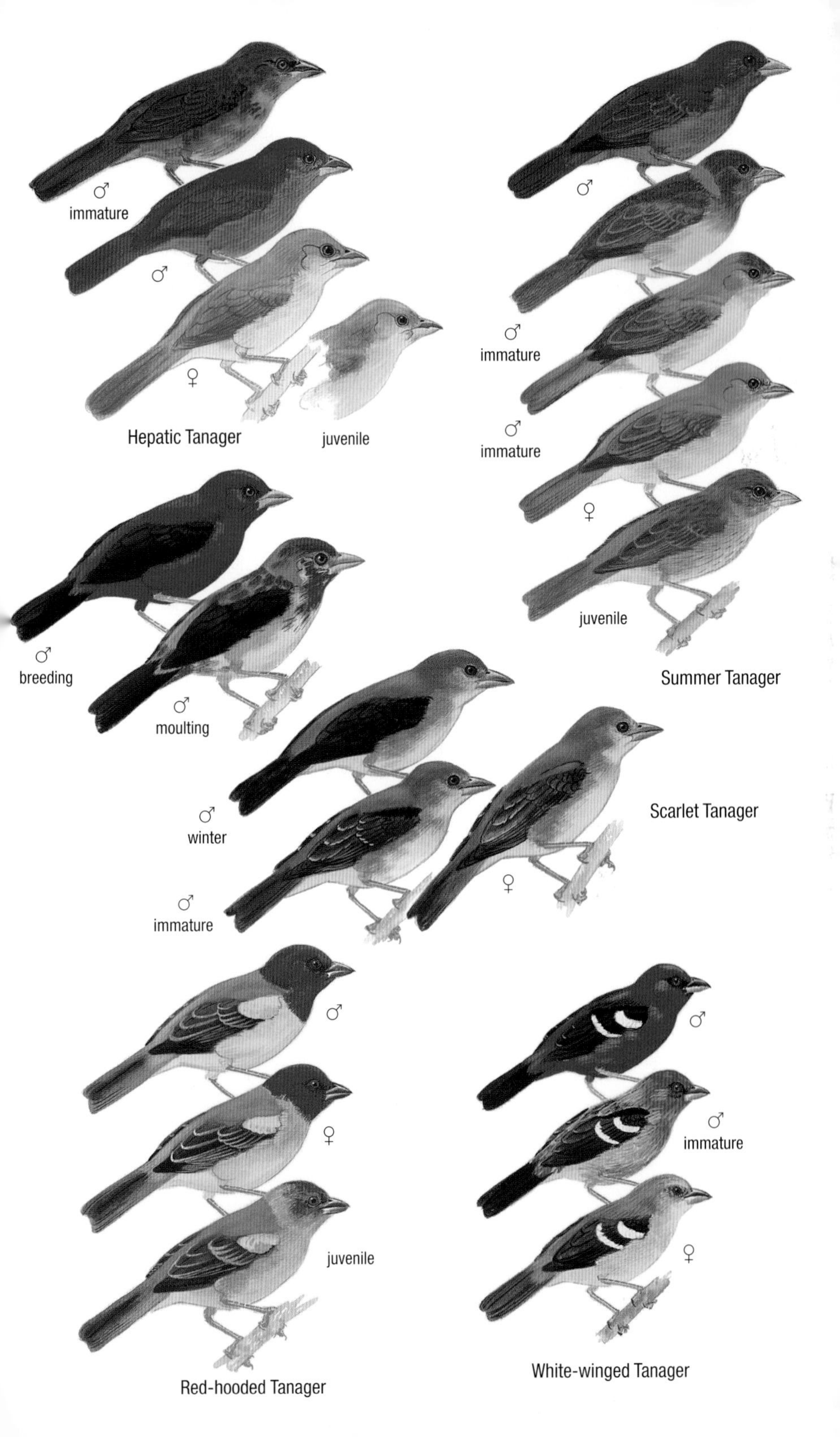
♂
immature
♂
♀
Hepatic Tanager
juvenile
♂
♂
immature
♂
immature
♀
juvenile
Summer Tanager
♂
breeding
♂
moulting
♂
winter
♂
immature
♀
Scarlet Tanager
♂
♀
juvenile
Red-hooded Tanager
♂
♂
immature
♀
White-winged Tanager

PLATE 282: GROSBEAKS

Mid-sized to large birds with heavy bills, ideal for cracking hard seeds, but also for picking and smashing fruit. Mainly in pairs or small groups in forests, borders, woodland and gardens. Precise affinities of Yellow-shouldered still unresolved.

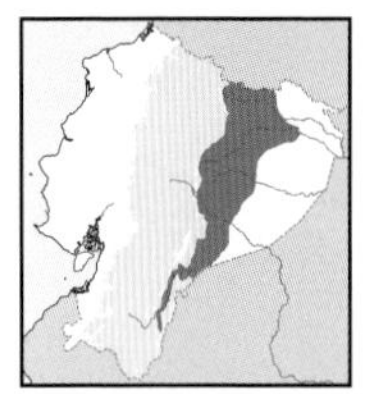

Yellow-shouldered Grosbeak *Parkerthraustes humeralis* 16–17cm

E lower foothills to adjacent lowlands (mainly below 600m, locally to 1,000m). Canopy and borders of *terra firme* forest. ***Heavy stout bill, red eyes. Grey crown, black mask, broad white malar, throat freckled white.*** Olive upperparts, grey underparts, ***bright yellow shoulders*** and vent. Monotypic. Singles, more often pairs, with mixed flocks, closer to ground at borders; regularly perches on exposed branches. **Voice** Sharp, high-pitched jumble: *sweE-sweE-sweesweé...*; piercing *tseét-swet.* **SS** Unmistakable if seen well, especially by face pattern and yellow shoulders. **Status** Rare.

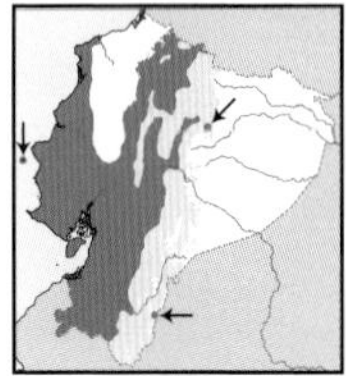

Golden-bellied Grosbeak *Pheucticus chrysogaster* 21cm

SW lowlands to subtropics, inter-Andean valleys to temperate zone (to 3,500m). Deciduous to semi-deciduous woodland, arid scrub, gardens, crops, montane forest borders. ***Golden-yellow*** with contrasting ***black wings and tail, white spots*** in wing-coverts, flight feathers and tail. ♀ and juv. ***yellow, coarsely streaked blackish above, dusky wings, two wingbars.*** Nominate. Singles, more often pairs or trios in various strata, mostly arboreal; vocal after light rain. **Voice** Rich, melodious *weerí-guoow, weerí-churí-churow*; loud *píck!* **SS** ♀ and juv. Black-backed Grosbeak darker above. **Status** Fairly common.

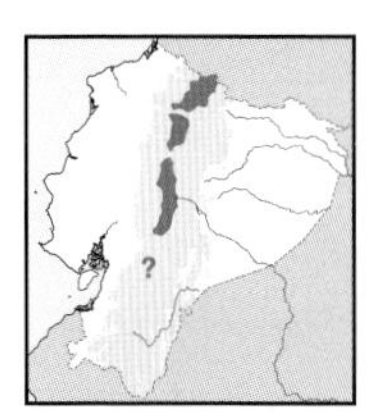

Black-backed Grosbeak *Pheucticus aureoventris* 21–22cm

Inter-Andean valleys to highlands S to Chimborazo (1,500–3,200m). Arid scrub, woodland, crops, gardens. ***Black hood and upperparts,*** yellow rump mottled dusky; ***yellow underparts, black wings, two white wingbars*** and spots in flight feathers, black tail tipped white. ♀ and juv. ***heavily mottled dusky above and on head,*** yellow below, wing as ♂. Race *crissalis*. Much like previous species, sometimes with mixed flocks; ecological segregation from Golden-bellied not clear. **Voice** Melodious carolling, faster than previous species: *chiri-weeri-chúu, wee-weeri-chuuow*; sharp *beek.* **SS** Head always darker than otherwise similar ♀ and juv. Golden-bellied. **Status** Uncommon.

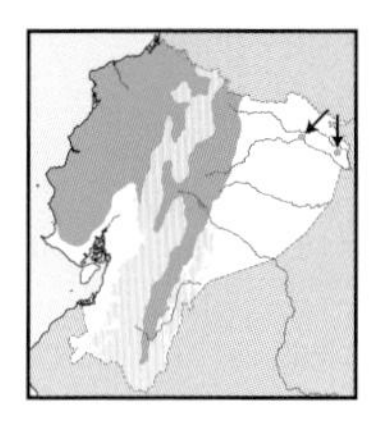

Rose-breasted Grosbeak *Pheucticus ludovicianus* 18–19cm

W & NE lowlands to subtropics, locally in inter-Andean valleys (to 2,000m). Montane to deciduous forest canopy, borders, woodland, clearings. Pale ***heavy bill, black hood and upperparts, rosy breast patch, white underparts,*** black wings with white bars and spots. ♀ and non-breeder have ***long white supercilium and crown-stripe, dusky cheeks,*** whitish underparts ***narrowly streaked dusky.*** Monotypic. Mostly singly in canopy, sometimes impassive on exposed perch. **Voice** Metallic *pínk!* **SS** Smaller, rarer Dickcissel and Bobolink, also smaller sparrows; heavy bill distinctive. **Status** Rare boreal winter visitor (Oct–Mar).

♀
♂
Yellow-shouldered Grosbeak
♂
immature
♀
♂
Golden-bellied Grosbeak
♀
♂
fresh
♂
worn
Black-backed Grosbeak
♂
immature
♀
♂
Rose-breasted Grosbeak

The small and rare Blue Seedeater, which is actually a 'cardinal', somewhat recalls species on Plates 274–277, but bill smaller, more pointed, and it is a forest bird. The boreal visitors Blue Grosbeak and Dickcissel have been recorded at a handful of sites between them, but might show up 'anywhere' and 'anytime' during migration.

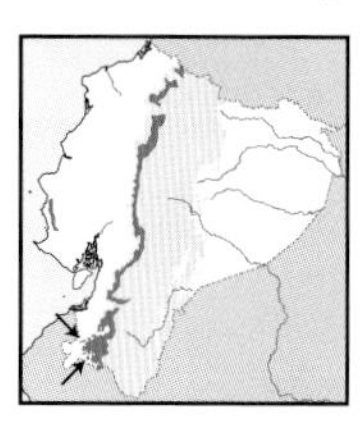

Blue Seedeater *Amaurospiza concolor* 12–13cm

Local in W Andean foothills to subtropics, also slopes of coastal mountains (1,100–2,300m, 400–600m in coastal cordilleras). Montane forest, borders, mature woodland. ***Entirely deep blue***, with darker cheeks, wings and tail. ♀: ***cinnamon-brown overall***, paler underparts. Race *aequatorialis*. Forest-based, shy pairs on ground or in undergrowth, often associated with bamboo. **Voice** Short, fast, melodious warbling *swee-swee-sweet-swee-sweet-sweea*.... **SS** No other blue seedeater; see glossier Blue-black Grassquit. ♀ recalls Thick-billed Seed-Finch but bill less heavy; more cinnamon or orange-washed than ♀ *Sporophila*. **Status** Rare, local, overlooked?

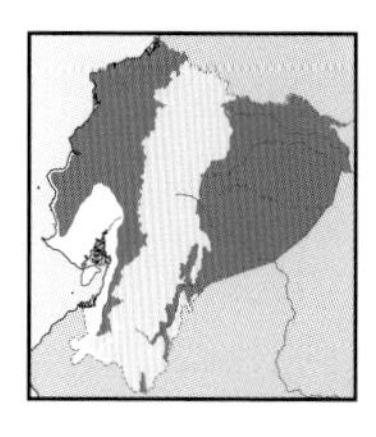

Blue-black Grosbeak *Cyanocompsa cyanoides* 16–17cm

E & W lowlands to foothills (to 1,000m). Humid forest undergrowth, borders. ***Heavy black bill. Dark metallic blue, shinier*** silvery-bluish ***forecrown, supercilium and shoulders*** (E). ♀: ***entirely chocolate-brown***. Nominate (W), *rothschildii* (E, possibly separate species). Pairs in dense thickets, vocal but difficult to see, rarely with mixed flocks. **Voice** Rich, musical, high-pitched, accelerating *dee-de, dee-de, deeyidiyideé?...*; piercing *chí-peénk, chí-peénk, chín* calls. **SS** No wingbar like accidental Blue Grosbeak; cf. ♀ with comparatively heavier-billed but smaller Large-billed and Black-billed Seed-Finches of semi-open areas. **Status** Fairly common.

Blue Grosbeak *Passerina caerulea* 16cm

Few records from E & W lowlands and slopes at 220–300m in shrubby clearings; once at 1600m. Entirely ***deep blue*** but non-breeders ***intermixed brown***, darker face, ***two prominent chestnut to cinnamon wingbars***. ♀: pale brownish bill, brown upperparts, paler below, ***cinnamon-buff wingbars***. Nominate. Does not winter in Ecuador; tends to remain atop shrubs and bushes, often spreads tail. **Voice** Metallic, very high-pitched *chink!* **SS** Wingbars separate from Blue-black Grosbeak. **Status** Accidental boreal winter visitor (Aug, Dec, Mar). **Note** Indigo Bunting (*Passerina cyanea*), recently recorded in Quito (a single, window crashed bird) is bluer, without wing bars, smaller, lighter-billed. Female shows some blue in wings.

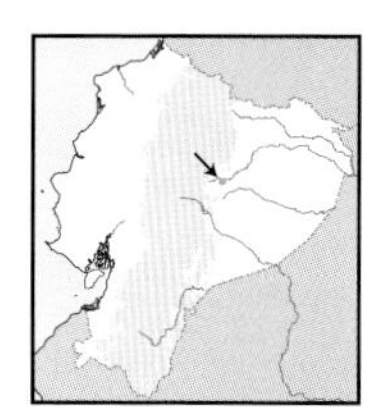

* Dickcissel *Spiza americana* 15–16cm

One record from E lowlands (400m) in roadside grassland and shrubs. Small ***conical pale horn bill. Grey to brownish-grey crown***, dusky-brown above ***streaked blackish***, long yellowish supercilium and ***broad malar, dusky sub-malar, white throat, V-shaped breast-band*** (breeder), ***yellowish breast***. Dusky wings, ***rufous shoulders***. Monotypic. Flocks with seed-eaters and sparrows low down; does not winter in Ecuador, elsewhere in large flocks. **Voice** Harsh, low-pitched *drrt*. **SS** Rose-breasted Grosbeak (no yellow breast, heavier bill), Bobolink (buff underparts, striped head, no rufous shoulders); juv. Saffron Finch (simpler face and wing patterns). **Status** Accidental boreal winter visitor (Jan).

♂
Blue Seedeater
♀
♂
♀
♂
♀
cyanoides
rothschildii
Blue-black Grosbeak
♂
♀
♂
♀
♂
immature
Blue Grosbeak
Dickcissel

Small finch-like birds, several species characterised by conspicuous wing patches and black hoods (♂♂). Identification keys still poorly known, especially for ♀♀; luckily range overlap is not extensive between most of the similar species. Yellow-bellied Siskin and Lesser Goldfinch are more distinctive. Voices similar among species.

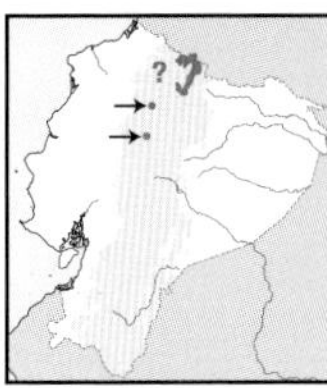

Andean Siskin *Sporagra spinescens* — 10–11cm

Local in N highlands, S to Pichincha (2,800–3,600m). Treeline, montane forest borders, *páramo*. ***Black crown***, dull olive above, black wings with bright yellow T, yellowish rump, dull olive-yellow underparts. ♀: ***dusky-olive*** crown to back, ***whitish mid belly***. Race *nigricauda*. Small flocks atop shrubs, *Espeletia* plants, ground bromeliads or grass tussocks; often with Hooded Siskin. **Voice** Sweet, lively twittering, squeakier than Hooded; calls thin *thsií*, longer than Hooded. **SS** Hooded has black hood (♂; but see imm.), more extensive yellow in tail (both sexes), little white below (♀). **Status** Locally uncommon.

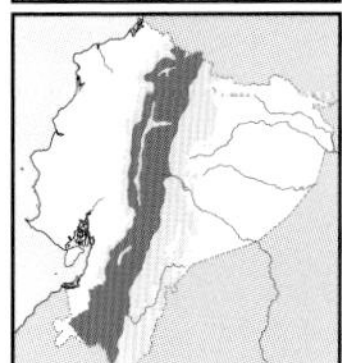

Hooded Siskin *Sporagra magellanica* — 10–11cm

Widespread in Andean highlands and inter-Andean valleys (1,000–3,500m). Open woodland, shrubby clearings, parks. ***Black hood***, olive to yellowish-olive above, ***rich yellow below, prominent*** yellow wing T, ***yellow tail base***. ♀: ***greyish-olive above, pale greyish to yellowish below***. Races *capitalis* (most of range), *paula* (S). Small flocks either at canopy level or near ground, noisy, undulating flight; restless, rather tame. **Voice** Variable, sweet twittering *dlee'tsii-dididi-glíglíglí-dlé-le-le…*; call thin *thíp, tiirp*. **SS** Very similar Olivaceous Siskin (rather heavier streaking above; ♀ duskier overall, upperparts streaked dusky, breast faintly mottled dusky-olive); Andean Siskin has black limited to crown; Saffron Siskin brighter overall. **Status** Common.

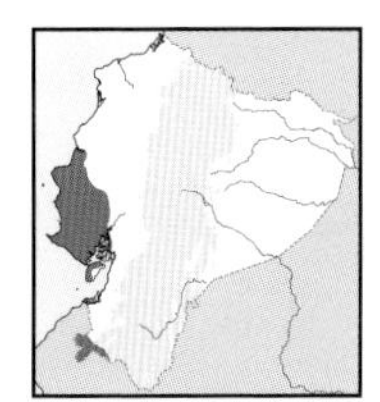

Saffron Siskin *Sporagra siemiradzkii* — 10–11cm

Local in SW lowlands to lower foothills (to 600m, locally to 1,300m). Deciduous to semi-deciduous forest, woodland, borders, xeric scrub. ***Black hood, bright olive-yellow above***, brighter rump, ***saffron-yellow below***. ♀: lacks black hood, throat to lower breast ***more olive*** than male. Monotypic. Few dozens may congregate at seeding bushes and herbs; possibly performs seasonal movements in search of food. **Voice** Variable, complex series of twitters, trills and sputters, very fast and rather melodious *tweep-weep-tlidididi…*; call thin *pee'*. **SS** Might overlap locally with similar Hooded Siskin in W Loja; Hooded less bright overall, ♀ more greyish-washed below. **Status** Uncommon, local, VU.

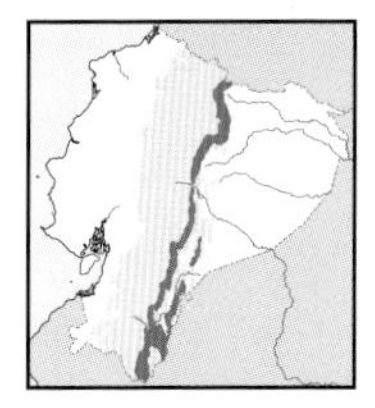

Olivaceous Siskin *Sporagra olivacea* — 10–11cm

E Andean foothills to subtropics (900–1,700m, locally to 2,000m). Montane forest, woodland, borders. ***Black hood, dull*** olive above ***finely streaked dusky***, underparts ***dull yellowish-olive***. ♀: lacks black hood, upperparts ***dull olive and streaked, throat faintly streaked dusky-olive***, dull yellowish-olive below. Monotypic. Small, restless and vocal groups primarily in forest canopy, lower at edges, seldom in open areas. **Voice** Resembles Hooded Siskin, melodious and sweet, various trills interspersed; call thin, high-pitched *tsiií*. **SS** Recalls Hooded, but more olive and more streaked above; ♀ Hooded often greyer below; Hooded tends to occur in more open situations. **Status** Uncommon.

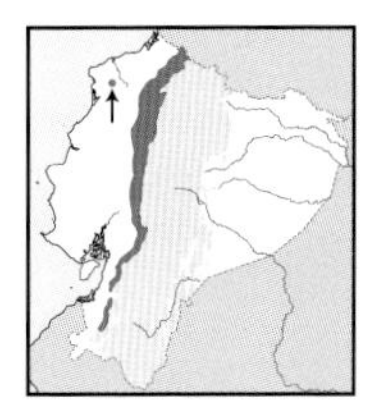

Yellow-bellied Siskin *Sporagra xanthogastra* — 11–12cm

Scattered records in W lowlands to subtropics (500–2,200m). Humid forest canopy, borders, woodland. ***Black head to breast and entire upperparts, bright yellow belly and large wing patch***. ♀: ***dusky-olive*** hood and upperparts, dull yellow belly, bold yellow wing patch. Nominate. Singles to small flocks, usually in wooded areas, seldom near ground; possibly performs seasonal movements. **Voice** Complex series of twitters, trills and sputters, sometimes prolonged; call thin, high-pitched *pee, pee'ee*. **SS** ♀ Hooded Siskin darker above, less striking wing patch; ♀ Saffron brighter overall. **Status** Rare.

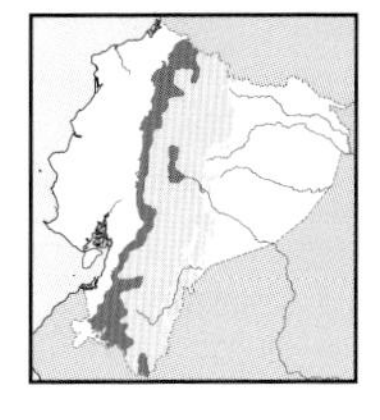

Lesser Goldfinch *Astragalinus psaltria* — 11–12cm

Scattered records in W Andean subtropics, locally to temperate zone, more local in E subtropics (mostly 400–2,700m). ***Black upperparts, bright yellow underparts, small white patch in wings***, white edges to tertials. ♀: ***dull olive above, yellower below, blacker wings with small white wing patch and fainter edges***. Race *columbiana*. Restless pairs to small flocks in various strata, erratic and apparently seasonal migrant. **Voice** Calls a sad, whistled *thee-thuee*, sometimes inflected *gleuu-gleuu?; gleeu-gleglegle, peeeu*; song a musical, scratchy, ascending, disjointed twittering. **SS** Confusion unlikely; no yellow wing patches. **Status** Rare, local, erratic?

♀
♂
uvenile
juvenile
Andean Siskin
♀
♂
paula
♂
Hooded Siskin
juvenile
♀
♂
Saffron Siskin
♀
♂
capitalis
juvenile
♀
♂
Olivaceous Siskin
immature
juvenile
♂
juvenile moult
♀
♂
worn
♂
fresh
Yellow-bellied
Siskin
Lesser
Goldfinch
juvenile
♂
juvenile
first moult
♀
fresh
♂
worn
♂
fresh

PLATE 285: EUPHONIAS

Euphonias are small, chunky and attractive forest birds, in appearance 'midway' between a tanager and a finch. Until recently, they were included with the Thraupidae (tanagers). Sexual dimorphism is strong, all ♂♂ on this plate combining steel purplish-blue above and deep yellow below. ♀♀ are puzzling as most are rather featureless. Voices complex, including some mimicry.

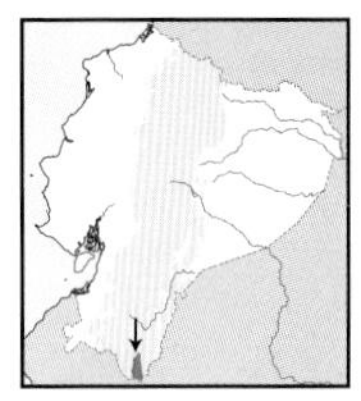

Purple-throated Euphonia *Euphonia chlorotica* 10–11cm

Extreme SE in Zumba region (650–1,100m). Semi-deciduous woodland, borders. ***Small yellow frontal patch***, steel-purple head and upperparts, ***yellow underparts***, white undertail patches. ♀: greyish-olive above, ***yellower front and most of face, greyish median underparts***. Race *taczanowskii*. Pairs to small groups mostly in canopy, with or without mixed flocks. **Voice** Whistled, warbled jerky song; calls plaintive, high-pitched *been-beé*, *ti-EE* and others. **SS** Orange-bellied (larger crown patch, orangey underparts; ♀ with more prominent orange forecrown, greyer crown); see ♀ Thick-billed (greyish lores, yellower underparts, heavier bill). **Status** Uncommon.

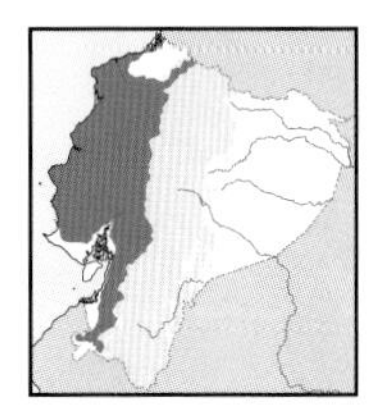

Orange-crowned Euphonia *Euphonia saturata* 10–11cm

W lowlands to subtropics (to 1,750m). Humid to semi-deciduous forest, borders, woody clearings. *Large orange-yellow crown patch, orange-yellow below, no white* in undertail. ♀: *plain olive above*, olive-yellow below, more yellowish belly. Monotypic. Mostly pairs at canopy level, often with mixed flocks. **Voice** High, rapidly accelerating *tsi't, tidididit*, with *di'cheew* whistles interspersed; call simple *bee-bee-bee-beé*. **SS** Orange-bellied (smaller crown patch, yellower underparts, white undertail); Fulvous-vented (richer rufescent vent). ♀: nondescript, plainer olive than sympatric euphonias, plainer face pattern. **Status** Uncommon.

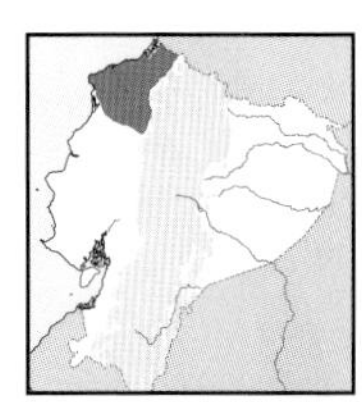

Fulvous-vented Euphonia *Euphonia fulvicrissa* 10–11cm

NW lowlands (below 500m). Humid forest, borders, woodland. Glossy purplish-blue with *small yellow frontal patch*; golden-yellow below, *rufescent vent*. ♀: *rufous front*, olive above, greener below, yellowish throat and central belly, *rufescent vent*. Race *purpurascens*. Pairs to small flocks, often in mid to lower strata, with mixed flocks. **Voice** Call brief, high rattling *p'rrrrt-p'rrrt* or high-pitched *whii-whii-whii...*; song fast, jumbled twittering and lower-pitched trills. **SS** No sympatric euphonia has fulvous to rufescent vent and dark undertail. **Status** Uncommon.

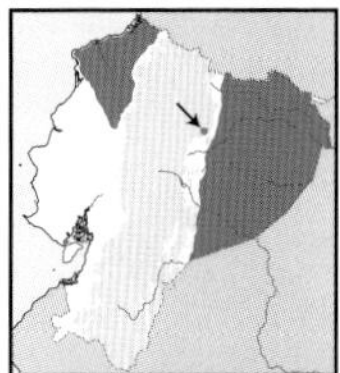

White-vented Euphonia *Euphonia minuta* 9–10cm

E & NW lowlands to foothills (to 700m). Forest borders, mature woodland. ***Very small yellow frontal patch***, yellow underparts, *white lower belly, vent and undertail*. ♀: greyish-olive above, yellower below, ***greyish-white throat, white lower belly to undertail***. Races *minuta* (E), *humilis* (NW). Singles or pairs in higher strata and emergent trees, regularly with mixed flocks. **Voice** High-pitched warbling and chipping: *vi-vree'it-vree-ti'i'i'i-chevit-ch-vreet...*; call a sharp *véEE-vréet*. **SS** Size and extensive white lower belly to undertail separate from congeners; voice rather siskin-like. **Status** Rare.

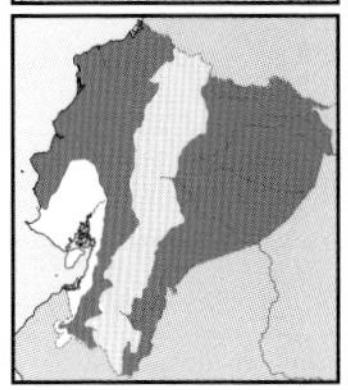

Orange-bellied Euphonia *Euphonia xanthogaster* 11–12cm

E & W lowlands to subtropics (to 2,000m). Forest canopy, borders, woodland, woody clearings. *Large orange-yellow crown patch, orange-yellow underparts, white undertail*. ♀: greyish-olive, dull ***orange forecrown, greyish nape and breast***, yellower below. Races *chocoensis* (NW), *quitensis* (CW–SW), *brevirostris* (E). Regular in understorey and canopy mixed flocks, bold. **Voice** Calls ringing *ding-ding-ding*, higher *cheé*, *duee-deeü*, *ding-ding-clé-clé-clé* and some mimicry; complex, semi-musical song with whistles and warbles. **SS** Orange-crowned (richer crown patch, deeper underparts); Fulvous-vented, Purple-throated and White-vented (smaller crown patches); ♀'s front, nape and breast rather distinctive but care needed. **Status** Common.

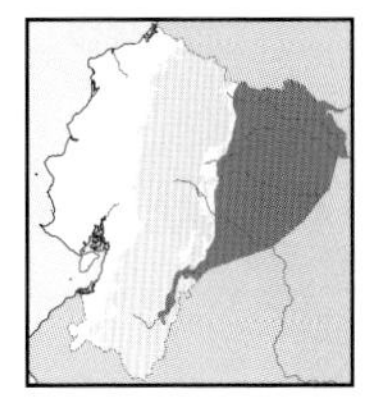

Rufous-bellied Euphonia *Euphonia rufiventris* 11–12cm

E lowlands, local into foothills (to 500m, locally higher). Mainly *terra firme* forest canopy, borders. ***Glossy purplish-blue hood and upperparts, orange to tawny-rufous below***. ♀: olive above, ***greyish nuchal patch***, olive-yellow sides to flanks, ***grey median underparts, orange to tawny vent***. Monotypic. Mostly pairs with mixed flocks, regular at epiphytes. **Voice** Fast, raspy chatter *drt'drt'drt'drt...*, ringing *dwi-dididí-dwít-dwít...*, *deeveet-dididídi*; short liquid *wëët* calls. **SS** ♂ lacks coronal patch of other euphonias, ♀ has fairly distinctive tawny-rufous vent, underparts mainly grey; see Orange-bellied. **Status** Uncommon.

♂
♀
Purple-throated Euphonia
♂
♀
Orange-crowned Euphonia
♀
♂
Fulvous-vented Euphonia
humilis
minuta
♂
♂
♀
White-vented Euphonia
chocoensis
brevirostris
♂
♂
♀
Orange-bellied Euphonia
♂
♀
Rufous-bellied Euphonia

PLATE 286: EUPHONIAS AND CHLOROPHONIAS

Two distinctive euphonias (Thick-billed and Golden-rumped) and two with less marked sexual dimorphism (White-lored and Bronze-green). Bronze-green is also the only euphonia in Ecuador confined to cloud forest. Chlorophonias are fantastically colourful versions of euphonias, with little sexual dimorphism: simply stunning. They appear to engage in seasonal movements.

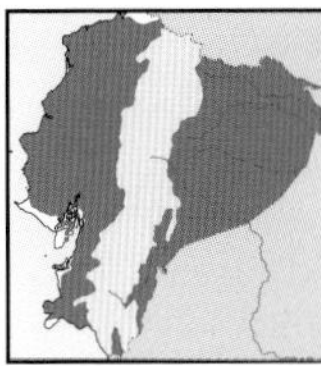

Thick-billed Euphonia *Euphonia laniirostris* 11–12cm

E & W lowlands to lower subtropics (to 1,500m, locally higher). Humid to deciduous forest borders, clearings. ***Bright yellow crown and underparts, steel-blue face and upperparts, white in undertail*** (W). ♀: greyish-olive above, ***greyish lores***. Races *hypoxantha* (W), *melanura* (E). Pairs to small groups with mixed flocks, not in forest. **Voice** Long, musical, complex series of twitters, sputters and imitations: *twii-wiwiwiwi-twi-ch'rrr-twi-fefefefe-twi…*, *twii-twiiti-twiititii..fé-fé*; calls *ch'weet*, *dueEE!..dEEeu* and others. **SS** All yellow underparts distinctive; ♀ has thicker bill, greyish lores. **Status** Common.

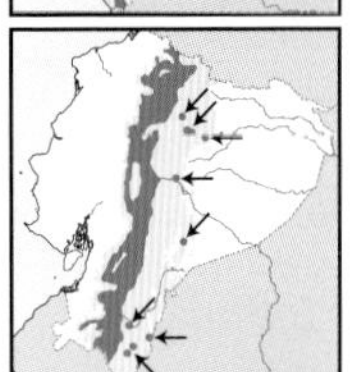

Golden-rumped Euphonia *Euphonia cyanocephala* 11–12cm

Inter-Andean valleys to W subtropics, local in E Andes (1,200–2,800m, locally higher). Montane forest borders, shrubby clearings, hedgerows. ***Black mask*** (orange front in S), ***blue crown to nape***, deep blue upperparts, ***golden-yellow rump and belly***. ♀: ***orange front, bluish crown***, olive above, yellowish-olive below. Races *pelzelni* (N), *insignis* (S). Pairs to small groups with or without mixed flocks; fond of mistletoes. **Voice** Fast, complex series of twitters, rattles and sputters, with long wheezy notes; call clear, far-carrying *tuee* or *wheeen*, also hoarser *weér*. **SS** Unmistakable. **Status** Uncommon.

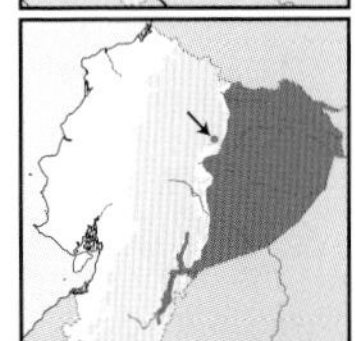

Golden-bellied Euphonia *Euphonia chrysopasta* 11–12cm

E lowlands, local in lower foothills (to 600m, locally higher). *Terra firme* and *várzea* forest canopy, borders. ***White foreface, bluish-grey nape***. ♀: ***white lores, bluish-grey nape, pale grey below***. Nominate. Mostly pairs, less often small groups, with mixed flocks, including other euphonias. **Voice** Song a fast, jumbled, short series of twitters, sputters, electric wheezes, given almost randomly; shorter song *veét, cheeteevéeu*; calls loud and explosive *píítz-week*, also clear *wheét* whistles. **SS** White in face distinctive even in duller and greyer ♀. **Status** Uncommon.

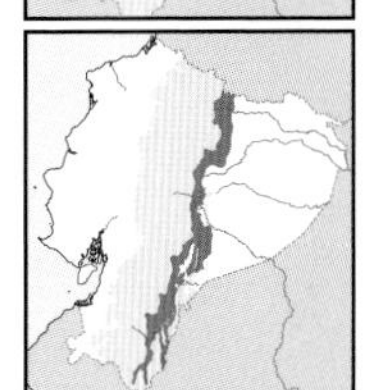

Bronze-green Euphonia *Euphonia mesochrysa* 10–11cm

E Andean foothills to subtropics (1,100–1,800m). Montane forest, borders, woodland. ***Small golden-yellow forecrown, bluish-grey nape***; mostly olive above, ***ochraceous-yellow central belly to vent***. ♀: olive above, ***bluish-grey nape***, underparts mostly dull olive, ***grey central belly, yellow undertail-coverts***. Nominate. Singles or pairs in higher strata, inconspicuous, often with mixed flocks. **Voice** Gives 2–3 whistles followed by short trill, *woo-teë-deeü-trrr, teë-deeü-trrr* or simpler single whistle and trill, *wooo-trrrr*; calls, a soft trill *t'rrr-tr-r-r-r* and sad *tree*. **SS** ♀ Orange-bellied (belly more orange yellow), Thick-billed (no grey below), Purple-throated (more extensive grey below). **Status** Uncommon.

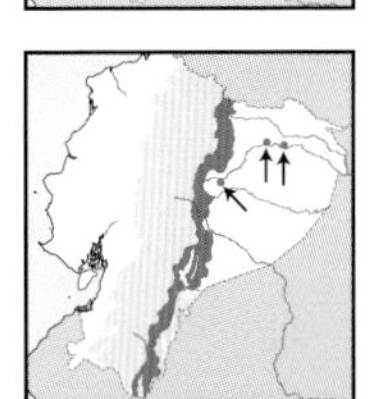

Blue-naped Chlorophonia *Chlorophonia cyanea* 11–12cm

E Andean foothills to subtropics, locally (seasonally?) in adjacent lowlands (800–2,000m). Montane forest canopy, borders. Superb. ***Bright grass-green hood, bluish nape, mantle and rump, blue eye-ring***. ♀: ***slightly duller***, blue limited to nape. Race *longipennis*. Pairs to small groups in higher strata, with or without mixed flocks, sluggish, fond of mistletoe. **Voice** Call a soft, far-carrying, plaintive *pleeeu*, somewhat downslurred; liquid short whistles and piercing notes (S *et al.*). **SS** Confusion unlikely; see very different ♀ Chestnut-breasted. **Status** Rare.

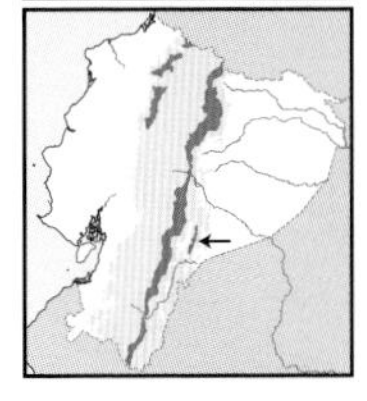

Chestnut-breasted Chlorophonia *Chlorophonia pyrrhophrys* 11–12cm

E & NW Andean subtropics to lower temperate zone (1,500–2,750 m). Montane forest, borders, woodland. ***Deep blue crown outlined black, bright green face to breast; orange-chestnut and yellow below***, yellow rump. Duller ♀ has ***chestnut crown-sides, no breast line***, olive-yellow below, ***no yellow rump***. Monotypic. Mainly in pairs in upper strata, erratic, seldom with mixed flocks. **Voice** Song a rather complex nasal *wee-neeh-nee-neä…*, with high *tseee* calls interspersed; more nasal *na-na..na-kEEyu, ke-na'na'na'na'na*; call drawn-out, downslurred *peeeea*. **SS** Confusion unlikely; see Blue-naped Chlorophonia at lower elevations; (♀ with plain green head). **Status** Rare.

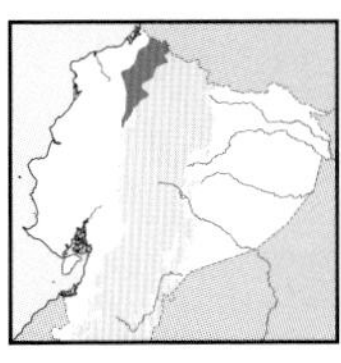

Yellow-collared Chlorophonia *Chlorophonia flavirostris* 10–11cm

NW foothills, lower subtropics, locally in adjacent lowlands (400–1,500m). Humid forest, borders, woodland. ***Bright orange bill, legs and eye-ring***, pale eyes. ***Mostly emerald-green, bold yellow spectacles, nuchal collar, rump*** and belly to vent. ♀: mainly green, with yellow belly and chin. Monotypic. Pairs to mid-sized groups with or without mixed flocks, mainly in upper strata, fond of mistletoes; apparently engages in seasonal or erratic wandering for food or following climatic shifts. **Voice** Thin, mournful, prolonged *peeee*, with *pik-pik* or *wik-wik* calls interspersed. **SS** Unmistakable. **Status** Rare.

pelzelni
insignis
pelzelni
juvenile
Golden-rumped Euphonia
immature
melanura
hypoxantha
Thick-billed Euphonia
Golden-bellied Euphonia
Bronze-green Euphonia
juvenile
Blue-naped Chlorophonia
Chestnut-breasted Chlorophonia
Yellow-collared Chlorophonia

Large passerines with long tails and long, pointed bills. Just one species in W lowlands, another on both Andean slopes, the others confined to E lowlands. Oropendolas are renowned for their fantastic liquid voices, spectacular displays on exposed perches and long pendant nests. Two species were recently reclassified as caciques.

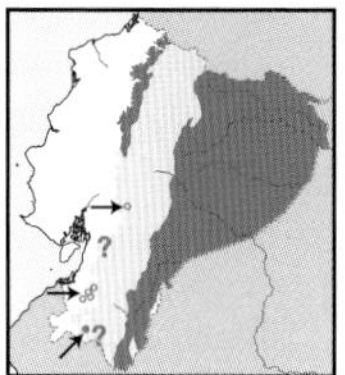

Russet-backed Oropendola *Psarocolius angustifrons* 35–48cm

E lowlands to subtropics, W foothills to subtropics (to 2,000m in E, 1,000–2000m in W). Riparian forest canopy, borders, adjacent clearings, river islands, montane forest. ***Yellow bill, yellow forecrown*** (W, SE) or ***black bill*** (E). Olive to brown head, ***brown to rufescent-brown overall.*** Races *angustifrons* (E), *atrocastaneus* (W), *alfredi* (SE slopes); possibly several species. Noisy, often in large aggregations, may form mixed flocks with other large passerines. **Voice** Accelerating, liquid *g-g-glukOOOk, glo-k-shrruUUk-glooo, gloo-gloo-glÖÖwk*, often starting with rattle; loud *cläk, chääk* in flight. **SS** Bill darker than other oropendolas in E; cf. Casqued. **Status** Common, declining in W?

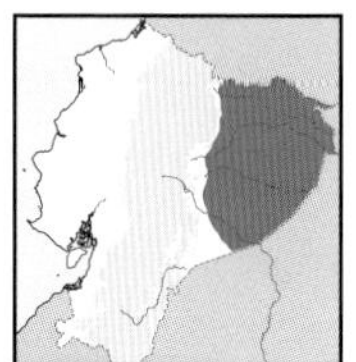

Green Oropendola *Psarocolius viridis* 37–50cm

E lowlands (to 600m). *Terra firme* canopy, borders. ***Greenish-yellow and red bill, pink orbital skin, bluish eyes.*** Mostly ***yellowish-olive***, rufescent rear parts. Monotypic. Flocks smaller than other oropendolas, including colonies; more forest-based than congeners. **Voice** Liquid, accelerating rolling *khek-'kkkk'll-kl'kl'kl'-cläclä-glooglák-gluü!, whyaa-whyaa-glglgl-gleglOgloo..., zweee-whowhe whowherüüp-wherüüp...* starting with screechy notes; harsh *cla'ká-kák* calls. **SS** Larger Olive Oropendola has chestnut wings, black bill tipped red; Casqued lacks red bill tip. **Status** Rare.

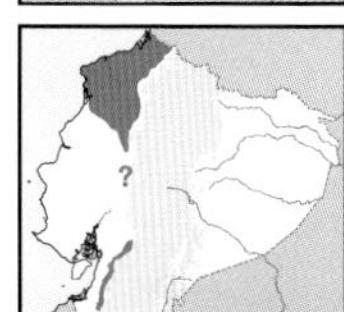

Chestnut-headed Oropendola *Psarocolius wagleri* 27–35cm

NW lowlands to foothills, S in Andean foothills to El Oro (to 700m). Humid forest canopy, woodland, borders. ***Ivory-yellow bill with large frontal shield. Dark chestnut head***, chestnut and black body, ***tail mostly yellow.*** Race *ridgwayi.* Noisy small to fairly large flocks, sometimes with other large passerines, but not frequently. **Voice** Explosive gurgling *glk-glk-glk-gloo-glAAAG!, gluu-gluu-phrrRRT'*; deep *glók-glók* calls. **SS** Confusion unlikely. **Status** Rare to locally uncommon, VU in Ecuador.

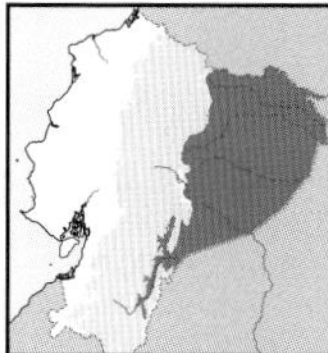

Crested Oropendola *Psarocolius decumanus* 35–48cm

E lowlands to foothills (to 1,000m). *Terra firme* and *várzea* canopy, borders, woodland, woody clearings. ***Ivory bill***, bluish eyes. ***Mostly black***, long hairy crest, chestnut rump and vent. ♀: slightly paler. Nominate. Noisy and conspicuous, often in large flocks with other large passerines for feeding and even roosting. **Voice** Loud, vibrating *tk'tk'tk'tk-gleeaák'oooo; gl'gl'gl-gl'-gleeeéeooou-goo-goop*; *clá* or *wág* calls. **SS** Blacker than other oropendolas, with distinctive ivory bill, voice more nasal; cf. rarer Band-tailed (smaller, little yellow in tail, darker maxilla, greenish wings). **Status** Fairly common.

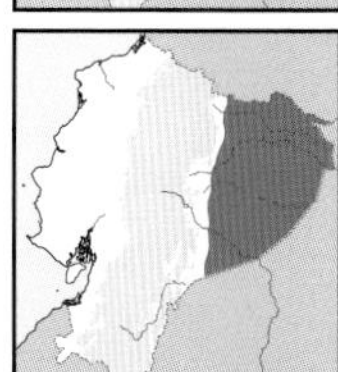

Olive Oropendola *Psarocolius bifasciatus* 41–52cm

E lowlands (to 300m). *Terra firme* and *várzea* canopy and borders, woodland, woody clearings. ***Black bill tipped red, pink facial skin***, dark eyes. Yellowish-***olive foreparts, chestnut body.*** Race *yuracares*, possibly valid species. Smaller flocks and nesting colonies than sympatric oropendolas, fond of nectar-rich large clumps of flowers. **Voice** Liquid gurgling, first notes rather metallic: *glöögu'lÖÖp, tkr'kr'kr'kr'-glööÖÖp...; teek-tek-tik-tktktk..wheeeeÖÖOpp*; also *chák!, drroop, dwát...* calls, like other oropendolas. **SS** Black-and-red bill, bare facial skin and chestnut wings separate from Green Oropendola. **Status** Rare.

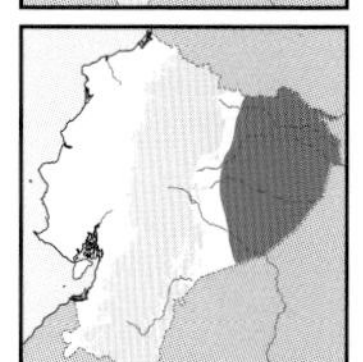

Casqued Oropendola *Cacicus oseryi* 28–38cm

E lowlands (to 300m, locally higher). *Terra firme* canopy, borders. Yellowish-ivory bill, ***large frontal shield.*** Rufescent-brown upperparts, ***yellowish-olive throat to lower breast.*** ♀: paler. Monotypic. Small flocks and nesting colonies usually in forest clearings (e.g. above forested rivers); flight more undulating than other oropendolas. **Voice** Harsh *khEEeeoou!*, with ringing quality; *cöö-EEk, shréeu-shréeu-shréuu, cöö-EEk, whéeu-whéeu-whéeu, googleé...*; harsh *waak, a'a'a'a-waak* calls. **SS** No 'casqued' oropendolas in range; see Crested Oropendola (blacker), Olive and Green (bicoloured bills). **Status** Rare.

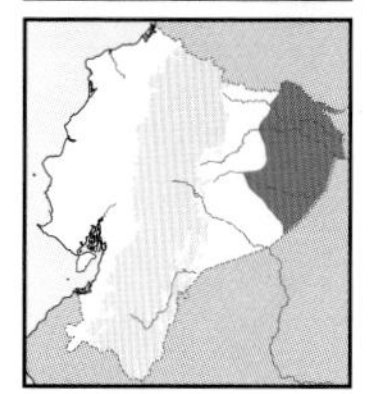

Band-tailed Oropendola *Cacicus latirostris* 24–34cm

E lowlands (below 300m). *Várzea* canopy, borders. ***Greyish and ivory bill, small frontal shield***, blue eyes. Dark chestnut crown, neck-sides and mantle grading into ***velvet black upperparts and underparts, wings glossed green, yellow in tail outlined black.*** Monotypic. Hardly known in Ecuador, elsewhere associates with other icterids, forming small flocks, not necessarily in canopy. **Voice** Emphatic chuckles and liquid, metallic chortles *chEoop!, ké-chOup!, skeedelop-chop..., kwá-tchOw!* **SS** Larger Crested Oropendola (more yellow in tail), Ecuadorian (slim bluish bill) and Solitary Black Caciques (all black, slimmer and paler bill, dark eyes). **Status** Very rare, few records.

♂
♀
♂
Russet-backed Oropendola
♂
alfredi
atrocastaneus
angustifrons
juvenile
Green Oropendola
adult
♂
♀
Crested Oropendola
Chestnut-headed Oropendola
♂
♀
Olive Oropendola
Casqued Oropendola
Band-tailed Oropendola

Fairly large, noisy birds, some in canopy, others in midstorey. Though primarily black, most are adorned with rich yellow or crimson around tail base; however, this is difficult to see when perched. Nests are hanging baskets in all *Cacicus*, but the monotypic Yellow-billed has a simple cup nest. Some species are conspicuous in open habitats.

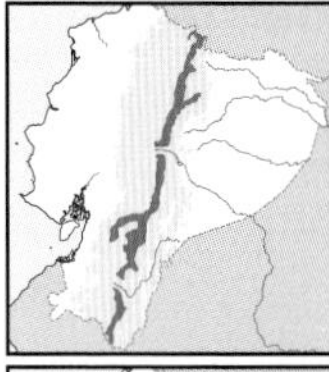

Mountain Cacique *Cacicus chrysonotus* 29–31cm

E Andean subtropics to temperate zone (2,000–3,100m). Montane forest canopy, borders, woody clearings. Pale bill, blue eyes. Velvet black, ***bright yellow shoulders and rump***. Juv. dark-billed, ***dark-eyed***. Race *leucoramphus*. Small noisy flocks often with other large passerines, acrobatic and lively. **Voice** Loud, metallic *wraa-weeweeweeweeé, wee-waa-wee-wee-wee'UU…*, *wheeaa*, nasal *yeëü-yeëü-yeëü…*, whining *wheenk* and sharp *tsueéé*. **SS** No overlap with Yellow-rumped; Yellow-billed (all black, pale yellow eyes); Scarlet-rumped lacks yellow in wings, heavier bill. **Status** Uncommon.

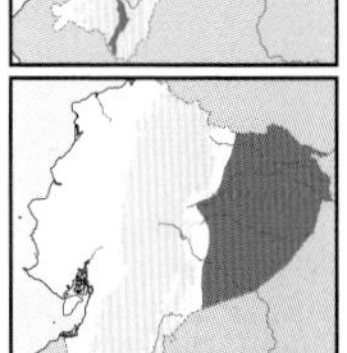

Ecuadorian Cacique *Cacicus sclateri* 23–24cm

E lowlands to lower foothills (to 800m locally). Canopy and borders of *várzea* and riparian forest. ***Bluish bill, blue eyes***. All black. Monotypic. Small flocks, not skulking like Solitary Black, may join other icterids. **Voice** Short *weep-weep, chééUU-chééUU* or *weep-weep, chééUU-chééUU'cheu'cheu…*, more ringing *k-cheeu', k'chow, k'chow*, softer than Yellow-rumped, quite melodic like Red-rumped. **SS** Larger Red-rumped looks dark when perched (but ivory bill is longer, curved); Solitary Black has dark eyes, ivory bill; behaviour differs. **Status** Rare, local, possibly overlooked.

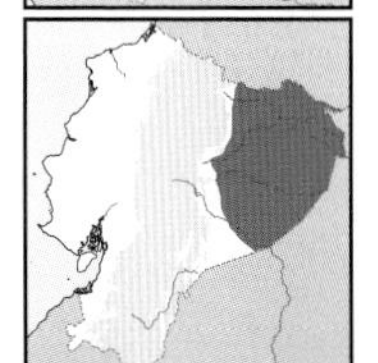

Solitary Black Cacique *Cacicus solitarius* 24–27cm

E lowlands to lower foothills (to 750 m). Undergrowth of dense scrub, woodland, riparian vegetation, river islands. ***Ivory bill, dark brown eyes***. Velvet black, ***long*** tail. Monotypic. Skulking pairs difficult to spot unless vocalising, sometimes follow mixed flocks, not colonial. **Voice** Strident, staccato *kheeoóp-kheeoóp, kEEY!-kEEY!-kEEY!*, faster *choychoychoychoychoy…*, more nasal *whe'whaaap!-whe'whaaap!-whe'whaaap!*; *weëp-weëp-weëp-weëp-pÿaaaau*. **SS** Dark brown eyes and ivory bill separate from slimmer Ecuadorian Cacique, juv. Ecuadorian is dark-eyed but bill slimmer; perched Red-rumped has blue eyes; behaviour and habitat differ from sympatric caciques. **Status** Uncommon, rather local.

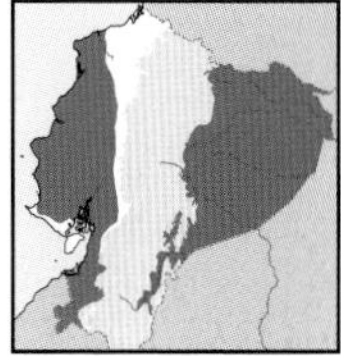

Yellow-rumped Cacique *Cacicus cela* 24–29cm

W lowlands to foothills, E lowlands (to 1,000m in W, to 300m in E). Forest canopy, borders, clearings. ***Ivory*** (E) to ***bluish*** (W) bill, blue eyes. Black, ***bright yellow shoulders and rear body***. Nominate (E), *flavicrissus* (W) possibly valid species. Noisy and conspicuous, often with other large passerines; nests close to oropendolas and wasps. **Voice** Variable, includes mimicry (in E) but more melodic (in W); loud *wóp-wóp-wóp-wóp…*, *cleecheéér*, liquid *clee-clee-cheewee'oo'uu, cheewee'oo'uu, sheek-sheek-clee'oo'ee-oouu…*, *claá-claá-claá…* **SS** No overlap with Mountain Cacique. **Status** Common.

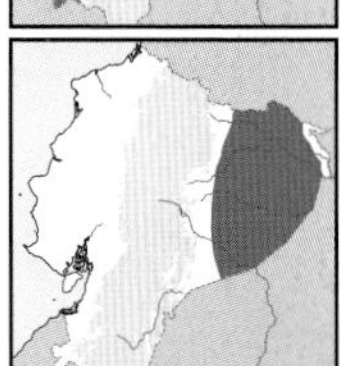

Red-rumped Cacique *Cacicus haemorrhous* 23–28cm

Local in E lowlands (below 400m). *Terra firme* canopy, borders. ***Ivory bill***, blue eyes. Velvet black, ***scarlet rump***. Dark-eyed juv. Nominate. As other caciques, small noisy flocks often with other icterids, nesting in colonies. **Voice** Guttural *däg-da-däng, zhooeeeeoo…*, *zhoowiíp, chaak-chaak…*, *klang-klang-weeu, weeu…* somewhat more melodic than sympatric caciques, no mimicry; harsh *tjak* calls. **SS** No overlap with higher-elevation Scarlet-rumped (smaller red rump patch, more curved, shallower bill); looks dark when perched, see Ecuadorian (bluish bill) and Solitary Black (brown eyes). **Status** Rare.

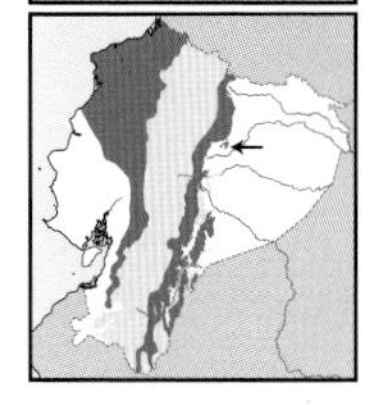

Scarlet-rumped Cacique *Cacicus uropygialis* 21–29cm

W lowlands to foothills, E foothills to subtropics (to 900m, locally higher in W, 1,000–2,100m E). Humid to montane forest canopy, borders. Black with ***scarlet rump*** (invisible when perched). Nominate (E), *pacificus* (W), possibly valid species. Small noisy flocks regularly with other large passerines; acrobatic and lively. **Voice** Loud bubbling *wheep-wheek-keeeyo-keeeyo-keeeyo…*, *whee'uüp-whee'uüp*, whistled *cleeo-cleeo-cleeo…*; *clee'ëëoop, clee'ëëoop…* (W), *kleee-kleee-kleee'eeoo-keee'eeoo*, ringing *kliklikliklikli, kleeyow-kleeyow…* (E). **SS** Yellow-billed Cacique (yellow eyes, deeper bill, dark rump); Red-rumped (larger, brighter red rump patch; no overlap). **Status** Uncommon, declining in W?

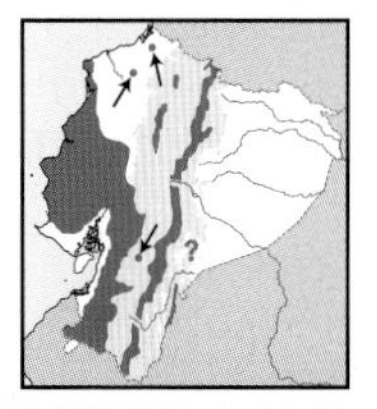

Yellow-billed Cacique *Amblycercus holosericeus* 22–24cm

W lowlands to subtropics in Loja; E subtropics to temperate zone (to 1,700m in SW, 1,900–3,100m in E). Deciduous and semi-deciduous forest, woodland (W), montane forest (E). ***Ivory bill, pale yellow eyes***. All black. Races *flavirostris* (W), *australis* (E). Skulking pairs in dense undergrowth, sometimes in dense shrubby clearings (W) or bamboo (E), often joins mixed flocks. **Voice** Loud, steady whistles: *peeo-peeo-peeo-peeo…* (W), more ringing *teee-teew, teee-teew, teee-tewe…* (E); also harsh *waa-waa-waa…*, *whew-whew…*and others. **SS** No other cacique in range has whitish bill and yellow eyes. **Status** Uncommon, rather local.

♂
juvenile
Mountain Cacique
juvenile
adult
Ecuadorian Cacique
Solitary Black Cacique
flavicrissus
adult
juvenile
cela
adult
Yellow-rumped Cacique
♂
♂
juvenile
♀
Red-rumped Cacique
♂
pacificus
♂
juvenile
♂
Yellow-billed Cacique
♂
adult
juvenile
♀
uropygialis
Scarlet-rumped Cacique

Long-tailed, elegant birds; pointed bill shorter than caciques. W species more tolerant of disturbance than in E. W species can be quite difficult to identify but they overlap little and segregate by habitat. One migrant (Baltimore) is rare in Ecuador. Spectacular voices in resident species, variable and rather melodious.

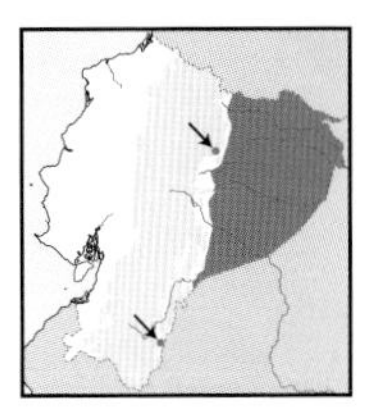

Orange-backed Troupial *Icterus croconotus* 23–24cm

E lowlands to foothills (to 750m). Dense second growth, regenerating scrub, lake and river edges, islands. ***Orange eyes, bright blue orbital skin.*** Black mask and bib, ***bright orange crown, upperparts and breast to vent.*** Black wings, ***white patch in secondaries.*** Nominate. Mostly in pairs, rather shy in dense lower growth, often near oxbow lakes; difficult to see; not with mixed flocks; uses cacique or oropendola nests. **Voice** Loud, musical *toooü'ee?, toooü'ee?, toooü'eeee?, tóow...* in long, leisurely repetitions; whistled *doo-dooEE-dooo?* **SS** Confusion unlikely; rare transient Baltimore Oriole has orange limited to lower underparts. **Status** Fairly common.

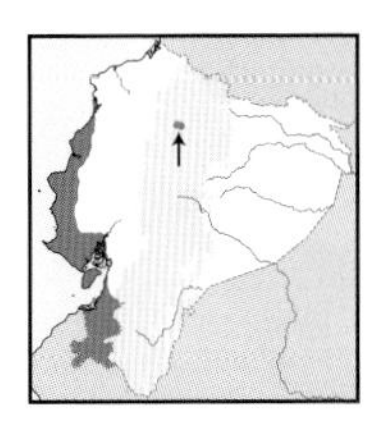

White-edged Oriole *Icterus graceannae* 20–21cm

SW lowlands, to subtropics in Loja (mainly below 500m, to 1,700m in Loja, locally higher). Xeric scrub, deciduous, semi-deciduous woodland, borders. Velvet black face and bib, ***bright yellow head, nape and underparts***, black back, yellow rump. Wings black, yellow shoulders and ***large white patch in tertials.*** Outer tail feathers ***edged white.*** Juv. duller overall, but ***white*** in wings and tail evident. Monotypic. Small family groups in lower growth, shier than Yellow-tailed, often loosely associated with mixed flocks. **Voice** Rapid, lively, musical *cheeu-cheeu-w'wree'u, cheërru-cheërru...weEE-weEE..., chwee-chwee-churr, ch'wheee-ch'wheee....* **SS** Yellow-tailed (little white in wings, extensive yellow in tail), usually segregated by habitat; more musical, higher-pitched voice. **Status** Uncommon.

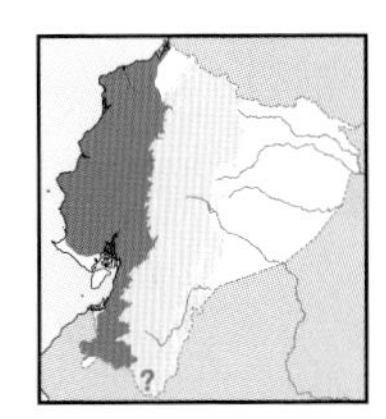

Yellow-tailed Oriole *Icterus mesomelas* 20–22cm

W lowlands to foothills, to subtropics in Loja (mainly to 900m, to 1,700m in Loja). Humid forest borders, woodland, clearings, gardens. Velvet black face and ***large*** bib, ***golden-yellow head, underparts and rump***, black back and wings. Yellow wing stripe, ***little white in tertials***, black tail, ***yellow outer feathers.*** Race *taczanowskii.* Conspicuous and noisy, often with mixed flocks and regular in open modified habitats. **Voice** Rich, melodious *cheed-chuu'peét, cheed-chuu'peét..., che-chu'peet-peet, che-chu'peet-peet..., chuu-pet'chuu-pet'chuu-pet'...; chee'kuupu, chee'kuupu...; wheeen'wa, wheeen'wa....* **SS** Yellower White-edged Oriole has more extensive white in wings, white in tail; lower-pitched, less melodious voice; see Yellow-backed in extreme NW. **Status** Fairly common.

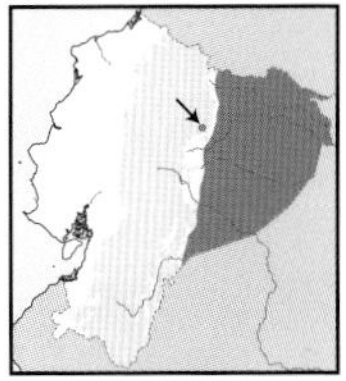

Epaulet Oriole *Icterus cayanensis* 20–22cm

E lowlands (to 500m). Canopy and borders, woody clearings. ***Bright yellow crown to nape, shoulders, rump and thighs*** contrasting with velvet black rest; long tail. Race *chrysocephalus*, possibly valid species. Mostly pairs to small groups, not regularly with mixed flocks; agilely searches foliage including palm fronds, fond of large nectar-rich flowers; jerks tail constantly. **Voice** Sweet, repetitive, variable *s'keey-s'keey, too-pyiu-toow, tee'ee'ee'ee; féet-véeu.* **SS** Confusion unlikely. **Status** Uncommon.

Yellow-backed Oriole *Icterus chrysater* 20–22cm

One record from extreme NW (L. Navarrete). Recalls Yellow-tailed but back all ***yellow***, tail ***black, no*** yellow bar in scapulars. Juv. has ***paler*** back, darker head, ***plainer*** wings than Yellow-backed. Race *giraudii*? Small family group in mangrove subcanopy; elsewhere conspicuous, often in pairs or small parties, with mixed-species flocks. **Voice** Sweet, clear, whistled series: *wheer-hee, who-hee... ha-heet, wita-wita-wita...* (HB); musical *chert* calls; *whirt-whirt-whirt...* in alarm, and others (JB). **SS** White-edged, of drier areas, has black back, patterned wings, pale undertail; not expected in sympatry. **Status** Possibly local breeder.

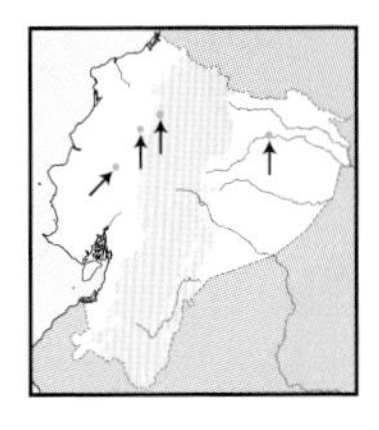

Baltimore Oriole *Icterus galbula* 17–20cm

Scarce records in E & W lowlands (below 800m). Humid forest borders, woodland, adjacent clearings. ***Black hood, orange to yellow-orange shoulders, lower back, rump and underparts, two white wingbars.*** ♀: dull olive-brown above mottled dusky. Blackish wings, ***two white wingbars, orange-buff underparts.*** Juv. similar but plainer above. Monotypic. Singly at flowering edge trees; elsewhere gathers in small flocks, often with mixed flocks. **Voice** Not very vocal in winter, slow whistles and chatters reported (JB). **SS** ♂ unmistakable; cf. ♀ and juv. with White-winged Tanager (richer yellow below, plain upperparts, shorter, less pointed bill). **Status** Rare boreal visitor (Nov–Apr).

juvenile
adult
Orange-backed Troupial
immature
adult
White-edged Oriole
juvenile
♀
♂
Yellow-tailed Oriole
juvenile
adult
Epaulet Oriole
juvenile
♀
♂
Yellow-backed Oriole
♀
juvenile
♂
immature
♂
Baltimore Oriole

With the notable exception of Oriole Blackbird along Amazonian rivers, all species on this plate are glossy black, but some more glossy than others. While W blackbirds are common and widespread in open and semi-open habitats, the two E species are rare and very local. Breeding system varies from parasitism to colonial.

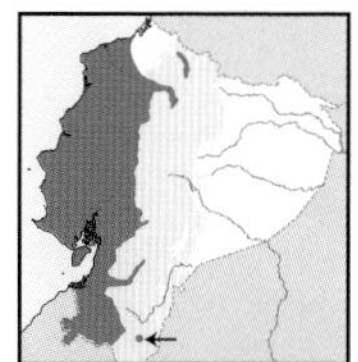

Scrub Blackbird *Dives warczewiczi* 22–24cm

W lowlands to foothills, in Loja to subtropics (to 2,100m, locally higher, and into inter-Andean valleys). Open and semi-open areas. Black *pointed* bill. Entirely glossy black, rather long tail. ♀ and imm. duller. Nominate. Conspicuous, noisy groups; bobs while singing. **Voice** Very loud, penetrating, whistled *wit-zeet'zeeeeét, teeét*; *woeep-cléo-cléo, cleé-cleé-cleé*; *cléo-weelee, glee'é*. **SS** Smaller ♂ Shiny Cowbird (more glossy, bill shorter and straighter, weaker legs); some vocalisations of Great-tailed Grackle are rather similar but higher-pitched. **Status** Common, spreading with deforestation.

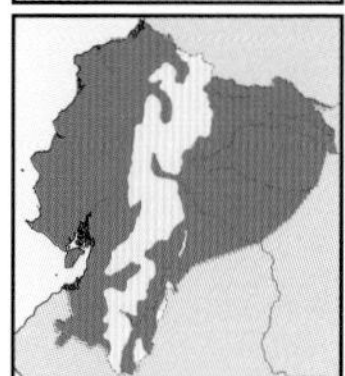

Shiny Cowbird *Molothrus bonariensis* 20–22cm

E & W lowlands to subtropics, locally in inter-Andean valleys (mostly to 1,500m, locally to 2,400m). Open areas. Short, *straight, pointed* bill. *Glossy purplish-black, wings and tail greener*. ♀: *dull brown* with *faint pale supercilium*; *duskier* in SW. Races *aequatorialis* (W), *occidentalis* (SW), *riparius* (E). Monospecific flocks of up 40+ birds; parasitises a wide variety of passerines. **Voice** Piercing *twéép* followed by complex ringing warbling and twittering; also *purr-purr-purr-tseeééEE*. **SS** Scrub Blackbird (less glossy, longer-billed, longer-legged), Velvet-fronted Grackle (not glossy, shorter-billed, long, fan-shaped tail), see White-lined Tanager. **Status** Common, spreading following deforestation.

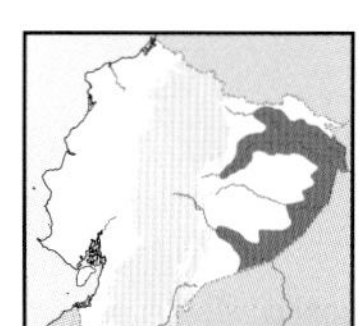

Oriole Blackbird *Gymnomystax mexicanus* 27–31cm

E lowlands (below 300m). Large rivers, sand bars, river islands. Large black pointed bill. Black skin on face; *bright golden-yellow head and underparts, black upperparts, wings and tail. Black crown* in duller juv. Monotypic. Pairs to small loose flocks (larger when not breeding) typically forage on ground or perch on logs and stumps. **Voice** Odd, harsh, metallic *zhiiirrr-zhiirr-zhíng...zhiirrr-zhrree*; liquid, ringing *cleén* and metallic *chínk!* **SS** Unmistakable. **Status** Uncommon.

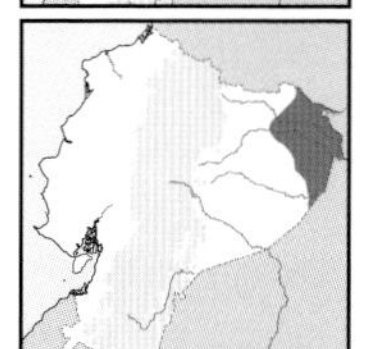

Velvet-fronted Grackle *Lampropsar tanagrinus* 19–22cm

E lowlands (to 300m). *Várzea* and riparian forest along large rivers and lakes. *Slim*, with *short pointed bill. Velvety black, plush front*, rather long, *fan-shaped tail*. Nominate. Compact, wary flocks at various heights, sometimes on ground or on floating vegetation but usually well concealed in dense lower growth. **Voice** Chuckling, fast *ch'k'k'k'k, chd'chd'chd'chd'chd* or gurgling *chuh-chuh-chuh, cheed-duu'duu...*; *trrrt-rrrrt* scolds. **SS** Shiny Cowbird is more glossed purplish or greenish overall, bill straighter and thicker; brown ♀ helps, voices differ considerably. **Status** Uncommon, rather local.

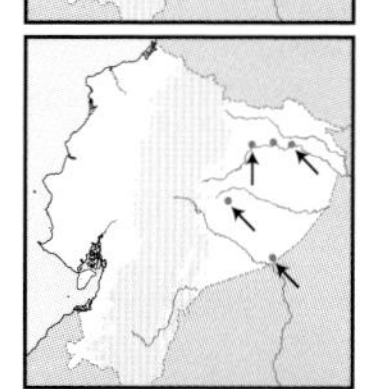

Pale-eyed Blackbird *Agelasticus xanthophthalmus* 20–21cm

Very local in E lowlands (below 300m). Tall swampy grass at lakesides. *Short pointed bill, pale yellow eyes*. Entirely *velvet black*, rather short tail. Monotypic. Pairs tend to remain well hidden in dense grasses near water level, but singing birds may perch atop tall grass. **Voice** Loud ringing whistles, *téw-téw-téw...*; harsh *chék* calls. **SS** Velvet-fronted Grackle has longer, fan-shaped tail, dark eyes; larger Solitary Black Cacique (longer, deeper-based pale bill, dark eyes). **Status** Very rare, possibly declining; VU in Ecuador.

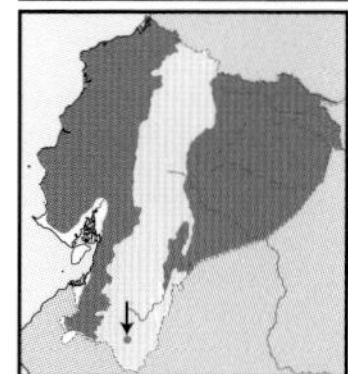

Giant Cowbird *Molothrus oryzivorus* 31–39cm

E & W lowlands to subtropics (to 2,000m). Forest borders, woodland, adjacent clearings. Massive black bill, *red* (♂) or *yellow eyes* (♀). *Glossy purple-black overall*, long tail, *hunchbacked, small-headed* silhouette. ♀: smaller. Nominate. Singles, pairs or small flocks often feed on ground or follow oropendolas in canopy borders; parasitises oropendola nests; flap-flap-flap-glide flight. **Voice** Piercing, unpleasant chattering *gla-díkdík!, dzk-dzk-dzk, cleég-dz'dz'dzí, chégg-chégg...*, but not very vocal. **SS** Great-tailed Grackle has longer, more slender bill, longer, creased tail; ♀ very different. **Status** Uncommon.

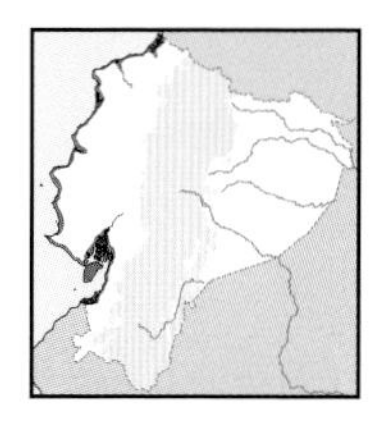

Great-tailed Grackle *Quiscalus mexicanus* 33–46cm

Coast S to C Manabí, Santa Elena Peninsula and Guayaquil Gulf (sea level). *Heavy black bill, yellow eyes*. Black overall, *glossed purple on head and underparts, bluish below; very long creased tail*. ♀: smaller, mostly *brown, paler eyebrow and underparts*; darker rear body, wing-coverts somewhat glossed green, *shorter tail*. Race *peruvianus*. Conspicuous scavenger, on ground at mangrove borders, tidal flats, estuaries, beaches. **Voice** Noisy, strident, fast *teeg, wíí-wíí-wíí-wíííi, wheeéék!...chá-chaak, wheeéé-chuk-chu'u'u'u...* and others. **SS** Confusion unlikely by habitat, size, tail shape. Giant Cowbird can look pale-eyed. **Status** Fairly common, spreading?

Scrub Blackbird
Shiny Cowbird
riparius
♂
occidentalis
♂
♀
aequatorialis
♀
occidentalis
♀
riparius
juvenile
juvenile
adult
Oriole Blackbird
Velvet-fronted Grackle
♀
♂
Pale-eyed Blackbird
♀
♂
Giant Cowbird
♀
♂
Great-tailed Grackle

PLATE 291: GRASSLAND ICTERIDS AND INTRODUCED PASSERINES

Three distinctive icterids of grassy fields, the two residents are common, widespread and spreading following deforestation. Bobolink is rare in Ecuador. Also, the only introduced passerines (to date). The sparrow has been established in Ecuador for nearly five decades, whilst the munias have recently colonised the W lowlands, where large-scale agriculture has dramatically changed the landscape. Small parties of escaped Red-crested Cardinal (*Paroaria coronata*) (not illustrated) seem likely to become established in Quito urban parks. More to come on this species.

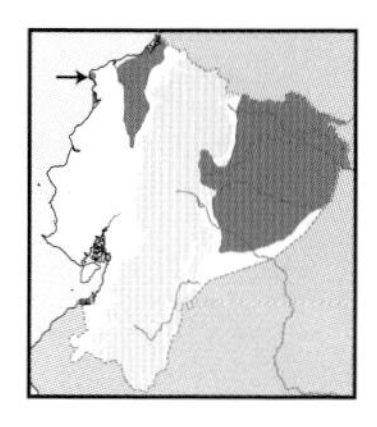

Red-breasted Meadowlark *Sturnella militaris* — 19–20cm

E & NW lowlands (below 400m in E, locally higher; below 400m in W). Open grassy areas. ***Velvet black***, with ***bright red throat to upper belly***. ♀: streaked blackish and buff above, ***long buff-white supercilium and coronal stripe***. Buff below, breast narrowly streaked, some ***red on belly***. Monotypic. Feeds near ground in tall grass; displaying ♂ perches on posts, stumps and other exposed perches. **Voice** High, metallic *tic-tic-péé* followed by long buzzy trill *bzeeeeee'e'e'e*; sputtering *pee-tititíti*; dry *djír*. **SS** In W, Peruvian Meadowlark (white supercilium, streaked upperparts, white underwing), ♀ duskier below, pink-washed breast, longer bill; see Bobolink. **Status** Uncommon, recently spreading.

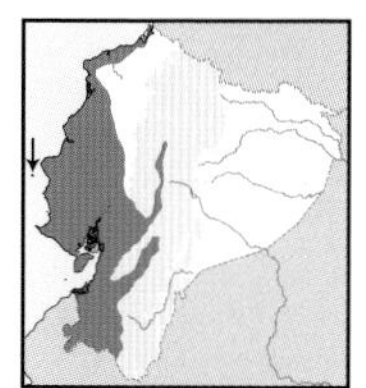

Peruvian Meadowlark *Sturnella bellicosa* — 20–21cm

W lowlands, to subtropics and inter-Andean valleys in S (to 2,500m). Open and semi-open grassy areas, arid scrub. ***Long*** pointed bill. ***Long whitish supercilium***, back streaked buff-brown. ***Bright red throat to upper belly, white underwing***. ♀: no red below, ***only faint pink wash on breast***, whitish throat. Nominate. Feeds on ground in cover but displaying ♂ sings from exposed perch; may form large aggregations when not breeding. **Voice** Fast *fifi'prr-bzzzzz*, sometimes more notes added, others shorter; sharp *cha'k* calls. **SS** See previous species. **Status** Fairly common.

Bobolink *Dolichonyx oryzivorus* — 17–18cm

E lowlands, few records in W lowlands and Andes (mostly below 300m). Tall grassland. ***Short conical bill. Velvet black head and underparts, golden nape, whitish scapulars and rump***. ♀: ***pinkish bill***, head boldly ***striped buff and dusky***, back markedly streaked tawny. Whitish chin, ***buff to tawny underparts***, dusky breast spotting. Short dusky, ***spiky tail***. Monotypic. Observed with seedeater flocks in tall grass, often low enough to be missed. **Voice** Metallic *fínk!* or *pink!* **SS** ♀ meadowlarks have longer, pointed bills and pink on breast; larger than most emberizids (Tumbes, Yellow-browed and Grasshopper Sparrows, ♀ Collared Warbling-Finch); see Dickcissel (no striped head, paler underparts, longer bill, etc.). **Status** Rare transient (Oct–Nov, Apr–May).

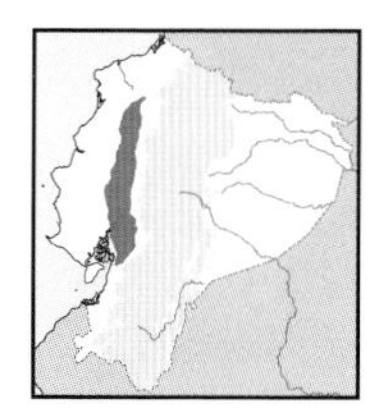

[i]Tricoloured Munia *Lonchura malacca* — 11–12cm

Several records from W lowlands, N to S Esmeraldas (to 600m). Grassy and scrubby, degraded areas. ***Stout, pale greyish bill. Black foreparts***, contrasting ***chestnut wings and upperparts***. Variable below, from ***white to cinnamon***, black central belly to vent. Juv. ***dull*** brownish-buff, paler below, ***wedge-shaped tail***. Monotypic? Small groups to several dozens reported (O. Carrión) feeding near ground in roadside rice paddies, grasses and crops; introduced from captive populations. **Voice** Clear *peet!*, squeaky song ending in clear *peeee* (R *et al.*). **SS** Confusion unlikely with resident seedeaters, only seed-finches have comparably large bills, but deeper-based. **Status** Uncertain, likely established and increasing; controls needed.

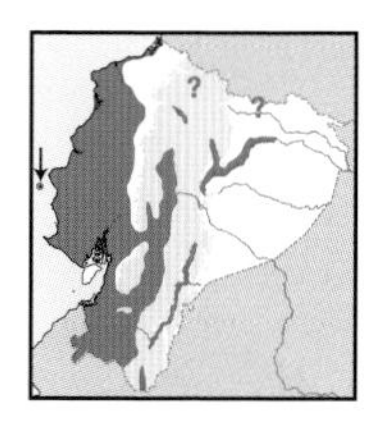

[i]House Sparrow *Passer domesticus* — 15cm

W lowlands, to S subtropics, inter-Andean valleys N to Pichincha, SE foothills (to 3,250m). Modified habitats. ***Rounded*** crown, ***black bib, chestnut upper face to neck-sides***. Brown above streaked dusky, white wingbar, ***buff underparts***. ♀: ***dusky-brown above***, back streaked darker, ***faint supercilium***, whitish chin, ***pale buff below***. Presumably nominate. Gregarious in parks and town squares, but not away from towns or crops; nests in palm or *Araucaria* fronds and leaves, apparently displaces Rufous-collared Sparrows. **Voice** Constant *pirrt-pirrt...* chattering, *piit-piit...*calls. **SS** Rufous-collared has crested head, no black bib, greyish underparts, etc. **Status** Locally common, spreading.

juvenile
♀
♂
immature
juvenile
♂
♀
♂
Red-breasted Meadowlark
Peruvian Meadowlark
♂
♂
♀
♀
Bobolink
cinnamon-bellied
morph
adult
juvenile
Tricoloured Munia
♂
breeding
♂
non-breeding
House Sparrow
♀
juvenile

APPENDIX 1: POTENTIAL SPECIES LIST

This list includes some of the species likely to be found in Ecuador, including its oceanic territory, in the near future. Some are migrants, others are breeding residents whose natural ranges lie very close to continental Ecuador. A few seabirds in this list have been recorded at the Galápagos Islands marine area.

English name	Scientific name	Expected region
Galápagos Penguin	*Spheniscus mendiculus*	offshore or SW coastline
Bulwer's Petrel	*Bulweria bulwerii*	offshore
Cook's Petrel	*Pterodroma cookii*	offshore
Kermadec Petrel	*Pterodroma neglecta*	offshore
Murphy's Petrel	*Pterodroma ultima*	offshore
Juan Fernández Petrel	*Pterodroma externa*	offshore
Grey Petrel	*Procellaria cinerea*	offshore
Black-bellied Storm-Petrel	*Fregetta tropica*	offshore
White-bellied Storm-Petrel	*Fregetta grallaria*	offshore
Leach's Storm-Petrel	*Oceanodroma leucorhoa*	offshore
Bare-throated Tiger-Heron	*Tigrisoma mexicanum*	mangroves of extreme SW
White-winged Guan	*Penelope albipennis*	dry forests of SW
Amazonian Pygmy-Owl	*Glaucidium hardyi*	far E lowlands
White-winged Potoo	*Nyctibius leucopterus*	extreme SE lowlands
Chuck-will's-widow	*Antrostomus carolinensis*	boreal winter visitor to N Andes
Black Swift	*Cypseloides niger*	boreal winter visitor to N
Steely-vented Hummingbird	*Amazilia saucerottei*	Andes of extreme NW
Rufous-necked Puffbird	*Malacoptila rufa*	far NE lowlands
Versicoloured Barbet	*Eubucco versicolor*	Andean slopes of extreme SE
Coastal Miner	*Geositta peruviana*	arid SW
Zimmer's Woodcreeper	*Dendroplex kienerii*	riparian areas of extreme E
Red-and-white Spinetail	*Certhiaxis mustelinus*	river islands of extreme E
Black-tailed Antbird	*Myrmoborus melanurus*	flooded areas of extreme E
White-masked Antbird	*Pithys castanea*	far SE lowlands
Brownish Elaenia	*Elaenia pelzelni*	islands of Rio Napo in far E
Helmeted Pygmy-Tyrant	*Lophotriccus galeatus*	lowlands of extreme NE

English name	Scientific name	Expected region
Olive Flycatcher	*Mitrephanes olivaceus*	extreme SE Andes and Cordillera del Cóndor
Rufous Flycatcher	*Myiarchus semirufus*	dry areas in extreme SW
Golden-breasted Fruiteater	*Pipreola auropectus*	extreme NW Andean slopes
Peruvian Plantcutter	*Phytotoma raimondii*	dry areas in extreme SW
Peruvian Martin	*Progne murphyi*	SW coastline
White-headed Wren	*Campylorhynchus albobrunneus*	extreme NW lowlands
Sooty-headed Wren	*Pheugopedius spadix*	Andean slopes of extreme NW
Orange-billed Nightingale-Thrush	*Catharus aurantiirostris*	Andean slopes of extreme N
Veery	*Catharus fuscescens*	E foothills and lowlands
Várzea Thrush	*Turdus sanchezorum*	riparian habitats of far E
Red-crested Cardinal	*Paroaria coronata*	feral in urban parks in Quito
Little Inca-Finch	*Incaspiza watkinsi*	extreme SE, in Zumba region

APPENDIX 2: CHECKLIST OF THE BIRDS OF ECUADOR

This checklist comprises all the species accepted for Ecuador, as validated by CERO. Species marked with an asterisk (*) are hypothetical species that require supporting evidence of occurrence in Ecuador. Taxonomy and sequence follow the SACC (December 2015). For the most recent version of the SACC list, see www.museum.lsu.edu/~Remsen/SACCBaseline.htm

Tinamidae

- ❑ Tawny-breasted Tinamou *Nothocercus julius*
- ❑ Highland Tinamou *Nothocercus bonapartei*
- ❑ Grey Tinamou *Tinamus tao*
- ❑ Black Tinamou *Tinamus osgoodi*
- ❑ Great Tinamou *Tinamus major*
- ❑ White-throated Tinamou *Tinamus guttatus*
- ❑ Berlepsch's Tinamou *Crypturellus berlepschi*
- ❑ Cinereous Tinamou *Crypturellus cinereus*
- ❑ Little Tinamou *Crypturellus soui*
- ❑ Brown Tinamou *Crypturellus obsoletus*
- ❑ Undulated Tinamou *Crypturellus undulatus*
- ❑ Pale-browed Tinamou *Crypturellus transfasciatus*
- ❑ Variegated Tinamou *Crypturellus variegatus*
- ❑ Bartlett's Tinamou *Crypturellus bartletti*
- ❑ Tataupa Tinamou *Crypturellus tataupa*
- ❑ Andean Tinamou *Nothoprocta pentlandii*
- ❑ Curve-billed Tinamou *Nothoprocta curvirostris*

Anhimidae

- ❑ Horned Screamer *Anhima cornuta*

Anatidae

- ❑ Fulvous Whistling-Duck *Dendrocygna bicolor*
- ❑ Black-bellied Whistling-Duck *Dendrocygna autumnalis*
- ❑ Orinoco Goose *Oressochen jubatus*
- ❑ Muscovy Duck *Cairina moschata*
- ❑ Comb Duck *Sarkidiornis melanotos*
- ❑ Brazilian Teal *Amazonetta brasiliensis*
- ❑ Torrent Duck *Merganetta armata*
- ❑ American Wigeon *Anas americana*
- ❑ Green-winged Teal *Anas crecca*
- ❑ Andean Teal *Anas andium*
- ❑ Northern Pintail *Anas acuta*
- ❑ Yellow-billed Pintail *Anas georgica*
- ❑ White-cheeked Pintail *Anas bahamensis*
- ❑ Blue-winged Teal *Anas discors*
- ❑ Cinnamon Teal *Anas cyanoptera*
- ❑ Northern Shoveler *Anas clypeata*
- ❑ Southern Pochard *Netta erythrophthalma*
- ❑ Ring-necked Duck *Aythya collaris*
- ❑ Lesser Scaup *Aythya affinis*
- ❑ Masked Duck *Nomonyx dominicus*
- ❑ Ruddy Duck *Oxyura jamaicensis*

Cracidae

- ❑ Sickle-winged Guan *Chamaepetes goudotii*
- ❑ Bearded Guan *Penelope barbata*
- ❑ Baudó Guan *Penelope ortoni*
- ❑ Andean Guan *Penelope montagnii*
- ❑ Spix's Guan *Penelope jacquacu*
- ❑ Crested Guan *Pipile purpurascens*
- ❑ Blue-throated Piping-Guan *Pipile cumanensis*
- ❑ Wattled Guan *Aburria aburri*
- ❑ Rufous-headed Chachalaca *Ortalis erythroptera*
- ❑ Speckled Chachalaca *Ortalis guttata*
- ❑ Nocturnal Curassow *Nothocrax urumutum*
- ❑ Great Curassow *Crax rubra*
- ❑ Wattled Curassow *Crax globulosa*
- ❑ Salvin's Curassow *Mitu salvini*

Odontophoridae

- ❑ Marbled Wood-Quail *Odontophorus gujanensis*
- ❑ Rufous-fronted Wood-Quail *Odontophorus erythrops*
- ❑ Dark-backed Wood-Quail *Odontophorus melanonotus*
- ❑ Rufous-breasted Wood-Quail *Odontophorus speciosus*
- ❑ Starred Wood-Quail *Odontophorus stellatus*
- ❑ Tawny-faced Quail *Rhynchortyx cinctus*

Podicipedidae

- ❑ Least Grebe *Tachybaptus dominicus*
- ❑ Pied-billed Grebe *Podilymbus podiceps*
- ❑ Great Grebe *Podiceps major*
- ❑ Silvery Grebe *Podiceps occipitalis*

Phoenicopteridae

- ❑ Chilean Flamingo *Phoenicopterus chilensis*

Spheniscidae

- ❑ Humboldt Penguin *Spheniscus humboldti*

Diomedeidae

- ❑ Waved Albatross *Phoebastria irrorata*
- ❑ Black-browed Albatross *Thalassarche melanophris*

- ❑ Buller's Albatross *Thalassarche bulleri**
- ❑ Salvin's Albatross *Thalassarche salvini**

Procellariidae

- ❑ Southern Giant Petrel *Macronectes giganteus*
- ❑ Southern Fulmar *Fulmarus glacialoides**
- ❑ Cape Petrel *Daption capense*
- ❑ Galápagos Petrel *Pterodroma phaeopygia*
- ❑ White-chinned Petrel *Procellaria aequinoctialis**
- ❑ Parkinson's Petrel *Procellaria parkinsoni*
- ❑ Wedge-tailed Shearwater *Ardenna pacifica*
- ❑ Buller's Shearwater *Ardenna bulleri*
- ❑ Sooty Shearwater *Ardenna grisea*
- ❑ Pink-footed Shearwater *Ardenna creatopus*
- ❑ Galápagos Shearwater *Puffinus subalaris**

Hydrobatidae

- ❑ Wilson's Storm-Petrel *Oceanites oceanicus*
- ❑ Elliot's Storm-Petrel *Oceanites gracilis*
- ❑ White-faced Storm-Petrel *Pelagodroma marina*
- ❑ Least Storm-Petrel *Oceanodroma microsoma*
- ❑ Wedge-rumped Storm-Petrel *Oceanodroma tethys*
- ❑ Band-rumped Storm-Petrel *Oceanodroma castro**
- ❑ Markham's Storm-Petrel *Oceanodroma markhami*
- ❑ Ringed Storm-Petrel *Oceanodroma hornbyi*
- ❑ Ashy Storm-Petrel *Oceanodroma homochroa**
- ❑ Black Storm-Petrel *Oceanodroma melania*

Phaethontidae

- ❑ Red-billed Tropicbird *Phaethon aethereus*

Ciconiidae

- ❑ Jabiru *Jabiru mycteria*
- ❑ Wood Stork *Mycteria americana*

Fregatidae

- ❑ Magnificent Frigatebird *Fregata magnificens*
- ❑ Great Frigatebird *Fregata minor**

Sulidae

- ❑ (Cape) Gannet cf. *Morus capensis**
- ❑ Blue-footed Booby *Sula nebouxii*
- ❑ Peruvian Booby *Sula variegata*
- ❑ Masked Booby *Sula dactylatra*
- ❑ Nazca Booby *Sula granti*
- ❑ Red-footed Booby *Sula sula*
- ❑ Brown Booby *Sula leucogaster*

Phalacrocoracidae

- ❑ Neotropic Cormorant *Phalacrocorax brasilianus*
- ❑ Guanay Cormorant *Phalacrocorax bougainvillii*

Anhingidae

- ❑ Anhinga *Anhinga anhinga*

Pelecanidae

- ❑ Brown Pelican *Pelecanus occidentalis*
- ❑ Peruvian Pelican *Pelecanus thagus*

Ardeidae

- ❑ Rufescent Tiger-Heron *Tigrisoma lineatum*
- ❑ Fasciated Tiger-Heron *Tigrisoma fasciatum*
- ❑ Agami Heron *Agamia agami*
- ❑ Boat-billed Heron *Cochlearius cochlearius*
- ❑ Zigzag Heron *Zebrilus undulatus*
- ❑ Pinnated Bittern *Botaurus pinnatus*
- ❑ Least Bittern *Ixobrychus exilis*
- ❑ Stripe-backed Bittern *Ixobrychus involucris**
- ❑ Black-crowned Night-Heron *Nycticorax nycticorax*
- ❑ Yellow-crowned Night-Heron *Nyctanassa violacea*
- ❑ Green Heron *Butorides virescens*
- ❑ Striated Heron *Butorides striata*
- ❑ Cattle Egret *Bubulcus ibis*
- ❑ Great Blue Heron *Ardea herodias*
- ❑ Cocoi Heron *Ardea cocoi*
- ❑ Great Egret *Ardea alba*
- ❑ Whistling Heron *Syrigma sibilatrix**
- ❑ Capped Heron *Pilherodius pileatus*
- ❑ Tricoloured Heron *Egretta tricolor*
- ❑ Reddish Egret *Egretta rufescens*
- ❑ Snowy Egret *Egretta thula*
- ❑ Little Blue Heron *Egretta caerulea*

Threskiornithidae

- ❑ White Ibis *Eudocimus albus*
- ❑ Scarlet Ibis *Eudocimus ruber*
- ❑ Glossy Ibis *Plegadis falcinellus*
- ❑ Puna Ibis *Plegadis ridgwayi**
- ❑ Green Ibis *Mesembrinibis cayennensis*
- ❑ Bare-faced Ibis *Phimosus infuscatus*
- ❑ Black-faced Ibis *Theristicus melanopis*
- ❑ Roseate Spoonbill *Platalea ajaja*

Cathartidae

- ❑ Turkey Vulture *Cathartes aura*
- ❑ Lesser Yellow-headed Vulture *Cathartes burrovianus**
- ❑ Greater Yellow-headed Vulture *Cathartes melambrotus*
- ❑ Black Vulture *Coragyps atratus*
- ❑ King Vulture *Sarcoramphus papa*
- ❑ Andean Condor *Vultur gryphus*

Pandionidae

- ❑ Osprey *Pandion haliaetus*

Accipitridae

- ❑ White-tailed Kite *Elanus leucurus*
- ❑ Pearl Kite *Gampsonyx swainsonii*
- ❑ Hook-billed Kite *Chondrohierax uncinatus*
- ❑ Grey-headed Kite *Leptodon cayanensis*
- ❑ Swallow-tailed Kite *Elanoides forficatus*
- ❑ Crested Eagle *Morphnus guianensis*
- ❑ Harpy Eagle *Harpia harpyja*
- ❑ Black Hawk-Eagle *Spizaetus tyrannus*
- ❑ Black-and-white Hawk-Eagle *Spizaetus melanoleucus*
- ❑ Ornate Hawk-Eagle *Spizaetus ornatus*
- ❑ Black-and chestnut Eagle *Spizaetus isidori*
- ❑ Black-collared Hawk *Busarellus nigricollis*
- ❑ Snail Kite *Rostrhamus sociabilis*
- ❑ Slender-billed Kite *Helicolestes hamatus*
- ❑ Double-toothed Kite *Harpagus bidentatus*
- ❑ Rufous-thighed Kite *Harpagus diodon**
- ❑ Mississippi Kite *Ictinia mississippiensis**
- ❑ Plumbeous Kite *Ictinia plumbea*
- ❑ Cinereous Harrier *Circus cinereus*
- ❑ Grey-bellied Hawk *Accipiter poliogaster*
- ❑ Tiny Hawk *Accipiter superciliosus*
- ❑ Semicollared Hawk *Accipiter collaris*
- ❑ Sharp-shinned Hawk *Accipiter striatus*
- ❑ Bicoloured Hawk *Accipiter bicolor*
- ❑ Crane Hawk *Geranospiza caerulescens*
- ❑ Plumbeous Hawk *Cryptoleucopteryx plumbea*
- ❑ Slate-coloured Hawk *Buteogallus schistaceus*
- ❑ Common Black Hawk *Buteogallus anthracinus*
- ❑ Savanna Hawk *Buteogallus meridionalis*
- ❑ Great Black Hawk *Buteogallus urubitinga*
- ❑ Solitary Eagle *Buteogallus solitarius*
- ❑ Barred Hawk *Morphnarchus princeps*
- ❑ Roadside Hawk *Rupornis magnirostris*
- ❑ Harris's Hawk *Parabuteo unicinctus*
- ❑ White-rumped Hawk *Parabuteo leucorrhous*
- ❑ Red-backed Hawk *Geranoaetus polyosoma*
- ❑ Black-chested Buzzard-Eagle *Geranoaetus melanoleucus*
- ❑ White Hawk *Pseudastur albicollis*
- ❑ Grey-backed Hawk *Pseudastur occidentalis*
- ❑ Semiplumbeous Hawk *Leucopternis semiplumbeus*
- ❑ Black-faced Hawk *Leucopternis melanops*
- ❑ Grey-lined Hawk *Buteo nitidus*
- ❑ Broad-winged Hawk *Buteo platypterus*
- ❑ White-throated Hawk *Buteo albigula*
- ❑ Short-tailed Hawk *Buteo brachyurus*
- ❑ Swainson's Hawk *Buteo swainsoni*
- ❑ Zone-tailed Hawk *Buteo albonotatus*

Aramidae

- ❑ Limpkin *Aramus guarauna*

Psophiidae

- ❑ Grey-winged Trumpeter *Psophia crepitans*

Rallidae

- ❑ Clapper Rail *Rallus longirostris*
- ❑ Virginia Rail *Rallus limicola*
- ❑ Brown Wood-Rail *Aramides wolfi*
- ❑ Grey-necked Wood-Rail *Aramides cajaneus*
- ❑ Rufous-necked Wood-Rail *Aramides axillaris*
- ❑ Red-winged Wood-Rail *Aramides calopterus*
- ❑ Uniform Crake *Amaurolimnas concolor*
- ❑ Chestnut-headed Crake *Anurolimnas castaneiceps*
- ❑ Russet-crowned Crake *Anurolimnas viridis*
- ❑ Black-banded Crake *Anurolimnas fasciatus*
- ❑ Rufous-sided Crake *Laterallus melanophaius*
- ❑ White-throated Crake *Laterallus albigularis*
- ❑ Grey-breasted Crake *Laterallus exilis*
- ❑ Yellow-breasted Crake *Porzana flaviventer**
- ❑ Sora *Porzana carolina*
- ❑ Ash-throated Crake *Mustelirallus albicollis*
- ❑ Colombian Crake *Mustelirallus colombianus*
- ❑ Paint-billed Crake *Mustelirallus erythrops*
- ❑ Spotted Rail *Pardirallus maculatus*
- ❑ Blackish Rail *Pardirallus nigricans*
- ❑ Plumbeous Rail *Pardirallus sanguinolentus*
- ❑ Common Gallinule *Gallinula galeata*
- ❑ Purple Gallinule *Porphyrio martinicus*
- ❑ Azure Gallinule *Porphyrio flavirostris*
- ❑ American Coot *Fulica americana*
- ❑ Slate-coloured Coot *Fulica ardesiaca*

Heliornithidae

- ❑ Sungrebe *Heliornis fulica*

Eurypygidae

- ❑ Sunbittern *Eurypyga helias*

Charadriidae

- ❑ American Golden Plover *Pluvialis dominica*
- ❑ Pacific Golden Plover *Pluvialis fulva**
- ❑ Black-bellied Plover *Pluvialis squatarola*
- ❑ Tawny-throated Dotterel *Oreopholus ruficollis*
- ❑ Pied Lapwing *Vanellus cayanus*
- ❑ Southern Lapwing *Vanellus chilensis*
- ❑ Andean Lapwing *Vanellus resplendens*
- ❑ Semipalmated Plover *Charadrius semipalmatus*
- ❑ Piping Plover *Charadrius melodus*

- ❑ Wilson's Plover *Charadrius wilsonia*
- ❑ Killdeer *Charadrius vociferus*
- ❑ Snowy Plover *Charadrius nivosus*
- ❑ Collared Plover *Charadrius collaris*

Haematopodidae

- ❑ American Oystercatcher *Haematopus palliatus*
- ❑ Blackish Oystercatcher *Haematopus ater*

Recurvirostridae

- ❑ Black-necked Stilt *Himantopus mexicanus*
- ❑ American Avocet *Recurvirostra americana*

Burhinidae

- ❑ Peruvian Thick-knee *Burhinus superciliaris*

Scolopacidae

- ❑ Upland Sandpiper *Bartramia longicauda*
- ❑ Whimbrel *Numenius phaeopus*
- ❑ Long-billed Curlew *Numenius americanus**
- ❑ Hudsonian Godwit *Limosa haemastica*
- ❑ Marbled Godwit *Limosa fedoa*
- ❑ Ruddy Turnstone *Arenaria interpres*
- ❑ Surfbird *Aphriza virgata*
- ❑ Red Knot *Calidris canutus*
- ❑ Sanderling *Calidris alba*
- ❑ Semipalmated Sandpiper *Calidris pusilla*
- ❑ Western Sandpiper *Calidris mauri*
- ❑ Least Sandpiper *Calidris minutilla*
- ❑ White-rumped Sandpiper *Calidris fuscicollis*
- ❑ Baird's Sandpiper *Calidris bairdii*
- ❑ Pectoral Sandpiper *Calidris melanotos*
- ❑ Dunlin *Calidris alpina*
- ❑ Sharp-tailed Sandpiper *Calidris acuminata**
- ❑ Curlew Sandpiper *Calidris ferruginea*
- ❑ Stilt Sandpiper *Calidris himantopus*
- ❑ Buff-breasted Sandpiper *Tryngites subruficollis*
- ❑ Ruff *Philomachus pugnax**
- ❑ Short-billed Dowitcher *Limnodromus griseus*
- ❑ Long-billed Dowitcher *Limnodromus scolopaceus*
- ❑ Imperial Snipe *Gallinago imperialis*
- ❑ Andean Snipe *Gallinago jamesoni*
- ❑ Noble Snipe *Gallinago nobilis*
- ❑ Wilson's Snipe *Gallinago delicata*
- ❑ South American Snipe *Gallinago paraguaiae*
- ❑ Puna Snipe *Gallinago andina**
- ❑ Wilson's Phalarope *Phalaropus tricolor*
- ❑ Red-necked Phalarope *Phalaropus lobatus*
- ❑ Red Phalarope *Phalaropus fulicarius*
- ❑ Spotted Sandpiper *Actitis macularius*
- ❑ Solitary Sandpiper *Tringa solitaria*
- ❑ Wandering Tattler *Tringa incana*
- ❑ Greater Yellowlegs *Tringa melanoleuca*
- ❑ Willet *Tringa semipalmata*
- ❑ Lesser Yellowlegs *Tringa flavipes*

Thinocoridae

- ❑ Rufous-bellied Seedsnipe *Attagis gayi*
- ❑ Least Seedsnipe *Thinocorus rumicivorus*

Jacanidae

- ❑ Wattled Jacana *Jacana jacana*

Stercorariidae

- ❑ Chilean Skua *Stercorarius chilensis**
- ❑ South Polar Skua *Stercorarius maccormicki**
- ❑ Pomarine Jaeger *Stercorarius pomarinus*
- ❑ Parasitic Jaeger *Stercorarius parasiticus*
- ❑ Long-tailed Jaeger *Stercorarius longicaudus*

Laridae

- ❑ Swallow-tailed Gull *Creagrus furcatus*
- ❑ Sabine's Gull *Xema sabini*
- ❑ Bonaparte's Gull *Chroicocephalus philadelphia*
- ❑ Andean Gull *Chroicocephalus serranus*
- ❑ Grey-hooded Gull *Chroicocephalus cirrocephalus*
- ❑ Grey Gull *Leucophaeus modestus*
- ❑ Laughing Gull *Leucophaeus atricilla*
- ❑ Franklin's Gull *Leucophaeus pipixcan*
- ❑ Belcher's Gull *Larus belcheri**
- ❑ Ring-billed Gull *Larus delawarensis*
- ❑ California Gull *Larus californicus*
- ❑ Kelp Gull *Larus dominicanus*
- ❑ Lesser Black-backed Gull *Larus fuscus*
- ❑ Herring Gull *Larus argentatus*
- ❑ Glaucous-winged Gull *Larus glaucescens*
- ❑ Brown Noddy *Anous stolidus*
- ❑ Black Noddy *Anous minutus**
- ❑ Bridled Tern *Onychoprion anaethetus*
- ❑ Least Tern *Sternula antillarum**
- ❑ Yellow-billed Tern *Sternula superciliaris*
- ❑ Peruvian Tern *Sternula lorata*
- ❑ Large-billed Tern *Phaetusa simplex*
- ❑ Gull-billed Tern *Gelochelidon nilotica*
- ❑ Caspian Tern *Hydroprogne caspia*
- ❑ Inca Tern *Larosterna inca*
- ❑ Black Tern *Chlidonias niger*
- ❑ Common Tern *Sterna hirundo*
- ❑ Arctic Tern *Sterna paradisaea*
- ❑ South American Tern *Sterna hirundinacea*

- ❑ Elegant Tern *Thalasseus elegans*
- ❑ Sandwich Tern *Thalasseus sandvicensis*
- ❑ Royal Tern *Thalasseus maximus*

Rynchopidae

- ❑ Black Skimmer *Rynchops niger*

Columbidae

- ❑ Rock Pigeon *Columba livia*
- ❑ Scaled Pigeon *Patagioenas speciosa*
- ❑ Band-tailed Pigeon *Patagioenas fasciata*
- ❑ Pale-vented Pigeon *Patagioenas cayennensis*
- ❑ Peruvian Pigeon *Patagioenas oenops**
- ❑ Plumbeous Pigeon *Patagioenas plumbea*
- ❑ Ruddy Pigeon *Patagioenas subvinacea*
- ❑ Dusky Pigeon *Patagioenas goodsoni*
- ❑ Purple Quail-Dove *Geotrygon purpurata*
- ❑ Sapphire Quail-Dove *Geotrygon saphirina*
- ❑ Ruddy Quail-Dove *Geotrygon montana*
- ❑ Olive-backed Quail-Dove *Leptotrygon veraguensis*
- ❑ White-tipped Dove *Leptotila verreauxi*
- ❑ Ochre-bellied Dove *Leptotila ochraceiventris*
- ❑ Grey-fronted Dove *Leptotila rufaxilla*
- ❑ Pallid Dove *Leptotila pallida*
- ❑ White-throated Quail-Dove *Zentrygon frenata*
- ❑ West Peruvian Dove *Zenaida meloda*
- ❑ Eared Dove *Zenaida auriculata*
- ❑ Common Ground Dove *Columbina passerina*
- ❑ Plain-breasted Ground Dove *Columbina minuta*
- ❑ Ruddy Ground Dove *Columbina talpacoti*
- ❑ Ecuadorian Ground Dove *Columbina buckleyi*
- ❑ Croaking Ground Dove *Columbina cruziana*
- ❑ Blue Ground Dove *Claravis pretiosa*
- ❑ Maroon-chested Ground Dove *Claravis mondetoura*
- ❑ Black-winged Ground Dove *Metriopelia melanoptera*

Opisthocomidae

- ❑ Hoatzin *Opisthocomus hoazin*

Cuculidae

- ❑ Little Cuckoo *Coccycua minuta*
- ❑ Dwarf Cuckoo *Coccycua pumila*
- ❑ Squirrel Cuckoo *Piaya cayana*
- ❑ Black-bellied Cuckoo *Piaya melanogaster*
- ❑ Dark-billed Cuckoo *Coccyzus melacoryphus*
- ❑ Yellow-billed Cuckoo *Coccyzus americanus*
- ❑ Pearly-breasted Cuckoo *Coccyzus euleri*
- ❑ Black-billed Cuckoo *Coccyzus erythropthalmus*
- ❑ Grey-capped Cuckoo *Coccyzus lansbergi*
- ❑ Greater Ani *Crotophaga major*
- ❑ Smooth-billed Ani *Crotophaga ani*
- ❑ Groove-billed Ani *Crotophaga sulcirostris*
- ❑ Striped Cuckoo *Tapera naevia*
- ❑ Pheasant Cuckoo *Dromococcyx phasianellus*
- ❑ Pavonine Cuckoo *Dromococcyx pavoninus*
- ❑ Rufous-vented Ground-Cuckoo *Neomorphus geoffroyi*
- ❑ Banded Ground-Cuckoo *Neomorphus radiolosus*
- ❑ Red-billed Ground-Cuckoo *Neomorphus pucheranii**

Tytonidae

- ❑ Barn Owl *Tyto alba*

Strigidae

- ❑ Tropical Screech-Owl *Megascops choliba*
- ❑ Peruvian Screech-Owl *Megascops roboratus*
- ❑ Colombian Screech-Owl *Megascops colombianus*
- ❑ Rufescent Screech-Owl *Megascops ingens*
- ❑ Cinnamon Screech-Owl *Megascops petersoni*
- ❑ Tawny-bellied Screech-Owl *Megascops watsonii*
- ❑ Vermiculated Screech-Owl *Megascops guatemalae*
- ❑ White-throated Screech-Owl *Megascops albogularis*
- ❑ Crested Owl *Lophostrix cristata*
- ❑ Spectacled Owl *Pulsatrix perspicillata*
- ❑ Band-bellied Owl *Pulsatrix melanota*
- ❑ Great Horned Owl *Bubo virginianus*
- ❑ Mottled Owl *Ciccaba virgata*
- ❑ Black-and-white Owl *Ciccaba nigrolineata*
- ❑ 'San Isidro' Owl *Ciccaba* sp.
- ❑ Black-banded Owl *Ciccaba huhula*
- ❑ Rufous-banded Owl *Ciccaba albitarsis*
- ❑ Cloud-forest Pygmy-Owl *Glaucidium nubicola*
- ❑ Andean Pygmy-Owl *Glaucidium jardinii*
- ❑ Subtropical Pygmy-Owl *Glaucidium parkeri*
- ❑ Central American Pygmy-Owl *Glaucidium griseiceps*
- ❑ Ferruginous Pygmy-Owl *Glaucidium brasilianum*
- ❑ Peruvian Pygmy-Owl *Glaucidium peruanum*
- ❑ Burrowing Owl *Athene cunicularia*
- ❑ Buff-fronted Owl *Aegolius harrisii*
- ❑ Striped Owl *Asio clamator*
- ❑ Stygian Owl *Asio stygius*
- ❑ Short-eared Owl *Asio flammeus*

Steatornithidae

- ❑ Oilbird *Steatornis caripensis*

Nyctibiidae

- ❑ Great Potoo *Nyctibius grandis*
- ❑ Long-tailed Potoo *Nyctibius aethereus*
- ❑ Common Potoo *Nyctibius griseus*
- ❑ Andean Potoo *Nyctibius maculosus*

- ❑ Rufous Potoo *Nyctibius bracteatus*

Caprimulgidae

- ❑ Nacunda Nighthawk *Chordeiles nacunda*
- ❑ Sand-coloured Nighthawk *Chordeiles rupestris*
- ❑ Lesser Nighthawk *Chordeiles acutipennis*
- ❑ Common Nighthawk *Chordeiles minor*
- ❑ Short-tailed Nighthawk *Lurocalis semitorquatus*
- ❑ Rufous-bellied Nighthawk *Lurocalis rufiventris*
- ❑ Band-tailed Nighthawk *Nyctiprogne leucopyga*
- ❑ Blackish Nightjar *Nyctipolus nigrescens*
- ❑ Band-winged Nightjar *Systellura longirostris*
- ❑ Common Pauraque *Nyctidromus albicollis*
- ❑ Scrub Nightjar *Nyctidromus anthonyi*
- ❑ Swallow-tailed Nightjar *Uropsalis segmentata*
- ❑ Lyre-tailed Nightjar *Uropsalis lyra*
- ❑ White-tailed Nightjar *Hydropsalis cayennensis*
- ❑ Spot-tailed Nightjar *Hydropsalis maculicaudus*
- ❑ Ladder-tailed Nightjar *Hydropsalis climacocerca*
- ❑ Chocó Poorwill *Nyctiphrynus rosenbergi*
- ❑ Ocellated Poorwill *Nyctiphrynus ocellatus*
- ❑ Rufous Nightjar *Antrostomus rufus*

Apodidae

- ❑ Spot-fronted Swift *Cypseloides cherriei*
- ❑ White-chinned Swift *Cypseloides cryptus*
- ❑ White-chested Swift *Cypseloides lemosi*
- ❑ Chestnut-collared Swift *Streptoprocne rutila*
- ❑ White-collared Swift *Streptoprocne zonaris*
- ❑ Band-rumped Swift *Chaetura spinicaudus*
- ❑ Grey-rumped Swift *Chaetura cinereiventris*
- ❑ Pale-rumped Swift *Chaetura egregia*
- ❑ Chimney Swift *Chaetura pelagica*
- ❑ Chapman's Swift *Chaetura chapmani**
- ❑ Short-tailed Swift *Chaetura brachyura*
- ❑ White-tipped Swift *Aeronautes montivagus*
- ❑ Fork-tailed Palm-Swift *Tachornis squamata*
- ❑ Lesser Swallow-tailed Swift *Panyptila cayennensis*

Trochilidae

- ❑ Fiery Topaz *Topaza pyra*
- ❑ White-necked Jacobin *Florisuga mellivora*
- ❑ White-tipped Sicklebill *Eutoxeres aquila*
- ❑ Buff-tailed Sicklebill *Eutoxeres condamini*
- ❑ Bronzy Hermit *Glaucis aeneus*
- ❑ Rufous-breasted Hermit *Glaucis hirsutus*
- ❑ Band-tailed Barbthroat *Threnetes ruckeri*
- ❑ Pale-tailed Barbthroat *Threnetes leucurus*
- ❑ Black-throated Hermit *Phaethornis atrimentalis*
- ❑ Stripe-throated Hermit *Phaethornis striigularis*
- ❑ Grey-chinned Hermit *Phaethornis griseogularis*
- ❑ Reddish Hermit *Phaethornis ruber*
- ❑ White-bearded Hermit *Phaethornis hispidus*
- ❑ White-whiskered Hermit *Phaethornis yaruqui*
- ❑ Green Hermit *Phaethornis guy*
- ❑ Tawny-bellied Hermit *Phaethornis syrmatophorus*
- ❑ Straight-billed Hermit *Phaethornis bourcieri*
- ❑ Long-billed Hermit *Phaethornis longirostris*
- ❑ Great-billed Hermit *Phaethornis malaris*
- ❑ Green-fronted Lancebill *Doryfera ludovicae*
- ❑ Blue-fronted Lancebill *Doryfera johannae*
- ❑ Wedge-billed Hummingbird *Schistes geoffroyi*
- ❑ Brown Violetear *Colibri delphinae*
- ❑ Green Violetear *Colibri thalassinus*
- ❑ Sparkling Violetear *Colibri coruscans*
- ❑ Tooth-billed Hummingbird *Androdon aequatorialis*
- ❑ Purple-crowned Fairy *Heliothryx barroti*
- ❑ Black-eared Fairy *Heliothryx auritus*
- ❑ Green-tailed Goldenthroat *Polytmus theresiae**
- ❑ Fiery-tailed Awlbill *Avocettula recurvirostris*
- ❑ Black-throated Mango *Anthracothorax nigricollis*
- ❑ Amethyst-throated Sunangel *Heliangelus amethysticollis*
- ❑ Gorgeted Sunangel *Heliangelus strophianus*
- ❑ Tourmaline Sunangel *Heliangelus exortis*
- ❑ Little Sunangel *Heliangelus micraster*
- ❑ Purple-throated Sunangel *Heliangelus viola*
- ❑ Royal Sunangel *Heliangelus regalis*
- ❑ Wire-crested Thorntail *Discosura popelairii*
- ❑ Green Thorntail *Discosura conversii*
- ❑ Black-bellied Thorntail *Discosura langsdorffi*
- ❑ Rufous-crested Coquette *Lophornis delattrei**
- ❑ Spangled Coquette *Lophornis stictolophus*
- ❑ Festive Coquette *Lophornis chalybeus*
- ❑ Ecuadorian Piedtail *Phlogophilus hemileucurus*
- ❑ Speckled Hummingbird *Adelomyia melanogenys*
- ❑ Long-tailed Sylph *Aglaiocercus kingii*
- ❑ Violet-tailed Sylph *Aglaiocercus coelestis*
- ❑ Ecuadorian Hillstar *Oreotrochilus chimborazo*
- ❑ Andean Hillstar *Oreotrochilus estella*
- ❑ Mountain Avocetbill *Opisthoprora euryptera*
- ❑ Black-tailed Trainbearer *Lesbia victoriae*
- ❑ Green-tailed Trainbearer *Lesbia nuna*
- ❑ Purple-backed Thornbill *Ramphomicron microrhynchum*
- ❑ Rufous-capped Thornbill *Chalcostigma ruficeps*
- ❑ Blue-mantled Thornbill *Chalcostigma stanleyi*
- ❑ Rainbow-bearded Thornbill *Chalcostigma herrani*
- ❑ Tyrian Metaltail *Metallura tyrianthina*

- ❑ Viridian Metaltail *Metallura williami*
- ❑ Violet-throated Metaltail *Metallura baroni*
- ❑ Neblina Metaltail *Metallura odomae*
- ❑ Greenish Puffleg *Haplophaedia aureliae*
- ❑ Hoary Puffleg *Haplophaedia lugens*
- ❑ Black-breasted Puffleg *Eriocnemis nigrivestis*
- ❑ Glowing Puffleg *Eriocnemis vestita*
- ❑ Black-thighed Puffleg *Eriocnemis derbyi*
- ❑ Turquoise-throated Puffleg *Eriocnemis godini*
- ❑ Sapphire-vented Puffleg *Eriocnemis luciani*
- ❑ Golden-breasted Puffleg *Eriocnemis mosquera*
- ❑ Emerald-bellied Puffleg *Eriocnemis alinae*
- ❑ Shining Sunbeam *Aglaeactis cupripennis*
- ❑ Bronzy Inca *Coeligena coeligena*
- ❑ Brown Inca *Coeligena wilsoni*
- ❑ Collared Inca *Coeligena torquata*
- ❑ Rainbow Starfrontlet *Coeligena iris*
- ❑ Buff-winged Starfrontlet *Coeligena lutetiae*
- ❑ Mountain Velvetbreast *Lafresnaya lafresnayi*
- ❑ Sword-billed Hummingbird *Ensifera ensifera*
- ❑ Great Sapphirewing *Pterophanes cyanopterus*
- ❑ Buff-tailed Coronet *Boissonneaua flavescens*
- ❑ Chestnut-breasted Coronet *Boissonneaua matthewsii*
- ❑ Velvet-purple Coronet *Boissonneaua jardini*
- ❑ Booted Racket-tail *Ocreatus underwoodii*
- ❑ White-tailed Hillstar *Urochroa bougueri*
- ❑ Purple-bibbed Whitetip *Urosticte benjamini*
- ❑ Rufous-vented Whitetip *Urosticte ruficrissa*
- ❑ Pink-throated Brilliant *Heliodoxa gularis*
- ❑ Black-throated Brilliant *Heliodoxa schreibersii*
- ❑ Gould's Jewelfront *Heliodoxa aurescens*
- ❑ Fawn-breasted Brilliant *Heliodoxa rubinoides*
- ❑ Green-crowned Brilliant *Heliodoxa jacula*
- ❑ Empress Brilliant *Heliodoxa imperatrix*
- ❑ Violet-fronted Brilliant *Heliodoxa leadbeateri*
- ❑ Giant Hummingbird *Patagona gigas*
- ❑ Long-billed Starthroat *Heliomaster longirostris*
- ❑ Blue-tufted Starthroat *Heliomaster furcifer**
- ❑ Purple-collared Woodstar *Myrtis fanny*
- ❑ Peruvian Sheartail *Thaumastura cora**
- ❑ White-bellied Woodstar *Chaetocercus mulsant*
- ❑ Little Woodstar *Chaetocercus bombus*
- ❑ Gorgeted Woodstar *Chaetocercus heliodor*
- ❑ Esmeraldas Woodstar *Chaetocercus berlepschi*
- ❑ Short-tailed Woodstar *Myrmia micrura*
- ❑ Amethyst Woodstar *Calliphlox amethystina*
- ❑ Purple-throated Woodstar *Calliphlox mitchellii*
- ❑ Western Emerald *Chlorostilbon melanorhynchus*
- ❑ Blue-tailed Emerald *Chlorostilbon mellisugus*
- ❑ Blue-chinned Sapphire *Chlorestes notata*
- ❑ Violet-headed Hummingbird *Klais guimeti*
- ❑ Grey-breasted Sabrewing *Campylopterus largipennis*
- ❑ Lazuline Sabrewing *Campylopterus falcatus*
- ❑ Napo Sabrewing *Campylopterus villaviscensio*
- ❑ White-vented Plumeleteer *Chalybura buffonii*
- ❑ Bronze-tailed Plumeleteer *Chalybura urochrysia*
- ❑ Crowned Woodnymph *Thalurania colombica*
- ❑ Fork-tailed Woodnymph *Thalurania furcata*
- ❑ Many-spotted Hummingbird *Taphrospilus hypostictus*
- ❑ Tumbes Hummingbird *Leucippus baeri*
- ❑ Spot-throated Hummingbird *Leucippus taczanowskii**
- ❑ Olive-spotted Hummingbird *Leucippus chlorocercus*
- ❑ Rufous-tailed Hummingbird *Amazilia tzacatl*
- ❑ Amazilia Hummingbird *Amazilia amazilia*
- ❑ Andean Emerald *Amazilia franciae*
- ❑ Glittering-throated Emerald *Amazilia fimbriata*
- ❑ Sapphire-spangled Emerald *Amazilia lactea**
- ❑ Blue-chested Hummingbird *Amazilia amabilis*
- ❑ Purple-chested Hummingbird *Amazilia rosenbergi*
- ❑ Golden-tailed Sapphire *Chrysuronia oenone*
- ❑ Violet-bellied Hummingbird *Damophila julie*
- ❑ Rufous-throated Sapphire *Hylocharis sapphirina*
- ❑ White-chinned Sapphire *Hylocharis cyanus*
- ❑ Humboldt's Sapphire *Hylocharis humboldtii*
- ❑ Blue-headed Sapphire *Hylocharis grayi*

Trogonidae

- ❑ Pavonine Quetzal *Pharomachrus pavoninus*
- ❑ Golden-headed Quetzal *Pharomachrus auriceps*
- ❑ Crested Quetzal *Pharomachrus antisianus*
- ❑ Slaty-tailed Trogon *Trogon massena*
- ❑ Blue-tailed Trogon *Trogon comptus*
- ❑ Ecuadorian Trogon *Trogon mesurus*
- ❑ Black-tailed Trogon *Trogon melanurus*
- ❑ White-tailed Trogon *Trogon chionurus*
- ❑ Green-backed Trogon *Trogon viridis*
- ❑ Gartered Trogon *Trogon caligatus*
- ❑ Amazonian Trogon *Trogon ramonianus*
- ❑ Blue-crowned Trogon *Trogon curucui*
- ❑ Black-throated Trogon *Trogon rufus*
- ❑ Collared Trogon *Trogon collaris*
- ❑ Masked Trogon *Trogon personatus*

Alcedinidae

- ❑ Ringed Kingfisher *Megaceryle torquata*
- ❑ Belted Kingfisher *Megaceryle alcyon**
- ❑ Amazon Kingfisher *Chloroceryle amazona*

- ❑ Green Kingfisher *Chloroceryle americana*
- ❑ Green-and-rufous Kingfisher *Chloroceryle inda*
- ❑ American Pygmy Kingfisher *Chloroceryle aenea*

Momotidae

- ❑ Broad-billed Motmot *Electron platyrhynchum*
- ❑ Rufous Motmot *Baryphthengus martii*
- ❑ Whooping Motmot *Momotus subrufescens*
- ❑ Amazonian Motmot *Momotus momota*
- ❑ Andean Motmot *Momotus aequatorialis*

Galbulidae

- ❑ White-eared Jacamar *Galbalcyrhynchus leucotis*
- ❑ Brown Jacamar *Brachygalba lugubris*
- ❑ Yellow-billed Jacamar *Galbula albirostris*
- ❑ Rufous-tailed Jacamar *Galbula ruficauda*
- ❑ White-chinned Jacamar *Galbula tombacea*
- ❑ Bluish-fronted Jacamar *Galbula cyanescens*
- ❑ Coppery-chested Jacamar *Galbula pastazae*
- ❑ Purplish Jacamar *Galbula chalcothorax*
- ❑ Paradise Jacamar *Galbula dea*
- ❑ Great Jacamar *Jacamerops aureus*

Bucconidae

- ❑ White-necked Puffbird *Notharchus hyperrhynchus*
- ❑ Black-breasted Puffbird *Notharchus pectoralis*
- ❑ Pied Puffbird *Notharchus tectus*
- ❑ Chestnut-capped Puffbird *Bucco macrodactylus*
- ❑ Spotted Puffbird *Bucco tamatia*
- ❑ Collared Puffbird *Bucco capensis*
- ❑ Barred Puffbird *Nystalus radiatus*
- ❑ Striolated Puffbird *Nystalus striolatus*
- ❑ White-chested Puffbird *Malacoptila fusca*
- ❑ White-whiskered Puffbird *Malacoptila panamensis*
- ❑ Black-streaked Puffbird *Malacoptila fulvogularis*
- ❑ Moustached Puffbird *Malacoptila mystacalis**
- ❑ Lanceolated Monklet *Micromonacha lanceolata*
- ❑ Rusty-breasted Nunlet *Nonnula rubecula*
- ❑ Brown Nunlet *Nonnula brunnea*
- ❑ White-faced Nunbird *Hapaloptila castanea*
- ❑ Black-fronted Nunbird *Monasa nigrifrons*
- ❑ White-fronted Nunbird *Monasa morphoeus*
- ❑ Yellow-billed Nunbird *Monasa flavirostris*
- ❑ Swallow-winged Puffbird *Chelidoptera tenebrosa*

Capitonidae

- ❑ Scarlet-crowned Barbet *Capito aurovirens*
- ❑ Orange-fronted Barbet *Capito squamatus*
- ❑ Five-coloured Barbet *Capito quinticolor*
- ❑ Gilded Barbet *Capito auratus*
- ❑ Lemon-throated Barbet *Eubucco richardsoni*
- ❑ Red-headed Barbet *Eubucco bourcierii*

Semnornithidae

- ❑ Toucan Barbet *Semnornis ramphastinus*

Ramphastidae

- ❑ Black-mandibled Toucan *Ramphastos ambiguus*
- ❑ White-throated Toucan *Ramphastos tucanus*
- ❑ Chocó Toucan *Ramphastos brevis*
- ❑ Channel-billed Toucan *Ramphastos vitellinus*
- ❑ Emerald Toucanet *Aulacorhynchus prasinus*
- ❑ Chestnut-tipped Toucanet *Aulacorhynchus derbianus*
- ❑ Crimson-rumped Toucanet *Aulacorhynchus haematopygus*
- ❑ Grey-breasted Mountain-Toucan *Andigena hypoglauca*
- ❑ Plate-billed Mountain-Toucan *Andigena laminirostris*
- ❑ Black-billed Mountain-Toucan *Andigena nigrirostris*
- ❑ Yellow-eared Toucanet *Selenidera spectabilis*
- ❑ Golden-collared Toucanet *Selenidera reinwardtii*
- ❑ Lettered Araçari *Pteroglossus inscriptus*
- ❑ Collared Araçari *Pteroglossus torquatus*
- ❑ Chestnut-eared Araçari *Pteroglossus castanotis*
- ❑ Many-banded Araçari *Pteroglossus pluricinctus*
- ❑ Ivory-billed Araçari *Pteroglossus azara*

Picidae

- ❑ Lafresnaye's Piculet *Picumnus lafresnayi*
- ❑ Ecuadorian Piculet *Picumnus sclateri*
- ❑ Rufous-breasted Piculet *Picumnus rufiventris*
- ❑ Olivaceous Piculet *Picumnus olivaceus*
- ❑ Yellow-tufted Woodpecker *Melanerpes cruentatus*
- ❑ Black-cheeked Woodpecker *Melanerpes pucherani*
- ❑ Smoky-brown Woodpecker *Picoides fumigatus*
- ❑ Red-rumped Woodpecker *Veniliornis kirkii*
- ❑ Little Woodpecker *Veniliornis passerinus*
- ❑ Scarlet-backed Woodpecker *Veniliornis callonotus*
- ❑ Yellow-vented Woodpecker *Veniliornis dignus*
- ❑ Bar-bellied Woodpecker *Veniliornis nigriceps*
- ❑ Red-stained Woodpecker *Veniliornis affinis*
- ❑ Chocó Woodpecker *Veniliornis chocoensis*
- ❑ White-throated Woodpecker *Piculus leucolaemus*
- ❑ Lita Woodpecker *Piculus litae*
- ❑ Yellow-throated Woodpecker *Piculus flavigula*
- ❑ Golden-green Woodpecker *Piculus chrysochloros*
- ❑ Golden-olive Woodpecker *Colaptes rubiginosus*
- ❑ Crimson-mantled Woodpecker *Colaptes rivolii*
- ❑ Spot-breasted Woodpecker *Colaptes punctigula*
- ❑ Andean Flicker *Colaptes rupicola*

- ❑ Cinnamon Woodpecker *Celeus loricatus*
- ❑ Scale-breasted Woodpecker *Celeus grammicus*
- ❑ Chestnut Woodpecker *Celeus elegans*
- ❑ Cream-coloured Woodpecker *Celeus flavus*
- ❑ Rufous-headed Woodpecker *Celeus spectabilis*
- ❑ Ringed Woodpecker *Celeus torquatus*
- ❑ Lineated Woodpecker *Dryocopus lineatus*
- ❑ Powerful Woodpecker *Campephilus pollens*
- ❑ Crimson-bellied Woodpecker *Campephilus haematogaster*
- ❑ Red-necked Woodpecker *Campephilus rubricollis*
- ❑ Crimson-crested Woodpecker *Campephilus melanoleucos*
- ❑ Guayaquil Woodpecker *Campephilus gayaquilensis*

Falconidae

- ❑ Laughing Falcon *Herpetotheres cachinnans*
- ❑ Barred Forest-Falcon *Micrastur ruficollis*
- ❑ Plumbeous Forest-Falcon *Micrastur plumbeus*
- ❑ Lined Forest-Falcon *Micrastur gilvicollis*
- ❑ Slaty-backed Forest-Falcon *Micrastur mirandollei*
- ❑ Collared Forest-Falcon *Micrastur semitorquatus*
- ❑ Buckley's Forest-Falcon *Micrastur buckleyi*
- ❑ Crested Caracara *Caracara cheriway*
- ❑ Red-throated Caracara *Ibycter americanus*
- ❑ Carunculated Caracara *Phalcoboenus carunculatus*
- ❑ Mountain Caracara *Phalcoboenus megalopterus*
- ❑ Black Caracara *Daptrius ater*
- ❑ Yellow-headed Caracara *Milvago chimachima*
- ❑ American Kestrel *Falco sparverius*
- ❑ Merlin *Falco columbarius*
- ❑ Bat Falcon *Falco rufigularis*
- ❑ Orange-breasted Falcon *Falco deiroleucus*
- ❑ Aplomado Falcon *Falco femoralis*
- ❑ Peregrine Falcon *Falco peregrinus*

Psittacidae

- ❑ Scarlet-shouldered Parrotlet *Touit huetii*
- ❑ Blue-fronted Parrotlet *Touit dilectissimus*
- ❑ Sapphire-rumped Parrotlet *Touit purpuratus*
- ❑ Spot-winged Parrotlet *Touit stictopterus*
- ❑ Barred Parakeet *Bolborhynchus lineola*
- ❑ Tui Parakeet *Brotogeris sanctithomae**
- ❑ Canary-winged Parakeet *Brotogeris versicolurus*
- ❑ Grey-cheeked Parakeet *Brotogeris pyrrhoptera*
- ❑ Cobalt-winged Parakeet *Brotogeris cyanoptera*
- ❑ Rusty-faced Parrot *Hapalopsittaca amazonina**
- ❑ Red-faced Parrot *Hapalopsittaca pyrrhops*
- ❑ Rose-faced Parrot *Pyrilia pulchra*
- ❑ Saffron-headed Parrot *Pyrilia pyrilia**
- ❑ Orange-cheeked Parrot *Pyrilia barrabandi*
- ❑ Red-billed Parrot *Pionus sordidus*
- ❑ Speckle-faced Parrot *Pionus tumultuosus*
- ❑ Blue-headed Parrot *Pionus menstruus*
- ❑ Bronze-winged Parrot *Pionus chalcopterus*
- ❑ Short-tailed Parrot *Graydidascalus brachyurus*
- ❑ Festive Amazon *Amazona festiva*
- ❑ Red-lored Amazon *Amazona autumnalis*
- ❑ Yellow-crowned Amazon *Amazona ochrocephala*
- ❑ Mealy Amazon *Amazona farinosa*
- ❑ Orange-winged Amazon *Amazona amazonica*
- ❑ Scaly-naped Amazon *Amazona mercenarius*
- ❑ Blue-winged Parrotlet *Forpus xanthopterygius*
- ❑ Dusky-billed Parrotlet *Forpus modestus*
- ❑ Pacific Parrotlet *Forpus coelestis*
- ❑ Black-headed Parrot *Pionites melanocephalus*
- ❑ Red-fan Parrot *Deroptyus accipitrinus*
- ❑ Rose-fronted Parakeet *Pyrrhura roseifrons*
- ❑ Maroon-tailed Parakeet *Pyrrhura melanura*
- ❑ El Oro Parakeet *Pyrrhura orcesi*
- ❑ White-necked Parakeet *Pyrrhura albipectus*
- ❑ Dusky-headed Parakeet *Aratinga weddellii*
- ❑ Red-bellied Macaw *Orthopsittaca manilatus*
- ❑ Blue-and-yellow Macaw *Ara ararauna*
- ❑ Military Macaw *Ara militaris*
- ❑ Great Green Macaw *Ara ambiguus*
- ❑ Scarlet Macaw *Ara macao*
- ❑ Red-and-green Macaw *Ara chloropterus*
- ❑ Chestnut-fronted Macaw *Ara severus*
- ❑ Golden-plumed Parakeet *Leptosittaca branickii*
- ❑ Yellow-eared Parrot *Ognorhynchus icterotis*
- ❑ Scarlet-fronted Parakeet *Psittacara wagleri*
- ❑ Red-masked Parakeet *Psittacara erythrogenys*
- ❑ White-eyed Parakeet *Psittacara leucophthalmus*

Sapayoidae

- ❑ Sapayoa *Sapayoa aenigma*

Thamnophilidae

- ❑ Rufous-rumped Antwren *Euchrepomis callinota*
- ❑ Chestnut-shouldered Antwren *Euchrepomis humeralis*
- ❑ Ash-winged Antwren *Euchrepomis spodioptila*
- ❑ Fasciated Antshrike *Cymbilaimus lineatus*
- ❑ Fulvous Antshrike *Frederickena fulva*
- ❑ Great Antshrike *Taraba major*
- ❑ Barred Antshrike *Thamnophilus doliatus*
- ❑ Chapman's Antshrike *Thamnophilus zarumae*
- ❑ Lined Antshrike *Thamnophilus tenuepunctatus*

- ❑ Collared Antshrike *Thamnophilus bernardi*
- ❑ Black-crowned Antshrike *Thamnophilus atrinucha*
- ❑ Plain-winged Antshrike *Thamnophilus schistaceus*
- ❑ Mouse-coloured Antshrike *Thamnophilus murinus*
- ❑ Cocha Antshrike *Thamnophilus praecox*
- ❑ Castelnau's Antshrike *Thamnophilus cryptoleucus*
- ❑ Northern Slaty Antshrike *Thamnophilus punctatus*
- ❑ Uniform Antshrike *Thamnophilus unicolor*
- ❑ White-shouldered Antshrike *Thamnophilus aethiops*
- ❑ Amazonian Antshrike *Thamnophilus amazonicus*
- ❑ Pearly Antshrike *Megastictus margaritatus*
- ❑ Black Bushbird *Neoctantes niger*
- ❑ Russet Antshrike *Thamnistes anabatinus*
- ❑ Plain Antvireo *Dysithamnus mentalis*
- ❑ Spot-crowned Antvireo *Dysithamnus puncticeps*
- ❑ Bicoloured Antvireo *Dysithamnus occidentalis*
- ❑ White-streaked Antvireo *Dysithamnus leucostictus*
- ❑ Dusky-throated Antshrike *Thamnomanes ardesiacus*
- ❑ Cinereous Antshrike *Thamnomanes caesius*
- ❑ Plain-throated Antwren *Isleria hauxwelli*
- ❑ Spot-winged Antshrike *Pygiptila stellaris*
- ❑ Checker-throated Antwren *Epinecrophylla fulviventris*
- ❑ Stipple-throated Antwren *Epinecrophylla haematonota*
- ❑ Brown-backed Antwren *Epinecrophylla fjeldsaai*
- ❑ Foothill Antwren *Epinecrophylla spodionota*
- ❑ Ornate Antwren *Epinecrophylla ornata*
- ❑ Rufous-tailed Antwren *Epinecrophylla erythrura*
- ❑ Pygmy Antwren *Myrmotherula brachyura*
- ❑ Moustached Antwren *Myrmotherula ignota*
- ❑ Amazonian Streaked Antwren *Myrmotherula multostriata*
- ❑ Pacific Antwren *Myrmotherula pacifica*
- ❑ Stripe-chested Antwren *Myrmotherula longicauda*
- ❑ White-flanked Antwren *Myrmotherula axillaris*
- ❑ Slaty Antwren *Myrmotherula schisticolor*
- ❑ Río Suno Antwren *Myrmotherula sunensis*
- ❑ Long-winged Antwren *Myrmotherula longipennis*
- ❑ Plain-winged Antwren *Myrmotherula behni*
- ❑ Grey Antwren *Myrmotherula menetriesii*
- ❑ Banded Antbird *Dichrozona cincta*
- ❑ Dugand's Antwren *Herpsilochmus dugandi*
- ❑ Ancient Antwren *Herpsilochmus gentryi*
- ❑ Yellow-breasted Antwren *Herpsilochmus axillaris*
- ❑ Rufous-winged Antwren *Herpsilochmus rufimarginatus*
- ❑ Dot-winged Antwren *Microrhopias quixensis*
- ❑ Striated Antbird *Drymophila devillei*
- ❑ Streak-headed Antbird *Drymophila striaticeps*
- ❑ Peruvian Warbling Antbird *Hypocnemis peruviana*
- ❑ Yellow-browed Antbird *Hypocnemis hypoxantha*
- ❑ Dusky Antbird *Cercomacroides tyrannina*
- ❑ Black Antbird *Cercomacroides serva*
- ❑ Blackish Antbird *Cercomacroides nigrescens*
- ❑ Riparian Antbird *Cercomacroides fuscicauda*
- ❑ Grey Antbird *Cercomacra cinerascens*
- ❑ Jet Antbird *Cercomacra nigricans*
- ❑ White-backed Fire-eye *Pyriglena leuconota*
- ❑ White-browed Antbird *Myrmoborus leucophrys*
- ❑ Ash-breasted Antbird *Myrmoborus lugubris*
- ❑ Black-faced Antbird *Myrmoborus myotherinus*
- ❑ Black-chinned Antbird *Hypocnemoides melanopogon*
- ❑ Black-and-white Antbird *Myrmochanes hemileucus*
- ❑ Silvered Antbird *Sclateria naevia*
- ❑ Slate-coloured Antbird *Schistocichla schistacea*
- ❑ Spot-winged Antbird *Schistocichla leucostigma*
- ❑ Chestnut-backed Antbird *Myrmeciza exsul*
- ❑ Esmeraldas Antbird *Myrmeciza nigricauda*
- ❑ Stub-tailed Antbird *Myrmeciza berlepschi*
- ❑ Zimmer's Antbird *Myrmeciza castanea*
- ❑ Black-throated Antbird *Myrmeciza atrothorax*
- ❑ White-shouldered Antbird *Myrmeciza melanoceps*
- ❑ Plumbeous Antbird *Myrmeciza hyperythra*
- ❑ Sooty Antbird *Myrmeciza fortis*
- ❑ Zeledon's Antbird *Myrmeciza zeledoni*
- ❑ Grey-headed Antbird *Myrmeciza griseiceps*
- ❑ Wing-banded Antbird *Myrmornis torquata*
- ❑ White-plumed Antbird *Pithys albifrons*
- ❑ Bicoloured Antbird *Gymnopithys bicolor*
- ❑ White-cheeked Antbird *Gymnopithys leucaspis*
- ❑ Lunulated Antbird *Gymnopithys lunulatus*
- ❑ Hairy-crested Antbird *Rhegmatorhina melanosticta*
- ❑ Spotted Antbird *Hylophylax naevioides*
- ❑ Spot-backed Antbird *Hylophylax naevius*
- ❑ Dot-backed Antbird *Hylophylax punctulatus*
- ❑ Scale-backed Antbird *Willisornis poecilinotus*
- ❑ Black-spotted Bare-eye *Phlegopsis nigromaculata*
- ❑ Reddish-winged Bare-eye *Phlegopsis erythroptera*
- ❑ Ocellated Antbird *Phaenostictus mcleannani*

Melanopareiidae

- ❑ Marañón Crescentchest *Melanopareia maranonica*
- ❑ Elegant Crescentchest *Melanopareia elegans*

Conopophagidae

- ❑ Chestnut-belted Gnateater *Conopophaga aurita*
- ❑ Ash-throated Gnateater *Conopophaga peruviana*
- ❑ Chestnut-crowned Gnateater *Conopophaga castaneiceps*
- ❑ Rufous-crowned Antpitta *Pittasoma rufopileatum*

Grallariidae

- ❑ Undulated Antpitta *Grallaria squamigera*
- ❑ Giant Antpitta *Grallaria gigantea*
- ❑ Moustached Antpitta *Grallaria alleni*
- ❑ Scaled Antpitta *Grallaria guatimalensis*
- ❑ Plain-backed Antpitta *Grallaria haplonota*
- ❑ Ochre-striped Antpitta *Grallaria dignissima*
- ❑ Chestnut-crowned Antpitta *Grallaria ruficapilla*
- ❑ Watkins's Antpitta *Grallaria watkinsi*
- ❑ Bicoloured Antpitta *Grallaria rufocinerea*
- ❑ Jocotoco Antpitta *Grallaria ridgelyi*
- ❑ Chestnut-naped Antpitta *Grallaria nuchalis*
- ❑ Yellow-breasted Antpitta *Grallaria flavotincta*
- ❑ White-bellied Antpitta *Grallaria hypoleuca*
- ❑ Rufous Antpitta *Grallaria rufula*
- ❑ Tawny Antpitta *Grallaria quitensis*
- ❑ Streak-chested Antpitta *Hylopezus perspicillatus*
- ❑ Thicket Antpitta *Hylopezus dives**
- ❑ White-lored Antpitta *Hylopezus fulviventris*
- ❑ Thrush-like Antpitta *Myrmothera campanisona*
- ❑ Ochre-breasted Antpitta *Grallaricula flavirostris*
- ❑ Peruvian Antpitta *Grallaricula peruviana*
- ❑ Rusty-breasted Antpitta *Grallaricula ferrugineipectus*
- ❑ Slate-crowned Antpitta *Grallaricula nana*
- ❑ Crescent-faced Antpitta *Grallaricula lineifrons*

Rhinocryptidae

- ❑ Rusty-belted Tapaculo *Liosceles thoracicus*
- ❑ Ocellated Tapaculo *Acropternis orthonyx*
- ❑ Ash-coloured Tapaculo *Myornis senilis*
- ❑ Blackish Tapaculo *Scytalopus latrans*
- ❑ Long-tailed Tapaculo *Scytalopus micropterus*
- ❑ White-crowned Tapaculo *Scytalopus atratus*
- ❑ Chocó Tapaculo *Scytalopus chocoensis*
- ❑ Ecuadorian Tapaculo *Scytalopus robbinsi*
- ❑ Nariño Tapaculo *Scytalopus vicinior*
- ❑ Spillmann's Tapaculo *Scytalopus spillmanni*
- ❑ Chusquea Tapaculo *Scytalopus parkeri*
- ❑ Páramo Tapaculo *Scytalopus opacus*

Formicariidae

- ❑ Rufous-capped Antthrush *Formicarius colma*
- ❑ Black-faced Antthrush *Formicarius analis*
- ❑ Black-headed Antthrush *Formicarius nigricapillus*
- ❑ Rufous-breasted Antthrush *Formicarius rufipectus*
- ❑ Short-tailed Antthrush *Chamaeza campanisona*
- ❑ Striated Antthrush *Chamaeza nobilis*
- ❑ Barred Antthrush *Chamaeza mollissima*

Furnariidae

- ❑ Tawny-throated Leaftosser *Sclerurus mexicanus*
- ❑ Short-billed Leaftosser *Sclerurus rufigularis*
- ❑ Scaly-throated Leaftosser *Sclerurus guatemalensis*
- ❑ Black-tailed Leaftosser *Sclerurus caudacutus*
- ❑ Grey-throated Leaftosser *Sclerurus albigularis*
- ❑ Slender-billed Miner *Geositta tenuirostris*
- ❑ Spot-throated Woodcreeper *Certhiasomus stictolaemus*
- ❑ Olivaceous Woodcreeper *Sittasomus griseicapillus*
- ❑ Long-tailed Woodcreeper *Deconychura longicauda*
- ❑ Tyrannine Woodcreeper *Dendrocincla tyrannina*
- ❑ Plain-brown Woodcreeper *Dendrocincla fuliginosa*
- ❑ White-chinned Woodcreeper *Dendrocincla merula*
- ❑ Wedge-billed Woodcreeper *Glyphorynchus spirurus*
- ❑ Cinnamon-throated Woodcreeper *Dendrexetastes rufigula*
- ❑ Long-billed Woodcreeper *Nasica longirostris*
- ❑ Northern Barred Woodcreeper *Dendrocolaptes sanctithomae*
- ❑ Amazonian Barred Woodcreeper *Dendrocolaptes certhia*
- ❑ Black-banded Woodcreeper *Dendrocolaptes picumnus*
- ❑ Bar-bellied Woodcreeper *Hylexetastes stresemanni*
- ❑ Strong-billed Woodcreeper *Xiphocolaptes promeropirhynchus*
- ❑ Striped Woodcreeper *Xiphorhynchus obsoletus*
- ❑ Ocellated Woodcreeper *Xiphorhynchus ocellatus*
- ❑ Elegant Woodcreeper *Xiphorhynchus elegans*
- ❑ Buff-throated Woodcreeper *Xiphorhynchus guttatus*
- ❑ Black-striped Woodcreeper *Xiphorhynchus lachrymosus*
- ❑ Spotted Woodcreeper *Xiphorhynchus erythropygius*
- ❑ Olive-backed Woodcreeper *Xiphorhynchus triangularis*
- ❑ Straight-billed Woodcreeper *Dendroplex picus*
- ❑ Red-billed Scythebill *Campylorhamphus trochilirostris*
- ❑ Curve-billed Scythebill *Campylorhamphus procurvoides*
- ❑ Brown-billed Scythebill *Campylorhamphus pusillus*
- ❑ Greater Scythebill *Drymotoxeres pucherani*
- ❑ Streak-headed Woodcreeper *Lepidocolaptes souleyetii*
- ❑ Montane Woodcreeper *Lepidocolaptes lacrymiger*
- ❑ Duida Woodcreeper *Lepidocolaptes duidae*
- ❑ Slender-billed Xenops *Xenops tenuirostris*
- ❑ Plain Xenops *Xenops minutus*
- ❑ Streaked Xenops *Xenops rutilans*
- ❑ Point-tailed Palmcreeper *Berlepschia rikeri*
- ❑ Rufous-tailed Xenops *Microxenops milleri*
- ❑ Buffy Tuftedcheek *Pseudocolaptes lawrencii*
- ❑ Streaked Tuftedcheek *Pseudocolaptes boissonneautii*
- ❑ Rusty-winged Barbtail *Premnornis guttuliger*
- ❑ Pale-legged Hornero *Furnarius leucopus*

- ❑ Pale-billed Hornero *Furnarius torridus**
- ❑ Lesser Hornero *Furnarius minor*
- ❑ Sharp-tailed Streamcreeper *Lochmias nematura*
- ❑ Chestnut-winged Cinclodes *Cinclodes albidiventris*
- ❑ Stout-billed Cinclodes *Cinclodes excelsior*
- ❑ Dusky-cheeked Foliage-gleaner *Anabazenops dorsalis*
- ❑ Slaty-winged Foliage-gleaner *Philydor fuscipenne*
- ❑ Rufous-rumped Foliage-gleaner *Philydor erythrocercum*
- ❑ Chestnut-winged Foliage-gleaner *Philydor erythropterum*
- ❑ Buff-fronted Foliage-gleaner *Philydor rufum*
- ❑ Cinnamon-rumped Foliage-gleaner *Philydor pyrrhodes*
- ❑ Montane Foliage-gleaner *Anabacerthia striaticollis*
- ❑ Scaly-throated Foliage-gleaner *Anabacerthia variegaticeps*
- ❑ Rufous-tailed Foliage-gleaner *Anabacerthia ruficaudata*
- ❑ Buff-browed Foliage-gleaner *Syndactyla rufosuperciliata*
- ❑ Lineated Foliage-gleaner *Syndactyla subalaris*
- ❑ Rufous-necked Foliage-gleaner *Syndactyla ruficollis*
- ❑ Chestnut-winged Hookbill *Ancistrops strigilatus*
- ❑ Henna-hooded Foliage-gleaner *Clibanornis erythrocephalus*
- ❑ Ruddy Foliage-gleaner *Clibanornis rubiginosus*
- ❑ Uniform Treehunter *Thripadectes ignobilis*
- ❑ Flammulated Treehunter *Thripadectes flammulatus*
- ❑ Striped Treehunter *Thripadectes holostictus*
- ❑ Streak-capped Treehunter *Thripadectes virgaticeps*
- ❑ Black-billed Treehunter *Thripadectes melanorhynchus*
- ❑ Chestnut-crowned Foliage-gleaner *Automolus rufipileatus*
- ❑ Brown-rumped Foliage-gleaner *Automolus melanopezus*
- ❑ Buff-throated Foliage-gleaner *Automolus ochrolaemus*
- ❑ Striped Woodhaunter *Automolus subulatus*
- ❑ Olive-backed Foliage-gleaner *Automolus infuscatus*
- ❑ Spotted Barbtail *Premnoplex brunnescens*
- ❑ Fulvous-dotted Treerunner *Margarornis stellatus*
- ❑ Pearled Treerunner *Margarornis squamiger*
- ❑ Andean Tit-Spinetail *Leptasthenura andicola*
- ❑ Rufous-fronted Thornbird *Phacellodomus rufifrons*
- ❑ White-browed Spinetail *Hellmayrea gularis*
- ❑ Many-striped Canastero *Asthenes flammulata*
- ❑ Streak-backed Canastero *Asthenes wyatti*
- ❑ White-chinned Thistletail *Asthenes fuliginosa*
- ❑ Mouse-coloured Thistletail *Asthenes griseomurina*
- ❑ Orange-fronted Plushcrown *Metopothrix aurantiaca*
- ❑ Double-banded Graytail *Xenerpestes minlosi*
- ❑ Equatorial Graytail *Xenerpestes singularis*
- ❑ Spectacled Prickletail *Siptornis striaticollis*
- ❑ Plain Softtail *Thripophaga fusciceps*
- ❑ Parker's Spinetail *Cranioleuca vulpecula*
- ❑ Red-faced Spinetail *Cranioleuca erythrops*
- ❑ Ash-browed Spinetail *Cranioleuca curtata*
- ❑ Line-cheeked Spinetail *Cranioleuca antisiensis*
- ❑ Speckled Spinetail *Cranioleuca gutturata*
- ❑ White-bellied Spinetail *Synallaxis propinqua*
- ❑ Plain-crowned Spinetail *Synallaxis gujanensis*
- ❑ Marañón Spinetail *Synallaxis maranonica*
- ❑ Necklaced Spinetail *Synallaxis stictothorax*
- ❑ Slaty Spinetail *Synallaxis brachyura*
- ❑ Dusky Spinetail *Synallaxis moesta*
- ❑ Dark-breasted Spinetail *Synallaxis albigularis*
- ❑ Azara's Spinetail *Synallaxis azarae*
- ❑ Blackish-headed Spinetail *Synallaxis tithys*
- ❑ Rufous Spinetail *Synallaxis unirufa*
- ❑ Ruddy Spinetail *Synallaxis rutilans*
- ❑ Chestnut-throated Spinetail *Synallaxis cherriei*

Tyrannidae

- ❑ Rough-legged Tyrannulet *Phyllomyias burmeisteri*
- ❑ Sooty-headed Tyrannulet *Phyllomyias griseiceps*
- ❑ Black-capped Tyrannulet *Phyllomyias nigrocapillus*
- ❑ Ashy-headed Tyrannulet *Phyllomyias cinereiceps*
- ❑ Tawny-rumped Tyrannulet *Phyllomyias uropygialis*
- ❑ Plumbeous-crowned Tyrannulet *Phyllomyias plumbeiceps*
- ❑ Yellow-crowned Tyrannulet *Tyrannulus elatus*
- ❑ Forest Elaenia *Myiopagis gaimardii*
- ❑ Grey Elaenia *Myiopagis caniceps*
- ❑ Foothill Elaenia *Myiopagis olallai*
- ❑ Pacific Elaenia *Myiopagis subplacens*
- ❑ Yellow-crowned Elaenia *Myiopagis flavivertex*
- ❑ Greenish Elaenia *Myiopagis viridicata*
- ❑ Yellow-bellied Elaenia *Elaenia flavogaster*
- ❑ Large Elaenia *Elaenia spectabilis*
- ❑ White-crested Elaenia *Elaenia albiceps*
- ❑ Small-billed Elaenia *Elaenia parvirostris*
- ❑ Slaty Elaenia *Elaenia strepera*
- ❑ Mottle-backed Elaenia *Elaenia gigas*
- ❑ Lesser Elaenia *Elaenia chiriquensis*
- ❑ Coopmans's Elaenia *Elaenia brachyptera*
- ❑ Highland Elaenia *Elaenia obscura*
- ❑ Sierran Elaenia *Elaenia pallatangae*
- ❑ Brown-capped Tyrannulet *Ornithion brunneicapillus*
- ❑ White-lored Tyrannulet *Ornithion inerme*
- ❑ Southern Beardless Tyrannulet *Camptostoma obsoletum*
- ❑ White-tailed Tyrannulet *Mecocerculus poecilocercus*
- ❑ White-banded Tyrannulet *Mecocerculus stictopterus*

- ❑ White-throated Tyrannulet *Mecocerculus leucophrys*
- ❑ Rufous-winged Tyrannulet *Mecocerculus calopterus*
- ❑ Sulphur-bellied Tyrannulet *Mecocerculus minor*
- ❑ Black-crested Tit-Tyrant *Anairetes nigrocristatus*
- ❑ Tufted Tit-Tyrant *Anairetes parulus*
- ❑ Agile Tit-Tyrant *Uromyias agilis*
- ❑ Torrent Tyrannulet *Serpophaga cinerea*
- ❑ River Tyrannulet *Serpophaga hypoleuca*
- ❑ Mouse-coloured Tyrannulet *Phaeomyias murina*
- ❑ Yellow Tyrannulet *Capsiempis flaveola*
- ❑ Subtropical Doradito *Pseudocolopteryx acutipennis*
- ❑ Bronze-olive Pygmy-Tyrant *Pseudotriccus pelzelni*
- ❑ Rufous-headed Pygmy-Tyrant *Pseudotriccus ruficeps*
- ❑ Ringed Antpipit *Corythopis torquatus*
- ❑ Tawny-crowned Pygmy-Tyrant *Euscarthmus meloryphus*
- ❑ Grey-and-white Tyrannulet *Pseudelaenia leucospodia*
- ❑ Lesser Wagtail-Tyrant *Stigmatura napensis*
- ❑ Chocó Tyrannulet *Zimmerius albigularis*
- ❑ Red-billed Tyrannulet *Zimmerius cinereicapilla*
- ❑ Slender-footed Tyrannulet *Zimmerius gracilipes*
- ❑ Golden-faced Tyrannulet *Zimmerius chrysops*
- ❑ Variegated Bristle-Tyrant *Phylloscartes poecilotis*
- ❑ Marble-faced Bristle-Tyrant *Phylloscartes ophthalmicus*
- ❑ Spectacled Bristle-Tyrant *Phylloscartes orbitalis*
- ❑ Ecuadorian Tyrannulet *Phylloscartes gualaquizae*
- ❑ Rufous-browed Tyrannulet *Phylloscartes superciliaris*
- ❑ Streak-necked Flycatcher *Mionectes striaticollis*
- ❑ Olive-striped Flycatcher *Mionectes olivaceus*
- ❑ Ochre-bellied Flycatcher *Mionectes oleagineus*
- ❑ Sepia-capped Flycatcher *Leptopogon amaurocephalus*
- ❑ Slaty-capped Flycatcher *Leptopogon superciliaris*
- ❑ Rufous-breasted Flycatcher *Leptopogon rufipectus*
- ❑ Amazonian Scrub Flycatcher *Sublegatus obscurior*
- ❑ Southern Scrub Flycatcher *Sublegatus modestus*
- ❑ Ornate Flycatcher *Myiotriccus ornatus*
- ❑ White-bellied Pygmy-Tyrant *Myiornis albiventris*
- ❑ Black-capped Pygmy-Tyrant *Myiornis atricapillus*
- ❑ Short-tailed Pygmy-Tyrant *Myiornis ecaudatus*
- ❑ Scale-crested Pygmy-Tyrant *Lophotriccus pileatus*
- ❑ Double-banded Pygmy-Tyrant *Lophotriccus vitiosus*
- ❑ White-eyed Tody-Tyrant *Hemitriccus zosterops*
- ❑ Johannes's Tody-Tyrant *Hemitriccus iohannis*
- ❑ Zimmer's Tody-Tyrant *Hemitriccus minimus*
- ❑ Black-throated Tody-Tyrant *Hemitriccus granadensis*
- ❑ Cinnamon-breasted Tody-Tyrant *Hemitriccus cinnamomeipectus*
- ❑ Buff-throated Tody-Tyrant *Hemitriccus rufigularis*
- ❑ Rufous-crowned Tody-Flycatcher *Poecilotriccus ruficeps*
- ❑ Black-and-white Tody-Flycatcher *Poecilotriccus capitalis*
- ❑ Rusty-fronted Tody-Flycatcher *Poecilotriccus latirostris*
- ❑ Golden-winged Tody-Flycatcher *Poecilotriccus calopterus*
- ❑ Spotted Tody-Flycatcher *Todirostrum maculatum*
- ❑ Common Tody-Flycatcher *Todirostrum cinereum*
- ❑ Black-headed Tody-Flycatcher *Todirostrum nigriceps*
- ❑ Yellow-browed Tody-Flycatcher *Todirostrum chrysocrotaphum*
- ❑ Brownish Twistwing *Cnipodectes subbrunneus*
- ❑ Olivaceous Flatbill *Rhynchocyclus olivaceus*
- ❑ Pacific Flatbill *Rhynchocyclus pacificus*
- ❑ Fulvous-breasted Flatbill *Rhynchocyclus fulvipectus*
- ❑ Yellow-olive Flycatcher *Tolmomyias sulphurescens*
- ❑ Orange-eyed Flycatcher *Tolmomyias traylori*
- ❑ Yellow-margined Flycatcher *Tolmomyias assimilis*
- ❑ Grey-crowned Flycatcher *Tolmomyias poliocephalus*
- ❑ Yellow-breasted Flycatcher *Tolmomyias flaviventris*
- ❑ Cinnamon-crested Spadebill *Platyrinchus saturatus*
- ❑ White-throated Spadebill *Platyrinchus mystaceus*
- ❑ Golden-crowned Spadebill *Platyrinchus coronatus*
- ❑ Yellow-throated Spadebill *Platyrinchus flavigularis*
- ❑ White-crested Spadebill *Platyrinchus platyrhynchos*
- ❑ Royal Flycatcher *Onychorhynchus coronatus*
- ❑ Flavescent Flycatcher *Myiophobus flavicans*
- ❑ Orange-crested Flycatcher *Myiophobus phoenicomitra*
- ❑ Roraiman Flycatcher *Myiophobus roraimae*
- ❑ Olive-chested Flycatcher *Myiophobus cryptoxanthus*
- ❑ Bran-coloured Flycatcher *Myiophobus fasciatus*
- ❑ Tawny-breasted Flycatcher *Myiobius villosus*
- ❑ Sulphur-rumped Flycatcher *Myiobius barbatus*
- ❑ Black-tailed Flycatcher *Myiobius atricaudus*
- ❑ Ruddy-tailed Flycatcher *Terenotriccus erythrurus*
- ❑ Cinnamon Manakin-Tyrant *Neopipo cinnamomea*
- ❑ Cinnamon Flycatcher *Pyrrhomyias cinnamomeus*
- ❑ Cliff Flycatcher *Hirundinea ferruginea*
- ❑ Handsome Flycatcher *Nephelomyias pulcher*
- ❑ Orange-banded Flycatcher *Nephelomyias lintoni*
- ❑ Euler's Flycatcher *Lathrotriccus euleri*
- ❑ Grey-breasted Flycatcher *Lathrotriccus griseipectus*
- ❑ Fuscous Flycatcher *Cnemotriccus fuscatus*
- ❑ Acadian Flycatcher *Empidonax virescens*
- ❑ Willow Flycatcher *Empidonax traillii*
- ❑ Alder Flycatcher *Empidonax alnorum*
- ❑ Olive-sided Flycatcher *Contopus cooperi*
- ❑ Smoke-coloured Pewee *Contopus fumigatus*
- ❑ Western Wood-Pewee *Contopus sordidulus*
- ❑ Eastern Wood-Pewee *Contopus virens*
- ❑ Tropical Pewee *Contopus cinereus*

- ❑ Blackish Pewee *Contopus nigrescens*
- ❑ Tufted Flycatcher *Mitrephanes phaeocercus*
- ❑ Black Phoebe *Sayornis nigricans*
- ❑ Vermilion Flycatcher *Pyrocephalus rubinus*
- ❑ Riverside Tyrant *Knipolegus orenocensis*
- ❑ Rufous-tailed Tyrant *Knipolegus poecilurus*
- ❑ Amazonian Black Tyrant *Knipolegus poecilocercus*
- ❑ Andean Tyrant *Knipolegus signatus*
- ❑ Drab Water-Tyrant *Ochthornis littoralis*
- ❑ Little Ground-Tyrant *Muscisaxicola fluviatilis*
- ❑ Spot-billed Ground-Tyrant *Muscisaxicola maculirostris*
- ❑ Dark-faced Ground-Tyrant *Muscisaxicola maclovianus**
- ❑ White-browed Ground-Tyrant *Muscisaxicola albilora*
- ❑ Plain-capped Ground-Tyrant *Muscisaxicola alpinus*
- ❑ Black-billed Shrike-Tyrant *Agriornis montanus*
- ❑ White-tailed Shrike-Tyrant *Agriornis albicauda*
- ❑ Streak-throated Bush-Tyrant *Myiotheretes striaticollis*
- ❑ Smoky Bush-Tyrant *Myiotheretes fumigatus*
- ❑ Red-rumped Bush-Tyrant *Cnemarchus erythropygius*
- ❑ Pied Water-Tyrant *Fluvicola pica*
- ❑ Masked Water-Tyrant *Fluvicola nengeta*
- ❑ White-headed Marsh-Tyrant *Arundinicola leucocephala**
- ❑ Tumbes Tyrant *Tumbezia salvini*
- ❑ Crowned Chat-Tyrant *Ochthoeca frontalis*
- ❑ Jelski's Chat-Tyrant *Ochthoeca jelskii*
- ❑ Yellow-bellied Chat-Tyrant *Ochthoeca diadema*
- ❑ Slaty-backed Chat-Tyrant *Ochthoeca cinnamomeiventris*
- ❑ Rufous-breasted Chat-Tyrant *Ochthoeca rufipectoralis*
- ❑ Brown-backed Chat-Tyrant *Ochthoeca fumicolor*
- ❑ White-browed Chat-Tyrant *Ochthoeca leucophrys*
- ❑ Long-tailed Tyrant *Colonia colonus*
- ❑ Short-tailed Field-Tyrant *Muscigralla brevicauda*
- ❑ Cattle Tyrant *Machetornis rixosa*
- ❑ Piratic Flycatcher *Legatus leucophaius*
- ❑ Rusty-margined Flycatcher *Myiozetetes cayanensis*
- ❑ Social Flycatcher *Myiozetetes similis*
- ❑ Grey-capped Flycatcher *Myiozetetes granadensis*
- ❑ Dusky-chested Flycatcher *Myiozetetes luteiventris*
- ❑ Great Kiskadee *Pitangus sulphuratus*
- ❑ Lesser Kiskadee *Pitangus lictor*
- ❑ White-ringed Flycatcher *Conopias albovittatus*
- ❑ Yellow-throated Flycatcher *Conopias parvus*
- ❑ Three-striped Flycatcher *Conopias trivirgatus*
- ❑ Lemon-browed Flycatcher *Conopias cinchoneti*
- ❑ Golden-crowned Flycatcher *Myiodynastes chrysocephalus*
- ❑ Baird's Flycatcher *Myiodynastes bairdii*
- ❑ Sulphur-bellied Flycatcher *Myiodynastes luteiventris*
- ❑ Streaked Flycatcher *Myiodynastes maculatus*
- ❑ Boat-billed Flycatcher *Megarynchus pitangua*
- ❑ Sulphury Flycatcher *Tyrannopsis sulphurea*
- ❑ Variegated Flycatcher *Empidonomus varius*
- ❑ Crowned Slaty Flycatcher *Empidonomus aurantioatrocristatus*
- ❑ Snowy-throated Kingbird *Tyrannus niveigularis*
- ❑ White-throated Kingbird *Tyrannus albogularis*
- ❑ Tropical Kingbird *Tyrannus melancholicus*
- ❑ Fork-tailed Flycatcher *Tyrannus savana*
- ❑ Eastern Kingbird *Tyrannus tyrannus*
- ❑ Grey Kingbird *Tyrannus dominicensis**
- ❑ Rufous Mourner *Rhytipterna holerythra*
- ❑ Greyish Mourner *Rhytipterna simplex*
- ❑ Chocó Sirystes *Sirystes albogriseus*
- ❑ White-rumped Sirystes *Sirystes albocinereus*
- ❑ Dusky-capped Flycatcher *Myiarchus tuberculifer*
- ❑ Swainson's Flycatcher *Myiarchus swainsoni*
- ❑ Panama Flycatcher *Myiarchus panamensis*
- ❑ Short-crested Flycatcher *Myiarchus ferox*
- ❑ Sooty-crowned Flycatcher *Myiarchus phaeocephalus*
- ❑ Pale-edged Flycatcher *Myiarchus cephalotes*
- ❑ Great Crested Flycatcher *Myiarchus crinitus*
- ❑ Brown-crested Flycatcher *Myiarchus tyrannulus**
- ❑ Large-headed Flatbill *Ramphotrigon megacephalum*
- ❑ Rufous-tailed Flatbill *Ramphotrigon ruficauda*
- ❑ Dusky-tailed Flatbill *Ramphotrigon fuscicauda*
- ❑ Cinnamon Attila *Attila cinnamomeus*
- ❑ Ochraceous Attila *Attila torridus*
- ❑ Citron-bellied Attila *Attila citriniventris*
- ❑ Bright-rumped Attila *Attila spadiceus*

Oxyruncidae

- ❑ Sharpbill *Oxyruncus cristatus*

Cotingidae

- ❑ Green-and-black Fruiteater *Pipreola riefferii*
- ❑ Barred Fruiteater *Pipreola arcuata*
- ❑ Orange-breasted Fruiteater *Pipreola jucunda*
- ❑ Black-chested Fruiteater *Pipreola lubomirskii*
- ❑ Scarlet-breasted Fruiteater *Pipreola frontalis*
- ❑ Fiery-throated Fruiteater *Pipreola chlorolepidota*
- ❑ Scaled Fruiteater *Ampelioides tschudii*
- ❑ Chestnut-bellied Cotinga *Doliornis remseni*
- ❑ Red-crested Cotinga *Ampelion rubrocristatus*
- ❑ Chestnut-crested Cotinga *Ampelion rufaxilla*
- ❑ Black-necked Red-Cotinga *Phoenicircus nigricollis*
- ❑ Andean Cock-of-the-rock *Rupicola peruvianus*
- ❑ Grey-tailed Piha *Snowornis subalaris*

- ❑ Olivaceous Piha *Snowornis cryptolophus*
- ❑ Purple-throated Fruitcrow *Querula purpurata*
- ❑ Red-ruffed Fruitcrow *Pyroderus scutatus*
- ❑ Amazonian Umbrellabird *Cephalopterus ornatus*
- ❑ Long-wattled Umbrellabird *Cephalopterus penduliger*
- ❑ Blue Cotinga *Cotinga nattererii*
- ❑ Plum-throated Cotinga *Cotinga maynana*
- ❑ Spangled Cotinga *Cotinga cayana*
- ❑ Dusky Piha *Lipaugus fuscocinereus*
- ❑ Rufous Piha *Lipaugus unirufus*
- ❑ Screaming Piha *Lipaugus vociferans*
- ❑ Purple-throated Cotinga *Porphyrolaema porphyrolaema*
- ❑ Black-tipped Cotinga *Carpodectes hopkei*
- ❑ Pompadour Cotinga *Xipholena punicea*
- ❑ Bare-necked Fruitcrow *Gymnoderus foetidus*

Pipridae

- ❑ Dwarf Tyrant-Manakin *Tyranneutes stolzmanni*
- ❑ Yellow-headed Manakin *Chloropipo flavicapilla*
- ❑ Jet Manakin *Chloropipo unicolor*
- ❑ Blue-backed Manakin *Chiroxiphia pareola*
- ❑ Golden-winged Manakin *Masius chrysopterus*
- ❑ Green Manakin *Cryptopipo holochlora*
- ❑ Blue-crowned Manakin *Lepidothrix coronata*
- ❑ Blue-rumped Manakin *Lepidothrix isidorei*
- ❑ Orange-crowned Manakin *Heterocercus aurantiivertex*
- ❑ White-bearded Manakin *Manacus manacus*
- ❑ Wire-tailed Manakin *Pipra filicauda*
- ❑ Club-winged Manakin *Machaeropterus deliciosus*
- ❑ Striped Manakin *Machaeropterus regulus*
- ❑ White-crowned Manakin *Dixiphia pipra*
- ❑ Red-capped Manakin *Ceratopipra mentalis*
- ❑ Golden-headed Manakin *Ceratopipra erythrocephala*

Tityridae

- ❑ Black-crowned Tityra *Tityra inquisitor*
- ❑ Black-tailed Tityra *Tityra cayana*
- ❑ Masked Tityra *Tityra semifasciata*
- ❑ Várzea Schiffornis *Schiffornis major*
- ❑ Northern Schiffornis *Schiffornis veraepacis*
- ❑ Foothill Schiffornis *Schiffornis aenea*
- ❑ Brown-winged Schiffornis *Schiffornis turdina*
- ❑ Speckled Mourner *Laniocera rufescens*
- ❑ Cinereous Mourner *Laniocera hypopyrra*
- ❑ White-browed Purpletuft *Iodopleura isabellae*
- ❑ Shrike-like Cotinga *Laniisoma elegans*
- ❑ Green-backed Becard *Pachyramphus viridis*
- ❑ Barred Becard *Pachyramphus versicolor*
- ❑ Slaty Becard *Pachyramphus spodiurus*
- ❑ Cinnamon Becard *Pachyramphus cinnamomeus*
- ❑ Chestnut-crowned Becard *Pachyramphus castaneus*
- ❑ White-winged Becard *Pachyramphus polychopterus*
- ❑ Black-and-white Becard *Pachyramphus albogriseus*
- ❑ Black-capped Becard *Pachyramphus marginatus*
- ❑ One-coloured Becard *Pachyramphus homochrous*
- ❑ Pink-throated Becard *Pachyramphus minor*

Insertae sedis

- ❑ Wing-barred Piprites *Piprites chloris*

Vireonidae

- ❑ Rufous-browed Peppershrike *Cyclarhis gujanensis*
- ❑ Black-billed Peppershrike *Cyclarhis nigrirostris*
- ❑ Slaty-capped Shrike-Vireo *Vireolanius leucotis*
- ❑ Yellow-throated Vireo *Vireo flavifrons*
- ❑ Chocó Vireo *Vireo masteri*
- ❑ Brown-capped Vireo *Vireo leucophrys*
- ❑ Philadelphia Vireo *Vireo philadelphicus*
- ❑ Warbling Vireo *Vireo gilvus**
- ❑ Red-eyed Vireo *Vireo olivaceus*
- ❑ Yellow-green Vireo *Vireo flavoviridis*
- ❑ Lemon-chested Greenlet *Hylophilus thoracicus*
- ❑ Olivaceous Greenlet *Hylophilus olivaceus*
- ❑ Tawny-crowned Greenlet *Tunchiornis ochraceiceps*
- ❑ Rufous-naped Greenlet *Pachysylvia semibrunnea*
- ❑ Dusky-capped Greenlet *Pachysylvia hypoxantha*
- ❑ Lesser Greenlet *Pachysylvia decurtata*

Corvidae

- ❑ Black-collared Jay *Cyanolyca armillata*
- ❑ Turquoise Jay *Cyanolyca turcosa*
- ❑ Beautiful Jay *Cyanolyca pulchra*
- ❑ Violaceous Jay *Cyanocorax violaceus*
- ❑ White-tailed Jay *Cyanocorax mystacalis*
- ❑ Green Jay *Cyanocorax yncas*

Hirundinidae

- ❑ Blue-and-white Swallow *Pygochelidon cyanoleuca*
- ❑ Brown-bellied Swallow *Orochelidon murina*
- ❑ Pale-footed Swallow *Orochelidon flavipes*
- ❑ White-banded Swallow *Atticora fasciata*
- ❑ White-thighed Swallow *Atticora tibialis*
- ❑ Southern Rough-winged Swallow *Stelgidopteryx ruficollis*
- ❑ Brown-chested Martin *Progne tapera*
- ❑ Purple Martin *Progne subis*
- ❑ Grey-breasted Martin *Progne chalybea*
- ❑ Southern Martin *Progne elegans*
- ❑ Tree Swallow *Tachycineta bicolor**

- ❑ Tumbes Swallow *Tachycineta stolzmanni*
- ❑ White-winged Swallow *Tachycineta albiventer*
- ❑ Bank Swallow *Riparia riparia*
- ❑ Barn Swallow *Hirundo rustica*
- ❑ Cliff Swallow *Petrochelidon pyrrhonota*
- ❑ Chestnut-collared Swallow *Petrochelidon rufocollaris*

Troglodytidae

- ❑ Scaly-breasted Wren *Microcerculus marginatus*
- ❑ Wing-banded Wren *Microcerculus bambla*
- ❑ Grey-mantled Wren *Odontorchilus branickii*
- ❑ House Wren *Troglodytes aedon*
- ❑ Mountain Wren *Troglodytes solstitialis*
- ❑ Sedge Wren *Cistothorus platensis*
- ❑ Band-backed Wren *Campylorhynchus zonatus*
- ❑ Fasciated Wren *Campylorhynchus fasciatus*
- ❑ Thrush-like Wren *Campylorhynchus turdinus*
- ❑ Plain-tailed Wren *Pheugopedius euophrys*
- ❑ Whiskered Wren *Pheugopedius mystacalis*
- ❑ Coraya Wren *Pheugopedius coraya*
- ❑ Speckle-breasted Wren *Pheugopedius sclateri*
- ❑ Stripe-throated Wren *Cantorchilus leucopogon*
- ❑ Bay Wren *Cantorchilus nigricapillus*
- ❑ Superciliated Wren *Cantorchilus superciliaris*
- ❑ Buff-breasted Wren *Cantorchilus leucotis*
- ❑ Rufous Wren *Cinnycerthia unirufa*
- ❑ Sharpe's Wren *Cinnycerthia olivascens*
- ❑ White-breasted Wood-Wren *Henicorhina leucosticta*
- ❑ Bar-winged Wood-Wren *Henicorhina leucoptera*
- ❑ Grey-breasted Wood-Wren *Henicorhina leucophrys*
- ❑ Chestnut-breasted Wren *Cyphorhinus thoracicus*
- ❑ Song Wren *Cyphorhinus phaeocephalus*
- ❑ Musician Wren *Cyphorhinus arada*

Polioptilidae

- ❑ Collared Gnatwren *Microbates collaris*
- ❑ Half-collared Gnatwren *Microbates cinereiventris*
- ❑ Long-billed Gnatwren *Ramphocaenus melanurus*
- ❑ Tropical Gnatcatcher *Polioptila plumbea*
- ❑ Slate-throated Gnatcatcher *Polioptila schistaceigula*

Donacobiidae

- ❑ Black-capped Donacobius *Donacobius atricapilla*

Cinclidae

- ❑ White-capped Dipper *Cinclus leucocephalus*

Turdidae

- ❑ Andean Solitaire *Myadestes ralloides*
- ❑ Slaty-backed Nightingale-Thrush *Catharus fuscater*
- ❑ Spotted Nightingale-Thrush *Catharus dryas*
- ❑ Grey-cheeked Thrush *Catharus minimus*
- ❑ Swainson's Thrush *Catharus ustulatus*
- ❑ Black Solitaire *Entomodestes coracinus*
- ❑ Rufous-brown Solitaire *Cichlopsis leucogenys*
- ❑ Pale-eyed Thrush *Turdus leucops*
- ❑ Plumbeous-backed Thrush *Turdus reevei*
- ❑ Hauxwell's Thrush *Turdus hauxwelli*
- ❑ Pale-vented Thrush *Turdus obsoletus*
- ❑ Ecuadorian Thrush *Turdus maculirostris*
- ❑ Lawrence's Thrush *Turdus lawrencii*
- ❑ Black-billed Thrush *Turdus ignobilis*
- ❑ Marañón Thrush *Turdus maranonicus*
- ❑ Chestnut-bellied Thrush *Turdus fulviventris*
- ❑ Slaty Thrush *Turdus nigriceps*
- ❑ Great Thrush *Turdus fuscater*
- ❑ Chiguanco Thrush *Turdus chiguanco*
- ❑ Glossy-black Thrush *Turdus serranus*
- ❑ White-throated Thrush *Turdus assimilis*
- ❑ White-necked Thrush *Turdus albicollis*

Mimidae

- ❑ Tropical Mockingbird *Mimus gilvus*
- ❑ Long-tailed Mockingbird *Mimus longicaudatus*

Motacillidae

- ❑ Red-throated Pipit *Anthus cervinus*
- ❑ Páramo Pipit *Anthus bogotensis*

Thraupidae

- ❑ Red-capped Cardinal *Paroaria gularis*
- ❑ Black-faced Tanager *Schistochlamys melanopis*
- ❑ Magpie Tanager *Cissopis leverianus*
- ❑ Black-and-white Tanager *Conothraupis speculigera*
- ❑ White-capped Tanager *Sericossypha albocristata*
- ❑ Rufous-crested Tanager *Creurgops verticalis*
- ❑ Black-capped Hemispingus *Hemispingus atropileus*
- ❑ Superciliaried Hemispingus *Hemispingus superciliaris*
- ❑ Oleaginous Hemispingus *Hemispingus frontalis*
- ❑ Black-eared Hemispingus *Hemispingus melanotis*
- ❑ Black-headed Hemispingus *Hemispingus verticalis*
- ❑ Grey-hooded Bush-Tanager *Cnemoscopus rubrirostris*
- ❑ Rufous-chested Tanager *Thlypopsis ornata*
- ❑ Orange-headed Tanager *Thlypopsis sordida*
- ❑ Buff-bellied Tanager *Thlypopsis inornata*
- ❑ Grey-headed Tanager *Eucometis penicillata*
- ❑ Flame-crested Tanager *Tachyphonus cristatus*
- ❑ Fulvous-crested Tanager *Tachyphonus surinamus*
- ❑ White-shouldered Tanager *Tachyphonus luctuosus*
- ❑ Tawny-crested Tanager *Tachyphonus delatrii*

- ❑ White-lined Tanager *Tachyphonus rufus*
- ❑ Fulvous Shrike-Tanager *Lanio fulvus*
- ❑ Masked Crimson Tanager *Ramphocelus nigrogularis*
- ❑ Silver-beaked Tanager *Ramphocelus carbo*
- ❑ Flame-rumped Tanager *Ramphocelus flammigerus*
- ❑ Vermilion Tanager *Calochaetes coccineus*
- ❑ Golden-chested Tanager *Bangsia rothschildi*
- ❑ Moss-backed Tanager *Bangsia edwardsi*
- ❑ Orange-throated Tanager *Wetmorethraupis sterrhopteron*
- ❑ Hooded Mountain-Tanager *Buthraupis montana*
- ❑ Masked Mountain-Tanager *Buthraupis wetmorei*
- ❑ Black-chested Mountain-Tanager *Cnemathraupis eximia*
- ❑ Grass-green Tanager *Chlorornis riefferii*
- ❑ Lacrimose Mountain-Tanager *Anisognathus lacrymosus*
- ❑ Scarlet-bellied Mountain-Tanager *Anisognathus igniventris*
- ❑ Blue-winged Mountain-Tanager *Anisognathus somptuosus*
- ❑ Black-chinned Mountain-Tanager *Anisognathus notabilis*
- ❑ Buff-breasted Mountain-Tanager *Dubusia taeniata*
- ❑ Purplish-mantled Tanager *Iridosornis porphyrocephalus*
- ❑ Yellow-throated Tanager *Iridosornis analis*
- ❑ Golden-crowned Tanager *Iridosornis rufivertex*
- ❑ Fawn-breasted Tanager *Pipraeidea melanonota*
- ❑ Blue-and-yellow Tanager *Pipraeidea bonariensis*
- ❑ Glistening-green Tanager *Chlorochrysa phoenicotis*
- ❑ Orange-eared Tanager *Chlorochrysa calliparaea*
- ❑ Blue-grey Tanager *Thraupis episcopus*
- ❑ Palm Tanager *Thraupis palmarum*
- ❑ Blue-capped Tanager *Thraupis cyanocephala*
- ❑ Golden-naped Tanager *Tangara ruficervix*
- ❑ Silvery Tanager *Tangara viridicollis*
- ❑ Black-capped Tanager *Tangara heinei*
- ❑ Green-throated Tanager *Tangara argyrofenges*
- ❑ Grey-and-gold Tanager *Tangara palmeri*
- ❑ Scrub Tanager *Tangara vitriolina*
- ❑ Masked Tanager *Tangara nigrocincta*
- ❑ Golden-hooded Tanager *Tangara larvata*
- ❑ Blue-necked Tanager *Tangara cyanicollis*
- ❑ Rufous-throated Tanager *Tangara rufigula*
- ❑ Speckled Tanager *Tangara guttata*
- ❑ Yellow-bellied Tanager *Tangara xanthogastra*
- ❑ Spotted Tanager *Tangara punctata*
- ❑ Blue-and-black Tanager *Tangara vassorii*
- ❑ Beryl-spangled Tanager *Tangara nigroviridis*
- ❑ Metallic-green Tanager *Tangara labradorides*
- ❑ Blue-browed Tanager *Tangara cyanotis*
- ❑ Turquoise Tanager *Tangara mexicana*
- ❑ Paradise Tanager *Tangara chilensis*
- ❑ Opal-rumped Tanager *Tangara velia*
- ❑ Opal-crowned Tanager *Tangara callophrys*
- ❑ Rufous-winged Tanager *Tangara lavinia*
- ❑ Bay-headed Tanager *Tangara gyrola*
- ❑ Golden-eared Tanager *Tangara chrysotis*
- ❑ Saffron-crowned Tanager *Tangara xanthocephala*
- ❑ Flame-faced Tanager *Tangara parzudakii*
- ❑ Green-and-gold Tanager *Tangara schrankii*
- ❑ Blue-whiskered Tanager *Tangara johannae*
- ❑ Golden Tanager *Tangara arthus*
- ❑ Emerald Tanager *Tangara florida*
- ❑ Silver-throated Tanager *Tangara icterocephala*
- ❑ Swallow Tanager *Tersina viridis*
- ❑ White-bellied Dacnis *Dacnis albiventris*
- ❑ Black-faced Dacnis *Dacnis lineata*
- ❑ Yellow-bellied Dacnis *Dacnis flaviventer*
- ❑ Scarlet-thighed Dacnis *Dacnis venusta*
- ❑ Blue Dacnis *Dacnis cayana*
- ❑ Scarlet-breasted Dacnis *Dacnis berlepschi*
- ❑ Short-billed Honeycreeper *Cyanerpes nitidus*
- ❑ Purple Honeycreeper *Cyanerpes caeruleus*
- ❑ Red-legged Honeycreeper *Cyanerpes cyaneus*
- ❑ Green Honeycreeper *Chlorophanes spiza*
- ❑ Golden-collared Honeycreeper *Iridophanes pulcherrimus*
- ❑ Scarlet-browed Tanager *Heterospingus xanthopygius*
- ❑ Guira Tanager *Hemithraupis guira*
- ❑ Yellow-backed Tanager *Hemithraupis flavicollis*
- ❑ Scarlet-and-white Tanager *Chrysothlypis salmoni*
- ❑ Chestnut-vented Conebill *Conirostrum speciosum*
- ❑ Bicoloured Conebill *Conirostrum bicolor*
- ❑ Cinereous Conebill *Conirostrum cinereum*
- ❑ Blue-backed Conebill *Conirostrum sitticolor*
- ❑ Capped Conebill *Conirostrum albifrons*
- ❑ Giant Conebill *Oreomanes fraseri*
- ❑ Tit-like Dacnis *Xenodacnis parina*
- ❑ Glossy Flowerpiercer *Diglossa lafresnayii*
- ❑ Black Flowerpiercer *Diglossa humeralis*
- ❑ White-sided Flowerpiercer *Diglossa albilatera*
- ❑ Indigo Flowerpiercer *Diglossa indigotica*
- ❑ Rusty Flowerpiercer *Diglossa sittoides*
- ❑ Deep-blue Flowerpiercer *Diglossa glauca*
- ❑ Bluish Flowerpiercer *Diglossa caerulescens*
- ❑ Masked Flowerpiercer *Diglossa cyanea*
- ❑ Plushcap *Catamblyrhynchus diadema*
- ❑ Black-backed Bush-Tanager *Urothraupis stolzmanni*
- ❑ Plumbeous Sierra-Finch *Phrygilus unicolor*

- ❑ Ash-breasted Sierra-Finch *Phrygilus plebejus*
- ❑ Band-tailed Sierra-Finch *Phrygilus alaudinus*
- ❑ Slaty Finch *Haplospiza rustica*
- ❑ Cinereous Finch *Piezorhina cinerea**
- ❑ Collared Warbling-Finch *Poospiza hispaniolensis*
- ❑ Saffron Finch *Sicalis flaveola*
- ❑ Grassland Yellow-Finch *Sicalis luteola*
- ❑ Sulphur-throated Finch *Sicalis taczanowskii*
- ❑ Wedge-tailed Grass-Finch *Emberizoides herbicola*
- ❑ Blue-black Grassquit *Volatinia jacarina*
- ❑ Lesson's Seedeater *Sporophila bouvronides*
- ❑ Lined Seedeater *Sporophila lineola*
- ❑ Parrot-billed Seedeater *Sporophila peruviana*
- ❑ Chestnut-throated Seedeater *Sporophila telasco*
- ❑ Drab Seedeater *Sporophila simplex*
- ❑ Chestnut-bellied Seedeater *Sporophila castaneiventris*
- ❑ Ruddy-breasted Seedeater *Sporophila minuta*
- ❑ Thick-billed Seed-Finch *Sporophila funerea*
- ❑ Chestnut-bellied Seed-Finch *Sporophila angolensis*
- ❑ Large-billed Seed-Finch *Sporophila crassirostris*
- ❑ Black-billed Seed-Finch *Sporophila atrirostris*
- ❑ Variable Seedeater *Sporophila corvina*
- ❑ Grey Seedeater *Sporophila intermedia**
- ❑ Caqueta Seedeater *Sporophila murallae*
- ❑ Black-and-white Seedeater *Sporophila luctuosa*
- ❑ Yellow-bellied Seedeater *Sporophila nigricollis*
- ❑ Slate-coloured Seedeater *Sporophila schistacea*
- ❑ Band-tailed Seedeater *Catamenia analis*
- ❑ Plain-coloured Seedeater *Catamenia inornata*
- ❑ Páramo Seedeater *Catamenia homochroa*
- ❑ Red-crested Finch *Coryphospingus cucullatus*
- ❑ Crimson-breasted Finch *Rhodospingus cruentus*
- ❑ Bananaquit *Coereba flaveola*
- ❑ Yellow-faced Grassquit *Tiaris olivaceus*
- ❑ Dull-coloured Grassquit *Tiaris obscurus*
- ❑ Yellow-shouldered Grosbeak *Parkerthraustes humeralis*

Incertae sedis

- ❑ Dusky-faced Tanager *Mitrospingus cassinii*
- ❑ Buff-throated Saltator *Saltator maximus*
- ❑ Black-winged Saltator *Saltator atripennis*
- ❑ Greyish Saltator *Saltator coerulescens*
- ❑ Streaked Saltator *Saltator striatipectus*
- ❑ Black-cowled Saltator *Saltator nigriceps*
- ❑ Masked Saltator *Saltator cinctus*
- ❑ Slate-coloured Grosbeak *Saltator grossus*

Emberizidae

- ❑ Tanager Finch *Oreothraupis arremonops*
- ❑ Yellow-throated Bush-Tanager *Chlorospingus flavigularis*
- ❑ Short-billed Bush-Tanager *Chlorospingus parvirostris*
- ❑ Ashy-throated Bush-Tanager *Chlorospingus canigularis*
- ❑ Common Bush-Tanager *Chlorospingus flavopectus*
- ❑ Dusky Bush-Tanager *Chlorospingus semifuscus*
- ❑ Yellow-green Bush-Tanager *Chlorospingus flavovirens*
- ❑ Tumbes Sparrow *Rhynchospiza stolzmanni*
- ❑ Grasshopper Sparrow *Ammodramus savannarum*
- ❑ Yellow-browed Sparrow *Ammodramus aurifrons*
- ❑ Black-striped Sparrow *Arremonops conirostris*
- ❑ Grey-browed Brush-Finch *Arremon assimilis*
- ❑ Orange-billed Sparrow *Arremon aurantiirostris*
- ❑ Black-capped Sparrow *Arremon abeillei*
- ❑ Chestnut-capped Brush-Finch *Arremon brunneinucha*
- ❑ Olive Finch *Arremon castaneiceps*
- ❑ Rufous-collared Sparrow *Zonotrichia capensis*
- ❑ White-naped Brush-Finch *Atlapetes albinucha*
- ❑ White-rimmed Brush-Finch *Atlapetes leucopis*
- ❑ White-headed Brush-Finch *Atlapetes albiceps*
- ❑ Tricoloured Brush-Finch *Atlapetes tricolor*
- ❑ Slaty Brush-Finch *Atlapetes schistaceus*
- ❑ Pale-naped Brush-Finch *Atlapetes pallidinucha*
- ❑ Yellow-breasted Brush-Finch *Atlapetes latinuchus*
- ❑ White-winged Brush-Finch *Atlapetes leucopterus*
- ❑ Pale-headed Brush-Finch *Atlapetes pallidiceps*
- ❑ Bay-crowned Brush-Finch *Atlapetes seebohmi*

Cardinalidae

- ❑ Hepatic Tanager *Piranga flava*
- ❑ Summer Tanager *Piranga rubra*
- ❑ Scarlet Tanager *Piranga olivacea*
- ❑ Red-hooded Tanager *Piranga rubriceps*
- ❑ White-winged Tanager *Piranga leucoptera*
- ❑ Red-crowned Ant-Tanager *Habia rubica*
- ❑ Carmiol's Tanager *Chlorothraupis carmioli*
- ❑ Lemon-spectacled Tanager *Chlorothraupis olivacea*
- ❑ Ochre-breasted Tanager *Chlorothraupis stolzmanni*
- ❑ Golden-bellied Grosbeak *Pheucticus chrysogaster*
- ❑ Black-backed Grosbeak *Pheucticus aureoventris*
- ❑ Rose-breasted Grosbeak *Pheucticus ludovicianus*
- ❑ Blue Seedeater *Amaurospiza concolor*
- ❑ Blue-black Grosbeak *Cyanocompsa cyanoides*
- ❑ Blue Grosbeak *Passerina caerulea*
- ❑ Indigo Bunting *Passerina cyanea**
- ❑ Dickcissel *Spiza americana**

Parulidae

- ❑ Ovenbird *Seiurus aurocapilla*
- ❑ Northern Waterthrush *Parkesia noveboracensis*

- ❑ Golden-winged Warbler *Vermivora chrysoptera*
- ❑ Black-and-white Warbler *Mniotilta varia*
- ❑ Prothonotary Warbler *Protonotaria citrea*
- ❑ Tennessee Warbler *Leiothlypis peregrina*
- ❑ Connecticut Warbler *Oporornis agilis*
- ❑ Masked Yellowthroat *Geothlypis aequinoctialis*
- ❑ Mourning Warbler *Geothlypis philadelphia*
- ❑ Olive-crowned Yellowthroat *Geothlypis semiflava*
- ❑ American Redstart *Setophaga ruticilla*
- ❑ Cerulean Warbler *Setophaga cerulea*
- ❑ Tropical Parula *Setophaga pitiayumi*
- ❑ Bay-breasted Warbler *Setophaga castanea*
- ❑ Blackburnian Warbler *Setophaga fusca*
- ❑ Yellow Warbler *Setophaga petechia*
- ❑ Chestnut-sided Warbler *Setophaga pensylvanica**
- ❑ Blackpoll Warbler *Setophaga striata*
- ❑ Black-throated Blue Warbler *Setophaga caerulescens*
- ❑ Black-throated Green Warbler *Setophaga virens**
- ❑ Citrine Warbler *Myiothlypis luteoviridis*
- ❑ Black-crested Warbler *Myiothlypis nigrocristata*
- ❑ Buff-rumped Warbler *Myiothlypis fulvicauda*
- ❑ Golden-bellied Warbler *Myiothlypis chrysogaster*
- ❑ Grey-and-gold Warbler *Myiothlypis fraseri*
- ❑ Russet-crowned Warbler *Myiothlypis coronata*
- ❑ Three-striped Warbler *Basileuterus tristriatus*
- ❑ Three-banded Warbler *Basileuterus trifasciatus*
- ❑ Canada Warbler *Cardellina canadensis*
- ❑ Wilson's Warbler *Cardellina pusilla*
- ❑ Slate-throated Redstart *Myioborus miniatus*
- ❑ Spectacled Redstart *Myioborus melanocephalus*

Icteridae

- ❑ Russet-backed Oropendola *Psarocolius angustifrons*
- ❑ Green Oropendola *Psarocolius viridis*
- ❑ Chestnut-headed Oropendola *Psarocolius wagleri*
- ❑ Crested Oropendola *Psarocolius decumanus*
- ❑ Olive Oropendola *Psarocolius bifasciatus*
- ❑ Solitary Black Cacique *Cacicus solitarius*
- ❑ Ecuadorian Cacique *Cacicus sclateri*
- ❑ Scarlet-rumped Cacique *Cacicus uropygialis*
- ❑ Yellow-rumped Cacique *Cacicus cela*
- ❑ Mountain Cacique *Cacicus chrysonotus*
- ❑ Band-tailed Oropendola *Cacicus latirostris*
- ❑ Red-rumped Cacique *Cacicus haemorrhous*
- ❑ Casqued Oropendola *Cacicus oseryi*
- ❑ Yellow-billed Cacique *Amblycercus holosericeus*
- ❑ Orange-backed Troupial *Icterus croconotus*
- ❑ White-edged Oriole *Icterus graceannae*
- ❑ Yellow-tailed Oriole *Icterus mesomelas*
- ❑ Epaulet Oriole *Icterus cayanensis*
- ❑ Yellow-backed Oriole *Icterus chrysater*
- ❑ Baltimore Oriole *Icterus galbula*
- ❑ Scrub Blackbird *Dives warczewiczi*
- ❑ Oriole Blackbird *Gymnomystax mexicanus*
- ❑ Velvet-fronted Grackle *Lampropsar tanagrinus*
- ❑ Pale-eyed Blackbird *Agelasticus xanthophthalmus*
- ❑ Giant Cowbird *Molothrus oryzivorus*
- ❑ Shiny Cowbird *Molothrus bonariensis*
- ❑ Great-tailed Grackle *Quiscalus mexicanus*
- ❑ Bobolink *Dolichonyx oryzivorus*
- ❑ Red-breasted Meadowlark *Sturnella militaris*
- ❑ Peruvian Meadowlark *Sturnella bellicosa*

Fringillidae

- ❑ Andean Siskin *Sporagra spinescens*
- ❑ Hooded Siskin *Sporagra magellanica*
- ❑ Saffron Siskin *Sporagra siemiradzkii*
- ❑ Olivaceous Siskin *Sporagra olivacea*
- ❑ Yellow-bellied Siskin *Sporagra xanthogastra*
- ❑ Lesser Goldfinch *Astragalinus psaltria*
- ❑ Purple-throated Euphonia *Euphonia chlorotica*
- ❑ Orange-crowned Euphonia *Euphonia saturata*
- ❑ Thick-billed Euphonia *Euphonia laniirostris*
- ❑ Golden-rumped Euphonia *Euphonia cyanocephala*
- ❑ Fulvous-vented Euphonia *Euphonia fulvicrissa*
- ❑ Golden-bellied Euphonia *Euphonia chrysopasta*
- ❑ Bronze-green Euphonia *Euphonia mesochrysa*
- ❑ White-vented Euphonia *Euphonia minuta*
- ❑ Orange-bellied Euphonia *Euphonia xanthogaster*
- ❑ Rufous-bellied Euphonia *Euphonia rufiventris*
- ❑ Blue-naped Chlorophonia *Chlorophonia cyanea*
- ❑ Chestnut-breasted Chlorophonia *Chlorophonia pyrrhophrys*
- ❑ Yellow-collared Chlorophonia *Chlorophonia flavirostris*

Estrildidae

- ❑ Tricoloured Munia *Lonchura malacca*[i]

Passeridae

- ❑ House Sparrow *Passer domesticus*[i]

REFERENCES

Some published sources of primary importance are marked with an asterisk (*). Among them are: Hilty & Brown (1986), Fjeldså & Krabbe (1990), Ortiz Crespo & Carrión (1991), Ridgely & Greenfield (2001, 2006), Krabbe & Nilsson (2003), Schulenberg *et al.* (2007), Haase (2011), McMullan & Navarrete (2013), the *Handbook of the Birds of the World* series, and the remarkable CD series published by John V. Moore and co-authors.

Alverson, W. S., Vriesendorp, C., del Campo, A., Moskovits, D. K., Stotz, D. F., García, M. & Borbor, L. A. (eds.) 2008. *Ecuador-Perú: Cuyabeno-Güeppí.* Rapid Biological and Social Inventories Report 20. Field Museum of Natural History, Chicago.

BirdLife International. 2016. IUCN Red List for birds. Version December 2016. http://www.birdlife.org/datazone.

Borman, R., Vriesendorp, C., Alverson, W. S., Moskovits, D. K., Stotz, D. F. & del Campo, A. (eds.) 2007. *Ecuador: Territorio Cofán-Dureno.* Rapid Biological and Social Inventories Report 19. Field Museum of Natural History, Chicago.

Boyla, K. A. & Estrada, A. (eds.) 2005. *Áreas Importantes para la Conservación de las Aves en los Andes Tropicales. Sitios Prioritarios para la Conservación de la Biodiversidad.* BirdLife International (Conservation Series 14), Quito.

Brinkhuizen, D. M., Carter, C., Lyons, J. A. & Albán, N. 2013. White-bellied Pygmy-Tyrant *Myiornis albiventris,* new to Ecuador. *Cotinga* 35: 131–132.

Brinkhuizen, D. M., Soldato, G., Lambeth, G., Lambeth, D., Albán, N. J. & Freile, J. F. 2015. Bluish-fronted Jacamar *Galbula cyanescens* in Ecuador. *Bull. Brit. Orn. Club* 135: 80–83.

*Brown, L. H. & Amadon, D. 1968. *Hawks, Eagles and Falcons of the World.* McGraw-Hill, London.

Buitrón-Jurado, G., Cabot, J. & de Vries, T. 2008. Patrón de distribución preliminar del Busardo dorsirrojo (*Buteo polyosoma*) en Ecuador. *Acta I Congr. Intern. Aves Rapaces y Conserv.:* 18–22.

Cabot, J. & de Vries, T. 2003. *Buteo polyosoma* and *Buteo poecilochrous* are two distinct species. *Bull. Brit. Orn. Club* 123: 190–207.

Cabot, J. & de Vries, T. 2004. Age- and sex-differentiated plumages in the two colour morphs of the Variable Buzzard *Buteo polyosoma*: a case of delayed maturation with subadult males disguised in definitive adult female plumage. *Bull. Brit. Orn. Club* 124: 272–285.

*Chantler, P. & Driessens, G. 1995. *Swifts: A Guide to the Swifts and Treeswifts of the World.* Pica Press, Robertsbridge.

*Cleere, N. & Nurney, D. 1998. *Nightjars: A Guide to the Nightjars and Related Nightbirds.* Pica Press, Robertsbridge.

Clements, J. F., Schulenberg, T. S., Iliff, M. J., Sullivan, B. L. & Wood, C. L. 2016. *The EBird Clements Checklist of Birds of the World: 2016.* Cornell University. www.birds.cornell.edu/clementschecklist/download

Coopmans, P. & Krabbe, N. 2000. A new species of flycatcher (Tyrannidae: *Myiopagis*) from eastern Ecuador and eastern Peru. *Wilson Bull.* 112: 305–312.

*Coopmans, P., Moore, J. V., Krabbe, N., Jahn, O., Berg, K. S., Lysinger, M., Navarrete, L. & Ridgely, R. S. 2004. *The Birds of Southwest Ecuador.* CDs. John V. Moore Nature Recordings, San Jose, CA.

*Curson, J., Quinn, D. & Beadle, D. 1994. *New World Warblers.* Christopher Helm, London.

*del Hoyo, J., Elliott, A. & Christie, D. (eds.) 2003–2011. *Handbook of the Birds of the World.* Vols. 8–16. Lynx Edicions, Barcelona.

*del Hoyo, J., Elliott, A. & Sargatal, J. (eds.) 1992–2002. *Handbook of the Birds of the World.* Vols. 1–7. Lynx Edicions, Barcelona.

*Ferguson-Lees, J. & Christie, D. A. 2001. *Raptors of the World.* Christopher Helm, London.

*Fjeldså, J. & Krabbe, N, 1990. *Birds of the High Andes.* Zoological Museum, Univ. of Copenhagen & Apollo Books, Svendborg.

Fogden, M. 2012. The non-breeding range of Esmeraldas Woodstar. *Neotrop. Birding* 10: 32–37.

*Forshaw, J. M. 2010. *Parrots of the World.* Christopher Helm, London.

Freile, J. F. 2005. Gustavo Orcés, Fernando Ortiz y el desarrollo de la ornitología hecha en Ecuador. *Orn. Neotrop.* 16: 321–336.

Freile, J. F. & Rodas, F. 2008. Conservación de aves en Ecuador: ¿Cómo estamos y qué necesitamos hacer? *Cotinga* 29: 48–55.

*Freile, J. F., Carrión, J. M., Prieto-Albuja, F. & Ortiz Crespo, F. 2005. *Listado Bibliográfico de las Aves del Ecuador.* EcoCiencia & Fundación Numashir, Quito.

Freile, J. F., Brinkhuizen, D. M., Solano-Ugalde, A., Greenfield, P. J., Ahlman, R., Navarrete, L. & Ridgely, R. S. 2013. Rare birds in Ecuador: first annual report of the Bird Committee of Ecuadorian Records in Ornithology (CERO). *Avances* 5: B24–B41.

Freile, J. F., Krabbe, N., Piedrahita, P., Buitrón-Jurado, G., Rodríguez-Saltos, C. A., Ahlman, F., Brinkhuizen, D. M. & Bonaccorso, E. 2014. Birds, Nangaritza River Valley, Zamora Chinchipe Province, southeast Ecuador: update and revision. *Check List* 10: 54–71.

Freile, J. F., Brinkhuizen, D. M., Greenfield, P. J., Lysinger, M., Navarrete, L., Nilsson, J., Ridgely, R. S., Solano-Ugalde, A., Ahlman, R. & Boyla, K. A. 2015–2017. Checklist of the Birds of Ecuador. Comité Ecuatoriano de Registros Ornitológicos. https://ceroecuador.wordpress.com/

*Granizo, T., Pacheco, C., Ribadeneira, M. B., Guerrero, M. & Suárez, L. (eds.) 2002. *Libro Rojo de las Aves del Ecuador.* Simbioe, Conservación Internacional, EcoCiencia, Ministerio del Ambiente & UICN, Quito.

Greeney, H. F. 2008. The Spotted Barbtail (*Premnoplex brunnescens*): a review of taxonomy, distribution, and breeding biology, with additional observations from northeastern Ecuador. *Bol. Soc. Antioqueña Orn.* 18: 1–9.

Guevara, E. A., Santander, T. & Duivenvoorden, J. F. 2012. Seasonal patterns in aquatic bird counts at five Andean lakes of Ecuador. *Waterbirds* 35: 636–641.

Guevara, E. A., Bonaccorso, E. & Duivenvoorden, J. F. 2015. Multi-scale habitat use analysis and interspecific ecology of the Critically Endangered Black-breasted Puffleg *Eriocnemis nigrivestis. Bird Conserv. Intern.* DOI:10.1017/S0959270914000367.

Gurney, M. 2006. Identification of Little Woodstar *Chaetocercus bombus* in Ecuador. *Neotrop. Birding* 1: 38–41.

*Haase, B. 2011. *Aves Marinas de Ecuador Continental y Acuáticas de las Piscinas Artificiales de Ecuasal.* Aves & Conservación, BirdLife International & Ecuasal S. A., Guayaquil.

*Hardy, J. W., Coffey, B. B. & Reynard, G. B. 1999. *Voices of New World Owls.* ARA Records, Gainesville, FL.

Harris, J. B. C., Ágreda, A. E., Juiña, M. E. & Freymann, B. P. 2009. Distribution, plumage, and conservation status of the endemic Esmeraldas Woodstar (*Chaetocercus berlepschi*) of western Ecuador. *Wilson Bull.* 121: 227–239.

*Harrison, P. 1987. *Seabirds of the World: A Photographic Guide.* Christopher Helm, London.

*Hayman, P., Marchant, J. & Prater, T. 1986. *Shorebirds: An Identification Guide.* Christopher Helm, London.

*Hilty, S. L. & Brown, W. L. 1986. *A Guide to the Birds of Colombia.* Princeton Univ. Press, Princeton, NJ.

Howell, S. N. G. 2011. Neotropical swifts—the final frontier? *Neotrop. Birding* 9: 18–22.

*Howell, S. N. G. & Dunn, J. 2007. *A Reference Guide to Gulls of the Americas.* Houghton Mifflin, Boston.

Howell, S. N. G., Jaramillo, A., Redman, N. & Ridgely, R. S.

2012. What's the point of field guides: taxonomy or utility? *Neotrop. Birding* 11: 16–21.
*Isler, M. L. & Isler, P. R. 1987. *The Tanagers: Natural History, Distribution and Identification.* Smithsonian Institution Press, Washington DC.
Jahn, O. 2011. Bird communities of the Ecuadorian Chocó: a case study in conservation. *Bonn. Zool. Monogr.* 56: 1–514.
*Jahn, O., Moore, J. V., Mena-Valenzuela, P., Krabbe, N., Coopmans, P., Lysinger, M. & Ridgely, R. S. 2002. *The Birds of Northwest Ecuador II, the Lowlands and Lower Foothills.* CDs. John V. Moore Nature Recordings, San Jose, CA.
Jahn, O., Palacios, B. & Mena-Valenzuela, P. 2007. Ecology, population and conservation status of the Chocó Vireo *Vireo masteri*, a species new to Ecuador. *Bull. Brit. Orn. Club* 127: 161–166.
Janni, O. & Pulcher, C. 2007. Reidentification of Ecuadorian specimens of *Pachyramphus rufus* as *P. castaneus. Bull. Brit. Orn. Club* 127: 246–247.
*Jaramillo, A. & Burke, P. 1999. *New World Blackbirds: The Icterids.* Christopher Helm, London.
*Kirwan, G. M. & Green, G. 2011. *Cotingas and Manakins.* Christopher Helm, London.
*König, C., Weick, F. & Becking, J.-H. 2008. *Owls of the World.* Second edn. Christopher Helm, London.
Krabbe, N. 2004. Pale-headed Brush-Finch *Atlapetes pallidiceps*: notes on population size, habitat, vocalizations, feeding, interference competition, and conservation. *Bird Conserv. Intern.* 14: 77–86.
*Krabbe, N. & Nilsson, J. 2003. *Birds of Ecuador: Sounds and Photographs.* Birdsongs International, Enschede.
Krabbe, N. & Ridgely, R. S. 2010. A new subspecies of Amazilia Hummingbird *Amazilia amazilia* from southern Ecuador. *Bull. Brit. Orn. Club* 130: 3–7.
Krabbe, N. & Schulenberg, T. S. 1997. Species limits and natural history of *Scytalopus* tapaculos (Rhinocryptidae), with descriptions of the Ecuadorian taxa, including three new species. *Orn. Monogr.* 48: 47–88.
Krabbe, N. & Stejskal, D. J. 2008. A new subspecies of Black-striped Sparrow *Arremonops conirostris* from south-eastern Ecuador. *Bull. Brit. Orn. Club* 128: 126–130.
Krabbe, N., Agro, D. J., Rice, N. H., Jácome, M., Navarrete, L. & Sornoza M., F. 1999a. New species of antpitta (Formicariidae: *Grallaria*) from the southern Ecuadorian Andes. *Auk* 116: 882–890.
Krabbe, N., Isler, M. L., Isler, P. R., Whitney, B. M., Alvarez A., J. & Greenfield, P. J. 1999b. A new species in the *Myrmotherula haematonota* superspecies (Aves: Thamnophilidae) from the western Amazonian lowlands of Ecuador and Peru. *Wilson Bull.* 111: 157–165.
*Krabbe, N., Moore, J. V., Coopmans, P., Lysinger, M. & Ridgely, R. S. 2001. *Birds of the Ecuadorian Highlands.* CDs. John V. Moore Nature Recordings, San Jose, CA.
Lee, C.-T., Birch, A. & Eubanks, T. L. 2008. Field identification of Western & Eastern Wood-Pewees. *Birding* 40(5): 34–40.
*Lysinger, M., Moore, J. V., Krabbe, N., Coopmans, P., Lane, D. F., Navarrete, L., Nilsson, J. & Ridgely, R. S. 2005. *The Birds of Eastern Ecuador I, the Foothills and Lower Subtropics.* CDs. John V. Moore Nature Recordings, San Jose, CA.
*Madge, S. & Burn, H. 1988. *Wildfowl: An Identification Guide to the Ducks, Geese and Swans of the World.* Christopher Helm, London.
*McMullan, M. & Navarrete, L. 2013. *Fieldbook of the Birds of Ecuador, including the Galápagos Islands.* Fundación Jocotoco, Quito.
*Moore, J. V., Coopmans, P., Ridgely, R. S. & Lysinger, M. 1999. *The Birds of Northwest Ecuador I, the Upper Foothills and Subtropics.* CDs. John V. Moore Nature Recordings, San Jose, CA.
*Moore, J. V., Krabbe, N., Lysinger, M., Lane, D. F., Coopmans, P., Rivadeneyra, J. & Ridgely, R. S. 2009. *The Birds of Eastern Ecuador. Volume II: the Lowlands.* CDs. John V. Moore Nature Recordings, San Jose, CA.
*Moore, J. V., Krabbe, N. & Jahn, O. 2013. *Bird Sounds of Ecuador, A Comprehensive Collection.* DVDs. John V. Moore Nature Recordings, San Jose, CA.
Nilsson, J., Freile, J. F., Ahlman, R., Brinkhuizen, D. M., Greenfield, P. J. & Solano-Ugalde, A. 2014. Rare birds in Ecuador: second annual report of the Committee for Ecuadorian Records in Ornithology (CERO). *Avances* 6: B38–B50.
*Olsen, K. M. & Larsson, H. 1997. *Skuas and Jaegers: A Guide to the Skuas and Jaegers of the World.* Pica Press, Robertsbridge..
*Onley, D. & Scofield, P. 2007. *Albatrosses, Petrels and Shearwaters of the World.* Christopher Helm, London.
Ortiz Crespo, F. 2003. *Los Colibríes: Historia Natural de unas Aves Casi Sobrenaturales.* Imprenta Mariscal, Quito.
Ortiz Crespo, F., Greenfield, P. & Matheus, J. C. 1990. *Aves del Ecuador, Continente y Archipiélago de Galápagos.* FEPROTUR, Quito.
Paynter, R. A. 1993. *Ornithological Gazetteer of Ecuador.* Second edn. Museum of Comparative Zoology, Harvard Univ., Cambridge, MA.
Pitman, N., Moskovits, D. K., Alverson, W. S. & Borman, R. (eds.) 2002. *Ecuador: Serranías Cofán–Bermejo, Sinangoe.* Rapid Biological Inventories Report 3. Field Museum of Natural History, Chicago
Putnam, C. G., Jones, A. W. & Ridgely, R. S. 2009. Two Long-billed Dowitcher *Limnodromus scolopaceus* specimens from Ecuador. *Cotinga* 31: 130–132.
*Ranft, R. & Cleere, N. 1998. *A Sound Guide to Nightjars and Related Nightbirds.* Pica Press, Robertsbridge & British Library of Wildlife Sounds, London.
*Remsen, J. V., Areta, J. I., Cadena, C. D., Jaramillo, A., Nores, M., Pacheco, J. F., Robbins, M. B., Stiles, F. G., Stotz, D. F. & Zimmer, K. J. 2015. A classification of the bird species of South America. http://www.museum.lsu.edu/~Remsen/SACCBaseline.html.
*Restall, R., Rodner, C. & Lentino, M. 2006. *Birds of Northern South America: An Identification Guide.* Christopher Helm, London.
*Ridgely, R. S. & Cooper, M. 2011. *Colibríes del Ecuador.* Fundación Jocotoco, Quito.
*Ridgely, R. S. & Greenfield, P. J. 2001. *The Birds of Ecuador.* Cornell Univ. Press, Ithaca, NY.
*Ridgely, R. S. & Greenfield, P. J. 2006. *Aves del Ecuador.* Academia de Ciencias de Philadelphia & Fundación Jocotoco, Quito.
Ridgely, R. S. & Robbins, M. B. 1988. *Pyrrhura orcesi*, a new parakeet from south western Ecuador, with systematic notes on the *P. melanura* complex. *Wilson Bull.* 100: 173–182.
*Ridgely, R. S. & Tudor, G. 2009. *Birds of South America. Passerines.* Christopher Helm, London.
Robbins, M. B., Rosenberg, G. H. & Sornoza M., F. 1994. A new species of cotinga (Cotingidae: *Doliornis*) from the Ecuadorian Andes, with comments on plumage sequences in *Doliornis* and *Ampelion. Auk* 111: 1–7.
Schulenberg, T. S. & Parker, T. A. 1997. A new species of tyrant-flycatcher (Tyrannidae: *Tolmomyias*) from the western Amazon basin. *Orn. Monogr.* 48: 723–731.
*Schulenberg, T. S., Stotz, D. F., Lane, D. F., O'Neill, J. P. & Parker, T. A. 2007. *Birds of Peru.* Christopher Helm, London.
Solano-Ugalde, A. & Freile, J. F. 2012. A decade of progress (2001–2010): overview of distributional records of birds in mainland Ecuador. *Orn. Neotrop.* 23: 29–35.
Tinoco, B. A., Astudillo, P. X., Latta, S. C. & Graham, C. H. 2009. Distribution, ecology and conservation of an endangered Andean hummingbird: the Violet-throated Metaltail (*Metallura baroni*). *Bird Conserv. Intern.* 19: 63–76.
*Turner, A. & Rose, C. 1989. *Swallows and Martins of the World.* Christopher Helm, London.
Vriesendorp, C., Alverson, W. S., del Campo, A., Stotz, D. F., Moskovits, D. K., Fuentes Cáceres, S., Coronel Tapia, B. & Anderson, E. P. (eds.) 2009. *Ecuador: Cabeceras Cofán-Chingual.* Rapid Biological and Social Inventories Report 21. Field Museum of Natural History, Chicago.
Whitney, B. M. & Alvarez-Alonso, J. 1998. A new *Herpsilochmus* antwren (Aves: Thamnophilidae) from northern Amazonian Peru and adjacent Ecuador: the role of edaphic heterogeneity of terra firme forest. *Auk* 115: 559–576.

INDEX